Biolubricants

Related titles:

Advances in biodiesel production
(ISBN 978-0-85709-117-8)
Biodiesel is one of the main biofuels capable of substituting fossil fuel usage in compression ignition vehicles, and is used in a variety of fuel blends worldwide. Advances in biodiesel production, aimed at more sustainable and effective feedstock utilisation alongside improved processing routes for increased production efficiency and fuel quality, are critical to the future both of this industry and of the continued use of biodiesel fuels in transportation. This book provides a timely reference on the advances in the development of biodiesel fuels, production processes and technologies.

Handbook of biofuels production
(ISBN 978-1-84569-679-5)
The global increase in the use of biofuels as substitute transportation fuels has prompted research and development into advanced chemical, biochemical and thermochemical biofuels production. This research and development is aimed at improving the quality and environmental impact of biofuels production, as well as the overall efficiency and output of biofuels production plants. The range of biofuels has also increased, with market developments leading to the increased production and utilisation of advanced biofuels such as biosyngas, biohydrogen and biobutanol. *Handbook of biofuels production* provides a thorough reference on the range of biomass conversion processes and technologies.

Bioalcohol production
(ISBN 978-1-84569-510-1)
Bioethanol is one of the main biofuels used as a petroleum-substitute in transportation. Second generation bioalcohol production technology, based on highly-efficient biochemical conversion, offers advantages over existing, energy-intensive bioethanol production processes. This book covers the process engineering, technology, modelling and integration of the entire production chain, from feedstock pretreatment on to hydrolysis, fermentation and purification. Primarily reviewing bioethanol production, coverage extends to the production of longer-chain bioalcohols which will be elemental to the future of the industry.

Details of these books and a complete list of titles from Woodhead Publishing can be obtained by:

- visiting our web site at www.woodheadpublishing.com
- contacting Customer Services (e-mail: sales@woodheadpublishing.com; fax: +44 (0) 1223 832819; tel.: +44 (0) 1223 499140 ext. 130; address: Woodhead Publishing Limited, 80, High Street, Sawston, Cambridge CB22 3HJ, UK)
- in North America, contacting our US office (e-mail: usmarketing@woodheadpublishing.com; tel.: (215) 928 9112; address: Woodhead Publishing, 1518 Walnut Street, Suite 1100, Philadelphia, PA 19102-3406, USA)

If you would like e-versions of our content, please visit our online platform: www.woodheadpublishingonline.com. Please recommend it to your librarian so that everyone in your institution can benefit from the wealth of content on the site.

Woodhead Publishing Series in Energy: Number 46

Biolubricants

Science and technology

Jan C. J. Bart, Emanuele Gucciardi
and Stefano Cavallaro

WP
WOODHEAD
PUBLISHING

Oxford Cambridge Philadelphia New Delhi

Published by Woodhead Publishing Limited,
80 High Street, Sawston, Cambridge CB22 3HJ, UK
www.woodheadpublishing.com
www.woodheadpublishingonline.com

Woodhead Publishing, 1518 Walnut Street, Suite 1100, Philadelphia,
PA 19102-3406, USA

Woodhead Publishing India Private Limited, G-2, Vardaan House, 7/28 Ansari Road, Daryaganj, New Delhi – 110002, India
www.woodheadpublishingindia.com

First published 2013, Woodhead Publishing Limited

British Library Cataloguing in Publication Data
A catalogue record for this book is available from the British Library.

Library of Congress Control Number: 2012946145

ISBN 978-0-85709-263-2 (print)
ISBN 978-0-85709-632-6 (online)
ISSN 2044-9364 Woodhead Publishing Series in Energy (print)
ISSN 2044-9372 Woodhead Publishing Series in Energy (online)

The publisher's policy is to use permanent paper from mills that operate a sustainable forestry policy, and which has been manufactured from pulp which is processed using acid-free and elemental chlorine-free practices. Furthermore, the publisher ensures that the text paper and cover board used have met acceptable environmental accreditation standards.

Typeset by Toppan Best-set Premedia Limited, Hong Kong
Printed and bound in the UK by MPG Books Group

Contents

Author contact details

Professor Jan C.J. Bart
E-mail: jbart@unime.it

Dr Emanuele Gucciardi
E-mail: gucciardi.emanuele@gmail.com

Professor Stefano Cavallaro
Dipartimento di Chimica Industriale
e Ingegneria dei Materiali dell'Università
Viale F. Stagno D'Alcontres, 31
I-98166 Messina
Italy
E-mail: stefano.cavallaro@unime.it

About the authors

Jan C.J. Bart (PhD Structural Chemistry, University of Amsterdam) is a senior scientist with a wide interest in materials science, heterogeneous catalysis and sustainable product development who spent a career in industrial R&D with Monsanto, Montedison and DSM Research in various technical functions. Dr Bart has held teaching assignments in several European universities, researched extensively both in industry and in academia, and published over 250 scientific papers, chapters in books and patents; he is also author of three monographs on polymer additive analytics, and recently co-authored a text on *Biodiesel science and technology: From soil to oil.* Dr Bart has acted as a Ramsay Memorial Fellow at the University of Leeds (Colour Chemistry) and Oxford (Material Science), a visiting scientist at IRC-Institut de Recherches sur la Catalyse (CNRS, Villeurbanne), and a Meyerhoff Visiting Professor at WIS-Weizmann Institute of Science (Rehovoth), and held an Invited Professorship at USTC-University of Science and Technology of China (Hefei). He is currently a Full Professor of Industrial Chemistry at the University of Messina (Italy).

Emanuele Gucciardi (PhD Industrial Chemistry, University of Messina) has interest in heterogeneous catalysis and renewable resources. He operated as a guest junior researcher at the Nicola Giordano CNR-TAE Institute of Advanced Technologies for Energy (Messina) and at the Boreskov Institute of Catalysis (BIC) in Novosibirsk (Russia). He has been involved in fuel cell research at the University of Milan (Italy) and, at present, he is a post-doctor researcher at the ZSW (Zentrum für Sonnenenergie-und Wasserstoff-Forschung) of Ulm, Germany.

Stefano Cavallaro graduated with an honours degree in Chemistry from the University of Messina, where he now acts as an Associate Professor of Industrial Chemistry. His research interests are applied and industrial chemistry, in particular heterogeneous catalysis, transformation and storage of energy, fuel cell research and new materials. Dr Cavallaro has a long-lasting co-operation with the CNR-TAE National Research Council Institute in Messina, has acted as a project leader of several Italian research projects, and is author of some 110 research papers and patents. He is also a co-author of *Biodiesel science and technology: From soil to oil* (Woodhead Publishing Series in Energy, 2010).

Woodhead Publishing Series in Energy

1 **Generating power at high efficiency: Combined cycle technology for sustainable energy production**
Eric Jeffs

2 **Advanced separation techniques for nuclear fuel reprocessing and radioactive waste treatment**
Edited by Kenneth L. Nash and Gregg J. Lumetta

3 **Bioalcohol production: Biochemical conversion of lignocellulosic biomass**
Edited by K. W. Waldron

4 **Understanding and mitigating ageing in nuclear power plants: Materials and operational aspects of plant life management (PLiM)**
Edited by Philip G. Tipping

5 **Advanced power plant materials, design and technology**
Edited by Dermot Roddy

6 **Stand-alone and hybrid wind energy systems: Technology, energy storage and applications**
Edited by J. K. Kaldellis

7 **Biodiesel science and technology: From soil to oil**
Jan C. J. Bart, Natale Palmeri and Stefano Cavallaro

8 **Developments and innovation in carbon dioxide (CO_2) capture and storage technology Volume 1: Carbon dioxide (CO_2) capture, transport and industrial applications**
Edited by M. Mercedes Maroto-Valer

9 **Geological repository systems for safe disposal of spent nuclear fuels and radioactive waste**
Edited by Joonhong Ahn and Michael J. Apted

10 **Wind energy systems: Optimising design and construction for safe and reliable operation**
Edited by John D. Sørensen and Jens N. Sørensen

11 **Solid oxide fuel cell technology: Principles, performance and operations**
Kevin Huang and John Bannister Goodenough

12 **Handbook of advanced radioactive waste conditioning technologies**
Edited by Michael I. Ojovan

13 **Membranes for clean and renewable power applications**
Edited by Annarosa Gugliuzza and Angelo Basile

14 **Materials for energy efficiency and thermal comfort in buildings**
Edited by Matthew R. Hall

15 **Handbook of biofuels production: Processes and technologies**
Edited by Rafael Luque, Juan Campelo and James Clark

16 **Developments and innovation in carbon dioxide (CO_2) capture and storage technology Volume 2: Carbon dioxide (CO_2) storage and utilisation**
Edited by M. Mercedes Maroto-Valer

17 **Oxy-fuel combustion for power generation and carbon dioxide (CO_2) capture**
Edited by Ligang Zheng

18 **Small and micro combined heat and power (CHP) systems: Advanced design, performance, materials and applications**
Edited by Robert Beith

19 **Advances in clean hydrocarbon fuel processing: Science and technology**
Edited by M. Rashid Khan

20 **Modern gas turbine systems: High efficiency, low emission, fuel flexible power generation**
Edited by Peter Jansohn

21 **Concentrating solar power (CSP) technology: Developments and applications**
Edited by Keith Lovegrove and Wes Stein

22 **Nuclear corrosion science and engineering**
Edited by Damien Féron

23 **Power plant life management and performance improvement**
Edited by John E. Oakey

24 **Electrical drives for direct drive renewable energy systems**
Edited by Markus Mueller and Henk Polinder

25 **Advanced membrane science and technology for sustainable energy and environmental applications**
Edited by Angelo Basile and Suzana Pereira Nunes

26 **Irradiation embrittlement of reactor pressure vessels (RPVs) in nuclear power plants**
Edited by Naoki Soneda

27 **High temperature superconductors (HTS) for energy applications**
Edited by Ziad Melhem

28 **Infrastructure and methodologies for the justification of nuclear power programmes**
Edited by Agustín Alonso

29 **Waste to energy (WtE) conversion technology**
Edited by Marco Castaldi

30 **Polymer electrolyte membrane and direct methanol fuel cell technology Volume 1: Fundamentals and performance of low temperature fuel cells**
Edited by Christoph Hartnig and Christina Roth

31 **Polymer electrolyte membrane and direct methanol fuel cell technology Volume 2: *In situ* characterization techniques for low temperature fuel cells**
Edited by Christoph Hartnig and Christina Roth

32 **Combined cycle systems for near-zero emission power generation**
Edited by Ashok D. Rao

33 **Modern earth buildings: Materials, engineering, construction and applications**
Edited by Matthew R. Hall, Rick Lindsay and Meror Krayenhoff

34 **Metropolitan sustainability: Understanding and improving the urban environment**
Edited by Frank Zeman

35 **Functional materials for energy applications**
Edited by John Kilner, Stephen Skinner, Stuart Irvine and Peter Edwards

36 **Nuclear decommissioning: Planning, execution and international experience**
Edited by Michele Laraia

37 **Nuclear fuel cycle science and engineering**
Edited by Ian Crossland

38 **Electricity transmission, distribution and storage systems**
Edited by Ziad Melhem

39 **Advances in biodiesel production: Processes and technologies**
Edited by Rafael Luque and Juan A. Melero

40 **Biomass combustion science, technology and engineering**
Edited by Lasse Rosendahl

41 **Ultra-supercritical coal power plant: Materials, technologies and optimisation**
Edited by Dongke Zhang

42 **Radionuclide behaviour in the natural environment: Science, impacts and lessons for the nuclear industry**
Edited by Christophe Poinssot and Horst Geckeis

43 **Calcium and chemical looping technology for power generation and carbon dioxide (CO_2) capture: Solid oxygen- and CO_2-carriers**
P. Fennell and E. J. Anthony

44 **Materials ageing and degradation in light water reactors: Mechanisms, modelling and mitigation**
Edited by K. L. Murty

45 **Structural alloys for power plants: Operational challenges and high-temperature materials**
Edited by Amir Shirzadi, Rob Wallach and Susan Jackson

46 **Biolubricants: Science and technology**
Jan C. J. Bart, Emanuele Gucciardi and Stefano Cavallaro

47 **Wind turbine blade design and materials: Improving reliability, cost and performance**
Edited by Povl Brøndsted and Rogier Nijssen

48 **Radioactive waste management and contaminated site clean-up: Processes, technologies and international experience**
Edited by William E. Lee, Michael I. Ojovan, Carol M. Jantzen

49 **Probabilistic safety assessment for optimum nuclear power plant life management (PLiM)**
Gennadij V. Arkadov, Alexander F. Getman and Andrei N. Rodionov

50 **Coal utilization in industry**
Edited by D. G. Osborne

51 **Coal power plant materials and life assessment: Developments and applications**
Edited by Ahmed Shibli

52 **The biogas handbook: Science, production and applications**
Edited by Arthur Wellinger and David Baxter

53 **Advances in biorefineries: Biomass and waste supply chain exploitation**
Edited by K. W. Waldron

54 **Geoscience of carbon dioxide (CO_2) storage**
Edited by Jon Gluyas and Simon Mathias

55 **Handbook of membrane reactors Volume 1: Fundamental materials science, design and optimisation**
Edited by Angelo Basile

56 **Handbook of membrane reactors Volume 2: Industrial applications and economics**
Edited by Angelo Basile

57 **Alternative fuels and advanced vehicle technologies: Towards zero carbon transportation**
Edited by Richard Folkson

58 **Handbook of microalgal bioprocess engineering**
Christopher Lan and Bei Wang

59 **Fluidized-bed technologies for near-zero emission combustion and gasification**
Edited by Fabrizio Scala

60 **Managing nuclear projects: A comprehensive management resource**
Edited by Jas Devgun

61 **Handbook of process integration: Energy, water, waste and emissions management in processing and power industries**
Edited by Jiří Klemeš

Preface

In the medium-term future (tens of years) fossil energy sources will still continue to satisfy most of the world's energy demand. According to the *World Energy Outlook 2010* (International Energy Agency), global energy use will further increase by 36% from 2008 to 2035, of which 50% is accounted for by fossil-based resources. This has enormous implications for the environment and will endanger climate control. Climate protection is one of the great challenges of the 21st century. In order to reduce CO_2 emissions by 2050, there will eventually have to be massive, much disputed and expensive carbon capture and storage (CCS) projects. More efficient processes are better than CCS. As the world's crude reserves are declining, heavy oil or bitumen reserves (about 53% of global total oil reserves) and coal (again) are called upon to play an ever-increasing role. However, for this to occur further development of oil sands heavy oil technology, bottom-of-the-barrel technology, offshore technology, green refining and clean coal technology are required. At the same time, primary sources urgently need to be developed that are neutral regarding the greenhouse effect in order to make the transition to a new and sustainable energy system for future generations. Primary objectives in the near future are clean energy, energy saving, an increase in the efficiency of energy-intensive processes and exploitation of unconventional resources (stranded gas, ultra-heavy oils, oil shales and gas hydrates).

Although energy generation may (partly) be replaced by various low- or zero-carbon alternative technologies (nuclear, photovoltaics, solar, wind, etc.), other solutions must be found to replace the tight fossil fuels as feedstock for the chemical industry. Raw material change and 'green technologies' are currently a hot topic in industry. Finding bio-based substances that can replace oil and gas as building blocks for chemicals is like chasing the Holy Grail. Some potential benefits of bio-based industrial products are diversification of petroleum feedstocks, national security, product quality, environmental quality and rural employment. The real motive for pushing bio-based materials is that they yield better products.

Currently, industry-wide efforts are underway to replace conventional petroleum-based chemical feedstocks with non-petroleum-based resources. There is a compelling need to consider renewable resources as future raw materials for our present industrial products. Today's production systems are largely unsustainable, being based on finite feedstocks, or wasteful production processes (both in terms of use of materials and energy) and resulting in degradation of the environment, including man-made global warming (0.2 °C from 1985 to 2010). Transition to a sustainable path is urgently needed. This book addresses moving towards a society in which bio-based feedstocks, processes and products are sustaining the economy. The search for environmentally friendly materials that have potential to substitute for mineral oil in various industrial applications is nowadays being considered a top priority research topic. The specific scope of this book concerns the prospects of replacement of mineral oil-based lubricants by biolubricants, especially as they impact value-in-use for a given application.

Lubricants are essential machine elements, which are vital for efficient transportation and industrial production, and contribute enormously to energy savings and the fuel economy. As an engineering fluid, a lubricant is an important construction element and is as significant as the material used itself and the surface treatment given. Most lubricants used today, in thousands of different technical applications, are traditionally still mineral oil-based. Advances in lubricant technology have brought green products to the forefront in many high-profile applications in environmentally sensitive areas. Over time, the ongoing development of next-generation lubricants has gradually shifted focus from biodegradable to renewable (bio-based) and to sustainable, with emphasis on energy economy. Lubrication saves energy. The adoption of biolubricants is based on performance, cost effectiveness and better environmental credentials. Environmentally friendly lubricants should be used whenever and wherever possible. At the same time, the quality and availability of raw materials should meet the potential demand.

Many of the most urgent global challenges, from the end of oil to climate change, pass through the intervention of chemical sciences and plant biotechnology. As not all technical innovations (e.g. conversion of vegetable oils to biodiesel) initially meet economic criteria, desirable technologies may temporarily need government assistance before becoming self-sustaining. The emerging bio-based economy can be encouraged by tax breaks or other incentives. The past decade has witnessed a strong trend to develop industrial applications from renewable materials, particularly from easily accessible feedstocks such as vegetable oils and animal fats. Development of energy-efficient, environmentally friendly, reliable, high-performance and affordable products, including biolubricants, enhances the global economy and is in the interest of the well-being of people worldwide.

Information on biolubricants is contained in classical lubricant textbooks, but often in a rather scattered or embedded mode. The present text describes advanced technologies for the production of lubricant base stocks from bio-based raw materials (mostly vegetable oils) more explicitly and gives some directions for the future. Methods to reduce the relatively high costs of biolubricants in comparison to mineral oil-based lubricants focus on minimising the raw materials cost and process costs (including catalyst). The role of (bio)catalysis is of utmost importance both technologically and economically. Emphasis is given to structure–property relations as a means for improved understanding and product design. Biolubricant formulations in relation to application engineering requirements are clearly outlined.

This book on bio-based lubricants gives a useful overview of biomass-derived raw materials, environmental advantages compared with the use of fossil fuels, lubricant design considerations, biolubricant chemistry and the major applications of biolubes. The text identifies potentially successful (and other less successful) products and technologies, promising vs less promising products and developments. It provides a comprehensive understanding for users and researchers in the lubrication industry alike interested in new products and technologies based on renewables. A substantial amount of literature, available in scientific journals, patents and on the web up to late 2011, has critically been reviewed and evaluated. In-depth analysis of technologies and applications makes this book a must read and a reference in the industry. The text is heavily referenced.

As we run out of fossil fuels, the 'green revolution' may be just as radical as the previous industrial and digital revolutions. The bio-economic society aims at responsible industrial development of bio-based products and bioenergy. Creating substitute competitive products and technologies founded on renewable raw materials is a challenging task. An analysis of state-of-the-art biolubricant technology is timely. This book offers chemists, engineers, agronomists, botanists and ecologists, an interdisciplinary introduction and orientation to biolubricant chemistry and its major applications, based on specific performance criteria. As industrial managers become more aware of what has been achieved they may be encouraged to further explore the application of biolubricants. Organisations and government policy makers may use this book as a basis for policy guidelines or for national programmes to underscore the expansion of biolubrication.

Jan C. J. Bart
Emanuele Gucciardi
Stefano Cavallaro
Messina

1
Renewable lubricants

Abstract: Renewable feedstocks are bound to gradually replace sources of fossil origin (oil, gas and coal), both as fuel and as raw materials for the chemical industry. Lubricants combine indispensable performance characteristics in transportation and industry with considerable economic value. The transition from mineral oil-based lubes to biodegradable lubricants from renewable raw materials is ongoing. Biolubricants originate from vegetable oils (triglycerides), wax esters and plant polymeric carbohydrates. Chemical sciences and plant biotechnology are crucial disciplines in designing environmentally acceptable lubricants for a wide variety of applications. Important actors in the biolubricant value chain are vegetable oil milling facilities, lubricant manufacturing industry, regulators and end users (for consumer acceptance).

Key words: bio-based economy, biolubricants, sustainability, renewable feedstocks, vegetable oils.

1.1 Introduction

The ever-increasing global population is faced with various future constraints in growth regarding its very basics of subsistence, such as the availability of fresh water, raw materials, energy and suitable land, abundant and affordable food production, coupled with the need for lower emissions and less waste, and mitigation of the anticipated climate change. This requires truly sustainable technologies and practices. The concept of sustainability dictates that modern technologies should be founded on renewable resources. The current petroleum-based economy cannot be sustained indefinitely, although naphtha is still likely to remain the main input for the chemical industry for the next few decades, despite the end of cheap conventional oil [1–3]. Prognoses of worldwide petroleum reserves are uncertain and are regularly being scaled up but generally predict maximum production levels around 2020–2030. Actually, peak oils were already expected in the past (IEA: 2006; BP: 2009). Proven global oil reserves amount to 1383 Bbbl (billion barrels), for 54% in the Middle East. At the same time, the potential of shale oil and gas deposits appears huge. The proven non-conventional oil reserves in oil sands and shale fields, available in many countries, amount to over 1000 Bbbl (with expectations up to 3000 Bbbl). Conventional gas reserves are of the order of 9150 EJ (for almost 50% in CIS), exceeded by non-conventional reserves of 62 150 EJ (for 29%

in Australia/Asia). At increasing oil prices new exploration and drilling technologies are being developed. However, exploitation of shale fields by hydraulic fracturing (fracking) is not without risks (groundwater pollution) and draws significant opposition.

At present, 86% of oil is used for transportation as gasoline, diesel and kerosene (55%), and energy and heating (31%), leaving only 14% to the chemical industry for transformation into a great variety of chemical products. This share might increase somewhat by more efficient engine and heating systems and increasing amounts of biofuels. A breakdown of the petroleum-derived products in the German market (totalling 104.8 Mt, 2010), as shown in Table 1.1, confirms the modest share of oil reserved for the chemical industry and for lubricant base oil production in particular. Deployments of renewable raw materials for the chemical industry and sustainable energy will require a concerted effort from scientists across a great number of disciplines to develop new technologies, and (most of all) political will.

Industrial production processes can reach higher levels of efficiency and products can be made more attractive through the added value lubricants can offer. A lubricant is a substance used to improve the ease of movement between surfaces [4]. Lubricants are used to reduce friction, reduce wear, and prevent overheating and corrosion. They are complex products, usually consisting of 70–99% base oils admixed with additives that modify the natural properties of the fluid to meet its intended requirements. Lubricants

Table 1.1 Breakdown of petroleum derivatives in Germany (Mt)

Main products	
• Diesel	30.1
• Petrol	21.5
• Heating oils	14.7/6.6/1.3[a]
• Refined oil[b]	8.0
• Other components	1.4
By-products	
• Liquid gas	2.6
• Refinery gas	3.8
• Heavy jet fuel	4.9
• Lubricants	2.5
• Bitumen	3.4
• Petrocoke	2.0
• Other residues	2.0

[a] Light fractions (households)/heavy fractions (industry)/heating oil components.
[b] Raw material for chemical industry.
Source: Mineralölwirtschaftsverband (2010).

represent considerable economic value, enabling high manufacturing speed, avoiding material failures in transportation and industry, and limiting downtime. Their cumulative benefits extend far beyond the tonnes produced.

Today, market demands for lubricants are increased environmental compatibility, reduced emissions, greater occupational safety and, above all, superior performance (including longer lifetime and energy efficiency). The global volume of lubricants used in various industrial applications, especially engine oils and hydraulic fluids, is considerable (about 40 Mt/yr). Demand for lubricants in the United States is about 2.8 billion gallons (2006), and mainly on account of process oils and high-performance lubricants. Most of these products (e.g. general industrial oils, engine oils, transmission and hydraulic fluids, gear oils and greases) are as yet still based on mineral oils. Concerns about environmental pollution caused by many chemicals, including lubricants, have created a growing worldwide trend of promoting new environmentally friendly products. In Europe, some 1.1 Mt/yr of used lubricating oils (representing 20% of total market) are not traceable. It has been estimated that about 600 kt of total-loss lubricants – including chainsaw oils, mould release oils, two-stroke engine oils, chassis and wheel flange greases – are released into the environment every year in the European Union alone [5]; for other release estimates, see ref. [6]. Whatever the precise figures, approximately 50% of all lubricants used worldwide end up somehow in the environment via total-loss applications, evaporation, spillage or accidents [7]. Estimates for the loss of hydraulic fluids are even higher (70–80%) [8, 9]. Currently, over 95% of these materials are mineral oil-based or synthetic oils. Mineral oils are toxic for mammals, fish and bacteria.

Replacing petroleum base oils with biodegradable products is one of the ways to reduce the adverse effects on the ecosystem caused by the use of lubricants. The potential use of renewable raw materials (be it starches from corn, tapioca, wheat or potatoes, cassava or sorghum, chitin, vegetable oils or animal fats, cellulose or algae biomass) in the lubrication industry is currently of great interest [10]. Growing concern about the use of mineral oils as lubricants has promoted extensive research into developing and using derivatives of vegetable oils as alternative base oils for environmentally benign lubricants. Interest in the production of biodegradable, environmentally acceptable esters for biodiesel, lubricating base oils, solvents, surfactants, etc., from vegetable oils has increased greatly in recent years. Utilisation of natural materials and green processes in the lubricant industry is especially attractive [11]. Biolubricants can typically be made from plant oils such as rapeseed, soybean, sunflower, palm and coconut, wax esters and plant polymeric carbohydrates. Since the cost of a lubricant is considerably higher than that of an equal volume of fuel, the use of oleochemicals as raw materials for lubricants makes good economic sense.

Replacement of petroleum by more widely distributed renewable raw materials in an agricultural product-based economy is an enormous challenge and requires integration of the biological and chemical sciences with development of the chemistry, modification, blending, processing, manufacturing and testing of agricultural products (Table 1.2). Brazil is a good example of valorisation of biomass. Non-food biomass is the only practical sustainable source of carbon in the long run [12, 13]. Biolubricants are usually considered to be lubricants that have high biodegradability and low human and environmental toxicity [4, 14]. Biolubricants that meet specific biodegradability and toxicity criteria can be based both on renewable, agricultural raw materials and on petroleum oils or synthetic esters [14, 15]. Criteria of consumer acceptance adopted in various countries are different. Some countries only require 50% of the oil to be renewable, others impose utilisation of biolubricants when near non-navigable waters.

Bio-based lubricants are part of an overall green chemistry solution for obtaining better, environmentally safer and more economical lubrication, and for creating efficiencies in manufacturing plants, end products, or in-house fleets and maintenance programmes. Development of bio-based high-performance lubricants requires insight in the relationship between physicochemical and tribological properties of the bio-based ingredients. While biodegradability nowadays is one of the most important design parameters for base oils and lubricant formulations, at the same time these

Table 1.2 Raw materials in fossil-derived and biomass-based economies

Fossil-derived	Biomass-based
• Mature technology	• Development of sustainable technologies
• Finite availability	• Renewable resource
• Geopolitically sensitive	• Universally available (national security)
• Unique resources	• Wide range of resources
• Concentrated raw material sourcing	• Widely dispersed raw material sources
• Limited logistics problems	• Considerable logistics problems
• Gaseous, liquid and solid raw materials; water free	• Liquid and solid raw materials; water-rich
• Monomeric to polymeric	• Mainly polymeric
• Crude source independent (similar) chemistry	• Highly source[a] and subsource[b] dependent chemistries
• C, H chemistry	• C, H, O chemistry
• Non-functionalised raw materials	• Highly functionalised, non-volatile product streams
• Hydrophobic compounds	• More hydrophilic

[a] E.g. plant, animal, fish.
[b] E.g. seeds, leaves, stems, roots.

desirable properties determine high susceptibility to tribochemical degradation processes. Therefore, an adequate balance of properties is needed. Another important objective of the lubricant industry is formulation of new environmentally friendly lubricants of high technical reliability with similar or better technical specifications than those based on mineral oils, in particular both improved kinematic viscosity values and adequate viscosity thermal susceptibility [16, 17].

Development of targeted synthetic biolubricants in chemical synthetic pathways alone is a lengthy and expensive process. Computer-based modelling, analysis of equilibrium energy (E_{min}), electron density distribution, steric structure and spatial orientation of functional groups can reveal important properties such as oxidation, molecular reactivity, high-temperature deposit formation, low-temperature fluidity, viscosity–temperature characteristics, solvency and lubricity. Technical, economic and legislative aspects play a role in the replacement of lubricants of petroleum origin by those based on renewable raw materials. Bio-based materials have clear advantages and are already economically competitive in a growing number of industrial sectors but further development work is required to achieve additional cost reduction.

1.2 Scope

The production of environmentally friendly, rapidly biodegradable fluids for lubricants based on petrochemicals (e.g. poly-α-olefins, polyalkylene glycols and synthetic esters) has been described elsewhere [18]. Plant oils are also attractive alternatives as raw materials for the chemical industry. Biolubricants are now available for numerous applications, with performance in competition with that of mineral oil-based products (ranging from worse to comparable and sometimes even better). Lubricant base fluids based on renewable raw materials have been reviewed in the past [7, 19–31]. In view of the rapid developments a critical update is timely. Therefore, the present text describes the state-of-the-art of advanced technologies for lubricant production from both virgin oils and waste oils as well as fats and other renewable resources on the basis of the technical, scientific and patent literature up to late 2011 and makes some recommendations for the future.

Consumer acceptance of renewable lubricants depends largely on how well they perform during high-temperature oxidation and low-temperature applications. A wide variety of (chemically modified) vegetable oils has been evaluated in terms of their thermal, oxidation and low-temperature behaviour for their potential use as base fluids for industrial and automotive lubrication. This text traces recent developments in lubricant base stocks of vegetable oil origin, putting emphasis on structure–property relations as a means for improved product design. Methods to reduce the high cost of biolubricants focus on minimising the raw material cost and process costs (in particular catalysts). The role of (bio)catalysis is an important one both technologically and economically. Advances in the plant biotechnology of vegetable oils and chemical derivatisations of natural esters have resulted in high-performance biolubricants at a lower cost compared with synthetic esters.

1.3 Chapter overview

The book focuses essentially mainly on biolubricant base stocks and on applications of biolubricant formulations (base stocks + additives). The text is divided into three main parts, introducing first general concepts of lubrication and renewable lubricants, and describing next raw materials and their chemical transformations to biolubricant base stocks, followed by applications and markets of commercial biolubricant formulations. The intent of these chapters is to convey the enormous opportunities renewable lubricants can offer both to government and industry. The chapters are structured to differentiate bio-based, renewable lubricants, often made from high-oleic base stocks, from petroleum-based lubricants made from non-renewable crude oil.

Chapter 2 examines lubrication in terms of friction, wear and corrosion and how these phenomena damage surfaces causing premature equipment failure. The mechanics of fluid film, boundary, extreme pressure and elastohydrodynamic lubrication to protect against surface deterioration are briefly outlined. Lubricants are then broadly classified chemically as mineral oil-derived and bio-based products (Chapter 3). Attention is given to the eco-designations for lubricants. Chapter 4 illustrates the consequences of the ongoing transition from fossil- to bio-based feedstocks for chemical transformation into commercially performing products. The potentials of various kinds of biomass are outlined as well as the role of plant biotechnology. Biomass limitations are also described. Chapter 5 introduces the main renewable and inherently biodegradable feedstocks for lubricant manufacturing, namely edible and non-edible vegetable oil sources, bio-based wax esters and plant polymeric carbohydrates. The importance of oil-crop engineering in providing high-oleic base stocks is stressed. Chemical functionalisation of vegetable oils and starch is described in necessary detail, clearly indicating in which case industrial application is possible or nowadays restricted for cost reasons (due to the absence of sufficiently performing catalysts; Chapter 6). Ample use is made of structure–performance relationships of plant oil-derived lubricant base fluids in rational product development.

Lubricant additives (partly similar to those in use in mineral oil base lubricants but with the extra dimension of biodegradability) are introduced without unnecessary detail in Chapter 7. Physical properties and quality requirements of renewable and environmentally acceptable lubricant formulations are given in Chapter 8. The description of testing of *bio*lubricant performance is restricted to the (extra) aspects related to the *bio* basis (biodegradability, ecotoxicity). Chapters 9 and 11 deal with product-focused legislation, the interaction of commercial lubricants with the environment, and life-cycle assessments of biolubricants. The scale and depth of international research and technological biolubricant development are illustrated in Chapter 10 by means of the scope and results of numerous framework and industrial programmes. Chapter 12 describes the main lubricant product groups (automotive, industrial, marine and aviation) in relation to the current and future prospects of technological applications for biolubricants, starting from the generally accepted, minimum performance criteria. The derived biolubricant market shares and prospectives are given in Chapter 13. Selection criteria for (renewable) lubricants are developed. Operational maintenance, lubricant monitoring, and (bio)lubricant disposal are the subjects of Chapter 14. The book concludes with advanced lubricant fluids (Chapter 15) and with an outlook for biolubricant science and technology (Chapter 16).

1.4 Sources of further information and advice

R.B. Laughlin, *Powering the Future*, Basic Books, New York, NY (2011).

Organization of Petroleum Exporting Countries, *World Oil Outlook 2010*, OPEC, Vienna (2010).

R. Höfer (ed.), *Sustainable Solutions for Modern Economies*, RSC Publishing, Cambridge (2009).

T. Volk, *CO_2 Rising: The World's Greatest Environmental Challenge*, MIT Press, Cambridge, MA (2008).

Intergovernmental Panel on Climate Change, *Summary for Policymakers*, IPCC, Geneva (2007); http://www.ipcc.ch

Th. Mang and W. Dresel (eds), *Lubricants and Lubrication*, 2nd edn, Wiley-VCH, Weinheim (2007).

J.S. Marth, *Renewable Lubricants Manual: Biobased Oils, Fluids & Greases*, United Bio Lube, Palo Alto, CA (2007); http://www.renewablelubricants.com

L.R. Rudnick (ed.), *Synthetics, Mineral Oils, and Bio-based Lubricants: Chemistry and Technology*, CRC Press/Taylor & Francis, Boca Raton, FL/New York, NY (2006).

A. Gore and Melcher Media, *An Inconvenient Truth: The Planetary Emergency of Global Warming and What We Can Do about It.* Rodale Press, Emmaus, PA (2006).

A. Igartua, J. Barriga and A. Aranzabe (eds), *Biodegradable Lubricants*, Virtual Tribology Institute Edition, Eibar (2005).

S.Z. Erhan (ed.), *Industrial Uses of Vegetable Oils*, CRC Press, Boca Raton, FL (2005).

S.Z. Erhan and J.M. Perez (eds), *Biobased Industrial Fluids and Lubricants*, AOCS Press, Champaign, IL (2003).

Committee on Biobiobased Industrial Products/National Research Council (ed.), *Biobased Industrial Products: Priorities for Research and Commercialization*, National Academies Press, Washington, DC (2000).

G. Knothe and J.T.P. Derksen (eds), *Recent Developments in the Synthesis of Fatty Acid Derivatives*, AOCS Press, Champaign, IL (1999).

F.D. Gunstone and F.B. Padley (eds), *Lipid Technologies and Applications*, CRC Press, Boca Raton, FL (1997).

W.J. Bartz (ed.), *Biologically Fast Degradable Lubricants and Operational Fluids*, Expert-Verlag GmbH, Ehningen (1993).

R.W. Johnson and E. Fritz (eds), *Fatty Acids in Industry*, Marcel Dekker, Inc., New York, NY (1989).

1.5 References

[1] C.J. Campbell and J.H. Laherrère, The end of cheap oil, *Scient. Amer.* (March 1998), pp. 78–83.

[2] D. Goodstein, *Out of Gas: The End of the Age of Oil,* W.W. Norton & Co., New York, NY (2004).

[3] International Energy Agency, *World Energy Outlook 2010*, IEA, Paris (2010).

[4] A. Lansdown, *Lubrication and Lubricant Selection: A Practical Guide*, 3rd edn, ASME Press, Champaign, IL (2004).

[5] E.M. Stempfel, H. Hostettler and H. Gasser, Practical experience with highly biodegradable lubricants, especially hydraulic oils and lubricating greases, Paper presented at Third German Schmierstoffforum, Bad Nauheim (March 1993).

[6] N.S. Battersby, The biodegradability and microbial toxicity testing of lubricants – some recommendations, *Chemosphere* **41** (2000) 1011–27.

[7] M. Schneider, Plant-oil-based lubricants and hydraulic fluids, *J. Sci. Food Agric.* **86** (2006) 1769–80.

[8] K. Carnes, Offroad hydraulic fluids: beyond biodegradability, *Tribol. Lubr. Technol.* **60**(9) (2004) 32–40.

[9] S. Miller, C. Scharf and M. Miller, Utilising new crops to grow the biobased market, in *Trends in New Crops and New Uses* (J. Janick and A. Whipkey, eds), ASHS Press, Alexandria, VA (2000), pp. 26–8.

[10] J. Hagstrom, USDA to Set Requirements for Agencies to Purchase Bio-based Products, in Govexec.com (2005).

[11] I. Gawrilow, Vegetable oil usage in lubricants, *INFORM* **15**(11) (2004) 702–5.

[12] J.H. Clark and F.E.I. Deswarte (eds), *Introduction to Chemicals from Biomass,* John Wiley & Sons, Ltd, Chichester (2008).

[13] R. Höfer (ed.), *Sustainable Solutions for Modern Economies*, RSC Publishing, Cambridge (2009).

[14] IENICA, *Biolubricants – Market Data Sheet* (2004), pp. 1–11.

[15] B.J. Bremmer and L. Plonsker, *Bio-based Lubricants: A Market Opportunity Study Update*, United Soybean Board, Chesterfield, MO (Nov. 2008).

[16] A. Igartua, X. Fernández, O. Areitioaurtena, R. Luther, C. Seyfert, J. Rausch, I. Illarramendi, M. Berg, H. Schultheiβ, B. Duffau, S. Plouseau and M. Woydt, Biolubricants and triboreactive materials for automotive applications, *Tribol. Intl.* **42** (2009) 561–8.

[17] L.A. Quinchia, M.A. Delgado, C. Valencia, J.M. Franco and C. Gallegos, Viscosity modification of high-oleic sunflower oil with polymeric additives for the design of new biolubricant formulations, *Environ. Sci. Technol.* **43** (2009) 2060–5.

[18] Th. Mang, Lubricants, in *Lipid Technologies and Applications* (F.D. Gunstone and F.B. Padley, eds), CRC Press, Boca Raton, FL (1997), pp. 737–58.

[19] A. Willing, Oleochemical esters, Environmentally compatible raw materials for oils and lubricants from renewable resources, *Fett/Lipid* **101** (1999) 192–8.

[20] F. Bongardt, Synthetic esters as multipurpose basestocks for lubricants, in *Jahrbuch für Praktiker*, Verlag für chemische Industrie H. Ziolkowsky GmbH, Augsburg (1996), pp. 348–61.

[21] A. Pavlovicová and J. Cvengroš, Lubricants based on vegetable oils, *Petrol. Coal* **41**(2) (1999) 99–102.

[22] D. Horner, Recent trends in environmental friendly lubricants, *Proc. 2nd Intl. Symp. Fuels & Lubricants* (S.P. Srivastava, ed.), New Delhi (10–12 March 2000), Vol. 2, pp. 753–66.

[23] H. Wagner, R. Luther and Th. Mang, Lubricant base fluids based on renewable raw materials. Their catalytic manufacture and modification, *Appl. Catal. A: General* **221** (2001) 429–42.

[24] S.Z. Erhan and J.M. Perez (eds), *Biobased Industrial Fluids and Lubricants*, AOCS Press, Champaign, IL (2003).

[25] S.Z. Erhan and A. Adhvaryu, Vegetable-based base stocks, in *Biobased Industrial Fluids and Lubricants* (S.Z. Erhan and J.M. Perez, eds), AOCS Press, Champaign, IL (2003), pp. 1–19.

[26] H.S. Hwang and S.Z. Erhan, Lubricant base stocks for modified soybean oil, in *Biobased Industrial Fluids and Lubricants* (S.Z. Erhan and J.M. Perez, eds), AOCS Press, Champaign, IL (2003), pp. 20–34.

[27] M. Pal and S. Singhal, Environmentally adapted lubricants, Part I. An overview, *Synth. Lubr.* **17**(2) (2000) 135–43.

[28] W. Klose, Significance of rapidly biodegradable lubricants in the German lubricant market, *Tribol. Schmierungstechn.* **50** (2003) 45–9.

[29] H. Murrenhoff and A. Remmelmann, Environmentally friendly oils, in *Fuels and Lubricants Handbook: Technology, Properties, Performance, and Testing* (G.E. Totten, S.R. Westbrook and R.J. Shah, eds), ASTM International, West Conshohocken, PA (2003), pp. 267–96.

[30] D.N. Bhomick and A.P. Pratap, Recent developments in biolubricants, *Chem. Ind. Digest* **18**(6) (2005) 53–8.

[31] A. Guo and Z. Petrović, Vegetable oils-based polyols, in *Industrial Uses of Vegetable Oils* (S.Z. Erhan, ed.), CRC Press, Boca Raton, FL (2005), Chapter 6.

2
Principles of lubrication

Abstract: Tribology is a highly interdisciplinary field. Tribosystems, composed of contacting surfaces with a lubricant in the interface, are present in all mechanisms in motion. Tribological variables comprise type of motion (sliding, rolling), applied load, sliding distance, speed, frequency, duration of applied stress, temperature, lubricating mode, nature of lubricant, etc. The primary objective of lubrication is reduction of friction, wear and heat between contacting surfaces in relative motion. Distinct lubrication regimes operate in different working conditions in various applications. The role of viscosity is stressed. Essential differences in interposition between metal surfaces of hydrocarbons and vegetable oils are briefly outlined.

Key words: tribology, friction, wear, lubrication, surface activity.

2.1 Introduction

Friction is the mechanical force which resists movement (dynamic or kinetic friction) or hinders movement (static friction) between two sliding or rolling surfaces in contact. These types of friction are also known as external friction. The causes of external friction are the microscopic contact points between the sliding surfaces. Depending on the application, friction may be desirable or undesirable. For example, tyre traction and braking rely on the beneficial effects of friction for their effectiveness. On the other hand, in applications such as operation of engines or equipment with bearings and gears, friction is undesirable because it causes wear, generates heat and frequently causes premature failure. The energy expended in overcoming friction is dispersed as heat and is essentially wasted. Waste heat is a major cause of excessive wear and premature failure of equipment. Friction occurs essentially in two different modes: sliding and rolling friction. The relationships between the frictional force and the load or weight of the sliding object differ for dry, or unlubricated, surfaces and lubricated surfaces. The coefficient of sliding friction is constant but depends on the type of material [1]. Adhesion occurs at the contacting points of surfaces and refers to the (cold) welding effect that occurs when two bodies are compressed against each other under pressure. A shearing force is required to separate cold-welded surfaces. Reduced wear and heat effects may be achieved by inserting a lower-viscosity (shear strength) material between wearing surfaces that have a relatively high coefficient of friction. Any material used to reduce friction in this way is a lubricant.

Much less force is required to roll an object than to slide or drag it. Unlike the coefficient of sliding friction, the coefficient of rolling friction varies with the rolling conditions. Rolling friction is very small compared with sliding friction. Housed ball or roller bearings are examples of practical applications which are subject to the laws of rolling friction. Roller bearings are the most important machine elements.

Wear is defined as progressive damage resulting in material loss due to contact between adjacent working parts. Although some wear may be expected during normal operation of equipment, excessive friction causes premature wear, which translates into significant economic loss due to equipment failure and downtime, and cost of spare parts. Friction and wear generate wasted energy that is not renewable. Therefore, wear can be held responsible for an overall loss in system efficiency.

The effects of wear are commonly detected by visual inspection of surfaces. Three types of surface damage are: (i) damage in the form of structural changes, plastic deformation or surface cracking, without exchange of material between the contacting parts; (ii) damage with loss of material (wear); and (iii) damage with gain of material (corrosion). Wear can result from abrasion, adhesion, fatigue or corrosion.

2.2 Purpose of lubrication

Tribology (Greek *tribos*, or rubbing) is the science and technology of interacting surfaces in relative motion and brings together all major disciplines involving friction, wear and lubrication. The term was originally coined by Jost in 1966 [2] who drew attention to its interdisciplinary nature and the large savings that proper application of tribology could produce. The aim of 'green tribology' is to save energy and materials and to enhance the quality of life and the environment.

A tribological system consists of four elements: the contacting partners, an interface with an enclosed medium (lubricant), and the environment. Tribosystems are present in all mobile mechanisms, from automotive, industrial and domestic to aerospace applications. In tribosystems, one material slides against another or against itself. For each combination of materials and contact configurations, the tribological variables include the type of movement, normal applied load, sliding distance, speed and frequency, duration of the applied stress, temperature (ambient and contact temperature), atmosphere, moisture, lubricating mode, etc. Tribometric parameters are friction, wear and temperature. Specific devices, called tribometers, are used to measure friction and wear. Tribology is a fast-growing discipline generating much interest in academia, industry and other research centres.

Because of its highly interdisciplinary nature, tribology requires efficient cooperation between mechanical engineers, materials scientists, metallurgists, chemists and chemical engineers, physicists, surface scientists, with contributions of specialists in environmental protection and toxicology.

The primary objective of lubrication is to reduce friction, wear and heat between contacting surfaces in relative motion. While wear and heat cannot be eliminated completely, both effects can be reduced to acceptable or even negligible levels. They can be minimised by introducing a material (the so-called lubricant) with low shear strength or coefficient of friction between the wearing surfaces. Lubricants are used to control friction and wear preventing direct contact between the surface asperities of the materials and lowering the contact temperature. Lubrication is meant to prevent scuffing, together with scoring, seizure and similar severe damage of friction surfaces [3]. Scuffing induces the heaviest damage in friction pairs, usually producing failure of the tribosystem. The lubricant must protect the component that it lubricates. In some cases, this protection is in the form of a fluid film that keeps opposing surfaces separated. In other cases, the lubricant provides wear protection by forming a chemical film on a surface. The lubricant transports protective chemicals to the sites where they are needed and transports waste products away from the sites where they are generated. Scuffing is caused by the breakdown of lubricant films on the rubbing surfaces. The rupture of films is due to mechanical and/or thermal loads, which in turn depend on the parameters of the tribosystem, such as sliding and rolling speeds, temperature and friction state. The geometry of the contact and the contact pressure change when one or both materials are worn during the sliding process. When wear debris is trapped at the contact zone, a third body is formed which can increase the surface damage by abrasion.

Tribosystems require adequate material selection, wear protection by special coatings and surface engineering, lubrication and lubricants, modern test equipment in laboratories and test fields. Scientific focus on lubrication technology and lubricants is relatively new. Lubrication is critical for wear protection of mechanical systems that should operate for extended periods of time. Lubrication may also be used for other purposes, namely to

- reduce oxidation and prevent rust;
- reduce energy consumption;
- provide insulation in transformer applications;
- transmit mechanical power in hydraulic fluid power applications; or
- seal against dust, dirt, and water (flushing out contaminants).

A major goal of current tribological research is to find effective lubricants for reactive light alloys such as magnesium, aluminium and titanium, which are of interest in the automotive industry in view of their lower densities and the ability to form corrosion-protective surface layers. Molecular inter-

Table 2.1 Benefits of lubrication

• Reduction of friction
• Wear and corrosion protection
• Shock cushioning
• Heat transfer
• Seal action
• Insulation
• Production efficiency
• Component lifetime (equipment durability)
• Energy economy

actions studied by nanotribology have impact on emerging uses in microelectromechanical systems as well as on traditional lubricant applications in the automotive and industrial sector [4]. Table 2.1 summarises the expected benefits of lubrication.

Commercial lubricants are mainly derived from petroleum, present environmental problems and are not suitable for many materials and conditions. Consequently, the need for new, effective, environmentally friendly lubricants is clear, especially as losses due to friction and wear caused by poor lubrication are estimated at about 1.5% of the gross domestic product (GDP) of industrialised countries [5].

2.3 Friction and lubrication conditions

For various applications in different operating conditions in tribological systems several forms of contact can exist between the moving partners, variously known as:

- dry friction or solid friction;
- fluid friction or full-film lubrication;
- hydrostatic lubrication;
- hydrodynamic or fluid-film lubrication;
- elastohydrodynamic (EHD) lubrication;
- mixed friction;
- boundary friction;
- extreme pressure (EP) lubrication.

It is important to differentiate between various lubrication regimes. Dry friction occurs when two solids have direct contact with each other without the presence of a separating layer. In these (lubricant-free) conditions, the coefficients of friction and wear rates are high. In the more common fluid friction or full-film lubrication, the surfaces are fully separated by a fluid lubricant film. This film is either formed hydrostatically or more commonly, hydrodynamically. Liquid or fluid friction is caused by the frictional resistance, because of the rheological properties of fluids. Compared with solid lubrication, fluid lubrication possesses a number of advantages, such as long-term endurance, low mechanical noise, promotion of thermal conductance and very low friction in the EHD regime. Gas lubrication refers to surfaces separated by a gas film.

The basic laws of friction, expressing that friction resistance is proportional to normal load P and independent on the contact area between solids, were first put forward by Leonardo da Vinci (1452–1519), later rediscovered by Amontons in 1699 [6] and verified by Coulomb in 1785 [7]. Coulomb made the distinction between static friction (force needed to start sliding) and kinetic friction (force required to maintain sliding), and expressed the friction coefficient μ as the ratio between the friction force and the normal load applied to the contacting bodies. In 1902 Stribeck introduced a diagram (Fig. 2.1) which differentiates friction in various lubrication regimes (boundary, mixed, EHD and hydrodynamic) as a function of the operating parameters, namely normal applied load P to the contact, sliding speed N and viscosity η of the lubricant [8]. The Hersey number (viscosity × sliding velocity/applied pressure) is proportional to the effective lubricant film thickness between rubbing surfaces [9]. Most recently, attempts are being made to develop a friction model to theoretically predict Stribeck curves as a function of operating conditions (applied load, slip, sum speed, surface topography, pressure and temperature distribution in the film) [10]. Friction is a very complicated matter.

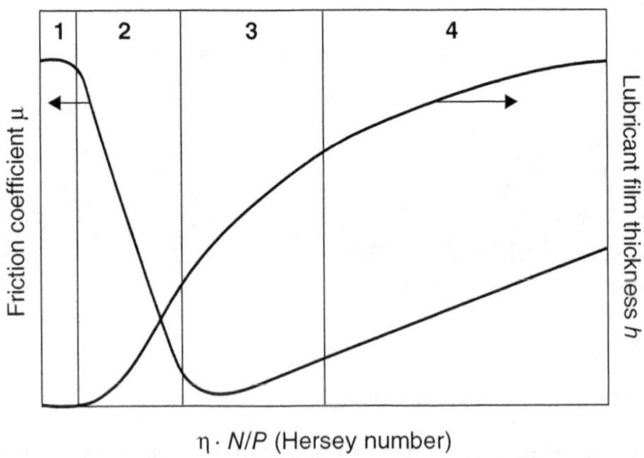

2.1 Schematic Stribeck's diagram (friction coefficient and film thickness) as a function of $\eta N/P$ (η, lubricant viscosity; N, relative sliding speed; P, normal applied load; μ, friction coefficient; h, lubricant film thickness; R, surface roughness): 1 boundary regime ($h \rightarrow 0$); 2 mixed film regime ($h \approx R$); 3 elastohydrodynamic lubrication ($h > R$); 4 hydrodynamic lubrication ($h >> R$). After ref. [11].

2.3.1 Lubricant film thickness

The lubricant film thickness (h) depends upon the constituent chemistry (viscosity of the formulated base oil) and operating (friction) conditions, in particular the applied load and sliding velocity, and should be compared with the surface roughness (R). At high speed, high viscosity or low load, sufficient hydrodynamic pressure is generated to fully separate the solid surfaces by a thin film of lubricant, i.e. hydrodynamic lubrication for $h >> R$ and EHD lubrication for $h > R$. In these conditions, friction is low and lifetime is high. As long as the lubricant film is sufficiently thick and resilient enough to prevent asperity–asperity contacts, the coefficient of friction tends to be very low. In this case, one talks about the film lubrication regime. In more severe operating conditions, such as higher contact pressures in combination with higher slip and lower viscosities, the surfaces may not fully be separated by a film and metal-to-metal contact will occur, i.e. mixed lubrication conditions ($h \approx R$) prevail. In this regime the friction is higher and system life will be shorter. At a sufficiently high load lubricant may be expelled from the friction zone ($h \rightarrow 0$), leaving the rubbing surfaces almost unlubricated (boundary conditions). In the boundary lubrication regime the average film thickness h is less than the elastically deformed surface roughness R and the surface asperities come into contact with each other under relative motion. In these conditions, the lubricant fluid viscosity has little effect on friction and wear. Physical collision of asperities causes deformation (elastic or plastic) and/or fracture, and produces high friction coefficients, severe wear and heat. Such collisions are generally accompanied by chemical interactions between surfaces and the lubricant molecules. Ionised vegetable oils function by postponing the onset of the boundary lubrication regime [12]; see also Sections 7.1 and 7.2.1.

2.3.2 Boundary lubrication

Boundary lubrication occurs when any of the essential factors that influence full-fluid film formation between contacting surfaces are lacking (e.g. when the fluid viscosity is insufficient to prevent surface contact). When thick, long-lasting lubricant films for separation of the two surfaces are technically impossible, a molecular layer of a substance whose specific properties can significantly influence the friction and wear characteristics is of great importance. One of the primary objectives of lubricant development is the creation of such boundary friction layers in a variety of geometric, dynamic and thermal conditions. Boundary lubricating layers are created from surface-active substances and their chemical reaction products. When a complete fluid film does not develop at the interface between potentially rubbing surfaces, the film thickness may be reduced to permit momentary dry contact between wear surface high points or asperities of the solid bodies. This condition is characteristic of boundary lubrication. Compared with hydrodynamic, EHD and mixed lubrication modes, friction and wear under boundary conditions are the most severe for the lubricant to control.

It is very important for a lubricant formulator to know in which regime a contact will operate. Boundary lubrication mostly occurs under high-load and low-speed conditions in bearings, gears, piston rings, pumps, transmissions, etc. It usually represents the critical regime that limits the life of components. In heavily loaded bearings, the fluid's viscosity alone is often not sufficient to maintain a film between the moving surfaces. In these cases higher fluid pressures are required to support the load until the fluid film is established. Hydrostatic lubrication refers to situations in which this pressure is supplied by an external source. If the pressure is generated internally, i.e. within the bearing by dynamic action, it is referred to as hydrodynamic or fluid-film lubrication. In hydrodynamic lubrication, the wearing surfaces are completely separated by a (thick) film of oil. As viscosity increases, the film thickness increases. Film thickness varies inversely with the load. Viscosity, sliding, velocity and operating temperature are also interrelated. At higher oil viscosity the operating temperature will increase, and this in turn, has the tendency to reduce viscosity. Velocity increases also cause temperature increases that subsequently reduce viscosity. Theoretically, hydrodynamic lubrication reduces wear to zero, but in practice, it reduces sliding friction and wear to acceptable levels. Essentially, fluid-film lubrication reduces friction between moving surfaces by substituting fluid friction for mechanical friction. In the case of liquid (hydrodynamic) lubrication, friction can be reduced by use of base oils with lower friction coefficient and lower viscosity. The principle of hydrodynamic lubrication is applied in thrust bearings used in the hydropower industry.

The term 'lubricity' refers to the slipperiness of lubricant films formed in boundary lubrication, a condition intermediate between unlubricated sliding and fluid-film lubrication, which is also defined as a condition in which the friction between the surfaces is determined by the properties of the surfaces and properties of the lubricant other than viscosity. Superlubricity is a special sliding regime of motion in which friction or resistance to sliding very nearly vanishes [13]. True zero friction is not thermodynamically permissible, but superlubricity gets close. Superlubricity may occur when two crystalline surfaces slide over each other in dry incommensurate contact (so-called structural lubricity) [14]. Energy can be conserved by reducing/removing friction in moving mechanical systems (both nano to micro-scale machines and large-scale engineering components). Superlubricity is interesting because it takes place in a dry environment rather than the wet environment normally associated with lubricants.

Hardy [15] has investigated the role of hydrocarbons on static friction (boundary regime), pointing out the role of molecular layers (boundary layers) adsorbed on the solid surfaces. Further research on boundary film physics and chemistry has resulted in the development of molecular theories of lubrication and in the design of lubricants, greases and lubricant additives with improved durability of mechanical devices by reducing both friction and wear. Zisman [16] published a systematic study of friction reduction action of monolayers for various molecules, in particular linear fatty acids. This work led to a monomolecular layer lubrication model. Vegetable oils are particularly effective as boundary lubricants since the high polarity of the base oil allows strong interactions with the lubricated surfaces. Boundary lubrication performance is affected by attraction of the lubricant molecules to the surface and also by possible reaction with the surface. The level of unsaturation of the fatty acids has only a minor effect on the lubrication performance. On the other hand, raising the concentrations of oxygen in the oil improves lubrication, denoting the importance of reactions on boundary lubrication performance [17]. The durability of boundary films composed of monomolecular fatty acids was also investigated [18]. The agglomerate observed during frictional contact under boundary lubrication conditions was composed of several substances resulting from the deterioration and recombination of fatty acid molecules with each other and the surrounding media. Upon friction the monolayers may be destroyed by bonds breaking or chemical reactions with the contacting surface leading to the built up of new interface materials with physical properties and thickness different from the initial monolayer, and finally progressive elimination from the sliding contact.

Boundary lubrication, which involves materials science, lubricant chemistry, fluid and contact mechanics, is a complicated subject of considerable industrial significance in the design of machinery, engines, gears, bearings, transmission, etc. Current understanding of the phenomena from a chemical perspective was reviewed recently by Hsu [19]. Freshly disturbed engineering surfaces produce a variety of reactive sites. The reactive surface and heat lead to tribological reactions ranging from oxidation of the metal surface to oxidation, degradation and polymerisation of the organic lubricant molecules, as well as organometallic chemistry. Tribochemistry can be defined as the chemical reactions that occur between the lubricant/environment and the surfaces under boundary lubrication conditions [20]. The surface is an integral part of the reacting system. The interface chemistry in the contact is complex and varied. Reactions at the surfaces depend on many variables including machining, defects, crystalline configurations, oxide layers, hydride layers, etc. There are oxidation and thermal reactions of lubricants, polymerisation to form high-molecular-weight (HMW) products, adsorption and corrosion reactions, and catalysis by metal surfaces. In addition, tribo-mechanical reactions, surface oxide reactions, mechano-chemical effects and phase transitions may take place. Organometallic compounds play an important role in the formation of boundary lubrication films [21]. However, the precise nature of the chemical reactions is still not well understood. This is partly due to the fact that the quantity of the reaction products involved is very small. It is speculated that charged particles and free electrons emanating from rubbing surfaces provide the energy for causing tribochemical reactions to take place. The intricate interactions between surfaces and molecules under rubbing conditions require further investigations, in particular as to the role of surface defects and dangling bonds [22]. Figure 2.2 is a schematical diagram of the formation of boundary films.

There appear to be several mechanisms by which boundary lubricating films can operate, namely sacrificial layer, low shear interlayer, friction modifying layer, shear resistant layer and load-bearing glasses. The operating mode of a sacrificial layer (e.g. oxide layer on steel surfaces) is that the tribological reaction products are easily renewable by rubbing [23]. Shear

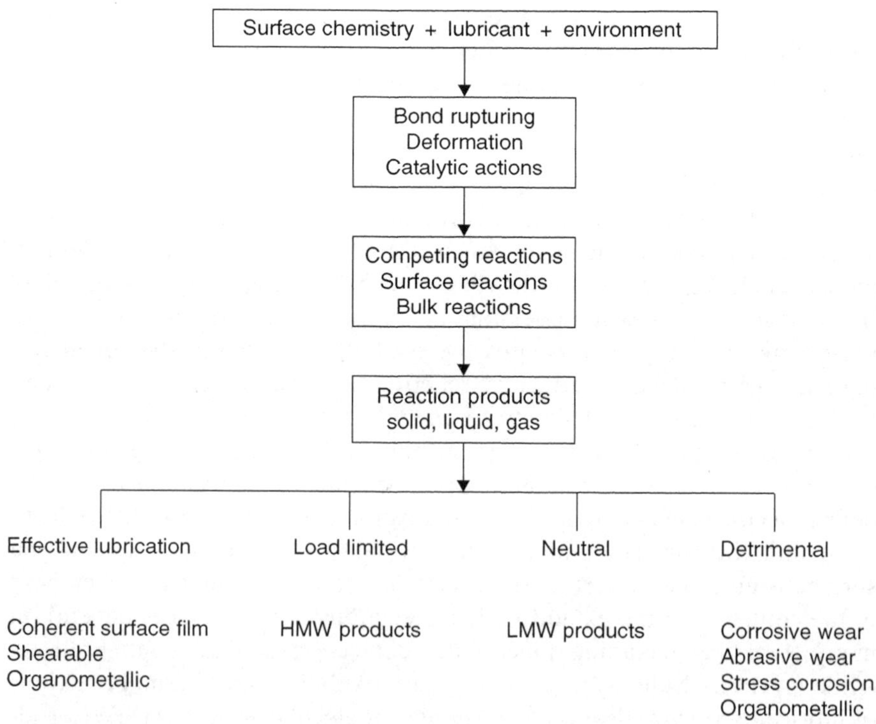

2.2 Boundary film formation. After ref. [19].

stresses remove the low shear interfacial layer rather than wear the surfaces. The low shear interlayer mechanism is typical of solid lubricants with weak interlattice forces between shear planes. Easy sliding of the lubricant molecules prevents the surfaces from contacting. In another mechanism (typical of friction modifiers such as fatty acids) reaction products form an ordered interface structure and sliding of the two surfaces is favoured by the two weakly bonded adsorbed layers. An alternative mechanism is to have a strongly adhered bonded layer which is shear resistant by itself.

The performance of lubricants under boundary conditions is affected by the chemical structure of base fluids and additives [24]. Lubricants required to operate under boundary conditions must possess the added quality of oiliness to lower the coefficient of friction of the oil between the rubbing surfaces. This oil enhancement property can be achieved by chemical additives known as anti-wear (AW) agents (Section 7.2.1). Most oils intended for use in heavier machine applications contain AW agents. Engine and compressor pistons rely exclusively on boundary lubrication. Bio greases containing additives to protect against EP are termed bio-EP greases (e.g. Bio Food Grade EP Grease-NLGI 2, United Bio Lube). EP lubrication is provided by a number of chemical compounds of boron, phosphorus, sulphur and chlorine. As the temperature increases, EP molecules become reactive and release derivatives of phosphorus, chlorine or sulphur to react with only the exposed metal surfaces to form a new compound (such as Fe choride or sulphide). The new compound forms a thin solid protective coating that fills the asperities on the exposed metal. AW agents in the EP oil continue to provide AW protection at sites where wear and temperature are insufficient to activate the EP agents. Bearings which normally operate with hydrodynamic or fluid-film lubrication commonly experience boundary lubricating conditions at starting up or switching off the equipment (mixed friction).

It is generally accepted that the tribological performance of an oil is closely related to film formation capability and the interaction strength with sliding surfaces at EHD and boundary lubrication. Most sliding pairs are in mixed EHD/boundary lubrication states. Therefore, complex processes may occur that include adhesion, deformation and ploughing. In the mixed friction model boundary friction combines with fluid friction. This form of friction requires sufficient load-bearing boundary layers to form. Under mixed friction conditions for roller bearings, formation of tribolayers is usually assured by using AW/EP lubricant formulations [25].

2.3.3 (Elasto)hydrodynamic lubrication

The lubrication principles applied to rolling bodies, such as ball or roller bearings, are typically described as EHD lubrication. Although lubrication of rolling objects operates on a considerably different principle from sliding objects, the principles of hydrodynamic lubrication can be applied, within limits, to explain lubrication of rolling elements. However, because the contact area is extremely small in a roller and ball bearing, the force per unit area, or load pressure, is extremely high. Roller bearing load pressure may reach 34 450 kPa and ball bearing load pressures even 690 000 kPa. Under these conditions, it would appear that the oil would be entirely squeezed from between the wearing surfaces. However, owing to viscosity increases under extremely high pressure, the oil is prevented from being entirely squeezed out. Consequently, a thin film of high-viscosity oil is maintained. Shear rates are very high (10^7 s^{-1}). The roughness of the wearing surfaces is an important consideration in EHD lubrication.

To minimise friction and wear, hydrodynamic lubrication is required. The efficiency of a lubricating oil at reducing friction and wear is greatly influenced by its viscosity. Clearly, for any given application, the optimum lubricant viscosity depends on the relative velocities of the surfaces, and on the load. It may seem logical to suppose that a thicker oil will give better lubrication than a thin one. However, provided an oil is within the correct viscosity range for the machinery to be lubricated, what matters is its performance rating. In some more extreme performance situations a thinner oil may perform better because it is 'wetter' and will tend to stay on the surface to be lubricated instead of flowing off and exposing the metal. A wide range of viscosities is required for different applications. Industrial lubricants in current use span a range of ISO grades [26], expressing the approximate viscosity in cSt at 40 °C from <10 to >1000. Car engine oils are usually classified according to the SAE grade system [27]; see also Section 3.3.

Surface activity is affected by several factors, including the interfacial properties of lube oil base stocks at oil/surface interfaces, lubricant solvency properties, additive interactions and variations in temperature. The activity of a surface also depends on the number of active sites on the surface. As loads, speeds and temperature increase, the number of active sites, the rate of diffusion of the lubricant molecules and their interaction with the surface increases. The chemistry of the molecules present in the lubricant formulation (base oil and additives) results in competition for these sites. The interposition of vegetable oil films is effective in preventing metal seizure under conditions of extreme pressure. Fatty oils are less easily displaced from metal surfaces by water than mineral oils, and hence are valuable ingredients for lubricants. Petroleum or mineral oils usually do not have the lubricity possessed by animal- and vegetable-type materials. The less chemically active petroleum oil (hydrocarbon) molecules adhere more loosely to the metal surfaces.

2.4 Sources of further information and advice

Th. Mang, K. Bobzin and Th. Bartels, *Industrial Tribology: Tribosystems, Wear and Surface Engineering, Lubrication,* Wiley-VCH, Weinheim (2010).

S.P. Srivastava, *Advances in Lubricant Additives and Tribology*, CRC Press, Boca Raton, FL (2010).

P.J. Blau, *Friction Science and Technology: From Concepts to Applications*, CRC Press, Boca Raton FL (2009).

Th. Mang and W. Dresel (eds), *Lubricants and Lubrication*, 2nd edn., Wiley-VCH, Weinheim (2007).

A. Erdemir and J.-M. Martin (eds), *Superlubricity*, Elsevier, Amsterdam (2007).

G.E. Totten (ed.), *Handbook of Lubrication and Tribology; Application and Maintenance*, 2nd edn, CRC Press, Boca Raton, FL (2006).

G.W. Stachowiak and A.W. Batchelor, *Engineering Tribology*, 3rd edn, Butterworth-Heinemann, Boston, MA (2005).

G.W. Stachowiak (ed.), *Wear – Materials, Mechanisms and Practice*, John Wiley & Sons, Ltd, Chichester (2005).

A.R. Lansdown, *Lubrication and Lubricant Selection: A Practical Guide*, 3rd edn, ASME Press, Champaign, IL (2004).

B.J. Hamrock, S.R. Schmid and B.O. Jacobson (eds), *Fundamentals of Fluid Film Lubrication*, 2nd edn, CRC Press, Boca Raton, FL (2004).

Z. Pawlak, *Tribochemistry of Lubricating Oils*, Elsevier Science, Ltd, Oxford (2003).

D.M. Pirro and A.A. Wessol (eds), *Lubrication Fundamentals*, 2nd edn, CRC Press, Boca Raton, FL (2001).

F.P. Bowden, *The Friction and Lubrication of Solids*, Oxford University Press, Oxford (2001).

M.J. Neale, *Lubrication: A Tribology Handbook*, Butterworth-Heinemann, Ltd, Oxford (2001).

D. Dowson, M. Priest, C.M. Taylor, P. Ehret, T.H.C. Childs, G. Dalmaz, Y. Berthier, L. Flamand, J.-M. Georges and A.A. Lubrecht (eds), *Lubrication at the Frontier: The Role of the Interface and Surface Layers in the Thin Film and Boundary Regime*, Elsevier, Amsterdam (1999).

D. Dowson, *History of Tribology*, 2nd edn, Professional Engineering Publishing, London (1998).

A.A. Seireg, *Friction and Lubrication in Mechanical Design*, Marcel Dekker, Inc., New York, NY (1998).

K.C. Ludema, *Friction, Wear, Lubrication: A Textbook in Tribology*, CRC Press, Boca Raton, FL (1996).

P.J. Blau, *Friction Science and Technology*, CRC Press, Boca Raton, FL (1995).

E. Rabinowicz, *Friction and Wear of Metals*, 2nd edn, Wiley-Interscience, New York, NY (1995).

A.R. Lansdown, *High Temperature Lubrication*, Mechanical Engineering Publications, Ann Arbor, MI (1994).

J.A. Williams, *Engineering Tribology*, Oxford University Press, New York, NY (1994).

E.R. Booser, *CRC Handbook of Lubrication and Tribology, Vol. III: Monitoring, Materials, Synthetic Lubricants, and Applications*, CRC Press, Boca Raton, FL (1993).

H. Czichos and K.-H. Habig, *Tribologie Handbuch*, Vieweg-Verlag, Wiesbaden (1992).

I.M. Hutchings, *Tribology: Friction and Wear of Engineering Materials*, Edward Arnold Publishing, London (1992).

J. Briant, J. Denis and G. Parc, *Rheological Properties of Lubricants*, Éditions Technip, Paris (1989).

E.S. Nachtman and S. Kalpakjian, *Lubricants and Lubrication in Metalworking Operations*, CRC Press, Boca Raton, FL (1985).

M.H. Jones and D. Scott (eds), *Industrial Tribology*, Elsevier, Amsterdam (1983).

J.A. Schey, *Tribology in Metalworking: Friction, Lubrication and Wear*, American Society of Metals, Metals Park, OH (1983).

A.Z. Szeri, *Tribology: Friction, Lubrication and Wear*, McGraw-Hill, New York, NY (1980).

A.S. Akhmatov, *Molecular Physics of Boundary Friction*, IPST, Jerusalem (1966).

British Petroleum Co., *Industrial Lubrication* (n.d.).

2.5 References

[1] E. Oberg and F.D. Jones, *Machinery's Handbook*, 28th edn, Industrial Press, Inc., New York, NY (2008).

[2] P.H. Jost, *Lubrication (Tribology) – A Report on the Present Position and Industry's Needs*, UK Dept. of Education and Science, HMSO (9 March 1966).

[3] M. Kozma, Investigation into the scuffing load capacity of environmentally-friendly lubricating oils, *J. Synth. Lubr.* **14**(3) (1997) 249–58.

[4] N. Canter, Tribology: the science of thinking small, *Tribol. Lubr. Technol.* **60**(6) (2004) 43–9.

[5] G.W. Stachowiak and A.W. Batchelor, *Engineering Tribology*, 3rd edn, Butterworth-Heinemann, Boston, MA (2005).

[6] G. Amontons, *Mém. Math. Phys. Acad. Roy. Sci.* (1699), p. 206.

[7] C.-A. de Coulomb, Theorie des machines simples, *Mém. Math. Phys. Acad. Roy. Sci.* (1785), p.161.

[8] R. Stribeck, Die wesentlichen Eigenschaften der Gleit- und Rollenlager, *Z. Verein. Deutsch. Ing.* **46**(38) (1902) 1341–8.

[9] M.D. Hersey, *Theory and Research in Lubrication*, John Wiley & Sons, Inc., New York, NY (1966), p. 137.
[10] Q. Liu, *Friction in Mixed and Elastohydrodynamic Lubricated Contacts Including Thermal Effects*, PhD Thesis, Twente University of Technology (2002).
[11] H. Czichos and K.-H. Habig, *Tribologie Handbuch*, Vieweg-Verlag, Wiesbaden (1992).
[12] M. Roegiers, H. Zhang and B. Zhmud, Elektrionized™ vegetable oils as lubricity and oiliness components in metalworking lubricants, *Proc. 6th Intl. Conf. on Tribology – Balkantrib'08,* Sozopol (12–14 June 2008), Paper BT-97; *FME Transactions* **36** (2008) 133–8.
[13] A. Erdemir and J.-M. Martin (eds), *Superlubricity*, Elsevier, Amsterdam (2007).
[14] M. Dienwiebel, G.S. Verhoeven, N. Pradeep, J.W.M. Frenken, J.A. Heimberg and H.W. Zandbergen, Superlubricity of graphite, *Phys. Rev. Lett.* **92**(12) (2004) 126101.
[15] W.B. Hardy, *Collected Works*, University Press, Cambridge (1936).
[16] W.A. Zisman, Durability and wettability properties of monomolecular films on solids, in *Friction and Wear* (R. Davies, ed.), Elsevier, Amsterdam (1959), pp. 110–48.
[17] T. Murakami and H. Sakamoto, Lubricating properties of vegetable oils and paraffinic oils with unsaturated fatty acids under high-contact-pressure conditions in four-ball tests, *J. Synth. Lubr.* **20**(3) (2003) 183.
[18] J.M. Georges, G. Meille, M. Jaquet, B. Lamy, T. Mathia, M. Cantarel and M. Pochard, A study of durability of boundary films, *Wear* **42**(2) (1977) 217–28.
[19] S.M. Hsu, Boundary lubrication: current understanding, *Tribol. Lett.* **3**(1) (1997) 1–11.
[20] S.M. Hsu, J. Zhang and Z. Yin, The nature and origin of tribochemistry, *Tribol. Lett.* **13**(2) (2002) 131–9.
[21] E.E. Klaus and E.J. Tewksbury, Microcorrosion studies with frictional fluids, *Lubric. Engng.* **29**(5) (1973) 205.
[22] P.M. Lenahan and S.E. Curry, First observation of the ^{29}Si hyperfine spectra of silicon dangling bond centres in silicon nitride, *Appl. Phys. Lett.* **56**(2) (1990) 157–9.
[23] T.F.J. Quinn, J.L. Sullivan and D.M. Rowson, Origins and development of oxidational wear at low ambient temperatures, *Wear* **94**(2) (1984) 175–91.
[24] W. Castro, D.E. Weller, K. Cheenkachorn and J.M. Perez, The effect of chemical structure of basefluids on antiwear effectiveness of additives, *Tribol. Intl.* **38** (2005) 32–6.
[25] R. Karbacher, Mixed film lubrication of roller bearings, *15th Intl. Colloquium Tribology, Automotive and Industrial Lubrication*, Esslingen (17–19 January 2006).
[26] ISO Standard 3448, *Industrial Liquid Lubricants, ISO Viscosity Classification*, International Organization for Standardization, Geneva (1992).
[27] *SAE J300, SAE Viscosity Grades for Engine Oils*, Society of Automotive Engineers, Warrendale, PA (2001).

3
Lubricants: properties and characteristics

Abstract: Being engineering fluids, lubricants play an essential role in industry and transportation, representing a considerable economic value as well as an ecological threat. Lubricant base stocks are highly diversified, including mineral, synthetic and bio-based oils. Fluid (oils and greases) and solid lubricants have been classified in many different ways using a variety of discriminating identifiers (base oil type, viscosity grade, application, etc.). There exists no general consensus in relation to the terminology 'bio' (in bio-based, biodegradable, biolubricant). Similarly, the lubricant industry uses a variety of subjective denominations for the concept of 'environmental lubricant'. General properties of lubricants are outlined.

Key words: lubricant base stocks, lubricant classifications, eco designations, environmentally acceptable lubricants, lubricant properties.

3.1 Introduction

A lubricant is a substance used to facilitate relative motion of solid bodies by minimising friction and wear between interacting surfaces. Lubricants can be considered as constructional elements or engineering fluids which have to be kept on the newest technological level because of the continuous development of tribological systems.

As a product group, lubricants play an essential role in industry and transportation: without lubrication, there is no transfer of energy. Reduction of friction and wear by use of lubricants enables machinery to work without premature failure, results in energy savings and controls emissions. Related functional fluids are used in applications where their primary purpose is not just friction and wear reduction, although some degree of lubricity may be required. Examples of such applications are hydraulic fluids, electrical transformer fluids, heat transfer fluids and metalworking coolants. Apart from high-volume applications such as engine, gear and hydraulic fluids, there are thousands of other specifically tailored lubricant or grease formulations that differ greatly in composition, both chemically and physically.

Lubricants represent considerable economic and ecological importance. Recently, in this context, saving resources and energy and reducing emissions have become essential matters. Lubricants have a particularly important contribution to make in relation to energy conservation, minimisation of waste and development of durable products. Correct lubricant selection can lead to improvements in energy efficiency of up to 10%. Minimising wear by efficient lubrication prolongs the useful life of machines, thereby decreasing the consumption of non-renewable resources. Industrial production processes can reach higher levels of efficiency and products can be made more attractive through the added value offered by lubricants. Truly green lubricants are those that optimise energy efficiency and minimise wear in the machinery they lubricate and have maximised service lifetimes in order to reduce the amount of lubricant required [1]. Adoption of high-performance lubricant technologies has led to a significantly lower lubricant consumption.

Modern high-performing commercial lubricants are usually complex materials composed of a lubricant base stock formulated with an additive package for specific property enhancement of the resulting full lubricant formulation. Traditional lubricant systems are highly diverse, ranging mainly from common lube oils (non-aqueous liquids) to oil-in-water emulsions (e.g. used in water-miscible cutting fluids), water-in-oil emulsions (as in metal-forming), oil-in-oil emulsions (applied in metalworking), water-based solutions (applied in chip-forming metalworking operations), greases and pastes, and solid lubricants. Water is a truly green lubricant or lubricant component. Water-based lubricants possess distinctive advantages over oil-based lubricants (such as environmental compatibility, biocompatibility, availability, cost effectiveness). Their environmental compatibility makes them suitable for a number of industrial applications, such as food processing or textile and pharmaceutical manufacturing, where the use of oil-based lubricants can be problematic due to contamination issues. For biomedical applications (such as lubricious coatings for catheters) aqueous lubrication is particularly important, since water is virtually the only acceptable base lubricant.

Lubricants can be classified in various ways using either a physical basis (aggregation state) or chemical criteria (chemical class or origin) (see Section 3.3). Accordingly, lubricants may be grouped as liquids (oils), semi-solids (greases), solids (dry lubricants) and gas (compressed air or other gases). The former two can be produced from mineral oil, plant oil, synthetic oil or re-refined oil. Mineral oil fractions are the most common lubricating oils utilised due to their availability, cost and compatibility with many engineering materials. It is less common to consider gases as lubricants. However, compressed air that actuates a pneumatic controlled device will separate the moving surfaces, reduce friction and permit easy movement, which are all characteristics typical of a lubricant. Air is used as a lubricant in the case of a dentist's drill. Graphite lubricates by the release of intercalated gases.

A further distinction between lubricant types is frequently based on the origin of the materials used, namely renewable and non-renewable. This broadly corresponds to oleochemical and petrochemical origins, respectively. Mineral-based oils are non-renewable, whereas vegetable oils are renewable. Lubricant formulations are also denoted as bio-based, biodegradable, or environmentally acceptable to express a certain degree of eco-sustainability (see Sections 3.4 and 3.5).

Among the more important properties for use as lubricants are viscosity, viscosity–temperature behaviour, lubricity, solvent power and oxidation stability (see Section 3.6). Low-temperature fluidity and behaviour and evaporation characteristics are equally of interest as well as compatibility with technical materials such as elastomers. The evaporation characteristics of lubricants are significant quality criteria. This stands in relation to undesired emissions as the lubricant evaporates and the concomitant change in composition. The evaporation behaviour is increasingly important for modern lower-viscosity, energy-saving oils.

3.2 Lubricant base stocks

Fatty acid-based products from vegetable and animal origins used to be the main source of lubricants up to and beyond the late 19th century when mineral oils began to become available in large volumes at increasingly competitive prices. For a time, blends of mineral oil and natural fatty acid products, referred to as fatty oils, were widely used. Since the 1950s and until recently the trend has been towards lubricants and functional fluids with solely petrochemical bases and containing additives from petrochemical or synthetic sources. With the overwhelming quantity of mineral lubricant base oils, lubricants are generally viewed as an integral part of the petroleum industry. Now industry and end-users are gradually turning again to more environmentally friendly products. Some recent high-performance lubricants no longer contain petroleum base oils. Increasing demands on lubricant performance and eco-compatibility have led to the development of more pure oils, such as hydrocracked products, poly-α-olefins and esters. Approximately 7–10% of base oils are now synthetic products (including hydrocracked oils). Also natural fatty oils are experiencing a come-back as their oleochemical derivatives because of rapid biodegradability. The recent use of renewable bio-resources for chemical product manufacturing is well documented [2–14].

Lubricants are broadly distinguishable on the basis of their origin: mineral, synthetic and fixed oils. Various lube base oils are being used such as virgin mineral oil, white oil, severely hydrotreated oil, re-refined mineral oil, poly-α-olefins (PAOs), synthetic esters, dibasic acid esters, polyol esters (POEs), alkylated aromatics, polyalkylene glycols (PAGs), phosphate esters, silicones, vegetable oils, animal fats or mixtures of some of these [15]. Today, few lubricant formulations contain only one base oil (see Section 7.1). Specialties such as silicone oils have been successfully used in aerospace due to their excellent fluidity in a wide temperature range and high viscosity index (VI, resistance of a lubricant to viscosity change with temperature), but are limited by their low extreme pressure capacity and easy migration properties.

3.2.1 Mineral oils

More than 90% of the lubricant market is dominated by mineral and related oils, which are heavily contaminating the environment but have a wide availability and low price. Development of lubricating oils (from crude oil to base stocks and full formulations) is a sophisticated technology. The process of converting crude oil into a finished mineral base oil is called refining (by distillation, de-asphalting, acid refining, solvent extraction, solvent dewaxing, etc.). Base oils obtained by processing of crude petroleum are classified as refined and residual compounded. Bright stocks are highly viscous base oils (viscosity > 45 cSt at 100 °C).

Mineral oils are complex mixtures of thousands of different compounds with a wide molecular weight distribution, but mainly C_{20}–C_{50} hydrocarbons including paraffins (linear alkanes or waxes, and branched alkanes), alicyclic and olefinic species, aromatics and polycyclic aromatics (PCAs) or polycyclic aromatic hydrocarbons (PAHs). The latter are environmentally harmful. This has limited the application of naphthenic base oils to less than 10% of total mineral oils. Non-renewable petroleum base oils, primarily composed of hydrocarbons (paraffinic, naphthenic and aromatic structures), lack chemical functionality. Mineral oils have low flash points, low viscosity indexes and high boundary friction coefficients compared with natural oils. The viscosity of the finished base stock is determined by the carbon chain length and processing temperature. Most refiners settle for three or four stocks from which a range of finished oils is prepared by blending.

Long-chain, high melting point paraffins are undesirable base oil components which negatively affect the cold-flow properties of lube oil distillates and lead to a high pour point. Dewaxing by crystallisation of paraffins at low temperatures and separation by filtration are the principal processes in traditional refining. Solvent refining is the separation of unwanted components from vacuum distillates. Paraffinic-rich oils are the most widely used base stock for lubricating oils because of their relatively good properties. They are preferentially used as engine oils, transmission fluids and gear oils. By convention, oils are labelled as naphthenic when the paraffinic carbon content is less than 55–60%. Naphthenic oils, with a high proportion of cycloalkanes, have different solvent properties from paraffinic oils and other highly desirable properties (see Table 3.1). Naphthenic oils are generally reserved for applications with narrow temperature ranges where a low

Table 3.1 General characteristics of mineral base oils

Paraffinic	Naphthenic	Aromatic
• Excellent oxidation stability	• Moderately good ageing behaviour	• Poor stability
• Higher PP	• Good PP (absence of wax)	• Varied PP
• Higher VI	• Lower VI	• Lowest VI
• Low volatility (high flash point)	• High volatility (lower flash point)	• Varied volatility
• Low specific gravities	• Higher specific gravities	• Highest specific gravities
• Poor solvency	• Outstanding solubility	• Excellent solvency

pour point (PP) is required, such as low-temperature hydraulic oils, refrigeration oils, rubber process oils, metalworking oils (coolants) and greases. They are also used when a not particularly high VI is required, e.g. in electrical transformer oils. The degradation products of naphthenic oils are soluble in the oils, leading to fewer problems with formation of sludges and deposits.

A two-stage hydrotreating process (hydrogenation and hydrocracking) in the manufacture of lubricant base oils significantly influences the chemical structures of mineral oils by giving rise to cracking to smaller molecules, saturation of (poly)aromatics, ring opening of naphthenics and isomerisation. The principal characteristics of the lube hydrocracking process are cracking of low-VI components and saturation of aromatics. The resulting products are high-VI and low-evaporation oils. Hydrocracked base oils (VI ~100) differ from solvent-extracted oils by their extremely low aromatic content and their chemical purity (only traces of hetero-atoms). By increasing the hydrotreating severity VI can be increased from about 100 to 125 to create very high viscosity index (VHVI) or unconventional base oils (UCBOs) [16]. VHVI mineral oil-based fluids have properties close to those of synthetic oils and are used in semi-synthetic automotive crankcase oils. Recently, changes in the dynamics of the base oil industry and the reduced availability of bright stock has created a shortage of high-viscosity base fluids.

New GTL (gas-to-liquid) base oils have premium characteristics, including ultra-high VIs (VI > 140), essentially no sulphur and nitrogen, very low evaporative losses and negligible aromatic content [17].

White oils are highly refined, odourless, chemically inert, low-unsaturated mineral oil raffinates preferably also without traces of polycyclic aromatics, which are produced by high-pressure hydrogenation and used for pharmaceutical, medical or food-processing applications, including citrus crop protection. The nearly colourless white oils are non-toxic and show poor biodegradability.

Petroleum-based lubricating oils, greases and hydraulic fluids are widely used in industry. However, these products suffer from a number of drawbacks, including biodegradability, ecotoxicity and dwindling reserves. Facilities using mineral oil-based lubricants such as hydropower plants, flood-control pumping plants and lock-and-dam sites are potentially highly polluting. Table 3.1 summarises the main general characteristics of mineral base oils.

3.2.2 Synthetic formulations

Even with additive chemistry the performance requirements of some applications cannot be met by mineral oil formulations. This has led to the development of a variety of synthetic base fluids [18]. Synthetic lubricants are composed of oil components that have been manufactured synthetically by reaction of a few well-defined chemical compounds (although often petroleum-based) rather than refined from existing petroleum crudes or vegetable oils. Synthetic processes enable matching precise oil properties. This base oil group is roughly equivalent to API Group IV. Note that the term 'synthetic' has led to heated legal arguments. Because the compositions of synthetic oils are much more well defined, synthetic oils are generally superior to refined oils in most circumstances, in particular in terms of:

- better oxidation stability;
- better viscosity index;
- much lower PP (as low as –78 °C);
- lower coefficient of friction.

The advantages offered by synthetic oils are most notable at very low and very high temperatures (see Table 3.5). The main disadvantage of synthetics is that they are inherently more expensive than mineral oils.

Many synthetic base fluids have been described, but the most commonly used in lubricant formulations are PAOs and synthetic (organic) esters (SEs) of different types (diesters and polyol esters). PAO/mineral oil blends are also widely used commercial semi-synthetic lubricants. Other synthetic fluids used in lubricant applications are PAGs, dialkylated benzenes, polyisobutylene (PIB), poly(internal olefins) (PIOs), and phosphate esters. Cost limits their use. PAOs, PIOs and PAGs are derived from petrochemicals, whereas SEs (monoesters, diesters, phthalate esters, polyol esters and complex esters) may also (partly) be of oleochemical origin. Alkylated aromatics have been described in detail elsewhere [19, 20]. The properties of synthetic alkylbenzenes vary widely and are difficult to generalise.

There are many different types of PAO oligomers, from PAO2 to PAO100, where the number denotes the viscosity (in cSt) at 100 °C. PAOs are made by polymerisation of linear alpha olefins (LAOs), such as 1-decene, but also of mixed C_8-C_{12} α-olefin fractions [22]. Table 3.2 lists the main characteristics of PAOs. PAOs are not generally produced in very high viscosity ranges. They are used as base fluids in many high-performance lubricants, synthetic motor oils, low-temperature hydraulic fluids, etc. Their low polarity can lead to seal shrinkage and loss of elasticity. PAOs were recently reviewed by Rudnick [23]. ExxonMobil's SpectraSyn Elite™ mPAO (metallocene poly-α-olefin) base stock has improved properties, notably higher VI for low- and high-temperature performance, improved shear stability for durability and lower PP for increased flow in cold environments. PIOs are fairly similar to PAOs and are made from cracked paraffinic base stocks. VIs are 10–20 units lower than the VIs of equiviscous PAOs.

Since their commercialisation over 50 years ago, PAG lubricants are generally considered as niche synthetic lubricants that are able to solve problems that mineral oil cannot [24]. Conventional PAGs, derived from (co)polymerisation of ethylene oxide (EO) and/or propylene oxide (PO), provide many performance benefits such as excellent friction control, good low-temperature properties, high viscosity indexes, and excellent deposit control. Polyglycols are used in applications such as gears, bearings and compressors for hydrocarbon gases. In some applications controlling deposits is a critical concern [25]; PAGs have been used extensively as rotary screw compressor lubricants and may be applied as engine oils.

Copolymers of EO and PO with a broad range of molecular weights and viscosities (up to 50 000 cSt at 40 °C) are the only major water-soluble lubricant base oil available today. This unique feature provides functional advantages for fire-resistant hydraulic fluids, metalworking fluids, quen-

Table 3.2 Main characteristics of PAOs

- Uniform distribution of molecular weight
- High VI (>130)
- Low volatility (narrow boiling range)
- High thermal stability
- Very low pour points
- Low polarity (high non-polarity index, NPI)
- Poor solvency for very polar additives
- Compatibility with mineral oils
- Slight compatibility with seals, elastomers
- Limited biodegradability[a]
- Use in demanding conditions (wind turbine, spacecraft)
- Relatively costly

[a] For low-molecular-weight PAOs only, e.g. PAO2.
After ref. [21].

chants, textile lubricants, etc. One known disadvantage of PAGs is their poor hydrocarbon oil (API Groups I–III) miscibility. This limits their use as co-base fluids (Section 7.1), in contrast to other synthetic lubricant chemistries such as PAOs, PIBs and esters which are used to form semi-synthetic lubricants that can meet higher performance specifications. PIBs are available in very high viscosity ranges and may substitute Group I bright stocks. Recently, Dow Europe has developed a new range of oil-soluble PAGs (OSPs) [26]. The design of OSPs can be accomplished by synthesising downstream derivatives of butylene oxide (BO) to form polybutoxylate homopolymers or by synthesising copolymers of PO and BO (Fig. 3.1). Butylene oxide has been largely unexplored as a building block of lubricant base oils. OSP base oils have higher VIs than mineral oils, excellent low-temperature properties with PPs as low as –57 °C and very low aniline points (< –30 °C). Synthetic OSPs provide opportunities to formulators to upgrade hydrocarbon oils and solve current lubrication challenges by leveraging the inherent benefits of PAGs. In particular, OSPs can help in boosting the lubricant viscosity and provide additional functionality. OSPs can be used as primary base oil, co-base oil and as a performance-enhancing additive.

A synthetic ester consists of a fatty acid and an alcohol moiety. In lubricant applications, esters normally contain two or more carboxylate groups. A large number of carboxylate groups in the ester molecule improves the thermal properties, and thereby a thicker lubricating film can be maintained in highly loaded, high-slip contacts [21]. SEs form a broad range of base fluids with properties varying greatly depending on the chemical structure. In principle, ester properties can be tailor-made to fit certain applications. This requires understanding of structure–property relationships. For a lubricant, the molecular structure is closely related to its properties, as shown for some selected ester structures (Table 3.3). Bis (*n*-octyl) adipate and bis (2-ethylhexyl) adipate just differ in branching only. Branching is beneficial for low-temperature properties, but also decreases VI. The increased fatty acid chain length of TMP oleate as compared to TMP C_8–C_{10} increases viscosity and VI, but PP is unaffected because the presence of double bonds

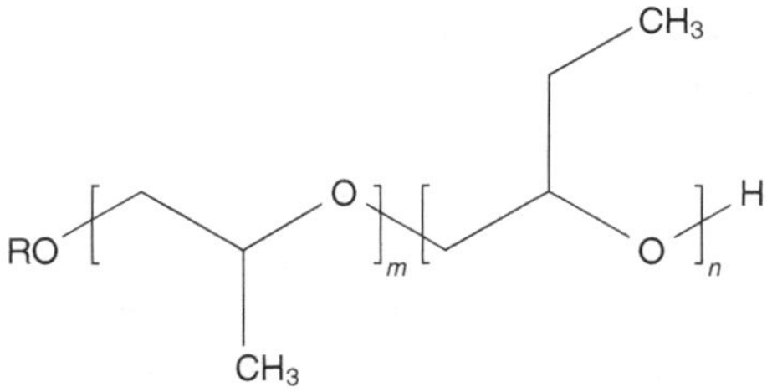

3.1 Structure of oil-soluble PAG. For BO homopolymers *m* = 0; for copolymers PO and BO are randomly distributed.

Table 3.3 Selected properties for synthetic ester structures

Synthetic ester	VI	Pour point (°C)	Viscosity @ 40 °C (cSt)
Bis(*n*-octyl)adipate	218	+4	8.29
Bis(2-ethylhexyl) adipate	123	−78	7.7
TMP C8-C10[a]	140	−53	20.0
TMP Oleate[a]	181	−54	43.8
NPG C16-C18[b]	158	−24	n.g.
NPG Iso C18[b]	148	−42	n.g.

[a] TMP, trimethylolpropane.
[b] NPG, neopentyl glycol.
n.g., not given.
After ref. [21].

in the carbon chain prevents crystallisation. The PP for the straight-chained neopentyl glycol (NPG) C_{16}–C_{18} is lowered from −24 °C to −42 °C for the branched chained NPG iso-C_{18} [21]. For structure-property relationships of POEs, see also Sections 10.5.3 and 12.7.1.

As SEs often have a high VI (see Table 3.3) there is no need for VI improvers in the formulation, which is particularly beneficial in highly loaded high-shear contacts. Synthetic esters are of great interest for formulation of environmentally adapted lubricants (EALs) (Section 3.5). SEs are non-toxic and generally show excellent biodegradability. High-viscosity ester base fluids (viscosity < 1100 cSt at 40 °C) with only readily biodegradable properties can be obtained by polymerisation of complex esters. High-performance ester base fluids were originally designed for lubrication of aircraft jet engines, but have found more general use in applications where biodegradability is required. Section 10.5.3 describes high-stability esters for lubricant applications. High-performance esters from natural oils as environmentally acceptable lubricants were reviewed by Kodali [27]. One of the major advantages of bio-based SEs is better performance at a lower cost compared with SEs.

Table 3.4 lists typical applications of various (not necessarily interchangeable) synthetic oil categories. Synthesised hydrocarbons are used in engine and turbine oils, hydraulic fluids, gear and bearing oils, and compressor oils. Esters are excellent substitutes for mineral oils [18]. Even before ecological aspects became a consideration in lubricant development, ester oils were used in specialty lubricants for technical reasons, e.g. as base fluids for aviation turbine oils and components for fuel economy oils, crankcase and compressor oils, and other applications of industrial oils and greases. Applications of phosphate esters are found where fire resistance is needed.

Synthetic oils, derived by synthetic chemical processes, have been developed to obtain intrinsic qualities such as lubricity and thermal stability.

Table 3.4 Applications of synthetic lubricant categories

Category	Applications
Synthesised hydrocarbons[a]	Engine and turbine oils (incl. transformer oils) Hydraulic fluids Gear and bearing oils Compressor oils (incl. refrigerator oils)
Organic esters[b]	Crankcase oils Compressor lubricants
Phosphate esters	Fire-resistance applications
Polyglycols	Gears Bearings Compressors for hydrocarbon gases
Silicones	Fire-resistance applications Industrial and military installations

[a] E.g. PAOs and dialkylated benzenes.
[b] E.g. dibasic acid and polyol esters.

Table 3.5 Characteristics of synthetic and vegetable-based base oils

Synthetic	Vegetable-based
• Better oxidation stability or resistance • Better VI • Much lower PP (as low as –78 °C) • Lower CoF	• Good boundary lubrication • General wear protection • High VI • High flash point • Low volatility

After ref. [29].

Synthetic products usually offer enhanced performance in comparison to vegetable or petroleum products (Table 3.5). Synthetic oil is used as a substitute for lubricant refined from petroleum when operating in extremes of temperatures. Some synthetic oils are readily biodegradable and non-toxic and can withstand operating temperatures above 200 °C. They are frequently designed for use in extreme conditions such as high temperature, vacuum, radiation or chemical environments. Aircraft turbines require the use of synthetic oils, whereas aircraft pistons engines do not. Synthetic lubricants are relatively costly (which is often compensated by increased lifetime), can be toxic and not readily biodegradable. Because of their cost (approximately three times higher than mineral-based refined oils), synthetic oils should not be considered for high leakage or total-loss applications. In 2004 about 7% of base oils were synthetic products (including severe hydrocracked oils), with a forecast growth to 10% by 2015 [28].

3.2.3 Bio-based oils

Fixed oils are fatty substances derived from animals, plants and fish. They are called fixed oils since they will not volatilise without decomposing. For example, some grease lubricants are made from animal fats. This subclassification not only identifies the lubricant source but also defines some of its characteristics and ability to perform its task. Vegetable oils provide another large source of lubricants. Fixed oils are generally composed of fatty acids and alcohols, the radicals of which are joined to form fatty acid esters.

Table 3.5 shows the main characteristics of vegetable-based base oils (mainly C_8 to C_{24}) compared with synthetic base oils. As to their tribological properties it appears that oils with longer carbon chains show lower friction and lower wear than shorter ones; saturated oils exhibit improved oxidative stability in high temperature/pressure applications in comparison with more unsaturated ones [30]. Vegetable oils are particularly effective as boundary lubricants (see Section 2.3). The fatty acid profile and the concentration of dissolved oxygen control the lubricating performance [31]. Free fatty acids with a high degree of polarity, which react readily with metal surfaces to form a protective layer, exhibit better tribological properties (improved boundary lubrication, lower friction and anti-wear (AW) abilities) compared to alcohols, methyl esters and less polar triacylglycerol (TAG) esters (see Section 10.5.5). Free fatty acids and/or sulphurised fatty acids can therefore be utilised as additives to improve the lubricating properties of hydrocarbon and vegetable-based oils [32, 33]. For the impact of vegetable oil oxidation compounds on lubrication, see Table 14.6. As at least 30% of lubricants used in the industry end up in the ecosystem, it is quite understandable that various countries are attracted by biolubricant alternatives. Although vegetable oils have a number of inherent qualities that give them advantages over petroleum oils, there are also several limitations to be overcome for their practical use at a wide scale. The similarity in all vegetable oil structures means that only a narrow range of viscosities is available for their potential use as lubricants. The strong intermolecular interactions whilst providing a durable lubricant film also result in poor low-temperature properties.

High-performance lubricating oils are often based on SEs (such as wax esters), sometimes with the fatty acid part from plant sources. Compounded oils, typically containing 3–10% fatty acid or synthetic oils, may be used to reduce the coefficient of friction in cases where extreme sliding friction occurs (e.g. in worm gear systems). Chevron-Phillips' SynFluid is an example of a renewable compounded oil composed of a high-oleic base stock (HOBS) and an ultimate, or readily biodegradable synthetic ester.

For a comparison of synthetic, mineral oil and bio-based lubricant fluids, see ref. [34]. Lubricant base fluids based on renewable raw materials have been reviewed in the past (for references, see Section 1.2).

3.3 Classifications for oils and lubricating greases

Various professional organisations and societies have established classifications for fluids and greases. The more widely encountered systems are as follows:

- API American Petroleum Institute (API Standard, 2002) [35]
- ISO International Organization for Standardization (ISO Standard, 1992) [36]
- NLGI National Lubricating Grease Institute (NLGI, 2010) [37]
- SAE Society of Automotive Engineers (SAE, 2010) [38]
- AGMA American Gear Manufacturers Association (AGMA 9005-D94) [39].

Standard oil classifications are based on base oil type, viscosity grade (VG), usage or additives. Other discriminating identifiers are producer brand name, product data and specifications [29]. A base fluid for use in engine oils is classified by the API [35], but a fluid for formulating a hydraulic fluid is only classified by the ISO VG system [36]. Table 3.6 shows the API/Association Technique de l'Industrie Européenne des Lubrifiants (ATIEL) classification of base oils according to their chemical composition. Even within the same group, base stocks may differ widely in composition and properties.

Groups I and II are universally referred to as mineral oils, whereas Group III is often considered as synthetic, except in some countries such as

Table 3.6 Lubricant base oil groups according to API/ATIEL

Base oil group	Base oil nature	Sulphur (wt%)		Saturates (wt%)	VI
Group I	Solvent dewaxed HVI oils[a]	>0.03	and/or	<90	80–111
Group II	Hydrogenated or hydrocracked oils[a]	≤0.03	and	≥90	80–111
Group III	VHVI oils[a,b]	≤0.03	and	≥90	≥120
Group IV	All poly-α-olefins (PAOs)				
Group V	All base oils not included in Groups I–IV or VI				
Group VI	All poly(internal olefins)				

[a] Mineral oil base stocks.
[b] Manufactured by severe hydrocracking and/or wax isomerisation.

Germany and Japan, where they must not be called synthetic. Groups IV/VI are synthetic oils. Group V base oils are highly diverse (synthetic esters, naphthenic oils and others). Group V base stocks are not often used in stand-alone mode, but are mostly used in small quantities as secondary base stocks in special situations to impart certain properties to lubricants. They also provide the raw material for additives.

The classification of mineral oil base stocks (Groups I to III) considers three parameters: sulphur content, saturates content and VI. Conventional crude oil-derived mineral oils, characterised by 10% aromatics and more than 300 ppm sulphur, are regarded as low-biodegradability API Group I products hazardous to the environment. Sulphur compounds and aromatic molecule structures are highly reactive and tend to react rapidly with oxygen to form various soluble and insoluble degradation products. Group II production requires more severe operations. Chevron developed hydrocracking in 1959 and successively (in 1993) invented the hydro-isomerisation process, where wax and paraffin molecules are reshaped into useful lubricants instead of simply broken up into smaller molecules. Base oils made with the Iso-DeWaxing process are called Group II, and are significantly more pure and have higher performance than Group I base oils. Hydrocracked Group II oils contain fewer aromatics and the majority of the reactive components from the solvent refining method are removed.

Base stock characteristics are mainly influenced by the requirements for automotive lubricants. Premium motor oils formulated with Group II base oils leave far fewer wax and tar deposits in the engine, and have much better low- and high-temperature performance than Group I oils (workhorse throughout Europe). Nevertheless, Group II base stocks leave something to be desired in terms of cold-start viscosity, extreme pressure durability and PP. Group II+ is based on raffinate hydroconversion (RHC) technology. Highly refined petroleum oils (Group III base oils) are even more severely hydrotreated (higher pressure, temperature, longer process times) and deliver higher performance. Group II/III plants have a low share in heavy viscosity grades. API Group III oils produced through hydroprocesses and GTL base oils (unofficial Group III+) are free from sulphur and nitrogen and have a higher VI. Highly stable Group III base oils are classified as highest top level mineral base stocks, are often labelled as semi-synthetic and sometimes even considered as fully synthetic oils. In the United States lubricants formulated with Group III base stocks are considered as synthetic motor oils. Group III-based lubricants are not allowed to be marketed as 'synthetic' in any market outside the United States. Within the United States, there are no official specifications, or standards as to which oils can be marketed as 'synthetic'. The cost of API Group III oils is about half that of traditional pure synthetics (Delvac-1, AMSOil, Redline and Motul 5100). Modern Group III oils can nearly match the performance of PAOs at a much lower cost. Because of this, PAO-based oils will gradually disappear unless new low-cost processes are developed. ConocoPhillips is one of the top producers of PAO Group IV base stocks. In 2005 (limited) production of GTL Group III base stocks started. Even though they are considered a synthetic product, they are still mineral base stocks and the mineral part of semi-synthetic lubricants. New GTL base oils, denoted as Group III+, have natural VI > 140 (extra high viscosity index, XHVI). Semi-synthetics often stand for blends of a petroleum oil and no more than 30% synthetic oil.

Synthetic oils are derived from either Group III, Groups IV/VI or some Group V bases. Synthetics include classes of lubricants like synthetic esters (Group V), as well as PAOs (Group IV), PIOs (unofficial Group VI), GTLs (Group III+), PAGs and OSPs (Group V). Higher purity and therefore better property control theoretically means a synthetic oil has better mechanical properties at extremes of high and low temperatures. Synthetic motor oils have a higher VI over the traditional petroleum base and often also a lower PP. With their improved VI, synthetic oils need lower levels of (polymeric) VI improvers, which are most vulnerable to thermal and mechanical degradation upon ageing.

Development of catalytic hydrotreatment technologies has created an abundant supply of highly refined base oils with extremely low sulphur content, low volatility and excellent antioxidant response. A trend in the base oil industry is the continuous transition of many lubricant formulations from Group I to Groups II/II+/III driven by PCMO (passenger car motor oil) and HDMO (heavy-duty motor oil) quality upgrades. The newer hydrocarbon base oils pose a challenge. Despite their aforementioned advantages over Group I oils, also in terms of VI and PP, severely hydrotreated and hydroisomerised base oils are slow in gaining market acceptance. Zhmud and Roegiers [40] have discussed the technological implications of the hydrotreated base oil interchange focusing particularly on solubility and lubricity issues (see also Section 3.6), whereas Greaves [26] has also pointed out formulation problems. These base oils lack a fundamental property, namely lubricity, as a result of their saturated character. The greater the degree of hydrotreatment, the lower also the solubility.

API Group III base oils, as well as GTL base stocks and PAOs, are regarded as 'dry' base oils because they contain fully saturated non-polar hydrocarbon (isoparaffinic) molecules, and have low sulphur levels. This can be seen by comparison of the aniline point values: higher aromatics, 20–40 °C; naphthenics, 70–100 °C; Group I paraffinics, 90–110 °C; Groups II–IV base oils, 100–130 °C. The aromatic and sulphur components in Group I base oils provide a degree of solvency which helps to impart solubility for performance-enhancing additives. The absence of such unwanted polar compounds in Groups II–IV base oils renders formulation with many conventional types of performance enhancing additives difficult. These problems can be overcome in several ways, namely by addition of ionised vegetable oils [40] or polar OPS base oils (aniline point < –30 °C) [26]. Elektrionized™ vegetable oils (typically 10%) added to the base stock significantly improve the lubricity of the mineral oil, which translates into higher processing speeds, reduced lubricant consumption, better production economy and reduced environmental load; see also Section 7.1.

The global nameplate capacity (2009) by petro-based base oil group qualities breaks down as follows, with strong regional differences reflecting diesel/gasoline demand: 60.2% Group I (40% Europe, 30% Asia-Pacific), 23.1% Groups II/II+ (60% N. America, 40% Asia-Pacific), 6.6% Group III (75% Asia-Pacific), 1.0% Groups IV/VI (50% W. Europe, 50% N. America), and 9.1% naphthenics (50% N. America, 30% Asia-Pacific) [41]. By moving into compliance with Euro VI emission standards Europe is expected to require more Group II base oils. The concept of Group II was born in the United States where crude oil is not suitable for base stock production. Group I oils produced in Western Europe are of a higher quality than in the United States. The virgin base oil nameplate capacity in Western Europe amounts to 6.8 Mt (utilised for 70%).

Regulatory initiatives (lower tailpipe emissions, better fuel economy, greater usage of low sulphur diesel) have had an impact on OEM decisions which in turn have influenced lubricant base stock selection (low viscosity/SAPS (sulphated ash, phosphorus and sulphur) oils, higher oxidation and VI oils) and additive fix (high ashless antioxidants). Lower sulphur contents determine increased usage of Group II over Group I oils. Low viscosity and improved oxidation stability determine a shift towards Group II+ and higher oils. Nevertheless, Group I base stocks (from 74% in 2004 to 57% in 2010) are expected to remain the workhorse category; a bright stock shortage is foreseen [42]. The effects of GTL base stock are still uncertain. Group III supply will rationalise in face of XHVI/GTL (Group III+) supply. Longer-term expectations are a decrease of Group I base oils to 50% or less, and an increase of Group II/II+/III/III+ to 40%; global nominal Group IV/VI capacity would stay unchanged at about 1% [41].

Some 10% of world's annual base oil production consists of (other) synthetic base oils, referred to as Group V base oils and several bio-based oils (triglycerides, unsaturated and saturated esters). According to assessments reported by Murphy *et al.* [43] approximately 80% of the synthetic base oil market is dominated by three groups of compounds: PAOs, organic esters and PAGs.

For most industrial applications, lubricating oils with viscosities ranging from 5 to 15 cSt at 100 °C cover the requirements. Classification of lubricants by internationally accepted viscosity grades is the most prevalent method of describing oils in relation to their application. The most common systems are those of ISO, SAE and AGMA. Each organisation uses a different kinematic viscosity range numbering system. The ISO 3448 classification distinguishes 18 viscosity grades (from 2 to 1500 mm^2/s), which apply to most industrial lubricants [36]. SAE viscosity gradings for passenger car motor oils (typically composed of C_{18}–C_{34} hydrocarbons) rank from 0 to 60 for low to high viscosity, respectively. The numbers are suffixed with the letter W, designating their 'winter' or cold-start viscosity, at lower temperature. Document SAE J300 (Engine Oil Viscosity Classification, July 2001), which has established 11 viscosity grades for monograde motor oils, of which 6 are winter grades (0W, 5W, 10W, 15W, 20W and 25W) and 5 are summer grades (20, 30, 40, 50 and 60), defines the viscometrics related to these grades.

Engine oils which only fit into one viscosity grade are known as monograde oils. The *V–T* behaviour of such oils corresponds to that of conventional mineral oils without VI improvers. Oils which cover two or more viscosity grades are called multigrade oils and are based on oils containing VI improvers or base oils with high natural VI. The SAE designation for multigrade (all-season) oils includes two viscosity grades. For instance, the common multigrade oil labelled as 10W-30 must pass the SAE J300 viscosity grade requirement for both 10W and 30. The high-temperature viscosity characteristics are indicated by the number following the W. The two numbers used are individually defined by SAE J300 for monograde oils. Monogrades dominated the engine oil market back in the 1950s. In the meantime, recommended viscosities for typical car engines have improved from the heavy 20W-50 oil in the 1960s to 10W-30 oils in the 1980s and to the modern 5W-20 low-viscosity oils. Engine designs are evolving to allow the use of lower-viscosity, fuel-efficient oils. A breakdown of the viscosity grade distribution for North American passenger car motor oils (PCMOs) is as follows: SAE 5W-30, 47%; SAE 5W-20, 18%; SAE 10W-30, 17%; SAE 10W-40, 6%; monograde, 2%; others (SAE 20W-50, SAE 5W-40, SAE 0W-20), 10%. In China SAE 15W-40 viscosity oils still hold most of the market.

SAE has a separate viscosity rating system for gear, axle and manual transmission oils (SAE J306: Automotive Gear Lubricant Viscosity Classification, 2005). Specific SAE viscosity grades (75W to 85W, 90–250) have been developed for automotive gearbox, axle and differential oils. AGMA has defined nine viscosity ranges for industrial gear oils (1–8, 8A). Viscosity grades for mineral base oils are traditionally expressed according to the Saybolt Universal Seconds (SUS) scale (100N to 650N, 150B).

Classification of oils on the basis of the additives used to improve the performance properties distinguishes categories such as: inhibited (rust and oxidation inhibited), AW, extreme pressure (EP) and residual. So-called residual compounds are heavy-grade straight mineral oils or EP oils, which are often used in admixture with a diluent to increase ease of application.

After application, the diluent evaporates, leaving behind a heavy adhesive lubricant coating. Residual oils – not to be confused with greases – are often used for open gear applications. Lower viscosity residual oil finds application in many closed-gear systems.

Base stocks and additives may be varied to provide formulations with desirable qualities for specific usage (Chapter 7). Lubricants are commonly classified into two major groups: automotive lubricants and industrial lubricants (oils and specialties such as greases, metalworking lubricants, solid lubricants, etc.). Table 12.1 subdivides lubricants in terms of engine and non-engine use. Some of the more common applications, which are a basis for lubricant classification, are engine oils (automotive, aircraft, marine, commercial), hydraulic fluids, bar and chain lubricants, cutting oils (coolants for metal cutting), insulating oils (transformers and circuit breakers), turbine oils, gear oils, slideway oils, wire rope lubricants, air tool oils, etc. (see Chapter 12). The occasional prefix 'bio' refers to the eco-sustainability quality of the oil and denotes biodegradable classes.

Some oils are classified as non-specialised industrial oils, i.e. oils not formulated for any specific application and are usually referred to as 'general purpose oils'. General-purpose (G-P) oils contain R&O (rust and oxidation) inhibitors, AW agents, antifoamants and demulsifiers. Examples are general-purpose biobased lubricants and EP gear oils. General-purpose biobased lubricants (ISO grade 32–460) are used in mechanical applications which do not require specialised oil. Such oils are commonly known as R&O oils or bio-hydraulic fluids (even though not developed for exclusive hydraulic use). EP gear oils are regarded as general purpose oils with EP additives and generally have higher viscosity. Bio-EP gear oils (ISO grades 68–1500) are essentially general-purpose bio-based lubricants with EP additives. They are commonly used in gear systems but can be used in any application which requires the specific viscosity range and additives.

For engine oils three internationally recognised sets of minimum performance requirements are in use: ACEA in Europe, API in USA, and ILSAC in Asia. API/SAE have defined gasoline and diesel engine oil classifications by specification. Current API service categories include SN, SM, SL and SJ for gasoline engines. There are six API diesel engine service designations. For the HDMO market API CJ-4 oils were developed in 2007. Today's high-tech, heavy-duty engines need enhanced wear protection. Chevron Group II/II+ base oils meet qualifications for ACEA, API, JAPA and major original equipment manufacturer (OEM) heavy-duty specs. It is possible for an oil to conform to both the gasoline and diesel standards.

Performance requirements for automotive gear lubricants depend on their intended use. GL-4 and GL-5 are standards set by the API for gear oils. GL-4 are intended for axles and GL-5 is indicated for high-speed/shock loads and low-speed/high-torque conditions. The premier standard in common use today, SAE J2360 is a new global quality standard that defines a level of performance equivalent to the US military standard MIL-PRF-2105E. It includes all of the most recent axle and transmission testing requirements identified in API GL-5, API MT-1, and MIL-PRF-2105E, and consequently defines a performance level beyond API GL-5.

Non-government specifications and standards provide a broad base of suppliers and promote competition, whereas producer-specific standards limit competition to a single supplier. New US federal specifications have been developed and existing specifications will be updated to establish requirements for commercial products only if specific design and performance are not covered adequately by non-government standards. Federal and military specifications are being replaced by specifications and industry standards developed by trade associations (SAE, AGMA, API) and professional organisations and technical societies (ISO, ASTM, ANSI). Military specifications are developed and updated for military-unique products or commercial products that require substantial modifications to satisfy military requirements. US military specifications are issued by the Department of Defense (DOD) and are listed in the DOD Index of Specifications.

Producer specifications amount to a certification that the product meets limited physical characteristics in terms of specific test values. Producers vary considerably in the information contained in the specifications. Kinematic viscosity (cSt) at 40 °C and 100 °C, SUS at 100 °F and 210 °F, API gravity, PP and flash point may generally be expected. Chemical impurities may also be indicated. Other physicochemical data may be included if relevant for the intended use.

Oil manufacturers often label products by brand names with or without any connection with standard classifications. Brand names are sometimes informative and may be indicative of their intended use. While Bio-Pro Bar & Chain Oil SAE 10W30 (Renewable Lubricants, Inc.; Hartville, OH) indicates a bio-based bar and chain oil with a specified SAE viscosity, the brand name Bio Fluid XTBO is much less informative and does not readily allow attribution to a specific class. Brand names, product data and producer specifications tend to vary with time.

Greases are classified by penetration number and by type of soap or other thickener. Table 3.7 lists the NLGI grease classification along with a description of the consistency of each class. A penetration number indicates how easily a grease can be fed to lubricated surfaces (i.e. pumpability) or how well it remains in place. ASTM D 217 and D 1403 are standards for performing penetration tests. No method exists to classify soap thickeners. The type of soap thickener indicates probable water resistance and maximum operating temperature and gives some indication of pumpability.

Table 3.7 NLGI grease classifications

NLGI grade	ASTM worked penetration 0.1 mm at 25 °C[a]	Consistency
000	445–475	Very fluid
00	400–430	Semi-fluid
0	355–385	Soft
1	310–340	Semi-soft
2	265–295	Multipurpose[b]
3	220–250	Semi-hard
4	175–205	Hard
5	130–160	Very hard
6	85–115	Solid

[a] After 60-stroke worked.
[b] Common grease.

3.3.1 Solid lubricants

A solid lubricant is a material used as powder or thin film which reduces friction and wear of contacting surfaces in relative motion and provides protection from damage. Solid lubricant friction is a special form of friction which occurs when using solid lubricants. It differs from friction models for liquid lubricants in view of the effects of particle shape, size, mobility and crystallographic characteristics of the particles. The main purpose of solid lubricants is to build up a continuous adherent soft or hard film in the rubbing surfaces. These films can be applied by mechanical, (electro)chemical or physical processes. ASTM D 2714 provides a test method for determination of the friction coefficient of a solid lubricant.

Solid lubricants fill a special niche in reducing wear in situations where the use of liquid lubricants is either impractical or inadequate, such as in vacuum, space technology or automotive transport. Solid lubricants are required for lubrication under extreme conditions where the bearing surfaces in tribological contact must still be effectively separated and where it is important for the lubricant to stay in place. Products containing solid lubricants are often used in applications when high specific loads are applied to sliding surfaces in the presence of boundary and mixed frictional regimes, at very low hydrodynamically effective speeds, or when the lubricant must perform over a wide temperature or under extreme temperature conditions (e.g. in aviation). Dry lubrication with solid lubricants is also required in aggressive environments, in high-vacuum applications, in nuclear reactors, and generally in applications where contamination by lubricating oils or greases must be avoided.

Many materials qualify as solid lubricants. Solid lubricants are usually classified as structural lubricants, mechanical lubricants, (reactive) soaps and chemically active lubricants. Lubricating properties of structural lubricants (e.g. graphite and metal dichalcogenides) are on account of their layered lattice structures. The class of mechanical lubricants comprises self-lubricating organic compounds (such as thermoplastics and thermosets) and naturally occurring metal oxide films (usually about 10 nm thick), chemical surface coatings (formed by chemical or electrochemical action on the metal surface), and glasses. Typical solid metal-forming lubricants are reactive soaps (salts of stearic, oleic and palmitic acid) in connection with a zinc phosphate coating. The main function of soaps (metal salts of fatty acids) in lubrication technology is in the preparation of greases (see Section 12.11). Soaps may also be formed *in situ* on a metal surface by chemical attack of a fatty acid on the metal. Chemically active lubricants include EP and AW additives and other chemicals which interact with the metal surface to produce a lubricating or protecting layer. Examples of common solid lubricants include the lamellar solids MoS_2 and graphite, as well as poly(tetrafluoro-ethylene) (PTFE) (admixed with organic and inorganic lubricating substances) [44] and other fluorine-containing polymers. Magnesium stearate dihydrate (MgSt-D) [45] and hydrogenated castor oil (m.p. 86 °C) [46] have also been used as solid lubricants.

Important material properties for solid lubricants include crystal structure, thermal and oxidation stability, volatility, chemical reactivity, melting point and hardness. Heat stability (in close relation to chemical stability) is required if thermal decomposition of the solid lubricant is to be avoided. Solid lubricants have a temperature range over which they are effective. For example, graphite can withstand 650 °C and moderate loads [47]. Starch is stable up to about 300 °C [48].

Very few solid lubricants have the required adhesive and cohesive properties which enable them to create an effective lubricating layer with low friction coefficients and sufficient lifetime. Most solid lubricants require a carrying medium, a bonding agent (such as oils, greases or water), and/or pretreatment of the material surface. Solid lubricant formulations are typically composed of a solid, a binder, and additives such as corrosion inhibitors or solvents. Products containing solid lubricants are in the form of

- powders;
- dispersions and suspensions;
- greases and grease pastes;
- pastes; and
- dry-film lubricants.

Solid lubricants are generally available as powder embedded in a resin or wax film or as dispersions in oil or water. Solid lubricants in the form of powder must have appropriate properties to enable them to create a film (cohesive properties) and adhere (adhesive properties). These requirements can be met by self-lubricating dry lubricants, but much less so by MoS_2, graphite and PTFE. Dispersions and suspensions of solid lubricants in water are used to coat mass elements for cold and hot forming. Dispersions and suspensions in oils are used as additions in gear- and oil-lubricating systems. Greases can contain solid lubricating compounds (typically graphite, MoS_2, ZnS, BN, PTFE, ultra-high-molecular-weight polyethylene, UHMWPE) in rather high concentrations. For most applications an average particle size of the micronised solids of between 5 and 15 μm is acceptable. Addition of solid lubricants to greases has a positive effect on the pressure absorption capacity, ability to withstand wear and friction. Pastes are solid lubricants in a carrying oil.

All drylubes have in common the solid state, or at least a waxy consistency at ambient temperature. The lubricant must have a fairly low viscosity for ease of application to the metal surface. Solids or waxes are difficult to apply and are usually thinned before application. Aqueous dispersions of wax or other solid organic compounds have the desired liquidity. Dry-film lubricants or lubricating varnishes are suspensions of solid lubricants (usually MoS_2, graphite, PTFE or starch) and additives in a solution of inorganic or organic binding agents. The main types of binding agent used are organic resins, cellulose, and inorganic silicates and phosphates. Hydrocarbons and water are used as solvents. Emulsifiers are needed to disperse the product in water. Dry-film lubricants can be used in a variety of ways (dipping, spraying). When the solvent (often water) has evaporated a wax or polymer film is formed. Heating can be used to accelerate the process. After hardening lubricating varnishes form a highly adhesive, dry film of lubricant. The proportion of solid lubricants in the layer created in this way can be as high as 70%.

Dry-film lubricants are applied on clean sheet metal surfaces for the purpose of protection of the part from corrosion, rust, stain and damage (scratches) and/or the environment during transportation, handling and storage [49]. Another important purpose of dry-film lubricants is to provide lubricity during the metalworking operation, where the pre-lubricated sheet metal is fabricated into various products. Dry-film lubricants can be used in a multiplicity of ways. Commercial dry-film lubricants are almost exclusively petroleum-based. In addition to liquid bio-based lubricants from unmodified and modified vegetable oils, dry-film bio-based lubricants are of interest. Successful application of dry-film bio-based lubricants in metalworking provides a number of benefits, including a clean and environmentally friendly work area, low health risk to workers due to the absence of vapours and mist since no sprayed lubricant is used, and significant cost savings due to the elimination of the costs associated with maintaining and operating a liquid lubricant system. Also, since dry-film lubricants can be efficiently distributed, less lubricant is typically needed [50].

Lubricants are available in liquid, solid and gaseous forms. Industrial machinery ordinarily uses oil or grease. Solid lubricants are more effective than fluid lubricants when the loading at contact points is high, or in high-speed conditions. Solid lubricants are used in applications not tolerated by more conventional lubricants (e.g. extreme pressure/temperature conditions). Reduced friction may also be achieved by plating the wearing surfaces of a material with a different metal.

Both liquid and dry-film lubricants can be used for the highly essential lubrication in sheet metal-forming in the automotive industry. Liquid lubricants are commonly applied in the press shop, whereas dry-film lubricants are applied in the rolling mill. Water-free dry-film lubricants (or hotmelts) are applied to sheet material at rolling mills to provide good drawing performance and excellent corrosion protection in automotive body construction. The benefits of dry lubricants in sheet metal forming are listed in Table 3.8. In aluminium sheet metal forming dry-film lubricants show advantages over conventional oil lubricants [50]. Solid lubricants play a significant role in metalworking, especially in non-cutting forming, cold and hot forging, pressure die-casting and drawing, and cold extrusion operations. Their main task is to generate an effective separating layer between tool and workpiece and to protect the sheet metal. E1 is a commercial drylube for automotive applications, produced by Zeller & Gmelin (Eislingen, Germany). Wheat flour (applied as an aqueous suspension) has been proposed as a non-polluting, adequately recyclable lubricant for the sheet metal-forming process [52, 53]. Also starch–oil composites are preferred ingredients for formulating bio-based dry-film lubricants because of their environmental and health characteristics (see Section 6.4.1).

Lubricant compositions comprising micronised magnesium stearate dihydrate (MgSt-D) may be used to lubricate a variety of solid materials (metal, glass, clay, ceramic or plastic surfaces) that are used in industrial or

Table 3.8 Benefits of dry lubricants in sheet metal forming

- Reduced lubricant usage
- Large process window (low friction coefficients)
- Elimination of die washing
- Uniform coating thickness and no runoff
- Good deep-drawing performance and corrosion protection
- Compatibility with assembly operations[a] and painting
- Minimal hazards and environmental compliance

[a] Welding, bonding, clinching and riveting.
After ref. [51].

consumer products [45]. MgSt-D has advantages over other lubricants because of its high melting temperature, high lubricity at a low concentration, large covering potential, general acceptance as safe, non-toxicity and excellent stability profile. MgSt is widely applied as a lubricant in the pharmaceutical and nutraceutical industry. MgSt lubricant compositions are used in powder blending applications for their lubricant activity (i.e. reduce friction, transfer heat, and prevent corrosion during the process), anti-adherent activity (i.e. prevent sticking to punch faces and die walls) and glidant activity (i.e. improve the flowability of the powder or granules). Such compositions may be used in the lubrication of dry paint products, as lubricants or dry binders in dry powder cosmetic formulations, as anti-caking agents in dry powder food products or as coating agents to extend the shelf-life of food products.

Surfaces of biological lubrication systems are often covered with sugar-rich layers. Poly(L-lysine)-*graft*-dextran (PLL-*g*-dex), adsorbed from aqueous solutions, is very effective for boundary lubrication of oxide-based tribosystems [54].

Solid lubricants and solid lubrication have been reviewed by Lancaster [55] and Busch [56].

3.4 Eco designations for lubricants

Many different terms are used to define and classify the effects of lubricants on the environment, which is utterly confusing. There are a variety of (partially misleading) designations used by lubricant oil suppliers such as 'natural', 'environmentally safe', 'biodegradable', 'readily biodegradable', 'bio-based', 'food grade', etc. In this context, the lamellar mineral MoS_2 (molybdenite) is to be considered as a natural lubricant, whereas α-glycoprotein, surface-active phospholipids (SAPLs) and cholesterol esters are examples of internal natural lubricants. Canola oil additivated with boric acid, which exhibits excellent lubrication properties in a brake valve assembly, forms a completely natural environmentally friendly lubricant formulation and does not require expensive disposal techniques [57]. Since there is no standard for environmentally preferable fluids, many different types of products claim to be environmentally safe. Table 3.9 outlines the performance of various base oil categories. True proof of performance is found in the field. Glycerides, vegetable oils, esters, glycols and polyoxyalkylenes are all considered as 'environmentally harmless' lubricants [58].

The often used prefix 'bio' refers to an association with a renewable resource of biological origin, such resources generally being exclusive of fossil fuels. However, it is necessary to distinguish the terms 'bio-based' and 'biodegradable'. The term bio-based does not have a unique, internationally accepted definition, although a generally accepted definition has come

Table 3.9 Fluid performance

Fluid type	Organic origin	Biodegradation time	Toxicity	Typical performance	Water tolerance	Elastomer compatibility
'Environmentally friendly'	Petroleum	Years	Low	Good	Good	Good
Mineral oil	Petroleum	Years	High	Good	Good	Good
'Inherently biodegradable'	Petroleum	Years	Variable	Good	Good	Excellent
'Food grade'	Petroleum	Years	Low	Weak	Weak	Excellent
Polyglycol (PAG)	Glycol	Variable	Medium	Variable[a]	Good	Poor[a]
Polyolefin (PAO)	Synthetic	Variable	Low	Excellent	Excellent	Low
Ester	Synthetic	Days	Low	Mixed	Very weak	Very poor
'Bio-based'	Vegetable	Days	Low	Variable	Medium	Mixed

[a] Better performance for new OSP polymers (see Section 3.2).

forward from the collaborative efforts between US government agencies and bio-based industry professionals. As defined by the US Department of Agriculture (USDA), 'bio-based' products are products determined by the US Secretary of Agriculture to be commercial or industrial goods (other than food or feed) composed in whole or in significant part of biological products, forestry materials or renewable domestic agricultural materials, including plant, animal or marine materials. Product groupings and minimum proposed bio-based content are shown in Table 9.8. These products are usually biodegradable and recyclable. Bio-based is not meant to identify a 100% pure vegetable base oil formula. Many commercial formulations include both bio-based components and petroleum mineral oils, PAOs, synthetic ester base oils and necessary additives. Made from renewable plant and animal sources, bio-based products are generally safer for the environment than their petroleum-based counterparts. A product may be biodegradable but *not* bio-based since biodegradable petroleum-based products can be produced (e.g. some PAOs). But a product can also be bio-based *and* biodegradable since most bio-based products have been shown to be biodegradable.

The terms bio-based and biodegradable hold some extent of confusion without the knowledge of the bio-based content or extent of biodegradability of the finished product. Biodegradation is the chemical breakdown or transformation of a substance caused by organisms or their enzymes. Biodegradability is thus not only a 'property' or characteristic of a substance, but is also a system's concept, i.e. a system with its conditions determines whether a substance within it is biodegraded. Mostly, when the term 'biodegradable' is used to describe a lubricant product, it means that the product is more biodegradable than petroleum base stocks or formulations.

'Biodegradable' or 'environmentally acceptable' lubricants are products characterised by rapid biodegradability and low eco-toxicity and typically contain a high proportion of renewable raw materials. The generic term for this product group of biodegradable and renewable base stocks is 'biolubricants'. However, this definition is not universally accepted. In some areas, only biodegradability is considered in the definition. For our present purposes, biolubricants are considered as being biodegradable *and* renewable materials.

Biolubricants do not have to be composed entirely of vegetable oil base stocks. They can be products derived from renewable oils, such as the fatty acids from fats and oils, reacted with synthetic alcohols or polyols to produce esters that can be considered biolubricants. Also, the natural vegetable oils can be treated to produce a modified product that is still biodegradable and renewable. Rapidly degradable lubricants, which minimise environmental stress, are chemically mainly composed of natural or synthetic esters and

PAOs. Bio-based lubricant compositions often comprise at least one bio-based natural oil or bio-based synthetic oil selected from the group consisting of natural or synthetic vegetable oil, natural or synthetic animal oil, genetically modified (GM) vegetable oil, natural or synthetic tree oil, and mixtures thereof. The vast majority of rapidly biodegradable lubricants currently on the market are based on saturated or unsaturated ester oils. SE lubricants having little or no unsaturation in the carbon to carbon bonds are used in premium quality motor oils owing to their desirable properties. However, the acids and alcohols used to make SEs usually are derived from petroleum distillates and are thus not from a renewable source. They are also more costly and less biodegradable than natural triglycerides.

The main chemistry of biodegradable lubricants consists of: (i) triglycerides (e.g. rapeseed oil, RSO; sunflower oil, SNO); (ii) esters of (modified) vegetable oils; (iii) semi-saturated, transesterified ester oils with natural fatty acids (e.g. trimethylolpropane trioleate); (iv) (partially bio-based) fully saturated, synthetic esters based on chemically modified vegetable oils or mineral oil (e.g. diisotridecyladipate); (v) polyglycols; and (vi) PAOs. Biodegradable lubricants have a long history. In fact, natural oils and fats, consisting of more or less unsaturated triglycerides, such as castor oil (CAS), palm oil (PMO), RSO, soybean oil (SBO), SNO, tallow (TLW), lard and sperm oil with excellent tribological qualities (low friction coefficient, good wear protection) are historically well documented for their lubrication properties. However, their range of use is generally limited by lower stability against thermal oxidative and hydrolytic stress and partly inferior cold-flow properties. These limits can be improved in several ways, e.g. by chemical or genetic modification or by additivation.

Vegetable oil esters and their derivatives can be used as a base stock for preparation of 'environmentally friendly', rapidly degradable lubricants. Again, there is no universal definition of environmental friendliness. Transesterified oils are frequently blended with petroleum oils to provide functionality to the latter. Although the fixed oils by themselves are biodegradable, once mixed with petroleum this property is lost.

Numerous applications exist for environmentally friendly lubricants, of which biolubricants are just one type, and they are slowly gaining market share from conventional mineral oil lubricants. Vegetable oil biolubricants are niche products with a smaller market share than synthetic esters derived from natural materials. There is in particular a need to develop biodegradable lubricants for use in applications which might result in the leakage of such lubricants into the soil and into waterways. The rate at which compounds biodegrade in the environment is limited mainly by their chemical structure, water solubility, the physicochemical parameters of the environment, and the qualitative and quantitative composition of the microorganisms taking part in the biodegradation process [59, 60]. Vegetable oils and animal fats, followed by esters of linear monocarboxylic acids and esters of dicarboxylic acids, are the most readily biodegradable. Synthetic phthalates and hydrocarbon-based oils are much more resistant to the action of microbes. The presence of aliphatic substituents causes the formation of biodegradation-resistant metabolites. Compounds that contain a quaternary carbon atom are most resistant to biodegradation. In the case of mineral and synthetic hydrocarbon-based oils biodegradability decreases rapidly with the increasing molecular weight of the oil. The biodegradation of SE oils is mainly an aerobic process [61].

3.4.1 Biodegradability classification

Industry has acknowledged the need for a standardised biodegradability test. Recognised tests to determine biodegradability of lubricants are listed in Table 8.2. Typical biodegradability values according to the CEC L-33-T-82 biodegradability test for common lubricant base oil categories are listed in Table 3.10. OECD testing is mostly adopted in the lubricant industry. For more details, see Section 8.2.2. Standardised biodegradability definitions according to the OECD and EPA (OPPTS 835.3110) guidelines are as follows:

- *Primary biodegradation* is the alteration in the chemical structure of a substance, brought about by biological action (microorganisms), resulting in the change of some measurable property of that substance.
- *Ultimate biodegradation* (aerobic) is the level of degradation achieved when the test compound is totally converted (100% biodegraded) by microorganisms resulting in the production of CO_2, CH_4, H_2O, mineral

Table 3.10 Biodegradability of base oils

Oil type	Primary biodegradability (%)[a]
Mineral oil	15–35
White oil (mineral base)	25–45
Alkylbenzene	5–20
PAO (synthetic)	5–30
Polyisobutylene (PIB)	0–25
Polyether	0–25
Phthalate and trimellitate esters	5–80
Polyol esters and diesters	55–100
Natural and vegetable oil	70–100

[a] According to CEC L-33-T-82 biodegradability test (now CEC L-33-A-93).
After refs. [1, 62].

salts, and new microbial cellular constituents (biomass). Vegetable oils are ultimate biodegradable. United Bio Lube's entire bio-based lubricant product line, based on RLI's Stabilized™ HOBS technology (see Section 10.5.2), is also rated ultimate biodegradable, the highest USDA rating attainable.

- *Readily biodegradable* is an arbitrary classification of chemicals which have passed certain specified screening tests for ultimate biodegradability; these tests are so stringent that it is assumed that such compounds will rapidly and completely biodegrade in aquatic environments under aerobic conditions. Readily biodegradable is usually taken to imply that 60% or more of the test material carbon has been converted to CO_2 in 28 days. More than 95% of the base oils used in manufacturing of lubricants fail to meet the ready biodegradability criterion. These materials can be made subject to inherent biodegradability tests, which are used to assess whether a substance has any potential for biodegradation.
- *Inherently biodegradable* is a classification of chemicals for which there is unequivocal evidence of biodegradation (primary or ultimate) in any test of biodegradability.

The most common test methods used by the lubricant industry for evaluating the acute toxicity of their products are EPA 560/6-82-002, Sections EG-9 and ES-6, and OECD 203. These tests determine the concentration of a substance that produces a toxic effect on a specified percentage of test organisms in 96 h. The acute toxicity test is normally conducted using rainbow trout. The ASTM D 6081 test method for determining the aquatic toxicity applies to lubricants and lubricant components with poor stability in water.

3.4.2 Advantages and disadvantages of biolubricants

Mineral oil-based lubricants are key elements for industrial, automotive and other transport applications. As they are polluting (by water and soil contamination and CO_2 emissions), there is a strong need to substitute current loss and high-risk lubricants by more environmentally friendly alternatives. Moreover, due to the high costs often associated with disposing of lubricant materials that are typically used to reduce friction and wear, efforts are being made to develop and implement lubricants that are environmentally benign.

Manufacture and use of biolubricants will result in a number of benefits to the environment, the health of citizens and the economy. Widespread use of biolubricants will reduce emission of greenhouse gases (GHGs), toxic waste streams into air, rivers and land, and reliance on imported petroleum.

In addition to reducing environmental impacts, biolubricants can result in a cleaner work environment, less skin problems, better safety due to flash point properties, constant viscosity, fewer oil mist and vapour emissions, and a competitive tool life. Moreover, manufacture of biolubricants will strengthen the farm-based domestic economy and generate new jobs. Although biolubricants have numerous benefits, several disadvantages have also been observed, particularly during the use phase. Some disadvantages include: temperature limitations, bad odour, metal discoloration, viscosity limitations, and poor thermal and oxidative stability [1]. These need to be corrected.

Traditionally, to achieve desired performance characteristics of mineral oils (no more than 40% biodegradable) for a given application, the benefit of additive technology is called in. However, as additives cannot produce a desired biodegradability, these attempts to modify mineral oils into a highly biodegradable base stock are doomed to be unsuccessful. Consequently, a switch to other base stocks is required to meet requirements for environmentally compatible lubricants [63].

Successful replacement of petroleum-based lubricants with biolubricants requires overcoming a number of technical, developmental and marketing hurdles (see Section 13.4). A manufacturer faced with the task of producing a biodegradable lubricant has several chemically different base stocks from which to make a selection. The production of environmentally friendly, rapidly biodegradable fluids for lubricants based on petrochemicals such as low-molecular-weight PAOs, PAGs, polyglycols, dialkyl adipates or polyol esters has been disclosed [64–68]. These synthetic base stocks, as well as vegetable oil-based fluids, all present specific performance limitations (Table 3.9). The lubricant industry has to overcome several challenges in increasing market share for biodegradable lubricants, in particular: (i) to achieve a lower temperature-dependence of the physical properties, i.e. to make either higher or lower temperature use accessible; (ii) to improve inertness, i.e. to prolong performance time; (iii) to improve ecological properties; and (iv) to reduce cost. In the future, super-clean GTL base oils with premium characteristics (extra high VI, very low evaporative losses) are expected to become more readily available. Fischer–Tropsch base oils are prospected as the next quantum leap in base oil quality [69] (see also Section 3.3).

3.5 Environmentally acceptable lubricants

The driving forces needed to reduce the problems caused by contacts of lubricants with the environment are environmental facts, public awareness, government directives and regulations, globalisation of markets and economic incentives. The preferred course of action is to reduce toxicity and to increase biodegradability. In its marketing literature the lubricant industry uses a great many of confusing denominations to classify lubricants with regard to their impairment of the environment. Some of these terms, all preceded by the common descriptor 'environmentally', include: acceptable, adapted, aware, benign, careful, compatible, conformable, considerate, favourable, friendly, harmless, justified, neutral, non-toxic, positive, preferable, protective, responsible, safe, sensitive, suitable. For instance, Mobil has introduced its Environmental Awareness Lubricants (EAL) line of hydraulic fluids. Pennzoil uses the terminology 'environmentally friendly' for its engine oils [70]. Similarly, Carl Bechem GmbH (Hagen, Germany) has developed a high-performance range of 'environmentally friendly' lubricants and additives (circulation oils, greases, rail lubricants, etc.). Environmentally acceptable lubricants are discussed in refs [71, 72]. Shell prefers the terms 'environmentally considerate products' for its lubricant product line based on carefully selected unsaturated and saturated SEs or advanced technology RSO with optimised additive packages to yield good AW/EP performance, corrosion protection and oxidative stability. Shell Naturelle HF-E is an advanced biodegradable hydraulic fluid [73]. Petro-Canada Lubricants (Mississauga, ON) has developed the EcoSia line of environmentally responsible products (inherently to readily biodegradable).

Bartz has rightly pointed out that no lubricant can be truly environmentally friendly (i.e. improving the environment); at best, a lubricant can remain neutral against the environment (i.e. harmless) [74]. The best one can hope for is that the lubricant affects the environment to an acceptable degree (minimal impact). Consideration of environmental aspects of lubricants should cover the full life cycle from production to disposal (see also Chapters 11 and 14). In particular, lubricant production and use should cause no depletion of resources, disposal problems, formation of waste or emissions, should have low energy consumption, give no health hazards, be ecotoxicologically acceptable and fast biologically degradable. The terminology used in connection with 'environmental compatibility' is both subjective (non-measurable) and objective (measurable). Subjective criteria are environmentally friendly and environmentally compatible. Table 3.11 lists some measurable or provable, objective criteria. Objective criteria for 'biolubricants' are biodegradability of at least 60% according to OECD 301 or 80% according to CEC L-33-A-93. Table 3.10 lists typical biodegradability results for various base oils.

There is no such thing as a wholly environmentally benign product. Environmentally friendly products are intended to have either lower or minimal impact. Environmentally acceptable (EA) is a commonly used term to address environmental lubricants. There is no precise definition of environmental acceptability of lubricants, which can encompass a broad range of potential environmental benefits, such as shown in Table 3.12. Any claim of

Table 3.11 Objective criteria for environmental compatibility

- Biodegradability
- Compatibility with conventional lubricants and materials
- Ecological toxicity and physiological safety
- Efficiency improvements, lower energy consumption
- Emission reduction in use
- Environmental labels
- Performance, approvals, oil change intervals
- Use of renewable raw materials
- Water solubility, water pollution

After ref. [75].

Table 3.12 Potential environmental benefits of biolubricants

• Renewable resource	• Reclamation
• Resource conservation	• Disposability
• Pollutant source reduction	• Biodegradability
• Low sulphur content	• Ecotoxicity
• Recycling	• Sustainability

environmental acceptability must be supported by appropriate technical evidence. Even many petroleum-based lubricants can claim a certain degree of environmental acceptability, as measured by various standards, e.g. extended oil drain capability. Of primary interest in selection and use of EA lubricants is defining and measuring the product attributes that could affect the environment. Obviously, the lubricants must provide the minimum performance requirements in key areas for the given application. In other words, EA lubricants must perform at levels equivalent to those achieved by conventional mineral- or synthetic-based lubricants in the equipment, while providing the additional characteristics that reduce the negative impact in the event of inadvertent release into the environment.

Commercial 'bio' hydraulic fluids are made from non-renewable synthetic polyol esters. Traditionally, this type of safer environmental fluid is known in industry as an EA fluid, as opposed to a bio-based fluid (see Section 12.3.1). There are no standards for EA lubricants or hydraulic fluids. Similarly, there are no industry or guide specifications for EA fluids and greases. Manufacturers and end-users agree that for a lubricant to be classified as an EA type it should be biodegradable and non-toxic, but not necessarily bio-based or renewable. Until specific standards and specifications are developed, qualification of fluids to be regarded as EA should rely on conformity with EPA 560/6-82-002 (LC50 > 1000) and ASTM D 5864 (60% conversion to CO_2 in 28 days) regarding toxicity and biodegradability, respectively. Some EA lubricants are rated as only 'readily biodegradable'; they are made from non-renewable resources.

In the context of environmentally friendly lubricants, other commonly used terms are 'food grade' and 'green'. Although some lubricants are identified as food grade, have been approved by the US Food and Drug Administration (FDA) and are subject to ASTM standard testing procedures, there is no worldwide standard definition or specification for environmental lubricants intended to replace standard lubricants. Food-grade lubricants are generally used in the food industry where incidental food contact may occur. Food-grade lubricants may or may not qualify as EA lubricants. Most food-grade lubricants are based on non-toxic US Pharmacopœia (USP) White Mineral Oil, which does not meet the biodegradability criteria applicable to EA lubricants. A green fluid mostly refers to vegetable oil-based lubricants.

The lubricant industry has made considerable efforts in developing and marketing biodegradable lubricants. The production of environmentally friendly, rapidly biodegradable fluids for lubricants based on petrochemicals (e.g. PAOs, PAGs and SEs) has been described elsewhere [64, 76]. The most popular EA base fluids are:

- ester oils;
- PAGs; and
- low-viscosity PAOs (PAO2).

EA formulations require entirely different additives from mineral oils.

Ester oils comprise:

- vegetable oils from harvestable raw materials;
- semi-saturated, transesterified ester oils with natural fatty acids (e.g. trimethylolpropane trioleate);
- fully saturated, synthetic esters based on chemically modified vegetable oils or mineral oil (e.g. diisotridecyladipate).

Natural fatty oils such as CAS, palm oil (PMO), RSO, SBO, SNO, TLW and sperm oil have been used in lubricants for years. Triglycerides (more or less unsaturated fatty esters) are biodegradable and show excellent tribological qualities (low coefficient of friction (CoF), good wear protection). Rapid Energy Services (Lafayette, LA) manufactures and supplies a range of environmentally friendly lubricant products, such as hydraulic fluids (ISO viscosity grades from 22 to 68), penetrating oil, chain oil and marine oil. All of Rapid's bio-based lubricant products, which are formulated from renewable agricultural (natural seed) plant base oils, are equal in performance to their petroleum-based counterparts while offering complete biodegradability. The base oils and additives in these products pass and exceed acute toxicity (LC50) criteria adopted by US EPA and US Fish and Wildlife.

In terms of properties, vegetable oil base stocks are more dissimilar in nature than paraffinic and naphthenic Group I, II or III base stocks. In the biolubricants industry, RSO and canola oil appear to be the base oils of choice for biodegradable hydraulic fluids (available since 1985). However, their range of applicability is limited in view of lower stability against thermal oxidation and hydrolytic stress in comparison to mineral oils, and partly inferior cold-flow properties. These limits can be improved with additives or by using HOBS based on canola, soy, corn or sunflower oils (see Section 5.3.1). Since the advent of the EA category, a new category of commercial high-quality performance lubricants has emerged, which is renewable, sustainable, recyclable and biodegradable, namely Renewable Lubricants, Inc. (RLI)'s Stabilized™ HOBS (see Section 10.5.2).

SEs cover a broad range of chemicals with different qualities and prices (see Section 6.2.1). For the development of EA lubricants, esters have to be selected which fulfil the ecological requirements and have more favourable properties than natural fatty oils. Chemistry offers a wide choice of possibilities in the area of SEs. The most important chemical reactions to improve the properties of esters are transesterification, (selective) hydrogenation, ozonolysis and dimerisation (see Chapter 6).

SEs have been in use longer than any other synthetic-based fluid. They were originally developed as aircraft jet engine lubricants in the 1950s and are still in use as such. For EA base lubricants, the most commonly used synthetic esters are trimethylolpropane (TMP) and pentaerythritol (PE) polyol esters. Many esters are susceptible to hydrolysis [77]. Hydrolytic stability of normal polyol esters differs little from RSO, but their oxidation resistance is much higher. Both properties are much improved in complex esters (see Section 6.2.1). However, usually an improvement in hydrolytic stability is detrimental to the base fluid's biodegradability. Several complex esters (medium chain, saturated fatty acids on TMP or other polyols) combine excellent thermol-oxidative characteristics with good hydrolysis resistance and good biodegradability. SEs are among the best biodegradable hydraulic fluids. Provided they are formulated with suitable additives they can be non-toxic. They exhibit excellent lubrication properties: high VI and low friction characteristics combined with good low- and high-temperature liquidity and ageing stability. They are applicable for extreme temperature range operations. By extending oil change intervals synthetic esters stay cost competitive.

PAGs were the first biodegradable oils on the market. However, not all PAGs degrade equally well. Rate and degree of biodegradation are controlled by the propylene to ethylene oxide (PO/EO) ratio, with polyethylene glycols (PEGs) being the more biodegradable. Polypropylene glycols (PPGs) are not easily biodegradable; the same probably holds true for polybutoxylates (OSPs). Not surprisingly, the rate and extent of biodegradability diminish with increasing molecular weight. PAG fluids have excellent stability in the –45 to 250 °C range and excel where fire hazards are a concern. Oil change intervals (2000 h or once a year) are similar to mineral oils. Not all polyglycols are water soluble. EO-based polyglycols are highly water-soluble, poorly miscible with mineral oils, and of high polarity. Polyglycols with a high PO content are not water-soluble, or only slightly so, to some extent miscible with mineral oil, and are significantly less polar than PEGs. As a result of their water solubility, polyglycol-based hydraulic fluids (HEPGs) may contain water.

PAGs can have very good technical properties and are well known in long-term practical use. The water-solubility of some polyglycols (PEGs) and their incompatibility with mineral oils has limited their acceptance. HEPGs are primarily used in the water-supply industry, in off-shore applications and canal lock hydraulics as rapidly biodegradable hydraulic fluids, in particular when the application leads to unavoidable contamination of the hydraulic fluid with ingressed water [78]. Polyglycol hydraulic fluids have been available for several decades and are still widely used, particularly in the food-processing industry, in construction machinery (excavators) and in various stationary installations. However, the use of polyglycols is declining due to their aquatic toxicity when mixed with lubricating additives and their incompatibility with mineral oils and seal materials. The chemical industry is making great strides to develop biodegradable, non-water soluble PAGs. These could be alternatives to esters.

Low viscous poly-α-olefins (PAO2) are biodegradable. However, these base stocks find only limited applicability in the formulation of lubricants. The global market share of PAOs is about 1% (see Section 3.3).

Properly formulated vegetable oil and SE-based fluids are both readily biodegradable. In contrast to mineral and synthetic based equivalents, EA lubricants are non-toxic and decompose into water and CO_2. Since EA fluids are biodegradable they will break down in the presence of water and bacteria. Conceptually, EA fluids should periodically be monitored to ensure that biodegradation is not occurring. While the base stocks of EA lubricants are non-toxic, additives in the formulations may lead to a wide range of toxicity.

The VI of most EA fluids meets or exceeds the VI of petroleum-based fluids. The pour point of non-stabilised vegetable oil-based hydraulic fluids and lubricants may be acceptable for many applications. EA lubricants normally produce good wear properties. Most formulated EA fluids do not have foaming problems. Some common paints used in fluid systems are incompatible with many EA fluids. Polyurethane seals should not be used with EA fluids and should be replaced by Viton and Buna N. EA fluids are compatible with steel and copper alloys and provide excellent rust protection. The fluid manufacturer should be consulted for specific compatibility data for each material used in the application. Disposal costs for EA oils are greater than for mineral oils due to reluctance of traditional recyclers to accept them (see Section 14.7).

EA (bio)lubricants are required in many applications such as chain saw oils, wire rope lubrication, marine oils, outboard engine lubricants, lubricants for the food industry, hydraulic fluids in forestry, agricultural equipment and building machinery, elevator oils, etc. In lubricant development the greatest problem for each product is to find the optimal compromise between the opposite requirements of long-term stability and rapid biological degradation. Successful application of bio-based materials in lubrication in various industrial sectors requires a thorough understanding of their tribochemical properties. For example, changes in the degree of unsaturation affect lateral interactions of adsorbate molecules, which change their adsorption and, hence, their boundary lubrication properties.

Formulation of biodegradable lubricants is a reverse of the present mineral-oil dominated lubricants market. It is accepted that over 90% of all lubricants could be made rapidly biodegradable. Nowadays, some Western European countries and Scandinavia are the greatest consumers of biodegradable lubricants. In Germany 5% of lubricant base oils were rapidly biodegradable (natural and synthetic) esters in 2005. Over the years the focus on lubricants has shifted from biodegradability to renewability. This influences the design of such products.

Descriptors such as environmentally friendly or compatible lubricants suggest there is no interaction with the natural surroundings. Since any oily substance stains the soil, the term 'environmentally adapted lubricants' has been suggested to express minimal harm and hazard to nature [79]. There are several qualifications that make an environmentally adapted lubricant (Table 3.13) [21, 79]. It appears that 'adapted' and 'acceptable' are interchangeable terms in this context. Future lubricants need to be more envi-

Table 3.13 Qualifications of environmentally adapted lubricants

- Good and stable properties
- High performance level
- Relative content of renewable raw material
- Biodegradable (ASTM D 5864)
- Non-toxic (EPA 560/6-82-002)
- No bioaccumulability and biomagnification
- Low total life-cycle cost (LCC)
- Energy saving and fuel economy
- Compliance with environmental regulations

After ref. [21].

ronmentally adapted, have a higher level of performance, and lower total life-cycle cost (LCC) than presently used lubricants [21]. As environmental regulations become more restrictive, finding EA lubricants that comply with manufacturers' specifications for turbines, pumps, gear drives and other equipment is becoming increasingly difficult. Accelerating research and development in the area of EA lubricants has been driven by public demand, industry concerns and governmental agencies.

3.6 Physicochemical properties of lubricants

Effective lubricants, either bio-based or petroleum-based, must meet several performance requirements. They should properly coat load-bearing surfaces to protect against wear and corrosion, reduce friction and not decompose under working conditions. Physical, chemical and film formation properties of base fluids influence the formulated lubricant performance [21]. Under boundary conditions, the friction coefficient is essentially independent of lubricant viscosity, and becomes dependent on the nature of the contacting metal surfaces. Surface additives can be used to modify the surface chemistry and reduce friction. A lubricant is vital for proper functioning of engines and machines.

Tribology provides the basis for understanding the factors influencing friction and lubrication. The most important, and obvious, property required of a lubrication fluid is the ability to reduce friction. This can be measured quantitatively by ball-on-disk friction and wear testing equipment. The instrument allows calculation of the CoF. Using the Langmuir model [80] or the Temkin adsorption isotherm model (eq. 3.1):

$$K = \exp\left[-\Delta H^* (1 - \alpha\theta)/RT\right] \qquad [3.1]$$

where K is the association equilibrium constant, ΔH^* is the adsorption enthalpy, $R = 8.317$ J mol^{-1} K^{-1}, T is the temperature (K), θ is the surface coverage, and α is a lateral interaction term, the effective adsorption energy of the lubricity additive is obtained. This may be used to derive an adsorption constant that is useful for comparison of materials [81–84].

The molecular structure of lubricants has a direct effect in elastohydrodynamic (EHD) and boundary lubrication. The effect of a lubricant's molecular structure on its friction behaviour is often described in terms of the so-called paper stack model [85]. The stack, composed of vigorously compressed lubricant molecules, is subject to a rate of shear. The resistance to shearing depends on the molecular shape, branching and the presence of active centres (such as oxygen, nitrogen or heteroatoms of functional groups). Minimising functional groups involved in intermolecular interactions facilitates molecular motion and decreases friction. The effects of molecular structure become more important at elevated pressures.

3.6.1 Viscosity

The ability to properly coat working surfaces is commonly quantified by measurements of the viscosity and VI. The properties of a lubricant are further determined by certain important characteristics such as compressibility, stability, PP, cloud point (CP), flash point and acid value (AV) [29]. Viscosity and VI are among the most important physical properties of any lubricant. Viscosity mainly controls the hydrodynamic behaviour of lubricants. Viscosity of an oil is essentially a measure of the oil resistance to shear. It is the friction between lubricant molecules or internal resistance to flow. High viscosity implies high resistance to flow. Viscosity varies inversely with temperature, is also affected by pressure and, in some cases, by the shear rate. The viscosity–pressure behaviour is described by the pressure–viscosity coefficient α, the value of which reflects the rate of oil viscosity increase under increasing pressure. Higher pressure enhances the load-carrying capacity of the oil. This property enables use of thin oils to lubricate heavy machinery. Load-carrying capacity also increases with the machine operating speed. Viscosity is commonly measured as shear stress and shear rate, or as the time required for an oil sample to flow through a standard orifice at a given temperature. Viscosity measured directly by shear, referred to as the dynamic viscosity η, is expressed in centiPoise (1 cP = 1 mPa s). The oil industry usually employs kinematic viscosity ν (dynamic viscosity divided by oil density), expressed in centiStokes (1 cSt = 1 mm^2/s), conventionally at 40 and 100 °C. The kinematic viscosity values (at 40 °C) are at the basis of the ISO grades for lubricating oils for industrial use. Viscosity as measured in time is expressed in SUS at 100 °F and 210 °F. The effective viscosity of a lubricant in a bearing may be different from the quoted viscosity measured by a standard test method, and the difference depends on the shear rate in the bearing [86].

With the modern VI improvers and other additives, kinematic viscosity (as measured by the time it takes a quantity of the oil to flow through an orifice at a specified temperature) is not the best measure of how the oil reacts in service under the influence of shearing effects. The high-temperature high-shear (HTHS) viscosity is a very different measure from the kinematic viscosity and is probably the best representation for the friction characteristics of an operating engine. The HTHS test is designed to simulate motor oil viscosity in operating crankshaft bearings. HTHS gives a much more realistic measure of what happens to the oil in an engine. Most vehicle engine components, including piston rings, operate in the hydrodynamic lubrication regime. Under the high shear rates encountered, multigrade oils have lower viscosity than monograde oils. This is due to (temporary) thinning of the lubricant, as the VI improvers align in the shear field in the lubricated components. Under high stress conditions within an engine polymeric VI improvers may even break down, leading to a (permanent) decrease in oil viscosity. These phenomena also explain why (VI improver-containing) multigrade lubricants produce lower fuel consumption than monograde oils (see Fig. 11.12).

Lubricant viscosity is a crucial property in application. Too high a viscosity of the lubricant for a given load means an excessive hydrodynamic film thickness, which translates into a waste of energy in shearing the thick film, heat development, and accelerated lubricant decomposition and mechanical failure of the surfaces. At too low a viscosity the system will operate in the boundary lubrication regime, leading to energy waste in overcoming high frictional forces, wear of metal components and reduced equipment lifetime.

VI is an arbitrary, dimensionless number which expresses the resistance of a lubricant to viscosity change with temperature. It is therefore considered a good indicator of the useful temperature range of a lubricant. Before the emergence of synthetic oil, the Dean and Davis viscosity index, defined as a function of Saybolt Universal viscosities of an oil at 100 °F and 210 °F, was on a scale of 0 to 100 [87]. With the advent of synthetic oils and additives, the scale now ranges from negative values to over 200 (super high VI), but VI still remains as an indicator of quality. A lubricating oil with high VI displays only small changes in viscosity with temperature, which denotes stable viscosity. The higher its VI value, the greater the resistance of a lubricant to thicken at low temperatures and thin out at high temperatures. The VI of an oil is crucial in selecting a lubricant for a given application. Lubricants with VI > 100 are often considered useful in many applications. Failure to use an oil with the proper VI when temperature extremes are expected may result in poor lubrication and equipment failure. Knowledge of VI is particularly relevant in extreme climatic conditions.

An ideal lubricant for most purposes would possess the same viscosity at all temperatures. All lubricants depart from this ideal, some more than others. For example, lubricating oils derived from highly paraffinic crudes have higher VI values than lubricating oils derived from highly naphthenic crudes. This difference was used, in fact, to fix the limits of 0 to 100 on the Dean and Davis scale, these values having been assigned to a poor naphthene-base oil and a good paraffin-base oil, respectively. Molecules having a high VI typically consist of long alkyl chains, have low aromatic content, no ethyl branching and no tertiary carbons. Bio-oils have a better VI than mineral oil products, without the need to add shear sensitive additives. The higher density of bio-oils also determines a higher dynamic to kinematic viscosity ratio. Dynamic viscosity is important for lubrication.

Although VI is empirically taken as a measure of the effect of temperature changes on the viscosity of an oil, it does not correlate with the well-defined flow activation energy E_a, which is the theoretically defined dependence of viscosity on temperature (slope of ln (viscosity) vs $1/T$) [88]. Therefore, the E_a should be preferred over VI as a measure of the influence of temperature on viscosity. Not surprisingly, the VI does not even always give a proper representation of the effect of temperature on the kinematic viscosity. However, despite the fact that VI is outdated and has no theoretical background, it continues to be widely used in the refining and lubricant industries.

In the absence of structural viscosity (i.e. shear rate dependency) lubricating oils are Newtonian and viscosity requirements vary with the industrial application. Typical requirements for the kinematic viscosity of lubricants at 40 °C range from 30 mm^2/s in the automation industry to 120 mm^2/s for bearings, and 240 mm^2/s for four-stroke engines and special gear assemblies [89]. The operational advantages offered by a lubricant having a high VI include principally less friction due to viscous 'drag' at low temperatures as well as reduced lubricant loss and lower wear at higher temperatures.

3.6.2 Temperature-related properties

The gelation index (GI) indicates an oil's tendency to form a gelated structure in the oil at lower temperature. GI is the largest rate of change of viscosity increase when an oil is slowly cooled from –5 °C to the lowest test temperature (ASTM D 5133).

The PP is the lowest temperature at which an oil will flow. This property is especially crucial for oils that must flow at low temperatures. A PP should be at least 10 °C lower than the lowest anticipated ambient temperature. A fundamental requirement of a lubricating oil is that it should be liquid over a broad temperature range. The usable liquid range is confined by PP (at low temperatures) and flash point (at high temperatures). For the most demanding applications, such as jet engine lubricants, an effective liquid range of over 300 °C may be required. Fats, especially animal fats, have a high pour point and generally exhibit a narrow viscosity range.

CP is the temperature at which dissolved particles and substances, such as paraffin wax, start solidifying and separating from the oil. The temperature drops, wax crystallises and becomes visible. Oils should generally be kept at temperatures above CP to prevent filter clogging.

Flash point is the lowest temperature to which a lubricant must be heated before its vapour, mixed with air, will ignite but not continue to burn. Fire point is the temperature at which lubricant combustion will be sustained. Flash and fire points are useful in determining volatility and fire resistance of a lubricant. In addition to its use for characterisation of base fluids and lubricants, the flash point is a very important characteristic for classification of products according to the international transport regulations. The flash point is commonly used also to determine the storage temperature requirements for lubricants. Auto-ignition temperature is the temperature at which a lubricant will ignite spontaneously without an external ignition source. The direct effect of an ester group on the physical properties of a lubricant is that it raises the flash point and lowers the lubricant volatility.

Volatility is a property of a liquid that defines its evaporation characteristics and is usually evaluated by tests for flash point, simulation distillation and volatility weight loss (NOACK). The NOACK volatility (ASTM D 5800) test is used to determine the evaporative losses of motor oil at high temperatures.

3.6.3 Acidity and alkalinity

The AV or neutralisation number is a measure of acidity of the (vegetable) oil. The total acid number (TAN) is the amount of KOH (in mg/g) needed for neutralisation of the acids in a lubricant or petroleum product. Hamblin *et al.* have described tests for TAN assessment and modifications thereof [90]. The TAN value can be deduced by a couple of different methods: according to non-aqueous potentiometric titration (ASTM D 664) or colorimetric titration (ASTM D 974), both with sample preparation, or directly – without sample preparation – by thermometric titration [91]. The acidity of unused mineral oils and fluids is normally derived from the type and concentration of specific additive material whereas the acidity of used oil is of interest as a measure of the degree of oxidation of a fluid. Acids are formed as oils oxidise due to age and service. The AV can be taken as an indication of ageing of the oil and can be used to determine when an oil must be changed. The TAN value indicates the potential of corrosion problems.

The total base number (TBN) is a measure of a lubricant's reserve alkalinity and is the reverse measurement of TAN. It is particularly used for engine oils where acidic combustion products use up the alkaline reserve. The higher the TBN value, the more effective is a lubricant in suspending wear-causing contaminants and reducing the corrosive effects of acids over an extended period of time. The associated measurements (ASTM D 2896 and ASTM D 4739–06) generally range from 6–80 mg KOH/g in modern lubricants to 7–10 mg KOH/g for general automotive use. For TBN < 2 mg KOH/g the lubricant is considered inadequate for engine protection and is at risk of corrosion.

3.6.4 Solvency

The aniline point (ASTM D 611, DIN 51775) measures the solubility temperature of aniline and a lubricant and is expressed as the temperature at which equal volumes of a lubricant and aniline separate in two phases. The aniline point depends on the products $V_M(\delta_{anil} - \delta_{oil})^2$, were V_M is the average molecular volume, and δ_{anil} and δ_{oil} are the Hildebrand solubility parameters for aniline and for the oil. The aniline point reflects the structure of base oils. Increasing molecular volumes raises the aniline point. The aniline point steadily increases with increasing viscosity for oils with identical polarity. In the series PAO2 to PAO40 it rises from ca 100 to 160 °C.

The aniline point can also be used to give an indication on the structure of mineral oils. The higher the temperature of separation the more paraffinic is the oil. It also gives an indication of the solvency behaviour of the oil: the lower the aniline point, the better the solvency for most types of additive. Because of their good solubility, aromatic structures give the lowest aniline values (see Section 3.3).

As a measure of solvency of a lubricant the aniline point is also profitably used as a good test for judging if a base stock will properly condition the most commonly used seals [92]. To reduce equipment leakage, it is important that elastomeric seals and gaskets maintain a slight positive swell and remain pliable. For a small positive swell the aniline point for Buna N and Neoprene should preferably be in the range of 65–75 °C, for silicone rubbers 90–100 °C. These figures do not necessarily apply to lubricants which contain seal conditioning additives. The term 'elastomer compatibility' refers to the tendency of a lubricant to alter the physical characteristics of elastomer seals with which it is in contact. To experimentally test and describe this tendency, static elastomer samples are stored in a container filled with the lubricant concerned under specified conditions (T, t), according to DIN ISO 1817. The change in volume, weight and mechanical characteristics is then determined. The ability of a lubricant to act as a seal is directly related to the consistency of the lubricant and its application. In applications requiring thick lubricants or grease (as in sleeve bearings) dirt and moisture are sealed out.

Lubricant base stocks influence lubricant performance through two main functions: solubility and oxidative stability. Synthetic base oils exhibit very specific additive solubility that is different from mineral oils. Solvent power of a lubricant is chiefly determined by the polarity of oil molecules. Non-polar base stocks (e.g. PAOs, paraffinics) have low solvent power. Polar base stocks (e.g. esters, alkylated aromatics) have high solvent power. Differences in polarity between various base stocks can be illustrated using the Teas diagram of Fig. 3.2. Groups II–IV base oils contain fully saturated hydrocarbon molecules, hence polar and hydrogen-bonding interactions are negligible. Esters reveal more significant polar interactions. Ionised vegetable oils have both polar interactions and hydrogen-bonded interactions.

Lubricity and solvent power are interrelated properties. However, solvent power alone does not guarantee good lubricity. Lubricity requires that polar and non-polar molecules be present simultaneously. Since metal surfaces

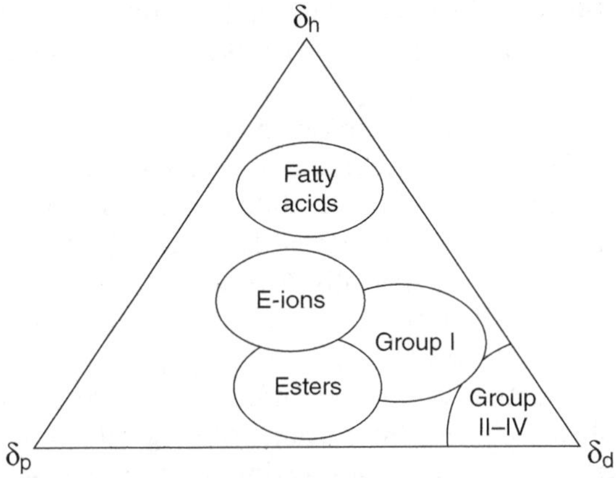

3.2 Teas diagram showing relative contributions of various types of intermolecular interactions to the intermolecular cohesive energy: δ, Hildebrand solubility parameter; d, dispersion interaction; p, polar interaction; and h, hydrogen-bonding interaction. After ref. [40].

are highly polar, polar oil molecules dissolved in non-polar ambient tend to adsorb to the metal surface, forming a protective surface film. Strength of the film and solvent power are linked to the same cohesion parameter. Group I base oils have sufficiently high content of polar species (heterocycles, alkylated aromatics) and demonstrate superior lubricity over Group II–IV base oils. Group II–IV base oils will benefit the greatest in terms of lubricity using boundary lubricity additives (or friction modifiers), such as fatty amides, or vegetable oils. Solubility and lubricity improvers help formulators address certain challenges brought about by 'dry' base stocks (see Section 3.3).

3.6.5 Adsorption, heat transfer and compressibility

At the level of the chemistry of lubrication, several factors affect the interactions occurring between the surfaces in motion and the interposed lubricant. A crucial property of lubricants is their ability to maintain a lubricating film of sufficient hydrodynamic stability at the metal contact zone. It is desirable to design lubricant molecules with optimum structure for effective metal adsorption and excellent boundary lubrication properties. The adsorption properties of friction-reducing additives, at monolayer coverage, operating in the boundary lubrication regime, have been reviewed [93]. The adsorption properties of typical friction reducing additives can be accounted for by hydrogen bonding to surfaces and the lateral interactions between adsorbate molecules. As long as a lubricant coats a surface it automatically reduces corrosion. Corrosion does not take place very quickly on parts that have been lubricated. Most lubricants are formulated with corrosion inhibitors and other chemical additives that improve their film consistency and reduce or limit corrosion from occurring.

A lubricant can remove heat from a surface by transferring it to a cooler surface; it thus acts as a heat transfer medium. Because of its chemical and physical composition, a lubricant has limited ability to retain heat compared with substances such as water. Absorbed heat leads to adverse effects on its lubricant characteristics. Even specially formulated high-temperature greases have heat-absorbing limitations before the oil components break down.

Temperature is an important parameter affecting the performance of lubricant ingredients. It affects such important tribological characteristics as viscosity, film thickness, adsorption, desorption, friction and wear. Temperature also promotes oxidation, polymerisation and degradation, which negatively impact the useful life of lubricant ingredients [94]. Qualifying thermal properties for lubricants are thermal conductivity and heat capacity. High thermal conductivity and high thermal capacity favour film formation, especially under high-slip and high-load conditions. SEs normally have higher thermal conductivity and heat capacity than mineral fluids.

Compressibility is a measure of the amount of volume reduction due to the pressure. It increases with temperature and pressure and is particularly relevant for hydraulic oils. According to existing theories of liquid structure, the more degrees of freedom in the molecular structure of a liquid, the higher is its compressibility [95]. Compressibility is sometimes expressed by the bulk modulus, which is the reciprocal of compressibility.

3.6.6 Quality and stability

The quality of a lubricant is the length of time that the fluid's essential properties will continue to perform as expected. Stability of a lubricant is an important property affecting service life. Foam resistance, water separation, rust prevention and physicochemical changes to the oil induced by processes such as oxidation, hydrolysis, cavitation, mechanical stress, etc., affect the performance of the lubricant. The lubricant must maintain its essential performance properties throughout long operating lifetimes (in some cases 10–20 yrs), during which time it may be exposed to atmospheric oxygen at high temperatures. The fluid physical properties are affected by thermal degradation and auto-oxidation. Consequently, a lubricant must possess extremely good thermo-oxidative stability in order to maintain performance. It must also be inert to potential chemical contact reactants, e.g. water, fuels and process gases. Under extremely high shear rates, typical of lubricated contacts, a large lubricant molecule must be stable to mechanical fragmentation.

A lubricant must also exhibit various application properties, such as AW activity, which is a function of the lubricant chemistry under extreme conditions of asperity contact. It is necessary to choose an optimal formulation (base oil and additives) to stabilise the fluid for a specific application. For the contribution of base stock and additives to lubricant properties, see Chapter 7.

Apart from physicochemical parameters, lubricant design also needs to consider economic, safety and environmental constraints.

3.7 Sources of further information and advice

L.Z. Pillon, *Surface Activity of Petroleum Derived Lubricants*, CRC Press, Boca Raton, FL (2010).

H.P. Bloch, *Practical Lubrication for Industrial Facilities*, The Fairmont Press, Inc., Lilburn, GA (2009).

T.R. Lynch, *Process Chemistry of Lubricant Base Stocks*, CRC Press, Boca Raton, FL (2008).

Th. Mang and W. Dresel (eds), *Lubricants and Lubrication*, 2nd edn, Wiley-VCH, Weinheim (2007).

J.S. Marth, *Renewable Lubricants Manual: Biobased Oils, Fluids & Greases*, United Bio Lube, Palo Alto, CA (2007); http://www.renewablelubricants.com

L.R. Rudnick (ed.), *Synthetics, Mineral Oils, and Bio-based Lubricants: Chemistry and Technology*, CRC Press/Taylor & Francis, Boca Raton, FL/New York, NY (2006).

A. Igartua, J. Barriga and A. Aranzabe (eds), *Biodegradable Lubricants*, Virtual Tribology Institute Edition, Eibar (2005).

S.Z. Erhan (ed.), *Industrial Uses of Vegetable Oils*, CRC Press, Boca Raton, FL (2005).

A. Lansdown, *Lubrication and Lubricant Selection: A Practical Guide*, 3rd edn, ASME Press, Champaign, IL (2004).

S.Z. Erhan and J.M. Perez (eds), *Biobased Industrial Fluids and Lubricants*, AOCS Press, Champaign, IL (2003).

G.E. Totten, S.R. Westbrook and R.J. Shah (eds), *Fuels and Lubricants Handbook: Technology, Properties, Performance, and Testing*, ASTM International, West Conshohocken, PA (2003).

V. Štěpina and V. Veselý, *Lubricants and Special Fluids*, Elsevier, Amsterdam (2002).

D.M. Pirro and A.A. Wessol (eds), *Lubrication Fundamentals*, 2nd edn, CRC Press, Boca Raton, FL (2001).

M.J. Neale, *Lubrication, A Tribology Handbook*, Butterworth-Heinemann, Ltd, Oxford (2001).

J. Denis, J. Briant and J.-C. Hipeaux, *Lubricant Properties Analysis and Testing*, Éditions Technip, Paris (2000).

L.R. Rudnick and R.L. Shubkin (eds), *Synthetic Lubricants and High-performance Functional Fluids*, 2nd edn, Marcel Dekker, Inc., New York, NY (1999).

G. Knothe and J.T.P. Derksen (eds), *Recent Developments in the Synthesis of Fatty Acid Derivatives*, AOCS Press, Champaign, IL (1999).

E.N. Frankel, *Lipid Oxidation*, The Oily Press, Dundee (1998).

F.D. Gunstone and F.B. Padley (eds), *Lipid Technologies and Applications*, CRC Press, Boca Raton, FL (1997).

E.R. Booser (ed.), *Tribology Data Handbook: An Excellent Friction, Lubrication, and Wear Resource*, CRC Press, Boca Raton, FL (1997).

A. Sequeira, *Lubricant Base Oil and Wax Processing*, Marcel Dekker, Inc., New York, NY (1994).

W.J. Bartz (ed.), *Biologisch schnell abbaubare Schmierstoffe und Arbeitsflüssigkeiten*, Expert-Verlag GmbH, Ehningen (1993).

R.L. Shubkin (ed.), *Synthetic Lubricants and High-performance Functional Fluids*, Marcel Dekker, Inc., New York, NY (1993).

W.H. Stadtmiller and A.N. Smith (eds), *Aspects of Lubricant Oxidation*, ASTM International, West Conshohocken, PA (1986).

A. Dorinson and K.C. Ludema, *Mechanics and Chemistry in Lubrication*, Elsevier, Amsterdam (1985).

D. Klamann, *Lubricants and Related Products*, VCH Verlagsgesellschaft, Weinheim (1984).

E.R. Braithwaite, *Solid Lubricants and Surfaces*, Pergamon Press, London (1964).

R.C. Gunderson and A.W. Hart (eds), *Synthetic Lubricants*, Reinhold, New York, NY (1962).

3.8 References

[1] S. Boyde, Green lubricants – Environmental benefits and impacts of lubrication, *Green Chem.* **4** (2002) 293–307.

[2] J.C.J. Bart, N. Palmeri and S. Cavallaro, *Biodiesel Science and Technology. From Soil to Oil*, Woodhead Publishing/CRC Press, Cambridge/Boca Raton, FL (2010).

[3] Committee on Biobased Industrial Products – National Research Council USA (ed.), *Biobased Industrial Products: Research and Commercialization Priorities*, National Academies Press, Washington, DC (2000).

[4] J.H. Clark and F.E.I. Deswarte (eds), *Introduction to Chemicals from Biomass*, John Wiley & Sons, Ltd, Chichester (2008).

[5] G. Centi and R.A. van Santen (eds), *Catalysis for Renewables*, Wiley-VCH, Weinheim (2007).

[6] B. Kamm, P.R. Gruber and M. Kamm (eds), *Biorefineries – Industrial Processes and Products. Status Quo and Future Directions*, Wiley-VCH, Weinheim (2006), 2 vols.
[7] M. Eggersdorfer, S. Warwel and G. Wulff (eds), *Nachwachsende Rohstoffe – Perspectiven für die Chemie*, VCH, Weinheim (1993).
[8] T. Werpy, G. Petersen and US Department of Energy (eds), *Top Value Added Chemicals from Biomass*, US DOE, Office of Scientific and Technical Information, Washington, DC (2004).
[9] H. Eierdanz (ed.), *Perspectiven nachwachsender Rohstoffe in der Chemie*, VCH, Weinheim (1996).
[10] M. Kjellin and I. Johansson (eds), *Surfactants for Renewable Resources*, John Wiley & Sons, Ltd, Chichester (2010).
[11] A. Vertès, N. Qureshi, H.P. Blaschek and H. Yukawa (eds), *Biomass to Biofuels*, John Wiley & Sons, Inc., Hoboken, NJ (2010).
[12] W. Soetaert and E. Vandamme (eds), *Biofuels*, John Wiley & Sons, Ltd, Chichester (2009).
[13] Th.F. McGowan, M.L. Brown, W.S. Bulpitt and J.L. Walsh Jr. (eds), *Biomass and Alternate Fuel Systems: An Engineering and Economic Guide*, John Wiley & Sons, Inc., Hoboken, NJ (2009).
[14] C.V. Stevens and R. Verhé (eds), *Renewable Bioresources: Scope and Modification for Non-food Applications*, John Wiley & Sons, Ltd, Chichester (2004).
[15] L.R. Rudnick (ed.), *Synthetics, Mineral Oils, and Bio-based Lubricants: Chemistry and Technology*, CRC Press/Taylor & Francis, Boca Raton, FL/New York, NY (2006).
[16] S.J. Miller, M.A. Shippey and G.M. Masada, Advances in lube base oil manufacture by catalytic hydroprocessing, NPRA National Fuels and Lubricants Meeting, Houston (1992).
[17] P.V. Snyder, GTL lubricants: The next stop, NPRA Lubricants and Waxes Meeting, Houston (1999).
[18] L.R. Rudnick and R.L. Shubkin (eds), *Synthetic Lubricants and High-performance Functional Fluids*, 2nd edn, Marcel Dekker, Inc., New York, NY (1999).
[19] H. Dressler, Alkylated aromatics, in *Synthetic Lubricants and High-performance Functional Fluids* (R.L. Shubkin, ed.), Marcel Dekker, Inc., New York, NY (1993), pp. 125–44.
[20] E.D. Sowle and T.M. Lachocki, Linear dialkylbenzenes as synthetic base oils, *Lubric. Engng.* **52** (1996) 116–20.
[21] A. Pettersson, High-performance base fluids for environmentally adapted lubricants, *Tribol. Intl.* **40** (2007) 638–45.
[22] R.L. Shubkin, Polyalphaolefins, in *Synthetic Lubricants and High-performance Functional Fluids* (R.L. Shubkin, ed.), Marcel Dekker, Inc., New York, NY (1993), pp. 1–40.
[23] L.R. Rudnick, Polyalphaolefins, in *Synthetics, Mineral Oils and Bio-based Lubricants: Chemistry and Tribology* (L.R. Rudnick, ed.), CRC Press, Boca Raton, FL (2006), pp. 3–36.
[24] S. Lawford, Polyalkylene glycols, in *Synthetics, Mineral Oils and Bio-based Lubricants* (L.R. Rudnick, ed.), CRC Press, Boca Raton, FL (2009), pp. 119–38.
[25] M.R. Greaves, *Compoundings Mag.* **59**(4) (2009) 25–6.

[26] M.R. Greaves, Oil soluble synthetic polyalkylene glycols. A new type of Group V base oil, *Lube Mag.* **104** (2011) 21–4.
[27] D.R. Kodali, High performance ester lubricants from natural oils, *Ind. Lubr. Tribol.* **54** (2002) 165–70.
[28] W.R. Downey Jr, Impact of GTL technology on the future of the lubes business, UEII World Congress, Rome (2005).
[29] US Army Corps of Engineers, *Engineering and Design, Lubricants and Hydraulic Fluids, Engineer Manual*, 1110-2-1424, Washington, DC (1999).
[30] S. Bhuyan, S. Sundarajan, L. Yao, E.G. Hammond and T. Wang, Boundary lubrication properties of lipid-based compounds evaluated using microtribological methods, *Tribol. Lett.* **22** (2006) 167–72.
[31] T. Murakami and H. Sakamoto, Lubricity properties of vegetable oils and paraffinic oils with unsaturated fatty acids under high-contact-pressure conditions in four-ball tests, *J. Synth. Lubr.* **20**(3) (2003) 183.
[32] Y. Cao, L. Yu and W. Liu, Study of the tribological behaviour of sulfurized fatty acids as additives in rapeseed oil, *Wear* **244** (2000) 126–31.
[33] N.J. Fox, B. Tyrer and G.W. Stachowiak, Boundary lubrication performance of free fatty acids in sunflower oil, *Tribol. Lett.* **16**(4) (2004) 275–81.
[34] L.R. Rudnick and W.J. Bartz, Comparison of synthetic, mineral oil, and bio-based lubricant fluids, in *Synthetics, Mineral Oils and Bio-based Lubricants: Chemistry and Technology* (L.R. Rudnick, ed.), CRC Press/Taylor & Francis, Boca Raton, FL/New York, NY (2006), pp. 331–49.
[35] API Standard 1509, *Engine Classification and Guide to Crankcase Oil Selection*, American Petroleum Institute, Washington, DC (2002).
[36] ISO Standard 3448, *Industrial Liquid Lubricants – ISO Viscosity Classification*, 2nd edn, International Organization for Standardization (ISO), Geneva (1992).
[37] NLGI homepage http://www.nlgi.com
[38] SAE homepage http://www.sae.org/servlets/techtrack?PROD_TYP=STD
[39] AGMA 9005-D94, *Industrial Gear Lubrication Standard*, American Gear Manufacturers Association, Alexandria, VA.
[40] B. Zhmud and M. Roegiers, New base oils pose a challenge, *Tribol. Lubr. Technol.* **65**(7) (2009) 34–9.
[41] A. Gosalia, The European base oils and lubricants industry. Trends in global context, *Lube Mag.* **95** (2010) 8–16.
[42] G.S. Agashe and M. Phadke, *Global Lubricant Baseoils Supply Overview*, Kline & Company, Inc. (18 February 2005); http://www.icislor.com; M. Phadke, Brightstock shortage, *Lube Mag.* **107** (2012) 8–9.
[43] W.R. Murphy, D.A. Blain and A.S. Galiano-Roth, Synthetic basics – benefits of synthetic lubricants in industrial applications, *J. Synth. Lubr.* **18**(4) (2002) 301–25.
[44] H. Beiβner and W. Schüler, Dry lubricants for bobbinet machines, DDR Patent No. DD 44,117 (23 December 1965).
[45] S.H. Wu, B.K. Cheng, G.A. Nichols and J.H. Park (to Mallinckrodt, Inc.), Use of magnesium stearate hydrate for lubrication of solid industrial or consumer products, PCT Int. Publ. No. WO 2009/114227 A1 (17 September 2009).
[46] E. Weiss, *Castor, Sesame and Safflower*, Leonard Hill, London (1971), Chapter 11.

[47] R.M. Gresham, Bonded solid film lubricants, in *Tribology Data Handbook* (E.R. Booser, ed.), CRC Press, Boca Raton, FL (1997), pp. 600–7.
[48] X. Zhang, J. Golding and I. Burgar, Thermal decomposition chemistry of starch studied by ^{13}C high-resolution solid-state NMR spectroscopy, *Polymer* **43** (2002) 5791–6.
[49] J.A. Schey, *Tribology in Metalworking: Friction, Lubrication and Wear*, American Society of Metals, Metals Park, OH (1983), pp. 27–130.
[50] M. Meiler, M. Pfestorf, M. Geiger and M. Merklein, The use of dry film lubricants in aluminum sheet metal forming, *Wear* **255** (2003) 1455–62.
[51] M. Meiler, M. Pfestorf, M. Merklein and M. Geiger, Tribological properties of dry film lubricants in aluminum sheet metal forming, *Proc. 2nd ICTMP*, Lyngby (2004).
[52] H. Yoshimura, S. Torikai, T. Nishihara, T. Nonishi and N. Inouchi, Application of wheat flour lubricants to the press-forming process, *J. Mat. Proc. Technol.* **125–6** (2002) 375–8.
[53] H. Yoshimura, What? Wheat flour can become a lubricant?, *Materia* **39**(2) (2000) 164–5.
[54] C. Perrino, S. Lee and N.D. Spencer, End-grafted sugar chains as aqueous lubricant additives: Synthesis and macrotribological tests of poly(L-lysine)-*graft*-dextran (PLL-*g*-dex) copolymers, *Tribol. Lett.* **33**(2) (2009) 83–96.
[55] J.K. Lancaster, Solid lubrication, in *CRC Handbook of Lubrication (Theory and Practice of Tribology)*, Vol. II, *Theory and Design* (E.R. Booser, ed.), CRC Press, Boca Raton, FL (1983), pp. 269–90.
[56] C. Busch, Solid lubrication, in *Lubricants and Lubrication* (Th. Mang and W. Dresel, eds), Wiley-VCH, Weinheim (2007), pp. 694–714.
[57] P. Deshmukh, M. Lovell, W.G. Sawyer and A. Mobley, On the friction and wear performance of boric acid lubricant combinations in extended duration operations, *Wear* **260** (2006) 1295–304.
[58] H. Ihrig, Environmentally harmless lubricants in the ninetees, *Tribol. Schmierungstechn.* **39** (1992) 121–5.
[59] J.F. Carpenter, Biodegradability and toxicity of polyalphaolefin (PAO) base stock, *J. Synth. Lubr.* **12** (1995) 13–20.
[60] N.J. Novick, P.G. Mehata and P.B. Goldrick, Assessment of the biodegradability of mineral oil and synthetic base stock, using CO_2 ultimate biodegradability tests and CEC L-33-T-82 test, *J. Synth. Lubr.* **13** (1996) 19–30.
[61] S.D. Haigh, Biodegradation of synthetic lubricants in soil, *J. Chem. Technol. Biotechnol.* **55** (1992) 406–9.
[62] S. Harold, *Biodegradability: Review of the Current Situation*, External report available from Lubrizol International Laboratory, PO Box 88, Belper, Derby DE56 1QN (1993).
[63] S.J. Asadauskas, J.M. Perez and J.L. Duda, Suitability of basestocks for biodegradable lubricants, *Prepr. ACS Div. Petr. Chem.* **42** (1997) 246–9.
[64] Th. Mang, Lubricants, in *Lipid Technologies and Applications* (F.D. Gunstone and F.B. Padley, eds), Marcel Dekker, Inc., New York, NY (1997), p. 737.
[65] R. van Voorst and F. Alam, Polyglycols as base fluids for environmentally-friendly lubricants, *J. Synth. Lubr.* **16**(4) (2000) 313–22.

[66] M. Rees and I. Macpherson (to Ethyl Petroleum Additives, Ltd.), Biodegradable lubricants and functional fluids, Eur. Patent No. EP 0,468,109 A1 (24 July 1990).

[67] C.B. Duncan (to ExxonMobil), Biodegradable synthetic ester base stocks formed from branched oxo acids, Eur. Patent No. EP 0,904,338 B1 (21 May 1997).

[68] R.J. Wittenbrink, D.F. Ryan and R.F. Bauman (to ExxonMobil Research and Engineering Co.), Process for the production of biodegradable high performance hydrocarbon base oils, Eur. Patent No. EP 0,876,446 B1 (15 November 1996).

[69] D. Kramer, Base oil supply. Demand and quality issues, *Proc. 8th Ann. Fuels & Lubes Asia Conference*, Singapore (29 January – 1 February 2002).

[70] I.-Ching Chiu, S. Gunsel and P. Lacey (to Pennzoil-Quaker State Co.), Environmentally friendly lubricants, PCT Int. Publ. No. WO 03/027212 A1 (3 April 2003).

[71] S.J. Randles, Formulation of environmentally acceptable lubricants, *Proc. 49th STLE Ann. Mtg*, Pittsburgh, PA (1–5 May 1994).

[72] B. Lehmann, Environmentally acceptable synthetic operational fluids and lubricating greases, in *Biologically Fast Degradable Lubricants and Operational Fluids* (W.J. Bartz, ed.), Expert-Verlag, Ehningen (1993) (in German).

[73] F. Farshchi, Why invest in environmentally considerate lubricants?, *World Pumps* **496** (2008) 36–7.

[74] W.J. Bartz, Lubricants and the environment, *Tribol. Intl.* **31**(1–3) (1998) 35–47.

[75] R. Luther, Lubricants in the environment, in *Lubricants and Lubrication* (Th. Mang and W. Dresel, eds), 2nd edn, Wiley-VCH, Weinheim (2007), pp. 119–82.

[76] D.M. Pirro and A.A. Wessol (eds), *Lubrication Fundamentals*, 2nd edn, CRC Press, Boca Raton, FL (2001), Chapter 6.

[77] M.R. Greaves and J. Knoell, A comparison of the performance of environmentally friendly anhydrous fire resistant hydraulic fluids, *J. ASTM Intl.* **6**(10) (2009) 1–10.

[78] W.J. Bartz (ed.), *Biologisch schnell abbaubare Schmierstoffe und Arbeitsflüssigkeiten*, Expert-Verlag GmbH, Ehningen (1993).

[79] T. Norrby, Environmentally adapted lubricants – where are the opportunities?, *Ind. Lubr. Tribol.* **55**(6) (2003) 268–74.

[80] I. Langmuir, The constitution and fundamental properties of solids and liquids. Liquids, *J. Am. Chem. Soc.* **39**(9) (1917) 1848–1906.

[81] G. Biresaw, A. Adhvaryu, S.Z. Erhan and C.J. Carriere, Friction and adsorption properties of normal and high-oleic soybean oils, *J. Am. Oil Chem. Soc.* **79**(1) (2002) 53–8.

[82] G. Biresaw, A. Adhvaryu and S.Z. Erhan, Friction properties of vegetable oils, *J. Am. Oil Chem. Soc.* **80**(7) (2003) 697–704.

[83] T.L. Kurth, G. Biresaw and A. Adhvaryu, Cooperative adsorption behaviour of fatty acid methyl esters from hexadecane via coefficient of friction measurements, *J. Am. Oil Chem. Soc.* **82**(4) (2005) 293–9.

[84] T.L. Kurt, J.A. Byars, S.C. Cermak, B.K. Sharma and G. Biresaw, Non-linear adsorption modeling of fatty esters and oleic estolide esters via boundary lubrication coefficient of friction measurements, *Wear* **262**(5–6) (2007) 536–44.

[85] K.-H. Hentschel, The influence of molecular structure on the frictional behaviour of lubricating fluids, Part 1, *J. Synth. Lubr.* **2**(2) (1985) 143–65; Part 2, *J. Synth. Lubr.* **2**(3) (1985) 239–53.
[86] M.J. Neale, *Lubrication and Reliability Handbook*, Newnes, Oxford (2000).
[87] E.W. Dean and G.B.H. Davis, Viscosity variations of oils with temperature, *Chem. Met. Engng.* **36** (1929) 618–19.
[88] S. Verdier, J.A.P. Coutinho, A.M.S. Silva, O.F. Alkilde and J.A. Hansen, A critical approach to viscosity index, *Fuel* **88** (2009) 2199–206.
[89] Th. Mang and W. Dresel (eds), *Lubricants and Lubrication*, 2nd edn, Wiley-VCH, Weinheim (2007).
[90] P. Hamblin, I. Rapenne-Jacob, J. Reyes-Gavilan and P. Rohrbach, Standard test methods for TAN assessment and modifications thereof, *Tribol. Lubr. Technol.* **60**(11) (2004) 40–6.
[91] Anon., *Thermometric Titration – The Ideal Complement to Potentiometric Titration*, Metrohm AG, Herisau (n.d.).
[92] J.Q. Griffith III, E.S. Williams and W.H. Reiland Jr (to Sun Oil Co. of Pennsylvania), Soap thickening hydraulic oil composition, US Patent No. 3,816,316 (11 June 1974).
[93] M. Beltzer and S. Jahanmir, Effect of additive molecular structure on friction, *Lubric. Sci.* **1** (1988) 3–26.
[94] G. Biresaw, Effect of temperature on lubrication with biobased oils, Program Guide 63rd STLE Ann. Meeting, Cleveland, OH (18–22 May 2008), p. 114.
[95] A. Dorinson and K.C. Ludema, *Mechanics and Chemistry in Lubrication*, Elsevier, Amsterdam (1985).

4

The transition from reliance on fossil resources to biomass valorisation

Abstract: In a future carbohydrate-based economy renewable feedstocks are bound to gradually replace fossil fuels of the present petro-based society. Life sciences and biotechnology are vital in addressing the growing global demands for crops for food, fodder, fibre and chemicals. As biomass has a limited food potential, development of a non-food agricultural industry for chemical applications has clear advantages. New feedstock opportunities vary significantly as to their potential impact in the market in terms of volume and timing to commercialisation. Development of biorefineries will mark the historic transition into a sustainable society in which biological feedstocks, processes and products constitute the main pillars of the economy.

Key words: renewable feedstock sources, transformation of biomass, non-food crops, biotechnology, biorefinery.

4.1 Introduction

The world is highly dependent on the utilisation of fossil resources (e.g. petroleum, natural gas and coal) to fulfil its energy needs. The European Union import dependency reached almost 54% of its energy requirements in 2006. Additionally, a wide range of commercial products such as polymers, resins and lubricants, etc., are also derived from the same fossil resources. Only a small fraction (approximately 14%) of mineral oil is used for the production of chemicals. The global consumption rate of conventional fossil resources increased by 50% in the 1980–2004 period. It is projected that the global needs for fossil resources in 2030 will amount to 600×10^{15} Btu [1].

In the last decades, the use of petrochemical products boomed, often at the expense of further development of renewable materials (vegetable oils, starches, cellulose, etc.). At the same time, a number of applications of classical petrochemical processes and products are being heavily criticised for their lack of sustainability. Recent major accidental oil spills (BP's 'Deepwater Horizon'/Gulf of Mexico and CNPC/Dalian Bay, both 2010, as well as frequent releases and blow-outs in Nigeria and Russia) highlight problems in safety management and environmental impact assessment in the petrochemical sector and are a reminder that our dependence on oil and derived energy can lead to massive catastrophes. The end of the availability of abundant conventional fossil fuels, without causing irreparable environmental damage, is approaching.

Clearly, in the long term the world can no longer afford to rely solely on fossil oil for its energy needs and oil-based materials. Finding a cure for humankind's oil (and carbon) addiction is urgently needed, in particular in consideration of the necessary long development times for alternatives. This requires energy efficiency awareness, political action and changes in consumer behaviour. With an increased awareness of health and the environment and a demand for sustainable products, the focus of consumer and industrial interests has recently begun to switch (back) to biomass-based raw materials. As higher oil prices, environmental issues and regulations begin to adversely affect the manufacture of chemicals from raw materials, the transition to a bio-based production system is unavoidable. During the 21st century the world is therefore likely to witness a reversal of the trends observed in the 20th century, whereby cheap (non-renewable) fossil-based hydrocarbons gradually replaced (renewable) biomass-based raw materials (both for chemical and fuel applications). This transition took several decades. The inverse process is now foreseen to happen in the transition from the petro-based to a bio-based economy (BBE) in which renewable bio-resources (agricultural, forestry and marine) and eco-efficient processes (including bioprocesses) will be used for non-food applications, such as sustainable chemicals, materials, transport fuels, electricity and heat. The forthcoming carbohydrate-based economy is driven by the demand for safer green chemistry and green energy. Future society's chemical products will be redesigned and built upon the Precautionary Principle. Nevertheless, both conventional and non-conventional oil and gas will still continue to play a significant role globally far into the 21st century. According to all major international oil companies, petroleum production in 2050 will be close to that of 2010, with output reaching a peak around 2030. Expectations are an increase in oil price, but relatively cheap natural gas [2].

Some alternatives to petrol and other fossil fuels have already been implemented in a variety of forms of transport. This is the case for bioethanol (since the 1970s) and biodiesel (since the 1990s). The 1973 Suez oil crisis led directly to Brazil establishing its much vaunted 'Proalcool' programme using sugar cane to produce ethanol as a transport fuel. The US corn ethanol programme was spurred on by similar concerns. Also biodiesel derived from oil crops as a liquid energy carrier for transport is a potential renewable and carbon neutral alternative to petroleum fuels. However, advanced biofuels based on edible energy crops and adequate technological development cannot be expected to supply more than some 10–15% of global demand in an environmentally responsible manner without affecting food production. Biodiesel is an example of such a product which has created a fundamentally new demand for agricultural commodities that was non-existent several years ago. However, current biofuel policies have already caused unwanted social and environmental problems (unjustified rising food prices, deforestation and soil degradation). At present, supply and demand of vegetable oils and fats for food, feed, fabrics and fuel are no longer in a healthy balance. Better prospects for farming biodiesel without undesirable side effects are based on the use of next-generation feedstock sources (under development).

Figure 4.1 illustrates the principal transitions in raw materials use and chemistry over the last 150 years. Fossil resources such as coal, natural gas and oil are distinctly hydrophobic, oxygen-free and devoid of functionality. During the petrochemical age the chemical industry has proficiently turned hydrocarbons into more functionalised, and therefore more versatile, chemicals. Thus, organic functional groups such as hydroxyl, amino, aldehyde, acid, ester or halo functionalities have been introduced, usually into olefinic (and paraffinic) hydrocarbons. Catalytic oxidation has been at the basis of these developments and has been the main (and most difficult) process step. This chemistry is no longer required for bio-based feedstocks with their inherent oxygen functionality. Renewable biomass is largely composed of lignocellulosics, a raw material source overfunctionalised with hydroxyl groups and more hydrophilic. The methods required for converting carbohydrates and lignin into viable industrial chemicals – reduction of oxygen content with introduction of C=C and C=O – are diametrically opposed to those developed and utilised by the petrochemical industry. We now need to understand how to catalyse conversion of the oxygenated molecules of biomass into high-quality, high-value chemicals. Carbohydrate conversion is expected to precede lignin valorisation.

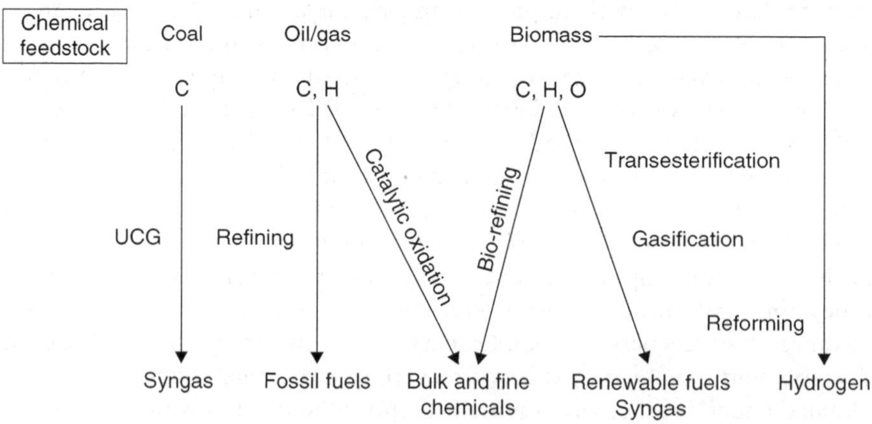

4.1 Raw materials and chemistry in transition (UCG = underground coal gasification).

Most of the world's chemicals are made from oil or natural gas. At present there is great interest in finding economical ways to produce chemicals from renewable feedstocks. Plant biomass is the only current sustainable source of organic carbon. Successful replacement of petroleum by agricultural product-based (ag-based) raw materials ('going green') requires overcoming a number of hurdles. Unlike petroleum, which has more or less similar chemistry relatively independent of its crude source, agricultural products encompass a wide variety of chemistries that are highly dependent on the source (e.g. plant, animal, fish) and sub-source (e.g. seed, leaves, stem, root). Successful use of ag-based raw materials requires the development of appropriate knowledge about the chemistry, modification, blending, processing, manufacturing and testing of agricultural-based products.

Recent innovations from oleochemistry have contributed to the sustainable development of environmentally compatible and powerful commercial products with high value added for industrial use. Further use of different types of vegetable raw materials, such as vegetable oils, carbohydrates, lignin and proteins to meet market and consumer needs constitutes a challenge for further development. Using carbohydrates and lignin in a wet medium (biomass) as a substrate instead of fossil hydrocarbons needs new catalyst materials, process and reactor technologies, all to be developed (and the sooner the better). For these processes, also catalytic technology based on enzymes, bacteria and other specialised microorganisms should be considered. There is a strong need for elaborating appropriate chemical and microbial processes to convert natural carbohydrate and mononuclear aromatic polymers into industrially useful products. There is also a need to improve fermentation technology, to develop larger-scale bioreactors, and to integrate these in biorefineries. Bioprocesses hold promise of cleaner and safer production as they can be used at atmospheric pressure and room temperature in aqueous environments.

Use of biomass instead of mineral oils as raw materials determines a shift away from the intermediate low-molecular-weight hydrocarbon platform chemicals of the petrochemical industry and their derivative products (e.g. polyolefins and engineering plastics) to (partially different) bio-based platform chemicals (see Fig. 5.11), intermediates and fine chemicals (vitamins, anti-infectants, biopolymers, etc.).

The non-food utilisation of inexpensive bulk scale-accessible, low-molecular-weight carbohydrates – sucrose, glucose, xylose and fructose being the most readily accessible – is at a rather modest level in terms of large-scale manufactured commodities currently on the market. However, a wide range of readily accessible products awaits further development, mainly for economic reasons [3]. Only about 20 of the 100 major organic chemicals in use today are aromatic compounds, invariably manufactured from fossil raw materials, mostly from the BTX (benzene–toluene–xylene) fraction derived from naphthas in refineries [4]. At present, there are few alternatives. The direct chemical conversion of biomass to an equivalent BTX product is fraught with difficulties as only small amounts of monocyclic aromatic hydrocarbons (phenols, cresols and catechol) are formed on pyrolysis or thermal cracking of woody feedstocks. However, a highly promising route from sugars to industrially relevant aromatic compounds is based on microbial conversions using genetically modified biocatalysts [3]. A more obvious route to aromatics is the valorisation of lignin.

A society utterly based on fossil fuels and raw materials for about 150 years cannot be expected to be suddenly turned into a bio-based society in which biomass and other renewable resources (wind, hydro, solar energy) are central for energy, transport and chemistry. Large-scale application of biomass for biofuels and bioproducts requires a transition in society at all levels, not dissimilar from the preceding switch from the coal-based society to oil and gas in the last century. This transition from a fossil-based to a bio-based economy will require macro changes, both locally, regionally, nationally and internationally in many branches of industrial activity. The transition process needs to be stimulated both by government (political support) and consumers/public opinion to master the learning curve (S curve) in due time. A true transition requires openness and debate in society as long-term economic advantages are expected. The slow transition to a bio-based economy will change the current international economic relations.

The sources of supply of fossil and bio resources already differ greatly. Petroleum, natural gas and coal are globally concentrated in a restricted number of well-defined regions, whereas biomass can be cultivated almost everywhere, provided certain minimum conditions are satisfied. The more centralised exploitation of fossil raw materials contrasts strongly with the rather diffuse and decentralised presence of biomass. Also the production processes and process technologies differ substantially as do the final products obtained from these resources. Biomass and fossil mass are not easily mutually exchangeable. For example, pyrolysis oil from biomass has a different composition from mineral oil.

The transition process from a petro-based to a bio-based economy requires considerable development time, high research and technology development (RT&D) costs and may occasionally benefit from a good fit with existing processes and structures. For example, bio-ethylene produced from biomass (i.e. from sugar cane and beet sugar bioethanol) can profitably use existing pipeline infrastructures. However, the likelihood for finding existing infrastructure for most bio-based products is rather limited. Industrial development and marketing of innovative bulk products (e.g. butanol, isobutanol and biomass-to-liquid (BTL) biofuels) requires huge investments at a time when the economics of these products is still questionable. Not surprisingly, without sufficient long-term governmental support technology developers are easily exposed to insolvency. Choren Industries (Germany) is a case in point.

Cooperation between the chemical and agro sectors is fairly new. It is not expected that the chemical industry will ever be fully biomass-based. Renewable resources represent especially attractive alternatives in applications where the synthetic chemistry of nature is (at least partially) retained in the final industrial end product. This is the case of hydraulic oils, gear oils, mould release agents and cooling lubricants based on fatty acid esters, or for phenolics from lignin. For economical industrial exploitation of plant sources at least about 10–20 dry wt% of matter should consist of the desirable component (or a derivative). Various technologies are needed for the (thermal or non-thermal) conversion of biomass to the desired end product. Other challenges are in the associated separation techniques, and such technologies need to be developed (see Section 4.3). However, many technologies for the conversion of residual streams from agriculture and forestry are already available. The bio-based economy also requires new reactor concepts and catalysts. The traditional catalysts used in the petrochemical industry, developed in the last century, may not be suitable for the conversion of biomass and need to be redesigned. Conversion of mineral oil takes place in a non-aqueous environment of hydrocarbons but conversion of biomass often occurs in aqueous acidic conditions.

Whereas there is a EU Directive on the Promotion of Use of Energy from Renewable Energy Sources (EU Directive 2009/28/EC) with national membership targets and a Biofuels Directive (EU Directive 2003/30/EC) on the Promotion and Use of Biofuels and Other Renewable Fuels for Transport (5.75% by 2010), there is as yet no EU Directive regarding the substitution of oil, coal and gas by biomass as the raw material for the chemical industry. Meanwhile, the United States, China, Canada and the EU have turned massively to biomass use with developments of second generation techniques for biofuels/biofeedstock, use of new raw materials (lignocellulose), biorefining, optimal use of agricultural land for biomass production, and emphasis on the food/feed/fuel/chemical feedstock equilibrium. It is expected that biomass/biofuels will be competitive with mineral oil by about 2020 [2].

At the present time, some 8–10% of the organic raw materials used by the chemical industries are renewable. *The Roadmap for Biomass Technologies* has predicted that by 2030 20% of transportation fuel and 25% of chemicals in the United States will be produced by biomass [5]. The Netherlands has set goals to have 20–45% of fossil-based raw materials substituted by biomass by 2040. The Dutch chemical industry aims at using 50% of its raw materials from bio-based platform chemicals by 2030. The size of global biomass production is sufficient to supply virtually all of the raw materials now required for the chemical industry. The transition to the carbohydrate economy is already under way with many companies, such as BP, Conoco-Phillips, Dow, DSM, DuPont, Petrobras, Shell, Unilever, UOP and others, developing the technology and infrastructure for biochemicals' production. Unilever strives at halving the environmental impact of its products by 2020 while using only agricultural raw materials that are produced in a sustainable manner [6]. The main technologies necessary for a carbohydrate economy are growth of biomass feedstock and biomass conversion to chemicals. In order to make the chemical industry more sustainable and competitive, access to renewable feedstocks should be improved. But little progress has been made in Europe since a CEFIC recommendation (2009) to increase use of bio-based materials to produce chemicals.

4.2 Biomass

The term 'biomass' comprises all organic matter originating from living organisms (excluding fossil resources and old-growth timber) that is available on a renewable basis, including dedicated energy crops and trees, aquatic plants, agro/forest by-products, wood pulp and paper residues, animal wastes and other (industrial and municipal) waste materials [7]. According to the EU Renewable Energy Directive (RED) 'biomass' means the biodegradable fraction of products, waste and residues from biological origin from agriculture (including vegetable and animal substances), forestry and related industries including fisheries and aquaculture, as well as the biodegradable fraction of industrial and municipal waste [8].

The biogeochemical carbon cycle (to/from atmosphere, plant matter, soil, oceans) is complicated. The net CO_2 fixation by land plants is approximately 56 Bt/yr [9]. Plants typically capture 0.1 to 1.0% of the solar energy which is used to convert CO_2 and water into a sugar building block $(CH_2O)_n$ and oxygen via photosynthesis, as follows:

$$nCO_2 + nH_2O + h\nu \xrightarrow{\text{chlorophyll}} (CH_2O)_n + nO_2 \qquad [4.1]$$

The sugar is stored in polymeric form as cellulose, starch, or hemicellulose. Approximately 75 wt% of biomass is composed of sugar polymer. Exceptionally high values (about 5%) for the biochemical conversion of solar into chemical energy are reported for corn, sugar cane and sorghum.

The annual production of biomass amounts to approximately 200 Bt [10] in the form of oils, sugars, polysaccharides and other biopolymers, representing an unprecedented resource for chemical production. Of this biomass production only 6 Bt (3.0%) are currently being used by people as a food and non-food resource, namely:

- 3.7 Bt (62%) for food/feed use;
- 2.0 Bt (33%) of wood for energy, timber, pulp and paper; and
- 0.3 Bt (5%) for technical (non-food) raw materials (clothing, detergents, chemicals, etc.).

Other biomass is used in natural ecosystems, is lost by burning or decays along natural pathways by mineralisation processes. There is no 'waste' in nature.

The term 'renewable resources' generally refers to all biomass produced from agriculture and forestry that is not used for food and animal feed. In the field of renewable resources the prefix 'bio' is short for 'bio-based' or 'biomass-based'. Producing chemicals and industrial products from biomass directly reduces the associated carbon released during the production of fossil-based products.

The organisms, processes, products and systems found in natural ecosystems have evolved over millions of years to become highly efficient; however, plants have generally not evolved to maximise biomass production. Plant growth rates range from 6 to 90 dry tonnes/ha/year. Plant breeding, biotechnology and genetic engineering are means to develop more efficient plant materials with faster growth rates, which require lower energy inputs. In nature, all bio-organic chemicals and materials are renewable, biodegradable and recycled. Fresh biomass reflects the biosynthesis performance of nature and has a different C:H:O:N ratio from petroleum (decayed biomass). The exact composition of each form of biomass varies depending on the plant species and residues collected (Table 4.1). Biomass has a complex chemical composition which strongly depends on its source. Generally, lignocellulosic (woody) biomass consists of approximately 38–50% cellulose, 25–35% hemicellulose and 15–25% lignin [11, 12]. On average, plant biomass is composed of 75% carbohydrates (mainly in the form of (hemi)cellulose, starch and saccharose), 20% lignin and only 5% other

Table 4.1 Composition (wt%) of typical biomass samples

Biomass	Variety	Lignin	Cellulose	Hemicellulose
Monterey pine	*Pinus radiata*	25.9	41.7	20.5
Poplar	DN-34 hybrid	24	40	22
Sugar cane bagasse	*Gramineae saccharum* var. 65–7052	24	43	25
Corn stover	*Zea mays*	18	35	22
Switchgrass	Alamo	18	31	24
Wheat straw	Thunderbird	17	33	23
Rice straw	*Oryza sativa* sp.	10	39	15

Source: http://www.eere.energy.gov/biomass/feedstock_database.html.

natural compounds such as lipids (oils and fats), proteins and other substances (antioxidants, vitamins, dyes, flavours, aromatic essences, terpenoids, nucleic acids, alkaloids) of very different structure (Fig. 4.2) [13]. Lipids are water-insoluble, hydrophobic substances. Cellulose is a chemically rather intractable, non-branched water-insoluble crystalline polysaccharide consisting of several hundreds up to tens of thousands of glucose units. Cellulose is the most abundant biopolymer synthesised by nature. Global generation of cellulose is equivalent in energy to 670 billion barrels of oil, or about 20 times the current annual global oil consumption. Cracking the major biofeedstock cellulose is one of the main industrial and biotechnological challenges of the 21st century. Hemicellulose is a complex amorphous polymer, whose major component is a xylose monomer unit (see also Section 5.5).

Lignin is among the most abundant biopolymers on earth. Lignins are found in arborescent plants and as the integral cell wall constituents in all vascular plants, including the herbaceous varieties. Lignin in the cell walls (more precisely termed 'protolignin') is intimately connected with the carbohydrate components. As a tough polymer lignin provides higher plants with rigidity, water-impermeability, and resistance to microbial decay. For lignin biosynthesis, see ref. [14]. The structure of lignin is complex and irregular. The lignin polymer with its dominant aromatic chemical structure is highly branched and substituted, and cross-linking occurs to some extent.

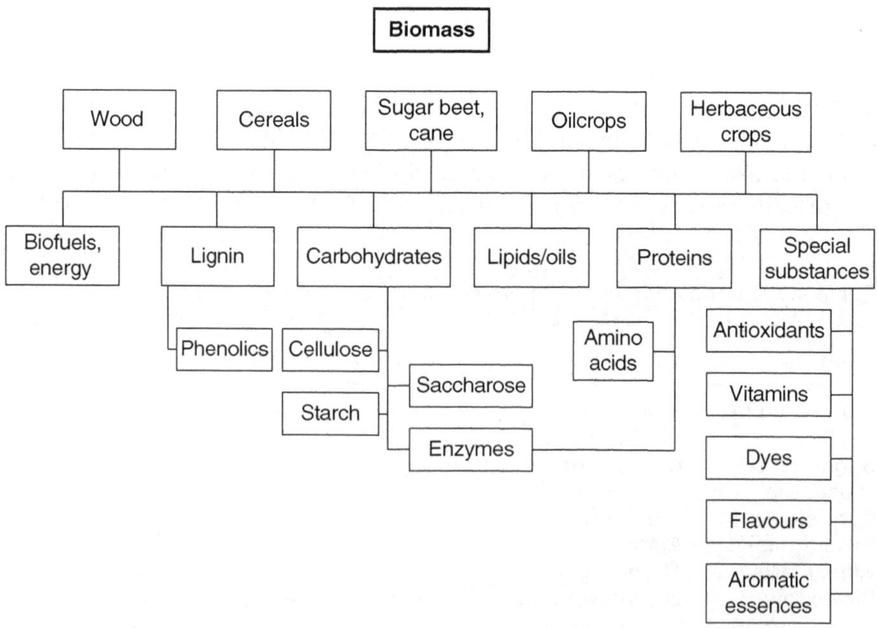

4.2 Basic components of biomass and their derivatives.

The nature of the primary building blocks (phenylpropanes) (Fig. 4.3), bonding pattern (via arylglycerol, phenylcoumaran, biphenyl and diaryl ether structures) and functional groups (primary and secondary hydroxyls, methoxy and phenolic groups) in lignins is now well established, at least qualitatively. Phenylpropane units (guaiacylpropane, 1; syringylpropane, 2; *p*-hydroxyphenylpropane, 3) constitute the main building blocks in lignins [15–17]; their proportion differs with the botanical origin of the lignin (Table 4.2). The structure of protolignins mainly consists of phenylpropane units with an oxygen atom at the *p*-position (as OH or O—C) and with methoxy groups in the positions *ortho* to this oxygen atom. These *ortho* positions may be C- or O-substituted with substituents other than methoxy. Biosynthesis of protolignin proceeds via oxidative polymerisation of three primary *p*-hydroxycinnamyl alcohols. The structural elements in protolignins are not linked to one another in any particular order. Linkages occur between protolignin and carbohydrates. Isolation of lignin from the other plant constituents is difficult. The chemical separation of lignin from

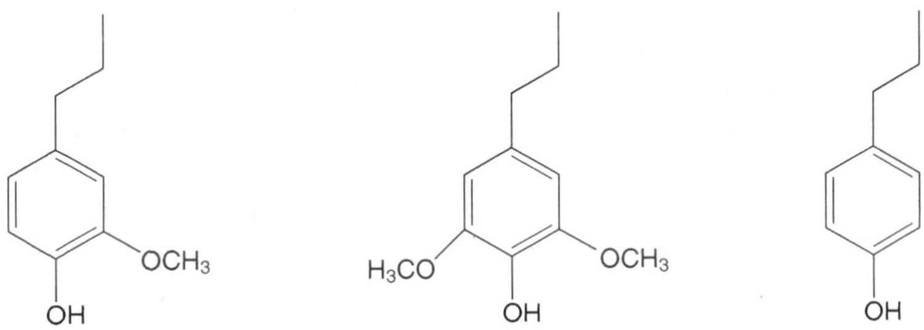

1 **2** **3**

4.3 Phenylpropane units of type 1 (guaiacylpropane), 2 (syringylpropane) and 3 (*p*-hydroxyphenylpropane) as main building blocks of lignins.

Table 4.2 Approximate composition (%) of some important classes of lignins

Lignin class	Phenylpropane units[a]		
	1	2	3
Softwood lignin	95	1	4
Hardwood lignin	50	50	2
Grass lignin	70	25	5

[a] 1, guaiacylpropane; 2, syringylpropane; 3, *p*-hydroxyphenylpropane. After ref. [15].

cellulose ('delignification') is a complex process and the products obtained greatly depend upon the process employed. Lignin represents the only viable source to produce renewable aromatic compounds. Current lignin valorisation processes were reviewed recently [18].

A wide variety of biomass is available for conversion and utilisation. Next-generation feedstocks for bioproducts are categorised as follows:

(a) the cellulosic components of municipal solid waste (MSW) (e.g. tree trimmings, yard waste, paper products);
(b) forest residues and thinnings (e.g. logging residues from commercial forests);
(c) annual crop residues (e.g. corn stover);
(d) dedicated herbaceous perennial crops (e.g. switchgrass, *Miscanthus*, native prairie grasses);
(e) short-rotation woody crops (e.g. hybrid poplar, willow shrubs, eucalyptus); and
(f) macro- and microalgae.

Feedstocks (a)–(c) are at present typically associated with the production of bioethanol, while (f) is generally associated with the production of biogas and biodiesel but wider application is desirable [19].

Biomass is expected to contribute significantly to renewable chemical feedstocks but at present the efficiency of integral biomass conversion is very limited (most advanced for corn and corn stover) and technologies are still evolving. Selection of the biomass feedstock is of paramount importance both technologically and socio-economically. Waste streams with a low or even negative value, such as agricultural waste, are preferred. When using dedicated crops, it is advantageous to select sources with a high growth rate per ha per year, that require limited amounts of fertiliser and water resources, that are available throughout the year and that are not prone to disease. For ethical reasons, the biomass feedstock should not compete directly with the food chain. Previous claims blaming biofuels for the 2006/08 food price increases have officially been deflated [20]; the price inflation was all about energy price inflation and the oil shock. There is no conflict in the use of biomass for food, feed, fuel and feedstock for the chemical industry in the 21st century.

Table 4.3 shows the current production figures and use of some important agricultural commodities, which are well up from the recent past (cf. wheat from 610 Mt in 2004/05 to 678 Mt in 2009/10, or oilseeds from 415 Mt in 2006/07 to 436 Mt in 2009/10). Crop forecasts of the International Grains Council (IGC, London) indicate a global corn (maize) production of 797 Mt in 2009/10. Record oilseed plantings for 2010 are being reported [21]. The aforementioned figures may be compared to the annual production of crude oil (4900 Mt, 2009) and coal (3800 Mt (as SKE), 2004).

Table 4.3 Projected 2011/12 world output and use for grains and oilseeds (Mt)

Commodity	Output	Trade[b]	Total use[d]
Total grains[a]	2300.38	303.91	2295.91
Wheat	694.64	146.74	694.35
Coarse grains[c]	1142.44	123.09	1143.71
Rice, milled	463.31	34.07	457.85
Oilseeds	437.24	106.52	388.29
Oilmeals	263.82	80.00	260.87
Vegetable oils	153.47	62.60	151.01

[a]Wheat, coarse grains and milled rice.
[b]Based on export estimate.
[c]Corn, sorghum, barley, oats, rye, millet and mixed grains.
[d]Crush only for oilseeds.
After ref. [21].

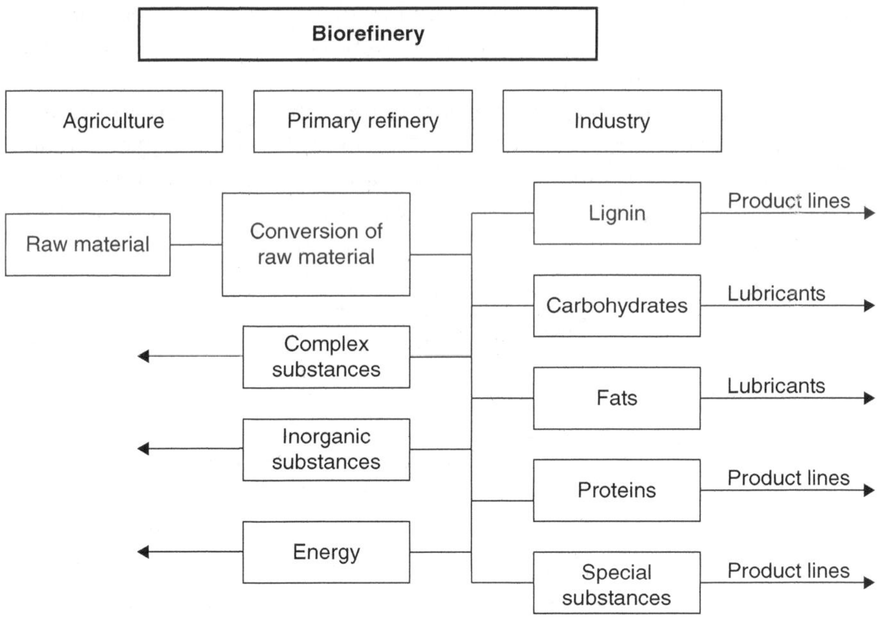

4.4 Conversion of biological raw materials into industrial intermediates and final products in a biorefinery.

The primary separation of biomass into its main substance groups (carbohydrates, lignin, lipids, proteins and special products) (Fig. 4.4) is of fundamental importance but not an easy matter. A technically feasible separation operation, which would enable separate use of all basic components, is still in its initial stages. Subsequent treatment and processing of these substances both chemically and biotechnologically leads to a whole range of derivatives. The composition of biomass determines the ease with which it can be converted to useful products and/or intermediates and affects the functionality of the final products. Biorefineries (see Section 4.3.2) combine the essential technologies which convert biological raw materials into the industrial intermediates and final products (see Fig. 4.8).

4.3 Transformation of biomass to bioproducts

Food and animal feed are the most important products of biological raw materials and should never be considered as a by-product of biomass-based chemical processes. Biomass denotes any (biodegradable) organic matter, available on a renewable or recurring basis that can be used in place of fossil-fuel sources to develop also value-added products such as power, heat, industrial chemicals and consumer goods [22]; see Figs 4.2 and 4.5. For top value-added chemicals from biomass, see. ref. [23]. Derivatives of natural, harvestable raw materials are finding increasing acceptance because of their environmental compatibility (biodegradability and ecotoxicity) and some technical advantages.

The production of renewable resources for industry by agriculture is of considerable economic importance for the farming community. A major political and social argument for the utilisation of renewable resources is the fact that these are not limited to a few restricted geographical areas, but

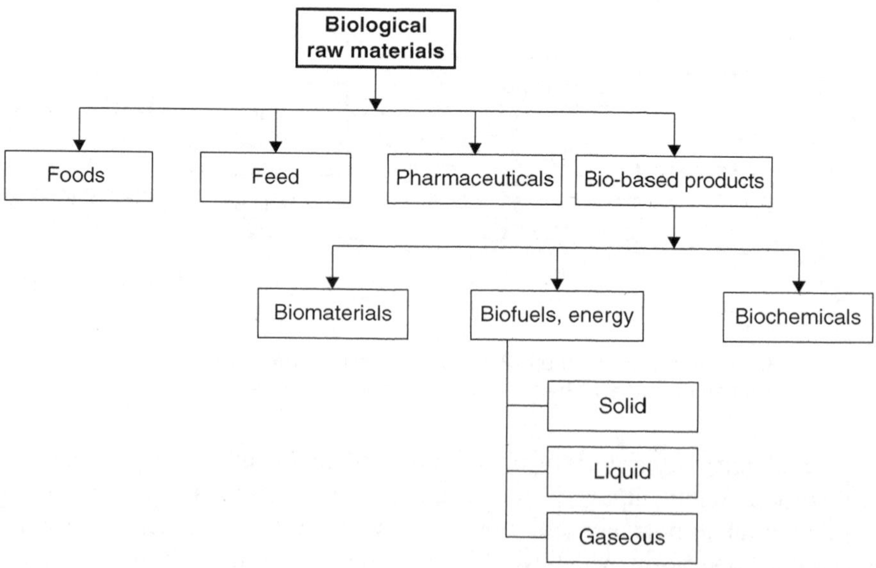

4.5 Products and product classes based on renewable resources.

are rather ubiquitous and annually available. However, use of renewable raw materials by industry is not quite without problems. Seasonal activities (not in the tropics) are problematic in comparison to continuous transformation of mineral oil supplies. Moreover, although industrial crops as sources of renewable materials do provide many valuable chemical compounds, these are always embedded in complex admixtures with broad distributions of physical and chemical properties of the constituting compounds. Also, biomass comprises a great variety of biomass types, arbitrarily divided into wet (>50%) and dry (<50%) biomass, roughly corresponding to agricultural and forestry crops. Elimination of water greatly affects the overall processing costs and total energy balance. Further, the quality of the matrix and of the individual components fluctuates greatly both geographically and temporally. Finally, sustainable and cost-effective processing of renewable resources requires total utilisation with little waste production.

Harvesting, treating, transporting, storing and delivering large volumes of biomass feedstock, at a desired quality, all year round, to a biorefinery (see Section 4.3.2) raises considerable supply chain issues and requires careful logistical analysis in order to ensure economic viability [24]. Biomass plants based on agricultural residues are to be operated decentrally (close to the source of production) in order to minimise transport costs. This holds equally true for areas with highly different population densities such as Germany, Australia or Oregon.

Renewable raw materials (biomass formed by photosynthesis in plants) are the only alternative to fossil raw materials for industrial use in the chemical industry. Plants as raw material suppliers represent one of the future-oriented fields that have been identified in several countries as being especially significant in terms of innovation policy. In the United States, a Billion-Ton Vision study has identified a potential 1.3 Bt biomass (1 t biomass = 0.43 tonne of oil equivalent, toe) that could be used as feedstock for a bioenergy and bioproducts industry [25]. In its 'Vision for Bioenergy and Biobased Products in the US' the Biomass Research & Development Technical Advisory Committee has put forward an ambitious goal for bioproducts [26]. The share of target chemicals that are bio-based is set at 25% by 2030.

The recent EU Biomass Action Plan [27] aims at increasing the biomass contribution to primary energy (1486 Mtoe/yr) from 57 Mtoe/yr in 2001 (or about 4%) to 188 Mtoe/yr in 2010 and 227 Mtoe/yr in 2020. There is sufficient biomass potential in the EU25 to support ambitious renewable energy and chemical raw material targets in an environmentally responsible way [28]. Compared to energy generation, material use of biomass still only plays a minor role. As little as 6.4 Mt of biomass are currently used as a raw material option for the production of chemicals (other than for energy use) in Europe, representing 8–10% of the industry's total feedstock.

In 2009, the Royal Society of Chemistry (RSC) roadmap for the chemical sciences defined ten top challenges, including agricultural productivity, conservation of scarce natural resources, conversion of biomass feedstocks, and sustainable product design, which are all themes of prime interest also to biolubricants [29]. The roadmap identifies where the chemical sciences can provide technological and sustainable solutions to these challenges. The key to addressing these problems is by advancing fundamental knowledge. Table 4.4 lists the trends and drivers for the conservation of scarce natural resources. Overarching principles are promotion of sustainable product design, cross-discipline collaboration, reduction, remanufacture, reuse and recycling (4Rs). In the near future, biomass will play an increased role as a source of fuel and chemicals. Establishing processes for converting biomass feedstocks, including agricultural, forestry and municipal wastes, into a range of valuable products will enable production of new sustainable fuels and chemical building blocks. Use of bio-based renewable chemicals instead of fossil fuel-based feedstocks is a great challenge for the chemical sciences and requires new technology to be developed (Table 4.5).

Germany is one of the leading countries in the world in the use of renewable resources. The 'Action Plan for the Industrial Use of Renewable Resources', which was adopted by the German Federal Government in 2009, promotes the industrial use of renewable resources, even apart from their use for energy generation. The Action Plan envisages a significant and sustainable increase in the proportion of biomass used in industry as well as an improvement in the efficiency of biomass use. In still another move

Table 4.4 Trends and drivers for the conservation of scarce natural resources

Social: • Population growth • Consumer demand • Skills availability	*Economic:* • Landfill tax • Scarce resources • Changing business models • Cost and performance
Technological: • Green chemistry • Life-cycle analysis • Rapid product development	*Political:* • Competition for resources • 80% cut CO_2 by 2050 • Product-focused legislation
Environmental: • Waste management • Climate change • Water shortages	

Source: www.rsc.org/roadmap.

Table 4.5 Key areas of technological development for biomass feedstock conversion to chemicals

Bioprocessing science
• Metabolic engineering for enhanced biomass feedstock properties
• Generation of homogeneous feedstocks
• Improved biocatalytic process design
• Advances in fermentation science
Separation technology
• Membrane and sorbent extraction of valuable components from biological media
• Pre-treatment methods for biomass component separation
(Bio)catalysis for biomass processing
• Techniques for lignocellulose and lignin breakdown
• Improved microorganisms by microbial genomics
• Pyrolysis and gasification techniques for biomass pre-treatment and densification
• Upgrading of pyrolysis oil
• Exploitation of biomethane from waste
• Oxygen and poison-tolerant catalysts and enzymes
Conversion of platform chemicals
• New synthetic approaches for high-value products
• New green and cost-effective processes

Source: www.rsc.org/roadmap.

away from oil, the German national strategy has recently (2010) been directed towards the global security of food supply, food safety, sustainable agricultural production, industrial use of renewable resources and to further development of biofuels [30].

In 2008 Germany used about 4.1% of the total fossil raw materials (456 Mt: petroleum, natural gas and coal) in the chemical industry. The German chemical industry is well under way to processing agricultural and forestry raw materials once again on a significant scale [31]. Approximately 15% of the German domestic agricultural raw materials are used for industry, whereas significant quantities of renewable raw materials are being imported. On the other hand, the country also exports a significant percentage of the products manufactured from renewable resources. The most important native plants cultivated for industrial use in Germany are oil-crops (120 kha rapeseed, 8.5 kha sunflower), cereals for starch (140 kha) and sugar plants (22 kha). Maize, wheat and potatoes are the primary raw materials used for starch production. These products contain up to 80% amylopectin, along with amylose. In Germany, sugar is produced exclusively from sugar beet. Fibre plants such as flax and hemp are currently being cultivated in Germany on a very small scale. Protein crops are rarely used as renewable industrial feedstocks. Their main use is as animal feed or green manure. Proteins are also obtained as by-products of oilseed processing or wheat starch production.

The German vegetable oil consumption (5.6 Mt, 2008) was as follows: rapeseed oil (RSO), 56%; palm oil (PMO), 16%; soybean oil (SBO), 11%; sunflower oil (SNO), 5%; palm kernel oil (PKO), 5%; coconut oil (CNO), 4%; other, 3% (*source:* OVID – see Appendix A.3). Only 1.1 Mt was used by industry. Not all oils are equally suitable for all oleochemical applications. Biolubricants (solid, semi-solid and liquid) accounted for only 2% of the industrial use of oils and fats in 2008 (*source:* FNR – see Appendix A.3). The German chemical industry is being hard-pressed to base its raw materials on sustainable bio-based resources by 2030 [30].

Green chemistry is actively being supported by the French Environment and Energy Management Agency (ADEME), including technologies in the field of plant chemistry: biofuels, biomolecules, and building blocks. Industries and Agro Resources (IAR) is a French competitiveness cluster of companies, associations and universities working on green chemistry in northern France. Brazil, China, and Australia are also focusing heavily on green chemistry initiatives to meet local needs.

4.3.1 Biomass generations in industrial processing

On the basis of the technology needed to convert biomass to bioproducts three generations of biomass are usually distinguished (Table 4.6). The first generation (G1) comprises the edible fractions of the common agricultural commodities which at present are also used as raw materials for the classical biofuels bioethanol (sugars, corn, starch) and biodiesel (transesterified vegetable oils). As many as 200 small- to medium-sized corn-to-ethanol refineries are operating in the United States. Some corn-based bioethanol plants have reached huge capacities (600–700 kt/yr, Jilin, China). Vegetable oils can also be hydrotreated to produce straight-chain alkanes that fit in the jet and diesel fuels range [32–37]. Hydroprocessing technology of vegetable oils is commercial (e.g. Neste's NExBTL, UOP LLC/Eni's Ecofining™ and Petrobras's H-BIO). Technologies have been developed but further investments are limited.

The second generation (G2) biomass refers to non-food agricultural products, which mainly consist of cellulose and hemicellulose. As this biomass material has a more complex structure than starch, more elaborate and costly methods are required to transform the different kinds of sugar (C_5 and C_6 sugars) in G2 bioethanol production technologies. Techniques and enzymes are needed for digestion of the less accessible sugar fractions of this type of biomass as well as genetically modified microorganisms for fermentation. Typical hydrolysis methods used nowadays include steam pretreatment, dilute sulphuric acid or hydrothermal treatment, and enzymes.

Table 4.6 Categorisation of biomass on the basis of technological requirements

	Biomass	Selected bioproducts
Main stream:		
First generation	Main (edible) products of agricultural commodities	Oils, sugars, starch, biofuels
Second generation	Non-food agriculture and forestry products (wood, herbaceous crops, etc.)	Cellulose, hemicellulose, (lignin)[a], C_5 and C_6 sugars
Third generation	Lignin, non-agricultural products	Carbohydrates, polysaccharides, aromatics
Rest streams:		
Primary	From cultivation: biomass residues from agriculture and forestry	Biogas, bio-coal, sugars, bioethanol
Secondary	From processing: residues from industrial processing	Xylose, galactose, animal feed
Tertiary	From use: residues from final product use	Biodiesel, animal feed

[a] For energy and electricity.

In the Plantrose™ process (Renmatix; Kennesaw, GA) woody biomass is converted to sugars through supercritical hydrolysis. Application of G2 biomass requires further development and huge investments. Depending on the cost effectiveness of the technology this might not be particularly well suited for third world countries! In G2 processes lignin will only be used for generation of heat and electricity.

The third generation (G3) biomass is usually considered as consisting of lignin and suberin as a chemical raw material (e.g. for production of aromatics such as phenol) and also comprises biomass which does not find its origin in agriculture or forestry, but refers to wet biomass such as micro- and macroalgae. Lignocellulosic biomass includes purpose-grown industrial crops such as vegetative grasses and short-rotation forests as well as cereal straw, bagasse and forest residues. Technology for the conversion of lignin is under development [18]. Breakthroughs are welcome. Lignocellulosic biomass and its derivatives can be processed in various ways to produce chemical intermediates [38, 39], liquid hydrocarbon fuels [40] including bio-oils [41], aromatic hydrocarbons [42], bioethanol [43–47] and hydrogen [48–51]. In the medium to long term biomass-to-liquid (BTL) fuel production will become more attractive than biodiesel from (non)edible vegetable oil sources.

The so-called primary rest streams of biomass consist of the low-value left-over (waste) products of agricultural and forestry cultivations (crop residues, crop straws, leaves, stalks, corn stovers, tubers, organic residues, forest slash and stumps) which are nowadays mostly simply abandoned on the fields, but have an important function as a soil enricher. Collection of these residues for industrial use leads to a considerable logistic and economic problem [53]. This problem can easily be alleviated by means of local fermentation installations close to the agricultural production site, which could produce biogas, electricity, heat and a concentrated soil fertiliser. Table 4.7 summarises the world production of crop residues. In all, approximately 1.55 Bt wasted crop residues are produced annually [52]. From a greenhouse gas (GHG) perspective, for ethanol production cellulose-containing materials are to be preferred over cornstarch. However, the lignin in woody plant material can prevent full conversion of cellulose into fermentable sugar. Iogen Corporation (Ottawa) has developed a process utilising cellulase enzymes that maximise conversion of wheat straw (after pretreatment to break down the fibrous structure) into fermentable sugar [54]. The yield and activity of the cellulase enzymes have been optimised using biotechnology. Genencor (now DuPont Industrial Biosciences) is planning a 100 ML/yr corn residue (cellulose)-to-ethanol plant in Iowa for 2012 using Accellerase® TRIO™. POET-DSM Advanced Biofuels LLC has developed a three-stage enzymatic (corn stover) cellulose-to-ethanol process, to be operational in Emmetsburg (IA) by 2013. The EU is operating three G2 bioethanol demo plants based on barley straw (Albengoa; Salamanca, Spain, 30–40 t/day^{-1}), softwood (Etek Etanolteknik; Ömsköldsvik, Sweden, 2 t/day^{-1}) and wheat straw/household waste (Inbicon; Fredricia, Denmark, 2.4–24 t/day^{-1}). A NGVA EU-funded project (FP7) 'Biowaste and Algae Knowledge for the Production of 2nd Generation Biofuels' (BioWALK4Biofuels) (2010–2014) aims at producing biogas (biomethane) by anaerobic digestion of biowaste and macroalgae rather than (mixed) cereal crops.

Table 4.7 Worldwide availability of crop residues (billion kg)

Residue	Main producer	Global production
Corn stover	North America (134)	204
Barley straw	Europe (44.2)	58.5
Oat straw	Europe (6.8)	10.6
Rice straw	Asia (668)	731
Wheat straw	Asia (145)	354
Sorghum straw	North America (6.9)	10.3
Bagasse	Asia (74.9)	181

After ref. [52].

Secondary rest streams are the by-products of industrial processes (such as sawdust, wood chips, pulp, black liquor, potato peels, soapstock, rendered oils and fats, sugar cane bagasse, waste liquid from distilling and spent grains, etc.). Traditionally, most of these products have already found some useful application (e.g. as animal feed) but some are still not being used. The logistic problem of their collection is limited as they are produced in industrial plants. Recently, a simple, integrated and economical way to produce jet and diesel fuel range (straight C_8–C_{13}) alkanes with high overall yield of 76% has been reported [55] using low-value waste hemicellulose-derived aqueous carbohydrate solutions, obtained from biomass power plants or composite wood manufacturing facilities [47, 56]. The overall stoichiometric reaction of the four-step process, comprising acid hydrolysis and xylose dehydration to furfural, aldol condensation with acetone, hydrogenation and hydrodeoxygenation, is as follows:

$$\underset{\text{xylose oligomers}}{2C_5H_{10}O_5} + \underset{\text{acetone}}{C_3H_6O} + 12H_2 \rightarrow \underset{\text{tridecane}}{C_{13}H_{28}} + 11H_2O \qquad [4.2]$$

A limitation of a large-scale commercial process is the need for hydrogen (eventually to be obtained from biomass). The process can be integrated into a pulp and paper refinery, or into a lignocellulosic ethanol refinery.

Tertiary rest streams are those originating after product use (such as used frying oils, animal manure, waste paper, etc.). These rather limited-volume residues are again highly dispersed at many locations and their use greatly depends on excellent logistics. Many stay unused and are wasted. Today's waste will be tomorrow's resource. Waste-derived syngas, obtained by steam reforming of agricultural and sanitary waste (mainly cellulose and lignin), municipal waste (70% carbohydrate, 10% plastic and 20% incombustibles) and industrial waste (about 60% combustibles) would be able to secure a renewable fuel production matching the present total oil supply.

4.3.2 Biomass separation and conversion techniques

A broader use of biolubricants will be possible when products and production processes are developed that are not based solely on oilseeds but on cheap and readily available renewable bulk plant materials such as sugars, cellulosic and waste materials. Given their positive environmental assessment, biolubricants from residues should be realised if technically and economically viable.

Any industrial use of renewable resources requires sophisticated technology. There are major technological challenges to overcome and the products must satisfy complex specifications. In the biomass-to-industrial product chain it is often not biomass which determines the end product, but rather the end product or application which determines the choice of biomass and the corresponding refining technique. A range of different technologies can be used industrially to convert the available biomass into renewable materials or energy carriers. Biomass energy in the form of traditional use (firing for cooking and heating) and modern use (producing electricity, steam and liquid biofuels) currently contributes 9–13% of global energy supply. Apart from biofuels today's bio-based products also include commodity and specialty chemicals, and materials. Industrial activity in renewable materials is very often linked to the food sector. Many of the renewable raw materials that can be considered for technical industrial applications (e.g. vegetable oils, starch, glucose, sugar) can be made in food-processing plants. The industrial sectors supplying the most important renewable raw materials are listed in Table 4.8. Quantitatively, the bio-based product groups, or fields of application, associated with the industrial use of renewable resources can roughly be sorted as follows:

- oleochemical applications and products (including biolubricants);
- bio-based materials;
- carbohydrate-based basic organic chemicals, intermediates and specialities; and
- industrial uses and products based on various other renewable resources.

Techniques for separation and conversion of biomass range from very low-value combustion (for generation of heat and electricity) to gasification (for syngas production), pyrolysis (for concentration of residual biomass to pyrolysis oil), mild pyrolysis and torrefaction (for bio-coal and coffee), fermentation (to biogas, bioalcohol), biorefining (e.g. of pyrolysis oil) and white biotechnology (for bio-based platform chemicals). White biotechnology is the application of selected (natural or genetically modified) microorganisms in the fermentation industry to produce high-value added

Table 4.8 Industrial sectors supplying the main renewable raw materials

Industrial sector	Feedstock	Product(s)
Sugar and starch	Plant materials[a]	Carbohydrates[b]
Oils and fats	Plant materials[c]	Oleochemical intermediates[d]
Cellulose and paper	Wood	Wood chemicals[e]

[a] Sugar cane, sugar beet, potatoes, wheat, corn, grain, rice, wood, etc.
[b] Sugar, glucose, starch and molasses.
[c] Rapeseed, soybean, palm, coconut, etc.
[d] Vegetable oils (triacylglycerols), fatty acids, fatty alcohols, glycerol.
[e] Cellulose and derivatives, lignin, tall oil, alkyd resins, rosins, pitch, fatty acids, turpentine.

Table 4.9 Primary conversion processes of biomass

Process	Major product(s)
Exsiccation, torrefaction (200–300 °C)	Biomass[a]
Carbonation (400 °C)	Charcoal
Pyrolysis (400 to >700 °C)[b]	Gas, bio-oil, chemicals, char, C
Thermolysis (450–800 °C)	Char, oil, gases
Gasification (650–900 °C)	CO, H_2, CO_2, CH_4, H_2O
Hydrothermolysis (250–600 °C)	Oil, char, gases (CO_2)
Combustion (900–1200 °C)	Heat, CO_2, H_2O
Fermentation	Ethanol, CO_2
Anaerobic digestion	CH_4, H_2O
Aerobic digestion	Heat, CO_2, H_2O
Extraction	Vegetable oil

[a] Without moisture and volatiles.
[b] Variable conditions, variable product compositions.

products from renewable raw materials (see Section 4.3.1). The primary conversion processes of biomass are summarised in Table 4.9. Most biomass technologies have not yet become competitive.

Many agricultural materials have a high water content (up to 50% in freshly cut wood, 7% in dried wood). This reduces their potential for combustion processes because the energy required to dry the materials exceeds the value of the energy recoverable through combustion. Anaerobic digestion reduces both the volume and mass of the waste materials and typically produces a product that is readily dewatered. Superheated steam drying is another biomass drying technology.

At variance to coal and oil, biomass contains a number of high-valued components (sugars, proteins, antioxidants, etc.) which deserve a better fate than just combustion. Pyrolysis and torrefaction are interesting techniques for the elimination of water from biomass, resulting in concentration and greater suitability for transportation and further processing in economic conditions. The pyrolysis technology that is closest to commercialisation is the pyrolysis of high lignin-containing lignocelluloses. Ensyn (Canada)'s Rapid Thermal Processing (RTP) can use sawmill residuals and construction and demolition wood waste as well as fast-growing wood species such as willow and poplar, liquefying 75% of the original in a fast pyrolysis process to 'bio-oil'. Some drawbacks of pyrolysis are:

- use of great amounts of heat;
- production of a wide range of compounds due to low selectivity;
- removal of oxygen from substrate molecules (thus eliminating one of the benefits of oxygenated products); and

- creation of solid residues of ash and carbon, thus requiring additional separation steps.

Pyrolysis oil, obtained by fast pyrolysis of lignocellulosic biomass, is the cheapest liquid fuel produced from biomass. Pyrolysis oils can be converted catalytically into industrial commodity chemicals (light olefins and aromatic hydrocarbons), polyols and synthetic gasoline [50, 57].

Fermentation is of interest not only for the production of (bio)ethanol but also for other bio-based chemicals using genetically modified microorganisms. Many of the top desirable chemicals from biomass (such as succinic, fumaric and itaconic acid, 3-HPA, PDO) can be produced by fermentation or enzymatic conversion of biomass (see Sections 5.5 and 6.2.7, p. 316).

First-generation conversion technologies of food crop biomass (to biofuels) have reached a high degree of maturity [58, 59]. 'Biofuels' are liquid or gaseous fuels for transport produced from biomass, such as pure vegetable oils (or derivatives), and bioethanol. Second-generation biomass is characterised by:

- use of non-food crops (agricultural and urban waste) and lignocellulosic mass; and
- development of innovative industrial crops.

The conversion of G2 biomass is relatively immature and still faces considerable technical barriers. The focus of the second generation is therefore on developing new technology. Second-generation technologies will maximise the use of low-cost and sustainable feedstocks, reduce dependence on food crops as feedstocks, and minimise the carbon footprint.

In even more advanced technologies conversion of all parts of the biomass, including the lignocellulosics, is considered as a source for energy, biofuel, chemicals and materials. Advanced bio-based platform chemicals are expected to be derived largely from industrial crops grown on marginal land (e.g. jatropha, switchgrass), short-rotation woody crops (e.g. willow, poplar), unused or new agricultural and forestry residues (wood chips, corn stover, wheat straw), as well as waste products (black liquor, animal waste) not required for food.

Figure 4.6 shows the main technologies used for conversion of the various biomass components. Biomass consists primarily of carbohydrates (cellulose and starch), and one method to convert these compounds is by fermentation to produce liquid alcohols. Biotechnology has had a positive impact on the cost efficiency of enzymatic conversion of biomass to fermentable sugars and has increased the range of products that can be produced by genetic engineering of fermentative organisms. Fermentation of glucose-based crops such as sugar cane (*Saccharinum officinarum*) and corn starch using *Saccharomyces* yeasts leads to ethanol. The technology to convert

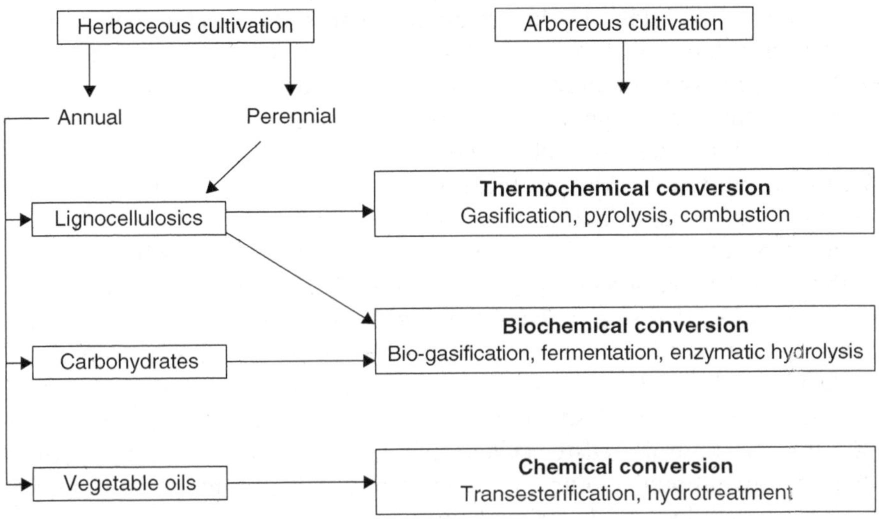

4.6 Technologies for conversion of biomass components.

grain-derived starches to ethanol by a combination of hydrolysis, fermentation and distillation is well established [60–62] and advances are being made in the cost-effective conversion of lignocellulosics to ethanol [63, 64]. The current high cost of cellulase enzymes is the key barrier to economical production of bioethanol from lignocellulosic material. The US Food, Conservation and Energy Act of 2008 promotes advanced biofuels (cellulosic ethanol, biobutanol and bio-based hydrocarbons) made from non-food crops and offers incentives to move away from corn ethanol production. Recently some robust microorganisms have been identified which enable crystalline cellulose deconstruction between 70 °C and 111 °C and resist denaturation in strong detergents, high salt concentrations, and ionic liquids [65].

In a different recent development a heterogeneous sulphonic acid functionalised catalyst has been described for the hydrolysis of the β-1,4-glycosidic bonds of cellobiose (a disaccharide) to form glucose in 96% yield [66].

Although biofuels are a tool to achieve national policy goals such as securing and diversifying the energy supply for road transport in an environmentally friendly way, biomass growth for the current vegetable oil-based diesel technologies (biodiesel/FAME (fatty acid methyl ester) and hydrotreated vegetable oil, HVO) is not able to keep pace with the growth in fuel demand. The production of next-generation feedstocks (e.g. MSW, forest residues, dedicated energy crops, microalgae) are expected to fare better than biofuel production from corn-grain or soybean [19]. Alternative biofuel production pathways are being investigated, such as BTL fuels, which aim at the integral valorisation of biomass. BTL technology permits the transformation of biomass (cellulosics, wood, organic waste, etc.) into biofuels by high-temperature gasification followed by Fischer–Tropsch. Producing biofuels from lignocellulosic material via gas-to-liquid (GTL) technologies – BTL fuels – could potentially overcome the aforementioned feedstock limitations. Various technologies are nowadays being proposed for processes to renewable diesels and other fuels, which differ in type of biomass used (vegetable oils, fats, agricultural waste or lignocellulosics) and in their conversion process [67]. All cellulosic BTL fuel technologies are at an early stage of development. BTL fuels have a greater potential than conventional biofuels to contribute to the national policy objectives of replacing traditional oil-based automotive fuels (petrol and diesel), owing to larger feasible land availability for woody and herbaceous biomass than that for agricultural feedstocks, and higher biofuel yield per hectare than that of conventional biofuels (see Section 4.4). These considerations for biofuels also hold for other bioproducts.

While the United States is favoured in terms of cellulosic feedstock availability (corn belt), Europe mainly needs to focus on waste feedstocks. This affects the technological options. As fermentation works best for bioproduct formation from single waste streams (such as corn stover), production of advanced bioproducts (mainly biofuels) in Europe is more likely to prefer thermochemical routes. Commercialisation of cellulosic technologies is as yet in its infancy and the technology is still unproven on a large scale. Several companies are aiming at becoming cellulosic ethanol producers; others have widened their focus to include the production of higher-value bio-based chemicals. Range Fuels, Inc. (Broomfield, CO) has recently (2010) produced the first batches of cellulosic methanol at its Soperton (GA) facility. A 198 kgal/yr cellulosic ethanol demo plant based on wood waste using Woodland's patented Catalyzed Pressure Reduction (CPR) gasification process has been on stream since mid-2011 in Sarnia, ON (Woodland Biofuels, Inc.) [68]. A pre-commercial demo-plant of forest raw material into biomethane is under construction by the Swedish Göteborg Energi AB – an energy company.

In the pulp and paper industry, lignin is removed or degraded chemically or mechanically. The sulphate or Kraft process produces the largest amount of pulp. Kraft lignin in spent pulping liquors can be recovered by acidification. An important factor is the cost of separating lignin from large volumes of aqueous industrial pulping liquors. After drying the product is not pure but contains ash, carbohydrates, and extraneous material of the spent liquor. The organic matter in the residual black liquor has substantial fuel value.

White-rot fungi degrade lignin in an essentially oxidative process and are the only organisms able to mineralise lignin to CO_2 and H_2O by processes initially catalysed by extracellular enzymes. Side chains of the biopolymer are oxidised with formation of carbonyl and carboxyl groups and aromatic nuclei are oxidatively cleaved after demethylation and introduction of hydroxyl groups in phenolic units to give dihydroxyphenyl moieties.

Much effort has been devoted to making use of lignin as feedstock in biomass conversion processes [15, 18]. Not surprisingly, in view of the complexity of lignin chemistry, there is only slow development towards products and applications. Lignin comprises a network of ring-like monomeric hydrocarbon units that are principally connected by C—O bonds. Cleaving these bonds without breaking up the individual ring structures is crucial to produce useful chemical building blocks, rather than a mixture of short-chain hydrocarbons. The challenge with lignin conversion is therefore the selective depolymerisation of the carbon–oxygen linkages: a catalyst should selectively cleave aromatic C—O bonds in various aromatic ethers without reduction of aromatic rings or cleavage of aliphatic C—O bonds. Recently, a homogeneous nickel–carbene-based catalyst has been described that breaks down lignin [69], although industrial applications are still some way off.

Solvolytic and reductive degradation of lignins yields products reflecting the structural units. Only a few low-molecular-weight chemicals (such as vanillin) are produced commercially from lignin. Polymeric lignin products are used for combustion, surface active properties of salts of a lignin derivative, and condensation [70]. For well-engineered bio-process technology, a preparation of lignocellulose containing plant matter is required which typically involves comminution of biomass to a particle size smaller than 250 μm. In favourable cases, specific enzymes may enable solubilisation, as occurs in the human body.

Valorisation of biomass and low-quality, low-price waste products into a high-quality, high-value feedstock for bulk and specialty chemicals is work in progress.

4.3.3 Biotechnology

Biotechnology is making it possible to sustainably increase biomass productivity in forestry and agriculture and to utilise that biomass and waste organic materials in a highly efficient and sustainable manner. Industrial or white biotechnology is that set of technologies that provides tools for adapting and modifying the biological organisms, products, processes and systems found in nature for developing processes and products that are eco-efficient, environment-friendly and profitable. Biotechnology mostly utilises renewable materials. Biotechnology enables the manufacture of new, innovative products or of existing products more effectively and efficiently, whose performance cannot be achieved using conventional chemical processes or petroleum as a feedstock. The development of bio-organic drilling fluids is an example in case (see Section 12.4.2).

Biotechnology has evolved over the last 35–40 years into a set of powerful tools for developing and optimising the efficiency of bioprocesses and the specific characteristics of bioproducts. In the past, industrial biotechnology has seen a number of failures but has equally well demonstrated its benefits [71]. The application of biotechnology in a wide range of industry sectors has led to both economic and environmental benefits by replacement of conventional processes. Adoption of more efficient bioprocesses and bioproducts with improved product quality/performance determines a more sustainable production.

Advances in biotechnology allow an economically viable and attractive way of replacing petroleum with biomass, a renewable feedstock derived mostly from plants. The Vision [72] and Technology Roadmap [73] for Plant/Crop Based Renewable Resources 2020 indicate the technological developments required for the near to long term:

- selection of value-added crops and trees for conventional and industrial requirements;
- high-yield, sustainable crop and tree production;
- eco-efficient harvesting and processing;
- stainable utilisation of the resulting products;
- preservation of the soil organic content and fertility.

The life sciences, and in particular biotechnology, will play a prominent role in meeting these challenges. A greater understanding of plant science could be exploited through biotechnology to generate crops with improved properties. The possibility of engineering the synthesis of large amounts of vegetable oil in crops has been envisaged [74]. The development of high-oleic vegetable oils from the seeds (as stable base stocks for lubricants) is another example (see Sections 5.3.1 and 10.5.2). Engineering the accumulation of triacylglycerols in non-seed tissues is an important target in plant biotechnology [75].

With advances in technology, environmental and economic benefits, and the need for innovation in the chemical industry the prospects of industrial biotech appear formidable. Biotechnology is now generally considered as a predominant driver for innovation in the chemical industry (see Sections 5.3, 5.4.2 and 6.2.7, p. 316) Compared with conventional chemical processes, biocatalytic processes usually consume fewer energy, produce less waste and use fewer organic solvents (see Table 6.22). Plant biomass can be processed and converted by fermentation and other processes into chemicals, fuels and materials that are renewable and result in no net GHG emissions. Biologically derived products ('bioproducts') are generally less toxic and less persistent than their petrochemical counterparts.

Today we are seeing the emergence of new bio-based building blocks (such as glucose, lactic acid, succinic acid, glycerol, furanics, propanediol

(PDO), 3-hydroxypropionic acid (3-HPA)) and green chemistry. This will stimulate the rekindling of chemical innovation. A broad spectrum of enzymes and fermentation systems is already available and the number is increasing rapidly.

For broad-based adoption, new products must be competitive with existing offerings. Few customers are willing to pay a green premium for the introduction of bio-based products. The cost differential between hydrocarbons (oil, natural gas) and carbohydrate (sugar, biomass) feedstocks is fluctuating. Apart from price increases of natural oil and gas, industrial biotech will benefit from the availability of low-cost waste biomass (such as wheat straw and corn stover) as an alternative feedstock to biomass. Bioethanol has been the first product to profit from such cheap residual biomass feedstocks. Advancements in biotechnology have allowed the conversion of cellulose into fermentable sugar utilising cellulase enzymes (see Section 4.3). Ethanol biorefineries based on waste biomass (cereal straw) are already operating (Iogen Corporation, Ottawa) [54] or are under development (Süd-Chemie AG/Linde Group, Munich). Biotech applications will be favoured by a decreasing price difference between fossil hydrocarbon and biological carbohydrate feedstocks. In 2007, 6% of total chemical sales depended on biotech. Expectations are 9% in 2012.

Apart from the Biochem project (initiated by SusChem), a European innovation scheme in biotechnology for chemical small and medium enterprises (SMEs), the EU has yet to implement any major initiatives to support industrial biotechnology, in contrast to the situation in the United States. Biotechnology is more a political than a scientific issue in Europe, where some genetically modified products have been held under review for more than six years, as compared to six to nine months in the United States, Canada and Japan [76]. The prospects of improvement remain dim. Austria, Denmark, France, Germany, Greece and Luxemburg have imposed marketing bans on some biotech products despite EU regulations (2001/18/EC, 1829/2003/EC, 1830/2003/EC) and approvals. Moreover, European carbohydrates are not competitive and funding for technological development is poor. Consumer acceptance is vital, as shown by the widespread rejection of agricultural biotech and genetically modified foods in Europe, despite their acceptance in the United States. Thus far, consumer movement against industrial biotech is limited, although UK supermarket chains have refused genetically modified biopolymer packaging materials. Genetic modification of microorganisms for use in fermentation is less controversial as they do not enter the food chain but stay confined to a bio-reactor. The aim of modifying the metabolic pathways of microorganisms by genetic engineering ('metabolic engineering') is to efficiently produce, in one process step and at high yield, what would take several steps with much lower yield in the conventional organic chemical fashion. Metabolic engineering of

Saccharomyces cerevisiae is used by Reverdia (DSM/Roquette Frères joint venture) to convert starch to biosuccinic acid at low pH. Similarly, Genomatica, Inc. (San Diego, CA) has demonstrated a commercially viable direct approach to turning sugar into 1,4-butanediol (bio-BDO) – at present produced exclusively through feedstock derived from oil and natural gas – using a metabolically engineered strain of *Escherichia coli* (Presidential Green Synthesis Award 2011, ACS/EPA) [77].

As to industrial sustainability, biotechnology has not yet developed into a generally profitable business. For case studies of implemented biotechnological processes over the last 35 years, see ref. [78]. There have been relatively few commercial success stories in the bioproduction of chemicals to date. For instance, the complex energy-intensive chemical process to produce vitamin C has now been replaced by one fermentation step. However, other projects such as 1,3-propanediol (Du Pont/Tate & Lyle), polylactic acid (Dow Chemical/Teijin/Cargill venture NatureWorks), polyhydroxybutyric acid (Metabolix/ADM) all have suffered problems. A new attempt is now being made with biosuccinic acid. Succinic acid (30 kt/yr) is currently produced from fossil-derived maleic anhydride at relatively high cost. Large-scale fermentation of renewable feedstocks (e.g. wheat-derived glucose, unrefined sugar, sorghum, glycerine) to biosuccinic acid is expected to create a 180 kt/yr lower cost platform molecule by 2015 for applications such as BDO (using DuPont technology), polybutylsuccinate (PBS), polyester polyols, polyurethanes and succinate-based plasticisers. Various industrial groups are developing commercial biosuccinic acid capacity (DNP Green Technology/ARD, DSM/Roquette Frères, Myriant Technologies/Sojitz, BASF/CSM, Mitsubishi Chemical Co./Ajinomoto). Myriant Technologies (Lake Province, LA) is constructing a 13.6 kt/yr biomass (sorghum)-to-biosuccinic acid plant. The Presidential Green Chemistry Challenge Award 2011 (ACS/EPA) was recently granted to BioAmber (Minneapolis, MN) for a biomass process route to biosuccinic acid. Two plants using this technology are under construction: 35 kt/yr in Sarnia, ON, and 70 kt/yr in Thailand (in cooperation with Mitsui). Using efficiency directed genome engineering (EDGE), Dow Chemical/OPX Biotechnologies are examining the viability of producing acrylic acid from renewable raw materials (corn) via 3-hydroxypropionic acid. For a global biotechnology industry outlook (including agricultural biotech), see ref. [79].

4.3.4 Biorefining

Biomass, typically in a low-density form, must be collected, transported and converted. The edible and non-edible parts of biomass and nutrients can be separated, and the non-edible fraction can then be converted into chemicals and fuels. As biomass is heterogeneous in nature, consisting of a minor percentage of high-valued components and a larger fraction of chemically less accessible to (currently) inaccessible material (for combustion), it is necessary to separate these fractions. Biorefineries are small to very large industrial complexes (yet smaller and more regional than petrorefineries) in which agricultural feedstocks are fractionated and processed into intermediates and partially converted into final products. Various definitions have been given of the concept of biorefinery [80–82]. According to the US Department of Energy (DOE) 'a biorefinery is an overall concept of a processing plant where biomass feedstocks are converted and extracted into a spectrum of valuable products' [80].

Biomass resources are widely available but are far more dispersed than oil. A key consideration for the production of bioproducts in a biorefinery is the continuous availability of sufficient biomass resources, preferably within 60 km of the refinery. For instance, Diesel Brewing (Salem, OR) is implementing a process for the gasification of waste biomass into bioalcohols and focuses on biobutanol in small-scale (125 t biomass per day) rural biorefineries. A biorefinery should ideally use all components of the biomass feedstock (waste free). Biorefining yields products greatly differing in volume and economic value (Fig. 4.7). Intermediate basic products may find their own applicability. The products often have little in common with the nature of the original plant feedstock. Rest streams (containing proteins, phosphates, potassium, lignocellulosics) may eventually serve as animal feed.

Biorefineries use physical, chemical and biotechnological processes. Fractionation technology is a preparation process intended to separate agricultural materials into separate substance families using both physical and chemical methods. Fermentation technology and biocatalysis using microorganisms and enzymes are particularly important. Thermochemical thermolysis, pyrolysis and gasification are already being used for the production of char, oil and gases. Thermochemical conversion processes, considered as having great potential, require further development. This holds in particular for catalytic thermochemical conversion technology for production of complex molecules from agricultural feedstocks. Industrial biotechnology, enzymatic technology and thermochemical conversion technology are complementary for full valorisation of renewable feedstocks into value-added chemicals and fuels. Figure 4.8 illustrates the biorefinery concept with a variety of biochemical and (thermo)chemical processing steps. Biorefineries may be designed to produce a range of co-products, some being *low-volume, high-value* products, such as intermediates and specialities, and others, like commodity chemicals and biofuels, being *high-volume, low-value.*

There are many types of biorefinery and the technology is in a stage of rapid development. In biorefineries, biomass is used as a feedstock for the production of food products, high added-value chemicals, solvents, materi-

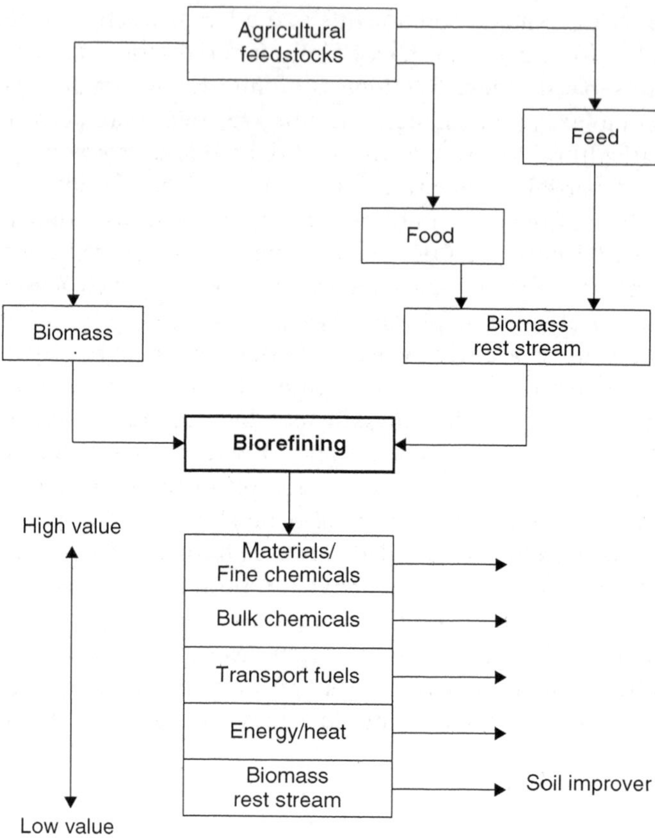

4.7 Separation of biomass in various fractions in a biorefinery.

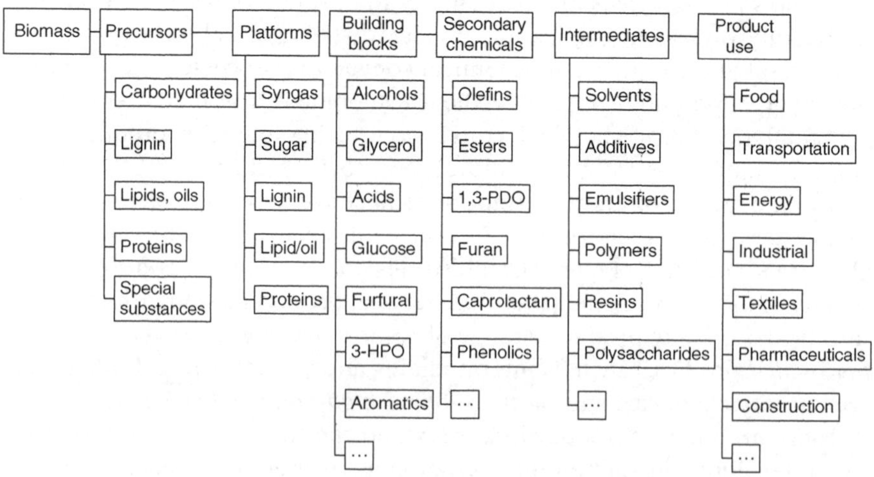

4.8 Product chart for biomass feedstock in a biorefinery.

als, intermediates and fine chemicals together with the production of energy carriers, preferably in the liquid phase (Fig. 4.8). Biorefineries can be compared with existing and fully integrated petrochemical refineries. Like the petroleum refinery, it is expected that the processes and products from a biorefinery will evolve to meet society's demand (in a bio-based economy). Although the primary focus of the biorefinery concept today is on fuels and energy production, large-scale implementation will stimulate development of a feedstock industry similar to that experienced for the petrochemical refinery. The final aim is to use dedicated crops in such a way that all the constituents of a whole plant (oils, proteins, fibres, cellulose, lignin, etc.) result in a total product mix with the highest economical return. Biorefineries and their associated crop production systems will be highly integrated.

Essential elements of a biorefinery are:

- multiple feedstock capability and flexibility;
- processing to fermentable sugars;
- biocatalysts for conversion of sugars to final products; and
- co-products (for recycling or sale).

Renewable carbon feedstocks for biorefineries are typically sugar cane/beet, starch/grains and lignocellulosics. Their primary products are sucrose, glucose/fructose and sugars, respectively.

Various complex multi-product biorefinery systems are under development:

- lignocellulose feedstock (LCF) biorefinery for conversion of 'nature-dry' raw materials (straw, corn stover, wood, paper waste, etc.);
- whole-crop biorefinery for conversion of cereals (rye, wheat, maize, etc.);
- green biorefinery for fractionation of 'nature-wet' biomass (green biomass); and
- two-platform (sugar and syngas) biorefinery for conversion of 75% carbohydrate containing biomass.

In an LCF biorefinery cellulose, hemicellulose and lignin are separated. However, there is still a need for further development and optimisation of the underlying fractionation techniques. In a whole-crop biorefinery the first step is mechanical separation into seed and straw (Fig. 4.9). Straw is a lignocellulosic feedstock and may further be processed in an LCF biorefinery with possibility of separation into cellulose, hemicellulose and lignin. Figure 4.9 shows the products of a whole-crop biorefinery based on cereals, where corn is either converted into starch and further derivatives or used directly after grinding to meal. Further processing may be conducted by four processes: breaking up, plasticisation, chemical modification (etherification into carboxymethyl starch; esterification and re-esterification into

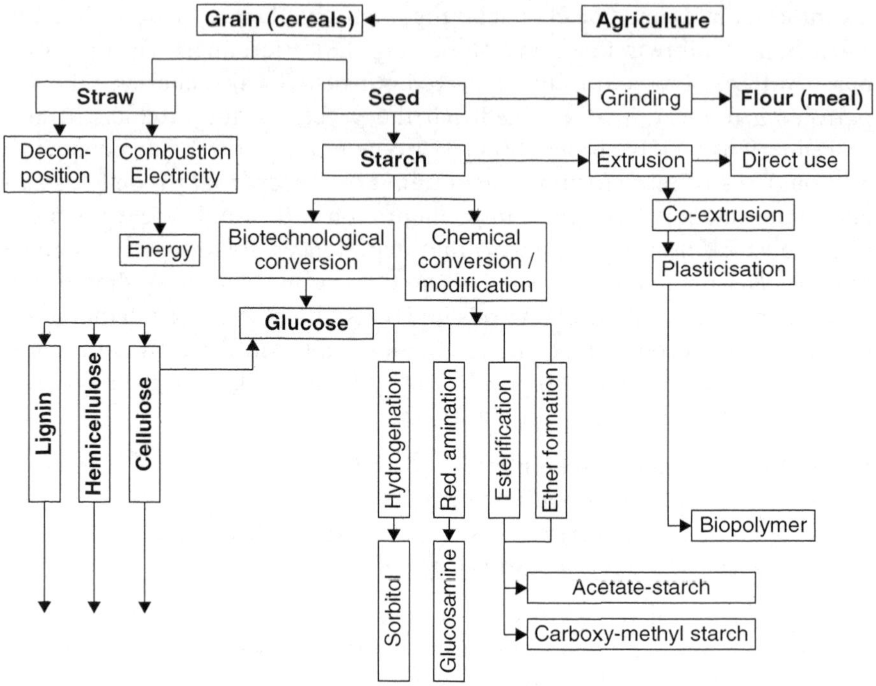

4.9 Products from a whole-crop biorefinery via dry fractionation.

fatty acid esters via acetate starch, etc.), or biotechnological conversion to glucose. Grain processing may take place not only by traditional dry fractionation of mature cereals but also by wet milling. Wet milling of corn yields corn oil, corn fibre and corn starch.

Green biorefineries represent complex (to fully integrated) systems of sustainable, environmentally and resource-friendly technologies for the comprehensive (holistic) material and energetic utilisation as well as exploitation of biological raw materials in the form of green and residue biomass from a targeted sustainable regional land utilisation [82]. Wet-fractionation of green biomass (green crop fractionation) in such a biorefinery can be used for the simultaneous manufacture of both food and non-food products [83]. In the so-called two-platform concept the sugar platform is based on biochemical conversions and focuses on fermentation of sugars extracted from the biomass feedstock; alternatively, in the syngas platform, based on thermochemical conversion, syngas and derivatives are produced.

Up to now, existing large-scale biorefineries have predominantly delivered food products such as soy oil and soy protein, wheat starch and gluten, potato starch and protein. A dry-milling ethanol plant, which uses grain as a feedstock and presents little processing flexibility, is an example of a first-generation biorefinery. In a second-generation biorefinery, using grain as feedstock and the current wet-milling technology, a variety of end products (such as starch, high-fructose corn syrup, ethanol, corn oil, corn gluten feed and meal) may be produced. Another example of a second-generation biorefinery is Iogen's wheat straw-to-ethanol plant [54]. In this process straw is chemically and/or physically pretreated. Enzymes then decouple the cellulose and hemicellulose chains, breaking them down into individual sugars, which will be used as feedstock for fermentation to ethanol by bacteria or yeast cells. The organisms can also be modified to produce vitamins, organic acids and other substances. Ineos Bio plans to build a biorefinery near Hull (UK) which will convert waste biomass into combustible gas. This gas can then be burned to provide electricity or fermented into bioethanol.

USDA operates a biorefinery assistance programme to promote the development, construction and/or retrofitting of commercial-scale biorefineries. In Europe, a Seventh Framework Programme (FP7) call concerns sustainable biorefineries for developing technologies to make biorefinery production cost-effective and coordinating biorefinery research [84]; see also Section 4.3.

Competition for resources between agriculture, urban areas and industry will only grow over the coming years. With the forthcoming integrated biorefineries using more than one feedstock the chemical industry will be in a position to play an essential role in producing an equivalent amount of industrial products (fuels, materials) as in the past, but relying on renewable biomass rather than petrochemical feedstocks, thereby achieving the much needed minimisation of waste, emissions and need for energy input. Oleaginous biomass-based renewable lubricant production may profitably be integrated in a biorefinery environment.

4.4 Biomass potentials and limitations

Hundreds of crops have been domesticated and cultivated by mankind during history, and are being utilised for food, forage, fibre, fuel, medicine and chemistry. However, only a small number provides the bulk of raw materials necessary for human survival. Bio raw materials are now increasingly becoming a need but for proper use of biomass certain boundary conditions should be satisfied both in relation to agriculture and in terms of socio-economic aspects. Cultivation of biomass should be sustainable. The EU Renewable Energy Directive (RED), which promotes the use of energy from renewable sources, specifies the land uses to be protected for (biofuel) feedstock production to meet expanded demand [8]. Developing a sustainable bio-based economy that uses eco-efficient bioprocesses and renewable bioresources (green economy) is one of the key strategic challenges of the 21st century. Ensuring environmental sustainability is one of the eight Millennium Development Goals (MDGs) defined by the UN in 2000 for the period up to 2015. The life sciences, and in particular biotechnology, will play a prominent role in meeting that challenge (see Section 4.3.1). In this context, sustainable development means development that meets the needs of the present without compromising the ability of future generations to meet their own needs. Sustainable economical activities, including the large-scale production of biomass, biofuels and bio-based materials, are those that do not negatively affect people and the environment, yet giving economic advantage (triple P: People, Planet, Profit). With proper management and technologies biomass feedstock can be produced sustainably. The process industry has defined a code for sustainable production of biomass. Examples are the Round Table on Sustainable Palm Oil and the Round Table on Responsible Soy Association. The socio-economic conditions for production of biomass by smallholders (i.e. local farmers) should not be dictated, monopolised and constrained by multinationals (Small is Beautiful vs Big Business). The triple P effects of agricultural production of GM crops (soybean, maize and cotton) have been evaluated [85].

For bioproducts including polymers, lubricants, solvents, adhesives, herbicides and pharmaceuticals potential markets are wide ranging and considerable. Current concerns regarding many first-generation biomass-derived bioproducts are that they:

- do not meet their claimed environmental benefits unless the biomass feedstock is produced sustainably;
- are accelerating deforestation with other negative indirect land-use change (ILUC) effects;
- have a potentially negative impact on biodiversity;
- provide only limited GHG reduction benefits (except for sugar cane ethanol) at high costs;
- are an expensive option for improving energy security; and
- contribute to higher food prices.

Development of a 'non-food' agricultural industry, where non-food refers to all biological materials and processes used in the production of industrial chemicals, polymers and fibres, servicing diverse industrial outlets, is a recent phenomenon.

The possibility to fully utilise not only the already naturally highly refined molecular structure of biogenous oils but also the 50–80% remainder of the plant for use (e.g. as co-generation fuel) will at least double the yield per hectare to 3–4 t (biodiesel + BTL rapeseed plant) and 8–10 t/ha (biodiesel + BTL from palm free fatty acids). It should be realised, however, that not all residuals of agricultural production (see Table 4.6) can be used for chemicals production; a considerable part needs to be re-invested in the soil to avoid gradual degradation.

The benefits of biomass as a raw material for the chemical industry are summarised in Table 4.10. Biomass is a convenient answer to an Inconvenient Truth [86]. Biomass as a feedstock for the chemical industry has a higher added value than for use as biofuels. If a non-food agricultural industry can be developed without the need for large processing units, such production facilities close to the source of the raw materials and end-users would strengthen the economy of rural areas.

At variance to solids like agricultural products, oil and gas remain easier to transport and are not affected by adverse weather conditions. Table 4.10 lists the main limitations to the use of biomass for non-food industrial applications. Major technical bottlenecks limiting the production and utilisation of industrial crops and products arise from the fact that the quality, quantity and price of agricultural raw materials are insufficient to supply industries with appropriate feedstocks for their processes [87]. More specifically:

Table 4.10 Potentials and limitations of biomass as raw material for the chemical industry

Advantages:
• Environmental benefits (CO_2 neutrality)
• Potentially sustainable
• Green economy
• Development of a new range of high added-value products
• Economically advantageous
• High potential (yet limited)
• Rural development
• New market outlets for agricultural products
• Sustained employment for farmers
• Strengthening of rural economy
• Decreasing dependence on foreign imports of raw materials (oil)
• Bio-based economy (stability of supply)
• Reduction of food surplus
Disadvantages:
• Agricultural restrictions
• Socio-economic implications
• Geopolitical, cultural aspects
• Threat on biodiversity
• Indirect land use change
• Complex chemical nature of raw materials
• Underdeveloped technology position
• Need for new reactor technology and catalysts
• Need for investments (irrigation, infrastructure, biorefineries)
• Transportation and production costs
• Food vs industrial use debate

- the batch-to-batch variable composition of biological raw materials is inconvenient for continuous industrial processing;
- continual supply of raw materials (except waste oils and fats) is difficult as agricultural production is essentially a seasonal activity (not valid for the tropics);
- plant size is limited by biomass collection capacity;
- limited knowledge base of biological raw materials;
- need for new technologies and sustainable (non-toxic) catalysts;
- traditional process technologies for biological materials do not fulfil current ecological requirements; and
- intensive use of fertilisers might determine a net loss of energy.

In particular, much technological development is still required for the second-generation bio raw materials (see Table 4.5).

It is easily appreciated that agricultural productivity is limited. Agricultural restrictions are the availability of land, water and solar energy (including good weather conditions), suitability of soil for certain crops, seed quality, need for nutrients, pesticides/insecticides, soil degradation, etc. The limited diversity of crops used for food and industrial use has increased the vulnerability of crops to adverse climatic conditions. Not every location is equally suitable for crop production. Essential requirements are the availability of energy (sun), water and micro-nutrients. Volumes of up to thousands of litres of water may be required to producte 1 kg of biomass. The agricultural productivity is strongly limited by the availability of water (compare a yield of 8.5–9 t/ha wheat in The Netherlands with 3 t/ha in Spain). Non-food crops, though not edible, compete with food crops for both land and water.

Chemistry has been instrumental to substantial advances in agricultural productivity, either directly (e.g. through the Haber–Bosch ammonia process and crop-protective chemical technology) or indirectly (e.g. by better conservation of perishable foods by refrigerants and flexible packaging). There exists a global need to increase agricultural productivity significantly and sustainably to meet the growing demands for food, feed, fibre, fuel and chemical feedstock. According to the UN Millennium Development goals on hunger food production will need to double by 2050 [88]. By 2025 one hectare of land will need to feed five people compared with only two people in 1960 [89]. This determines a wide variety of challenges (Table 4.11) and calls for the application of new technologies across the entire food supply chain, partly to be provided by the chemical sciences (Table 4.12). Without effective use of crop protection chemicals, up to 40% of the agricultural productivity would be lost [90]. As agriculture is facing emerging and resistant strains of pest, the development of new crop protection strategies is essential. Understanding soil structure and science is important to ensure high productivity.

Table 4.11 Agricultural productivity

Category	Challenge
Effective farming	Adopt best agronomic practices
Livestock and aquaculture	Technologies for containment of environmental impact and waste
Pest control	New crop protection strategies
Plant science	More efficient nutrient uptake and utilisation in plants
Soil science	Improved soil structure for high productivity
Water	Maintaining adequate water supply and quality

Source: www.rsc.org/roadmap.

Table 4.12 Potential contributions of the chemical sciences to enhanced agricultural productivity

- Biosensors for monitoring of soil quality, crop condition and water availability
- Analysis of climate change parameters
- Improved engineering tools for on-farm practices
- Aquaculture production for food and industrial use (e.g. algae)
- Formulation engineering for delivery and minor component release
- More targeted, high-potency agrochemicals
- Better pest control strategies
- Tailored pesticides
- Understand plant growth regulators
- Understand the impact of nutrients
- Improve fertiliser formulations
- Understand soil structure and biochemistry of soil ecosystems

Source: www.rsc.org/roadmap.

The food/feed/fuel dispute, which raged several years ago on the unwanted competition between food and fuel crops, obviously applies equally well to the application of biomass for chemical use. The production of biomass for industrial use should not adversely affect biodiversity and be in competition with food production. Food/fuel competition not only needs to be avoided in view of higher food prices but also because of indirect land use change (with indirect hidden CO_2 emissions). A strategy based on the use of agricultural areas for food production and development of robust (non-edible) industrial crops of the second generation (with high drought resistance, high photosynthesis capacity and possibilities for fixation of nitrogen and CO_2) for marginal areas limits competition for food and energy/fuels/chemistry.

The Brazilian bioethanol production (using only 4% of the present agricultural area) shows that it is possible to combine biofuels with food production, adequate socio-economic conditions and environmental protection [91]. The food-famine problem in the world is not so much a matter of production but more one of politics, corruption, economy and distribution. There is sufficient food to feed 9 billion people by 2050 (even without application of biotechnology). Use of biomass waste for sustainable chemistry circumvents the whole issue.

4.5 Sources of further information and advice

R.B. Laughlin, *Powering the Future*, Basic Books, New York, NY (2011).

Wetenschappelijke & Technologische Commissie (WTC) Biobased Economy (Dutch Ministry of Economic Affairs, Agriculture and Innovation), *Naar een Groene Chemie en Groene Materialen,* http://www.biobasedeconomy.nl/Pages/Naargroenechemieengroenematerialen.aspx (21 April 2011).

Commission of the European Communities, *Roadmap to a Resource Efficient Europe*, Brussels (20 September 2011).

Nuffield Council on Bioethics, *Biofuels: Ethical Issues*, Nuffield Press, Abingdon (UK) (2011); http://www.nuffieldbioethics.org

B.J.P. Lovell, *Challenged by Carbon, the Oil Industry and Climate Change*, Cambridge University Press, Cambridge (2010).

J.C.J. Bart, N. Palmeri and S. Cavallaro, *Biodiesel Science and Technology. From Soil to Oil,* Woodhead Publishing Ltd/CRC Press, Cambridge/Boca Raton, FL (2010).

K. Waldron (ed.), *Bioalcohol Production: Biochemical Conversion of Lignocellulosic Biomass*, Woodhead Publishing Ltd/CRC Press, Cambridge/Boca Raton, FL (2010).

A. Hoogendoorn and H.J.M.N.V. van Kasteren, *Transportation Biofuels. Novel Pathways for the Production of Ethanol, Biogas and Biodiesel*, RSC Publishing, Cambridge (2010).

J. Emsley, *A Healthy, Wealthy, Sustainable World*, RSC Publishing, Cambridge (2010).

J. Sanders, E. Heeres, F. van Oranje and M. Evenblij, *Biogrondstoffen. Van Fossiel naar Biomassa*, Stichting BWM, The Hague, 29 (2010) Cahier 1 (in Dutch).

W. Soetaert and E.J. Vandamme (eds), *Industrial Biotechnology. Sustainable Growth and Economic Success*, Wiley-VCH, Weinheim (2010).

A.A. Vertès, N. Qureshi, H.P. Blaschek and H. Yukawa (eds), *Biomass to Biofuels*, John Wiley & Sons, Inc., Hoboken, NJ (2010).

J. Cheng (ed.), *Biomass to Renewable Energy Processes*, CRC Press, Boca Raton, FL (2010).

D. Deublein and A. Steinhauser, *Biogas from Waste and Renewable Resources*, 2nd edn, Wiley-VCH, Weinheim (2010).

H. Langeveld, M. Meeusen and J. Sanders (eds), *The Biobased Economy*, Routledge, New York, NY (2010).

J. Harmsen and J.B. Powell (eds.), *Sustainable Development in the Process Industries. Cases and Impact*, John Wiley & Sons, Inc./AIChE, Hoboken, NJ/New York, NY (2010).

R. Höfer (ed.), *Sustainable Solutions for Modern Economics*, RSC Publishing, Cambridge (2009).

F. Cavani, G. Centi, S. Perathoner and F. Trifirò (eds), *Sustainable Industrial Chemistry*, Wiley-VCH, Weinheim (2009).

Global Industry Analysts, Inc., *Biotechnology, A Global Industry Outlook*, San José, CA (2008).

D.M. Mousdale, *Biofuels. Biotechnology, Chemistry and Sustainable Development*, CRC Press, Boca Raton, FL (2008).

J.H. Clark and F.E.I. Deswarte (eds), *Introduction to Chemicals from Biomass*, John Wiley & Sons, Ltd, Chichester (2008).

C. Hou and J.-F. Shaw (eds), *Biocatalysis and Biotechnology for Functional Foods and Industrial Products*, AOCS Press, Urbana, IL (2007).

Biofuels Research Advisory Council, *Biofuels in the European Union: A Vision for 2010 and Beyond*, EU Report, Brussels (2006).

G.A. Olah, A. Goeppert and G.K. Surya Prakash, *Beyond Oil and Gas: The Methanol Economy*, Wiley-VCH, Weinheim (2006).

B. Kamm, P.R. Gruber and M. Kamm (eds), *Biorefineries – Industrial Processes and Products: Status Quo and Future Directions*, Wiley-VCH, Weinheim (2006), 2 vols.

R.D. Perlack, L.L. Wright, A.F. Turhollow, R.L. Graham, B.J. Stokes and D.C. Erbach, *Biomass as Feedstock for a Bioenergy and Bioproducts Industry: The Technical Feasibility of a Billion-Ton Annual Supply*, DOE/USDA, Washington, DC (2005).

B. Kavalov and S.D. Peteves, *Status and Perspectives of Biomass-to-Liquid Fuels in the European Union*, EUR 21745 EN, European Commission, Directorate-General Joint Research Centre, Institute for Energy, Petten (2005).

R.C. Ramesh (ed.), *Microbial Biotechnology in Agriculture and Aquaculture*, Science Publishers, Inc., Enfield, NH (2005).

Commission of the European Communities, *Biomass Action Plan*, Brussels (2005), COM 628.

Wissenschaftlicher Beirat der Bundesregierung, *Globale Umweltveränderungen, Welt im Wandel: Energiewende zur Nachhaltigkeit*, Springer-Verlag, Berlin (2003).

R.C. Brown, *Biorenewable Resources*, Iowa State University Press, Ames, IA (2003).

Biomass R&D Technical Advisory Committee, *Roadmap for Biomass Technologies in the United States*, Washington, DC (2002); http://www.eere.energy.gov/biomass/publications.html

M. Griffiths, A. Gram, W. Treffenfeldt, U. Lange, T. McIntyre and O. Wolf (eds), *The Application of Biotechnology to Industrial Sustainability – A Primer*, OECD, Paris (2001).

H. Zoebelein (ed.), *Dictionary of Renewable Resources*, 2nd edn, Wiley-VCH, Weinheim (2001).

National Agricultural Biotechnology Council, *The Biobased Economy of the 21st Century: Agriculture Expanding into Health, Energy, Chemicals and Materials*, NABC Report 12, Ithaca, NY (2000).

R.P. Overend and E. Chornet (ed.), *Biomass: A Growth Opportunity in Green Energy and Value-added Products*, Pergamon Press, Oxford (1999).

D.L. Klass (ed.), *Biomass for Renewable Energy, Fuels and Chemicals*, Academic Press, San Diego, CA (1998).

K. Soyez, B. Kamm and M. Kamm (eds), *The Green Biorefinery*, Verlag GÖT, Berlin (1998).

A.V. Bridgwater and D.G.B. Boocock (eds), *Developments in Thermal Biomass Conversion*, Blackie Academic and Professional, London (1997).

C.E. Wyman, *Handbook on Bioethanol Production and Utilization*, Taylor & Francis, Washington, DC (1996).

H. Eierdanz (ed.), *Perspektiven nachwachsender Rohstoffe in der Chemie*, VCH, Weinheim (1996).

A. Wiseman, *Handbook of Enzyme Biotechnology*, 3rd edn, Ellis Horwood, London (1995).

E.-C. Oerke, H.-W. Dehne, F. Schönbeck and A. Weber, *Crop Production and Crop Protection*, Elsevier Science, Amsterdam (1994).

W.J. Bartz (ed.), *Biologically Fast Degradable Lubricants and Operational Fluids*, Expert-Verlag GmbH, Ehningen (1993).

G. Kitani and C.W. Hall, *Biomass Handbook*, Gordon & Breach, New York, NY (1989).

D.O. Hall and R.P. Overend (eds), *Biomass*, John Wiley & Sons, Ltd, Chichester (1987).

R.P. Overend, T.A. Milne and L.K. Mudge (eds), *Fundamentals of Thermochemical Biomass Conversion*, Elsevier, London (1985).

D. Fengel and G. Wegener, *Wood: Chemistry, Ultrastructure, Reactions*, Walter de Gruyter, Berlin (1984).

4.6 References

[1] Energy Information Administration (EIA), *International Energy Outlook 2007*, EIA Washington, DC; http://www.eia.doe.gov/oiaf/ieo/index.html

[2] International Energy Agency (IEA), *World Energy Outlook 2010*, IEA, Paris (2010).
[3] F.W. Lichtenthaler, The key sugars of biomass: availability, present non-food uses and potential future development lines, in *Biorefineries – Industrial Processes and Products: Status Quo and Future Directions* (B. Kamm, P.R. Gruber and M. Kamm, eds), Wiley-VCH, Weinheim (2006), Vol. 2, pp. 3–59.
[4] D.H. Klass, Organic commodity chemicals from biomass, in *Biomass for Renewable Energy, Fuels and Chemicals* (D.H. Klass, ed.), Academic Press, San Diego, CA (1998), pp. 495–546.
[5] *Biomass Research and Development Technical Advisory Committee Roadmap for Biomass Technologies in the US*, US Government (2002); http://www.eere.energy.gov/biomass/publications.html
[6] *Unilever Sustainable Living Plan*, http://www.unilever.com/images/UnileverSustainableLivingPlan_tcm13–239379.pdf (accessed 20 September 2011).
[7] B. Kamm, M. Kamm, P.R. Gruber and S. Kromus, Biorefinery systems: An overview, in *Biorefineries – Industrial Processes and Products: Status Quo and Future Directions* (B. Kamm, P.R. Gruber and M. Kamm, eds), Wiley-VCH, Weinheim (2006), Vol. 1, pp. 3–40.
[8] Directive 2009/28/EC of the European Parliament and of the Council of 23 April 2009 on the promotion of the use of energy from renewable sources and amending and subsequently repealing Directives 2001/77/EC and 2003/30/EC, *Off. J. Eur. Un.* **L 140/**16 (5 June 2009).
[9] C.B. Field, M.J. Behrenfeld, J.T. Randerson and P. Falkowski, Primary production of the biosphere: integrating terrestrial and oceanic components, *Science* **281** (1998) 237–40.
[10] H. Lieth, Primary production of the major vegetation units of the world, in *Primary Production of the Biosphere* (H. Lieth and R.H. Whittaker, eds), Springer-Verlag, Berlin (1975), pp. 203–15.
[11] Energy Efficiency and Renewable Energy (EERE), *Biomass Program: Biomass Basics*, US Department of Energy, Washington, DC; http://www.eere.energy.gov/biomass
[12] M. Pauly and K. Keegstra, Cell-wall carbohydrates and their modification as a resonance for biofuels, *Plant J.* **54** (2008) 559–68.
[13] H. Röper, Perspectiven der industriellen Nutzung nachwachsender Rohstoffe, insbesondere von Stärke und Zucker, *Mitt. Fachgruppe Umweltchemie und Ökotoxicologie der Gesellschaft Deutscher Chemiker* **7**(2) (2001) 6–12.
[14] W. Boerjan, J. Ralph and M. Baucher, Lignin biosynthesis, *Annu. Rev. Plant Biol.* **54** (2003) 519–46.
[15] G. Brunow, Lignin line and lignin-based product family trees, in *Biorefineries – Industrial Processes and Products. Status Quo and Future Directions* (B. Kamm, P.R. Gruber and M. Kamm, eds), Wiley-VCH Verlag, Weinheim (2006), Vol. 2, pp. 151–63.
[16] G. Brunow, K. Lundquist and G. Gellerstedt, Lignin, in *Analytical Methods in Wood Chemistry, Pulping and Papermaking* (E. Sjöström and R. Alén, eds), Springer-Verlag, Berlin (1998), pp. 77–124.
[17] K. Freudenberg, The constitution and biosynthesis of lignin, in *Constitution and Biosynthesis of Lignin* (K.K. Freudenberg, and A.C. Neish, eds), Springer-Verlag, Berlin-Heidelberg (1968), pp. 47–122.

[18] J. Zakzeski, P.C.A. Bruijnincx, A.L. Jongerius and B.M. Weckhuijsen, The catalytic valorization of lignin for the production of renewable chemicals, *Chem. Rev.* **110** (2010) 3552–99.
[19] P.R.D. Williams, D. Inman, A. Aden and G.A. Heath, Environmental and sustainability factors associated with next-generation biofuels in the US: What do we really know?, *Environm. Sci. Technol.* **43**(13) (2009) 4763–75.
[20] World Bank, *Placing the 2006/08 Commodity Price Boom into Perspective*, Washington, DC (2010).
[21] Agricultural Marketing Service (WABO), *World Agricultural Supply and Demand Estimates*, USDA, Washington, DC (10 May 2012).
[22] B. Kamm and M. Kamm, Principles of biorefineries, *Appl. Microbiol. Biotechnol.* **64** (2004) 137–45.
[23] T. Werpy and G. Petersen (eds), *Top Value Added Chemicals from Biomass*, Vol. 1, National Renewable Energy Laboratory and Pacific Northwest National Laboratory, Golden, CO (August 2004).
[24] E. Henrich, N. Dahmen and E. Dinjus, Cost estimate for biofuel production via biosyncrude gasification, *Biofuels Bioprod. Bioref.* **3** (2009) 28–41.
[25] R.D. Perlack, L.L Wright, A.F. Turhollow, R.L. Graham, B.J. Stokes and D.C. Erbach, *Biomass as Feedstock for a Bioenergy and Bioproducts Industry: The Technical Feasibility of a Billion-ton Annual Supply*, DOE/USDA Rept. DOE-GO-102995–2135 and ORNL/TM 2005–66 (April 2005).
[26] Biomass Research & Development Technical Advisory Committee, *Vision for Bioenergy & Biobased Products in the US*, Washington, DC (October 2002); http://www.bioproducts-bioenergy.gov/pdfs/BioVision_03_-Web.pdf
[27] Commission of the European Communities, *Biomass Action Plan*, Brussels (2005), COM/2005 628.
[28] European Environmental Agency (EEA), *How Much Biomass can Europe Use Without Harming the Environment?*, Copenhagen (2005), EEA Briefing 02; http://www.eea.eu.int
[29] Royal Society of Chemistry, *Chemistry for Tomorrow's World: A Roadmap for the Chemical Sciences*, RSC, London (2009); www.rsc.org/roadmap
[30] German Forschungsministerium (Bundesregierung), *Nationaler Forschungsstrategie Bioökonomie 2030*, Berlin (9 November 2010).
[31] D. Peters, N. Holst, B. Herrmann, D. Lulies and H. Stolte, *Renewable Resources in Industry*, 3rd edn, Fachagentur Nachwachsende Rohstoffe e.V., Gülzow (2010).
[32] G.W. Huber, P. O'Connor and A. Corma, Processing biomass in conventional oil refineries: Production of high quality diesel by hydrotreating vegetable oils in heavy vacuum oil mixtures, *Appl. Catal. A: Gen.* **329** (2007) 120–9.
[33] S. Mikkonen, Second-generation renewable diesel offers advantages, *Hydrocarb. Proc.* (February 2008) 63–8.
[34] L. Rantanen, R. Linnaila, P. Aakko and T. Harju, NExBTL – biodiesel fuel of the second generation, *SAE Techn. Paper Series* 2005-01-3771 (2005).
[35] J. Holmgren, C. Gosling, R. Marinangeli, T. Marker, G. Faraci and C. Perego, New developments in renewable fuels offer more choices, *Hydrocarb. Proc.* (2007) 67–8.
[36] J. Holmgren, C. Gosling, T. Marker, P. Kokayeff, G. Faraci and C. Perego, Green diesel production from vegetable oil, *Proc. AIChE Spring Natl. Mtg.*, Houston, TX (22–27 April 2007).

[37] T. Kalnes, T. Marker and D.R. Shonnard, Green diesel: a second generation biofuel, *Int. J. Chem. Reactor Engng.* **5** (2007) Art. A48.

[38] F. Carrasco, Production of furfural by dilute-acid hydrolysis of wood: methods for calculating furfural yield, *Wood Fiber Sci.* **25** (1993) 91–102.

[39] G. Vially, R. Marchal and N. Guilbert, L(+) lactate production from carbohydrates and lignocellolosic materials by *Rhizopus oryzae* UMIP 4.77, *World J. Microbiol. Biotechnol.* **26** (2010) 607–14.

[40] P.B. Weisz, W.O. Haag and P.G. Rodewald, Catalytic production of high-grade fuel (gasoline) from biomass compounds by shape-selective catalysts, *Science* **206** (1979) 57–8.

[41] D.C. Elliott, D. Beckman, A.V. Bridgwater, J.P. Diebold, S.B. Gevert and Y. Solantausta, Developments in direct thermochemical liquefaction of biomass: 1983–1990, *Energy Fuels* **5** (1991) 399–410.

[42] T.R. Carlson, G.A. Tompsett, W.C. Conner and G.W. Huber, Aromatic production from catalytic fast pyrolysis of biomass-derived feedstocks, *Topics Catal.* **52** (2009) 241–52.

[43] M.H. Thomsen, A. Thygesen, H. Jørgensen, J. Larsen, B.H. Christensen and A.B. Thomsen, Preliminary results on optimization of pilot scale pretreatment of wheat straw used in coproduction of bioethanol and electricity, *Appl. Biochem. Biotechnol.* **129–132** (2006) 448–60.

[44] H. Mao, J.M. Genco, S.-H. Yoon, A. van Heiningen and H. Pendse, Technical economic evaluation of a hardwood biorefinery using the 'near-neutral' hemicelluloses pre-extraction process, *J. Biobased Mater. Bioenergy* **2** (2008) 177–85.

[45] L.R. Lynd, R.T. Elander and C.E. Wyman, Likely features and costs of mature biomass ethanol technology, *Appl. Biochem. Biotechnol.* **57–8** (1996) 741–61.

[46] H. Mao, J.M. Genco, A. van Heiningen and H. Pendse, Kraft mill biorefinery to produce acetic acid and ethanol: technical economic analysis, *Bioresources* **5** (2010) 525–44.

[47] S. Walton, A. van Heiningen and P. van Walsum, Inhibition effects on fermentation of hardwood extracted hemicelluloses by acetic acid and sodium, *Bioresourc. Technol.* **101** (2010) 1935–40.

[48] G.W. Huber, S. Iborra and A. Corma, Synthesis of transportation fuels from biomass: chemistry, catalysts and engineering, *Chem. Rev.* **106** (2006) 4044–98.

[49] G.W. Huber and J.A. Dumesic, An overview of aqueous-phase catalytic processes for production of hydrogen and alkanes in a biorefinery, *Catal. Today* **111** (2006) 119–32.

[50] T.P. Vispute and G.W. Huber, Production of hydrogen, alkanes and polyols by aqueous phase processing of wood-derived pyrolysis oils, *Green Chem.* **11** (2009) 1433–45.

[51] S. Szernik, R. French, C. Feik and E. Chornet, Hydrogen by catalytic steam reforming of liquid byproducts from biomass thermoconversion processes, *I&EC Res.* **41** (2002) 4209–15.

[52] B.E. Dale and S. Kim, Biomass refining global impact – the biobased economy of the 21st century, in *Biorefineries – Industrial Processes and Products: Status Quo and Future Directions* (B. Kamm, P.R. Gruber and M. Kamm, eds), Wiley-VCH, Weinheim (2006), Vol. 1, pp. 41–66.

[53] S. Kim and B.E. Dale, Global potential bioethanol production from wasted crops and crop residues, *Biomass Bioen.* **26** (2004) 361–75.

[54] J.S. Tolan, Iogen's demonstration process for producing ethanol from cellulosic biomass, in *Biorefineries – Industrial Processes and Products: Status Quo and Future Directions* (B. Kamm, P.R. Gruber and M. Kamm, eds), Wiley-VCH, Weinheim (2006), Vol 1, pp. 193–208.

[55] R. Xing, A.V. Subrahmanyan, H. Olcay, W. Qi, G.P. van Walsum, H. Pendse and G.W. Huber, Production of jet and diesel fuel range alkanes from waste hemicellulose-derived aqueous solutions, *Green Chem.* **12** (2010) 1933–46.

[56] B.-H. Um and G.P. van Walsum, Evaluation of enzyme mixtures in releasing fermentable sugars from pre-pulping extract of mixed Northeast hardwood, *Appl. Biochem. Biotechnol.* **161** (2010) 432–47.

[57] (a) T.P. Vispute, H. Zhang, A. Sanna, R. Xiao and G.W. Huber, Renewable chemical commodity feedstocks from integrated catalytic processing of pyrolysis oils, *Science* **330** (6008) (2010) 1222–7; (b) S. Crossley, J. Faria, M. Shen and D.E. Resasco, Solid nanoparticles that catalyze biofuel upgrade reactions at the water/oil interface, *Science* **327** (2010) 68.

[58] J.C.J. Bart, N. Palmeri and S. Cavallaro, *Biodiesel Science and Technology. From Soil to Oil*, Woodhead Publishing, Ltd/CRC Press, Cambridge/Boca Raton, FL (2010).

[59] D.M. Mousdale, *Biofuels. Biotechnology, Chemistry and Sustainable Development*, CRC Press, Boca Raton, FL (2008).

[60] A.E. Farrell, R.J. Plevin, B.T. Turner, A.D. Jones, M. O'Hare and D.M. Kammen, Ethanol can contribute to energy and environmental goals, *Science* **311** (2006) 506–8.

[61] D.R. Grüll, F. Jetzinger, M. Kozich, M.M. Wastyn and R. Wittenberger, Industrial starch platform – Status quo of production, modification and application, in *Biorefineries – Industrial Processes and Products: Status Quo and Future Directions* (B. Kamm, P.R. Gruber and M. Kamm, eds), Wiley-VCH, Weinheim (2006), Vol. 2, pp. 61–95.

[62] H. Shapouri, J.A. Duffield and M. Wang, The energy balance of corn ethanol: An update, *Agricultural Economic Report* No. **814** (2002) 1–15.

[63] P.O. Ward and A. Singh, Microbial technology for bioethanol production from agricultural and forestry wastes, in *Microbial Biotechnology in Agriculture and Aquaculture* (R.C. Ramesh, ed.), Science Publishers, Inc., Enfield, NH (2005).

[64] K.W. Waldron (ed.), *Bioalcohol Production: Biochemical Conversion of Lignocellulosic Biomass*, Woodhead Publishing, Ltd/CRC Press, Cambridge/Boca Raton, FL (2010).

[65] J.E. Graham, M.E. Clark, D.C. Nadler, S. Huffer, H.A. Chokhawala, S.E. Rowland, H.W. Blanch, D.S. Clark and F.T. Robb, Identification and characterization of a multidomain hyperthermophilic cellulase from an archaeal enrichment, *Nature Comm.* **2** (2011) 375.

[66] D.-M. Lai, L. Deng, J. Li, B. Liao, Q.-X. Guo and Y. Fu, Hydrolysis of cellulose into glucose by magnetic solid acid, *ChemSusChem.* **4** (2011) 55–8.

[67] J.C.J. Bart, N. Palmeri and S. Cavallaro, *Biodiesel Science and Technology. From Soil to Oil*, Woodhead Publishing, Ltd/CRC Press, Cambridge/Boca Raton, FL (2010), p. 761.

[68] L. Gibson, Woodland develops demo cellulosic ethanol plant, *Abstr. Pacific West Biomass Conference & Trade Show*, San Francisco, CA (16–18 January 2012).

[69] A.G. Sergeev and J. Hartwig, Selective, nickel-catalysed hydrogenolysis of amyl ethers, *Science* **332** (22 April 2011) 439–43.

[70] C.H. Hoyt and D.W. Goheen, Polymeric products, in *Lignins–Occurrence, Formation, Structure and Reactions* (K.V. Sarkanen and C.H. Ludwig, eds), Wiley-Interscience, New York, NY (1971), pp. 833–65.

[71] Organisation for Economic Co-operation and Development (OECD), *The Application of Biotechnology to Industrial Sustainability*, OECD, Paris (2001), www.oecd.org/sti/biotechnology

[72] US Department of Energy (DOE), *Vision for Plant/Crop Based Renewable Resources 2020*, DOE, Washington, DC (1998); www.oit.doe.gov. agriculture/pdfs/vision2020.pdf

[73] US Department of Energy (DOE), *The Technology Roadmap for Plant/Crop Based Renewable Resources 2020*, DOE, Washington, DC (1999); www.oit.doe.gov/agricultural/pdfs/ag25945.pdf

[74] J.A. Napier and I.A. Graham, Tailoring plant lipid composition: designer oilseeds come of age, *Curr. Opin. Plant Biol.* **13** (2010) 329–37.

[75] S.P. Slocombe, J. Cornah, H. Pinfield-Wells, K. Soady, Q. Zhang, A. Gilday, J.M. Dyer and I.A. Graham, Oil accumulation in leaves directed by modification of fatty acid breakdown and lipid synthesis pathways, *Plant Biotechnol. J.* **7** (2009) 694–703.

[76] US Mission to the European Union, *Foreign Agricultural Science*, http://www.fas.usda.gov/posthome/useu/GMOs.html (31 August 2010).

[77] H. Yim, R. Haselbeck, W. Niu, C. Pujol-Baxley, A. Burgard, J. Boldt, J. Khandurina, J.D. Trawick, R.E. Osterhout, R. Stephen, J. Estadilla, S. Teisan, H. Brett Schreyer, S. Andrae, Tae Hoon Yang, Sang Yup Lee, M.J. Burk and S. Van Dien, Metabolic engineering of *Escherichia coli* for direct production of 1,4-butanediol, *Nature Chem. Biol.* **7** (2011) 445.

[78] M. Griffiths, A. Gram, W. Treffenfeldt, U. Lange, T. McIntyre and O. Wolf (eds), *The Application of Biotechnology to Industrial Sustainability – A Primer*, OECD, Paris (2001).

[79] Global Industry Analysts, Inc., *Biotechnology, A Global Industry Outlook*, San José, CA (2008).

[80] US Department of Energy (DOE), *Energy, Environmental and Economics (E^3) Handbook*, 1st edn, DOE, Washington, DC (1997); http://www.oit.doe.gov/e3handbook

[81] National Renewable Energy Laboratory (NREL), http://www.nrel.gov/biomass/biorefinery.html

[82] K. Soyez, B. Kamm and M. Kamm (eds), *The Green Biorefinery*, Verlag GÖT, Berlin (1998).

[83] R. Carlsson, Sustainable primary production – Green crop fractionation: effects of species, growth conditions, and physiological development, in *Handbook of Plant and Crop Physiology* (M. Pessarakli, ed.), Marcel Dekker, Inc., New York, NY (1994), pp. 941–63.

[84] FP7-2009-biorefinery_cp

[85] A.C. Franke, M.L.H. Breukers, W. Broer, F. Bunte, O. Dolstra, F.M. d'Engelbronner-Kolff, L.A.P. Lotz, J. van Montfort, J. Nikoloyuk, M.M. Rutten,

M.J.M. Smulders, C.C.M. van de Wiel and M. van Zijl, *Sustainability of Current GM Crop Cultivation*, DLO Foundation, Wageningen UR, Report **386** (2011).

[86] A. Gore and Melcher Media, *An Inconvenient Truth: The Planetary Emergency of Global Warming and What We Can Do About It*, Rodale Press, Emmaus, PA (2006).

[87] B. Kerckow, C. Mangan and L. Breslin, Industrial crops and products and European Union research policy, *Ind. Crops Prod.* **6** (1997) 325–31.

[88] United Nations Environment Programme, *Global Environment Outlook GE 04*, United Nations, New York, NY (2007).

[89] Food and Agriculture Organization of the United Nations Statistics, FAOSTAT, Rome (2004).

[90] E.-C. Oerke, H.-W. Dehne, F. Schönbeck and A. Weber, *Crop Production and Crop Protection*, Elsevier Science, Amsterdam (1994).

[91] L.I. da Silva, *FAO Conference on Food Safety, Climate Change and Bioenergy*, New York, NY (2007).

5 Renewable feedstocks for lubricant production

Abstract: The traditional role of agriculture has recently been extended to non-food applications, including vegetable oil-, wax ester- and carbohydrate-based biolubricants. Rapeseed, soy and sunflower oils are increasingly finding industrial applications for biolubricant production beyond their oleochemical use. Selection criteria for vegetable crops for potential application in lubrication include chemical structure and various quality parameters of the oil. Increasing pressure on globally limited edible oil commodities poses commercial and ethical threats. As biomass has a limited edible oil potential, development of a non-food agricultural chemistry for chemical applications has clear advantages. Biotechnology is vital in addressing the growing global demands for crops for chemical industrial use.

Key words: renewable feedstock sources, non-food crops, advanced agro-energy crops, plant breeding, genetic engineering.

5.1 Introduction

The application of biomass for the synthesis of chemicals is not new: about 5–10% of all chemicals on the market today are from biomass resources [1]. Renewable resources as industrial raw materials are polysaccharides and sugars, plant oils and animal fats, mostly used for nutrition. Sugar (145 Mt, 2004) from cane and beets is an important commodity for food and fuel (bioethanol). Other important examples are the oleochemical and starch industries. The oleochemical industry uses vegetable oils as the feedstock and is a global business with a volume of about 18 Mt (2004) [2]. Products are used as lubricants, surfactants, soaps, surface coatings, solvents, polymers (ingredients) and plasticisers.

Use of renewable raw materials is in line with the development of sustainable chemistry. Oils and fats of vegetable and animal origin make up the greatest portion of the current consumption of raw materials in the chemical industry. Oleochemistry (based on CHO feedstocks) offers applications which cannot easily be met by petrochemistry (essentially CH feedstocks only). However, the chemical possibilities of renewable oils and fats are still far from being fully exploited. Substitution of mineral oil with biodegradable lubricant base oils could significantly reduce environmental pollution. Vegetable oils are a major source of these base fluids. Vegetable oils basically have been used for lubrication at all times and have an advantage over mineral-based oils in being biodegradable and non-toxic. Because of health, economic and environmental issues, a revival in the development of bio-based lubricants for renewable resources is ongoing as agriculturally based materials are generally considered non-toxic, easily biodegradable and abundant.

A vast range of renewable resources is available, differing locally. Agricultural products provide a viable alternative to resource-limited petroleum in a number of applications. Renewable, agricultural raw materials have been used for many years in the chemical-technical sector. Practically all common crop plants are already being exploited as raw materials for industrial processes. Other types of renewable resources are wild plants, animal products and biogenetic materials, wastes and residues that originate from agricultural and forestry production (such as straw, hemp shives, molasses, glycerol from fats and oils processing, wood residues, sawmill by-products, whey, liquid manure and slaughterhouse wastes). In addition to lipids, carbohydrates (starch and sugar from agricultural raw materials as well as cellulose from fibre plants or in the form of dissolving pulp from wood) are processed in significant quantities. Starch is one of the most abundantly available agricultural products that has been developed for various applications. Starch is used for the production of several high-added value non-food products [3]. Together, oils/fats and carbohydrates account for about 80% of the industrially processed renewable resources. Various other renewable raw materials are used on a smaller scale (especially proteins, as well as other plant constituents and exudates, such as plant waxes).

Surplus corn, soybeans, wheat, barley and other cereals have suppressed prices that farmers get for their crops. One way of overcoming this problem is to develop new uses for starches, proteins and oils, which are ingredients of most cereals. An important application area for agricultural products is in lubrication, which is currently almost exclusively dominated by petroleum-based products. Lubricants represent a large non-food product area in which plant oils and other biomass can be increasingly utilised. In addition to liquid bio-based lubricants derived from unmodified and modified vegetable oils, dry-film bio-based lubricants are equally of interest. Successful application of bio-based materials in lubrication requires a thorough understanding of the tribochemical properties of these agricultural products.

5.2 Natural vegetable oils and animal fats in lubrication

Bio-renewables, such as plant-derived oils, are a sustainable means of providing the essential products needed by society. Plant oils (Fig. 5.1) are

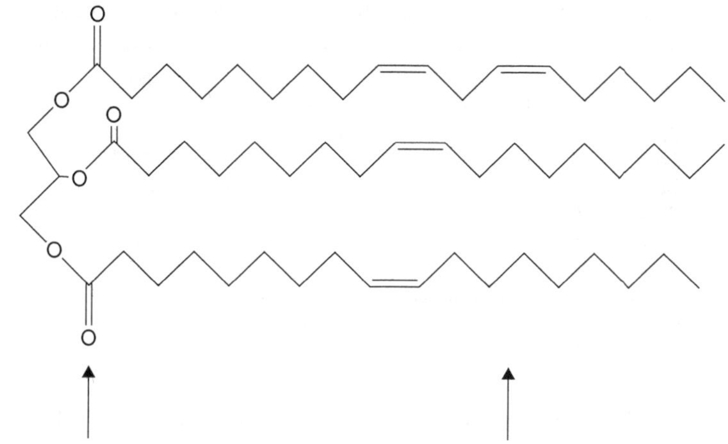

5.1 Structure of a representative triacylglycerol (plant oil).

already major agricultural commodities with around 20% by value used for non-food applications. The value and application of an oil are determined largely by its fatty acid (FA) composition. Vegetable oils and fats have different physicochemical properties, which leads to various areas of application (in competition with petroleum products), depending on the carbon chain distribution. While the chain length range of C_{12} (e.g. lauric acid) provides important raw materials for detergents and surfactants, chain lengths of C_{18}–C_{22} (e.g. oleic and erucic acid) are used mainly in industrial applications such as lubrication and polymer additives). Alteration in fatty acid composition of vegetable oils is desirable for meeting specific food and industrial uses. Historically, cost has been the major bottleneck limiting the development of new plant-derived oils. However, with the escalating costs of crude oil and concerns about security of supply, there is an increased strategic need to develop additional renewable products from vegetable oil resources.

Oils, fats and waxes from renewable origin constitute a huge resource. The annual production of animal fats (tallow, lard and butter) is approximately 22 Mt, while fish oils contribute 1 Mt [4]. It is reported that the yearly world production (2009/10) of the main plant oils is approximately 138 Mt, consumed in nutritional as well as in industrial sectors [5]. Table 5.1 shows the impressive increase in the global production of oils and fats in the last decennia. The significant rise in volume and value in the world market for vegetable oils is continuing also in the recent past (Table 5.2). While the EU and North America dominated the world production of vegetable oils in the early 1990s with a global share of 35%, this has now been reduced to 20% due to increased production in China, Brazil, Argen-

Table 5.1 Average annual production of oils and fats (Mt)

Oil type	1976–80	2000	2016–20[a]
World total	52.6	114.5	175.8
Palm	4.2	21.7	45.8
Soybean	11.2	25.5	37.1
Rapeseed/sunflower	7.2	23.8	38.8
Other vegetable oils	12.8	22.8	26.2
Animal fats	17.2	20.7	27.9

Source: *Lipid Technology* (June 1999).
[a] Forecast.

Table 5.2 World vegetable oil production (Mt)

Vegetable oil	2005/06	2009/10
Coconut	3.46	3.67
Cottonseed	4.90	4.66
Olive	2.66	2.99
Palm	35.83	45.88
Palm kernel	4.40	5.50
Peanut	4.97	4.56
Rapeseed	17.30	22.12
Soybean	34.62	37.88
Sunflower	10.62	11.31
Total	118.72	138.57

After ref. [6].

tina (for soy oil), Malaysia and Indonesia (for palm oil). The share of tropical oils on the world market has risen from 61% in 2005 to over 65% in 2010. The 15% appreciation of oilseed prices in 2009 is primarily on account of the increased demand for palm oil. The Asian market is currently dominating from both production and consumption standpoints.

Oils and waxes are commonly found in many plant species (Table 5.3). Oils are abundant in seeds, while waxes are normally abundant on the surface of leaves or stems. Oil-yielding crop plants are very important for the agricultural sector. Oil volumes and values vary widely. In the past, the world's vegetable oil and fat outputs have been more than enough to satisfy the needs for human and animal nutrition (food and feed, respectively). At the same time, the oleochemical and health-related industries were assured of a reliable supply of raw materials (14% of oils and fats) for chemicals manufacture [7]. FAs are used to produce soaps, lubricants and greases, surfactants, detergents, biodiesel and chemical intermediates. Higher demands for human nutrition from highly populated developing countries

(China, India) and for fuel applications (biodiesel) have recently determined a change in global use of vegetable oils (2007: 84% foodstuff, 10% oleochemical and 6% biodiesel).

Natural lipid biosynthesis is regulated by enzymatic activity and leads to a broad variety of oils in seeds and fruits in which a restricted subset of FA compositions predominates and accounts for the traditional use of oils and fats in food and oleochemical applications. Both genetic and environmental factors affect triacylglycerol (TAG) biochemistry. An increased understanding of how plants synthesise fatty acids and TAGs will ultimately allow the development of novel (energy) crops. For example, knowledge of the regulation of oil synthesis has suggested ways to produce TAGs in abundant non-seed tissues.

The end products of FA synthase activities are usually 16- and 18-carbon FAs. The seed oils of most plants, including the major domesticated, edible oil-producing species, contain the same limited set of the five most important FAs found in the structural lipids of plant tissues, namely C16:0 (palmitic), C18:0 (stearic), C18:1 (oleic), C18:2Δ9,12 (linoleic) and C18:3Δ9,12,15 (linolenic) (see Table 5.3 and Fig. 5.2). A wide variety of structurally diverse FAs occurs in the seed oils of wild plant species and many 'unusual' FAs represent potentially outstanding feedstocks for industry. They include unusual monounsaturated fatty acids (MUFAs), short, medium or very-long-chain FAs, FAs with additional functional groups such as epoxy and hydrogen groups, or FAs with conjugated or acetylenic bonds. There are several plant species that store 8- to 14-carbon (medium-chain) FAs in their oilseeds. Among the medium-chain fatty acids (MCFAs), caprylic (8:0) and capric (10:0) are minor components of coconut oil, which are used in many industrial, nutritional and pharmaceutical products. There are also plant families that contain unusual FAs present almost exclusively in their seed oils while their structural lipids have conventional FA composition. Among unusual fatty acids are hydroxy fatty acids (HFAs). The fatty acid composition of a lipid determines its chemical and physical properties and hence the type of application.

Despite a large genetic variation for the different quality traits of vegetable oil plants, a restricted subset of FA compositions predominates. This has largely determined the value and use of oils and fats in food and oleochemical applications [7–11]. Lubricant application sets specific requirements to the oily feedstocks, both in terms of composition (See Table 5.4) and chemical and physical properties. Globally dominating feedstocks for lubricant production nowadays are rapeseed (canola) oil (RSO), soybean oil (SBO) and sunflower oil (SNO). Alteration of the FA synthesis for the purpose of usability for both industrial and nutritional applications can be achieved by chemical modification of the oil products (see Chapter 6) as well as by natural breeding methods and genetic engineering techniques of the seeds

Table 5.3a Fatty acid profiles of common vegetable oils and animal fats

Oil and fat source	Fatty acid profile[a], % by weight						
	≤10	12:0	14:0	16:0	16:1	18:0	18:1
Canola	0.0	0.0	0.0	4.0–5.6	0.0	2.4	60.0–75.0
Castor[b]	0.0	0.0	0.0	0.9–1.4	0.0	0.9–1.0	2.6–3.7
Coconut[c]	14.0–21.0	41.5–53.2	16.0–21.0	6.0–11.3	0.0	1.5–4.2	4.0–8.7
Corn (maize)	0.0	0.0	0.5–2.0	10.0–12.0	0.1–1.5	1.6–4.5	20.0–32.5
Cottonseed	0.0	0.1–0.2	0.7–1.4	17.0–31.0	0.4–2.0	0.9–4.0	13.3–22.9
Crambe	0.0	0.0	0.0	1.8–2.2	0.0	0.7–0.9	15.9–18.8
Jatropha	0.0	0.0	0.1–1.4	12.8–17.7	0.0–1.3	3.7–9.8	37.8–45.8
Karanja	0.0	0.0	0.0	3.7–7.9	0.0	2.4–8.9	44.5–71.3
Linseed	0.0	0.0	<0.1	4.0–7.0	0.1	2.0–5.0	18.0–24.0
Olive	0.0	0.0	0.4–1.2	6.9–15.6	0.5–1.1	1.4–4.4	62.8–84.4
African palm	0.0	0.1–0.5	0.6–2.4	32.3–47.9	0.2–1.4	3.7–6.3	37.0–52.4
Palm kernel	5.1–13.0	42.4–55.0	12.0–19.0	3.0–11.4	0.2	1.3–8.0	8.7–21.0
Palm stearin (HO)	0.0	0.2	0.7	26.8	0.0	10.1	48.2
Peanut	0.0	0.0	0.0	8.0–12.6	1.4	1.0–4.5	41.0–56.0
Rapeseed (LE)[d]	0.0	0.0	0.0–0.5	3.0–5.7	0.0	0.8–1.5	55.0–64.4
Rice bran	0.0	0.0	0.3–1.0	12.0–22.1	0.2–0.4	0.8–3.0	38.9–50.0
Safflower	0.0	0.0	0.0–0.1	5.2–9.0	0.0	1.0–3.0	10.0–16.2
Safflower (HO)[e]	0.0	0.0	0.0	4.8	0.0	2.0	77.4–79.0
Sesame	0.0	0.0	0.0	7.8–13.0	0.1	3.7–7.0	36.8–53.0
Soybean	0.0	0.1	0.1–0.4	7.0–12.7	0.1–1.0	1.4–5.9	11.5–27.2
Soybean (HO)	0.0	0.0	0.0	5.4	0.0	4.1	81.3
Soybean (MO)	0.0	0.0	0.0	4.3–11.1	0.0	3.5–5.0	43.7–55.0
Sunflower	0.0	0.1	0.1–0.2	2.0–9.0	0.3	1.6–6.5	14.0–39.4
Sunflower (MO)[f]	0.0	0.0	0.4–0.8	4.0–5.5	0.1	2.1–5.0	43.1–71.8
Sunflower (HO)[g]	0.0	0.0	0.1	2.6–5.9	0.1	2.8–6.2	75.0–90.7
Jojoba oil-wax[h]	0.0	0.0	0.0	1.1–1.9	0.1–0.3	<0.1	10.1–12.2
Beef tallow	0.0	0.1–0.7	2.0–8.0	23.0–38.0	0.0–4.7	14.0–29.0	34.9–50.0
Lard	0.0	0.1	1.4–2.3	17.3–30.0	1.9–3.0	12.0–20.0	42.5–50.0
Poultry fat	0.0	0.0	0.0–0.5	22.2–23.0	8.4	5.1–5.4	42.4
Yellow grease	0.0	0.0	2.4	23.2	3.8	13.0	44.3

[a] Acids: caproic, caprylic, capric (<10); lauric (12:0); myristic (14:0); palmitic (16:0); palmitoleic (16:1); stearic (18:0); oleic (18:1); linoleic (18:2); linolenic (18:3); arachidic (20:0); gadoleic (20:1); behenic (22:0); erucic (22:1); lignoceric (24:0); methyl 15 – tetracosenoate (24:1).
[b] Contains 87–88 % of ricinoleic acid.
[c] Contains <4% free fatty acids.
[d] Zero–Zero.
[e] High-oleic safflower.
[f] Mid-oleic sunflower, NuSun.
[g] High-oleic sunflower.
[h] Contains about 97% wax esters.

18:2	18:3	20:0	20:1	22:0	22:1	24:0	24:1
12.0–23.4	3.2–9.1	0.5–1.3	<0.5	0.0	<0.2	0.0	0.0
4.1–4.9	0.3–0.4	0.2	0.0	2.1	0.0	0.0	0.0
0.8–2.6	0.0	0.1	0.0	0.0	0.0	0.0	0.0
52.0–62.0	0.5–1.7	0.2–1.0	0.2–0.6	0.1–0.5	0.1–0.3	0.1–0.5	0.0
47.8–58.2	0.1–2.1	0.2–1.3	0.1–0.5	0.1–0.6	0.1–0.3	0.1	0.0
8.7–9.3	4.8–8.7	1.0–2.0	0.0–4.7	0.8–2.0	56.2–62.5	0.7–1.1	1.6
29.0–45.0	0.0–0.3	0.1–0.3	0.0	0.2	0.0	0.0	0.0
10.8–18.3	0.0	0.0	0.0	0.0	0.0	1.1–3.5	0.0
14.0–19.0	50.0–58.0	0.1	0.3–0.4	0.2	0.0	0.0	0.0
3.0–20.2	0.5–4.0	0.1–0.6	0.2–0.4	0.1–0.2	0.0	0.4	0.0
6.4–12.5	0.2–0.6	0.1–1.0	0.1–0.4	<0.2	0.0	0.0	0.0
0.5–3.5	0.2	0.2	0.2	0.2	0.0	0.0	0.0
12.8	0.3	0.7	0.0	0.0	0.0	0.0	0.0
26.0–31.9	0.9	0.8–4.2	0.6–2.6	1.5–5.4	0.0	0.5–2.5	1.0
20.0–26.0	8.2–12.0	0.0	1.0–2.0	0.5	0.5–2.0	0.0	0.0
29.4–35.2	0.4–1.1	0.5–1.2	0.0	<1.0	0.0	0.4	<0.1
76.3–82.0	0.1–0.4	0.0	0.0	0.0	0.0	0.0	0.0
13.0–13.8	0.5	0.6	0.0	0.0	0.0	0.0	0.0
35.0–53.0	0.3	0.4–0.6	0.2	0.1	0.0	0.1	0.0
51.5–63.1	2.9–12.1	0.3–4.0	0.1–0.4	0.1–0.8	0.4	0.0	0.0
3.8	5.3	0.0	0.0	0.0	0.0	0.0	0.0
24.0–35.5	<3.1	0.6	0.0	0.9	0.0	0.0	0.0
48.3–74.0	0.2–0.7	0.1–1.2	0.4	0.3–1.5	0.3	0.5	0.0
18.7–45.3	0.1–0.7	0.2–2.9	0.2–0.3	0.6–6.8	0.0	0.3–0.4	0.0
2.1–17.0	0.1–0.3	0.2–0.5	0.1–0.5	0.5–2.0	0.3	0.5	0.0
0.2	0.1	0.1–0.4	63.5–73.9	0.5	12.7–18.9	0.0	0.9–1.9
1.0–5.0	0.5	0.1	0.1	0.0	0.0	0.0	0.0
7.0–13.0	0.4–1.0	0.3	1.5	0.0	0.0	0.0	0.0
17.0–19.3	1.0	0.0	1.0–2.0	0.0	0.0	0.0	0.0
7.0	0.7	0.0	0.0	0.0	0.0	0.0	0.0

Table 5.3b Some collective properties of common vegetable oils and animal fats

Oil and fat source	Iodine number	Saturated fatty acids (SFA)[a]%	Monounsaturated fatty acids (MUFA)[a]%	Polyunsaturated fatty acids (PUFA)[a]%
Canola	100–127	7.0	69.0	24.0
Castor[b]	82–88	2.7	92.5[i]	4.8
Coconut[c]	6–17	92.0	6.0	2.0
Corn (maize)	109–140	15.0	26.0	59.0
Cottonseed	90–117	29.0	19.0	52.0
Crambe	85–93	6.5	78.5	15.0
Jatropha	92–110	22.0	41.0	37.0
Karanja		15.8	66.6	16.7
Linseed	170–211	10.0	20.0	70.0
Olive	78–94	14.0	71.0	15.0
African palm	51–58	47.0	44.0	9.0
Palm kernel	12–19.7	84.5	15.0	0.5
Palm stearin (HO)		20.0	51.0	29.0
Peanut	73–107	19.0	51.0	30.0
Rapeseed (LE)[d]	96–121	5.5	62.0	32.5
Rice bran		20.0	44.5	35.5
Safflower	126–150	8.0	13.0	79.0
Safflower (HO)[e]	80–100	7.0	78.0	15.0
Sesame	82–120	15.0	43.0	42.0
Soybean	121–143	16.0	21.0	63.0
Soybean (HO)	90.0	9.5	81.5	9.0
Soybean (MO)	107.0	17.5	43.5	39.0
Sunflower	118–141	11.5	27.0	61.5
Sunflower (MO)[f]	94–122	14.0	53.5	32.5
Sunflower (HO)[g]	78–90	10.0	81.0	9.0
Jojoba oil-wax[h]	81–89	2.3	97.5	0.2
Beef tallow	35–48	55.0	42.5	2.5
Lard	55–71	40.0	49.5	10.5
Poultry fat	75–85	28.2	52.2	19.5
Yellow grease	90	40.0	51.0	9.0

[a] Average values.
[b–h] see Table 5.3a.
[i] Including ricinoleic acid (12-hydroxyoleic acid).

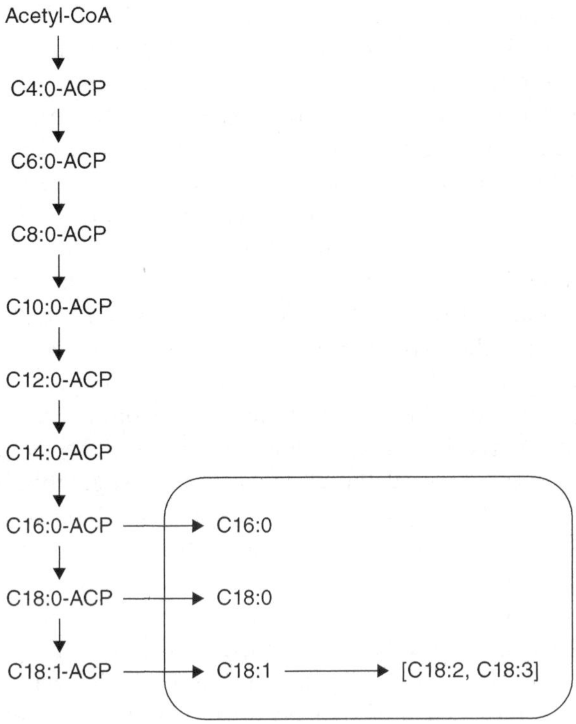

5.2 Simplified scheme of plastid fatty acid synthesis leading to the accumulation of most common oils.

Table 5.4 Fatty acid specificity and lubricant applications

Fatty acid composition	Industrial crop source	Applicability
Lauric acids (C12)	Coconut, palm kernel	Lubricant base stocks
Oleic/linoleic acids (C18)	Rapeseed, soybean, sunflower	Lubricant base stocks
Epoxy fatty acids (C18)	Euphorbia	Potential lubricant base stocks
Hydroxy fatty acids (C18-OH, C20-OH)	Castor, *Lesquerella*	Lubricity enhancers
Long-chain fatty acids (C20–C22)	Meadowfoam, high-erucic acid rapeseed	Specialty lubricants

(see Section 5.3). At present, oil chemists have at their disposal a new generation of vegetable oils, the characteristics of which are particularly suitable for lubricant applications. Use of plant TAGs as feedstock for the production of biolubricants has recently been reviewed [12–15].

FA-based products from vegetable and animal origins were once used as the main source of lubricants but were displaced by mineral oils in the late 19th century. More recently, industry and the general public are becoming much more environmentally conscious. Vegetable-based lubricants are practical and feasible for use in many applications. Vegetable-based oils have substantial benefits over petroleum, or mineral-based oils for use as lubricant base stocks: higher viscosity index (VI), lower evaporation loss and a potential to enhance lubricity, which could lead to improved energy efficiency. The worldwide trend to the use of more environmentally friendly lubricants is also motivated by the strict legislation to reduce the environmental impact of lubricants (see Chapter 9). The use of vegetable oils is an obvious choice because of their excellent biological degradability and non-toxicity, and the fact that they can be readily extracted from renewable resources. Yet, despite their obvious advantages, of the 70 or so available vegetable oils, only a few are currently being used in technical applications, such as hydraulics, gears, motor oils, namely rapeseed, castor, sunflower and soybean oils. However, other oils (e.g. sesame and wheat germ oil) and derivatives of new crop oils (*Crambe*, meadowfoam and *Lesquerella*) have technical potential yet to be exploited.

A number of vegetable oils and their derivatives have been tested for lubricant application, either as base stock or as lubricant additive (see Chapter 12). In order to broaden the feedstock basis, suitable (edible and non-edible) oils need to be identified. Although no unique, ideal FA profile for lubricant application can be defined, some vegetable oils are more desirable than others, both in terms of composition and from an agronomic point of view. In fact, while chemical composition data are important in relation to lubricant potential, other considerations are equally essential. Important criteria for crop production include:

- genotype (seed yield, seed weight, oil content, oil/ha, oil properties);
- ecology (climatological tolerance for poor soil, drought and frost; seed deterioration time);
- cultivation (growth conditions, crop season, fertiliser need, ease of machine harvesting);
- biotic factors (resistance to pests and diseases);
- maturation (harvesting date, storage); and
- oil extraction performance (temperature, time).

Important aspects of a field crop include crop physiology (crop development, growth and yield), agronomy, weeds and their control, diseases, insect pests, plant breeding, biotechnology and impact of production on the surrounding environment. Plant species capable of giving oil-bearing seeds to produce lubricants need to be selected and identified. These should preferably be able to thrive on any type of soil, need minimum input and post-plantation management, and have low moisture demand. A successful oilseed crop for lubricant application should be easy to propagate, precocious, rustic, productive, easy to harvest, and be able to be put into current farm practice. Getting both food/feed and fuel/chemical feedstock from a single crop is equally most desirable. The crop should also have high supply potential, high oil yields and low oil costs. Although oil yields will differ in different agroclimatic zones, some agricultural commodity crops for lubricants (RSO, SNO, SBO) are not among the oil yield champions.

In terms of utilisation, the main considerations are processing the seed and oil, seed chemistry and analysis, oil properties in human nutrition and industrial use, meal and by-product utilisation in animal nutrition. Lubricant oilseed feedstock requires a market for the post-crush meal. This presents a limitation for castor beans and jatropha. Desirable crop technology changes include yield increases, more efficient harvest technology, and using less wasteful tillage practices.

Nature has generated an enormous variety of FAs, differing in chain length, number and position of double bonds and functional groups. Vegetable oils fall into two broad chemical categories: triesters and monoesters. Most seed oils are triesters of glycerol with FAs (triglycerides) whose characteristics are dependent on the chemistry and composition of the fatty acid residues. Some seed oils consist of monoesters of long-chain FAs and fatty alcohols of varying chain length and degree of unsaturation. Triacylglycerol isomers are TAGs including the same esterified acid residues but with variation in the location of the moieties in the TAG. The isomers may be symmetrical or unsymmetrical. Unsaturated TAGs can contain *trans* as well as *cis* isomers. The fatty acid profile largely determines quality and use of a vegetable oil and fat (see Table 5.5). Optimum applicability requires tailored FA profiles. For industrial non-food applications such as oleochemicals, high concentrations of a single fatty acid composition in vegetable oils is of considerable economic value. Certain plant species accumulate mainly

Table 5.5 Classification of oils on the basis of the degree of saturation

Saturated: non-drying oils (IV 100–80 g I_2/hg)
Palm, peanut, olive, rape, castor
Moderately unsaturated: semi-drying oils (IV 130–100 g I_2/hg)
Cottonseed, soybean, sunflower, sesame, corn, curcas
Highly unsaturated: drying oils (IV 200–130 g I_2/hg)
Linseed, tung, lallemantia, hempseed, safflower

one such triglyceride. Some of the Brassicaceae, e.g. *Crambe abyssinica*, *Brassica napus* L., *Nasturtium* (watercress) and a specially bred high-erucic acid rapeseed (HEAR) are rich sources (40–75%) of erucic acid (C22:1Δ13). The high level of erucic acid makes HEAR unsuitable for human consumption. On the other hand, oils with high oleic content, such as olive (64.6–84.4%), canola (60.0–75.0%) and NuSun (43.1–71.8%) oils [16], are highly valued for their nutritional qualities. Foodstuff oils are quite expensive. Rapeseed oil (*B. napus*) with its high content of oleic acid (60%), low saturated fatty acid (SFA) level (5.5%) and sufficiently low polyunsaturated fatty acid (PUFA) level (32.5%) is currently a raw material of preference for some lubricant applications.

The oleic (C18:1)/linoleic (C18:2) group comprises crops such as sunflower (*Helianthus annuus* L., C18:1 + C18:2 > 85%), sesame (*Sesamum indicum* L., > 85%), corn (*Zea mays* L., > 80%), safflower (*Carthamus tinctorius* L., > 90%), rapeseed (*B. napus* L. ssp. *oleifera*, > 80%) and high-oleic soybean (*Glycine max*, > 85%) oils. Safflower is the highest source of linoleic acid (79%). The increase of oleic content is at the expense of linoleic acid.

The linolenic acid (C18:3) group has semi-drying properties and is widely used industrially in paints and varnishes. As all tend to oxidise readily, they are regarded as low-quality oils. The highest contents of C18:3 (68%) in plant species is found in Iberian dragonhead (*Lallemantia iberica* L.). Soybean oil from current commercial cultivars typically contains roughly 8% linolenic acid but breeding programmes have allowed the development of germplasm with lower than normal levels of C18:3, namely 3.3% [17, 18]. The linolenic level is most vulnerable to environmental effects in soybean seed oil [19]. Conventional cold-pressed flaxseed (linseed, *Linum usitatissimum* L.) oil is an excellent source of linolenic acid, containing approximately 50 wt% C18:3. In oils containing conjugated FAs, such as tung oil (containing 80% of *cis, trans, trans*-9,11,13-octadecatrienoic acid or C18:3c, where 'c' stands for conjugation), the position of double bonds favours oxidation and polymerisation.

Seed oils, which comprise a mixture of saturated and unsaturated FA esters, provide a promising source of renewable non-petroleum-based feedstocks for industrial utilisation. Vegetable oils are the main biodegradable lubricant base stocks used worldwide. However, consumer acceptance of economical vegetable oil-based lubricants for universal application requires overcoming the major inconveniences of vegetable oils, notably thermal and oxidative instabilities and limited viscosity range, which prevent their uses in some of the more extreme environments. Approaches to correct these negative properties are several: (i) development of (new) crops that provide vegetable oils that are less susceptible to oxidation and other degradation reactions; (ii) transesterification; (iii) winterisation; (iv) functionalisation; (v) use of synthetic vegetable-based lubricant oils; or (vi) incorporation of appropriate inhibitors in the lubricant formulation (additivation). On average, lubricating oils consist of 93% base oil(s) and 7% additives (from 0.5 to 40%), depending on the specific application. With additives the properties limiting application in the marketplace can sometimes be improved, but only at the sacrifice of biodegradability, toxicity and cost. By functionalisation a wide variety of chemical derivatives may be obtained with varied properties. A compromise is needed between the performance based on chemical structure and the desired biodegradability and ecotoxicity. Modified plant-based lubricants have been developed that meet or even exceed the performance expectations of petroleum lubricants.

When it comes to lubricant production, the raw materials from the petrochemical industry are well defined and standardised. However, vegetable oil composition and quality are highly diversified. The primary type of vegetable oil used for biolubricants, at least in the European market, is rapeseed. Biolubricant production in the United States is mainly based on canola oil (85%), sunflower and soy. Other crops, such as castor, corn and safflower only play a minor role. Palm oil (IV 50–60) is a semi-solid at room temperature and is generally not useful as a lubricant despite its relatively good oxidative stability. Esters of castor oil show improved lubricity over other oils with similar carbon chain-length (C_{18}) fatty acids [20].

Animal fat products, reviewed by Haas [21], are another potential source of TAGs. As a result of the bovine spongiform encephalopathy (BSE) problem, which has created alarm for all products of animal origin, a significant amount of animal fat from so-called high-risk material (tallow, yellow grease, etc.), possibly contaminated with infectious prions, has now become available at an attractive price for industrial purposes. Analysis by the EC [22], FDA [23] and WHO [24] has indicated that rendered animal fat is not an agent of transmission of BSE. Community legislation acts as a guidance for minimising the risk of transmitting animal spongiform encephalopathy (TSE) agents [25]. The BSE problem has led to an increased EU biodiesel production based on used frying oils (UFOs). VALUIL is a European Craft project (2002–04) dealing with non-nutritious valorisation of low-cost UFOs as biolubricant base oils for loss applications and hydraulic fluids [26]. Unlike their derivatives, neat fats are more suitable in application as lubricating greases than as liquid lubricant base stocks. In ancient history, pig fat has been used as a lubricating grease [27]. The first greases of the industrial age were made from tallow or olive oil [28, 29].

5.2.1 Edible vegetable oil sources

Triglyceride oils suitable for use as lubricants are both conventional vegetable oils and modified (biotech) vegetable oils (Table 5.6). The vegetable oil triglycerides are naturally occurring oils. By 'naturally occurring' is meant

Table 5.6 Vegetable oils as potential lubricant feedstocks

Conventional oils		Biotech oils
Edible	Non-edible	
Coconut	Castor	Canola
Corn	HERO[a]	*Crambe*
Cottonseed	*Jatropha*	High-oleic[b]
Groundnut	*Lesquerella*	
Moringa	Meadowfoam	
Oil palm	Mustard	
Rapeseed		
Sesame		
Soybean		
Sunflower		
Wheat germ		

[a] High-erucic rapeseed oil.
[b] HOSNO, HOSBO, HOSFO, HORSO, HOCAS, HOPNO, high-oleic canola, palm olein, cottonseed, meadowfoam and *Lesquerella* oil.

that the seeds from which the oils are obtained have not been subjected to any genetic altering. Further, by 'naturally occurring' it is meant that the oils obtained are not subjected to hydrogenation or any chemical treatment that alters the di- and tri-unsaturation character. Naturally occurring vegetable oils (edible or not) having most potential utility as lubricant base stocks are SBO, RSO and canola oil (CO), SNO, coconut oil (CNO), *Lesquerella* oil, peanut oil (PNO), corn oil (CRO), cottonseed oil (CSO), palm oil (PMO), safflower oil (SFO), meadowfoam oil (MFO), and castor oil (CAS). Particularly useful are vegetable oils with a MUFA content of at least about 50%, based on total fatty acid content: rapeseed (*Brassica*), sunflower (*Helianthus*), soybean (*G. max*), *Crambe* and meadowfoam (*Limnanthes*). The MUFA content is usually composed of oleic acid (C18:1), eicosenoic acid (C20:1), erucic acid (C22:1) or combinations thereof.

Oilseeds of the Brassica oleracea family, which include rapeseed, canola, *Crambe*, mustard and *Camelina*, are among the few edible oil crops that can be cultivated in the temperate zones of the globe, at higher elevations and as winter crops. Both summer and winter species are available, allowing cultivation in climatically different regions. Rapeseed (*B. napus* L. ssp. *oleifera*) is the third most important source of vegetable oil in the world (main producers: Europe, China and Canada). European production was heavily supported through the Common Agricultural Policy. In 2004–05 approximately one-third of the EU rapeseed crop was used by the non-food sector.

Rapeseed oil is characterised by a high content of monounsaturated oleic acid (*c.* 60%) and low levels of saturated fatty acids (*c.* 5.5%). The major drawback of some rapeseed oils is their high content of linoleic (C18:2) and linolenic (C18:3) fatty acids, with two and three double bonds, respectively. These determine poor oxidative stability for use as a lubricant base stock.

High-erucic rapeseed oils (HERO; erucic acid content > 45% detrimental to food quality) constitute the classic source of erucic acid and are important oils for oleochemical and lubricating purposes [30]. Although rapeseed (colza or coleseed) oil and meal used to contain high levels of erucic acid and glucosinolates, nowadays most rapeseed cultivars – developed by classical breeding methods – belong to the so-called 'zero' (0) and 'double zero' (00) varieties with reduced levels of both erucic acid (<2%) in the oil and glucosinolates (<30 μmol/g) in the meal, also referred to as edible rapeseed or canola (CANadian Oil Low Acid), in accordance with FDA standards. Canola is not a botanical term, but a trademark used to describe compositional qualities. From an agronomic standpoint, canola offers the advantage of being both a spring and a winter crop. Canola was cultivated on some 16.2 million acres of Canadian farmland in 2009. Producing biolubricants from canola is a profitable venture for Canadian farmers. A typical canola oil (Puritan™, Procter & Gamble) consists of 6% SFA (C16:0, C18:0), 62% C18:1, 22% C18:2, 10% C18:3 and < 1% C22:1 (see Table 5.3). Europe has concentrated on 'single-low' types (low in C22:1).

Rapeseed is highly accessible to biotechnological methods. The manipulation of oil synthesis in rapeseed to produce specific, desired, chemically altered triglycerides has already been implemented on a large scale [31], see also Sections 5.3 and 5.3.1. Whereas rapeseed traditionally refers to any HERO variety, the low-erucic (LE) acid high-oleic varieties (with C18:1 > 60%) are of interest to the food industry in view of their high nutritional qualities and a low level of saturated fat.

Soybeans (*G. max*) are native to north-east Asia (Japan, Korea, north-east China) and are a temperate-climate crop, sensitive to temperature changes and requiring four distinct seasons. Global soybean production (2009/10) is estimated at being 257.5 Mt. The main soybean producing countries (the United States, Argentina) have temperate climates. However, genetically modified (GM) soybean has also been developed for the Brazilean cerrado (see Section 5.3). During the 2002–07 period soya provided about 50% of the growth in world oilseed crush (about 31 Mt). The soybean market has historically been driven by the demand for meal, which is not a very storable product, as opposed to oil (making up 35–40% of soybean value). Although refined soybean oil is a good lubricant, it oxidises too readily for use in many applications. The problem of low resistance to oxidation and limited shelf-life can be overcome in many ways, such as chemical modification (notably partial hydrogenation and epoxidation) [32–35], use of additives [36], interesterification with palm oil, genetic modification with C18:3 reduction [37–39] (see Section 5.3.1), or blending with jojoba oil (JO) [40]. The use of the epoxidation route is particularly interesting since it yields a material suitable for other use that can be further modified by a ring opening reaction [41]. JO was effective in reducing the formation of hydroperoxides and volatile compounds in SBO. All SBO/JO blends were more stable than pure SBO. The 80:20 SBO/JO blend had the largest impact on the oxidative stability performance of SBO. Soybean breeding lines with low linolenic acid content have yet to be commercialised (see Table 5.20); interesterification greatly increases processing costs. Therefore, it is of interest to consider non-conventional natural lipid sources [42], as well as simple processes such as mixing [40, 43, 44]. The SBO pour point (PP) problem can be solved by blending with other fluids such as synthetic oils with lower PP (compromising the biodegradability) or by means of chemical addition. Mixing only lessens our dependency on petro-based lubricants and is therefore only a short-term solution. High-oleic crops represent interesting but limited feedstock sources for biolubricants.

Sunflower (*H. annuus* L.) yields (about 800 kg oil/ha) are lower than rapeseed yields (about 1000 kg oil/ha); yet, SNO represents a useful vegetable oil source in countries with warm and dry climatic conditions and rich soil. Although seed oil of the standard cultivated sunflower is considered to be of good quality for edible purposes, it has more limited use for fuel or lubricant production in view of its high linoleic acid content. In this respect, high-oleic varieties with higher oxidative stability are of more interest (see Section 5.3.1). Sunflower oil offers environmentally friendly biolubricants.

Corn (maize, *Z. mays* L.) is a native grain crop of the Americas and is the third most important grain in the world after wheat and rice. Since only 7–8% of the grain is oil, corn oil is a high-cost bonus by-product with high value as an edible oil. Corn oil is characterised by high levels of unsaturated fatty acids (MUFA, 26%; PUFA, 59%), low levels of saturated fatty acids (15%), very low levels of linolenic acid (0.5–1.7%), high levels of unsaponifiables (including phytosterols and tocopherols) and stability during frying. The food industry relies heavily on (hydrogenated) corn oils [45].

Owing to their wide availability, in the United States soybean and corn are the main vegetable oils used for food and industrial purposes, including lubricant production, while rapeseed and sunflower oils are mostly utilised in Europe [46, 47]. Despite the fact that RSO and SNO have superior flow properties compared to other vegetable oils, their oxidative, hydrolytic and thermal stabilities are not sufficient for use in a circulating system.

The oil palm (*Elaeis guineensis)* is the highest oil-yielding crop and has the potential of becoming the major supplier of both edible oil and renewable industrial feedstock. Domesticated oil palm is regarded as the most cost-effective vegetable oil crop with typical yields of about 5.0 t/ha/yr. Through breeding and selection, the oil yield of commercial plantations could reach as much as 8 t/ha/yr. Production of palm oil is increasing most rapidly, from a modest 7.9% market share in 1980 to 19.0% by 2000, with highest growth figures expected until 2020 (production forecast of 49.4 Mt). *E. guineensis* is the major supplier of both edible oil and renewable industrial feedstocks since 2003. Oil palms differ from most other oil crops in that both the fruit pulp and seed are used. Two totally different kinds of oil are produced from the oil palm, namely orange red palm oil (PMO) and colourless palm kernel oil (PKO). While PMO contains 47% SFA (mainly C16:0), 44% MUFA (mainly C18:1) and 9% PUFA, PKOs are far more saturated with 84.5% SFA (mainly C12:0), 15% MUFA and 0.5% PUFA. Biotechnological efforts are directed towards production of palm oil with increased oil yields, high IV and high MUFA (C18:1) contents for edible purposes and industrial uses [48]. Palm oil is emerging as a new source for biolubricants.

At tropical ambient temperature (27 °C) palm oil consists of two phases which can be cooled and separated into olein and stearin. Palm olein is the liquid fraction whereas palm stearin is the solid fraction. Palm olein consists mainly of low melting triacylglycerols and is used as cooking oil and suitable for frying. Palm stearin is dominated by high melting triacylglycerols and is used for margarines and shortenings. In normal practice, palm stearin is blended with other soft oils which contain more PUFAs, such as SBO, CNO, RSO (canola) and SNO to impart the required plasticity to the end product. As palm stearin contains no *trans* fatty acids it is a good replacement for hydrogenated fats in food applications. A mandatory labelling of *trans* FA content in every food product was introduced by the United States in 2007.

Palm oil with high SFA content (see Table 5.3b) is often considered not suitable as a lubricant because of its poor low-temperature flow properties (PP 8–15 °C). However, when palm oil is used for lubricant applications in tropical regions, its high SFA content may be considered advantageous because it determines better oxidative stability. In those areas, the limitation in low-temperature properties may be ignored. High-oleic palm trimethylolpropane (TMP) esters have improved low-temperature properties (PP of –33 °C) compared with palm oil and palm kernel oil esters (see Table 10.18).

High-oleic palm oil is produced by the High Oleic Pilot Plant, Malaysian Palm Oil Board Head Quarters [49]. High-oleic palm stearin with an oleic content of 48.2% (FA composition: C12:0, 0.2%; C14:0, 0.7%; C16:0, 26.8%; C18:0, 10.1%; C18:1c, 48.2%; C18:2c, 12.8%; C18:3, 0.3%; C20:0, 0.7%) was obtained from fractionation of high-oleic palm oil [50]. High-oleic palm oil TMP esters, with adequate PP and lubrication properties, have potential as base stocks for biodegradable lubricants [51].

Coconut (*Cocos nucifera*) oil is another primary source of oleochemicals (soaps, detergents, lubricants, plasticisers, surfactants, detergents) [52]. The main coconut producers are Indonesia, Philippines and India (73.2% market share, 2005). CNO and PKO are so-called lauric oils, both containing high contents of C12:0 (approximately 50%) and saturated short-chain fatty acids (≈85–90%). CNO shows a high PP (24 °C) due to the predominantly saturated nature of its FA constituents [53]. Additives fail to improve the PP, whereas chemical modification procedures (estolide formation) are more effective. CNO is used as a two-stroke engine lubricant in southern India [54].

Peanut (*Arachis hypogaea*) cultivars are characterised by a slightly lower oil yield (890 kg oil/ha) than rapeseed (1000 kg oil/ha) but much higher than soybean (375 kg oil/ha). Peanuts are native to South America. Peanut oil has excellent stability, is extremely durable and is less prone to oxidation than other vegetable oils in frying applications; the oil price is high. High-oleic peanut cultivars, developed using conventional breeding procedures, contain up the 81% oleic acid [55].

Sesame (*S. indicum* L.) is a herbaceous oilseed crop of the Pedaliaceae family. The economically important crop is widely cultivated in many parts of the world (5 million acres), mainly in (sub)tropical areas (main producers: India, China, Myanmar) and has been adapted to semi-arid regions. Sesame is currently being developed as a major oilseed crop in Turkey (Southeastern Anatolia Project). Global sesame production amounts to 3.3 Mt/yr. Sesame is widely used in food, and for nutraceutical and pharmaceutical purposes because of its high contents in oil (37–63%), protein (18–25%) and antioxidants (sesamin and sesaminol lignans) [56]. The FA composition consists of C16:0 (11%), C18:0 (7%), C18:1 (43%) and C18:2 (35%). Sesame oil shows high oxidation stability and has been indicated as a potential lubricant base stock [57]; see Section 5.2.3.

Safflower or false saffron (*Carthamus tinctorius* L., Asteraceae family) is an annual drought-resistant niche crop, which grows in the temperate zone where wheat and barley do well. India is the main producer of safflower, which is grown exclusively for its food oil, which is high in essential unsaturated fatty acids. Traditional plant breeding and plant biotechnology play an important role in developing safflower. High-oleic (HO) safflower with 77.4% C18:1 (SFA 7.0%, MUFA 78.0%, PUFA 15.0%) is characterised by a low iodine value (80–100 g I_2/hg) and high cetane number (CN, 52.2) [58].

CSO is a vegetable oil extracted from the seeds of the cotton plant (*Gossypium herbaceum*) after the cotton lint has been removed. It must be refined to remove the naturally occurring toxin gossypol (a polyphenol); gossypol induces male infertility. Cottonseed typically contains about 15 wt% oleic acid; increased seed oleic acid content in transgenic cotton plants ranges from 21 to 30 wt% [59]. The increase in oleic acid content is at the expense of linoleic acid. CSO is one of the few oils considered acceptable for reducing saturated fat intake. It is among the most unsaturated oils (PUFA, 52%); others include safflower (PUFA, 79%), corn (PUFA, 59%) and sunflower seed oils (PUFA, 61.5%). CSO has many food applications (salad oil, cooking oil, shortenings or margarine, baking and frying fats) [60]. Only a small amount (18%) is used for other industrial purposes. CSO has been described as an environmentally accepted lubricant additive [61]. Cotton is also one of the main four (soy, corn, rapeseed/canola and cotton) GM crops grown around the world. Oil contents and yields are low.

Wheat germ oil is extracted from the germ of the wheat kernel, which makes up only 2.5 wt% of the kernel. Wheat germ oil is particularly high in octacosanol, a C_{28} long-chain saturated primary alcohol. Wheat germ oil contains the following FAs: C16:0, 16 wt%; C18:1, 14 wt%; C18:2, 55 wt%; C18:3, 7 wt%. The product is expensive and easily perishable.

The development of new and alternative agricultural crops (Table 5.7) provides a basis for farm diversity and a source of novel compounds for chemistry and industry. Some emerging new oil crops also present interesting characteristics for potential application as lubricant base stocks, notably *Crambe*, *Jatropha*, *Lesquerella*, meadowfoam, *Moringa* and mustard. Recently, several of these new oilseed plants have progressed to the stage of commercial production, such as *Crambe*, jojoba, *Lesquerella*, meadowfoam and mustard.

Moringa oleifera of the Moringaceae family is an exceptionally nutritious vegetable tree growing rapidly in semi-arid (sub)tropical areas [62]. It is considered as one of the world's most useful trees, as almost every part can be used for food. *Moringa* seeds contains 38–40% oleic acid-rich oil (C18:1,

Table 5.7 Main characteristics of emerging new oil crops

Crop	Oil content (%)	Main component
Jatropha curcas (physic nut)	40	C18:1 (40%)
Crambe abyssinica (Crambe)	35[a]–46[b]	C22:1 (58%)
Brassica juncea (mustard)	25–50[c]	C22:1 (46%)
Limnantes alba (meadowfoam)	28	C20:1 (62%)
Lesquerella fendleri	30	C20:2-OH (56%)
Camelina sativa	30–43	C18:3 (38%)
Moringa oleifera	38–40	C18:1 (71%)
Eruca sativa	29	C22:1 (45%)

[a] Whole seed.
[b] Dehulled seed.
[c] Yellow mustard (27%), brown mustard (36%), some oriental mustards (50%).

71.2–74.0%; SFA, 22%; MUFA, 77%; PUFA, 1%; SN 186 mg KOH/g; IV 65.6 g I_2/hg). *Moringa* deserves further development and attention for prospective industrial use (see also Section 5.2.3, p. 163). *Moringa* is a natural high-oleic oil.

5.2.2 Non-edible oilseeds for lubricants

Alternative feedstocks for lubricants are non-food crops and food crop residues. Non-food feedstocks are playing a vital role in the market, both for economic and environmental reasons. Oils that contain toxic substances in their seeds and therefore do not conflict with food resources may eventually be good candidates for supply of plant oils for the lubricant industry. These include castor oil, curcas oil (physic nut oil) from *Jatropha curcas*, and *Crambe*. Although it is generally considered that non-edible oils should be exploited as far as possible in order to allot edible oils to human consumption, large-scale cultivation of non-food oil crops for industrial application is not quite a panacea as large amounts of toxic meal are co-produced, as in case of castor beans, *Jatropha*, etc.

Oxygenated fatty acids, also termed oxylipins [63], are contained in high quantities only in castor oil. Castor bean (*Ricinus communis*) oil (world production volume of about 600 kt/yr, mainly originating from India, China and Brazil), which is an inedible, inexpensive, non-volatile oil with good shelf-life, is rich in ricinoleic or 12-hydroxy-9-*cis*-octadecenoic acid (85–95%), which is the only hydroxy acid oil of commercial value. In the developing endosperm of castor, ricinoleate is synthesised by direct hydroxyl substitution of an oleic acid moiety rather than via an unsaturated, keto or epoxy intermediate [64]. Enzymatic hydrolysis of castor oil with lipase yields ricinoleic acid [65]. However, in view of its agronomic characteristics castor bean is not an ideal source for this HFA. Castor beans grow on marginal lands, require little rainfall (15–18 in./yr), and can withstand long periods of drought. The poisonous plant provides beans with a high oil content (47–49%); cake contains toxins (ricin, ricinine and certain allergens). Castor bean oil is an option for lubricant production because it does not compete with food crops for arable land space. Castor oil has once been considered as a strategic material by Brazil.

Castor oil, obtained by a combination of mechanical pressing and solvent extraction, is one of the most promising renewable raw materials for the chemical industry, giving access to the platform chemicals sebacic acid (by oxidation) and 10-undecenoic acid (by pyrolysis) [66, 67]. EcoPaXX™ (DSM) is a partially (70%) renewable nylon (polyamide 410) based on diaminobutane (DAB) and sebacic acid (from castor oil). Being classified as a non-drying oil, castor oil is also used in the preparation of brake fluids, as an ingredient of specialty lubricants, inks, paints and varnishes, soap, and as the main ingredient in motor oils for high-speed automobile engines. The cost of castor oil is relatively high.

Castor oil is a vegetable oil with a long history in lubrication. In the past, this oil has been used as a lubricant in carts and wheels in Persia (now Iran) for a very long time [68]. Nowadays, castor oil is used for many applications including base oil formulations for lubricants, functional fluids and process oils. Certain characteristics of castor oil, such as high lubricity, high viscosity over a wide range of temperatures, and insolubility in aliphatic petrochemical fuels and solvents, make it directly applicable as a lubricant for equipment operating under extreme conditions. It was recognised as early as 1937 by the German Reichsluftfahrtministerium (RLM) that high output aircraft engines needed to use castor oil because of its outstanding lubricating qualities, notwithstanding its gumming propensities. In this context, work by Zorn [69] (see also ref. [70]) needs to be mentioned (see also Section 10.5.3). Castor oil-based fluids were also used some 50 years ago by London Bus and other vehicle companies in their rear axles, thereby obtaining major advantages in fuel consumption owing to the excellently low coefficient of friction [71]. More recently, Brazilian electric power utilities have worked hard to develop insulating vegetable oil based on castor oil as a specialty lubricant [72]. Several other vegetable oil-based products are on the transformer insulating fluids market nowadays (see Section 12.8.2). Ricinoleate esters and amides are used as lubricant additives.

A satisfactory VI (≈90) and the extraordinary high kinematic viscosity of castor oil (297 mm^2/s at 38 °C) can be related to hydrogen bonding of hydroxy monounsaturated triglycerides. This property makes the oil useful as a component in blending lubricants [73]. The presence of the hydroxyl group in castor oil (as ricinoleic acid) adds extra stability to the oil and its derivatives by preventing the formation of hydroperoxides. Although castor oil is a unique naturally occurring polyhydroxy compound, a limitation of the oil is the slight reduction of its hydroxyl value and acid value on storage (about 10% in 90 days). The reduction of these values is due to the reaction between hydroxyl and carboxyl groups in the fatty acid molecule to form estolides (see Section 6.2.2).

Castor oil shows lower deposit forming tendencies than high-oleic sunflower oil (HOSNO) but much higher than super-refined mineral oil (SRMO). Castor oil solubilises higher concentrations of antioxidants, but the oxidative stability of the formulation is lower when compared with HOSNO. Without additives castor oil is superior to SRMO in lubrication performance on the four-ball wear tester [74]. Despite high viscosity, wear properties of castor oil with a high hydroxyl content are inferior to vegetable oils without the hydroxyl groups [75]. Rudnick and Morehouse [76] have reacted the hydroxyl functionality in castor oil to form the methyl ether, which modifies the properties (viscosity, PP) and performance (wear) relative to unmodified castor oil. Ricinoleic acid can be hydrogenated to give 12-hydroxystearic acid, which is extensively used as a gellant in lubricating greases in the form of its alkali metal salts [77]. Castor oil can be modified by reduction with hydrogen to produce hydrogenated castor oil (HCO), which is a wax-like material with melting point of 86 °C. HCO may be used as a solid lubricant [78]. Selective hydrogenation of castor oil to a high-oleic oil would be of greater interest. Lubricity analysis by a high-frequency reciprocating rig (HFRR) shows that castor oil methyl ester performs quite well as a lubricity enhancer for diesel fuel at concentrations less than 1.0% [79]. Dioctylsebacate (from castor oil) is used as a jet lubricant and in air-cooled combustion motors.

At present, the market is mainly investing in new, non-edible raw materials such as *Jatropha*, *Camelina*, mustard and *Crambe*. *Crambe*, also known as Ethiopian mustard or colewart, is a member of the Brassicaceae family. It is a drought-tolerant, low-input, industrial oilseed, and has been grown and processed on a commercial scale in the United States (North Dakota) since 1990 on a limited production area (30 000 to 60 000 acres) [80]. *Crambe*, with an oil content of about 30%, has a higher concentration of crucic acid (>55% C22:1Δ13) than most other species and low free fatty acid (FFA) contents (<0.5%). Refined *Crambe* oil (IV, 90.5 g I_2/hg; viscosity, 51.1 mm^2/s at 40 °C; VI, 204; PP, –9 °C) is mainly a source of erucic acid but it also shows promise of becoming a new industrial oil crop with great potential [81–83]. This is illustrated by the efforts for the production of wax esters in *Crambe* in the framework of the EPOBIO project for the realisation of economic potential of sustainable resources, in particular bioproducts from non-food crops (see Section 5.4.2). *Crambe* oil and HERO are among the more important vegetable oils for lubricating purposes. However, both contain significant amounts of linoleic and linolenic acids that negatively influence mainly the oxidative stability. The same holds for wild mustard (*Brassica juncea* L.) with C18:2, 14.2%; C18:3, 13.0%; C22:1, 45.7% [84]. The high erucic acid content is responsible for high viscosity values, whereas it negatively affects the cold parameters because of the high melting point [85]. The main derivative of erucic acid is erucamide, which finds use as a slip and antiblocking agent in polymeric materials. Other derivatives are behenic acid and its esters, brassylic acid and pelargonic acid (see Fig. 6.45). Brassylic acid may be used in lubricant chemistry. Development of high-fat content (25–40%), low cost (US$ 0.10/lb), non-food grade mustard seed is under way.

Jatropha is a genus of about 175 succulent plants, shrubs and trees (including *Jatropha curcas*) of the Euphorbiaceae family growing in tropical and subtropical countries. *Jatropha curcas* (physic nut), native to the Caribbean, Central and South America, is a non-food crop now being actively developed for fuel applications (*Jatropha* biodiesel) and is considered as having one of the best potentials among a large number of oil yielding species. *Jatropha* is a weed, wildly growing in arid and semi-arid regions on degraded wastelands having low fertility and moisture. When grown in marginal, non-agricultural areas, *Jatropha* does not compete with existing agricultural resources. A benefit of *Jatropha* is the long-term yield. It can be harvested almost continuously over a 30-year period. If irrigated and fertilised, one hectare of *Jatropha* can produce some 2.5 t of seed oil. While its composition (C16:0, 12.8%; C18:0, 7.3%; C18:1, 44.8%; C18:2, 34.0%) is similar to other oils which are used for edible purposes (see Table 5.3a), the presence of some antinutritional factors such as toxic phorbol esters renders this oil unsuitable for use in cooking [86]. *Jatropha* oil is indicated as a raw material for many potential applications: biodiesel [87], illumination, soaps, cosmetics, medical uses, biopesticides (phorbol esters), and also has good potential as a lubricant feedstock [88]. The production of a biolubricant from high-FFA containing jatropha oil (up to 30%) has been reported [89]. A drawback of *Jatropha* is its high C18:2 content. At present, large-scale *Jatropha* plantation (over 1.75 Mha) is under way in South-East Asia, Africa, the United States and Central and South America [90]. However, as domestication is still in its infancy some considerable challenges lie ahead for this crop, such as tree management, uniform maturation rates, mechanical harvesting, consistency of oil yields, etc. Consequently, the impact of large-scale production of *Jatropha* is expected to require some more 6–8 years. While *Jatropha* plantations are quite successful in Colombia, many other *Jatropha* projects have failed due to diseases, unsuitable soil conditions or disappointing low yields. Large-scale cultivations of toxic, wild-type *Jatropha* has recently severely been criticised [91].

Hydroxyl functionality is rare in plant oils, although the HFA ricinoleic acid has been identified as a constituent of the seed storage oil in at least 12 genera from 10 families of higher plants [92]. HFAs are nowadays classified as strategic materials by the US government. Special properties of HFAs compared to other fatty acids include higher viscosity and reactivity, caused by the presence of the hydroxyl group. Although various plants synthesise C18:1-OH (e.g. *Crotolaria striata*, *Ochrocarpus africanus* and *Alternanthera triandra* with 16–22% HFA), none can match castor bean (*R. communis*) for seed oil content and degree of HFA purity. The uniquely high level of ricinoleic acid in castor oil (88–90%) imparts increased lubricity to the oil esters compared with normal vegetable oil esters with similar carbon chain length. By engineering ricinoleic acid into a high-oleic seed background, a superior base stock can be supplied having all the advantages of a vegetable oil without the poor oxidative stability. Similarly, the developing oilseed *Lesquerella* is also of interest for lubricant application. The primary fatty acid in the non-commercial crop *Lesquerella fendleri* (Fendler's bladderpod), a species in the mustard family and perennial in nature but cultivated as a winter annual, is lesquerolic acid, a 14-hydroxy-*cis*-11-eicosenoic acid (C20:1-OH, 55–69%), a homologue to ricinoleic acid with high kinematic oil viscosity (127 mm^2/s at 40 °C). Hydrogen bonding is at the origin of the high viscosities of hydroxy fatty acids [93]. *Heliophila amplexicaulis* L.f (Cruciferae/Brassicaceae family) oil contains about 30% lesquerolic acid [94]. *Lesquerella* oil applications are targeted at high-value products such as lubricants, coatings, cosmetics and additives. Intensive research efforts currently aim at successful introduction of *L. fendleri* into agriculture.

Vernonia (*Vernonia galamensis*) and vernin spurge (*Euphorbia lagascae*) are both poor producers of the epoxy FA vernolic acid. The manufacture of these oils has not yet been commercialised. The occurrence of a wide variety of oxygenated fatty acids (oxylipins) has been reported in mammals, plants, fungi and bacteria. Biotechnological routes for the production of epoxy fatty acids and HFAs are being investigated (see Section 6.2.7, p. 316).

There are numerous plant breeding efforts to domesticate wild plants, such as species of the genera *Lesquerella, Crambe* or *Limnanthes*, in order to develop useful plants that may be more productive [95–97]. Among the underexploited temperate climate industrial crops with potential new oilseeds, meadowfoam (*Limnanthes* spp.) has already gained active crop status, whereas *Lesquerella*, which has the capacity to produce an oil containing a high proportion of unusual components, requires further breeding, agronomy or processing research. Lubricity analysis by HFRR has shown that *Lesquerella* oil methyl ester enhances lubricity of diesel fuel to acceptable levels already at concentrations as low as 0.25% (cf. 1.0% for castor oil methyl ester) [79]. Non-hydroxylated methyl esters, such as soybean and rapeseed methyl esters, require higher concentrations than 1% before a significant improvement in lubricity can be observed.

Crude meadowfoam oil (*Limnanthes alba*) with long-chain FAs (97% ≥ C20) is very stable (see Section 5.2.3, p. 169). Meadowfoam is a rich source of unique unsaturated long-chain fatty acids (LCFAs), in particular *cis*-5 unsaturated FAs (20:1Δ5, 22:1Δ5 and 22:2Δ5Δ13). Meadowfoam seed oil thus contains some 95% of FAs with chain lengths of 20 carbon atoms or longer (C20 and C22 monoenes and dienes) [98]. The unique chemistry of meadowfoam (SFA 1.0%, MUFA 82.0%, PUFA 17.0%) has stimulated its development as an industrial oilseed crop [99]. Being mainly composed of monoenoic FAs, meadowfoam oil has many interesting physical properties such as an unusually high oxidative stability index (OSI) compared with other vegetable oils (see Table 5.13). The oil chemistry of meadowfoam results in commercial oil values exceeding those of SBO; initial target markets are therefore primarily high-value applications.

Development of high-fat content (25–40%), low cost (US$ 0.10/lb), non-food grade mustard seed is under way in the United States. The key development that makes industrial mustard so promising is that the defatted meal (after the oil is removed) can be used as an organic pesticide without further processing.

The fringe oilseed crop *Camelina sativa* L. (Brassicaceae), commonly known as gold-of-pleasure, false flax or linseed dodder, is also considered as a promising potential oilcrop for non-food industrial applications [100]. The under-exploited canola relative is a short-season crop (90 to 105 days) with favourable agronomical characteristics, high oil content (29–41%) and high yields (1200 to 1500 lbs/acre). *Camelina* is fit for cool regions such as Montana where canola production is difficult. *Camelina* does not have generally recognised as safe (GRAS) status from federal US agencies and the oil is nowadays used for biodiesel production [101, 102]. *Camelina*'s high percentage of PUFAs (>50%) makes it oxidation susceptible, thus requiring additional antioxidants, but also lowers the PP. *Camelina* is prospected as a future crop for lubricants, to be produced in Montana on a small scale [103, 104].

The Mediterranean crop eruca or rocket (*Eruca sativa* ssp. *oleifera*), newly developed for industrial use (non-food oil) in the Po Valley pedoclimatic conditions within the framework of the European Common Agricultural Policy, contains 29% oil and a high level (53–55%) of long-chain fatty acids (mainly C22:1, 45.3%) [105].

5.2.3 Application-relevant vegetable oil properties

Biodegradable vegetable oils offer specific environmental benefits over mineral-based lubricants. In addition, vegetable oils also have certain performance advantages over conventional mineral oil base stocks. These include low volatility, high flash points, high VI, high polarity and excellent lubricity. New vegetable oil-based lubricant product development requires understanding of the relationship between chemical structure and physical properties. An in-depth understanding of the molecular origin of the structure–property–functionality relationship is needed for application development and creation of market value (Fig. 5.3).

Most vegetable oils are liquids at room temperature with easy miscibility with other fluids and high solubilising power for both contaminants and a variety of substances used in lubricant formulations, thus making them good candidates for use as base oils. Variability in composition of vegetable oils can be expected from one crop to another and from one year to the next, depending on weather conditions, soil, plant characteristics, storage and treatment conditions. This makes it hard to establish structure–property relationships and affects consistency in lubricant quality. High lubricant quality is a prerequisite for market acceptance. Impurities in vegetable oils can be in the form of natural minor ingredients (oil specific), degradation

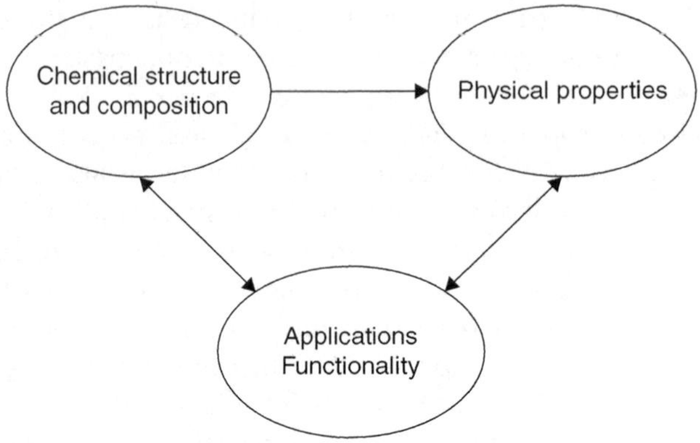

5.3 Properties vs functionality of chemical structures for lubricant application.

products (from seed, processing or oil storage) and remaining chemicals (growing and processing).

Vegetable oils are rapidly biodegradable and therefore promising candidates for use as base fluids in formulation of environment friendly lubricants. Under standard conditions (e.g. OECD 301D test method), a typical vegetable oil can biodegrade up to 80% into CO_2 and H_2O in 28 days, compared with 25% or less for typical petroleum-based lubricating fluids. The biodegradability also of plant oil-based lubricants has been tested under operating conditions [106–108]. It seems that biodegradability is not affected by usage. On the other hand, the ease of biodegradability also determines high susceptibility to undesired tribochemical degrading reactions in severe conditions of high temperature, pressure and shear. In contrast to mineral oil-based fluids, most vegetable oils undergo cloudiness, precipitation (causing lower filterability), poor flow and solidification at –10 °C upon prolonged exposure. Combination with water in the form of emulsions must be prevented at every stage of production. The presence of C22:1 (erucic acid) and SFAs having high melting points leads to poor cold flow in lubricants.

Vegetable oils still have a number of other inherent qualities that give them advantages over petroleum oils as the feedstock for lubricants (Table 5.8). Because vegetable oils are derived from a renewable resource with excellent biodegradability and ecotoxicity characteristics, their use avoids the upstream pollution associated with petroleum extraction and refining. Similarly, their disposal requires minimal expense. From a worker safety perspective, plant-based lubricants are more attractive than their petroleum counterparts because of their relatively low toxicity, skin compatibility, high flash point and low volatile organic compound (VOC) emissions; they constitute a reduced fire hazard.

Table 5.8 Main characteristics of vegetable oil-based biolubricant base stocks

Advantages:
• Good viscosity and very high viscosity index
• Excellent boundary lubricating properties
• High bulk modulus
• Superior thin film strength
• Low friction coefficients
• Improved fuel economy
• High tool-life
• High cleanliness at working place
• Compatibility with mineral oil and lubricant additives
• Applicability as base stock and as additive
• Easy miscibility with other fluids
• Very low volatility (reduced emissions); high flash point (*c.* 300 °C)
• Oil mist and vapour reduction
• Excellent (worker) safety characteristics
• Skin compatibility
• Excellent biodegradability
• Low ecotoxicity, non-water polluting
• Aromatic – free
• Reduced engine emissions
• Environmentally friendly, renewable
• Liquids at room temperature
• Water insoluble
• Disposal at minimal expense
• Wide production facilities, low cost
Disadvantages:
• Operating temperature limitations
• Low thermo-oxidative stability (in service and storage)
• Hydrolytic instability
• Ageing
• Poor low-temperature fluidity
• High pour point
• Lack of broad viscosity range
• Susceptibility to tribochemical degradation processes
• Limited additive technology
• Easy formation of sludge
• Filter-clogging tendency[a]
• Poor air release, foaming and corrosion properties[a]
• Poor tolerance for alkalinity
• Yellow metal discoloration[a]
• Swelling and softening of elastomeric seals
• Product cost

[a] Depending on vegetable oil.

Major differences in chemical composition of vegetable oils to other lubricant base stocks are: (i) much higher molecular weight (866–894 g/mole for high-oleic varieties) than mineral oil or synthetic lubricants of similar viscosity grades; and (ii) unsaturation levels of triglycerides. Vegetable oils have some limitations in key lubrication areas. Most of them fall into a narrow range of inherent kinematic viscosities. In terms of ISO viscosity grades the vegetable oils cover only ISO VG 32 and 46, whereas in many applications higher viscosities are required; this problem cannot always be solved through the use of thickeners. Other restrictions in application of vegetable oils to lubrication relate to oxidative instability and poor low-temperature performance.

Some of the qualities of vegetable oils, excellent biodegradability and low ecotoxicity, are particularly important for lubricants used in environmentally sensitive areas such as marine ecosystems, and for those with a high potential of being lost to the surrounding environment. R&D efforts of vegetable-based industrial lubricants have mainly been concentrated on rapeseed/canola and sunflower oils and derived oleochemical products. These oils are naturally more stable than soybean oil. As lubricant base oils RSO and SNO are being preferred because their oxidation stability is acceptable in some technical applications and performance is combined with good environmental properties. However, the oxidation, hydrolytic and thermal stability of pure vegetable oils are not adequate for their use in circulating systems. Soy lubricants are not widely used in commerce, again due to their limited hydrolytic and thermo-oxidative stability; additional processing is required to achieve an optional glyceride composition. Optimised soybean base oils containing >83% C18:1 and low C18:2 content (<3%) [109] are excellent for hydraulic fluids [110], gear oils, metalworking and other applications where the product is subject to long-term machinery exposure. The chemical fatty acid structures of the high-oleic vegetable oils provide unique natural corrosion protection, inherent detergent and solubility properties. Even though fresh vegetable oils and mineral oils may present similar PPs, the confidence level in these two oils differs; compare the PP variations in high-oleic RSO from –27 °C (fresh) to –20 °C and –10 °C (after 3 and 10 days, respectively). This problem can (partially) be overcome by additivation.

As the use of vegetable oils is still relatively new and uncommon in many lubricant applications, it is often difficult to obtain comparable performance data and conflicting opinions have been reported. This is partly due to the fact that additive packages for vegetable oil and mineral oil formulations differ (necessarily), which invalidates comparison of the base oil performance. A few studies have directly compared the use of mineral and biodegradable oils in hydraulic systems [111, 112]. In recent years the use of RSO has become more widespread in mobile hydraulic systems. Contrary to reasonable expectations that systems running on biorenewable RSO should automatically be better for the environment, this does not necessarily hold true [112]. The disappointing environmental performance of RSO as a lubricant is on account of its unsatisfactory technical qualifications. As further promotion of the use of mineral oil is unsustainable because of its non-renewable origin, it is necessary to improve the performance characteristics of vegetable oils where weaknesses have been identified. According to Eichenberg [113] RSO is an acceptable hydraulic fluid, but its high-temperature stability is critical. This limits its applicability. RSO degrades faster than mineral oil.

Gerbig *et al.* [57] have compared the lubrication-relevant properties of 16 commercially available neat vegetable oils (without additives) with those of mineral oil and synthetic esters (Table 5.9). Physical properties, chemical characteristics and oxidative stability of the oils were evaluated in relation to typical lubricant requirements. The tribological performance, lubrication and wear protection of the oils were judged under adhesive as well as abrasive conditions. The comparative study of the test vegetable oils showed great differences in the properties of the oils. All the vegetable oils tested revealed weak points during testing, including rapeseed and castor oils, which have established uses in industry; none of the oils excelled in all the tests. The mineral oil and synthetic esters showed markedly superior performance when tested under adhesive conditions, but comparable behaviour to the other oils under abrasive conditions.

The functionality of the oils depends strongly on the tribosystem. For the combination chrome steel 100Cr6 against 100Cr6 (DIN W.13505, AISI 52100), representative of adhesive wear, only linseed, olive, walnut and safflower oils showed acceptable behaviour. However, the friction and wear performance of the mineral oil and the synthetic esters was far superior to any of the vegetable oils tested. Among the vegetable oils, the wheat germ and olive oils best lubricated the 100Cr6/100Cr6 system. Under abrasive wear conditions (Al_2O_3/100Cr6), the castor, sesame and soybean oils provided the best results of the vegetable oils. In contrast to the clear superiority of the mineral and synthetic esters under adhesive wear conditions, the friction and wear performance of the vegetable oils was comparable to or better than that of the mineral oil or the synthetic esters under abrasive conditions. In these tests RSO, which is already being used in technical applications, was not as good as the aforementioned oils.

Wheat germ, olive, peanut, sesame and soybean oils can be considered as viable alternatives to rapeseed oil. Wheat germ and sesame oils in particular showed relatively good oxidative stability and may have potential in selected technical applications. The industrial suitability of these oils should further be evaluated, in particular as to hydrolytic stability and compatibility with additives [57].

Table 5.9 Basic characteristics of selected vegetable oils, mineral oil and synthetic esters[a]

Vegetable oil	MUFA (%)	PUFA (%)	AV (mg KOH/g)	SN (mg KOH/g)	IV (g I_2/hg)	ρ (g/cm^3)	FP (°C)	FIP (°C)	CP (°C)	PP (°C)
Almond oil	65	21	0.3	173	96 n	0.911	328	365	–24	–29
Castor oil	92	5	0.8	179	90 n	0.957	300	358	<0[b]	–27
Corn oil	27	53	0.3	176	109 s	0.916	324	362	–9	–14
Grapeseed oil	16	68	0.7	198	150 d	0.918	324	362	–11	–15
Hazelnut oil	71	11	2.0	204	83 n	0.912	320	360	–6	–13
Linseed oil	16	70	1.4	194	192 d	0.925	320	362	–14	–15
Olive oil	76	7	1.1	186	89 n	0.909	318	360	6	–3
Peanut oil	45	32	0.3	199	105 s	0.912	326	364	–3	–12
Pumpkinseed oil	25	55	6.1	202	100 s	0.915	302	360	<0[b]	–6
Rapeseed oil	56	31	1.5	180	114 s	0.913	320	362	–12	–29
Safflower oil	12	74	0.4	244	154 d	0.918	328	364	–12	–18
Sesame oil	42	44	2.0	193	111 s	0.915	316	362	–2	–5
Soybean oil	20	67	0.3	195	143 d	0.917	326	364	–4	–9
Sunflower oil	13	58	0.3	195	130 s	0.918	325	360	–6	–14
Walnut oil	25	63	0.4	197	120 s	0.918	326	360	–14	–18
Wheat germ oil	16	66	4.9	184	130 s	0.922	316	360	–6	–12
Mineral oil	–	–	–	–	–	0.874	227	232	–7	–12
Diester oil	–	–	–	–	–	0.897	273	281	–23	–28
Triester oil	–	–	–	–	–	0.916	231	239	–21	–38

[a] MUFA, monounsaturated fatty acids; PUFA, polyunsaturated fatty acids; AV, acid value; SN, saponification number; IV, iodine value; ρ, density; FP, flash point; FIP, fire point; CP, cloud point; PP, pour point; d, drying; n, non-drying; s, semi-drying.
[b] Could not be determined.
After ref. [57].

FAs are used as additives in compounded petroleum-based lubricants, as they provide lower friction and add to the load-carrying capacity of the petroleum oil [114]. Among the FAs of SNO, the saturated stearic acid plays a major role in protecting sliding interfaces and has the most significant influence on wear and friction reduction [115].

Physical properties of vegetable oils

The effects of vegetable oil composition and structure on their physico-chemical properties have widely been explored. A great variety of vegetable oils has been evaluated in terms of their thermal, oxidation and low-temperature behaviour for their potential use as base fluids for industrial and automotive lubrication. Understanding the structure–properties–functionality relationship is very important in order to arrive at an oil composition which best suits a particular application. Table 5.10 lists the molecular properties and derived functionalities of vegetable oils.

Vegetable oils possess most of the specific properties required for lubricants (see Table 5.8), such as excellent frictional properties with high lubricity, high VI and low volatility (high flash point due to the high molecular weight of the triglyceride molecules). The VIs of triglycerides (typically 180–275) are higher than those of petroleum-based mineral oils with no additives (typically 50–120), so that triglycerides are to their nature so-called multigrade oils. This is of considerable importance under conditions in which the operating temperature may vary within rather wide limits [116]. The better oil viscosity control decreases the need for expensive VI improver additives. High VIs of triglyceride oils allow the use of less viscous

Table 5.10 Molecular properties and derived functionalities of vegetable oils

Molecular property	*Derived functionality*
• Chain length	• Viscosity (flow properties)
• Conjugation	• Lubricity
• Molecular weight	• Low-temperature behaviour
• Heterogeneity	• Oxidative stability (shelf-life)
• Molecular packing	• Volatility (VOC)
• Reactivity	• Drying (film formation)
• Iodine value	• Adhesion
• Saponification value	• Tack/rub-off
• Acid value	• Biodegradability
• Peroxide value	• Compatibility
• Polarity	• Appearence/colour
• Solvency	
• Hydrophobicity	

base stocks to meet higher temperature requirements in top ring and groove zones of pistons. Reduced friction of triglyceride oils results in improved fuel economy. Detergent and solubility properties help keep moving parts free from sludge and deposits.

As shown in Table 5.9 the density of most oils ranges from 0.874 to 0.925 g/cm^3, except for castor oil (0.957 g/cm^3). The lowest densities are observed for mineral oil and the diester. Castor oil is also exceptional in its dynamic viscosity, namely η = 0.230 Pa s at a shear gradient of D = 124 s^{-1}; other oils range from 0.0458 (walnut oil) to 0.0703 Pa s (sunflower oil) at the same shear gradient. The similarity in most vegetable oil structures means that only a narrow range of viscosities is available for their potential use as lubricants. All the vegetable oils of Table 5.9 exhibit viscosities that are adequate to sustain the basic demands made on a lubricant, which is to have a kinematic viscosity of at least 25 mm^2/s [117]. Also, an inverse proportionality was observed between the shear gradient and the viscosity, which is indicative of structure-based viscosity [118]. While the resulting decrease in viscosity in response to an increase in the shear gradient results in lower friction, this same effect can also lead to wear. Although providing a durable lubricant film the strong intermolecular interactions also result in poor low-temperature properties.

The fire points (FIPs) of the vegetable oils of Table 5.9 lie within a close temperature interval (358–365 °C). Flash points (FPs) differ more widely, with castor oil having the lowest flash point at 300 °C; almond and safflower oil develop an inflammable air–gas mixture at 328 °C. However, all flash points are well above 150 °C. This indicates a low tendency to evaporation, which fulfils yet another basic requirement for possible application as a lubricant [117]. Also the fire and flash point values of mineral oil and synthetic esters meet typical lubricant requirements.

Cloud points (CPs) and PPs vary widely for different oils with extreme CP values for olive oil (6 °C) and almond oil (–24 °C); the latter value is similar to the CPs of di- and triesters (–23 and –21 °C, respectively). However, for most lubricants, acceptable behaviour at low temperatures is defined as an oil solidification point that lies below 0 °C [117]. Among the vegetable oils of Table 5.9 all but olive oil fulfil this criterion. Almond, rapeseed and castor oils show the lowest CP and PP temperatures and thus (in relative terms) the best low-temperature behaviour. Similarly, good low-temperature behaviour was exhibited by the synthetic esters, while the behaviour of the mineral oil was comparable to the other vegetable oils.

The PP is the most important low-temperature property of any oil used as a lubricant. As evident from Table 7.1, and indicated elsewhere (Section 3.6), the cold-flow properties (as expressed by pour points) of most vegetable oil-based lubricants are limiting their applicability as industrial oils unless corrected. For example, HERO has a PP of –16 °C, but undergoes a significant increase in viscosity with increasing temperature. Possible corrective actions are several: pour point depressants (PPDs), chemical or physical modification. It has been shown that microwave-irradiation (MI) and heat treatments such as heat-bodying (HB) affect structure, thermal properties and lubricity of SBO (Table 5.11) [119]. HB and MI oils form a cyclic ring structure with polymerisation which increases viscosity. PP decreases from –9 °C (SBO) to –15 °C (HB-SBO) and –18 °C (MI-SBO) despite higher viscosities. MI oil shows reduced potential as a lubricant. Thermally polymerised oils (heat-bodied or stand oils) are produced mainly from refined linseed and soybean oils or tung oil at 280 °C and high vacuum.

Because of the predominantly saturated nature of its fatty constituents the PP of CNO is very high (24 °C) compared with other vegetable oils, which precludes its use as a base oil for lubricants in temperate and cold climatic conditions. This can eventually be corrected to some extent by using PPDs and/or subjecting CNO to chemical modification (branching) [120]. However, addition of PPDs does not greatly alter the PP of CNO [53, 121]. The PP of CNO can also be brought down to some extent by introducing branched or aromatic hydrocarbon molecules (e.g. styrenated phenol) as additives. Chemical modification is more successful. Esterification with 2-ethylhexyl alcohol of a mixture of 90% CNO and 10% CAS reduces the PP to –3 °C. This improvement is to be ascribed to estolide formation (see Section 6.2.2) [53].

Functional properties of vegetable oils such as stability and viscosity can be altered by chemical and genetic modification. Chemical modification of the fatty chains of SBO affects its lubricant and physical properties [32, 33, 122, 123].

Table 5.11 Effects of heat treatment and microwave irradiation on properties and lubricity of soybean oil[a]

Property[b]	HB-SBO	MI-SBO
Viscosity	+	+
Pour point	–15 °C	–18 °C
Oxidative stability	+	++
CoF	+	+
Wear	0	–
Cold-flow	0	+
Lubricity	0	–

[a] HB, heat-bodied; MI, microwave-irradiated.
[b] Compared with untreated SBO: +, increased; 0, equal; –, decreased.
After ref. [119].

Structure and quality parameters

The FA profile largely determines properties and uses of vegetable oils (see Table 5.4). Optimum applicability requires tailored FA profiles. Screening vegetable oil feedstocks for application in lubrication needs to consider various chemical and physical restrictions (hydrolytic and oxidative stability of oil, winter operability, etc.). Analytical methods in FA chemistry are categorised in terms of structure and quality parameters. Common structure indexes, such as iodine value (IV), oil stability index (OSI), saponification number (SN) and hydroxyl value (HV), are widely used despite the fact that modern analytical methods yield more reliable information. Quality parameters describe components in oils and fats arising from processing, storage and naturally occurring, non-fatty materials. Quality indexes are the FFA, peroxide, anisidine, phosphor and other similar values. Quality indexes were formerly termed processing-related parameters.

The acid value (AV) of a vegetable oil or animal fat is a measure of the FFA level and is determined by titrating FFA (e.g. according to ASTM D 664, ASTM D 974 or AOCS Official Method Cd 3d-63). Commercial grade oils and fats always contain relatively important quantities of FFAs, water and impurities such as polypeptides and phospholipids. Crude vegetable oils and fats frequently contain more than 3 wt% free fatty acids. The AV of good quality vegetable oils is less than 10 mg KOH/g, but amounts to 20–25 mg KOH/g for lower quality oils. The acid number of natural oils and fats, and hence their FFA content, may vary within wide limits even for the same crop, also in relation to oil extraction and storage. Prime quality cottonseed should contain no more than 1.8% FFAs and less than 1.0% foreign matters. Cottonseed, having more than 12.5% FFA, over 20% moisture, and more than 10% foreign matter, is classified as an off-quality product. Table 5.12 lists indicative FFA values for some vegetable oil and fat feedstocks. Waste oils are characterised by a very high AV.

Table 5.12 Free fatty acid content in vegetable oils and animal fats[a]

Feedstock	Approximate wt% FFA	Feedstock	Approximate wt% FFA
Soybean oil	1	*Jatropha curcas* oil	4–10
Rapeseed oil	3	Corn oil	9
Sunflower oil	3.5	Yellow grease	5–20
Crude palm oil	4.5	Inedible tallow	2–35
Coconut oil	5	Animal fats	2–50[b]
Used cooking oils	2–7	Crude rice bran oil	15–60

[a] Indicative values only.
[b] E.g. pig fat, 1.4; low-grade chicken fat, 6; duck fat, 23.

As to chemical characteristics of vegetable oils for use in lubrication, the AV should be as low as possible as otherwise unwanted chemical reactions (corrosion) can occur between the lubricant and components of the technical system. As indicated in Table 5.9, pumpkinseed, wheat germ, sesame and hazelnut oil show high AV (from 6.1 to 2.0 mg KOH/g) compared with other vegetable oils.

The SN pertains to all FAs present in the sample (free and esterified) and is usually determined according to ref. [124] or AOCS Official Methods Cd 3c-9 and Cd 3b-76. To determine the SN, the sample is completely saponified with an excess of alkali, which excess is then determined by titration (in mg KOH/g). The SN depends on the molecular weight and the percentage concentration of fatty acid components present in the oil. Lauric oils, with a higher percentage of ester bonds than longer chain oils, have a higher SN (240–250 mg KOH/g for CNO compared with 190–195 mg KOH/g for SBO). The distinction is quite relevant for lubricant properties of oils. High SN values point to increased ester bonding indicative of a lower hydrolytic stability. In oils with high saponification values, at raised temperatures water can readily split the ester groups from the glycerol ester. Oils with high SN (about 250 mg KOH/g) impart high foamability (e.g. palm kernel, coconut and babassu oils). It is noticed that the reported high SN value (244 mg KOH/g) for safflower oil in Table 5.9 is beyond the usually observed range for this oil [125]. Jojoba oil is an example of an oil characterised by a very low SN value (92–95 mg KOH/g). The difference between saponification number and acid value (SN–AV) is referred to as the ester value (EV, mg KOH/g) and is indicative of the number of ester bonds in the sample.

The SN and IV of an FA composition can be calculated as follows:

$$\mathrm{SN} = \sum\left(\frac{560 \times A_i}{\mathrm{MW}_i}\right) \tag{5.1}$$

$$\mathrm{IV} = \sum\left(\frac{254 \times D \times A_i}{\mathrm{MW}_i}\right) \tag{5.2}$$

where A_i is the percentage, D is the number of double bonds and MW_i is the molecular mass of each component.

The IV, expressed as grams of I_2 absorbed/100 g sample under standard conditions, is taken as a measure of the unsaturation of oils and fats, and can be determined in different ways, including AOCS method Cd 1d-92 [126]. The IV depends mainly on the percentage concentration of unsaturated FA components and the number of double bonds D present in the structure. Determination of the IV gives a reasonable quantification of unsaturation if the double bonds are not conjugated with each other or with carbonyl oxygen [127]. Theoretical IVs calculated from the total number of double bonds using the FA composition tend to be slightly higher (5–10%)

than the IVs determined by titration, due to the presence of unsaponifiables in the crude oils. The IV stands in some relation to oxidative stability and reflects both the propensity of an oil or fat to oxidise and to polymerise and form engine deposits. However, many FA profiles and different FA structures can yield the same IV [128]. As the (old-fashioned) IV index does not take into account the positions of the double bonds, it is nowadays considered as being an unsatisfactory structure index and not necessarily the best indicator of a fuel's oxidation stability. Several new structural indexes have been proposed that better relate structure and amount of common component FAs in vegetable oils to observed properties [128].

On the basis of their IVs, vegetable oils are commonly categorised as follows: drying (IV, 130–200 g I_2/hg), semi-drying (IV, 100–130 g I_2/hg) and non-drying (IV, 80–100 g I_2/hg). Drying oils have the property of forming a solid, elastic substance when exposed to air in thin layers. The drying power is proportional to the amount of unsaturated FAs present. Generally, it is assumed that drying oils are not suitable as lubricants because they harden as a result of reactions between the polyunsaturated bonds and atmospheric oxygen [129]. They are mainly used in paints and coatings, inks and resins. Vegetable semi-drying oils are characterised by forming a skin when exposed to air at somewhat elevated temperatures. Non-drying oils thicken at elevated temperature but do not dry to a skin. They are used as lubricants, heat transfer fluids, etc. (see Table 5.5). The oils of Table 5.9 have been classified according to their IVs. The relatively high IVs indicate a high amount of unsaturated bonding. For non-drying oils, such as castor, hazelnut and olive oils, the PUFA content is only 5–21%, while for drying oils this value lies within the range of 67–74%. The IV and SN values not only provide information about the oxidative and hydrolytic stability of the vegetable oils, but can also be used to calculate the heating value of the oil [130].

The HV (mg KOH/g) is used to determine the hydroxyl content according to AOCS Official Method Cd 13–60.

Thermo-oxidative stability

Stability of plant oils depends on the FA profile, presence of naturally occurring stabilisers (tocopherols) and metal contaminants. Vegetable oils are prone to oxidative degradation and exhibit poor low-temperature fluidity. Oxidation leads both to formation of volatile small-chain FAs, as well as polymers which can develop into deposits and residues.

Oxidation stability covers both storage and thermal stability. The oxidation stability of a vegetable oil depends on its source (structure), processing (extraction) conditions, contaminants (particularly trace metals, water, radicals and peroxides), natural antioxidants and storage conditions. Oxidation processes start immediately during and after production. Storage stability can be influenced by humidity, sunlight, microorganisms, temperature and oxygen. Thermal stability additionally generates gum and solid deposits.

The structure of the FAs determines the lubricant properties and operational stabilities. Although many vegetable oils have excellent lubricity, they possess poor thermo-oxidative, cold-flow and hydrolytic stability (as compared to conventional mineral oil-based fluids). In a performance ranking of base oils [131] vegetable oils score best in terms of evaporation loss and toxicity, but poorest for high-temperature performance and ageing. Consequently, use of neat vegetable oils in many industrial lubrication applications is faced with several problems. In conditions of extreme heat, vegetable oils are subjected to degradative oxidation. Most neat vegetable-based lubricants have a maximum operating temperature of 70 °C, in formulations up to 120 °C [116]. The hydrolytic instability is due to the ester linkage, while C=C polyunsaturation leads to rapid oxidation, and the presence of saturated long alkyl chain results in poor cold-flow properties. The physicochemical properties can be improved by additivation or by chemical modification.

Fundamental knowledge of the oxidative properties of lubricants is necessary to predict the long-term thermal stability of these fluids, which is a critically important lubricant property. The reason for the poor thermo-oxidative properties of vegetable oils lies in their molecular structure, in particular the central β-CH group on the glycerol moiety and the double bonds on the alkyl chains. Oxidative attack on unsaturated fatty acids is initiated by a one-electron transfer process, leading to a radical intermediate which reacts with molecular oxygen in a radical chain reaction; see Fig. 5.4 [132]. The resulting hydroperoxides are further degraded with the production of small molecules (water, alcohols, aldehydes, ketones, carboxylic acids) and condensation products, which increase the viscosity and may even form sediments by polycondensation. Degradation results in volatile low-molecular weight fragments and corrosive breakdown products which impair the properties of the lubricants [133]. Carboxylic acid groups gener-

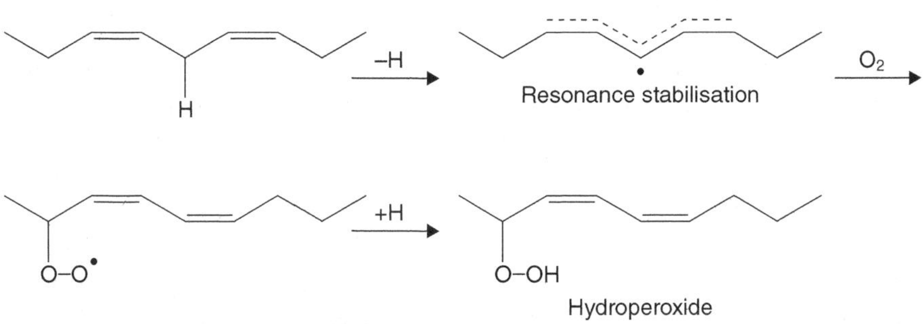

5.4 Oxidation mechanism of vegetable oils.

ated from the oxidation of a triglyceride make the lubricant acidic. These acidic groups have attraction for oxidised metals and can solubilise them in oil promoting metal removal from some surfaces of lubricated metal parts. The amount and nature of unsaturation in the fatty acid affect the susceptibility of the triglyceride to oxidise. Oxidation can include polymerisation reactions that link two or more triglycerides together through reactions of atoms near the unsaturation. These reactions may form higher molecular weight material which can become insoluble deposits (sludge). The increase in oil acidity and viscosity reduce the lubrication functionality. The β-hydrogen atom of triglycerides is also easily eliminated from the molecular structure through oxidation. This leads to weakening of the centrally positioned ester linkage, which will break by heat alone creating a carboxylic acid and a glycerol moiety and further degradation of the oil. The extent of oxidation and formation of oxidation products is influenced by the presence of antioxidants and metals. Additives and blending are frequently used to improve a vegetable oil's oxidative stability [134] and PP [121] (see also Section 7.2.2), but usually with some potential sacrifice to biodegradability and toxicity. Other conventional methods to improve the oxidative stability of vegetable oils are hydrogenation of the fatty acids (see Section 6.2.3). The structure of highly monounsaturated oils can also be modified chemically to improve the functional properties of the oil (see Section 6.2). Finally, advanced plant breeding and genetic engineering are used to develop high-oleic (HO) varieties with improved oxidative stability (see Section 5.3.1).

The high degree of multiple C=C unsaturations in the FA chain of many vegetable oils confines their use as lubricants to a modest range of temperature. The limited viscosity range restricts application as lubricants. Neat vegetable oils usually show unsatisfactory performance in a wide range of lubricant applications, in particular as machine lubricants [135].

Low-temperature properties evaluated experimentally are often used to predict actual lubricant service life at high temperature and in other extreme applications. The more resistant a lubricant is to oxidation, less is the tendency to form deposits, sludge and corrosive by-products in grease, engine oil and industrial oil applications. It is also more resistant to undesirable viscosity increases during use. There are various chemical reaction mechanisms based on free radicals that are involved in the oxidative degradation of vegetable oil-based lubricants and engine oils [136, 137]. ASTM D 943 is a widely used test method to assess storage and long-term service oxidation stability of oils in the presence of oxygen, water, copper and iron at elevated temperature (95 °C). As oxidative degradation starts at a lower temperature than thermal degradation, determination and modification of oxidative stability is more important when vegetable oils are used as base oil for industrial lubricants.

A high degree of polyunsaturates in vegetable oils (e.g. in SBO) significantly diminishes the stability. Autoxidation of unsaturated fatty compounds proceeds at different rates depending on the number and position of double bonds [138]. A study of oxidative stability with respect to the olefinic position of monoenoic fatty methyl esters shows that the Δ5 double bond is the most stable by more than an order of magnitude compared with Δ6, Δ9 and Δ13 FAs. Meadowfoam oil is unique in that the TAG mixture consists of > 95% of FAs with carbon chain lengths greater than 18. The high contribution of *cis* Δ5 FAs (≈66%) determines an exceptionally high OSI value [44]; see Table 5.13. The positions allylic to double bonds are especially susceptible to oxidation. Bis-allylic positions in polyunsaturated FAs (at C11 in linoleic acid and at C11 and C14 in linolenic acid) are even more prone to autoxidation than allylic positions. Relative rates of oxidation are: stearate s:oleates:linoleates:linolenates = 1:10:100:200 (Fig. 5.5).

The presence of various structural units in the FA molecule, such as isolated and conjugated unsaturations, methylene chains, bis-allylic methylene groups, etc., and their relative abundance, makes that oxidation is kinetically and thermodynamically a highly complex process leading to numerous oxidation products involving various intermediates. Different structural parameters participate in the reaction at different stages of oxidation. It is generally observed that the overall activation energy E_a for oxidation, which typically varies between 63 and 89 kJ/mol, and represents the cumulative effect of all the activation energies available in the system during the period of oxidation, is influenced by the degree of polyunsaturation of the vegetable oils. High C18:2 and C18:3 contents in the chain lower E_a, whereas

Table 5.13 OSI values of vegetable oils[a]

Vegetable oil	OSI at 110 °C
Crude meadowfoam oil (CMFO)	246.9
Refined meadowfoam oil (RMFO)	67.3
Cold pressed jojoba oil (PJO)	55.9
Refined high-oleic sunflower oil (HOSNO)	49.8
Refined castor oil (RCAS)	56.1
Crude jojoba oil (CJO)	34.5
Refined jojoba oil (RJO)	31.4
Deodorised jojoba oil (DJO)	23.5
Refined soybean oil (RSBO)	19.9
Reagent grade triolein	8.5
High-erucic acid rapeseed oil (HERO)	8.1
Triolein 99+%	2.7
Triolein 95%	1.8

[a] After ref. [44].

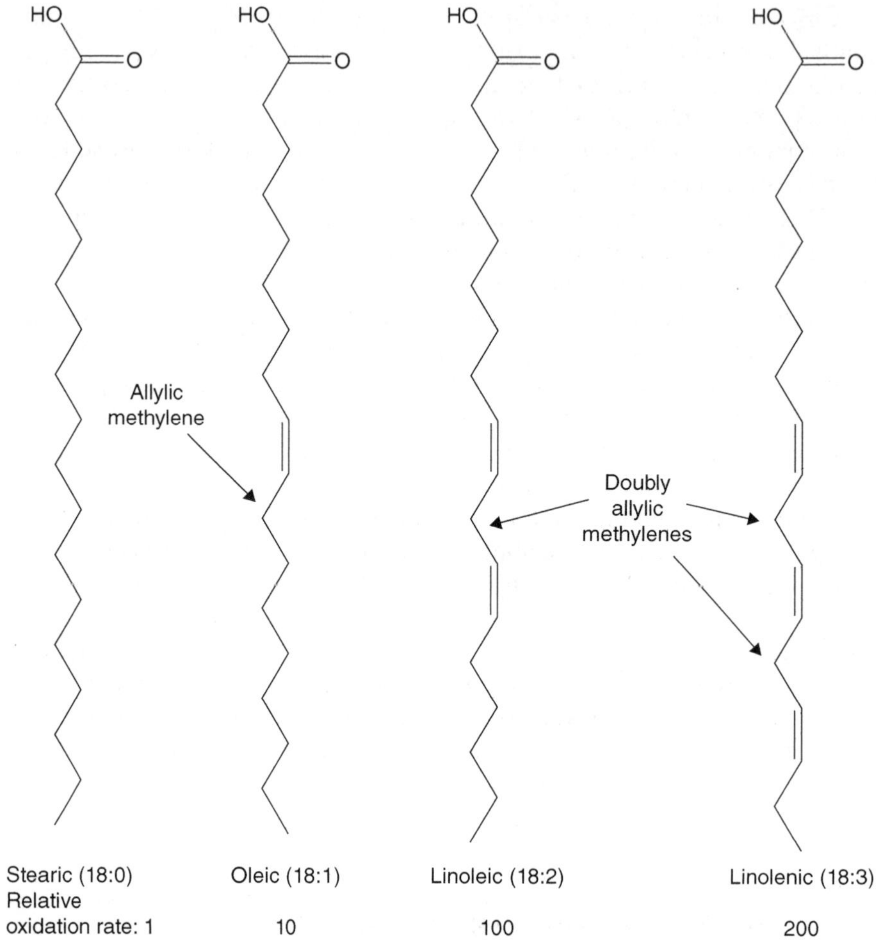

5.5 Relative oxidation rates of stearic, oleic, linoleic and linolenic acid.

high C18:1 (oleic) content increases E_a for oxidation. Higher fractions of saturated —CH_2 carbons improve the resistance to initial thermal breakdown and delay the onset of initial oxidation processes (T_S). However, the presence and abundance of polyunsaturation and saturated methylene chain length of the FAs do not conclusively explain the relative variation of E_a and other kinetic data.

General wisdom holds that high E_a is associated with high percentage of saturated –CH_2 carbons and low percentage of bis-allylic —CH_2 carbons. An increase in the methylene carbons of the FA chains, without being interrupted by —C=C— bonds, increases the thermal and oxidative stability. Protons on bis-allylic methylene groups are highly susceptible to rapid radical attack and initiate oxidative degradation of the FA molecule. An increase in the percentage of olefin proton in the FA chain decreases the E_a of oxidation.

The effect of several structural moieties (as quantified experimentally by compositional gas chromatography (GC) and ^{1}H and ^{13}C nuclear maynetic resonance (NMR) data [139]) on thermal and oxidative properties of fatty acids has been investigated in more detail [140]. The relative abundance of olefin protons, bis-allylic CH_2 groups, allylic CH_2, α-CH_2 to C=O and —CH_2 protons on saturated carbons all influence oxidation and kinetic data of a variety of vegetable oils (CSO, CRO, SFO, high-oleic safflower oil (HOSFO), high-linoleic safflower oil (HLSFO), high-oleic sunflower oil (HOSNO), SBO and SNO) [140]. Owing to the complexity of the thermo-oxidative process, which involves several short-lived species and various intermediate steps, there are many limitations of using a single structural parameter in rationalisation of the observed behaviour. Various structural parameters are needed for correlation with start and onset temperatures of vegetable oil oxidation and kinetic parameters. The oxidation start temperature T_S is the temperature where bond scission takes place to form primary oxidation products in the vegetable oil matrix, whereas the onset temperature T_O represents the temperature of rapid increase in the oxidation rate. Lower T_S and T_O values indicate a thermally unstable matrix. Multicomponent statistical analyses have established the following correlations:

$$T_S\,(^\circ\mathrm{C}) = 54.8 - 13.5A + 1.37B - 3.93C + 21.7D \qquad [5.3]$$

(statistically valid for $100 < T_S < 200\,^\circ$C); and

$$T_O\,(^\circ\mathrm{C}) = 103 - 12.4A - 4.84C + 24.7D + 5.93E \qquad [5.4]$$

(statistically valid for $140 < T_O < 190\,^\circ$C); where A percentage of olefin carbons; B saturated —CH_2; C bis-allylic —CH_2; D ω-2 carbons of saturated, mono- and n-6 polyunsaturated fatty acids; and E terminal methyl carbons [140]. A and D exert maximum influence on both T_S and T_O. Higher monounsaturation improves the thermal/oxidative stability as expressed by T_S and T_O. The effect of terminal methyl groups on T_O is noticed by improvement of oxidative behaviour when branching sites are introduced in the FA chain.

For the kinetic parameters (E_a and k) the following correlations (based on ^{1}H NMR data) have been derived [140]:

$$E_a\,(\mathrm{kJ/mol}) = 112 - 2.12\,A - 3.29\,B + 1.43\,C \qquad [5.5]$$

(statistically valid for $63 < E_a < 89$ kJ/mol); and

$$k\,(\mathrm{min}^{-1}) = 0.0485\,B - 0.0228\,D + 0.102\,E - 0.602 \qquad [5.6]$$

(statistically valid for $0.3 < k < 0.55$ min^{-1}); where A percentage of olefin proton; B allyl —CH_2 protons; C —CH_2 on saturated carbons; D bis-allylic

CH_2 protons; and E α-CH_2 protons to C=O. Equations 5.3–5.6 thus indicate that different structural parameters contribute in different magnitude to the thermal and oxidative stability of vegetable oils. E_a is significantly affected by the allylic —CH_2 protons in the FA chain (B); other influences are on account of unsaturation (A) and chain length of saturated —CH_2 moieties (C). Allylic-CH_2 (B) and bis-allylic CH_2 (D) protons influence the rate constant significantly with some contribution from —CH_2 protons α to > C=O groups (E) (Eqn 5.6). As polyunsaturation in FAs is associated with bis-allylic methylene protons and low percentage of saturated CH_2 carbons (short methylene chain length), it is not surprising that the thermal and oxidative stability of such oils is compromised. Pressurised differential scanning calorimetry (PDSC) experiments have been shown to generate linear log b (heating rate) vs. T^{-1} (inverse of peak height temperature) for all vegetable oils studied (CSO, CRO, SFO, HOSFO, HLSFO, HOSNO, SBO, SNO) [140].

Saturated FAs, such as CNO (coconut oil) and PKO (palm kernel oil), have relatively high oxidation stability, which decreases with increasing unsaturation in the molecule. The oxidative stability of vegetable oils increases with decreasing PUFA content.

Jayadas [141] has compared the oxidative characteristics of nine Indian vegetable oils and a commercial automotive lubricant of SAE 20W-50 grade using thermogravimetric differential thermal analysis (TG/DTA), infrared (IR) and NMR techniques. ^{1}H NMR can be effectively used to predict the oxidative performance of vegetable oils. Results for multicomponent analysis of ΔW (weight gain) and H_{exo} (exothermic heat flow, mW/g) are

$$\Delta W(\%) = -0.037 - 0.005\ (\%\ \text{olefin H}) + 0.0117\ (\%\ \text{allylic H}) + 0.193\ (\text{bis-allylic H}) \quad [5.7]$$

and

$$H_{exo} = -0.803 - 0.205\ (\%\ \text{olefin H}) + 0.137\ (\%\ \text{allylic H}) + 0.448\ (\%\ \text{bis-allylic H}) \quad [5.8]$$

with R^2 = 75.3% and 94.3%, respectively.

As shown above, vegetable oil oxidation is a complex process. Different test methods may be used to get a global view of the oxidative stability of a vegetable oil or derived product, such as oxygen bomb, PDSC and OSI (see Section 8.5.1). The OSI, which determines the oxidative stability of an oil by passing air through a sample under stringent temperature control, is an accepted method for evaluating the stability of vegetable oils [142]. Various other analytical techniques provide insight into the oxidative process.

OSI values of various vegetable oils determined at 110 °C were compared (Table 5.13) [44]. Meadowfoam oil (*L. alba*) is the most stable oil with an

OSI time of 67.3 h for refined oil and 246.9 h for crude oil. Other oils with good oxidative stabilities are refined HOSNO and crude jojoba oil (CJO) with OSI times of 49.8 and 34.5 h, respectively. Several new oil crops such as meadowfoam have superior stability compared to traditional vegetable oils. Factors influencing the OSI time of lipids are the inherent natural antioxidant content, degree of polyunsaturation and the presence of Δ5 double bond entities [143]. Natural antioxidants increase the oxidative stability of vegetable oils. Refined vegetable oils with reduced tocopherol content result in lower oxidative stability (see Table 5.13). Linear relationships of the ratio of total tocopherol/iodine value vs. OSI time at 110 °C are generally reported [44, 144].

The thermal stability of oils is important in defining the operability conditions for lubricants prepared using such vegetable oils [146–149]. In this respect, *Moringa* oil, commercially also known as 'ben oil' or 'behen oil' because of its content of behenic (docosanoic) acid (6.0%), is of considerable interest as it possesses favourable oxidation and thermal stability characteristics in comparison to other oils, including canola oil, CSO, *Jatropha* oil and SNO [145, 150]; see also Table 5.14. Its PDSC onset temperature (OT), which is a measure of oxidative stability, is high. This also holds for OSI. The higher oxidative stability of *Moringa* oil is ascribed to its high MUFA content (75.8%), low unsaturation number (1 × [MUFA] + 2 × [C18:2] + 3 × [C18:3]) compared with other oils and the presence of naturally occurring antioxidants [150]. In comparison, the highest unsaturation number of Table 5.14 for sunflower seed oil makes it least oxidative stable with low OT and OSI values. Canola oil demonstrated superior low-temperature stability compared with *Moringa* oil, as shown by cryogenic dif-

Table 5.14 Oxidation and thermal stability of vegetable oils measured using PDSC, Rancimat and TG

Vegetable oil	OT[a] (°C)	OSI[b] (h)	T_{onset}[c] (°C)	T_{end}[d] (°C)	UN[e]
Moringa	191 ± 0.4	15.3 ± 1.3	347	393	78
Jatropha	169 ± 0.3	2.6 ± 0.05	322	413	115
Cottonseed	159 ± 0.4	1.9 ± 0.04	343	406	120
Canola	164 ± 1.2	3.4 ± 0.01	339	391	124
Sunflower	153 ± 0.7	1.1 ± 0.01	342	403	142

Values are mean ± SD for triplicate determinations.
[a] PDSC onset temperature.
[b] Rancimat oxidative stability index.
[c] T_{onset} = onset temperature of thermal mass loss transition (TG).
[d] T_{end} = end temperature of thermal mass loss transition (TG).
[e] UN = unsaturation number (1 × [MUFA] + 2 × [C18:2] + 3 × [C18:3]).
After ref. [145].

ferential scanning calorimetry (DSC), PP and CP measurements [145]. The high PP (4 °C) of *Moringa* oil is due to the presence of about 23% SFA. More SFA in an oil results in close packing of the TAG molecules during cooling, leading to gel-like structures entrapping low melting molecules also and thus results in high PP and CP. These properties require improvement; a PPD is recommended. The lubricity properties of *Moringa* oil are on a par with other vegetable oils [145]. *Moringa* oil has kinematic viscosity in the range suitable for formulating a lubricant of ISO viscosity grade 32. The VI of *Moringa* oil (239) is much higher than for many other vegetable oils (e.g. CO 219, CSO 219, JCO 213, SNO 218) [145], making it suitable for use as multigrade lubricant. *Moringa* oil has high potential in formulation of industrial fluids for high-temperature applications.

Table 5.15 and Fig. 5.6 show the inverse relationship between oxidative stability and fluidity of plant oils. Vegetable oils with high levels of saturates (SFAs) have poorer cold-weather operability but better oxidative or storage stability. Whereas petroleum-based hydraulic fluids function satisfactory at −25 °C, in cold weather vegetable oils have the tendency to solidify more readily, as expressed by a higher PP. For vegetable oil based lubricants cloudiness, precipitation, poor flowability and pumpability, and even solidification at −15 °C may occur. Low-temperature behaviour of TAGs relates first of all to their (complex) crystallisation kinetics. It is firmly established that the presence of *cis* unsaturation, lower molecular weights, and diverse chemical structures of TAGs favour lower temperatures of solidification. In the industry one major characteristic of the low-temperature properties of lubricating fluids is PP. Effects of dilution of vegetable oils with major biodegradable fluids, namely poly-α-olefin (PAO2), diisodecyl adipate (DIDA), and oleates, as well as PPDs, were investigated [121]. Dilution with oleates is less effective than with PAO2 and DIDA. Addition of 1 wt% PPD depressed PPs down to −33 °C for canola oil and to −24 °C for HOSNO. No further depression could be achieved. Low-temperature performance of

Table 5.15 Low-temperature properties vs oxidative stability of fatty acids and triacylglycerols

Property	Structure			
	C18:0	C18:1	C18:2	C18:3
Iodine value (TAG)	0	86	173	262
Molecular weight (TAG)	892	886	880	874
Melt temperature (FA)[a] (°C)	71	16	−5	−11
Relative oxidation rate (FA)	1	10	100	200

[a] Ref. [151].

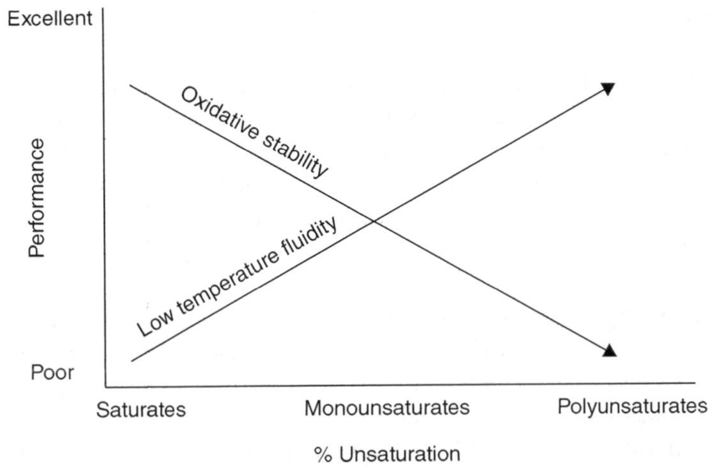

5.6 Oxidative stability vs fluidity of plant oils.

vegetable oils thus continues to limit their prospects as biodegradable lubricants, although adequate use of diluents and PPDs can deliver some improvements.

Since at least one *cis*-(*Z*) double bond is essential for the required low-temperature behaviour, a high oleic acid (C18:1) content is optimal. Standard RSO and SBO do not meet this qualification and require either extensive chemical modification or additivation. While polyunsaturation in the FA chain accelerates oxidative degradation it reduces evaporative loss. Synthetic or PAO-based fluids are considered an option at lower temperatures.

Upon high thermal loading, a consistency change was observed for all the vegetable oils of Table 5.9 except for safflower, sesame, walnut and wheat germ oils which remained in their fluid state. The other oils solidified. No significant viscosity changes for wheat germ oil were measured before or after thermal loading. Walnut, sesame and safflower oils showed a distinct viscosity increase (from 5-fold to 14-fold) upon ageing. Surprisingly, the semi-drying or drying oils passed the ageing tests while, as expected, the non-drying oils did not. The mineral oil showed the same performance as wheat germ oil, with no significant viscosity change. In contrast, the diester showed a 2-fold viscosity increase after the ageing test, while the triester showed a marked consistency change during testing.

Ageing of bio-based fluids or loss of performance is related to three dominating processes: (i) oxidative attack at the bis-allylic C—H positions, mainly of PUFAs; (ii) hydrolytic degradation of ester bonds by water; and (iii) thermal degradation.

Tribological performance

All the bio-based fats, oils and their derivatives, have better lubricities than #2 diesel fuel. All bio-based oils and fats also show a better lubricity than a turbine drip mineral oil (control) [152]. Crude vegetable oils exhibit significantly better lubricities than their refined or transesterified (methylated or ethylated) forms. The type of fatty acid (saturated or unsaturated) or its concentrations do not correlate with the lubricity value of a particular oil, fat or their derivatives.

Whereas vegetable oils without additives outperform mineral base oils in anti-wear and friction [74, 153], scuffing load capacity [154] and fatigue resistance [155], native vegetable oils are not fully satisfactory for lubrication purposes, especially as machine lubricants. In comparison to mineral oil formulations fully formulated vegetable oil lubricants display a lower coefficient of friction (CoF), equivalent scuffing load capacity and better pitting resistance, but also poorer thermal and oxidative stability. At extreme loads vegetable oil-based lubricants become significantly less effective [156]. In short, vegetable oils have very good lubrication properties but these need improvement for applications in the bulk lubricant market.

Seed oils are amphiphilic with a hydrophilic polar head and a non-polar hydrophobic long hydrocarbon chain (see Fig. 5.1). The polar end constitutes the ester functional group(s). Because of their high dipole moments, carboxylic groups reduce the volatility and increase the flash point of lubricating oils, enhance the thermal stability, solvency, lubricity and biodegradability; however, they negatively influence the hydrolytic stability and reactivity with copper or lead-containing metals [157]. The non-polar tails are hydrocarbons of varying chain lengths, degrees of unsaturation and stereochemistry. Depending on the type of seed oil, various functional groups (hydroxide, epoxide, keto, furanoid, etc.), may be present in the hydrocarbon chain. The tribological and other important physicochemical properties of seed oils such as oxidation, low-temperature stability and rheology are highly dependent on their FA distribution, composition and additional functional groups in the chain structure. The chemical structure is the predominant factor that determines the relative polar attraction of the molecules for the active sites on the surfaces in motion and how they interact with the surface materials. This polarity of the molecules affects their migration or rate of diffusion to the surface; all components of the lubricant formulation compete for these active surface sites. The amphiphilic character of seed oil molecules allows for adsorption on metal surfaces and the separation between the dynamic mechanical surfaces. Amphiphilic properties affect the boundary lubrication or additive properties while fluid or rheological properties affect the hydrodynamic properties of seed oils. Most vegetable oils are functional fluids with at least one functional group (an ester) and are liquid at room temperature, at variance to animal fats. These oils are essentially water insoluble.

The excellent lubricity provided by vegetable oils is due to their ester functionality [158]. The polar heads of the fatty acid chains attach to the metal surfaces by a chemical process that allows monolayer film formation with the non-polar end of fatty acids sticking away from the metal surface. The fatty acid —CH_2— chains offer a sliding surface that prevents direct metal-to-metal contact. The polar nature of triglycerides accounts for their superb ability to be adsorbed on metal faces as a very thin adhering film. The adhering nature of the film ensures lubrication while the thin nature allows for parts to be designed with less intervening space for lubricant. Without film formation, contact may result in rising temperature at the contact zones of moving parts causing adhesion, scuffing or even metal-to-metal welding. During this rubbing process under lubricated conditions at high load and low speed, bond cleavage of fatty acid molecules might take place. Film-formation properties of triglycerides are particularly advantageous in hydraulic systems. In addition, water cannot force an adhering triglyceride oil film off a metal face as easily as a hydrocarbon film.

Limitations on the use of vegetable oil in its natural form as an industrial base fluid or as an additive relate to poor thermal/oxidation stability, poor low-temperature behaviour and other tribochemical degrading processes that occur under severe conditions of temperature, pressure, shear stress, metal surface and environment. To meet the increasing demands for stability during various tribochemical processes, the oil structure has to withstand extremes of temperature variations, shear degradation and maintain excellent boundary lubricating properties through strong physical and chemical adsorption with the metal. The film-forming properties of triacylglycerol molecules inhibit metal-to-metal contact and progression of pits and asperities on the metal surface. Strength of the protective fluid film and extent of adsorption on the metal surface dictate the efficiency of a lubricant's performance. Friction coefficient and wear rate are dependent on the adsorption energy of the lubricant [159].

Vegetable oils can be used in most lubricating regimes: boundary, hydrodynamic and mixed film lubrication [160–162] albeit with some major limitations. Understanding of their chemistry in relation to fluid and boundary properties is essential in successful application of vegetable oils in lubricant formulations. The effects of seed oil chemistry on their boundary lubrication properties have been investigated [163–165]. Vegetable oils are particularly effective as boundary lubricants as the high polarity of the entire base oil allows strong interactions with the lubricated surfaces. Boundary lubrication performance is affected by attraction of the lubricant molecules to the surface and also by possible reaction with the surface. Adsorption requires interaction of the polar ester functional groups of the vegetable oils with the friction surfaces, as quantified by the free energy of adsorption (ΔG_{ads}). When amphiphiles adsorb on surfaces, two types of interaction occur: (i) adhesive interaction between polar groups and surface (depending on type and number of functional groups); and (ii) lateral dipole–dipole interactions. As shown in Table 5.16, triglycerides have consistently lower ΔG_{ads} values than monoesters. Consequently, the former are superior to the latter for effective metal adsorption and boundary lubrication properties. PUFAs display better lubrication properties at higher temperatures and loads than less unsaturated counterparts. Higher oxygen concentrations in the oil improve the lubrication properties, denoting the importance of reactions on boundary lubrication performance [166].

Also several additive classes show high polarity, not unlike base oils consisting of vegetable oils or synthetic esters. This can result in a competitive reaction on the metal surface. Several studies have been made of the role of the base fluid (vegetable oils, esters, hydrocarbon fluids) and additives in the boundary regime [75, 167, 168]. It is generally observed that friction is reduced by increasing the effective chain length of the base oil molecules. On the other hand, some additives functioned better in high unsaturated vegetable oils while others are more effective in fully saturated base fluids.

Poorly understood tribological reactions refer to chemical interactions between oil and the environment (metal, oxygen, moisture) in the interface or friction zone due to shear, high temperature and pressure. Such interac-

Table 5.16 Free energies of adsorption, ΔG_{ads}, of seed oils, computed from friction-derived adsorption isotherms using the Langmuir adsorption models[a]

Seed oil	ΔG_{ads} (kcal/mol), Langmuir model
Canola[b]	–3.81
Cottonseed[b]	–3.71
Olive[b]	–3.98
Meadowfoam[b]	–3.57
Safflower[c]	–3.65
Soybean[d]	–3.60
Methyl oleate[c]	–2.91
Methyl palmitate[c]	–2.70
Methyl laurate[d]	–1.90

[a] Data obtained from friction measurement on steel/steel friction surface using ball-on-disc test geometry.
[b] Ref. [163].
[c] Ref. [165].
[d] Ref. [164].

tions may lead to oxidation, hydrolysis, degradation of the oil, generation of friction polymers, etc.

The tribological performance of vegetable-based lubricants was also investigated in terms of the influence of the fatty acid composition [169]. Friction and abrasion rate under sliding wear at ambient conditions are less severe for SBO than for SNO. This has been ascribed to the lower contents of linoleic and oleic acids in SBO.

Tribological testing of the oils of Table 5.9 using a reciprocating tribometer according to the established standard DIN 51834 were performed under mixed lubrication conditions, where solid–solid contact is possible if oil lubrication is inadequate [57]. Distinct differences were observed with regard to lubricity of the various vegetable oils under adhesive wear conditions (100 Cr6/100Cr6 material combination). Only linseed, walnut and wheat germ oil showed a stable low friction coefficient ($\mu \sim 0.11$). The remaining oils exhibited unstable friction behaviour versus time. Amongst these oils, less friction peaks occurred with safflower and olive oils, which should lead to lower wear than the other oils. Pumpkinseed oil showed the lowest wear protection with a total wear volume W of 19.7×10^{-3} mm^3; olive oil unexpectedly exhibited the best wear protection ($W = 9.3 \times 10^{-3}$ mm^3). Neither the friction nor the wear performance of the various vegetable oils was as good as that of the mineral oil or the synthetic esters, which showed superior overall performance. Only the friction behaviour of linseed oil was comparable. The total wear performance of the triester oil was far better than that of any of the vegetable oils, followed by that of the mineral oil and the diester oil. Thus, under adhesive wear conditions the performance of natural vegetable oils, without additives, cannot match that of mineral oil and synthetic esters [57].

Under abrasive wear conditions (Al_2O_3/100Cr6 material combination) all oils showed steady and low friction values ranging from $\mu = 0.11$ to 0.13. However, significant differences were observed in wear behaviour. Walnut and linseed oils yielded 2- to 2.5-fold greater wear volumes than sesame and castor oils, which showed the least abrasive wear damage. The abrasive behaviour in terms of friction and wear performance of the vegetable oils was comparable to that of mineral oil and synthetic oils. While the diester oil still showed the lowest wear, it was not significantly less than the sesame, castor, soybean and peanut oils. When lubricated with triester and mineral oils, wear actually exceeded that of these vegetable oils. The tribological functionality of a vegetable oil (friction decrease, wear protection effect) is strongly dependent upon the material combination (100Cr6/100Cr6 *vs.* Al_2O_3/100Cr6).

The tribological properties of CNO were evaluated by Jayadas *et al.* [54] using a four-ball tester and a test rig to test the (extensive) wear on two-stroke engines. Addition of anti-wear/extreme pressure (AW/EP) additives effects a considerable reduction in wear with coconut oil as 2T oil. Certain vegetable oils, such as sesame oil and wheat germ oil may have potential for use in selected technical applications [57].

Being amphiphilic fluids, vegetable oils can be used both as base oils and as film strength additives in lubricant formulations. However, for some lubricant applications neat vegetable oils are not adequate to provide all the necessary needs. Various methods are being pursued to overcome the shortcomings of plant-based oils for various lubrication applications. One approach is to blend plant-based oils with non-polar synthetic or petroleum-based oils (e.g. PAOs) that have complementary properties [170]. Blending with non-polar fluids strongly affects the surface and interfacial properties and therefore various lubricant functions [171]. Such an approach could result in a lubricant with acceptable lubrication properties, which may or may not be fully biodegradable. Blending allows vegetable oils to be used in lubrication applications that until now were possible only with synthetic lubricants.

In mineral oils, FFAs are considered as one of the classical boundary lubrication improvers [172, 173]. The boundary lubrication performance of FFAs in vegetable oils has been extensively studied [115, 156, 174–177]. Cao *et al.* [174, 176] assessed the lubrication properties of RSO formulations containing alcohols, FAs and sulphurised FAs. Minami *et al.* [177] compared the performance of a range of organic compounds, including acids, amines and thiols, in HOSNO in various conditions. Vižintin *et al.* [156] tested fatty acids and sulphur-phosphorus anti-wear additives in RSO-based hydraulic fluids. It was generally concluded that FAs are the most effective boundary lubrication improvers in vegetable oils at low to medium loads. Fox *et al.* [115] have examined the influence of FFAs (C18:0, C18:1 and C18:2) on the boundary lubrication performance of SNO. Stearic acid was the most effective boundary lubrication additive in SNO, reducing wear and providing a steady reduction of the coefficient of friction; however, the performance is limited above 150 °C. Formulation of RSO and SNO with stearic acid as an additive improves the AW properties of these vegetable oils under boundary conditions at different temperatures [175].

Structure–performance relationships of plant oil-derived lubricant base fluids

The molecular structure of base fluids affects important operational properties of lubricants such as viscosity, VI, hydrolytic and oxidative stability, and low-temperature performance such as PP. In summary, viscosity of the base fluid generally increases with chain length of the carboxylic acid and the alcohol, but is also affected by the presence of hydroxyl functions. Castor oil (>88 % 12-hydroxyoleic acid) is characterised by much higher viscosity than native oils without hydroxyl groups. Viscosity also depends on the employed polyol with pentaerythritol (PE) > TMP > glycerol (GL) > neopentylglycol (NPG). The VI, which describes the dependence of viscosity on temperature, increases with chain length of carboxylic acid and alcohol, and decreases with branching. Base oils based on natural fatty acids generally have high VI and are considered as multigrade oils.

Although viscosity is the most commonly used parameter to assess the effectiveness of a lubricant, it alone does not predict functionality sufficiently, especially in field conditions. Inherent differences of the chemical composition of bio-based oils greatly affect how they perform. Lubricity is another important parameter to consider in the determination of a lubricant's effectiveness. Using the method of optical interferometry, Biresaw and Bantchev [178] have investigated the effect of chemical structure on film-forming properties of seed oils. In the low entrainment speed region, film thickness of seed oils (i) rarely correlates with viscosity as proposed by the Hamrock-Dowson (H-D) equation, (ii) increases with decreasing polarity of the oil, and (iii) increases with decreasing degree of unsaturation of the oils. On the other hand, the H-D equation adequately describes film thickness in the high entrainment speed region.

Good low-temperature properties require a low saturated FA (C16:0, C18:0) content and/or short chains. Unsaturated FAs display excellent low-temperature properties. PPs of simple esters derived from saturated and unsaturated fatty acids of the same carbon number differ dramatically. PP is improved (lowered) with increased shortening and branching of the chain.

Crystallisation onset and wax appearance temperatures for a series of vegetable oils (natural, genetically and chemically modified), as determined by DSC, relate as follows:

$$T\,(^\circ\mathrm{C}) = 32.9 - 10.9\,(\%\ \text{olefin H}) + 13.6\,(\%\ \text{bis-allylic } CH_2) - 1.2\,(\%\ \text{allylic } CH_2) \qquad [5.9]$$

and

$$T\,(^\circ\mathrm{C}) = 91.2 - 15.8\,(\%\ \text{olefin H}) + 17.9\,(\%\ \text{bis-allylic } CH_2) - 1.5\,(\%\ \text{allylic } CH_2) \qquad [5.10]$$

respectively, with coefficients of determination (R^2) of 0.98 and 0.93 [179].

Hydrolytic stability strongly depends on the ester moiety. Saturated esters with straight-chain components are generally more stable than unsaturated or branched structures. Oxidative stability is favoured by a low PUFA content (notably C18:2 and C18:3). One double bond (as in C18:1) is essential for good low-temperature properties without affecting the oxidative stability. Fully saturated esters exhibit excellent oxidative stability. Partially unsaturated systems need improvement of oxidative stability performance for applications in automotive (engine, transmission fluids), diesel and industrial (hydraulic, compressor) lubricants.

Vegetable oils with particularly good tribological properties, namely good boundary friction lubricity, general wear protection with stable viscosity–temperature behaviour and very low evaporation, and high biodegradability, are excellent raw materials for the formulation of ecologically friendly lubricants. A major limitation is their inadequate ageing resistance. Consequently, they are perfect products for total-loss applications in which the lubricant enters the environmental cycle (e.g. chain saw oils), but less so for circulation lubricant systems.

Bio-based synthetic fluids can be chemically custom designed for a specific application. Computer-generated structural modelling of a representative molecule and subsequent computation of equilibrium energy can assist in the design of new lubricants at the molecular level and often provides helpful information prior to actual synthesis. In particular, molecular modelling of the effects of chemical modification in desired compounds and computation of their minimum energy profile, steric environment, electron charge density distribution and quantitative structure–property relations can give valuable information on the physicochemical performance parameters prior to expensive chemical synthesis. Modelling TAGs in an effort to develop synthetic fluids is difficult because of the high molecular weight of vegetable oil molecules and conformations of the fatty acid chain structures [180].

Data from molecular modelling can be used to predict important properties of bio-based derivatives. Tan *et al.* [181] reported on the use of molecular orbital indexes criteria to study interaction between lubricant polar end groups and metal surface. Jabbarzadeh *et al.* [182] investigated the effects of branching on rheological properties and behaviour of molecularly thin liquid films of alkanes in a thin film lubrication regime. The dynamics of the molecules and their orientation are affected by the degree of branching. Molecular dynamics was also used to predict the tribological properties of coconut oil in a qualitative manner on the basis of the chemical and physical characteristics of the constituent FAs [183]. Konno *et al.* [184] have used computational chemistry to predict the viscosity of lubricants.

Synthesis efforts of new seed oil derivatives for product development have taken advantage of molecular modelling [185]. Efforts were made to understand the high- and the low-temperature behaviour of seed oils. Ring opening of the TAG epoxy group (in epoxidised soybean oil, ESBO) and subsequent derivatisation of the epoxy carbons can improve the oxidation and low-temperature stability of soybean-based lubricant base oil [180].

Boundary lubrication phenomena are usually associated with adsorption and tribochemical reaction occurring on the metal surface [186–189]. Adsorption refers to the ability of lubricant molecules to attach to the friction surface and therefore prevent their contact during a tribological process. Adsorption occurs mainly through the polar head groups of the molecules, as quantified by the free energy of adsorption (ΔG_{ads}). Tribochemical reactions with formation of a tribo film result from a complex interaction of the lubricant with other materials (oxygen, moisture and metal) in the friction zone in conditions of high temperature, pressure and shear. Oil degradation and generation of friction polymers are often responsible for mechanical failures.

The effects of seed oil chemistry on their boundary properties have been investigated [164, 165]. The effects of degree of unsaturation of the FA residues [164] and degree of functionalisation [165] of the seed oils on free energy of adsorption have been described. Whereas the degree of functionalisation greatly influences the adsorption properties of seed oils, the degree of unsaturation only has a minimal effect [164, 165]. Chain length effects on boundary lubrication properties have also been described [163, 187]. In particular, the adsorption properties of CSO, canola, olive and MFO oils, which have different average chain length, were investigated in terms of free energy of adsorption (ΔG_{ads}) for lubricant formulations consisting of seed oils dissolved in non-polar hexadecane as base oil using the Langmuir model as a function of average chain length of the triglyceride structure [163]. The effect of seed oil is observed in drastically lowering the CoF between the metal surfaces in relative motion with increasing concentration reaching a steady state at about 0.1 M; beyond this point the CoF is independent of additive concentration. The polar heads of triglycerides (ester groups) are adsorbed to the metal surface through adhesive interaction making a unimolecular barrier, resulting in a sharp decrease in CoF. Friction-derived adsorption isotherms can be obtained from boundary CoF versus concentration data, which can be analysed according to Langmuir. A prediction model for ΔG_{ads}, based on a limited data set on adsorption behaviour of triglycerides (canola, CSO, olive, MFO, SFO, SBO) (see Table 5.16) was obtained as follows:

$$\Delta G_{ads} = -3.56 - (0.00918 \times \text{C16:0}) + (0.0462 \times \text{C18:0}) - (0.00553 \times \text{C18:1}) \qquad [5.11]$$

The triglyceride chemistry affects the ΔG_{ads} values of seed oils in several ways:

- the degree of functionality or molecular polarity (mono- to triesters) exerts a strong inverse effect;
- the degree of unsaturation has a strong direct effect;
- average C_{18} chain lengths favour strong adsorption on the metal surface (lower ΔG_{ads}); and
- oleic acid content has a strong inverse effect.

Equation 5.11 may be used for design of suitable lubricant molecules with optimum structure for effective metal adsorption and excellent boundary lubrication properties. For over 150 vegetable oil and animal fat compositions [190, 191] optimal ΔG_{ads} values (<–3.80), calculated according to Eqn 5.11, were noticed for high-oleics (PNO, RSO, SBO, SFO, SNO), low-linolein and low-erucic RSO, but also for a wide variety of natural oils, ranging from coriander (–3.96) to buriti palm, avocado and hazelnut (*C. avellana*) (–3.90), Macadamian nut (–3.86), algae, olive and tigernut (*C. esculentus*) (–3.84), and canola, bacuri and pecan oil (–3.80). For structure–property relationships of ester base fluids, see Section 10.5.3.

5.3 Industrial oil-crop engineering

Biomass production needs improvement as well as development of technologies to obtain the maximum benefit at an acceptable cost. Advances in plant sciences and genetics are required to develop vegetable oilcrops having increased yield and utility. Plant breeding is a core technology traditionally mainly aimed at improvement in the yield performance of useful plants. Although many annual crops have been subjected to continuous domestication, this is much less so for perennial species. In recent years, knowledge of the biochemical relationships of plant metabolism – in particular, of the biosynthesis of the storage fats in commercial use – has increased considerably [192]. Efforts may now be directed to meet the demands of industry for tailor-made oils and fats. Breeding allows propagation of superior plant varieties which are vital to ensure the sustainability of food/feed/fuel and oleochemical supplies. While in the past mutation breeding has frequently proved to be successful in oil plants, the advent of agricultural biotechnology, including genetic engineering, has now further extended the range of tools available. Gene technology is a universal approach for the introduction of new genetic variation in cultivated plants, allowing the breeder not only to change the amount and composition of the stored oil but also to introduce totally new qualities [193]. In designing oilcrops, of particular interest are medium to very long chain fatty acids (VLCFAs) as well as FAs with unusual functionality resulting from the number and position of double bonds or the presence of hydroxyl, oxo or epoxy groups [194]. Transferring a novel oil quality into a high yielding, agronomically adapted plant species is a challenging task. Manipulation of fatty acid and TAG synthesis for the purpose of usability for both nutritional and industrial applications by means of natural breeding methods and genetic engineering techniques rely on detailed knowledge of biochemical pathways.

Alternatively, plant oil fatty acid compositions can be modified chemically (see Chapter 6) to make value-added products useful as lubricants.

Modern synthetic methods together with enzymatic microbiological methods have led to an extraordinary expansion in the potential for the synthesis of novel FAs, which are selectively modified. These deserve further attention for their action, properties and new applications.

Breeding and biotechnology are the primary methods used to increase crop yields. Yield gains are considerable and range from 10 (1925) to 40 bushels/acre (2005) for soybean and from 1500 (1930) to 9500 kg/ha (2005) for corn. Soy production in Argentina has almost tripled in the last 15 years, increasing sharply after the introduction of GM soy in the 1996–97 season [195]. At present, 100% of the country's soy production (used for food/feed/fuel) is genetically modified. In the United States approximately 80% of soy and corn are GM crops. Genetically modified crops can ultimately decrease the cost of feedstocks.

Genetic modification of crops is a highly controversial issue in some circles. There are certain barriers to the realisation and use of gene technology (gentech) because of poor acceptance by some end-users. This applies especially to the nutritional and feed areas. However, there exists overwhelming scientific evidence and positive commercial experience of biotech crops around the world with no fewer than 125 Mha of land in 2008 in 22 countries (mainly US 62.5 Mha, Argentina 21 Mha, Brazil 15.8 Mha, Canada 7.6 Mha, India 7.6 Mha, and China 3.8 Mha) producing genetically modified crops, in particular soy (58.6 Mha), corn (25.2 Mha), cotton (13.4 Mha) and rapeseed (4.8 Mha) (2006 data) [196]. GM crops represent 71% of global soy and 29% of global corn. While the United States has a substantial list of authorised oil-bearing GM crops (soybean, rape, cottonseed, maize), which are being widely grown and mainly modified in view of the need for sturdier crops (resistance to pests, floods, drought), few GM crops are being grown in the EU (with France, Germany, Italy, Austria and Luxemburg being notorious anti-GM countries). Germany has banned GM corn as from 2009.

The EU is over-cautious with its regulatory regime (2001/18/EC, 2003/1829/EC, 2003/1830/EC and 2003/1946/EC). In Europe, honey with traces of GM materials is considered technically as food produced from GM organisms, even if the GM material got there by accident! Progress on GM approvals and crop growing in the EU is slow. Current authorisation licences for cultivating GM crops are valid across the whole of the EU. The EU is presently aiming for more GM crop freedom. Under the proposals, crops would still have to be approved by the EU on health and environmental grounds but then it would be left to individual countries to decide whether to allow them to be grown at a national level. Recently, the European Commission has approved the release of some GM products, including MON801 maize. The EU policy as to GM organisms is disproportionate and damages European competitiveness. European gentech suffers from an unfavourable socio-political climate. Genetic modification of plants in

Europe will get new chances when used not for food products but to offset high energy costs. Genetic modification is especially important for increasing the share of second and third generation bio-feedstocks.

First generation GM crops have been developed for various purposes such as insect resistance. The reason that a farmer chooses for GM crops is not necessarily an increase in production but may also be a reduction of risk and/or an increased flexibility of operations. The sustainability of current GM and non-GM crop production (soybean, maize and cotton) has recently been evaluated [197]. As the term 'GM crops' encompasses a broad diversity of traits and crops with different goals and accompanying effects no single value of sustainability can be given that is valid for all GM crops under all conditions. The contribution of GM crop production to sustainability (in 'Profit' and 'People' terms) is highly dependent on local legal and institutional systems. Bt maize and Bt cotton, cultivated according to Good Agricultural Practice (GAP), generally contribute to sustainability in the 'Planet' sense.

The concept of identity preservation (IP) stands for traceability of agricultural products from seed and farm field to delivery to the consumer in order to preserve some unique characteristics of added value that must be identified and maintained [198]. Clearly, IP has been accorded greater importance in the world trade of agricultural commodities with the introduction of hybrid and GM organisms in crop growth, in relation to their segregation from non-GM commodities. Table 5.17 lists the requirements for IP oil development. Costs of IP GM materials will be higher in view of the special handling and isolation from the unmodified products. The timeline for IP development is about 7 years, and includes various processes such as transportation, regeneration, testing for gene expression, genetic analysis, seed increase, environmental stability, yield crosses, microscope culture, yield testing, etc. Genetic work is slow.

Biotechnologies are vital in addressing the growing demand for crops for food, feed, fibre, fuel and feedstock for the chemical industry, their primary production and processing. If the production of industrially important FAs in high-yielding oilcrops can be developed successfully through genetic

Table 5.17 Requirements of identity preserved (IP) oil development

- Genetic analyses
- Trait stability testing
- Variety development
- Hybrid development
- Disease resistance
- Oil yield
- Herbicide tolerance
- Fungal tolerance

engineering approaches, then there is a huge market volume available. Given the apparent worldwide limitation on land suitable for oilseed production, genetic enhancement of oilseed productivity is a means of achieving the quantity and quality of oils needed to accommodate both food markets and the potential demand for oleochemicals and biofuels. However, GM oils as lubricant base stocks could meet the same public resistance in the future as have food products.

The synthesis of novel fatty compounds based on oils and fats has recently made important progress. With the breeding of new oil plants – including the use of gene technology – numerous fatty compounds of adequate purity are now available, which makes them attractive for synthesis. By modifications of existing oilcrops or through selection of new crops higher yields of one fatty acid as well as fatty acids with uncommon structures are often obtained. The genotype of the biological source does not change the chemical structures of individual molecules – only their relative abundance. Gene technology is being used to modify nature's traditional oilcrops into advanced agroenergy crops with increased oilseed productivity and customised fatty acid profile [206, 207]. Initial applications of modern bio- and gene technology are more readily introduced in the technical and chemical domains, as opposed to poor acceptance in the nutritional and feed areas. Table 5.18 shows some commercially available and extreme fatty acid variants in breeding material from important oilseeds, with special reference to high-oleics. Rapeseed is highly accessible to biotechnological methods. The manipulation of oil synthesis in rapeseed to produce specific, desired, chemically altered triglycerides has already been implemented on a large scale [31]. For industrial use, erucic acid must be obtained in high concentration in order to reduce purification costs. Thus, increasing the erucic acid content in *B. napus* (rapeseed) by genetic engineering has been a goal for several groups. *Crambe abyssinica* and HEAR (*B. napus*) oilseed, which contain 59.5% and 42% erucic acid, respectively, are used for the production of erucamide. The breeding of high-oleic (>80%) rapeseed forms has been achieved by induced mutation [200] and genetically by inhibition of the inherent 12- or 15-desaturase genes [201].

Whereas industrial rapeseed traditionally refers to any HERO variety, the low-erucic (LE) acid high-oleic (HO) varieties (with some 60% C18:1) are of interest to the food industry in view of their high nutritional qualities and low level of saturated fat. Since commodity LE rapeseed is now widely available with supplies beyond that needed for food purposes, LE RSO (canola) is now the source of oil for some non-food applications, including biodiesel and biolubricants. Roughly 60% of RSO production finds non-food applications.

Genetic modification for obtaining products for the chemical industry is still at the beginning. A recent Home-Grown Cereals Authority (HGCA)-

Table 5.18 Conventional and some extreme fatty acid mutants in breeding materials from important oilseeds

Type	Mutant	Origin	Composition						Source
			16:0	18:0	18:1	18:2	18:3	Others	
Rapeseed	Conventional (HERO)	Natural	3	1	11	12	9	64	[199]
	0 or 00 (Canola)	Natural mutation	4	2	60	21	10	3	[199]
	High oleic (HO)	Mutagenesis	4.2	2.2	80.2	4.5	5.2	3.7	[200]
	High oleic (HO)	Gene technology	4.3	1.4	84.1	5.2	2.9	2.1	[201]
Sunflower	Conventional	Natural	7	5	19	68	–	1	[199]
	High oleic (HO)	Mutagenesis	3	4	83	10	–	–	[199]
	High oleic/low saturated	Mutagenesis	3.2	2.4	92.1	2.3	–	–	[202]
Soybean	Conventional	Natural	11	4	23	54	8	–	[199]
	High oleic (HO)	Gene technology	6.6	3.6	84.9	0.6	1.9	2.4	[203]
Peanut	Conventional	Natural	12	4	47	31	–	6	[204]
	High oleic (HO)	Natural mutation	6	2	81	3	–	8	[55, 204]
Palm olein	Conventional	Fractionation	41.1	4.2	41.5	11.2	0.3	1.7	[205]

coordinated project (project no. 3039; partners: HGCA, Monsanto, Fuchs Lubricants, NNFCC, Scottish Agricultural College; period 2004–2008) has aimed at improved winter oilseed rape cultivars that meet the requirements for biolubricants and other specific uses as closely as possible [208, 209]. As a result, Splendor was developed as the first commercial winter HOLL (high-oleic, low-linolenic) oilseed rape variety (with improved C18:1, and lower C18:2 and C18:3 contents) that can be produced competitively in the UK. Splendor is most useful as a frying oil, but is not directly suitable for the key range of biolubricant applications (though more suitable than other conventional or hybrid varieties) and requires further increase in oleic acid content (necessitating GM). It appears that crop management has little influence on the fatty acid profile of the oilseed rape but there is site to site variation.

LE acid varieties (canola types) were developed by classical breeding methods. High-oleic canola oil can be obtained from *Brassica napus* or *B. campestris* (see Tables 5.19 and 5.20). Special FA variants with short- to medium-chain SFAs (C_8–C_{14}), developed by gene technology, are of special interest for oleochemistry [210, 211]. Most advanced is the development of high-lauric acid rapeseed (40–50% C12:0) by Calgene, Inc. (Davis, CA), which has been already commercially planted.

Research on genetic modification of soybean oil and other vegetable oils is an ongoing effort for many years. Major life science companies (Monsanto/Asgrow and DuPont/Pioneer) are active players in this field. Much of the research efforts have focused on reducing both saturated and polyunsaturated fatty acids. Some varieties are being marketed with improved cold-flow properties and increased stability. Other goals include reduction of chemical crop protection strategies through GM crops.

Conventional sunflower oil is characterised by a high linoleic acid (C18:2) content (see Table 5.3a). HO types were developed by mutagenesis [212, 213]. In recent years, the development of cultivars with high oil content and HO acid concentration is an important breeding objective for this crop [214]. Industrial use of HOSNO is favoured by a low SFA content [202]. Breeding has reduced the stearic acid content to 1.5%, which affects the solidification temperature and cloud point. For non-food purposes economic reasons demand at least 83% oleic acid. Sunflower is the only major crop native to western North America. While the United States is concentrating on the production of nutritionally attractive mid-range oleic NuSun [16, 215], Argentina produces regular and HO sunflower oils; Europe has focused on HO sunflower. For a recent review, see ref. [216].

The two most promising vegetable oils being developed for lubricants are HOSNO and low-erucic acid rapeseed oil (LEAR). In 1996 about 870 kt of LEAR were grown in the UK, mainly for use as foodstuff, for detergent manufacture and for export to France and Austria in biodiesel production.

Soybean (*G. max*) oil is a complex mixture of triglycerides with a high proportion of PUFA (63.0%), a characteristic that renders SBO very susceptible to oxidation. There are several thousand sub-species of soybean. As a result of intensive quality breeding, the FA pattern of soybean is now remarkably variable [217]. In addition to low-linolenic acid varieties, which should contribute considerably to the improvement in oxidative stability (mainly in the food oil area), HO soybeans have been produced by routes based on genetic engineering [203]. A transgenic soybean seed with an oil composition comprising high levels of oleic and linoleic acid, and low levels of linolenic acid and SFAs has been claimed [218].

Major US seed companies (Cargill, Monsanto, Pioneer) are in the process of commercialising new soybean varieties that will increase oilseed yields by 9–12% on current acreage. A yield increase of 10% corresponds to an additional 250 million bushels of soybeans. These technologies are set to have an impact as from 2010. Further research is under way to increase the oil yield of soybean by more than 20% [219]. In this respect, the long-term supply of soy as a chemical raw material looks positive. Although soybean oils have been modified by various breeding methods to create benefits for specific markets, a soybean oil that is broadly beneficial to major users in the food *and* industrial markets is not readily available. An optimised SBO composition for both food and fuel use is retained to contain about 24 wt% PUFA [220].

Almost all of Argentina's soy production is based on GM material. In the past, Brazil has been importing GM soya seeds and is now the world's third largest user of GM after the United States and Argentina. Embrapa (Empresa Brasileira de Pesquisa Agropecuária; Brazilean Agricultural Research Corporation) has turned soybean into a tropical crop. Embrapa has created varieties of soya that are more tolerant of acid soils than usual and allow short-cycle, no-till, high-yield cultivation in the hot, acidic Brazilean backlands (Brazil's cerrado or savannah in Mato Grosso and Goiás States).

Peanut (*A. hypogaea*) oil has excellent stability, is extremely durable and less prone to oxidation than other vegetable oils. HO mutants provide oil with high oxidative stability [221, 222].

Substitution of the synthetic bases of fossil origin used in the lubricant industry by environment-respecting compounds has become a central question. In relation to lubrication, conventional plant breeding and genetic modification of oil-producing crops is of interest in view of: (i) increasing the monounsaturated level; (ii) introduction of branching; (iii) heterogeneous chain length oils; and (iv) accumulation of unusual FAs (such as hydroxylated FAs). Lowering the levels of both SFAs and PUFAs while increasing the amount of MUFAs, such as oleate (C18:1) or palmitoleate (C16:1), are important targets for biolubricants as well as for optimal FAME production in order to improve the thermo-oxidative stability [223]. This can be achieved by manipulation of TAG biosynthesis. In soybean, reduced levels of the saturated FA palmitate were obtained by downregulation of FATB, an acyl-ACP thioesterase, causing the accumulation of oleic acid up to 85% from 18% in the wild type [224]. In PCT Int. Publ. No. WO 00/07433 (to Cargill, Inc.), Kodali *et al.* [225] describe *Brassica* plants, seeds and oils having a total long-chain MUFA (C18:1, C20:1, C22:1) content of 82 to 90% and an erucic acid content of at least 15%. Lubricant formulations with an additive package composed of an antioxidant, rust inhibitor, corrosion inhibitor, PPD, antifoam additive, colorant and detergent are suitable as hydraulic fluids. Table 5.19 lists several commercial vegetable oil cultivars with high MUFA content.

One of the many unusual FAs known to occur in nature, namely branched-chain fatty acids (BCFAs), could advantageously replace synthetic bases. In contrast with other – straight-chain – vegetable oils used as lubricant bases, BCFA-containing oil has both excellent oxidative resistance and thermal stability, thereby making it a candidate substitute for high-temperature applications such as motor lubricants. In French Patent No. 2,769,320 Duhot *et al.* (to Total) describe transgenic plants producing branched fatty acids for use as hydraulic fluids or motor oils [235]. Only modest performance improvements were reported. The European research project REFLAX (Rational Engineering of lipid metabolism of FLAX; project no. QLK3-2000-00349; period 2001–2004) has acted as a feasibility programme aiming at the integration of molecular, cellular and physiological investigations to provide a rational strategy for the engineering of oilseeds – and more specifically flax – towards the induced production of BCFAs [236]. The capability of genetically engineered oilseeds to produce BCFAs has been assessed with rapeseed (proof-of-concept). The final transformation with optimal genes has concerned flax, an oilseed with limited risks of gene dissemination, as the producer. However, it was not an objective of REFLAX to deliver high BCFA-producing oilseeds, but only to demonstrate the feasibility. The general conclusion of the REFLAX project for the development of a bio-based production of BCFAs by engineering of oilseeds is that is seems feasible to promote their synthesis in plants. However, more research is needed to achieve higher, commercially interesting, levels of these compounds.

TAGs containing FAs with heterogeneous chain lengths and with high monounsaturate levels can provide useful traits for industrial purposes. Plants with FA compositions that have high monounsaturate levels and heterogeneous chain lengths provide a source of industrial oils for uses such as lubrication. Heterogeneous oils having a high 1,3-dierucoyl 2-oleoyl-*sn*-

Table 5.19 Selection of vegetable oil cultivars with high monounsaturated fatty acid content (>50%)

Vegetable oil	Monounsaturated content (%)	Trade name	Distributor	Reference(s)
	C18:1			
Canola (*B. napus* L.)	71–80[a]	IMC-130	Cargill, Inc. (Minneapolis, MN)	[226, 227]
Canola (*B. napus* L.)	Unspecified[b]	InVigor® Health Hybrids	Bayer CropScience/Cargill, Inc. (USA)	–
Rapeseed	80	RS80	SVO Enterprises (Eastlake, OH)	[228]
Soybean	83.6	–	Pioneer Hi-Bred Intl. (Johnston, IA)	[179]
Sunflower	77–81, 86–92	Sunyl®80, Sunyl®90	SVO Enterprises (Eastlake, OH)	[228]
Sunflower	80.3	–	Intl. Flora Tech (Gilbert, AZ)	[179]
Sunflower	87–92	Sigco 41B	Lubrizol Corp. (Wickliffe, OH)	[229, 230]
Safflower	77.5	–	Oilseeds International, Ltd (San Francisco, CA)	[179]
Olive	62.8–84.4	(Native plant oil)	–	–
	C20:1/C22:1			
Meadowfoam	60–65/16	Fancor® Meadowfoam Seed Oil	Elementis Specialties, Inc. (Hightstown, NJ)	[231]
	C22:1			
Rapeseed	51.2–53.3	HERO	CanAmera Foods (Winnipeg, MB)	[232, 233]
Summer rape (*B. napus* L.)	53.0	Venus	idem	[232]
Summer rape (*B. napus* L.)	54.1	Mercury	idem	[234]
Summer rape (*B. napus* L.)	53.5	Neptune	idem	[233]
Crambe	56.2–62.5	(Native plant oil)	Ag Grow Oils LLC (Carrington, ND)	[80]

[a] PUFA (C18:2 and C18:3) content of about 14%.
[b] For high-heat food-processing applications.

glycerol (EOE) content (at least 50%) are of considerable interest. These oils can be synthetic or produced by plants. In US Patent No. 6,281,375 B1 (to Cargill, Inc.) Kodali *et al.* [237] have disclosed high-oxidative stability, canola-structured biodegradable oils having an EOE content of about 75–90%, i.e. TAG oils having an erucic acid moiety at the *sn-1* and *sn-3* positions and an oleic acid moiety at the *sn-2* position of glycerol (Fig. 5.7). Such chemical feedstock-derived, heterogeneous chain length oils exhibit a high oxidative stability of about 80–300 active oxygen method (AOM) hours in the absence of added antioxidants, have good low-temperature properties, and show good lubrication performance with a low friction coefficient and a viscosity index exceeding 195. The proportions of TAGs in an oil that are EOE can be readily determined according to AOCS Official Method Ce 5b-89. The oxidative stability, viscosity and AW properties of high-EOE oils make them suitable for industrial uses such as hydraulic oils or lubrication additives. In another aspect, the aforementioned invention features plants having a seed-specific reduction in Δ-12 desaturase activity in comparison with a corresponding wild-type plant. Plants that naturally produce erucic acid can be manipulated to produce high EOE levels through genetic engineering, mutagenesis or combinations thereof. Suitable plants for such manipulation include *Brassica* species such as *B. napus*, *B. juncea* and *B. rapa*, *Crambe* species such as *C. abyssinica* and *C. hispanica*, and *Limnanthes* species such as *L. alba alba* and *L. douglasii* (meadowfoam). Such modified plants produce seeds yielding an oil comprising from about 50% to 70% erucic acid (C22:1) and from about 25% to 35% oleic acid (C18:1). Other transgenic plants have a seed-specific reduction in palmitoyl ACP thioesterase activity and a seed-specific increase in Δ9 desaturase activity in comparison with corresponding wild-type plants. The plants also can have a seed-specific reduction in Δ15 desaturase activity.

In another recent development Taylor *et al.* [238] have reported the significant accumulation of the unusual VLCFA nervonic acid (C24:1Δ15) in *Brassica* oilseeds by metabolic engineering. Annual honesty (*Lunaria annua*

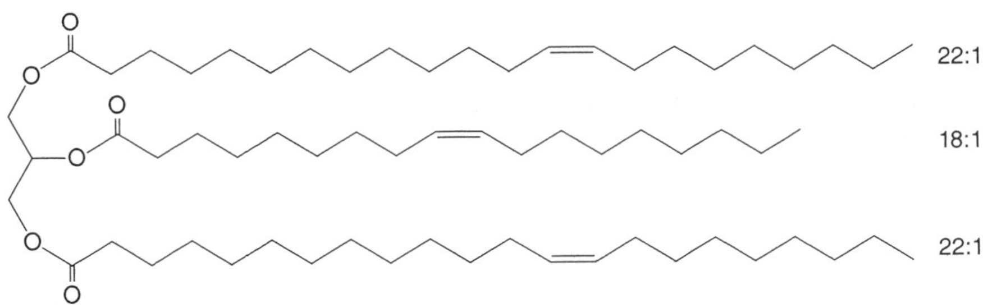

5.7 Canola-structured 1,3-dierucoyl-2-oleoyl-*sn*-glycerol (EOE) oil. After ref. [237].

L.) is a nervonic acid-containing oilseed [239]. Recently, also real progress has been made in the synthesis of industrially useful HFAs in transgenic plants [240]. Ricinoleic acid is produced in castor bean by the direct hydroxylation of the common fatty acid, oleic acid (C18:1). The hydroxylation reaction is catalysed by a single, highly efficient enzyme, the fatty acid hydroxylase. This enzyme has been cloned and expressed in genetically engineered plants to produce novel seed oils containing ricinoleic acid. Linnaeus Plant Sciences, Inc. (Vancouver, BC) aims at engineering an existing oilseed crop species to produce HFAs in its seed oil by transfer of genes from castor oil, for use as high-performance biolubricants [241]. The oil is used in engine oils and two-strokes (cooperation with AtoFina). Biotechnology offers many opportunities to alter the composition of castor oil fatty acids or to create entirely new, toxic-free alternatives of castor oil [242]. A natural mutant of castor seed oil characterised by HO acid and low-ricinoleic acid content has recently been developed [243]. New oilcrops synthesising unusual fat molecules in high contents are coriander (75% petroselinic acid, C18:1), calendula (64% C18:3 conjugated) and meadowfoam (65% C20:1). For transgenic oilcrops rich in oxygenated FAs, see Section 6.2.7, p. 316. A review on transgenic oils is available [244].

Although applications of modern bio- and gene technology are quite likely to provide vegetable raw materials of improved quality and yield, industrial demand requires that these new materials of vegetable origin are made available in sufficient quantities at competitive prices. Transgenics are seen as imperative for biofuel crops [91]. The same might apply for use as chemical feedstock for other industrial applications. There are divergent opinions on the need for the GM approach in relation to biolubricants. Apart from public concern over the environmental and human health effects of genetic modification, some lubricant developers are of the opinion that GM oils are not necessary to achieve high-performance goals. To enhance performance traits, GEMTEK (Phoenix, AZ) uses a mechanically expelled soy oil, which is more highly clarified and oxidatively stable than chemically extracted oil, and applies its plant-based additive package.

5.3.1 High-oleic base stocks (HOBS)

Recent developments in gene segment transfer and modification techniques suggest an alternative route for development of lubricants from plant sources. Major technology goals of genetic manipulation of vegetable oils for lubricants are increased productivity and economic return while providing environmental sustainability (decreased greenhouse gas (GHG) emissions), reduced agricultural inputs (nutrient requirements and pesticide reduction), tolerance for biotic and abiotic stresses, and control of chemical composition (oil concentration and profiles). Benefits sought by the lubricant industry are improved oxidative stability and improved cold-flow properties. These properties are interconnected.

Genetic modification occurs in the seed stock. Crucial criteria for success in genetic engineering of oil crops are, firstly, the seed crop must be amenable to genetic transformation, and secondly, the seed crop should contain high levels of TAGs to be considered suitable to achieve quantities of modified oil sufficient to constitute an economically viable product. Soybean, canola and other vegetable oil crops meet both criteria. To engineer a defined change in seed oil composition, it is essential to understand seed oil metabolism and have knowledge of the enzyme(s) catalysing the reaction of interest.

Today, hybrid breeding and genetic engineering have produced soy, corn, rapeseed, canola and sunflower oils with very high concentrations of oleic acid for industrial and military applications requiring better oxidative stability, thermal stability and load carrying capacity. Such HO vegetable oils contain at least 60% oleic acid, which exceeds the normal C18:1 contents (except for the natural HO olive oil with 65–85% C18:1). The HO contents of GM vegetable oils is at the expense of di- and tri-unsaturated acids. GM vegetable oils generate a much lower linoleic and linolenic acid profile. For compositions of several HO vegetable oils, see Tables 5.18 and 5.19. It is possible to produce HO acid profiles (up to over 90% C18:1) without compromising on the agronomic performance (yield, maturity, plant height, lodging, seed weight or oil content) [245]. The generation of transgenic plant lines with HO acid content represents a contribution of plant biotechnology to the improvement of lubricants.

The primary drawback of conventional vegetable oils is their lower oxidation stability relative to mineral oils and certain synthetic esters whereas low volatility, high flash points, VI and lubricity are their assets. With recent advances in hybrid breeding technology it is now possible to alter the physical properties of conventional vegetable oils by changing FA profiles. Benefits sought by the lubricant industry are improved oxidative stability by increasing the oleic acid (C18:1) content of various vegetable oils and their low-temperature properties. Oleate content that impacts oxidative stability, while maintaining liquidity of the oil, can be increased through genetic modification. Nevertheless, uninhibited HO vegetable oils are still much poorer in oxidative stability than neat mineral oils, but show an advantage of low volatility [153]. Using the AOM, Naegely [246] has indicated that oils with FA compositions with higher values show better oxidative stability.

Renewable, bio-based high-oleic base stocks (HOBS) are rated ultimate biodegradable and show performance improvements over petroleum-based oils in nearly all categories of lubricant properties. These properties include a super high VI (>200) compared with mineral oils (<100). The high VI, combined with an excellent mechanical shear stability, provides film protection (viscosity) on metal surfaces. The unique FA composition produces natural corrosion protection and detergency factors. According to NOACK (an international standardised test for volatility; ASTM D 5800), HOBS have significantly less volatility over solvent refined petroleum (HOBS < 1% compared with petroleum > 15%). Consequently, Stabilized™ HOBS lubricants (Renewable Lubricants, Inc.) (see also Section 10.5.2) will perform at higher temperature with less evaporation than mineral and synthetic base stocks. Application of HOBS lubricants requires no modification or engineering changes to the equipment. HOBS lubricants are compatible with the same seals and filters as are petroleum lubricants. By producing oils more usable as lubricant base stocks there is also reduced need for additional additives and/or manufacturing steps.

HOBS comply with the White House Executive Orders (see Section 9.3) in an effort to achieve major objectives for the environment, agriculture and US national security. HOBS products help industry meet compliance with worker safety regulations and environmental laws, and gain enhanced machine performance and productivity.

In the aforementioned chemical modifications oleic acid or oleic acid esters constitute the main starting point. The oleic acid content in common RSO is typically about 55–65%. Clearly, a >90% technical oleic content in tailor-made raw materials would produce fewer by-products. Many complex biochemical pathways have been manipulated genetically to produce FA compositions with increased oleate (C18:1) level [218, 247]. HO acids seem to be the best compromise between performance, price and biodegradability. Several oilcrops which in their present form do not possess traits ideal for lubricant production have been subject to genetic modification. Main biotech crops are soy, corn, cotton and rapeseed, but also sunflower, safflower, peanut, coconut, palm, linseed and castor bean have been genetically engineered or obtained by selective breeding [248, 249]. Recently, progress has been made in engineering high-performance biolubricants in crop plants. In high-oleic vegetable oils (HOSNO, high-oleic soybean oil (HOSBO), HOSFO, high-oleic rapeseed oil (HORSO), high-oleic castor oil (HOCAS), high-oleic peanut oil (HOPNO), HO canola, palm olein, cottonseed, meadowfoam and *Lesquerella* oil), all with a higher ratio of monounsaturate oleic to diunsaturate linoleic residues, a reduced tendency to oxidation is provided as well as improved heat stability [164, 243, 250–257].

In recent years, HO varieties of rapeseed, sunflower and soybean with oleic acid levels of up to 94% have been developed [202, 230, 258]. Table 5.20 lists some trait-enhanced commercial vegetable oils, mainly developed for the food consumer market. In particular, HO oil from sources such as canola and safflower have lucrative nutritional markets. The improvement in oxidative stability achieved in low-linolenic (LL) oils is limited. For that

Table 5.20 Trait-enhanced commercial vegetable oils

Product	Company	Description[a]	Application	Supply	Reference
Vistive™ I	Monsanto	LL (<3%) SBO	Frying, baking	1.5 billion lb (2007)	[b]
Vistive™ III	Monsanto	HO/LL SBO	Frying, baking	Pending FDA approvals	[b]
TReUS® (low-linolenic)	Pioneer	LL (<2.8%) SBO	Most	250 million lb (2007)	[c]
TReUS® (high-oleic)	Pioneer	HO (80–90%)/LL (<3%) SBO	Heavy frying, industrial use	Full availability >2011	[c]
Omega-9 canola	Dow	HO canola	Food service	2.5 billion lb (2008)	[d]
Omega-9 sunflower	Dow	HOSNO	Edible oil	Small	[d]

[a] HO, high-oleic; LL, low-linolenic.
[b] http://www.monsanto.com/monsanto/content/farmprogress/pdf/vistive.pdf.
[c] http://www.pioneer.com/llsoy/lowlin.htm.
[d] http://www.dowagro.com/omega9oils.

purpose better alternatives are HO oils, oil blends and/or use of antioxidants. HO oils are more oxidative stable than LL oils. Other commercial HOSNO with over 80% oleic acid are Sunyl® 80 (SVO Enterprises; Eastlake, OH) and TriSun™ 80 (A.C. Humko; Cordova, TN). High-oleic (HOSNO) oils have also been developed by Dow, DuPont, Instituto de la Grasa (CSIC, Sevilla). Other high-oleics with over 80% oleic acid content are HORSO (RS®-80; SVO Enterprises), high-oleic canola oil (Cargill, DuPont), HOSBO (DuPont, Monsanto), HOCRO (DuPont) and HOPNO (Mycogen). InVigor®Health canola (Bayer CropScience/Cargill) is a high-oleic acid hybrid for the food market. New rapeseed varieties (e.g. LZ7632) are reaching high levels of up to 87% C18:1.

High-oil sunflower hybrids (40–45% total oil) were developed in the USSR in the mid-1960s, and were followed by HO (64–79%) sunflower cultivars ('Pervenets') with proportionally less linoleic acid (C18:2) in the late 1970s [259]. In a further development, US Patent No. 4,627,192 discloses a Pervenets-derived sunflower hybrid Sigco 41B with high-oleic acid content (87–92%) and low C18:2 content (1.5–4%) [229, 230]. Table 5.21 shows the effect of linoleic acid concentration of sunflower and safflower oils on AOM values, and therefore on shelf-life. Duda *et al.* [153] have investigated the oxidative stability and anti-wear properties of HO vegeta-

Table 5.21 Effect of linoleic acid content of safflower and sunflower oils on AOM value

Vegetable oil	C18:2/C18:1 ratio	AOM (h)
SFO	79/12	10
HOSFO	15/80	35
HOSNO	1.5/92	100

After ref. [230].

ble oils. HOSNO is comparable with or better than mineral oils as a boundary lubricant. While neat HOSNO already shows superior lubricity over neat mineral oil, properly formulated HOSNO (antioxidant (AO), zinc dialkyldithiophosphate (ZDDP)) is better than a commercial crankcase lubricant in AW properties. Deposit formation in HOSNO can be inhibited by 1 wt% AO, resulting in oxidative stability comparable to that of the conventional SAE 10W-30 SG-grade crankcase lubricant. However, addition of ZDDP to a HO vegetable oil reduces the antioxidant effect, denoting a ZDDP–AO interaction [153].

HOSNO and polyol esters derived therefrom have considerably higher stabilities as compared to materials that contain multiply unsaturated fatty acids. HOSNO with an oleic acid content of >90% and very low stearic acid (C18:0) content (1.0–1.5%) is extremely oxidation- and ageing-stable when compared to RSO and TMP oleate synthesised from rapeseed oil [260]. The low stearic acid content determines the product's excellent low-temperature behaviour. Even unmodified, HOSNO is suitable as a base fluid for various biodegradable lubricants. The economicity of its industrial application needs further attention. Agri-Pure^{TM}-560 (Cargill, Inc.; Minneapolis, MN) is a transesterified high-oleic SNO with short saturated fatty acid esters, with specifications as shown in Table 10.22.

Rapeseed has also been engineered. US Patent No. 5,638,637 to Pioneer Hi-Bred International, Inc. (Des Moines, IA) describes improved rape cultivars (i.e. *B. napus* and *B. campestris*) that have the ability to yield an endogenous vegetable oil of increased heat stability on account of an unusually high oleic acid content (80–85%), low α-linoleic acid (<3.5%) and erucic acid content (<2%) [261]. The oil of the improved cultivar is particularly well suited for use as frying oil. Genetically modified *B. napus* oils with a total long-chain MUFA content of over 82% (of which up to the 56% C18:1) and an erucic acid (C22:1) content of at least 15% have very high oxidative stability and desirable lubricant or hydraulic fluid characteristics [225].

For Cargill's HO canola grades AP65, AP75 and AP85 (with oleic contents of 65, 75 and 85%, respectively) the oxidative stability (expressed as

Table 5.22 Fatty acid composition (%) of rapeseed and sunflower seed oil varieties

Fatty acids	Rapeseed varieties								Sunflower varieties	
	Caracas	Contact	Cabriolet	Calida	Spiral	MSP05	MSP11	MSP13	Elansol	Aurasol
C16:0 (palmitic)	5.0	3.9	3.8	3.6	4.5	4.3	3.4	3.0	3.2	3.5
C16:1 (palmitoleic)	–	–	0.2	0.1	–	0.3	0.3	–	–	–
C18:0 (stearic)	1.6	1.6	1.6	1.6	1.9	1.6	2.1	1.5	4.6	4.1
C18:1 (oleic)	63.2	73.4	76.9	63.2	64.1	75.9	84.9	81.9	88.0	89.6
C18:2 (linoleic)	17.7	9.3	8.0	28.7	25.0	12.8	5.5	9.6	2.5	1.8
C18:3 (n-3)	10.6	9.8	7.9	0.8	2.4	2.8	2.0	1.9	–	–
≥C20	1.9	2.1	1.7	2.1	2.3	2.3	1.9	2.1	1.7	1.0
Saturated acid (%)	7.5	6.3	6.0	5.9	7.4	6.9	6.1	5.3	8.8	8.4
Monounsaturated acid (%)	64.2	74.6	78.1	64.6	65.2	77.6	86.3	83.2	88.1	89.6
Polyunsaturated acid (%)	28.4	19.1	15.9	29.5	27.3	15.5	7.5	11.5	2.5	1.8

After ref. [262].

AOM hours) increases from 30 to 34 and 60 h, respectively, in comparison to 12 h for canola and 25 h for hydro canola.

In US Patent Appl. No. 2009/0286704 A1 to Monsanto [262] Despeghel relates to the use of blends of high-oleic (>85%), low-linolenic (<4%) (HOLL) rapeseed oil (extracted from varieties such as Caracas, Contact, Cabriolet, Calida, Spiral, MSP05, MSP11 or MSP13) and sunflower oil (extracted from seeds of Aurasol and/or Elansol varieties) or soybean oil (Table 5.22). The disclosure also provides for the use of mono-alkyl ester compositions as biolubricant base stocks. Table 5.23 lists the physicochemical characteristics of Elansol (SBO)/Caracas (RSO) blends.

HOSBO, which is significantly more saturated than SBO (SFA/MUFA values of 9.5/81.5% and 16.0/21.0%, respectively), shows considerable improvement in thermal and oxidative stability over SBO. At the same time, reduced unsaturated content may also affect the boundary and hydrodynamic lubrication properties of HOSBO [164]. DuPont/Pioneer have recently announced evaluation and sales of high-oleic (>80%), low-linolenic (<3%) soybean oil (brand name TReUS®), mainly for food applications; however, this product could also constitute a boost to the lubricant industry. Monsanto is marketing a low C18:3 (<3%), low saturate (about 9%) soy oil (Vistive™ I) that has been developed by conventional breeding. Vistive III (73% C18:1 and <3% C18:3 and saturates) is being developed by breeding Vistive I with a transgenic variety to achieve a high-oleic IP basestock that could be more useful in lubricant applications. The products are being evaluated by Valvoline for use as high-temperature engine oils and hydraulic fluids. Graef *et al.* [245] have also described an HO (>85%), low-palmitic acid (<5%) soybean with increased oxidative stability and cold-flow oil properties, which can be used for biolubricant applications. An HO variety of soybeans, originally modified for food purposes, is already in use for

Table 5.23 Physicochemical characteristics of Elansol/Caracas blends

Parameter	Blend ratio				
	95/5	75/25	50/50	25/75	5/95
Viscosity, 40 °C (mm^2/s)	40.0	39.0	37.8	36.6	35.6
Viscosity, 100 °C (mm^2/s)	8.6	8.5	8.4	8.2	8.1
VI	201	204	207	211	213
PP (°C)	−11.7	−14.5	−18.0	−21.5	−24.3
Rancimat[a] (h)	52.1	42.7	31.0	19.2	9.8
IV (g I_2/hg)	84.2	90.4	98.3	106.1	112.3

[a] 98 °C, 20 L/h.
After ref. [262].

lubricant production. Lubricant base stocks from modified soybean oil have recently been reviewed [263]. The biotechnological enhancement of soybean oil for lubricant applications has also been reviewed [264].

Domestication of oil palm has led to its improvement, especially through breeding. Biotechnological efforts at MPOB (Malaysian Palm Oil Board, Kuala Lumpur) are directed toward the production of palm oil with high IV and high MUFA (C18:1) content for edible purposes and industrial uses [48]. The aim is to increase the efficiency of conversion of palmitate (C16:0) to oleate (C18:1) levels by switching off the palmitoyl acyl carrier protein (ACP) thioesterase gene. HO palm oil TMP esters with excellent PP and lubricating properties have been reported [51].

Successful use of vegetable oils as environmentally friendly, biodegradable, base fluids in industrial applications is contingent upon improving their low-temperature viscometries. HOSNO containing 80% oleic acid has a PP of –12 °C, whereas many applications require a PP of less than –25 °C. This can be achieved by blending (see Table 5.23) or by formulation of HOSNO (eventually in the presence of an additional oil) with a PPD [265].

High-performance base fluids may also be based on native high-oleic oils (mainly SNO, SBO), selectively hydrogenated RSO, all with replacement of the GL moiety by TMP. Lawate and Lal (to Lubrizol) have disclosed HO (75–95%) polyol esters [266]. HO TMP trioleate (PP –45 °C) displays excellent low-temperature properties. Table 5.24 shows the properties of HO (90%+) TMP trioleate. TMP esters of oleic acid are widely applied for hydraulic fluids.

Many HO acid applications have been described. US Patent No. 5,338,471 (to Lubrizol Corp.) discloses industrial lubricants containing mixtures of FA esters and HO vegetable oils (HOSNO or HORSO) as base oils [267]. In US Patent No. 5,538,654 Lawate *et al.* (to Lubrizol Corp.) describe environmentally friendly food-grade lubricant compositions, useful as hydraulic

Table 5.24 Properties of high-oleic (90%+) TMP trioleate

Property	Value
Viscosity, 40 °C (cSt)	47
Viscosity, 100 °C (cSt)	9.5
Viscosity index (VI)	190
Pour point (°C)	–45
Flame point (°C)	300
Evaporation rate (%)	1.3
Oxidation stability (Baader)	Fair to good
Biodegradation CEC (%)	95
Biodegradation OECD (%)	90

oil, gear oil and compressor oil, composed of GM triglycerides and containing FDA-approved, conventional performance, additives [268]. In US Patent No. 6,300,292 A1 to Nippon Oil Corp. Konishi and Kikuchi [269] describe a hydraulic oil composition consisting of a base oil composed of a mixture of at least 20% mineral oil and synthetic oil (diester, polyol ester) and at most 80% of an HO acid content (>80%) vegetable oil having a low degree of unsaturation (<0.2 according to JIS 1557-1970) [269]. US Patent No. 6,312,623 to Oommen and Claiborne [270] discloses an HO acid triglyceride composition (>75% C18:1) for application as a biodegradable electrical insulation fluid; see also ref. [271]. HOBS are particularly interesting for crankcase applications where stability is a major problem. Oils from GM plants are preferred for applications where the use temperature exceeds 250 °C, such as internal combustion engines. In PCT Int. Publ. No. WO 2006/116502 A1 (to Renewable Lubricants, Inc.) Garmier describes an HO acid application in a bio-based high-temperature engine lubricant formulation [272]. For HOSBO-base crankcase oil, see also Section 13.3.

Industrial demand requires that these new materials of vegetable origin are made available in sufficient quantities at competitive prices, allow economically viable isolation of the relevant components of adequate purity and present advantages over alternatives provided by the petrochemical industry. Continuous development ensures increased use of oils and fats as renewable raw materials in the chemical industry. The generation of transgenic sunflower lines with HO acid content is an example in which plant biotechnology has already provided new opportunities for bio-based lubricants and greases [255]. Despite its somewhat limited availability HOSNO (with >90 % C18:1) is considered as a commodity oil and is currently the best choice. With its low viscosity value (38–42 mm^2/s^{-1} at 40 °C) HOSNO is suitable for applications where no particularly high viscosity is required.

5.4 Bio-based wax esters

Wax esters constitute another molecular variation by which properties of lipids can be improved making them targets for lubricant application. Wax esters are oxo-esters of long-chain fatty acids (LCFAs) and long-chain fatty alcohols. Natural waxes are mixtures of esters and usually contain hydrocarbons as well. Of the various potential biolubricants, wax esters are regarded as having excellent performance properties. Wax esters have lubrication properties that are superior to ordinary vegetable oils (TAGs), due to their high oxidation stabilities and resistance to hydrolysis. The high linearity of wax esters enhances the VI of the oil and imparts specific characteristics such as antirust, antifoam, AW and friction reduction properties to the lubricant [273, 274]. These properties make wax esters excellent feedstocks for production of high-temperature and -pressure lubricants as well as hydraulic fluids. Therefore, wax esters can in principle be used both in high-volume/low-price base oils and in the high-price/low-volume additive segments of the market. However, owing to the high cost of obtaining waxes from existing sources, they are currently limited in use to specialised and low-volume product areas such as cosmetics and for specialty lubricants.

Similar to TAGs, wax esters are neutral lipids. They are solid at room temperature and liquid at higher temperatures. Wax esters are naturally present in different organisms where they fulfil diverse and important biological functions. In particular, these abundant lipids are found coating the surfaces of plants, insects and animals (surface wax). The only natural sources of wax esters of commercial interest have been sperm whale (*Physeter macrocephalus*) oil and the seeds of the slow-growing desert shrub jojoba. Winterised sperm oil, derived by a process that removes the more saturated lipid components, and composed of C_{24}–C_{42} wax esters, was the raw material formerly employed in the lubricant industry [275]. After the Endangered Species Act (USA) banned the killing of whales, a suitable replacement product was not readily available that would yield the lubricating and heat-dissipating properties spermaceti oil provided as an automatic transmission fluid. Replacements of sperm whale oil such as phosphorised lard oils and sulphurised polyisobutylene have met with limited success. Better lubricating properties (i.e. friction and reduced wear) have been achieved with a natural wax ester, such as jojoba oil and derivatives [276].

Jojoba (*Simmondsia chinensis* L. Schneider) is a perennial shrub that grows naturally in the Sonora desert (Mexico) and in south-west USA [277]. Commercial jojoba plantations are found worldwide (in Argentina, Australia, Chile, Egypt, Israel, Mexico, Peru and the United States) covering some 8000 ha [278]. In the seeds of the jojoba plant wax esters serve as energy storage materials [279]. At variance to most seed-bearing plants, the light yellow oil of the jojoba seed (50% of dry weight) [280] is composed almost entirely (97%) of wax esters (mainly C_{40}–C_{42}) and does not contain triacylglycerols, which are instead the constituents of other vegetable oils and animal fats. This odourless monoester wax, composed of a narrow mixture of straight-chain esters of long-chain monounsaturated acids and long-chain primary alcohols, has typically an average molecular weight of 606–610 Da [281, 282], see Fig. 5.8. Many factors affect wax ester, FA and fatty alcohol molecular weight [283]. JO is composed of 5.4% C_{38}, 24.1% C_{40}, 55.1% C_{42}, 13.1% C_{44} and 2.0% C_{46} wax esters [40]. Jojoba molecules are quite symmetric with double bonds more or less equidistant from the central ester linkage (see Fig. 5.8). The approximate FA composition of JO oil, consisting for 98% of monounsaturated ω-9 FAs, is 66–71% C20:1 (*cis*-11-eicosenoic acid), 14–20% C22:1 and 10–13% C18:1 [284]. The final composition of this

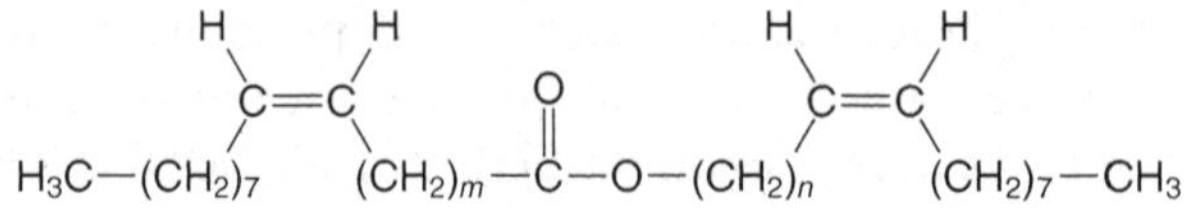

m = 7–13, n = 8–14

5.8 A typical jojoba ester molecule.

unique storage-lipid wax is a result of activities and affinities of three main enzymes: a fatty acyl-CoA elongase (β-ketoacyl-CoA synthase), which produces 20:1, 22:1 and 24:1 acyl CoA as precursors to the wax; fatty alcohol reductases (acyl CoA reductases); and acyl:alcohol transacylase [285].

Environmental conditions prevailing at the time of wax synthesis affect the chain lengths of the FA and fatty alcohols. Wax composition in jojoba is highly variable. The composition of the wax affects its quality and suitability for different applications. Chain length influences properties such as viscosity and melting point. JO, with a melting point about 10 °C, has outstanding qualities for lubrication applications. As a lubricant, JO is superior to vegetable oils such as SBO, SNO and CAS [274]. JO is characterised by high flash and fire points, high thermal stability, high PP (+9 °C) and high VI; the wax maintains its viscosity at high temperatures. Jojoba is recognised as one of the more oxidatively stable oils with an OSI value of 55.9 h at 110 °C for cold-pressed JO. JO is more shelf-stable than safflower or canola oil, but less than castor or coconut oil. Jojoba seed production is still quite small (<5 kt/yr) [286] – mainly in Mexico and the United States – and insufficient to meet global demand. Consequently, this expensive commodity is mainly used commercially for production of value added products, as in the cosmetics and pharmaceutical industries [287]. However, jojoba liquid wax ester from second pressing is most often used for lubricant purposes. Interest in JO stems from its unusual properties differing from all other known seed oils, in particular pronounced thermal and oxidative stability [288, 289]. JO is a low-viscosity oil of 6 cSt at 100 °C, which limits its utility as a base stock. For most industrial applications, lubricating oils with viscosities ranging from 5 to 15 cSt at 100 °C are required. JO displays good lubricity and can be utilised as a component in lubricating oil formulations [290, 291]. However, PP, AV and oxidative stability, which do not match those of mineral oil, are limiting factors in its use as a base stock. These properties can be modified by physical and chemical methods and also by additive treatment [290].

In applications where PP is not a critical requirement, JO has a potential to replace mineral oil. It has proven to be an excellent lubricant for mechanical applications, in particular at high pressure [292]. In Indian conditions, a JO base stock may serve as a two-stroke gasoline engine lubricant [293].

Jojoba wax ester has also been indicated as an essential component in a general purpose base oil formulation for internal combustion engines and total-loss lubrication developed by Agro Management Group, Inc. (Colorado Springs, CO) [294]. US Patent No. 5,888,947 to Lambert and Johnson [294] relates to a composition with three main components: a base oil (primarily triglycerols), an oil source containing HFAs (5–20%) and one composed of vegetable or animal waxes (5–10%).

The unique JO has been subjected to a variety of transformations. The PP of JO (+9 °C) can be reduced to –9 °C by blending with mineral oil. The acid content of JO (1.1 mg KOH/g) can be brought down to 0.1 mg KOH/g (acidity normally observed in mineral oils) by washing with $NaCO_3$ solutions. Refining reduces the stability of JO due to the removal of natural antioxidants (see Table 5.13). To improve the oxidative stability, the effect of refining, chemical modification by sulphurisation, blending with mineral oils of high stability and additive treatment have been considered [290]. Jojoba's wax ester structure, with a *cis* stereochemistry (Fig. 5.8), makes it reactive in a variety of chemical transformations, including sulphur chlorination, sulphurisation, hydrogenation, isomerisation and epoxidation. Sulphurised jojoba oil (SJO) finds application for lubrication in automotive transmissions. SJO has also been described as being an extreme pressure additive for lubricants [295]. Partial sulphurisation of jojoba oil to levels of 0.2–2.0% sulphur leads to viscosities at 100 °C of 7.1 and 10.3 cSt, respectively, which greatly extends the scope of its utilisation in developing lubricant formulations [290].

As new industrial applications are on the horizon the demand for this natural product is expected to increase dramatically. The production of waxes with different compositions (i.e. different chain lengths), and hence changed wax properties such as viscosity, melting point, and thermal stability, may be of importance for future requirements of the jojoba industry. Price is a limiting factor for application in the lubricant industry. The IJEC (International Jojoba Export Council) Quality Certification Seal is the only international validation of JO quality available worldwide.

Natural waxes differ widely in biological significance and chemical composition [287]. Waxes in the cuticle of plants are mainly composed of alkanes, fatty alcohols, secondary alcohols, ketones and FAs. Plant waxes consist predominantly of C_{20}–C_{35} carbon chains and are derived from VLCFAs [296]. Rice bran (*Oryza sativa*) wax (RBX), mainly composed of aliphatic acids (C_{16}–C_{24} wax acids) and higher alcohol esters (C_{26}–C_{30}), may also be used as a lubricant. Carnauba wax (also called Brazil or palm wax), the most valuable of the natural waxes, is obtained from the coating on the leaves of the Brazilian palm *Copernicia prunifera*, and consists of 80–85% esters of C_{24}–C_{28} fatty acids with C_{32}–C_{34} alcohols, 10–16% fatty alcohols, 3–6% acids and 1–3% hydrocarbons. There is a considerable amount of ω-hydroxy fatty acids, HO—$(CH_2)_x$COOH, where $x = 17–29$. These ω-hydroxy fatty acids can form long polymer esters, which gives this wax its unique properties of being hard and impervious (m.p. 80–87 °C). Carnauba wax has been used as a mould release agent. Wool wax (wool grease, degras) is recovered from the scouring of wool and is unusual because it forms a stable, semi-solid emulsion containing up to 80% water. A purified product is known as lanolin. The 'wax' consists mainly of fatty acid esters of cholesterol, lanosterol and fatty alcohols.

A mammalian alternative to jojoba esters is found in the accumulation of liquid wax esters (C_{28}–C_{36}) and TAGs (66 and 30 wt%) in sperm whales as spermaceti oil [297]. The main wax ester of spermaceti oil is cetyl palmitate (the ester of cetyl alcohol and palmitic acid, $C_{15}H_{31}COOC_{16}H_{33}$), m.p. 42–47 °C. Sulphurised spermaceti oil served as an excellent additive in many critical lubricant applications until 1971, when sperm whale hunting was banned. Wax esters are also produced in the sebaceous and meibomian glands of some other mammals [298]. For mammalian wax biosynthesis, see refs [299, 300].

Unlike inshore/midwater fish, wax esters are also found at rather high levels (95% of the lipid fraction) in deep-sea fish species such as orange roughy (*Hoplostethus atlanticus*), smooth oreo (*Pseudocyttus maculatur*) and black oreo (*Allocyttus* sp.) [297]. The range of wax esters in deep-sea fishes is mainly C_{34}–C_{42} (orange roughy). Natural wax esters differ greatly in chain length of the fatty acid and alcohol moieties as well as in the degree of branching. Also specialised tissues in some birds such as mallard ducks (*Anus platyrhynchos*) produce liquids rich in wax esters [301]. Insect wax esters are found as surface waxes. The red harvester ant (*Pogonomyrmex barbatus*) produces C_{19}–C_{31} wax esters, some of which are branched [302]. The honey bee (*Apis mellifera*) produces cuticular wax consisting of 23% wax esters (C_{36}–C_{50}) and 55% hydrocarbons [303]. Beeswax used for the honeycomb is composed of 35% wax esters (mainly C_{40} with methyl-branched side chains) and 15% hydrocarbons (mainly C_{31}) [304]. TAGs and wax esters are rather uncommon storage lipids in bacteria, but some bacteria do accumulate some fairly high levels of wax ester as a common storage lipid [305]. In particular, bacteria of the genus *Acinetobacter* store wax esters as intracellular storage lipids as C_{30}–C_{36} mono esters composed of saturated and monounsaturated fatty acid and fatty alcohol moieties [306].

In principle, production of natural wax esters could be increased if the cultivation of jojoba could be substantially extended, but the species is unsuitable as a mainstream agricultural crop in Europe. Outside Europe there is significant potential to develop alternative sources of wax esters, in particular jojoba varieties. Alternatively, wax esters can also be produced by chemical catalytic or enzymatic processes from many different plant oils

[307–309]. This requires long-chain fatty alcohols that can be transesterified with LCFAs to produce a mixture of C_{40}–C_{42} jojoba-type wax esters. *Crambe* and *Limnanthes* seed oils, two promising agricultural raw materials, as well as honesty (*Lunaria annua* L.), can be converted to liquid wax esters similar to those of sperm oil [275, 310, 311]. Synthetic wax esters and diesters were derived from *Limnanthes* and partially hydrogenated *Crambe* seed oils using a *p*TSA catalyst [307].

Synthetic wax esters can be made by esterifying an unsaturated FA and an unsaturated fatty alcohol. Sulphurised and phosphor derivatives improve friction, wear and EP properties of such fluids. Sulphurised vegetable wax esters, made from HERO by a complex and expensive process, are described in US Patent No. 4,152,278 [312]. Liquid wax esters made from HERO are a substitute for sperm whale oil or a natural wax ester, such as JO. Sulphurised synthetic wax esters often display excellent lubricating properties. However, the cost of a process to manufacture a synthetic wax ester is high and comparable with the cost of natural wax esters. Phosphite adducts of synthetic vegetable oil wax esters are described in US Patent No. 4,970,010 [313].

Industrial Lubricants, Inc. (ILI; Seattle, WA) has introduced three patented lubricant technologies – Liquid Wax Esters (LXE®), Synergol® Lubricity and Anti-Wear Technology. Synergol products are non-sacrificial AW additives, diesel fuel lubricity and general lubricant lubricity additives. Lubegard™, a highly successful biodegradable automatic transmission fluid supplement (not formulated with ZDDP) with improved thermo-oxidative properties, is based on the unique LXE® technology in which the FA of HERO is linked with an alcohol to form a linear liquid wax. The synthetic Lubegard™ wax ester is a molecular replacement for sperm whale oil, or more specifically, spermaceti oil. Lubegard® lubricants substantially reduce fuel consumption and increase fuel economy.

Lipase-catalysed reactions offer great potentials for the preparation of a wide variety of wax esters under mild reaction conditions [314]. Thus, wax esters have been obtained in high yield by esterification of FAs with long-chain alcohols [308, 315–324] and in moderate to good yields by interesterification (alcoholysis) of TAGs or natural fats and oils with long-chain alcohols [318, 320, 322, 325–327]. For example, wax esters having compositions approaching that of jojoba were prepared in high yield via lipase (Novozym 435 and papaya)-catalysed esterification of *Crambe* FAs with *Camelina* alcohols or *Camelina* FAs with *Crambe* alcohols [308]. Wax esters have also been prepared by lipase-catalysed alcoholysis of alkyl esters of fatty acids with long-chain alcohols [323, 328, 329].

Wax esters produced by bacteria [305] or enzymatically from *n*-alkanes [330] or from alcohols and FAs (bio-based or mineral-based) [308] are too costly for the bulk lubricant market. It follows that other sources of wax esters need to be investigated and explored. In short, there is a compelling need to establish a new cost-effective supply of plant-derived wax esters.

5.4.1 Wax biosynthesis pathways

Wax esters are esters between an FA and a fatty alcohol, and their biosynthesis includes reduction of an FA to a primary alcohol, catalysed by an FA reductase, and then esterification of an FA from acyl CoA to this alcohol, which is performed by a wax synthase. The capability for biosynthesis of neutral lipids such as TAGs and wax esters is widely distributed in nature and is found in animals and plants as well as microorganisms. In plant biology plant cuticular wax is of considerable importance. Cuticular wax is mainly composed of long-chain aliphatic compounds derived from VLCFAs. Wax biosynthesis in plants starts with fatty acid synthesis in the plastid, followed by fatty acid elongation (FAE) to very long chains (C_{24}–C_{34}), and the subsequent processing of these elongated products into aldehydes, alkanes, secondary alcohols and ketones (decarbonylation pathway) and/or primary alcohols and wax esters (acyl reduction pathway); see Fig. 5.9 [296, 331]. Little is known about the intercellular transport of the hydrophobic wax components.

Wax biosynthesis requires the coordinated activities of a large number of enzymes that carry out the elongation of FA wax precursors and catalyse the formation of a multitude of aliphatic compounds. The reactions of wax production are localised in the epidermal cells [332]. Formation of saturated VLCFA wax precursors is a complex process that is accomplished in two stages in different cellular compartments, namely *de novo* fatty acid synthesis of C_{16} and C_{18} acyl chains and FAE. The first stage occurs in the stroma of plastids by soluble enzymes forming the fatty acid synthase (FAS) complex [333]. During synthesis, the elongating acyl chains are esterified to acyl carrier proteins (ACP) [334]. Fatty acid synthesis proceeds through four consecutive steps: (i) condensation of a C_2 moiety originating from malonyl-ACP to acetyl-CoA; (ii) reduction of β-ketoacyl-ACP; (iii) dehydration of β-hydroxyacyl-ACP; and (iv) reduction of *trans*-Δ^2-enoyl-ACP. This sequence of reactions results in a fully reduced acyl chain extended by two carbon atoms. Three different types of FAS complexes are involved in the synthesis of C_{18} fatty acids in the plastid, which differ in their condensing enzymes with strict acyl chain length specificities, namely KAS III (C_2–C_4), KAS I (C_4–C_{16}) and KAS II (C_{16}–C_{18}). Following liberation from ACP, the LCFAs (C_{16} or C_{18}) are thiol-esterified to CoA by acyl-CoA synthetases in the chloroplast envelope [335].

During the second stage of FAE, extension of C_{16} and/or C_{18} fatty acids to VLCFA chains is catalysed in consecutive two-carbon extensions by extra-plastidial (endoplasmic reticulum, ER) membrane-associated multi-

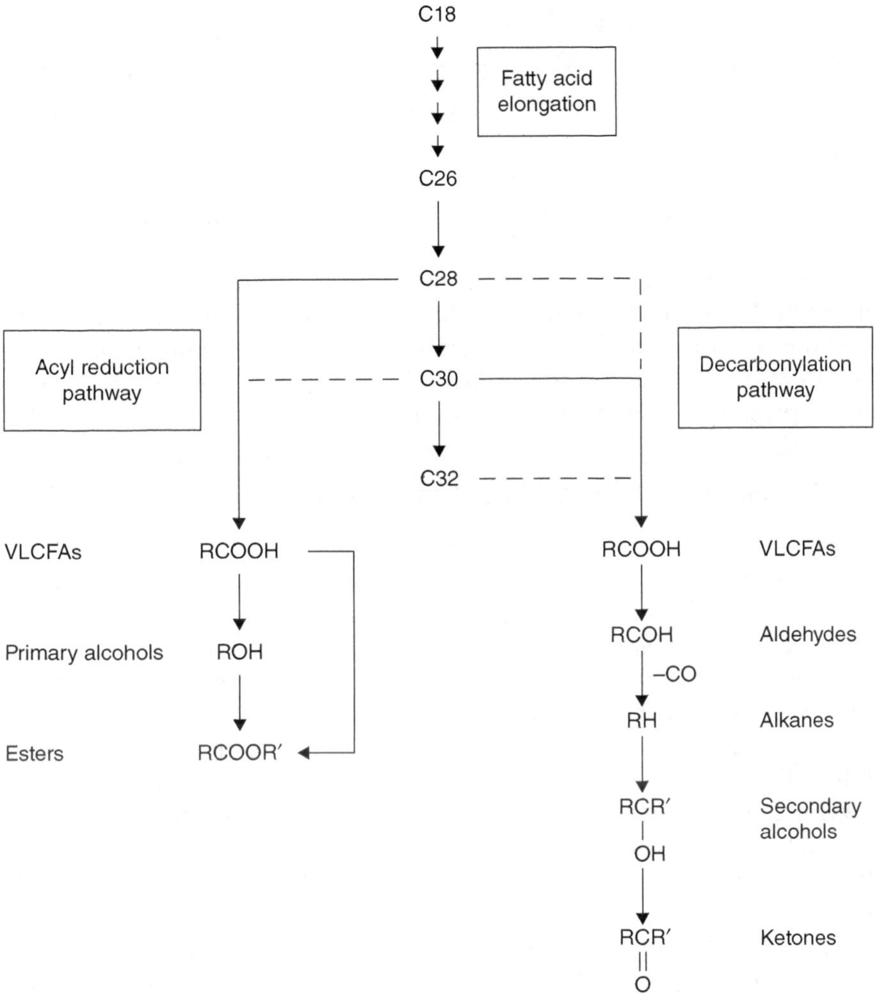

5.9 Wax biosynthetic pathways in *Arabidopsis*. After ref. [336].

enzyme complexes, fatty acid elongases. Since chain length of wax components are typically in the range of 20–34 carbons, multiple elongations are needed to extend the acyl chain to its final desired length.

In most plants, including *Arabidopsis*, there are two principal wax biosynthetic pathways: acyl reduction (to primary alcohols and wax esters) and decarbonylation (to aldehydes, alkanes, secondary alcohols and ketones), as indicated schematically in Fig. 5.9 [336]. VLCFAs in the epidermal cells are reduced to primary fatty alcohols by a single fatty acyl-CoA reductase (FAR) of the acyl-reduction pathway, which appears to be associated with the ER [337]. The key enzymatic step of the acyl reduction pathway consists

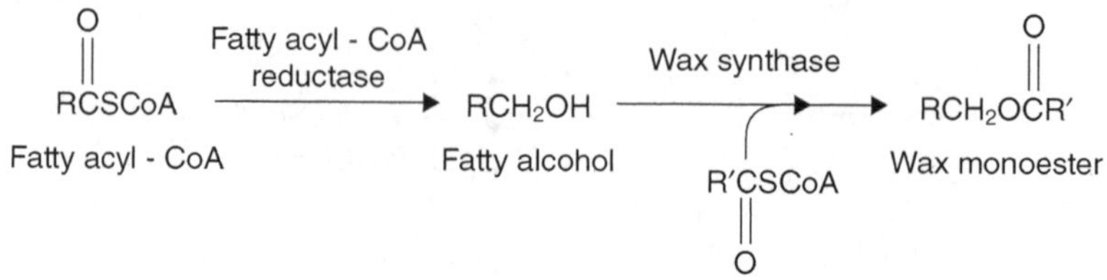

5.10 Catalytic steps involved in wax monoester production.

in the synthesis of wax monoesters by condensation of fatty acyl-CoA and primary fatty alcohols, a reaction catalysed by acyl-CoA: fatty alcohol acyl-transferase (wax ester synthase, WS) (Fig. 5.10). WS is an integral membrane protein [338], but its site of action is unknown.

5.4.2 Genetically modified *Crambe* oil

The current price and availability of wax esters make them unsuitable for widespread use for the lubricant market. An international project, called EPOBIO (Realising the Economic Potential of Sustainable Resources – Bioproducts from Non-Food Crops) and supported by the European Commission (FP6 action SSPE-CT-2005-022681) and the United States Department of Agriculture (USDA), has opted for development of wax esters as target for a bulk lubricant base oil in view of their outstanding lubrication properties at a currently unfavourable cost. In particular, EPOBIO has identified lubricant product development in the non-food oilcrop *Crambe* (*Crambe abyssinica*) as the first target to consider for the biotechnological production in Europe of wax esters at an acceptable price [83].

The production of plant-derived wax esters in Europe will necessitate the use of a GM crop. Use of a crop which cannot be used for food or feed is considered essential from a regulatory perspective, given that the agricultural infrastructure cannot ensure full separation of different varieties of the same crop species. To minimise risks in using a genetic engineering strategy for the production of wax esters the preferred strategy is establishing this production in a non-food crop in order to avoid mixing an industrial feedstock into the food chain. Risks can further be mitigated for crops for which interspecies crosses with the closest-related species give sterile offspring.

The potential of *Crambe* as a platform for production of bio-based wax esters within the framework of EPOBIO is based on arguments summarised in Table 5.25. *Crambe* (60% C22:1) already has a proven record as a producer of an industrial feedstock. Erucic acid is a suitable LCFA for conversion into the fatty alcohols needed for production of wax esters. The high

Table 5.25 Potential of non-food oilcrop *Crambe* for wax ester production

• Specific non-food oilcrop required for industrial feedstock • Risk of development of industrial feedstock production in food-use crops • High yielding oilcrop • Proven industrial resource (erucic acid) • No outcrossing to wild relatives • Distinct seed morphology (identity preservation) • Self-pollinating • Low environmental burden

erucic FA content of *Crambe* excludes its use in food applications. This fulfils an essential condition since the manufacture of wax esters in *Crambe* can only be achieved through genetic modification of the plant. Considerable risk is involved in developing industrial feedstock production in the same crop that is also used for the food market [339, 340]. Industrial oils to be produced in genetically engineered plants are not intended for food use (although not necessarily toxic) and must therefore not enter the food or feed chain. Dominating and high yielding oilcrops in Europe such as rapeseed, sunflower and flax are all more or less used for food production and do therefore not qualify as candidate oilcrops for wax ester production. It is of interest to notice that pre-market wax ester production in oilseed rape was developed by Monsanto in the 1980s/1990s and subsequently abandoned for strategic reasons.

Crambe is a high-yielding oilcrop with high oil content (35%) which can grow wherever rapeseed is grown [341]. Overcrossing to wild indigenous relatives or other Cruciferae species is considered unlikely [342]. As *Crambe* is a self-pollinating crop outcrossing of the different traits among varieties is also reduced. When growing non-food GM *Crambe* the same precautions are recommended as in growing GM foodcrops (such as rapeseed) in order to help ensure identity preservation of the crop. The seed morphology of *Crambe* is distinct from other established oilcrops which simplifies identity preservation. With regard to cultivation, *Crambe* is a low-input crop (meaning reduced environmental burden in terms of fertilisers and water) compared with the many other oilcrops that can be cultivated in Europe. *Crambe* for oil production can be grown on maincrop land and also on set-aside land for as long as that cultivation option is retained within the EU Common Agricultural Policy (CAP).

Wax esters provide an as-yet unexplored potential for accumulation of unusual fatty acids in combination with industrially useful alcohols. EPOBIO has identified three different classes of wax esters to be potentially produced by the GM *Crambe* oilcrop for lubricating formulations. Table 5.26

Table 5.26 EPOBIO target wax oils from GM *Crambe*

Target wax	Target oil structure	
	Fatty acid moiety[a]	Alcohol moiety
Long-chain wax esters, C_{40}–C_{44}[b]	Monounsaturated LCFA C_{18}–C_{22}	Long-chain C_{20}–C_{22}
Branched wax esters, C_{22}–C_{31}	Monounsaturated MCFA C_8–C_{12}	Long-chain C_{18} with none or one methyl branch
Hydroxy wax esters, C_{38}–C_{40}–OH	C_{18}–OH	Long-chain C_{20}–C_{22}

[a] LCFA = long-chain fatty acid; MCFA = medium-chain fatty acid.
[b] Jojoba-type.

shows the targets for delivery of specific wax esters from *Crambe* using genetic engineering, namely long-chain (jojoba-type) and novel medium-branched wax esters and hydroxy wax esters. Similar compounds already exist in nature. Wax esters containing branched FAs are found in gland waxes of birds [301]. HFAs are produced by castor bean (*R. communis* L.) and *Lesquerella* sp. [343].

Crambe contains 60% of erucic acid in its seed oil and thus has an ideal fatty acid composition for production of jojoba type wax esters. While the world production of JO is only 3.5 kt (2002) potential demand is much higher (estimates ranging from 64 to 200 kt). Once produced with increased volumes at an affordable price level from a high-yielding oilcrop these types of wax esters could profitably be considered for market segments such as automotive, transmission and hydraulic fluids. Addition of JO (15–20%) to SBO effectively improves the oxidation stability of the vegetable oil [40]. However, owing to the high price of jojoba wax esters at present the improved soy oil does not compete with synthetic oils and even less with petroleum-based oils.

The proposed branched chain wax esters mimic wax esters used by birds to confer high water repellent properties to the plumage. The presence of C_4 and C_5 branched FAs and a double bond irregularity in the hydrocarbon chain (preferably internally) disrupts the lipid packing of the wax and reduces the melting point [344]. The effect of lowering melting temperature and consequently the pour point of base oils without adding oxidative instability renders BCFAs very useful in lubricant formulations.

Bio-based HFAs are synthetically produced by introducing hydroxy groups in hydrocarbon chains of common plant oils by using a chemical catalyst. Among unusual FAs, a natural source of HFAs is the seed oil of castor bean (up to 90% C18:1-OH). However, castor bean cultivation is cumbersome and requires laborious production methods. Moreover, its seeds are highly toxic due to the presence of ricin and allergenic proteins. Transgenic production of HFAs as wax esters in a high-yielding agricultural crop is therefore of considerable interest. HFAs are presently used in lubricant formulations as additives in order to improve VI, low temperature and oxidative stability properties in a way similar to branched fatty acids. The hydroxy wax esters from GM *Crambe* are proposed as a value-added wax ester base oil. Moreover, the HFAs can be isolated from the wax ester and used as an additive in lubricant formulations. At high temperatures in the presence of alkali castor oil forms capryl alcohol and sebacic acid. Sebacic acid, a 10-carbon dicarboxylic acid, is used in jet lubricants and in air-cooled combustion motors [67]. EPOBIO assumes that the proposed wax ester types will have different material characteristics and properties (still to be assessed). They should have significant potential utility in various lubricant applications both as base oils and as additives.

A transformation protocol is critical in the process of providing *Crambe* with the necessary genes for wax ester production and needs to be developed [83]. Various other members of the Brassicaceae family, such as *Arabidopsis thaliana* and *B. napus* are routinely transformed [258, 337, 338, 342, 343, 345–350]. A gene discovery programme has been recommended to identify the relevant enzymes for production of the various wax ester classes in high yields [83]. Target genes to modify and redirect the flow of carbon chain in GM *Crambe* to produce jojoba-type wax esters include fatty acid reductase (FAR) with specificity towards C22:1 and acyl-CoA: fatty alcohol acyltransferase (WS) with specificity towards C22:1 alcohol and C_{18} or C_{20} fatty acyl-CoA substrates. The C22:1 alcohol is used by WS for production of wax esters with a C_{18} or C_{20} fatty acid. Table 5.27 summarises the proposed biosynthesis in GM *Crambe* for the other wax esters of interest.

Waxes are synthesised from a long-chain FA such as erucic acid (C22:1) and a fatty alcohol. A high concentration of C22:1 in the oil may increase the amount of wax produced in transgenic *Crambe*. Selection of elevated erucic acid levels by breeding seems to be possible by selecting for a low content of the SFAs. Alternatively, molecular mutation breeding or gene silencing may raise the C18:1 and C22:1 levels and reduce the level of unwanted polyunsaturated fatty acids (14%) in *Crambe* oil which makes the wax esters more vulnerable to oxidation [85]. A reduction in PUFA content can be achieved by abolishing the Δ12-desaturase activity in the *Crambe* seeds. This technique has already been used to drastically reduce C18:2 and C18:3 acids in *Arabidopsis* seeds [347].

The wax ester-accumulating jojoba plant contains only wax esters in the seed oil and no triacylglycerols. Lardizabal *et al.* [338] have shown that insertion of two genes from jojoba and one from honesty (*L. annua* L.) in thale cress (*A. thaliana*) resulted in transgenic plants producing seed oil with up to 70% of wax esters instead of the common TAGs. The value of

Table 5.27 Biosynthesis of wax esters in GM *Crambe*

Enzyme	Specificity
Jojoba type wax esters:	
FAR	C22:1
WS	C22:1 alcohol and C18 or C20 fatty acyl-CoA
Branched medium-chained fatty acids:	
FAR	C18
WS	Methyl branched MCFAs
Thioesterase	C8, C10 or C12
Methyltransferase	Fatty acid chain
Short chained and branched wax esters:	
FAR	C18
WS	Branched short-chained fatty acids
Aminotransferase/oxoacid dehydrogenase	C_4 and C_5 branched fatty acids[a]
Hydroxy wax esters:	
FAR	C22
WS	Hydroxy fatty acids
Δ12-desaturase	Fatty acid chain

[a] From leucine or valine.

such transgenic oils would increase if the proportion of TAGs could be lowered even more. As most TAGs in seeds are synthesised from diacylglycerols (DAGs) by two different DGAT (acyl-CoA: diacylglycerol acyltransferase) enzymes eliminating these enzyme activities in *Crambe* seeds in combination with the introduction of FAR and WS enzymes might increase the percentage of wax in the oil to near 100%.

The potential of bifunctional WS/DGAT enzymes to establish novel processes for biotechnological production of jojoba-like wax esters was demonstrated by heterologous expression in recombinant *Pseudomonas citronellolis* if a long-chain fatty alcohol was provided as a carbon source [351]. By varying the fatty alcohol used as the carbon source it should be possible to vary the composition of the produced wax esters.

Production of high levels of wax esters in *Crambe* will need efficient and tissue-specific promoters. Seed-specific promoters such as napin or FAE have been used for expression in *Arabidopsis* and *B. napus* and are therefore quite likely also a good choice for *Crambe* [346, 348].

Potential agronomic impacts of increasing wax ester levels in *Crambe* seeds, such as the possibility of reduced germination efficiency, need also to be investigated and reconciled for widespread cultivation of the transgenic crop. An agronomy programme is to ensure a robust, mainstream agricultural crop. Oil yields of non-food crops generally need to be improved as they have not yet been optimised for mainstream agriculture.

Processing of *Crambe* seed with high levels of wax esters is best integrated in a biorefining process for increasing the industrial value of non-food/non-feed by-products. Viability for the manufacture of lubricants from the non-food oilcrop *C. abyssinica* is enhanced in scenarios where co-products such as the hulls and meal remaining after oil extraction are used to produce bioenergy in the form of heat and/or electricity.

Production of plant-derived wax ester oils in GM *Crambe* has a considerable economic potential compared with wax esters from jojoba and/or an enzymatic or chemical process. It has been shown that wax ester production for bulk consumption by an enzymatically or chemical catalytic process is less economically viable than via GM *Crambe* [83]. The wax esters also have the potential to compete on a price basis with synthetic base oils and mineral base oils.

Current end-uses of jojoba-type wax esters are very limited due to low production and high price and consist mainly of low-volume, high-cost specialties (cosmetics, pharmaceuticals). At present, the JO lubricant market does not exceed 100 t/yr and is directed to special applications such as high-pressure, high-temperature lubricants in gearboxes, differentials, crankcases and cutting oils. Wax esters are also used in oil-in-water (O/W) emulsions for lubricating conveyor belt systems in the food industry [352]. However, the synthetic ester segment of the lubricant market – which is the target of *Crambe* wax esters – is substantial (500 kt/yr PAOs and 50 kt/yr organic esters). Wax esters constitute a target lubricant base oil but can also function as an additive in vegetable-base oils and improve their stability. The production costs of GM *Crambe* wax ester oils have been estimated for various yield scenarios [83], based on previous *Crambe* field trials [341, 353–355].

Current GM regulations in Europe determine implications for the use of GM plants. Considering also the associated substantial development and regulatory compliance costs exploitation of GM plants is likely to be undertaken only by multinationals. It is not expected that wax esters from GM *Crambe* will reach the marketplace before 2015–2020.

5.5 Plant polymeric carbohydrates

Most biomass consists of natural polymers and 75% of annually renewable biomass of 200 billion tons are carbohydrate in nature. Virtually all plants contain carbohydrates as (hemi)cellulose, starch and saccharose (see Fig. 4.2). Cellulose is found in association with other polymers, hemicellulose and lignin. Cellulose is a crystalline material which is broken down into glucose by acid hydrolysis with great difficulty [356, 357]. Hemicellulose is a complex polysaccharide that consists of a large fraction of pentose sugars (mainly xylose and arabinose) and some hexoses (galactose, glucose and mannose). The most abundant building block of hemicellulose is xylan (a xylose polymer linked at the 1- and 4-positions). Hemicellulose is amorphous and more easily hydrolysed to its monomer sugars than cellulose. Starch is an energy-storage compound found in the seeds of many plants. Starches, which are glucose polysaccharides with both α-(1–4) and α-(1–6) glycoside linkages, are readily hydrolysed enzymatically. Starch and cellulose are both polymers made up of glucose units, but the units are linked together in a slightly different way in the two polymers. In cellulose the extended polymer chains are essentially linear, whereas in starch the chains have a helical structure. Hydrogen bonds maintain the flat, linear conformation of the cellulose chains. The sides of the cellulose chains are hydrophilic. These structural differences give the two polymers vastly different properties; for example, cellulose is the water-insoluble, fibrous building material of plants, but starch has no fibrous structure and has different solubility properties.

As carbohydrates represent the most important class of organic material in terms of volume produced by biomass, it is necessary to focus special attention on efficient access to these biomass constituents. As a single class of natural products carbohydrates are the major biofeedstocks from which to develop industrially and commercially viable organic chemicals and materials as a replacement of petrochemically derived products in a future carbohydrate economy [358]. Cellulosic material can potentially meet our fuel and chemical feedstock needs if it can be efficiently converted into sugar molecules such as glucose with higher energy densities than the parent biomass [359]. Carbohydrate-based products have the potential to improve the sustainability of natural resources, environmental quality and national security while expanding the world industrial base. Bio-based products have a wide range of uses in energy and intermediate chemicals for food, industrial, consumer and pharmaceutical applications. Agricultural crop producing rural areas are well positioned to support regional processing facilities dependent on locally grown crops.

Cellulose, by far the most abundant of the carbohydrate polysaccharides, is the principal component of the cell walls of all higher plants. Saccharides and polysaccharides are hydroxycarbons; their basic chemical structure is $(CH_2O)_n$. Most hydroxycarbons occur naturally as either five- or six-membered ring structures. The ring structure may include only one or two connected rings (sugars) or they may be very long polymer chains (cellulose, hemicellulose, starch). The basic six-sided saccharide structure is exemplified by glucose. Attempts to fit carbohydrate chemistry within a hydrocarbon-based chemistry framework are doomed to fail.

Current utilisation of carbohydrates as a feedstock for the chemical industry (either bulk, commodity, intermediate, fine or high value-added specialty chemicals) is modest in comparison to their ready availability at low cost. Carbohydrates constitute a largely unexploited, huge potential. Cellulosic plant materials are used as fuel, lumber, mechanical pulp and textiles. The estimated world production of cellulose amounts to 320 Mt/yr (2006). Purified cellulose is currently used in many applications, including wood-free paper, membranes, organic-solvent-soluble polymers, etc. The non-food utilisation of polysaccharides in the textile, paper and coating industries is mostly in the form of simple esters and ethers. Dextrans are commercially produced higher molar mass polysaccharides. The principal cellulose derivative is cellulose acetate, which can be used to make acetate rayon and a variety of thermoplastic products (membranes, films). Cellulose acetate products are biodegradable. Other high value added cellulose derivatives are methylcellulose (MC) and cellulose ether (DWC; Midland, MI). A carbohydrate-to-biodiesel process is under development [360]. The technology utilises an engineered microbe to convert carbohydrate feedstocks (including sugars, glucose, glycerol, hydrolysates from biomass and algae) into lipids.

Low molar mass saccharides are very common in food such as sweets, fruits and honey. Low-molecular-weight organic commodity chemicals are more expediently acquired from low-molecular-weight carbohydrates (mono- and disaccharides) than from polysaccharides. Sucrose is world's most abundantly produced organic compound (approximately 140 Mt/yr; 2004/2005). Bulk scale accessible are also the component sugars D-glucose (30 Mt/yr), produced by hydrolysis of starch, and D-fructose (60 Mt/yr). Despite their large-scale accessibility at comparatively low cost, the chemical industry currently utilises mono- and disaccharides only to a minor extent as feedstock for organic chemicals (mainly for the production of ethanol, sorbitol, citric acid, lysine and glutamic acid). This can be attributed to the still more economic fossil raw materials and exceedingly well-developed process technology for conversion of petrochemical feedstock.

Fermentation ethanol with a production of about 37 Mt in 2005 is the largest-volume bio-based chemical for non-food use today. Actually, 'bioethanol' obtained by fermentation and simple distillation is a mixture consisting of methanol, ethanol, fusel oil (*n*-propyl, *n*-butyl, *sec*-butyl, isobutyl, *n*-amyl and isoamyl alcohol) and water [361]. A DuPont/Genencor joint venture is developing a low-cost technology to make cellulosic ethanol from corn and sugar cane waste (pilot 2009; commercial production 2012).

With an annual production of approximately 250 kt, furfural is the only unsaturated large-volume organic chemical prepared from carbohydrate sources. Glucose, accessible by microbial or chemical methods from starch, sugar or cellulose, holds a key position as a platform chemical because a broad range of biotechnological or chemical products is accessible from glucose. For starch, the advantage of enzymatic compared with chemical hydrolysis is well known [362]. For cellulose this is not yet realised. Cellulose-hydrolysing enzymes are only effective after energy-intensive pretreatment (thermal, thermomechanical or thermochemical) to break up the lignin/(hemi)cellulose composites.

Recently, 12 potential bio-based building block chemicals have been selected which can be produced from sugar by biological and chemical conversions and can subsequently be converted to several high-value bio-based chemicals or materials [363]; see Fig. 5.11. Building block chemicals are molecules with multiple functional groups with high potential to be transformed into new families of useful molecules. The 12 sugar-based building blocks to be obtained by white biotechnology are 1,4 diacids (succinic, fumaric and malic), 2,5-furan-dicarboxylic acid (FDCA), 3-hydroxypropionic acid (3-HPA), aspartic acid (ASP), glucaric acid, glutamic acid (GLU), itaconic acid, levulinic acid, 3-hydroxybutyrolactone (3-HBL), glycerol (GL), sorbitol and xylitol/arabinotol [363]. Succinic acid on the basis of starch is now reaching its commercial stage [364].

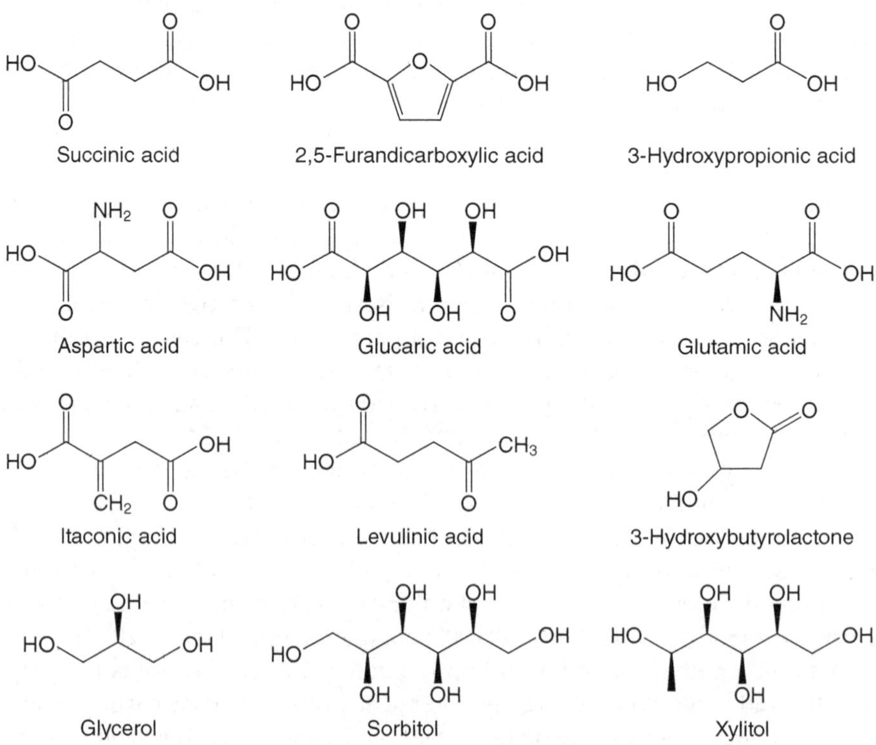

5.11 Top-12 molecules for carbohydrate-based bulk chemistry.

Cellulose, which is chemically rather intractable and insoluble in water and common organic solvents, can be solubilised in some ionic liquids (ILs); see Section 15.2. In the absence of added acid catalysts typically employed in biomass conversion, cellulose dissolved in certain ILs has been converted under relatively mild conditions (<140 °C, 1 atm) into water-soluble reducing sugars with high yield (97%), or directly into the biomass platform chemical 5-hydroxymethylfurfural (HMF) in high conversion (89%) when $CrCl_2$ is added [365]. Recently, direct enzymatic degradation of cellulose to cellodextrin products has been demonstrated [366].

5.5.1 Starch production

Starch is one of the most commonly available agricultural products [367]. Plants assimilate CO_2 from the atmosphere and starch, the most abundant storage carbohydrate, is generated in many plants by photosynthesis. The source of carbon for starch synthesis in storage organs such as cereal endosperms and potato tubers is generally believed to be sucrose. Sucrose is converted to glucose, the base unit for starch production, and finally to starch by a series of enzyme-catalysed reactions [3, 368]. The starch synthases can be divided into two classes, the granule-bound (GBSS I and II) and the soluble starch synthases (SSS I and II). The enzyme responsible for the synthesis of amylose is thought to be ADP-glucose-dependent GBSS [369]. Starch branching enzyme catalyses the formation of amylopectin by the introduction of α-(1–6)-linkages.

Starch occurs as granules in the chloroplasts of green leaves and in the amyloplasts of storage organs such as seeds, tubers, roots and fruits. Amyloplast starch is deposited over a period of days to weeks, stored and then remobilised during the germination of seeds and the sprouting of tubers. The small starch granules (from about 1 to over 100 μm) are insoluble in cold water. Plants that store these carbohydrates in the form of starch granules, which can be extracted, constitute the most useful raw materials for starch production. While starch is mainly derived from cereals, it is also obtained from tubers (such as potatoes) and roots (e.g. manioc or cassava). The main sources for starch (worldwide production of 68 Mt in 2009 with a yearly global growth of 2–3% [370]) for both the food and non-food markets are met by a restricted selection among the major commodity crops, namely maize (80.9%), wheat (8.6%), potatoes (5.3%), and rice and cassava for tapioca starch (5.1%). The quantities derived from other crops (sorghum, sweet potato, amaranth, arrowroot, taro, barley, millet, oat, rye, peas, beans, lentils and yam) are small. Many cellulosic materials, including straw and grass, contain up to 10% starch. Wheat straw contains 3–4% starch.

The US starch production (24.6 Mt; 2000) is dominated by maize starch (99%), with only 1% of wheat starch and minor amounts of potato starch.

There are only a few starch processors in the US (including ADM, Cargill, National Starch). The EU starch industry (9.4 Mt, 2006) is more widely diversified in its raw material sources: maize (46%), wheat (33%), potatoes (21%), including minor amounts of barley and rice. Table 5.28 shows the mean composition of the main raw materials used as sources of starch. In modern production plants the amount of contaminants (proteins, lipids and salts) in commercially produced starch is usually less than 1%.

Because of some basic differences between cereals, tubers and roots, starch production technologies depend on the type of raw materials. Whereas cereal-starch may be produced throughout the year, tuber- and root-starches require production immediately after harvesting. Starches are isolated by wet grinding of the grain or tuber, followed by wet sifting, centrifugation and drying. Wheat starch is an exception in that it may also be extracted from flour.

The process of corn dry milling is traditionally used for food consumption. In this process moist corn granules are physically turned into composite products such as flakes, meal and flour. The dry milling process is also used for industrial products and for fermentation alcohol. Corn refining is distinguished from corn milling in that the refining process separates the corn grain into its components, starch, fibre, protein and oil, and further processes the starch into a variety of products. Corn 'wet milling' is an aqueous slurry process by which the corn grain is separated into its component parts [372]. By wet milling in a corn refinery a 99.5+% pure 'refined' starch may be obtained. The greatest portion of the wet milled cornstarch is converted to sweeteners or ethanol. Starch production from yellow maize is technically well developed [371]. The production of waxy maize starch, a special form of yellow maize containing more than 99% amylopectin starch, is similar to that of normal yellow maize starch.

Industrial production of wheat starch is almost exclusively from wheat flour but may also be based on wet milling of wheat grain. Wheat starch is

Table 5.28 Mean composition of starch raw materials

Component (%)	Maize	Wheat	Potato	Tapioca	Rice
Starch	62	60	19	24	77
Moisture	16	13	75	70	14
Protein	8.2	13	2	1.5	7
Minerals (ash)	1.2	1.7	1.2	2	0.5
Sugars	2.2	8	1.1	0.5	0.3
Fibres	2.2	1.3	1.6	0.7	0.3
Fat	4.2	3	0.1	0.5	0.4

After ref. [371].

characterised by two fractions of different granule size which are obtained separately: prime starch (20–50 μm diameter, 90% starch, <0.3% residual protein) and second-grade starch (2–15 μm diameter, with higher protein content as well as pentosans and hemicelluloses).

Potato is the most important tuber used for industrial starch production. The extraction process is simpler than that used for cereals [371]. However, tubers have a low starch content and produce a large quantity of fruit water, causing a sewage problem. Starch yields in modern potato starch plants exceed 97%. Sweet potatoes are used for starch production in Japan.

Tapioca starch is a root starch produced from the roots of manioc or cassava plants. The starch content is as low as that of potato. As the roots are not storable processing should take place within 24 h from harvesting.

Other cereal starches are used for industrial production regionally (e.g. rice and millet); rye and barley are rarely utilised because of their minor availability, low starch content and difficult processing. The production of tropical starches (from banana, shoti roots and curcuma) as well as starch from different kinds of peas and beans is still in the development stage.

5.5.2 Structure and properties of native starches

All starches are high-molecular-weight biopolymers composed of glucose sub-units linked in various ways and stored as water-insoluble granules in different parts of the plant. During starch biosynthesis two types of polysaccharides are produced, namely amylose and amylopectin (usually in approximately 1:3 ratio) which differ in the arrangement of the glucose units. Amylose starch is composed of largely linear molecules of 1,4 linked α-D-glucopyranosyl units (without branching, see Fig. 5.12) with an average molecular mass of 10^5–10^6. To be more precise, amylose is a mixture of unbranched and randomly, limitedly branched polymers [373, 374]. In amylopectin starch, some glucose units are connected to three glucose units, resulting in a branched structure (Fig. 5.12). Amylopectin thus consists of short α-(1–4) linked chains which are connected by α-(1–6) bonds; the molecular weight of amylopectin (10^7–10^9) is several orders of magnitude higher than that of amylose. The molecular mass determined varies substantially depending on the method of isolation of starch and also on the analytical method (usually size-exclusion chromatography, SEC). The amylose contents can be determined colorimetrically [375] or by means of SEC [376]. Amylose can be extracted from starch by leaching in hot (50 °C) water. As the resulting paste cools, aggregation (retrogradation) occurs.

Starches from most plants comprise a mixture of 75–80% amylopectin and 20–25% amylose (see Table 5.29) [367]. Mutants of yellow maize, so-called waxy maize, produce a starch composed of pure amylopectin. Plant species which produce starches having an unusually high amylose content

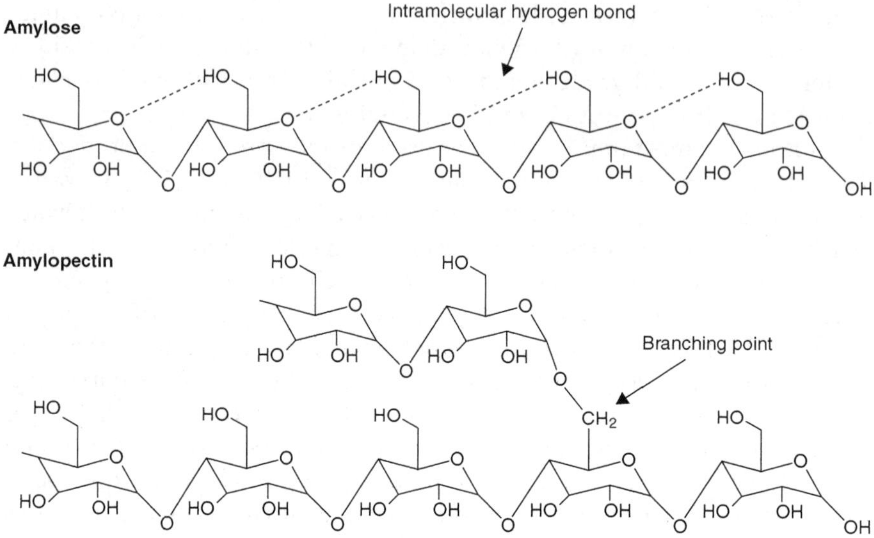

5.12 Structures of starch amylose and amylopectin molecules.

Table 5.29 Composition and properties of commercial (native) starches

Property	Starch type					
	Maize	Waxy maize	Potato	Manioc	Wheat	Rice
Amylose (%)	20–31	<1	16–24	16–18	27–31	15–25
Amylopectin (%)	69–80	>99	70–80	84	69–73	75–80
Moisture (%)	<14	<14	13–21	<17	<14	<14
Crystal structure	A	A	B	C	A	A
Particle size range (μm)	5–25	2–25	15–100	3–25	2–35[a]	2–10
Gelatinisation temperature (°C)	62–80	63–72	56–66	52–65	52–85	61–78
Swelling power at 95 °C	24	64	>1000	71	21	20–30
Paste viscosity	Medium	Medium–high	Very high	High	Medium–low	n.d.
Retrogradation rate	High	Very low	Medium-low	Low	High	n.d.

[a] Bimodal distribution.
After refs [3, 371].

(70%+) have been bred (so-called amylomaize) [377]. Cerestar® (Cerestar Co.; Hammond, IN) is a commercial amylomaize variety. Also potatoes are being developed which produce high-amylose starch.

The physicochemical properties of starches synthesised in plants depend on their origin. Quite obviously, this stands in direct relation to the different specific properties of the two starch fractions. Amylopectin is responsible for the partial crystallinity of starch. X-ray diffraction (XRD) allows the identification of different types of starch and further classification of the material. Cereal starches have been assigned to type A, root and tuber starches to type B, and mixed forms (e.g. manioc starch) to type C. In type B, the double helices are less closely packed than in type A. The molecular structure (linear vs branched) and molecular weight of starch molecules in a crop greatly influence the processability for various applications. Yet starches with similar amylose: amylopectin ratios can differ in physical properties such as viscosity.

Starch in its native state is highly crystalline with helical structure and strong intramolecular H-bonding between the free hydroxyl groups of its glucose units. The free hydroxyl groups are not available to participate in intermolecular H-bonding with other molecules. As a result, starch is insoluble in water at room temperature. However, starch is soluble in solvents (e.g. some ionic liquids) that are capable of disrupting the H-bonding arrangement. In lubrication, water is the most preferred (green) solvent. Chemical modification of starch also enables disruption of the intramolecular hydrogen bonding by the free hydroxyl groups which makes them available for intermolecular H-bonding. The availability of free hydroxyl groups for intermolecular H-bonding causes the starch surface to become highly polar and hydrophilic, and explains the observed high water solubility and high CoF of chemically modified starches. Water-soluble starch can be used as the base to formulate and apply dry-film lubrication.

Native starches only dissolve if heated in an aqueous suspension above a certain temperature, the so-called gelatinisation temperature. The gelatinisation or pasting temperature is a qualitative index of crystalline structure. Gelatinisation is the collapse (disruption) of molecular order within the starch granule manifested in irreversible changes in properties such as granular swelling, native crystallite melting, loss of birefringence and starch solubilisation [378]. Gelatinisation is accompanied by irreversible swelling of starch granules, because disruption of hydrogen bonds of starch helices allows storage of water. The volume of starch can thus increase up to a thousand-fold (Table 5.29), with consequences for the rheological behaviour and practical applications. The gelatinisation temperature range of starch is a characteristic of the genotype of the plant but is affected by the environmental conditions (especially temperature). Gelatinisation is required for particular processes, e.g. textile sizing and industrial starch hydrolysis, as it affects the rheology and viscosity properties of the system and makes the starch more accessible to enzymic action. Gelatinisation effects a transition from a suspension to a paste. On an industrial scale the energy input necessary for starch gelatinisation is considerable, and is a significant part of the processing costs.

Another specific property of starch paste is retrogradation, i.e. a recrystallisation process caused by formation of hydrogen bonds between parallel starch molecules, which reduces the water binding ability and enhances viscosity. Retrogradation involves the alignment of linear segments of amylose chains which associate into a more thermodynamically stable form through hydrogen bonding. Retrogradation can be reduced by chemical modification.

Starch granules in the crop plant range in size (Table 5.29) from those of canna (100 μm) to those of the small-granule starch amaranth (0.5–2.0 μm) [379]. Starch granule shape can be characteristic of a genus and species. The shape and size of starch granules, as measured microscopically, allow identification of the origin of the starch. The mechanical (extraction) processes necessary for the industrial isolation of starch granules can damage the granule structure. The composition and structure of individual starch components determine many starch properties, especially in solution, but the arrangement of these components within the granule determine its structure for applications requiring intact or modified granules.

There are considerable differences in the properties of the starches from various species. These differences depend not only on differences in the relative propositions of amylose and amylopectin and the characteristics of these molecules, but also on the differences in the non-starch components of starch granules such as lipids, proteins and phosphate groups.

Viscosity is an important starch quality parameter. Obtaining a consistent viscosity of starch solutions is a major aim in industrial processes that involve starch pastes. Most native starches do not maintain a stable viscosity when their pastes are subjected to high shear rates and heated for an extended period but chemically modified starches perform appropriately under these conditions.

There are considerable gaps in our knowledge of the relationship between starch chemistry and starch properties. This is because the architecture of the starch granule is inherently complex and methods of analysis not sufficiently sensitive for detailed probing.

The chemistry and properties of starch granules are under genetic control and are also significantly affected by the environmental conditions experienced by the plant during starch deposition. In order to produce starches of specified and novel chemical composition and properties, it will be necessary to alter the genome of starch-producing crop plants. Attempts have been made to develop plants which produce new starches having specific properties (amylopectin/amylose ratio, particle size distribution, degree of branching or amylopectin sidechain length, viscosity, etc.).

5.5.3 Starch modification

The major application of starch is as food and feed. However, new non-food products from starch have been developed. Starch is now to be considered as an industrial raw material, in (partial) replacement of petroleum [380]. Native starch with its granular structure is used as such only in a few applications. For many food and non-food applications the property profiles of starch products require complex chemical, physical or enzymatic modifications of native starch. Development of new applications for starch requires that it be soluble in a suitable solvent, preferably in a benign solvent such as water. However, starch in its native state is insoluble in room temperature water. This is attributed to its highly crystalline structure as a result of a strong intramolecular H-bonding between the free hydroxyl groups of its glucose units. Usually starch is converted into a hydrated or gelatinised form either by the user or the producer. Processing makes modified starches more expensive than native starches.

Figure 5.13 illustrates the main technologies for starch modification. A corn refinery allows downstream processing of the primary product, the carbohydrate. The granular starch slurry can be dried to produce native corn starch, or processed into modified starch products. In slurry technology starches are usually suspended in aqueous solution and are chemically, physically or enzymatically modified. The grain structure of starch remains substantially unaltered. The process yields almost pure starch derivatives. These starch products are insoluble in cold water (cook up starches). In dry reactions the starches are dried under vacuum and then subjected to thermal treatment at temperatures of up to 180 °C with or without the presence of chemicals. The grain structure changes only slightly. Conversion paste reactions, often in autoclaves, aim at obtaining a high degree of substitution.

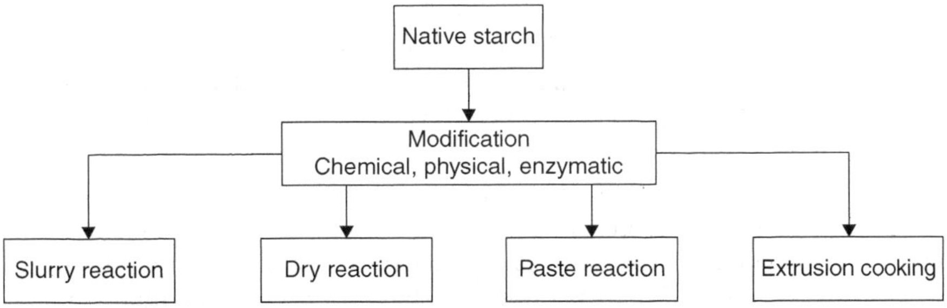

5.13 Starch modification technology.

Reaction conditions are more severe than in slurry processes. The starches are converted into an amorphous structure. The resulting product is very soluble in cold water. Extrusion cooking also yields products that are soluble in cold water. The process allows only limited chemical modification and low degrees of substitution. Extrusion cooking is predominantly used for physical modification of starch.

As there is a need for starch products having better or different properties this has led to the development of new technologies for the production and modification of starch products. For example, ultrasonics are useful for starch extraction, whereas new treatment methods, such as high-pressure or cryo treatment, lead to starches with special rheological characteristics. Reactive extrusion processing requires further development.

Using the aforementioned technologies and chemical, physical or enzymatic treatments leads to starch molecules which are degraded, chemically modified by the introduction of functional groups, or changed by hydrothermal technology (annealing and heat-moisture treatment). Hydrothermal technology modifies the physicochemical properties of starch without destroying the granular structure. Degraded starches are produced by mechanical, thermal, acidic, oxidative and enzymatic degradations. Starches are obtained with different degrees of reduced polymerisation. Usually the amylopectin is degraded and the amylose only slightly split. Enzymatically degraded starch products, using enzymes such as α- and β-amylases, glucoamylases and oligosaccharide hydrolases, are mainly used in food applications, as well as for adhesives and paper surface treatment.

Dry reactions may be used to produce acetylated, phosphated and carbamate starches. Already low degrees of substitution of starch (degree of substitution 0.01–0.1) will result in major changes in the properties. Esterification (e.g. to phosphates and acetylated starches) and etherification are important kinds of modification. Hydroxyalkylation of starch, performed by reaction with propylene oxide, is certainly the most important reaction. Etherification also allows introduction of ionic groups into starch. Anionic (e.g. carboxymethyl) starches are used in sizings and textile printing. Cationic starches (i.e. positively charged polysaccharides) are usually produced by reacting starch with tertiary or quaternary ammonium compounds. Cationic starch ethers have a special affinity to negatively charged celluloses, fibres, sludges, etc. and are used in the paper industry and as flocculants. The modification of starch (Fig. 5.14) varies depending on the chosen application.

Improvements in the properties of starches for industrial use can also be achieved by the manipulation of starch biosynthesis in the plant itself by conventional breeding techniques or through the use of genetic modification. Growers try to breed varieties of starch-producing plants of higher agricultural performance (higher yield per hectare, better pest resistance, etc.) or starch characteristics. A mutant of yellow maize (so-called waxy

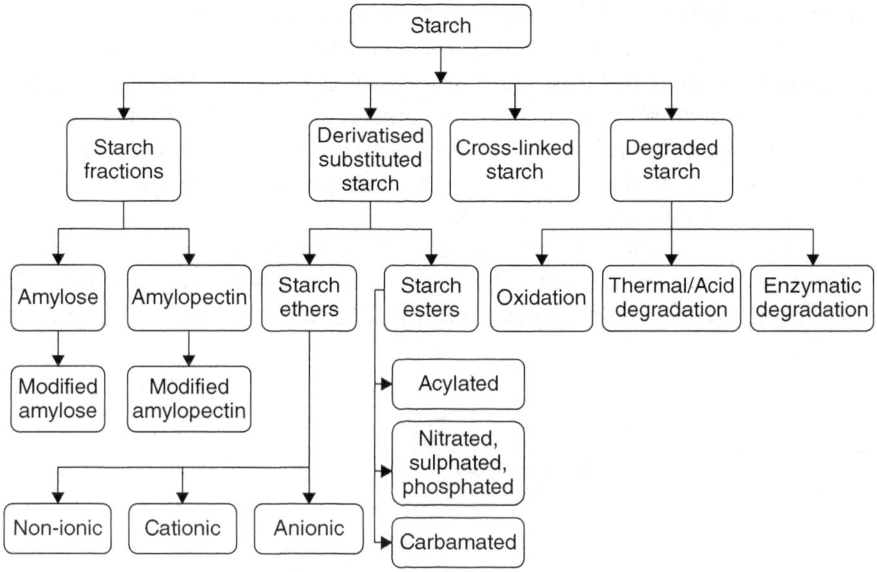

5.14 Types of modified starch.

maize), which produces starch made up of pure amylopectin, overcomes the need for the complicated fractionation of conventional starch for the production of amylopectin. Also mutants or GM wheat, rice, tapioca and potatoes already have their specific industrial uses. Plant species which produce high amylose content starches have been bred in the conventional way (amylomaize). Potatoes are also being developed for the production of high-amylose starch. It is to be expected that other starches having special functions (e.g. altered particle size distribution, higher viscosity, different degrees of branching) will be introduced to the market. Transgenic plants produced by new biotechnological tools are expected to produce new tailor-made starches.

5.5.4 Industrial use of starch and starch derivatives

Starch use has been recorded in early history, e.g. to cement strips of papyrus, to stiffen and whiten cloth, and as a wholesome dietary product. With the diversity of native starches and the numerous modifications many different starches are available nowadays for selected applications. Native starches, modified starches and starch derivatives are used in the manufacture of a wide variety of both food [381] and non-food products [371, 382, 383]. This is usually accomplished by blending of starch with ingredients such as sugars, oils or fatty acids.

The starch industry produces over 600 products, from native starches to physically or chemically modified starches, through to liquid and solid sweeteners. Starch products are used as ingredients and functional supplements in a vast array of food, non-food and feed applications (Fig. 5.15). Thousands of food products use corn starch derivatives. While the major applications of starch are as food, new non-food products from starch are being developed [380]. Application areas have been opened up in particular by starch modification, resulting in new properties, and by biotechnologically produced tailor-made starches.

The global demand for starch is growing steadily. Its use in the largest markets, the United States and the EU, is quite different, as shown in Table 5.30. In the US the main applications are for the production of isoglucose,

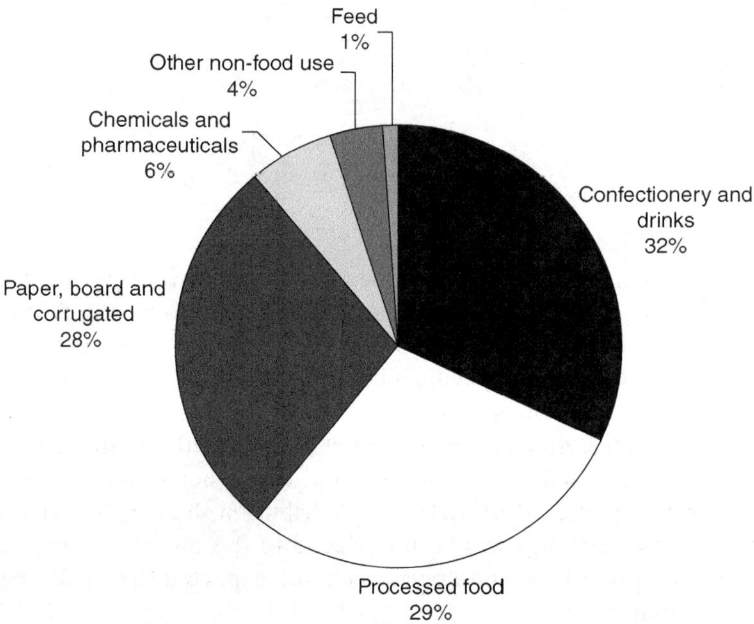

5.15 Sectors of application of starch products in the EU in 2009.

Table 5.30 Starch use in US and EU in 2000 (Mt)

Product	US	EU
Ethanol	10.1	0.1
Isoglucose/HFCS	7.9	0.3
Other syrup-based products	3.2	3.6
Native and modified starches	3.6	4.5

After ref. [371].

high-fructose corn syrup (HFCS) and ethanol. Production of starch and starch derivatives from maize, wheat and potato in the EU market reached 8.5 Mt in 2000 and 9.4 Mt in 2009. Consumption of these products in Europe (excluding co-products) amounted to 8.5 Mt in 2009 with 62% total food/feed and 38% total non-food applications. In the latter category, the paper, board and corrugating industries are the largest user of starch and starch derivatives (Fig. 5.15). Other important fields are textiles, adhesives, pharmaceuticals, cosmetics, construction, paints and agrochemicals. The use as a lubricant filler or extender is only marginal (see Section 3.3.1).

The main industrial uses of starch are summarised in Table 5.31. The ability of starch to adjust the viscosity of solutions and pastes is used in the food and lubricant industries. This property is also exploited in the formulation of starch-based drilling fluids used to drill wells into geological formations for exploitation of oil, gas or minerals [384]. Starch in suspension may act as a liquid lubricant [385].

The dimensions of small starch granules make them suitable as a paper coating [386]. Granular starches have also been used as biodegradable fillers in polymers. Biodegradable plastic production is still in its infancy when compared to petrochemical plastic production.

Demands on sizing agents are high solubility, good adhesion to the fibre, formation of a tough, abrasion-resistant and elastic film, and good desizing. Originally, native products were used for sizings; they have now been replaced by degraded modified starch esters and starch ethers. Starch derivatives also act as excipients in pharmaceutical products, are used as flocculants and in the textile glass-fibre industry.

Table 5.31 Industrial uses of starch

Industry	Use of starch/modified starch
Adhesive	Adhesive production
Agrochemical	Pesticide delivery, seed coatings
Cosmetics	Face powder
Detergent	Surfactants, builders, bleaching agents
Fermentation	Enzymes
Fine chemicals	Glucose
Food	Viscosity modifier, glazing agent
Lubricant	Dry film
Medical	Transplant organ preservation, absorbents
Oil drilling	Viscosity modifier
Paper and board	Binder, sizing, coating
Pharmaceuticals	Diluent, binder, drug delivery
Polymer	Biodegradable filler
Separation	Flocculant
Textile	Sizing, finishing, fire resistance

As a consequence of a large surplus over the demand for starch-bearing crops, which has depressed the price of starch-producing agricultural products and, hence, the income of farmers growing such crops, there has been a considerable effort at developing new starch-based products [387, 388]. Non-food applications of starch that lately have been investigated include biodegradable polymers [389], composites [390], elastomers [391] and lubricants [392]. USDA has developed a starch-based product called Fantesk™ that has opened up new applications for starch-based products in food and non-food industries [393, 394].

It is expected that extending the range of the chemical composition of the native starches produced in crop plants will greatly increase the opportunities for applications requiring specific starch properties. The industrial starch platform has recently been reviewed by Grüll *et al.* [371].

Industrial biotechnology and the application of starch as a source of carbon in the fermentation industry are becoming increasingly important, especially as fossil energy resources are becoming scarce. Starch processing and refining plants are highly sophisticated bio-process operations. Hydrolysis of starch accounts for a significant use of starch [395]. In particular, starch may serve as a feedstock for glucose. Starch-based or sucrose-based processes are already widely used to make ethanol. The production of bioethanol from starch-containing raw materials, in particular maize in the United States, is booming. Usually, starch is saccharified enzymatically (by α-amylase and amyloglucosidase) and the glucose-containing organic hydrolysate is fermented to alcohol by means of yeast. Bioethanol thus produced is mostly used in admixture to mineral fuels but also for the synthesis of ethyl tertiary butyl ether (ETBE). The residues from fermentation may be made into highly nutritious animal feed. Sucrose-based feedstocks for bioethanol are also sugar cane (Brazil) and sugar beets (Europe). Other feedstocks used to make small quantities of bioethanol in some regions include potatoes.

Starch hydrolysate also serves as a carbon source in many other fermentations, e.g. for the production of many organic acids such as citric, acetic, gluconic and itaconic acid. Other organic acids, such as malic, fumaric and L-tartaric acid, can be produced biotechnically from glucose syrup. Processing of the hydrolysis products of starch by microbial fermentations can also lead to conversion into alcohols, ketones, polyols, amino acids, nucleotides, biopolymers, lipids, proteins, vitamins (e.g. vitamin C), antibiotics and hormones [396]. In these applications starch thus acts as a feedstock chemical for numerous fine chemicals.

Most of the enzymes which are widely used in food and non-food applications are produced in biotechnical processes using glucose of starch as the carbon source. α-Amylases and amyloglucosidases are produced on an industrial scale by fermenting hydrolysed maize starches.

There is also growing interest in alternative economical processes for producing organic chemicals, such as propanediol, butanediol and 2-propanol, by fermentation of glucose syrups.

Starch offers the important product property of sustainability because of biodegradability. Another great advantage of starch products is their low cost. The increasing need for replacement of crude oil as a carbon source will trigger starch use in new application areas. In many fields of application new solutions are being investigated that are based on starch. Intensive research has already been carried out in the field of thermoplastic starches, and specially modified starches for inks, coatings, paints and lubricants (see Section 6.4). New product introductions may be expected.

5.6 Sources of further information and advice

Nuffield Council of Bioethics, *Biofuels, Ethical Issues*, Nuffield Press, Abingdon (UK) (2011); http://www.nuffieldbioethics.org

L.A. Johnson, P.J. White and R. Galloway, *Soybeans: Chemistry, Processing and Utilization*, AOCS Publ., Urbana, IL (2010).

J.H. Clark and F.E.I. Deswarte (eds), *Introduction to Chemicals from Biomass*, John Wiley & Sons, Ltd, Chichester (2008).

D.E. Vance and J.E. Vance (eds), *Biochemistry of Lipids, Lipoproteins, and Membranes*, 5th edn, Elsevier, Amsterdam (2008).

M.K. Gupta, *Practical Guide to Vegetable Oil Processing*, AOCS Press, Urbana, IL (2007).

J.S. Marth, *Renewable Lubricants Manual: Biobased Oils, Fluids & Greases*, United Bio Lube, Palo Alto, CA (2007); http://www.renewablelubricants.com

A.S. Carlsson, *Production of Wax Esters in* Crambe, CPL Press, Newbury (2006); http://www.epobio.net

B. Kamm, P.R. Gruber and M. Kamm (eds), *Biorefineries – Industrial Processes and Products*, Wiley-VCH, Weinheim (2006), 2 vols.

R.A. Della Porta (ed.), *Edible Oils Manual*, AOCS Press, Urbana, IL (2006).

K. Larsson, P. Quinn, K. Sato and F. Tiberg, *Lipids: Structure, Physical Properties and Functionality*, Vol. 19, The Oily Press Lipid Library, Bridgwater (2006).

F. Shahidi (ed.), *Bailey's Industrial Oils and Fat Products*, 6th edn, Wiley-Interscience, Hoboken, NJ (2005), 6 vols.

S.Z. Erhan (ed.), *Industrial Uses of Vegetable Oils*, AOCS Press, Urbana IL (2005).

R.D. Perlack, L.L. Wright, A.F. Turhollow, R.L. Graham, B.J. Stokes and D.C. Erbach, *Biomass as Feedstock for a Bioenergy and Bioproducts Industry: The Technical Feasibility of a Billion Ton Annual Supply*, DOE/USDA, Washington, DC (2005).

A. Steinbüchel and S.K. Rhee (eds), *Polysaccharides and Polyamides in the Food Industry*, Wiley-VCH, Weinheim (2005), 2 vols.
Commission of the European Communities, *Biomass Action Plan*, Brussels (2005), COM 628.
R.M. Goodman (ed.), *Encyclopedia of Plant and Crop Science*, Marcel Dekker, Inc., New York, NY (2004).
O.B. Würzburg (ed.), *Modified Starches: Properties and Uses*, CRC Press, Boca Raton, FL (2004).
A.-C. Eliasson (ed.), *Starch in Food: Structure, Function and Applications*, Woodhead Publishing, Ltd, Cambridge (2004).
G. Tegge (ed.), *Stärke und Stärkederivative*, 3rd edn, Behr's Verlag, Hamburg (2004).
T. Werpy and G. Petersen (eds), *Top-value Added Chemicals from Biomass*, Vol. 1, US DOE, Washington, DC (2004).
H.R. Boerma and J.E. Specht (eds), *Soybeans: Improvement, Production and Uses*, 3rd edn, Amer. Soc. Agronomy, Madison, WI (2004).
R.C. Brown, *Biorenewable Resources*, Iowa State University Press, Ames, IA (2003).
Nuffield Council on Bioethics, *The Use of Genetically Modified Crops in Developing Countries: A Follow up*, Nuffield Council on Bioethics, London (2003).
J.A. Duke, *Handbook of Energy Crops*, Purdue University Center for New Crops and Plant Products, West Lafayette, IN (2003); http://www.hort.purdue.edu./newcrop
T.M. Kuo and H.W. Gardner (eds), *Lipid Technology*, Marcel Dekker, Inc., New York, NY (2002).
Biomass R&D Technical Advisory Committee, *Roadmap for Biomass Technologies in the United States*, Washington, DC (2002).
F.D. Gunstone (ed.), *Vegetable Oils in Food Technology: Composition, Properties and Uses*, CRC Press, Boca Raton, FL (2002).
D. Strayer, *Identity-Preserved Systems. A Reference Handbook*, CRC Press, Boca Raton, FL (2002).
F.D. Gunstone and R.J. Hamilton (eds), *Oleochemical Manufacture and Applications*, Sheffield Academic Press and CRC Press, Sheffield/Boca Raton, FL (2001).
Organisation for Economic Co-operation and Development (OECD), *The Application of Biotechnology to Industrial Sustainability – A Primer*, OECD Publishing, Paris (2001).
H. Zoebelein (ed.), *Dictionary of Renewable Resources*, 2nd edn, Wiley-VCH, Weinheim (2001).
R.D. O'Brien, W. Farr and P.J. Wan (eds), *Introduction to Fats and Oils Technology*, 2nd edn, AOCS Press, Urbana, IL (2000).

L. Roth and K. Kormann, *Ölpflanzen – Pflanzenöle*, Ecomed-Verlagsgesellschaft, Landsberg (2000).
D. Zohary and M. Hopf, *Domestication of Plants in the Old World*, Oxford University Press, Oxford (2000).
U.T. Bornscheuer (ed.), *Enzymes in Lipid Modification*, Wiley-VCH, Weinheim (2000).
National Agricultural Biotechnological Council, *The Biodiesel Economy of the 21st Century: Agriculture Expanding into Health, Energy, Chemicals and Materials*, NABC Report 12, Ithaca, NY (2000).
V. Moses, R.E. Cape and D.G. Springham (eds), *Biotechnology: The Science and the Business*, Harwood, New York, NY (1999).
Nuffield Council on Bioethics, *Genetically Modified Crops: The Ethical and Social Issues*, Nuffield Council on Bioethics, London (1999).
M.C. Flickinger and S.W. Drew (eds), *Encyclopedia of Bioprocess Technology: Fermentation, Biocatalysis and Bioseparation*, John Wiley & Sons, Inc., New York, NY (1999).
G.T.S. Mugnozza, E. Porceddu and M.A. Pagnotta (eds), *Genetics and Breeding for Crop Quality and Resistance*, Kluwer Academic Publishers, Dordrecht (1999).
W. Hamm and R.J. Hamilton, *Edible Oil Processing*, Sheffield Academic Press and CRC Press, Sheffield/Boca Raton, FL (1999).
R.P. Overend and E. Chornet (ed.), *Biomass: A Growth Opportunity in Green Energy and Value-added Products*, Pergamon Press, Oxford (1999).
K. Liu, *Soybeans: Chemistry, Technology and Utilization*, Aspen Publishers, Inc., Gaithersburg, MD (1999).
R.D. O'Brien, *Fats and Oils: Formulating and Processing for Applications*, Technomic Publ. Co., Lancaster, PA (1998).
D.L. Klass (ed.), *Biomass for Renewable Energy, Fuels and Chemicals*, Academic Press, San Diego, CA (1998).
Fachagentur Nachwachsende Rohstoffe (ed.), *Chemische Nutzung heimischer Pflanzenöle*, Landwirtschaftsverlag, Münster (1998).
J.L. Harwood (ed.), *Plant Lipid Biosynthesis: Fundamentals and Agricultural Applications*, Cambridge University Press, Cambridge (1998).
E.N. Frankel, *Lipid Oxidation*, The Oily Press, Dundee (1998).
M. Bockisch, *Fats and Oils Handbook*, AOCS Press, Urbana, IL (1998).
J. Smartt and N. Haq (eds), *Domestication, Production and Utilization of New Crops*, Colorline Printers, Dhaka (1997).
F.D. Gunstone and F.B. Padley (eds.), *Lipid Technologies and Applications*, CRC Press, Boca Raton, FL (1997).
J.K. Setlow (ed.), *Genetic Engineering*, Plenum Press, New York, NY (1997).
J. Janick (ed.), *Progress in New Crops*, ASHS Press, Alexandria, VA (1996).
H. Eierdanz (ed.), *Perspektiven nachwachsender Rohstoffe in der Chemie*, VCH, Weinheim (1996).

K. Warner and N.A.M. Eskin (eds), *Methods to Assess Quality and Stability of Oils and Fat-containing Foods*, AOCS Press, Champaign, IL (1995).

R.J. Hamilton (ed.), *Waxes: Chemistry, Molecular Biology and Functions*, The Oily Press, Dundee (1995).

D.J. Murphy (ed.), *Designer Oil Crops: Breeding, Processing and Biotechnology*, VCH, Weinheim (1994).

B.S. Kamel and Y. Kakudo (eds), *Technological Advances in Improved and Alternative Sources of Lipids*, Chapman & Hall, London (1994).

F.D. Gunstone, J.L. Harwood and F.B. Padley, *The Lipid Handbook*, 2nd edn, Chapman & Hall, London (1994).

K.A. Larson, *Lipids: Molecular Organization, Physical Functions and Technical Applications*, The Oily Press, Dundee (1994).

W.J. Bartz, *Additive für Schmierstoffe*, Expert-Verlag, Renningen-Malmsheim (1994).

W.J. Bartz (ed.), *Biologisch schnell abbaubare Schmierstoffe und Arbeitsflüssigkeiten speziell auf pflanzlicher Basis*, Expert-Verlag, Ehningen (1993).

P.R. Shewry and A.K. Stobart (eds), *Seed Storage Compounds: Biosynthesis, Interactions and Manipulation*, Clarendon Press, Oxford (1993).

S.L. Mackenzie and D.C. Taylor (eds), *Seed Oils for the Future*, AOCS Press, Champaign, IL (1993).

J. Janick and J.E. Simon (eds), *New Crops*, John Wiley & Sons, Inc., New York, NY (1993).

K.R.M Anthony, J. Meadly and G. Röbbelen (eds), *New Crops for Temperate Regions*, Chapman & Hall, London (1993).

T.S. Moore Jr (ed.), *Lipid Metabolism in Plants*, CRC Press, Boca Raton, FL (1993).

F.C. Naughton (ed.), *The Chemistry of Castor Oil and Its Derivatives and Their Applications*, ICOA Technical Bulletin No. 2, ICOA, Westfield, NJ (1992).

F.D. Gunstone and B.G. Herslöf, *A Lipid Glossary*, The Oily Press, Dundee (1992).

A. Cherif, D.B. Miled-Daoud, B. Marzouk, A. Smaoui and M. Zarrouk (eds), *Metabolism, Structure and Utilisation of Plant Lipids*, Centre Pédagogique, Tunis (1992).

S.L. Mackenzie and D.C. Taylor (eds), *Seed Oils for the Future*, AOCS Press, Champaign, IL (1992).

H. Niewiadomski, *Rapeseed Chemistry and Technology*, Elsevier, New York, NY (1990).

Alternative Field Crops Manual, University of Wisconsin Cooperative Extension Service/University of Minnesota Extension Service and Center for Alternative Plant and Animal Products (1990).

O. Kitani and C.W. Hall, *Biomass Handbook*, Gordon & Breach, New York, NY (1990).

G. Hoffman, *The Chemistry and Technology of Edible Oils and Fats and Their High Fat Products*, Academic Press, London (1989).

G. Röbbelen, R.K. Downey and A. Ashri (eds), *Oil Crops of the World*, McGraw-Hill, New York, NY (1989).

P.A. Biacs, K. Gruiz and T. Kremmer (eds), *Biological Role of Plant Lipids*, Akadémiai Kiadó, Budapest and Plenum Publishing, New York, NY (1989).

D.J. Sessa and J.L. Willet (eds), *Paradigm for Successful Utilization of Renewable Resources*, AOCS Press, Champaign, IL (1988).

T. Galliard (ed.), *Starch Properties and Potential: Critical Reports on Applied Chemistry*, John Wiley & Sons, Ltd, Chichester (1987), Vol. 13.

J. Wisniak, *The Chemistry and Technology of Jojoba Oil*, American Oil Chemists' Society, Champaign, IL (1987).

W.H. Stadtmiller and A.N. Smith (eds), *Aspects of Lubricant Oxidation*, ASTM International, West Conshohocken, PA (1986).

E.S. Pattison (ed.), *Fatty Acids and Their Industrial Application*, Marcel Dekker, Inc., New York, NY (1986).

G. Kunkel, *Plants for Human Consumption*, Koeltz Scientific Books, Koenigstein (1984).

R.L. Whistler, J.N. Bemiller and E.F. Paschall (eds), *Starch: Chemistry and Technology*, 2nd edn, Academic Press, New York, NY (1984).

C. Ratledge, P. Dawson and J. Rattray (eds), *Biotechnology for the Oils and Fats Industry*, AOCS, Champaign, IL (1984).

J.K.G. Kramer, F.D. Sauer and W.J. Pigden (eds), *High and Low Erucic Acid Rapeseed Oils*, Academic Press, San Diego, CA (1983).

J.A. Schey, *Tribology in Metalworking Friction, Lubrication and Wear*, American Society of Metals, Metals Park, OH (1983).

E.A. Weiss, *Oilseed Crops*, Longman, Inc., New York, NY (1983).

S. Tsunada, K. Hinata and G. Campo (eds), *Brassica Crops and Wild Allies. Biology and Breeding*, Japan Scientific Press, Tokyo (1980).

C.N. Williams, W.Y. Chew and J.H. Rajaratnam, *Trees and Field Crops of the Wetter Regions of the Tropics*, Longman, Inc., London (1980).

C. Paquot, *Standard Methods for the Analysis of Oils, Fats and Derivatives*, IUPAC Commission on Oils, Fats and Derivatives, Pergamon Press, London (1979).

J.F. Carter (ed.), *Sunflower Science and Technology*, American Society of Agronomy, Madison, WI (1978).

J.G. Vaughan, *The Structure and Utilisation of Oil Seeds*, Chapman & Hall, London (1976).

J.G. Vaughan, A.J. Macleod and B.M.G. Jones (eds), *The Biology and Chemistry of the* Cruciferae, Academic Press, New York, NY (1976).

5.7 References

[1] J.H. Clark and F.E.I. Deswarte (eds), *Introduction to Chemicals from Biomass*, John Wiley & Sons, Ltd, Chichester (2008).

[2] R. Verhé, Dilemma: Petrochemistry and oleochemistry as resources for fuel and oleochemicals, *Eur. J. Lipid Sci. Technol.* **112** (2010) 427.

[3] R.P. Ellis, M.P. Cochrane, M.F.B. Dale, C.M. Duffins, A. Lynn, I.M. Morrison. R.D.M. Prentice, J.S. Swanston and S.A. Tiller, Starch production and industrial use, *J. Sci. Food Agric.* **77** (1998) 289–311.

[4] F.D. Gunstone and J.L. Harwood, Occurrence and characterisation of oils and fats, in *The Lipid Handbook* (F.D. Gunstone, J.L. Harwood and A.J. Dijkstra, eds), 3rd edn, Taylor & Francis, Boca Raton, FL (2006), pp. 37–142.

[5] Agricultural Marketing Service (WAOB), *World Agricultural Supply and Demand Estimates*, USDA, Washington, DC (9 April 2010).

[6] Foreign Agricultural Service, *Oilseeds, World Markets and Trade*, USDA, Washington, DC (yearly).

[7] F. Shahidi (ed.), *Bailey's Industrial Oil and Fat Products*, 6th edn, John Wiley & Sons, Inc., Hoboken, NJ (2005), 6 vols.

[8] R. Verhé, Industrial products from lipids and proteins, in *Renewable Bioresources. Scope and Modification for Non-food Applications* (C.V. Stevens and R. Verhé, eds), John Wiley & Sons, Ltd, Chichester (2004), pp. 208–50.

[9] M. Svensson, Surfactants based on natural fatty acids, in *Surfactants from Renewable Resources* (M. Kjellin and I. Johansson, eds), John Wiley & Sons, Ltd, Chichester (2010), pp. 3–20.

[10] J.M. Dyer, S. Stymne, A.G. Green and A.S. Carlsson, High-value oils from plants, *Plant J.* **54** (2008) 640–55.

[11] S.Z. Erhan (ed.), *Industrial Uses of Vegetable Oil*, CRC Press, Boca Raton, FL (2005).

[12] S.Z. Erhan and A. Adhvaryu, Vegetable-based base stocks, in *Biobased Industrial Fluids and Lubricants* (S.Z. Erhan and J.M. Perez, eds), AOCS Press, Champaign, IL (2003), pp. 1–19.

[13] S.J. Asadauskas and J.M. Perez, Utilisation of vegetable oils in metalworking fluids, in *Biobased Industrial Fluids and Lubricants* (S.Z. Erhan and J.M. Perez, eds), AOCS Press, Champaign, IL (2003), pp. 59–64.

[14] B.N. Rhodes and D. Johnson, Vegetable-based motor oils, in *Biobased Industrial Fluids and Lubricants* (S.Z. Erhan and J.M. Perez, eds), AOCS Press, Champaign, IL (2003), pp. 85–109.

[15] L.R. Rudnick and S.Z. Erhan, Natural oils as lubricants, in *Synthetics, Mineral Oils and Bio-Based Lubricants: Chemistry and Technology* (L.R. Rudnick, ed.), CRC Press, Boca Raton, FL (2006), pp. 353–60.

[16] L.W. Kleingartner, NuSun sunflower oil: redirection of an industry, in *Trends in New Crops and New Uses* (J. Janick and A. Whipkey, eds), ASHS Press, Alexandria, VA (2002), pp. 135–8.

[17] J.W. Burton, R.F. Wilson, C.A. Brim and R.W. Rinne, Registration of soybean germplasm lines with modified fatty acid composition of seed oil, *Crop Sci.* **29** (1989) 1583.

[18] R.C. Leffel, Registration of BARC-12, a low linolenic acid soybean germplasm line, *Crop Sci.* **34** (1994) 1426–7.

[19] G. Hou, G.R. Ablett, K.R. Pauls and I. Rajcan, Environmental effects on fatty acid levels in soybean seed oil, *J. Am. Oil Chem. Soc.* **83**(9) (2006) 759–63.
[20] D.C. Drown, K. Harper and E. Frame, Screening vegetable oil alcohol esters as fuel lubricity enhancers, *J. Am. Oil Chem. Soc.* **78** (2001) 579–84.
[21] M.J. Haas, Animal fats, in *Bailey's Industrial Oil and Fat Products*, 6th edn, Vol. 1 (F. Shahidi, ed.), John Wiley & Sons, Inc., Hoboken, NJ (2005), pp. 161–212.
[22] European Commission, *Preliminary Report on Quantitative Risk Assessment on the Use of the Vertebral Column for the Production of Gelatine and Tallow*, submitted to the Scientific Steering Committee, Brussels (13–14 April 2000).
[23] Department of Health and Human Service, US Food and Drug Administration, 21 CFR Part 589, *Substances Prohibited from Use in Animal Food or Feed*, US Fed. Regist. 62, 30935-78.
[24] World Health Organization, Report of a WHO Consultation on Medicinal and Other Products in Relation to Human and Animal Transmissible Spongiform Encephalopaties, *Inform* **12** (2001) 588.
[25] Note for guidance on minimising the risk of transmitting animal spongiform encephalopathy agents via human and veterinary medicinal products (EMEA/410/01 Rev. 2 – October 2003) adopted by the Committee for Proprietary Medicinal Products (CPMP) and by the Committee for Veterinary Medicinal Products (CVMP), *Off. J. Eur. Union* **C 24**/6 (28 January 2004).
[26] A.V. Avellan and C. Alfos, Valorisation non alimentaire des huiles de friture usagées en tant que lubrifiants biodegradables, *Oléag. Corps Gras Lipides* **12**(4) (2005) 290–4.
[27] R. König (ed.), *C. Plini* Secundi Naturalis Historiae Liber XXVIII, Artemis, Munich (1988), par. 141.
[28] British Patent 6,945 (7 December 1835).
[29] British Patent 12,571 (16 April 1849).
[30] C.E. Temple-Heald, Sourcing erucic acid for the industrial market, *Lipid Technol. Newsl.* **6** (2000) 59–63.
[31] T.A. Voelker, T.R. Hayes, A.N. Cranmer, J.C. Turner and H.N. Davies, Genetic engineering of a quantitative trait: metabolic and genetic parameters influencing the accumulation of laurate in rapeseed, *Plant J.* **9** (1996) 229–41.
[32] T.L. Kurth, B.K. Sharma, K.M. Doll and S.Z. Erhan, Adsorption behavior of epoxidized fatty esters via boundary lubrication coefficient of friction measurements, *Chem. Eng. Commun.* **194**(8) (2007) 1065–77.
[33] B.K. Sharma, K.M. Doll and S.Z. Erhan, Ester hydroxy derivatives of methyl oleate: Tribological, oxidation and low temperature properties, *Bioresour. Technol.* **99** (15) (2008) 7333–40.
[34] S.Z. Erhan, A. Adhvaryu and B.K. Sharma, Chemically functionalized oils, in *Synthetics, Mineral Oils and Bio-based Lubricants: Chemistry and Technology* (L.R. Rudnick, ed.), CRC Press, Boca Raton, FL (2006), pp. 361–87.
[35] S.Z. Erhan, B.K. Sharma and J.M. Perez, Oxidation and low temperature stability of vegetable oil-based lubricants, *Ind. Crops Prod.* **24**(3) (2006) 292–9.
[36] B.K. Sharma, J.M. Perez and S.Z. Erhan, Soybean oil-based lubricants: a search for synergistic antioxidants, *En. Fuels* **21**(4) (2007) 2408–14.

[37] K. Liu, Chemistry and nutritional value of soybean components, in *Soybeans: Chemistry, Technology and Utilization*, Chapman & Hall, New York, NY (1997), pp. 25–113.
[38] W.E. Neff, M.A. El-Agaimy and T.L. Mounts, Oxidative stability of blends and interesterified blends of soybean and palm olein, *J. Am. Oil Chem. Soc.* **71** (1994) 1111–16.
[39] N. Shen, W. Fehr, L. Johnson and P. White, Oxidative stabilities of soybean oils with elevated palmitate and reduced linolenate contents, *J. Am. Oil Chem. Soc.* **74** (1997) 299–302.
[40] M. Torres, C. Lioret, M. Sosa and D. Maestri, Composition and oxidative stability of soybean oil in mixtures with jojoba oil, *Eur. J. Lipid Sci. Technol.* **168** (2006) 513–20.
[41] K.M. Doll, B.R. Moser, B.K. Sharma and S.Z. Erhan, Current uses of vegetable oil in the surfactant, fuel and lubrication industries, *Chimica Oggi/ Chem. Today* **24**(6) (2006) 41–4.
[42] L.G. Angelini, E. Moscheni, G. Colonna, P. Belloni and E. Bonari, Variation in agronomic characteristics and seed oil composition of new oilseed crops in Central Italy, *Ind. Crops Prod.* **6** (1997) 313–23.
[43] Y.H. Chu and Y.L. Kung, A study on vegetable oil blends, *Food Chem.* **62** (1998) 191–5.
[44] T.A. Isbell, T.P. Abbott and K.D. Carlson, Oxidative stability index of vegetable oils in binary mixtures with meadowfoam oil, *Ind. Crops Prod.* **9** (1999) 115–23.
[45] R.A. Moreau, Corn oil, in *Bailey's Industrial Oil and Fat Products*, 6th edn, Vol. 2 (F. Shahidi, ed.), John Wiley & Sons, Inc., Hoboken, NJ (2005), pp. 149–72.
[46] B.J. Bremmer and L. Plonsker, *Bio-based Lubricants: A Market Opportunity Study Update*, United Soybean Board, Chesterfield, MO (November 2008).
[47] Th. Mang and W. Dresel (eds), *Lubricants and Lubrication*, 2nd edn, Wiley-VCH, Weinheim (2007).
[48] B.S. Jalani, S.C. Cheah, N. Rajanaidu and A. Darus, Improvement of palm oil through breeding and biotechnology, *J. Am. Oil Chem. Soc.* **74** (1997) 1451–5.
[49] M.R. Ramli, W.L. Siew, K.Y. Cheah and E. Hishamuddin, Pilot plant production of high-oleic palm oil, *MPOB VIVA Report* No. 386/2007(2), MPOB, Bangi, Malaysia (2007).
[50] M.R. Ramli, W.L. Siew, K.Y. Cheah, I. Nor Aini and M.M. Sahri, Physicochemical properties and performance of high oleic and palm-based shortenings, *J. Oleo Sci.* **57**(11) (2008) 605–12.
[51] R. Yunus, A. Fakhru'l-Razi, T.L. Ooi, S.E. Iyuke and J.M. Perez, Lubrication properties of trimethylolpropane esters of palm oil and palm kernel oils, *Eur. J. Lipid Sci. Technol.* **106** (2004) 52–60.
[52] E.C. Canapi, Y.T.V. Agustin, E.A. Moro, E. Pedroso Jr and M.L.J. Bendaño, Coconut oil, in *Bailey's Industrial Oil and Fat Products*, 6th edn, Vol. 2 (F. Shahidi, ed.), John Wiley & Sons, Inc., Hoboken, NJ (2005), pp. 123–47.
[53] A. Govindapillai, N.H. Jayadas and M. Bhasi, Analysis of the pour point of coconut oil as a lubricant base stock using differential scanning calorimetry, *Lubr. Sci.* **21**(1) (2009) 13–26.

[54] N.H. Jayadas, K. Prabhakaran Nair and G. Ajithkumar, Tribological evaluation of coconut oil as an environment-friendly lubricant, *Tribol. Int.* **40**(2) (2007) 350–4.

[55] R.S. Jonnala, N.T. Dunford and K.E. Dashiell, New high-oleic peanut cultivars grown in the Southwestern United States, *J. Am. Oil Chem. Soc.* **82**(2) (2005) 125–8.

[56] L.S. Hwang, Sesame oil, in *Bailey's Industrial Oil and Fat Products*, 6th edn, Vol. 2 (F. Shahidi, ed.), John Wiley & Sons, Inc., Hoboken, NJ (2005), pp. 537–76.

[57] Y. Gerbig, S.I.-U. Ahmed, F.A. Gerbig and H. Haefke, Suitability of vegetable oils as industrial lubricants, *Synth. Lubric.* **21**(3) (2004) 177–91.

[58] J. Smith, Safflower oil, in *Bailey's Industrial Oil and Fat Products*, 6th edn, Vol. 2 (F. Shahidi, ed.), John Wiley & Sons, Inc., Hoboken, NJ (2005), pp. 491–536.

[59] K.D. Chapman, S. Austin-Brown, S.A. Sparace, A.J. Kinney, K.G. Ripp, I.L. Pirtle and R.M. Pirtle, Transgenic cotton plants with increased seed oleic acid content, *J. Am. Oil Chem. Soc.* **78**(9) (2001) 941–7.

[60] R.D. O'Brien, L.A. Jones, C.C. King, P.J. Wakelyn and P.J. Wan, Cottonseed oil, in *Bailey's Industrial Oil and Fat Products*, 6th edn, Vol. 2 (F. Shahidi, ed.), John Wiley & Sons, Inc., Hoboken, NJ (2005), pp. 173–279.

[61] E. Durak and F. Karaosmanoğlu, Using a cottonseed oil as environmentally accepted lubricant additive, *En. Sources* **26**(7) (2004) 611–25.

[62] J.F. Morton, The Horseradish tree, *Moringa pterigosperma* (Moringaceae) – a boon to arid lands?, *Econ. Bot.* **45** (1991) 318–33.

[63] E. Blee, Biosynthesis of phytooxylipins. The peroxygenase pathway, *Fett/Lipid* **100** (1998) 121–7.

[64] L.J. Morris, The mechanism of ricinoleic acid biosynthesis in *Ricinus communis* seeds, *Biochem. Biophys. Res. Commun.* **29** (1967) 311–15.

[65] M.S. Puthli, V.K. Rathod and A.B. Pandit, Enzymatic hydrolysis of castor oil: process intensification studies, *Biochem. Engng. J.* **31** (2006) 31–41.

[66] H. Mutlu and M.A.R. Meier, Castor oil as a renewable resource for the chemical industry, *Eur. J. Lipid Sci. Technol.* **112** (2010) 10–30.

[67] D.S. Ogunniyi, Castor oil: a vital industrial raw material, *Bioresour. Technol.* **97**(9) (2005) 1086–91.

[68] M. Brink and G. Belay, Alphabetical treatment of vegetable oils, in *Plant Resources of Tropical Africa. I. Cereals and Pulses* (M. Brink and G. Belay, eds), Backhuys Publishers, Leiden (2006), p. 144.

[69] H. Zorn, *Esters as Lubricants*, U.S. Air Force translation Report F-TS-957-RE (1946–47).

[70] H.L. West, *Major Developments in Synthetic Lubricants and Additives in Germany*, Section B1, Technical Information and Documents Unit, London (1948), pp. 44–60.

[71] C.W. Lea, European development of lubricants derived from renewable resources, *Ind. Lubr. Tribol.* **54**(6) (2002) 268–74.

[72] J.R. Marinho, E.G. Sampaio and M.M. Monteiro, Castor oil as an insulating fluid, *CIGRÉ 500-06, Symposium 05-87*, Vienna (1987).

[73] R.E. Kirk and D.F. Othmer, *Kirk-Othmer Encyclopedia of Chemical Technology*, 3rd edn, John Wiley & Sons, New York, NY (1979).

[74] S. Asadauskas, J.M. Perez and J.L. Duda, Lubrication properties of castor oil – potential basestock for biodegradable lubricants, *Lubr. Eng.* **53** (1997) 35–41.

[75] D.E. Weller Jr and J.M. Perez, A study of the effect of chemical structure on piston and wear. Part 2. Vegetable oils and ester, *Lubr. Eng.* **57**(5) (2001) 20–6.

[76] L.R. Rudnick and J. Morehouse, Castor oil derivatives: effect on lubricant performance, *Abstr. 221st ACS Natl. Mtg*, San Diego, CA (1–5 April 2001), CHED-117.

[77] R.M. Mortier and S.T. Orzulik (eds), *Chemistry and Technology of Lubricants*, 2nd edn, Blackie Academic and Professional, London (1997).

[78] E. Weiss, *Castor, Sesame and Safflower*, Leonard Hill, London (1971), Chapter 11.

[79] J.W. Goodrum and D.P. Geller, Influence of fatty acid methyl esters from hydroxylated vegetable oils on diesel fuel lubricity, *Bioresour. Technol.* **96** (2005) 851–5.

[80] S.E. Knights, Crambe. *A North Dakota Case Study*, RIRDC Publ. WO2/005, Kingston, ND (2002).

[81] Y.P. Wang, J.S. Tang, C.Q. Chu and J. Tian, A preliminary study on the introduction and cultivation of *Crambe abyssinica* in China, an oil plant for industrial uses, *Ind. Crops Prod.* **12** (2000) 47–52.

[82] C. Watkins, *Crambe*, ready to be a commercial success, *Int. News Fats, Oil Relat. Mater.* **10** (1999) 828–36.

[83] A.S. Carlsson, *Production of Wax Esters in* Crambe, CPL Press, Newbury (2006); http://www.epobio.net.

[84] G.N. Jham, B.R. Moser, S.N. Shah, R.A. Holser, O.D. Dhingra, S.F. Vaughn, M.A. Berhow, J.K. Winkler-Moser, T.A. Isbell, R.K. Holloway, E.L. Walter, R. Natalino, J.C. Anderson and D.M. Stelly, Wild Brazilian mustard (*Brassica juncea* L.) seed oil methyl esters as biodiesel fuel, *J. Am. Oil Chem. Soc.* **86** (2009) 917–26.

[85] P. Bondioli, L. Folegatti, L. Lazzeri and S. Palmieri, Native *Crambe abyssinica* oil and its derivatives as renewable lubricants: an approach to improve its quality by chemical and biotechnological processes, *Ind. Crops Prod.* **7** (1998) 231–8.

[86] W. Haas, H. Sterk and M. Mittelbach, Novel 12-desoxy-16-hydroxyphorbol diesters isolated from the seed oil of *Jatropha curcas*, *J. Nat. Prod.* **65** (2002) 1434–40.

[87] J.C.J. Bart, N. Palmieri and S. Cavallaro, *Biodiesel Science and Technology. From Soil to Oil*, Woodhead Publishing Ltd/CRC Press, Cambridge/Boca Raton, FL (2010), pp. 193–7, 412–13.

[88] T.I. Mohd. Ghazi, M.F.M. Gunam Resul and A. Idris, Bioenergy II: Production of biodegradable lubricant from *Jatropha curcas* and trimethylolpropane, *Int. J. Chem. Reactor Engng.* **7** (2009) Art. A68.

[89] M.F.M. Gunam Resul, T.I. Mohd. Ghazi, A.M. Syam and A. Idris, Synthesis of biodegradable lubricant from Jatropha oil with high content of free fatty acids, *Proc. AIChE Ann. Mtg.*, Philadelphia, PA (16–21 November 2008), pp. 667/1–667/7.

[90] W. Thurmond, Feedstock trends: jatropha, *Biofuels Int.* **1**(4) (2007) 62–3.

[91] J. Gressel, Transgenics are imperative for biofuel crops, *Plant. Sci.* **174** (2008) 246–63.

[92] F.J. van de Loo, B.G. Fox and C.R. Somerville, Unusual fatty acids, in *Lipid Metabolism in Plants* (T.S. Moore Jr, ed.), CRC Press, Boca Raton, FL (1993), pp. 91–126.
[93] D. Firestone, *Physical and Chemical Characteristics of Oils, Fats and Waxes*, 2nd edn, AOCS Press, Champaign, IL (2006).
[94] R.D. Plattner, K. Payne-Wahl, L.W. Tjarks and R. Kleiman, Hydroxy acids and estolide triglycerides of *Heliophila amplexicaulis* L.f. seed oil, *Lipids* **14** (1979) 576–9.
[95] F. Hirsinger, *Oilseed Crops: Present Situation and Future Developments for Oleochemical Application*, Comm. Eur. Communities, Brussel (1990), EUR 11583.
[96] J. Janick (ed.), *Progress in New Crops, Proc. 3rd Nat. Symp. New Crops*, ASHS Press, Alexandria, VA (1996).
[97] J. Smartt and N. Haq (eds), *Domestication, Production and Utilization of New Crops*, Centre for Underutilized Crops, Southampton University (1997).
[98] C.R. Smith, M.O. Bagby, T.K. Miwa, R.L. Lohmar and I.A. Wolff, Unique fatty acids from *Limnanthes douglasii* seed oil, *J. Org. Chem.* **25** (1960) 1770–4.
[99] S.J. Knapp and J.M. Crane, Fatty acid diversity of Section Inflexae *Limnanthes* (meadowfoam), *Ind. Crops Prod.* **4** (1995) 219–27.
[100] A.L. Pilgeram, D.C. Sands, D. Boss, N. Dale, D. Wichman, P. Lamb, C. Lu, R. Barrows, M. Kirkpatrick, B. Thompson and D.L. Johnson, *Camelia sativa*: a Montana omega-3 and fuel crop, in *Issues in New Crops and New Uses* (J. Janick and A. Whipkey, eds), ASHS Press, Alexandria, VA (2007), pp. 129–31.
[101] K. Sawyer, Is there room for *Camelina*?, *Biodiesel Mag.* **5**(7) (2008) 82–7.
[102] D. Johnson, Developing *Camelina* as a source of renewable biofuels, *Proc. Harvesting Clean Energy Conference IX*, Billings, MT (27 January 2009).
[103] S. Burkdoll, Biolubricant market could compliment Montana's interest in biodiesel production, *The Prairie Star* (5 July 2006).
[104] D. Johnson, Future crops for bio-diesel and bio-lubricants, *Proc. Harvesting Clean Energy Conference V*, Great Falls, MT (20–21 January 2005); http://www.harvestcleanenergy.org/conference/HCE5/HCE5_PPTs/Johnson.pdf
[105] L. Lazzeri, M. Errani, O. Leoni and G. Venturi, *Eruca sativa* opp. *oleifera*: a new non-food crop, *Ind. Crops Prod.* **20** (2004) 67–73.
[106] A. Eisenträger, M. Schmidt, H. Murrenhoff, W. Dott and S. Hahn, Biodegradability testing of synthetic ester lubricants: effects of additives and usage, *Chemosphere* **48** (2002) 89–96.
[107] E. Remmele and B. Widmann, Hydraulic fluids based on rapeseed oil in agricultural machinery – suitability and environmental impact during use, in *Industrial and Automotive Lubrication* (W.J. Bartz, ed.), Vol. 1, *11th Intl. Colloquium*, TA Esslingen (13–15 January 1998).
[108] N.S. Battersby, S.E. Pack and R.J. Watkinson, A correlation between the biodegradability of oils products in the CEC L-33-T-82 and modified Sturm tests, *Chemosphere* **24** (2000) 1989–2000.
[109] L.A.T. Honary, An investigation of the use of soybean oil in hydraulic systems, *Bioresour. Technol.* **56** (1996) 41–7.
[110] L.A.T. Honary, Soybean based hydraulic fluid, US Patent No. 5,972,855 A (26 October 1999).

[111] J. Hudson, Life performance of hydraulic pumps with various hydraulic fluids, *Proc. IMechE S673: Environmental Impact of Fluid Power Systems*, Institute of Mechanical Engineering, London (1999).

[112] M.C. McManus, G.P. Hammond and C.R. Burrows, Life-cycle assessment of mineral and rapeseed oil in mobile hydraulic systems, *J. Ind. Ecol.* **7** (2003) 163–77.

[113] H.F. Eichenberg, Biodegradable hydraulic lubricant: An overview of current developments in Central Europe, *Proc. 42nd Earthmoving Industrial Conference*, Peoria, IL (9–10 April 1991), *SAE Techn. Paper Ser.* No. 910962.

[114] A.Z. Szeri, *Tribology: Friction, Lubrication and Wear*, McGraw-Hill, New York, NY (1980).

[115] N.J. Fox, B. Tyrer and G.W. Stachowiak, Boundary lubrication performance of free fatty acids in sunflower oil, *Tribol. Lett.* **16**(4) (2004) 275–81.

[116] M.A. Maleque, H.H. Masjuki and S.M. Sapuan, Vegetable-based biodegradable lubricating oil additives, *Ind. Lubr. Tribol.* **55** (2003) 137–43.

[117] GfT-Arbeitsblatt 4, *Pflanzenöle als Schmiermittel*, GfT, Moers (1993).

[118] H. Czichos and K.-H. Habig, *Tribologie-Handbuch Reibung und Verschleiß*, Vieweg, Braunschweig (1992).

[119] A. Biswas, A. Adhvaryu, D.G. Stevenson, B.K. Sharma, J.L. Willett and S.Z. Erhan, Microwave irradiation effects on the structure, viscosity, thermal properties and lubricity of soybean oil, *Ind. Crops Prod.* **25** (2007) 1–7.

[120] N.H. Jayadas and K.P. Nair, Coconut oil as base oil for industrial lubricants. Evaluation and modification of thermal, oxidative and low temperature properties, *Tribol. Int.* **39** (2006) 873–8.

[121] S. Asadauskas and S.Z. Erhan, Depression of pour points of vegetable oils by blending with diluents used for biodegradable lubricants, *J. Am. Oil Chem. Soc.* **76**(3) (1999) 313–16.

[122] S.Z. Erhan, A. Adhvaryu and Z. Liu (to USDA), Chemically modified vegetable oil-based industrial fluid, US Patent No 6,583,302 B1 (24 June 2003).

[123] K.M. Doll and S.Z. Erhan, Synthesis of carbonated fatty methyl esters using supercritical carbon dioxide, *J. Agr. Food Chem.* **53** (2005) 9608–14.

[124] C. Paquot, *Standard Methods for the Analysis of Oils, Fats and Derivatives*, IUPAC Commission on Oils, Fats and Derivatives, Pergamon Press, London (1979), p. 52.

[125] J.C.J. Bart, N. Palmieri and S. Cavallaro, *Biodiesel Science and Technology. From Soil to Oil*, Woodhead Publishing Ltd/CRC Press, Cambridge/Boca Raton, FL (2010), p. 234.

[126] *Official Methods and Recommended Practices of the American Oil Chemists' Society*, AOCS Press, Champaign, IL (1999), Method Cd 1d-92.

[127] R.R. Allen, Determination of unsaturation, *J. Am. Oil Chem. Soc.* **32** (1955) 671–4.

[128] G. Knothe, Structure indices in FA chemistry. How relevant is the iodine value?, *J. Am. Oil Chem. Soc.* **79** (2002) 847–54.

[129] A. Hubmann, Chemie pflanzlicher Grundöle, in *Biologisch schnell abbaubare Schmierstoffe und Arbeitsflüssigkeiten speziell auf pflanzlicher Basis* (W.J. Bartz, ed.), Expert-Verlag. Ehningen (1993), p. 41.

[130] A. Demirbaş, Fuel properties and calculation of higher heating values of vegetable oils, *Fuel* **77** (1988) 1117.

[131] G. Lingg, Unconventional base oils for liquid and semisolid lubricants, *Proc. 14th Intl. Colloquium Esslingen* (2004), Vol. I, pp. 1–4.
[132] N.A. Porter, L.S. Lehmann, B.A. Weber and K.J. Smith, Unified mechanism of polyunsaturated fatty acid autoxidation: competition of peroxy radical hydrogen abstraction, beta-scission, and cyclization, *J. Am. Oil Chem. Soc.* **103** (1981) 6447–55.
[133] D.R. Kodali, High performance ester lubricants form natural oils, *Ind. Lubr. Tribol.* **54** (2002) 165–70.
[134] R. Becker and A. Knorr, An evaluation of antioxidants for vegetable oils at elevated temperatures, *Lubr. Sci.* **8** (1996) 95–117.
[135] E.S. Nachtman and S. Kalpakjian, *Lubricants and Lubrication in Metalworking Operations*, CRC Press, Boca Raton, FL (1985).
[136] J. Igarashi, Oxidative degradation of engine oils, *Jpn. J. Tribol.* **35** (1990) 1095–104.
[137] N.J. Fox and G.W. Stachowiak, Vegetable oil-based lubricants – A review of oxidation, *Tribol. Intl.* **40** (2007) 1035–46.
[138] E.N. Frankel, *Lipid Oxidation*, The Oily Press, Dundee (1998).
[139] Y. Miyake, K. Yokomizo and N. Matsuzaki, Determination of unsaturated fatty acid composition by high-resolution nuclear magnetic resonance spectroscopy, *J. Am. Oil Chem. Soc.* **75** (1998) 1091.
[140] A. Adhvaryu, S.Z. Erhan, Z.S. Liu and J.M. Perez, Oxidation kinetic studies of oils derived from unmodified and genetically modified vegetables using pressurized differential scanning calorimetry and nuclear magnetic resonance spectroscopy, *Thermochim. Acta* **364** (2000) 87–97.
[141] N.H. Jayadas, Evaluation of the oxidative properties of vegetable oils as base stocks for industrial lubricants using spectroscopic and thermogravimetric analysis, *J. Synth. Lubr.* **25**(3) (2008) 105–13.
[142] D. Firestone (ed.), *Official Methods and Recommended Practices of American Oil Chemists' Society*, 4th edn, AOCS Press, Champaign, IL (1992), Oxidative stability index (OSI), Method Cd 12b-92.
[143] K.M. Schaich, Lipid oxidation: theoretical aspects, in *Bailey's Industrial Oil and Fat Products*, 6th edn, Vol. 1 (F. Shahidi, ed.), John Wiley & Sons, Inc., Hoboken, NJ (2005), pp. 269–355.
[144] B.W. Matthäus, Determination of the oxidative stability of vegetable oils by Rancimat and conductivity and chemiluminescence measurements, *J. Am. Oil Chem. Soc.* **73** (1996) 1039–43.
[145] B.K. Sharma, U. Rashid, F. Anwar and S.Z. Erhan, Lubricant properties of *Moringa* oil using thermal and tribological techniques, *J. Therm. Anal. Calorim.* **96** (2009) 999–1008.
[146] J.C.O. Santos, A.D. Oliveira, C.C. Silva, J.D.S. Silva, A.G. Souza and L.N. Lima, Kinetic and activation thermodynamic parameters on thermal decomposition of synthetic lubricant oils, *J. Therm. Anal. Calorim.* **87** (2007) 823–9.
[147] C.C. Garcia, P.I.B.M. Franco, T.O. Zappa, N.R.A. Filho and M.I.G. Leles, Thermal stability studies of some cerrado plant oils, *J. Therm. Anal. Calorim.* **87** (2007) 645–8.
[148] S. Vecchio, L. Campanella, A. Nuccilli and M. Tomassetti, Kinetic study of thermal breakdown of triglycerides contained in extra-virgin olive oil, *J. Therm. Anal. Calorim.* **91** (2008) 51–6.

[149] B. Lin, L. Yang, H. Dai, Q. Hou and L. Zhang, Thermal analysis of soybean oil based polyols, *J. Therm. Anal. Calorim.* **95** (2009) 977–83.
[150] S. Lalas and J. Tsaknis, Extraction and identification of natural antioxidant from the seeds of the *Moringa oleifera* tree variety of Malawi, *J. Am. Oil Chem. Soc.* **79** (2002) 677–83.
[151] G. Knothe, J. Van Gerpen and J. Krahl, *The Biodiesel Handbook*, AOCS Press, Urbana, IL (2005).
[152] S. Fernando, M. Hanna and S. Adhikari, Lubricity characteristics of selected vegetable oils, animal fats, and their derivatives, *Appl. Eng. Agric.* **23**(1) (2007) 5–11.
[153] S. Asadauskas, J.M. Perez and J.L. Duda, Oxidative stability and antiwear properties of high oleic vegetable oils, *Lubr. Eng.* **52**(12) (1996) 877–82.
[154] M. Kozma, Investigation into the scuffing load capacity of environmentally-friendly lubricating oils, *J. Synth. Lubr.* **14**(3) (1997) 249–58.
[155] S. Odi-Owei, Tribological properties of some vegetable oils and fats, *Lubr. Eng.* **45**(11) (1988) 685–90.
[156] J. Vižintin, A. Arnšek and T. Ploj, Lubricating properties of rapeseed oils compared to mineral oils under a high load oscillating movement, *J. Synth. Lubr.* **17**(3) (2000) 201–17.
[157] S.J. Randles, Esters, in *Synthetic Lubricants and High-performance Functional Fluids* (R.L. Shubkin, ed.), Marcel Dekker, Inc., New York, NY (1993), pp. 41–65.
[158] O.O. Steren, Trobological properties of some vegetable oils and fats, *Lubr. Eng.* **45** (1989) 685–90.
[159] E.P. Kingsbury, The heat of adsorption of a boundary lubricant, *ASLE Trans.* **3** (1960) 30–3.
[160] A. Adhvaryu, S.Z. Erhan and J.M. Perez, Tribological studies of thermally and chemically modified vegetable oils for use as environmentally friendly lubricants, *Wear* **257** (2004) 359–67.
[161] F.A. Zaher and H.M. Nomany, Vegetable oils and lubricants, *Grasas Aceites* (Sevilla) **39** (1988) 235–8.
[162] A. Willing, Lubricants based on renewable resources – An environmentally compatible alternative to mineral oil products, *Chemosphere* **43** (2001) 89–98.
[163] A. Adhvaryu, G. Biresaw, B.K. Sharma and S.Z. Erhan, Friction behaviour of some seed oils: Biodiesel lubricant applications, *I&EC Res.* **45** (2006) 3735–40.
[164] G. Biresaw, A. Adhvaryu, S.Z. Erhan and C.J. Carriere, Friction and adsorption properties of normal and high-oleic soybean oils, *J. Am. Oil Chem. Soc.* **79** (2002) 53–8.
[165] G. Biresaw, A. Adhvaryu and S.Z. Erhan, Friction properties of vegetable oils, *J. Am. Oil Chem. Soc.* **80** (2003) 697–704.
[166] T. Murakami and H. Sakamoto, Lubricating properties of vegetable oils and paraffinic oils with unsaturated fatty acids under high contact-pressure conditions in four-ball tests, *J. Synth. Lubr.* **20**(3) (2003) 183–201.
[167] W. Castro, D.E. Weller, K. Cheenkachorn and J.M. Perez, The effect of chemical structure of basefluids on antiwear effectiveness of additives, *Tribol. Intl.* **38** (2005) 32–6.

[168] D.E. Weller Jr and J.M. Perez, A study of the effect of chemical structure on friction and wear. Part 1. Synthetic ester base fluids, *Lubr. Eng.* **56**(11) (2000) 39–44.
[169] M.T. Siniawski, N. Saniei, B. Adhikari and L.A. Doezema, Influence of fatty acid composition on the tribological performance of two vegetable-based lubricants, *J. Synth. Lubr.* **24** (2007) 101–10.
[170] G. Bantchev and G. Biresaw, Elastohydrodynamic study of vegetable oil–poly-alphaolefin blends, *Lubr. Sci.* **20** (2008) 283–97.
[171] G. Biresaw, Lubricant formulation with vegetable oils. Interfacial aspects, *59th STLE Ann. Meeting & Exhibition*, Toronto (16–20 May 2004), p. 172.
[172] J.A. Williams, *Engineering Tribology*, Oxford University Press, New York, NY (1994).
[173] G.W. Stachowiak and A.W. Batchelor, *Engineering Tribology*, Elsevier, Amsterdam (1993).
[174] Y. Cao, L. Yu and W. Liu, Study of the tribological behavior of sulfurized fatty acids as additives in rapeseed oil, *Wear* **244** (2000) 126–31.
[175] I.-ul-Haq, M. Farooq and N. Muhammd, Some studies on the use of vegetable oils as environmentally-friendly lubricants *Tribol. Online* **6**(5) (2011) 230–4.
[176] Y. Cao and L. Yu, Effect of alcohols and fatty acids as additives on the anti-wear and extreme pressure properties of rapeseed oil, *Mocaxue Xuebao/Trib.* **20**(4) (2000) 288–91.
[177] I. Minami, H.-S. Hong and N.C. Mathur, Lubrication performance of model organic compounds in high oleic sunflower oil, *J. Synth. Lubr.* **16**(1) (1999) 3–12.
[178] G. Biresaw and G. Bantchev, Effect of chemical structure on film-forming properties of seed oils, *J. Synth. Lubr.* **25**(4) (2008) 159–83.
[179] A. Adhvaryu, S.Z. Erhan and J.M. Perez, Wax appearance temperatures of vegetable oils determined by differential scanning calorimetry: effect of tria-cylglycerol structure and its modification, *Thermochim. Acta* **395** (2003) 191–200.
[180] A. Adhvaryu, B.K. Sharma, H.S. Hwang, S.Z. Erhan and J.M. Perez, Development of biobased synthetic fluids: Application of molecular modeling to structure–physical property relationship, *I&C Res.* **45** (2006) 928–33.
[181] Y. Tan, W. Huang and X. Wang, Molecular orbital indexes criteria for friction modifiers in boundary lubrication, *Tribol. Intl.* **35** (2002) 381–4.
[182] A. Jabbarzadeh, J.D. Atkinson and R.I. Tanner, The effect of branching on slip and rheological properties of lubricants in molecular dynamics simulation of Couette shear flow, *Tribol. Intl.* **35** (2002) 35–46.
[183] N.H. Jayadas and K.P. Nair, Coconut oil as biolubricant: study of the anti-wear properties using quantum chemical calculations and tribological tests, *Proc. World Tribology Congress III*, Washington, DC (12–15 September 2005), pp. 513–14.
[184] K. Konno, D. Kamei, T. Yokosuka, S. Takami, M. Kubo and A. Miyamoto, The development of computational chemistry approach to predict the viscosity of lubricants, *Tribol. Intl.* **36** (2003) 455–8.
[185] K.V. Camarda, P. Sunderesan, S. Siddhaye, G.J. Suppes and J. Heppert, An optimization approach to the design of value-added soybean oil products, *Eur.*

Symp. Computer Aided Process Engng-11 (R. Gani and S.B. Jorgensen, eds), Elsevier Science, Amsterdam (2001) pp. 369–74.
[186] S. Jahanmir and M. Beltzer, An adsorption model for friction in boundary lubrication, *ASLE Trans.* **29** (1986) 423–30.
[187] S. Jahanmir, Chain length effects in boundary lubrication, *Wear* **102** (1985) 331–49.
[188] M. Beltzer and S. Jahanmir, Role of dispersion interactions between hydrocarbon chains in boundary lubrication, *ASLE Trans.* **30** (1987) 47–54.
[189] J.A. Schey, *Tribology in Metalworking: Friction, Lubrication and Wear*, American Society of Metals, Metals Park, OH (1983), pp. 27–130.
[190] J.C.J. Bart, N. Palmeri and S. Cavallaro, *Biodiesel Science and Technology, From Soil to Oil*, Woodhead Publishing, Ltd/CRC Press, Cambridge/Boca Raton, FL (2010), Chapters 5 and 6.
[191] S.D. Sanford, J.M. While, P.S. Shah, C. Wee, M.A. Valverde and G.R. Meier, *Feedstock and Biodiesel Characteristics Report*, Renewable Energy Group, Inc., Ames, IA (2009) http://www.regfuel.com
[192] T. Voelker and A.J. Kinney, Variations in the biosynthesis of seed-storage lipids, *Annu. Rev. Plant Physiol. Plant Mol. Biol.* **52** (2001) 335–61.
[193] D.J. Murphy, Transgenic plants – a future source of novel edible and industrial oils, *Lipid Technol.* **6**(4) (1994) 84–91.
[194] W. Lühs and W. Friedt, Non-food uses of vegetable oils and fatty acids, in *Designer Oil Crops – Breeding, Processing and Biotechnology* (D.J. Murphy, ed.), VCH, Weinheim (1994), pp. 73–130.
[195] P. Lamers, K. McCormick and J.A. Hilbert, The emerging liquid biofuel market in Argentina: implications for domestic demand and international trade, *En. Policy* **36** (2008) 1479–90.
[196] International Service for the Acquisition of Agri-Biotech Applications (ISAAA), Manila/Nairobi/Ithaca (2006).
[197] A.C. Franke, M.L.H. Breukers, W. Broer, F. Bunte, O. Dolstra, F.M. d'Engelbronner – Kolff, L.A.P. Lotz, J. Van Montfort, J. Nikoloyuk, M.M. Rutten, M.J.M. Smulders, C.C.M. van de Wiel and M. Van Zijl, *Sustainability of Current GM Crop Cultivation*, DLO Foundation, Wageningen UR, Report 386 (2011).
[198] D. Strayer, *Identity-preserved Systems, A Reference Handbook*, CRC Press, Boca Raton, FL (2002).
[199] D.J. Murphy (ed.), *Designer Oil Crops: Breeding, Processing and Biotechnology*, VCH-Verlag, Weinheim (1994).
[200] D.L. Auld, M.K. Heikkinen, D.A. Erickson, J.L. Sernyk and J.E. Romero, Rapeseed mutants with reduced levels of polyunsaturated fatty acids and increased levels of oleic acid, *Crop Sci.* **32** (1992) 657–62.
[201] W.D. Hitz, N.S. Yadav, R.S. Reiter, C.J. Mauvais and A.J. Kinney, Reducing polyunsaturation in oils of transgenic canola and soybean, in *Plant Lipid Metabolism* (J.-C. Kader and P. Mazliak, eds), Kluwer Academic Publishers, Dordrecht (1995), pp. 506–8.
[202] G. Cole, G.S. Coughlan, N. Frey, J. Hazebroek and C. Jennings, New sunflower and soybean cultivars for novel vegetable oil types, *Fett/Lipid* **100** (1998) 177–81.
[203] J.R. Booth Jr, R.M. Broglie, W.D. Hitz, A.J. Kinney, S. Knowlton and S.A. Sebastian (to E.I. Du Pont de Nemours & Co.), Novel gene combinations that

alter the lipid profile of soybean oil, resulting in oils useful for margarine and other spreads, PCT Int. Publ. No. WO 99/58689 (18 November 1999).
[204] J.B. Mugendi, C.A. Sims, D.W. Gorbet and S.F. O'Keefe, Flavor stability of high-oleic peanuts stored at low humidity, *J. Am. Oil Chem. Soc.* **75**(2) (1998) 21–5.
[205] J.C.J. Bart, N. Palmeri and S. Cavallaro, *Biodiesel Science and Technology. From Soil to Oil*, Woodhead Publishing, Ltd/CRC Press, Cambridge/Boca Raton, FL (2010), p. 154.
[206] T.A. McKeon, Genetic modification of seed oils for industrial applications, in *Industrial Uses of Vegetable Oils* (S.Z. Erhan, ed.), CRC Press, Boca Raton, FL (2005), Chapter 1.
[207] W. Lühs, K.J. Dehmer, R. Bergmann and W. Friedt, Molekulargenetische Methoden – Werkzeuge für die Züchtung von Ölpflanzen als Quelle nachwachsender oleochemische Rohstoffe, in *Perspektiven nachwachsender Rohstoffe in der Chemie* (H. Eierdanz, ed.) VCH, Weinheim (1996), pp. 232–9.
[208] Home-Grown Cereals Authority (HGCA), *Improved Oilseed Rape Cultivars for the Biolubricant Market*, Annual Report (2004–2008); http://www.hgca.com
[209] J. Orson, E. Booth, C. Merritt and C. Lea, Growing 'high oleic low linolenic' (HOLL) oilseed rape for specialised markets, HGCA Project Report No. 442 (Nov. 2008).
[210] W. Friedt and W. Lühs, Recent developments and perspectives of industrial rapeseed breeding, *Fett/Lipid* **100** (1998) 219–26.
[211] C. Stoll, W. Lühs, M.K. Zarhloul and W. Friedt, Genetic modification of saturated fatty acids in oilseed rape (*Brassica napus*), *Eur. J. Lipid Sci. Technol.* **107** (2005) 244–8.
[212] W. Lühs and W. Friedt, Non-food use of vegetable oils and fatty acids, in *Designer Oil Crops, Breeding, Processing and Biotechnology* (D.J. Murphy, ed.), Verlag Chemie, Weinheim (1994), p. 74–130.
[213] J.M. Martínez-Rivas, P. Sperling, W. Lühs and E. Heinz, Spatial and temporal regulation of three different microsomal oleate desaturase genes (FAD2) from normal-type and high-oleic varieties of sunflower (*Helianthus annuus* L.), *Mol. Breed.*, **8** (2001) 159–68.
[214] Z. Flagella, T. Rotunno, E. Tarantino, R. Di Caterina and A. De Caro, Changes in seed yield and oil fatty acid composition of high oleic sunflower (*Helianthus annuus* L.) hybrids in relation to the sowing date and the water regime, *Eur. J. Agron.* **17** (2002) 221–30.
[215] M.K. Gupta, NuSun oil – success and challenges, *Inform* **18**(9) (2007) 639–41.
[216] M.A. Grompone, Sunflower oil, in *Bailey's Industrial Oil and Fat Products*, 6th edn, Vol. 2 (F. Shahidi, ed.), John Wiley & Sons, Inc., Hoboken, NJ (2005), pp. 655–730.
[217] E.G. Hammond, L.A. Johnson, C. Su, T. Wang and P.J. White, Soybean oil, in *Bailey's Industrial Oil and Fat Products*, 6th edn, Vol. 2 (F. Shahidi, ed.), John Wiley & Sons, Inc., Hoboken, NJ (2005), pp. 577–653.
[218] J.J. Fillatti, N.A. Bringe and K. Dehesh (to Monsanto Technology LLC), Nucleic acid constructs and methods for producing altered seed oil composition, US Patent Appl. No. 2004/0107460 A1 (3 June 2004).
[219] http://www.danforthcenter.org.

[220] N.A. Bringe, Soybean oil composition for biodiesel, in *The Biodiesel Handbook* (G. Knothe, J. Van Gerpen and J. Krahl, eds), AOCS Press, Urbana, IL (2005), pp. 161–4.
[221] A.J. Norden, D.W. Gorbet, D.A. Knauft and C.T. Young, Variability in oil quality among peanut genotypes in the Florida breeding program, *Peanut Sci.* **14**(1) (1987) 7–11.
[222] T.G. Isleib, C.T. Young and D.A. Knauft, Fatty acid genotypes of five Virginia-type peanut cultivars, *Crop Sci.* **36**(3) (1996) 556–8.
[223] T.P. Durrett, C. Benning and J. Ohlrogge, Plant triacylglycerols as feedstocks for the production of biofuels, *Plant J.* **54** (2008) 593–607.
[224] T. Buhr, S. Sato, F. Ebrahim, A. Xing, Y. Zhou, M. Mathiesen, B. Schweiger, A. Kinney, P. Staswick and T. Clemente, Ribozyme termination of RNA transcripts down-regulate seed fatty acid genes in transgenic soybean, *Plant J.* **30**(2) (2002) 155–63.
[225] D. Kodali, Z. Fan and L.R. DeBonte (to Cargill, Inc.), Plants, seeds and oils having an elevated total monounsaturated fatty acid content, PCT Int. Publ. No. WO 00/007433 (17 February 2000).
[226] Z. Fan (to Cargill, Inc.), *Brassica napus* plants which produce non-hydrogenated canola oil for food applications, US Patent No. 5,767,338 (16 June 1998).
[227] L.R. DeBonte, Z. Fan and W.H.-T. Loh (to Cargill, Inc.), Oil from canola seed with altered fatty acid profiles and a method of producing oil, US Patent No. 5,861,187 (19 January 1999).
[228] S.S. Lawate and K. Lal (to Lubrizol Corp.), High oleic polyol esters, compositions and lubricants, functional fluids and greases containing the same, Eur. Patent No. EP 0,712,834 A1 (14 November 1995).
[229] G.N. Fick (to Sigco Research, Inc.), Sunflower products and methods for their production, US Patent No. 4,627,192 (9 December 1986).
[230] G.N. Fick (to Sigco Research, Inc.), Novel sunflower products and methods for their production, US Patent No. 4,743,402 (10 May 1988).
[231] D.R. Kodali and S.C. Nivens (to Cargill, Inc.), Oils with heterogeneous chain lengths, PCT Int. Publ. No. WO 00/41515 (20 July 2000).
[232] P.B.E. McVetty, R. Scarth, S.R. Rimmer and C.G.J. Van den Berg, Venus high erucic acid, low glucosinolate summer rape, *Can. J. Plant Sci.* **76** (1996) 341–2.
[233] P.B.E. McVetty, S.R. Rimmer, R. Scarth and C.G.J. Van den Berg, Neptune high erucic acid, low glucosinolate summer rape, *Can. J. Plant Sci.* **76** (1996) 343–4.
[234] R. Scarth, P.B.E. McVetty and S.R. Rimmer, Mercury high erucic low glucosinolate summer rape, *Can. J. Plant Sci.* **75** (1995) 205–6.
[235] P. Duhot, E. Gontier, D. Thomas and B. Thomasset (to Total), Procédé de production d'acides gras ramifiés au moyen de plantes génétiquement modifiées, Fr. Patent No. FR 2,769,320 A (3 October 1997).
[236] REFLAX – Rational engineering of lipid metabolism in flax to produce branched-chain fatty acids, unusual fatty acids with potent biolubricant properties, Proj. QLK3-2000-00349 (2001–2004); http://ec.europa.eu/research/quality-of-life/cell-factory/volume1/projects/qlk3-2000-00349_en.html
[237] D.R. Kodali, Z. Fan and L.R. DeBonte (to Cargill, Inc.), Biodegradable high oxidative stability oils, US Patent No. 6,281,375 B1 (28 August 2001).

[238] D.C. Taylor, T. Francis, Y. Guo, J.M. Brost, V. Katavic, E. Mietkiewska, E.M. Giblin, S. Lozinsky and T. Hoffman, Molecular cloning and characterization of a KCS gene from *Cardamine graeca* and its heterologous expression in Brassica oilseeds to engineer high nervonic acid oils for potential medical and industrial use, *Plant Biotechn. J.* **7** (2009) 925–38.

[239] Y. Guo, E. Mietkiewska, T. Francis, V. Katavic, J.M. Brost, M. Giblin, D.L. Barton and D.C. Taylor, Increase in nervonic acid content in transformed yeast and transgenic plants by introduction of a *Lunaria annua* L., 3-keto-acyl-CoA synthase (KCS) gene, *Plant Mol. Biol.* **69** (2009) 565–75.

[240] J. Burgal, J. Shockley, C. Lu, J. Dyer, T. Larson, I. Graham and J. Browse, Metabolic engineering of hydroxy fatty acid production in plants: RcDGAT2 drives dramatic increases in ricinoleate levels in seed oil, *Plant Biotechnol J.* **6** (2008) 819–31.

[241] J. Grushcow and M. Smith, Engineering high-performance biolubricants in crop plants, *Ind. Biotechnol.* **2** (2006) 48–50.

[242] T.A. McKeon, Engineering plants for industrial uses, in *Biocatalysis and Agricultural Technology* (C.T. Hou and J.F. Shaw, eds), CRC Press, Boca Raton, FL (2009), pp. 89–97.

[243] P. Rojas-Barros, A. De Haro, J. Muñoz and J.M. Fernández-Martínez, Isolation of a natural mutant in castor (*Ricinus communis* L.) with high-oleic/low ricinoleic content in the oil, *Crop Sci.* **44** (2004) 76–80.

[244] T.A. McKeon, Transgenic oils, in *Bailey's Industrial Oil and Fat Products*, 6th edn, Vol. 3 (F. Shahidi, ed.), John Wiley & Sons, Inc., Hoboken, NJ (2005), pp. 154–74.

[245] G. Graef, B.J. LaVallee, P. Tenopir, M. Tat, B. Schweiger, A.J. Kinney, J.H. Van Gerpen and T.E. Clemente, A high-oleic-acid and low-palmitic-acid soybean: agronomic performance and evaluation as a feedstock for biodiesel, *Plant Biotechnol. J.* **7** (2009) 411–21.

[246] P.C. Naegely, *Environmentally Acceptable Lubricants*, Lubrizol Corporation, Wickliffe, OH (1992).

[247] W.D. Hitz (to E.I. du Pont de Nemours & Co.), Nucleotide sequences of canola and soybean palmitoyl-ACP thioesterase genes and their use in the regulation of fatty acid content of the oils of soybean and canola plants, US Patent No. 5,955,650 (21 Sept. 1999).

[248] D.N. Duvick, Utilization of biotechnology in plant breeding in North America: A status report, in *Perspektiven nachwachsender Rohstoffe in der Chemie* (H. Eierdanz, ed.), VCH Verlag, Weinheim (1996), pp. 3–10.

[249] W. Friedt and W. Lühs, Die Zukunft der transgenen Pflanzen für europäische Entwicklungen, in *Perspektiven nachwachsender Rohstoffe in der Chemie* (H. Eierdanz, ed.), VCH Verlag, Weinheim (1996), pp. 11–20.

[250] S.Z. Erhan, A. Adhvaryu and Z. Liu, Chemical modification of vegetable oils for lubricant basestocks, *Proc. 2003 Spring Techn. Conf. ASME Internal Combustion Engine Division*, American Society of Mechanical Engineers, Salzburg (2003), pp. 369–80.

[251] M. Schneider, Plant-oil-based lubricants and hydraulic fluids, *J. Sci. Food Agric.* **86** (2006) 1769–80.

[252] L.A. Quinchia, M.A. Delgado, C. Valencia, J.M. Franco and C. Gallegos, Viscosity modification of high-oleic sunflower oil with polymer additives for the

design of new biolubricant formulations, *Environm. Sci. Technol.* **43** (2009) 2060–5.

[253] R. Garcés, E. Martínez-Force, J.J. Salas and M. Venegas-Calerón, Current advances in sunflower oil and its applications, *Lipid Technol.* **21** (2009) 79–82.

[254] L. Velasco Varo, J.M. Fernández-Martínez and B. Perez Vich, Sunflower oil with high heat stability, PCT Int. Publ. No. WO 2009/080858 A1 (2 July 2009).

[255] L. Lazzeri, M. Mazzoncini, A. Rossi, E. Balducci, G. Bartolini, L. Giovannelli, R. Pedriali, R. Petroselli, G. Patalano, G. Agnoletti, A. Borgioli, B. Croce and L. D'Avino, Biolubricants for the textile and tannery industries as an alternative to conventional mineral oils: an application experience in the Tuscany province, *Ind. Crops Prod.* **24** (2006) 280–91.

[256] S. Smith, R. King and D. Min, Oxidative and thermal stabilities of genetically modified high oleic sunflower oil, *Food Chem.* **102** (2007) 1208–13.

[257] O.H. Botz, High oleic sunflower as new base oil – properties and scope of application, *Mineralöltechn.* **50**(9) (2005) 1–17.

[258] P.A. Stoutjesdijk, C. Hurlestone, S.P. Singh and A.G. Green, High-oleic acid Australian *Brassica napus* and *B. juncea* varieties produced by co-suppression of endogenous D12-desaturases, *Biochem. Soc. Trans.* **28** (2000) 938–40.

[259] K.I. Soldatov, Chemical mutagenesis in sunflower breeding, *Proc. 7th Intl. Sunflower Conference*, Krasnodar (1976), pp. 352–7.

[260] J. Seemann, B. Olbrich-Deußner and R. Groß-Lannert, High-oleic sunflower oil 90plus: sunflowers are capable of more than expected, *SÖFW J.* **126**(10) (2000) 50–4.

[261] R.S.C. Wong, W.D. Beversdorf, J.R. Castagno, I. Grant and J.D. Patel (to Pioneer Hi-bred Intl., Inc.), Production of improved rapeseed exhibiting an enhanced oleic acid content, US Patent No. 5,638,637 (17 June 1997).

[262] J.-P. Despeghel (to Monsanto S.A.S.), Use of rapeseed in biolubricants, US Patent Appl. No. 2009/0286704 A1 (19 November 2009).

[263] H.S. Hwang and S.Z. Erhan, Lubricant base stocks for modified soybean oil, in *Biobased Industrial Fluids and Lubricants* (S.Z. Erhan and J.M. Perez, eds), AOCS Press, Champaign, IL (2003), pp. 20–34.

[264] M.A. Schmidt, C.R. Dietrich and E.B. Cahoon, Biotechnological enhancement of soybean oil for lubricant applications, in *Synthetics, Mineral Oils and Bio-based Lubricants: Chemistry and Technology* (L.R. Rudnick, ed.), CRC Press, Boca Raton, FL (2006), pp. 389–98.

[265] K. Lal, D.M. Dishong and J.G. Dietz (to Lubrizol Corp.), Pour point depressants for high monounsaturated vegetable oils and for high monounsaturated vegetable oils/biodegradable base and fluid mixtures, US Patent No. 5,413,725 (9 May 1995).

[266] S.S. Lawate and K. Lal (to Lubrizol Corp.), High oleic polyol esters, compositions and lubricants, functional fluids and greases containing the same, Eur. Patent No. EP 0,712,834 A1 (14 November 1995).

[267] K. Lal (to Lubrizol Corp.), Pour point depressants for industrial lubricants containing mixtures of fatty acid esters and vegetable oils, US Patent No. 5,338,471 (16 Aug. 1994).

[268] S.S. Lawate, P.C. Naegely and V.A. Carrick (to Lubrizol Corp.), Environmental friendly food grade lubricants from edible triglycerides containing FDA approved additives, US Patent No. 5,538,654 (23 July 1996).

[269] T. Konishi and Y. Kikuchi (to Nippon Mitsubishi Oil Corp.), Hydraulic oil composition, US Patent No. 6,300,292 A1 (9 October 2001).
[270] T.V. Oommen and C.C. Claiborne (to ABB Power T&D Company, Inc.), High oleic acid compositions and methods of making electrical insulation fluids and devices comprising the same, US Patent No. 6,312,623 B1 (6 November 2001).
[271] T.V. Oommen and C.C. Claiborne, Biodegradable insulating fluid from high oleic vegetable oils, *CIGRÉ 15-302*, Paris (1998).
[272] W.W. Garmier (to Renewable Lubricants, Inc.), High temperature biobased lubricant compositions comprising boron nitride, PCT Int. Publ. No. WO 2006/116502 A1 (2 November 2006).
[273] R.P.S. Bisht, G.A. Sivasankaran and V.K. Bhatia, Additive properties of jojoba oil for lubricating oil formulations, *Wear* **161** (1993) 193–7.
[274] O. El Kinawy, Comparison between jojoba oil and other vegetable oils as substitutes to lubricating oils, *En. Sources* **26** (2004) 639–45.
[275] G.F. Spencer and W.H. Tallent, Sperm whale oil analysis by gas chromatography and mass spectrometry, *J. Am. Oil Chem. Soc.* **50** (1973) 202–6.
[276] P.S. Landis and F. Erickson, Jojoba oil and jojoba oil derivative lubricant compositions, US Patent No. 4,873,008 (10 October 1989).
[277] G.J. Ash, A. Albistion and E.J. Cother, Aspects of jojoba agronomy and management, *Adv. Agronomy* **85** (2005) 409–37.
[278] International Jojoba Export Council (Scottsdale, AZ); http://www.ijec.net.
[279] J.B. Ohlrogge, M.R. Pollard and P.K. Stumpf, Studies on biosynthesis of waxes by developing jojoba seed tissue, *Lipids* **13** (1978) 203–10.
[280] D.J. Undersander, E.A. Oelke, A.R. Kaminski, J.D. Doll, D.H. Putnam, S.M. Combs and C.V. Hanson, Jojoba, in *Alternative Field Crops Manual*; http://www.hort.purdue.edu/newcrop/afcem/jojoba.html.
[281] J. Busson-Breysse, M. Farines and J. Soulier, Jojoba wax: its esters and some of its minor components, *J. Am. Oil Chem. Soc.* **71** (1994) 999–1002.
[282] L. Tobares, M. Frati, C.A. Guzmán and D.M. Maestri, Agronomical and chemical traits as descriptors for discrimination and selection of jojoba (*Simmondsia chinensis*) clones, *Ind. Crops Prod.* **19** (2004) 107–14.
[283] P. Cappillino, R. Kleiman and C. Botti, Composition of Chilean jojoba seeds, *Ind. Crops Prod.* **17** (2003) 177–82.
[284] G.F. Spencer, R.D. Plattner and T. Miwa, Jojoba oil analysis by high pressure liquid chromatography and gas chromatography/mass spectrometry, *J. Am. Oil Chem. Soc.* **54** (1997) 187–9.
[285] A. Benzioni and Y. Vaknin, Effect of female and male genotypes and environment on wax composition in jojoba, *J. Am. Oil Chem. Soc.* **79** (2002) 297–302.
[286] L. Canoira, R. Alcántara, M.J. García-Martínez and J. Carrasco, Biodiesel from jojoba oil-wax: transesterification with methanol and properties as a fuel, *Biomass Bioen.* **30** (2006) 76–81.
[287] R.J. Hamilton (ed.), *Waxes: Chemistry, Molecular Biology and Functions*, The Oily Press, Dundee (1995).
[288] L. Tobares, C.A. Guzmán and D.M. Maestri, Effect of the extraction method and bleaching on jojoba (*Simmondsia chinensis*) wax quality, *Eur. J. Lipid Sci. Technol.* **105** (2003) 749–53.

[289] J. Wisniak, *The Chemistry and Technology of Jojoba Oil*, American Oil Chemists' Society, Champaign, IL (1987).

[290] V.K. Bathia, A. Chaudhry, G.A. Sivasankaran, R.P.S. Bisht and M. Kashyap, Modification of jojoba oil for lubricant formulations, *J. Am. Oil Chem. Soc.* **67** (1990) 1–7.

[291] M.Y.E. Selim, M.S. Radwan and S.M.S. Elkefy, Combustion of jojoba methyl ester in an indirect injection diesel engine, *Renewable Energy* **28** (2003) 1401–20.

[292] G.K. Sandha and V.K. Swami, Study of quality parameters of jojoba oil important for production of value added products, *Intl. J. Chem. Sci.* **6** (2008) 959–74.

[293] G.A. Sivasankaran, R.P.S. Bisht, V.K. Jain, M. Gupta, A. Sethuramiah and V.K. Bhatia, Jojoba oil-based two-stroke gasoline lubricant, *Tribol. Intl.* **21**(6) (1988) 327–33.

[294] J.W. Lambert and D.L. Johnson (to Agro Management Group, Inc.), Vegetable oil lubricants for internal combustion engines and total loss lubrication, US Patent No. 5,888,947 (30 March 1999).

[295] G.A. Sivasankaran, An improved process for the sulfurisation of jojoba oil for use as an extreme pressure additive for lubricants, Indian Patent No. IN 169,857 A1 (4 January 1992).

[296] L. Kunst and A.L. Samuels, Biosynthesis and secretion of plant cuticular wax, *Progr. Lipid Res.* **42** (2003) 51–80.

[297] D. Buisson and S.F. Hannon, Studies on wax esters in fish V-E-1–6; http://www.nzic.org.nz (accessed 05 October 2006).

[298] J.P. McCulley and W.E. Shine, Meibomian gland and tear film lipids: Structure, function and control, *Adv. Exp. Med. Biol.* **506** (2002) 373–8.

[299] J.B. Cheng and D.W. Russell, Mammalian wax biosynthesis. I. Identification of two fatty acyl-coenzyme A reductases with different substrate specificities and tissue distributions, *J. Biol. Chem.* **279**(3) (2004) 37789–97.

[300] J.B. Cheng and D.W. Russell, Mammalian wax biosynthesis. II. Expression cloning of wax synthase cDNAs encoding a member of the acyltransferase enzyme family, *J. Biol. Chem.* **279**(3) (2004) 37798–807.

[301] P.E. Kolattukudy, S. Bohnet and L. Rogers, Disappearance of short chain acids from the preen gland wax of male mallard ducks during eclipse, *J. Lipid Res.* **23** (1982) 989–93.

[302] D.R. Nelson, M. Tissot, L.J. Nelson, C.L. Fatland and D.M. Gordon, Novel wax esters and hydrocarbons in the cuticular surface lipids of the red harvester ant, *Pogonomyrmex barbatus, Comp. Biochem. Physiol. Part B* **128** (2001) 575–95.

[303] G.J. Blomquist, J.C. Arthur and S. Remaley, Biosynthesis of wax in the honeybee, *Apis mellifera* L., *Insect Biochem.* **10** (1980) 313–21.

[304] I. Basson and E.C. Reynhardt, An investigation of the structures and molecular dynamics of natural waxes: I. Beeswax, *J. Phys. D: Appl. Phys.* **21** (1988) 1421–8.

[305] T. Ishige, A. Tani, Y. Sakai and N. Kato, Wax ester production by bacteria, *Curr. Opin. Microbiol.* **6** (2003) 244–50.

[306] L.M. Fixter, M. Nagi, J.G. McCormack and C.A. Fewson, Structure, distribution and function of wax esters in *Acinetobacter calcoaceticus, J. Gen. Microbiol.* **132** (1986) 3147–57.

[307] H. Nieschlag, G.F. Spencer, R.V. Madrigal and J.A. Rothfus, Synthetic wax esters and diesters from *Crambe* and *Limnanthes* seed oils, *I&EC Prod. Res. Dev.* **16**(3) (1977) 203–7.

[308] G. Steinke, P. Weitkamp, E. Klein and K.D. Mukherjee, High-yield preparation of wax esters via lipase-catalyzed esterification using fatty acids and alcohols from crambe and camelina oils, *J. Agric. Food Chem.* **49** (2001) 647–51.

[309] G. Hills, Industrial use of lipases to produce fatty acid esters, *Eur. J. Lipid Sci. Technol.* **105** (2003) 601–7.

[310] T.K. Miwa and I.A. Wolff, Fatty acids, fatty alcohols, and wax esters from *Limnanthes douglasii* (meadowfoam) seed oil, *J. Am. Oil Chem. Soc.* **39** (1962) 320–2.

[311] T.K. Miwa and I.A. Wolff, Fatty acids, fatty alcohols, wax esters, and methyl esters from *Crambe abyssinica* and *Lunaria annua* seed oils, *J. Am. Oil Chem. Soc.* **40** (1963) 742–4.

[312] E.W. Bell (to USDA), Wax esters of vegetable oil fatty acids useful as lubricants, US Patent No. 4,152,278 (1 May 1979).

[313] F.L. Erickson, R.E. Anderson and P.S. Landis (to International Lubricants, Inc.), Vegetable oil derivatives as lubricant additives, US Patent No. 4,970,010 (13 November 1990).

[314] K.D. Mukherjee, Lipid biotechnology, in *Food Lipids: Chemistry, Nutrition, and Biotechnology* (C.C. Akoh and D.B. Min, eds), Marcel Dekker, Inc., New York, NY (1998), pp. 589–640.

[315] S. Bloomer, P. Adlercreutz and B. Mattiasson, Facile synthesis of fatty acid esters in high yields, *Enzyme Microb. Technol* **14** (1992) 546–52.

[316] T. García, M. Martínez and J. Aracil, Enzymatic synthesis of an analogue of jojoba oil: optimization by statistical analysis, *Enzyme Microb. Technol.* **15** (1993) 607–11.

[317] M. Habulin, V. Krmelj and Z. Knez, Synthesis of oleic acid esters catalyzed by immobilized lipase, *J. Agric. Food Chem.* **44** (1996) 338–42.

[318] D.G. Hayes and R. Kleiman, Lipase-catalysed synthesis of lesquerolic acid wax and diol esters and their properties, *J. Am. Oil Chem. Soc.* **73** (1996) 1385–92.

[319] C. Miller, H. Austin, L. Posorske and J. Gonzalez, Characteristics of an immobilized lipase for the commercial synthesis of esters, *J. Am. Oil Chem. Soc.* **65** (1988) 927–31.

[320] K.D. Mukherjee and I. Kiewitt, Preparation of esters resembling natural waxes by lipase-catalyzed reactions, *J. Agric. Food Chem.* **36** (1988) 1333–6.

[321] R. Multzsch, W. Lokotsch, B. Steffen, S. Lang, J.O. Metzger, H.J. Schäfer, S. Warwel and F. Wagner, Enzymatic production and physiochemical characterization of uncommon wax esters and monoglycerides, *J. Am. Oil Chem. Soc.* **71** (1994) 721–5.

[322] M. Trani, F. Ergan and G. André, Lipase-catalyzed production of wax esters, *J. Am. Oil Chem. Soc.* **68** (1991) 20–2.

[323] E. Ucciani, M. Schmitt-Rozières, A. Debal and L.C. Comeau, Enzymatic synthesis of some wax esters, *Fett/Lipid* **98** (1996) 206–10.

[324] E. Wehtje, D. Costes and P. Adlercreutz, Continuous lipase-catalyzed production of wax ester using silicone tubing, *J. Am. Oil Chem. Soc.* **76** (1999) 1489–93.

[325] B.K. De, D.K. Bhattacharyya and C. Bandhu, Enzymatic synthesis of fatty alcohol esters by alcoholysis, *J. Am. Oil Chem. Soc.* **76** (1999) 451–3.

[326] B. Decagny, S. Jan, J.C. Vuillemard, C. Sarazin, J.P. Séguin, C. Gosselin, J.N. Barbotin and F. Ergan, Synthesis of wax ester through triolein alcoholysis: Choice of the lipase and study of the mechanism, *Enzyme Microb. Technol.* **22** (1998) 578–82.

[327] R. Schuch and K.D. Mukherjee, Interesterification of lipids using an immobilized *sn*-1,3- specific triacylglycerol lipase, *J. Agric. Food Chem.* **35** (1987) 1005–8.

[328] N. Goma-Doncescu and M.D. Legoy, An original transesterification cation route for fatty acid ester production from vegetable oils in a solvent-free system, *J. Am. Oil Chem. Soc.* **74** (1997) 1137–43.

[329] M.L. Hallberg, D. Wang and M. Härröd, Enzymatic synthesis of wax esters from rapeseed fatty acid methyl esters and a fatty alcohol, *J. Am. Oil Chem. Soc.* **76** (1999) 183–7.

[330] T. Ishige, A. Tani, K. Takabe, K. Kawasaki, Y. Sakai and N. Kato, Wax ester production from *n*-alkanes by *Acinetobacter* sp. strain M-1: Ultrastructure of cellular inclusions and role of acyl coenzyme A reductase, *Appl. Environ. Microb.* **68** (2002) 1192–5.

[331] M.A. Jenks, S.D. Eigenbrode and B. Lemieux, Cuticular waxes in *Arabidopsis*, in *The* Arabidopsis *Book* (C.R. Somerville and E.M. Meyerowits, eds), American Society of Plant Biologists, Rockville, MD (2002); http://www.aspb.org/publications/arabidopsis.

[332] P.E. Kolattukudy, Further evidence for an elongation-decarboxylation mechanism in the biosynthesis of paraffins in leaves, *Plant Physiol.* **43** (1968) 375–83.

[333] J.B. Ohlrogge and J. Browse, Lipid biosynthesis, *Plant Cell* **7** (1995) 957–70.

[334] J.B. Ohlrogge, D.N. Kuhn and P.K. Stumpf, Subcellular localization of acyl carrier protein in leaf protoplasts of *Spinacia oleracea*, *Proc. Natl. Acad. Sci. USA* **76**(3) (1979) 1174–5.

[335] J. Joyard and P.K. Stumpf, Synthesis of long-chain acyl-CoA in chloroplast envelope membranes, *Plant Physiol.* **67** (1981) 250–6.

[336] A.A. Millar, S. Clemens, S. Zachgo, E.M. Giblin, D.C. Taylor and L. Kunst, *CUT1*, an *Arabidopsis* gene required for cuticular wax biosynthesis and pollen fertility, encodes a very-long-chain fatty acid condensing enzyme, *Plant Cell* **11** (1999) 825–38.

[337] J.G. Metz, M.R. Pollard, L. Anderson, T.R. Hayes and M.W. Lassner, Purification of a jojoba embryo fatty acyl-coenzyme A reductase and expression of its cDNA in high erucic rapeseed, *Plant Physiol.* **122** (2000) 635–44.

[338] K.D. Lardizabal, J.G. Metz, T. Sakamoto, W.C. Hutton, M.R. Pollard and M.W. Lassner, Purification of a jojoba embryo wax synthase, cloning of its cDNA, and production of high levels of wax in seeds of transgenic *Arabidopsis*, *Plant Physiol.* **122** (2000) 645–55.

[339] L. Bucchini and L.R. Goldman, Starlink corn: a risk analysis, *Environ. Health Perspect.* **110** (2002) 5–13.

[340] H.A. Kuiper and G.A. Kleter, The scientific basis for risk assessment and regulation of genetically modified foods, *Trends Food Sci. Technol.* **14** (2003) 277–93.

[341] G. Laghetti, A.R. Piergiovanni and P. Perrino, Yield and oil quality in selective lines of *Crambe abyssinica* and *C. hispanica* grown in Italy, *Ind. Crops Prod.* **4** (1995) 203–12.

[342] Y. Wang and P. Peng, Intergeneric hybridization between *Brassica* species and *Crambe abyssinica*, *Euphytica* **101** (1998) 1–7.

[343] P. Broun and C. Somerville, Accumulation of ricinoleic, lesquerolic, and densipolic acids in seeds of transgenic *Arabidopsis* plants that express a fatty acyl hydroxylase cDNA from castor bean, *Plant Physiol.* **113** (1997) 933–42.

[344] A. Gibbs and J.G. Pomonis, Physical properties of insect cuticular hydrocarbons: The effects of chain length, methyl branching and unsaturation, *Comp. Biochem. Physiol.* **112B**(2) (1995) 243–9.

[345] M.A. Smith, H. Moon, G. Chowrira and L. Kunst, Heterologous expression of a fatty acid hydroxylase gene in developing seeds of *Arabidopsis thaliana*, *Planta* **217** (2003) 507–16.

[346] M. Rossak, M. Smith and L. Kunst, Expression of the FAE1 gene and FAE1 promoter activity in developing seeds of *Arabidopsis thaliana*, *Plant Mol. Biol.* **46** (2001) 717–25.

[347] P.A. Stoutjesdijk, S.P. Singh, Q. Liu, C.J. Hurlstone, P.A. Waterhouse and A.G. Green, hpRNA-mediated targeting of the *Arabidopsis* FAD2 gene gives highly efficient and stable silencing, *Plant Physiol.* **129** (2002) 1723–31.

[348] K. Ståhlberg, M. Ellerström, L.G. Joseffson and L. Rask, Deletion analysis of a 2S seed storage protein promoter of *Brassica napus* in transgenic tobacco, *Plant Mol. Biol.* **23** (1993) 671–83.

[349] S. Thomæus, A.S. Carlsson and S. Stymne, Distribution of fatty acids in polar and neutral lipids during seed development in *Arabidopsis thaliana* genetically engineered to produce acetylenic, epoxy and hydroxy fatty acids, *Plant Sci.* **161** (2001) 997–1003.

[350] E. Wiberg, P. Edwards, J. Byrne, S. Stymne and K. Dehesh, The distribution of caprylate, caprate and laurate in lipids from developing and mature seeds of transgenic *Brassica napus* L., *Planta* **212** (2000) 33–40.

[351] R. Kalscheuer and A. Steinbüchel, A novel bifunctional wax ester synthase/acyl-CoA: diacylglycerol acyltransferase mediates wax ester and triacylglycerol biosynthesis in *Acinetobacter calco-aceticus* ADP1, *J. Biol. Chem.* **278** (2003) 8075–82.

[352] S. Küpper, C. Kohlstedde and M. Schneider (to Ecolab, Inc.), Use of O/W emulsions for chain lubrication, US Patent Appl. No. 2008/108532 A1 (8 May 2008).

[353] F. Fontana, L. Lazzeri, L. Malagutti and S. Galletti, Agronomic characterization of some *Crambe abyssinica* genotypes in a locality of the Po Valley, *Eur. J. Agron.* **9** (1998) 117–26.

[354] T. Troegel, *Abschlußbericht über das Forschungsprojekt 'Modellvorhaben Krambe-Anbau 1996–1999'*, Gülzow (2000).

[355] G. Fila, F. Fontana, C. Maestrini and G. Bellocchi, Field evaluation of crambe cultivars in Northern Italy, *Proc. VII ESA (European Society for Agronomy) Congress*, Cordoba (15–18 July 2002).

[356] T.-A. Hsu, M.R. Ladisch and G.T. Tsao, Alcohol from cellulose, *ChemTech* **10**(5) (1980) 315–19.

[357] G.T. Tsao, M.R. Ladisch, C.M. Ladisch and T.-A. Hsu, Process for treating cellulosic materials and obtaining glucose therefrom, US Patent No. 4,281,063 (28 July 1981).
[358] F.W. Lichtenhaler, The key sugars of biomass: availability, present non-food uses and potential future development lines, in *Biorefineries – Industrial Processes and Products* (B. Kamm, P.R. Gruber and M. Kamm, eds), Wiley-VCH, Weinheim (2006), Vol. 2, pp. 3–59.
[359] G.W. Huber (ed.), *Breaking the Chemical and Engineering Barriers to Lignocellulosic Biofuels: Next Generation Hydrocarbon Biorefineries*, National Science Foundation Workshop, Washington, DC, 25–26 June 2007, (2008).
[360] G. Stephanopoulos, unpublished results (2011).
[361] N. Kosaric and F. Varder-Sukan, Potential source of energy and chemical products, in *The Biotechnology of Ethanol Classical and Future Applications* (M. Roehr, ed.), Wiley-VCH, Weinheim (2001), p. 182.
[362] B. Kamm, M. Kamm and K. Richter, Entwicklung eines Verfahrens zur Konversion von hexosenhaltigen Rohstoffen zu biogenen Wirk- und Werkstoffen – Polylactid aus fermentiertem Roggenschrot über organische Aluminiumlactate als alternative Kuppler biotechnischer und chemischer Stoffwandlungen, in *Chemie nachwachsender Rohstoffe* (P.B. Czedik-Eysenberg, ed.), Österreichisches Bundesministerium für Umwelt, Vienna (1997), pp. 83–7.
[363] T. Werpy and G. Petersen (eds), *Top Value Added Chemicals from Biomass*, US Dept. of Energy, Washington, DC (2004), No. DOE/GO-102004–1992; http://www.osti.gov/bridge
[364] Anon., Succinic acid plans, *Chem. World* **8**(6) (2011) 14.
[365] Y. Zhang, H. Du, X. Qian and E.Y.-X. Chen, Ionic liquid – water mixtures: Enhanced K_w for efficient cellulosic biomass conversion, *En. Fuels* **24** (2010) 2410–17.
[366] R.J. Quinlan, M.D. Sweeney, L. Lo Leggio, H. Otten, J.-C.N. Poulsen, K.S. Johansen, K.B.R.M. Krogh, C.I. Jørgensen, M. Tovborg, A. Anthonsen, T. Tryfona, C.P. Walter, P. Dupree, F. Xu, G.J. Davies and P.H. Walton, Insights into the oxidative degradation of cellulose by a copper metalloenzyme that exploits biomass components, *Proc. Natl Acad. Sci.* (USA) **108**(37) (2011) 15079–84.
[367] R.L. Whistler, J.N. Bemiller and E.F. Paschall, *Starch: Chemistry and Technology*, 2nd edn, Academic Press, New York, NY (1984).
[368] C.M. Duffus, Starch synthesis and deposition in developing cereal endosperms, in *Seed Storage Compounds: Biosynthesis, Interactions and Manipulation* (P.R. Shewry and A.K. Stobart, eds), Clarendon Press, Oxford (1993), pp. 191–209.
[369] J. Preiss, Biology and molecular biology of starch synthesis and its regulation, in *Oxford Surveys of Plant Molecular and Cell Biology* (B.J. Miflin, ed.), Oxford University Press, Oxford (1991), Vol. 7.
[370] S.K. Patil, *Strategic Review of Global Corn Starch Industry and Markets – Starch Derivatives, Sweeteners and Co-products*, S.V. Patil and Associates, Inc., Munster, IL (Aug. 2010).
[371] D.R. Grüll, F. Jetzinger, M. Kozich, M.M. Wastyn and R. Wittenberger, Industrial starch platform – Status quo of production, modification and application,

in *Biorefineries – Industrial Processes and Products* (B. Kamm, P.R. Gruber and M. Kamm, eds), Wiley-VCH, Weinheim (2006), Vol. 2, pp. 61–95.
[372] D.L. Johnson, The corn wet milling and corn dry milling industry – A base for biorefinery technology developments, in *Biorefineries – Industrial Processes and Products* (B. Kamm, P.R. Gruber and M. Kamm, eds), Wiley-VCH, Weinheim (2006), Vol. 1, pp. 345–53.
[373] Y. Takeda, N. Muruta and S. Hizukuri, Examination of the structure of amylose by tritium labelling of the reduced terminal, *Carbohydr. Res.* **227** (1992) 113–20.
[374] K. Shibanuma, Y. Takeda, S. Hizukuri and S. Shibata, Molecular structures of some wheat starches, *Carbohydr. Polym.* **25** (1994) 111–16.
[375] C. Martinez and J. Prodolliet, Determination of amylose in cereal and non-cereal starches by a colorimetric assay. Collaborative study, *Starch/Stärke* **48** (1996) 81–5.
[376] A.-C. Salomonsson and B. Sundberg, Amylose content and chain profile from normal, high amylose and waxy barleys, *Starch/Stärke* **46** (1994) 325–8.
[377] B. Müller-Röber and J. Koßmann, Approaches to influence starch quantity and starch quality in transgenic plants, *Plant Cell Environm.* **17** (1994) 601–13.
[378] W.A. Atwell, L.F. Hood, D.R. Lineback, E. Varriano-Marston and H.F. Zobel, The terminology and methodology associated with basic starch phenomena, *Cereal Foods World* **33** (1988) 306–11.
[379] J. Jane, T. Kasemsuwan, S. Leas, H. Zobel and J.F. Robyt, Anthology of starch granule morphology by scanning electron microscopy, *Starch/Stärke* **46** (1994) 121–9.
[380] W.M. Doane, Starch, industrial raw material, in *Cereals: A Renewable Resource, Theory and Practice* (Y. Pomeranz and L. Munch, eds), American Association of Cereal Chemists, St. Paul, MN (1981), pp. 265–90.
[381] A.-C. Eliasson (ed.), *Starch in Food: Structure, Function and Applications*, Woodhead Publishing, Ltd, Cambridge (2004).
[382] H. Röper, Renewable raw materials in Europe – Industrial utilization of starch and sugar, *Starch/Stärke* **54** (2002) 89–99.
[383] O.B. Würzburg (ed.), *Modified Starches: Properties and Uses*, CRC Press, Boca Raton, FL (1986).
[384] L. Munck, F. Rexen and L. Haastrup Pedersen, Cereal starches within the European Community – Agricultural production, dry and wet milling and potential use in industry, *Starch/Stärke* **40** (1988) 81–7.
[385] H. Yoshimura, T. Nonishi, N. Inouchi and J. Nakakigashi (to Nippon Sangyo KK), Vegetable lubricant, Jap. Patent No. JP 2008/056881 (13 March 2008).
[386] E.C. Wilhelm, Properties of small starch granules and their application in paper coatings, in *Plant Polymeric Carbohydrates* (F. Meuser, D.J. Manners and W. Seibel, eds), The Royal Society of Chemistry, Cambridge (1993), pp. 180–90.
[387] D.L. Kaplan (ed.), *Biopolymers from Renewable Resources*, Springer-Verlag, Berlin (1998).
[388] K. Eskins and G.F. Fanta (to USDA), Non-separable compositions of starch and water-immiscible organic materials, US Patent No. 5,882,713 (16 March 1999).

[389] J.W. Lawton, Nonfood uses of cereals, in *Handbook of Cereal Science and Technology* (K. Kulp and J.G. Ponte, eds), Marcel Dekker, Inc., New York, NY (2000), pp. 725–40.

[390] C. Bastioli, Global status of the production of biobased packaging materials, *Starch/Stärke* **53** (2001) 351–5.

[391] L. Belard, P. Dole and L. Averous, Current progress on biodegradable materials based on plasticized starch, *Austr. J. Chem.* **58**(6) (2005) 457–60.

[392] G. Biresaw and S.Z. Erhan, Solid lubricant formulations containing starch-soybean oil composites, *J. Am. Oil Chem. Soc.* **79** (2002) 291–6.

[393] G.F. Fanta and K. Eskins, Stable starch – lipid compositions prepared by steam jet cooking, *Carbohydr. Polym.* **28** (1995) 171–5.

[394] G.F. Fanta and K. Eskins, Starch-oil composites prepared by jet cooking: Preparation, properties and commercial applications, in *Paradigm for Successful Utilization of Renewable Resources* (D.J. Sessa and J.L. Willet, eds), AOCS Press, Champagne, IL (1988), pp. 266–86.

[395] H. Gusmán-Maldonado and O. Paredes-López, Amylolytic enzymes and products derived from starch: A review, *Crit. Rev. Food Sci. Nutr.* **5** (1995) 373–403.

[396] W.M. Doane, New and potential markets for wheat starch, in *What is Unique* (Y. Pomeranz, ed.), Am. Assoc. Cereal Chemists, St. Paul, MN (1989), pp. 615–31.

6
Chemical transformations of renewable lubricant feedstocks

Abstract: Chemical modification is a means of improving tribological and physicochemical performance of renewable feedstocks (vegetable oils, starches) as lubricant base stocks. Most oleochemical transformations concern the carboxyl moiety of fatty acids; less attention has been paid to functionalisation of the (unsaturated) fatty acid chains. Important transformations of triglycerides leading to enhanced performance are (trans)esterification, selective hydrogenation, condensation, branching and oxyfunctionalisation (mainly epoxidation, hydroxylation, maleation). Industrial applications of modification reactions of vegetable oils are frequently hampered by lack of selective catalysts. Enzymatic catalysis is being explored. Added processing cost is another commercial hurdle. Also modified starches have been evaluated. Starch–oil composites are applied in dry-film lubricant technology.

Key words: chemical transformations, vegetable oils, (trans)esterification, oxyfunctionalisation, modified starches.

6.1 Introduction

Chemistry has laid the foundation for many of the products in use today. As mentioned in Section 3.2, the lubricant industry nowadays uses many different base oil chemistries, including oligomeric/polymeric materials (poly-α-olefins (PAOs), poly(internal olefin)s (PIOs), polyalkylene glycols (PAGs), oil-soluble polyalkylene glycols (OSPs), polyisobutylenes (PIBs), etc.), mostly petro-based. Despite best efforts, it will take decades before we shall see a real shift away from the convenient liquid hydrocarbons to plant matter as the predominant feedstock for the chemical industry and fuel source. The pace of changeover will be driven by technological innovation as much as by unpredictable geopolitical factors and legislation. (Bio) chemistry and chemical engineering are to make significant contributions to a sustainable world by developing substitute technologies (including fermentation) founded on exploitation of renewable raw materials from agricultural and forestal production in a low-carbon economy. The race is on to produce industrial materials from renewable sources. The production of biolubricants from green feedstocks differs essentially from that of conventional lubricants from mineral oil.

Some qualifying requirements for a lubricating oil comprise: (i) satisfactory lubricity in load-carrying or anti-wear (AW) capacity; (ii) high oxidative stability; (iii) satisfactory fluidity at low temperatures; and (iv) high viscosity index (VI). From an environmental preservation standpoint it is desirable to replace the present poorly biodegradable mineral oils by more highly biodegradable base oils. Vegetable oils and polyol esters (POEs) are such highly biodegradable materials.

Vegetable oils are convenient raw materials for use as a renewable resource and are being applied in various industry sectors (e.g. lubricants, fuels, detergents, solvents, composites). Although the use of vegetable oils and fats as lubricants and greases is ancient, their application (in chemically modified form) as biofuels goes back only 25 years. The threat of gradually depleting petrochemical resources, increased environmental awareness and governmental regulations (such as promotion of fallow land practice) have recently resulted in a revival of lipochemistry with an acceleration in innovation and development of new vegetable oils applications [1]. As shown in this chapter, vegetable oil chemistry opens up a great potential for innovation to the lubricant sector. Vegetable fatty oils have advantages such as high VIs, excellent lubricity in extreme pressure, low volatility and good compatibility with additives. However, many of the native oils also have deficiencies for specific applications based on their structures. In particular, the fatty acid (FA) compositions of most of the vegetable oils that are readily available and inexpensive are not directly suitable for lubrication application due to their high saturated or polyunsaturated fatty acid (SFA/PUFA) content. The properties of triglyceride oils of interest to lubrication can be improved in various ways, such as blending (see Section 7.1), genetic modification (see Section 5.3), additivation (see Section 7.2) or by chemical means (this chapter). Selective hydrogenation improves ageing resistance, introduction of branching affects the low-temperature behaviour (pour point, PP) and hydrolytic stability, and hetero-atoms alter the polarity. Any structural modification of vegetable oils is accompanied by economic implications and adds to the overall environmental footprint of utilising these oils for industrial purposes.

Figure 6.1 shows the vulnerable positions of triglyceride molecules. Under high temperature oxidation conditions (not unusual in lubrication applications) vegetable oils can degrade. There are two distinct modes of decomposition, namely thermolysis and autoxidation. Thermolysis occurs by β-elimination at the ester group. Autoxidation is initiated by hydrogen abstraction from the carbon atoms vicinal to double bonds. Esters, such as triglycerides, are also subject to hydrolysis. Consideration of the mechanisms of chemical decomposition of natural oils can be used in molecular design of synthetic esters having improved performance characteristics (see Sections 6.2.1, 10.5.3 and 10.5.4). Functional properties of vegetable oils such as stability and viscosity can be adjusted by chemical modification.

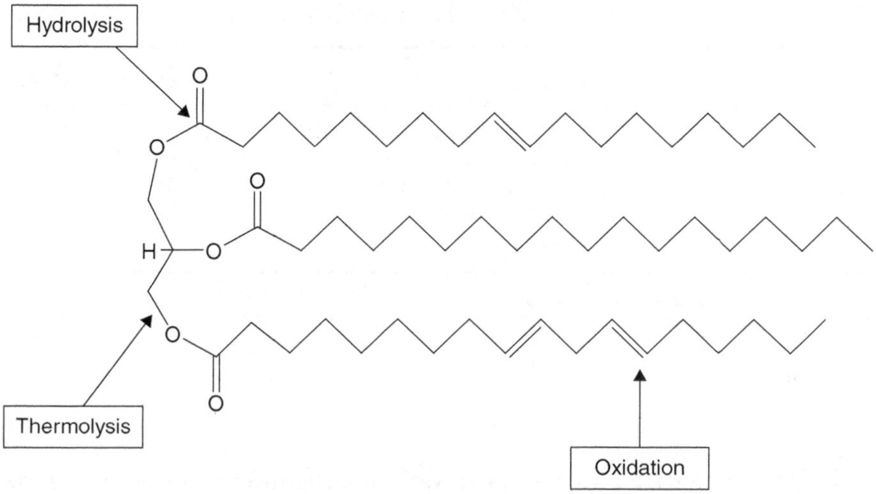

6.1 Schematic structure of triglyceride natural oil, showing positions vulnerable to chemical degradation.

Chemical modification is necessary to improve certain performance limitations of ester-based lubricant base stocks without impairing their excellent tribological and environmentally relevant properties. Optimisation of natural oils refers specifically to properties such as hydrolytic stability, viscosity, evaporation (flash point), oxidation stability, foaming, etc. Polyunsaturation in the FA chains leads to poor oxidation stability, whereas complete saturation of the double bonds results in poor low-temperature fluidity. Therefore, a limiting situation is reached beyond which further improvement in high- and low-temperature stability cannot be obtained. This restricts the use of triglyceride molecules in industrial applications operating under a broad temperature range. Chemical modification of triglycerides will help build molecules with desirable properties for lubricant applications, as all the physical and chemical properties of seed oil are solely based on the molecular structure and their relative distribution. Designing alterations in the FA composition of vegetable oils for a specific industrial application is a challenging task [2]. Table 6.1 shows the effect of various commercial chemical modifications of oils (Cargill processes) on a selected lubricant property (viscosity).

6.2 Chemically modified fatty compounds in lubrication

The linear and long-chain structure of (un)saturated fatty compounds is quite suitable for some applications (e.g. detergents), but is frequently less

Table 6.1 Effect of chemical modification of oil on viscosity

Process[a]	Product[a]	Viscosity @ 25 °C (cP)
Transesterified oils	FAME of soybean, linseed	15
Heat-bodied oils	Soybean, linseed oil	400–50 000
Oxidised oils	Soybean, linseed oil	100–3000
Maleinised oils	Falkowood-51	2000–3000
CPD condensed oils[b]	Dilulin[c]	300–600

[a] Cargill, Inc. processes/products.
[b] CPD, cyclopentadienyl.
[c] Reactive diluent.
After ref. [3].

ideal for lubricants. For this purpose, various structural modifications of the monofunctional FAs have been suggested, for example leading to structures with longer or shorter carbon lengths or bi- or multifunctional substrates (see Sections 6.2.1 to 6.2.9), or otherwise to branched structures (see also Sections 6.3 and 10.5.4).

There are basically three positions in triglycerides where a chemical modification can most easily be effected (Fig. 6.1), namely the ester moieties, β-hydrogen position, and the double bonds along the FA chains. Unsaturated fatty compounds are alkenes and contain an electron-rich double bond that can be functionalised by reactions with electrophilic reagents. Most oleochemical modifications (90%) concern the carboxyl moiety of the FAs (see Sections 6.2.1 and 6.2.2), whereas reactions of the unsaturated or saturated FA chain (Sections 6.2.3 to 6.2.9) account for only 10% [4]. Focus is on eliminating the critical β-CH-group (relative to the ester group) and bis-allylic hydrogen functionalities in methylene interrupted polyunsaturation (for better oxidative stability) and optimal extent of structural alteration for improved low-temperature performance [5].

The glycerol moiety of triglycerides is readily destructible at high temperatures, namely by partial fragmentation of the molecule with formation of unsaturated compounds (Fig. 6.2). The compounds formed may undergo polymerisation, increasing the viscosity and finally resulting in precipitates. Replacing glycerol with polyhydric alcohols such as neopentyl glycol (NPG), trimethylolpropane (TMP) or pentaerythritol (PE) (see Fig. 6.7), which do not contain β-hydrogens, obviates this problem (see Section 6.2.1). Although such synthetic esters (SEs) also do decompose at high temperatures, their thermal decomposition has a radical character and proceeds slowly (Fig. 6.2). Simple SEs produced from natural acids cannot be used at very high temperatures, but they are very suitable for less extreme applications such as two-stroke engines, chain bar oils, and concrete mould release agents.

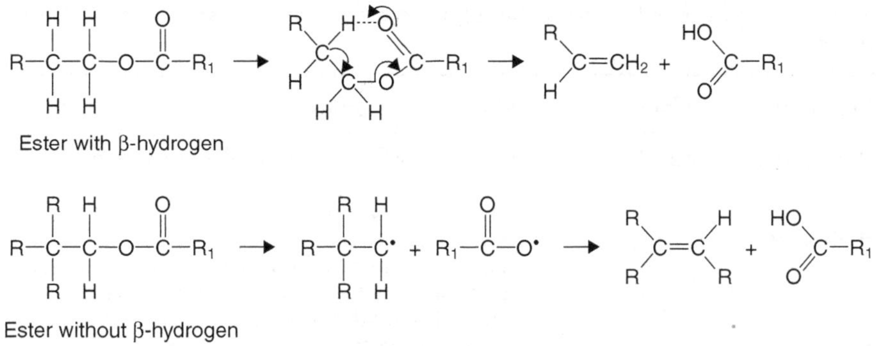

6.2 Thermal decomposition of esters.

With appropriate chemical modifications, SEs can be adapted to the most demanding applications (see Sections 10.5.3 and 10.5.4).

In oleochemistry, there are numerous catalytic conversions (often homogeneous), including catalysed reactions at the double bonds of the unsaturated FA chains (see Sections 6.2.3–6.2.7). Functionalisations, i.e. additions to the double bond, oligomerisation of alkenes, oxidations to epoxides, acids, aldehydes or ketones, or metathesis reactions to different substituted olefins are important examples of these reactions. Chemically functionalised vegetable oils were reviewed by Erhan *et al.* [6]. It should be considered that catalysis of short-chained apolar petrochemical alkenes and of long-chained polar unsaturated fatty compounds differs considerably. While simple alkenes coordinate easily to catalysts, double bonds of mostly balled long-chain fatty molecules interact only with some difficulty with a catalyst site because of steric interactions. Moreover, petrochemical alkenes are mostly devoid of substituents, whereas unsaturated FAs always contain a functional group, e.g. carboxyl, carboxylic ester or alcohol functions. These groups are able to interact strongly with transition metal catalysts and can cause deactivation of the catalyst. Therefore, not all petrochemical alkene reactions can be transferred automatically to the chemistry of unsaturated oleochemicals.

Unsaturation improves the low-temperature properties of vegetable oils, but the increased chemical reactivity is harmful to oxidative stability and is likely to influence many other lubricating properties. The double bonds of unsaturated FAs and their derivatives are easy targets for chemical transformation such as hydrogenation, di/oligomerisation, formation of C—C and C—X bonds (X = hetero-atom), metathesis and oxidation (see Table 6.3). Many of these chemical functionalisations lead to materials showing interesting properties for practical applications. The influence of chemical structure on physical properties is summarised in Table 6.2. Reactions at the double bonds are used to increase the product stability. At the same

Table 6.2 Influence of chemical structure of modified triglycerides on physical properties

Chemical modification	Physical properties
High degree of branching	Excellent low-temperature characteristics High hydrolytic stability Low viscosity index
High linearity	High viscosity index Relatively poor low-temperature characteristics
Low saturation	Excellent low-temperature characteristics Limited oxidation stability
High saturation	Excellent oxidation stability Poor low-temperature characteristics

time, the introduction of functional groups and branches changes the physical properties of the resulting fluids considerably. Ageing resistance can be improved by selectively increasing the degree of saturation. Introduction of hetero-atoms alters the polarity and branching improves low-temperature behaviour and hydrolytic stability. The fluidity characteristics of unsaturated triacylglycerols can be improved by structural modification with 'spacer groups', i.e. (bulky) side groups to the long acyl chains (see Figs 6.18, 6.35 and 6.37). 'Spacer groups' disrupt the close packing ability of the fatty acyl chains and lower the viscosity. Increasing branching reduces pour point and melting point.

Modification of the FA chain entails making a compromise between the following criteria: performance (tribological properties), stability towards oxidation, low-temperature behaviour, biodegradability and cost. At present, only part of the known modification reactions of FA esters are industrially applicable, mainly as a result of lack of sufficient highly selective catalysts and for cost reasons. This offers prospects for further developments (see Section 6.2.7, p. 316).

A variety of structural modifications of (unsaturated) vegetable oils have been suggested (Table 6.3), which appreciably alter overall lubricant properties of vegetable oils. Possible chemical reactions of vegetable oils are numerous. Ester reactions comprise hydrolysis, (trans/inter)esterification, alcoholysis, saponification, reduction, amidation, etc. Double bond reactions comprise oxidation, polymerisation, hydrogenation, epoxidation, halogenation, addition reactions, sulphonation, metathesis and others. Some reported changes that address the polyunsaturated problem include alkylation, acylation, hydroformylation, hydrogenation, oligomerisation (polymerisation) and epoxidation. Moreover, in case of hydroxy fatty acids (HFAs) hydroxyl group reactions are dehydration, hydrolysis, pyrolysis, alkoxylation, esterification, halogenation, sulphonation, etc. Unfortunately, in many cases

Table 6.3 Chemical modifications of (unsaturated) fatty acids

Modification	Catalyst(s)	Reference(s)
Modification of the carboxyl moiety:		
• Esterification/transesterification	Acids, bases	[7–18]
• Estolides	$HClO_4$	[19–24]
Modification of the (unsaturated) fatty acid chain:		
• Selective hydrogenation	Copper catalysts	[7, 25–31]
• Dimerisation/oligomerisation	Aluminosilicates	[4, 7]
• Diels–Alder cycloaddition	–	[32]
• Telomerisation	H_2O vapour	[33]
• Co-oligomerisation	$RhCl_3 \cdot 3H_2O$	[34, 35]
• Hydroformylation (oxo-synthesis)	$Rh(Ph_3P)_3$	[7, 36–41]
• Friedel–Crafts alkylation	$Et_3Al_2Cl_3$	[7, 42]
• Aminoalkylation	$[Rh(COD)Cl]_2$	[43]
• Friedel–Crafts acylation	$EtAlCl_2$	[7, 44, 45]
• Ene-reaction	$EtAlCl_2$, SnO_4	[7, 46, 47]
• Radical addition	$Mn(OAc)_3$, Cu	[7, 48–51]
• Acyloxylation	Nafion/SiO_2	[7, 52–54]
• Cyclopropanation	CH_2I_2, Zn(Cu), Et_2Zn	[55, 56]
• Metathesis	Grubbs' catalyst	[7, 57–63]
• Epoxidation	Peracids, chemo-enzymatic	[4, 64–77]
• Ozonation	–	[78–80]
• Oxidative cleavage	Pb_3O_4	[81, 82]

industrial application of such modification reactions of vegetable oils is still being hampered by lack of suitable catalysts.

Not unexpectedly, saturated fatty compounds are far less reactive. Reactions of saturated fatty compounds are far fewer than those of the more reactive unsaturated systems, namely radical C–C coupling reactions and functionalisation of C–H bonds (see Sections 6.2.8 and 6.2.9).

Oleic acid and oleic acid esters are usually the prototype starting point for chemical modification. The oleic acid content in rapeseed oil (RSO) (canola) is about 60–70%. A >90% technical oleic acid content would produce much purer end-products. In this respect, cultivation of high-oleic vegetable oils such as high-oleic sunflower oil (HOSNO) (C18:1 >90%) is of great interest [83]; see also Section 5.3.1. HOSNO is extremely oxidation- and ageing-stable compared with RSO and TMP-oleate synthesised from rapeseed oil. Even unmodified, HOSNO is suitable as a base fluid for various biodegradable lubricant applications (see also Section 10.5.2).

Several commercially available products based on these modifications have been identified (see Table 6.1). Although it is possible that the end product oils are performing well in the intended applications for some of these reported transformations, the added processing cost is likely to be a hurdle limiting broad commercial adoption.

6.2.1 Esterification/transesterification

In addition to direct environmental advantages, synthetic esters have technical advantages for lubricants in comparison with conventional mineral base oils. Oleochemical esters are being used increasingly because their thermal oxidation stability opens up a wider field of application than that of vegetable oils. The extensive use of esters as lubricants started during World War II to meet the low-temperature requirements of military aviation applications [84]. Today, many industrial lubricating applications use esters [85]. Some esters have a PP below −30 °C, possess high VI and biodegradability. FA esters and amides (mainly RSO-based) are also used as polymer additives (internal lubricants, release agents, slip agents).

A broad variety of bio-based ester oil stocks can be synthesised from the available carboxylic acids (mono-, di-, polycarboxylic acids, aliphatic and aromatic carboxylic acids) and alcohols (diols, polyols, linear and branched alcohols). Oleochemical esters for lubricant applications are commonly chemically classified as monocarboxylic FA esters (monoesters), dicarboxylic acid esters (diesters), glycerol esters (triglycerides), polyol esters, complex and oligomeric esters (Fig. 6.3). The collective name 'synthetic esters' therefore covers a broad range of chemicals with greatly different qualities and prices. The basis of these are mainly alcohols from petrochemical and oleochemical industries and FAs derived from natural oils. FA esters which are obtained from FAs (mono-, diacids) and alcohols (mono-, polyols) are gaining increasing importance as biodegradable replacements for mineral oils. They have more favourable properties than natural fatty acids. There is a need for improved ester-based lubricants with higher kinematic viscos-

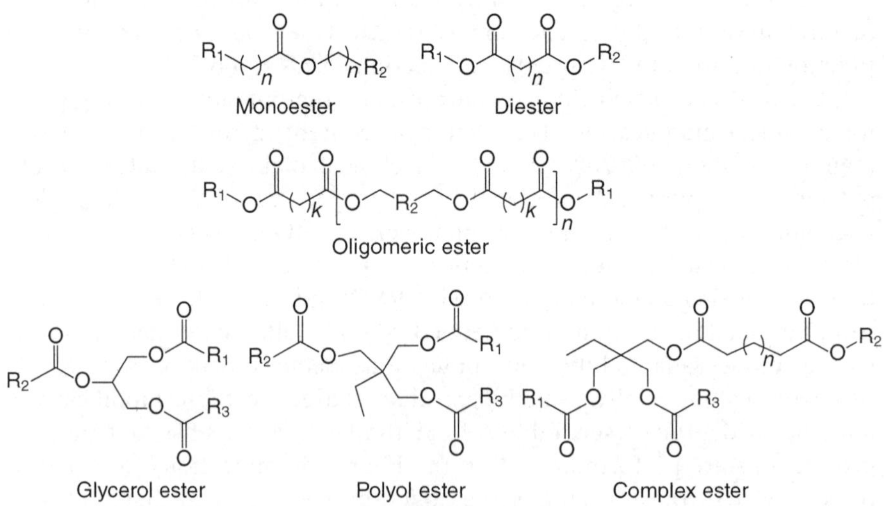

6.3 Structure of different types of esters.

ity (>50 mm^2/s at 40 °C) and other desirable properties such as: PP < −40 °C, VI > 140, low volatility and high flash point (>250 °C), high thermo-oxidative stability and biodegradability. Such properties can be attained by esters composed of branched primary alcohols (C_{18}–C_{22} Guerbet alcohols) and oligomeric (dimeric, trimeric), linear or branched monocarboxylic (C_6–C_{22}) acids (such as oleic, ricinoleic or tall oil acids) with acid value of 195–198 mg KOH/g [86]. POEs such as trimethylolpropane esters (TMP-esters) dominate. Synthetic ester lubricants are used in premium quality motor oils. A variety of companies produce synthetic ester base oils such as BASF (Glissofluid®), Ciba-Geigy (Reolube®), Henkel & Co./Emery Group (Empol®), Oleofina SA (Radialube®), Uniqema (Emkarate®, Pripol®, Priolube®).

Ester-based lubricants exhibit lubrication properties due to the polarity of the ester molecules. The high polarity determines high surface affinity. The polar ester groups adhere to positively charged metal surfaces, creating protective films which slow down the wear and tear of the metal surfaces. Such lubricants are less volatile than the traditional lubricants and tend to have much higher flash points and much lower vapour pressure. Ester lubricants are also excellent solvents and dispersants, and can readily solvate and disperse the degradation by-products of oils. This greatly reduces sludge build up. Ester lubricants are stable to thermal and oxidative processes and show increased biodegradability in comparison to mineral oil-based analogues. However, the preparation of esters is more complex and costly than that of PAOs. Table 6.4 lists the main characteristics of ester oils.

Biodegradation of ester oils is impeded by the presence of aliphatic substituents, cyclic structures, aromatic rings, heteroatoms and unsaturated systems in their molecules. Esters with a linear unbranched chains are most readily biodegradable. Because of their chemical composition, the esters

Table 6.4 Main characteristics of ester oils

Advantages:
- Excellent lubricity
- High biodegradability
- Medium viscosity indexes
- High thermo-oxidative stability
- Low pour points
- Good hydrolysis resistance
- Low volatility
- High flash points
- Compatibility with additives

Disadvantages:
- High production costs
- Limited raw materials for production[a]

[a] For natural products only.

can undergo biological oxidation according to the classical scheme of β-oxidation [87], proceeding by hydrolysis into an FA and an alcohol. The final products of these biological processes are carbon dioxide and water.

Alkyl esters of natural FAs are usually classified as SEs. These products show a relatively good hydraulic stability in comparison to natural esters. Such alkyl esters may be obtained by esterification of FAs with the corresponding alcohols. For that purpose native vegetable oils or animal fats can first be decomposed with water at high temperature and pressure (250 °C, 50 bar) to form FAs and glycerol. The resulting fatty acid alkyl esters (FAAEs), obtained by esterification at high temperature and pressure (e.g. 240 °C, 100 bar) according to Fig. 6.4, are relatively expensive and not cost competitive to mineral oil lubricants.

However, the biodiesel industry nowadays offers fatty acid methyl esters (FAMEs) at competitive costs, in particular when using animal or used nutrient fats or oils as raw materials. Biodegradable lubricant base oils can be produced by re-esterification of the palmitin or stearin fraction of FAME, as obtained by crystallisation fractionation (winterisation) according to Fig. 6.5, with higher alcohols (preferably branched C_3–C_8) [10]. Typical properties of various such FAAE base oils with improved oxidation stability are given in Table 6.5.

Monocarboxylic acid esters are usually derivatives of linear FAs with a C_8–C_{22} alkyl chain length esterified with branched and/or linear monoalcohols. Methyl esters of normal saturated FAs are biodegradable and have excellent oxidative stability, but their melting points are too high for application as lubricants. The melting points of FAs are affected by chain length, molecular structure and polymorphism. Branches on the carbon chain prevent close chain packing and greatly lower the melting point. Utilisation of branched FAs can expand the use of stable saturated FAs in biolubricants (see Section 6.3).

An ester-type biolubricant has also been produced enzymatically from oleic acid and fusel oil (a distillery by-product, typically composed of 10%

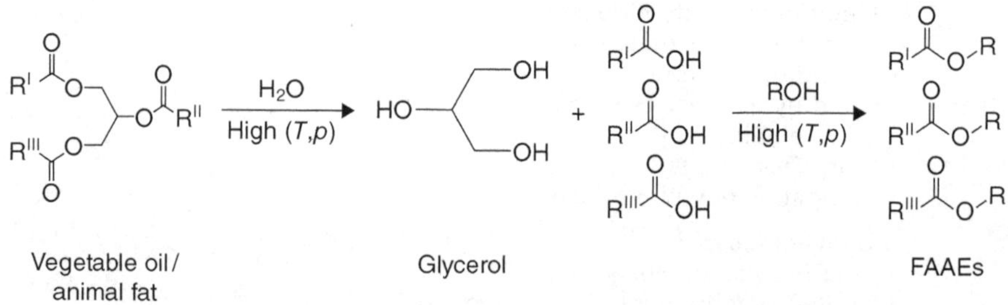

6.4 Transformation of triglycerides to fatty acid alkyl esters (FAAEs) via fatty acids.

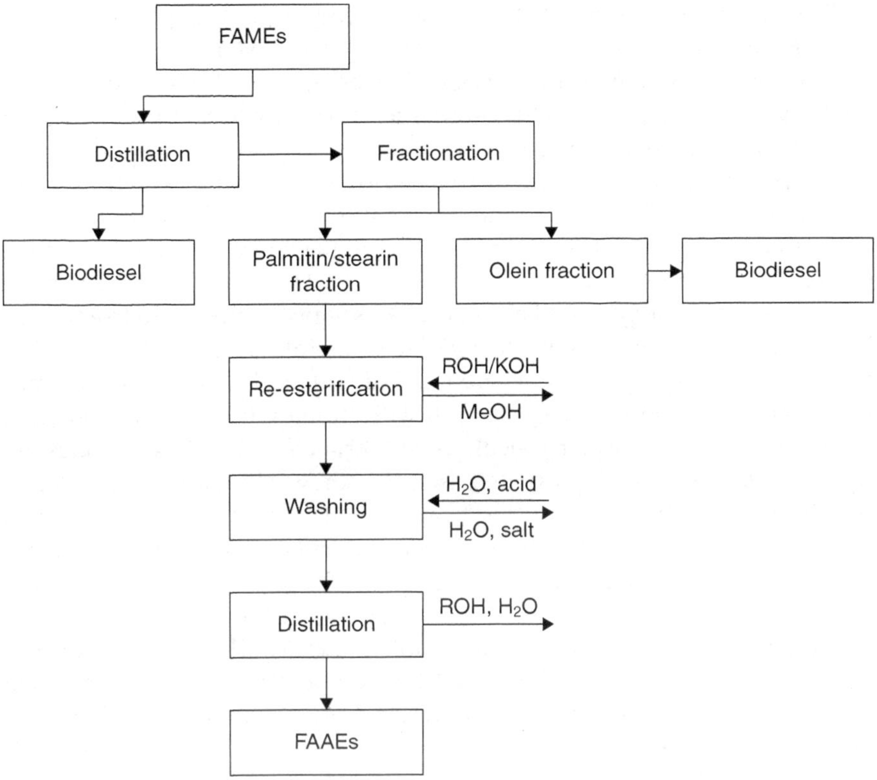

6.5 Re-esterification of the palmitin/stearin fraction of FAME (biodiesel). After ref. [10].

Table 6.5 Properties of fatty acid alkyl ester base oils

FAAE[a]	Flash point (°C)	Viscosity (cSt)[b]	Pour point (°C)	Solubility in isohexane
2-Ethyl-1-hexanol	180–220	7.5–8.5	0–6	yes
1-Butanol	170–190	5.5–6.5	10–14	yes
1-Propanol	160–180	5.3–6.3	10–16	yes

[a] Alcohol.
[b] Temperature not specified.
After ref. [10].

EtOH, 13% *n*-PrOH, 15% *i*-BuOH, 51% *i*-amylalcohol, 11% miscellaneous alcohols and water) [8, 88]. The resulting bioester (PP –27 °C, VI 207, viscosity at 40 °C 6.39 cSt, flash point 211 °C) qualifies as a low-viscosity, low-flash point lubricant, which can be applied at low temperatures. Its thermo-oxidative stability, exceeding that of a commercial, synthetic dicarboxylic acid ester reference, suggests use in high speed/low load regimes, e.g. as a cooling lubricant. Immobilised-lipase (Lipozyme RM IM)-mediated production of *n*-octyl oleate by esterification of oleic acid with 1-octanol in $scCO_2$ has been reported [89]. Biosynthesis for large-scale production in continuous mode is envisaged.

The process of transforming a triglyceride oil (such as high-erucic acid rapeseed oil, HERO) into a long wax mono-ester form is a difficult and expensive process with little, if any, cost advantage over direct use of jojoba oil (see also Section 5.4).

Synthetic dicarboxylic acid esters (diesters) may be prepared from dicarboxylic acids and monovalent alcohols (see general formula of Fig. 6.3). Two types of dicarboxylic acid esters are most suitable as lubricant components: (i) esters of branched primary alcohols with straight dicarboxylic acids; and (ii) esters of straight primary alcohols with branched dicarboxylic acids [9]. Synthetic esters may be partially based on renewable material. The dicarboxylic acids can be derived either from natural sources such as azelaic acid (obtained by ozonolysis of oleic acid), sebacic acid (from oxidation of castor oil) or dimeric FA, isostearic acid or from purely petrochemical sources such as adipic or maleic acid. Consequently, dicarboxylic acid esters are eventually only partly based on renewable materials. The alcohols are generally branched to achieve improved low-temperature performance, such as isooctanol (2-ethylhexanol), isodecanol, isotridecanol or Guerbet alcohols (Fig. 6.6). Esters of straight-chain dicarboxylic acids have higher VI values than mineral oils. Also branched FAs, e.g. 12-hydroxystearic acid (derived from ricinoleic acid by hydrogenation), can be employed. Diesters from oleic acid have been reported [90].

Esters of dicarboxylic acids have found widespread application as high-quality synthetic oils for over 50 years in view of their desirable properties (Table 6.6). Adipic and sebacic acid esters are used mainly as components of modern engine oils and as compressor oils, hydraulic fluids and white oils in the textile and food industry. Branched diesters have been applied as lubricants and hydraulic fluids over a wide range of temperatures [92, 93]. Although diesters have good lubricating properties, their thermal stability is surpassed by polyol esters.

One of the most important modifications of the carboxyl group of the FA chain is (acid or base catalysed) transesterification of the glycerol esters or direct (acid catalysed) esterification of alcohols by the FAs obtained by cleavage of such esters; enzyme-catalysed reactions are also known [94]. The features of the selected substrates influence the properties of the ester manufactured. Catalytic transesterification (homogeneous, heterogeneous or enzymatic) of triglycerides with simple alcohols yields lubricity enhancing products. Low-erucic RSO monoalkyl esters were claimed as base fluids in biolubricants [95]. Biodiesel, FAMEs, obtained by catalytic transesterifi-

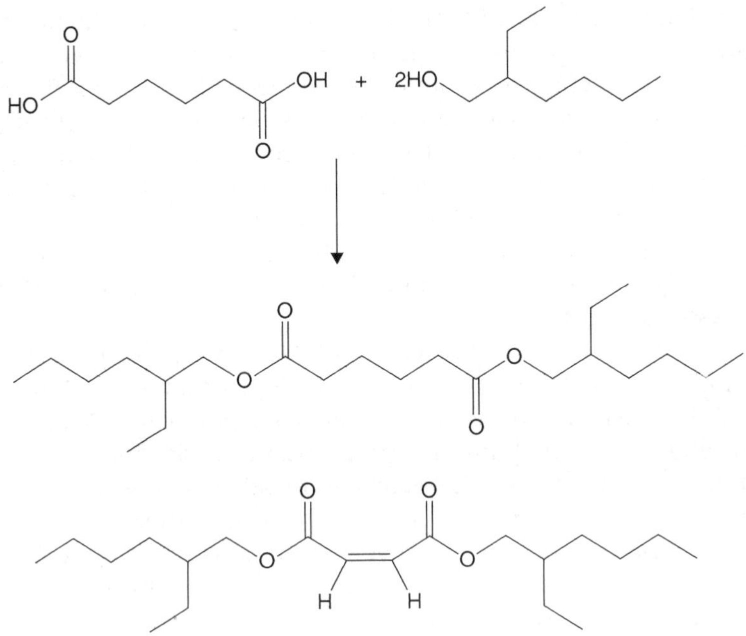

6.6 Diesters from adipic and maleic acids and 2-ethylhexanol. After ref. [91].

Table 6.6 Characteristics of dicarboxylic acid esters as lubricant base stock

- Low freeze point
- Good thermo-oxidative stability
- Low volatility at high temperatures
- Good lubricating properties
- High viscosity index
- Non-toxic
- Readily biodegradable
- Relatively cheap

cation of vegetable oils or animal fats with methanol [94], may be used as an additive to improve the lubricity of petroleum fuels. For the origin of the lubrication properties of FAME, see Section 10.5.5. Ethyl esters of vegetable oils show noticeable improvement in wear properties compared with FAMEs [96]. While the fatty acid methyl, ethyl, propyl and butyl esters belong to the category of biofuel [94], hexyl and octyl esters are constituents of biolubricants [97]; see also Table 6.5 [10]. The physical properties of *n*-octyl esters as premium biolubricant base oils, obtained from sunflower oil by transesterification using NCL (National Chemical Laboratory, Pune)

Fe-Zn double metal cyanide (DMC) complexes as hydrophobic, solid acid catalysts [98], are listed in Table 6.7. Biodegradable lubricants were also obtained by reaction of various Indian vegetable oils (neem, rice bran, linseed, castor, pongamia and jatropha oils) and long-chain (C_5–C_{10}) alcohols [99]. In the context of cost-effective production and refinement routes of lubricating oils, membrane-added separation procedures of unsaturated FAMEs were reported, resulting in enrichment of saturated acid esters and reduction of free fatty acids (FFAs) [100]. Transesterification of castor oil with C_1–C_{10} diols gives diol esters and reaction of these esters with NaOMe gives estolides [101].

The oxidative stability of rapeseed oil, which contains a large quantity of unsaturated FAs, is unsatisfactory. Technology of improved low-temperature fluidity has introduced a medium-chain saturated FA into RSO by ester interchange [102]. Intramolecular transesterification keeps the glycerol structure intact. A biodegradable lubricating oil in which vegetable oil is made the base, and which is a base oil concurrently endowed with satisfactory lubricity, improved oxidative stability, low-temperature fluidity and a low cloud point, has been developed by Fuji Oil Co. by a lipase-catalysed ester-interchange reaction of monoene FAs, *trans*-isomeric FAs and medium-chain FAs [103]. For example, the long-chain acidic residues of RSO have been exchanged for medium-chain length residues from palm oil. Interesterification products have also been prepared from castor oil and

Table 6.7 Physical properties of fatty acid-*n*-octylesters (biolubricants) obtained by transesterification of sunflower oil with *n*-octanol

Test	Method	Value
Kinematic viscosity at 40 °C (cSt)	IS:1448	7.93
Kinematic viscosity at 100 °C (cSt)	IS:1448	2.74
Viscosity index	IS:1448	226
Density at 15 °C (g/cm^3)	ASTM D 4052	0.8813
Pour point (°C)	IS:1448	–3
Copper strip corrosion at 100 °C for 3 h	IS:1448	No. 1
Total acid number (mg KOH/g)	IS:1448	0.76
RBOT, induction period in min at 100 °C	ASTM D 4742	23
Lubricity on HFRR, friction coefficient, load 2 N, 60 °C	ISO 12156	280 μm
Lubricity on HFRR, wear scar dia., load 2 N, 60 °C	ISO 12156	180 μm
Lubricity on HFRR, contact potential, load 2 N, 60 °C	ISO 12156	44–48 mV

RBOT, rotary bomb oxidation test; HFRR, high-frequency reciprocating rig.
After ref. [98].

substantially non-hydroxyl-containing naturally occurring triglyceride oils [104]. However, the critical secondary hydrogen in the β-position in triglycerides, with glycerol as the alcohol component, renders these substances still sensitive to thermal degradation by elimination and cleavage of the esters into acid and olefin.

Although diesters already possess excellent lubricating properties, their thermal stability is surpassed by the polyol esters. The term 'polyol ester' (POE) means a substantially fully esterified product obtained when a polyol (or polyhydric aliphatic alcohol) having at least two hydroxyl groups is used (Fig. 6.7). The polyol esters include esters derived from hindered neopentyl alcohols such as NPG, trimethylolethane (TME), TMP, higher trimethylol-alkanes, PE, dipentaerythritol or tripentaerythritol; see Fig. 6.8. A variety of polyols are commercially available, e.g. from Perstorp Polyols, Inc. (Toledo, OH). For instance, Perstorp Polyol TP30 is an ethoxylated TMP with a molar ratio of 1:3 (TMP:ethylene oxide). Polyol ester stocks are usually made up of long-chain linear FAs (C_8–C_{18}) of plant or animal origin esterified with multifunctional alcohols which include two to six hydroxy groups. Cramail *et al.* [105] have reported the preparation of polyols by transesterification and thiolation.

Figure 6.9 shows the transesterification of TMP with FAMEs to biodegradable base stocks. The forward reaction of Fig. 6.9 is favoured by either

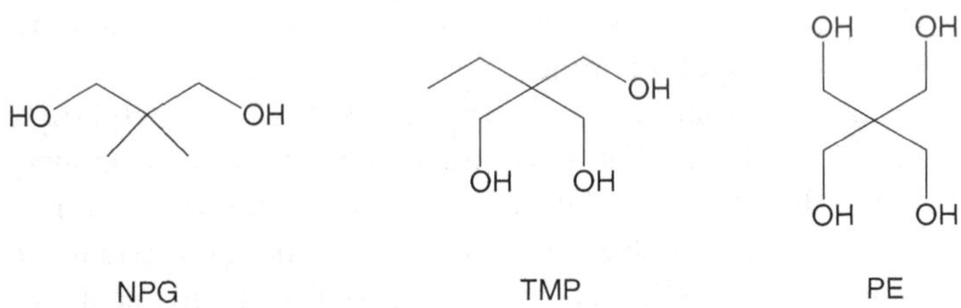

6.7 Polyols: neopentyl glycol (NPG), trimethylolpropane (TMP) and pentaerythritol (PE).

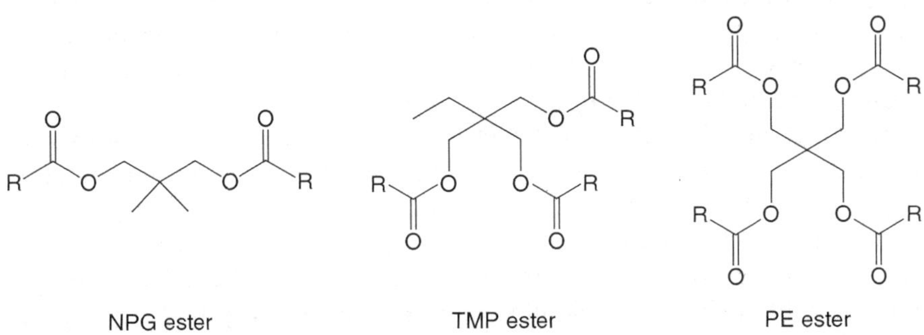

6.8 Esters of neopentyl glycol (NPG), trimethylolpropane (TMP) and pentaerythritol (PE). After ref. [91].

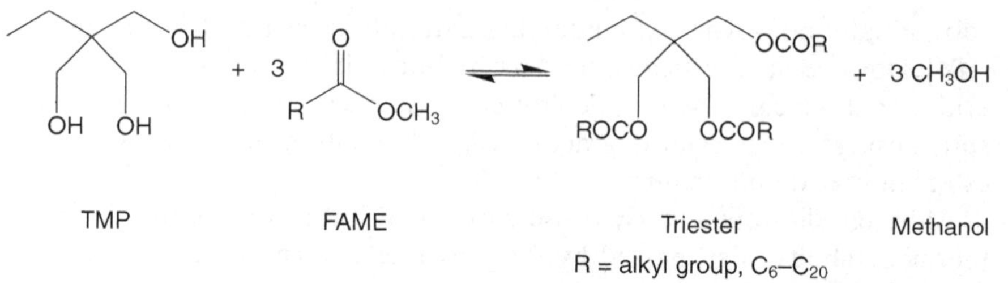

6.9 Transesterification of trimethylolpropane (TMP) with fatty acid methyl esters (FAMEs) to biodegradable polyester base stocks.

continuous removal of methanol or the use of excess FAME. Removal of methanol can be achieved by the application of a vacuum [106], high-temperature system under nitrogen [107], and the use of an azeotropic agent such as isooctane [108]. At atmospheric pressure, an operating temperature of at least 200 °C is required to produce a reasonable conversion to POEs [107]. An optimal choice of vacuum conditions is crucial for the product yield. If the vacuum is too high, it tends to entrain the methyl esters away with methanol, thus lowering the product yield. On the other hand, if the vacuum is too low (e.g. 100 mbar) not all the methyl esters are converted to TMP esters.

Glycerol esters are considered as a separate class of products as compared with polyol esters with a quaternary carbon atom. In synthetic products in which the glycerol moiety has been replaced by polyols with a quaternary carbon atom in the centre, namely NPG [7, 109, 110], TMP [15, 106] or PE [7, 110] (see Fig. 6.8), much improved performance towards hydrolysis and thermal degradation is achieved. Table 10.21 summarises the main properties of POEs. Beside excellent thermal and oxidation stability, the esters have good *V–T* and lubricating properties and excellent viscosity characteristics at low temperatures. Bio-based POEs show very good biodegradability, moderately good oxidative stability and moderate price level. They display good shear stability. Yet, being esters, POEs are hygroscopic lubricants (i.e. they absorb water) which may hydrolyse over longer periods of time for water content >200 ppm. These materials are considered to be semi-synthetics, not being entirely renewable.

Recently, alkali-catalysed production of a *Jatropha*-based lubricant was reported by transesterification at 65 °C of *Jatropha* oil to a *Jatropha* methyl ester (JME), followed by transesterification (at 150 °C, 10 mbar) of JME with TMP (at JME/TMP molar ratio of 4:1) at a maximum yield of 50% [111]. The basic properties of the jatropha biolubricant were found comparable to other plant-based biolubricants (kinetic viscosity at 40 °C, 43.9 cSt; VI, 180; PP, –6 °C). The PP could eventually be improved by blending with other plant-based lubricants.

Relationships between structure (functionality of the alcohol, length of the FA hydrocarbon chain, degree of unsaturation of the acyl group, FA profile, and amount of diacid) and lubricating properties (viscosity at 40 °C, VI, flash point and PP) of neopentyl polyol esters have been reported [11]. Qualitative and quantitative relationships were derived for esters from alcohols including TMP, PE and NPG hydroxypivalate and from C5:0 to C22:1 acids. SEs derived from vegetable base stocks (e.g. TMP trioleate) significantly improve the cold-flow properties of biodegradable base stocks. Typically, vegetable oils have PPs of –10 to –20 °C [112] compared with TMP trioleates with PPs of –20 to –45 °C [12], which are considered to meet most field lubricant applications.

POE lubricant base stocks from C_5 polyols and C_6–C_{14} carboxylic acids were proposed as biodegradable automatic transmission fluids (ATFs) [113], others were used for environmentally sensitive (marine) applications [114]. With some additivation POEs are an excellent solution for medium to heavy applications. Some of these esters display extraordinary thermal stability. TMP esters of low viscosity and high temperature stability have long been used as lubricants for jet turbine engines (see Section 10.5.3). Also NPG esters are applied as lubricants in jet engines. Other derivatives are used in arctic conditions. PE derivatives of C_5–C_9 carboxylic acids are used in gas turbine engines. TMP esters of oleic acid are presently the most widely applied materials for hydraulic fluids. Hydraulic fluids based on TMP esters of RSO or sunflower oil (SNO) have good cold stability, friction and wear characteristics, and resistance against oxidation at elevated temperatures. Transesterification of TMP and RME (rapeseed oil methyl ester) has been catalysed by both enzymatic and chemical means [106]. Pentaerythritol oleate has been described as an environmentally acceptable and renewable lubricant base oil [13].

Transesterification can also be performed by adding a short-chain FA ester to a vegetable oil in the presence of a suitable catalyst and heating the mixture. Esters of short-chain FAs include methyl esters and polyol esters. Transesterification of a short-chained polyol ester with a vegetable oil results in the short FA chains of the polyol and the long FA chains of the triacylglycerol (TAG) being randomly distributed among both the polyol and glycerol backbones (Fig. 6.10). Chemical methods to produce such transesterified vegetable oils with heterogeneous chain lengths are disclosed in US Patents Nos 6,278,006 B1 [115], 6,281,375 B1 [116] and 6,465,401 B1 [14]. In US Patent No. 6,465,401 B1 (to Cargill, Inc.) Kodali and Nivens [14] disclose lubricating oils with improved oxidative stability, low-temperature properties and low friction coefficients, based on polyhydric alcohols with heterogeneous FA chain lengths obtained by interesterification of vegetable oils with short-chain (C_6–C_{10}) FA esters. A statistical model based on a random distribution was developed to determine how

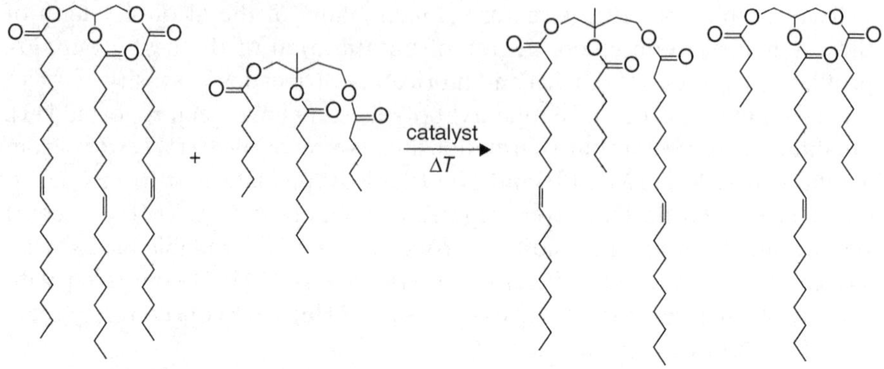

6.10 Transesterification of a vegetable oil and a short-chained polyol ester to produce oils with heterogeneous chain lengths. After ref. [115].

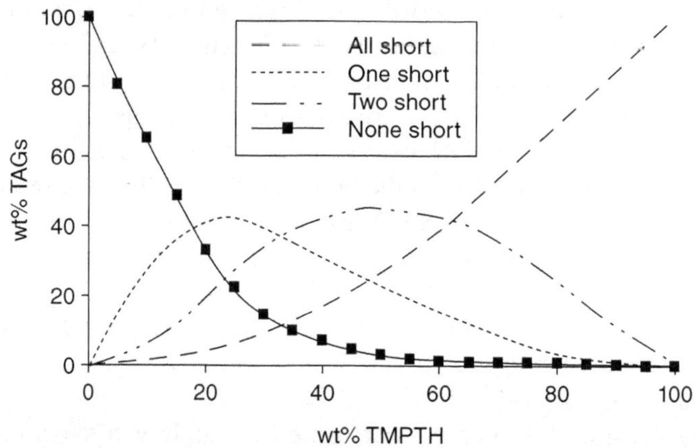

6.11 Predicted fatty acid distribution of triacylglycerols of interesterified products of canola oil and trimethylolpropane triheptanoate (TMPTH). After ref. [14].

the long-chain FAs of canola oil TAGs and the short-chain FAs of the non-glycerol ester would be distributed when short-chain FA esters were transesterified with canola oil at different concentrations (see Fig. 6.11). Interesterifying about 20–25 wt% TMPTH (trimethylolpropane triheptanoate) with canola oil yields a large number of TAGs with one short chain, and modifies over 70% of the original TAGs found in canola oil. See also Section 10.5.3, p. 513. Interesterification of natural triglycerides with short-chain carboxylic acid esters leads to products with low-viscosity, fuel-like characteristics (TBK-biodiesel) [117].

Polyhydroxy acids and their derivatives qualify as biolubricants. Pentaerythritol-tetra [poly(ricinoleic acid)] is a ricinoleic acid star polymer possessing desirable physical properties for bio-based lubricants with high VI (155) and low melting point (<–15 °C) [118]. A targeted viscosity value can be readily achieved by controlling the product's M_n (number average molecular weight), since viscosity vs. M_n for star polymers is a linear relationship. Castor oil contains 87% ricinoleic acid, which is a renewable resource. Ricinoleate based polyols are available from Caschem and Spencer-Kellogg.

POEs differ only slightly from RSO in regard to hydrolytic stability, but oxidation resistance is much greater. Both characteristics are significantly improved with complex esters (see Fig. 6.3). Improvement in hydrolytic stability usually leads to lower biodegradability. The term 'complex esters' is usually meant to imply an ester formed from the reaction of three or more of the following compounds: monohydric aliphatic alcohols, monobasic aliphatic acids, aliphatic glycols or polyglycols, polyhydric aliphatic alcohols, dibasic aliphatic acids and polybasic aliphatic acids, where at least one polyfunctional alcohol and at least one polyfunctional acid are employed. This definition includes glycol centred complex esters, dibasic acid centred complex esters and alcohol acid terminated complex esters [119]. Examples of aliphatic polyols used for manufacturing complex esters include glycerol, alkylene glycols (e.g. ethylene glycol, diethylene glycol, triethylene glycol and NPG), PE, TME, TMP and sorbitol. Alicyclic polyols include cyclohexanediols and inositol as well as natural cyclic polyols such as glucose, galactose and sorbose. Complex esters (or oligomer ester mixtures) are disclosed in various patents [120–123]. Typical complex esters were prepared using low-molecular-weight saturated FAs, TMP and a dicarboxylic acid to obtain products having high viscosity, high shear stability and good lubricating activity [15]. The products, based on renewable resources, show additional advantages over other synthetic esters in having superior oxidative stability due to the presence of SFAs, and very low melting points due to the high degree of branching. While the toxicity is rather low and comparable to normal POEs, the biodegradability is significantly influenced by the degree of cross-linking in the molecule. High viscosity types (>300 cSt) do not surpass the OECD limiting values for ready biodegradability. Only medium viscosity products up to 140 cSt qualify for WGK 0.

Complex esters that contain straight or branched diols or polyalkylene glycols and straight or branched dicarboxylic acids as well as (mono)carboxylic acids and monofunctional alcohols have higher molecular weights and higher viscosities than the common esters. Higher complex metalworking lubricants (of unknown composition, resulting from interchange and condensation reactions) of increased viscosity were derived by reacting natural oils and fats with a hindered polyol (e.g. NPG, TMP, PE) and a (cyclic or acyclic) dicarboxylic acid, eventually followed by chlorination or sulphurisation [124].

High-viscosity complex alcohol esters, derived from a polyhydroxyl compound (e.g. NPG, TMP, PE), a polybasic acid (e.g. adipic acid) or an anhydride of a polybasic acid, and a monohydric alcohol (such as isodecyl alcohol) exhibit PPs of less than –20 °C and a viscosity of 100–700 cSt at 40 °C. They are useful as a base stock for various lubricants [123]. The ratio of polybasic acid to polyol is critical in the formation of a complex alcohol ester. Such complex alcohol esters are characterised by excellent lubricity, seal swell, biodegradability, low toxicity, friction modification, low-temperature properties, high viscosity index, thermal and oxidative stability, and polarity. The higher-viscosity oils are desirable in certain end-use applications such as greases, heavy-duty engine oils, certain hydraulic fluids, drilling fluids, aircraft and other turbine oils, compressor oils and the like. These complex esters are particularly well suited for use as the base stock in engine oils for reducing wear and increasing fuel efficiency. Conventional complex alcohol esters, which are formed with adipates, result in much lower viscosity (<100 cSt) and poor seal swell properties. In PCT Int. Publ. No. WO 00/29521 (to Exxon Chemical Co.) Turner *et al.* describe complex alcohol ester compounds and compositions which exhibit continuous solubility up to 15 wt% in mineral oil [120]. In PCT Int. Publ. No. WO 2005/019395 A1 (to Cognis Corp.) Zehler and Costello [16] disclose complex esters, namely polyol (NPG, TMP, PE) esters of long-chain saturated or unsaturated dicarboxylic acid residues (from C_9 to C_{22}) which show improved performance (biodegradability >73%) as compared to esters containing short-chain dicarboxylic acid residues such as adipic acid. Complex esters are particularly useful in high-flying aircrafts where very low temperatures are encountered for prolonged periods of time.

A drawback of vegetable oil-based or esterified lower molecular weight polyglycerol-based compounds as lubricants is their low viscosity profile, which limits their range of usefulness. The viscosity of vegetable oils may be increased in several ways, including sulphurisation. Another possibility is to use synthetic polyol esters of polyhydric alcohols (e.g. NPG, TMP, PE). Although biodegradable, such esters are still relatively low in viscosity. Higher viscosities may be achieved by using dibasic acids to increase molecular weight, but biodegradability suffers. Such products are also insoluble in functional fluids. Polyglycerols (for $n \geq 8$) are highly viscous and highly polar, and therefore oil insoluble. Esterification with fatty acids or transesterification with methyl esters of fatty acids or triglycerides alters their polarity, hydrophilic–lipophilic characteristics and improves the solubility of polyglycerols in oil. Esterification can take place at any or all of the hydroxy groups of the polyglycerol chain (usually at least 25%) depending upon the reaction conditions. Viscosity increases and hydroxyl value decreases with increasing polyglycerol chain length. US Patent No. 5,380,469 to Calgene Chemical, Inc. [17] describes high-viscosity functional fluids prepared by reacting a polyglycerol with a triglyceride oil (preferably canola) or fat. The completely biodegradable products (viscosity > 1000 cSt at 25 °C) can be used as high-viscosity lubricants or blended with triglyceride-based functional fluids in mud drilling, as low-temperature lubricants, cutting fluids, hydraulic fluids, food-grade lubricants or as additives for other functional fluids. Decaglyceryl monostearate is used as a food-grade lubricant [125]. It is noticed that the maximum degree of polymerisation of glycerol in polyglycerol esters of fatty acids which are permitted by law as food additives is 10.

Common oleochemical esters (i.e. monoesters, triglycerides, polyol esters and complex esters) for lubricants are only scarcely toxic or non-toxic (LD50 ≈ 1000–10 000 mg/L) due to their very low water solubility. They also exhibit high degradation rates and low bioaccumulation potential. As a matter of fact, biodegradability is mainly determined by chemical structure and less by water solubility. FA esters (monoesters, glycerol and polyol esters) are classified as WGK 0, whereas mineral oil with its limited biodegradability is generally classified as WGK 1. In addition, fatty acid esters are biodegradable also under anaerobic conditions, whereas mineral oils are not [126].

Ester oils having an oligomeric structure are less well known. Oligomeric esters can be derived from dicarboxylic acids and dihydric alcohols. Transesterification of dimethyl adipate and dimethyl sebacate by neopentyl glycol (Fig. 6.12) has been reported [127]. Such esters can be designed in

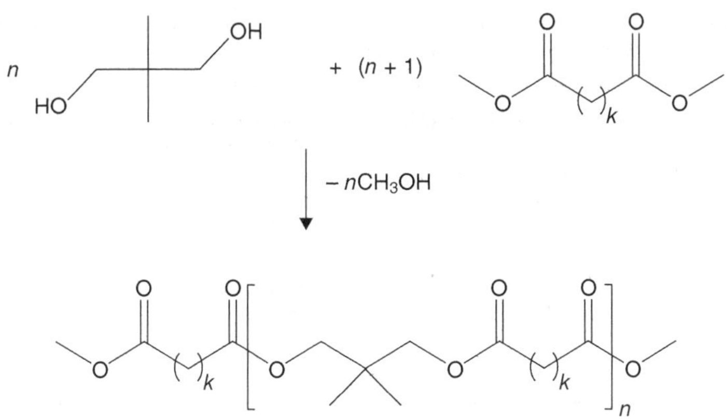

k = 4, adipic acid; k = 8, sebacic acid;
n = degree of oligomerisation

6.12 Transesterification of dimethyl adipate and dimethyl sebacate by neopentyl glycol. After ref. [127].

different ways in order to obtain different properties in terms of viscosity (medium to high) and polarity. As the molecular weight of the esters increases, so does the viscosity.

In US Patent No. 5,916,854 (to Kao Corp.) Inaya *et al.* disclose a biodegradable lubricating oil obtained by addition of an alkylene oxide to a polyhydric alcohol, followed by transesterification between the addition product and oils or fats [128]. The general formula of the (oligomeric) reaction product is shown in Fig. 6.13. Table 6.8 lists lubrication-related parameters of some typical reaction products.

Most vegetable oils, SEs and glycols display excellent lubricity in boundary lubricant conditions. The high degree of polarity of these lubricants results in superiority against mineral oil-based lubricants. The friction coef-

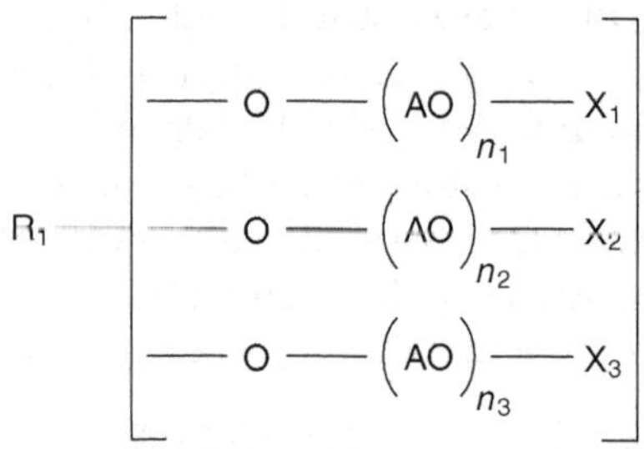

6.13 Transesterified reaction product with oils and fats of the adduct of a polyhydric alcohol and an alkylene oxide. R_1, hydrocarbon residue of polyhydric alcohol; AO, alkylene oxide; X_i, H or R′CO; R′, long-chain alkyl group.

Table 6.8 Characteristics of transesterified reaction addition products of a polyhydric alcohol and an alkylene oxide

Composition[a]	Biodegradability	Kinematic viscosity @ 100 °C (mm^2/s)	Pour point (°C)
RSO/GL/EO	72[b]	7.4	−22.5
SBO/GL/EO	76[b]	16.3	−22.5
PKO/GL/EO	70[b]	10.7	−12.5
CNO/GL/PO	92[c]	10.2	−32.5
CNO/EG/bEO-PO	82[c]	9.2	−27.0
CNO/EG/rEO-PO	70[c]	2.1	−35.0
CNO/TMP/EO	74[b]	10.7	−2.5

[a] CNO, coconut oil; PKO, palm kernel oil; RSO, rapeseed oil; EG, ethylene glycol; GL, glycerol; TMP, trimethylolpropane; EO, ethylene oxide; PO, propylene oxide; b, block; r, random.
[b] OECD 301B.
[c] CEC L-33-A-93.
After ref. [128].

ficient of vegetable oils, SEs (such as TMP-ester) and glycols is half that of mineral oils. Weller and Perez [129] have compared the effect of chemical structure of two ester-based and three vegetable-based fluids (jojoba, corn and castor oil) on friction and wear characteristics. The ester chemical structure (alcohol structure, acid chain length, acid chain branching) influences friction and wear with ester base stocks [130]. The number of ester linkages and chain length affects the measured friction. An increasing number of ester linkages results in higher friction coefficients. As the acid chain length increases, the friction coefficient and delta wear show large improvements. A mono ester structure shows the best wear with the base fluids. Branching of the acid group used in the esterification process and the polarity of the ester molecule play major roles in the wear and scuffing results [131].

All ester oils can hydrolyse more or less rapidly in contact with water. It is therefore essential that these products are shielded from water, and moisture in general, during storage and use. Ester oils are ultra-dried and are filled into air-tight metal drums with a water content of <30–100 ppm in a nitrogen atmosphere. Ester oils have higher VI than mineral oils. The higher VI translates into a wider temperature range of application and into multigrade lubricants with improved shear stability. Esters and vegetable oils show excellent evaporation loss: RSO < POEs < hydrocracked oils < paraffinic solvent refined oils < alkylbenzenes. The lower evaporation loss of ester oils has great advantages in metalworking and in internal combustion engines. Compared with mineral oil-based products, the selection of low-evaporation ester and vegetable oil-based lubricants can allow lower viscosities to be used. This enables faster machining, thus increasing productivity. Rapidly biodegradable total-loss lubricants based on esters, polyglycols and natural oils decrease the lubricant-based pollution.

Typical examples of highly biodegradable base oils are vegetable oils (biodegradation rate of 95–100% according to CEC L-33-A-93), polyol esters (ditto, 80–100%) and dicarboxylic acid esters (ditto, 70–95%) [132]. Biodegradable lubricants were first developed explicitly for two-stroke outboard engines in the early 1980s, with the main base fluid composed of NPG esters of branched-chain fatty acids. Developments in environmentally friendly lubricating oils based on neopentyl polyol esters were reviewed by Eychenne *et al.* [133].

Ester-based formulations derived from natural products are able to satisfy both technical requirements and environmental/health issues [134]. A specially developed FA ester (Petrofree®; Halliburton/Baroid) for offshore drilling applications not only fulfils the requirements of biodegradability and low toxicity but also has a better lubricating effect than products based on mineral oils (see also Section 12.4.2). For high-stability esters for lubricant applications, see Section 10.5.3. Advanced bio-based polyol esters are described in Section 10.5.3, p. 510; for hydroxyl esters, see Section 10.5.3, p. 513. For more information on branched biodegradable ester base stocks, see Section 10.5.4. One of the major advantages achieved by bio-based SEs is better performance at a lower cost compared with SEs [135]. This is possible due to recent advantages in the biotechnology of vegetable oils and chemical modifications that can be applied to convert these natural esters into high-performance biolubricants. Wax esters constitute another molecular variation by which properties of lipids can be improved, making them targets for lubricant application (see Section 5.4). Modifications of the FA chain also comprise secondary ester formation (See Section 6.2.2).

6.2.2 Estolide ester base stocks

Estolides are formed when the carboxylic acid group of one FA molecule forms an ester link at a site of unsaturation of another FA. More generally, estolides are unique oligomeric FAs formed by condensation of two or more FAs, (Fig. 6.14). Estolides are thus a class of esters with secondary ester linkages between FA chains [19–21, 24]. Monoestolides contain one ester link. Estolide formation may also involve a hydroxy functionality as a site for esterification (see Fig. 6.18). Estolides occur in nature [136–138]. *Heliophila amplexicaulis* L.f. seed oil contains about 30% lesquerolic acid and estolide triglycerides [136]. The kernel oil of Chinese tallow (*Sapium sebiferum*) seed contains tetraester triglycerides composed of *trans*-2,*cis*-4-decadienoic acid joined in an estolide linkage to 8-hydroxy-5,6-octadienoic acid [137]. *Trewia nudiflora* (false white teak) seed oil contains glycerides of α-kamlolenic (18-hydroxy-9,11,13-octadecatrienoic) acid, largely as estolide triglycerides [138].

Estolides are oligomeric esters of FAs resulting from a cationic dimerisation of two fatty groups in the presence of an acid catalyst. Various Brönsted acids can catalyse this oligomerisation [19], which may also be induced by lipase catalysis [139, 140]. Technically, optimal conditions, in terms of properties, yields and cost, for the synthesis of a complex estolide are often 0.05 equivalents of $HClO_4$ at 60 °C for 24 h followed by the direct conversion to the corresponding esters [22, 23]. FAs can be readily converted into estolides

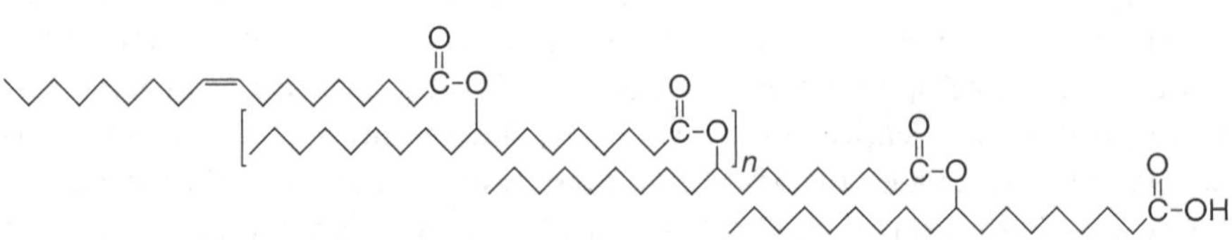

6.14 Oleic acid polyestolide. After ref. [139].

either as triglycerides in the presence of an FFA or from homopolymerisation of split FAs.

The mechanism of acid-catalysed estolide (oxyfatty acid oligomer) formation of a monoenoic FA involves carbocationic oligomerisation across the double bond [19]; see Fig. 6.15. This condensation can continue, resulting in polyestolides, where the average extent of oligomerisation is defined as the estolide number EN (EN = n + 1, Fig. 6.16), which represents the average number of FAs added to the base FA [23]. For monoestolides in the structure of Fig. 6.16 n = 0.

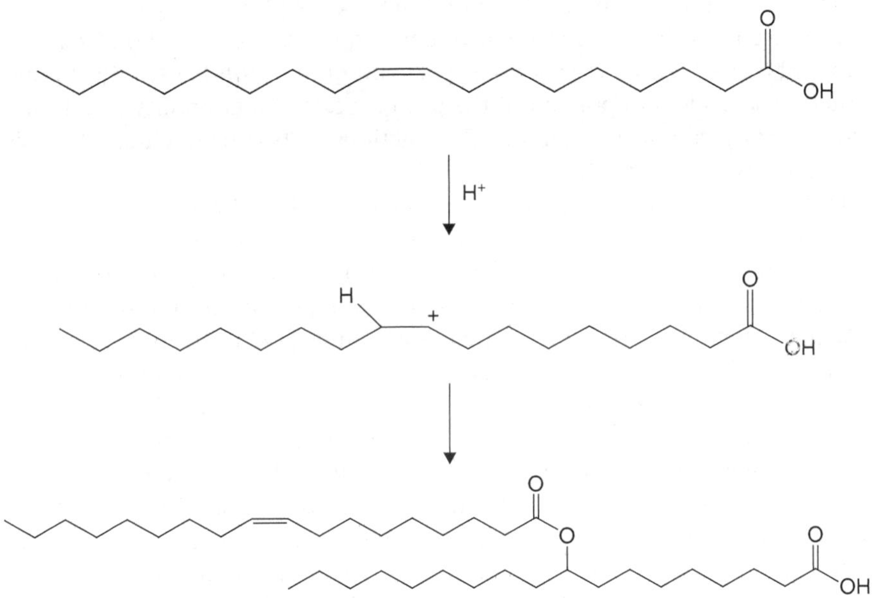

6.15 Acid-catalysed condensation of oleic acid into oleic estolide. After ref. [19].

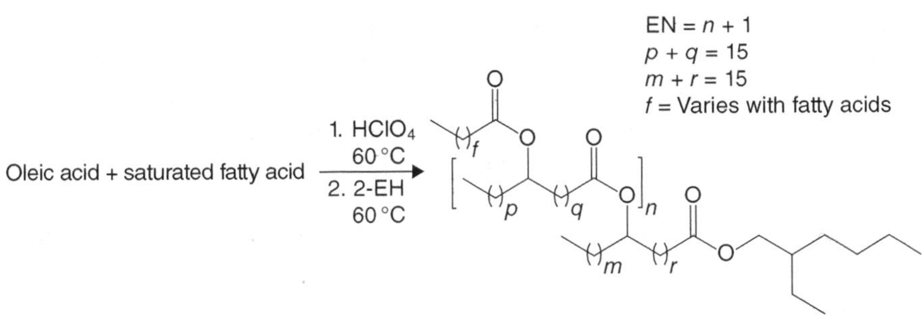

6.16 Reaction scheme for the formation of (poly)estolide 2-ethylhexyl esters.

By using (un)saturated FAs as precursors to estolides, a new class of oxygenated FAs has been disclosed. Estolide formation is based on the double bond sites in triacylglycerols. Since coconut oil is predominantly saturated (SFA, 92%), the double bond sites for reaction are very few in number. For increased estolide formation (via double bonds and hydroxyl groups) castor oil may be included in the reactant mixture [141]. Conversely, when SFAs are added to an olefin-containing reaction mixture, the oligomerisation terminates because the saturate provides no additional reaction site to further the reaction. Consequently, the estolide is stopped from further growth and is termed as being 'capped' [142]. During the estolide reaction migration of the double bond may occur [22].

The actual estolides are to be distinguished from estolide esters (see Figs 6.15 and 6.16). Figure 6.16 shows a specimen of an estolide ester backbone structure. The alcohol portion of the primary ester functionality in estolide esters plays a significant role in PP reductions as branched chain alcohols dramatically lower the PP [143].

Different molecular parameters (molecular weight, branching, etc.) and physical properties (e.g. viscosity and density) are obtained by modifying the nature of the base material and reaction conditions. Estolides have certain physical characteristics that could help eliminate common problems associated with vegetable oils as functional fluids [142]. The main concerns with vegetable oils are their stability toward oxidation and their low-temperature characteristics. The unique structure of estolides and their esters results in materials with far superior physical properties compared with vegetable and mineral oils [142]. Estolides have potential as base stocks for lubricants without the need for fortifying additives normally required to improve the lubricating properties of base stocks (see Table 6.9). The superior properties of estolide esters are dictated less by the positions of linkage and site of unsaturation and more by the combination of the degree of oligomerisation, decrease in level of unsaturation, the virtual absence of hydroxyl functionalities on the estolide backbone and the nature of the specific ester moiety. The condensation process inherently introduces a distribution of secondary linkage positions in the estolide, which in general affects low temperature and viscometric behaviour very favourably. Viscosities of estolide esters are typically <100 cSt at 40 °C and depend on the degree of oligomerisation. The free acid estolides are generally more viscous due to hydrogen bonding of the carboxylate functionality. Preferred estolide esters exhibit a viscosity at 40 °C of >30 cSt, VI > 150, PP < –25 °C and volatility of less than 10% at 175 °C. The secondary ester linkages of the estolide are more resistant to hydrolysis than those of triglycerides and impart branch points to the molecule. The structural aspect of the estolide results in superior low-temperature properties in addition to the properties already inherent to the FA structure [19]. Cold weather properties of estolides are

Table 6.9 Properties of common base stocks

Property (test method)	Canola oil	TMP trioleate	Polyol ester	PAO2	Tridecyl adipate	Oleic estolide 2-EH
Biodegradability, % (OECD 301B)	>85	70	<40	>70	<30	>80[a]
Viscosity @ 40 °C (ASTM D 445)	38.5	49	78.3	5.55	27	53.6
VI (ASTM D 2270)	207	190	147	n.a.	135	161
Pour point, °C (ASTM D 97)	−18	−24	−21	−72	−54	−27
Cold storage @ −25 °C, days	<1	<1	<1	7+	7+	7+
Flash point, °C (ASTM D 92)	162	315	n.a.	160	221	250
Evaporation, wt% (30 min at 150 °C)	1	1	n.a.	98	10	3

[a] Estolide.
n.a., not available.
After ref. [24].

superior to materials currently on the market. Furthermore, estolides can improve thermal oxidative stability [144] and are readily biodegradable [145].

Estolides and estolide esters of oleic acid [19, 21, 24, 143, 144, 146, 147], meadowfoam oil [23, 143], castor oil [148–152], or any source of monoenoic FAs or HFAs [148, 152] show promise as biolubricants. Isbell *et al.* [143] have compared a series of estolides and estolide esters from oleic, meadowfoam and *Crambe* FAs (see Table 6.10). Minor components other than oleic acid, such as linoleic acid or stearic acid, may lead to variations in the basic estolide structure shown in Fig. 6.15.

Oleic acid can be used to obtain estolides with remarkable viscosities, a fact that makes them promising compounds for biolubricant applications [146, 147]. Estolides from HOSNO oleins (split soapstock) were evaluated as viscosity modifiers for biolubricant applications [146]. The largest viscosity increments were observed within the first 3 hours of processing, due to a dramatic increase in molecular weight. Maximum viscosities (430 cSt at 40 °C) and estolide molecular weight (4300 a.m.u.) were obtained after 10 h of reaction. Estolide viscosity is controlled by the degree of oligomerisation [21].

Table 6.10 Some physical properties of estolides

Compound	Melting point (°C)	Viscosity index	Viscosity (cSt)	
			40 °C	100 °C
Crambe estolide (EN = 2.3)	0	151	679.0	58.6
Crambe estolide EH[a]	−12	177	184.4	26.1
Meadowfoam estolide (EN = 1.7)	6	154	229.8	27.4
Meadowfoam estolide EH[a]	−1	172	104.2	16.5
Oleic estolide (EN = 2.3)	−31	148	404.9	40.0
Oleic estolide EH[a]	−34	167	161.2	22.5

[a] EH, 2-ethylhexyl ester.
After ref. [143].

Estolides from oleic acid have been explored most intensively. When formulated with a small amount of oxidative stability package, simple (unsaturated) homo-oleic estolides (oligomers of oleic acid) show better oxidative stability than both petroleum and vegetable oil-based fluids [21]. Oleic estolides provided the lowest melting points of various estolides compared (Table 6.11) [143]. Oleic estolide esters have the lowest PPs, namely −43 °C for Guerbet ester and −34 °C for 2-ethylhexyl (EH) ester. The oxidative stability of homo-oleic estolides can further be improved by partial hydrogenation [143]. However, while hydrogenated (completely saturated) oleic estolides are oxidatively more stable than unsaturated estolides, their unfavourable cold weather properties (PP ≈ −9 °C) make them useless as bio-based functional fluids [143].

Oleic acid and SFAs, butyric through stearic, were transformed into complex estolides by treatment with $HClO_4$ at about 50 °C [21, 142]. The corresponding estolide-2EH esters (see Fig. 6.16) with VI 172–196 displayed desirable properties as a base stock for biodegradable lubricants and functional fluids. By varying the capping material on the estolide, the crystal lattice structure of the material could be disrupted as it approached its PP. This leads to superior PPs, cloud points, and other low-temperature properties. As the chain length of the saturated FA component increases from C_4 to C_{10}, the PPs of the estolide-2EH esters decrease to −39 °C; with chain length increase to C_{18} the PPs increased to −15 °C. The viscosity range of oleic estolides and estolide-2EH esters are useful for numerous applications. Saturated FA capped estolides and their corresponding esters, synthesised according to Fig. 6.16, have superior biodegradability and lubricating properties with excellent physical properties, PPs, cloud points, viscosities and oxidative stabilities [20–22, 144]. Table 6.11 shows that an unformulated C_8 saturated estolide 2-ethylhexyl ester (OO-EH) outperforms various formulated commercial products such as petroleum-based hydraulic fluids,

Table 6.11 Comparison of physical properties of estolide esters and commercial lubricants

Lubricant	Pour point (°C)	Cloud point (°C)	Viscosity @ 40°C (cSt)	Viscosity index	RBOT (min)
Commercial petroleum oil[a]	–27	2	66.0	152	228
Commercial synthetic oil[b]	–21	–10	60.5	174	256
Commercial soy-based oil[c]	–18	1	49.6	220	28
Commercial hydraulic fluid[d]	–33	1	56.6	146	464
Commercial aviation oil	–33	–30	20.4	170	552
O-EH[e]	–34	n.a.	161.2	167	8.5, 426[j]
OO-EH[f]	–36	–41	104.4	175	–
LO-EH[g]	–41	–40	52.2	176	–
CO-EH[h]	–33	–26	55.5	162	17, 477[j]
Oleic-Cas[i]	–54	<–54	34.5	196	159[j]
Oleic-H-Cas[i]	–36	<–36	68.3	178	403[j]
2-EH-Les[i]	–54	<–54	51.1	189	–

[a] Mobil® 10 W-30.
[b] Castrol Synthetic® 10 W-30.
[c] BioSoy®.
[d] Traveler Universal Hydraulic Fluid®.
[e] Unformulated oleic estolide 2-EH ester.
[f] Unformulated oleic-octanoate estolide 2-EH ester.
[g] Unformulated lauric-oleic estolide 2-EH ester.
[h] Unformulated coconut-oleic estolide 2-EH ester.
[i] 2-Ethylhexyl esters.
[j] Formulated with 3.5% Lubrizol® 7652 oxidative stability package.
After refs. [22, 142, 143, 148].

soy-based fluids and petroleum oils. All the commercial products of Table 6.11 have functional PPs except the soy-based oil. The PPs of soy-based products are too high for cold weather climates [5].

Estolides can also be synthesised from meadowfoam oil (62% 5-eicosenoic acid) (see Fig. 6.10). In contrast to oleic estolides and estolide esters, meadowfoam estolides and the parent triglyceride have nearly identical PPs (6 and 1 °C, respectively). Meadowfoam estolides are more suitable for the cosmetics industry.

Chemical modification of coconut oil to 2-ethylhexyl ester of coconut oil estolides brings the PP down to –3 °C [141], which is much less than in case of 2-ethylhexyl esters of oleic acid (–34 °C) [22, 143]. This is not surprising as highly saturated CNO offers only few double bond sites for estolide formation.

Unformulated oleic estolide-2EH shows a low rotating bomb oxidation test (RBOT) time of 8.5 min, as typical of vegetable oils. RBOT times of various formulated estolide esters compare favourably with a wide range of petroleum, vegetable-based and synthetic materials (Table 6.12). The RBOT value reaches a maximum of 426 min with a 3.5% oxidative stability package (Lubrizol® 7652), which compares favourably with most premium petroleum hydraulic fluids (200–460 min) [144].

Unformulated coco-oleic estolide esters (Fig. 6.16, $f = 10$), such as CO-EH, display superior low-temperature properties (not unlike the simple oligomeric oleic estolide O-EH; see Table 6.11) because the random short chains disrupt the stacking interactions and produce a low PP (–33 °C) [22, 144]. However, CO-EH shows an increased oxidative stability. Taken together with their relatively low cost, superior biodegradability and lubricating properties, coco-oleic estolide esters are more suitable as a base stock for

Table 6.12 RBOT times of various functional fluids

Sample	Nature	Average time (min)
Crambe oil[a]	Vegetable oil	13
Soybean oil[a]	Vegetable oil	13
Meadowfoam oil[a] (crude)	Vegetable oil	20
BioSoy®	Soy-based oil	28
Valvoline®	Petroleum crankcase oil	228
Castrol Synthetic® 10W-30	Synthetic motor oil	256
Traveller® All Season H.F. IVG-46	Hydraulic fluid	274
AeroShell® W 15W-50	Multigrade aviation oil[b]	552

[a] Unformulated (no oxidative stability package).
[b] Mineral/synthetic oil blend.
After ref. [144].

biolubricants and functional fluids than current vegetable oil-based commercial materials. Table 6.11 shows that unformulated estolide esters outperform (formulated) commercial products in terms of cold weather performance. All of the commercial products listed had higher cloud points than the estolide-2EH esters. A high cloud point could lead to filter plugging and poor pumpability in cold weather applications. The high cloud point of commercially available oils demonstrates the need for a better cold weather performing oil. Coconut-oleic and oleic estolide 2-ethylhexyl esters can be formulated to meet commercial crankcase requirements (RBOT ≈ 200 min) with as little as 1.0 and 1.5% oxidative stability package (Lubrizol® 7652); see Fig. 6.17. The maximum value of 504 min for 3.0% Lubrizol® 7652 for coco-oleic estolide EH compares favourably with aviation oil for single-engine propeller planes as a top commercial performer. Not unexpectedly, with a small amount of additives coco-oleic estolide esters display better RBOT values than the corresponding oleic esters. The viscosity indexes for these two formulated estolide esters ranged from 164–174 and 179–190, respectively [144]. Both have potential as a functional fluid with properties better than current commercial products at a reasonable cost.

While (poly)estolides are usually formed by carbocationic homo-oligomerisation of unsaturated FAs from the addition of an FA carboxyl across the olefin [19], estolide formation may alternatively also involve a hydroxyl functionality as a site for esterification [148, 153, 154]; see Fig. 6.18. Conversion of the hydroxyl group into an ester functionality is a strikingly different method from conversion of the FA double bond. Castor oil estolides demonstrate that this reaction can be run in the absence of a catalyst at high

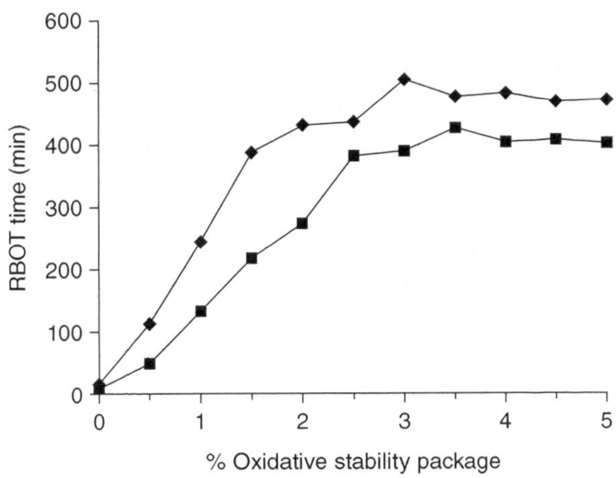

6.17 RBOT times of oleic estolide (■) and coconut-oleic estolide 2-ethylhexyl esters (◆) with varying amounts of Lubrizol® 7652 additive. After ref. [144].

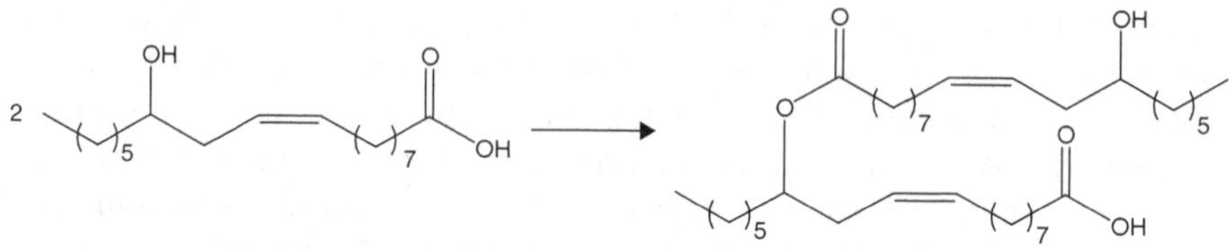

6.18 Monoestolide (diricinoleic acid) formation from ricinoleic acid.

temperature under vacuum or a blanket of carbon dioxide [155]. The esterification to form the estolide as the product of a hydroxyl-containing FA and a carboxylic acid occurs at ambient temperature up to the decomposition temperature of any reactant or product. Usually the temperature does not exceed 120 °C. To shift the equilibrium in the direction of estolide formation, it is necessary to use either a large excess of carboxylic acid, or else remove water as it is formed. In either case, excess carboxylic acid or formed water can be removed by distillation.

Triglycerides containing —OH groups occur in nature as castor oil (with 80–89% ricinoleic acid) and *Lesquerella* oil (with 55–60% lesquerolic acid). Another fatty acid similar to ricinoleic acid, isoricinoleic acid [(*S*)-9-hydroxy-*cis*-12-octadecenoic acid] is a major component of the seed oils of *Wrightia tinctoria* and *W. coccinea* [156]. As an example, under proper conditions the —OH from one ricinoleic acid molecule can react with the —COOH of another ricinoleic acid molecule to give an estolide (Fig. 6.18). The estolide would continue to cross-link or react linearly at the unreacted —OH and —COOH sites to form a polyestolide. Estolides have been synthesised by homopolymerisation of castor oil FAs [149, 150] or 12-hydroxystearic acid [157] under thermal or acid-catalysed conditions. Also the enzymatic production of estolides from HFAs in castor oil (mainly ricinoleic acid or 12-hydroxy-*cis*-9-octadecenoic acid) has been reported [140, 151]. Triglyceride estolides of castor oil [152, 153, 158] and *Lesquerella* oil [152, 153] with various acids (such as heptanoic and oleic acid) were reported. HFAs such as *Lesquerella* can be readily converted to estolides using clays [159]. Similarly, estolides from free lesquerolic acid (or 14-hydroxy-*cis*-11-eicosenoic acid, a homologue to ricinoleic acid) and oleic acid were synthesised using a lipase catalyst [154]. *Lesquerella* oil, which contains 53 % lesquerolic acid, is a good candidate for development into a biodegradable lubricant. Castor and *Lesquerella* FA esters were also combined with different types of saturated, unsaturated, and branched FAs to produce estolides and estolide esters [148].

In the absence of a domestic commercial source of HFAs, Burg *et al.* (to USDA) have described formation of estolides by catalytic hydrolysis of unsaturated FAs, in particular erucic acid (C22:1Δ13), palmitoleic (C16:1Δ9) and oleic (C18:1Δ9) acids, petroselinic acid (C16:1Δ6), and especially Δ5 unsaturated FAs (C18:1Δ5, C20:1Δ5, C22:1Δ5 and C22:2Δ5,13) [159]. Some of these unsaturated FAs occur naturally in a variety of plants (such as castor bean, meadowfoam).

As shown in Fig. 6.18, estolides derived from the aforementioned HFAs are composed of esters at the 12 carbon of the FAs and have a residual hydroxyl group on the estolide backbone. Several oleic-castor and *Lesquerella* estolide esters, which can be produced at reasonable costs, display far superior low-temperature properties (PP –54 °C) than any other estolides reported to date (see Tables 6.11 and 6.13). All of the commercial products listed have higher cloud points than the castor and *Lesquerella* 2-ethylhexyl esters. Whereas the oxidative stability of vegetable oils is very poor and normal estolides increase the oxidative stability by reducing the amount of unsaturation in the molecule [144], estolides synthesised from HFAs do not decrease the unsaturation of the starting molecule, and, therefore, do not improve the oxidative stability.

Limitations of estolide technology reside in the synthesis details, i.e. number of synthetic steps, product work up, yield, viscosities, colour, etc., and the breadth of the product line and applications. The estolide reaction is non-specific and there exists limited control allowing production of relatively homogeneous materials (although this does not preclude useful applications, as in case of other technical materials). The reaction product invariably contains a range of oligomers with different molecular weights, with implications as to viscosity.

Mono- and polyestolides have a variety of potential applications as lubricants, greases, printing inks, cosmetics and surfactants. Biodegradability is an advantage for many of these industrial applications. The biodegradability of mono- and polyestolides, made from oleic acid, meadowfoam oil FAs and erucic acid, has been assessed [145]. Estolides and, particularly, estolide esters may find application as lubricant base stocks [160] or as lubricant

Table 6.13 Physical properties of unsaturated castor and lesquerella-based estolide 2-ethylhexyl esters

Hydroxy FA ester	Capping FA	Product name	Pour point (°C)	Cloud point (°C)	Viscosity @ 40 °C (cSt)	Viscosity index
Castor	Oleic	Oleic-Cas	–54	<–54	34.5	196
Castor	2-Ethylhexanoic	2-EH-Cas	–51	<–51	70.6	164
Lesquerella	Oleic	Oleic-Les	–48	–35	35.4	200
Lesquerella	2-Ethylhexanoic	2-EH-Les	–54	<–54	51.1	189

[a] 2-EH, 2-ethylhexanoic acid.
After ref. [148].

additives [154, 161]. Both HFAs and derived estolide triglycerides find useful industrial applications as lubricants, greases and plasticisers, as well as in cosmetics [159]. Biodegradable lubricating oils have been disclosed which contain 2-ethylhexyl or Guerbet esters from hindered alcohols (TMP, NPG, PE) and estolides of castor oil or (hydrogenated) castor oil fatty acids [162]. In US Patent No. 4,428,850 (to Texaco, Inc.) Zoleski and Gaetani [163] disclose a low foaming lubricating oil composition comprising a mineral oil and an estolide of 12-hydroxystearic acid.

In many industrial applications oils are too thin to be of value. Successful use of oils in industrial applications and also as a fuel additive to liquid fuels may be contingent upon increasing the viscosity of the oils (thickening). Thickened oils find utility in particular in high-temperature applications. Lawate (to Lubrizol Corp.) has disclosed oils (synthetic ester base oil, mineral oil or PAO) thickened with estolides of hydroxy-containing triglycerides [161]. Viscosity improvement can be achieved by admixture of a triglyceride oil with an estolide of a hydroxy-containing triglyceride fatty acid or by catalytic interesterification of the two components. Figure 6.19 shows the various interesterification products.

Successful use of estolides of hydroxy-containing triglycerides as environmentally friendly, biodegradable base fluids in industrial applications depends on increasing various crucial performance values such as improved AW/EP, oxidation inhibition, rust/corrosion inhibition, metal passivation, friction modification, viscosity modification, foam inhibition, emulsification, demulsification, lubricity, dispersancy, detergency and the like by additivation. In many applications these values of the estolide base stock are too poor to be of practical use. In order to take advantage of the biodegradability of estolides, it is necessary to increase the performance values [160].

6.2.3 Selective hydrogenation

Hydrogenation, a process called 'hardening' in oleochemistry, is very well known. Hydrogenation of naturally occurring triglycerides is the primary means of chemical modification. By chemically modifying a vegetable oil by hydrogenation it is meant that hydrogen is permitted to react with the

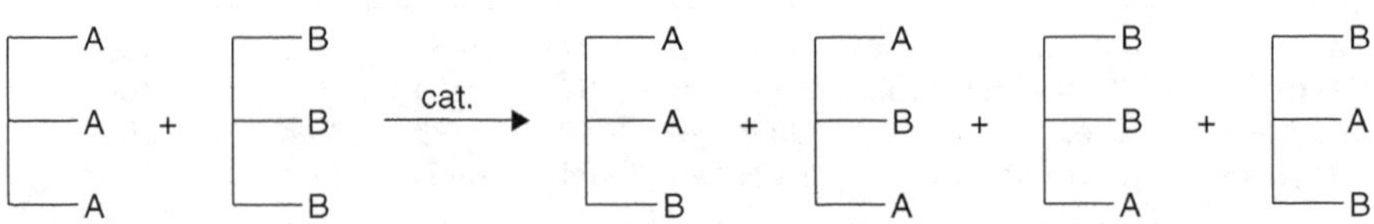

6.19 Interesterification of a triglyceride oil (A) and an estolide of a hydroxy-containing triglyceride (B).

usaturated FA profile present such as oleic acid (C18:1), linoleic acid (C18:2), and linolenic acid (C18:3) to improve the ageing resistance. The object is not to remove all the unsaturations, but to reduce or convert a substantial portion of the C18:2 and C18:3 fractions to an oleic acid profile. Catalytic hydrogenation of vegetable oils is a major processing step in the fats and oils industry, which changes a liquid oil of PUFAs into a (semi-) solid product containing mostly monounsaturated fatty acids (MUFAs) [164]. Properties of concern in this process are the final iodine value (IV), FA composition, melting point and solid fat content. Improving the shelf-life of soybean oil for food and feed applications historically has been addressed via partial hydrogenation. This process introduces unwanted *trans* FAs into the oil, which has undesirable effects on human health and nutrition [165]. Catalytic hydrogenation of double bonds in unsaturated compounds serves primarily to improve the stability as well as to increase the melting point of unsaturated fatty derivatives. Hydrogenation is generally accompanied by a concomitant and undesirable increase in PP, i.e. reduction in fluidity (cf. melting points of trilinolein, triolein and tristearin of –11, 5, and 74 °C, respectively). In the past, hydrogenated olive and castor oils have been used as lubricants in steam cylinders.

The hydrogenation of fats and oils is very complicated. Besides the addition of hydrogen to double bonds, dehydrogenation processes take place so that not only position isomeric acids but also stereoisomeric acids may be formed in the selective hydrogenation of polyunsaturated FAs. It is difficult to control the entire reaction uniformly and selectively either by using various catalysts or by varying the reaction conditions such as the amount of hydrogen used, the pressure, hydrogenation temperature and time.

The aim of partial hydrogenation of FA esters is thus a certain reduction in the amount of polyunsaturates, while generally avoiding formation of saturates or *trans* products with high melting points. Selective hydrogenation of high IV FAAEs for the forced removal of conjugated polyene, triene and diene systems and reduction of the linoleic acid alkyl ester content is meant to lead to formation of monoenes, without in any way achieving complete saturation of the molecule [25–28]. Selective hydrogenation of vegetable oils serves many purposes and, in particular, may be carried out with the purpose of obtaining low *trans* isomer formation [166], high conjugated linoleic acids (CLA) formation [167], reduction of undesirable flavour properties (in relation to linolenic constituents) [168, 169], production of biodiesel quality [170–172]. In lubricant chemistry, selective hydrogenation in which the fatty residue is not fully saturated is of great interest. Here selective hydrogenation aims at transforming multiple unsaturated FAs into single unsaturated FAs (one isolated *cis* double bond) improving the oxidative stability without increasing the saturated part of the substance in order to avoid deterioration of low-temperature behaviour such as PP.

Table 6.14 shows the oleic acid (C18:1), linoleic acid (C18:2) and linolenic acid (C18:3) profiles of selected naturally occurring vegetable oils. The linoleic acid content of naturally occurring SNO is about 67.5%. It is a goal of chemical modification to hydrogenate such that C18:2 is reduced to about 25%. At the same time, the oleic acid fraction increases to about 61%.

Hydrogenated oils with different physical and chemical properties, whenever needed, are produced by applying various hydrogenation conditions. Reaction conditions (catalyst types and amount, reaction temperature, agitation rate, hydrogen pressure, etc.) greatly affect the mix of fatty acids in hydrogenated vegetable oils [29, 173, 174]. One of the most important hydrogenation rate-determining factors is the mass transfer of hydrogen gas to liquid oil near the catalyst surface. Selective hydrogenation, i.e. a hydrogenation of only one of the two double bonds of linoleic acid (C18:2) to yield *cis*- and *trans*-C18:1 fatty acids, is obviously more difficult than total hydrogenation.

On the basis of the hydrogenation mechanism, catalytic partial hydrogenation of triacylglycerols will invariably generate some *trans* isomers. For selective partial hydrogenation leading to a moderate increase in SFAs and also a low *trans*-isomer content, various solutions are open, namely use of: (i) zeolites [175–177]; (ii) precious metal catalysts [178]; or (iii) copper catalysts [179–181]. It is essential to use a hydrogenation catalyst which reduces the degree of unsaturation as much as possible but without increasing the stearic acid content and limiting *cis–trans* and functional isomerisation. Precious metal catalysts are active at the low temperature of 70 °C. With a colloidal solution of a palladium nano-catalyst (4 nm) linoleic acid can be mono-hydrogenated nearly quantitatively [26]. The selectivity depends on the stabilising solvent used (Table 6.15). Selective hydrogenation was also investigated extensively using homogeneous transition metal catalysts. Copper catalysts are active in hydrogenation reactions of conjugated double

Table 6.14 Oleic (C18:1), linoleic (C18:2) and linolenic (C18:3) profiles of selected naturally occurring vegetable oils[a]

Oil	C18:1	C18:2	C18:3
Corn	25.4	59.6	1.2
Cottonseed	18.6	54.4	0.7
Peanut	46.7	32.0	–
Rapeseed (LE)[b]	59.7	23.0	10.1
Safflower	12.0	77.7	0.4
Soybean	23.2	53.7	7.6
Sunflower	18.7	67.5	0.8

[a]Average values.
[b]Low erucic.

Table 6.15 Selective hydrogenation of a mixture of sunflower fatty acids in different stabilising solvents

Solvent	Composition			
	C18:0 (%)	C18:1 (%)	C18:2 (%)	*cis/trans* C18:1
Substrate	2	22	70	100/0
Dimethyl acetamide	2	84	2	74/26
Propylene carbonate	2	93	0	55/45

After ref. [186].

bond systems [182] but do not catalyse the saturation of oleic acid to stearic acid [183]. Catalyst performance of copper catalysts has been studied mainly in relation to the high linolenic acid selectivity [180, 181] and thus as a means to increase soybean oil stability. Recent patents [170, 184] reflect renewed interest in the use of copper catalysts. A basic copper carbonate catalyst may be used in a selective hydrogenation process of unsaturated vegetable oils producing high monoene compositions [185].

Heterogeneous hydrogenation of vegetable oils has repeatedly been reviewed [29, 167, 187–189]. Partially hydrogenated soybean oil (SBO) methyl esters (PHSME, IV = 116 g I_2/hg) have superior oxidative stability, but inferior low-temperature performance, kinematic viscosity, and lubricity in comparison to soybean oil methyl esters (SBME, IV = 120 g I_2/hg) [190]. The University of Northern Iowa's Ag-Based Industrial Lubricants (ABIL) programme uses partial hydrogenation and other chemical processes to increase the oxidative stability of SBO [30, 191]. Through a proprietary technology (called XBO) Terresolve Technologies, Ltd. (Eastlake, OH) chemically modifies soy and canola oils to produce commercially available hydraulic fluids and two-stroke engine oils. PCT Int. Publ. No. WO 2008/043454 A1 (to Cognis) describes a process for the partial hydrogenation of fatty esters of high (C18:2 + C18:3) acyl group content to mainly *cis*-C18:1 glycerides where the hydrogenation is conducted in an aqueous/organic two-phase system in the presence of a water soluble catalyst consisting of a Group VIII metal and a hydrophilic ligand, such as trisulphonated triphenylphosphine (TPPTS) [31]. Lubricant base oils were also produced by esterification of several Indian vegetable oils (neem, rice bran, linseed, castor, *Pongamia* and *Jatropha*) with long-chain (C_5–C_{10}) alcohols [99]; after partial hydrogenation biolubricants were obtained with IV > 160 g I_2/hg [192]. For light lubrication applications the liquid oils obtained from brush hydrogenation of polyunsaturated vegetable oils followed by fractionation can be used. Siew has described the properties of products of palm kernel oil, produced from fractionation and hydrogenation (conditions not specified) [193]. The products are mainly of interest to the confectionery industry.

A fatty oil mixture composed of hardened palm fractionated oil containing >20 wt% *trans*-isomeric FAs and <12 wt% di-ene fatty acids, HOSNO (C18:1) with <12 wt% di-ene fatty acids, and medium-chain saturated triglycerides (MCTs, C_8–C_{10}) in approximate ratios of 30–60:5–35:15–45 wt%, respectively, subjected to 1,3-specific lipase (*Rhizopus delemar, Mucor miehei*)-catalysed ester interchange, has been claimed to present optimum lubricant properties with good low-temperature fluidity, low cloud point and good oxidative stability [103]. The synthetic biolubricant is a mixed-acid group triglyceride composed of 40–58% mono-ene fatty acid, 10–20 wt% *trans*-acid and 20–40 wt% medium-chain SFA.

Mutation breeding and gene-technological improvements of plants with maximum potential for desirable oil qualities would eliminate the need for selective catalytic hydrogenation of unsaturated oils. A genetic approach has been used to develop soybean germplasm low in linolenic acid as a means to augment oxidative stability.

6.2.4 Dimerisation/oligomerisation

Dimerisation and oligomerisation are technologically feasible modifications of the double bonds of unsaturated FAs. In the presence of montmorillonite at 210–250 °C C_{18} FAs with one or more double bonds may form a complex mixture of C_{36} dicarboxylic acids (dimeric FAs), C_{54} trimer FAs and C_{18} monomer FAs [4]. Complex dimer and trimer FAs may be used in lubricants [194]. It is assumed that reaction to dimeric FAs may take place via a head/head or head/tail Diels–Alder (DA) addition in which one molecule acts as the diene (after double bond shifting) and the other as the dienophile; see Fig. 6.20 [4]. Parallel to this reaction other mechanisms can occur with the formation of other dimer products or ene-reactions. The Diels–Alder reaction is a cornerstone in organic synthesis, forming two

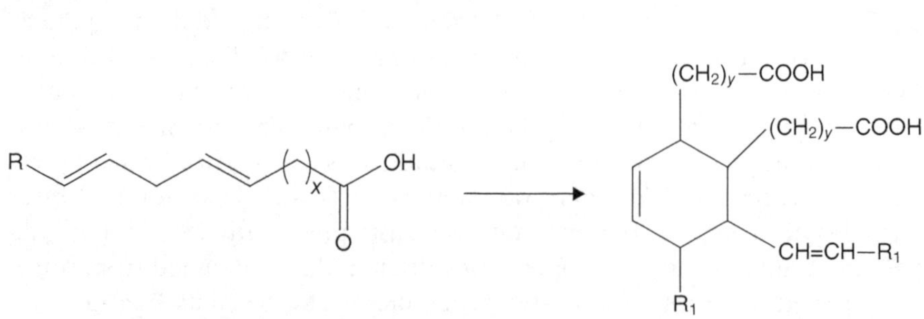

6.20 Dimerisation of C_{18}-fatty acids via (head/head) Diels–Alder reaction. In the product $y = x$, $R_1 = RCH_2$.

carbon–carbon bonds and up to four new stereogenic centres in one step. Quite noticeably, there are no naturally occurring enzymes known to catalyse bimolecular Diels–Alder reactions. However, recently, *de novo* computational enzyme design has resulted in a 'Diels-Alderase' for a stereoselective DA reaction [195].

Additional by-products of the dimerisation are C_{18}-monomeric FAs which consist of a mixture of straight-chain, branched, saturated and unsaturated FAs. Hydrogenation of the unsaturated C_{18}-monomers results in a solid phase (straight-chain FAs) and a liquid phase (technical isostearic acid). The main components of isostearic acid are mono- and poly-branched C_{18}-FAs with branching points mainly in the centre of the molecule. Base fluids synthesised from branched FAs display excellent low-temperature behaviour such as low PP, low viscosity, high chemical stability and high flash point. With a low PP similar to oleic acids and good thermal and oxidation resistance, isostearic acid and its derivatives are ideal raw materials for lubricants. However, branched substituents typically cause a lowering of the biodegradability.

6.2.5 Formation of C—C and C—O bonds

Branching of the fatty acid residues is initiated by C–C and C–O linkages. Reactions leading to chains are several, including co-oligomerisation, telomerisation, hydroformylation, Friedel–Crafts alkylation and acylation, alkoxylation, ane and ene reactions, radical addition, cyclopropanation and acyloxidation (*cfr.* Table 6.3 and Fig. 6.21). For oxyfunctionalisation of fatty acids, see also Section 6.2.7.

Saturated, branched fatty acids are highly desirable base fluids for lubricants because of their physical properties (low PP, low viscosity, high chemical stability and high flash point). Branching significantly lowers the PP [196]. By adding sterically hindered branches, hydrolytic stability can be increased. HFAs, which may be formed using oxygenase enzymes, are equally highly desirable materials. Hydroxyl groups are very reactive and help prevent the breakdown of oils under extreme conditions of heat and friction, forming dimers. HFAs and liquid wax esters (C_{24}–C_{48}) are particularly well suited to bonding to metals, coating the wear surfaces and thus ensuring less metal-to-metal contact and reducing wear.

Unsaturated FA esters (e.g. monoesters, triglycerides) may be reacted with olefinic hydrocarbons to form cycloaddition products [32]. The olefinic hydrocarbons used to form the adduct may be either a dienic or monoolefinic hydrocarbon. The dienic hydrocarbon (e.g. cyclopentadiene and derivatives or acyclic dienes such as isoprene and dimethylbutadiene) can react with a double bond in the fatty acyl chain of an FA ester molecule (the 'dienophile') to form a [4 + 2] Diels–Alder cycloadduct (Fig. 6.22). Cycloaddition

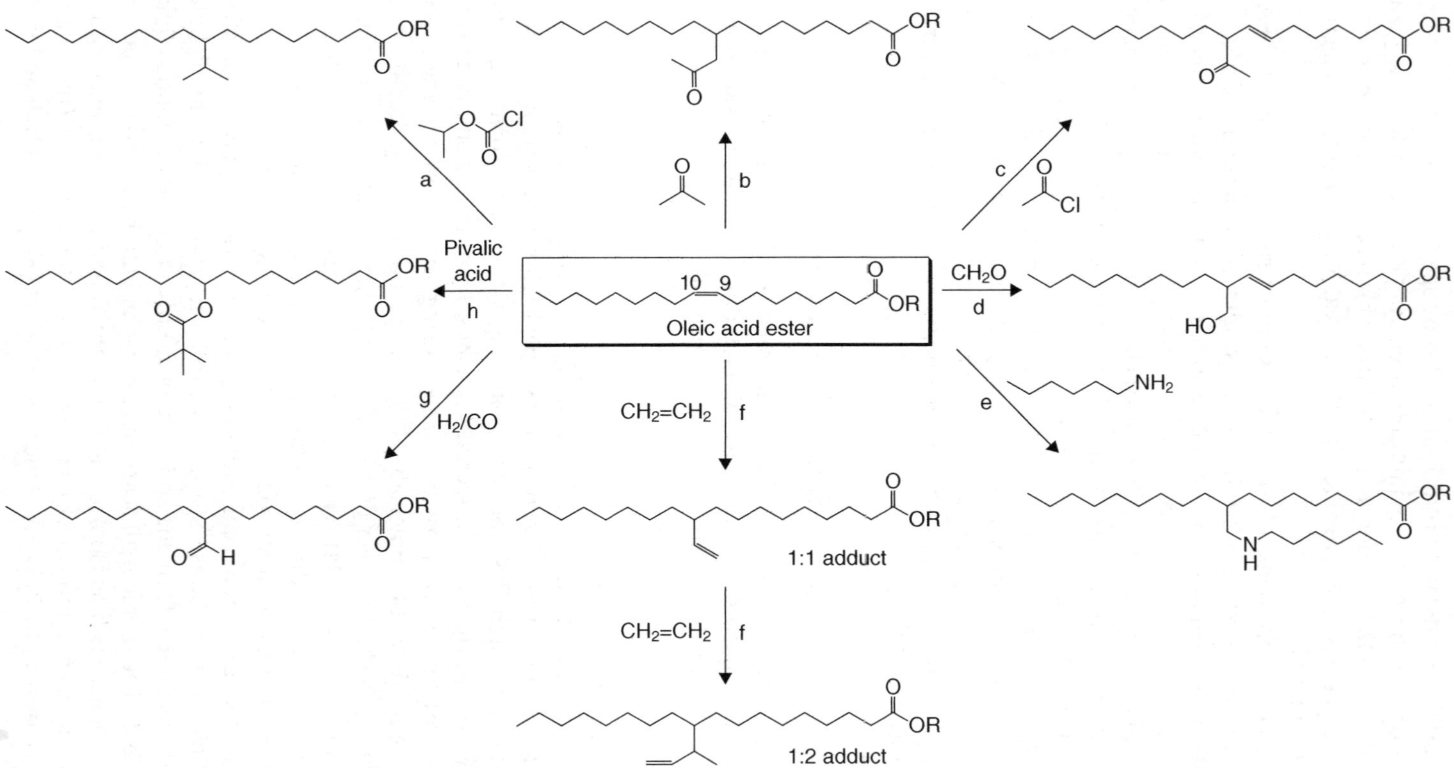

6.21 Some chemical modifications of oleic acid derivatives: (a) alkylation; (b) radical addition; (c) acylation; (d) ene reaction; (e) aminoalkylation; (f) co-oligomerisation; (g) hydroformylation; (h) acyloxylation. Regioisomers not shown.

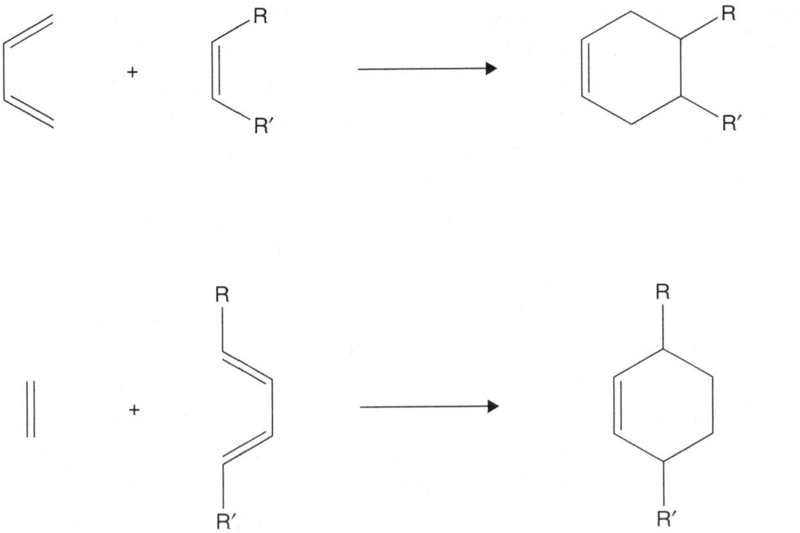

(a)

(b)

6.22 Cycloaddition of a diolefinic hydrocarbon to a monounsaturated fatty acid ester (a) and of a monoolefinic hydrocarbon to a polyunsaturated fatty acid ester (b).

of these dienes results in the introduction of only hydrophobic (alkyl) substitution in fatty acyl chains. Similarly, monoolefinic compounds (e.g. cyclohexene, propylene, butylene, maleic anhydride) can act as a dienophile and react with FA ester molecules having a fatty acyl chain with a diene moiety (Fig. 6.22).

The reaction of maleic anhydride with soy oil to yield maleated soybean oil (MSO) was reported a long time ago [197, 198]. More recently, a single-step, free radical-initiated maleation procedure has been developed that proceeds smoothly and gives high product yields [199]. Typical initiators used are 2,5-bis-(*t*-butylperoxy)-2,5-dimethylhexane peroxide and di-*t*-butylperoxide. The peroxide-initiated functionalisation of SBO with maleic anhydride in a closed vessel at elevated pressure and temperature was found to proceed by a Diels–Alder mechanism. The extent of the maleation reaction can be determined by the acid number test (ASTM D 1980) and iodine number (ASTM D 1959). MSO is a viscous liquid which is used primarily in cosmetic applications [200] and in biocomposites [201]. Chemically modified soy oils with (maleic) anhydride and hydroxyl functionalities were also evaluated as lubricants [202]. Maleation allows introduction of an anhydride functionality into a triglyceride.

Maleated products (primarily mono-maleated) are often called succinated triglyceride oils because maleic anhydride loses one double bond in the thermal coupling reaction (i.e. converting the maleic acid or anhydride to a succinic acid or succinic anhydride substituent). The terms 'succination' and 'maleation' can be used interchangeably. The degree of anhydride or unsaturated acid coupling to a triglyceride, which consists in thermal condensation at 175–250 °C for 1–5 hs at the unsaturated C=C sites of the FA moieties ('ene' mechanism or Alder reaction) to form alkenylsuccinic anhydride moieties, is a function of the oil composition and molar charge ratio of anhydride or unsaturated acid (dienophile) to triglyceride oil. Triglyceride oils comprising high levels of linoleic acid esters can condense with maleic anhydride in a Diels–Alder manner to form cyclohexene–dicarboxylic anhydride moieties. The maleated intermediate is an anhydride (or carboxylic acid)-grafted vegetable oil which may be used directly as a self-emulsifiable lubricity agent in aqueous emulsions, or can be converted to other derivatives (esters, amidic acids, amidic esters, imides) that are useful in metalworking compositions, as lubricants, lubricity agents, friction modifiers, and anti-wear agents [203]. Examples are reaction products of SBO/maleic anhydride/triethanolamine and RSO/maleic anhydride/monoethanolamine (1:1:2 molar ratio). Emulsifiers may be formed by reaction of succinated triglycerides with water, Group IA and IIA metals, ammonium hydroxide or amines. These functionalised triglycerides can self-emulsify, i.e. form an instantaneous emulsion when added to water or other aqueous-based formulations. Thus, the reaction product of maleating a triglyceride oil is particularly useful for oil-in-water emulsions. Yahiro *et al.* have disclosed maleated fish oil for metalworking lubricating oils [204]. Maleic anhydride functionalised vegetable oils, such as Falkowood 51 (Cargill, Inc.) and maleinised linseed oil, dehydrated castor oil, SBO or tung oil and the like, are also used as components of grease compatible dielectric encapsulant compositions of use with signal transmission devices [205].

When it is intended that the triglycerol is to act as diene component in the cycloaddition reaction, the starting unsaturated TAG oil may first be subjected to an isomerisation reaction to convert the polyunsaturated fatty acyl chains into conjugated isomers. Cycloaddition has several beneficial effects, namely reduction of the susceptibility to oxidation (by a decreased number of bis-allylic hydrogens) and decreased ease of packing. In particular, certain vegetable oil-cyclopentadiene (CPD) cycloaddition adducts (see Fig. 6.23) are useful lubricant base stocks with enhanced fluidity and/or oxidative stability [32].

Formation of cycloadducts through the reaction of a dienic hydrocarbon with an FA ester does not alter iodine value or average unsaturation content of the FA ester, i.e. the number of double bonds present in the FA ester molecule remains unchanged. However, the number of active methylene groups present can be reduced by cycloaddition of an olefinic hydrocarbon to a polyunsaturated fatty acyl chain. For example, reaction of CPD with

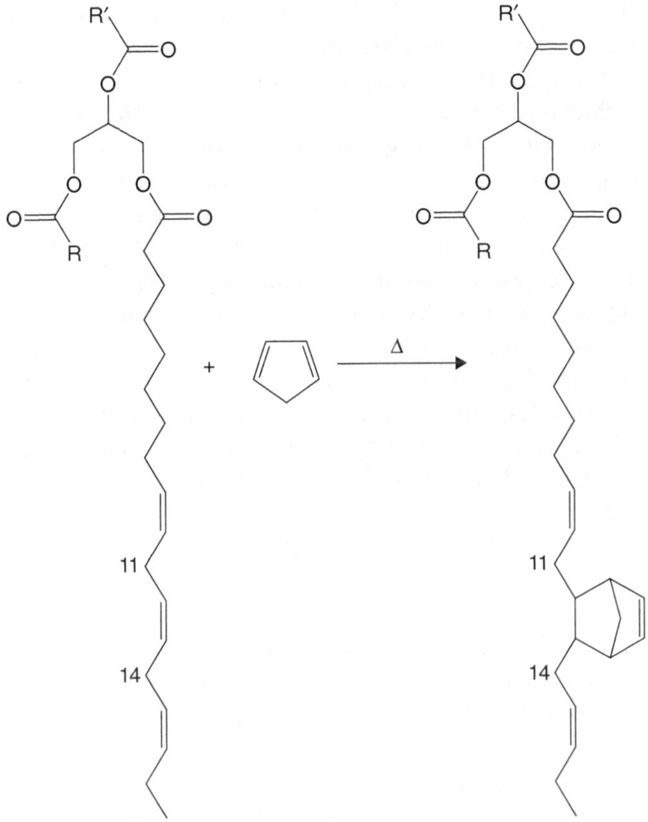

6.23 Cycloaddition of cyclopentadiene to a linolenic (C18:3) ester chain.

the C_{12} double bond of a linolenic ester destroys the bis-allylic character of two methylene groups at the 11 and 14 positions (Fig. 6.23), thus improving the oxidative stability. Moreover, introduction of the bulky bicyclic ring onto acyl chains of TAGs leads to lower viscosity. It is noticed that where a fatty acyl chain acts as the diene in a [4 + 2] cycloaddition reaction, e.g. via reaction with a mono-olefin (Fig. 6.22b), the number of double bonds (or degree of unsaturation) in the altered fatty acyl chain is decreased by one. Cycloaddition of triglycerides with mono- or diene hydrocarbons does not greatly change the viscosity index [32]. Moreover, as shown in Fig. 6.23, a spacer group generated via Diels–Alder reaction moves a double bond from the acyl chain backbone to a remote position (i.e. into a portion of a ring extending from the acyl chain backbone). While this decreases the possibility of double bond migration to generate a less stable polyunsaturated chain, the cyclic nature of the adduct results in non-volatile oxidation products of the remote double bond. CPD-modified linseed oils are useful for coating binder applications [206, 207].

Further improvement in the oxidation stability of the cycloaddition products can be achieved by partial hydrogenation to sufficiently reduce the active methylene content. Hydrogenation is conducted until the stability of the resulting oil, with respect to oxidation, has increased to a desired level (IV 90–110). The preferred partially hydrogenated base stocks (adducts) typically have VI > 150 and a viscosity of <100 cP at 40 °C. The PP of the hydrogenated cycloaddition adduct is generally lower than that of the product obtained by a corresponding partial hydrogenation of the unsaturated FA ester stock (see Table 6.16).

In PCT Int. Publ. No. WO 92/07051 (to International Lubricants, Inc.) Landis [33] discloses telomerised vegetable oils as substitute lubricant additives with high pressure performance. A telomer is a highly viscous polymerised triglyceride oil that has thermo-oxidative stability and viscosity improvement characteristics that make the telomer an essential component or ingredient of a large variety of lubricant formulations. Telomerised triglycerides can be produced by heating a polyunsaturated vegetable oil at 200–400 °C for 6–40 h in a reducing atmosphere in the presence of water vapour as a catalyst. Telomerisation results in a reduced iodine number due to a decrease in unsaturation, enhanced oxidation resistance and an increase in viscosity due to higher molecular weight (up to 2000–5000 Da) and formation of MUFA residues connected by intra- and intermolecular six-membered rings; see Fig. 6.24.

In US Patent No. 5,985,806 A (to Lambent Technologies, Inc.) O'Lenick [208] relates to the telomerisation of a triglyceride vegetable oil with a high conjugated diene concentration (typically linoleyl groupings), with complex esters having multiple single double bonds (such as pentaerythritol

Table 6.16 Characteristics of soybean oil and its cyclopentadiene adducts

Adduct[a,b]	Reactant ratio	Pour point (°C)	Iodine value (g I_2/hg)	Viscosity @ 40 °C (cP)
SBO	–	−7	130	30
SBO (H)	–	+15	90	n.a.
SBO/CPD	93/7	−12	127	55
SBO/CPD(H)[c]	93/7	−5	76	n.a.

[a] H, (partially) hydrogenated.
[b] CPD, cyclopentadiene.
[c] 1 h.
n.a., not available.

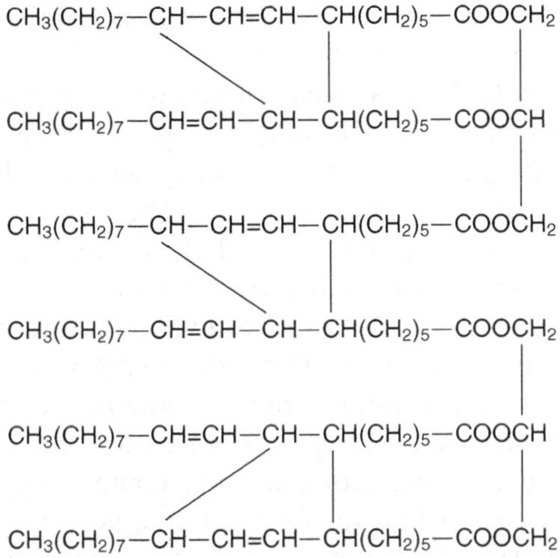

6.24 Reaction product chemical configuration of a telomerised vegetable oil.

tetraoleate). This results in a telomerised complex ester triglyceride. The reaction is a Diels–Alder coupling between a reactive conjugated diene and a dienophile group. Monitoring of the telomerisation reaction can be made by measuring the iodine number, which decreases with time, or viscosity, which increases. The advantage of these telomerised products is that they are relatively inexpensive, very high molecular weight (hence low volatility), highly branched (hence liquid to low temperature), and very stable to oxidation. Addition of a telomerised complex ester triglyceride (or a sulphurised or phosphorus derivative) improves the viscosity of a lubricant. Telomer esters are soluble in organic solvents and in hydrocarbon oils and esters. Telomerised oils have been proposed as jojoba replacements.

Most natural FAs are linear, unbranched chains. This linearity confers them characteristic physical properties, including high melting points. However, for use as lubricants, products with a lower melting point are often desirable. These substrates can be generated almost quantitatively by catalytic co-oligomerisation of unsaturated fatty compounds with short chain olefins (ethylene, propylene) in mild reaction conditions ($T = 70\,°C$, $p = 30$ bar) using $RhCl_3 \cdot 3H_2O$ as a catalyst; see Fig. 6.21. In the case of reaction of an oleic acid ester with ethylene a 1:1 adduct is formed first, whereas a second addition may take place at the existing branch [209]. In the case of methyl linoleate, after isomerisation to 10,12- or 9,11-octadecadienoic acid methyl ester, ethylene addition takes place on one of the double bonds, thus forming the corresponding vinyloctadecenoic acid methylester. Depending on the control of oligomerisation 1:1, 2:1 or 3:1 adducts of ethylene and an FA ester are produced. Further hydrogenation of these co-oligomerates over Pd/C catalysts generates saturated derivatives which are also of interest to the lubricants industry [210].

Instead of hydrogen, many H–Y substrates can add to unsaturated fatty compounds and a new functional group Y is then inserted, which can change the physical and chemical properties of the substrate considerably. Some examples are hydroformylation (Y = —CHO), hydrosilylation (Y = —SiX_3) and hydroaminomethylation (Y = —CH_2—NR_2). Hydroformylation or oxo synthesis is the reaction of unsaturated substrates (such as oleyl alcohol) with synthesis gas (H_2/CO); see Fig. 6.21. Hydroformylation of unsaturated fatty compounds leads to the introduction of an aldehyde group in the FA chains with formation of fatty formyl esters. Frankel and coworkers [37, 40, 41] have reported hydroformylation of vegetable oils using Rh catalysts. The preferred catalyst for hydroformylation of fatty compounds in mild conditions is based on rhodium, [Rh(acac)$(CO)_2$]/Biphephos [211]. In US Patent No. 5,177,228 (to Mitsubishi Kasei Corp.) Sato *et al.* disclose the hydroformylation of an unsaturated FA ester, such as methyl oleate, to a monoformyl FA ester, such as methylformyl stearate [212]. Hydroformylation of methyl oleate forms approximately 60% of the isomeric mixture of formyl stearic acid methyl esters, but also the undesired hydrogenation by-product methyl stearate. Apart from hydrogenation of the olefinic unsaturations, side reactions also include isomerisations. Subsequent hydrogenation of the aldehydes with Raney nickel yields the corresponding alcohols; again, a considerable degree of hydrogenation side reactions may occur.

Hydroformylation also allows branching of unsaturated FA chains. Dow Chemical has recently disclosed novel aldehyde compositions derived by hydroformylation of unsaturated transesterified seed oils (mixtures of mono-, di- and tri-formyl substituted FAs and FA esters) using a stabilised Group VIII transition metal–organophosphine metal salt ligand complex catalyst [213]. The corresponding (poly)alcohols derived by hydrogenation were mainly intended for application in manufacturing of polyurethanes. In PCT Int. Publ. No. WO 2008/124265 A2 (to Dow Chemical Co.) Sonnenschein *et al.* describe polyester polyol lubricant compositions, some of which are capped, that include two or more chemically linked ester moieties [214]. An alkanolysed, hydroformylated and reduced vegetable oil was used as the renewable raw material source. The capping agent was a short-chain (C_6–C_{12}) FA or its methyl ester. The resulting compositions, with PP –10 °C and a viscosity of 40–2000 cP at 25 °C, have utility in the industrial segment (as a hydraulic fluid) and automotive segment (transmission fluid, gear oil).

Alkylated (branched-chain) FAs have interesting properties [215] and an effective synthesis of these products is important. The ane reaction is the thermally initiated radical addition of alkanes to alkenes at elevated temperatures (200–450 °C) and pressures (200–250 bar) [216]. Figure 6.25 shows

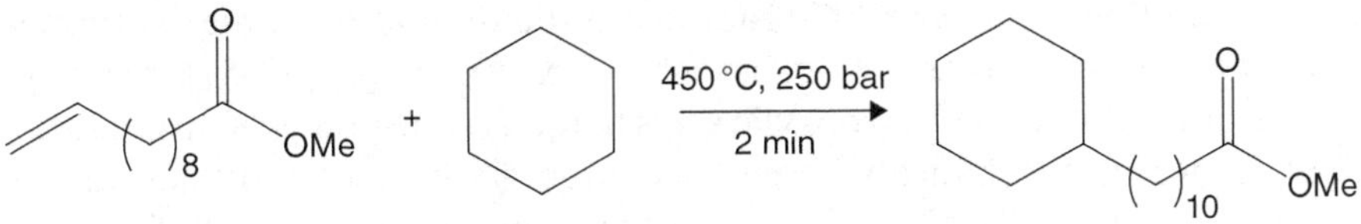

6.25 Thermally initiated addition of cyclohexane to methyl 10-undecenoate at high *p, T*. After ref. [217].

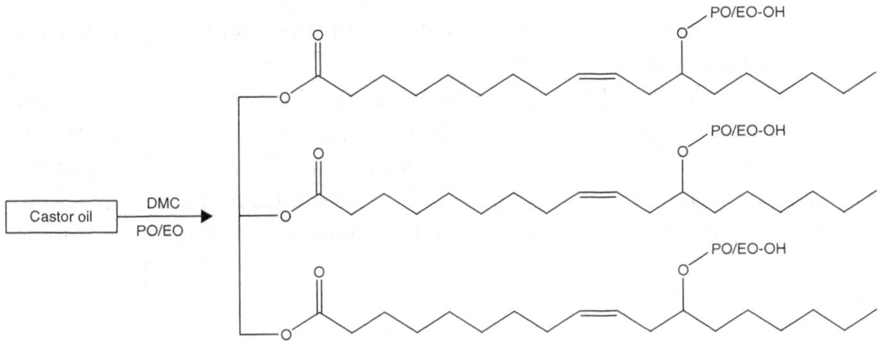

6.26 DMC-catalysed alkoxylation of castor oil to form polyetherols. EO, ethylene oxide; PO, propylene oxide.

the addition of cyclohexane to methyl-10-undecenoate at supercritical fluid conditions [217]. Highly alkylated products formed by Friedel–Crafts alkylation exhibit low pour point, good oxidation resistance and good hydrolytic stability [42]. Alkylation is technologically only of interest with high-performance catalysts. Friedel–Crafts acylation is another reliable method for functionalisation of unsaturated fatty substances and allows the introduction of a keto group in the side chain [44]. Also this process is currently not economically viable because of high catalyst consumption. Lipases are flexible biocatalysts for acylation.

Triglycerides are also used in the manufacture of polyetherols (Fig. 6.26). Alkoxylation of castor oil with a solid double-metal (commonly Zn, Co) cyanide (DMC) catalyst is the basis of the BASF polyol (Lupranol Balance 50) production technology [218].

The ene-reaction, defined as the reaction of an alkene with an allylic hydrogen (the 'ene' compound) with a substance containing a double or triple bond (the 'enophile') to form a new C–C bond with migration of the ene double bond and with a 1,5-hydrogen shift [219], allows introduction of the hetero functions such as —COOR, —OH, —CN in the unsaturated side chain of fatty substances. A special case of ene reactions is addition of formaldehyde to vegetable oils with formation of di- and tri-functionalised triglycerides [47]. Practical application requires high-performance catalysts.

The addition of electrophilic radicals to the double bond allows formation of functionalised and short-branched FA derivatives [48–51]. Acyloxylation, or the addition of carboxylic acids to the double bonds of unsaturated FAs, improves the hydrolytic and oxidation stability of fatty substances when using sterically hindered acids such as neo acids [52, 220]. Esters of pivalic acid with sterically hindered alcohols give very stable synthetic lubricants [54]. Keller *et al.* [53] reported formation of the biodegradable lubricant base oils dimethylpropionyloxymethylstearate and formyloxymethylstearate by addition of pivalic and formic acid to methyl oleate using Nafion/SiO_2 catalysts.

US Patents Nos. 6,051,539 [221] and 6,291,409 [56] to Cargill, Inc. describe the formation of cyclopropanated oils. Cyclopropanation may be effected using the Simmons–Smith reagent diiodomethane [222]. In US Patent No. 6,291,409 (to Cargill, Inc.) Kodali *et al.* disclose a modified Simmons–Smith process with an active Zn(Cu) catalyst for modifying unsaturated polyol FA ester stock to form a cyclopropanated product [56]. Cyclopropanation of the starting ester stock is not necessarily uniform in the mixture. Low toxicity and readily biodegradable cyclopropanated polyol FA ester stocks exhibit an enhanced oxidative stability with respect to that of the corresponding unmodified unsaturated polyol FA ester stock. In most instances, the PP is also reduced.

6.2.6 Oleochemical metathesis

The metathesis reaction, discovered almost 50 years ago, has introduced a breakthrough in C—C bond formation and is one of the most powerful synthetic tools applicable both to petroleum-based and renewable raw materials [59, 60, 223, 224]. Olefin metathesis ('change places', essentially $AB + CD \rightarrow AD + CB$) results in the exchange of substituents of two reacting olefins in an equilibrium reaction, transalkylidenation, according to the catalytic Chauvin mechanism, which consists of a [2 + 2] cycloaddition to form a metallacyclobutane intermediate followed by cycloelimination. Solvent-free metathesis is a green reaction being catalytic under mild reaction conditions with high atom efficiency. Olefin metathesis, used in the petrochemical industry to vary the olefin chain lengths, can be extended to vegetable oil chemistry [57, 59]. Accordingly, the double bonds present in plant oils and fatty acid derivatives can be modified to introduce chemical functionality or to alter the properties of the oils and fatty acid derivatives.

Technical realisation of metathesis in oleochemistry is within reach owing to the development of new active 'Grubbs-type' ruthenium carbene catalysts [186]. At high turnover number and low catalyst concentrations (25 to 50 ppm) metathesis can be performed economically. Oleochemical metathesis is distinguished in two different types: in self- or homo-metathesis the same fatty substrate reacts with itself, whereas in cross-metathesis a fatty substrate reacts with an alkene. Metathesis cannot only be realised with oleochemical base chemicals, but also directly with unsaturated triglycerides, as shown in Figs 6.27 and 6.28. Self-metathesis of MUFAs leads to monounsaturated α,ω-dicarboxylic acids and long-chain unsaturated hydrocarbons [225]. Self-metathesis of unsaturated FA esters provides a convenient route to unsaturated diesters. Methyl oleate may be converted into equimolar amounts of dimethyl 9-octadecenedioate and 9-octadecene in the presence of $WCl_6/Sn(CH_3)_4$ (Fig. 6.27). Grubbs' ruthenium-based catalysts are much more effective [62]. Metathesis of vegetable oils, such as glycerol trioleate (Fig. 6.28), results in viscous oils [60]. The reaction brings about *cis–trans* isomerisation.

Cross-metathesis of fatty oils consisting of long-chain unsaturated FA triglycerides with lower olefins allows transformation of natural fatty oils into fatty oils of lower molecular weight. Typical cross-metathesis reaction partners are ethylene or 2-butene [58, 60] and methyl acrylate [226]. Figure 6.29 shows the cross-metathesis of methyl oleate and ethylene (ethenolysis) to produce methyl 9-decenoate and 1-decene, both key intermediates in lubricant manufacture. Instead of simple alkenes, the cross-metathesis of fatty compounds can also be carried out with substituted alkenes. In one reaction step, very different multifunctional products can be produced with this method. Similarly, cross-metathesis of triolein and ethylene in the presence of bis(tricyclohexylphosphine) benzylidene ruthenium (IV) dichloride (Grubbs catalyst) leads to tridecenoic acid with terminal (9, 10) double bonds and 1-decene (Fig. 6.30) [58]. Ethenolysis of olive oil followed by hydrogenation yields tricaprin [63]. Self-metathesis or co-metathesis of fatty substances often leads to an unselective mixture of diesters and alkenes. The catalyst choice is crucial.

Metathesis reactions of SBO [227] and soy, rapeseed, tall and linseed FAs [225] have been reported. However, technical utilisation of this reaction in oleochemistry has not yet been achieved despite recent promising catalyst developments [57]. Elevance Renewable Sciences, Inc. (Chicago, IL), a Materia, Inc./Cargill, Inc. partnership, is to develop functional oils, lubri-

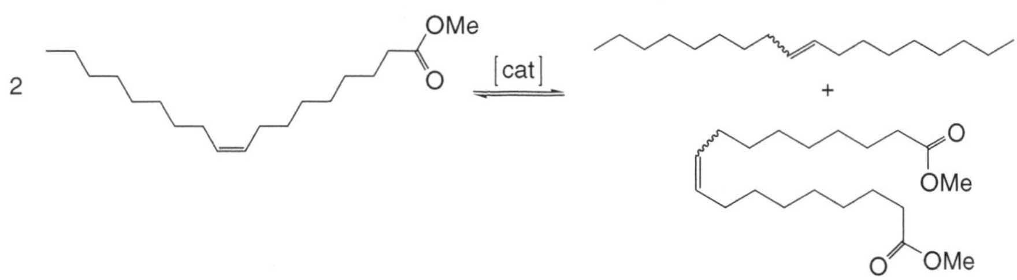

6.27 Homo-metathesis of methyl oleate. After ref. [186].

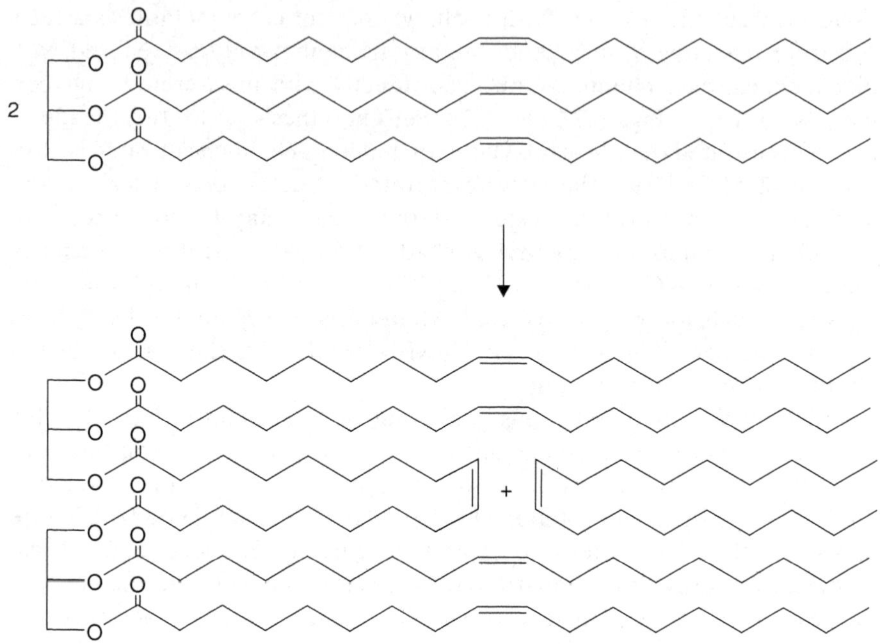

6.28 Self-metathesis of glycerol trioleate. After ref. [60].

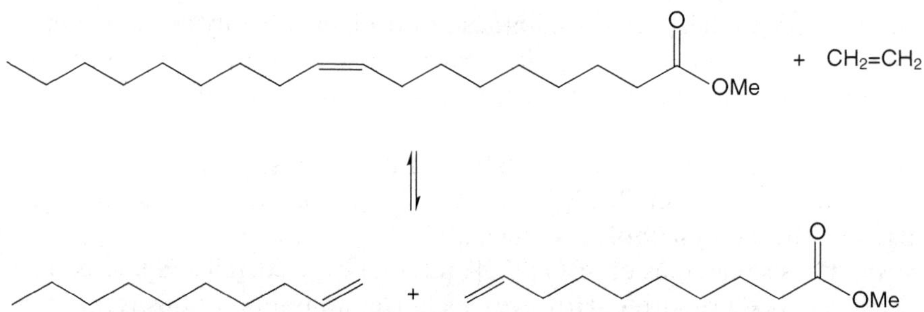

6.29 Cross-metathesis of methyl oleate and ethylene (ethenolysis) to 1-decene and methyl 9-decenoate.

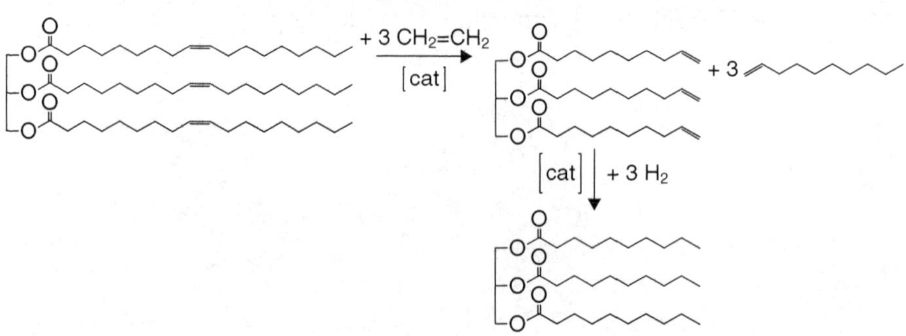

6.30 Cross-metathesis of glycerol trioleate with ethylene.

cants, industrial chemicals, feedstocks, and consumer products with high added value by applying metathesis technology to low-cost renewable oils and fats (soy, canola, corn, etc.) for improving their stability by reacting out the undesirable parts of the oil [228]. Questions remain as to the biodegradability of the new potential products and whether they should be considered renewable or synthetic.

Homogeneous olefin metathesis catalysts such as tricyclophosphine [1,3-bis (2,4,6-trimethylphenyl) imidazol-2-ylidene] [2-thienylmethylene] ruthenium(II) dichloride are being produced by Evonik Industries (formerly Degussa) for a broad spectrum of reactions in cross-metathesis, ring-closing and ring-opening metathesis (catMETium® RF2-4; RF, royalty-free) [229].

6.2.7 Catalytic oxidations of unsaturated oils and fats

Oxyfunctionalisation of fatty acids (oxidation reactions at the alkyl chain of triglycerides) represents less than 10% of today's oleochemistry. Methods available today generally lack selectivity and require harsh conditions. Notable exceptions are the epoxidation of unsaturated plant oils and the synthesis and use of a few HFAs.

Unsaturated oils and fats can react in various ways with oxidants such as hydrogen peroxide, organic hydroperoxides or peracids (see Fig. 6.31), resulting in:

- oxidation of the double bond to an epoxide (oxirane);
- addition of H_2O_2 with formation of a vicinal diol;
- double bond cleavage with formation of mono- and dicarboxylic acid.

As shown in Section 10.5.5 lubricity greatly benefits from various oxygen functionalities in a molecule.

As a result of biodiesel production FAMEs have become extensively available and are now being produced with high purity. This has opened up

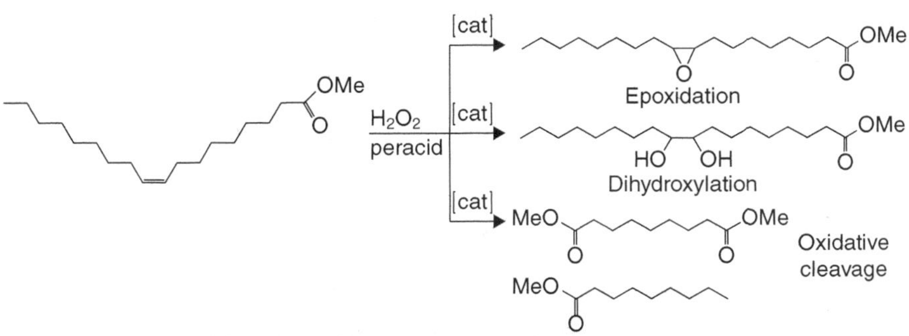

6.31 Possible oxidations of unsaturated fatty acids. After ref. [186].

new pathways for the synthesis of oleochemical products, such as epoxidised FAMEs [230] and lubricants based on branched-chain derivatives [231], partially hydrogenated methyl esters [232], diesters [90] and α-hydroxy ethers [233].

Epoxidation of unsaturated FA esters with classical chemocatalytic methods (using peracids) [4, 66, 73] or enzymatically [67, 68, 74] is industrially performed on a large scale. In the Prileshajev plant oil epoxidation process, peroxy acids such as peracetic acid are generated from the corresponding acid and hydrogen peroxide in the presence of a strong mineral acid (Fig. 6.32). Owing to the potential danger of handling peroxy acids, the intermediate peroxy acids are not isolated during large-scale epoxidation of unsaturated triglycerides (Fig. 6.33). Other drawbacks of the Prileshajev epoxidation are a relatively low selectivity (<80%) in view of a multitude of side reactions via oxirane opening, leading to diols, hydroxyesters, estolides and other dimers, which are catalysed by strong mineral acids. The

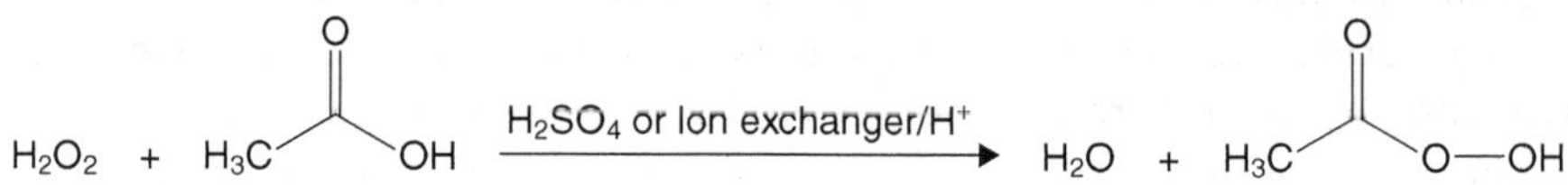

6.32 Prileshajev epoxidation: generation of short-chain peroxy acids.

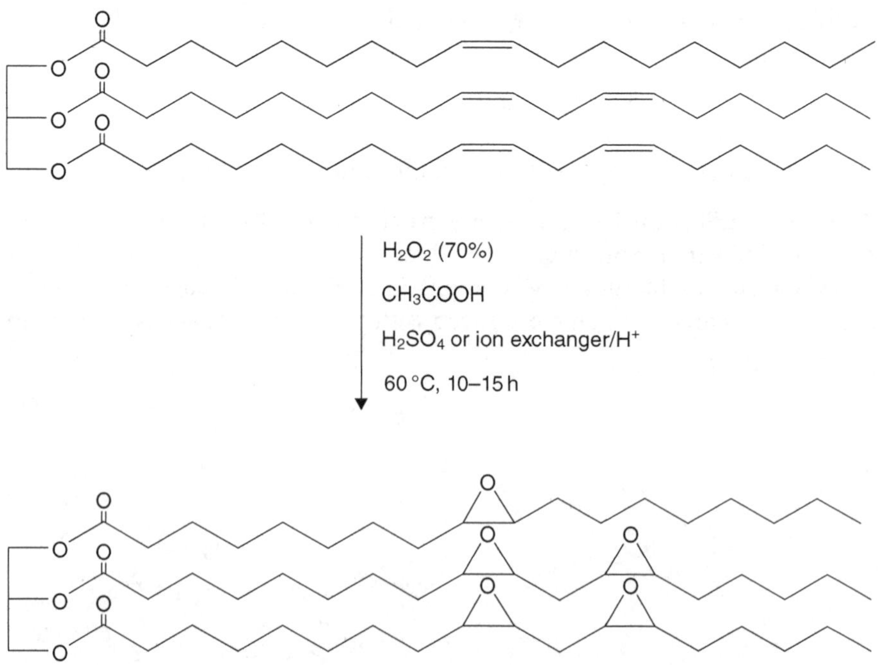

6.33 Prileshajev epoxidation: use of short-chain peroxy acids for epoxidation of unsaturated plant oils.

presence of a strong acid in an oxidative environment causes corrosion of the equipment. The Prileshajev epoxidation of unsaturated plant oils (predominantly SBO) is used worldwide for the production of epoxidised soybean oil (ESBO). FA epoxides are mainly used as poly (vinyl chloride) (PVC) plasticisers and stabilisers.

Sobczak and Ziółkowski [234] reported epoxidation of oleic acid with organic hydroperoxides to 9,19-epoxystearic acid with a selectivity of 87% using molybdenum complexes (e.g. $Mo(CO)_6$) as catalysts; also the highly selective epoxidation of HERO was described. Chemo-enzymatic epoxidation is of considerable interest because this method totally suppresses ring opening of the epoxide. Initially, the unsaturated FA or ester is converted into an unsaturated percarboxylic acid by a lipase-catalysed reaction with H_2O_2 and is then self-epoxidised in an essentially intermolecular reaction (Fig. 6.34) [74]. The second reaction step occurs without involvement of the enzyme, following the rules of the Prileshajev epoxidation. Excellent stability and activity is shown by Novozym 435® (*Candida antarctica* lipase B immobilised on polyacryl). This allows production of fully saturated esters with improved stability against oxidative attack and conversion of native oils, including ESBO, into potential base fluids for lubricants (Fig. 6.35) [70, 91]. Whereas the stereochemistry of triglyceride molecules with various FA (oleic, linoleic, and linolenic) chains differs in orientation, steric bulkiness, distances, etc., epoxidation restores uniformity to the chain and low steric hindrance. Such molecules readily undergo crystallisation upon cooling (cf. PPs of SBO and ESBO of –6 and 3 °C, respectively). Epoxy oils are more thermally stable than the parent triglycerides [71]. However, when exposed to higher temperature over extended time, epoxy oils undergo rapid oxidative degradation. Vegetable oil epoxides may be used to improve lubricity. Pure epoxidised rapeseed oil (ERSO) can be used as a lubricant base fluid with improved oxidation stability and lubricity (friction-reducing and extreme pressure abilities) as compared to RSO [69]. As epoxidation increases the PP of the oil, epoxidised vegetable oils are limited to higher-temperature applications. The epoxidation treatment has no adverse effect on the biodegradability of the base stock.

Natural sources of epoxy oil (vernolic acid, C18:1 > O) include *Vernonia galamensis* (>74%), *V. anthelmintica* (67%) and *Euphorbia lagascae* (60%).

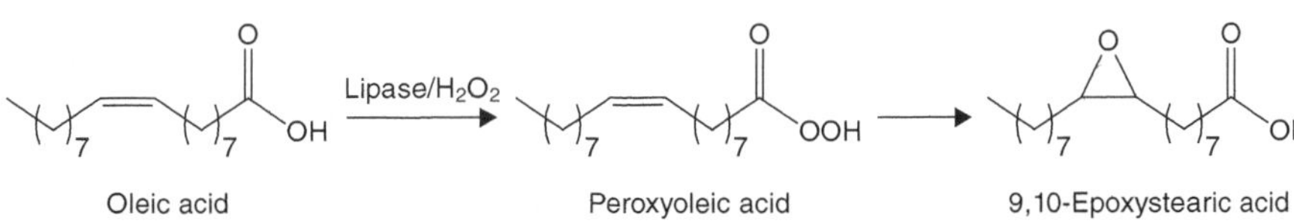

6.34 Reaction principle of the chemo-enzymatic self-epoxidation of oleic acid with intermediate enzymatic formation of peroxyoleic acid.

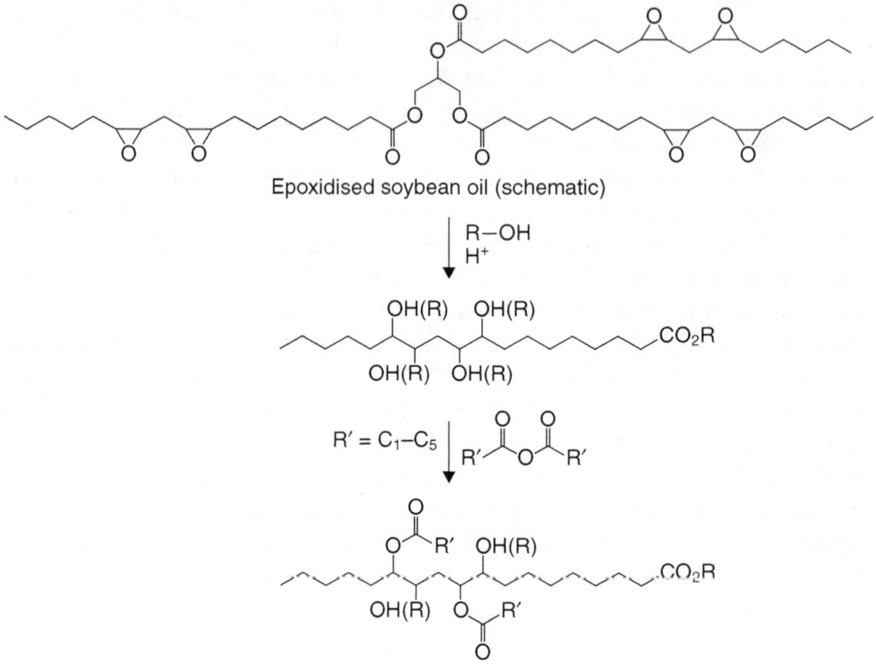

6.35 Base fluids from epoxidised soybean oil. After ref. [91].

The naturally occurring epoxidised oil of *Vernonia* and the chemically expoxidised SBO and linseed oil are similar, but there is an important difference. Expoxidised soybean oil and linseed oil are highly viscous and are non-pourable below 0 °C. *Vernonia* oil, however, has low viscosity and can be stored below 0 °C. Methyl vernoleate is very stable towards oxidation.

Currently, the most common chemical modification of soybean oil is the introduction of epoxy groups. ESBO is used primarily as a process aid in PVC. It is a potential candidate for high-temperature lubrication, which outperforms HOSBO and SBO in certain applications [71]. Conversion of C=C unsaturations in the FA chain into epoxy groups influences the thermo-oxidative behaviour of ESBO. The olefin and bis-allylic protons in SBO and HOSBO have high susceptibility to thermal and oxidative degradation. Removal of these reactive protons in ESBO determines an enhanced oxidative stability. More precisely, ESBO shows considerable improvement in oxidative stability over SBO and HOSBO at low additive concentrations (0.5% AO) with pressure differential scanning calorimetry (PDSC) onset temperatures of 243, 179 and 203 °C, respectively. Similarly, also in terms of oxidation induction times (measured under isothermal conditions) ESBO (at 195 °C) showed greater stability than SBO (at 150 °C) and HOSBO (at 160 °C), namely 52.2, 7.8 and 24.7 min, respectively [71]. Oxidative stability is rated about six times higher for ESBO than for HOSBO. Deposit formation (thin film micro-oxidation, TFMO, test) increases in the order ESBO < HOSBO < SBO due to the increased presence of conjugated unsaturation in the FA chain. The viscosity of ESBO at 40 °C is significantly higher than for SBO and HOSBO (170, 33 and 41 cSt, respectively). The higher molecular weight and more polar structure compared with SBO and HOSBO results in stronger intermolecular interaction in ESBO. This property of ESBO translates into enhanced lubricity in a dynamic system. The frictional behaviour of these three vegetable oil-based lubricants is similar. ESBO, SBO and HOSBO have excellent coefficient of friction (CoF) reducing properties due to formation of a stable polymeric film on the metal surface during boundary lubrication. Such film formation in ESBO is achieved through the —O— cross-linking on the metal surface, thus reducing friction between the moving parts [71].

Doll *et al.* [235, 236] have examined the properties of SBO-based olefins and the epoxided methyl esters of oleic, linoleic and linolenic acid (EMO, EMLO and EMLEN, respectively) relevant to lubricant application such as oxidative stability, PP, VI and CoF (see Table 6.17). Neat SBO has a very high VI (>200), and methyl oleate is nearly as good. The other olefinic materials perform less well. Epoxidation increases the viscosity of the oleo-

Table 6.17 Properties of soybean oil-based olefins and epoxides for lubricant application

Sample	OT (°C)[a]	PP (°C)	VI	E[e]
Olefins[b]				
SBO	155	−9	225	−15.1
MO	177	−27	199	−9.6
MLO	139	−48	Undefined[d]	−10.0
MLEN	117	−60	Undefined[d]	−10.5
Epoxides[c]				
ESBO	199	3	142	−13.8
EMO	190	0	151	−11.3
EMLO	180	−1.5	132	−10.9
EMLEN	131	−7.5	63	−9.2
E2EHS	193	−4.5	140	−12.6

[a] PDSC, oxidation onset temperature.
[b] SBO, soybean oil; MO, methyl oleate; MLO, methyl linoleate; MLEN, methyl linolenate.
[c] ESBO, epoxidised soybean oil; EMO, epoxidised methyl oleate; EMLO, epoxidised methyl linoleate; EMLEN, epoxidised methyl linolenate; E2EHS, epoxidised 2-ethylhexyl soyate.
[d] Viscosity below 2.0 mm^2/s at 100 °C.
[e] Calculated adsorption energy (kJ/mol) from Temkin model.
After ref. [235].

chemical at 100 °C to above 2.0 mm^2/s, enabling proper definition of their VIs. The epoxides show acceptable VI values, although inferior to unmodified SBO. Unmodified olefinic oleochemicals of Table 6.17 all have PPs suitable for use in most lubrication applications. Compounds of increased unsaturation display even lower PPs, which are adversely affected by epoxidation [235]. However, these materials will still flow above 3 °C and, therefore, are suitable for operation at room temperature. However, their PPs are a limiting factor for use in outdoor winter conditions.

Compounds containing larger headgroups, such as a glycerol backbone in SBO or a 2-ethylhexyl moiety in epoxidised 2-ethylhexyl soyate (E2EHS) show stronger adsorption to the steel surface. As the level of unsaturation increases, the adsorption energy becomes stronger (see Table 6.17). The opposite trend is observed by the increased level of epoxidation. The less epoxidised EMO has the best adsorption energy of the epoxy methyl esters. As to wear scar, the best performing epoxidised lubricant is ESBO and the worst lubricant is EMLEN. It is thus shown that epoxides have superior oxidative stability (PDSC) and stronger metal adsorption compared to olefins, which have better pour points and viscosity indexes.

Epoxidation has received special attention and is of great synthetic interest because it opens up a wide range of feasible reactions that can be carried out under moderate reaction conditions [237, 238]. Epoxidation of native plant oils transforms multiply unsaturated functions into highly reactive oxiranes, which are susceptible to ring opening reactions. Figure 6.36 shows various possibilities of oxirane ring opening [239]. Ring opening of the

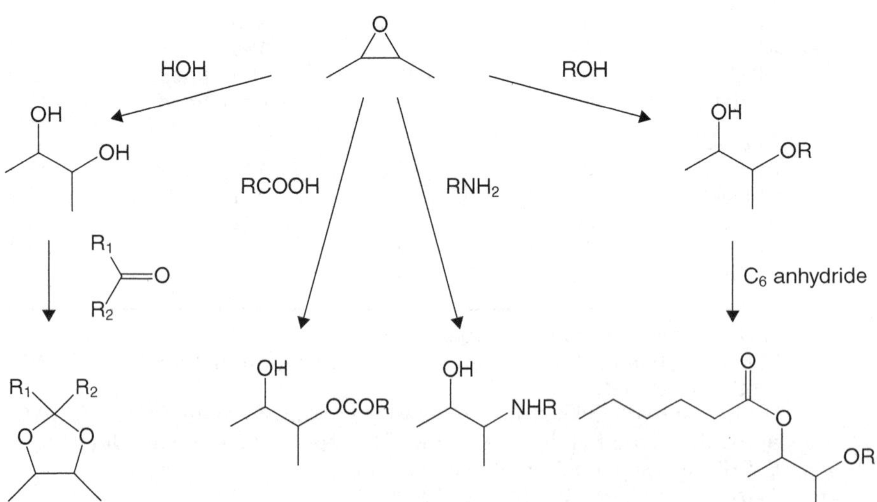

6.36 Reactivity of the oxirane group.

epoxy group and subsequent derivatisation of the epoxy carbons can improve the oxidative and low-temperature stability of the molecule. Acid-catalysed ring opening of epoxidised oils with low-molecular-weight mono- or polyfunctional acids or alcohols leads to low-molecular-weight liquid epoxy polyol esters or ethers [65, 240]. Depending on the reaction conditions either polyols with high OH-functionality (complete reaction) or epoxy polyol esters with remaining epoxy groups (partial conversion) are obtained. Oleochemical polyols have a relatively low viscosity. Polyols are widely used in the production of foams, dispersants and fluid polyurethane resins [240].

The oxirane ring opening reaction of epoxy-HOSME, produced from high-oleic sunflower methyl esters (HOSME) using H_2O_2 and formic acid with epoxidation efficiency of 90%, with an excess of oleic acid under atmospheric pressure without any catalyst or solvent, leads to complete estolide formation (mixtures of C_{36} oleates of methyl hydroxystearate and C_{54} dioleates of methyldihydroxystearate) [238]. Oxirane ring opening with alcohols may be carried out under acid catalysis at 100 °C and atmospheric pressure, leading to ether alcohols and secondary products resulting from dehydration, transesterification and dimerisation side reactions. Neutral Zn, Co double-metal cyanide complexes catalyse the ring opening reactions of epoxides.

Cleavage of the epoxy ring with other nucleophiles allows also introduction of hetero-atoms and a variety of functional groups producing mono-alcohols, diols, alkoxyalcohols, hydroxyesters, *N*-hydroxyalkylamides, mercaptoalcohols, aminoalcohols, hydroxynitriles, etc. [4, 77, 241]. The cleavage reaction of epoxy-HOSME with butylamine was conducted under pressure, at high temperature (180–200 °C) [238]. Reaction products are composed of amides formed by transesterification and a mixture of fatty amines/imines obtained by ring opening.

Vicinal diols of unsaturated fatty compounds may be prepared by epoxidation and subsequent nucleophilic ring opening of the epoxide. Since ring opening of fatty epoxides requires harsh reaction conditions [242], the direct synthesis of *vic*-dihydroxy FAs is of interest. Catalytic hydroxylation of oleic acid by H_2O_2 affords syn diols via the epoxy intermediate [243, 244]. Ring opening of the oxirane ring with HCl or HBr (with formation of chloro- and bromohydrins) introduces halogens in the structure which increase the polarity of the compounds [245]. The presence of free —OH groups determines high viscosity of such fluids as a result of intermolecular H-bonding. Products obtained by ring opening of epoxidised FA esters followed by esterification of the resulting hydroxyl group show good performance for lower temperature applications [65]. Reaction with an anhydride yields an ester ether compound. Incorporation of a long-chain substitution improves the low-temperature fluidity and viscosity characteristics of the fluid. It is challenging to improve both oxidation and low-temperature properties. However, in most cases improving one property destroys the other (Murphy's Law).

Epoxidised vegetable oils are promising intermediates for the utilisation of vegetable oils as the oxirane group can be easily functionalised. Gast *et al.* [246] have reported low PP polyhydroxystearic acid ester lubricants derived from 9,10-epoxystearic acid and 9,10-dihydroxystearic acid. The products obtained by ring-opening reaction of epoxidised FA esters by esterification of the resulting hydroxyl group show good performance for lower temperature lubricant applications. More recently, a high-yield (100%) process has been developed to lower the PP of ESBO to –15 °C by ring-opening reaction with branched alcohols (*iso*-amyl alcohol and 2-ethylhexanol) in the presence of the solid catalyst Amberlyst® 15 Dry [65]. Longer branched alcohols result in lubricants with lower PPs. The catalyst was reusable for an estimated five reactions, which allows for greater economic viability. Lubricant base stocks have also been produced from acid-catalysed oxirane ring-opening reactions using ESBO, employing several alcohols and by successive esterification of the resulting hydroxyl group with an acid anhydride [247–249].

Figure 6.37 shows ring opening reactions of ESBO with acid anhydrides to form chemically modified soybean oil (CMSO) derivatives [250]. As shown in Table 6.18 the PPs of these products decrease with an increasing chain length of the ester branching, on account of steric interaction counteracting easy stacking of the molecules. The CMSO products also show higher oxidation stability than unmodified SBO, which is due to removal of unsaturation.

Biolubricants were also synthesised from di-hydroxylated soybean oil, obtained by acid hydrolysis of the epoxide group followed by esterification of the hydroxyl groups [251, 252]. More recently, acyl derivatives of ricinoleic/oleic acid have been described as obtained by the opening of the oxirane ring by the nucleophilic addition of FAs in the presence of a *p*-TSA catalyst. These products appear to be promising biolubricants [253–255]. Salimon *et al.* [253] have investigated various biolubricant base stocks

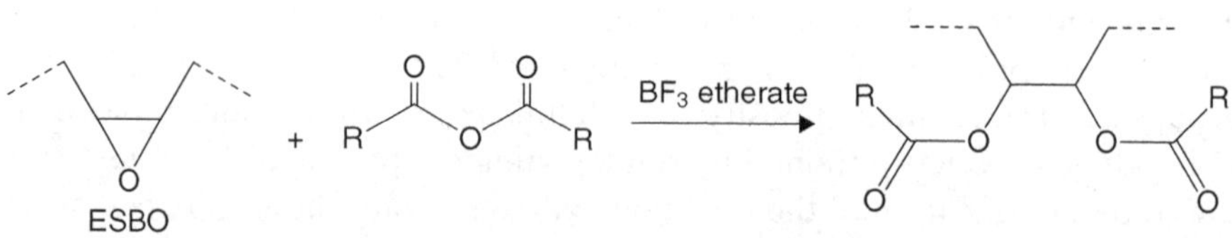

6.37 Ring-opening reaction of the oxirane ring of epoxidised soybean oil (ESBO) using acid anhydrides with formation of acyl derivatives of soybean oil.

Table 6.18 Pour points and onset temperatures of neat chemically modified soybean oil (CMSO) derivatives

CMSO	R[a]	Pour point (°C)[b]	OT (°C)[c]
SBO	–	−9	172.2
ESBO	–	0	–
ASO	CH_3	−3	181.0
PSO	C_2H_5	−9	185.0
BSO	$n\text{-}C_3H_7$	−15	182.8
VSO	$n\text{-}C_4H_9$	−18	182.7
HxSO	$n\text{-}C_5H_{11}$	−21	183.5
HpSO	$n\text{-}C_6H_{13}$	−21	181.9

[a] See Fig. 6.37.
[b] Using diluent (dibutyl adipate)/oil ratio of 0.1.
[c] Onset temperature, obtained by PDSC at 10 °C/min in air under 200 psi pressure.
After ref. [250].

from epoxidised ricinoleic acid (ERA) with improved low-temperature properties (Fig. 6.38 and Table 6.19). Butyl-10,12-dihydroxy-9-behenoxy-stearate (BDHBS) with a bulky ester behenyl mid-chain moiety exhibits the most favourable low-temperature performance (PP, −47 °C) and highest VI (243). On the other hand, butyl-10,12-dihydroxy-9-octyloxystearate (BDHOS) shows the highest oxidation stability (onset temperature, OT, 221 °C). Improved oxidation stability is achieved when the chain length of the mid-chain ester decreases. It is evident that hydrogen bonding is a critical parameter influencing the low-temperature properties and oxidation stability of synthetic esters. Increased hydrogen bonding determines an increased pour point and decreased oxidation stability of the compounds of Fig. 6.38. The VI and flash point of ERA, DHASA (10,12-dihydroxy-9-acyloxystearic acid) and BDHAS (butyl-10,12-dihydroxy-9-acyloxystearate) are significantly influenced by compound structure.

It is very important to design an optimised structure that will offer the best performance properties over a broad spectrum. Short-chain substitution in hydroxyl ether (I) of Fig. 6.39 results in a higher PP, whereas a bulky and asymmetrically branched chain in hydroxyl ether (II) delays solidification during the cooling process. Substitution of a bulkier and branched alcohol (2-ethylhexanol) in hydroxyl ether (II) shows a significant lowering of PP (−18 °C) compared with methanol in hydroxyl ether (I) (3 °C) [256]. The bulkier molecule (II) is characterised by low oxidation stability. Free hydroxyl groups in the synthetic oils of Figs 6.35 and 6.39 make a strong chemical interaction with the more reactive steel surface. This interaction is a result of iron atoms being able to offer coordination sites to polar

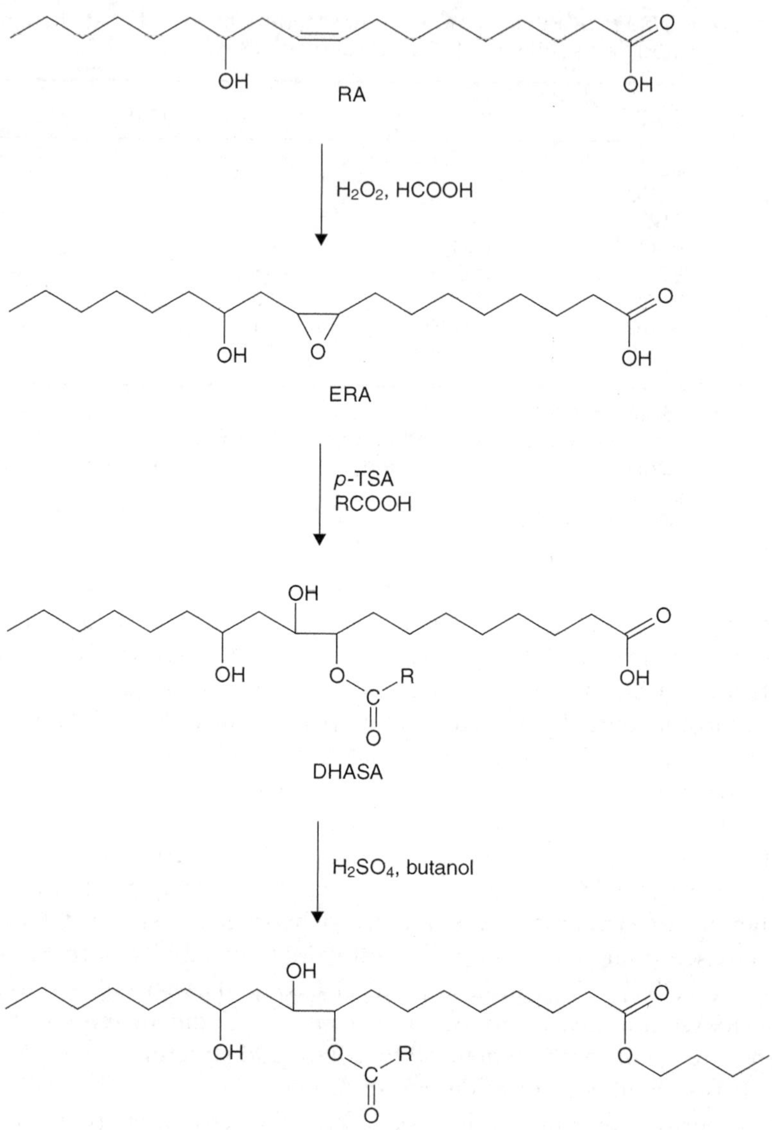

6.38 Schematic reactions of epoxidation, ring opening and esterification of ricinoleic acid (RA) and its derivatives; ERA, epoxidised ricinoleic acid; *p*-TSA, *p*-toluenesulphonic acid; RCOOH: octanoic, nonanoic, myristic, palmitic, stearic, and behenic acids; DHASA, 10,12-dihydroxy-9-acyloxystearic acid; BDHAS, butyl-10,12-dihydroxy-9-acyloxystearate. After ref. [253].

Table 6.19 Physicochemical properties of epoxidised ricinoleic acid derivatives

Compound[a]	Pour point (°C)	Flash point (°C)	Viscosity index	OT (°C)
ERA	9	80	67	60
DHOSA	−10	63	110	105
DHBSA	−23	192	171	64
BDHOS	−30	163	180	221
BDHBS	−47	180	243	155

[a] ERA: epoxidised ricinoleic acid; DHOSA: 10,12-dihydroxy-9-octyloxystearic acid; DHBSA: 10,12-dihydroxy-9-behenoxystearic acid; BDHOS: butyl 10,12-dihydroxy-9-octyloxystearate; BDHBS; butyl 10,12-dihydroxy-9-behenoxystearate.
After ref. [253].

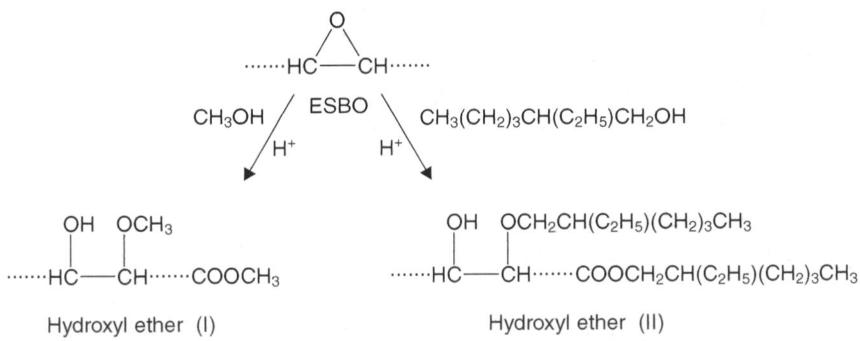

6.39 Short-chain and highly branched hydroxyl ethers. After ref. [256].

oxygen atoms. OT increases as the free —OH is substituted with —$COCH(CH_3)_2$ or —$CO(CH_2)_4CH_3$.

Sharma *et al.* [257] have described branched ester hydroxy derivatives of methyl oleate, obtained by epoxidation and ring opening acylation using organic acids (propionic, levulinic, hexanoic, octanoic and 2-ethylhexanoic acid) (Fig. 6.40). Thermo-oxidative stability, low-temperature fluidity and lubricant properties of the hydroxy ester products outperform their methyl ester analogues. These products show the dual function of improving low-temperature fluidity as well as an effective friction modifier and AW additive in various lubrication systems.

In PCT Int. Publ. No. WO 2008/100822 A2 (to Chevron U.S.A., Inc.) Miller and Elomari [259] disclose high-value diester based (bio)lubricants prepared from epoxidising an olefin followed by oxirane ring opening to a diol and subsequent (di)esterification with carboxylic acid or acid chloride. A follow-up patent describes an alternative synthetic procedure [258]. Formation of (vicinal) diester species may take place via *direct* diesterification

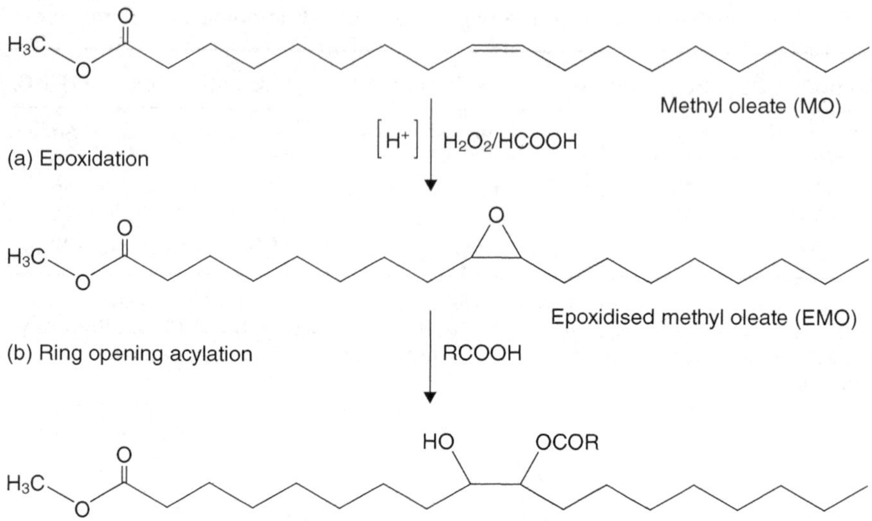

6.40 Epoxidation of methyl oleate (a), followed by ring opening acylation of EMO using an organic acid (b). After ref. [257].

of the epoxide (Fig. 6.41), i.e. without epoxide ring-opening to a diol intermediate. The procedure utilises a biomass precursor and/or low value Fischer–Tropsch (FT) olefins (C_8–C_{16}) and/or alcohols, or FT olefins and fatty acids. The fatty acid may be from a bio-based source (i.e. biomass, renewable source) or can be derived from FT alcohols via oxidation. It is necessary to convert α-olefins to internal olefins because diesters of α-olefins, especially if short chain, tend to be solids or waxes. 'Internalising' α-olefins, followed by transformation to the diester functionalities introduces branching along the chain, which reduces the PP of the intended lubricants. It is noticed that the diester-based lubricant compositions obtained via direct or indirect esterification of epoxides differ essentially from estolides.

Substituted triesters of stearic acid were prepared by epoxidation of methyl oleate, oxirane ring opening and esterification of free hydroxyl groups (Fig. 6.42) [260]. The fully degradable triesters are potential lubricants. Table 6.20 shows some relevant physicochemical properties.

Epoxidised unsaturated FAs have been used as antirust additives in water-based cutting fluids [261] and to eliminate corrosion from chlorine-containing compounds [262]. Adhvaryu and Erhan [71] reported ESBO as a potential source for high-temperature lubricant applications. Epoxidised unsaturated fatty esters and glycerides have been claimed as lubricating oil stabilisers for internal combustion engines (ICEs) [263]. Epoxidation of new crop oils of *Lesquerella* and *Limnanthes* species yields materials with

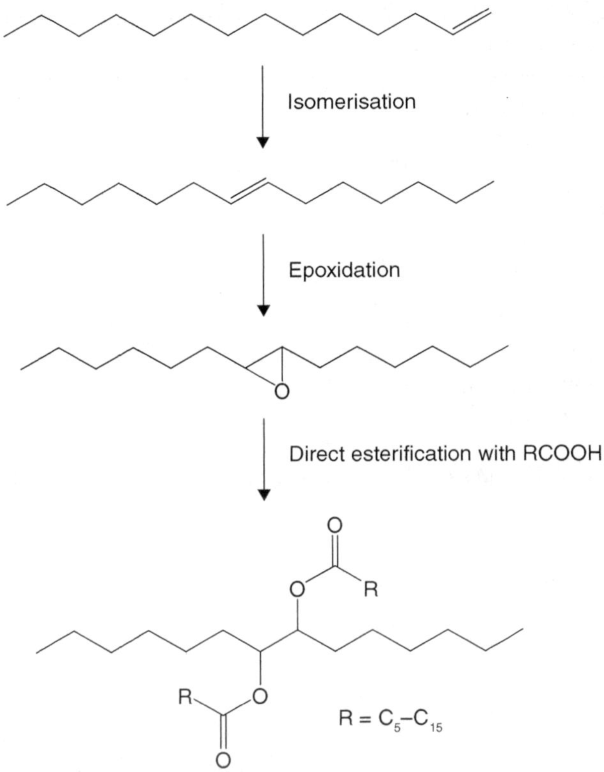

6.41 *a*-Olefin isomerisation, epoxidation and direct diesterification with a (bio-derived) carboxylic acid to form a diester species. After ref. [258].

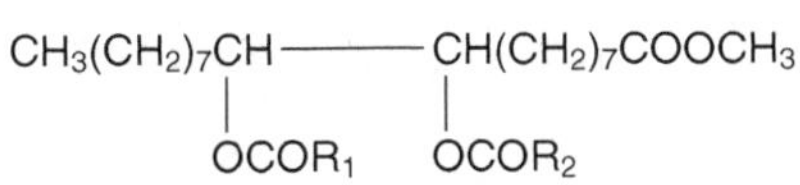

6.42 Substituted triester derivatives of stearic acid ($R_1 = C_{11}H_{23}$, $C_{15}H_{31}$; $R_2 = CH_3$, C_5H_{11}).

significant oxirane values and properties that are potentially useful in lubricants, plasticiser-stabilisers and coatings [264].

Potential new bio-based oxygenated lubricants and fuel lubricity enhancers have been prepared from methyl 9,10-dihydroxystearate and long-chain aldehydes (>C_7) to form cyclic acetals in an acid-catalysed process according to Fig. 6.43 [265]. These materials are oils down to low temperatures, unlike symmetric cyclic ketals derived from the same diol, which are waxes at room temperature. Acetals generally are stable in alkaline media, but are sensitive to acidic hydrolysis.

Table 6.20 Physicochemical properties of triesters compared to di(2-ethylhexyl)sebacate (EHS)

Triester formulation					
R_1	R_2	VI	CP (°C)	PP (°C)	OT (°C)[a]
$C_{11}H_{23}$	CH_3	127	<–30	<–30	176.6
$C_{15}H_{31}$	CH_3	171	4	–20	176.6
$C_{11}H_{23}$	C_5H_{11}	137	5	<–30	176.3
$C_{15}H_{31}$	C_5H_{11}	135	–2	–20	188.5
EHS		126	–	–60	188.4

[a] Onset temperature.
After ref. [260].

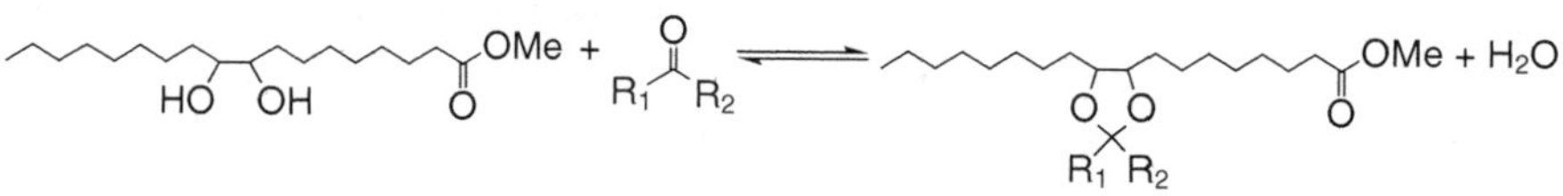

6.43 Cyclic acetal formation from methyl 9,10-dihydroxystearate. After ref. [265].

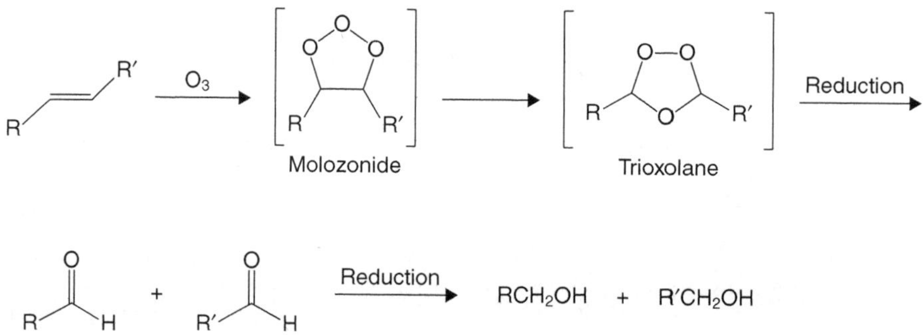

6.44 Ozonation and reduction mechanism.

Apart from epoxidation, oxidative functionalisation of vegetable oils may also be effected industrially in other ways, such as ozonation (see Fig. 6.44). The chemistry of ozonation of alkenes has been studied extensively [78, 79, 266–268]. Ozonolysis of olefins is typically performed at moderate to elevated temperatures whereby the initially formed molozonide (a 1,2,3-trioxolane) dissociates to carbonyl oxide and an aldehyde or ketone that recombine to form a secondary ozonide (a 1,2,4-trioxolane) [269].

The oxidative cleavage of unsaturated fats is an interesting means of selectively generating short-chain mono- or dicarboxylic acids from longer chain carboxylic acids. Ozonation of oleic acid has been successfully applied on a commercial scale to the production of pelargonic and azelaic acids.

Reductive cleavage leads to fatty aldehydes as viscous oils. Azelaic acid is used for the manufacture of high-performance lubricants [4, 270]. Similarly, ozonolytic cleavage of erucic acid in acetic acid followed by oxidation of the resultant aldehyde by oxygen at 100 °C leads to the saturated diacids brassylic acid and pelargonic acid (Fig. 6.45) [271]. Both products are useful in a variety of industrial applications. Brassylic acid is used as a plasticiser, biodegradable solvent and lubricant; pelargonic acid is a herbicide.

Highly selective ozonation of unsaturated FAs (such as SBO, castor, canola, RSO, SNO) at ambient temperature forms ozonides (1,2,4-trioxolane rings) and peroxy hemi-acetals which are usually further transformed by ozonolysis (defined as the breakdown of ozonation products to carbonyl compounds). While oxidative cleavage of the products of ozonation gives polyfunctional acids or ketones, reductive decomposition (in a variety of reducing conditions) gives aldehydes and ketones or primary and secondary alcohols [272]; see Fig. 6.44.

As shown in Fig. 6.46 for soy oil, ozonation may convert all the double bonds to ester groups in a single-step chemical attack in the absence of any catalyst. As the modified vegetable oil products from this process do not contain any tertiary carbons or unsaturation, the modified oil has much higher thermal and oxidative stability. The triglyceride ester linkage is not affected in the process and the fragments of the unsaturated fatty acids beyond the double bonds are also oxidised to form mono-esters and diesters. The structure of the triglyceride lubricant after the ozonolysis reaction is

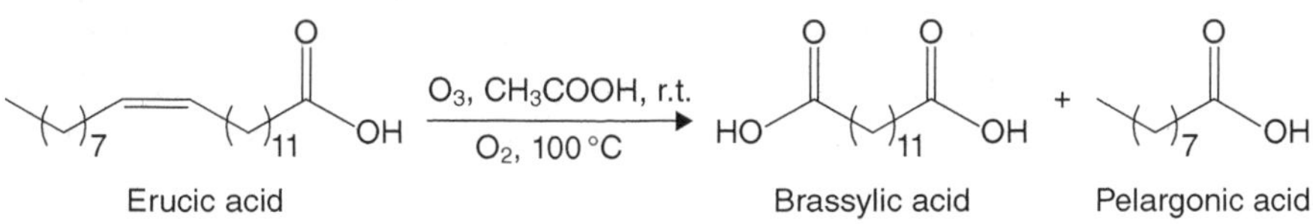

6.45 Ozonolytic cleavage of erucic acid. After ref. [271].

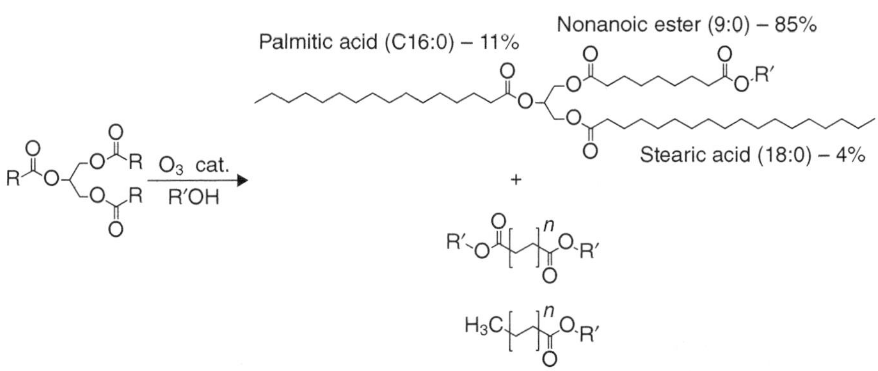

6.46 Ozonolysis of soy-oil (R = 11% C16:0, 4% C18:0, 23% C18:1, 51% C18:2, 7% C18:3).

more uniform than the original soy-oil, with only three FA components (palmitic, stearic and nonanoic).

Ozonation followed by reduction is a very efficient technique for carbon–carbon double bond cleavage in unsaturated oils for the formation of triglyceride polyols and low-molecular-weight diols and monols of well-defined structure [78, 79]. The oxidative cleavage products of unsaturated oils formed by the oxidation process can be in the form of oils or solids at ambient temperature.

Ozonolysis of triglycerides leads to oleochemical polyols with terminal primary hydroxyl groups (Fig. 6.47). As SFAs are left unaffected by ozonolysis, their presence decreases the functionality and hydroxyl number of the polyol. While epoxidised and hydroformylated SBO polyols may have a hydroxyl functionality of 4.5 or higher, ozonolysis produces polyols with a maximal functionality of 3 (see Fig. 6.47a).

Graiver *et al.* [80] have described a novel single-step catalytic ozonolysis and *in situ* reaction which proceeds rapidly and efficiently at room temperature in the absence of solvents. The reaction has been used to reduce the unsaturation in SBO and for the preparation of a number of potentially useful materials such as a biolubricants with good thermo-oxidative stability. The soy-based materials are competitive both in cost and performance to petroleum-based materials, besides being environmentally friendly and sustainable. Polyols were prepared by passing ozone through a solution of SBO and ethylene glycol in the presence of an alkaline catalyst. Unlike conventional ozonolysis that yields aldehydes and carboxylic acids by spontaneous decomposition of the ozonide intermediates, the ozonides formed reacted with the hydroxyl group of the glycerol to form an ester linkage with a terminal hydroxyl group. Statistical analysis of the product mixture

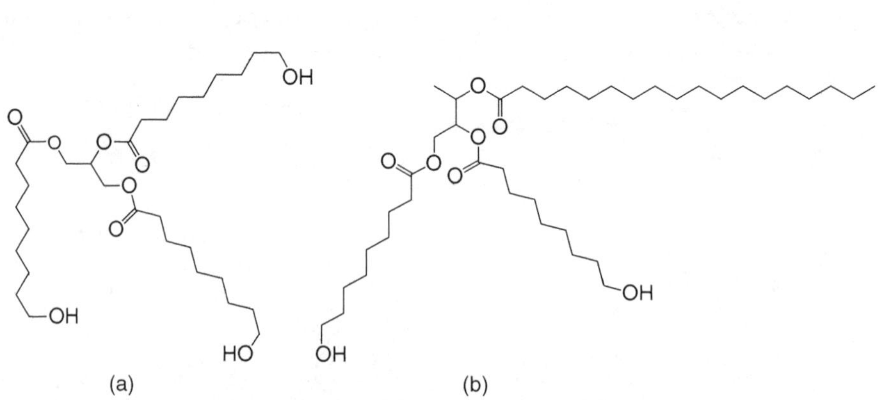

6.47 Triol obtained by ozonolysis of trilinolein (triglyceride of 9-hydroxy-nonanoic acid) (a) and a diol from soybean (ozo-soy-polyol) or canola oil (ozo-canola-polyol) containing saturated fatty acids (b). After ref. [79].

indicates that the resulting polyol mixture is more uniform than the original triglyceride mixture, having (2-hydroxy) nonanoate as the major component of the new hydroxyl functional triglyceride [273].

As shown in Fig. 6.47, the structures of the triglyceride polyols are dependent on the chemical composition of the starting material. Trilinolein gives only one polyol structure, while SBO yields various species, including the one in Fig. 6.47b. Polyols obtained by ozonation of trilinolein, canola and SBO are waxy solids at room temperature with melting points comparable to paraffins (Table 6.21). Due to the lower molecular weight of ozonolysis polyols, their viscosities in the liquid state at 25 °C are an order of magnitude lower than those of triolein and soybean polyols obtained by epoxidation [245]. However, epoxidation polyols are liquid at room temperature.

Ozonised vegetable oils with reduced viscosity compared with their parent triglycerides [274, 275] are used as pour point depressants (PPDs) of biodiesel fuel [266, 276]. Methanolysis of ozo-canola-polyol leads to 9-hydroxynonanoic acid methyl ester [78], while ozonolysis of methyl oleate yields the corresponding acid [277]. ω-HFAs can be used for the preparation of mixed diesters.

There are many ways to functionalise unsaturated vegetable oils to form polyols. Hydroxylation of oleic acid to produce ricinoleic acid is catalysed by castor hydroxylase enzyme [278]. By dihydroxylation, vicinal diols can be generated from alkenes. For dihydroxylation of unsaturated fats, tungstic acid (H_2WO_4) or rhenium catalysts (e.g. CH_3ReO_3) are appropriate [243, 279]. Epoxidation followed by oxirane ring opening gives polyols with secondary hydroxyl groups [58, 245, 280]. The resulting products differ in the number of hydroxyl groups, their position and distribution in the FA chain. Their properties depend on the type of vegetable oil [280]. Hydroformylation and subsequent reduction of the aldehydes formed gives polyols with primary hydroxyls [39]. In both cases hydroxyl groups are located along the FA chains. However, ozonolysis gives polyols with primary, terminal hydrox-

Table 6.21 Properties of triglyceride polyols by ozonolysis

Polyol	OH no. (mg KOH/g)	f[a]	Melting point (°C)	Viscosity at 25 °C (Pa s)
Ozo-trilinolein-polyol	298	3.0	50	0.89
Ozo-canola-polyol	260	2.8	37	0.81
Ozo-soy-polyol	228	2.55	39	0.68

[a] Hydroxyl functionality.
After ref. [79].

yls. This process very efficiently cleaves the carbon–carbon double bond and leads to the formation of carboxylic acid, aldehyde, or hydroxyl groups depending on the reaction conditions. High-quality bio-based polyols obtained by hydroformylation or ozonolysis of vegetable oils such as SBO find many polyurethane applications [281].

Short-chain linear α,ω-dicarboxylic acids such as suberic acid (C_8), azelaic acid (C_9) or sebacic acid (C_{10}) can be derived from oils and fats by chemical oxidation (alkali fission) of unsaturated fatty acids, which constitute particularly important building blocks for the synthesis of lubricants [81, 82].

Allylic C—H bonds of unsaturated FAs are activated C—H bonds, which may be functionalised by various oxidising agents. SeO_2/*t*BuOOH is suitable for allylic oxidation of methyl oleate and methyl 10-undecenoate [282]. The reaction with singlet oxygen has proved suitable for the preparation of α,β-unsaturated ketones from methyl oleate (Fig. 6.48) [283].

Biocatalytic oxyfunctionalisation of FAs

Apart from the epoxidation of unsaturated oils and the synthesis of a few HFAs, the chemical methods available nowadays for oxidation reactions at the alkyl or alkenyl chains lack selectivity and/or require harsh conditions. Even the Prileshajev epoxidation reaction, carried out industrially at more than a 200 kt/yr level to yield ESBO, is far from ideal and actually quite problematic. As organic chemistry currently does not dispose of efficient tools for (stereo)selective oxygenation of FA carbon chains, the potential of enzymatic oxyfunctionalisation deserves attention [284].

Biocatalysis is slowly making progress in organic chemistry. There are no natural catalysts for every conceivable chemical reaction that chemists wish to promote. Consequently, there is a need to re-engineer enzymes (protein engineering) to adapt their specificity to the reaction and reaction conditions (e.g. extreme pH values) at stake [285]. Table 6.22 shows the many advantages of enzymes over classical catalysis of specific organic reactions. Enzyme-based chemical transformations typically proceed with high selectivity under mild conditions and are becoming increasingly important industrially. However, as substrates and products from biotransformations

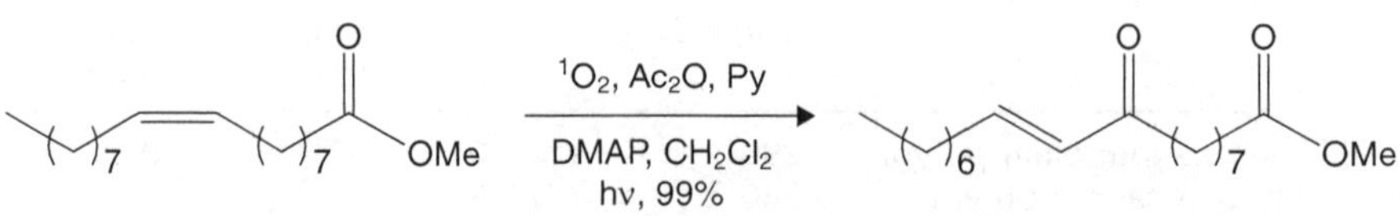

6.48 Photo-oxygenation of methyl oleate with singlet oxygen and tetraphenylporphyrin as a sensitiser to yield enone fatty acids. DMAP = 4-dimethylaminopyridine.

Table 6.22 Comparison of enzymatic and conventional chemical catalysis

Advantages of enzymes:
- High efficiency in simple and complex chemical transformations
- High chemo-, enantio- and regio-selectivity
- High substrate specificity
- Allowance for cheap feedstocks (high tolerance for FFAs and water content)
- Green chemistry (aqueous media)
- Metal-free or heavy metal-free
- Mild reaction conditions (low *T*, *p*)
- High purity products (no by-products)
- No production of chemical waste
- Fewer reaction steps
- No need for tedious (de)blocking steps
- Simple production systems, reduced production costs
- Suitable for continuous and batch processing
- Heterogeneous processes (immobilised enzymes or cells)
- Fully biocompatible and biodegradable

Disadvantages of enzymes:
- Expensive enzymes (few providers)
- Relatively low lipase activity
- Long reaction times
- Enzyme deactivation
- Generally low thermal stability
- Limited tolerance for organic solvents (maturing non-aqueous enzymology)
- Low catalytic activities in non-aqueous conditions
- Low product concentrations
- Difficult recyclability
- Need for specific reactor and hardware configurations
- Requirement for special production extraction methods
- Long process development time

are often poorly soluble or unstable in aqueous media and/or are toxic to living cells, usually low product concentrations result. Industry demands the development of thermally stable, salt-, poison- and acid/base-resistant synthetic enzymes, because naturally existing enzymes are not robust. Biocatalysis results in sustainable technology.

New biocatalytic processes will often be determined by the availability of the biocatalyst. There now exist several specialised suppliers of biocatalytic reagents, such as Novozymes (Bagsværd, Denmark), Genencor/DuPont, Maxygen Corp. (Redwood City, CA) and Diversa/Verenium (San Diego, CA). Many enzymes have become commercially available and cheaper due to recombinant expression systems. Enzymes are increasingly being used as biocatalysts in the generation of products that have until now been derived using traditional chemical processes. The use of biocatalysis for industrial chemical synthesis is becoming easier and applications of enzymes in the chemical industry are already well established. Recent advances in enzymatic catalysis have extended its use in the food, pharma and chemical industries to the synthesis of specialty chemicals (amino acids, vitamins, sweeteners, antibiotics) and even of some bulk chemicals (e.g. adipic acid, acrylamide). Certik and Shimizu [286] have developed efficient fungal-based systems for the production of PUFAs.

Lipase-catalysed and microbial transformations allow a variety of selective syntheses of fatty compounds. For instance, the lipase-catalysed synthesis of carbohydrate esters has been reported [287]. Typical microbial transformations are the hydration of unsaturated FAs, the ω- and β-oxidation of FAs and the conversion of oils/fats and glucose into glycolipids. *Candida* yeasts have been used for FA oxygenation in fermentation processes. Apart from yeasts, several bacteria such as pseudomonades, bacilli or rhodococci are able to hydroxylate FAs in a terminal or subterminal manner [288, 289]. The chemical addition of water to unsaturated fatty compounds is neither regioselective nor stereoselective, in contrast to microbial hydration. Microbial enantioselective and regioselective addition of water to oleic, linoleic and linolenic acids gives 10-HFAs [290, 291]. Hydratases from a number of bacteria and yeasts convert (9*Z*)-fatty acids into 10-HFAs [292]. Microbial ω-oxidation of fatty acids leads to dicarboxylic acids. For example, oleic acid can be converted into the corresponding unsaturated dicarboxylic acid with yeast of the genus *Candida tropicalis* [293]. Palmitic, stearic and erucic acid may be converted similarly. Oleic acid can also be transformed into the unsaturated hydroxy diacid (*Z*)-3-hydroxy-9-octadecendioic acid with the yeast mutant *C. tropicalis* M25 [294]. However, these processes are not sufficiently cheap to make the preparation of oxy-functionalised FAs commercially interesting.

Enzymes possess considerable potential in the area of asymmetric oxidation [295–299] and biocatalytic epoxidations [300–302]. Among the industrial enzyme applications in the exploratory phase are oxidations [302, 303] and carbon–carbon bond formations [304], processes often limited by product inhibition. Xylase monooxygenase is used in *Escherichia coli* for sequential hydroxylation of xylenes [303]. Recombinant *E. coli* strains have been used to produce fine chemicals by redox reactions, such as dihydrodiols [305, 306] and the oxidation of styrene to (*S*)-styrene oxide at an enantiomeric excess greater than 99% [302]. This demonstrates the possibility of whole cell biotransformations in an organic/aqueous reaction system with toxic substrates and products.

Cytochrome P450 monooxygenases (P450s) constitute a large family of enzymes of particular interest in view of the ability to catalyse the selective insertion of oxygen into a wide variety of compounds [307]. Practical use of P450s as isolated enzymes is limited by low turnover rates (particularly towards non-natural substrates), low stability and the need for electron-donating cofactors.

Selective biological hydroxylation and epoxidation reactions of fatty acids are being explored along three major biotechnological routes: (i) transgenic oil crops; (ii) fermentation using genetically engineered micro-organisms; and (iii) bioreactors using isolated enzymes. Genetic engineering of oil crops producing oxygenated FAs is a developing discipline in which the model plant *Arabidopsis thaliana* has frequently served as a host system for seed-specific expression of hydroxylases or epoxidases [308–310]. Developing seeds of *Arabidopsis* were used for the expression of castor bean oleate Δ12-hydroxylase and the corresponding enzyme from *Lesquerella fendleri* [308]. In an attempt to design new oil crops yielding epoxy FAs, CYP726A1 from the Euphorbiacea *E. lagascae* was expressed in the seeds of various plant hosts; low yields (up to 15%) of fatty acid epoxides in the seed oils were obtained [311]. Soybean was successfully engineered for the production of FA epoxides. Interest in using plants expressing P450 genes for production of oxygenated FAs is growing [312].

The C—H bond of an FA methylene group is one of the most stable chemical entities whose breakage requires approximately 98 kcal/mol. This energy is beyond the range of simple enzyme reactions and requires either a sequence of enzyme reactions or a metal cofactor. Enzymatic FA oxygenation essentially takes places by three main reaction types, namely: (i) oxidation by P450 monooxygenases; (ii) oxidation by diiron centre oxygenases; and (iii) oxidation by lipoxygenases (LOXs). With respect to unsaturated FAs, cytochrome P450 enzymes can act both as hydroxylases and as epoxidases [313]. The peroxide-mediated cytochrome P450 aromatic hydroxylation of naphthalene in the absence of cofactors was reported [314]. Euphorbiaceae species use a P450 monooxygenase for fatty acid epoxidation. Diiron-centre oxygenases (mainly in plants and bacteria) use a diiron-centre for reductive oxygen activation, hydrogen abstraction and oxygen insertion to yield an epoxy or hydroxy compound [315]. Catalysis by proteins containing binuclear non-haem iron clusters is currently under extensive investigation as these enzymes catalyse a diverse set of reactions including hydroxylation, desaturation and epoxidation [316]. Despite their occurrence in most plant species producing oils enriched in epoxy acids, diiron cluster-containing FA monooxygenases in biotechnology have found little application.

Lipoxygenases are limited to eukaryotes for FA oxyfunctionalisation [317]. They catalyse the regio- and stereo-selective dioxygenation of polyenoic FAs and yield a vast range of compounds including epoxides, mono-, di- and trihydroxy derivatives, allene oxides, epoxyalcohols, aldehydes, oxoacids, etc. Lipid hydroperoxides formed by LOXs can either be reduced to hydroxyl compounds or processed further by enzymes of the lipoxygenase pathway.

Biocatalytic hydroxylation and epoxidation of FAs using isolated enzymes in a bioreactor is still in its infancy and limited to either LOXs or lipase-catalysed perhydrolysis with subsequent self-epoxidation. Cognis has recently developed a new and efficient method to produce FA hydroperoxides based on LOX [318]. Hydrogen peroxide is also used as an oxidant in the chemoenzymatic synthesis of fatty epoxides by lipase-catalysed perhydrolysis. Unbranched SFAs (C_4–C_{22}) can be converted to peroxy FAs using H_2O_2 and Novozym 435 [319]. If an unsaturated FA or its ester is treated with H_2O_2 in the presence of Novozym 435, self-epoxidation occurs [74]. Depending on the chain length, peroxy FA yields from 70 to 95% were obtained. Provided successful scale-up is possible, this could substitute the problematic chemical Prileshajev epoxidation process for lipase-catalysed perhydrolysis. Oleic acid, linoleic acid and arachidonic acid have been converted nearly quantitatively to their corresponding epoxides using *t*-butylhydroperoxide as oxidant in the presence of oat *(Avena sativa)* peroxygenase [320].

FA-oxygenating enzymes are a versatile and useful class of biocatalysts. However, industrial applications of this class of enzymes are still to be developed. More R&D effort is needed to overcome the problems in biocatalytic oxidation.

Nowadays the development time for enzymes by protein engineering is still considerable and the present trial-and-error methods need to be replaced by a more (semi)rational approach. In the medium term (>10 years) the availability of enzymes will likely be considerably improved by progress in synthetic biology. The long term aims at novo-enzyme design. Also, the so-called extremophiles – enzymes produced by organisms that inhabit extreme environments – might be able to catalyse reactions far outside the long-established temperature, pressure and pH norms, suiting them to a range of biotechnological applications [321, 322]. It is not expected that enzymes will completely outplace the classical catalysts. A mix of traditional and biocatalysis of organic reactions is most probable. Widespread application of enzymes in the chemical industry will depend on the ability to couple enzymatic and chemical steps (hybrid chemical/biocatalytic processes). Chemoenzymatic reaction sequences (such as in Fig. 6.34, in Tosoh/DSM's aspartame process or in DuPont's glycolic acid synthesis) profit from the high technical development level of both chemical and enzymatic reactions. Advances in biocatalyst development need to be complemented by improvement in process design. Today, the market for basic, intermediate, fine and specialty chemicals and polymers made by bioprocesses accounts for only a few percent of the chemical market. Industrial biocatalysis has been reviewed by Schmid *et al.* [298, 323].

6.2.8 Radical and catalytic C—C coupling reactions

C—C coupling, with formation of symmetrical products, may be achieved by dimerisation of two radicals. Radicals may be formed selectively under mild conditions in high concentrations by the oxidation of anionic precursors. In particular, the α-positioned C—H bond of the FA ester group possesses comparably high acidity and is suitable for anionisation and subsequent reactions (Fig. 6.49) [324].

Anodic decarboxylation of aliphatic carboxylic acids provides a rapid and technically useful path to radicals for dimerisation and coupling (Kolbe electrolysis) [325–327]. This efficient synthetic method is used extensively in homocouplings with natural and modified FAs, such as in the preparation of specifically functionalised alkanes or in the preparation of long-chain diesters. Through heterocoupling, i.e. electrolysis of two different carboxylic acids, new unsaturated FAs are formed (Fig. 6.50).

An interesting reaction to transform carboxylic acids into symmetrical ketones is their coupling or ketonic decarboxylation [328]. The mechanism of this reaction remains under debate. The products obtained from the condensation of FAs (Fig. 6.51), also called fatty ketones (or their derivatives), can make excellent lubricants [329].

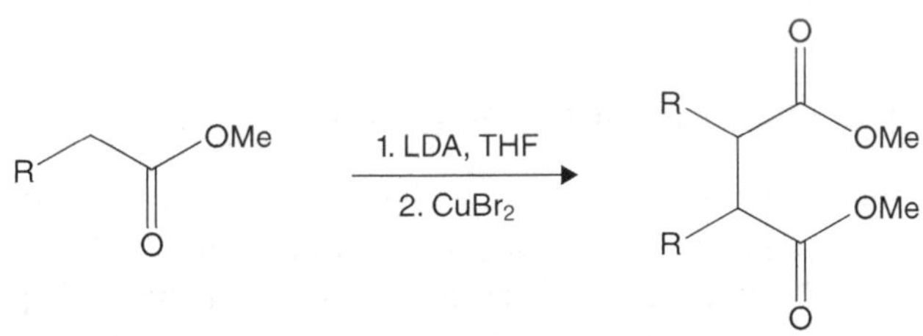

6.49 Radical α, α′ dimerisation of fatty acid methyl esters (FAMEs) by anionisation with LDA (lithium diisopropylamide) and oxidation with $CuBr_2$.

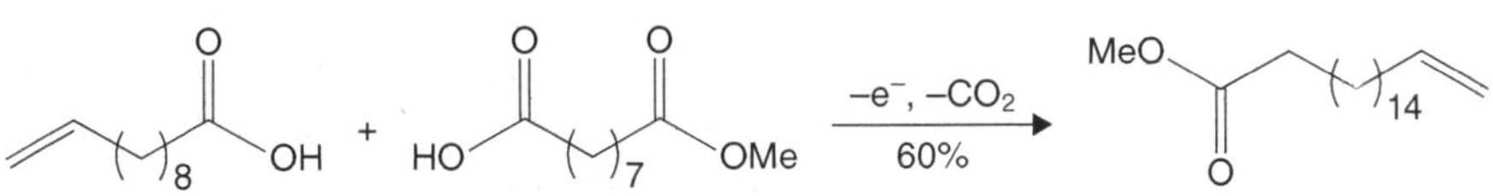

6.50 Kolbe electrolysis of two different fatty acids forming a new ω-unsaturated fatty acid. After ref. [327].

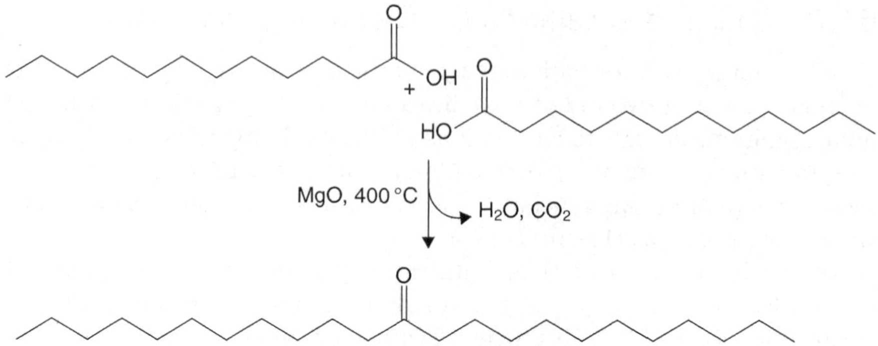

6.51 Ketonic decarboxylation of lauric acid. After ref. [329].

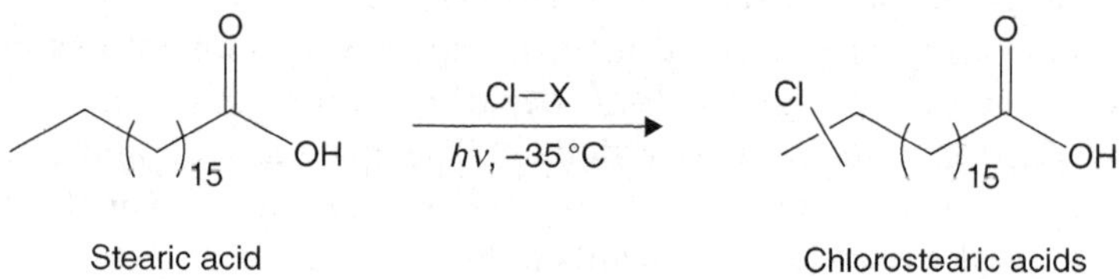

6.52 Photochemical gas-phase chlorination of stearic acid to chlorostearic acids. X = Cl, *t*BuO. After ref. [332].

6.2.9 Functionalisation of C—H bonds

Selective reactions at the saturated alkyl chain of FAs are few but of considerable interest, in particular oxidation of non-activated C—H bonds. Non-activated C—H bonds may be functionalised chemically [330] or enzymatically [331]. Photochemical gas-phase chlorination of FAs adsorbed onto Al_2O_3 with chlorine or *t*BuOCl has been reported (Fig. 6.52) [332].

6.3 Branched-chain fatty acids (BCFAs) in lubrication

Utilisation of branched-chain fatty acids (BCFAs) can expand the use of stable SFAs in biolubricants. The improvement of lubricity properties of triglycerides upon the introduction of branching has been demonstrated [333]. BCFA oils are characterised by good thermo-oxidative stability, viscosity and biodegradability, and low melting point. The irregularity imparted by BCFAs or hydroxyl groups disrupts the lipid packing ability of the hydrocarbon chains, thereby reducing the melting point of the oil. For example, branched oleochemicals such as 9(10)-hydroxy-10(9)-ester derivatives of methyl oleate show improved low-temperature properties (PP) over olefinic oleochemicals and increased thermo-oxidative stability, and have potential as a lubricating oil or fuel additive [257]. Introduction of branching also affects the hydrolytic stability.

In view of the desirable properties of branched FAs, it is worthwhile considering their natural occurrence. Nature offers an abundance of unusual FAs in plant oils with various functionalities in the FA chain, like acetylenic, allenic, conjugated, epoxide, keto, hydroxy, furanoid, fluoro, cyclopropene and cyclopentenyl FAs [334]. Other unusual FAs are short- and long-chain length triglycerides. Over 300 different unusual FAs have been identified in seed oils from various plants [335–339]. Some of these occur in very high amounts (up to 95%) in the seed oil of certain species. Of these plants only relatively few, e.g. castor bean, are nowadays being used for any significant commercial production. However, many other unsaturated FAs have considerable potential as renewable industrial feedstocks.

Oxygenated FAs are mainly mono- and dihydroxy FAs, keto and epoxy FAs. Hydroxyl functionality is rare in plant oils (i.e. castor oil). Natural sources of epoxy oil (vernolic acid, C18:1 > 0) are *V. galamensis* (>74%), *V. anthelmintica* (67%) and *E. lagascae* (>60%). Currently, no oilseed crop has been commercialised as a source of natural epoxidised oils. However, *V. galamensis* is considered having potential for domestication as a new industrial oilseed source of natural epoxy FAs. Epoxidation of SBO is carried out on an industrial scale.

Cyclopropenoid FAs occur most commonly in the seed oils of the order Malvales (Sterculiaceae, Malvaceae, Bombaceae and Tiliaceae), but in low amounts (8.2–22.0%). Various cyclopentenyl fatty acids constitute up to 90% of the total FAs of the seed oils of *Hydnocarpus*, a medicinal oil [340].

Available sources of branched FAs and property data of branched fatty esters are rather few. BCFAs of various kinds can be found in limited amounts in many animals (as surface fats and wool wax), bacteria and in some plants. The most common types of branched acids are iso-acids, usually with an even number of total carbon atoms, and the anteiso-acids, usually with an odd number of carbon atoms. Usually they occur as saturated iso (*i*)- and anteiso (*ai*)-methyl-branched and saturated mid-chain methyl-branched FAs. The branch can consist of a methyl group attached to a carbon chain of 10 to more than 30 carbons, usually C_{14}–C_{18}, but ethyl and propyl branches are also found. The biosynthesis and elongation of short- and medium-chain-length FAs has been described [341]. An enzyme that can introduce methyl groups on an FA chain is referred to as a methyltransferase.

Saturated branched FAs and their derivatives, which are both biodegradable and stable to oxidation, exhibit lower melting points than their straight-chain analogues, which makes them potential candidates for biolubricants. There is growing interest in using the derivatives of iso- and anteiso-methyl branched FAs in lubricants. Branched-chain methyl and isopropyl esters (C_{14}–C_{18}) obtained from lanolin were explored [342]. Some BCFA methyl esters of lanolin and oleate esters of oleyl alcohol and normal alcohols (up to C_{12}) have been suggested as biolubricants on the basis of their melting point and viscosities [343]. BCFAs have been proposed as low-temperature stabilisers of vegetable oils and derivatives [344].

In the absence of any significant commercial natural source of BCFAs considerable effort has been devoted to their synthesis (see Sections 6.2.2, 6.2.4–6.2.8). Here we mention a few illustrative examples. Conversion of methyl oleate to branched-chain alkyl and aryl derivatives (ethyl, *n*-butyl, *n*-hexyl, phenyl) by bromination in the allylic position, followed by reaction with cuprate reagents, has been reported by Dailey *et al.* [345]. Potential use of these materials, in which the double bond is retained, is as additives to biodiesel at low temperatures. In US Patent No. 6,051,539 (to Cargill, Inc.) Kodali and Li disclose manufacturing of cyclopropanated TAG oils [221]. Sharma *et al.* [346] have described the synthesis of acyl derivatives of vegetable oils using ESBO and hexanoic anhydride (see Fig. 6.36). The products are diesters at double bond sites of the FA chains of triacylglycerol. These branched SBOs may find use in some lubricant end-use applications as such or when formulated with functional components, such as extreme-pressure (EP) additives, AW additives, PPDs, other base stocks, diluents and the like. The tribological behaviour of such CMSOs has been reported [250].

Unusual FAs are of particular interest for various industrial applications. Consequently, considerable efforts have been made to alter the genome of traditional oilseed crops in such a way that they accumulate new economically valuable oils [347]. The EU funded REFLAX (Rational Engineering of lipid metabolism of FLAX) project (project no. QLK3-2000-00349; 2001–2004) has aimed at developing a bio-based production of branched FAs by the engineering of oilseeds [348]. Although it seems feasible to promote the biosynthesis of BCFAs in plants, more research is needed to achieve high concentration levels of these compounds. It was not an objective of REFLAX to deliver high BCFA-producing oilseeds, but merely to demonstrate the feasibility.

Production of BCFAs synthetically from bio-based products (vegetable oils and animal fats) has also been one of the objectives of a US ARS project [349].

6.4 Modified starch-based lubricants

Starch possesses lubricating ability and may act as a liquid lubricant in suspension [350]. Development of new products from starch for various industrial applications (including lubrication) is proceeding along various directions: (i) modification to diversify the properties; (ii) blending with other materials, such as polymers, sugars, oils and FAs; and (iii) modification and blending.

Formulation of starch-based lubricants requires solubilising starch and various lubricant ingredients in water. Solubilisation of starch in water can be achieved through chemical modification [351]. This is usually carried out by reacting granular starch in aqueous suspension with anhydrides (forming esters, see Fig. 6.53) or other reagents that can react with the hydroxyl groups of starch. By esterification of a portion of the hydroxyl groups of starch the number of free hydroxyl groups available for participation in the intramolecular H-loading required for the formation of the helical crystalline structure of native starch is reduced. The net effect is disruption of the crystalline network and improved water solubility of chemically modified starch.

Application of dry-film lubricant formulations requires dissolving the formulation in a suitable solvent, preferably water. Accomplishing both of these goals, i.e. formulation and application, requires disrupting the crystalline structure of starch and enhancing its stability in water. The most widely practised method for solubilising starch in water is by cooking it with steam under high pressure and temperature [352]. This procedure successfully disrupts the crystalline structure and produces water-soluble starch that can then be used for formulation by blending. In an extension of this process, called Fantesk™, a slurry of starch pre-blended with the appropriate oil-phase is processed with high pressure and temperature steam to produce the final product directly [352, 353]. The resulting dispersion is dried, milled and used for the intended application. The continuous steam-jet cooking process can be used to produce water-soluble starch composites with dispersed and encapsulated ingredients for use in various applications [354]. Ingredients that have been blended with starch and encapsulated in Fantesk™ composites as micrometre-sized droplets include vegetable oils, synthetic oils and formulated products for use in food, medicine, cosmetics, agriculture, coatings, adhesives and lubricants [355]. Some of the Fantesk™ products have been developed commercially.

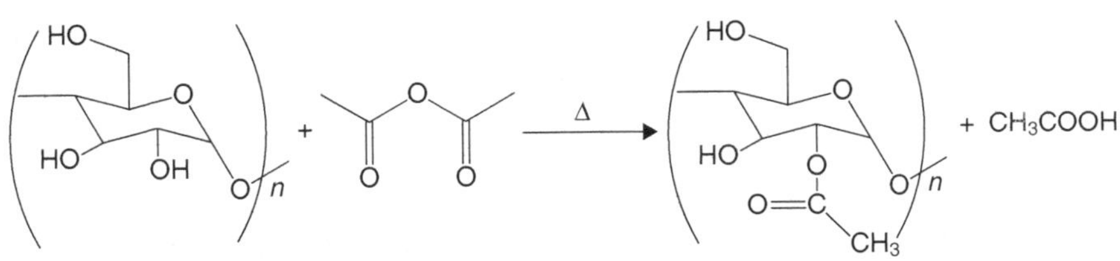

6.53 Esterification reaction between glucose units of starch and an anhydride.

Fantesk™ composites have been evaluated in lubrication, in particular as ingredients of water-based dry-film lubricant formulations for sheet metal forming applications for delivery of biocides [356, 357]. Dry-film lubricants are pre-applied to sheet metals for use in sheet metal forming and/or to protect sheet metal surfaces from scratches and/or the environment during transformation and storage [358]. Fantesk™ compositions have also been evaluated for encapsulation of lubricant formulations for application in oil drilling mud [359].

6.4.1 Starch–oil composites in bio-based dry-film lubricant technology

Bio-based lubricants comprising starch and vegetable oils have promising lubrication properties. Stable starch–lipid composites are water dispersable over a wide range of dilution [360]. Starch–oil composites are preferred ingredients for formulating solid lubricants because they are based on renewable and abundantly available raw materials and also have superior environmental and health characteristics.

Biresaw and Erhan [356, 361] have investigated the effect of film thickness on the friction properties of dry-film lubricants formulated from purified food-grade corn starch (PFGS) – SBO composites. Steel sheets coated with the PFGS-SBO containing solid lubricant were evaluated for boundary CoF and wear properties using ball-on-flat test geometry. The CoF was highly dependent on the SBO to PFGS ratio in the composite. In the absence of SBO, the CoF was high, and decreased sharply with increasing SBO content. Water-based solid lubricants formulated with the PFGS–SBO starch–oil composite have acceptable friction and wear properties and should be evaluated further for use in sheet metal forming.

Starches with various degrees of amylose content, namely waxy maize (1% amylose), normal purified food-grade corn starch (27% amylose) and high (70%+) amylose corn starch (Hylon 7; National Starch, Bridgewater, NJ) were also modified by reaction with anhydrides (forming starch esters, see Fig. 6.53) to improve their water solubility [362]. The chemically modified starches were then dissolved in water and formulated into dry-film lubricants by incorporation of canola oil. Specific starch–oil interactions between the starch and oil molecules consist of hydrogen bonding between free hydroxyl groups of the glucose units of the starch macromolecule and the ester linkage in the triglyceride oil. The resulting formulation was investigated for its friction properties as a function of the starch chemical structure, degree of chemical modification and the concentration of canola in the formulation. The results indicated that dry-film lubricants with acceptable friction properties can be developed using all farm-based ingredients.

Chemically modified starch without added vegetable oil displays a very high CoF on account of the high polarity of the starch surface due to the presence of free hydroxyl groups that do not react with the anhydride and also do not precipitate in intramolecular H-bonding. Addition of canola oil lowers the CoF, which is attributed to H-bonding between the ester groups of the canola oil triglyceride and the free hydroxyl groups of the starch glucose units. The H-bonding reduces the polarity of the starch surface and, hence, its CoF.

An important criterion for commercialisation of aqueous starch–oil composites as dry-film lubricants is their application to metal surfaces in a cost-effective manner. Aqueous dry-film lubricant formulations containing starch–SBO composites and sucrose were applied to steel plates as thick films, where sucrose is added to promote composite adherence to the metal surface. In these formulations sucrose not only acts as a promoter for adhesion to the metal substrates, but also to plasticise the thick films, and prevent cracking. However, the presence of sucrose in the formulation is undesirable as it may promote moisture retention in the coating that could promote oxidation.

It has been shown that the adsorption of vegetable oils on steam-jet cooked starch (Fantesk™), which does not have any of its free hydroxyl groups converted to esters, is stronger than that of chemically modified starch, where some of the hydroxyl groups are converted to esters [363]. Aqueous dispersions of starch-SBO and starch-JO composites, prepared by excess steam jet cooking, can effectively be used in bio-based dry-film lubricant technology. However, dry-film lubricants prepared by air-assisted spray application of aqueous dispersions of waxy maize starch–oil composites allow formation of thinner, sucrose-free coatings [364]. Spray applications of such composites render bio-based dry-film lubricant technology more efficient.

6.5 Sources of further information and advice

C. Jiménez-González and D.J.C. Constable, *Green Chemistry and Engineering. A Practical Design Approach*, Wiley-Blackwell, Oxford (2011).

J.C.J. Bart, N. Palmeri and S. Cavallaro, *Biodiesel Science and Technology. From Soil to Oil.* Woodhead Publishing Ltd/CRC Press, Cambridge/Boca Raton, FL (2010).

A. Matlack, *Introduction to Green Chemistry*, 2nd edn, CRC Press, Boca Raton, FL (2010).

M. Lancaster, *Green Chemistry. An Introductory Text*, RSC Publishing, Cambridge (2010).

R.M. Mortier, M.F. Fox and S.T. Orszulik (eds), *Chemistry and Technology of Lubricants*, 3rd edn, Springer, Heidelberg (2009).

G. Grogan, *Practical Biotransformations: A Beginner's Guide*, Wiley-Blackwell, Oxford (2009).

G. Rothenberg, *Catalysis. Concepts and Green Applications*, Wiley-VCH, Weinheim (2008).

G. Centi and R.A. van Santen (eds), *Catalysis for Renewables*, Wiley-VCH, Weinheim (2007).

J.G. de Vries and C.J. Elsevier (eds), *The Handbook of Homogeneous Hydrogenation*, Wiley-VCH, Weinheim (2006), 3 vols.

G.P. Chiusoli and P. Maitlis, *Metal-catalysis in Industrial Organic Processes. Fundamentals and Applications*, The Royal Society of Chemistry, Cambridge (2006).

R.A. Sheldon, I. Arends and F. van Rantwijk, *Green Chemistry and Catalysis*, Wiley-VCH, Weinheim (2006).

L.R. Rudnick (ed.), *Synthesis, Mineral Oils, and Bio-Based Lubricants: Chemistry and Technology*, CRC Press/Taylor & Francis, Boca Raton, FL/ New York, NY (2006).

A. Igartua, J. Barraga and A. Aranzabe (eds), *Biodegradable Lubricants*, Virtual Tribology Institute Edition, Eibar (2005).

K. Buchholz, V. Kasche and U.T. Bornscheuer, *Biocatalysts and Enzyme Technology*, Wiley-VCH, Weinheim (2005).

R.H. Crabtree, *The Organometallic Chemistry of the Transition Metals*, Wiley-Interscience, New York, NY (2005).

G. Knothe, J. Van Gerpen and J. Krahl (eds), *The Biodiesel Handbook*, AOCS Press, Urbana, IL (2005).

C.H. Hou (ed.), *Handbook of Industrial Biocatalysts*, CRC Press, Boca Raton, FL (2005).

A.S. Bommarius and B.R. Riebel, *Biocatalysis*, Wiley-VCH, Weinheim (2004).

P.W.N.M. van Leeuwen, *Homogeneous Catalysis*, Kluwer Academic Publishers, Dordrecht (2004).

S.Z. Erhan and J.M. Perez (eds), *Biobased Industrial Fluids and Lubricants*, AOCS Press, Champaign, IL (2003).

K. Faber, *Biotransformations in Organic Chemistry*, Springer-Verlag, Berlin (2000).

A. Liese, K. Seelbach and C. Wandrey, *Industrial Biotransformations*, Wiley-VCH, Weinheim (2000).

U.T. Bornscheuer (ed.), *Enzymes in Lipid Modification*, Wiley-VCH, Weinheim, New York, NY (2000).

J. Hagen, *Industrial Catalysis*, Wiley-VCH, Weinheim (1999).

L.R. Rudnick and R.L. Shubkin (eds), *Synthetic Lubricants and High-performance Functional Fluids,* 2nd edn, Marcel Dekker, Inc., New York, NY (1999).

G. Knothe and J.T.P. Derksen (eds), *Recent Developments in the Synthesis of Fatty Acid Derivatives*, AOCS Press, Urbana, IL (1999).
P.T. Anastas and J.C. Warner, *Green Chemistry: Theory and Practice*, Oxford University Press, New York, NY (1998).
H. Buchholz and V. Kasche, *Biokatalysatoren und Enzymtechnologie*, VCH, New York, NY (1997).
M. Sinott (ed.), *Comprehensive Biological Catalysis*, Academic Press, London (1997).
T. Bugg, *An Introduction to Enzyme and Coenzyme Chemistry*, Blackwell Science, Oxford (1997).
F.D. Gunstone and F.B. Padley (eds), *Lipid Technologies and Applications*, CRC Press, Boca Raton, FL (1997).
P.R. Ortiz de Montellano (ed.), *Cytochrome P450: Structure, Mechanism and Biochemistry*, 2nd edn, Plenum Press, New York, NY (1995).
A. Dorinson and K.C. Ludema, *Mechanics and Chemistry in Lubrication*, Elsevier, Amsterdam (1985).

6.6 References

[1] C. Alfos, Lipochemistry and innovations, *Oléag., Corps Gras, Lipides* **10**(5–6) (2003) 344–7.
[2] G. Knothe and J.T.P. Derksen (eds), *Recent Developments in the Synthesis of Fatty Acid Derivatives*, AOCS Press, Champaign, IL (1999).
[3] D.R. Kodali, New developments in seed oils and their industrial applications, *Green Tech. 2002 – 5th European Symposium on Industrial Crops and Products*, Floriade/Amsterdam (24–26 April 2002).
[4] H. Baumann, M. Bühler, H. Fochem, F. Hirsinger, H. Zoebelein and J. Falbe, Natürliche Fette und Öle – nachwachsende Rohstoffe für die chemische Industrie, *Angew. Chem.* **100** (1988) 41–62; *Angew. Chem. Int. Ed.* **27** (1988) 41–62.
[5] S.Z. Erhan and S. Asadauskas, Lubricant basestocks from vegetable oils, *Ind. Crops Prod.* **11** (2000) 277–82.
[6] S.Z. Erhan, A. Adhvaryu and B.K. Sharma, Chemically functionalized vegetable oils, in *Synthetics, Mineral Oils and Bio-Based Lubricants: Chemistry and Technology* (L.R. Rudnick, ed.), CRC Press/Taylor & Francis, Boca Raton, FL /New York, NY (2006), pp. 361–88.
[7] H. Wagner, R. Luther and Th. Mang, Lubricant base fluids based on renewable materials: their catalytic manufacture and modification, *Appl. Catal. A: General* **221** (2001) 429–42.
[8] N. Dörmő, K. Bélafi-Bakó, L. Bartha, U. Ehrenstein and L. Gubicza, Manufacture of an environmental-safe biolubricant from fusel oil by enzymatic esterification in solvent-free system, *Biochem. Engng. J.* **21** (2004) 229–34.
[9] W.G. Dudek and A.H. Popkin, Dibasic acid esters, in *Synthetic Lubricants* (R.C. Gunderson and A.W. Hart, eds), Reinhold, New York, NY (1962), pp. 151–245.

[10] O. Falk and R. Meyer-Pittroff, Biologisch schnell abbaubares Schmierstoffgrundöl, Germ. Gebrauchsmusterschrift No. DE 20,305,164 U1 (9 October 2003).
[11] V. Eychenne and Z. Mouloungui, Relationships between structure and lubricating properties of neopentyl polyol esters, *I&EC Res.* **37** (1998) 4836–44.
[12] Th. Mang, Lubricants, in *Lipid Technologies and Applications* (F.D. Gunstone and F.B. Padley, eds), Marcel Dekker, Inc., New York, NY (1997), pp. 737–58.
[13] K. Li, X. Wang, L. Shan, Q. Jin, X. Niu and Y. Liu, Synthesis of pentaerythritol oleate, *Zhongguo Youzhi* **32** (2007) 53–6.
[14] D.R. Kodali and S.C. Nivens (to Cargill, Inc.), Oils with heterogeneous chain lengths, US Patent No. 6,465,401 B1 (15 October 2002).
[15] P. Bondioli, L. Della Bella and A. Manglavati, Synthesis of biolubricants with high viscosity and high oxidation stability, *Oléag., Corps Gras, Lipides* **10** (2003) 150–4.
[16] E. Zehler and C. Costello (to Cognis Corp.), Complex polyol esters with improved performance, PCT Int. Publ. No. WO 2005/019395 A1 (3 March 2005).
[17] F.J. Flider (to Calgene Chemical, Inc.), Polyglycerol esters as functional fluids and functional fluid modifiers, US Patent No. 5,380,469 (10 January 1995).
[18] A. Willing, Oleochemical esters. Environmentally compatible raw materials for oils and lubricants from renewable resources, *Fett/Lipid* **101** (1999) 192–8.
[19] T.A. Isbell, R. Kleiman and B.A. Plattner, Acid-catalyzed condensation of oleic acid into estolides and polyestolides, *J. Am. Oil Chem. Soc.* **71** (1994) 169–74.
[20] S.C. Cermak and T.A. Isbell (to USDA), Biodegradable oleic estolide ester having saturated fatty acid end group useful as lubricant base stock, US Patent No. 6,316,649 B1 (13 November 2001).
[21] S.C. Cermak and T.A. Isbell, Synthesis of estolides from oleic and saturated fatty acids, *J. Am. Oil Chem. Soc.* **78** (2001) 557–65.
[22] S.C. Cermak and T.A. Isbell, Synthesis and physical properties of estolide-based functional fluids, *Ind. Crops Prod.* **18** (2003) 183–96.
[23] T.A. Isbell and R. Kleiman, Mineral acid-catalysed condensation of meadowfoam fatty acids into estolides, *J. Am. Oil Chem. Soc.* **73** (1996) 1097–107.
[24] T.A. Isbell, T.P. Abbott, S. Asaudaskas and J.E. Lohr Jr. (to USDA), Biodegradable oleic estolide ester base stocks and lubricants, US Patent No. 6,018,063 (25 January 2000).
[25] L.E. Johansson and S.T. Lundin, Copper catalysts in the selective hydrogenation of soybean and rapeseed oils, 1. Activity of the copper chromite catalyst, *J. Am. Oil Chem. Soc.* **56** (1979) 974–80.
[26] A. Behr, N. Döring, S. Durowicz-Heil, B. Ellenberg, C. Kozik, C. Lohr and H. Schmidke, Selective hydrogenation of multi-unsaturated fatty acids in the liquid phase, *Fat Sci. Technol.* **95** (1993) 2–12.
[27] A. Behr, Homogeneous transition metal catalysis in oleochemistry, *Fat Sci. Technol.* **92** (1990) 375–88.
[28] B. Fell and W. Schäfer, Selective hydrogenation of fats and derivatives using Ziegler-type organometallic catalysts. 1. Selective hydrogenation of methyl-linoleate and other dienic compounds with isolated double bonds, *Fat Sci. Technol.* **92** (1990) 264–72.

[29] A.J. Dijkstra, Revisiting the formation of *trans* isomers during partial hydrogenation of triacylglycerol oils, *Eur. J. Lipid Sci. Technol.* **108** (2006) 249–64.
[30] L.A.T. Honary, Soybean based hydraulic fluid, US Patent No. 5,972,855 A (26 October 1999).
[31] G. Papadogianakis, A. Bouriazos, K. Mouratidis and N. Psaroudakis (to Cognis), Process for the partial hydrogenation of fatty acid esters, PCT Int. Publ. No. WO 2008/043454 A1 (17 April 2008).
[32] D.R. Kodali (to Cargill, Inc.), Process for modifying unsaturated fatty acid ester stocks, resulting products and uses thereof, PCT Int. Publ. No. WO 98/47855 A1 (29 October 1998).
[33] P.S. Landis (to International Lubricants, Inc.), Telomerised vegetable oil for lubricant additives, PCT Int. Publ. No. WO 92/07051 (30 April 1992).
[34] A. Laufenberg and A. Behr (to Henkel KGaA), Olefinisch ungesättigte Addukte von Ethylen an mehrfach ungesättigte Fettsäuren bzw. Fettsäureester, Verfahren zu ihrer Herstellung und ihre Verwendung, German Patent No. DE 4,002,012 A1 (24 January 1990).
[35] A. Behr and A. Laufenberg, Synthesis of new branched fatty acids by rhodium catalyzed homogeneous oligomerization (in German), *Fat Sci. Technol.* **93**(1) (1991) 20–4.
[36] E.H. Pryde, Hydroformylation of unsaturated fatty acids, *J. Am. Oil Chem. Soc.* **61** (1984) 419–25.
[37] E.N. Frankel and E.H. Pryde, Catalytic hydroformylation and hydroxycarboxylation of unsaturated fatty compounds, *J. Am. Oil Chem. Soc.* **54** (1977) A873–81.
[38] Z. Xia, U. Kloeckner and B. Fell, Hydroformylation of mono- and multiple unsaturated fatty substances with heterogenized cobaltcarbonyl and rhodiumcarbonyl catalysts (in German), *Fett/Lipid* **98** (1996) 313–21.
[39] A. Guo, D. Demydov, W. Zhang and Z.S. Petrović, Polyols and polyurethanes from hydroformylation of soybean oil, *J. Polym. Environm.* **10** (2002) 49–52.
[40] E.N. Frankel (to USDA), Selective hydroformylation of unsaturated fatty compounds, US Patent No. 3,787,459 (22 January 1974).
[41] E.N. Frankel, F.L. Thomas and W.L. Rohwedder, Hydroformylation of methyl linoleate and linolenate with rhodium-triphenylphosphine catalyst, *I&EC Prod. Res. Dev.* **12** (1973) 47–53.
[42] U. Biermann and J.O. Metzger, Friedel–Crafts alkylation of alkenes: ethylaluminium-sesquichloride induced alkylations with alkyl chloroformates, *Angew. Chem. Int. Ed.* **38** (1999) 3675–7.
[43] A. Behr, M. Fience, C. Buβ and P. Eilbracht, Hydroaminomethylation of fatty acids and secondary amines: a new route to interesting surfactant substrates, *Eur. J. Lipid Sci. Technol.* **102** (2000) 467–71.
[44] J.O. Metzger and U. Biermann, Alkylaluminium dichloride induced Friedel–Crafts acylation of unsaturated acids and alcohols, *Liebigs Ann. Chem.* (1993) 645–50.
[45] U. Biermann and J.O. Metzger, Reaction of unsaturated fatty compounds, in *Recent Developments in the Synthesis of Fatty Acid Derivatives* (G. Knothe and J.T.P. Derksen, eds), AOCS Press, Champaign, IL (1999), pp. 80–9.

[46] U. Biermann and J.O. Metzger, Lewis acid catalyzed additions to unsaturated fatty compounds II. Alkylaluminium halide catalyzed ene reactions of unsaturated fatty compounds and formaldehyde, *Fat Sci. Technol.* **93** (1991) 282–4.

[47] U. Biermann, W. Friedt, S. Lang, W. Lühs, G. Machmüller, J.O. Metzger, M. Rüsch gen. Klaas, H.J. Schäfer and M.P. Schneider, New syntheses with oils and fats as renewable raw materials for the chemical industry, *Angew. Chem. Intl. Ed.* **39** (2000) 2206–24.

[48] J.O. Metzger, U. Biermann and R. Mahler, Neuartige funktionalisierte und verzweigte Fettstoffe, in *Perspektiven nachwachsender Rohstoffe in der Chemie* (H. Eierdanz, ed.), VCH, Weinheim (1996), pp. 255–8.

[49] J.O. Metzger and U. Riedner, Free radical additions to unsaturated fatty acids, *Fat Sci. Technol.* **91** (1989) 18–23 (in German).

[50] J.O. Metzger and U. Linker, New results on free radical additions to unsaturated fatty compounds, *Fat Sci. Technol.* **93** (1991) 244–9.

[51] J.O. Metzger, R. Mahler, G. Francke and A. Hayen, Synthesis of new oleochemicals: Functionalization of fatty compounds using radical reactions, in *Recent Developments in the Synthesis of Fatty Acid Derivatives* (G. Knothe and J.T.P. Derksen, eds), AOCS Press, Champaign, IL (1999), pp. 90–9.

[52] U. Keller, J. Fischer, W.F. Hölderich, H. Murrenhoff, M. Schmidt, W. Dott, A. Eisenträger, S. Hahn and G. Maxam, New lubricants from renewable resources: ecotoxicities and oxidative characteristics, *Ölhydr. Pneum.* **4** (2000) 240–5.

[53] U. Keller, W.F. Hölderich, M. Schmidt and H. Murrenhoff, New environmentally friendly lubricants by chemical modification of oleochemicals, *Agro-Food-Industry Hi-Tech* **12**(4) (2001) 17–21.

[54] W.M. Weigert, in *Ullmann's Enzyklopädie der Industriellen Chemie* (E. Bartholomé, E. Biekert, H. Hellmann, H. Ley and W. Weigert, eds), 4th edn, Verlag Chemie, Weinheim (1975), Vol. 9, p. 144.

[55] J.W. Lambert and D.L. Johnson (to Agro Management Group, Inc.), Vegetable oil lubricants for internal combustion engines and total loss lubrication, US Patent No. 5,888,947 (30 March 1999).

[56] D.R. Kodali and K. Li (to Cargill, Inc.), Process for modifying unsaturated triacylglycerol oils, resulting products and uses thereof, US Patent No. 6,291,409 B1 (18 September 2001).

[57] S. Warwel, P. Bavaj, M. Rüsch gen. Klaas and B. Wolff, Polymerbausteine aus Pfanzenölen durch katalytische Reaktionen, in *Perspektiven nachwachsender Rohstoffe in der Chemie* (H. Eierdanz, ed.), VCH, Weinheim (1996), pp. 119–36.

[58] A. Zlatanić, Z.S. Petrović and K. Dušek, Structure and properties of triolein-based polyurethane networks, *Biomacromol.* **3** (2002) 1048–56.

[59] A. Rybak, P.A. Fokou and M.A.R. Meier, Metathesis as a versatile tool in oleochemistry, *Eur. J. Lipid Sci. Technol.* **110** (2008) 797–804.

[60] J.C. Mol, Catalytic metathesis of unsaturated fatty acid esters and oils, *Topics Catal.* **27** (2004) 97–104.

[61] P.B. Van Dam, M.C. Mittelmeijer and C. Boelhouwer, Metathesis of unsaturated fatty acid esters by a homogeneous tungsten hexachloride-tetramethyltin catalyst, *J.C.S. Chem. Commun.* **78** (1972) 1221–2.

[62] M. Scholl, T.M. Trnka, J.P. Morgan and R.H. Grubbs, Increased ring closing metathesis activity of ruthenium-based olefin metathesis catalysts coordinated with imidazolin-2-ylidene ligands, *Tetrah. Lett.* **40** (1999) 2247–50.
[63] J.C. Mol, Metathesis of unsaturated fatty acid esters and fatty oils, *J. Mol. Catal.* **90** (1994) 185–99.
[64] S.J. Miller (to Chevron), Production of biofuels and biolubricants from common feedstock, US Patent Appl. No. 2009/0084026 A1 (2 April 2009).
[65] P.S. Lathi and B. Mattiasson, Green approach for the preparation of biodegradable lubricant base stock from epoxidized vegetable oil, *Appl. Catal. B: Environm.* **69** (2007) 207–12.
[66] A. Westfechtel and W. Giede (to Cognis Deutschland GmbH), Method for producing polyolesters, PCT Int. Publ. No. WO 99/57092 (11 November 1999).
[67] M. Rüsch gen. Klaas and S. Warwel, Chemoenzymatic epoxidation of unsaturated fatty acid esters and plant oils, *J. Am. Oil Chem. Soc.* **73** (1996) 1453–7.
[68] M. Rüsch gen. Klaas and S. Warwel, Complete and partial epoxidation of plant oils by lipase-catalyzed perhydrolysis, *Ind. Crops Prod.* **9** (1999) 125–32.
[69] X. Wu, X. Chang, S. Yang, H. Chen and D. Wang, The study of epoxidized rapeseed oil used as a potential biodegradable lubricant, *J. Am. Oil Chem. Soc.* **77** (2000) 561–3.
[70] H.-S. Wang and S.Z. Erhan, Lubricant base stocks from modified soybean oil, in *Biobased Industrial Fluids and Lubricants* (S.Z. Erhan and J.M. Perez, eds), AOCS Press, Champaign, IL (2002) pp. 20–34.
[71] A. Adhvaryu and S.Z. Erhan, Epoxidized soybean oil as a potential source of high temperature lubricants, *Ind. Crops Prod.* **15** (2002) 247–54.
[72] S. Sinadinović-Fišer, M. Janković and Z.S. Petrović, Kinetics of *in situ* epoxidation of soybean oil in bulk catalyzed by ion exchange resin, *J. Am. Oil Chem. Soc.* **78** (2001) 725–31.
[73] Z.S. Petrović, A. Zlatanić, C.C. Lava and S. Sinadinović- Fišer, Epoxidation of soybean oil in toluene with peroxoacetic and peroxoformic acids – kinetics and side reactions, *Eur. J. Lipid Sci. Technol.* **104** (2002) 293–9.
[74] S. Warwel and M. Rüsch gen. Klaas, Chemo-enzymatic epoxidation of unsaturated carboxylic acids, *J. Mol. Cat. B: Enzymatic* **1** (1995) 29–35.
[75] S. Watanabe, T. Fujita and M. Sakamota, Epoxides of unsaturated fatty acids as antirust additives in water based cutting fluids, *J. Am. Oil Chem. Soc.* **65** (1988) 1311–12.
[76] D. Tao, H.L. Zhu and Z.M. Hu, Free chlorine fixation and the protective film in boundary lubrication, *Lubric. Sci.* **8** (1996) 397–407.
[77] A. Pavlovicová and J. Cvengroš, Lubricants based on vegetable oils, *Petrol. Coal* **41**(2) (1999) 99–102.
[78] I. Cvetković, J. Milić, M. Ionescu and Z.S. Petrović, Preparation of 9-hydroxynonanoic acid methyl ester by ozonolysis of vegetable oils and its polycondensation, *Hem. Ind.* **62**(6) (2008) 319–28.
[79] Z.S. Petrović, W. Zhang and I. Javni, Structure and properties of polyurethanes prepared from triglyceride polyols by ozonolysis, *Biomacromol.* **6** (2005) 713–19.
[80] D. Graiver, P. Tran, L. Patrick, K. Farminer and R. Narayan, Modifications of soybean oil using novel ozone-based chemistry, *ACS Symp. Ser.* **939** (2006) 76–100.

[81] M.J. Diamond, R.G. Binder and T.H. Applewhite, Alkaline cleavage of hydroxy unsaturated fatty acids. I. Ricinoleic acid and lesquerolic acid, *J. Am. Oil Chem. Soc.* **42** (1965) 882–4.
[82] H. Mutlu and M.A.R. Meier, Castor oil as a renewable resource for the chemical industry, *Eur. J. Lipid Sci. Technol.* **112** (2010) 10–30.
[83] W. Lühs, K.J. Dehmer, R. Bergmann and W. Friedt, Molekular-genetische Methoden – Werkzeuge für die Züchtung von Ölpflanzen als Quelle nachwachsender oleochemische Rohstoffe, in *Perspektiven nachwachsender Rohstoffe in der Chemie* (H. Eierdanz, ed.), VCH, Weinheim (1996), pp. 232–9.
[84] W. Zorn, Esters as lubricants, *U.S. Air Force Trans. Rep.* F-T-S957RE (1946–7).
[85] G. van der Waal and D. Kenbeek, Testing, application and future development of environmentally friendly ester base fluids, *J. Synth. Lubr.* **10**(1) (1993) 67–83.
[86] S. Busch and J. Röder (to Cognis), Esters and mixtures thereof and their application as lubricants or in hydraulic oils, Eur. Patent No. EP 1,958,932 A1 (20 August 2008).
[87] C. Ratledge, Microbial oxidations of fatty alcohols and fatty acids, *J. Chem. Technol. Biotechnol.* **55** (1992) 399–400.
[88] N. Dörmő, D. Bucsu, S. Kovacs, B. Belafine and L. Gubicza, Preparation of an environmentally acceptable biolubricant by enzymatic esterification based on a fatty acid in a solvent-free system, *Olaj, Szappan, Kozm.* **53** (2004) 105–9.
[89] C.G. Laudani, M. Habulin, Ž. Knez, G. Della Porta and E. Reverchon, Immobilized lipase-mediated long-chain fatty acid esterification in dense carbon dioxide: bench-scale packed-bed reactor study, *J. Supercrit. Fuids* **41** (2007) 74–81.
[90] B.R. Moser, B.K. Sharma, K.M. Doll and S.Z. Erhan, Diesters from oleic acid: synthesis, low temperature properties and oxidation stability, *J. Am. Oil Chem. Soc.* **84** (2007) 675–80.
[91] M. Schneider, Plant-oil-based lubricants and hydraulic fluids, *J. Sci. Food Agric.* **86** (2006) 1769–80.
[92] D.C. Kenneth, V.E. Johns, R.E. Perkins Jr and E.L. Huffman, Brassylic acid: chemical intermediate from high erucic oils, *I&EC Prod. Res. Dev.* **16** (1977) 95–100.
[93] R.G. Kadesch, Fat based dibasic acids, *J. Am. Oil Chem. Soc.* **56** (1979) 845A-9A.
[94] J.C.J. Bart, N. Palmeri and S. Cavallaro, *Biodiesel Science and Technology. From Soil to Oil*, Woodhead Publishing Ltd/CRC Press, Cambridge/Boca Raton, FL (2010).
[95] J.-P. Despeghel (to Monsanto S.A.S.), Use of rapeseed oil in biolubricants, PCT Int. Publ. No. WO 2007/034336 A3 (29 March 2007).
[96] D.C. Drown, K. Harper and E. Frame, Screening vegetable oil alcohol esters as fuel lubricity enhancers, *J. Am. Oil Chem. Soc.* **78** (2001) 579–84.
[97] V.V. Bokade and G.D. Yadov, Synthesis of biodiesel and biolubricant by transesterification of vegetable oil with lower and higher alcohols over heteropolyacids supported by clay (K-10) *Process Safety Environm. Prot.* **85** (2007) 372–7.

[98] P.S. Sreeprasanth, R. Srivastava, D. Srinivas and P. Ratnasamy, Hydrophobic, solid acid catalysts for production of biofuels and lubricants, *Appl. Catal. A: General* **314** (2006) 148–59.
[99] V.K. Chhibber, A.K. Gupta, O.N. Anand, J. Mehta, M. Pal and I.P. Kaur (to CSIR), A process for the preparation of vegetable oil based biodegradable liquid lubricants, Ind. Patent No. IN 2004-DE 00362 A (5 March 2004).
[100] H. Wichmann, T. Sahlabji, M. Ohnesorge, R. Vogt and M. Bahadir, Feasibility study on membrane-aided cleanup and fractionation of fatty acid esters produced from waste fats, *Clean* **36** (10–11) (2008) 840–4.
[101] H. Cramail, A. Boyer, E. Cloutet and C. Alfos, Castor oil derivatives and method for their preparation, Fr. Patent No. FR 2,950,052 A1 (18 March 2011); PCT Int. Publ. No. WO 2011/030075 A1 (17 March 2011).
[102] M. Nakazato and T. Murata (to Nisshin Oil Mills), Antifreezing oil or fat and its production, Jap. Patent No. JP 7179882 A (18 July 1995).
[103] S. Nagaoka and M. Ibuki (to Fuji Oil Co.), Biodegradable lubricant base oil and its manufacturing process, US Patent No. 6,117,827 (12 September 2000).
[104] S. Kusakawa, Y. Itoh, S. Morooka, K. Komazawa and K. Ueda (to Itoh Seiya KK and Sanyu Resin UK), Curable urethane composition, US Patent No. 4,603,188 (29 July 1986).
[105] H. Cramail, A. Boyer, E. Cloutet and C. Alfos, Novel method for preparing polyols by transesterification and thiolation, Fr. Patent No. FR 2,951,168 A1 (15 April 2011); PCT Int. Publ. No. WO 2011/045536 A1 (21 April 2011).
[106] E. Uosukainen, Y.-Y. Linko, M. Lämsä, T. Tervakangas and P. Linko, Transesterification of trimethylolpropane and rapeseed oil methyl ester to environmentally acceptable lubricants, *J. Am. Oil Chem. Soc.* **75** (1998) 1557–63.
[107] V. Eychenne, Z. Mouloungui and A. Gaset, Total and partial erucate of pentaerythrytol. Infrared spectroscopy study of relationship between structure, reactivity and thermal properties, *J. Am. Oil Chem. Soc.* **75** (1998) 293–9.
[108] S. Gryglewicz, W. Piechocki and G. Gryglewicz, Preparation of polyol esters based on vegetable and animal fats, *Bioresour. Technol.* **87** (2003) 35–9.
[109] T.G. Smith, Neopentyl polyol esters, in *Synthetic Lubricants* (R.C. Gunderson and A.W. Hart, eds), Reinhold, New York, NY (1962), pp. 388–401.
[110] D.R. Kodali, Biobased lubricants: chemical modification of vegetable oils, *Inform* **14** (2003) 121–3.
[111] T.I. Mohd. Ghazi, M.F.M. Gunam Resul and A. Idris, Bioenergy II: Production of biodegradable lubricant from *Jatropha curcas* and trimethylolpropane, *Intl. J. Chem. Reactor Engng.* **7** (2009) Art. A68.
[112] S. Asadauskas and S.Z. Erhan, Depression of pour points of vegetable oils by blending with diluents for biodegradable lubricants, *J. Am. Oil Chem. Soc.* **76** (1999) 313–16.
[113] R.P.S. Bisht, S. Kaul, P. Nagendramma, V.K. Bhatia and A.K. Gupta, Eco-friendly base fluids for lubricant oil formulations, *Synth. Lubr.* **19** (2002) 243–8.
[114] J.M. Kurosky and Z.A. Tahir, Biodegradable lubricants, US Patent Appl. No. 2004/235679 A1 (25 November 2004).
[115] D.R. Kodali and S.C. Nivens (to Cargill, Inc.), Transesterified oils, US Patent No. 6,278,006 B1 (21 August 2001).
[116] D.R. Kodali, Z. Fan and L.R. DeBonte (to Cargill, Inc.), Biodegradable high oxidative stability oils, US Patent No. 6,281,375 B1 (28 August 2001).

[117] J. Thész, B.L. Boros and Z. Király, Use of fuels or fuel additives based on triglycerides of modified structure and process for their preparation, PCT Int. Publ. No. WO 2008/096187 A1 (14 August 2008).
[118] (a) D.G. Hayes, Biocatalytic synthesis of ricinoleic acid star polymers: 'Green' manufacturing of potentially valuable lubricant additives and drug delivery materials, *ACS Symp. Ser.* 939 (2006) 126–39; (b) A.R. Kelley and D.G. Hayes, Lipase-catalysed synthesis of polyhydric alcohol-poly(ricinoleic acid) ester star polymers, *J. Appl. Polym. Sci.* **101** (2006) 1646–56.
[119] S.J. Metro and D.D. Carr (to Exxon Research & Engineering Co.), Synthetic ester lubricating oil composition containing particular *t*-butylphenyl substituted phosphates and stabilized hydrolytically with particular long chain alkyl amines, US Patent No. 4,440,657 (3 April 1984).
[120] D.W. Turner, G.A. Knudsen, R.H. Schlosberg, C.B. Duncan, J.W. Chu, J.S. Szobota, H.S. Aldrich, C.P. Fitzpatrick, J.M. Brown and K.E. Edwards (to Exxon Chemical Co.), Soluble complex alcohol ester compounds and compositions, PCT Int. Publ. No. WO 00/29521 (25 May 2000).
[121] P.M. Kerschner and B.W. Greenwald (to Cities Service R&D Co.), Method of preparing complex diesters of a dibasic acid with a diol and an excess of a monohydric alcohol, US Patent No. 3,086,044 (16 April 1983).
[122] H.D. Kluge, H. Krug and P.A. Cantor (to Texaco, Inc.), Complex ester lubricant, US Patent No. 3,377,377 (9 April 1968).
[123] C.B. Duncan, P.R. Geissler, D.W. Turner, W.J. Munley Jr. and M.A. Krevalis (to Exxon Chemical Co.), High viscosity complex alcohol esters, US Patent No. 5,750,750 (12 May 1998).
[124] R.J. Sturwold (to Cincinnati-Vulcan Co.), Metalworking lubricants derived from natural fats and oils, US Patent No. 4,885,104 (5 December 1989).
[125] R. Ogake, M. Okada, H. Takashima, T. Shimizu, N. Yamamoto and H. Uenoyama (to Nippon Oil Co., Ltd), Greasy oil and fat composition for food processing machines, US Patent No. 5,185,091 (9 February 1993).
[126] J. Steber, C.-H. Herold and J.M. Limia, Comparative evaluation of anaerobic biodegradability of hydrocarbons and fatty derivatives currently used as drilling fluids, *Chemosphere* **31** (1995) 3105–18.
[127] B. Kolwzan and S. Gryglewicz, Synthesis and biodegradability of some adipic and sebacic esters, *J. Synth. Lubr.* **20** (2003) 99–107.
[128] S. Inaya, H. Sawada, Y. Kobayashi and T. Hagihara (to Kao Corp.), Biodegradable lubricating base oil, lubricating oil composition containing the same and the use thereof, US Patent No. 5,916,854 (29 June 1999).
[129] D.E. Weller Jr. and J.M. Perez, A study of the effect of chemical structure on friction and wear, Part I. Synthetic ester base fluids, *Lubr. Engng.* **56**(11) (2000) 39–44.
[130] C.B. Duncan and L.K. Meade (to Exxon Chemical Co.), Biodegradable branched synthetic ester base stocks and lubricants formed therefrom, US Patent No. 5,817,607 (6 October 1998).
[131] T. Konishi and J.M. Perez, Properties of polyol esters – Lubrication of an aluminum silicon alloy, *Tribol. Trans.* **40**(3) (1997) 500–506.
[132] N.S. Battersby, The biodegradability and microbial toxicity testing of lubricants – some recommendations, *Chemosphere* **41** (2000) 1011–27.

[133] V. Eychenne, Z. Mouloungui and A. Gaset, Neopentylpolyol ester-based lubricants, *Oléag., Corps Gras, Lipides* **3** (1996) 57–63.

[134] S.J. Randles, Esters, in *Synthetics, Mineral Oils and Bio-Based Lubricants: Chemistry and Tribology* (L.R. Rudnick, ed.), CRC Press/Taylor & Francis, Boca Raton, FL/New York, NY (2006), pp. 47–74.

[135] D.R. Kodali, High performance ester lubricants from natural oils, *Ind. Lubr. Tribol.* **54** (2002) 165–70.

[136] R.D. Plattner, K. Payne-Wahl, L.W. Tjarks and R. Kleiman, Hydroxy acids and estolide triglycerides of *Heliophila amplexicaulis* L.f. seed oil, *Lipids* **14** (1979) 576–9.

[137] K. Payne-Wahl and R. Kleiman, Quantitation of estolide triglycerides in *Sapium* seeds by high performance liquid chromatography with infrared detection, *J. Am. Oil Chem. Soc.* **60** (1983) 1011–12.

[138] R.V. Madrigal and C.R. Smith, Estolide triglycerides of *Trewia nudiflora* seed oil, *Lipids* **17** (1982) 650–5.

[139] D.G. Hayes and R. Kleiman, Lipase-catalyzed synthesis and properties of estolides and their esters, *J. Am. Oil Chem. Soc.* **72** (1995) 1309–16.

[140] S. Okumura, M. Iwai and Y. Tsujisaka, Synthesis of estolides during hydrolysis of castor oil by *Geotrichum candidum* lipase, *Yakagaku* **32** (1983) 271–3.

[141] A. Govindapillai, N.H. Jayadas and M. Bhasi, Analysis of the pour point of coconut oil as a lubricant base stock using differential scanning calorimetry, *Lubric. Sci.* **21**(1) (2009) 13–26.

[142] S.C. Cermak and T.A. Isbell, Physical properties of saturated estolides and their 2-ethylhexyl esters, *Ind. Crops Prod.* **16** (2002) 119–27.

[143] T.A. Isbell, M.R. Edgcomb and B.A. Lowery, Physical properties of estolides and their ester derivatives, *Ind. Crops Prod.* **13** (2001) 11–20.

[144] S.C. Cermak and T.A. Isbell, Improved oxidative stability of estolide esters, *Ind. Crops Prod.* **18** (2003) 223–30.

[145] S.M. Erhan and R. Kleiman, Biodegradation of estolides, *J. Am. Oil Chem. Soc.* **74** (1997) 605–7.

[146] L.A. García-Zapateiro, M.A. Delgado, J.M. Franco, C. Valencia, M.V. Ruiz-Méndez, R. Garcés and C. Gallegos, Oleins as a source of estolides for biolubricant applications, *Grasas Aceites* **61** (2010) 171–4.

[147] L.A. García-Zapateiro, J.M. Franco, C. Valencia and C. Gallegos, Polyestolides as viscosity modifiers of vegetable oils for lubricant applications, in *Rheology in Product Design and Engineering* (A. Guerrero, J. Muñoz and J.M. Franco, eds), *Ibereo 2008*, Madrid (11–12 September 2008).

[148] S.C. Cermak, K.B. Brandon and T.A. Isbell, Synthesis and physical properties of estolides from lesquerella and castor fatty acid esters, *Ind. Crops Prod.* **23** (2006) 54–64.

[149] S.N. Modak and J.G. Kane, Studies in estolides. I. Kinetics of estolide formation and decomposition, *J. Am. Oil Chem. Soc.* **42** (1965) 428.

[150] R. Neissner, TLC separation of estolides of castor oil fatty acids, *Fette Seifen Anstrichm.* **82** (1980) 183–6.

[151] C. Yamaguchi, M. Akita, S. Asaoka and F. Osada (to Itoh Oil), Production of castor oil fatty acid estolide, Jap. Patent No. JP 1,016,591 A (20 January 1989).

[152] T.A. Isbell and S.C. Cermak, Synthesis of triglyceride estolides from lesquerella and castor oils, *J. Am. Oil Chem. Soc.* **79** (2002) 1227–33.

[153] S.S. Lawate (to Lubrizol Corp.), Triglyceride oils thickened with estolides of hydroxy-containing triglycerides, US Patent No. 5,427,704 (27 June 1985).
[154] D.G. Hayes and R. Kleiman, Lipase-catalyzed synthesis and properties of estolides and their esters, *J. Am. Oil Chem. Soc.* **72** (1995) 1309–16.
[155] K.T. Achaya, Chemical derivatives of castor oil, *J. Am. Oil Chem. Soc.* **48** (1971) 758–63.
[156] I. Ahmad, Oleochemicals from isoricinoleic acid (*Wrightia tinctoria* seed oil), *I&EC Res.* **47** (2008) 2091–5.
[157] M.W. Raynor, K.D. Bartle, A.A. Clifford, J.M. Chalmers, T. Katase, C.A. Rouse, K.E. Markides and M.L. Lee, Analysis of aliphatic and phenol carboxylic acids by capillary supercritical fluid chromatography-Fourier transform infrared microspectrometry, *J. Chromatogr.* **505** (1990) 179–90.
[158] A.T. Erciyes, L. Dandik and F.S. Erkal, The decomposition of secondary esters of castor oil with fatty acids, *J. Am. Oil Chem. Soc.* **68** (1991) 642–5.
[159] D.A. Burg, R. Kleiman and S.M. Erhan (to USDA), Production of hydroxy acids and estolide intermediates, US Patent No. 5,380,894 (10 January 1995).
[160] S.S. Lawate (to Lubrizol Corp.), Estolides of hydroxy-containing triglycerides that contain a performance additive, US Patent Appl. No. 08/188,263 (28 January 1994).
[161] S.S. Lawate (to Lubrizol Corp.), Triglyceride oils thickened with estolides of hydroxy-containing triglycerides, US Patent No. 5,427,704 (27 June 1995).
[162] T. Kusakawa, Y. Ito, Y. Yamada and Y. Taketsuji (to Itoh Oil), Biodegradable lubricants, Jpn. Kokai Tokkyo Koho JP 08027473 A (30 January 1996).
[163] B.H. Zoleski and F.J. Gaetani (to Texaco, Inc.), Low foaming railroad diesel engine lubricating oil composition, US Patent No. 4,428,850 (31 January 1984).
[164] M.W. Balakos and E.E. Hernandez, Catalyst characteristics and performance in edible oil hydrogenation, *Catal. Today* **35** (1997) 415–25.
[165] A. Archerio, *Trans* fatty acids and blood lipids, *Artherosclerosis Suppl.* **7** (2006) 25–7.
[166] C.H. Riesz and H.S. Weber, Catalyst for selective hydrogenation of soybean oil. III. Hydrogenation catalyst prepared on molecular sieves and other supports, *J. Am. Oil Chem. Soc.* **41** (1970) 464–8.
[167] E.S. Jang, M.Y. Jung and D.B. Min, Hydrogenation for low *trans* and high conjugated fatty acids, *Compr. Revs. Food Sci. Food Safety* **4** (2005) 22–30.
[168] C.H. Riesz and H.S. Weber, Catalysts for selective hydrogenation of soybean oil. I. An experimental method for evaluating selectivity, *J. Am. Oil Chem. Soc.* **41** (1970) 380–3.
[169] C.H. Riesz and H.S. Weber, Catalysts for selective hydrogenation of soybean oil. II. Commercial catalysts, *J. Am. Oil Chem. Soc.* **41** (1970) 400–6.
[170] P.F. Bondioli, M.N. Ravasio and F. Zaccheria (to CNR and Università degli Studi di Milano), A method for the production of biodiesel, starting from high iodine number fatty substances, PCT Int. Publ. No. WO 2006/111997 A1 (26 October 2006).
[171] N. Ravasio, Workshop on *Bioraffineria, Carburanti e Materie Prime da Fonti Rinnovabili per uno Sviluppo Sostenibile*, Milano (19 November 2008).
[172] M.N. Ravasio, F. Zaccheria, P.F. Bondioli and L. Della Bella, Selective hydrogenation as a value tool for the production of high quality biodiesel, *Proc. Intl. Congress on Biodiesel*, Vienna (5–7 November 2007).

[173] M.O. Jung, S.H. Yoon and M.Y. Jung, Effect of temperature and agitation rate on the formation of conjugated linoleic acids in soybean oil during hydrogenation process, *J. Agric. Food Chem.* **49** (2001) 3010–16.
[174] M.O. Jung, J.W. Ju, D.S. Choi, S.H. Yoon and M.Y. Jung, CLA formation in oils during hydrogenation process as affected by catalyst types, catalyst contents, hydrogen pressure, and oil species, *J. Am. Oil Chem. Soc.* **79** (2002) 501–10.
[175] P.A. Jacobs, P.J.A. Maes, S.J. Paulussen, M. Tielen, D.F.E. Van Steenkiste and L.K. Van Looveren, Elimination of *trans*-unsaturated fatty acid compounds by selective adsorption with zeolites, PCT Int. Publ. No. WO 98/54275 (3 December 1998); US Patent No. 6,229,032 B1 (8 May 2001).
[176] A.F. Pérez-Cadenas, F. Kapteijn, M.M.P. Zieverink and J.A. Moulijn, Selective hydrogenation of fatty acid methyl esters over palladium on carbon-based monoliths. Structural control of activity and selectivity, *Catal. Today* **128** (2007) 13–17.
[177] T. Boger, M.M.P. Zieverink, M.T. Kreutzer, F. Kapteijn, J.A. Moulijn and W.P. Addiego, Monolithic catalysts as an alternative to slurry systems: hydrogenation of edible oil, *I & EC Res.* **43** (2004) 2337–44.
[178] A.E.W. Beers and G. Mangnus, Hydrogenation of edible oils for reduced *trans*-fatty acid content, *Inform* **15** (2004) 404–5.
[179] L.E. Johansson and S.T. Lundin, Copper catalyst in the selective hydrogenation of soybean and rapeseed oils: I. The activity of the copper chromite catalyst, *J. Am. Oil Chem. Soc.* **56** (1979) 974–80.
[180] A. de Jonge, J.W.E. Coenen and C. Okkerse, Selective hydrogenation of linolenate groups in soybean oil, *Nature* **206** (1965) 573–4.
[181] S. Koritala and H.J. Dutton, Selective hydrogenation of soybean oil. II. Copper–chromium catalysts, *J. Am. Oil Chem. Soc.* **43** (1966) 556–8.
[182] S. Koritala, Hydrogenation of unsaturated fatty esters with copper-chromite catalyst: kinetics, mechanism and isomerization, *J. Am. Oil Chem. Soc.* **47** (1970) 463–6.
[183] C. Okkerse, A. de Jonge, J.W.E. Coenen and A. Rozendaal, Selective hydrogenation of soybean oil in the presence of copper catalysts, *J. Am. Oil Chem. Soc.* **44** (1967) 152–6.
[184] R.T. Sleeter (to Archer Daniels Midland Co.), Copper-chromium catalyzed hydrogenation of polyunsaturated oils, US Patent Appl. No. 2003/0158433 A1 (21 August 2003).
[185] R. Sleeter (to Archer Daniels Midland Co.), Hydrogenation process and high monoene compositions obtained therefro, PCT Int. Publ. No. WO 2008/048495 (24 April 2008).
[186] A. Behr, A. Westfechtel and J. Pérez Gomes, Catalytic processes for the technical use of natural fats and oils, *Chem. Engng. Technol.* **31**(5) (2008) 700–14.
[187] J.W. Veldsink, M.J. Bouma, N.-H. Schöön and A.A.C.M. Beenackers, Heterogeneous hydrogenation of vegetable oils: a literature review, *Catal. Rev. Sci. Engng.* **39** (1997) 253–318.
[188] W.T. Koetsier, Hydrogenation of edible oils. Technology and applications, in *Lipid Technologies and Applications* (F.D. Gunstone and F.B. Padley), Marcel Dekker, Inc., New York, NY (1997), pp. 265–303.

[189] W.E. Farr, Hydrogenation: processing technologies, in *Bailey's Industrial Oil and Fat Products* (F. Shahidi, ed.), 6th edn, John Wiley & Sons, Inc., Hoboken, NJ (2005), Vol. 5, pp. 385–96.
[190] B.R. Moser, M.J. Haas, M.A. Jackson, S.Z. Erhan and G.R. List, Evaluation of partially hydrogenated methyl esters of soybean oil as biodiesel, *Eur. J. Lipid Sci. Technol.* **109** (2007) 17–24.
[191] L.A.T. Honary, An investigation of the use of soybean oil in hydraulic systems, *Bioresour. Technol.* **56** (1996) 41–7.
[192] V.K. Chhibber, A.K. Gupta, O.N. Anand, J. Mehta, M. Pal and A. Sharma (to CSIR), A process for the preparation of lubricants based on hydrogenated vegetable oils, Ind. Patent No. IN 2004-DE 00361 A (5 March 2004).
[193] W.L. Siew, Crystallisation and melting behaviour of palm kernel oil and related products by differential scanning calorimetry, *Eur. J. Lipid Sci. Technol.* **103** (2001) 729–34.
[194] R. Höfer, Anwendungstechnische Aspekte der Verwendung natürlicher Öle und ihrer Derivate in der Polymer-Synthese und -Verarbeitung, in *Perspektiven nachwachsender Rohstoffe in der Chemie* (H. Eierdanz, ed.), VCH, Weinheim (1996), pp. 91–106.
[195] J.B. Siegel, A. Zanghellini, H.M. Lovick, G. Kiss, A.R. Lambert, J.L. St. Clair, J.L. Gallaher, D. Hilvert, M.H. Gelb, B.L. Stoddard, K.N. Houk, F.E. Michael and D. Baker, Computational design of an enzyme catalyst for a stereoselective bimolecular Diels–Alder reaction, *Science* **329** (2010) 309.
[196] I. Lee, L. Johnson and E. Hammond, Use of branched chain esters to reduce the crystallization temperature of biodiesel, *J. Am. Oil Chem. Soc.* **72**(10) (1995) 1155–60.
[197] F.B. Root (to Ellis-Foster Co.), Drying oil products and process of making, US Patent No. 2,374,381 (24 April 1945).
[198] M. Nagakura and K. Yoshitomi, Reaction of maleic anhydride with unsaturated fatty acids in the presence of organic peroxides, *Yukagaku* **21**(2) (1972) 83–91; *C.A.* **76** (1972) 139717b.
[199] P. Tran, K. Seybold, D. Graiver and R. Narayan, Free radical maleation of soybean oil via a single-step process, *J. Am. Oil Chem. Soc.* **82**(3) (2005) 189–94.
[200] A.A. Gripp and D.C. Steinberg, Maleated soybean oil and its multifunctional properties, *Proc. In-Cosmetics Exhibit. & Conf.*, Barcelona (22–24 March 1994), pp. 293–310.
[201] P. Tran, D. Graiver and R. Narayan, Biocomposites synthesized from chemically modified soy oil and bio fillers, *J. Appl. Polym. Sci.* **102**(1) (2006) 69–75.
[202] P.T. Tran, *Engineered Soy Oils for New Value Added Applications*, PhD Thesis, Michigan State University (2005).
[203] R.M. Lange, S.L. Bartley, C.G. Ollinger and J.M. Hogan (to Lubrizol Corp.), Maleated vegetable oils and derivatives, as self-emulsifying lubricants in metalworking, PCT Int. Publ. No. WO 2005/071050 A1 (4 August 2005).
[204] T. Yahiro, T. Murakami, T. Okita, A. Nakajima, Y. Hashiguchi and M. Takai, Maleated fish oil for metalworking lubricating oils, Jpn. Kokai Tokkyo Koho JP 61060793 A (28 March 1986).

[205] T.S. Croft and H.A. Haugen (to Minnesota Mining and Manufacturing Company), Encapsulant compositions for use in signal transmission devices, US Patent No. 4,857,563 (15 August 1989).
[206] D.R. Kodali (to Cargill, Inc.), Linseed oil based copolymer coating composition, US Patent No. 5,693,715 (2 December 1997).
[207] D.R. Kodali (to Cargill, Inc.), Linseed oil based polymeric vehicle, US Patent No. 5,288,805 (22 February 1994).
[208] A.J. O'Lenick Jr. (to Lambent Technologies, Inc.), Telomerized complex ester triglycerides, US Patent No. 5,985,806 A (16 November 1999).
[209] (a) A. Behr and A. Laufenberg, Synthesis of new branched fatty acids by rhodium catalyzed homogeneous oligomerization, *Fat Sci. Technol.* **93** (1991) 20–24 (in German); (b) A. Behr and Q. Miao, A new temperature-dependent solvent system based on polyethylene glycol 1000 and its use in rhodium catalyzed cooligomerization, *J. Mol. Catal. A: Chemical* **222** (2004) 127–32.
[210] A. Laufenberg and A. Behr (to Henkel KGaA), Gesättigte, verweigte Fettsäuren mit 20 bis 28 Kohlenstoffatomen bzw. Ester derselben mit C_1–C_{36}, Alkanolen, Verfahren zu ihrer Herstellung und ihre Verwendung, Germ. Offenl. No. DE 4,002,011 A1 (25 July 1991).
[211] A. Behr, D. Obst and A. Westfechtel, Isomerizing hydroformylation of fatty acid esters. Formation of ω-aldehydes, *Eur. J. Lipid Sci. Technol.* **107** (2005) 213–19.
[212] K. Sato, Y. Kawaragi and M. Takai (to Mitsubishi Kasai Corp.), Process for hydroformylation, US Patent No. 5,177,228 (5 January 1993).
[213] Z. Lysenko, D.L. Morrison, D.A. Babb, D.L. Bunning, C.W. Derstine, J.H. Gilchrist, H.R. Jouett, J.S. Kanel, K.D. Olson, W.-J. Peng, J.D. Phillips, B.M. Roesch, A.W. Sanders, A.K. Schrock and P.J. Thomas (to Dow Chemical), Aldehyde and alcohol compositions derived from seed oils, US Patent Appl. No. 2006/0193802 A1 (31 August 2006).
[214] M.F. Sonnenschein, M.R. Greaves, A.W. Sanders, Z. Lysenko, G.E. Spilman, G.J. Frycek, J.D. Phillips, A.K. Schrock and S.J. Martin (to Dow Global Technologies, Inc.), Capped polyol lubricant composition, PCT Int. Publ. No. WO 2008/124265 A2 (16 October 2008).
[215] D.V. Kinsman, Branched-chain fatty acids, in *Fatty Acids in Industry* (R.W. Johnson and E. Fritz, eds), Marcel Dekker, Inc., New York, NY (1989), pp. 233–76.
[216] J. Hartmanns, K. Klenke and J.O. Metzger, Thermally initiated addition of alkanes to alkenes. II. Addition of cyclohexane to acrylate in a free-radical chain reaction, *Chem. Ber.* **199** (1986) 488–99.
[217] J.O. Metzger and F. Bangert, Ane additions to unsaturated fatty compounds: thermally initiated additions of alkanes to methyl 10-undecenoate, *Fat Sci. Technol.* **97** (1995) 7–9.
[218] J. Müller, V. Manea, S. Quaiser, P. Saling and J.E. Maloney, Lupranol balance – eco-efficiency analysis, *PU Mag.* **5**(4) (2008) 2–5.
[219] W. Oppolzer and V. Snieckus, Intramolekulare En-Reaktionen in der organischen Synthese, *Angew. Chem.* **90** (1978) 506–16.
[220] J. Fischer and W.F. Hölderich, Heterogeneously catalysed synthesis of new 3-alkoxypivalic acids from substituted 1,3-dioxanes, *J. Catal.* **194** (2000) 294–300.

[221] D. Kodali and K. Li (to Cargill, Inc.), Process for modifying unsaturated triacylglycerol oils. Resulting products and uses thereof, US Patent No. 6,051,539 (18 April 2000).

[222] H.E. Simmons and R.D. Smith, A new synthesis of cyclopropanes, *J. Am. Oil Chem. Soc.* **81** (1959) 4256.

[223] R.L. Banks and G.C. Bailey, Olefin disproportionation. A new catalytic process, *I&EC Prod. Res. Dev.* **3** (1964) 170–3.

[224] K.J. Ivin and J.C. Mol, *Olefin Metathesis and Metathesis Polymerization*, Academic Press, London (1997).

[225] H.L. Ngo, K. Jones and T.A. Foglia, Metathesis of unsaturated fatty acids. Synthesis of long-chain unsaturated α,ω-dicarboxylic acids, *J. Am. Oil Chem. Soc.* **83** (2006) 629–34.

[226] A. Rybak and M.A.R. Meier, Cross-metathesis of fatty acid derivatives with methyl acrylate. Renewable raw materials for the chemical industry, *Green Chem.* **9** (2007) 1356–61.

[227] S.Z. Erhan, M.O. Bagby and T.C. Nelsen, Drying properties of metathesized soybean oil, *J. Am. Oil Chem. Soc.* **74** (1997) 703–6.

[228] http://www.materia-inc.com/News/Cargill082005.html

[229] W.A. Herrmann, W. Schattenmann and T. Weskamp (to Degussa AG), Alkylidene complexes of ruthenium containing *N*-heterocyclic carbene ligands; use as highly active, selective catalysts for olefin metathesis, US Patent No. 6,635,768 B1 (21 October 2003).

[230] A. Campanella, C. Fontanini and M.A. Baltanás, High yield epoxidation of fatty acid methyl esters with performic acids generated in-situ, *Chem. Engng. J.* **144** (2008) 466–75.

[231] B.R. Moser and S.Z. Erhan, Branched chain derivatives of alkyl oleates: tribological, rheological, oxidation, and low temperature properties, *Fuel* **87** (2008) 2253–7.

[232] B.R. Moser, M.J. Haas, J.K. Winkler, M.A. Jackson, S.Z. Erhan and G.R. List, Evaluation of partially hydrogenated methyl esters of soybean oil as biodiesel, *Eur. J. Lipid Sci. Technol.* **109** (2007) 17–24.

[233] B.R. Moser and S.Z. Erhan, Preparation and evaluation of a series of α-hydroxy ethers from 9,10-epoxystearates, *Eur. J. Lipid Sci. Technol.* **109** (2007) 206–13.

[234] J.M. Sobczak and J.J. Ziołkowski, Molybdenum complex-catalysed epoxidation of unsaturated fatty acids by organic hydroperoxides, *Appl. Catal. A: General* **248** (2003) 261–8.

[235] K.M. Doll, B.K. Sharma and S.Z. Erhan, Friction reducing properties and stability of epoxidized oleochemicals, *Clean – Soil, Air, Water* **36**(8) (2008) 700–5.

[236] K.M. Doll and S.Z. Erhan, Stability and friction-reducing properties of epoxidized oleochemical methyl esters, *Abstr. 234th ACS Natl. Mtg*, Boston (19–23 August 2007), IEC-016.

[237] A. Padwa and S.S. Murphree, Epoxides and aziridines – a minor review, *ARKIVOC* **3** (2006) 6–33.

[238] X. Pages and C. Alfos, Synthesis of new derivatives from vegetable sunflower oil methyl ester via epoxidation and oxirane opening, *Oléag., Corps Gras, Lipides* **8**(2) (2001) 122–5.

[239] X. Pagès-Xatart-Parès, C. Bonnet and O. Morin, Synthesis of new derivatives from vegetable oil methyl esters via epoxidation and oxirane opening, in *Recent Developments in the Synthesis of Fatty Acid Derivatives* (G. Knothe and J.T.P. Derksen, eds), AOCS Press, Champaign, IL (1999), pp. 141–56.

[240] B. Gruber, R. Höfer, H. Kluth and A. Meffert, Polyols auf Basis fettchemischer Rohstoffe, *Lipid/Fett* **89**(4) (1987) 147–51.

[241] L.A. Rios, P.P. Weckes, H. Schuster and W.F. Hölderich, Resin catalyzed alcoholysis of epoxidised fatty esters: effect of the alcohol and the resin structures, *Appl. Catal. A: General* **284** (2005) 155–61.

[242] B. Dahlke, S. Hellbardt, M. Paetow and W.H. Zech, Polyhydroxy fatty acids and their derivatives from plant oils, *J. Am. Oil Chem. Soc.* **72** (1995) 349–53.

[243] T.M. Luong, H. Schriftmann and D. Swern, Direct hydroxylation of fats and derivatives with a hydrogen peroxide tungstic acid system, *J. Am. Oil Chem. Soc.* **44** (1967) 316–20.

[244] S. Warwel, M. Rüsch gen. Klaas and M. Sojka, Formation of vicinal diols by Re_2O_7-catalysed hydroxylation of alkenes with hydrogen peroxide, *Chem. Commun.* (1991) 1578–9.

[245] A. Guo, Y.-J. Cho and Z.S. Petrović, Structure and properties of halogenated and nonhalogenated soy-based polyols, *J. Polym. Sci., Part A: Polym. Chem.* **38** (2000) 3900–10.

[246] L.E. Gast, C.B. Croston, W.J. Schneider and H.M. Teeter, Synthetic lubricants from polyhydroxystearic acids, *J. I&EC* **46** (1954) 2205–8.

[247] S.Z. Erhan and J.M. Perez (eds), *Biobased Industrial Fluids and Lubricants*, AOCS Press, Champaign, IL (2003), pp. 14–30.

[248] H.-S. Hwang and S.Z. Erhan, Modification of epoxidized soybean oil for lubricant formulations with improved oxidative stability and low pour point, *J. Am. Oil Chem. Soc.* **78** (2001) 1179–84.

[249] H.-S. Hwang, A. Adhvaryu and S.Z. Erhan, Preparation and properties of lubricant basestocks from epoxidized soybean oil and 2-ethylhexanol, *J. Am. Oil Chem. Soc.* **80** (2003) 811–15.

[250] S.Z. Erhan, B.K. Sharma, Z. Liu and A. Adhvaryu, Lubricant base stock potential of chemically modified vegetable oils, *J. Agric. Food Chem.* **56** (2008) 8919–25.

[251] A. Adhvaryu, Z. Liu and S.Z. Erhan, Synthesis of novel alkoxylated triacylglycerols and their lubricant base oil properties, *Ind. Crops Prod.* **21** (2005) 113–19.

[252] B.K. Sharma, A. Adhvaryu, Z. Liu and S.Z. Erhan, Chemical modification of vegetable oils for lubricant applications, *J. Am. Oil Chem. Soc.* **83** (2006) 129–36.

[253] J. Salimon, N. Salih and E. Yousif, Synthetic biolubricant basestocks from epoxidized ricinoleic acid: improved temperature properties, *Kem. Ind.* **60**(3) (2011) 127–34.

[254] J. Salimon, N. Salih and E. Yousif, Biolubricants: raw materials, chemical modifications and environmental benefits, *Eur. J. Lipid Sci. Technol.* **112** (2010) 519–30.

[255] J. Salimon and N. Salih, Modification of epoxidized ricinoleic acid for biolubricant base oil with improved flash and pour points, *Asian J. Chem.* **22**(7) (2010) 5468–76.

[256] A. Adhvaryu, B.K. Sharma, H.-S. Hwang, S.Z. Erhan and J.M. Perez, Development of biobased synthetic fluids: application of molecular modeling to structure-physical property relationships, *I&EC Res.* **45** (2006) 928–33.
[257] B.K. Sharma, K.M. Doll and S.Z. Erhan, Ester hydroxy derivatives of methyl oleate: tribological, oxidation and low temperature properties, *Bioresour. Technol.* **99**(15) (2008) 7333–40.
[258] S.J. Miller, Z. Zhou and S.A. Elomari (to Chevron U.S.A., Inc.), Synthesis of diester-based biolubricants from epoxides, US Patent Appl. No. 2009/198075 A1 (6 August 2009).
[259] S. Miller and S.A. Elomari (to Chevron U.S.A., Inc.), Diester based lubricants and methods making same, PCT Int. Publ. No. WO 2008/100822 A2 (21 August 2008).
[260] A. Kleinová, P. Fodran, L. Brnčalová and J. Cvengroš, Substituted esters of stearic acid as potential lubricants, *Biomass Bioen.* **32** (2008) 366–71.
[261] S. Watanabe, T. Fujita and M. Sakamoto, Epoxides of unsaturated fatty acids as antirust additives in water based cutting fluids, *J. Am. Oil Chem. Soc.* **65** (1988) 1311–12.
[262] D. Tao, H.L. Zhu and Z.M. Hu, Free chlorine fixation and the protective film in boundary lubrication, *Lubric. Sci.* **8** (1996) 397–407.
[263] V. Schäfer and O.H. Botz (to Natoil AG), Epoxidized unsaturated fatty esters and glycerides as lubricating oil stabilizers, Eur. Patent No EP 2,138,556 A1 (30 December 2009).
[264] K.D. Carlson, R. Kleiman and M.O. Bagby, Epoxidation of *Lesquerella* and *Limnanthes* (meadowfoam) oil, *J. Am. Oil Chem. Soc.* **71** (1994) 175–82.
[265] J. Filley, New lubricants from vegetable oil: cyclic acetals of methyl 9,10-dihydroxystearate, *Bioresour. Technol.* **96** (2005) 551–5.
[266] M. Matsumura (to Sun Care Fuels Corp.), Modifier for biodiesel fuel, fuel, methods relating to those, Eur. Patent No. EP 1,681,338 A1 (29 September 2004).
[267] H. Eierdanz and F. Hirsinger, Neue Fettrohstoffe für oleochemische Reaktionen, *Fett Wiss. Technol.* **92** (1990) 463–7.
[268] E.H. Pryde, D.E. Anders, H.M. Teeter and J.C. Cowan, Ozonation of soybean oil. The preparation of some properties of aldehyde oils, *J. Am. Oil Chem. Soc.* **38** (1961) 375–9.
[269] R. Criegee, Mechanism of ozonolysis, *Angew. Chem. Int. Ed.* **14** (1975) 745.
[270] R.G. Fayter, Technical reactions for production of oleochemical monomers, in *Perspectiven nachwachsender Rohstoffe in der Chemie* (H. Eierdanz, ed.), VCH, Weinheim (1996), pp. 107–18.
[271] T.A. Isbell, Derivatives of new oil crops, in *Recent Developments in the Synthesis of Fatty Acid Derivatives* (G. Knothe and J.T.P. Derksen, eds), AOCS Press, Champaign, IL (1999), pp. 44–58.
[272] F.D. Gunstone, Uses of ozone in organic chemistry, *Educ. Chem.* **5**(4) (1968) 166–70.
[273] P. Tran, D. Graiver and R. Narayan, Ozone-mediated polyol synthesis from soybean oil, *J. Am. Oil Chem. Soc.* **82**(9) (2005) 653–9.
[274] M. Matsumura and S. Murakami, Method and equipment of refining plant oil and waste vegetable oil into diesel engine fuel, US Patent No. 6,364,917 B1 (2 April 2002).

[275] N.U. Soriano, V.P. Migo and M. Matsumura, Ozonation of sunflower oil: spectroscopic monitoring of the degree of unsaturation, *J. Am. Oil Chem. Soc.* **80** (2003) 997–1001.

[276] N.U. Soriano Jr., V.P. Migo and M. Matsumura, Ozonised vegetable oil as pour point depressant for neat biodiesel, *Fuel* **85** (2006) 25–31.

[277] G. Liu, X. Kong, H. Wan and S. Narine, Production of 9-hydroxynonanoic acid from methyl oleate and conversion into lactone monomers for the synthesis of biodegradable polylactones, *Biomacromolecules* **9** (2008) 949–53.

[278] J. Grushcow and M. Smith, Engineering high-performance biolubricants in crop plants, *Ind. Biotechnol.* **2** (2006) 48–50.

[279] W.A. Herrmann, Essays on organometallic chemistry V. State of and outlook for rhenium chemistry in catalysis, *J. Organomet. Chem.* **382**(1–2) (1990) 1–18.

[280] A. Zlatanić, C. Lava, W. Zhang and Z.S. Petrović, Effect of structure on properties of polyols and polyurethanes based on different vegetable oils, *J. Polym. Sci., Polym. Phys.* **42** (2004) 809–19.

[281] H.P. Benecke, B.R. Vijayendran, D.B. Garbark and K.P. Mitchell, Low cost and highly reactive biobased polyols: a co-product of the emerging biorefinery economy, *Clean – Soil, Air, Water* **36**(8) (2008) 694–9.

[282] H.J. Schäfer, M. aus dem Kahmen, L. Hinkamp and R. Maletz, Neue Oleochemikalien aus ungesättigten Fettsäuren durch Cycloaddition, nukleophile Addition und anodische Oxidation, in *3. Symposium Nachwachsende Rohstoffe, Perspectiven für die Chemie* (Bundesministerium für Ernährung, Landwirtschaft und Forsten, ed.), Landwirtschaftsverlag, Münster (1994), pp. 217–34.

[283] M. Zobel and H.J. Schäfer, Nukleophile Additionen und Substitutionen an ungesättigten Fettsäuren, in *Schriftenreihe Nachwachsende Rohstoffe,* Bd. 12, *Chemische Nutzung heimischer Pflanzenöle* - Abschlußkolloquium des BML, Landwirtschaftsverlag, Münster (1998), pp. 75–110.

[284] S. Maurer and R. Schmid, Biocatalysts for the epoxidation and hydroxylation of fatty acids and fatty alcohols, in *Handbook of Industrial Biocatalysis* (C.T. Hou, ed.), Taylor & Francis, New York, NY (2005); http://elib.uni-stuttgart.de/opus/volltexte/2006/2655/

[285] N.J. Turner, Direct evolution drives the next generation of biocatalysts, *Nature Chem. Biol.* **5** (2009) 567–73.

[286] M. Certik and S. Shimizu, Biosynthesis and regulation of microbial polyunsaturated fatty acid production, *J. Biosci. Bioengng.* **87**(1) (1999) 1–14.

[287] R.D. Schmid and R. Verger, Lipases: Interfacial enzymes with attractive applications, *Angew. Chem. Intl. Ed.* **37** (1998) 1608–33.

[288] T.M. Kuo, A.C. Lanser, L.K. Nakamura and C.T. Hou, Production of 10-ketostearic acid and 10-hydroxystearic acid by strains of *Sphingobacterium thalpophilum* isolated from composted manure, *Curr. Microbiol.* **40** (2000) 105–9.

[289] T.M. Kuo, L.K. Nakamura and A.C. Lanser, Conversion of fatty acids by *Bacillus sphaericus*-like organisms, *Curr. Microbiol.* **45** (2002) 265–71.

[290] Y. Yamada, H. Uemura, H. Nakaya, K. Sakata, T. Takatori, M. Nagao, H. Iwase and K. Iwadate, Production of hydroxy fatty acid (10-hydroxy-12(*Z*)-octadecenoic acid) by *Lactobacillus plantarum* from linoleic acid and its cardiac effects by guinea pig papillary muscles, *Biochem. Biophys. Res. Commun.* **226** (1996) 391–5.

[291] S.H. El-Sharkawy, W. Yang, L. Dostal and J.P.N. Rosazza, Microbial oxidation of oleic acid, *Appl. Environ. Microbiol.* **58**(7) (1992) 2116–22.

[292] C.T. Hou, Microbial oxidation of unsaturated fatty acids, in *Advances in Applied Microbiology* (S.L. Neidleman and A.I. Laskin, eds), Academic Press, San Diego, CA (1995), Vol. 41, pp. 1–23.

[293] Z.-H. Yi and H.-J. Rehm, Identification and production of Δ^9-*cis*-1,18-octadecenedioic acid by *Candida tropicalis*, *Appl. Microbiol. Biotechnol.* **30**(4) (1988) 327–31.

[294] D. Fabritius, H.J. Schäfer and A. Steinbüchel, Identification and production of 3-hydroxy-Δ^9-*cis*-1,18-octadecenedioic acid by mutants of *Candida tropicalis*, *Appl. Microbiol. Biotechnol.* **45** (1996) 432–8.

[295] M. Wubbolts, P. Reuvekamp and B. Witholt, TOL plasmid-specified xylene oxygenase is a wide substrate range monooxygenase capable of olefin epoxidation, *Enz. Microbiol. Technol.* **16** (1994) 608–15.

[296] P. Besse and H. Veschambre, Chemical and biological synthesis of chiral epoxides, *Tetrah.* **50** (1994) 8885–927.

[297] K. Furuhashi, Biological routes to optically active epoxides, in *Chirality in Industry* (A.N. Collins, G.N. Sheldrake and J. Crosby, eds), John Wiley & Sons, Ltd, Chichester (1992), pp. 167–86.

[298] A. Schmid, J.S. Dordick, B. Hauer, A. Kiener, M. Wubbolts and B. Witholt, Industrial biocatalysis today and tomorrow, *Nature* **409** (2001) 258–68.

[299] G.R. Stephenson, Asymmetric oxidation, in *Advanced Asymmetric Synthesis* (G.R. Stephenson, ed.), Chapman & Hall, London (1996), pp. 367–91.

[300] S.D. Doig, A.T. Boam, A.G. Livingstone and D.C. Stuckey, Epoxidation of 1,7-octadiene by *Pseudomonas oleovorans* in a membrane bioreactor, *Biotechnol. Bioengng.* **63** (1999) 601–11.

[301] K. Furuhashi, M. Shintani and M. Takagi, Effects of solvents on the production of epoxides by *Nocardia corallina* B-276, *Appl. Microbiol. Biotech.* **23** (1986) 218–23.

[302] S. Panke, M. Held, M.G. Wubbolts, B. Witholt and A. Schmid, Pilot-scale production of (*S*)-styrene oxide from styrene by recombinant *Escherichia coli* synthesizing styrene monooxygenase, *Biotechnol. Bioengng.* **80** (2002) 33–41.

[303] B. Bühler, B. Witholt, B. Hauer and A. Schmid, Characterization and application of xylene monooxygenase for multistep biocatalysis, *Appl. Environ. Microbiol.* **68** (2002) 560–8.

[304] W.-D. Fessner and V. Helaine, Biocatalytic synthesis of hydroxylated natural products using aldolases and related enzymes, *Curr. Opin. Biotechnol.* **12** (2001) 574–86.

[305] J. Reddy, C. Lee, M. Neeper, R. Greasham and J. Zhang, Development of a bioconversion process for production of *cis*-1*S*, 2*R*-indanediol from indene by recombinant *Escherichia coli* constructs, *Appl. Microbiol. Biotechnol.* **51** (1999) 614–20.

[306] O. Favre-Bulle, T. Schouten, J. Kingma and B. Witholt, Bioconversion of *n*-octane to octanoic acid by a recombinant *Escherichia coli* cultured in a two liquid phase bioreactor, *Bio/Technol.* **9** (1991) 367–71.

[307] D.R. Nelson, L. Koymans, T. Kamataki, J.J. Stegeman, R. Feyereisen, D.J. Waxman, M.R. Waterman, O. Gotoh, M.J. Coon, R.W. Estabrook, I.C. Gunsalus and D.W. Nebert, P450 superfamily: update on new sequences, gene

mapping, accession numbers and nomenclature, *Pharmacogenetics* **6**(1) (1996) 1–42.
[308] M.A. Smith, H. Moon and L. Kunst, Production of hydroxy fatty acids in the seeds of *Arabidopsis thaliana*, *Biochem. Soc. Trans.* **28** (2000) 947–50.
[309] M.A. Smith, H. Moon, G. Chowrira and L. Kunst, Heterologous expression of a fatty acid hydroxylase gene in developing seeds of *Arabidopsis thaliana*, *Planta* **217** (2003) 507–16.
[310] S. Singh, S. Thomaeus, M. Lee, A. Green and S. Stymne, Inhibition of polyunsaturated fatty acid accumulation in plants expressing a fatty acid epoxygenase, *Biochem. Soc. Trans.* **28** (2000) 940–2.
[311] E.B. Cahoon, K.G. Ripp, S.E. Hall and B. McGonigle, Transgenic production of epoxy fatty acids by expression of a cytochrome P450 enzyme from *Euphorbia lagascae* seed, *Plant Physiol.* **128** (2002) 615–24.
[312] N. Tijet, F. Pinot, I. Benveniste, R. Le Bouquin, C. Helvig, Y. Batard, F. Cabello-Huartado, D. Werck-Reichhart, J.-P. Salaun and F. Durst (to CNRS), Plant fatty acid hydroxylase genes, PCT Int. Publ. No. WO 99/18224 (15 April 1999).
[313] D. Werck-Reichhart and F. Feyereisen, Cytochrome P450: A success story, *Genome Biol.* **1** (2000) 3001–12.
[314] H. Joo, Z. Lin and F.H. Arnold, Laboratory evolution of peroxide-mediated cytochrome P450 hydroxylation, *Nature* **399** (1999) 670–3.
[315] J. Shanklin and E.B. Cahoon, Desaturation and related modifications of fatty acids, *Ann. Rev. Plant Phys. Plant Mol. Biol.* **49** (1998) 611–41.
[316] B.G. Fox, Catalysis by non-heme iron, in *Comprehensive Biological Catalysis* (M. Sinott, ed.), Academic Press, London (1997), pp. 261–348.
[317] A.R. Brash, Lipoxygenases: Occurrence, functions, catalysis and acquisition of substrate, *J. Biol. Chem.* **274** (1999) 23679–82.
[318] A. Weiss, U. Schörken, V. Candar, Y. Eryasa, M. Wunderlich, N. Buyukuslu and C. Beverungen (to Cognis), Process for the enzymic generation and recovery of fatty acid hydroperoxides, Eur. Patent No. EP 1,336,959 A2 (7 February 2003).
[319] F. Björkling, S.E. Godtfredsen and O. Kirk, Lipase-mediated formation of peroxy carboxylic acids used in catalytic epoxidation of alkenes, *J. Chem. Soc. Chem. Commun.* (1990), pp. 1301–3.
[320] G.J. Piazza, A. Nuñez and T.A. Foglia, Epoxidation of fatty acids, fatty methyl esters, and alkenes by immobilized oat seed peroxygenase, *J. Mol. Catal. B: Enzym.* **21** (2003) 143–51.
[321] B. Perks, Extreme potential, *Chem. World* **8**(6) (June 2011) 48–51.
[322] J.E. Graham, M.E. Clark, D.C. Nadler, S. Huffer, H.A. Chokhawala, S.E. Rowland, H.W. Blanch, D.S. Clark and F.T. Robb, Identification and characterization of a multidomain hyperthermophilic cellulase from an archaeal enrichment, *Nature Comm.* (July 2011) doi:10.1038/ncomms1373; ISSN (online) 2041–1723.
[323] A. Schmid, F. Hollmann, J.B. Park and B. Bühler, The use of enzymes in the chemical industry in Europe, *Curr. Opin. Biotechnol.* **13** (2002) 359–66.
[324] R. Quermann, R. Maletz and H.J. Schäfer, Conversion of fatty acids and derivatives. IV. Conversion of fatty acid methyl esters into dialkylated succinic esters by oxidative coupling of their enolates, *Liebigs Ann. Chem.* (1993) 1219–23.

[325] H.J. Schäfer, Recent contributions of Kolbe electrolysis to organic synthesis, *Top. Curr. Chem.* **152** (1990) 91–151.

[326] D. Degner, Organic electrosynthesis in industry, *Top. Curr. Chem.* **148** (1988) 1–95.

[327] H.J. Schäfer, A. Weiper, M. aus dem Kahmen and A. Malzeit, Spezialchemikalien durch Kolbe-Elektrolyse von Fettsäuren, in *Nachwachsende Rohstoffe – Perspektiven für die Chemie* (M. Eggersdorfer, S. Warwel and G. Wulff, eds), Verlag Chemie, Weinheim (1993), pp. 97–108.

[328] M. Renz, Ketonization of carboxylic acids by decarboxylation: mechanism and scope, *Eur. J. Org. Chem.* (2005), 979–98.

[329] A. Corma, M. Renz and C. Schaverien, Coupling fatty acids by ketonic decarboxylation using solid catalysts for the direct production of diesel, lubricants and chemicals, *ChemSusChem.* **1** (2008) 739–41.

[330] J.A. Davies, P.L. Watson, J.F. Liebman and A. Greenberg, *Selective Hydrocarbon Activation, Principles and Progress*, VCH, Weinheim (1990).

[331] M. Bühler and J. Schindler, Aliphatic hydrocarbons, in *Biotechnology* (J. Rehm and G. Reed, eds), VCH, Weinheim (1984), Vol. 6a, pp. 329–85.

[332] L. Hinkamp, H.J. Schäfer, B. Wippich and H. Luftmann, Regioselective conversion of nonactivated CH bonds. VI. Selective ω- to (ω – 2)-chlorination of fatty acids by way of adsorption on alumina, *Liebigs Ann. Chem.* (1992) 559–63.

[333] S.Z. Erhan, A. Adhvaryu and Z. Liu (to USDA), Chemically modified vegetable oil-based industrial fluid, US Patent No. 6,583,302 B1 (24 June 2003).

[334] V. Spitzer, Screening analysis of unknown seed oils, *Fett/Lipid* **101** (1999) 2–19.

[335] J.L. Harwood, Plant acyl lipids: structure, distribution and analysis, in *Biochemistry of Plants* (P.K. Stumpf, ed.), Academic Press, New York, NY (1980), Vol. 4, pp. 2–58.

[336] F.J. van de Loo, B.G. Fox and C. Somerville, Unusual fatty acids, in *Lipid Metabolism in Plants* (T.S. Moore Jr., ed.), CRC Press, Boca Raton, FL (1993), pp. 91–126.

[337] C.R. Smith Jr., Occurrence of unusual fatty acids in plants, *Prog. Chem. Fats Other Lipids* **11** (1971) 137–77.

[338] R.C. Badami and K.B. Patil, Structure and occurrence of unusual fatty acids in minor seed oils, *Prog. Lipid Res.* **19** (1980) 119–53.

[339] J. Jaworski and E.B. Cahoon, Industrial oils from transgenic plants, *Curr. Opin. Plant Biol.* **6** (2003) 178–84.

[340] P. Blaise, M. Farines and J. Soulier, Identification of cyclopentenyl fatty acids by ^{1}H and ^{13}C nuclear magnetic resonance, *J. Am. Oil Chem. Soc.* **74**(6) (1997) 727–30.

[341] R.S. Van der Hoeven and J.C. Steffens, Biosynthesis and elongation of short- and medium-chain-length fatty acids, *Plant Physiol.* **122** (2000) 275–82.

[342] L. Yao and E.G. Hammond, Isolation and melting properties of branched-chain esters from lanolin, *J. Am. Oil Chem. Soc.* **83** (2006) 547–52.

[343] L. Yao. E. Hammond and T. Wang, Melting points and viscosities of fatty acid esters that are potential targets for engineered oilseed, *J. Am. Oil Chem. Soc.* **85** (2008) 77–82.

[344] D. Muller, P. Lopes, M. Brewer and E. Kelderman (to Arizona Chemical Co.), Compositions containing fatty acids and/or derivatives thereof and a low

temperature stabilizer, PCT Int. Publ. No WO 2006/105306 A2 (5 October 2006).
[345] O.D. Dailey Jr., N.T. Prevost and G.D. Strahan, Synthesis and characterization of branched-chain derivatives of methyl oleate, *Clean-Soil, Air, Water* **36**(8) (2008) 687–93.
[346] B.K. Sharma, Z. Liu, A. Adhvaryu and S.Z. Erhan, One-pot synthesis of chemically modified vegetable oils, *J. Agric. Food Chem.* **56** (2008) 3049–56.
[347] J.A. Napier, The production of unusual fatty acids in transgenic plants, *Ann. Rev. Plant Biol.* **58** (2007) 295–319.
[348] REFLAX – Rational engineering of lipid metabolism in flax to produce branched-chain fatty acids, unusual fatty acids with potent biolubricant properties, Proj. QLK3-2000-00349 (2001–2004); http://ec.europa.eu/research/quality-of-life/cell-factory/volume1/projects/qlk3-2000-00349en.html
[349] T. Foglia, G. Piazza, V. Wyatt, M. Haas and W. Marmer, Production of value-added lipids, biofuels, and biobased products from fats and oils, ARS Project # 1935-41000-066-00 (2004–2009); http://www.ars.usda.gov/research/projects/projects.htm?ACCN_NO=408597
[350] H. Yoshimura, T. Nonishi, N. Inouchi and J. Nakakigashi (to Nippon Sangyo KK), Vegetable lubricant, Jap. Patent No. JP 2008/056881 (13 March 2008).
[351] A. Viswanathan, Effect of degree of substitution of octenyl succinate starch on the emulsification activity on different oil phases, *J. Environm. Polym. Degrad.* **7** (1999) 191–6.
[352] G.F. Fanta and K. Eskins, Stable starch–lipid compositions prepared by steam jet cooking, *Carbohydrate Polym.* **28** (1995) 171–5.
[353] Fantesk™ is a registered trademark of the National Center for Agricultural Utilization Research, Agricultural Research Service, USDA, Peoria, IL.
[354] K. Eskins and G.F. Fanta, Fantesk: Carbohydrate-oil composites useful in low-fat foods, adhesives, seed coatings, cosmetics and drug delivery systems, *Lipid Technol.* **8**(3) (1996) 53–7.
[355] G.F. Fanta and K. Eskins, Starch–oil composites prepared by jet cooking: preparation, properties and commercial applications, in *Paradigm for Successful Utilization of Renewable Resources* (D.J. Sessa and J.L. Willet, eds), AOCS Press, Champaign, IL (1988), pp. 266–86.
[356] G. Biresaw and S.M. Erhan, Solid lubricant formulations containing starch-soybean oil composites, *J. Am. Oil Chem. Soc.* **79** (2002) 291–6.
[357] S.M. Erhan, G.F. Fanta, F.C. Felker, D.C. Ciccognani and L.A. Nelson, The use of biocides in metalworking fluids containing starch–oil composites, *Program Guide 55th Ann. Mtg. STLE*, Nashville, TN (7–11 May 2000), p. 70.
[358] J.A. Schey, *Tribology in Metalworking: Friction, Lubrication and Wear*, American Society of Metals, Metals Park, OH (1983), pp. 27–130.
[359] S.M. Erhan, G.F. Fanta, K. Eskins, F.C. Felker and H. Muys, Starch–oil composites in water-based drilling mud formulations, *Program Guide 54th Ann. Mtg. STLE*, Las Vegas, AZ (22–27 May 1999), pp. 143–4.
[360] K. Eskins and G.F. Fanta (to USDA), Non-separable compositions of starch and water-immiscible organic materials, US Patent No. 5,882,713 (16 March 1999).
[361] G. Biresaw, Biobased dry-film metalworking lubricants, *J. Synth. Lubr.* **21**(1) (2004) 43–57.

[362] G. Biresaw and R.L. Shogren, Friction properties of chemically modified starch, *J. Synth. Lubr.* **25** (2008) 17–30.
[363] G. Biresaw, J.A. Kenar, T.L. Kurth, F.C. Felker and S.M. Erhan, Investigation of the mechanism of lubrication in starch–oil composite dry film lubricants, *Lubric. Sci.* **19** (2007) 41–55.
[364] J.A. Kenar, F.C. Felker, G. Biresaw and T.L. Kurth, Properties of dry film lubricants prepared by spray application of aqueous starch–oil composites, *Ind. Crops Prod.* **29** (2009) 45–52.

7
Formulating lubricating oils

Abstract: High-performance lubricants are formulated from a base stock (often an oil blend) and property-enhancing additives. Additives can improve the intrinsic lubrication properties of the base oil (surface additives), develop new properties (performance-enhancing additives), or extend lifetime (fluid protective additives). Additive design needs to take into account the chemical and physical properties of base oils, notably their polarity. Additive packages for mineral-based oils require redesign for application in more polar biolubricant formulations. Most additives specifically enhance one property; multifunctional additives combine different functionalities. Additive chemistry should avoid interaction between components of a fully formulated lube oil. Development of more highly performing additive systems is an active area of research.

Key words: base oil blends, lubricant additive technology, additive design, bio-based additives, biolubricant formulations.

7.1 Introduction

All lubricant systems are materials formulated from one or more base fluids (native oils, synthetic esters, etc.) and property-enhancing chemical substances, commonly called additives. The percentage of additives in a formulation differs greatly, from hundreds of ppm level for antifoam agents and biocides to 1% in simple compressor oils and up to 30% in metalworking fluids and gear lubricants. On average, lubricating oils consist of 93% base oil(s) and 7% additives [1]. A base fluid means a lubricating fluid whose flow, ageing, lubricity and anti-wear (AW) properties, as well as its properties regarding contaminant suspension, have not been altered by the inclusion of additives. The formulation of high-performance lubricants requires excellent high-quality base fluids and is in line with the observed trends in increased use of severely hydrotreated highly refined mineral oils, synthetic esters and poly-α-olefins (PAOs). Table 7.1 compares physicochemical properties of biodegradable base fluids with mineral oil.

Technical and environmental pressures (such as ACEA 2008 and 2010, REACH, API ILSAC GF-5 and dexos™) continue to force improvements in lube specifications. Today, few commercial lubricants contain only one base oil; there are usually two or three. A wide variety of different characteristics required by lubricants necessitate special performance which often cannot be satisfied by just one base stock. This can easily be rationalised.

Table 7.1 Comparison of physicochemical properties of biodegradable base fluids with mineral oil (average values)

Property	Vegetable oil	Jojoba oil	Synthetic ester	PAG[a]	Mineral oil
Density at 20 °C (kg/L)	0.940	0.866	0.930	1.100	0.880
Viscosity index	100–200	210–233	120–220	100–200	100
Shear stability	Good	Very good	Good	Good	Good
Pour point (°C)	–20 to +10	+9	–60 to –20	–40 to +20	–15
Cold flow	Poor	Poor	Very good	Very good	Good
Miscibility with mineral oil	Good	Good	Good	Not miscible	–
Solubility in water	Not soluble	Not soluble	Not soluble	Very good to not soluble	Not soluble
Compatibility with paints/coatings	Good	Good	Good to poor	Poor	Good
Biodegradability (%)	70–100	70–100	10–100	10–100	10–35
Oxidation stability	Fair	Good	Good	Good	Good
Hydrolytic stability	Poor	Good	Fair	–	Good
Cost	2–3	10	4–20	2–4	1

[a] PAG, polyalkylene glycol.
After ref. [2].

As mentioned in Section 3.6, lubricity and solvent power are interrelated properties. Lubricity requires the simultaneous presence of polar and non-polar molecules. This can readily be achieved by appropriate blending. A good reason for the popularity of synthetic and mineral oil blends is that the additives attached to the mineral oil structure become more soluble with the addition of synthetic oil. Blends and additives can improve and complement the technical performance of a given base oil. In lubricant design, individual lube oils have mutual effects on each other and these are frequently nearly additive. Formulators use blending charts to identify optimal viscometrics and NOACK volatility requirements for a specific additive system. Using a blend of base stocks can sometimes achieve desired properties with only a minimal increase in cost.

A mineral oil blended with vegetable oil to improve its performance as a lubricant is denominated a compounded oil. Compounded oils are base stocks with excellent lubricity and AW properties, thermal stability, high specific heat, high flash point and improved health, safety and the environment (HSE) profile. The chief disadvantage is insufficient oxidation stability. The friction modifier performance of such a compounded oil, namely rapeseed oil (RSO) added to mineral oil (base stock for SAE 20W-50), was examined by Durak [3]. RSO reduces considerably the friction coefficient of the mineral oil both in mixed boundary and fluid film lubrication regimes, and more so at low (25 °C) than at high (100 °C) temperatures.

Elektrionization™ is a proprietary process whereby feedstocks undergo electroionising treatment resulting in an increase in viscosity, polarity and viscosity index (VI) [4]; see also Section 7.3. Increasing the polar functionality of the triglyceride molecules of vegetable oil has a positive impact on wear protection. Featuring a unique combination of viscosity and polarity, ionised vegetable oils form sufficiently thick and resilient protective layers by adsorption to rubbing surfaces. Blends of mineral oils and ionised vegetable oils have unique properties, such as high lubricity and AW efficiency (from vegetable oil) and added oxidation stability and antisludge capability (from mineral oil) [5, 6]. Addition of 10% elektrionised vegetable oils to a mineral base stock significantly improves the lubricity, which translates into higher processing speeds, reduced lubricant consumption, better production economy and reduced environmental load.

In some lubricant applications, specific vegetable oils might not be adequate enough to provide all the necessary needs. In PCT Int. Publ. No. WO 2007/034336 A3 (to Monsanto) Despeghel relates to blends of RSO with another oleaginous oil (sunflower oil, SNO; high-oleic sunflower oil, HOSNO; or soybean oil, SBO) as co-base fluids [7]. In other cases, the vegetable oils have been blended with synthetic and/or mineral oils, which are non-polar. Such blending of amphiphilic fluids with non-polar oils has a profound effect on various properties of the blend, and most notably on its surface and interfacial properties. Surface and interfacial properties are particularly important because they affect various lubricant functions, including the formulation of water-based lubricants from such neat blends. PCT Int. Publ. No. WO 03/027212 A (to Pennzoil-Quaker State Co.) describes environmentally friendly crankcase lubricant base oils composed of vegetable oil/synthetic ester/mineral oil blends and a suitable additive package [8]; see also Section 10.5.1. Engine oils are usually composed of blends of lubricating base oils [9].

Mixtures of base oils (e.g: mineral hydrocarbon oil/ester; transesterified triglyceride base oil/synthetic ester) are generally meant to improve specific base oil properties, such as to enhance the low-temperature properties of vegetable oils or the biodegradability of mineral hydrocarbon oils [8, 10, 11]. The main classes of synthetic material to blend lubricants for a variety of applications (aeronautic, AE; automotive, AU; industrial, I) include:

- olefin oligomers (AU, I);
- dibasic acid esters (AE, AU);
- polyol esters (AE, AU);
- alkylated aromatics (AU, I);
- polyalkylene glycols (I);
- phosphate esters (I).

Synthetic base fluids are frequently blended with mineral oils. As indicated in Section 3.2, oil-soluble polyalkylene glycols (OSPs) as co-base oils provide opportunities for formulators to upgrade Group II and III base oils by improving lubricant viscosity, low-temperature properties (pour points, PP), lubricity and polarity, thus favouring performance-enhancing additive formulation [12]. Many of the more polar synthetic base oils serve as additives to less polar oils, e.g. esters in hydrocarbons and vice versa. An ester oil/hydrocarbon oil blend with a kinematic viscosity of 1–20 mm^2/s at 40 °C is particularly indicated for metalworking [13]. In US Patent Appl. No. 2010/0105583 A1 (to Renewable Lubricants, Inc.) Garmier describes the use of a bio-based blend composed of HOSNO and a base oil such as a synthetic ester [14]. Also low-cost mixtures of corn oil and synthetic diesters, in particular di-2-ethylhexyladipate (DOA) and di-2-ethylhexylsebacate (DOS), have been described as base stock oils for lubricants [15]. Binary blends of PAOs with vegetable oils (canola, SBO) have been studied in relation to their elastohydrodynamic (EHD) properties [16]. In PCT Int. Publ. No. WO 2008/134179 Thoen *et al.* [17] describe fully miscible lubricant compositions comprising a polyether and a renewable raw material such as an unsaturated seed or vegetable oil.

Synesstic™ alkylated naphthalene (AN) base stocks (ExxonMobil Chemical) help formulators improve the performance of PAO and mineral oil-based lubricants used in many automotive and industrial applications, including applications with incidental food contact. AN base stocks provide thermal and oxidative stability for improved lubricant life, hydrolytic stability for use in high-moisture environments, seal compatibility enhancement and improved additive effectiveness as an ester replacement. Blending dry base oils (which contain fully saturated non-polar hydrocarbons) with naphthenic base stocks or alkylaromates leads to an improvement in solvency but also to a lower VI. In contrast to alkylaromates, naphthenic bases have a higher tendency to form sludge, which limits their applications. Blends with synthetic esters improve solvency and lubricity but deteriorate stability. Some linear-chain esters and their decomposition products passivate the surface against reaction with extreme pressure (EP) additives, undermining AW protection at high loads.

In US Patent Appl. No. 2007/0232506 A1 (to ExxonMobil) Gao and Habeeb [18] describe blends of mineral oil base stocks with synthetic polyol (neopentyl glycol, NPG; trimethylolpropane, TMP; pentaerythritol, PE, etc.) esters that have a viscosity lower than that of the individual components. US Patent No. 7,465,696 B2 (to Chevron Oronite Corp.) discloses a blend composed of a Fischer–Tropsch derived lubricating base oil and at least one polyol ester, which exhibits improved fuel economy in an internal combustion engine [19]. US Patent Appl. No. 2008/0255011 A1 (to ExxonMobil) relates to fuel economy lubricant compositions comprising a blend of a gas-to-liquid (GTL) lubricating base oil and PAO, and about 0.5 wt% of a friction modifier consisting essentially of oil soluble fatty acid (FA) esters of a polyol (such as glycerol monooctadecanoate) [20]. Such lubricating compositions have reductions in their friction coefficients exceeding those of similar compositions formulated with Group III or PAO base oils. GTL base oils have high VI (>140) and very low sulphur and nitrogen contents (<10 ppm). When lubricating compositions are based on a GTL base oil, or Group I–IV petroleum base oils, the overall formulation may not be readily biodegradable. Tools for blending control and formulation management, including blend optimisation, are important for the lubricants industry.

Base stocks usually make up more than 80% of the lubricant formulation and thus greatly determine the overall properties. Nevertheless, base oils or even base oil blends are mostly unable to meet the modern high-performance lubrication requirements of an application. Therefore, a great variety of additives are necessarily employed to improve lubrication by modifying the properties of the base oil and/or those of the metal surfaces. Lubricants frequently contain some 0–20 wt% additives to tailor their viscometric properties (flow), low-temperature behaviour, oxidative stability (ageing), corrosion protection, EP behaviour, demulsibility and water rejection, friction coefficients, lubricity, wear protection, air release, colour and other properties [21]. The quality and performance of a lubricant formulation depend on the quality of the base fluid and the combination of additives or additive package used. Although additives can improve lubricant performance, high performance is achieved only with the use of high-quality base oils.

Table 7.2 lists general requirements of lubricant formulations, with some properties being mainly attributable to the base stock oil and others mainly to the additives. Some lubricant properties are exclusively base stock dependent, such as biodegradability, effective metal adsorption, volatility (flash point), solvency, miscibility, etc.; other properties are greatly determined by the base stock, such as viscosity, VI, low-temperature fluidity (PP), thermal and oxidative stability. To some extent, additives can improve the intrinsic properties of the base oil (VI improvers, antifoam agents/demulsifiers, pour point depressants, PPDs), develop new properties (AW/EP, detergents, dispersants, anticorrosion agents) or extend lifetime (antioxidants). Some properties cannot be influenced by additives, e.g. volatility, air release prop-

Table 7.2 Contributions of base stock and additives to lubricant properties

Property	Dependency	
	Base stock	Additives
• Viscosity for film maintainance	+	+
• Viscosity–temperature characteristics/Viscosity Index (VI)	+	+
• Low-temperature fluidity	+	+
• Effective metal adsorption	+	
• Satisfactory lubricity in load-carrying capacity		+
• Heat transfer	+	+
• Thermal and oxidative stability	+	+
• Hydrolytic stability/water tolerance	+	+
• Low volatility	+	
• Anti-wear protection	+	+
• Corrosion protection	+	+
• Metal passivation		+
• Deposit forming tendencies	+	
• Acidity	+	+
• Ash content		+
• Cleanliness and filterability	+	
• Demulsibility		+
• Foam inhibition		+
• High solvency	+	
• Miscibility or compatibility with other system fluids	+	
• Biodegradability	+	
• Ecotoxicity	+	+
• Colour	+	

erties, water separation, thermal conductivity, compressibility and, boiling point. In particular, biodegradability of a base stock cannot be improved by using additives and is even compromised by the use of non-biodegradable substances as additives. Additives are particularly effective in establishing satisfactory lubricity in load-carrying capacity (AW and EP additives), rust and corrosion inhibition, foam inhibition. Oils have the ability to emulsify. An emulsion (mixture of oil and water) is often undesirable as water is not a good lubricant for machined surfaces. Moreover, the foaming action allows moisture to collect on surfaces that otherwise should be free of moisture. Emulsified oils have a tendency to trap dirt. If a lubricant emulsifies it should have the ability to demulsify (separation of oil and water); additives are used for this purpose.

In other circumstances, additives may turn out to be ineffective, e.g. the PP of jojoba oil (+9 °C) cannot be improved using the classical PPDs polyisobutylene (PIB) or poly(methylacrylate) (PMA) [22]. Additive action is dependent upon the nature of the base oil (see Section 7.3). Additives can react in different ways depending on the type of base fluid being employed. For example, the differing chemistry of mineral oils and synthetic esters means that, over the long term, changes in properties may differ markedly between the two lubricant types.

Main components of lubricant formulations are substances accounting for more than 5 wt% of the oil. Standard mineral oil hydraulic fluids and compressor oils are typically composed of 98% base fluid and 2% chemical additives (AW, corrosion inhibitor, AO, PPD, copper passivator, antifoam). Many high-performance lubricants no longer contain petroleum base oils; synthetic ester and/or vegetable oil-based hydraulic fluids usually contain a higher proportion (2–5 wt%) of additives to achieve the required performance. The higher cost of both the base fluid(s) and additives means that environmentally acceptable (EA) hydraulic fluids are usually more expensive than conventional mineral oil-based fluids. While the amount of additives in hydraulic fluids is quite limited, in motor oils typically 10% of additives are present, whereas for special applications (e.g. some metalworking fluids, gear lubricants, transmission oils and greases) up to 30% additives may be needed.

Vegetable oils are amphiphilic fluids since their molecules comprise distinctly separated polar and non-polar groups. As a result, vegetable oils and derivatives can be used both as base oils and as film strength additives in lubricant formulation. For example, epoxidised unsaturated FAs find application as metalworking fluids and as lubricating additives to eliminate corrosion from chlorine containing compounds [23, 24]. ESBO (epoxidised soybean oil) is also used as a plasticiser and stabiliser in polymeric materials. Similarly, sulphurised products have found extensive commercial use as additives for metalworking and industrial oils [25]. Various other lubricant classes (FAs, waxes, wax esters, polyol esters, complex esters, metal soaps, etc.) are in use to improve dispersion, release and flow properties in thermoplastics processing [26]. Oleamide, erucamide and waxes are well-known slip agents. One of the disadvantages of vegetable oils as lubricant base stocks is their poor compatibility with customary additives used for mineral oil formulation (see Section 7.3).

Dialkylcarbonates, the diesters of carbonic acid, $R_1OC(O)OR_2$, are synthetic fluids that have been proposed as base oils for lubricants, but only find niche applications. The performance of dialkylcarbonates is similar to that of common dicarboxylic acid esters but are preferable with regard to toxicity and seal compatibility [27, 28]. Owing to their amphiphilic nature and their decomposition to non-corrosive simple alcohols and carbon dioxide, dialkylcarbonates represent both attractive lubricants and additives [29]. Dialkylcarbonates are useful as AW additives, but fatty acid methyl esters (FAMEs) of similar chain length show better performance.

7.2 Lubricant additive technology

Additives improve certain characteristics of the lubricant formulation which the base fluid cannot provide, or cannot provide to a sufficient extent. Most of the conventional synthetic lubrication oils are not good EHD or boundary lubricants due to their low load-carrying capacity, which may be due to reduced polarity and hence poor film formation ability and less reactivity to form AW layers. For successful lubrication, additives are usually applied to these oils to achieve a comprehensive performance that includes enhancement of AW and friction reduction and prevention of oil degradation and corrosion. Key issues with additives must be resolved before severely hydrotreated and hydroisomerised base oils will gain widespread acceptance [5]; see also Section 3.3.

Generally, lubricant base oil modification by additives provides a cost-effective and flexible means to alter lubricant properties. A base oil is usually additivated by customary lubricant additives and may also contain solvents, FA derivatives, in particular their metal salts, organics and inorganics, natural or synthetic polymers. The components of an additive package should remain substantially solubilised or stabilised in the base oil without the need for additional solubilising additives so that the additive package is substantially free from additive drop-out in the long run. There are limits to the amounts of additives that can be introduced in base oils to improve performance. Beyond these limits, the benefits in performance are minimal. The cost of additives is a strong economical deterrent for abuse. Moreover, additives tend to thicken the oil, which reduces fuel economy. An increasing trend is the preference to develop lubricants that do not contain additives (or at least in the least possible amounts). Ideally, additives should not be needed with properly formulated base lubricants. Experiences such as decreased lubricant stability, miscibility/solubility (at all temperatures), sludge formation and component deposits, are concerns related to the use of additives. At the same time, additives can provide the advantage of improved lubricity, and may be necessary in particularly demanding applications. Motor oils are being formulated with reduced ash build-up tendency (see also Section 7.3).

Additives are substances formulated for the improvement of antifriction, chemical and physical properties of base oils (mineral, synthetic, vegetable or animal), which results in enhancing the lubricant performance and extending the equipment life. Combination of different additives and their quantities are determined by the lubricant type (engine oils, gear oils, hydraulic fluids, cutting fluids, way lubricants, compressor oils, etc.) and the specific operating conditions (temperature, loads, machine parts materials, environment). Industrial oils are formulated with different additives from engine oils. Apart from meeting the minimum required performance in given applications (e.g. protection of metal surfaces from wear), selected additives should preferably improve the efficiency, exhibit low toxicity, be stable and avoid microbial degradation of lubricants. For eco-requirements for lubricant additives, see ref. [30].

Additin® RC 9330 (Rhein Chemie, Mannheim) is a new metal-free multipurpose additive package for both hydraulic oils and industrial gear oils. It features outstanding oxidation stability, filtration and anti-wear performance and excellent solubility in various base fluids including PAOs and white oils. Depending on the base fluid, Additin® RC 9330 is certified by NSF and can be used by the food-processing industry for H1 lubricants.

Synthetic lubricant base fluids (e.g. PAOs, synthetic esters) are frequently used in formulations containing the same types of functional additives as used in mineral oils. This is less appropriate for the more polar ester oils, and for vegetable oil base stocks. Ageing of lubricant formulations is usually on account of additive depletion (resulting from degradation due to shearing, oxidation, adverse chemical reactions, drop-out); see also Sections 14.4 and 14.5. In general, lubricant additives are not recyclable owing to degradation in service use [31]. Their degradation products are therefore removed as sludge during lubricant re-refining and contribute to waste generation. Most additives are combustible.

Most additives are specifically enhancing one property. However, additive components can either complement or counter each other. The benefits of lubricant additives are not marginal. They are not simply operational extras but essential ingredients which make all the difference in lubricant performance. The bottom line on the use of any additive is a desired level of performance. The additive package formulation needs to achieve the minimum requirements for a given application cost effectively. Although the overall performance of an oil can be improved by introducing additives, a poor quality oil cannot be converted into a premium quality oil simply by introducing additives. In some cases, additives may show internal or external incompatibility. Additive chemistry should avoid interaction between the components of a fully formulated lube oil (base stock and other additives), or between additives and other materials in the application (metals and seals, etc.); see also Section 7.2.4.

Most additives depend on a competitive reaction mechanism, i.e. the additive reacts with an undesirable substance that occurs in the lubrication system that, if unchecked, could lead to premature failure of either the base lubricant or component. For instance, an antioxidant reacts with the initial free radicals formed early in the oxidation process by delaying or preventing the attack on the base oil. It is usually important that additives are dormant and stable during normal lubricant function and act only on demand. Antioxidants become active and reactive only when oxidative free radicals are present. An additive may work by more than one mechanism, for example by both blocking and trapping water.

Most currently used high-performance synthetic oils, such as perfluoropolyether (PFPE), multiply-alkylated cyclopentanes (MACs) and silicone oils, have very poor miscibility with most additives. Low solubility makes it difficult to dissolve some needed additives and compromises essential parameters such as dispersancy and seal compatibility. PAOs with favourable PP and volatility have the lowest lubricity and solubility ranking. This can be compensated by using solubility improvers in finished lubricants. The high polarity and low aniline points of oil-soluble polyalkylene glycols (OSPs) will help formulators to solubilise critical additives in newer Group II and Group III base oils which are more isoparaffinic and less polar in nature [12].

Use of additives also brings along some potential disadvantages. Many people have been influenced by a widespread public suspicion of chemicals in general (and additives in particular, whether in foods, plastics or otherwise). Some materials used historically as lubricant additives have found to present potential hazards. Use of these materials, such as lead naphthenates, chlorinated naphthenates and tris(*o*-cresyl) phosphate, has generally been discontinued [32]. Some additives are classified as irritant (e.g. zinc dialkyldithiophosphate ZDDP) or sensitising [33]. Technological actions must take place within an increasingly (and understandably) strict environment, which regulates the potential hazards of chemicals in the workplace, the use of lubricants in incidental contact with foodstuffs, the long-term influence of additives on the environment when the product is recycled or otherwise comes to final disposal. Nowadays, many customers are looking for 'green additives' that are non-metallic, non-chlorinated, non-sulphur additives, and are also looking for natural products with low cost relative to the conventional additives.

Concerns are expressed by legislation and regulations, such as:

- General health and safety – fitness for purpose.
- EU Directives – disposal of waste oils, recycling, biocidal products.
- Water Management Acts – toxic effects in surface waters.
- Restrictions on Hazardous Substances (RoHS) – chemical hazards.
- Registration, Evaluation, Authorisation and Restriction of Chemicals (REACH), Globally Harmonised System (GHS) – European chemicals regulation system.
- Eco-labels – ecological criteria.
- Life-cycle analysis (LCA) – sustainability, evaluation of product use.
- BioPreferred^{SM} – green procurement.

All additives are subject to some form of regulatory control through general health and safety at work legislation. From an environmental and legislative point of view several additive types and elements in particular experience pressure, e.g. biocides, plasticisers and chlorine-containing compounds. Incineration of additive-containing materials, which recovers considerable energy for further use, leads to concern about the effects of any additives on the emissions produced [34]. European emissions legislation has forced the development of new additive technology. Similarly, the toxicity of certain additives has been a driving force towards new, safer, more biocompatible additives. For instance, European waste oil legislation [34], with restrictions on re-refining and disposal of used oils containing more than 0.2% Cl, has promoted the development of chlorine-free lubricants (base oils and additives). Where food contacts are an issue, base oil and additives used must be rigorously tested to avoid any contamination [35]. On the whole, the benefits of additives far outweigh their disadvantages.

Lubricant formulations usually contain a significant number of additives which are used to enhance an already existing useful property and suppress deficiencies, or to impart desirable new characteristics up to a level where they meet the specific requirements for a particular application. This is especially important in the case of vegetable oils. They perform many and varied functions such as friction reduction, protection of metal surfaces, heat transfer, engine seal protection, and separation and suspension of contaminants. The principal functional classes of additives are (natural or synthetic) antioxidants, detergents and dispersants, neutralising agents, PPDs, viscosity modifiers, friction modifiers and antifoam agents. Other property enhancing additives are typically selected from the group consisting of corrosion and rust inhibitors, metal deactivators, dyes, hydrolysis inhibitors, AW agents, thickeners, (d)emulsifiers, EP agents, etc., depending on the characteristics of the base oil and requirements of the particular application. Most industrial oils have an additive package containing rust and oxidation inhibitors, defoamant, PPD and sometimes AW and/or EP additives. For long-term trends in industrial lubricant additives, see ref. [36].

Additives can broadly be classified into two types that influence chemical and physical properties of base fluids (e.g. *V–T* characteristics, demulsibility, low-temperature properties, oxidation stability) and affect the metal surfaces (e.g. reduction of friction, increase in EP behaviour, wear protection, corrosion inhibition). Additives used in lubricant base oils are normally classified according to their intended performance, rather than on a chemical or physical basis. For ease of survey it is convenient to classify them into groups with similar functions. Additives then serve three main functions [37]:

- protection of the lubricated surface (surface additives);
- improvement of the performance of the lubricant (performance-enhancing additives);
- protection of the lubricant base oil (fluid protective additives).

The additives impart special properties, such as reduced friction, reduced wear, increased viscosity, improved viscosity index, resistance to corrosion, oxidation, ageing and/or contamination, to the lubricant formulation. Table 7.3 gives a short overview of the main types of additives used. For more details, see refs. [38–41].

Additives can enhance one property while adversely affecting another. Compatibility of additives is very important to minimise adverse chemical reactions that may destroy essential properties. The effectiveness of an additive package depends on the chemistry of the base oil and the additives, the properties of the material surfaces and the wear regime [42]. Oils for similar applications but produced by different manufacturers may be incompatible owing to the additives used. By mixing such incompatible fluids, additives may be consumed by adverse chemical reactions, resulting in undesired additive depletion. Similarly, some petroleum-based oils, such as those used in turbine, hydraulic, motor and gear applications are acidic, whereas other oils, such as motor oils and transmission fluids, are basic. Acid and alkaline lubricants are incompatible. When servicing oil lubricating systems, it is important to ascertain that the old and new oils are compatible. For additive–additive interactions, see also Section 7.2.4.

A great deal of interest exists in multifunctional additives, which combine different additive functionalities (e.g. antioxidant and AW agent, antioxidant and thickening agent, lubricity additive and emulsifier, rust and oxidation inhibitor) and are therefore capable of imparting more than one desirable property to lubricating compositions. ZDDP is the most widely used multifunctional additive in engine oil formulations, where it plays a role both as an antioxidant and as an anti-wear additive [43]. For other

Table 7.3 Types of additive

Additive class	Function	Example(s)[a]
Surface active additives		
Anti-wear agents	Reduce wear of loaded contact surfaces	Ethyl stearate, ZDDPs
Extreme pressure agents	Prevent microwelding	Overbased sulphonates, MoS_2
Friction modifiers	Reduce frictional forces	Fatty acid esters, organomolybdenum compounds
Rust inhibitors	Prevent corrosion of ferrous materials	Alkylbenzene sulphonates
Corrosion inhibitors	Protect against chemical attack	Long-chain carboxylic acid derivatives
Detergents	Keep metal surfaces clean	Sulphonates, phenolates, salicylates
Dispersants	Disperse particulate matter (debris)	Polybutenes, metal soaps
Tackiness agents	Increase surface adhesiveness	Long-chain fatty acid soaps
Performance-enhancing additives		
Viscosity index improvers	Improve *V–T* relation	Polymers (PIB, EVA, SBS)
Pour point depressants	Affect fluidity by controlling crystal formation	Polymethacrylates
Emulsifiers	Disperse water in base oil	Naphthenic acids
Demulsifiers	Separate water contamination from lube oil	Polyethylene glycols
Deposit control additives	Control deposits	OSPs
Thickeners	Convert oil into solid or semi-solid lube	Calcium soaps
Lubricant protective additives		
Antioxidants	Improve oxidation resistance	Phenolics, aromatic amines, phosphites
Foam inhibitors	Prevent foaming	Polydimethylsiloxanes
Antileak agents	Prevent leakage; seal conditioning	Polymers, soaps
Biocides	Regulate microbial activity	Phenol derivatives, isothiazoles

[a] ZDDP, zinc dialkyldithiophosphate; PIB, polyisobutylene; EVA, ethylene–vinylacetate copolymer; SBS, styrene hydrogenated butadiene–styrene terpolymer; OSPs, oil-soluble polyalkylene glycols.

examples of multifunctional additives for lubricating oils and cutting fluids, see refs [44–46]. One of the problems associated with a multifunctioning additive is the difficulty in achieving a proper balance of its various characteristics, so that each desirable property can fully manifest itself under actual service conditions and yet complement other desirable properties. For example, one might wish a VI improver also to have other properties such as AW and anti-sludge properties. One solution is to incorporate into the molecular structure of such additive certain polar groups which are capable of imparting detergent or dispersant properties. Yet, the incorporation of such polar groups necessarily affects the balance of the limited but sufficient solubility characteristics. In US Patent No. 3,702,300 (to Lubrizol Corp.) Coleman [47] discloses nitrogen-containing esters of carboxy-containing anhydride interpolymers which act as VI improvers and anti-sludge agents.

New base oils and new applications determine complex additivation challenges and drive the development of new lubricant additives. Infineum International, Ltd (ExxonMobil/Shell jv; Abingdon, UK) is a world leader in thc formulation, manufacture and marketing of additives for lubricants and fuels. Other well-known high-performance lubricant additive companies include Lubrizol Corp., Chevron Oronite Co., Ethyl Petroleum Additives Ltd, Afton Chemical Corp., Croda Lubricants, Rhein Chemie (Lanxess Group). For instance, Croda's technology platform includes multifunctional esters (Priolube™) polyalkylene glycols (Emkarox™), polymerised FAs (Pripol™, Prisorine™) as well as specially additives (e.g. Perfad™, Crodacor™). As to production, more severe HSE regulations continue to be placed on the handling of (solid) additives. The use of non-dusting polymeric bound additives for lubricants, such as Additin® RC8400 (Rhein Chemie, Mannheim), outperforms traditional solid alternatives in EP greases.

Traditionally, lubricant producers do not specify which compounds are used to enhance the lubricant quality, but specify only the generic functions (e.g. AW, EP, rust and oxidation, R&O). Their composition is usually trade secret and classified as confidential business information (CBI). The user must assume that the appropriate agents have been added in sufficient quantities to obtain the necessary degree of protection. Additive packages are also subject to change (without notice). For selection of lubricant additives for the food-processing industry, see ref. [30].

A typical vegetable oil lubricating composition with improved thermal and oxidative stability, corrosion resistance, and anti-wear properties contains at least 90 wt% of a vegetable oil and an additive composition including 1.5–2.0 wt% triphenylphosphorothionate (TPPT), 0.1–3.0 wt% hindered phenolic antioxidant, 0.05–0.25 wt% 1-[di(phenyl)aminomethyl]tolutriazole, and about 0.05–0.5 wt% alkyl succinic acid half ester as a rust inhibitor [48]. Extensive lists of many types of additives for environmentally friendly lubricant compositions are found in Eur. Patent No. EP 2,017,329 A1 [40] and elsewhere. US Patent No. 5,413,725 details additive packages for biodegradable base fluids (vegetable oils) [49].

Grease additives are usually not stated explicitly except for solid additives (such as MoS_2 or graphite), or that an EP additive is included. Alcell™ organosolv lignin (Repap Technologies, Inc. Valley Forge, PA) is a competitive, green and multifunctional grease additive [50]. Additives for grease applications have been recently reviewed [51].

Customary additives for lubrication systems are usually not adequate for aerospace (aircraft and spacecraft) applications (gas turbine engine oils, greases and hydraulic fluids, etc.), which are called upon to operate in extreme conditions of temperature and space environment, and require more development [52].

In view of their miscibility with polar and non-polar solvents, ionic liquids have been proposed as advanced lubricating additives both in water and in lubricating oils (see Section 15.2.2). For further technological progress, it is necessary to develop a better understanding of how additives affect the interactions between lubricants and rubbing surfaces. It is also of interest to gain better insight in the rate of degradation of oil additives.

ASTM D 6443 is a standard test method (based on WDXRF) for a variety of elements (typically Ca, Cl, Cu, Mg, P, S and Zn) in lubricating oils and additives and can be used to verify that additive packages meet specifications with respect to the added elements, and used oils with respect to chlorine contamination [53]; see also Section 8.7.2.

7.2.1 Surface additives

Surface additives are meant to protect (bio)lubricated surfaces. Additives such as AW and EP agents, boundary lubricity additives, oiliness additives, (bio)rust inhibitors, (bio)corrosion inhibitors, tackiness agents and others act by coating the lubricated surfaces and preventing wear, rust, corrosion, etc. In the same way as the phenomenon of lubrication is complex (see Chapter 2) so are the processes which take place in the presence of lubricity additives.

AW agents

AW agents are polar molecules capable of binding or attaching themselves to metal surfaces. In this way they form a stable layer that protects the metal surface from the wear that can occur during boundary lubricating conditions, e.g. direct contact between asperities. This reduces friction and wear up to 250 °C. The AW (and antifriction) properties of most commonly used commercial additives are on account of functional groups containing a variety of elements (B, P, S, N, Cu, Zn, Pb) capable of reacting with the metal surface. Traditionally, wear protection in lubricants is provided by ZDDP or other phosphorus compounds, which act through formation of a complex glassy polyphosphate AW film. However, zinc and phosphate residues from ZDDP breakdown, along with sulphur and sulphated ash from lubricant detergents and dispersants, disrupt automotive catalytic converters. Consequently, the automotive industry has phased out the use of phosphorus and requires reduced sulphated ash, phosphorus and sulphur (low SAPS) lubricants (see also Section 7.3). This has been achieved commercially by introducing catalyst-friendly, titanium-based, non-phosphorus-containing AW additives (Afton Chemical Corp.; Richmond, VA) [54]. It is postulated that such additives form protective $FeTiO_3$ films on the engine surfaces.

VanLube® 289 (R.T. Vanderbilt Company, Inc.; Norwalk, CT) is a boron-containing AW (SAPS) additive. Sulphurised FAs added to base oils can equally achieve AW improvements [55]. Poly(hydroxy thioether) derivatives, chemically modified vegetable oils prepared by reacting epoxidised triglyceride oils with thiols, also have utility as AW/antifriction additives for industrial oils and automotive applications [56]; see Section 7.3.

Other related antifriction/AW lubricant additives were disclosed by Caterpillar, Inc. [57]. Industrial Lubricants, Inc. (ILI; Seattle, WA) has introduced three patented lubricant technologies – Liquid Wax Esters (LXE®), Synergol® Lubricity and Anti-Wear Technology. Synergol products are non-sacrificial AW additives, diesel fuel lubricity and general lubricant lubricity additives. Lubricant oil and grease compositions have recently been disclosed containing an additive package comprising wear-resistant additives in the form of nanoparticles [58]. AW additives are particularly important for oils used in racing vehicles, as those engines are operated at extremely high temperatures and pressures.

EP agents

EP additives alleviate the dramatic effect of 'dry' friction by preventing destructive metal-to-metal contact in lubrication at high temperature and pressure conditions such as those found in certain gear elements in automotive vehicles and various industrial machines where high pressure can cause a film of lubricant to rupture. When the pressure between two moving surfaces becomes so high that most of the film of lubricating oil is squeezed out (a 'boundary lubricant' situation), some other kind of lubrication is required. Lubricating compositions composed only of relatively high-molecular-weight hydrocarbon base stock tend to fail and the parts become damaged. This problem can be addressed by additivation to increase high-pressure performance. EP additives, which are often composed of sulphur, phosphorus, chlorine or molybdenum derivatives, react with the rubbing surfaces to form a surface layer with lower shear strength than the metal. This protective coating, which acts as a solid lubricant when the rubbing surfaces come into direct contact, will reduce friction, wear and surface scoring.

EP agents are used in particular in applications that may experience heavy loading or shock, such as turbines, gears, and ball and roller bearings. EP additives are not as necessary in engine oils as they are in many gear oils. AW/EP lubricants should have good lubricity, good cooling properties, high film strength, good load-bearing ability and miscibility with the usual types of base oils. Typical products qualifying are natural wax esters, e.g. jojoba oil (in short supply), sulphurised vegetable oils [25, 59], synthetic wax esters (formed by esterifying an unsaturated FA and an unsaturated fatty alcohol) [60], or telomerised triglyceride vegetable oils [61–64]. Applications are found in continuous casting of steels. Sulphurised jojoba oil and sulphurised FAs (oleic, C18:1 and erucic, C22:1) have proven excellent performance as EP additives in vegetable-based oils [55]. Botanical additive packages allow formulation of 'load-carrying' capabilities into individual products for specific applications while optimising performance (equal or better than EP agents) and economics without pollution concerns. Use of chlorinated paraffins is under scrutiny in North America. For trends in EP additives, see ref. [65].

Boundary lubricity additives

While conventional AW/EP additives such as ZDDP reduce friction and wear by chemically reacting with the metal surface under boundary contact conditions once asperity–asperity contacts occur to yield a reaction product which prevents cold welding, they do not prevent the occurrence of such direct contacts. Boundary lubricity additives reduce friction and wear by forming adsorbed surface layers (fatty amides, esters) or slippery surface deposits (graphite, Teflon), which physically separate the rubbing surfaces from one another. Boundary lubricity additives keep their lubricity-enhancing effect even if there is no reciprocal motion between the rubbing surfaces. They are particularly efficient in controlling stick-slip and chatter phenomena.

Friction modifiers

Friction modifiers are variously known as lubricity or fatty oiliness additives. Unlike EP additives, which act when a direct asperity-to-asperity contact occurs in the boundary lubrication regime, lubricity additives work by postponing the onset of the boundary lubrication regime, as illustrated in Fig. 7.1. In other words: in a loading cycle the film lubrication regime is expanded, whereas in an unloading cycle the change from boundary to film lubrication will occur earlier. In the area of low slide velocities, moderately increased loads and low viscosities at higher temperatures fluid lubrication can easily turn into mixed friction conditions. While lubricity additives elevate the threshold pressure at which the dry contact will occur they do not provide protection beyond that point. Therefore, for maximum protection under various tribological scenarios, it is important that EP additives be deployed as well in finished formulations. A good lubricity additive dissolved in a base oil combines a sufficiently high surface affinity with the ability to form a sufficiently thick and resilient protective film. The protective film should not inhibit reactions of conventional EP additives. Surface adsorptivity normally increases with increasing polarity of the additive, while thickness of the adsorbed film increases with increasing its viscosity (related to average molecular weight). However, excessive polarity and viscosity undermine miscibility of the additive with (hydrocarbon) base stocks. An amphiphilic molecular structure is to be preferred.

Friction modifiers prevent stick–slip oscillations and noises by reducing frictional forces. They can be regarded as mild AW or EP additives working at moderate temperatures and loads in the area of beginning mixed friction. Friction modifiers (as well as detergents) are highly polar molecules. Using a highly polar base fluid with these kinds of additives leads to competition between the base fluid and the additives for access to the surface. A small amount of vegetable oil or animal fat added to a mineral oil (to form a compounded oil) reduces the coefficient of friction (CoF) without affecting the viscosity. Compounded oils with improved oiliness are generally used in worm gears. Friction modifiers are typically used in engine oils to improve fuel efficiency. Oil-soluble friction modifiers effective in engine oils are organomolybdenum compounds (molybdenum dithiophosphate/dialkyldithiocarbamate – MoDTP and MoDTC) and ashless organic ester- or acid-based additives like stearic acid or octadecanol. AW/EP and boundary lubricity additives are often both referred to as friction modifiers, despite their differences.

Superlubricity additives

As shown in Fig. 7.1, superlubricity additives (e.g. E-ION® additives) expand the range of operating conditions under which film lubrication is sustained. High pressure and low sliding velocities force the tribosystem into the boundary lubrication regime in which most intense friction and wear occur. AW/EP additives, as well as regular boundary lubricity additives, shift the Stribeck curve down, reducing friction in the boundary lubrication regime. Superlubricity additives shift the Stribeck curve to the left, maintaining the

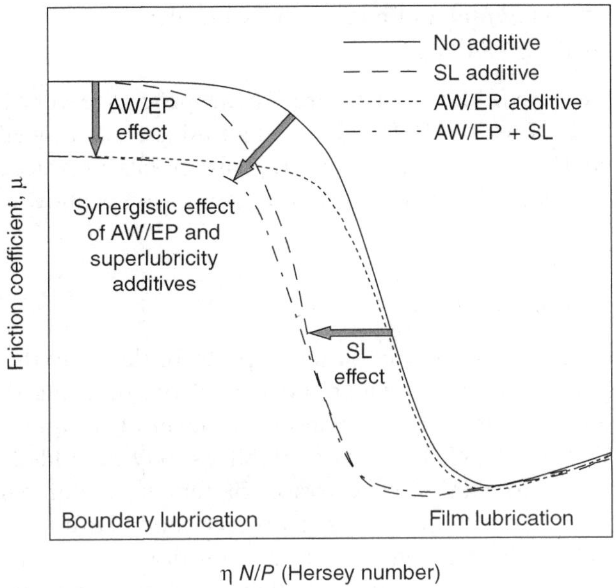

7.1 Stribeck diagram comparing the tribological effect of conventional AW/EP and boundary lubricity additives with the effect of superlubricity (SL) additives (μ, the coefficient of friction; η, viscosity; *N*, relative sliding velocity; *P*, applied pressure). After ref. [5].

film lubrication regime over a broader range of tribological conditions [5]. They are a special class of boundary lubricity additives. Examples are certain amphiphilic ester-based comb-copolymers and Elektrionized™ vegetable oils. Roegiers and Zhmud [6] have described how such additives function in various tribosystems and which parameters control lubricity of finished formulations. Electrionisation of vegetable feedstocks produces outstanding tribological performance of bio-based lubricity and fatty oiliness additives. Featuring a unique combination of viscosity and polarity, ionised vegetable oils form sufficiently thick and resilient protective layers by adsorption to rubbing surfaces. Unlike EP additives, which act when a direct asperity–asperity contact occurs in the boundary lubrication regime, ionised vegetable oils function by postponing the onset of the boundary lubrication regime (see Fig. 7.1).

Superlubricity additives provide the following benefits [66]:

- enhanced wear protection and extended tool life;
- clean running and excellent surface finish;
- better process runnability and ability to operate at higher speeds;
- reduced lubricant consumption;
- better emulsion stability in the case of soluble oils;

- improved environmental and health safety profile;
- excellent antisludge activity.

In combination with synthetic and severely hydroprocessed very high and extra-high viscosity index (VHVI, XHVI) mineral-base oils, as well as with emerging Group III+, or 'Super-Group III' base stocks produced by the Fischer–Tropsch process, lubricity and solubility improvers allow formulating top-quality lubricants.

Antirust and corrosion inhibitors

These additives bind to the surface of metal parts. In that way they create a protective layer that prevents water/chemicals from reaching the metal and rust/corrosion from starting to develop. This layer also suppresses oxidation and formation of acids. Bio-rust inhibitors may be added to most industrial lubricants to minimise corrosion of ferrous metal parts. Bio-corrosion inhibitors are used primarily to protect alloy bearings and other metals in internal combustion engines from deterioration by oxidation or other chemical action. Epoxidised 2-ethylhexyl tallate is effective as an anticorrosion, antifatigue and AW additive [67]. An ashless, calcium compatible corrosion inhibitor for use in industrial lubricating oils has been described [68]. Many chemistries used for other primary functions such as detergency, emulsification and AW properties can prevent corrosion or rust from forming.

Detergents and dispersants

Detergents (such as metal sulphonates, salicylates and thiophosphonates) and dispersants (such as FAs, metal soaps, waxes, fatty alcohols, polyamine succinimides, hydroxybenzyl polyamines) are polar substances that serve a cleaning function. Most fuel additives are detergents aimed at keeping engines clean to maintain optimal running conditions. Added to a lubricant base oil, they keep compounds and substances dissolved in the liquid and avoid deposition onto metal surfaces. Polyalkenyl succinimide and polybutenes are metal-free, ashless dispersants. Dispersing agents aid in the uniform dispersion of additives. Bio-detergents and dispersants are used primarily in hydraulic systems and internal combustion engines to keep metal surfaces clean by preventing deposition of oxidation products. Commercial dispersant/inhibitor packages are only available for the non-polar mineral oils, but are lacking for the more polar vegetable oils. Section 10.5.1 illustrates the consequences. With the present trend toward smaller but equally powerful engines, lubricants experience more stressful conditions. This requires better rather than more detergents and dispersants in the lubricant formulations.

Tackiness agents

For applications where the oil needs to adhere to surfaces extremely well, tackiness agents are added such as aluminium soaps of long-chain fatty acids.

7.2.2 Performance-enhancing additives

VI improvers, PPDs, emulsifiers, demulsifiers, deposit control additives and thickeners are added to (bio)lubricants to improve their quality performance up to the desired level for the specific application. More in particular, their actions are as follows.

VI improvers

VI improvers are chemicals which are added to lubricating oils to make them conform more closely to the ideal lubricant (with an almost constant viscosity at all temperatures). All commercially important VI improvers are oil-soluble organic polymers. Chemical classes of VI additives are polymeric hydrocarbons (polyisobutene, ethylene and propylene copolymers and butadiene and styrene copolymers), ester-containing polymers (polyacrylates and polymethylacrylates), and polymers modified with dispersant properties (by introduction of monomers containing polar groups such as tertiary amines, imidazoles, pyrrolidone and pyridines). Polymer viscosity modifiers can keep oil working over a wider temperature range. For a polymeric substance to function as a VI improver, it must have sufficient but limited solubility in the lubricating oil at ordinary temperatures. Sufficient solubility determines a desired oil concentration, however without being able to exert full viscosity-modifying potential until service temperature. VI improvers exert little effect on oil viscosity at low temperature. Owing to the increased solubility at higher temperature the VI improver displays a greater thickening effect and thus prevents excessive thinning of the lubricant oil at the desired operating conditions. The dual requirements of limited solubility and sufficient stability are difficult to meet. The quality of VI improvers is mostly appreciated in the application of motor oils. Commercial VI improvers are olefin copolymers (for engine and hydraulic oils), polyalkylacrylates and PIB (for gear oils), and hydrogenated styrene–isoprene copolymers and hydrogenated styrene–butadiene (for engine oils). Polymeric viscosity improver additives can boost vehicle fuel economy [69].

PPDs

Oils with high PP render start-up of machinery difficult or impossible. Paraffin waxes tend to solidify and stiffen at low temperatures. PPDs reduce the size and cohesiveness of crystal structures, resulting in reduced PP and increased cold flow at reduced temperatures. Most of the commercially available PPDs are polymethacrylate-based or alkylated polystyrenes. US Patent No. 5,413,725 to Lal *et al.* (to Lubrizol Corp.) has described PPDs for high monounsaturated (>60%), non-genetically modified vegetable oils (i.e. high-oleic cottonseed oil (HOCSO), high-oleic rapeseed oil (HORSO), high-oleic soybean oil (HOSBO), high-oleic safflower oil (HOSFO), high-oleic sunflower oil (HOSNO), high-oleic corn oil, high-oleic lesquerella oil, high-oleic meadowfoam oil and high-oleic palm oils) [49]. The influence by any particular polymer on the PP of an oil is rather unpredictable. PPDs are most valuable if they can be used in small concentrations yet be effective over a wide range of viscosities.

Emulsifiers and demulsifiers

Emulsifiers fall under the chemical class of surfactants or surface-active substances, more specifically water-soluble surface-active substances. Emulsifiers, which reduce the interface tension and counteract droplet enlargement, are divided into ionic products (mostly anionic; soaps) and non-ionic products (ethoxylates). Soluble oils require that water and oil are emulsified rapidly and form stable emulsions. Emulsifiers (e.g. alkali salts of FAs) are used in lubricants for cutting, grinding and drilling applications in machine shops. Self-emulsifying surfactants combine lubrication and emulsification in one product. Demulsifiers promote separation of water and oil in lubricants exposed to water of contamination, such as industrial oils in circulation systems (hydraulic, gear, turbine and compressor oils). Demulsifiers and emulsifiers use the same class of chemical substances (surfactants). Special polyethylene glycols are highly efficient demulsifiers.

Deposit control additives

In some applications (e.g. gas turbine oils) controlling deposits and varnish formation is a critical concern [70]. OSP polymers are excellent deposit control additives; see Section 3.2 [12]. Deposit control is tested by modified ASTM D 2893B test.

Thickeners

Thickeners transform liquid lubricants into consistent lubricants (lubricating greases) and affect the properties. Most commercially important thickeners (including simple soaps) are comparably competitive.

7.2.3 Fluid protective additives

Fluid protective additives are employed to protect the (bio)lubricant instead of the equipment.

Oxidation inhibitors

Hydrocarbon compositions tend to oxidise during use and over time. Oxidation products of lubricating oils (carboxylic acids, ketones and alcohols) may damage the metal parts and condense to form polymeric materials, leading to the formation of varnish and sludge and to an increase in viscosity. Oxidation inhibitors function by reducing the quantity of oxygens reacting with oil by forming inactive soluble compounds. As (synthetic or natural) antioxidants are consumed during use, lubricant performance for the specific application is reduced and equipment failure may result under continued use. Oil monitoring programmes should be implemented to periodically test oils and verify that the essential additives have not been depleted to unacceptable levels. Oxidation inhibitors are used in most industrial lubricant applications where oil is continuously circulated or contained in a housing. There are several types of oxidation inhibitors used in lubricants, which are classified as:

- UV absorbers;
- peroxide decomposers;
- chain breaking – electron donors;
- chain breaking – electron acceptors.

Antioxidants for lubricant protection are disclosed in many patents, e.g. US Patents No. 5,336,420 [71], 5,451,334 [72] and 5,773,391 [73], and PCT Int. Publ. No. WO 03/027212 [8]. The performance characteristics of different lubricants formulated with low- and high-molecular-weight phenolic antioxidants have been reported [74]. Solubility and lubricity improvers help formulators address certain challenges brought about by 'dry' base stocks. However, the antioxidant response of vegetable-based products improvers is different from that of hydrocarbon bases and, therefore, the antioxidant package may need to be redesigned accordingly. Phosphites and some conjugated dual antioxidant systems such as tocopherol–disulphide are known to be efficient antioxidants for vegetable oils.

Foam inhibitors

Air and other gases mixed up in the lubricating oil during action in an engine or machine may produce foam. Foaming of lubricants is a very undesirable effect that can cause enhanced oxidation as well as insufficient oil transport in circulation systems, thus affecting lubrication. Foaming can result in insufficient oil delivery to bearings, causing premature failure. Foams may thus interfere with proper operation of equipment. Antifoaming additives prevent reservoir overflow, overheating and contact between adjacent metal surfaces. Antifoaming additives also eliminate sponginess in hydraulic systems and prevent damage caused by pump cavitation. Antifoaming agents added to lubricant oils act both as fluid protective and performance-enhancing additives. Silicon-free defoamers are special polyethylene glycols (PEGs), polyethers and polymethacrylates.

Anti-leak or seal conditioning additives

Prevent the undesirable leakage of fluid past seals or worn tolerances under high-pressure conditions, reducing lubricant consumption and pollution. To reduce equipment leakage, elastomeric seals and gaskets should maintain a slight positive swell and remain pliable. This is determined by the solvency of the lubricant base oil (as expressed by the aniline point; see Section 3.6). Obstruction of small leaks with retardant materials requires polymers which have an oil thickening effect, or gelling agents (such as Li or Al soaps), which barricade small openings, preventing fluid flow. The types of leak retardants which can be used are limited by other performance characteristics. Anti-leak components should not plug filters. There are no known products which completely stop leakage.

Biocides

Antimicrobials are used to regulate microbial activity (preservation of coolants). The growth of bacteria and fungi can be kept under control by the use of biocides. There is no consensus on the level of bacteria or fungi that demands a biocide response. Experienced lubricant manufacturers usually can recommend to users the microbial level required of biocidal addition to determine the best prophylactic level. In Europe, the handling of biocides is regulated by the Biocidal Products Directive (BPD), Guideline 98/8/EC [75]. The BPD is similar to the US Environmental Protection Agency (EPA) Federal Insecticide, Fungicide, and Rodenticide Act (FIFRA)-based regulations [76]. The EC Commission has published a Final List of Notified and Identified Biocides, comprising product-type 13 'metalworking fluid preservatives' [77]. Numerous chemical classes exhibit antimicrobial activity, including alkanolamines [78] and alkylphenols [79]. Biocides are generally added in small amounts. However, biocides have a limited lifetime and raise issues regarding working safety. Some biocides are highly toxicologically active and hazardous. International Specialty Products (ISP)/Ashland Specialty Ingredients (Wayne, NJ) operates an effective line of biocides such as Fungitrol® and Nuosept®. As biofouling is a major problem in metalworking shops, where workers are exposed to skin contact, inhalation or swallowing, biocides are added to metalworking fluids (MWFs). When lubricant manufacturers develop and formulate MWFs, toxicological, medical and industrial hygiene aspects are a high priority. Some 80 active ingredients are currently approved by the EPA for use in MWFs, but only few account for more than 90% of total MWF microbicide use, in particular hexahydro-1,3,5-tris(2-hydroxyethyl)-*S*-triazine (HTHT) and various isothiazolines (5-chloro-2-methyl-4-isothiazolin-3-one, CIT, 2-methyl-3(2H)-isothiazolin-3-one, MIT).

7.2.4 Additive–additive interactions

Lube oil stability is strongly influenced by the particular combinations and concentrations of transition metal or lube additives used. Many additives, being polar in nature, have been found to interact with each other in solution and at surfaces, thereby exhibiting both antagonistic and synergistic effects compared with the performance of the individual additives [80]. Additive–additive interactions may have important consequences for the development of oil formulations. For example, AW agents, friction modifiers, and detergents and dispersants in engine oil lubricants have been found to interact with each other, thereby affecting the overall performance of a lubricant system [80, 81]. The stability of the additive package is quite different from that of the additives on their own. ZDDP is well known to interact with most other additives employed in engine oil formulations, such as detergents (metal sulphonates/phenates) [43, 80, 82], AW/EP ashless dialkyldithiophosphate or its amine salts [83], succinimide-based dispersants (PIBS) [83] and MoDTC [81]. In particular, ZDDPs, once used extensively as anti-wear additives and antioxidants in automotive lubricants, strongly inhibit iron-catalysed oxidations, but their activity is greatly reduced in the presence of basic additives (e.g. overbased detergents) [84]. Additive–additive interactions in the ZDDP–PIBS–OBS [83] and zinc dialkyldithiocarbamate (ZDTC)-ZDDP systems [85] in mineral base oils have been investigated. Synergy of AW additives (ZDTC) and phenolic antioxidants has been reported [86] as well as the synergistic effect on tribological behaviour of two EP and AW additives [87, 88]. Rounds [83] has examined changes in lubrication performance caused by interactions among surface-active engine oil additives.

With the introduction of better quality and environmentally friendly base oils (API Groups II to IV), the study of additive–additive interactions has gained in practical importance due to the entirely different types of behaviour shown by formulations prepared using these base oils [43, 82, 83]. More systematic studies on the effect of base oils on interaction behaviour is important in relation to the development of additives that will have better compatibility and show improved performance, in particular in the more polar vegetable based oils. The role of analytical techniques is significant in understanding the mechanism of additive–additive interactions leading to various adducts or complexes being formed (see Section 8.7.2). These results will have impact on the selection and optimisation of additive dosages and the production of thermodynamic stable formulations.

Decomposition of additives, eventually accompanied by auto-oxidation of vegetable oils, is another phenomenon found to occur. For example, anti-wear additives may degrade in vegetable oils and give less oxidative stability. It has also been observed in engine tests that ZDTP decomposes in zinc-containing compounds and phosphorus-containing intermediate products such as tetraalkylthioperoxidiphosphate [89]. However, in this case the AW performance of the oils used was not greatly affected, being mainly attributable to the phosphorus-containing compounds. Other forms of breakdown of additives are observed for deposit control additives (oxidation inhibitors, detergents and dispersants), film-forming additives (friction modifiers), AW/EP additives, viscosity regulators (viscosity modifiers and PPDs) and miscellaneous additives (tackifiers, seal swell additives, antimicrobial additives, surfactants, corrosion inhibitors and rust preventatives) [30].

7.3 Additive design for renewable lubricants

For a variety of reasons, nowadays renewable or biodegradable lubricant additives are attracting considerable attention. Most additives currently used for the traditional mineral-based oils are non-biodegradable. Therefore, they cannot be used for EA fluids (see Section 3.5). When additive packages are based on petrochemical sources a (bio)lubricant formulation is no longer totally renewable or eventually not even totally biodegradable. The next challenge is therefore to move from petrochemical to renewable additive feedstocks.

It is paramount for the industry to examine the use of tribologically functional additive components that are compatible with environmental regulations. In particular, substances that are persistent in the environment, bioaccumulative and toxic should be banned from use (see refs [75, 77]). Hence, there is a need for additive compositions and lubricant formulations containing such compositions that incorporate more environmentally acceptable materials. Not all additives pass the requirements for low ecotoxicity. Biocides present an inherent ecotoxicity problem. The use of microbiocides is regulated (BPD, FIFRA); see Sections 7.2.3 and 12.9. Clearly, a formulated oil consisting of a highly biodegradable base fluid in combination with toxic additives is an environmentally unacceptable lubricant.

According to the Additive Technical Committee (ATC) the mammalian toxicological potential of most commonly used lubricant additives is low [90]. Consequently, they are not classified as dangerous. Of the most important additive types only ZDDP and long-chain calcium alkaryl sulphonates are classified as irritant. The widely used commercial antioxidant/AW agent ZDDP faces strong environmental challenges and cannot be used in biodegradable lubricants because it leads to toxicity to water bodies and land soils by decomposing and generating S- and P-containing toxic compounds [91]. *p*-Toluidine, an effective oxidation inhibitor for vegetable oils, improves the thermo-oxidative properties of jojoba oil, as an alternative to additives such as ZDDP and 2,6-di-*t*-butyl-*p*-cresol (DBPC), as used in mineral oil-based lubricants [22]. Additives containing chlorine pose another considerable threat to human health and to the environment. Chloroparaffins (once used as EP additives in cutting fluids) are now banned from use. Contact with chlorine-containing cutting fluids can lead to chlorine acne.

In the past, it was (erroneously) assumed that additives designed for mineral oil lubricants perform similarly when used with vegetable oils. This led to the production of technically inferior plant-based lubricants. Many of the industry's negative opinions of plant-based lubricants stem from these early products. As the physicochemical properties of EA lubricants (such as vegetable oils) differ considerably from mineral oils, optimised lube oil performance can be achieved only by specific additive design (additive reformulation). Bio-based lubricants have a higher lubricity and therefore a much lower CoF than mineral-based lubricants. Enhanced lubricity reduces friction and wear, which reduces the necessity for many additives, including those for AW and EP. Owing to environmental concerns, development of vegetable- or plant-based biodegradable lubricating oil additives and additive packages with specific properties and applications is of interest. Recently, industry has started reformulating and/or designing additives specifically for vegetable oils with successful results. Highly soluble antioxidants, suited to vegetable oils for application as hydraulic fluids with improved biodegradable properties, were described by Konishi and Kikuchi in Germ. Patent DE 60,122,400 T (to Nippon Mitsubishi Oil Corp.) [92]. Other companies have developed purely plant-based additive packages. GEMTEK (Phoenix, AZ), which produces a full line of total-loss lubricants and hydraulic fluids, uses vegetable seed oil telomers, other seed esters, and rarified waxy substances as additives. OLEOTEK (Québec, QC) aims at developing bio-based additives with increased lubricating properties or performance in cold conditions. Badger Oil (Spooner, WI) has licensed a patented technology from the German company TESSOL GmbH (Stuttgart) that also produces additives from plant sources. Eur. Patent No. EP 2,017,329 A1 (to Afton Chemical Corp.) [40] discloses lubricant additive compositions having environmentally compatible characteristics as defined by EC Directive 67/548/EEC in which the components are selected from non-persistent, non-bioaccumulative materials, or non-toxic materials. Additive requirements for biodegradable lubricants (RSO, esters, PAGs) were discussed [93, 94]. Additives are frequently used to improve a vegetable oil's oxidative stability [95] and PP [96], but usually with some sacrifice in terms of biodegradability and toxicity.

Usage issues limit the application of vegetable oil-based lubricants, in particular formulation challenges. Attention should be paid especially to hydrolysis, oxidation and emulsification. Emulsification is more difficult with natural oils, which require a lower hydrophilic–lipophilic balance (HLB), resulting in the need for more and stronger emulsifiers. Canola oil, containing primarily monounsaturated fatty acid (MUFA) chains, is the easiest vegetable oil for formulation; rapeseed oil is slightly more challenging to emulsify. SBO, with high levels of polyunsaturation, is best reserved for applications where expectations for long emulsion service are low. Emulsification is easier for synthetic ester oils. Drawbacks to vegetable emulsifier technologies are: (i) high treat rates (up to 35–40%; cf. 15 to 20% for mineral oil formulations); and (ii) intolerance to alkalinity. This translates in higher costs, higher foaming tendency and the need for 'two-phase' systems. Lubrizol® MC944V overcomes many of the issues of the vegetable oil emulsifier technology [97].

Elektrionization™ or Voltolization is a process in which a lubricating oil is bombarded with ions formed by a high-frequency, high-voltage alternating current (Elektrion S.A process) [98]. The process of condensation and polymerisation of electro-ionising treatment of a mixture of a relatively light mineral oil and a refined fatty oil by means of an electric discharge in hydrogen atmosphere leads to a considerable increase in molecular weight, viscosity and VI, and a decrease in the CoF. The reaction products, which feature an optimum balance between polarity – the property essential for adsorptivity – and miscibility with base oils, may be used for blending with lighter lubricating oils. Electrionised vegetable oils allow formulating top-quality lubricants. By carefully selecting constituents of the initial blends and processing conditions, a variety of products with varying spectrum of properties may be produced, including pure vegetable oil-based bioadditives (E-ION® additives used for metalworking fluids, specialty lubricants and greases). E-Ion S.A. (Brussels) produces Elektrionized™ vegetable oils as lubricity additives. The products may be used as friction modifiers in metalworking lubricants [66].

In contrast with traditional lubricants, biolubricants can be ultimate biodegradable and may have a low toxicity as long as the additives utilised also possess these characteristics. In preceding sections (7.2.1–7.2.3) we have already indicated various examples of biodegradable additives. For instance, succinic acid derivatives are used to produce ashless (metal-free) corrosion protection additives. Fully biodegradable bio-based additives typically consist of alkanolamines, amino acids, plant and vegetable extracts, and non-ionic surfactants. Additives that are more than 80% biodegradable (CEC L-33-T-82) are available. Further development work is under way to produce environmentally suitable additives for improving the properties of EA base fluids.

Use of additives to improve the lubricating capacity and durability of oil plays an important role in the wear and friction process of materials. Organic straight-chain compounds with polar end-groups, such as oleic acid amide, are widely used as friction-modifying additives in lubricating oils. Fatty acid esters of a polyol have been used as a friction modifier in a GTL lubricating base oil [20]. Palm oil methyl ester (PME) has potential as an AW additive for lubrication [99]. Branched oleochemicals were used as additives in SBO and in PAO for lubrication enhancement [100]. Feng [101] has described a modified vegetable oil-based water-soluble antirust lubricity additive for use in a water-based metal cutting fluid. Vegetable-based biodegradable lubricant oil additives with specific properties and application have been recently reviewed [99]. Although borate esters have been indicated as new types of relatively non-toxic friction modifier and anti-wear additives with good oxidation stability [102], their use in lubricants is restricted because of their susceptibility to hydrolysis, which leads to formation of oil-insoluble and abrasive boric acid [103]. For low-toxicity and ashless additives for bioderived and biodegradable lubricants, see also ref. [30].

Efforts to replace sulphurised sperm whale oils as AW/antifriction lubricants by sulphurised vegetable oils have often resulted in products with too high a level of intermolecular cross-linking, and therefore unacceptable viscosities and thickening. This was typically the case of sulphurisation of unrefined jojoba oil [104] and meadowfoam oil (MFO) [105]. Rubbery and resinous product formation has been reported [25, 106]. In US Patent No. 4,970,010 sulphurised vegetable oil derivatives are described as additives in a mineral oil base [107]. Successively, in US Patent No. 5,282,989 (to International Lubricants, Inc.) Erickson *et al.* [59] have described AW/antifriction lubricants composed of blends of a vegetable oil (MFO, RSO or *Crambe*) or wax ester, and a sulphurised or phosphite adduct of these oils or waxes. More recently, US Patent No. 7,279,448 to Erhan *et al.* [56] has disclosed chemically modified vegetable oils prepared by reacting epoxidised triglyceride oils with thiols (Fig. 7.2). The resulting poly(hydroxy thioether) derivatives have utility as AW/antifriction additives for environmentally friendly industrial oils and automotive applications [108]. For environmentally friendly hydraulic oil compositions, see refs [92, 109]. The extreme pressure additive Additin® RC 2519 (Rhein Chemie, Mannheim) is a new reduced viscous sulphur carrier derived from vegetable-based fatty

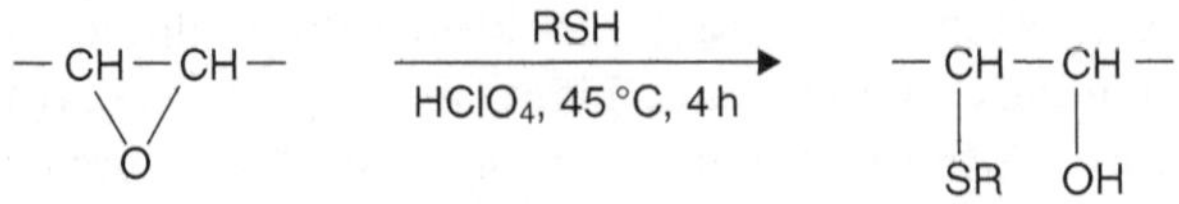

R = C_1 to C_{22}

7.2 Formation of poly(hydroxy thioether) vegetable oil derivatives from epoxidised oils.

acid esters and olefins for use in metalworking processes as a replacement of polysulphides.

In US Patent No. 5,229,023 Landis describes telomerised vegetable oils for lubricant additives containing no more than 4% polyunsaturated fatty acids (PUFAs) in the telomerised product, which is obtained from lower polyunsaturated oils such as RSO or higher polyunsaturated oils such as corn or safflower oil [61]. A telomer is a highly viscous polymerised triglyceride oil that has thermal oxidative stability and viscosity improvement characteristics that make the telomer an essential component or ingredient of a large variety of lubricant formulations (see Section 6.2.5). The lower-range viscosity telomer oils of ref. [61] are plagued by high acid number (about 50 mg KOH/g) due to cleavage of some free fatty acids (FFAs) during processing which excludes use of such products as hydraulic or gear fluids in view of oxidation and corrosion associated with the free acids. Improved (much thicker) telomer oils may be obtained by adding 30–50% of a more expensive conjugated triglyceride oil [62]. These products are useful as AW and thickening agents in lubricant compositions. In another disclosure to International Lubricants, Inc. (ILI; Seattle, WA), lower viscosity range telomer oil products are described which can be used as an AW agent and a thickening agent in hydraulic fluid and gear fluid formulations without causing problems related to corrosion or oxidation [63]. Specifically, such low acid number (<20 mg KOH/g) telomer products were made from a mixture of triglyceride oils wherein only 5–15% consist of conjugated triglycerides (e.g. tung oil, oiticica oil, fish oils) and 85–95% of non-conjugated unsaturated triglycerides (e.g. RSO, SBO, SNO). These high-performance vegetable oil-based thickeners and VI improvers are marketed by ILI under the brand-name Erucichem®.

Novitas Chem Solutions LLC (Bellaire, TX) has developed an environmentally friendly lignin-based AW/EP additive (Desilube™ AEP) for use in lithium soap and aluminium complex greases [110, 111]. The EP performance is due to the ability of the oxygen-containing organic functionalities to form a natural barrier between metal surfaces under extreme pressure conditions. Alcell™ organosolv lignin (Repap Technologies, Inc.; Valley Forge, PA) is a green and multifunctional grease additive [50].

A major challenge in the formulation of motor oils is reduction in AW agents and metal-organic additives (antioxidants and engine cleanliness additives) in order to reduce ash build-up. As additives capable of friction reduction, sulphur-containing molybdenum compositions (such as MoDTC) have been in use. However, as such metal-containing compounds may be harmful to exhaust-abatement catalysts, ashless lubricating oil additives are more desirable. Moreover, although sulphur and phosphorus are capable of preventing wear, reduction of their use is demanded. The trend towards low-ash lubricating compositions has focused oil formulators' efforts on using ashless friction modifiers. Ashless friction modifiers typically include FA esters, FA amides, organic dithiocarbamates or dithiophosphates. Eur. Patent No. EP 2,172,535 A1 to Sekiguchi and Okada (to Idemitsu Kosan Co., Ltd) describes lanolin alcohol and sterol derivatives as ashless lubricating oil additives capable of providing sufficient lubrication without use of metals (such as Mo), sulphur or phosphorus [112]. Lanolin is a light-yellow waxen substance, obtained by refining 'wool grease' (i.e. collected wash with which wool has been washed) through deacidification, discoloration and deodorisation [113]. Lubricant formulations should contain the least possible amounts of metal-containing compounds, sulphur and phosphorus compounds. Various low-ash biodegradable lubricants have been formulated within the European EREBIO project (see Section 10.5). The aforementioned European patent gives examples of many suitable additives acting as VI improvers, PPDs, antioxidants, ashless dispersants, friction modifiers, detergents, AW agents, rust inhibitors, metal deactivators, antifoaming agents and antiemulsifiers. For long-term trends in crankcase additives, see ref. [114].

Objectives of the collective European IBIOLAB research project (FP6 Contract no. 030307) were to generate a database describing the actual additives used (bioadditives and mineral ones) for several industrial applications, to develop new bioadditives (e.g. standoils and functional partial glycerides) and to use bioadditives in biolubricant formulations. See also Section 10.5.

Renewable Lubricants, Inc. (RLI; Hartville, OH) has developed advanced proprietary additives (many bio-based) with exceptional performance for high-oleic base stocks (HOBS). Successful high-performance bio-based lubricant formulations, based on HOBS and RLI's patented Stabilized™ additive technology, are applicable for high- and low-temperature engine, transmission, hydraulic, gear and grease applications. RLI's Bio Hydraulic Fluids are compounded with an environmentally friendly zinc-free additive package. Similarly, the Bio EP Gear Oils are *not* formulated with poly(tetrafluoro-ethylene) (PTFE) or WS_2, as these chemical additives are non-renewable, toxic and hazardous. Bio Turbine R&O Fluids are fortified with Bio Corrosion Inhibitors (BCI™) to protect bearing surfaces from rust and corrosion. Similarly, the corrosion resistance of Bio EP Greases is enhanced by patented bio corrosion inhibitors. In PCT Int. Publ. No. WO 03/106599 A1 (to RLI) Garmier discloses typical additive packages for penetrating lubricants [115]. Extensive government, military and independent laboratory testing has demonstrated that Stabilized™ HOBS have all the required physical properties lubricants must have in order to meet standards set by industry, and some of the new GL-4 and GL-5 requirements set by API for gear oils. Fatty base oils from HOBS adhere more strongly to natural than mineral oils and provide more protective thin boundary and fluid type films. Bio-based lubricants, fluids, and greases can be tailor designed for any equipment under any conditions by using the right combination of HOBS and performance-boosting additives. See also Section 10.5.2. Lubrizol Corp. (Cleveland, OH) has developed PPDs for high monounsaturated (>60%) vegetable oils [49]. Lubrizol's VEG-ESTER™ vegetable oil additive technology simplifies the challenging task of stabilising vegetable-based MWFs.

7.4 Biolubricant formulations

Lubricant formulation is based on a mineral or synthetic base oil or grease with a mixture of different additives to control viscosity, surface interactions and corrosion and to increase load-carrying capacity, thermal diffusivity and service life. Commercial lubricants derived from petroleum present environmental problems and are not suitable for all materials and all conditions. Base stocks for biodegradable lubricant applications should typically meet five criteria:

- solubility with dispersants and other additives;
- good cold-flow properties (PP < –40 °C, viscosity < 7500 mPa s at –25 °C);
- sufficient biodegradability;
- good lubricity without the aid of wear additives;
- high flash point (>260 °C).

Early naturally occurring vegetable oil compositions were formulated using the customary additives for lubricants [49, 109, 116–121]. In US Patent No. 5,863,872 to Renewable Lubricants, Inc. (Hartville, OH) Garmier [122] discloses lubricant compositions consisting of a triglyceride oil and an oil-soluble copper compound antioxidant, making them suitable for use in a variety of applications demanding higher temperatures such as lubricating engines, transmissions, gear boxes and for hydraulic applications. Similarly, US Patent No. 5,990,055 (to RLI) discloses an oil-soluble antimony compound, such as antimony dialkyldithiocarbamate (VanLube®73) or antimony dialkylphosphorodithioate (VanLube®622), as an antioxidant for a high-oleic triglyceride base stock for uses such as motor oils [123]. Clearly, true biolubricants are based on a renewable base stock (usually vegetable oils) and a botanical additive package (e.g. Safe Lube™ industrial and metalworking fluids, GEMTEK®, Phoenix, AZ). Such products, which do not contain petroleum fractions, synthetics, sulphur, chlorine and heavy metals, are readily biodegradable, non-toxic, free of volatile organic compounds (VOCs), non-carcinogenic, non-reactive to skin, compatible with all metals, safe to use, to store and dispose of, or may replace EP additives.

Oil formulations, whether they are bio-based or mineral oil, are generally regarded as proprietary information. Unless additive analysis is carried out it is usually difficult to know the identity of the additives added to a base oil in formulations that have been developed. Patent literature indicates that stable products have been based on additive technology.

The formulated lubricant performance of base fluids is influenced by their chemical, physical and film formation properties. As indicated elsewhere (Chapters 2 and 3), the rheological properties of a base fluid are mainly determined by the viscosity, which is influenced by temperature, pressure and shear rate. Important chemical properties are molecular formula, molecular weight (MW), and polarity. Using the wrong polarity in the base fluid can lead to problems with seal swell [124]. According to van der Waals [124], a non-polarity index (NPI) is expressed as follows:

$$\text{NPI} = \frac{\text{total number of C atoms} \times \text{MW}}{\text{number of carboxylic groups} \times 100} \qquad [7.1]$$

NPI rates the polarity for a molecule, where a low number corresponds to a highly polar molecule. NPIs for lubricants range from about 50 for short synthetic esters up to over 300 for PAO2. Similarly, a polarity index (PI) [125] has been expressed as:

$$\text{PI} = \frac{\text{number of ester groups} \times 10\,000}{\text{total carbon number} \times \text{MW}} \qquad [7.2]$$

There are various means of increasing the polarity of organic molecules, such as the introduction of polar functional groups into an appropriate position of the molecule, and removal of oleophilic group(s) such as hydrocarbon chains from a molecule. Additive technology for synthetic fluids needs to take into account the effect of polarity. Conventional AW additives developed for mineral oils cannot be applied in low-viscous TMP esters. In polar base oils, polar anti-wear additives are effective. Hydroxyalkyl phosphates/phosphonates reduce wear in polar synthetic esters [126]. The AW properties of such phosphorus-containing additives depend on the polarity of the base oil [125], see Fig. 7.3.

Biolubricant base stocks generally benefit from mild refining processes of the raw materials. Despite their wide availability, attractive properties

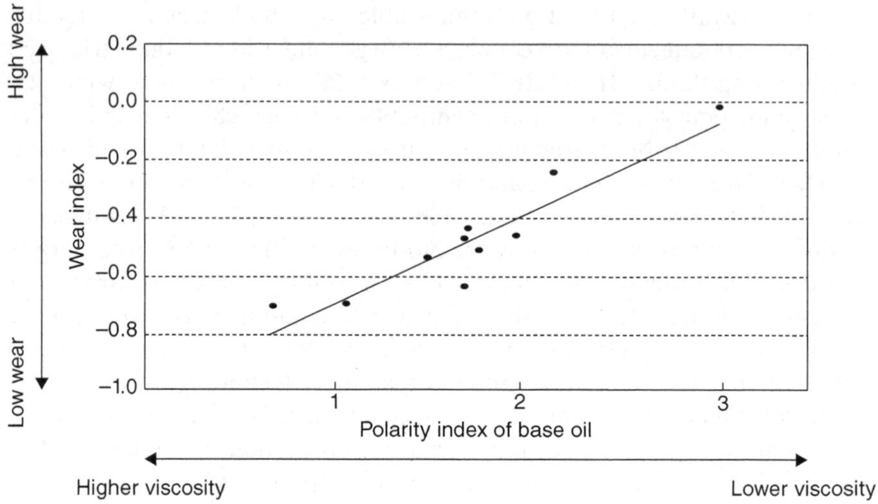

7.3 Effect of synthetic ester base oil polarity on anti-wear properties of dibutylphosphonate, $(BuO)_2P(O)H$. After ref. [125].

and biodegradability, pure native plant oils have not widely been used as high-performance lubricants, although RSO and castor bean oil have found some specific applications. Some of the restrictions of vegetable oils (see Table 5.8) can be overcome by using high-oleic varieties of vegetable oils, eventually in combination with available additives, diluents or functional fluids (see Sections 5.3.1 and 10.5.2). In particular, vegetable oils may be additivated in order to reduce or eliminate undesired reactivity of the labile structural groups [127, 128].

Apart from virgin oils also cheaper used frying oils (UFOs) – considered as a waste – have been proposed for use in lubricant formulations [129]. The potential European UFO market is about 700 kt/yr and is nowadays directed mainly towards oleochemical applications such as production of fatty acids and biodiesel [130]. Used oils show fluctuations in chemical composition. Compared with virgin oils, the main difference with UFOs is the presence of triglyceride polymers and FFAs. This affects the nature and quantity of additives to be needed in the base lubricant formulation. FFAs have a positive lubricating effect (see Section 10.5.5).

Minor components, e.g. tocopherols, are able to modify oil properties such as thermo-oxidative resistance. Natural antioxidants contribute to the oxidative stability of vegetable oils. Consequently, oil processing techniques affect the oxidative stability of natural oils. Mechanically expelled soy oil is more oxidatively stable than chemically extracted oil. Refined vegetable oils with reduced tocopherol content result in lower oxidative stability (see Table 5.13). As natural antioxidants are typically unstable under the harsh operating conditions of machinery such as hydraulic systems, synthetic antioxidants may be needed to improve the oxidation stability of vegetable oils and reduce ageing phenomena. Special phenolic and aminic materials are suitable antioxidants for the formulation of biodegradable lubricants.

As one of the most relevant physical properties of lubricant base oils is viscosity (and VI) it is not surprising that also various VI improvers (e.g. polyisobutylenes or polymethacrylates) have been added to enhance the viscosity of plant oils. For example, high-viscosity biolubricant formulations were obtained by additivation of vegetable oils (castor oil, SNO, SBO, RSO, HOSNO) with 1–4 wt% viscosity improvers (ethylene–vinylacetate copolymer, EVA, styrene hydrogenated butadiene–styrene copolymer, SBS, EVA/SBS): e.g. for castor oil from 210 cSt at 40 °C to 411 cSt and for HOSNO from 39 cSt at 40 °C to 165 cSt [131, 132]. A high-VI fluid will enable cooler operating temperatures; it will also enable efficient and smooth operation at higher temperatures and enable start-up at lower temperatures. AW additives (e.g. PTFE) may be used to improve the cold-flow behaviour of vegetable oils. Diluents and PPDs may exert a positive influence on cold-storage stability of vegetable oils. However, cold-storage properties of vegetable oils do not respond to PPDs to the same extent as in case of mineral oils. Acid scavengers are added to neutralise acids and reduce wear.

In Japanese Patent No. JP 5,209,187 A Shoji *et al.* [133] disclose an improved cold-resistant lubricant formulation for frozen food applications obtained by additivation of a vegetable oil with a polyglycerol ester of an FA, a sucrose ester of an FA and lecithin. In using RSO the oxidative stability remains unsatisfactory.

Asadauskas and Erhan [96] have studied the pour point depression of vegetable oils by use of commercial PPDs and by blending with various diluents. Both provide only limited improvement in low-temperature performance. Owing to the predominantly saturated nature of its fatty acid constituents (see Table 5.3), coconut oil (CNO) has a very high congelation temperature precluding its use as a base oil for industrial lubricants in temperate and cold climatic conditions. Vegetable oils with a high degree of polyunsaturation have comparatively better low-temperature properties. Govindapillai *et al.* [134] have modified CNO by means of a styrenated phenol (SP) PPD, which is also an effective antioxidant [135], and by admixture with castor oil with formation of an esterified estolide. As shown in Table 7.4, chemical addition of PPDs did not greatly improve the pour point of CNO, at variance to the chemical modification with 10 wt% castor oil and formation of a 2-ethylhexyl esterified estolide according to ref. [136].

In US Patent No. 5,888,947 (to Agro Management Group, Inc.) Lambert and Johnson [137] describe a biodegradable, non-toxic, liquid lubricant that

Table 7.4 Pour points and DSC endotherms of various neat and modified vegetable oils

Vegetable oil	Pour point (°C)[a,b]	Endothermal peak (°C)
Coconut oil	24	25.3 ± 0.7
Castor oil	−33	−26.2 ± 2.2
CNO (90 wt%)/CAS (10 wt%)	24	23.1
Sesame oil	−12	−17 ± 0.6
Groundnut oil	3	−14.5 ± 0.9[c]
Linseed oil	n.g.	−28.9 ± 0.4
Mustard oil	−18	−15.3 ± 1.5
Olive oil	−9	−3.8 ± 0.6
Palm oil	n.g.	6 ± 1.2[c]
Sunflower oil	−18	−17.2 ± 1.3
Coconut oil – SP[d]	24 to 12	25 to 21
Coconut oil – PMA[e]	24 to 18	n.g.
Coconut estolide – 2 EH	−3	+5, −22[f]

[a] According to ASTM D 97.
[b] Repeatability 2.9 °C at 95% confidence.
[c] Multiple peaks.
[d] CNO with styrenated phenol-based commercial PPD (1 to 20%).
[e] CNO with polymethacrylate-based commercial PPD (0.1 to 0.5%).
[f] Ascribed to 2-EH ester of oleic acid estolide.
n.g., not given.
After ref. [134].

is essentially entirely composed of vegetable-based components, namely an unrefined base oil (about 85 vol%), an oil source containing hydroxy fatty acids (HFA; 10 vol%), an oil source containing vegetable or animal waxes (5 vol%), and natural antioxidants (when needed). The base oil consists of primarily long-chain triglycerides (C_{16}–C_{22}, preferably canola, but also soybean, sunflower and other oils, including their blends), mono- and diglycerides and FFAs. HFA sources were castor and *Lesquerella*; waxes were derived from jojoba, meadowfoam or lanolin.

Additive technology is often fundamental in the commercial development of (bio)lubricants. This is illustrated quite well in case of John Deere's second generation universal tractor transmission oil BIO-HY-GARD II (see Section 12.2.1, p. 593). In PCT Int. Publ. No. WO 03/027212 A1 (to Pennzoil-Quaker State Co.) Chiu *et al.* [8] formulate a biodegradable lubricant (>60% biodegradability) based on a transesterified triglyceride base oil and a synthetic ester (with ester and olefin copolymer VI improvers). By blending with mineral oils it is possible to prepare a full range of SAE grade engine oils for gasoline-fuelled and diesel-fuelled engines (see Section 10.5.1). Improving the fuel efficiency of internal combustion engines through additive technology is a key objective of the automotive industry [138]. Development of improved additive systems offering higher activity and improved oxidative stability is an active area of research; see also Section 12.2.

The Cortec® Corporation (St. Paul, MN) is a pioneer of environmentally friendly corrosion protection, and has developed innovative Vapor phase Corrosion Inhibitor (VpCI™) and Migratory Corrosion Inhibitor (MCI™) technologies for application in various industries (including metalworking, construction, water treatment, oil and gas). EcoLine® Bearing, Chain & Roller Lubricant is a bio-based, low-environmental impact, rust preventive lubricant formulated with soy oil that meets USDA's Biobased product definition for EO 13101. As a sustainable product, it contains 86% less mineral oil than traditional lubricants. Formulated with Vapor phase Corrosion Inhibitors (VpCI™) for indoor and outdoor conditions, that form a thin, tight layer on the metal surface, completely sealing off the metal against moisture and air, it offers corrosion protection of ferrous and non-ferrous metals in the most aggressive environments. The product is excellent for conveyor and lubricating drive chains.

Utilising a lipase-catalysed ester-interchange reaction on oil blends typically consisting of HOSNO (C18:1), hardened palm fractionated oil (C16:1) and medium-chain saturated fatty acids (SFAs) in appropriate ratios, US Patent No. 6,117,827 to Fuji Oil Co. [139] discloses a biodegradable lubricant base oil consisting of 20–40 wt% medium-chain (C_6–C_{12}) SFAs in the 1,3 positions of glyceride, 40–58 wt% mono-ene FAs (such as C18:1) and about 10–20% *trans*-acid content with satisfactory low-temperature fluidity, oxidative stability and lubricity, and low cloud point. The lubricant base oil is suitable in rolling mill applications, in two- and four-cycle engine lubricating oil applications, as well as in cutting oil applications.

7.5 Sources of further information and advice

S. Srivastava, *Advances in Lubricant Additives and Tribology*, CRC Press, Boca Raton, FL (2010).

L.R. Rudnick (ed.), *Lubricant Additives: Chemistry and Applications*, 2nd edn, CRC Press, Boca Raton, FL (2009).

R.M. Mortier, M.F. Fox and S.T. Orszulik (eds), *Chemistry and Technology of Lubricants*, 3rd edn, Springer, Heidelberg (2009).

E.N. Frankel, *Antioxidants in Food and Biology: Fats and Friction*, The Oily Press Lipid Library, Vol. 20, Bridgwater (2007).

Th. Mang and W. Dresel (eds), *Lubricants and Lubrication*, 2nd edn, Wiley-VCH, Weinheim (2007).

J.S. Marth, *Renewable Lubricants Manual: Biobased Oils, Fluids & Greases*, United Bio Lube, Palo Alto (CA) (2007); http://www.renewablelubricants.com

A. Lansdown, *Lubrication and Lubricant Selection: A Practical Guide*, 3rd edn, ASME Press, Champaign, IL (2004).

G.E. Totten, S.R. Westbrook and R.J. Shah, *Fuels and Lubricants Handbook: Technology, Properties, Performance and Testing*, ASTM International, West Conshohocken, PA (2003), Vol. I.

V. Štěpina and V. Veselý, *Lubricants and Special Fluids*, Elsevier, Amsterdam (2002).

7.6 References

[1] Th. Mang and W. Dresel (eds), *Lubricants and Lubrication*, 2nd edn, Wiley-VCH, Weinheim (2007).

[2] J. Toscano, Biodegradable lubricants, *Science Tech. Entrepreneur* (December 2005).

[3] E. Durak, A study on friction behavior of rapeseed oil as an environmentally friendly additive in lubricating oil, *Ind. Lubr. Tribol.* **56** (2004) 23–37.

[4] M. Roegiers, *Le Procédé Elektrion, son Historique, son Mécanisme, son Action sur les Huiles de Graissage*, De Cavel & Roegiers, Ghent (1952).

[5] B. Zhmud and M. Roegiers, New base oils pose a challenge for solubility and lubricity, *Tribol. Lubr. Technol.* **65**(7) (2009) 34–9.

[6] M. Roegiers and B. Zhmud, Tribological performance of ionised vegetable oils as lubricity and fatty oiliness additives in lubricants and fuels, *Lube Mag.* **88** (2008) 19–26; *Lubric. Sci.* **21**(5) (2009) 169–82.

[7] J.-P. Despeghel (to Monsanto S.A.S.), Use of rapeseed oil in biolubricants, PCT Int. Publ. No. WO 2007/034336 A3 (29 March 2007).

[8] I.-Ching Chiu, S. Gunsel and P. Lacey (to Pennzoil-Quaker State Co.), Environmentally friendly lubricants, PCT Int. Publ. No. WO 03/027212 A1 (3 April 2003).

[9] K. Haniyuda and E. Nagatomi (to Showa Shell Sekiyu), Lubricating oil composition for engine, Jap. Patent No. JP 2008/031289 (14 February 2008).

[10] B. Genuyt, M. Janssen, R. Reguerre, J. Cassiers and T. Breye (to Total Raffinage Distribution), Biodegradable lubricating composition and its use in a drilling fluid, PCT Int. Publ. No. WO 2001/083640 A1 (8 November 2001).

[11] P.W. Gold, C. Assmann and U. Fassbender, Miscibility of biodegradable lubricants with mineral oil, *Erdöl, Erdgas, Kohle* **14** (1998) 336–7.

[12] M.R. Greaves, Oil soluble synthetic polyalkylene glycols. A new type of Group V base oil, *Lube Mag.* **104** (2011) 21–4.

[13] S. Suda, H. Yokota and M. Ibi (to Nippon Oil Corp.), Metal working fluid, US Patent Appl. No. 2008/026967 A1 (31 January 2008).

[14] W.W. Garmier (to Renewable Lubricants, Inc.), High temperature biobased lubricant compositions from boron nitride, US Patent Appl. No. 2010/0105583 A1 (29 April 2010).

[15] L. Pop, C. Puşcaş, G. Bandur, G. Vlase and R. Nuţiu, Basestock oils for lubricants from mixtures of corn oil and synthetic diesters, *J. Am. Oil Chem. Soc.* **85**(1) (2008) 71–6.

[16] G. Bantchev and G. Biresaw, Elastohydrodynamic study of vegetable oil-polyalphaolefin blends, *Lubric. Sci.* **20** (2008) 283–97.

[17] J.A. Thoen, R. Geiger, M.R. Greaves and D.C. Busby (to Dow Global Technologies, Inc.), Lubricant blend composition, PCT Int. Publ. No. WO 2008/134179 (6 November 2008).
[18] J.Z. Gao and J.J. Habeeb (to ExxonMobil Res. & Engng. Co.), Blends of bio-lubricant basestocks with polyol esters, US Patent Appl. No. 2007/0232506 A1 (4 October 2007).
[19] S.H. Roby and S.G. Ruelas (to Chevron Oronite Corp.), Lubricating base oil compositions and methods for improving fuel economy in an internal combustion engine using same, US Patent No. 7,465,696 B2 (16 December 2008).
[20] M.A. Poirier and O. Sutton (to ExxonMobil), Fuel economy lubricant compositions, US Patent Appl. No. 2008/0255011 A1 (16 October 2008).
[21] A. Lansdown, *Lubrication and Lubricant Selection: A Practical Guide*, 3rd edn, ASME Press, Champaign, IL (2004).
[22] V.W. Bathia, A. Chaudhry, G.A. Sivasankaran, R.P.S. Bisht and M. Kashyap, Modification of jojoba oil for lubricant formulations, *J. Am. Oil Chem. Soc.* **67** (1990) 1–7.
[23] S. Watanabe, T. Fujita and M. Sakamoto, Epoxides of unsaturated fatty acids as antirust additives in water based cutting fluids, *J. Am. Oil Chem. Soc.* **65** (1998) 1311–12.
[24] D. Tao, H.L. Zhu and Z.M. Hu, Free chlorine fixation and the protective film in boundary lubrication, *Lubric. Sci.* **8** (1996) 397–407.
[25] K.P. Kammann Jr. and A.I. Phillips, Sulfurized vegetable oil products as lubricant additives, *J. Am. Oil Chem. Soc.* **62** (1985) 917–23.
[26] E. Richter, Lubricants, in *Plastics Additives Handbook* (H. Zweifel, ed.), 5th edn, Hanser Verlag, Munich (2000), pp. 511–52.
[27] G. Fisicaro and G. Gerbaz, Dialkylcarbonates, in *Synthetic Lubricants and High-performance Functional Fluids* (R.L. Shubkin, ed.), Marcel Dekker, Inc., New York, NY (1993), pp. 229–39.
[28] L.R. Rudnick, Dialkylcarbonates, in *Synthetics, Mineral Oils and Bio-Based Lubricants: Chemistry and Technology* (L.R. Rudnick, ed.), CRC Press/Taylor & Francis, Boca Raton, FL/New York, NY (2006), pp. 235–45.
[29] J.A. Kenar and T.L. Kurth, Dialkyl carbonates as lubricant additives, *Proc. 97th AOCS Ann. Mtg. & Expo*, St. Louis, MO (30 April–3 May 2006), p. 74.
[30] L.R. Rudnick (ed.), *Lubricant Additives: Chemistry and Applications*, 2nd edn, CRC Press, Boca Raton, FL (2009).
[31] The Technical Committee of Petroleum Additive Manufacturers in Europe (ATC), *Lubricant Additives and the Environment*, ATC Document 49, CEFIC, Brussels (1993).
[32] A.R. Eyres, Health and safety aspects of lubricants, in *Industrial Tribology* (M.H. Jones and D. Scott, eds), Elsevier, Amsterdam (1983), p. 275.
[33] J.A. Henry, Composition and toxicity of petroleum products and their additives, *Hum. Exp. Toxicol.* **17** (1998) 111.
[34] Council Directive 87/101/EEC of 22 December 1986 amending Directive 75/439/EEC on the disposal of waste oils, *Off. J. Eur. Comm.* **L 042** (12 February 1987), 43–7.
[35] K. Grob, A. Artho, M. Biedemann and J. Egli, Food contamination by hydrocarbons from lubricating oils and release agents: determination by coupled LC-GC, *Food Add. Contam.* **8**(4) (1991) 437–46.

[36] F. Linn Lee and J.W. Harris, Long-term trends in industrial lubricant additives, in *Lubricant Additives: Chemistry and Applications* (L.R. Rudnick, ed.), 2nd edn, CRC Press, Boca Raton, FL (2009), pp. 609–39.
[37] US Army Corps of Engineers, *Engineering and Design, Lubricants and Hydraulic Fluids*, Engineer Manual 1110-2-1424, Washington, DC (1999).
[38] J. Braun, Additives, in *Lubricants and Lubrication* (Th. Mang and W. Dresel, eds), 2nd edn, Wiley-VCH, Weinheim (2007), pp. 88–118.
[39] H. Zweifel, R.D. Maier, M. Schiller and St. E. Amos (eds), *Plastics Additives Handbook*, 6th edn, Hanser Gardner, Cincinnati, OH (2009).
[40] I. MacPherson and A. Pittigrew (to Afton Chemical Corp.), Environmentally-friendly lubricant compositions, Eur. Patent No. EP 2,017,329 A1 (21 January 2009).
[41] L.R. Rudnick, Additives for industrial lubricant applications, in *Lubricant Additives: Chemistry and Applications* (L.R. Rudnick, ed.), 2nd edn, CRC Press, Boca Raton, FL (2009), pp. 493–509.
[42] W. Castro, D.E. Weller, K. Cheenkachorn and J.M. Perez, The effect of chemical structure of basefluids on antiwear effectiveness of additives, *Tribol. Intl.* **38** (2005) 32–6.
[43] A.M. Barnes, K.D. Bartle and V.R.A. Tribon, A review of zinc dialkyldithiophosphates (ZDDPs): characterisation and role in the lubricating oil, *Tribol. Intl.* **34**(6) (2001) 389–95.
[44] A.N. Agaev, S.M. Velieva and I.D. Kulaliev, Multifunctional additives for lubricating oils, *Azarb. Neft. Tasar.* **7** (2007) 64–5.
[45] P.Sh. Mamedova, S.A. Gamzayeva and B.M. Aminova, Multifunctional additives for oils and cutting fluids, *Azarb. Neft. Tasar.* **12** (2005) 34–8.
[46] S.B. Borshchevskii, E.V. Shabanova, A.A. Fufaev, F.N. Merzilkin and O.V. Belyaeva, Multifunctional additive for lubricant oils, *Khim. Tekhnol. Topl. Mas.* (4) (1991) 5–6.
[47] L.E. Coleman (to Lubrizol Corp.), Lubricant containing nitrogen-containing ester, US Patent No. 3,702,300 (7 November 1972).
[48] B. Stunkel and G.A. Aguilar (to R.T. Vanderbilt Co., Inc.), Vegetable oil lubricating composition, US Patent Appl. No. 2008/132434 A1 (5 June 2008).
[49] K. Lal, D.M. Dishong and J.G. Dietz (to Lubrizol Corp.), Pour point depressants for high monounsaturated vegetable oils and for high monounsaturated vegetable oils/biodegradable base and fluid mixtures, US Patent No. 5,413,725 (9 May 1995).
[50] E.K. Pye and J.H. Lora, The Alcell™ process. A proven alternative to Kraft pulping, *Tappi J.* **74**(3) (1991) 113–18.
[51] R. Silverstein and L.R. Rudnick, Additives for grease applications, in *Lubricant Additives: Chemistry and Applications* (L.R. Rudnick, ed.), Marcel Dekker, Inc., New York, NY (2003), pp. 513–39.
[52] C.E. Snyder, L.J. Gschwender and S.K. Sharma, Long-term additive trends in aerospace applications, in *Lubricant Additives: Chemistry and Applications* (L.R. Rudnick, ed.), 2nd edn, CRC Press, Boca Raton, FL (2009), pp. 637–46.
[53] K. Behrens, A. Bühler and L. Arias, ASTM D 6443 standard test method for determination of Ca, Cl, Cu, Mg, P, S and Zn in unused lubricating oils and additives, *Petro Ind. News* **12**(2) (April/May 2011) 38–9.

[54] M. Guevremont, G.H. Guinther, D. Szemenyi, M.T. Devlin, T.-C. Jao, C. Jaye, J. Woicik and D.A. Fischer, Enhancement of engine oil wear and friction control performance through titanium additive chemistry, *Tribol. Trans.* **51**(3) (2008) 324–31.
[55] Y. Cao, L. Yu and W. Liu, Study of the tribological behaviour of sulfurized fatty acids as additives in rapeseed oil, *Wear* **244** (2000) 126–31.
[56] S.Z. Erhan, A. Adhvaryu and B.K. Sharma (to USDA), Poly(hydroxy thioether) vegetable oil derivatives useful as lubricant additives, US Patent No. 7,279,448 (10 September 2007).
[57] A. Adhvaryu (to Caterpillar, Inc.), Anti-friction and anti-wear lubricant additive, PCT Int. Publ. No. WO 2008/054625 A2 (8 May 2008).
[58] J.A. Waynick (to South-West Research Institute), Lubricant oils and greases containing nanoparticle additives, PCT Int. Publ. No. WO 2006/119502 A2 (9 November 2006).
[59] F.L. Erickson, R.E. Anderson and P.S. Landis (to International Lubricants, Inc.), Vegetable oil derivatives as lubricant additives, US Patent No. 5,282,989 (1 February 1994).
[60] E.W. Bell (to USDA), Wax esters of vegetable oil fatty acids useful as lubricants, US Patent No. 4,152,278 (1 May 1979).
[61] P.S. Landis (to International Lubricants, Inc.), Telomerized vegetable oil for lubricant additives, PCT Int. Publ. No. WO 92/07051 (30 April 1992); US Patent No. 5,229,023 (20 July 1993).
[62] A. Shanahan and P.S. Landis (to International Lubricants, Inc.), Telomerized triglyceride oil product, US Patent No. 5,454,965 (3 October 1995).
[63] P.S. Landis (to International Lubricants, Inc.), Lower viscosity telomer oil, US Patent No. 5,567,345 (22 October 1996).
[64] A.J. O'Lenick Jr. (to Lambent Technologies, Inc.), Telomerized complex ester triglycerides, US Patent No. 5,985,806 A (16 November 1999).
[65] N. Canter, Trends in extreme pressure additives, *Tribol. Lubr. Technol.* **63**(9) (2007) 10–18.
[66] M. Roegiers, H. Zhang and B. Zhmud, Elektrionized™ vegetable oils as lubricity and oiliness components in metalworking lubricants, *Proc. 6th Intl. Conf. on Tribology, Balkantrib'08,* Sozopol (12–14 June 2008), Paper BT-97; *FME Transactions* **36** (2008) 133–8.
[67] R.G. Rowland and C.A. Migdal (to Crompton Corp.), Epoxidized ester additives for reducing lead corrosion in lubricants and fuels, US Patent Appl. No. 2006/0090393 A1 (4 May 2006).
[68] M. Fletchinger, P. Hamblin, F. Lichtenberg, J. Reyes-Gavilan and P. Rohrbach, New ashless, calcium compatible corrosion inhibitor for use in industrial lubricating oils, *Tribol. Lubr. Technol.* **64**(4) (2004) 38–50.
[69] M.J. Souza de Carvalho, P.R. Seidl, C.R. Pereira Belchior and J.R. Sodré, Lubricant viscosity and viscosity improver additive effects on diesel fuel economy, *Tribol. Intl.* **43** (2010) 2298–302.
[70] G. Livingstone, Varnish: looking ahead, *Turbomach. Intl.* (September/October 2008), no page number.
[71] A.-G. Horodysky, S.-Y. Hsu, A. Jeng and L.R. Rudnick (to Mobil Oil Corp.), Antioxidants for functional fluids, US Patent No. 5,336,420 A (9 August 1994).

[72] F. Bongardt, K. Schmid and R. Wüst (to Henkel K.g.A.), Environment-friendly basic oil for formulating hydraulic fluids, US Patent No. 5,451,334 (19 September 1995).
[73] S.S. Lawate and K. Lal (to Lubrizol Corp.), High oleic polyol esters, compositions and lubricants, functional fluids and greases containing the same, US Patent No. 5,773,391 (30 June 1998).
[74] P. Rohrbach, P. Hamblin and J. Reyes-Gavilan, The benefits of a high molecular weight phenolic antioxidant compared with BHT and 2,6-DTBP, *Tribol. Lubr. Technol.* **60**(6) (2004) 56–63.
[75] Directive 98/8/EC of the European Parliament and of the Council of 16 February 1998 concerning the placing of biocidal products on the market, *Off. J. Eur. Comm.* **L 123**/1 (24 April 1998).
[76] Federal Insecticide, Fungicide, and Rodenticide Act (FIFRA), http://www.epa.gov/agriculture/lfra.html
[77] EC Commission Regulation No. 1451/2007 of 4 December 2007 on the second phase of the 10-year work programme referred to in Article 16(2) of Directive 98/8/EC of the European Parliament and of the Council concerning the placing of biocidal products on the market, *Off. J. Eur. Comm.* **L 325**/3 (11 December 2007).
[78] E.O. Bennett, Corrosion inhibitors as preservatives for metalworking fluids – ethanolamines, *Lubr. Engng.* **3** (1979) 137–44.
[79] A.M. Magerramov, M.R. Bayramov, O.N. Javadova, S.Q. Aliyeva and S.A. Hasanova, Study of sulfur- and nitrogen-containing alkylphenols as antimicrobial additives for lubricating oils, *Kim. Problem.* **3** (2009) 528–30.
[80] S.S.V. Ramakumar, N. Aggarwal, A.M. Rao, A.S. Sarpal, S.P. Srivastava and A.K. Bhatnagar, Studies on additive–additive interactions: effect of dispersant and antioxidant additives on the synergistic combination of overbased sulphonate and ZDDP, *Lubric. Sci.* **7**(1) (1995) 25–38.
[81] A.S. Sarpal, J. Christopher, S. Mukherjee, M.B. Patel and G.S. Kapur, Study of additive–additive interactions in a lubricant system by NMR, ESCA, and thermal techniques, *Lubric. Sci.* **17**(3) (2005) 319–45.
[82] A.S. Sarpal, V. Bansal, M.I.S. Sastry, S. Mukherjee and G.S. Kapur, Molecular spectroscopic studies of the effect of base oils on additive-additive interactions, *Lubric. Sci.* **16**(1) (2004) 29–45.
[83] F. Rounds, Changes in friction and wear performance caused by interactions among lubricant additives, *Lubric. Sci.* **1**(4) (1989) 333–63.
[84] T. Colclough, Role of additives and transmission metals in lubricating oil oxidation, *I&EC Res.* **26** (1987) 1888–95.
[85] D.-C. Du, S.-S. Kim, J.-S. Chun, C.-M. Suh and W.-S. Kwon, Antioxidation synergy between ZnDTC and ZnDDP in mineral oil, *Tribol. Lett.* **13**(1) (2002) 21–7.
[86] I. Minami and K. Mimura, Synergistic effect of antiwear additives and antioxidants in vegetable oil, *J. Synth. Lubr.* **21** (2004) 193.
[87] Y.-S. Ma, J.-J. Liu, L.-Q. Zheng, D.-P. Wei, R.-L. Wang, X.-S. Zhang and H. Liu, The synergistic effects of EP and AW additives with a sulphurized steel surface under boundary lubrication, *Lubric. Sci.* **7**(2) (1995) 171–9.
[88] J.-Q. Hu, D.-P. Wei and J.-B. Yao, Synergistic antiwear effects of organic borate ester containing nitrogen as lubricating oil additive, *Runhuayon* **21**(4) (2006) 26–8.

[89] K. Hosonuma, K. Yoshida and A. Matsunaga, The decomposition products of zinc dialkyldithiophosphate in an engine and their interaction with diesel soot, *Wear* **103** (1985) 297–309.

[90] J.H. Raddatz and Ch. von Eberan-Eberhart, Lubricant additives and the environment, *TAE Seminar 21591 'Ökonomische Aspekte von Schmierstoffe'*, Ostfildern (November 1996).

[91] J. Li, Y. Zhang, T. Ren, W. Liu and D. Wang, A study of the tribological behaviour of dialkyldithiophosphate esters as additives in rapeseed oil, *Synth. Lubric.* **19** (2002) 99–108.

[92] T. Konishi and Y. Kikuchi (to Nippon Mitsubishi Oil Corp.), Hydraulic oil composition with improved biodegradable properties, Germ. Patent No. DE 60,122,400 T (8 November 2007).

[93] J. Korff and A. Fessenbecker, Additives for biodegradable lubricants, *NLGI Spokesman* **57** (1993) 107–12.

[94] M. Miller, Additives for bioderived and biodegradable lubricants, in *Lubricant Additives: Chemistry and Applications* (L.R. Rudnick, ed.), 2nd edn, CRC Press, Boca Raton, FL (2009), pp. 445–56.

[95] R. Becker and A. Knorr, An evaluation of antioxidants for vegetable oils at elevated temperatures, *Lubric. Sci.* **8** (1996) 95–117.

[96] S. Asadauskas and S.Z. Erhan, Depression of pour points of vegetable oils by blending with diluents for biodegradable lubricants, *J. Am. Oil Chem. Soc.* **76** (1999) 313–16.

[97] J. Ineman, Advanced emulsifier technology for renewable resource-based metalworking fluids, *Tribol. Lubr. Technol.* **67**(11) (2011) 46–8.

[98] M. Roegiers, Treatment of oils by electrical discharges ('Elektrion' process), *Chim. Ind. (Paris)* (1931), Special No. 443–7; The Elektrion process, *Petroleum (London)* **15** (1952) 179–85.

[99] M.A. Maleque, H.H. Masjuki and S.M. Sapuan, Vegetable-based biodegradable lubricating oil additives, *Ind. Lubr. Tribol.* **55** (2003) 137–43.

[100] B.K. Sharma, K.M. Doll and S.Z. Erhan, Ester hydroxy derivatives of methyl oleate: Tribological, oxidation and low temperature properties, *Bioresour. Technol.* **99** (2008) 7333–40.

[101] L. Feng (to Changzhou Xialei Chemical Co.), Method for preparing water-soluble anti-rust additive, Chin. Patent No. CN 101265435 (17 September 2008).

[102] A.G. Paper, Anti-wear and extreme pressure additives in lubricants, *Proc. 10th Intl. Colloq. Tribology, Tribology – Solving Friction and Wear Problems*, Technische Akademie Esslingen (9–11 January 1996), Vol. 2, pp. 1093–9.

[103] Y. Junbing, *Tribological Research on Borate Esters Used as Lubricant Additives*, Beijing Aviation Oil Research Institute, Beijing (1999).

[104] T.K. Miwa, J.A. Rothfus and E. Dimitroff, Extreme-pressure lubricant tests on jojoba and sperm whale oils, *J. Am. Oil Chem. Soc.* **56**(8) (1979) 765–70.

[105] L.H. Princen and J.A. Rothfus, Development of new crops for industrial raw materials, *J. Am. Oil Chem. Soc.* **61**(2) (1984) 281–9.

[106] J.M. Wakim (to Shell Oil Co.), Cosulfurized products of high iodine value triglyceride and nonwax ester of monoethenoid fatty acid as lubricant additives, US Patent No. 3,986,966 (19 October 1976).

[107] F.L. Erickson, R.E. Anderson and P.S. Landis (to International Lubricants, Inc.), Vegetable oil derivatives as lubricant additives, US Patent No. 4,970,010 (13 November 1990).
[108] B.K. Sharma, A. Adhvaryu and S.Z. Erhan, Synthesis of hydroxy thio-ether derivatives of vegetable oil, *J. Agric. Food Chem.* **54** (2006) 9866–72.
[109] T. Konishi and Y. Kikuchi (to Nippon Mitsubishi Oil Corp.), US Patent No. 6,300,292 (9 October 2001).
[110] N. Canter, Environmentally friendly extreme pressure additive, *Tribol. Lubr. Technol.* **67**(10) (2011) 10–11.
[111] L. Wei and J. King, Novel lignin based grease additives, *Presented at the STLE 2011 Ann. Meeting & Exhibit.*, Atlanta, GA (16–20 May 2011).
[112] H. Sekiguchi and T. Okada (to Idemitsu Kosan Co., Ltd), Lubricant additive and lubricant composition, PCT Int. Publ. No. WO 2009/004893 (8 January 2009); Eur. Patent No. EP 2,172,535 A1 (6 June 2008).
[113] U. Hoppe (ed.), *The Lanolin Book*, Beiersdorf AG, Hamburg (1999).
[114] S.E. Schwartz, S. Papasavva and L.R. Rudnick, Long-term trends in crankcase additives: Lubrication for the future, in *Lubricant Additives: Chemistry and Applications* (L.R. Rudnick, ed.), Marcel Dekker, Inc., New York, NY (2003), pp. 541–56.
[115] W.W. Garmier (to Renewable Lubricants, Inc.), Biodegradable penetrating lubricant, US Patent Appl. No. 2003/0069146 A1 (10 April 2003); PCT Int. Publ. No. WO 03/106599 A1 (24 December 2003).
[116] K.V.J. Jokinen, H.K. Kerkkonen, E.A. Leppämäki and E.I. Piirilä (to Oy Kasviöljy-Växtolje Ab), Hydraulic fluids, US Patent No. 4,783,274 (8 November 1988).
[117] J.A. Stoffa (to Lubrizol Corp.), Functional fluid with triglycerides, detergent-inhibitor additives and viscosity modifying additives, US Patent No. 5,298,177 (29 March 1994).
[118] K. Lal (to Lubrizol Corp.), Pour point depressants for industrial lubricants containing mixtures of fatty acid esters and vegetable oils, US Patent No. 5,338,471 (16 August 1994).
[119] R.M. Lange, J.G. Seebauer and C.A. Mamajek (to Lubrizol Corp.), Environmentally friendly viscosity index improving compositions, US Patent No. 5,399,275 (21 March 1995).
[120] P.C. Naegely (to Lubrizol Corp.), Biodegradable chain bar lubricant composition for chain saws, US Patent No. 5,641,734 (24 July 1997).
[121] D.E. Chasan and P.R. Wilson (to Ciba-Geigy Corp.), Stabilized lubricant compositions, US Patent No. 5,580,482 (3 December 1996).
[122] W.W. Garmier (to Renewable Lubricants, Inc.), Biodegradable lubricant composition from triglycerides and oil soluble copper, US Patent No. 5,863,872 (26 January 1999).
[123] W.W. Garmier (to Renewable Lubricants, Inc.), Biodegradable lubricant composition from triglycerides and oil soluble antimony, US Patent No. 5,990,055 (23 November 1999).
[124] G. van der Waals, The relationship between the chemical structure of ester base fluids and their influence on elastomer seals, and wear characteristics, *J. Synth. Lubr.* **1**(4) (1984) 280–301.
[125] I. Minami and S. Mori, Anti-wear additives for ester oils, *J. Synth. Lubr.* **22** (2005) 105–21.

[126] I. Minami, K. Hirao, M. Memita and S. Mori, Investigation of anti-wear additives for low viscous synthetic esters: hydroxyalkyl phosphonates, *Tribol. Intl.* **40** (2007) 626–31.

[127] A. Fessenbecker and J. Korff, Additive für ökologisch unbedenklichere Schmierstoffe, *Tribol. Schmierungstechn.* **42** (1995) 26–30.

[128] R. Becker and A. Knorr, Antioxidantien für pflanzliche Öle, *Tribol. Schmierungstechn.* **42** (1995) 272–6.

[129] A.V. Avellan and C. Alfos, Valorisation non alimentaire des huiles de friture usagées en tant que lubrifiants biodegradables, *Oléag., Corps Gras, Lipides* **12**(4) (2005) 290–4.

[130] J.C.J. Bart, N. Palmeri and S. Cavallaro, *Biodiesel Science and Technology, From Soil to Oil*, Woodhead Publishing Ltd/CRC Press, Cambridge/Boca Raton, FL (2010), pp. 198–204, 336.

[131] L.A. Quinchia, M.A. Delgado, C. Valencia, J.M. Franco and C. Gallegos, Viscosity modification of high-oleic sunflower oil with polymeric additives for the design of new biolubricant formulations, *Environm. Sci. Technol.* **43** (2009) 2060–5.

[132] L.A. Quinchia, C. Gallegos, J.M. Franco, C. Valencia and M.A. Delgado (to University of Huelva), Additives that modify the viscosity of vegetable oils for use as biolubricants, PCT Int. Publ. No. WO 2009/118435 A1 (1 October 2009).

[133] S. Shoji, T. Yoshida and K. Kano (to Tsukishima Shokuhin Kogyo Kk), Cold resistant vegetable oil, Jap. Patent No. JP 5,209,187 A (20 August 1993).

[134] A. Govindapillai, N.H. Jayadas and M. Bhasi, Analysis of the pour point of coconut oil as a lubricant base stock using differential scanning calorimetry, *Lubric. Sci.* **21**(1) (2009) 13–26.

[135] N.H. Jayadas, Evaluation of the oxidative properties of vegetable oils as base stocks for industrial lubricants using spectroscopic and thermogravimetric analyses, *J. Synth. Lubr.* **25**(3) (2008) 105–13.

[136] S.C. Cermak and T.A. Isbell, Synthesis and physical properties of estolide-based functional fluids, *Ind. Crops Prod.* **18** (2003) 183–96.

[137] J.W. Lambert and D.L. Johnson (to Agro Management Group, Inc.), Vegetable oil lubricants for internal combustion engines and total loss lubrication, US Patent No. 5,888,947 (30 March 1999).

[138] S. Boyde, Green lubricants – Environmental benefits and impacts of lubrication, *Green Chem.* **4** (2002) 293–307.

[139] S. Nagaoka and M. Ibuki (to Fuji Oil Co.), Biodegradable lubricant base oil and its manufacturing process, US Patent No. 6,117, 827 (12 September 2000).

8
Quality assurance of biolubricants

Abstract: Biolubricant quality depends on the raw materials and chemicals used in processing, production process parameters and post-production conditions. Standard biolubricant analytical and performance methods are outlined. Considerable analytical attention is devoted to the assessment of lubricant biodegradability and ecotoxicity. The current status of international biolubricant specifications is given. Biolubricant quality management sets different analytical requirements at various stages of the product life cycle, involving production, storage, in-service use and disposal. Off- and on-line analytical measurements from feedstock to final bio-based end product are crucial to ensure reliable product quality. Field measurements of oils and greases are illustrated. The analysis of formulated biolubricants is complex.

Key words: biolubricant quality requirements, product quality management, process control, standardised testing procedures, analytical methodology.

8.1 Introduction

High (bio)lubricant quality is a prerequisite for market acceptance. Various parameters influence the quality of the biolubricant product. These include in the first place the renewable source (e.g. vegetable oil) and other materials used in the production process, the production process conditions as well as post-production parameters, including storage stability. Since the quality of vegetable oil raw material depends on the nature of the source (e.g. rapeseed, soy; recycled or not), location, season, etc., the various fatty acid (FA) profiles (chain length, degree of unsaturation and branching of the chain do influence the ultimate biolubricant properties (see Section 5.2.3). Important lubricant properties that are influenced by the FA profile (see Table 5.3) and by the structural features of the various fatty esters are viscosity and wear, and ultimately cold flow, oxidative stability and emissions. Production technology should always lead to the same quality standard, independent of the oils or fats used. Flexible multi-feedstock use (including non-food oils) at the highest yield levels is a technological challenge. The possibility of a qualitative and quantitative analytical characterisation of all components of the fluids is very helpful during development and application.

Accreditation of a modern analytical laboratory passes through the core EN-ISO Standard [1] and associated standards and guides, ISO Guides 30 and 35 (Certified Reference Materials, CRMs) and ISO Standard 17043 (Proficiency Testing). Standard setting bodies such as the AOAC, ASTM and RSC develop standard methods and certified reference materials, and validate commercial test kits.

The analytical issues connected with formulated biolubricants are numerous. Analytical measurements (from off- to on-line) are important from feedstock to final bio-based end product. Some vital analyses in the determination of the quality of a biolubricant refer to thermo-oxidative and hydrolytic stability, the absence of moisture and residual catalyst (ash content), etc. Reliable production and minimisation of pollution depend upon knowledge of the feedstock components and evaluation of the final product quality. Production facilities need to ensure lubricant quality.

Important analyses in use of lubricants are measurement of viscosity, moisture in oil, oil in cooling water, additive depletion, lubricant oxidation, foreign oil contamination, etc. (see also Section 8.8). For on-line monitoring of wear in running conditions sophisticated analytical tools such as a radionuclide technique (RNT) may be required.

As a very large proportion of the overall lubricant production is dispersed into the environment on first use, considerable attention has been given to lubricant biodegradability and persistence in the environment [2]. For eco designations of lubricants, see Section 3.4.

8.2 Biolubricant quality requirements

The quality of a (bio-based) lubricant is an indication of the time the lubricant may be expected to perform its essential properties. Primary properties of formulated lubricants affecting quality are lubricity (IP 450/2000), friction (ASTM D 5183), viscosity index (VI) (ASTM D 2270), anti-wear (AW) properties (ASTM D 4172), pour point (PP) (ASTM D 97), oxidation stability (ASTM D 943), rust prevention (ASTM D 665), foam resistance (ASTM D 892), water separability (demulsibility) (ASTM D 1401), paint and elastomeric seal compatibility, and degradability. Many of these properties are also on account of the use of chemical additives.

Specifications for industrial lubricants have been defined in relation to the demands of application. The determination of the physical, chemical and tribological characteristics of lubricants by standardised test methods and product specifications (AFNOR, ASTM, CEC, DIN, FTM, IP, ISO, JPI, MIL, etc.) enables assessment of their suitability for specific use. For laboratory methods for testing lubricants, see ref. [3]. For example, the mini-rotary viscometer (MRV) (ASTM D 3829, D 4684) measures an apparent yield stress and viscosity under low shear rates and is used to predict low-temperature pumpability of motor and hydraulic oils in field service. Mechanical-dynamic test methods for lubricants are described by ref. [4].

Frequently, no correlation has been established between mechanical-dynamic tests and field service. For instance, this is the case of the standard test method (ASTM D 2266) for wear preventive characteristics of lubricating grease (four-ball method). Field tests in real-world conditions (performance testing) are by far the most expensive and lengthy tests in the development of a new lubricant; see also Sections 10.4 and 10.5.1. In most industrial applications the actual conditions of a lubricant in use are also monitored (see Sections 8.8 and 14.4).

In principle, quality and performance requirements for biolubricants, as set by original equipment manufacturers and end-users, do not differ from those of conventional lubricants, with the provision that ecotoxicity and biodegradability are additional qualifiers. In this respect, it is noticed that the nature of the feedstock used for biolubricants (such as vegetable oils and derivatives) determines enhanced oxidative instability with respect to mineral oil-based lubricants, which needs to be taken care of.

8.2.1 Toxicological evaluation

Volumes of lost lubricants (30–50%) represent a threat for the environment and for human health. Lubricating oils are generally mobile liquids which are insoluble and less dense than water. They therefore tend to spread over surface waters in the form of very thin oil films, are visually polluting and have a damaging effect on aquatic life. The spread of lubricants in the environment affects different parts of the natural medium (soils, underground waters, rivers, etc.), which explains the difficulty of selecting analytical methods to assess the real environmental impact of these products. The most usual analyses provide degradation rates of the substance in contact with microorganisms, doses of non-toxicity or measures regarding the oxygenation rates of waters. It is important to select the most suitable method according to the context. Vegetable-based lubricants are considered as a reliable solution to combine technical performances and ecocompatibility characteristics.

Ecotoxicological tests can be divided into acute and chronic tests (Table 8.1). Acute tests measure the effects of high concentration of chemicals during a relatively short exposure time. Toxicity is expressed as concentration (ppm) of the test material that results in a 50% mortality rate after 96 h (lethal dose, LC50, or other relevant effects, EC50). A lubricant is generally considered acceptable if a concentration exceeding 1000 ppm of material in an aqueous solution is needed to achieve a 50% mortality rate in the test organism. Sub-chronic or chronic tests investigate long-term effects at sublethal concentrations (determination of No Observed Effect Concentration, NOEC). The environmental classification of substances is usually based on data from acute tests. Aquatic ecotoxicity tests measure

Table 8.1 Standard ecotoxicity tests

Type	Test method	Test organism	Reference
Acute tests	OECD 201	Algae	[5]
	OECD 202/1	Daphnia	[6]
	OECD 203	Fish	[7]
	OECD 209	Bacteria	[8]
Chronic tests	OECD 202/2	Daphnia	[6]
	OECD 204	Fish	[9]

the extent to which a fluid will poison environmental species such as algae (OECD 201) [5], daphnia (OECD 202) [6], fish (typically Gold orfe, *Leuciscus idus,* or zebrafish, *Brachydanio rerio*) (OECD 203) [7] and bacteria (*Pseudomonas putida*) (OECD 209) [8]. The minimum requirements for testing of aquatic toxicity are OECD 201 and 202. Testing on fish (OECD 203) is not necessary since fish is less sensitive than algae and daphnia. OECD test guideline 401 for mammalian toxicities was deleted in 2002.

The most common test methods used by the lubricant industry for evaluating the acute toxicity of their products are EPA 560/6-82-002, Sections EG-9 and ES-6, and OECD 203. These tests determine the concentration of a substance that produces a toxic effect on a specified percentage of test organisms in 96 h. The acute toxicity test is normally conducted using rainbow trout. Standard acute and chronic aquatic toxicity procedures are most appropriate for lubricants with compositions that are completely soluble at the prescribed test treat rates.

Most fully formulated lubricants and lubricant compositions are difficult to evaluate in toxicity tests because they are mixtures of chemical compounds with varying and usually poor solubility in water. Some of the substances that are normally added to base fluids in order to enhance their applicability may possess a high toxicological potential. Lubricants and lubricant component mixtures should not be added directly to aquatic systems for toxicity testing because the mode of the addition procedure has a large impact on the results of the toxicity test. Use of mechanical dispersion techniques described in ASTM D 6081-98 (Standard Practice for Aquatic Toxicity Testing of Lubricants: Sample Preparation and Results Interpretation) will produce well-characterised test systems that will give reproducible and meaningful results for acute and chronic aquatic toxicity tests. Samples prepared in accordance with ASTM D 6081 may be used in acute or chronic aquatic toxicity tests conducted in fresh water with fish, large invertebrates or algae.

The toxicity of mixtures of poorly soluble components cannot be expressed in the usual terms of lethal concentration (or the similar terms of effect concentration or inhibition concentration) because the mixtures may not be completely soluble at treat levels that lead to toxic effects. The test results are then best expressed in terms of loading rate.

Ecotoxicological risk classification of lubricant oils is usually carried out with new oil fluids (base stocks) on the basis of aqueous bioassays. Environmentally acceptable lubricants can undergo a change of their ecotoxicological potential during use. Friction and ageing influence the toxic and ecotoxic behaviour of used bio-based lubricants [10]. Used oils present a more complex toxicological picture than fresh oils. Their hazards are harder to quantify since used oils may have a wide range of compositions, reflecting the potential range of service conditions and possibilities of contamination (from fuel combustion by-products, tribochemical or thermochemical decomposition of the lubricant). A toxicological evaluation of age-dependent change of biogenous lubricating oils has been reported [11]. Photocatalytic treatment of effluents from metalworking fluids and coolants can be used to reduce toxicity to meet ever-tightening regulatory standards [12].

For proper ecotoxicological evaluation of synthetic ester lubricants it is appropriate to take into account the changes in the chemical composition caused by service use, as well as the effects of additives [13]. Because of polycyclic aromatic hydrocarbon (PAH) formation in service, used engine oil is recognised as a potential carcinogen. Careful monitoring of re-refining processes is required in order to control the level of PAH in recycled oils.

For general information on the toxicological behaviour of (mineral) base oils, see refs. [14, 15]; for toxicity and eco-requirements of lubricant additives, see refs. [16–18].

8.2.2 Biodegradability: definitions and tests

Biodegradability describes the environmental fate of a substance as a result of biochemical breakdown by microorganisms. There is no uniform definition of biodegradability or of environmental safety. Internationally there is a wide range of definitions of 'environmentally preferable'. ASTM D 5864 defines biodegradation as 'the process of chemical breakdown or transformation of a substance caused by organisms or their enzymes'. Ultimate biodegradation is defined as 'degradation achieved when the substance is totally utilized by microorganisms resulting in the production of CO_2, water, inorganic compounds and new microbial cellular constituents (biomass or secretions, or both)'. ASTM D 6006 describes biodegradability as a function of the degree of degradation, time and test methodology. Readily biodegradable is referred to as 'breaking down rapidly in the environment by a defined amount in a specific time frame'. Inherently biodegradable means 'has the propensity to break down with no defined amount or time frame'.

Harmonisation of the term biolubricants is highly desirable (see also Section 3.4).

In terms of chemical processes, biodegradation (usually aerobic) occurs in two steps, namely first disappearance of the original substance (primary biodegradability) and next complete mineralisation to CO_2 and H_2O (secondary or ultimate biodegradability) (Fig. 8.1). Primary degradation stands for the alteration in the chemical structure of the substance with consequent loss of specific properties. While primary degradation describes the disappearance of the parent organic compound (as determined by compound-specific analysis) it does not necessarily indicate that the substance is completely biodegradable [19]. Functional groups or fragments with a positive effect on biodegradation are alcohols, acids, aldehydes, esters and amides.

Table 8.2 lists the most established standardised biodegradability tests. The biodegradability of organic substances can be evaluated using a wide variety of testing procedures, but the most commonly adopted are the standard tests specified in the OECD Guidelines for the testing of chemicals and in the ISO standardised methods [20]. The biodegradation process can be analysed under aerobic and anaerobic conditions, in fresh water or seawater, or in soil. In most cases, biodegradability is evaluated by measuring the aerobic biodegradability of a substance in an aqueous medium. For the principles of evaluating the biodegradability of organic substances in aerobic aquatic media by means of screening tests, see ref. [21].

The choice of an appropriate test is an essential prerequisite for the determination of the biodegradability of lubricants [22]. Testing of lubricants for biodegradability raises considerable problems [23]. Most of the lubricating base oils are poorly water-soluble (e.g. lower than 1 mg/L for synthetic ester oils). The recommended concentrations in test solutions often exceed the ester-solubility values for oils. Moreover, lubricants generally constitute a mixture of compounds (or even blends). Also, not all tests for biodegradability are created equal.

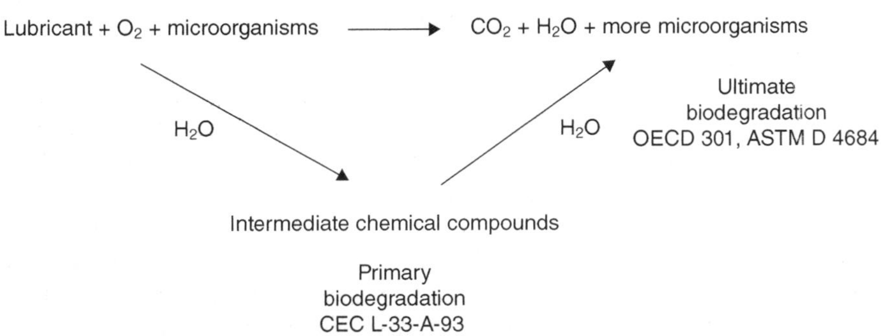

8.1 Primary vs ultimate biodegradation.

Table 8.2 Most established test methods for biodegradability

Method	Organisation
CEC L-33-A-93[a]	Coordinating European Council (CEC)
OECD 301B CO_2 Evaluation Test (Modified Sturm) OECD 302B Zahn-Wellens/EMPA Test OECD 302D (draft) CONCAWE Test OECD 310	Organisation for Economic Co-operation and Development (OECD)
ISO 6060, 9439, 9888, 10634, 10707, 10708, 14239, 14593	International Organization for Standardization (ISO)
EPA 560/6-82-003, No. CG-2000 (Shake Flask Test)	US Environmental Protection Agency (EPA)
Environmental Persistance Classification	US military
ASTM D 5864-05, ASTM D 6006	American Society for Testing and Materials (ASTM)

[a] Formerly CEC L-33-T-82.

OECD guidelines 301A–F [24] and ISO standards 10707 [25] and 10708 [26] are standardised test methods for assessment of ultimate biodegradability of tested material in water by microbial action within 28 days. OECD 301B (Modified Sturm) test [27] is adequate for soluble and insoluble organic, non-volatile materials, and measures only complete oxidation. Compared with OECD 310 [28], the OECD 301B procedure is labour-consuming. The OECD 310 test seems to be better suited for determining the ultimate aerobic biodegradability of poorly water-soluble substances [23]. The procedure of ISO 14593 [29] in combination with the conditions of the OECD 310 testing method permits the ready biodegradability of oils to be evaluated in a convenient way. Using the same apparatus and reagents, and taking into account the conditions and criteria of the OECD 302D (draft) CONCAWE testing method [30] allows assessment of the inherent biodegradability of oils.

The approved test method CEC L-33-A-93 (CEC, Coordinating European Council) for two-stroke engine oils [31], which measures the disappearance of IR bands of C-H bonds (CH_3: 2958 cm^{-1}; CH_2: 2928 cm^{-1} of lipophilic molecules extractable into chloroalkane solvent), does not clearly differentiate between primary and ultimate degradation but is essentially a measure of the former [32, 33]. Typical values for biodegradability of a range of a base fluids by CEC L-33-A-93 are listed in Table 3.9. The method applies to most organic compounds, whether soluble or insoluble in water. It measures the overall biodegradability of hydrocarbons, or similar compounds containing methylene groups, recording all transformations that the starting material undergoes, including oxidation and hydrolysis. The CEC test, which was developed to characterise the biodegradability of outboard engine oils, is accepted in a number of Blue Angel Environmental Labels, and requires 80% or greater biodegradability. The CEC L-33-A-93 test for 'inherent degradation' is becoming obsolete. Moreover, as the CEC test deals only with the aquatic environment, assessment of the relative biodegradability in soil of lubricants for agricultural machinery requires a modification of the method [34].

ASTM D 5864-05 (Standard Test Method for Determining Aerobic Aquatic Biodegradation of Lubricants or Their Components) has adopted the OECD 301B Modified Sturm procedure and covers the degree of aerobic aquatic biodegradation of fully formulated lubricants or their components to an inoculum under laboratory conditions. The inoculum may be the activated sewage sludge from a domestic sewage treatment plant, or derived from soil or natural surface waters. The test method is intended to specifically address the difficulties associated with testing water insoluble materials and complex mixtures such as are found in many lubricants. According to ASTM D 5864 a lubricant is classified as readily biodegradable when 60% or more of the test material is converted to CO_2 in 28 days. ASTM D 5864 is only a test method, not a hierarchy of biodegradability. The Environmental Persistence Classification (aerobic fresh water) is used by US military.

ASTM D 6006 is a guide for assessing biodegradability of hydraulic fluids, whereas ASTM 6046-02 describes the standard classification of hydraulic fluids for environmental impact. Criteria for environmentally acceptable hydraulic fluids according to ISO 15380 classification are much less stringent [35]. No defined regulation has been drawn up for disposal of oils. Laboratory tests have shown that the degradation rates may vary widely among the different test methods indicated above. It is therefore necessary to always make reference to the specific biodegradability test used.

Beran [23] recommends use of CEC L-33-A-93 for the evaluation of primary biodegradability of lubricating oils, for evaluation of the ultimate biodegradability in an aerobic environment use of testing methods OECD 301B and OECD 310 for ready biodegradability, and OECD 302D (draft) CONCAWE test [30] and OECD 302B Zahn-Wellens/EMPA test [36] for inherent biodegradability.

Evaluation of the biodegradability of lubricating oils by standard screening tests is useful not only for the assessment of the potential environmental impact of substances but also in order to acquire better insight in the relationship of chemical structure and biodegradability of lubricants. Figure 8.2 shows the assessment of primary and ultimate biodegradability of various lubricants [23]. Rapeseed oil (RSO), the polyester oils NYCOBASE 3118

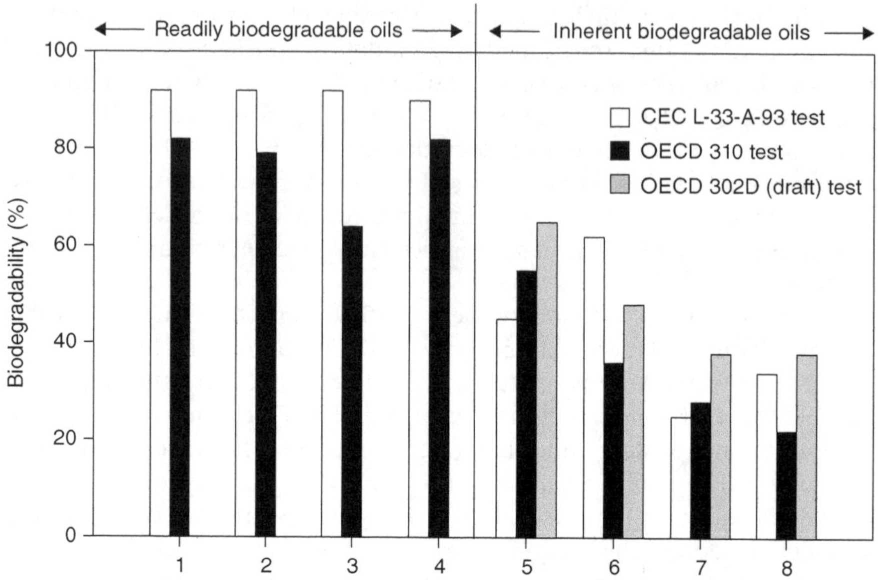

8.2 Biodegradability of lubricating base oils differing in chemical structure: rapeseed oil (1), NYCOBASE 3118 oil (2), PRIOLUBE 3999 oil (3), PE tetracapronate oil (4), PE tetra(sec-capronate) oil (5), PAO 4 oil (6), PAO 6 oil (7), and mineral oil ISO VG (8). After ref. [23].

(TMP trioleate; NYCO), PRIOLUBE 3999 (Uniqema), and PE tetracapronate (ester of pentaerythritol and *n*-hexanoic acid [37]) can be classified as readily biodegradable substances. The inherent biodegradability of the other oils is as follows: PE tetra(sec-capronate) > PAO 4 (Nexbase 2004, Fortum Oil N.V.) > PAO 6 (Nexbase 2006, Fortum Oil N.V.) > conventional mineral oil ISO VG 32 (SAE 10/95 type). The results of Fig. 8.2 indicate that the CEC test fails to reveal the relationship between chemical structure and biodegradability in the case of readily biodegradable ester oils. On the other hand, the OECD 310 (ISO 14593) test discloses the effect of chemical structure on the biodegradability of the oil, as can be inferred from the lowest ultimate degradability value for PRIOLUBE 3999 with branching in the structure of the acid residue of the polyester molecule. It is well known that branched alkyl groups, particularly those containing quaternary branching, are more resistant to biodegradation. Alkylbenzenes having a high degree of branching are generally of low biodegradability (see Table 3.9).

Biodegradability may vary noticeably from one ester oil to another according to their chemical structure. For example, the ultimate biodegradability of PE tetra(sec-capronate) is higher than the primary biodegradability. Chemical hydrolysis can support the process of biological degradation of PE tetra (sec-capronate) oil and can be responsible for the higher extent of ultimate than of primary biodegradation [23]. Contrary to common belief, not all ester oils are readily biodegradable. The determination of the inherent biodegradability of oils 5–8 in Fig. 8.2 provides the data needed for the assessment of their potential environmental impact.

Typical criteria for environmentally friendly base fluids are: (aerobic) biodegradability (CEC L-33-A-93) > 80% after 28 days, (OECD 301, ASTM D 5864) > 60% after 28 days; aquatic toxicity LD50 > 1000 ppm. The biodegradability of most of the fresh biodegradable oils is usually well established [38], but on-line monitoring of biodegradability during ageing is less well developed. During oil ageing, the chemical structure of a biolubricant changes, and its biodegradability tends to decrease [39]. The biodegradability of the aged lubricant in a diesel vehicle is dependent on the applied fuel. As the OECD 301 'Manometric respiratory test' for measuring biodegradability is very time consuming, near-infrared (NIR) spectroscopy combined with statistical techniques (PLS) has been used for rapid assessment of the biodegradability of fully formulated ester oils and a biodegradable base oil without additives during ageing at 140 °C in air flux [40]. Oxidation processes of the ester-based lubricant hydroxy-isobutoxy-stearic acid methyl ester (HISM) have influenced the ecotoxic and genotoxic behaviour of the oil [11]. Consequently, there appears to be a need for bioassays of used fluids alongside with chemical analysis. Owing to wide and different types of applications of lubricants it is difficult to assess their harmfulness to the environment and control their effective distribution on territory [41]. For environmental characteristics of lubricants, see also ref. [42].

The recognised biodegradability tests used today, such as the OECD 301 series or CEC L-33-A-93 test, assess biodegradability of chemicals in an aquatic environment. Also the ecotoxicity tests used as a reference (OECD Guidelines 201, 202 and 203) refer to the environmental impact of a substance in an aqueous environment by measuring its impacts on algae, daphnia and fish. Soil is a non-homogeneous medium with composition varying with depth. Aerobic processes will only be possible within the first centimetres of the ground's surface. Soil microorganisms differ from those used in aqueous medium tests. In case of assessment of the environmental impact of a substance deposited on the soil, it should firstly be determined whether it is carried along by surface water or if it penetrates the ground. In the latter case, migration through the various soil layers needs to be considered (with monitoring in time and space by means of a lysimeter). Experiments in lysimeters are relatively complex and costly to implement. ISO standard 14239 lists different methods available to assess biodegradability in soil [43]. A simplified protocol based on the principle of the Modified Sturm test has been proposed for the assessment of the ultimate biodegradability of substances deposited in a bioreactor containing natural soil with its original microorganisms [44]. Biodegradability tests performed in soil are the best approaches for the evaluation of the environmental impact of organic compounds suspect of pollution. If a chemical remains adsorbed by the soil particles, as long as it is biodegradable, it will be transformed by the soil's microorganisms and cause no environmental damage. Vegetable-based lubricants and synthetic products can be considered readily biodegradable. An inherently biodegradable product (typically a conventional petroleum-based lubricant), which breaks down very slowly over time (years), can cause substantial damage. They require long-term remediation due to environmental persistence.

Various studies have assessed the properties of vegetable oil-based lubricants towards the environment and users [38, 45–50]. Cecutti and Agius [38] have compared migration and biodegradability of several hydraulic fluids in two forest soils and in a liquid medium, namely of the HEES trimethylolpropane ester Biohydran TMP 46 (FINA), the RSO- and SNO-based Biolube HETG 346 (IGOL), the SNO-based HETG Hélianthe TRF (TECHNOL), and the classic, mainly paraffinic oil-based hydraulic Mobilfluide (Mobil), widely used by lumberjacks. The experimental methodology followed consisted in use of a lysimeter for assessment of the migration and progressive primary biodegradation of the lubricants, integrated by ultimate biodegradation tests in soil and in aqueous medium. Evaluation of the hydraulic fluids' biodegradability and ecotoxicity was carried out with new and used fluids, i.e. fluids that had been used for 1000 h in a logging machine. Figure 8.3 shows the biodegradation kinetics of used hydraulic fluids in sandy oil. The three biolubricants, with very similar kinetics, degraded more rapidly than the lubricant of mineral origin. As shown in Table 8.3, biodegradation in a liquid medium is more rapid and complete than in a soil environment, in accordance with similar findings by other authors [51, 52].

Ecotoxicity tests were carried out with both fresh and used fluids and the test results on algae are shown in Table 8.4. None of the biolubricants was toxic to fish (OECD 203), as opposed to the fluid of mineral origin that was lethal already at relatively low concentrations (380 mg/L). The test on algae (OECD 201) was the most sensitive and enabled classification and comparison of biolubricants, and underscored an increase in toxicity after use for Hélianthe TRF and Mobilfluide. The toxicity levels of the biolubricants were never high however, even after use, unlike the petroleum-based fluid.

Use of biolubricants in forestry is an excellent way of protecting the environment as these easily biodegradable products entail no risk of pollution of soils or subsurface waters. Tests in Finland have compared the biodegradability of mineral and vegetable oils in the soil. Oil concentrations were monitored to a depth of 15 cm for 16 weeks. RSO had completely

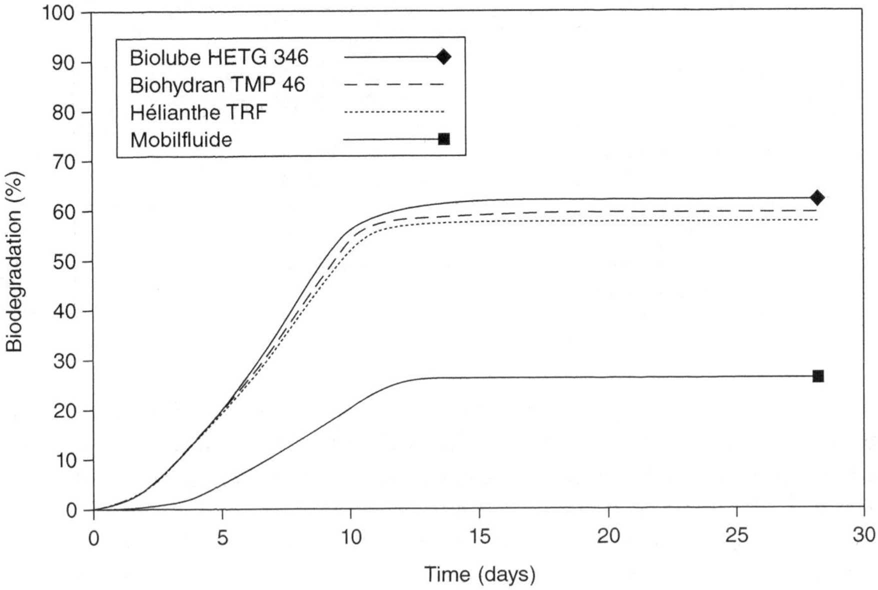

8.3 Ultimate biodegradation of used hydraulic fluids in sandy oil. After ref. [38].

Table 8.3 Ultimate biodegradability rate (%) of hydraulic fluids in two forest soils and in aqueous medium[a]

Lubricant	Biodegradation rate		Biodegradation OECD 301B test
	Clayey soil	Sandy soil	
Biolube HETG 346	64/62	64/61	93/79
Biohydran TMP 46	63/61	62/60	78/75
Hélianthe TRF	61/57	60/57	85/69
Mobilfluide	29/27	27/25	93/78

[a] Before/after use.
After ref. [38].

disappeared after 6–7 weeks but 20% of the mineral oil still remained in the ground at the end of the four-month test period [53].

Haigh examined biodegradation and effect on crops of a range of lubricants, including Priolube 3999, dibasic acid esters based on dodecanoic acid, sunflower oil (SNO), white spirit and mineral oils, in field studies [54]. Factors affecting the degradation of Priolube 3999 (TMP ester of 'isostearic acid') in natural soil are nitrogen availability (fertiliser), temperature and water saturation/oxygen availability. Similar findings have been reported

Table 8.4 Hydraulic fluid ecotoxicity values[a]

Lubricant	Algae EC50-72h OECD 201 (mg/L)
Biolube HETG 346	5400/5600
Biohydran TMP 46	2800/1800
Hélianthe TRF	4800/100
Mobilfluide	1300/790

[a] Before/after use.
After ref. [38].

for the degradation of crude and waste oils in soil [55, 56]. Seasonal variation is one of the most important factors influencing the degradation of crude oil in soil [57]. Temperature is known to affect the biodegradation of lubricants not only by controlling the rate of microbial metabolism, but also by its effect on the physical nature of the oil [58]. Synthetic ester-based lubricants tested were degraded more rapidly and extensively in soil than mineral oil-based lubricants but not as rapidly nor as extensively as vegetable oil. Surprisingly, the rate of degradation of the oils, including natural vegetable oil, was slow in natural soil (>1 year), and considerably longer than in laboratory studies (40 days) [59]. The different rates and extents of biodegradation between laboratory and field situations are largely due to the influence of temperature and availability of oxygen and water. Even at low levels of vegetable oil (the fastest degrading oil) in the soil still significant effects on the yield of wheat crop were observed. Oil is known to retard plant growth by inhibition of the germination of seeds, suffocation of roots, physical interface with water and nutrient uptake capability, etc. In general, the smaller the molecule the greater is its ability to penetrate plant tissue and hence the more toxic the oil is to plants [60].

Use of biodegradable synthetic lubricants may help to reduce the impact of such molecules on the environment. The persistence of a lubricant must be assessed by taking into account the conditions under which a lubricant interacts with the soil. Care must be taken that reduced persistence is not accompanied by increased toxicity. Both biodegradability and low toxicity are characteristics of an environmentally acceptable lubricant and any assessment of the potential impact of a lubricant on the environment must consider both.

8.2.3 Bioaccumulability and biomagnification

Bioaccumulation, or bioconcentration, occurs when a substance accumulates in an organism. Bioaccumulation is defined as the increase in concentration of a test substance in or on an organism or specified tissues thereof relative to the concentration of the test substance in the surrounding medium. A predictive indicator for bioaccumulation is a substance's octanol–water partition coefficient. The criterion for acceptable bioaccumulation is defined in European Council Directive 67/548/EEC Annex XIII section 1.2, as a substance having a bioconcentration factor BCF (concentration of test substance in organism divided by concentration of the substance in the surrounding medium) of 2000; test procedures are set forth by EC Directive 76/548/EEC Annex V, section C.13 (flow-through fish test). None of commonly used base fluids for lubricants is bioaccumulative.

Biomagnification is when the accumulated substances concentrate higher up in the food chain. For biomagnification to occur, the pollutant must be

- long-lived (display persistence, resistance to biodegradation);
- mobile;
- soluble in fatty tissue; and
- biologically active.

Biomagnification can be high for certain substances (perchlorinated biphenyls (PCBs), dichlorodiphenyltrichloroethane (DDT), etc.) in some ecosystems. Bioaccumulation and biomagnification may cause severe damage to an ecosystem during prolonged exposure. One of the objectives of the OSPAR Convention (Oslo and Paris Convention for the Protection of the Marine Environment of the North East Atlantic, ratified in March 1998) is to prevent and eliminate any marine pollution in the convention area by hazardous substances, that means persistent, toxic and bioaccumulative substances.

Within the framework of the European REACH (Registration, Evaluation, Authorisation and Restriction of Chemicals) regulation [61, 62], companies that produce, import and use chemicals have the duty to assess the risks arising from their use, and to take the necessary measures to manage any risks identified. The registration requirements vary, depending on the volume of a substance produced and on the likelihood of exposure of humans or the environment. Tight controls have been introduced for the chemicals of highest concern, such as carcinogens, mutagens and reproductive toxicants (CMRs), persistent, bioaccumulative and toxic substances (PBTs), very persistent and very bioaccumulative substances (vPvBs), and persistent organic pollutants (POPs).

8.3 Biolubricant quality management

Reliable product output is essential, as is adherence to (inter)national norms and standards (see Chapter 9). Biolubricant quality control is challenging, especially if we consider the need for extensive analytical and testing facilities (see Tables 10.4 and 10.10).

The quality of lubricant is to be guaranteed by a quality management (QM) system such as according to the automotive standard ISO/TS 16949 [63]. A QM system needs to cover the whole life cycle of a biolubricant from raw material to disposal (and beyond):

- production (including feedstock selection and quality, production control, laboratory testing, and product quality);
- transport and storage conditions;
- consumers (complaints, problem solving);
- original equipment manufacturers (technical working groups); and
- disposal.

The quality of a biolubricant depends on several factors, including feedstock quality, the FA composition of the (modified) parent oil or animal fat, the production process and other materials used in the process. Biolubricants obtained by process chemistry of renewable feedstocks should only contain detrimental impurities such as moisture, unreacted triglycerides, free fatty acids (FFAs) and trace metals within prescribed limits. Commercial manufacturers must ensure that the manufacturing process quality is well controlled in order to produce on-spec biolubricants. Biolubricants are more sensitive to environmental factors than petroleum products. Biodegradable products are inherently more susceptible to biodegradation and microbial contamination than non-biodegradable material.

Certified lubricant laboratory testing services are available worldwide or locally (e.g. JOAP TSC, Oelcheck GmbH) for lubricants produced from various sources, as well as for experimental lubricant formulations. In the development phase of a biolubricant it is not unusual to have recourse to specialised laboratories, such as PerkinElmer Automotive Research Laboratory (San Antonio, TX) for high-temperature TEOST MHT-4 thermo-oxidation engine oil simulation test, homogeneity and miscibility (H&M) test, foam sequence testing, high-temperature foam test, engine oil filterability (EOFT), engine oil filterability/water tolerance test (EOWTT), gelation index, NOACK volatility, volatility loss, phosphorus and ball rust test in case of experimental engine oils [64]. In the H&M test, the tested oil shall remain homogeneous and miscible when mixed with SAE reference oils. Table 10.10 shows the results of testing for the experimental engine oil formation of Table 10.9. According to ASTM D 4485-99b, this oil meets the bench test of the ILSAC GF-3/API superlubricity minimum performance standard and passes all API SL bench test requirements.

Producers usually dispose of their own quality control laboratories for analysing (bio)lubricants according to the standards or specifications. Quality is directly related to liability issues. Bad product can cause harm to equipment. On the whole, the lubricant industry presents a favourable control history.

8.4 Quality control of biolubricant feedstocks

The need for analytical control of the quality of biolubricant feedstocks is apparent when considering the wide range of oil- and fat-containing products which may be used as a raw material. Oil sources vary regionally and may also vary throughout the growing season, resulting in changes in oil composition. The quality of vegetable oil or fat may be assessed using methods of analysis in the *Official Methods and Recommended Practices of the AOCS* [65]. Table 8.5 lists AOCS methods for oil products. It is quite reasonable to defer feedstock analysis to the supplier and carry out random checks. Consistent quality control (QC) of incoming feedstocks is required for the optimal regulation of the production process.

FFA content in oils and fats can be determined by titration methods or by ^{1}H nuclear magnetic resonance (NMR) spectroscopy [66], and fatty acid compositions by gas chromatography–flame ionisation detection (GC-FID) after derivatisation as methyl esters or by quantitative ^{1}H and ^{13}C NMR [67].

Almost all lubricants are multicomponent mixtures. The oils used as starting materials for many biolubricants are of natural origin and therefore not 'pure' in terms of molecular structure. Differential scanning calorimetry (DSC) is a useful tool for analysing the melting/crystallisation behaviour of vegetable oils with different FA compositions, e.g. in case of palm kernel oil [68]. Unlike most other compounds, vegetable oils and fats, which are

Table 8.5 AOCS methods for vegetable oil quality parameters

Chemical tests:	
1. Acid value (AV)	Cd 3d-63
2. Free fatty acids (FFA)	Ca 5a-40
3. Oil content	Ba 3-38
4. Iodine value (IV)	Cd 1d-92
5. Moisture content (KF)	Ca 2e-84
6. Peroxide value (PV)	Cd 8b-90
7. Saponification value (SV)	Cd 3b-76
8. Soap content	Cc 17-95
9. Free gossypol	Ba 7-58
Physical tests:	
1. Cloud point (CP)	Cc 6-25
2. Colour	Cc 13c-92
3. Flash point	Cc 9a-48
4. Specific gravity (SG)	AOCS 10a-25

not pure organic compounds but mixtures of different FA, may show multiple melting peaks depending on the chain length of the triacylglycerols (TAGs), the configuration of the unsaturations (*cis*- or *trans*-olefinic), and the number and relative positions of the unsaturated centres and crystalline form. DSC was also used for the analysis of PP of coconut oil as a lubricant base stock [69]. In cases of vegetable oils with complex FA constitution (e.g. groundnut oil), the endothermic peaks fail to indicate the PP because of the multiple peaks.

8.5 Standardised methods for testing lubricating fluids and greases

Lubricants and greases can be tested by a variety of methods, and should be evaluated for lubricity, friction and wear properties, low-temperature behaviour, oxidative stability, rheology, biodegradability, ecotoxicity and many other properties. Table 8.6 lists the main ASTM test methods [70] for lubricants. For an overview of equivalent (inter)national standardised laboratory methods (AFNOR [71], DIN [72], IP [73], and DIN ISO [74]) for testing lubricants and greases, see ref. [3].

A wide variety of analytical methods is needed in order to guarantee a high level of (bio)lubricant manufacturing and final product quality (Table 8.6). The choice of equipment needed for a specific analysis depends on regulatory requirements, or time, cost and labour requirements. The standard analytical methods are slow (requiring hours or days) and often require multiple physical analyses per sample. Most traditional reference instrumentation must be kept in a stable laboratory environment, away from the process line. Some analyses for process control or quality verification are quick, while others require a more complete test according to the ASTM approved methods. The parameters for lubricant quality require different types of instruments. No analyser will do it all.

The high-frequency reciprocating rig (HFRR) lubricity test (ASTM D 6079) is used to determine the lubricity or the ability of a fluid to affect friction and wear between the surfaces in relative motion under load. It has been stressed, however, that standard lubricity tests such as HFRR and the ball-on-cylinder lubricity evaluator (BOCLE) (ASTM D 6078) do not really test lubricity, but rather AW/extreme pressure (EP) functionality [75]. For other mechanical-dynamic test methods for lubricants, see Section 10.4.1. The viscosity-pressure behaviour of a lubricant is principally responsible for the load-carrying properties of a lubricating film. The dynamic viscosity of fluid media increases with pressure. Vibrational fork viscometer technology allows the direct on-line measurement of both viscosity and density [76]. Low-temperature flow properties can be evaluated by means of cryogenic DSC (for wax disappearance temperature, WDT, and signal

Table 8.6 ASTM standardised methods for testing lubricating fluids and greases

Additive depletation	D 7412
Ageing/oxidation stability	D 942, D 943, D 2112, D 2272, D 2893, D 4310, D 4742, D 6186, D 7214, D 7414
Air release	D 3427
Aniline point	D 611
Ash content	D 128, D 482, D 874
Atomic spectroscopy	D 4628, D 5185, D 6443, D 6595
Biodegradability	D 4684, D 5864, D 6006, D 6064
Bio-oil content	D 6866
Boiling range/distillation	D 86, D 2887
Carbon distribution	D 2140, D 3238
Carbon residue	D 189, D 524
Cloud point	D 2500, D 5551, D 5771, D 5773
Condition monitoring	D 7412, D 7414, D 7415, D 7418, E 2412
Corrosion tests	B 117, D 130, D 665, D 1748, D 4310, D 4627, D 4683
Demulsibility (water separability)	D 1401, D 2711
Density	D 1298, D 7042
Dropping point/melting point	D 127, D 566
Ecotoxicity	D 6046
Evaporation loss	D 972, D 5480, D 5800
Flash point	D 56, D 92, D 93, D 6417, D 6450, D 7094
Foaming characteristics	D 892
Friction	D 5183
Hydrolytic stability	D 2619
Insolubles, soot	D 893, D 7686
Neutralisation number	D 128, D 974
Oil separation	D 1742
Penetration	D 217, D 937, D 1321
Pour point	D 97, D 5949, D 5950, D 5985, D 6892
Refractive index	D 1218
Roll stability	D 1831
Rust prevention	D 665, D 1748
Saponification number	D 94
Shear stability	D 3945, D 5293
Specific gravity	D 287
Sulphur	D 3120, D 6443
Total acid/base number	D 664, D 974, D 2896, D 4739
Viscosity (kinematic, dynamic, CCS apparent)	D 445, D 446, D 1092, D 2602, D 2983, D 3829, D 4683, D 4684, D 4741, D 5133, D 5293, D 6074, D 7042, D 7279, D 7483
Viscosity index	D 2270, D 7483
Water content	D 95, D 1533, D 1744, D 4377, D 4928, D 6304
Water resistance	D 1264
Wear	D 1947, D 2266, D 2271, D 2596, D 2782, D 2783, D 2882, D 4172, D 7043
Wear metals and additive elements	D 5185, D 6443, D 6595

maxima temperature, SMT), PP and cloud point. The PP and the boundary viscosity define the lowest temperature at which an oil can normally be used. In industry the low-temperature properties of lubricating fluids are mostly evaluated in terms of PP, determined according to ASTM D 97. The method is extremely time-consuming and interlaboratory reproducibility is poor (repeatability of 2.87 °C at 95% confidence level) [77].

The evaporation characteristics of lubricants are other important quality criteria. This is easily understood by the emissions created upon evaporation and the accompanying change in lubricant composition. Evaporation is dependent on the vapour pressure of the base oil components. The evaporative loss of rapidly biodegradable lubricants is generally lower than that of conventional oils. In most applications of energy-saving oils low evaporation behaviour and low viscosity are increasingly being appreciated.

8.5.1 Oxidative stability tests

Oxidation is a general term used to denote chemical degradation of lubricants. Lubricant oxidation is a three-stage process, consisting of: (i) initiation, or formation of a highly reactive, short-lived free radical; (ii) propagation, i.e. reaction of hydroperoxy radicals with base oil or additives to regenerate an alkyl radical or an alcohol and water; and (iii) termination when the designated oxidation inhibitor (antioxidant) performs its function. The most significant contributors to oxidation include air, moisture, heat/temperature (by oxidation rate), pressure (by variation in fluid viscosity, causing frictional heat generation), contaminants (by catalysis) and certain metals (e.g. Cu, by oxidation in the presence of water). Metal deactivators and chain breakers may be added as oxidation inhibitors to reduce the potential for oxidation. Additives are selected for the chosen base stocks to help stop, or at least postpone, oxidation. Once antioxidant additives are consumed, oxidation can continue unabated. The inherent chemical degradation reactions result in the formation of aldehydes and acids, which can negatively influence the fluid viscosity and lubricity and can cause corrosion. Condensation and polymerisation produce insoluble sludge and gum that can increase wear, reduce clearances, and plug lines and valves. The degradation compounds may be removed with specially designed electrophoresis and electrophysical filtration systems [78].

Traditional base stocks, containing high levels of aromatics and sulphur (Group I oils), have been substantially displaced by more highly refined base stocks, which have very low levels of aromatics and no sulphur (Group II/III oils). The oxidation performance of these various base stock classes is quite different. This development determines the need to adapt current bench tests to adequately mimic oxidation processes in modern lubricant formulations [79].

Lipid oxidation (also called autoxidation) promoted by heat, light, certain metals (iron and copper) or lipoxygenases, is one of the chief processes by which lipids degrade. Analytical protocols are required to measure the quality of lipids. Lipid oxidation can be determined by a variety of techniques, including redox titrimetry (peroxide value) and headspace volatile analysis (HSGC). During the early stages of the autocatalytic free-radical chain reaction, the positions of double bonds in unsaturated FAs are changed and hydroperoxides are produced (see Fig. 5.4). These alterations can be monitored by simple analytical (spectrophotometric, iodometric and colorimetric) protocols. The conjugated dienes of oil samples can be determined according to a IUPAC method [80]. UV absorbance at 233 and 268 nm is employed to monitor the formation of conjugated dienes (CDs) and trienes (CTs) of polyunsaturated fatty acids (PUFAs), respectively. For CD formation to occur, unsaturated FAs with at least two double bonds are required. Likewise, more than two double bonds are a prerequisite for CT formation. Increasing absorption values are an indication that oxidation is proceeding. Owing to the sensitivity of UV spectrophotometry only small quantities of lipid are required. The oxidative stability of vegetable oils can also be evaluated using AOCS method Cd 19-90 for measurement of 2-thiobarbituric acid reactive substances (TBARS) [81].

The oxidative status of vegetable oils as base stocks for industrial lubricants can also be assessed by a combination of Fourier transform infrared (FTIR) and NMR spectroscopy, and thermogravimetric/differential thermal analysis (TGA/DTA) techniques [82]. FTIR spectroscopy can be used as a quick method to measure the degree of unsaturation in vegetable oils in a qualitative manner. NMR spectroscopy can give this a quantitative basis. TGA/DTA thermograms show weight changes and thermal activities in terms of exo- or endothermic heat flows during the oxidative reactions. The oxidative degradation can be evaluated in terms of the weight gain by the respective vegetable oils due to the formation of oxidative polymerisation products and the value of exothermic peaks as observed by DTA.

Evaluation of the oxidative stability of lubricant oils is generally complex and no universally accepted protocol is available [83]. Oxidation tests can be differentiated into two groups: (i) tests which describe the condition of the lubricant after a defined test period (as expressed by acid number, viscosity change and sludge formation); and (ii) tests measuring the oxidation induction time. There are several ways in which the oxidative stability of an oil (a measure of the expected lifetime of a lubricant) may be measured, including oxidative stability index (OSI) [84], rotating bomb oxidation test (RBOT) (ASTM D 2272-98), DSC [85], and others. RBOT measures the time in which the oxygen in a pressurised autoclave has been used up by the lubricant. Synthetic oils have good oxidation stability. Comparative tests have shown that some synthetic ester oils have much greater oxidation stability than mineral oil-base products. Several test methods were developed more specifically for the determination of the oxidation stability of mineral oils, such as ASTM D 943 and the PSU (Penn State University) micro-oxidation test (for heavy-duty diesel engine oils) [86, 87]. The latter thin film test requires relatively short test durations [88].

Methods used to monitor and analyse the products of vegetable oil oxidation have been reviewed by Fox and Stachowiak [89]. Screening tools for oxidative stability of vegetable oils are numerous. The determination of peroxide value (PV) is carried out by iodometric titration. In the presence of excess iodide, iodine is liberated by hydroperoxides. The amount of iodine is determined by titration with sodium thiosulphate using a starch indicator. The PV is an estimate of peroxide content and can be used as an index of oxidation if the peroxides formed are stable, which is usually not the case. Indeed, the activation energy of decomposition of lipid peroxides (typically 84–185 kJ/mol) is less than that for peroxide formation (146–272 kJ/mol). In the active oxygen method (AOM) [90], the oxidation of a test oil heated at 100 °C is measured until PV = 100 meq/kg, which determines the AOM end point. This end point might not be reached in samples that form unstable peroxides. In the Rancimat method, volatile acids formed during oxidation are used in end point detection [91, 92]. A good correlation has been noticed between the results of the Rancimat induction times at 100 °C and oil stability as measured by peroxide development during storage at 20 °C [93]. In yet another test, the so-called Sylvaner test [94], the pressure drop due to O_2 consumption of a sample heated to 100 °C in a closed vessel is recorded. In its automated version (Oxidograph) an induction period is determined from the sudden decrease in O_2 pressure.

Photochemiluminescence (CL) is another highly selective method for measurement of lipid oxidation [95], which correlates linearly with the iodometric peroxide determination according to DGF Method C-IV 6a [96, 97]. Values of an induction period, as measured by the OSI method [98] generally correspond well with AOM values for PV $\geq$ 100 meq/kg [92]. While the automated OSI method has numerous advantages over the AOM procedure, the inability to differentiate between small changes in vegetable oil matrix and large experimental errors associated with small changes in O_2/air flow rate are the main disadvantages [99]. In particular, the OSI (AOCS Cd 12b-92) is unreliable for higher stability oils due to poor reproducibility. Chemiluminescence was also used to assess the oxidation stability of lubricating oils [100].

Pressurised differential scanning calorimetry (PDSC) is also frequently used for the determination of thermal and oxidative stabilities of vegetable oils [101–105]. High-pressure differential scanning calorimetry (HPDSC) has been used to evaluate the thermo-oxidative stabilities of synthetic lubricating oils, polyol ester derived lubricating oils, as well as of formulated automotive lubricating oils [106]. PDSC–oxidation induction time (OIT) measurements are highly reproducible and versatile and may be applied to oils of low to very high oxidative stability [107]. PDSC allows calculation of the oxidation energy and other kinetic parameters. NMR research into the oxidative stability of vegetable oils generally concerns fatty acids and their changes [108], but may equally well focus on aldehydes [109]. Correlation with qualitative ^{1}H and ^{13}C NMR spectroscopy reveals the effect of differences in the structural parameters of the FA chains of the oils in determining the relative oxidation behaviour [105]. GC compositional data and/or ^{1}H and ^{13}C NMR-derived structural information can be used to rationalise the thermal and kinetic behaviour of pure vegetable oils. PDSC according to ASTM D 6186-98 [110] and bulk oil oxidation stability tests according to ASTM D 943 [111] are frequently used to assess the quality of a lubricant.

Industry usually dictates the method or instrument to be used for a particular test. OSI is a milder test than RBOT. Most food applications require the use of OSI. The petroleum industry requires testing with much harsher conditions than the food industry; here, RBOT and micro-oxidation are recommended tests in view of the environment in which they are used. For ageing tests, see also Section 10.4.2.

8.6 Biolubricant process and product quality control

The petroleum industry is highly regulated (i.e. tested to specification from production to transportation to market); it is subject to relatively small variations during its production processes. Biolubricants are produced in smaller quantities than lubricants (thus rendering testing more expensive) and subject to various and different quality-affecting factors including the feedstock used.

Process control, optimised product output and biolubricant quality require adequate accuracy and reliability, matched with economy, accessibility and ease-of-use. Biolubricant processing and quality are closely related. The processes used to refine the feedstock and convert it to biolubricant determine whether the product will meet the applicable specifications. The production process must be resilient enough to accommodate a wide range of operating parameters to adjust the feed variations. Monitoring the process parameters enables the controlled use of expensive substances.

Biolubricant process and quality control bring about a distinct set of challenges. Producers of biolubricants need to be able to ensure the quality of the incoming feedstock at the time of delivery, optimise the ratio of process chemicals used, refine and troubleshoot the manufacturing process and guarantee the quality and batch-to-batch consistency of the end product.

Although any kind of vegetable oil – virgin, crude, refined or used – may qualify as a raw material, the final product should be within (inter)national quality specifications. Monitoring the process parameters enables the controlled use of expensive substances. On-site concentration measurement allows the exact dosage of raw materials.

Biolubricant monitoring technology may take advantage of a wide range of laboratory, plant-based and portable instruments for numerous test methods including analytical instruments for in-process analysis equipment and quality control. Continuous monitoring of a biolubricant plant avoids the need for (expensive) manual batch testing for quality assurance. Options for quality testing are several (Table 8.7).

Product quality is closely linked to measurement quality. Test kits (e.g. Castrol's CAREMAX™ Onboard), designed to complement laboratory test results, allow rapid determination of several key quality parameters, such as viscosity, density, water content, total acid number (TAN), microbial contamination, and a visual quality check. L-Vis 510 is an in-line viscometer allowing 24-hour monitoring of lubricants in non-hazardous areas whereas L-Dens 427 (both Anton Paar GmbH, Graz, Austria) is an on-line density measurement sensor.

Small portable instruments are often typically designed for measurements of one particular product quality characteristic [112]. Several multifunctional instruments use NIR spectroscopy. Fully-fledged laboratory instruments are the ultimate testing option allowing for high-quality results derived according to official ASTM protocols.

The need for streamlined production processes has spiked interest in more efficient and less costly on-line analytical methods which provide real-time information, instead of the traditional, slow and expensive wet-chemical reference methods and physical testing. The NIR qualification method provides a fairly complete quality assurance program for lubricant

Table 8.7 Options for quality testing

Characteristics	Instrumentation			
	Test kits	Portable	Multifunctional	Laboratory
Cost	Inexpensive	Cheap	Medium	High
Ease of use	Simple	More difficult	Simple	Technical staff
Quality indicators	Several	One	Several	One
Results	Go/no-go	Nearly lab-quality	Accurate	High quality

production and has the power to replace many conventional methods in lubricant analysis. NIR technology has potential for on-line lubricant oil production process control, as shown by the determination of lubricant base oil properties (density, molecular weight, flash point, temperature 50% recovered; aromatic, naphthenic and paraffinic carbon content) for mineral-based lubricants [113]. NIR spectroscopy enables carrying out at-line biolubricant production quality control and analysing the production process in real time from start to finish, measuring some key point indicators, such as water content, TAN, density and viscosity to give an early indication of off-spec batches or production trends. Manufacturers can optimise the ratio of chemicals used, gain real-time batch-to-batch data, check quality of base stocks, refine and troubleshoot production and ensure end-product quality and consistency. For reasons beyond meeting the requirements of a modern quality assurance (QA) system (product specifications), demonstrated testing of lubricant performance is of fundamental importance for end-users. End-users can ensure quality of delivered lubricant, monitor batch-to-batch consistency, check stored biolubricant for deterioration over time and gain confidence in the producer. Methods are available for the determination of dynamic viscosity and water content [114–116], density, flash point and base oil composition [113], various contaminants [117], carcinogenic potential [118] and biodegradability [40].

At-line NIR spectroscopy monitoring of a water-soluble cooling metal-working oil and of oil impurities has been reported [119]. On the whole, quantitative at-line process control for (bio)lubricants is still rather underdeveloped.

In order to ensure the highest level of QA, CRMs from a suitably qualified source are highly desirable. CRM-FCLU-2 and CRM-PMLU-2 are lubricant flash point standards (available from Tamson Instruments, Bleiswijk, NL) with certified flash points of 265 °C (ASTM D 92) and 213 °C (ASTM D 93), respectively. CRM-DELU-2 is certified for density (0.8837 g/mL @ 15 °C) according to test methods ASTM D 4052/ISO 12185. Certified total base number (TBN) standards from 6 to 70 mg KOH/g to ASTM D 2896/IP 276 are manufactured by Paragon Scientific, Ltd (Birkenhead, UK). The VHG Labs (Manchester, NH) Sulphur and Metals in Lube Oil Performance Testing Program (PTP) is designed to provide analytical laboratories with a method of monitoring their analytical performance as measured against CRMs. Currently, the types of instrumentation included in the programme are inductively coupled plasma (ICP), viscosity, and particle sizing systems, sulphur, and sulphur and metals in lube oil. PTPUBEMO-25 is a Performance Testing Program Lube Oil Test Standard useful as an internal standard in elemental analysis (see Section 8.7.2). NIST Standard Reference Material 2806 (Medium Test Dust in Hydraulic Fluid) is used for particle counting in quality control. ISO Guide 34:2000 and ISO/IEC 17025 (the standard for laboratory competence) are the premier standards for CRM producers.

8.7 Biolubricant analytical methodology

Bio-based content in a base oil can be measured by the radiocarbon analysis method according to ASTM D 6866-11 [120]; see Fig. 8.4. More practically, sensors have been developed for law enforcement in areas where (more expensive) bio-oils are mandatory. Portable oil sensors measuring the capacitance of a fluid may be used to differentiate between the dielectric behaviour of polarised molecules, such as esters (present in biodegradable oils) and the less polar mineral oils and poly-α-olefins (PAOs) [121].

Lubricant formulations are complex chemical mixtures with a composition which may undergo changes in practical use. Qualitative and quantitative analytical characterisation of all components of a lubricant formulation may be helpful during research and development as well as during application. Lubricant deformulation, i.e. qualitative and quantitative compositional analysis of the lubricant components and their reaction products generated during service life in a given application, is a highly complex matter, requiring more than simple routine analytics [123]. Table 8.8 shows the basic needs for lubricant deformulation and Table 8.9 lists some factors affecting lubricant analysis. Published literature on systematic lubricant deformulation is rather scarce. A complete separation scheme of the base oil and additives using dialysis, column chromatography and ion exchange

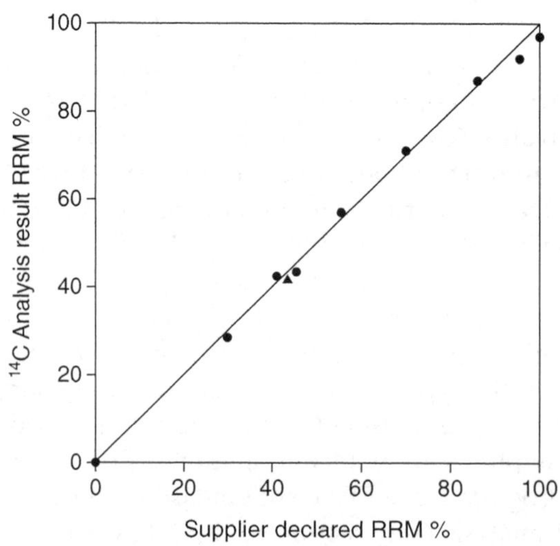

8.4 Test of radiocarbon method for renewable raw material (RRM) detection. After ref. [122].

Table 8.8 Basic needs for lubricant deformulation

- Separation of components (base oil(s), additives and reaction products)
- Qualitative identification and quantitative determination of base stock components
- Qualitative identification of components
- Quantitative determination of additive concentrations
- Identification of reaction (degradation and interaction) products
- Reliability, accuracy
- Sensitivity (down to 0.01 wt% or less, component dependent)
- Short analysis time (e.g. simultaneous analysis, automation)
- Low cost

Table 8.9 Factors affecting lubricant analysis

- Wide coverage of chemical materials (both organic and inorganic, varying greatly in molecular weight, volatility and polarity)
- Complex mixtures of compounds (up to completely unknown type and concentration in an unknown base stock)
- Wide additive concentration range (restricting the analytical choice of method)
- Purity (many base stocks and additives are technical grade substances or (isomeric) mixtures)
- High reactivity and low thermal stability of many additives (especially antioxidants)
- Mutual interferences of the components (presence of transformation and degradation products)

techniques has recently been described for a model lubricant system [124]. Possibilities for the performance of screening analyses were outlined. Quantification of the components is possible.

Analytical measurements provide not only a method of quality control but also support fundamental mechanistic studies or can be used for competitor analysis. There are many analytical methods available for the characterisation of non-additivated mineral oils. Analysis of polyol esters is frequently carried out by means of GC [125], but ^{1}H NMR analysis is to be preferred as a quick analysis procedure [126]. For detailed characterisation chromatographic and spectroscopic methods as well as element analyses of the narrow distillate fractions have to be carried out. Antioxidants in various synthetic polyol esters for turbine-powered aircraft engines (military specifications: MIL-L-23699 and MIL-L-7808) were identified by gradient-elution high-performance liquid chromatography (HPLC) separation (with UV detection), thin-layer chromatography (TLC), and FTIR and NMR spectroscopy [127]. Typical antioxidants identified were triphenylphosphate, tris(alkylphenyl)phosphates, *N*-phenyl-*N*-α-naphthylamine and *p,p′*-

dioctyldiphenylamine. A GC-MS (mass spectrometry) method was developed to elucidate the base oil composition of various formulated synthetic polyols (MIL-L-23699 and MIL-L-7808) [128]. The products were PE, trimethylolpropane (TMP), neopentylglycol (NPG) and dipentaerythritol (DPE) based, (partially) esterified with a variety of linear or branched C_5–C_{10} acyl substituents. As it turns out, many foods are contaminated with mineral oil products used as lubricating oils, greases or as release agents. The mineral oil base can be determined by coupled LC-GC [129].

Elemental analysis of fully formulated (additivated) lubricants is of interest mainly for two reasons: (i) quantitative analysis of additives; and (ii) study of wear mechanism. Elemental analysis of lubricants and greases can be carried out by the usual techniques, such as high- and low-power wavelength dispersive x-ray fluorescence (WDXRF) [130] and bench-top energy dispersive x-ray fluorescence (EDXRF) [131], ICP-AES (atomic emission spectroscopy) [132], scanning electron microscopy (SEM/EDAX) [133], spectrographic analysis and electrochemical methods [134]. Zinc (II) in lubricant oils has been determined by means of stripping chronopotentiometry (detection limit: 0.02 mg/g) [134]. The results were not significantly different from those obtained by atomic absorption spectroscopy (AAS). Quantitative spectrographic analysis allows detection of trace metals (from wear and silica).

Among the most widely used additive types are those applied for AW. These additives typically contain high concentrations of phosphorus, sulphur and zinc, and are used to reduce premature wear of vehicle engines. Accurate and reliable elemental analysis of oil additives has been carried out by means of high-dispersion ICP spectrometry, which allows determination of up to 70 elements in a sample in under 30 seconds [132]; see Section 8.7.2. Cabrera *et al.* have reported a study of wear mechanisms and characterisation of coating films on the Falex block-on-ring tester using SEM and EDAX [135].

8.7.1 Spectroscopic methods

Spectroscopic analysis is often a quicker method than chromatography for production monitoring and verification of product quality. The main spectroscopic methods used are mid-IR and NIR spectroscopy; ^{1}H and ^{13}C NMR spectroscopy and Raman spectroscopy have been used less frequently. Electron paramagnetic resonance (EPR) has only been used occasionally, e.g. for the evaluation of the efficiency of petrolatum (vaselina), transformer oils and industrial oils [136]. Metals had a marked effect on radical formation in the lubricant.

FTIR spectroscopy is an important and commonly used method to sense the quality of a lubricant [137]. The speed of IR measurement and the value and diversity of the results it provides have led to becoming a routine method for oil analysis laboratories. IR measurements are sensitive to a number of additives (applicable for monitoring depletion of AW additives or phenolic antioxidants), degradation products (such as oxidation products, nitric oxides, acidic sulphur compounds, ester breakdown residues) and potential contaminants (soot, water, glycol, unburned fuel) that can be found in oil. Most recently, ASTM International introduced a series of FTIR standard practices and test methods (ASTM E 2412, D 7412, D 7414, D 7415, D 7418). Many crucial measurements in the biolubricant industry can rely on mid-IR analyses. A rapid, practical and accurate FTIR method was reported for the determination of free fatty acids, in edible oils [138]. FTIR (with or without chemometric modelling) for rapid feedstock and process check can determine viscosity, acid number, flash point, cloud point and water (ppm). FTIR offers rapid multiproperty analysis. FTIR can also be used during the production process by monitoring the conversion or detection of the end of the reaction. The speed of spectroscopic procedures allows a considerable increase in reactor throughput.

FTIR has been indicated as an alternative to Karl Fischer (KF) titration for the determination of moisture in mineral- and ester-based lubricants [112]. Portable iPAL FTIR analyser technology (A2 Technologies, Danbury, CT) for moisture determination in lubricating oils is an advancement over Karl Fischer titration (Table 8.10). Speed, ease of use, on-site measurement capability and accuracy are significantly improved with this FTIR device and additionally no reagents are required. The level of water in lubricating

Table 8.10 Measurement of water in lubricating fluids by FTIR and KF methods

FTIR	Karl Fischer (KF)
• Speed (3–5 min)	• Analysis time (5–30 min) dependent on moisture level
• Reagent free, no sample preparation	• Potentially toxic reagents (coulometric titration)
• Single portable analyser	• Need for multiple KF titrators
• Pre-programmed; virtually no training	• Operator training required
• Ease of use	• Need for skilled technician
• More accurate than KF at high moisture level	• Analytically accurate
• Detection limit: 65 ppm	• Detection limit: 20 ppm
• Real-time, on-site measurement capability	• Less time effective
• High confidence level	• Low confidence level (off-site testing)

oil is critical to measure and control since it can accelerate oxidation and affect lubrication performance. In some applications such as conventional and nuclear power stations determination of water in lubricating oils in massive turbines is considered highly critical [139]. The FTIR system can also act as a check on off-site testing lab measurements. FTIR analysis technology can further be used for measuring oil in cooling water, additive depletion of oils and lubricant oxidation. FTIR was also applied in condition monitoring of in-service lubricants [140] and for strategic oil analysis by estimating the remaining lubricant life [78].

ASTM D 7066-04e1 covers the determination of oil and grease and non-polar material in water and wastewater by IR determination of dimer/trimer of chlorotrifluoroethylene (S-316) extractable substances from an acidified sample. In accordance with ASTM D 7066-4, portable infrared analysers (e.g. Wilks' InfraCal TOG/TPH Analyser) can be used for rapid (10–15 min) measurement of total oil and grease (TOG), or fats, oils and grease (FOG) levels in water in plant discharges for compliance with regulatory limits. See also Section 8.8.1.

NIR spectroscopy

NIR spectroscopy is a suitable secondary analytical technique for rapid evaluation of a great many critical chemical *and* physical properties of substances and can be used for characterisation and identification of complex sample material. The greatest advantage of NIR spectroscopy is the ease of sample handling (intact sampling). It allows various measurement modes: in reflection (for solids, powders, granulate), transflection (for slurries, semi-solids, liquids, films, emulsions) or transmission (for clean liquids). For lubricants, transflection or transmission mode is the most indicated. NIR spectroscopy instrumentation is rugged, flexible and even portable, and can be used in a process environment.

NIR spectra contain a wealth of information about the chemical and physical properties of a sample, which however require considerable effort to be deducted from the spectra. Because NIR is composed of overlapping overtones and combinations of bands originating in the mid-IR, chemometric data analysis algorithms are used. Although multiple component quantitations are now routinely being performed, NIR spectroscopy is not a technique without development problems. Each specific application needs a relatively high number of samples for robust calibration and development and maintenance of a calibration routine. NIR spectroscopy is not a technique for analysis of products beyond the training set.

NIR spectroscopy allows simultaneous qualitative and accurate quantitative analysis of different parameters, high sample throughput, real-time monitoring, and is a non-destructive, cost-effective and user-friendly quality and process tool. It is not a trace analytical method. The detection limits are in the low percentage range, in favourable cases in the high per mil range for analyses in liquid media. Table 8.11 summarises the main characteristics of NIR spectroscopy.

NIR analysis was born of a need to solve practical quality control problems rather than the desire to perform high-resolution molecular structure

Table 8.11 Main characteristics of near-infrared spectroscopy

Advantages:

- Flexible sample presentation
- All sample types (of any convenient sample size or shape, from transparent to totally opaque)
- Representative sampling (10–100 g of material)
- Large dynamic range of sample thicknesses
- Non-invasive, non-destructive
- Small absorption coefficient(s)
- (Ultra)fast measurement (<1 min)
- Various measurement configurations; excellent diffuse reflectance spectra
- Multicomponent analysis
- Medium sensitivity
- Favourable S/N ratio (10^5:1)
- Mature technology
- Wavelength and ordinate accuracy and precision
- Simple and robust instrumentation (favourable hardware cost)
- High energy sources
- Fibre-optic probes
- Long pathlength cells
- Low maintainance costs
- Qualitative and quantitative applications
- Ideal for quality control and manufacturing environment (off-, at-, on-line)
- Limited operator training needed
- 2D correlation spectroscopy; imaging

Disadvantages:

- Secondary method (requires calibration against reference method)
- Dependence on a large reference set
- Influence of sample morphology
- Slow and costly method development
- Need for quantitative calibration model
- Troublesome calibration transfer
- Strict sample temperature control required
- Spectroscopic complexity (lack of specificity; no characteristic absorption bands)
- Lack in structural interpretative value (difficult to identify unknowns)
- Lack of reference data
- Need for sophisticated data evaluation algorithms (heavy computation load)
- Weak sensitivity to minor components: minimum concentration >0.1% (no trace analysis)

analysis in the laboratory. The samples subjected to NIR spectroscopy are often very complex mixtures (such as lubricants) and are studied without any sample preparation.

NIR spectroscopy can be used for product identification, classification and quality control, as well as for the determination of product properties (chemical and physical) and component concentrations in process applications, all with the object of rapid analysis. Its primary advantage for industrial applications is the ability to employ robust, inexpensive optical fibres to form a convenient optical probe for gathering the spectral data on which the material identification is based. It is used not only for identification of materials anywhere in the formulation process, from raw materials to intermediates and finished products, but also for quantitative analysis. Quantitative analysis of materials is inherently more complex than identification and requires a more elaborate calibration set on which to base the analysis of new samples.

As lubricants are formulated products (usually blends of base oils containing a great variety of additives), it is important to use the right calibration model. The concentration of the components in the 'standards' used for model development (validation set) must span the values of the samples to be analysed successively. Nonetheless, once calibrated, these methods (which are labour intensive in development) can be of great economic value in maintaining consistent product quality, in reducing spoilage, waste and off-spec material, and in staying as close as possible to regulatory requirements.

Table 8.12 gives an overview of the applications of NIR spectroscopy to lubricants. Non-destructive analysis of the samples makes NIR spectroscopy unique in that physical and chemical properties can be measured simultaneously. Spectra have been correlated to a variety of physical properties such as viscosity and density. It is equally of interest to notice that NIR spectroscopy can be used for the evaluation of the carcinogenic potential and biodegradability of lubricants. Competitor analysis, such as the determination of an unknown additive package, is not possible.

NIR spectroscopy is recognised worldwide as technology appropriate for industrial quality and process control. Since only one instrument is all that is needed in using FT-NIR spectroscopy for multiple measurements in a biolubricant product line, the testing protocol for process samples is simplified, multiple pieces of equipment can be replaced and laboratory supply costs are reduced. Moreover, the ability to quickly analyse samples for multiple components using FT-NIR spectroscopy much more quickly (30 s) than traditional primary analytical laboratory methods (up to 1 h) results in considerable time savings.

NIR spectroscopy is used not only for quantitative analysis but also for identification of materials used in the manufacturing process: raw materials,

Table 8.12 Applications of NIRS to lubricants

Chemical properties:
• Identification of raw material, in-process stock and finished products
• Composition determinations (total oil contents, impurities)
• Concentrations of (in)organic chemical constituents (e.g. additives: >0.5%)
• Volatiles and water content
• Mineral oil compositional analysis (aromatics, naphthenics, paraffinics)
• Degree of degradation
Physical properties:
• Density
• Viscosity
• Molecular weight
• Flash point
• Cloud point
• Transparency
• Colour
• Total acid number (TAN)
Ecotoxicological properties:
• Carcinogenic potential
• Biodegradability

intermediates, and finished products. From the characterisation of oil content in oilseeds to on-line quantification of key parameters in lubricant processing and quality analyses of the product, its use is extending in the lubricant industry provided a sufficient number of samples is available as reference for calibration. The popularity of NIR spectroscopy is directly attributable to the technique's convenience and ease of use. Some of the additives for lubricants are readily distinguished (for concentrations exceeding 0.5%) because their NIR spectra are different. Similar compounds with nearly identical NIR spectra require special chemometric treatment in order to identify them as a proper substance. These same chemometric methods can be used to qualify the material as the specified grade and quality.

Timm *et al.* [119] used NIR spectroscopy for qualification of lubricants and other mineral oil products for QA and examined 22 samples of six different additive packages (viscosity improvers, polymethacrylate) in transmission mode (pathlength: 0.5 cm). Other applications concerned discrimination of hydraulic oils and quantification of the oil content in water-soluble cooling lubricants. The durability of such lubricants in the metalworking industry depends on the amount of impurities accumulated as reaction products and hydraulic or other process oils (foreign oil). NIR spectroscopy can be used in continuous mode as a quick and effective at-line method to determine the total oil content and oil impurities within seconds, thus replacing the more inaccurate and time-consuming reference method with the oil tester [119]. NIR spectroscopy (1300–1700 nm) has also been used for the analysis of gasoline and ethylene glycol contaminants in lubricant oil [117].

The determination of a variety of mineral oil-based lubricant base oil properties (notably density, molecular weight, flash point, temperature 50% recovered; aromatic, naphthenic and paraffinic carbon content) has also taken advantage of NIR spectroscopy [113]. Work was based on 88 samples from three Brazilian refineries (Petrobas). Results obtained were well within the reproducibility values defined by ASTM methods (except for relative density). These results indicate the potential of NIR spectroscopy and chemometric techniques for on-line lubricant oil production process control by accurate quantification of key product parameters.

VIS-NIR spectroscopy was also applied to fast and non-invasive measurement of the dynamic viscosity and water content of an engine lubricant [114–116]. For the determination of dynamic viscosity of automobile lubricants a calibration set of 125 samples was used and 25 samples in the validation set collected from five brands [116].

Several biolubricant synthesis procedures are complex catalytic dynamic equilibrium processes, with no guaranteed end point. Target analytical properties for biolubricant production monitoring (such as viscosity, TAN, flash point, cloud point), as specified in relevant ASTM norms, can rapidly be carried out using off-, at- or on-line FT-NIR methods. Use of an at-line monitoring tool avoids the need for expensive, slow and frequently off-site wet-chemical quality screening measurements for production process monitoring, thereby improving the reliability and throughput of both batch and continuous reactors. Very time-consuming analytical procedures of biolubricant quality control are an obstacle to quickly address problems arising during production.

It is of considerable interest that NIR spectroscopy allows determination of the carcinogenic potential of lubricant base oils [118]. NIR spectroscopy combined with advanced statistical analysis (PLS) was also used for the assessment of biodegradability during use [40]. UV-VIS-NIR spectroscopy was used to evaluate the degree of degradation of lubricating oils [141].

In manufacturing of synthetic fibres it is common practice to add a small amount of lubricating material ('finish') to the fibre to assist its performance and renewability in downstream textile manufacturing processes. The amount of finish on the fibre surface, called finish-on-fibre (FOF), is of critical importance in textile manufacturing processes. Blanco *et al.* [142] have reported quality control analyses of finishing oils in acrylic fibres using NIR diffuse reflectance spectroscopy and PLS regression methods.

NMR spectroscopy

In comparison with spectroscopic screening techniques such as FTIR or NIR spectroscopy, NMR provides much more significant and selective qualitative and quantitative information. Proton NMR has been used to measure iodine value [143], FA composition [144] and positional distribution of FAs [145], and provides a valuable tool for quantitation of oxidation of vegetable oils [108]. NMR also provides a method to estimate the overall changes in the primary and secondary oxidation products of lipids.

Although NMR spectroscopy is not as sensitive as other spectroscopic methods (for ^{1}H NMR concentrations > 100 μM), the detail and lack of ambiguity of the information makes it the most effective tool for structural identification and elucidation in the liquid state. ^{1}H NMR is a technique that is capable of measuring a wide range of different organic compounds in a simple sample without the need for any complex sample preparation. ^{1}H NMR is quantitatively reliable. Depending on the strength of the magnetic field used it is possible to measure compounds down to a level of about 1 ppm. High-resolution NMR is a laboratory-based technique.

In principle, the ^{13}C nucleus is the most informative probe for organic structure determination by means of FTNMR; however, ^{13}C NMR has an intrinsic low relative sensitivity (natural abundance of ^{13}C of only 1.108%). ^{1}H NMR and ^{13}C NMR spectroscopy have been used, often in combination, to sense the quality of lubricants [146–148].

The high sensitivity of ^{31}P NMR spectroscopy has allowed its application to the study of phosphorus in lubricants containing commercially used levels of phosphorus-based antioxidants and other additives. In particular, ^{31}P NMR spectroscopy has found wide application for the study of degradation (oxidation, decomposition, hydrolysis) of stabilisers, additive–additive interactions and the identification of conversion products (see also Section 8.7.2). Informative spectra can be obtained even at low levels of phosphorus (<1 ppm).

The chemistry of additive components used in lubricants is quite diverse, encompassing several categories of organic chemicals, organometallics and specialty polymers. ^{1}H, ^{13}C and ^{31}P NMR (as well as specialised NMR techniques such as attached proton test (APT), distortionless enhancement of polarisation transfer (DEPT), pulsed-gradient spin-echo (PGSE)) have been employed extensively in the field of lubricants, both for characterisation of lubricants and additives, as well as for the study of their degradation mechanisms under use and the interaction mechanisms between polar additive components [149]. Recently, NMR relaxometry has been used to analyse lubricant oils degradation [150].

Low-resolution NMR (LR-NMR) technology has been making inroads into near-line product quality control applications since the beginning of the 1990s. Process NMR, which discriminates in mobility between various components of a sample, may be used for the determination of oil content in oilseeds or in grease, droplet size in oil-in-water emulsions, total moisture content in seeds, process oils in rubber, etc. These are some advantages of LR-NMR which have particular significance for process and product quality control, namely accuracy, reproducibility and speed of measurement. The samples are measured 'as is' without any treatment. If we consider that the prime objective of the quality control engineer is to devise a scheme which will enable a statistically valid assessment of production, the advantages of LR-NMR are crucial. LR-NMR measurements take only 20 s, and an appropriate statistical quality control scheme may be set up while maintaining a high degree of accuracy. LR-NMR is an ideal, robust bench-top analysis tool for routine operation by non-specialist production workers.

8.7.2 Additive analysis

Whereas additive analysis in polymers is a highly mature field [151–153], the analytics of additives in lubricants appears to be less extensively developed [154]. Additive analysis is notoriously a difficult matter; see Table 8.9. However, on the basis of the chemical nature of mineral oil-based lubricants and polyolefins (both hydrocarbons) it would not be surprising if the additives in commercial use for these non-polar systems were essentially very similar. In analogy, also comparable analytical methods might be expected for both systems, obviously with the provision that there is no need for extractive processes from solids (polymers) in the case of lubricant oils. This would translate into better quantitation for additives in mineral oil-based lubricants. For (polar) biolubricants commercial additive development is lacking far behind that of conventional oils. For some additive categories (e.g. dispersant/inhibitor packages) even no valid commercial equivalent is on the market. Fully formulated biolubricants represent more reactive systems than traditional Group I lubricants. This also means that more biolubricant–additive and additive–additive interactions may be expected.

Critical end-users (military, aeronautics) may be interested in a screening analysis of proprietary additive packages. The screening analysis and separation scheme of additives in a model lubricating oil have been described [124, 155]. Identification of antioxidant additives in synthetic polyol-based lubricant oils for aircraft engines was reported by Garofalo *et al.* [127]. Separation was carried out by gradient-elution HPLC-UV and TLC, and identification of triphenylphosphate, various tris(alkylphenyl) phosphates, *N*-phenyl-*N*-α-naphthylamine and *p,p′*-dioctyldiphenylamine by means of FTIR and NMR. Analysis of dithiophosphate lubricating oil antioxidants was carried out by separation by TLC and identification by means of ^{1}H, ^{13}C and ^{31}P NMR [156].

The quantitative determination of additive content in lubricants requires an even more demanding analytical approach than qualitative identification. Quantitative analytical measurements prove a means of establishing the loss of additives as a function of time. FTIR is a suitable technique for monitoring depletion of specific additives such as phenolic antioxidants and AW additives. However, in favourable cases general elemental analysis is a sufficient indicator.

The ASTM Standard Test Method D 6443-04 can be used to determine if the unused lubricating oils, additives, and additive packages meet specifications with respect to the added elements (typically Ca, Cu, Mg, P, S and Zn), and with respect to chlorine contamination. Liquid sample analysis can be carried out by means of WDXRF with lower limits of detection (LLD of 0.3 ppm for Zn to 6.6 ppm for S) [130, 157]. Accurate and reliable elemental analysis of oil additives and wear metals (for particles $< 8\ \mu m$) can also be carried out using high dispersion ICP excitation. Motor oil additive formulations (containing 0.5–4.0% Ca, Mg, Na, P, S and Zn, and 10–22% P, S and Zn, respectively) have been analysed in high-purity kerosene dilution by high dispersion ICP-AES using a radially viewed plasma configuration [132]. A Performance Testing Program Lube Oil Test Standard (PTPLUBEMO-25; VHG Labs, Manchester, NH) containing Ba, Ca, Cl, Mg, Mo, P, S, Si and Zn was used as an internal standard. Use of an IS minimises differences in nebulisation efficiencies between samples which results in improved precision values. Table 8.13 shows the emission wavelengths chosen for analysis and the instrument's detection limits in radial view mode. As the most sensitive wavelengths for Ca, Mg and Zn produce intense emission, which results in short linear dynamic ranges, weaker emission lines were employed for those elements to increase the linear working range over a wide concentration range. Results of the performance testing standard in comparison to the certified concentrations are given in Table 8.14. ICP-AES can provide accurate and reliable analysis of elements that are present in a wide range of concentrations in viscous sample matrices.

Table 8.13 Wavelengths and detection limits in kerosene matrix

Element	Wavelength (nm)	Detection limit (ppb)
Ca	422.673	13
Mg	277.982	13
Na	589.592	43
P	213.618	24
S	182.034	82
Zn	481.053	1700

After ref. [132].

Table 8.14 ICP-AES analysis of the Performance Testing Program Lube Oil Standard (PTPLUBEMO-25)

Element	Wavelength (nm)	Avg. measured conc. (%)	Certified conc. (%)	Difference (%)
Ca	422.673	0.315	0.318	0.82
Mg	277.982	0.053	0.049	7.6
P	213.618	0.084	0.086	2.7
S	182.034	1.7	1.8	5.6
Zn	481.053	0.1346	0.141	4.5

After ref. [132].

Additive packages often contain reactive compounds. Additive–additive interactions in lubricant systems may be synergistic or antagonistic in nature and are therefore of considerable interest or concern, respectively (see Section 7.2.4). Various studies on additive–additive interactions have been reported [158–166] using techniques such as IR [160], NMR [149, 158, 160, 162], x-ray photoelectron spectroscopy (XPS) [158] or TGA [158]. The interaction between ZDDP (zinc dialkyldithiophosphate) with PIBS (polyisobutylene succinimide) in solution has been studied by viscosimetry, IR, NMR and DSC/TGA techniques [159, 164, 166]. Analytical studies using ^{31}P NMR, XPS and TGA in the ZDDP–PIBS–MoDTC (molybdenum dithiocarbamate) system provide strong evidence for interactions between PIBS and ZDDP and between ZDDP and MoDTC [158].

8.8 Quality of in-service lubricants

Suppliers cannot guarantee that each stage of the handling process from manufacturer to local marketer and end-user is completely free of chemical, solid and moisture contaminants. Consequently, incoming lubricant quality control testing is good practice. As in the case of any quality control testing programme, sample collection practices should be defined in a standardised operating procedure (SOP). The test methods should be rigorously defined and repeatable, and the instruments routinely calibrated to a recognised standard. For bulk deliveries oil tests that can be conducted on-site would be the most useful for quality control analysis (see also Section 14.5.1).

Present market trends require lubricants to function under demanding conditions for extended periods. To avoid damage to the equipment or unscheduled maintenance calls, monitoring of in-service lubricants is essential. Measurable consequences of lubricant degradation are, for example, viscosity increase, discoloration, sludge and acid formation. Lubricant monitoring aims at utilising the useful life of a lubricant to the full extent without discarding the material too soon, and at avoiding using the fluid too long to the detriment of the host machine or engine. Strategic oil analysis estimates the remaining lubricant life [78]. The analysis of used (in-service) oil is an effective tool to ensure the control of the lubricating function of the oil. When warranted by the nature of the application, laboratory analysis of lubricants may be conducted. Lubricants are non-unique, complex chemical entities, changing their composition more or less in practical use, while also creating reaction products. This complexity mostly prevents use of simple routine analytics for lubricants [123].

Many studies have been undertaken to determine degradation mechanisms and the effects of the degradation of lubricants in service use [110, 128, 140, 141, 146–150, 167–172], including assessment of biodegradability during use [40]. For practical applications sensorial systems would be ideal [171–174]; see Section 14.5.1.

A lubricating oil may become unsuitable for its intended use as a result of oxidation, water and/or air contamination, additive depletion, accumulation of contaminants, thermal or biological degradation (see Sections 14.5 and 14.5.1). These processes are indicative of the analytical tools useful for monitoring of oil quality. Several test methods for lubricants in use have been developed. Analysis may include spectrographic or ferrographic analysis, acid number, viscosity and water content. Viscosity is a good indicator of oil quality. The presence of water, fuel and other contaminants causes viscosity of the oil to decrease. The build-up of soot and sludge in the oil leads to an increase in viscosity. Therefore, monitoring oil viscosity can provide timely warnings of deteriorating oil quality. It is common practice in refineries to monitor lubricant viscosity in each major compressor once a month by external laboratory testing. As shown in Section 12.7 this may be inadequate as lubricant viscosity could change drastically much more quickly. The unpredictability of viscosity changes means monthly viscosity checks are insufficient to prevent compressor failure [175].

It is equally very important to carefully monitor the oxidation of oils (ASTM D 943 and D 2272) and to determine the remaining useful life of the oil (oxidation stability reserve) [78]; see Sections 14.4 and 14.5. GC-MS was proposed for monitoring the thermooxidative degradation of synthetic polyol ester lubricant oils for turbine-powered aircraft engines (base components and some additives) [128]. Additive depletion (e.g. of antioxidants) is a limiting factor in lubricant life. Marshall [168] has reported the qualitative and quantitative changes in P-based additives in lubricating oils in service by ^{31}P NMR monitoring. In some favourable cases additive decomposition does not greatly affect performance. For instance, ZDDP decomposition in engine tests leads to zinc compounds (adsorbed by diesel soot) and non-adsorbed phosphorus-containing intermediate products such as tetraalkylthioperoxidiphosphate which are responsible for maintaining AW performance over a long period [170].

ASTM D 6224-09 (Standard Practice for In-service Monitoring of Lubricating Oil for Auxiliary Power Plant Equipment) provides a list of proposed tests recommended for quality control testing of lubricants. ASTM D 7418-07 describes the standard practice for set-up and operation of FTIR spectrometers for in-service oil condition monitoring. Various standard methods concern condition monitoring of phosphate AW additives (ASTM D 7412-10), oxidation (ASTM D 7414-09) and sulphate by-products (ASTM D 7415-09) in in-service petroleum and hydrocarbon-based lubricants by trend analysis by means of FTIR. Condition monitoring of in-service lubricants by FTIR has been reviewed and placed in the context of the recently approved ASTM Practice E 2412-04, developed by the Joint Oil Analysis Program (JOAP) as a standardised means of trending oil/lubricant condition [140]. Table 14.13 illustrates the usefulness of infrared for generic used-oil analysis, in particular for lubricant fluid analysis (oxidation, process contaminants, additive depletion) and contaminant monitoring (soot, water, glycol).

Several optical fibre spectroscopies (UV-VIS-NIR spectroscopy, fluorescence spectroscopy and scattering measurements) were used for measuring various quality parameters (TAN, JOAP index, water content and phosphorus) of lubricant oils, indicating the degree of degradation of the oil [141]. Degradation mechanisms of lubricants under use were studied by NMR [149]. For moisture analysis it is more advantageous to use portable on-site FTIR technology than off-site Karl Fischer analysis. See Section 8.7.1. Acid formation is quantitatively assessed by titration and reported as TAN (in mg KOH/g). TAN is indicative of the degree of oxidation and hydrolysis of the lubricant in use, warning limits for which have been stipulated in industry standards.

Lubricating oils are used in very difficult conditions in engines reaching high temperatures and varying pressure conditions in contact with many metallic parts. ASTM D 6923-10a (Caterpillar 1R test procedure) is a laboratory test method for stressing an engine oil under modern high-speed diesel operating conditions which measures the oil's deposit control lubrication ability and resistance to oil consumption [176]. It is carried out using a standardised high-speed, single-cylinder diesel engine. The lubricating function of the oil can be modified according to the amount and nature of the metals present. Wear debris analysis (WDA) can be performed by means of ICP-OES following the recommendations of ASTM D 5185 [177]. However, elemental analysis has its limitations (see Section 14.5). Wear measurement based on the RNT is another, unique way to determine the tribological performance of a system continuously and in real-time [169]. The advantages of RNT lie in the on-line monitoring of wear in defined running conditions and in selective examination of critical engine components [178]. The method is especially suited to mechanical systems showing low wear rates. The RNT wear measurement can be applied to the analysis of wear protection of lubricating oils and the effect of additives.

Johnson and Spurlock have addressed strategic oil analysis in terms of remaining lubricant life [78], on-site oil analysis with and without instruments [179, 180], machine and lubricant condition [181], and analytical methods and practices for use in selecting an oil analysis supplier [182]. For effective industrial plant oil analysis practices, see also Section 14.5.

Oil analysis is the 'process of analysing oil and other fluids used to lubricate or operate mechanical equipment, evaluating the condition of the fluid or the equipment from which the fluid originated, and recommending maintenance actions to the equipment operating activity' [183]. An oil analysis programme (OAP) provides rapid and accurate oil analysis support through standardised procedures, data elements, analytical equipment and diagnostic techniques. The JOAP is a combined US military effort to set up and maintain a standard programme for conducting aviation and non-aviation machine and oil analysis [184]. This widely used programme aims at ensuring timely and accurate oil and other fluid-wetted component analysis support for equipment of the US Army, Navy and Air Force through the consolidation and strategic location of all oil analysis laboratories and through the standardisation of analytical instrumentation and diagnostic techniques and customer procedures. JOAP was instituted to continually maximise use of oil analysis with the object of detecting oil-wetted equipment failures before serious malfunction or secondary damage occurs.

US Army Regulation 700-132 (28 July 2008) defines the JOAP policy, goals and responsibilities. US Army documents TM 38* 301-1,-2,-3 and -4 illustrate why the programme exists, what tests are to be run, how instrument management is to be performed and how the data is to be evaluated. The specific goals of JOAP comprise:

- improvement of the operational readiness and economy of equipment through the use of oil analysis on the basis of a condition-monitoring concept that relies primarily on the detection and measurement of wear metals in lubricants and to a very limited extent on physical property testing (viscosity, water/fuel contamination, additive levels, etc.);
- cost effective, inter-service determination of the condition of lubricants, fluids, and mechanical systems through the use of various analytical techniques;
- improvement of wear metal and oil analysis instruments, materials and technologies.

The JOAP TSC (Technical Support Center) develops and maintains SOPs for the manufacture and quality verification of reference/calibration fluid standards, ensures (external) validation of such standards on a periodic basis, controls the distribution of reference standards, and conducts laboratory instrument certification programmes. The JOAP TSC also evaluates available technologies from military, industrial and academic sources for future oil/fluid analysis monitoring requirements. The determination of the applicability of oil analysis techniques and related evaluation criteria for equipment and cognisant lubricant products falls within the scope of JOAP.

Selection of an (external) oil analysis laboratory should be based on technical and quality issues: logistics, testing capability and accuracy, interpretation quality and information management [182]. There should be a balance between timeliness and quality. Labs with highly standardised practices (e.g. JOAP TSC, Oelcheck GmbH) will have better quality and efficiency. Oil analysis services should be evaluated in terms of instrument set, calibration and data accuracy, interpretation accuracy, reporting effectiveness and cost. Qualifying a lab for the quality of its rendered evaluations is difficult to measure, requiring large amounts of experience and perspective by the evaluator. Sample tracking, turnaround and retention are important logistical concerns. Sample results and report quality are both influenced by the lab's testing capabilities and effectiveness for managing its instruments. Instrument calibration and repeatability are keys to producing quality results. Once the right tests are selected and accurate results produced, the data should be interpreted. Analyst experience and depth of perspective about the operating context of the equipment, sump and sample point, is critical. Oil analysis services provide a critical part of condition-based maintenance (see Section 14.5.1). Used oil analysis is a service often provided by lubricant suppliers. Evaluation of oil analysis information is complex matter.

Regardless of a lab's test methods, there should be a process to manage instrument operation. ISO 17025 was developed as a guideline for labs to use in developing their management systems for quality and technical operations [1]. As an independent standard that qualifies competence, it is also used to grade competence.

During operations using biodegradable oil for lubrication, the chemical structure of the biolubricant changes and, as a consequence, its biodegradability tends to decrease. Despite the fact that the biodegradability of most of the fresh biodegradable oils has been well determined, monitoring of the biodegradability of oil during ageing has not been studied extensively. NIR spectroscopy was used for the assessment of biodegradability during use (air flux at 140 °C) of three fully formulated ester oils and a biodegradable base oil without additives [40].

There is a general need to better understand the modifications occurring during in-use operation; suitable analytical methods should be developed. On-line monitoring of lubricant biodegradability provides a significant advance in lubrication and environmental management. An advanced system for monitoring biolubes is under development (BIOMON project).

8.8.1 Field measurement of oils and greases

SKF has developed a new methodology to evaluate grease condition directly in the field [185]. Three different tests are included in the protocol: consistency, oil bleeding, and contamination evaluation. An SKF Grease Test Kit TKGT 1 is provided.

Tightening wastewater regulations require more frequent testing from the industries producing high oil and grease levels. A quick and easy field analysis method for determining oil and grease concentration levels is important in many instances, such as offshore oil platforms, industrial wastewater and soil remediation sites. Measurements of FOG (fats, oils and grease) are not simple by any means. Because oil and grease are not made up of a unique chemical entity, analysis is fraught with difficulties. The analytical definition of the oil and grease entity depends on the procedure and solvents used. As different testing methods analyse different chemical or physical properties of oil and grease, it is not surprising that analyses may end up showing different results. Solvent extraction is another source of problems as each solvent has a different extraction efficiency. These analytical problems are illustrated here for the case of waste streams from offshore oil rigs.

On offshore oil rigs the produced water must constantly be monitored to ensure the waste stream is in compliance with discharge regulations. ASTM D 3921 uses Freon 113 as an extraction solvent and IR for quantitative analysis. EPA Methods 413.2 and 418.1 use IR analysis. Following up the Montreal Protocol in 1995, which banned Freon 113, the US EPA Method 1664A (1999) uses hexane rather than freon for the extraction process in combination with gravimetric analysis. The gravimetric EPA Method 1664A measures the weight of the oil that is removed from the sample with hexane after the hexane is evaporated off. Hexane is not a suitable solvent for IR analysis as the hydrocarbons in hexane will contribute to the oil and grease concentration. Whether the IR intensity correlates well with the total weight of the material depends on the composition. As it turns out, EPA method 1664A and hexane/IR analysis generally correlate well with each other.

Gravimetric analysis is a laboratory method. On-site testing can lower laboratory costs, avoid out-of-compliance fees, and improve remediation efficiency. According to ASTM D 7066-04e1 oil and grease in water and wastewater is defined as that which is extractable, and measured by IR absorption at 2930 cm^{-1} (or 3.4 μm). IR determination of oil and grease concentration levels is quick and accurate and the ability to do analysis on-site greatly expands its usefulness. With the new ASTM D 7066-04 method, on-site oil and grease analysis options are increased. ASTM D

7066-04 uses the more ozone friendly solvent S-316 (dimer/trimer of chlorotrifluoroethylene) for extraction and IR analysis [186]. S-316 is comparable to the earlier freon/IR extraction efficiency and what is extracted.

In the past, laboratory FTIR oil and grease condition monitoring systems had not been well suited to on-site analysis, suffering either from a lack of robustness, poor portability or complex user interfaces requiring expert operation. Modern instrumentation now allows obtaining lab-quality results with portable systems. PerkinElmer's In-Service Lubricants FT-IR Analysis System is an on-site oil analyser which provides immediate feedback about the state of the oil [187]. Fixed filter portable IR analysers (e.g. InfraCal Soot Meter and InfraCal TOG/TPH Analyser, Wilks Enterprise, Inc.) allow on-site field condition monitoring according to ASTM D 7686 and also provide the quick and accurate measurements necessary to ensure that TOG and TPH (total petroleum hydrocarbons) concentration levels comply with various regulations. Application specific, portable fixed filter IR analysers allow ease of use in the field or in-plant by relatively unskilled personnel [188]. For portable, on-site FTIR technology, see also Section 8.7.1. As already mentioned in Section 8.7 portable oil sensors are also used for the detection of bio-oils in base oil blends.

Eralytics GmbH (Vienna) has developed portable stand-alone QCL-IR (Quantum Cascade Laser Infra-Red) analyser technology (ERACHECK™) for chlorofluorocarbon (CFC)-free testing of TOG and TPH in water and soil samples at sub-ppm concentrations according to ASTM D 7678-11. The technology allows field analysis at offshore oil platforms and can also be operated flexibly at plant operational sites, typically prior to and after water purification systems. With the manufacture of Freon 113 banned by the Montreal Protocol, older methods like ASTM D 3921 and DIN 38409 that used IR analysis and freon as the extraction solvent had to be replaced. ERACHECK uses a very similar extraction method with cyclohexane or cyclopentane. Correlation with ASTM D 3921 is excellent. ITT Corporation (White Plains, NY)/UNC (Chapel Hill, NC) and IFP (Rueil Malmaison)/Néréïdes (Courtabœuf) have developed sensor arrays for real-time detection of light hydrocarbons and oxygen depletion associated with offshore oil and gas monitoring (oil spill prevention). The Offshore Chemical Notification Scheme (OCNS) manages chemical use and discharge by the UK and Netherlands offshore petroleum industries within the OSPAR framework [189].

8.9 Sources of further information and advice

H.P. Bloch, *Practical Lubrication for Industrial Facilities*, The Fairmont Press, Inc., Lilburn, GA (2009).

R. Thomas, *Practical Guide to ICP-MS*, CRC Press, Boca Raton, FL (2008).

Th. Mang and W. Dresel (eds), *Lubricants and Lubrication*, 2nd edn, Wiley-VCH, Weinheim (2007).

J.S. Marth, *Renewable Lubricants Manual: Biobased Oils, Fluids & Greases*, United Bio Lube, Palo Alto, CA (2007); http://www.renewablelubricants.com.

D. Firestone (ed.), *Physical and Chemical Characteristics of Oils, Fats, and Waxes*, AOCS Press, Urbana, IL (2006).

M.M. Mossoba, J.K.G. Kramer, J.T. Brenna and R.E. McDonald (eds), *Lipid Analysis and Lipidomics: New Techniques and Applications*, AOCS Press, Urbana, IL (2006).

W.C. Byrdwell (ed.), *Modern Methods for Lipid Analysis by Liquid Chromatography/Mass Spectrometry and Related Techniques*, AOCS Press, Urbana, IL (2005).

A.K.-Eldin and J. Pokorný (eds), *Analysis of Lipid Oxidation*, AOCS Press, Urbana, IL (2005).

Official Methods and Recommended Practices of the AOCS, 5th edn, AOCS Press, Champaign, IL (2004).

W.W. Christie, *Lipid Analysis*, The Oily Press Lipid Library, Vol. 15, Bridgwater (2003).

G.E. Totten, S.R. Westbrook and R.J. Shah (eds), *Fuels and Lubricants Handbook: Technology, Properties, Performance and Testing*, ASTM International, West Conshohocken, PA (2003), Vol. I.

H.W. Siesler, Y. Ozaki, S. Kawata and H.M. Heise (eds), *Near-Infrared Spectroscopy, Principles, Instruments, Applications*, Wiley-VCH, Weinheim (2002).

D.A. Burns and E.W. Ciurczak (eds), *Handbook of Near-infrared Analysis*, Marcel Dekker, Inc., New York, NY (2001).

J.C. Lindon, G.E. Tranter and J.L. Holmes (eds), *Encyclopedia of Spectroscopy and Spectrometry*, Academic Press, San Diego, CA (2000).

J. Denis, J. Briant and J.-C. Hipeaux, *Lubricant Properties Analysis & Testing*, Éditions Technip, Paris (2000).

K.H. Koch, *Process Analytical Chemistry*, Springer-Verlag, Berlin (1999).

R. Kramer, *Chemometric Techniques for Quantitative Analysis*, Marcel Dekker, Inc., New York, NY (1998).

E.N. Frankel, *Lipid Oxidation*, The Oily Press, Dundee (1998).

W.W. Christie, *Gas Chromatography and Lipids – A Practical Guide*, The Oily Press, Ayr (1992).

R.G. Ackman, *Analysis of Fats, Oils and Lipoproteins*, AOCS, Champaign, IL (1991).

H.W. Rossmore (ed.), *Biodegradation and Biodeterioration*, Elsevier, Amsterdam (1991).

W.H. Stadtmiller and A.N. Smith (eds), *Aspects of Lubricant Oxidation*, ASTM International, West Conshohocken, PA (1986).

8.10 References

[1] EN/ISO Standard 17025, *General Requirements for the Competence of Testing and Calibration Laboratories*, International Organization for Standardization, Geneva (2005).

[2] R.B. Cain, Biodegradation of lubricants, in *Biodegradation and Biodeterioration* 8 (H.W. Rossmoor, ed.), 2nd edn, Elsevier Applied Science, London (1991), pp. 249–75.

[3] S. Noll and R. Müller, Laboratory methods for testing lubricants, in *Lubricants and Lubrication* (Th. Mang and W. Dresel, eds), 2nd edn, Wiley-VCH, Weinheim (2007), pp. 715–35.

[4] F. Bartels, Mechanical-dynamic test methods for lubricants, in *Lubricants and Lubrication* (Th. Mang and W. Dresel, eds), 2nd edn, Wiley-VCH, Weinheim (2007), pp. 736–85.

[5] OECD 201, *Guideline for Testing of Chemicals: Algae, Growth Inhibition Test*, Organisation for Economic Co-operation and Development, Paris (1984).

[6] OECD 202, *Guideline for Testing of Chemicals:* Daphnia *sp., Acute Immobilisation Test and Reproduction Test*, Organisation for Economic Co-operation and Development, Paris (1984).

[7] OECD 203, *Guideline for Testing of Chemicals: Fish, Acute Toxicity Test*, Organisation for Economic Co-operation and Development, Paris (1992).

[8] OECD 209, *Guideline for Testing of Chemicals. Inhibition of Respiration of Activated Sludge (Proposed Revised Guideline)*, Organisation for Economic Co-operation and Development, Paris (2009).

[9] OECD 204, *Guideline for Testing of Chemicals: Fish, Prolonged Toxicity Test: 14-Day Study*, Organisation for Economic Co-operation and Development, Paris (1984).

[10] J. Bressling, O.-C. Goehler, C. Bugiel, W. Dott and B. Erlenkämper, How friction and ageing influence the toxic and ecotoxic behaviour of used, bio-based lubricants, in *Friction, Wear and Wear Protection* (A. Fischer and K. Bobzin, eds), *Intl. Symposium on Friction, Wear and Wear Protection*, Aachen, 9–11 April 2008 (2009), pp. 515–20.

[11] J. Bressling, W. Dott, H. Murrenhoff and J. Schumacher, Toxicological evaluation of age-dependent change of biogenous lubricating oils, *Tribol. Schmierungstechn.* **56**(4) (2009) 5–9.

[12] http://www.catalysystems.com

[13] G. Maxam, S. Hahn, W. Dott and A. Eisenträger, Assessment of the influence of use on ecotoxicological characteristics of synthetic ester lubricants, *Ecotoxicology* **11**(5) (2002) 349–55.

[14] CONCAWE, *Health Aspects of Lubricants*, CONCAWE Rept. 5/87, Brussels (1987).

[15] J. Pocklington, Zur Frage der Toxicität von Schmierstoffgrundölen, in *Ökonomische Aspekte von Schmierstoffen, TAE Seminar 21591*, Ostfildern (November 1996).

[16] J.H. Raddatz and Ch. von Eberan-Eberhart, Lubricant additives and the environment, in *Ökonomische Aspekte von Schmierstoffen, TAE Seminar 21591*, Ostfildern (November 1996).

[17] A. Cains and R. Haycock, *Automotive Lubricants Reference Book*, SAE, Warrendale, PA (1996).

[18] T. Habereder, D. Moore and M. Lang, Eco requirements for lubricant additives, in *Lubricant Additives: Chemistry and Applications*, 2nd edn (L.R. Rudnick, ed.), CRC Press, Boca Raton, FL (2009), pp. 647–66.
[19] L. Tocci, What does biodegradable really mean? *Lubes 'n' Greases* **10** (2004) 40–3.
[20] U. Pagga, Testing biodegradability with standardized methods, *Chemosphere* **35**(12) (1997) 2953–72.
[21] OECD, *Introduction to the OECD Guidelines for Testing of Chemicals*. Section 3, part 1: *Principles and Strategies to the Testing of Degradation of Organic Chemicals*, Organisation for Economic Co-operation and Development, Paris (2003).
[22] N.S. Battersby, The biodegradability and microbial toxicity testing of lubricants – some recommendations, *Chemosphere* **41**(7) (2000) 1011–27.
[23] E. Beran, Experience with evaluating biodegradability of lubricating base oils, *Tribol. Intl.* **41**(12) (2008) 1212–18.
[24] OECD 301A-F, *Guidelines for Testing of Chemicals. Ready Biodegradability*, Organisation for Economic Co-operation and Development, Paris (1992).
[25] ISO Standard 10707, *Water Quality – Evaluation in an Aqueous Medium of the 'Ultimate' Aerobic Biodegradability of Organic Compounds – Methods by Analysis of Biochemical Oxygen Demand (Closed Bottle Test)*, International Organization for Standardization, Geneva (1994).
[26] ISO Standard 10708, *Water Quality – Evaluation in an Aqueous Medium of the Ultimate Aerobic Biodegradability of Organic Compounds – Determination of Biochemical Oxygen Demand in a Two-phase Closed Bottle Test*, International Organization for Standardization, Geneva (1997).
[27] OECD 301B, *Guidelines for Testing of Chemicals. Ready Biodegradability: CO_2 Evolution Test (Modified Sturm Test)*, Organisation for Economic Co-operation and Development, Paris (1992).
[28] OECD 310, *Guidelines for Testing of Chemicals. Ready Biodegradability: CO_2 in Sealed Vessels (Headspace Test)*, Organisation for Economic Co-operation and Development, Paris (2006).
[29] ISO Standard 14593, *Water Quality – Evaluation of Ultimate Aerobic Biodegradability of Organic Compounds in Aqueous Medium – Method of Analysis of Inorganic Carbon in Sealed Vessels (CO_2 Headspace Test)*, International Organization for Standardization, Geneva (1999).
[30] OECD, *Guideline for Testing of Chemicals. Draft Document. Proposal for a New Guideline 302D. Inherent Biodegradability – CONCAWE Test*, Organisation for Economic Co-operation and Development, Paris (2001).
[31] CEC L-33-A-93, *Test Method: Biodegradability of Two-stroke Cycle Outboard Engine Oils in Water*, Co-ordinating European Council for the Development of Performance Tests for Lubricants and Engine Fuels, Brussels (1995).
[32] W. Buchser and H.W. Jäckle, Bestimmung der Bioabbaubarkeit von nicht wasserlöslichen Flüssigkeite (Schmieröle, etc.) nach CEC L-33-A-94, *Chimia* **48** (1994) 227–31.
[33] N.S. Battersby, P.A. Fieldwick, T. Ablitt, S.A. Lee and G.R. Moys, The interpretation of CEC L-33-T-82 biodegradability test data, *Chemosphere* **28** (1994) 787–800.
[34] S.D. Haigh, Determination of synthetic lubricant concentrations in soil during laboratory-based biodegradation studies, *J. Synth. Lubr.* **11**(2) (1994) 83–98.

[35] ISO Standard 15380, *Lubricants, Industrial Oils and Related Products (Class L) – Family H (Hydraulic Systems) – Specification for Categories HETG, HEPG, HEES and HEPR,* International Organization for Standardization, Geneva (2002).
[36] OECD 302B, *Guidelines for Testing of Chemicals. Inherent Biodegradability: Zahn-Wellens/EMPA Test*, Organisation for Economic Co-operation and Development, Paris (1992).
[37] E. Beran, Application of waste carboxylic acids to the manufacture of biodegradable polyol ester base oils, *J. Synth. Lubr.* **18** (2001) 39–50.
[38] C. Cecutti and D. Agius, Ecotoxicity and biodegradability in soil and aqueous media of lubricants used in forestry applications, *Bioresour. Technol.* **99** (2008) 8492–6.
[39] J. Schramm, Application of biodegradable lubricant in a diesel vehicle, *SAE Spec. Publ.* SP-1810 (2003), pp. 17–28.
[40] E. Aranzabe, A. Marcaide, A. Arnaiz, E. Gorritxategi and M. Hernaiz, New method proposed for the assessment of lubricant biodegradability during its use, *J. New Infrared Spectrosc.* **16** (2008) 291–6.
[41] B. Cunningham, N. Battersby, W. Wehrmeyer and C. Fothergill, Sustainability assessment of a biolubricant, *J. Ind. Ecol.* **7** (2003) 179–92.
[42] M.L. Hinman, Environmental characteristics of fuels and lubricants, in *Fuels and Lubricants Handbook: Technology, Properties, Performance, and Testing* (G.E. Totten, S.R. Westbrook and R.J. Shah, eds), ASTM International, West Conshohocken, PA (2003), pp. 885–908.
[43] ISO Standard 14239, *Soil Quality – Methods for Measuring the Mineralization of Organic Chemicals in Soil under Aerobic Conditions Using Laboratory Incubation Systems*, International Organization for Standardization, Geneva (1995).
[44] C. Cecutti, Un nouveau test d'évaluation de la biodégradabilité dans le sol, Travaux de normalisation, *Oléag., Corps Gras, Lipides* **10**(5–6) (2003) 354–9.
[45] E. Remmele and B. Widmann, Biodegradability and ecotoxicity of hydraulic fluids based on rapeseed oil used in agricultural machinery, *Biomass for Energy and Bioindustry*, Würzburg (8–11 June 1998).
[46] A. Willing, Oleochemical esters – environmentally compatible raw material for oils and lubricants from renewable resources, *Fett/Lipid* **6** (1999) 192–8.
[47] A. Willing, Lubricants based on renewable resources – an environmentally compatible alternative to mineral oil products, *Chemosphere* **42** (2001) 89–98.
[48] S.M. Mudge and G. Pereira, Stimulating the biodegradation of crude oil with biodiesel. Preliminary results, *Spill Sci. Technol. Bull.* **5** (1999) 353–5.
[49] P. Gateau, F. van Dievoet, G. Vermeersch, S. Claude and F. Staat, Environmentally friendly properties of vegetable oil methyl esters, *J. Am. Oil Chem. Soc.* **12** (2005) 308–13.
[50] P. De Caro, A. Gaset and N. Nguyen The, Vegetable based lubricants in forestry: results from a test campaign, *Agro-food Industry Hi-tech* (July/August 2001), pp. 26–30.
[51] C.H. Chaineau, J. Morel, J. Dupont, E. Bury and J. Oudot, Comparison of the fuel oil biodegradation potential of hydrocarbon assimilating microorganisms isolated from a temperate agricultural soil, *Sci. Total Environ.* **227** (1999) 237–47.

[52] C.L. Sabourin, J.C. Carpenter, T.K. Leib and J.L. Spivack, Mineralization of dimethylsilanediol by microorganisms isolated of soil, *Environ. Toxicol. Chem.* **18** (1999) 1913–19.
[53] B. Wilson, Lubricants and functional fluids from renewable sources, *Ind. Lubr. Tribol.* **50**(1) (1998) 6–15.
[54] S.D. Haigh, Fate and effects of synthetic lubricants in soil: biodegradation and effect on crops in field studies, *Sci. Total Environ.* **168**(1) (1995) 71–83.
[55] I. Bossert and R. Bartha, The fate of petroleum in soil ecosystems, in *Petroleum Microbiology* (R.M. Atlas, eds), MacMillan, New York, NY (1984), Chapter 10.
[56] S. Sandvik, A. Lode and T.A. Pedersen, Biodegradation of oily sludge in Norwegian soils, *Appl. Microbiol. Biotechnol.* **23** (1986) 297–301.
[57] J.T. Dibble and R. Bartha, Leaching aspects of oil sludge biodegradation in soil, *Soil Sci.* **127** (1979) 365–70.
[58] J.G. Leahy and R.R. Colwell, Microbial degradation of hydrocarbons in the environment, *Microbiol. Rev.* **54** (1990) 305–15.
[59] S.D. Haigh, Biodegradation of synthetic lubricants in soil, *J. Chem. Technol. Biotechnol.* **55** (1992) 406–9.
[60] J.M. Baker, The effects of oils on plants, *Environ. Pollut.* **1** (1970) 27–44.
[61] Regulation 06/1907/EC (18 December 2006), Registration, Evaluation, Authorisation and Restriction of Chemicals (REACH), establishing a European Chemicals Agency, amending Directive 1999/45/EC and repealing Council Regulation (EEC) No. 793/93 and Commission Regulation (EC) No. 1488/94 as well as Council Directive 76/769/EEC and Commission Directives 91/155/EEC, 93/67/EEC, 93/105/EC and 2000/21/EC, *Off. J. Eur. Comm.* L396 (2006).
[62] The Royal Society of Chemistry (RSC), *A Brief Guide to REACH: What You Need to Know*, RSC, Cambridge (2008); http://www.rsc.org/ScienceAndTechnology/EHSCGuidance.asp
[63] ISO/TS 16949, *Quality Management Systems – Particular Requirements for the Application of ISO 9001:2008 for Automotive Production and Relevant Service Part Organizations*, International Organization for Standardization, Geneva (2009).
[64] I.-Ching Chiu, S. Gunsel and P. Lacey (to Pennzoil-Quaker State Co.), Environmentally friendly lubricants, PCT Int. Publ. No. WO 03/027212 A1 (3 April 2003).
[65] *Official Methods and Recommended Practices of the AOCS*, 5th edn, AOCS Press, Champaign, IL (2004).
[66] J.K. Satyarthi, D. Srinivas and P. Ratnasamy, Estimation of free fatty acid content in oils, fats and biodiesel by ^{1}H NMR spectroscopy, *En. Fuels* **23** (2009) 2273–7.
[67] Y. Miyake, K. Yokomizo and N. Matsuzaki, Determination of unsaturated fatty acid composition by high-resolution nuclear magnetic resonance spectroscopy, *J. Am. Oil Chem. Soc.* **75** (1998) 1091.
[68] W.L. Siew, Crystallisation and melting behaviour of palm kernel oil and related products by differential scanning calorimetry *Eur. J. Lipid Sci. Technol.* **103** (2001) 729–34.

[69] A. Govindapillai, N.H. Jayadas and M. Bhasi, Analysis of the pour point of coconut oil as a lubricant base stock using differential scanning calorimetry, *Lubr. Sci.* **21**(1) (2009) 13–26.

[70] *Annual Book of ASTM Standards*, Vols. 5.01, 5.02, 5.03, 5.04, American Society for Testing and Materials, West Conshohocken, PA.

[71] Association Française de Normalisation (AFNOR), *Produits Pétroliers et Lubrifiants,* Vols. 1 and 2, L'Imprimerie Chirat, St.-Just-La-Pendue (1992).

[72] Deutsches Institut für Normung (DIN), *Handbuch 2-1, 2-2*, Beuth Verlag GmbH, Berlin (1997).

[73] Institute of Petroleum (IP), *Standard Methods for Analysis and Testing of Petroleum and Related Products*, John Wiley & Sons, Ltd, Chichester (1999).

[74] Deutsches Institut für Normung (DIN), *Handbuch 2-3,* Beuth Verlag GmbH, Berlin (1997).

[75] B. Zhmud and M. Roegiers, New base oils pose a challenge for solubility and lubricity, *Tribol. Lubr. Technol.* **65**(7) (2009) 34–9.

[76] K. Durdag and J. Andle, Real-time viscosity measurement for condition-based monitoring using solid-state viscosity sensor, *Tribol. Lubr. Technol.* **66**(2) (2010) 40–7.

[77] A. Adhvaryu, S.Z. Erhan and J. Perez, Wax appearance temperatures of vegetable oils determined by differential scanning calorimetry: effect of triacylglycerol structure and its modification, *Thermochim. Acta* **395** (2003) 191–200.

[78] M. Johnson and M. Spurlock, Strategic oil analysis: estimating remaining lubricant life, *Tribol. Lubr. Technol.* **66**(1) (2010) 22–24, 26–30.

[79] V.J. Gatto, W.E. Moehle, T.W. Cobb and E.R. Schneller, Oxidation fundamentals and its application to turbine oil testing, *J. ASTM Intl.* **3**(4) (April 2006), JAI 13498.

[80] International Union of Pure and Applied Chemistry (IUPAC), *Standard Methods for the Analysis of Oils and Fats and Derivatives*, 7th edn, Blackwell Scientific Publ., Ltd, Oxford (1987).

[81] AOCS Method Cd 19-90 (2-Thiobarbituric acid value. Direct method), in *Official Methods and Recommended Practices of the American Oil Chemists' Society*, 5th edn, AOCS Press, Champaign, IL (2000).

[82] N.H. Jayadas, Evaluation of the oxidative properties of vegetable oils as base stocks for industrial lubricants using spectroscopic and thermogravimetric analysis, *J. Synth. Lubr.* **25**(3) (2008) 105–13.

[83] G.J. Cochrac and S.Q.A. Rizvi, Oxidation of lubricants and fuels, in *Fuels and Lubricants Handbook: Technology, Properties, Performance and Testing* (G.E. Totten, S.R. Westbrook and R.J. Shah, eds), ASTM International, West Conshohocken, PA (2003), pp. 787–824.

[84] C.C. Akoh, Oxidation stability of fat substitutes and vegetable oils by the oxidative stability index method, *J. Am. Oil Chem. Soc.* **71** (1994) 211–16.

[85] W.F. Bowman and G.W. Stachowiak, Application of sealed capsule differential scanning calorimetry. Part I: Predicting the remaining useful life of industry-used turbine oils, *Lubr. Eng.* **54** (1998) 19–24.

[86] J.M. Perez, F.A. Kelley, E.E. Klaus and V. Bagrodia, Development and use of the PSU microoxidation test for diesel engine oils, *SAE Techn. Paper Series* No. 872028 (1987).

[87] W. Castro, J.M. Perez, S.Z. Erhan and F. Caputo, A study of the oxidation and wear properties of vegetable oils: soybean oil without additives, *J. Am. Oil Chem. Soc.* **83** (2006) 47–52.
[88] J.M. Perez, P.T. Pei, Y. Zhang and S.M. Hsu, Diesel deposit forming tendencies – microanalysis methods, *SAE Techn. Paper Series* No. 910750 (1991).
[89] N.J. Fox and G.W. Stachowiak, Vegetable oil-based lubricants. A review of oxidation, *Tribol. Intl.* **40** (2007) 1035–46.
[90] AOCS Official Method Cd 12-57, *Fat Stability, Active Oxygen Method*, American Oil Chemists' Society, Champaign, IL (2007).
[91] Metrohm Application Bulletin No. 204/1d, *Oxidationsstabilität von Ölen und Fetten – Rancimatmethode*, Metrohm AG, Herisau (1994).
[92] M.W. Läubli and P.A. Bruttel, Determination of the oxidative stability of fats and oils: comparison between the active oxygen method (AOCS Cd 12-57) and the Rancimat method, *J. Am. Oil Chem. Soc.* **63** (1988) 792.
[93] M.H. Gordon and E. Mursi, A comparison of oil stability based on the Metrohm Rancimat with storage at 20 °C, *J. Am. Oil Chem. Soc.* **71** (1994) 649.
[94] A.R. Wewala, Prediction of oxidative stability of lipids based on the early stage oxygen consumption rate, in *Natural Antioxidants. Chemistry, Health Effects and Applications* (F. Shahidi, ed.), AOCS Press, Champaign, IL (1997), pp. 331–45.
[95] B.M. Matthäus, C. Wiezorek and K. Eichner, Fast luminescence method for the detection of oxidized lipids, *Fat Sci. Technol.* **96** (1994) 95.
[96] B.M. Matthäus, C. Wiezorek and K. Eichner, Bestimmung von Hydroperoxides in Fetten und Ölen durch Chemiluminescenz, *Lebensmittelchemie* **47** (1993) 85.
[97] *DGF Einheitsmethoden zur Untersuchung von Fetten, Fettprodukten, Tensiden und verwandten Stoffen*, Deutsche Gesellschaft fur Fettwissenschaften e.V., Münster, Wissenschaftliche Verlagsgesellschaft GmbH, Stuttgart, Method C-IV 6a (1984).
[98] AOCS Official Method Cd 12b-92, *Fat Stability, Oil Stability Index (OSI)*, American Oil Chemists' Society, Champaign, IL (2009).
[99] S.E. Hill and E.G. Perkins, Determination of oxidation stability of soybean oil with the oxidation stability instrument: operation parameter effects, *J. Am. Oil Chem. Soc.* **72** (1995) 741–3.
[100] M.A.S. Aráujo and M. Carvalho Coutinho, The use of chemiluminescence to assess the oxidation stability of lubricating oils, *Ciencia-Tecnica-Petroleo, Sec. Quimica* **3** (2001) 255–61.
[101] B. Kowalski, Evaluation of activities of antioxidants in rapeseed oil matrix by pressure differential scanning calorimetry, *Thermochim. Acta* **213** (1993) 135–46.
[102] G. Litwinienko, A. Daniluk and T. Kasprzycka-Guttman, A differential scanning calorimetry study on the oxidation of C_{12}–C_{18} saturated fatty acids and their esters, *J. Am. Oil Chem. Soc.* **76** (1999) 655–7.
[103] B. Kowalski, Oxidative stabilities of engine oils contaminated by vegetable oil, *Thermochim. Acta* **250** (1995) 55–63.
[104] G. Litwinienko, T. Kasprzycka-Guttman and M. Jarosz-Jarszewska, Dynamic and isothermal DSC investigation of the kinetics of thermooxidative decomposition of some edible oils, *J. Therm. Anal.* **45** (1995) 741–50.

[105] A. Adhvaryu, S.Z. Erhan, Z.S. Liu and J.M. Perez, Oxidation kinetic studies of oils derived from unmodified and genetically modified vegetables using pressurized differential scanning calorimetry and nuclear magnetic resonance spectroscopy, *Thermochim. Acta* **364** (2000) 87–97.
[106] J.A Walker and W. Tsang, Characterization of lubricating oils by differential scanning calorimetry, *SAE Techn. Paper Series* No. 801383 (1980).
[107] D.R. Kodali, Oxidative stability measurement of high-stability oils by pressure differential scanning calorimeter (PDSC), *J. Agric. Food Chem.* **53** (2005) 7649–53.
[108] S.P.J.N. Senanayake and F. Shahidi, Measuring oxidative stability of structured lipids by proton nuclear magnetic resonance, *J. Food Lipids* **14** (2007) 217–31.
[109] D.W. Lachenmeier, M. Gary, Y.B. Monakhova, T. Kuballa and G. Mildau, Rapid NMR screening of total aldehydes to detect oxidative rancidity in vegetable oils and decorative cosmetics, *Spectrosc. Europe* **22**(6) (2010/11), 11–14.
[110] B.K. Sharma and A.J. Stipanovic, Development of new oxidation stability test method for lubricating oils using high-pressure differential scanning calorimetry, *Thermochim. Acta* **402** (2003) 1–18.
[111] T. Colclough, Role of additives and transition metals in lubricating oil oxidation, *I&EC Res.* **26** (1987) 1888–95.
[112] F.R. van der Voort, J. Sedman, R. Cocciardi and S. Juneau, An automated FTIR method for the routine quantitative determination of moisture in lubricants: an alternative to Karl Fischer titration, *Talanta* **72** (2007) 289–95.
[113] F.S.G. Lima, M.A.S. Aráujo and L.E.P. Borges, Determination of lubricant base oil properties by near infrared spectroscopy using different sample and variable selection methods, *J. Near Infrared Spectrosc.* **12**(3) (2004) 159–66.
[114] Y. Zhang, D. Wu, L.-L. Jiang and L.-H. Tan, Fast determination of qualities of engine lubricant by visible and near infrared spectroscopy, *Guangpu Shiyanshi* **27**(4) (2010) 1629–32.
[115] Y. Zhang, L.-L. Jiang, D. Wu, L.-H. Tan and Y. He, Non-invasive measurement of water content in engine lubricant using visible and near infrared spectroscopy, *Guangpuxue Yu Guangpu Fenxi* **30**(8) (2010) 2111–14.
[116] Y. Zhao, L.-L. Jiang, Y. Zhang, L.-H. Tan and Y. He, Determination of dynamic viscosity of automobile lubricant using visible and near infrared spectroscopy, *Guangpuxue Yu Guangpu Fenxi* **30**(9) (2010) 2498–9.
[117] J. Paschoal, F.D. Barboza and R.J. Poppi, Analysis of contaminants in lubricant oil by near infrared spectroscopy and interval partial least-squares, *J. Near Infrared Spectrosc.* **11** (2003) 211–18.
[118] F.S.G. Lima, M.A.S. Aráujo and L.E.P. Borges, Determination of carcinogenic potential of lubricant base oil using near infrared spectroscopy and chemometrics, *Tribol. Intl.* **36**(9) (2003) 691–6.
[119] D. Timm, C. Kurowski, U. Grummisch, U. Meyhack and H. Grunewald, Applications of near infrared spectroscopy for qualification of lubricants and other mineral oil products, *J. Near Infrared Spectrosc.* **6**(1–4) (1998) A243–6.
[120] ASTM D 6866-11, Standard Test Methods for Determining Biobased Content of Solid, Liquid, and Gaseous Samples Using Radiocarbon Analysis, *Annual Book of ASTM Standards*, ASTM International, West Conshohocken, PA (2011).

[121] T. Meindorf and H. Theissen, Fast recognition of bio-oil, *o+p Ölhydraulik Pneumatik* **53**(9) (2009) 390–4 (in German).
[122] H. Theissen, The German Market Introduction Program for Biobased Lubricants, *Tribol. Online* **5**(5) (2010) 225–9.
[123] W. Holweger, Simple routine analysis of lubricant components, *Tribol. Schmierungstechn.* **56**(6) (2009) 37–44.
[124] M. Ohnesorge, H. Wichmann, R. Vogt and M. Bahadir, Analysis of additives in lubricating oils. Part I: Characterisation of all components of a model lubricating oil and screening analysis, *Tribol. Schmierungstechn.* **56**(3) (2009) 18–24.
[125] R. Yunus, T.L. Ooi, A. Fakhru'l-Razi and S. Basri, Simple capillary GC method for analysis of palm oil based polyol ester, *J. Am. Oil Chem. Soc.* **79**(11) (2002) 1075–80.
[126] B. Nebel, M. Mittelbach and G. Uray, Determination of the composition of acetylglycerol mixtures by ^{1}H NMR followed by GC investigation, *Anal. Chem.* **80** (2008) 8712–16.
[127] F. Garofalo, G. Sechi and G. Seghieri, Identification of antioxidant additives in neopentyl ester-based synthetic lubricant oils, *Rassegna Chim.* **44**(5) (1992) 243–50 (in Italian).
[128] F. Garofalo, G. Sechi and G. Seghieri, New analytical methodology for identification and quality control of neopentyl ester synthetic lubricants, *Rassegna Chim.* **46**(1) (1994) 21–8 (in Italian).
[129] K. Grob, A. Artho, M. Biedemann and J. Egli, Food contamination by hydrocarbons from lubricating oils and release agents: determination by coupled LC-GC, *Food Add. Contam.* **8**(4) (1994) 437–46.
[130] K. Behrens, A. Bühler and L. Arias, ASTM D 6443 standard test method for determination of Ca, Cl, Cu, Mg, P, S, and Zn in unused lubricating oils and additives, *Petro Ind. News* **12**(2) (April/May 2011) 38–9; L. Arias, K. Behrens and A. Bühler, S8 TIGER ASTM D 6443 standard test method for determination of Ca, Cl, Cu, Mg, P, S, Zn in unused lubricating oils and additives, *Petro Ind. News* **13**(1) (February/March 2012) 21–2.
[131] J. Wolska, B. Vrebos and P. Brouwer, Analysis of fuels, lubricants, and greases using x-ray fluorescence spectrometry, *J. ASTM Intl.* **3**(1) (2006), no pp. given; *ASTM Spec. Techn. Publ. STP 1468 (Elemental Analysis of Fuels and Lubricants)* (2005), pp. 98–107.
[132] Teledyne Leeman Laboratories, Analysis of Oil Additives Using the Teledyne Leeman Labs Prodigy High Dispersion ICP, *Appl. Note #1060* (2010); http://www.leemanlabs.com
[133] M.A. Maleque, H.H. Masjuki and S.M. Sapuan, Vegetable-based biodegradable lubricating oil additives, *Ind. Lubr. Tribol.* **55** (2003) 137–43.
[134] F. Lo Coco, S. Rizzotti, C. Locatelli, V. Novelli and L. Ceccon, Determination of zinc(II) in lubricant oils by stripping chronopotentiometry, *Ann. Chim.* (Roma) **93** (2003) 291–5.
[135] C.R. Cabrera, H.J. Franco and N. Torres, Study of wear mechanisms and characteristics of coating films on the Falex block-on-ring tester using SEM and EDAX, *SAE Spec. Publ. SP-936 (Rheology and Tribology of Engine Oils)* (1992), pp. 191–212.
[136] V.N. Latyshev and N.A. Mozhin, Activity of lubricants studied by electron paramagnetic resonance, *Stank. Instrum.* **5** (1975) 31–2.

[137] H.E. Henderson, M. Fefer and A. Legzdins, Fundamental oxidation studies of polyalphaolefins and VHVI specialty base fluids, *Preprints Am. Chem. Soc. Div. Petr. Chem.* **44**(3) (1999) 284–8.

[138] A. Al-Alawi, F.R. van der Voort and J. Sedman, New FTIR method for determination of FFA in oils, *J. Am. Oil Chem. Soc.* **81** (2004) 441–6.

[139] R.C.J. Wilson, iPAL FTIR analyser for moisture determination in lubricating oils – an advancement over Karl Fischer titration; http://www.a2technologies.com/downloads/ferrybridgedownload.pdf (September 2010).

[140] F.R. van der Voort, J. Sedman, R.A. Cocciardi and D. Pinchuk, FTIR condition monitoring of in-service lubricants: ongoing developments and future perspectives, *Tribol. Trans.* **49**(3) (2006) 410–18.

[141] A.G. Mignani, L. Ciaccheri, N. Diaz-Herrera, A.A. Mencaglia, H. Oltevaere, H. Thienpont, S. Francalanci, A. Paccagnini and P. Pavone, Optical fiber spectroscopy for measuring quality indicators for lubricant oils, *Proc. SPIE 7004 (Pt. 2, 19th Intl. Conf. on Optical Fiber Sensors)* (2008) 70045 R/1-70045 R/4.

[142] M. Blanco, J. Coello, J.M. García Fraga, H. Iturriaga, S. Maspoch and J. Pagès, Determination of finishing oils in acrylic fibers by near-infrared reflectance spectroscopy, *Analyst* **122**(8) (1997) 777–81.

[143] Y. Miyake, K. Yokomizo and N. Matsuzaki, Rapid determination of iodine value by ^{1}H NMR nuclear magnetic resonance spectroscopy, *J. Am. Oil Chem. Soc.* **75** (1998) 15–19.

[144] Y. Miyake, K. Yokomizo and N. Matsuzaki, Determination of unsaturated fatty acid composition by high-resolution nuclear magnetic resonance spectroscopy, *J. Am. Oil Chem. Soc.* **75** (1998) 1091–4.

[145] K.F. Wollenberg, Quantitative high resolution ^{13}C nuclear magnetic resonance of olefinic and carbonyl carbons of edible vegetable oils, *J. Am. Oil Chem. Soc.* **67** (1990) 487–94.

[146] B. Basu, M.P. Singh, J.S. Kapur, A. Nakazat, M.I.S. Sastry, S.K. Jain, S.P. Srivastava and A.K. Bhatnagar, Prediction of biodegradability of mineral base oils from chemical composition using artificial neural networks, *Tribol. Intl.* **31** (1998) 159–68.

[147] F. Owrang, H. Mattsson, J. Olsson and J. Pedersen, Investigation of oxidation of a mineral and a synthetic engine oil, *Thermochim. Acta* **413** (2004) 241–8.

[148] J.C.O. Santos, I.M.G. dos Santos A.G. Souza, E.V. Sobrinho, V.J. Fernandez Jr and A.J.N. Silva, Thermoanalytical and rheological characterization of automotive mineral lubricants after thermal degradation, *Fuel* **83** (2004) 2393–9.

[149] A.S. Sarma and N.C. Joshi, Application of NMR spectroscopy in lubricant and lubricant additive fields, *J. Sci. Industr. Res.* **57** (1998) 503–14.

[150] M. Ballari, F. Bonetto and E. Anoardo, NMR relaxometry analysis of lubricant oils degradation, *J. Phys. D: Appl. Phys.* **38**(19) (2005) 3746–50.

[151] J.C.J. Bart, *Additives in Polymers. Industrial Analysis and Applications*, John Wiley & Sons, Ltd, Chichester (2005).

[152] J.C.J. Bart, *Plastics Additives. Advanced Industrial Analysis*, IOS Press, Amsterdam (2006).

[153] J.C.J. Bart, *Polymer Additive Analytics. Industrial Practice and Case Studies*, Firenze University Press (2006).

[154] L.R. Rudnick (ed.), *Lubricant Additives: Chemistry and Applications,* 2nd edn, CRC Press, Boca Raton, FL (2009).

[155] M. Ohnesorge, H. Wichmann, R. Vogt and M. Bahadir, Analysis of additives in lubricating oils. Part 2: Separation scheme for the complete analysis of a model lubricant, *Tribol. Schmierungstechn.* **56**(4) (2009) 26–31.

[156] Y. Cui, Separation and determination of antioxidant by TLC, IR and NMR, *Shiyou Lianzhi* **24**(2) (1993) 59–63.

[157] ASTM D 6443-04, Test Method for Determination of Calcium, Chlorine, Copper, Magnesium, Phosphorus, Sulfur, and Zinc in Unused Lubricating Oils and Additives by Wavelength Dispersive X-ray Fluorescence Spectrometry (Mathematical Correction Procedure), *Annual Book of ASTM Standards*, ASTM International, West Conshohocken, PA (2004).

[158] A.S. Serpal, J. Christopher, S. Mukherjee, M.B. Patel and G.S. Kapur, Study of additive–additive interactions in a lubricant system by NMR, ESCA, and thermal techniques, *Lubr. Sci.* **17**(3) (2005) 319–45.

[159] S.S.V. Ramakumar, N. Aggarwal, A.M. Rao, A.S. Sarpal, S.P. Srivastava and A.K. Bhatnagar, Studies on additive–additive interactions: effect of dispersant and antioxidant additives on the synergistic combination of overbased sulphonate and ZDDP, *Lubr. Sci.* **7**(1) (1996) 25–38.

[160] G.S. Kapur, A. Chopra, A.S. Sarpal, S.S.V. Ramakumar and S.K. Jain, Studies on competitive interactions and blending order of engine oil additives by variable temperature ^{31}P-NMR and IR spectroscopy, *Trib. Trans.* **42**(4) (1999) 807–12.

[161] A.S. Sarpal, V. Bansal, M.I.S. Sastry, S. Mukherjee and G.S. Kapur, Molecular spectroscopic studies of the effect of base oils on additive–additive interactions, *Lubr. Sci.* **16**(1) (2004) 29–45.

[162] A.S. Sarpal, S. Mukherjee, R. Sarin, A.K. Mehtha and A.K. Bhatnagar, NMR studies of additive interaction of zinc phosphorodithioate and ashless phosphorodithioate system, *Lubr. Engng.* **58**(5) (2002) 28–32.

[163] G.S. Kapur, A. Chopra, S.S.V. Ramakumar and A.S. Sarpal, Molecular spectroscopic studies of ZDDP–PIBS interactions, *Lubr. Sci.* **10**(4) (1998) 309–21.

[164] L. Bartha, G. Kis, J. Auer and Z. Kocsis, Thickening effects of interactions between engine oil additives, *Lubr. Sci.* **13**(3) (2001) 231–49.

[165] H.A. Spikes, Additive–additive and additive–surface interactions in lubrication, *Lubr. Sci.* **2**(1) (1989) 3–23.

[166] F. Rounds, Changes in friction and wear performance caused by interactions among lubricant additives, *Lubr. Sci.* **1**(4) (1989) 333–63.

[167] M.F. Fox, The degradation of lubricants in service use, in *Handbook of Lubrication and Tribology: Application and Maintenance* (G.E. Totten, ed.), CRC Press, Boca Raton, FL (2006), Vol 1, pp. 29/3–29/40.

[168] G.L. Marshall, The characterisation of oil additives and associated products using phosphorus-31 Fourier transform nuclear magnetic spectroscopy, *Proc. Inst. Petr.* (London) (*2, Petroanal. '81*) (1982) pp. 409–21.

[169] M. Scherge, K. Pöhlmann and A. Gervé, Wear measurement using radionuclide-technique, *Wear* **254** (2003) 801–17.

[170] K. Hosonuma, K. Yoshida and A. Matsunaga, The decomposition products of zinc dialkyldithiophosphate in an engine and their interaction with soot, *Wear* **103** (1985) 297–309.

[171] H.S. Lee, S.S. Wang, D.J. Smolenski, M.B. Viola and E. Klusendorf, *In-situ* monitoring of high-temperature degraded oil condition with microsensors, *Sens. Actuators* **B20** (1994) 49–54.

[172] M.F. Smiechowski and V. F. Lvovich, Iridium oxide sensors for acidity and basicity detection in industrial lubricants, *Sens. Actuators* **B96** (2003) 261–7.
[173] S.S. Wang and H.-S. Lee, The development of *in-situ* electrochemical oil condition sensors, *Sens. Actuators* **B17** (1994) 179.
[174] I.C. Halalay, E.W. Schneider and M.W. Rogers (to General Motors Co.), Determining quality of lubricating oils in use, US Patent Appl. No. 2006/0232267 A1 (19 October 2006).
[175] R. Kasameyer, Lubricant viscosity control for gas compressors, *Petro Ind. News* **11**(6) (2011) 22.
[176] ASTM D 6923-10a, Standard Test Method for Evaluation of Engine Oils in a High Speed, Single-Cylinders Diesel Engine – Caterpillar 1R Test Procedure, *Annual Book of ASTM Standards*, ASTM International, West Conshohocken, PA (2010).
[177] M. Chausseau, S. Lebouil-Arlettaz and G. Thomas, Wear metals and additive elements analysis in used lubricating oils using ICP-OES, *Petro Ind. News* **11**(2) (2010) 6–7.
[178] W. Mach and A. Gervé, IAVF Information Brochure, Institut f. angewandte Verschleissforschung, Karlsruhe (2000).
[179] M. Johnson, Strategic oil analysis: onsite practices using sensory inspection, *Tribol. Lubr. Technol.* **66**(4) (2010), 18–23.
[180] M. Johnson and M. Spurlock, Strategic oil analysis: instrument-based on-site lubricant analysis, *Tribol. Lubr. Technol.* **66**(7) (2010), 26–33.
[181] M. Johnson, Strategic oil analysis: time-dependent alarms for extended lubricant lifecycles, *Tribol. Lubr. Technol.* **66**(8) (2010), 26–9.
[182] M. Johnson, Selecting an oil analysis laboratory, *Tribol. Lubr. Technol.* **66**(12) (2010), 22.
[183] Air Force Instruction AFI 21-124, *Oil Analysis Program,* Department of the Air Force, Washington, DC (2003); http://www.e-publishing.af.mil
[184] Joint Oil Analysis Program, *JOAP Manual* (*NAVAIR 17-15-50/TM 38-301/T.O. 33-1-37/CGTO 33-1-37*), Headquarters Department of the Army, the Navy, and the Air Force, Washington, DC (2008) (unclassified).
[185] P.M. Lugt, A. v.d. Kommen, H. Lindgren and C. Roth, The R0F+ methodology for grease life testing, *ELGI 23rd Ann. General Meeting*, Paris (30 April–3 May 2011).
[186] ASTM D 7066-4, Standard Test Method for Dimer/Trimer of Chlorotrifluoroethylene (S-316) Recoverable Oil and Grease and Nonpolar Material by Infrared Determination, *Annual Book of ASTM Standards*, ASTM International, West Conshohocken, PA (2004).
[187] B. Perston, Raising the bar in FT-IR getting the most out of on-site oil monitoring, *Lube Mag.* **104** (2011) 16–20.
[188] S. Rintoul, New ASTM test method offers quick and easy oil and grease measurement for water and oil samples, http://www.WilksIR.com/pdf/ASTM
[189] http://www.cefas.defra.gov.uk/industry-information/offshore-chemical-notification-sc

9
Legislation of relevance to lubricants

Abstract: In industrialised countries various regulatory measures have been taken to protect the environment from regular use of chemicals, including lubricants. Most recently, the European legislative framework regulating chemicals has been reviewed. Regulations are becoming ever more restrictive with regard to contents, use and disposal of lubricants. There is mounting governmental pressure to establish environmentally sustainable products. Governmental policies and regulations facilitating the introduction of biolubricants avail themselves of subsidies, mandates, preferred procurement, market introduction schemes and other promotional instruments. Voluntary (supra-)national ecolabel schemes focusing on toxicity, ecotoxicity and biodegradability assist the consumer in making a rational product choice.

Key words: product-focused legislation, environmental regulations, biolubricant standards, ecolabels, incentive policies.

9.1 Introduction

Over a sixth of all pollution incidents affecting the environment (caused by negligence or vandalism) involve oil, mostly from tanks and pipelines. Control of Pollution (Oil Storage) Regulations 2001 (OSR England) aim at reducing the impact of such incidents [1]. About 1% of the total mineral oil consumption is used to formulate lubricants. Environmental hazards connected with lubricants fall within the same category of general oil pollution, and despite their lower volumes are equally insidious because of their wide spreading. Between 13% (EC countries) and 32% (USA) of all used lubricants return into the environment more or less changed in properties and appearance [2]. The total amount of lubricants in Germany released to the environment has been estimated at 300 kt/yr, worldwide 12 Mt/yr [3]. Spain consumes 500 kt/yr of lubricants, which cause up 40% of the river pollution. Product safety for the consumer and the environment requires careful evaluation by exposure analysis, biodegradation and ecotoxicity tests. Current health issues with lubricants primarily relate to inhalation of mists and to skin contact. The production, application and disposal of lubricants has to cover the requirements of the best possible protection of the environment.

Lubricants have lately attracted increasing public attention because of pollution problems resulting from high material losses, leakages, spillages and evaporation. Lubricants impact on the environment in all stages of production, application and disposal. The increased public awareness for a pollution-free environment has determined greater use of bio-based consumer products such as lubricants and ecofriendly technology.

It is the aim of various measures taken by legislators and regulatory authorities to protect the environment and to increase the market share of biodegradable lubricants. Research and the use of lubricants prepared from vegetable oils have greatly advanced in the last years by the enforcement of strict governmental regulations and mandates in various parts of the world that stimulate a broader use of renewable resources, in particular for sensitive areas such as forestry, mining, marine, agriculture, heavy industry, transportation, rail and shipping, pulp and paper mills, saw mills, draglines, motorcycle chains, etc. Technological progress though has been more restricted [4, 5]. Notably, the European Union has supported a growing number of projects that propose research activities as well as disseminating knowledge in the field (see Section 10.5).

By any objective measure, there are significant economic and environmental benefits associated with displacing petroleum-based lubricants by bio-based lubes. Table 9.1 lists pros and cons for substitution of mineral oil-based lubricants by equally performing biolubricants. The possibilities of supporting the substitution of mineral lubricants are manifold and include: (i) economic incentives to crop production; (ii) subsidies for manufacturing lubricants from crops; (iii) subsidies for developing innovative products and processes; (iv) creation of a regulatory environment; and (v) country-specific market and legal instruments to favour application of the finished products. The biodiesel history with its effects of incentives in various countries (Germany, USA, Canada) has clearly demonstrated that the success or failure of various areas of the energy sector, particularly nascent industries, is directly affected by the legislative, regulatory and legal framework. Without such incentives, the biodiesel market would be almost

Table 9.1 Substitution of mineral oil-based lubricants by high-performance biolubricants

Favourable:
• Risk of environmental contamination
• Use of renewable resources (CO_2 neutrality)
• Energy security
• Public awareness
• Agricultural support
Unfavourable:
• Production costs of biolubricants
• Substitution costs

non-existent (Germany/USA, 2009–2010). At this point it is to be noticed that if the biodiesel industry is subsidised, the petroleum industry is really super-subsidised.

As shown in Fig. 9.1, various countries (Austria, Germany and Switzerland) have pursued different political strategies to promote the substitution of mineral lubricants by biodegradable lubricants. Since the 1980s Germany has granted continuous support for R&D activities related to industrial use of agricultural raw materials, including the development of biodegradable lubricants. The German 'Market Introduction Programme' (MIP) has been operative from 2001 to 2008 [7, 8] and is an example of rollout subsidies. This has increased the biodegradable lubricant market share in a large number of application fields without setting priorities. Austria has introduced a legal obligation in 1991 for use of biodegradable lubricants on forestry saw chains (BGBl. No. 647/1990 ST0255), which has been broadly adhered to and has been quite effective. Similar regional and local regulatory measures (e.g. in national parks) are now being enforced most frequently. France's economic incentive policy is based on differential taxation according to results of ranking (e.g. award of ecolabel).

Mann [6] has argued that only a policy that prioritises full replacement of mineral lubricants by biodegradable substitutes in the applications with the greatest benefits can be efficient. Regulations requiring substitution in

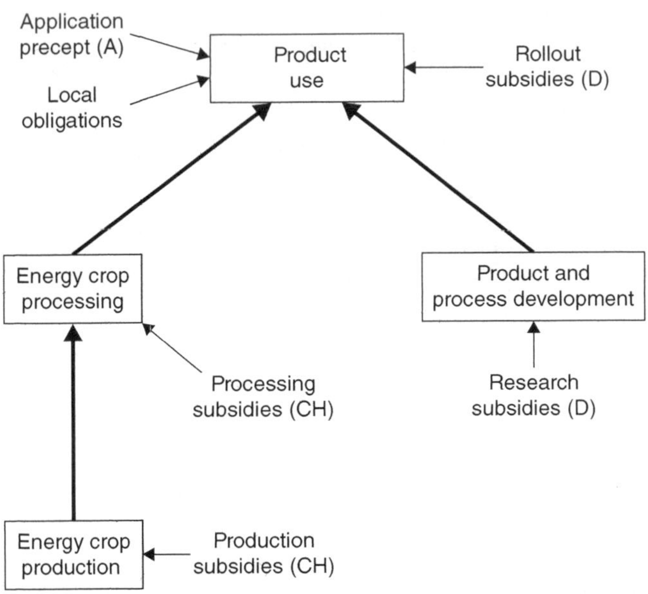

9.1 Product-focused political strategies in favour of biodegradable lubricants. A, Austria; CH, Switzerland; D, Germany. After ref. [6].

certain fields are likely to fulfil this criterion, while subsidies for production, processing, market introduction and research usually fail to meet the efficiency criteria. Industry needs clarity in terms of reliable long-term governmental policies on industrial crops instead of the present rather volatile and *ad hoc* measures, and stable supply of vegetable oils. The biodiesel experience is emblematic.

9.2 Chemicals policy initiatives

Given the problems associated with non-renewable fossil fuels, together with efforts to clean up industrial supply chains of hazardous chemicals, the international chemicals market is bound to undergo fundamental changes. Making the transition from fossil-based to bio-based industrial products requires a strategic vision, a supportive infrastructure and considerable (long-term) investments before reaping the expected benefits (see Table 9.2).

Green chemical innovation is quite likely to act as a key competitive advantage (not unlike quality) for companies manufacturing chemicals

Table 9.2 Making the transition to bio-based industrial products

Vision
• Raw material resource base
• Opportunities: range of bio-based products
• Milestones for the future
Supportive infrastructure
• Education of politicians; public awareness
• Legal obligations
• Information and databases
• Technical training
• Research priorities
• Processing advances (biorefinery concept)
• Environmental research
• Economic feasibility
Investments
• Governmental cooperation
• R&D subsidies
• Production and processing subsidies
• Rollout incentives
Potential benefits
• Diversification of feedstock
• Competitive advantage
• Environmental quality
• Rural development

used in consumer products. Green chemistry processes and products using bio-based materials are becoming increasingly important. In 2000, the US National Academy of Sciences (NAS) evaluated bio-based materials in the USA and proposed national targets (Table 9.3). It has been argued that the NAS timeline might be conservative [10].

In an increasingly competitive global economy, and with a global chemical production expected to double in the next 25 years, the need for a modern chemicals policy is recognised worldwide. Only developing a chemicals policy that addresses *all* the relevant issues associated with the management, design of new chemicals and new chemical processes, manufacturing and use, as mentioned in Table 9.4, will provide long-term solutions to the many chemical problems facing public and environmental health, business, industry and government. It is time to acknowledge the need for a green chemistry technology transition in the chemical industry. A gradual chemical transition to green materials from fossil to vegetable feedstocks is foreseen from now to 2050 [11].

Table 9.3 National Academy of Sciences targets for bio-based industrial materials, as percentage derived from bio-based feedstock material

Bio-based product		US national targets	
	2000 level	2020 level	2090 level
Liquid fuels	1–2%	10%	50%
Organic chemicals	10%	25%	90%
Materials	90%	95%	99%

After ref. [9].

Table 9.4 Essential chemicals policy issues

- Reduce work-related diseases from chemical exposures
- Prevent proliferation of hazardous waste and pollution of air, water and land resources
- Improve information about chemical toxicity and ecotoxicity
- Reduce or eliminate use of the most hazardous chemicals
- Reinforce the scientific and technological basis for safer chemicals
- Motivate industry to invest proactively in the design and use of safer chemicals
- Build productive capacity in ways that improve environmental health
- Be a global leader in green chemistry technology innovation
- Harmonise requirements for chemicals worldwide

Broad chemical registration status is an integral part of product development. Lubricant base stocks and additives should be registered for use (e.g. SEPA Registered, China) or listed on chemical control inventories such as TSCA (USA), DSL List (Canada), KECI (Japan, Korea and Taiwan), and being REACH compliant in Europe. For food-grade lubricants NSF registration is required (see Section 12.10).

Chemical notification systems require chemical producers to generate and disclose standardised information on chemical toxicity, ecotoxicity, use, basic measures of exposure, etc. Chemical use reporting improves accountability in chemical management and often leads to strategies to reduce chemical throughput and use. In this respect, a number of chemical policy initiatives worldwide should be mentioned, ranging from the Massachusetts Toxic Use Reduction Act (TURA) of 1989 to EPA's High Production Volume (HPV) chemical programme, launched in 1997, which concerns over 99 vol% of chemicals in commercial circulation in the United States. Other green chemistry initiatives are Europe's Restriction on Hazardous Substances (RoHS) and Registration, Evaluation, Authorisation and Restriction of Chemicals (REACH), California's Framework for Leadership in Chemicals Policy and Innovation [12], Japan's Green Procurement Survey Standardization Initiative (JGPSSI), and P.R. China's Cleaner Production Promotion Law. The purpose of JGPSSI, a voluntary Japanese organisation for green compliance, established in 2001, is to promote the standardisation of chemical management practices for E&E equipment products. Some of the provisions within China's Cleaner Production (pollution prevention) law (effective as of 1 January 2003) are elimination of obsolete or obsolescent production technologies, processes, equipment and products gravely hazardous to environments and wasteful of resources (art. 12), and preferred procurement (art. 16). Recently, the Chinese Ministry of Environmental Protection (MEP) has issued more specific measures on environmental management of new chemical substances [13]. OECD member countries as well as other countries (e.g. Philippines) are implementing a New Chemical Notification Scheme (NCNS), as a means to assess new chemicals within the framework of an overall program of chemical management. NCNSs represent a high level of environmental protection activity.

Many countries have also adopted the Globally Harmonised System (GHS) of classification and labelling of chemicals into their regulatory systems. CLP (Classification, Labelling and Packaging) Regulation EC No. 1272/2008 implements GHS into the European Union. The CLP regulation replaces the Dangerous Substances Directive (DSD) 67/548/EEC and the Dangerous Preparations Directive (DPD) 1999/45/EC.

The European Union's RoHS Directive (2002/95/EC) restricts hazardous substances in E&E equipment (mainly metals and polybrominated flame retardants, FRs). The directive is currently under review [14]. More importantly (certainly for the lubricant sector), the European industry must comply with the highly comprehensive REACH regulations, which came into force across the EU on 1 June 2007 [15, 16]. REACH intends to streamline and improve the former (scattered) legislative framework on chemicals in the European Union. In the United States, REACH is (incorrectly) classified as protectionist legislation. REACH makes industries bear most responsibilities to manage the risks posed by chemicals and provide appropriate safety information to the users. The controversial REACH regulation (to be reviewed in 2012) is placing a heavy burden on the lubricants sector, particularly the lubricating grease and metalworking fluid manufacturers [17]. The full costs of compliance are still not apparent. Regulatory approval mostly incurs in very substantial expense. The ambitious EU Directive aims at ensuring that all chemicals used commercially in Europe shall be registered and the toxicity, biotoxicity and persistence properly documented. As from 2010 bulk chemicals (>1000 t/yr) and very hazardous chemicals may be supplied only if they are covered by registration dossiers (currently about 25 000). Deadlines in 2013 and 2018 will apply to chemicals supplied in smaller quantities. Restrictions to the use of current mineral oil-based lubricant formulations could eventually derive from this Directive, finalised to reduce the hazard of chemicals currently in use. The EU Biocidal Products Directive (BPD) [18] has regulated the use of metalworking fluid (MWF) microbicides; BPD is similar to the US Environmental Protection Agency's (EPA) Federal Insecticide, Fungicide, and Rodenticide Act (FIFRA)-based regulations [19].

Precautions have long been taken by industry to comply with the (now replaced) EU DSD 67/548/EEC, which establishes criteria for a product's potential hazards to the aquatic environment. The 'Global Automotive Declarable Substance List' of the automotive industry is a corporate instrument for the management of problematic substances [20]. The chemical industry needs finding alternatives to reduce its environmental impact and employment of bio-based lubricants is then a sustainable and technically acceptable solution.

The US regulations are becoming all more restrictive with regard to the contents, use and disposal of lubricants. Four regulations are of particular interest at the federal level: Toxic Substance Control Act (TSCA), Resource Conservation and Recovery Act (RCRA), Comprehensive Environmental Response Compensation and Liability Act (CERCLA) and Superfund Amendments and Reauthorization Act (SARA); see Section 14.7.1. Increasing legislative action is likely to promote further development of the biolubricant market and help overcoming the cost vs environment dilemma [21]. However, the technological level of lubricants will decrease if unnecessary restrictions are imposed.

New VOC (volatile organic compound) regulation requirements worldwide represent environmental and regulatory challenges for solvent-based protectants and for materials with low flash point.

9.3 (Bio)lubricant regulations

In view of the above, it is no longer a matter of deciding *if* we need to switch away from conventional lubricants, but *how* the transition can best be accomplished both technically and economically. A legislative framework is quite necessary, although regulation alone is not considered to be sufficient to improve the biolubricant market [21].

All industrialised countries have laws which are designed to protect waters, land, air and working places from pollution. If chemicals such as lubricants are released into the environment during or after use, surface waters, soil and air are endangered. Although no EU legislation addresses the regular use of biolubricants (vegetable oil-based as well as some synthetics), nowadays many initiatives and national level legislations are in vogue across Europe to enforce replacement of mineral oils. Only few countries have imposed early compulsory legislative measures in favour of plant oil-based lubricants, such as Austria (for chain bar saws) and Portugal (for outboard two-stroke engine oils), both in 1991. In other European countries regulations now exist regarding biolubricants. For example, regulations in Germany, Austria and Switzerland forbid the use of mineral oil-based lubricants around inland waterways and in forest areas. UKWAS (UK Woodland Assurance Standard) requires that 'biodegradable lubricants are used where practical' and all chainsaws and harvesters working on Forestry Commission land use biodegradable oils. In 2004/05 80 t of mineral oils were thus replaced. The biodegradable hydraulic oil currently used by the Forestry Commission is BP Biohyd SE-S, which is a synthetic ester that degrades at over 85% as measured by CEC L-33-A-93. The chain oil is rapeseed oil-based BP Bioforst 2000, which degrades at over 99%. French legislation, enforced as from 1 January 2008, which bans the use of non-biodegradable lubricants when they can be replaced by Euro ecolabel alternatives [22], is expected to push demand for biolubricants in France. The new Forest Code adopted on 15 July 2008 in Wallonia provides for the compulsory use of vegetable oil for chainsaw and forest exploitation equipment.

Germany has set up some of the most extensive guidelines for environmentally acceptable lubricants. The production, application and disposal of lubricants have to be performed without impairment of human beings or of the environment. For this purpose, the German legislation has enacted an extensive framework of laws [3, 23]. Table 9.5 lists the relevant ecological laws and regulations applicable to lubricants in Germany. The German

Table 9.5 Ecological laws and regulations pertaining to regular use of lubricants in Germany

- Water Management Act
- Drinking Water Directive
- Soil Protection Law
- Federal Emissions Law
- Waste and Recycling Act
- Chemical Substances Act
- Environmental Liability Law
- Transport regulations
- German Parliament initiatives
- 'Blue Angel' environmental seal

Table 9.6 Allocation of lubricants within the German water hazard classification scheme for chemicals[a,b]

Water hazard class[c]	Ecological properties	Application to lubricants
'nwh'/WGK 0	Not water hazardous	Biodegradable lubricants, vegetable oils
WHC 1/WGK 1	Slightly water polluting	Plain lubricating oils, base oils, white oils[d]
WHC 2/WGK 2	Water polluting	Additivated lubricating oils, engine and industrial oils
WHC 3/WGK 3	Highly water polluting	Additivated water-miscible cutting fluids, lubricants with emulsifiers

[a] Criteria: acute oral toxicity (rat), fish toxicity, bacterial toxicity and biodegradability.
[b] According to VwVwS of 18 April 1996 (ref. [24]), revised on 1 June 1999.
[c] Water hazard class (WHC); Wassergefährdungsklasse (WGK); status: 1 March 2010.
[d] Mineral oils without additives.

Water Management Act (Wasserhaushaltsgesetz) has led to a classification of substances (Table 9.6), such as lubricants, in Water Hazard Classes (Wassergefährdungsklassen, WGK 0-3) based on their mammalian toxicity, toxic effects on aquatic organisms (bacteria and fish), persistence/biodegradability, and mobility in soil and groundwater [25]. The 1996 provisions were revised on 1 June 1999. It is an objective for ester based products, especially in metalworking, to be defined as 'not water hazardous'.

In Scandinavian countries a tax exemption on biolubricants is in place. The Gothenburg 'Ren Smörja' (Clean Lubricants) project encourages the Swedish manufacturing industry to switch to biolubricants. In fact, classification in the best category of 'Clean Lubricants' requires an origin of the product – at least partly – from renewable resources. In the past, most countries have made recommendations only, as in the UK [26, 27] and Canada [28], or relied on community regulations regarding the use of lubricants for special applications.

Europe is some ten years ahead of the US in terms of acceptance and use of vegetable oil-based lubricants. This can mainly be attributed to the environmental policy prohibiting the use of mineral oil lubricants in certain applications. Strict specifications on various environmental matters, such as biodegradability, toxicity, occupational health and safety, and emissions, have also become mandatory in certain markets. For example, increasingly strict regulations in the field of environment, product compliance and safety are of great interest for MWF producers. EU regulation 67/548 EWG and German TRGS 611 are influential in the production of water-miscible cooling lubricants.

In the United States responsibility for developing rules or regulations to implement the requirements of Acts of Congress is given to various federal government agencies, such as the EPA. Proposed regulations developed by these agencies are published in the Federal Register for review. The final rules from the Federal Register are compiled annually in the Code of Federal Regulations (CFR), which is divided in 50 broad areas or titles. Regulations concerning environmental protection are contained in Title 40. The EPA has developed water pollution and quality regulations within the framework of the Federal Water Pollution Control Act (or Clean Water Act). Relevant parts of the regulation addressing oil pollution are 40 CFR 110, 40 CFR 112 and 40 CFR 113. Criteria for harmful oil discharges are defined in 40 CFR 110 ('Discharge of Oil'). Subchapter 40 CFR 112 ('Oil Pollution Prevention') prescribes a plan for spill prevention control and countermeasures (SPCC) in relation to discharging harmful oil into navigable US waters or adjoining shorelines. 40 CFR 113 ('Liability Limits for Small Onshore Storage Facilities') establishes size classifications and associated liabilities for onshore storage facilities (with a maximum capacity of 1000 barrels) that discharge oil into US waters.

Regulations regarding oil contamination of soil and waters vary from state to state. For instance, the Great Lakes Water Quality Initiative (GLWQI) aims at restoring, protecting and maintaining the quality of the unique Great Lakes water resource. GLWQI essentially bans zinc compounds in the Great Lakes Basin. For guidelines on preventing and handling soil contamination from oil spills state (and local) laws and regulations should be reviewed.

US governmental regulatory action is lagging behind Europe. However, manufacturers expect a shift may emerge in 5–10 years. There are no federal regulations specifically pertaining to the production and usage of biodegradable lubricants in place of mineral oil. US regulations favour the use of biolubricants to meet the government's goal of reducing dependence on petroleum products [29]. Rather than restricting the use of petroleum-based lubricants both federal and several state policy initiatives call for expanding the production and use of bio-based products, including lubricants. There is mounting pressure from the US Federal Government to utilise environmentally sustainable products. The US Farm Security and Rural Investment Act (FSRIA, known as the 'Farm Bill'; G.W. Bush, 13 May 2002) includes a mandate to the US Department of Agriculture (USDA) to develop and implement a comprehensive programme for designating bio-based products, and a directive to all federal agencies to increase their purchase and use of 'preferred' products. As defined by the Federal Biobased Product Preferred Procurement Program (FB4P), Section 9002, 'bio-based' products are products determined by the US Secretary of Agriculture to be commercial or industrial goods (other than food or feed) composed in whole or in significant part of biological products, forestry materials, or renewable domestic agricultural materials, including plant, animal, or marine materials [30]. Current USDA product categories under FB4P comprise adhesives; construction materials and composites; fibres; paper and packaging (P&P); fuel additives; inks; landscaping materials, compost and fertilisers; lubricants, rust inhibitors and functional fluids; paints and coatings; plastics; solvents and cleaners; and sorbents. Table 9.7 lists the lubricant-related categories considered under USDA's BioPreferred^{SM} programme. Made from renewable plant and animal sources, bio-based products are generally safer for the environment than their petroleum-based counterparts. They are usually biodegradable or recyclable.

The US Government has taken action to determine what percentage of bio-based material will be allowed in bio-based products. The USDA has established Biobased Content Guidelines (listed in law H.R. 2646 SEC. 9001) under a mandate from FB4P. Table 9.8 shows the percentage of minimum required renewable content for different product categories for preferred procurement.

Various US Presidential Executive Orders (1993–2007) have dealt with sustainable development (EO 12852); Greening the Government through waste prevention, recycling and federal acquisition (EO 12873, EO 13101), through efficient energy management (EO 13123) and through federal fleet and transportation efficiency (EO 13149); developing and promoting bio-based products and bioenergy (EO 13134); and strengthening federal environmental, energy, and transportation management (EO 13423). Executive Order No. 13423 establishes that agencies are required to acquire sustainable products and services that are bio-based, environmentally preferable, energy-efficient, water-efficient, and have recycled-content, reduce the use of toxic and hazardous materials, and reduce petroleum consumption in vehicle fleets, among other requirements. The White House Executive

Table 9.7 Lubricant-related categories considered under USDA's BioPreferred^{SM} Programme

Air tool lubricants
Chain and cable lubricants
Concrete and asphalt release fluids
Engine crankcase oils
Forming lubricants
Gear lubricants
Greases
Greases (food grade)
Greases (multipurpose)
Greases (rail track)
Greases (truck)
Hydraulic fluids (mobile equipment)
Hydraulic fluids (stationary equipment)
Machine oils
Metal foundry mold release
Metalworking fluids
Metalworking fluids – straight oils
Multipurpose lubricants
Penetrating lubricants
Rock drill oil
Slide way lubricants
Soybean oil
Surfactants and emulsifiers
Total-loss lubricants
Transmission fluids
Turbine drip oils
Two-cycle engine oils
Water turbine bearing oils

Orders (EOs), in combination with several other US government policies, laws, and initiatives have intended to stimulate the emergence of a US bio-based economy by boosting purchase and use of environmentally compatible products by government agencies.

As for lubricants, the USDA manages the BioPreferred Biobased Products Catalog. BioPreferred^{SM} is a source of information on bio-based products designated for preferred procurement. In September 2008 this database of available and acceptable products that meet the US government's need included more than 200 lubricants in the following categories: chain and cable lubricants, firearm lubricants, forming lubricants, gear lubricants, greases, hydraulic fluids, penetrating lubricants, metalworking fluids, and two-cycle engine oils.

Federal agencies are required by FSRIA to purchase bio-based products over their petroleum-based counterparts, as long as the bio-based materials are reasonably available, reasonably priced and comparable in performance. As the single largest consumer in the United States, purchasing roughly

Table 9.8 Minimum proposed bio-based content required per product grouping[a]

Product grouping	Minimum bio-based content (%)
• Hydraulic fluids for mobile equipment	44
• Hydraulic fluids for stationary equipment	46
• Penetrating lubricants	68
• Fluid-filled transformers	
Synthetic ester-based	66
Vegetable oil-based	95
• Metalworking fluids	
Straight oils	66
General-purpose soluble, semi-synthetic and synthetic oils	57
High-performance soluble, semi-synthetic and synthetic oils	40
• Two-cycle engine oils	34
• Chain and cable lubricants	77
• Concrete and asphalt release fluids	87
• Firearm lubricants	49
• Gear lubricants	58
• Greases	
Food-grade grease	42
Multipurpose grease	72
Rail track grease	30
Truck grease	71
Other greases	75

[a] Under the USDA BioPreferred^{SM} programme (www.biopreferred.gov).

$400 billion annually in goods and services, the Federal Government asked to use bio-based resources. However, the BioPreferred^{SM} programme states the each agency shall consider life-cycle costs and performance prior to investing in bio-based products. Also, re-refined products have a priority over bio-based products [29]. Therefore, bio-based content is not necessarily the sole criterion for selecting lubricants; an organisation may justify the purchase of non-biobased lubricants. Unfortunately, some lubricants have higher costs and sometimes are considered to have inferior performance compared with mineral-based products, which limits their competitiveness. Use of BioPreferred^{SM} lubricants in government agencies has not been quantified. As the US Federal Government is a very large purchaser of lubricants and fluids, once the programme is fully operative and implemented (with annual reporting to the Office of Federal Procurement Policy), this should be a major driver for sales of biolubricants in the United States. As a result of federal action, a number of states are also beginning to implement their own purchasing preference programmes.

To aid federal agencies in their selection of bio-based products, the BioPreferred website (http://www.biopreferred.gov/aboutus.aspx) collects voluntary manufacturing and product information from producers and vendors of bio-based materials, including indications of bio content and other key information. There are four main sources of information used to gather data about items for designation in the FB4P, namely Internet search engines (such as Google, Yahoo, WebCrawler, MetaCrawler, Vivisimo, Alta Vista), the Thomas Register, industry associations (Table 9.9), and bio-based industry contacts. When investigating products through the industry associations and bio-based industry contacts, the results tend to be product or company specific. A thorough list of bio-based companies which produce a specific designation item can be developed best by utilising all four information sources. This information is then evaluated against programme criteria established collaboratively. Results from the aforementioned four information sources have indicated that 13 manufacturing companies in the United States currently producing mobile hydraulic fluids might be considered for inclusion in FB4P.

Thanks to the variety of governmental actions, many US companies are now developing bio-based lubricants. Increased federal procurement is a valuable first step in allowing plant-based lubricant companies to increase their share in the consumer market. In January 2008, 18 manufacturing companies in the United States were producing biogreases, 20 were producing stationary hydraulic fluids and 11 were producing two-cycle engine oils that may be considered in the FB4P. The product catalogue (BioPreferred Website) enlisting 'preferred' bio-based lubricants is a useful source of information on bio-based products (see Table 9.10).

Table 9.9 Selected list of US industry associations[a]

- Biobased Manufacturers Association
- National Fluid Power Association
- Fluid Power Society
- Fluid Power Distributor's Association
- Association for Manufacturing Technology
- Material Handling Industry of America
- United Soybean Board
- Consolidated Fluid Power Institute
- US Grains Council
- Independent Lubricant Manufacturers Association
- Society of Tribologists and Lubrication Engineers
- Institute of Electrical and Electronics Engineers
- American Public Power Association

[a] Relevant to mobile hydraulic fluids.

Table 9.10 Bio-based lubricant items under FB4P

Bio high-pressure oil barrier lubricants
Bio two-cycle engine oils
Bio high-performance motor oils
Bio heavy-duty truck grease
Bio 5th wheel truck grease
Bio industrial lubricants
Bio multipurpose machine shop grease
Bio chain lubricant
Bio wire cable corrosion inhibitors
Bio slideway lubricants
Bio rail curve lubricants with lithium thickeners
Bio penetrating oils
Bio turbine lubricants
Bio vacuum pump oils
Bio mobile hydraulic fluids
Bio stationary hydraulic fluids

Current US government policy encourages use and adoption of non-government specifications and standards instead of developing new or updating existing federal and military specifications. Types of specifications, in order of usage preference, are: (i) non-government specifications; (ii) commercial item descriptions; (iii) federal and military specifications; and (iv) proprietary.

Environmental legislation has economic repercussions. On the one hand, the implementation of these regulations may require upgrading or renewal of old machinery. For example, the action of adopting emission standards for several groups of non-road engines as of 7 January 2003 in the United States has led to the improvement of equipment (such as spark-ignition industrial engines as used in forklifts and compressors; airport ground-service equipment, off-highway motorcycles and recreational marine diesel engines) with substantially reduced fuel consumption [31, 32]. On the other hand, the selection of low-evaporation ester and vegetable oil-based lubricants instead of mineral oil-based products does allow use of lower viscosities. This enables faster machining, which in turn increases productivity, generates product optimisation and stimulates the economy. As environmentally harmless products are generally non-toxic and more skin compatible than mineral oils, health and safety at work is promoted, again with a positive economic impact.

9.4 Ecolabels and international standards

While environmental classification schemes place emphasis on the dangerous properties of given substances, 'eco-labels' do exactly the opposite and stress their ecological virtues. Several (supra)national eco-labels/schemes and one international standard (ISO 15380) [33] have been developed for identifying lubricants with reduced environmental impact [34–37].

All ecolabelling is in principle voluntary (consumer choice) and not mandatory by law [38]. The Global Ecolabelling Network (GEN) consists of 26 of the world's leading ecolabelling programmes. In Germany, the Institute for Quality Assurance and Labelling (RAL) grants an environmental label called 'Blue Angel', according to regulations laid down by the Federal Environment Agency (UBA). The Blue Angel is granted for a period of three years. In the framework of the Blue Angel scheme, worldwide the first environmental label (1978), optional criteria are in place for hydraulic fluids, rapidly biodegradable chain lubricants for motor saws, and lubricating oils and greases (including concrete release agents) [39–42]. The 'Blue Angel' combines the performance level of the requirements of ISO 15380 with environmental criteria, which are not compatible with the requirements of the European Ecolabel. In Nordic countries Nordic Swan (Eco-labelling of Lubricating Oils, 1997) sets high requirements for renewability [43]. Separate requirements are imposed to chain oil, hydraulic oil, two-stroke oil, lubricating grease, metal cutting fluid and transmission/gear oil. In the case of chain oil, mould oil and metal cutting fluids only products based on renewable resources will qualify. The Nordic Swan requirements not only cover environmental and health aspects but also technical performance of various lubricant product groups [44]. Criteria for Swedish Standards (legal requirements) were developed for hydraulic fluids (SS 155434) and greases (SS 155470) in close cooperation between manufacturers and users [45, 46]. The environmental criteria for lubricating grease compiled within the Clean Lubricants project are now part of SS 155470. France and Austria have established ecolabels for chainsaw oils. Green Seal is the largest US-based ecolabelling organisation with certification programs for re-refined engine oil. Canada's 'Environmental Choice Program' (Maple Leaf) also includes lubricants and has a separate category for vegetable oil-based and synthetic industrial lubricants. Various national ecolabels, which all have formulated different sets of criteria to define environmentally friendly lubricants, do not contribute to global transparency.

EC regulation 1980/2000 (17 July 2000) has established a voluntary European ecolabel (Euromarguerite, Fig. 9.2), which sets criteria for products and services in order to meet high environmental and performance objectives [47]. The original EU ecolabel scheme according to EC regulation No. 880/92 (23 March 1992) had been largely ignored by industry because of its very strict criteria. In 2005 the EU Eco-labelling Board (EUEB) has defined ecological criteria for the product subgroups of total-loss and high-risk lubricants [48, 49]. Criteria of the European ecolabel for lubricants (EEL) aim at promoting the reduction of water and soil pollution and savings of

9.2 European ecolabel 'Euromarguerite'.

exhaustive energy resources. EEL requires ultimate aerobic biodegradability of >95% for hydraulic fluids, chainsaw oils, concrete release agents and other total-loss lubricants, and >75% for greases and two-stroke engine oils. Additives must fulfil the requirements for R(risk)-phrases, aquatic toxicity, biodegradability/bioaccumulation and the OSPAR (Oslo/Paris) List of Substances of Possible Concern to the marine environment (latest revision: 2009). EEL has been set as the official standard for the Dutch VAMIL regulation (see Section 13.2), whereas France has prescribed the use of EEL for environmentally sensitive zones (January 2008). A list of additives that satisfy the ecolabel criteria is being drafted by Ciba [50]. Revised rules for the establishment and application of the voluntary EU Ecolabel scheme are contained in EC Regulation No. 66/2010 (25 November 2009) [51], replacing the earlier EC Regulation No. 1980/2000. Further measures to establish specific EU ecolabel criteria for each product group are foreseen. It is necessary to ensure coordination between the EU ecolabel scheme and the establishment of the requirements in the context of Directive 2009/125/EC establishing a framework for the setting of ecodesign requirements for energy-related products [52]. Recommendations for revision of EU ecolabel criteria further comprise compliance with new EU regulations on chemicals policy (REACH), increasing harmonisation with other environmentally relevant schemes and ecolabels (especially those that apply to the marine environment) and simplification of the application process.

One of the targets of the EU sponsored IBIOLAB project (see Section 10.5) has been promoting the diffusion of the Euro ecolabel for hydraulic oils, chainsaw oils, demoulding agents, two-stroke oils and greases. The number of licenced lubricant suppliers or products (status: February 2009) varies considerably per ecolabel, which can also be interpreted as a measure of success of the label: SS 155434 Hydraulic oils, 37/84; SS 155470 Lubricating greases, 13/19; Nordic Swan, 0/0; Blue Angel Hydraulic fluids RAL-UZ 79, 25/77; Blue Angel Total-loss lubricants RAL-UZ 64, 25/53; European Ecolabel, 14/57; and German positive list, 40/232. It is noticed that the number of entries varies considerably in time (often in relation to changing regulatory policies).

Ecolabelling schemes focus on toxicity, ecotoxicity and biodegradability but ignore the potential environmental benefits of optimised performance in use. Ecolabels play a valuable role in aiding decision making and influencing market activities but do not consider the relative importance of the various potential environmental impacts of a product or process and could therefore be misleading [53]. Life-cycle assessments (LCAs) provide more objective and quantitative information (see Section 11.2.1). Being based on the properties of unused lubricants, most ecolabels also do not take into account the significant changes in toxicity or ecotoxicity resulting from ageing of oils in use [54, 55]. Future revisions of ecolabel criteria for lubricants should consider inclusion of used oils.

Section 9002 of FSRIA, as amended by the Energy Policy Act of 2005 and the Food, Conservation, and Energy Act of 2008, requires establishment by the USDA of a voluntary programme and criteria for producers of bio-based products to use a 'USDA Certified Biobased Product' label. Among the issues under consideration are: bio-based content levels, testing procedures, performance requirements, life-cycle information, environmental and health benefits, recertification processes and compliance audits. The intent of the USDA logo is to support product identification and consumer awareness. The 'USDA Certified Biobased Product' label is expected to provide a strong third party validation of bio-based products.

9.5 Sources of further information and advice

Umweltbundesamt (ed.), *Guide on Sustainable Chemicals*, Federal Environment Agency, Dessau – Roßlau (2011).

Union Indépendante de l'Industrie Européenne del Lubrifiants, *Purchase and Use of Lubricants within the European Union*, UEIL, Brussels (2011); http//:www.ueil.org

J. Paull, *The Value of Eco-Labelling*, VDM Verlag, Saarbrücken (2009).

The Royal Society of Chemistry, *A Brief Guide to REACH: What You Need to Know*, RSC, Cambridge (2008).

M.P. Wilson, D.A. Chia and B.C. Ehlers (for The California Senate Environmental Committee)/The California Assembly Committee on Environmental Safety and Toxic Materials), *Green Chemistry in California; A Framework for Leadership in Chemicals Policy and Innovation*, California Policy Research Centre, University of California (2006).

Committee on Biobased Industrial Products/National Research Council (eds), *Biobased Industrial Products. Priorities for Research and Commercialization*, National Academies Press, Washington, DC (2000).

9.6 References

[1] http://www.environment-agency.gov.uk/osr

[2] P. Hamblin, Environmentally compatible lubricants: Trends, standards and terms, *Proc. Environmental Aspects in Production and Utilization of Lubricants*, Sopron (13–15 September 1995).

[3] W.J. Bartz, Lubricants and the environment, *Tribol. Intl.* **31**(1–3) (1998) 35–47.

[4] A. Igartua, X. Fernández, O. Areitioaurtena, R. Luther, C. Seyferth, J. Rausch, I. Illarramendi, M. Berg, H. Schultheiß, B. Duffau, S. Plouseau and M. Woydt, Biolubricants and triboreactive materials for automotive applications, *Tribol. Intl.* **42** (2009) 561–8.

[5] A.E.F. Fontes, W.S. Nogueira, F.A.D. Sanitago and J.R. Gomes (to Petrobras), Process to obtain biolubricants and bioparaffins by hydroprocessing mixtures of wax obtained from renewable resources and waxes of mineral origin, Eur. Pat. Appl. No. EP 1,927,642 A2 (4 June 2008).

[6] S. Mann, Ranking without valuing in the face of major uncertainty – The case of promotion of biodegradable lubricants, *J. Environ. Managm.* **85** (2007) 198–203.

[7] H. Murrenhoff, *Markteinführungsprogramm Biogene Treib- und Schmierstoffe – Positivliste*, RWTH, Aachen (2006).

[8] H. Theissen, Experiences with the German Market Introduction Program for biolubricants, *Proc. European VTI-LLINCWA Workshop on Biolubricants* (The Virtual Tribology Institute/Mol and Fundacion Tekniker/Eibar, eds), Mol (15 January 2003).

[9] Committee on Biobased Industrial Products/National Research Council (ed.), *Biobased Industrial Products. Priorities for Research and Commercialization*, National Academies Press, Washington, DC (2000).

[10] C.H. Deutsch, DuPont looking to displace fossil fuels as building blocks of chemicals, *The New York Times* (28 February 2006).

[11] Wetenschappelijke & Technologische Commissie (WTC), Biobased Economy (Dutch Ministry of Economic Affairs, Agriculture and Innovation), *Naar een Groene Chemie en Groene Materialen* (accessed 21 April 2011); http://www.biobasedeconomy.nl/Pages/Naargroenechemieengroenematerialen.aspx

[12] M.P. Wilson, D.A. Chia and B.C. Ehlers, *Green Chemistry in California. A Framework for Leadership in Chemicals Policy and Innovation*, California Policy Center, University of California, Berkeley, CA (2006).

[13] Ministry of Environmental Protection (MEP), *Measures on Environmental Management of New Chemical Substances* (Order 7), Beijing (2010).
[14] J. Calder, *The New EU RoHS Directive: Key Topics Under Review*, http://www.intertek.com/rohs (2010).
[15] Regulation (EC) No. 1907/2006 of the European Parliament and of the Council of 18 December 2006 concerning the Registration, Evaluation, Authorisation and Restriction of Chemicals (REACH), establishing a European Chemicals Agency, amending Directive 1999/45/EC and repealing Council Regulation (EEC) No. 793/93 and Commission Regulation (EC) No. 1488/94 as well as Council Directive 76/769/EEC and Commission Directives 91/155/EEC, 93/67/EEC, 93/105/EC and 2000/21/EC, *Off. J. Eur. Comm.* **L 396** (2006).
[16] The Royal Society of Chemistry, *A Brief Guide to REACH: What You Need to Know*, RSC, Cambridge (2008); http://www.rsc.org/ScienceAndTechnology/EHSCGuidance.asp
[17] S. Hausmann, REACH registration progress, *ELGI 23rd Ann. General Meeting*, Paris (30 April–3 May 2011).
[18] Directive 98/8/EC of the European Parliament and of the Council of 16 Febrary 1998 concerning the placing of biocidal products on the market, *Off. J. Eur. Comm.* **L 123** (24 April 98).
[19] Federal Insecticide, Fungicide and Rodenticide Act (FIFRA), http://www.epa.gov/agriculture/lfra.html
[20] http://www.gadsl.org
[21] R.D. Whitby, Will regulation help improve the biolubricant market?, *Tribol. Lubr. Technol.* **60**(3) (2004) 64.
[22] French Law no. 2006–11 (05.01.06), Loi d'orientation agricole, *J. Off.* no. 5 (6 January 2006).
[23] M. Schmoltzi, Framework of laws and economic aspects of rapeseed oils as lubricants (in German), in *Biologically Fast Degradable Lubricants and Operational Fluids* (W.J. Bartz, ed.), Expert-Verlag, Ehningen (1993).
[24] *Verwaltungsvorschrift wassergefährdender Stoffe* (VwVwS) (18 April 1996); http://www.umweltbundesamt.de/wgs/vwvws.htm
[25] Federal Environment Agency Secretariat of the Commission for the Evaluation of Substances Hazardous in Water (KBWS), *Classification of Substances in Water Hazard Classes. Current State and Opportunities for Further Developments*, Umweltbundesamt, Berlin (1997).
[26] Britain: Environmental Agency regarding use of chain saw lubricants (1996).
[27] Britain: Environmental Agency regarding use of biodegradable hydraulic fluids (1996).
[28] Canada: Canadian Environmental Choice Programme (ECP-05-94); Rapidly biodegradable chain saw lubricants; Rapidly biodegradable biodegradable fluids (1994).
[29] B.J. Bremmer and L. Plonsker, *Bio-based Lubricants: A Market Opportunity Study Update*, United Soybean Board, Chesterfield, MO (November 2008).
[30] USDA, *Establishing the USDA Biobased Products Procurement Program*, Secretary's Memorandum 1042-003, Washington, DC (19 January 2005).

[31] Control of Emissions from Nonroad Large Spark-Ignition Engines and Recreational Engines (Marine and Land-Based), *Federal Register*, **67**(217) (8 November 2002).
[32] http://www.epa.gov/EPA-AIR/2002/November/Day-08/a23801.htm
[33] ISO 15380, *Lubricants, Industrial Oils and Related Products (Class L) – Family H (Hydraulic Systems) – Specification for Categories HETG, HEPG, HEES and HEPR*, International Organization for Standardization, Geneva (2002).
[34] F. Bongardt and A. Willing, European ecolabels for biodegradable hydraulic oils – A challenge to material producers and formulators, *J. Synth. Lubr.* **20-1** (2003) 53.
[35] A. Willing, Lubricants based on renewable resources; an environmentally comparative alternative to mineral oil products. *Chemosphere* **43** (2001) 89–98.
[36] S. Miller, C. Scharf and M. Miller, Utilising new crops to grow the biobased market, in *Trends in New Crops and New Uses* (J. Janick and A. Whipkey, eds), ASHS Press, Alexandria, VA (2000), pp. 26–8.
[37] T. Norrby, M. Torbacke and M. Kopp, Environmentally adapted lubricants in the Nordic marketplace: recent developments, *Ind. Lubr. Tribol.* **54** (2002) 109–16.
[38] J. Paull, *The Value of Eco-Labelling*, VDM Verlag, Saarbrücken (2009).
[39] RAL-UZ 48, *Rapidly Biodegradable Chain Lubricants for Power Saws*, Umweltbundesamt, Berlin (1988).
[40] RAL-UZ 64, *Rapidly Biodegradable Total Loss Lubricants and Shuttering Oils*, Umweltbundesamt, Berlin (1991).
[41] RAL-UZ 79, *Rapidly Biodegradable Hydraulic Fluids*, Umweltbundesamt, Berlin (1995).
[42] http://www.blauer-engel.de/en/index.php
[43] *Nordic Ecolabelling, Ecolabelling of Lubricants.* Criteria document version 4.4 (6 October 2000–30 June 2011); http://www.svanen.nu/eng/default.asp
[44] M. Lämsä, Nordic environmental label for lubricants, *Tribol. Schmierungstechn.* **49** (2002) 40–3.
[45] Svensk Standard SS 155434, *Hydraulic Fluids: Requirements and Test Methods*, Swedish Standards Institute, Stockholm (2003).
[46] Svensk Standard SS 155470, *Lubricating Grease: Requirements and Test Methods*, Swedish Standards Institute, Stockholm (2004).
[47] EC Regulation No. 1980/2000 of 17 July 2000 on a revised Community eco-label award scheme, *Off. J. Eur. Comm.* **L 237** (21 September 2000), 1–12.
[48] Commission Directive 2005/360/EC of 26 Apr. 2005 establishing ecological criteria and the related assessment and verification requirements for the award of the Community eco-label to lubricants, *Off. J. Eur. Comm.* **L 118** (5 May 2005), 26–34.
[49] http://ec.europa.eu/comm/environment/ecolabel/product/pg_lubricants_en.htm
[50] E. Nehls and D. Moore, Are you eco-ready?, *Lubes 'n' Greases Mag.* (November 2007), Issue 6.
[51] EC Regulation No. 66/2010 of 25 Nov. 2009 on the EU ecolabel, *Off. J. Eur. Comm.* **L 27** (30 January 2010), 1–14.

[52] Directive 2009/125/EC of the European Parliament and of the Council of 21 Oct. 2009 establishing a framework for the setting of ecodesign requirements for energy-related products, *Off. J. Eur. Comm.* **L 285** (31 October 2009), 10–35.
[53] I. Boustead, Eco-labels and eco-indices. Do they make sense?, *Proc. Fourth Intl. Ryder Transpak Conf.*, Brussels (19–20 May 1999).
[54] R.P.H. Schmitz, P.H. Roland, A. Eisenträger, T. Lindvogt, F. Möller and W. Dott, Increase of the toxic potential of synthetic ester lubricant oils by usage, *Chemosphere* **36** (1998) 1513.
[55] R.P.H. Schmitz, P.H. Roland, A. Eisenträger, T. Lindvogt, F. Möller and W. Dott, Operation-caused impairment of the environmental compatibility of lubricating oils based on synthetic esters, *Tribol. Schmierungstechn.* **45** (1998) 10.

10
Biolubricant product development

Abstract: Industrial biolubricant product development starts from setting minimum requirements and performances for given applications and consists of defining the most suitable raw materials and processes at the lowest possible cost. Performance testing of the proposed biolubricant formulations by standardised methods is an integral part of the technological development process. Various industry–university consortia have joined R&D efforts. Many collective (inter)national research projects are illustrated. In particular, the development of environmentally friendly lubricant compositions for internal combustion engines, stabilised high-oleic base oil formulations, high-stability bio-based polyol esters, hydroxyl esters and branched biodegradable lubricants are highlighted. The origin of biodiesel lubricity is attributed.

Key words: original equipment manufacturer (OEM) specifications, product development, biolubricant R&T, high-stability esters, branched biolubricants.

10.1 Introduction

Lubricant development should be based on the latest tribological knowledge, thereby duly taking into account the advances of special chemistry (see Chapter 6). Development of (bio)lubricating oils requires sophisticated technology. Industrial research into improving biolubricants has been underway for only about 20 years and interesting results have been obtained, such as very good lubricity characteristics and corrosion-inhibiting properties, thermo-oxidative stability and lower volatility. Applications for environmentally friendly lubricants include machinery operating in sectors such as forestry, agriculture, mining, construction, earth-moving, marine and water treatment.

Within the oil and petroleum industry, as well as additive industries, the impressive wave of increased consolidation in the last decennium has resulted in fewer competing companies along with a reduction in research and technology development (RTD) personnel. Of significant importance is the consideration that a more consolidated branch of industry – in need for greater and longer-term returns on investments (ROIs) – may be less apt at responding to the requirements of original equipment manufacturers (OEMs) for rapid specification changes.

Industrial lubricant users define specifications which are derived from the demands of applications. Starting from the minimum requirements and performances, biolubricant development consists in working out the most suitable processes and raw materials at the lowest possible cost. Testing of lubricant performance is an essential part in the development of lubricant formulations by manufacturers. Lubricant testing can be divided into two major areas:

- physical, chemical and tribological measurements and specifications to characterise the lubricants; and
- lubricant testing in formulations under processing conditions to evaluate their application in industry.

The properties of lubricants are usually determined by use of standardised test methods. Laboratory testing is an indispensable part of research and development (R&D) of lubricants. In most industrial applications, moreover, it is necessary to monitor the actual condition of a lubricant in use with the object of prolonging its lifetime and sustaining the proper functionality of machines and lubricating systems (see Sections 14.5 and 14.5.1). The test results obtained from these methods enable characterisation of the physical, chemical and tribological proprieties of lubricants and assessment of their suitability for specific applications. Testing of lubricant performance is important for manufacturers and end users alike.

The last two decades have witnessed many initiatives aimed at technology transfer in the biolubricant sector, both in Europe, the United States and elsewhere. This chapter describes some of these joint projects (involving manufacturers, RTD institutions and end-users) in more detail (e.g. IBIOLAB), while others are described elsewhere in the text.

Industrial developments have been impressive and are illustrated by means of several case studies. A description of the industrial development of environmentally friendly motor oils by Pennzoil-Quaker State Co. (Houston, TX) is particularly revealing. The various steps in the process are outlined in sufficient detail. It is illustrative to observe how the lack of specific commercial additive packages formulated for the more polar vegetable oils necessitates correction of the composition of the proposed base stock blend.

By combining high-oleic base stocks (HOBS; see Section 5.3.1) and proprietary Stabilized™ (additive) technology, Renewable Lubricants, Inc. (Hartville, OH) has created an impressive product line of renewable and functional oils, fluids and greases which outperform leading petroleum lubricants for many applications.

High-stability esters are particularly necessary in a variety of high-tech lubricant applications. Starting from outstanding early work (Zorn, Zisman) the important line of industrial research on synthetic esters is followed up to the present time with particular emphasis on bio-based polyol esters and hydroxyl esters. Moreover, Exxon Chemical's efforts are illustrated to preserve the desirable properties of branched lubricants without sacrificing biodegradability. Finally, the origin of the lubricity characteristics of biodiesel and vegetable oils is attributed to the presence of free fatty acids.

10.2 Original equipment manufacturer (OEM) specifications for lubricants

Regulatory/environmental initiatives (see Chapter 9) have an impact on OEM decisions, which in turn impact the lubricant base stock and additive selection. The performance of machinery and equipment is greatly dependent on lubrication. Minimum specifications for lubricants proposed for given applications are set by the OEM in close cooperation with the lubricant industry. For example, development of engine oils has been focused on specifications issued by the international automobile industry. In principle, over time all specifications have reflected the adaptation of oil qualities to developments in engine design. Monogrades dominated the engine oil market in the 1950s (distinct summer and winter oils). Mineral oil-based multigrades (i.e. combined summer and winter oils) were introduced in the 1960s. Successively, new qualities with comparatively low evaporation losses and optimised cold flowing properties have been introduced. More recently, fully synthetic oils dominate the premium-quality market in the passenger car sector. Meanwhile, a niche for environmentally friendly, biodegradable products has been created [1]; see also Section 10.5.1. The ecological demands on these oils have also increased and include:

- longer service life despite higher thermal and mechanical loads [2];
- improved emission characteristics by reduced fuel consumption [3];
- lower oil-related particulate emissions [4]; and
- improved wear protection even in severe conditions.

Similar developments may be noticed in other lubricant sectors; Chapter 12.

10.3 Biolubricant standardisation

Standards are vital for commercialisation and market introduction. Original equipment manufacturers rely on specifications for trouble-free operation. A standard, which is comprised of both chemical and physical tests, is meant to guarantee that all (bio)lubricant manufactured for use conforms to a specified quality. This ensures production of a product that has no adverse effects on the equipment or engine, is going to run with no long-term components degradation, is contaminate-free and is not going to pollute the environment with any toxic gases or particulates.

If biolubricants are to compete with fossil-derived lubricants and gain acceptance in the market, they must be manufactured to the highest performance standards. The only way this can be guaranteed is to ensure that the process is well-controlled by manufacturing to strict quality specifications.

Specific legislation promoting and regulating the use of biolubricants is in force in various countries (see Chapter 9). Standardisation bodies, such as the American Society for Testing and Materials (ASTM International), the European Committee for Standardisation (CEN) and others, have designed metrics for lubricants. By using appropriate instrumental techniques to monitor both the physical and chemical properties, it will help to establish biolubricants as high-quality, environmentally safe and viable alternatives to traditional lubricants.

Certain quality parameters (e.g. iodine value) are strictly related to the type of feedstock, while others can be influenced by pre-treatment of the oil or fat (S-, P-content), the process chemistry (ester content, acid value, mono-, di- and triglycerides, estolide number) or any post-treatment of the biolubricant base oil (property enhancement by additives). The standard specifications place restrictions in particular on several contaminants such as water, free fatty acids (FFAs) (acid value) and residual catalyst (ash content). Metal-free additives are mostly being preferred.

Lubricant standards have been developed in close cooperation by the OEMs and lubricant oil producers. It is important that users of biolubricants be conversant with issues related to performance in specific applications, and with impact upon engine or machinery warranties. Manufacturer warranties extend to biolubricants as long as the products sold meet the specifications required for the particular application.

Development of worldwide lubricant standards and international harmonisation is actively being pursued. Global standards offer better protection to original equipment manufacturers. The increasing importance of ecologically harmless lubricants is reflected in the international standard ISO 15380 ('Lubricants, industrial oils and related products (class L) – Family H (Hydraulic systems) – Specification for categories HETG, HEPG, HEES and HEPR') [5]. The rapidly biodegradable hydraulic fluids (formulated end products) according to ISO 15380 are HETG (water-insoluble triglycerides), HEPG (water-soluble polyalkylene glycols), HEES (water-insoluble synthetic esters) and HEPR (water-insoluble poly-α-olefins (PAOs) and related hydrocarbons), see Fig. 12.1. The ecological requirements for this standard cover biodegradability, aquatic toxicity and renewability. The technical criteria of EC Directive 2005/360/EC ('EC Eco-label for Lubricants'), RAL-UZ 79 ('Blue Angel') and Nordic Ecolabelling for Lubricants (Version 4.4, 'White Swan') for hydraulic fluids refer to ISO 15380. Note that definitions for biolubricants in the aforementioned different regulations are not equal. There are also different views on the use of renewable raw materials.

10.4 Performance tests for lubricants and lubricating greases

Lubricant product development requires extensive testing of the proposed experimental formulations, ranging from the determination of physico-chemical and tribological properties by standardised laboratory methods [6] to mechanical-dynamic test methods (ranging from experiments with model systems to customer or field trials) [7] and ageing experiments. Tables 10.1 and 10.2 list the properties of lubricants and lubricating greases which are usually determined by standardised laboratory methods.

Because realistic engine oil tests cannot be performed only by long-lasting field trials, various international committees have created methods of testing engine oils in defined test engines operated in reproducible and practically relevant conditions. The non-profit organisations CRC (Coordinating Research Council) and CEC (Coordinating European Council for the Development of Performance Tests for Lubricants and Engine Fuels)

Table 10.1 Properties of lubricants as determined by the most common standardised laboratory methods

• Acidity/alkalinity	• Hydrolytic stability
• Ageing	• Oil compatibility[a]
• Aniline point	• Pour point
• Ash content	• Refractive index
• Cloud point	• Structural analysis
• Corrosion	• Surface phenomena[b]
• Density	• Viscosity
• Evaporation loss	• Water content
• Flash point	

[a]With seal materials.
[b]Air release, water separation and demulsibility, foaming characteristics.

Table 10.2 Properties of lubricating greases as determined by common standardised laboratory methods

• Consistency	• Oxidation stability
• Corrosion /rust-preventive characteristics	• Shear stability
• Dropping point	• Water resistance
• High-temperature performance	• Wheel-bearing leakage
• Oil separation	• Wheel-bearing life

direct engineering and environmental studies on the interaction between automotive equipment and petroleum products. In Europe, CEC represents the automotive fuels, lubricants, additives, and allied industries in the development of suitable standardisation oils, calibration oils or reference oils (CEC Reference Oil System) and standardised performance test procedures, and is responsible for testing and approval [8]. Performance requirements are set up in the form of ACEA (Association des Constructeurs Européens d'Automobiles) oil sequences which are decided together by the lubricant and additive industry. In the USA, the American Petroleum Institute (API), in cooperation with the automobile industry, lays down test procedures and limits. The Asian ILSAC largely conforms to American automobile specifications. The test procedures focus on the following general performance criteria:

- oxidation and thermal stability [9];
- dispersion of soot and sludge particles;
- protection against wear and corrosion [10]; and
- foaming and shear stability.

For some of the important test methods used to evaluate the performance of formulated oils, see also refs [11, 12].

10.4.1 Mechanical-dynamic test methods for lubricants

Mechanical-dynamic lubricant testing is an essential element in the development of modern lubricants [7]. The wide range of tribological, mechanical-dynamic test machines and test methods stretches from field tests under real conditions to small laboratory instruments. Table 10.3 shows the various tribological system categories within lubricant testing.

Simple mechanical-dynamic lubricant test machines are the four-ball apparatus, Reichert's friction-wear balance, Brugger rig, Falex test machines, Timken test machine, high-frequency reciprocating test machines, low-velocity friction apparatus (LVFA) and diesel injector apparatus. For details concerning performance tests for gear applications, roller bearing applications, synchroniser applications, automatic transmissions, continuously variable transmissions (CVT) and hydraulic fluid applications, see Bartels [7]. High repeatability is a basic prerequisite for satisfactory reproducibility of a test result. Comparison of existing test conditions with real conditions is important in assessing the relevance of results obtained in tribological lubricant tests.

Standardised tribological, mechanical-dynamic test rigs and test methods play a decisive role in the development of lubricants and additives. Standardised test methods are the basic means of achieving the required performance of OEM specifications for lubricants. A variety of additional

Table 10.3 Tribological system categories within lubricant testing[a]

Category	Test		
	Type	Operating conditions	System
I[b]	Field trial	Normal or extended	Original equipment
II[b]	Test laboratory with complete equipment (vehicle) or plant	Close to normal or extended	Original equipment
III[b]	Test laboratory with plant or construction elements	Normal or extended	Original equipment
IV[c]	Standard construction elements or scaled down plant	Normal or extended	Original equipment; experimental samples
V[c]	Test equipment	Close to normal	Experimental sample
VI[c]	Simple laboratory test equipment	Bench test	Experimental sample

[a] Classification according to DIN 50322 [13].
[b] Customer or field trial or trial under similar conditions.
[c] Experiment with model system.

in-house methods complete the range of modern tribological, mechanical-dynamic testing. This is the case for the CEC engine tests which do not allow comprehensive testing of all required oil properties. OEM-specific, long-term performance tests for developing oils are part of these engine tests. The engine tests have been supplemented by radionuclide techniques (RNT) [14, 15]. The advantages of this technology lie in the on-line monitoring of wear in defined running conditions and in selective examination of critical engine components. For details on engine oil tests, see ref. [16].

10.4.2 Ageing tests

Ageing of lubricants is a function of temperature and other operating parameters. Ageing tests are carried out on base oils as well as on fully formulated products in order to assess the efficiency of the additive package. The most commonly used methods for determining the ageing characteristics of lubricants are based on oxidation stability testing. Standardised test methods may vary in duration of the testing period (from hours to months). In most cases, the test fluids are exposed to oxygen or air at relatively high temperatures in the presence of a catalyst to enhance the oxidation rates and reduce the testing time. Oxidation stability may be evaluated by the quantitative determination of oxidation products, oxygen absorption, viscosity changes, changes in acidity, or formation of sludge.

The turbine oxidation stability test (TOST; ASTM D 943) is used to evaluate the oxidation stability of inhibited hydraulic, turbine and rust and oxidation (R&O) oils in the presence of oxygen, water, copper and iron catalyst at 95 °C. The best result is expressed as the number of hours required to reach a total acid number (TAN) of 2.0 mg KOH g^{-1}.

The rotary pressure vessel oxidation test (RPVOT), formerly RBOT, is used to determine the oxidation stability of hydraulic, gear, R&O, turbine and transformer oils in the presence of water, oxygen and copper as a metal catalyst. Testing consists of heating a rotary vessel, pressurised at 620 kPa with oxygen at ambient temperature, up to 150 °C whereby the pressure increases to about 1300 kPa. As oxidation occurs, the oxygen volume decreases and also the pressure inside the vessel. The test termination criterion is the time (induction period, in minutes) to reach a pressure drop of 166 kPa. The rotating bomb oxidation test (RBOT; ASTM D 2272-98) is an analogous rapid method of comparing the oxidation life of lubricants in similar formulations. In this case, the results are reported as the number of minutes to a 25 psi loss. The longer the time, the greater is the stability of the fluid.

Baader's ageing test (DIN 51554) is an oxidation test using atmospheric air and intermittent immersion of a copper spiral at test temperature (95 °C). After a given time the saponification number (SN, mg KOH g^{-1}) is measured.

10.5 Biolubricant research and technology development (RTD)

Research on biodegradable functional fluids emerged as one of the top priorities in lubrication in the early 1990s. Environmentally adapted lubricants (EALs) have been repeatedly heralded as one of the few future growth segments of the lubricants business, hence the relatively large increase in R&D activities (see also Chapter 6). Nevertheless, in sales terms, growth has been slow (see Chapter 13). Industry has a special responsibility for developing and implementing innovative environment-friendly products and manufacturing processes and also many industrial-academic consortia have been formed for that purpose. Notably, the European Union has been supporting a growing number of projects that propose research activities as well as disseminating knowledge in the field. Many European or joint projects (e.g. BIOGREASE, BIOMON, CTVO-Net, EPOBIO, EREBIO, IBIOLAB, IENICA, LLINCWA, LUBRICOAT, LUVE, MEFOLUB, REFLAX, SOILCY, SUNOIL, VALUIL and VOSULUB) and several national projects aimed at technology transfer (e.g. Canada: OLEOTEK; USA: AMfRR, ARS, HOBS programme; Germany: UNICUT; Sweden: GREENCHEM; Italy, BIOVIT; Spain, BIOVESIN; France, initiatives by ADEME and ONIDOL [17]) have promoted development of ecologically harmless lubricants since 1985. Government research sponsorship is offered for projects aimed at improving and demonstrating the performance of biodegradable lubricants, for example producing lubricants which are thermally and hydrolytically stable, and stable to oxidation, without significantly altering their biodegradability. Joint research projects which combine the disciplines of engineering, materials science, chemistry, health and safety, with work by lubricant manufacturers offer the best guarantee for practical results.

CTVO-Net: Chemical-technical Utilisation of Vegetable Oils (FAIR-CT97-3884) has defined the elements to be considered when developing an environmentally friendly lubricant, namely: (i) chemical (performance and cost); (ii) technological (compatibility with the working fluids in the process); (iii) shop floor (health and safety); and (iv) environmental (disposal of waste fluids, recycling of exhausted fluids), and has promoted several workshops on lubricants and hydraulic fluids [18]. The IENICA (Interactive European Network for Industrial Crops and their Applications) project (1997–2005), sponsored within FP4 (FAIR-CT96-1495) and FP5 (QLK5-CT-2000-00111), has addressed the status of non-food crop development within the EU (production-supply-processing-market chain), including attention to biolubricants [19]. Within the framework of the FP5 LUBRICOAT programme the problem of limited lifetime of vegetable oil-based lubricants by friction heating was faced by use of advanced low friction surface coatings. VALUIL is a European Craft project (2002–04) dealing with non-nutritious valorisation of low-cost used frying oils (UFOs) as biolubricant base oils for loss applications (demoulding agents), industrial gear oils, cutting oils and hydraulic fluids [20]. The environmentally friendly lubricant base oils were obtained by methanolysis of UFOs, followed by transesterification in the presence of trimethylolpropane (TMP). Within the SUNOIL European project the Eibar (Gipuzkoa)-based technology centre Tekniker-IK4 developed a sunflower oil-based lubricant for use in hydraulic parts and gears of diggers. The EU-funded REFLAX (Rational Engineering of lipid metabolism of FLAX) project (proj. no. QLK3-2000-00349; 2001–04) has aimed at developing a bio-based production of branched fatty acids by the engineering of oilseeds [21]; see also Sections 5.3 and 6.3. Production of branched-chain fatty acids (FAs) synthetically from bio-based products (vegetable oils and animal fats) has also been one of the objectives of a US ARS project [22].

The collective IBIOLAB research project (FP6 Contract no. 030307; 2006–2009), composed of a consortium of four associations (representing manufacturers and users of oils and fats), 12 SMEs and five RTD performers was aimed at the implementation of several process and product innovations:

- development of specific mild refining processes of raw materials;
- production of vegetable oils directly emulsionable; and
- development of new lubricant bases and bio-additives.

The IBIOLAB project intended to assist small and medium enterprises (SMEs) of the biolubricant sector in developing new biolubricants, to enhance the market share of biolubricants and reinforce their competitiveness. Particular attention was directed at developing a new cost-effective generation of biolubricants for a variety of applications: automotive and industrial lubes, including two-stroke engine oils, hydraulic fluids, chainsaw oils, metalworking fluids, demoulding agents, and lubricating greases. In particular, the formulation of special lubricants for extreme applications (very high and low temperatures, extreme heat variations, high and low rotation speeds, heavy loads, aggressive atmospheres) was being pursued. Targeted vegetable oils were rapeseed oil (RSO), sunflower oil (SNO), high-oleic sunflower oil (HOSNO) and UFO.

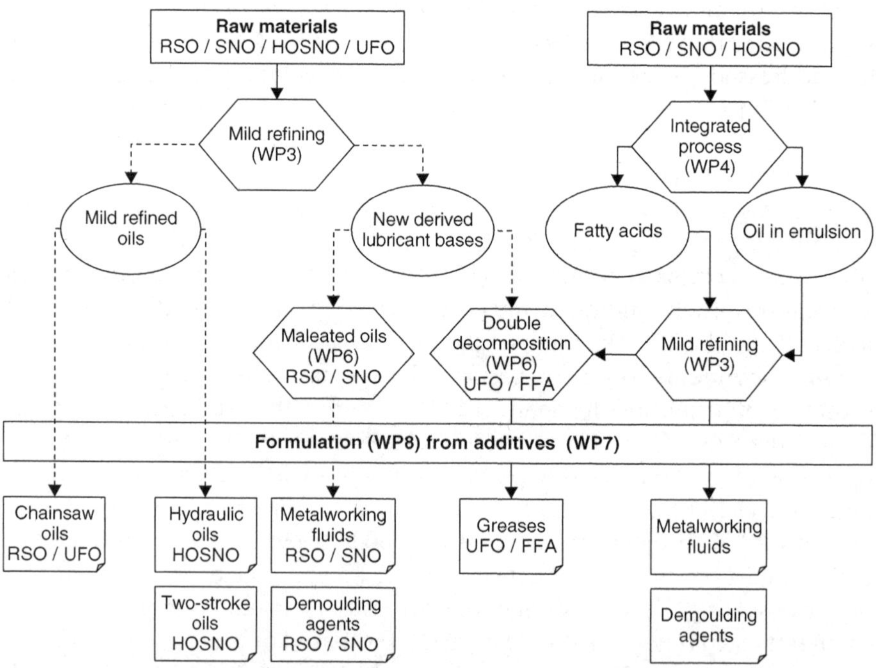

10.1 Production of biolubricants for applications targeted by the IBIOLAB project. FFA, free fatty acids; HOSNO, high-oleic sunflower oil; RSO, rapeseed oil; SNO, sunflower oil; UFO, used frying oil; WP, workpackage.

With reference to Fig. 10.1, mild refining of raw materials (WP3) aims at defining a process of purification of raw materials which is suited to the targeted applications and allows obtaining oils with high stability and lowest impact on the environment at low cost. The main general objective of an integrated process of emulsion production from whole seeds (WP4) is to produce new biolubricant base stocks, namely: (i) O/W emulsions from oil seeds; and (ii) direct, inverse or multiple emulsions of FFAs via *in situ* lipolyse of the seed storage triacylglycerols. Objectives of WP5 are to produce new greases by a double decomposition process for saponification and to characterise the grease bases obtained from UFO fractions and FFAs. The use of maleated oils was proposed as a routeing towards several biolubricant applications (WP6). The objective of WP7 was to develop new bioadditives for formulation with biolubricant bases. Finally, WP8 (vegetable oil lubricant formulation and use) aimed at developing EU ecolabel compliant formulations of biolubricants for industrial testing in various applications. For the biolubricants developed within the framework of the IBIOLAB project, see Section 13.2.

As from 1983 to 2006 some 1354 relevant patents concerning fatty oil fractions as base material for lubricants are found in the European Patent database (esp@cenet). Figure 10.2 shows 196 patents selected on the basis of IBIOLAB's keywords, which clearly denotes an increasing trend. Main applicants among 92 US and 55 German patents (out of 196) were Henkel/Cognis, Exxon Chemical and Lubrizol.

The European Renewable Resources and Materials Association (ERRMA) is a representative body promoting the use of renewable raw materials in technical applications, including biolubricants. As to national

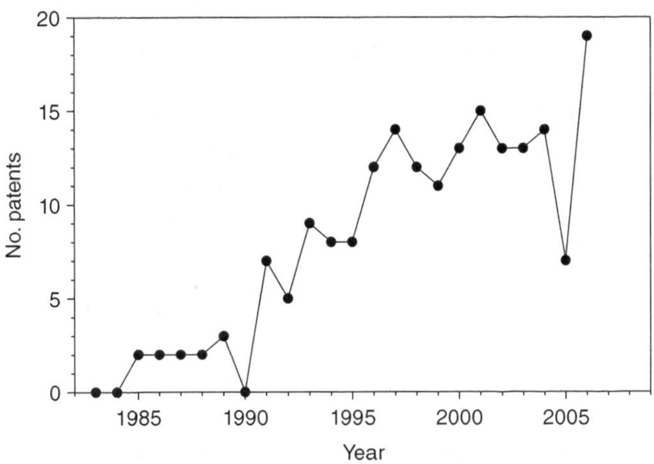

10.2 Number of patent applications out of class C10M (IBIOLAB's selection).

biolube programmes, a green process for the production of biodegradable lubricant-based stocks from epoxidised vegetable oils with lower pour points (–5 to –15 °C) via a cationic ion-exchange resin (Amberlyst 15 Dry) as catalyst was developed within the framework of the MISTRA (Swedish Strategic Fund for Environmental Research)-sponsored GREENCHEM programme [23]; see also Section 6.2.7.

The objectives of the Spanish BIOVESIN project (acronym in Spanish for 'Biolubricants based on vegetable oils and their synthetic derivatives') (PSE 42000-2008-004, co-financed with FEDER funds, 2006–2012) – a partnership of 14 Spanish institutions – are the generation of environmentally friendly lubricating oils and greases based on the last generation modified vegetable oils (HOSNO, new castor oil varieties) and biodegradable additives to attain utmost performance ratings while featuring a good cost–performance ratio and compliance with the Eco-label reference framework. Project developments concern hydraulic oils (with ISO 15380 performance requirements), two-stroke oils (with NMMA TCW-3 performance requirements), gear oils and lubricating greases for open gears. Two hydraulic prototype oils have ultimately been selected for field tests, conducted in wind turbines (Gamesa Eólica; Zamudio, Vizcaya) and agricultural tractors (Agria Hispania S.A.; Amorebieta, Vizcaya). Two-stroke oils are being tested in Abamotor Energía's (Astola, Vizcaya) test engines. Gear-oil evaluation is in progress. The characteristics of a lubricating grease for wind turbine guide rings have been defined.

Target areas of the US ARS research project (2004–2009) on Production of Value-Added Lipids, Biofuels, and Biobased Products from Fats and Oils (Project no. 1935-41000-066-00) include producing lipids with improved physical and/or nutritional properties; introducing branching into the linear fatty acids common to natural fats and oils; producing polyol FAs, amides, as functional or metalworking fluids; and using natural and/or modified phospholipids as metalworking fluids and/or lubricant additives [22]. Production of alkyl (C—C) and alkoxy (C—O) branched FAs aims at producing materials with enhanced lubricity while preserving biodegradability. Incorporation of oxygen or phosphate functionalities into fats and oils or their constituent FAs is to increase their hydrophilic character. Within the framework of ARS unsaturated FAs from vegetable oils (soybean, rapeseed, tall and linseed) were converted in high yields using mild reaction conditions into a mixture of long-chain unsaturated dicarboxylic acids which can be important intermediates in the production of plasticisers.

In the framework of an assistance programme for the intensification of technology in business sponsored by the Canadian Ministry of Economic Development, Innovation and Export Trade, OLEOTEK (Québec, QC) is to: (i) develop bio-based additives with increased lubricating properties or performance in cold conditions; (ii) formulate emulsified total-loss lubricants based on frying oils; (iii) develop renewable biodegradable synthetic lubricants with high oxidative stability; and (iv) develop a metalworking fluid based on 'structured' castor oil.

Renewable Lubricants, Inc. (RLI; Hartville, OH) operates one of the largest known bio-based lubricant programmes in industry. Since 1991, RLI has pioneered an extensive R&D programme of vegetable oil-based lubricants, focused initially primarily on formulating vegetable oil-based automotive engine and hydraulic oils for transportation use. The programme has directed R&D also toward food-grade lubricants, industrial fluids, penetrating lubricants, corrosion inhibitors/preservative oils and transformer oils. Like petroleum oils, vegetable oils have different traits that are revealed in their performance in numerous formulations with various additives. RLI has developed a base oil read-across database formulating vegetable base oils in combination with other vegetable, synthetic, petroleum and re-refined oils. RLI and Penn State University have participated in the USDA-DOD Advanced Materials from Renewable Resources Program (AMfRR), entitled 'Formulating Functional Vegetable Base Motor and Hydraulic Oils' (1993–1996). During this R&D programme more than 15 different types of vegetable-base oils were evaluated, including many different additives and vegetable base formulations. The main areas of interest were oxidation and wear. Joint R&D programmes with the United Soybean Board (USB) on 'Commercializing Super High VI Soybean Vegetable Lubricants for High Temperature Engine and Hydraulic Applications' (Projects No. 7430 and 7431) have led to technology which has been commercialised. Other joint projects have involved organisations such as Chevron-Phillips, DOD, DOE, DOI, Dow AgroSciences, Lubrizol (now Berkshire Hathaway) and NCGA.

In US Patents No. 6,383,992 and 6,534,454 Garmier and Rotondo (to Renewable Lubricants, Inc.) disclose biodegradable (genetically modified) vegetable oil triglycerides (in particular high-oleics) formulated with a pour point depressant (such as the alkylated polystyrene Keil-Flo™150 or the polyalkyl methacrylate RohMax EF171) and an amine antioxidant [24, 25]. These lubricant compositions, eventually as blends with 25–50% other – environmentally friendly – oils (synthetic esters, PAOs), provide oxidative stability as well as cold-temperature performance and can be used for lubricating engines, transmissions, gear boxes, hydraulic applications, penetrating and metalworking compositions and as a base stock for biodegradable greases.

RLI's product range comprises: Bio Hydraulic Fluids, Bio Trans-Hydraulic Fluids, Bio Hydrostatic Fluids, BPL™-Bio Penetrating Lubricants, Bio Bar & Chain Oils, Bio Two-Cycle Engine Oils, and BioPower™ Diesel Fuel Conditioners (see also Table 10.12). Renewable Lubricants, Inc. has successfully developed many vegetable hydraulic and engine oil formulations in different viscosities (SAE 20 to SAE 70). RLI's hydraulic oils were formulated to meet MIL Specifications as described in the US Army's TARDEC report #13640 (46 cSt @ 40 °C). RLI's sunflower-based 10W-30 motor oil has received recognition with a Product of the Year Award (1993) in *Lubricants World Magazine*. Later, its SAE 70 High Performance Corn Base Racing Oil has set track records in corn oil lubrication in the engine transmission and differential of a racing car. A soy oil-based 10W-40 motor oil formulation, developed with the Ohio Soybean Council and the United Soybean Board performs as good as or better than mineral-based formulas. RLI has also been instrumental in developing fire-resistant Bio Hydraulic Fluids to meet military specifications for tactical equipment.

In US Patent Appl. No. 2009/0084026 A1 to Chevron [26] Miller discloses an integrated method for the production of biolubricants and biofuels by the hydrolysis of a common triglyceride feedstock, successive separation on the basis of saturation type using urea complexation or fractional crystallisation of the resulting FFAs to yield two fractions (monounsaturated fatty acids (MUFAs) and saturated/polyunsaturated fatty acids (SFAs/PUFAs)), followed by esterification of the monounsaturated fatty acids (MUFAs) to an ester-based biolubricant (e.g. comprising triesters); the SFA/PUFA fraction can be transformed into monoester-based fuels (biodiesel) or alternatively hydroprocessed into a paraffinic-based transportation fuel (hydrodiesel); see Fig. 10.3. The resulting biolubricant can be used either neat or as an additive.

Formulated with high-quality base stocks refined from collected used oil, the Pure Green™ product line (North American Lubricants, Co., NAL;

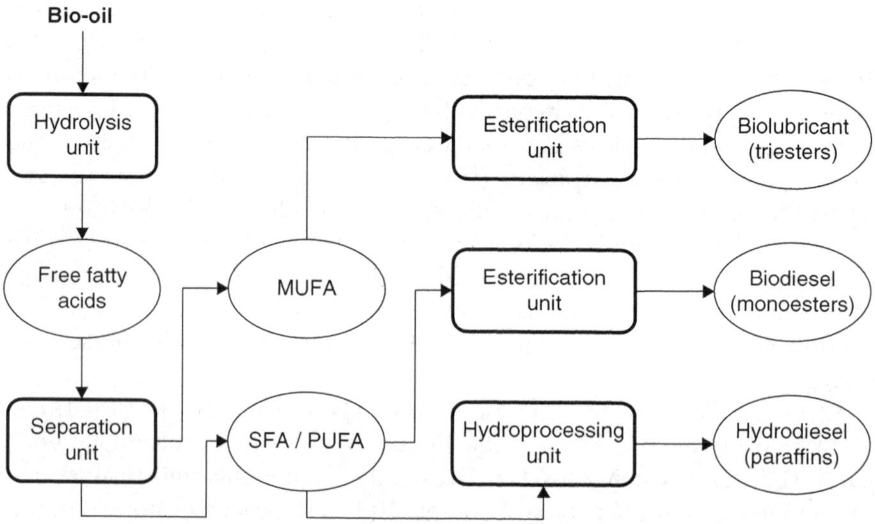

10.3 Production of biolubricants and biofuels from a common triglyceride feedstock.

Scottsdale, AZ) includes passenger car motor oil, heavy-duty engine oil, hydraulic oil, universal tractor fluid and other automotive and industrial lubricants. Terresolve™ Envirologic® Products (Eastlake, OH) has formulated a great variety of non-toxic, biodegradable lubricants for industrial and consumer applications, including hydraulic fluids, gear oils, bar and chain, general purpose and two-cycle engine oils, a dust suppressant, universal tractor fluid, grease, shock absorber and asphalt release agent, etc. [27]. Envirologic® 146H is a vegetable-based hydraulic fluid. Rapid Energy Services (Lafayette, LA) manufactures a range of environmentally friendly lubricant products for a number of industrial applications: hydraulic fluids, penetrating oil, chain oil and marine oil. These high-performance bio-based lubricants from plant oils are equal in performance to their petroleum-based counterparts.

Petrobras (Brazil) has recently described a process to obtain basic lubricating oils ('food grade' biolubricants and bioparaffins) by hydrorefinery technology (hydrotreatment, HDT; hydroisodewaxing, HIDW; and hydrofinishing, HF) applied to waxes obtained from renewable sources (vegetable waxes such as carnauba, jojoba, ouricuri/licuri, or from animal origin), eventually in admixture with waxes of mineral origin [28]. Vegetable waxes typically consist mainly of esters of long chain fatty acids and fatty alcohols and small amounts of FAs, free alcohols and resins [29]; see Section 5.4. Low-cost carnauba wax (composition: 80–85% FA esters, 10–16% fatty alcohols, 3–6% acids) is obtained from leaves of the native Brazilean palm *Copernica cerifera.* Neat carnauba waxes offer lubrication in conformity with US Food and Drug Administration (FDA) regulations. Jojoba oil (JO) is a monoester vegetable oil, mainly composed of C_{40}–C_{44} wax esters (see Section 5.4). Ouricuri (licuri) is a cheaper substitute for carnauba wax with high resin content. Beeswax and lanolin (wool grease) are examples of waxes of animal origin. In the combined Petrobras process, hydrocracking reactions are used to reduce the number of carbon atoms in the chain, hydrotreatment is used for removing oxygenated compounds and hydrogenation of unsaturated fats eliminates double bonds; finally, isomerisation removes paraffins. The resulting high-performance products are biodegradable, cost effective and characterised by a high viscosity index and low fluidity point, surpassing lubricants of mineral origin. The synthetic vegetable-based lubricant oil can be used in production of diesel with elevated cetane number.

Advances in formulations technology is providing new opportunities for bio-based lubricants and greases. Within the European EREBIO project (GROWTH Project no. GRD2-2001-50119, Contract no. G3RD-CT-2002-00796-EREBIO) various low-ash biodegradable prototype lubricants have recently been formulated by FUCHS-Mannheim (blends of hydrocarbons and esters, HCE) and BAM-Berlin (water-soluble polyalkylene glycols –

PAGs – and oil-soluble polypropylene glycols – PPGs) for heavy-duty engines and passenger cars. Application of low or no sulphur, ash and phosphorus (low SAP) ester- and polyglycol-based oils for passenger car engine lubricants as substitutes for hydrocarbon-based oils requires the formulation of lubricants with comparable tribological and functional properties [30, 31]. The main challenge in the formulation of passenger car motor oils (PCMOs) is reduction in crucial anti-wear agents and metal-organic additives (antioxidants and engine cleanliness additives) in order to reduce ash build-up. Additionally, these lubricants need to be biodegradable (according to OECD 301F) and non-toxic (according to OECD 202 Daphnia Magna test) to the aqueous environment in compliance with Directive EC/1999/45.

Frost & Sullivan have recently given an overview of the global technology advancements of biolubricants [32].

10.5.1 Environmentally friendly lubricant compositions for internal combustion engines

Technically and commercially, engine oils are the best among lubricants and functional fluids and account for more than 60% of the global market (see Section 12.2.1). Most commercial motor oils are formulated from petroleum distillates. These distillates, which are resistant to biodegradation, require additives to reduce wear and oxidation, lower the pour point and modify the viscosity index (VI) (to adjust either the high- or low-temperature viscosity). The additive package (often containing metals and reactive compounds), used to adjust certain characteristics, further detracts from the biodegradability of the spent lubricant.

Synthetic ester lubricants having little or no carbon–carbon saturation are used in premium quality motor oils. The acids and alcohols used to make synthetic esters are usually derived from petroleum distillation and thus are not renewable. Synthetic lubricants are also more costly and less biodegradable than natural triglycerides. Hubmann *et al.* have indicated the advantages of a synthetic carboxylic acid ester engine oil (SAE 5W-30) in terms of biodegradability, low particle emissions, reduced oil consumption and high engine cleanliness (low deposition) [1].

There is a need for a vegetable oil-based lubricant that can be used as internal combustion engine oils of varying SAE viscosities that are biodegradable (minimum 60%) and meet the current requirements of the ILSAC GF-3/API SL minimum performance standards, in particular: gelation index (GI) < 12; high-temperature thermo-oxidation engine oil simulation test (TEOST) deposits, 45 mg max; low volatility; P level, 0.1% max; pass foam, filterability and ball rust tests; and remain homogeneous and miscible when mixed with SAE reference mineral oils. In vegetable-based oils it is particularly difficult to meet the SAE low-temperature viscosity requirement. The GI indicates the oil's tendency to form a gelated structure at lower temperature (ASTM D 5133). GI is the largest rate of change of viscosity increase when an oil is slowly cooled from –5 °C to the lowest test temperature. To engine manufacturers GI > 12 is of concern; GI > 15 means field-failing oils.

In PCT Int. Publ. No. WO 03/027212 A1 Chiu *et al.* (to Pennzoil-Quaker State Co., Houston, TX) describe engine oil formulations based on blends of transesterified triglyceride base oil, synthetic ester and mineral oil, additivated with a combination of an ester and an olefin copolymer VI improver and other additives [33]. Minimum targets set for these environmentally friendly motor oils are formulations which are at least 60% biodegradable (ASTM D 5864-95), have GI < 12 and also meet all ILSAC GF-3/API SL bench tests. The testing methods used in the development of the environmentally friendly engine oils are listed in Table 10.4.

The transesterified vegetable oils used in the proposed formulations may be mixtures of transesterified corn oil, RSO, soybean oil (SBO) and SNO. Preferably, high-oleic vegetable oils are used (see Section 5.3.1). Vegetable oils show a tendency to crystallise at low temperature because the triacyl structures tend to be quite regular and subject to packing. This causes the viscosity to abruptly increase at lower temperatures, resulting in failure of the gelation index tests. To meet the GI requirement (GI < 12), in accordance with the ILSAC GF-3/API SL bench test specification, and to improve viscosity, a low gelation saturated synthetic ester (different from

Table 10.4 Testing methods for environmentally friendly engine oils

Property	Test method
Biodegradability	ASTM D 5864
Scanning Brookfield viscosity	ASTM D 5133
Cold cranking simulator (CCS)	ASTM D 5293
Gelation index	ASTM D 5133-99
Gelation temperature	ASTM D 5133-99
Kinematic viscosity	ASTM D 445
Determination of yield stress and apparent viscosity of engine oils at low temperature (MRV TP-1)	ASTM D 4684-98
Pour point	ASTM D 97
Viscosity at high shear rate and high temperature by tapered bearing simulator (TBS)	ASTM D 4683
Viscosity index	ASTM D 2270
Viscosity	ASTM D 445, D 5293
Volatility at 371 °C (simulated distillation, flash point)	ASTM D 6417
Evaporation loss (NOACK) (wt%)	ASTM D 972, D 5800

the vegetable oil) is added to the transesterified oil. In practice, some 10–30% of saturated synthetic ester lowers GI significantly. The synthetic ester (with low GI) may be a dibasic ester such as an adipate or sebacate ester, a tribasic ester such as trimethylolethane (TME) or TMP ester, or a polyol ester, such as pentaerythritol (PE) ester. Several of these products are manufactured by Uniqema Performance Chemicals (New Castle, DE), such as Emkarate 1130™ (diester of a C_{10} alcohol with sebacic acid), Emkarate 1550™ (tribasic ester), Emkarate 1700™ (PE ester of a C_5–C_7 alcohol); Priolube 1831™, Priolube 3939™ and Priolube 3960™ are polymers made from a dimer acid with a di-alcohol.

To increase the viscosity at higher temperatures, VI improvers were added to the blend formulation. There are two types of VI improvers: relatively polar ester types and non-polar hydrogenated olefin copolymer (OCP) type viscosity modifiers. Typical polar ester VI improvers are composed of long-chain esters of maleic anhydride styrene copolymers or polymethacrylate copolymers, such as Lubrizol™ 7671, 7764, 7776, 7783, 7785, 7786 (Lubrizol; Wickliffe, OH). Lubrizol™ 7671 is a thickener which also acts as a pour point depressant (PPD) for vegetable oils. Non-polar hydrogenated olefin copolymer type viscosity modifiers include Lubrizol™ 7070, 7075, 7075D, 7077, 7740 (Lubrizol, OH); Infineum™ SV140, SV145, SV150, SV200, SV205, SV300, SV305 (ExxonMobil, TX); and Paratone™ 8900 series (Chevron, CA). Cost-effective hydrogenated olefin copolymers are the most widely used type of viscosity modifier for PCMOs and heavy-duty diesel engine oils. Table 10.5 shows typical properties of polar ester-type and non-polar hydrogenated olefin copolymer-type viscosity modifiers. Both types of viscosity modifiers were tested in the formulations.

The ester-type viscosity modifiers contribute to the lowering of the gelation index. Figure 10.4 shows the relation between GI and VI improver content for two Lubrizol™ grades for various proposed motor oil formulations [33]. Using Lubrizol™ 7764 and Lubrizol™ 7785, the maximum amount of ester viscosity modifiers allowable in transesterified glyceride

Table 10.5 Properties of polar ester-type (A) and non-polar hydrogenated olefin copolymer-type (B) viscosity modifiers

Property	Value	
	A	B
Flash point (°C)	165	190
Specific gravity (g/cm^3)	0.90	0.87
Viscosity (cSt)	8500 @ 40 °C	–
	1500 @ 100 °C	825 @ 100 °C

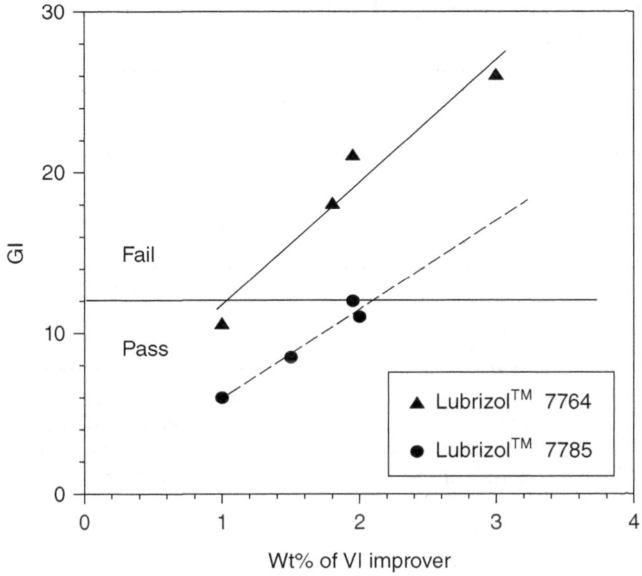

10.4 Gelation index (GI) of formulated environmentally friendly motor oils composed of transesterified glyceride oil/synthetic ester blends as a function of VI improver content. After ref. [33].

oil/synthetic ester formulations without failing the GI specification is from about 1.7% to 2.0% (see Fig. 10.4). At this low concentration of the ester type viscosity modifier alone, the viscosity grade of the formulated oil corresponds to SAE 30 grade or lower. When using a combination of two types of viscosity modifiers a wide range of viscosity grades of motor oils can be produced.

Detergents are generally combined with dispersant additives in crankcase oils. Detergents chemically neutralise acidic contaminants in the oil before they become insoluble and fall out of the oil, forming a sludge. Neutral or basic compounds are created which can remain in suspension in the oil. Lubricating oils typically contain about 2–5% detergent. The compositions of dispersant/inhibitor (DI) packages are generally a proprietary secret, but they usually contain an anti-wear agent, antioxidant, detergent, dispersant, corrosion inhibitor, rust inhibitor, friction modifier and anti-foaming agent. Typical commercial DI packages for gasoline and diesel engine oils are Lubrizol™ 9850 and 9850U (Lubrizol, OH), Oronite™ (Chevron, CA) and Infineum™ (ExxonMobil, TX). ILSAC GF-3 DI packages include OLOA™ 55007 and 59029 (Chevron, CA), Infineum™ 3421, 3422, 5063 (ExxonMobil, TX) and Lubrizol™ 20000, 20000A (Lubrizol, OH). Typical characteristics of dispersant additive packages are shown in Table 10.6.

Table 10.6 Characteristics of typical dispersant/inhibitor additive packages

Property	Value
Flash point (°C)	146–167
Specific gravity (g/cm^3)	0.96–0.97
Viscosity (cSt)	1350–1400 @ 40 °C
	100–125 @ 100 °C

Table 10.7 Group II and III mineral oils

Brand name	Producer	API Group	Kinematic viscosity (cSt) @ 40/100 °C	VI
EXCEL 100-HC™	Pennzoil	II	20.85/4.2	101
EXCEL 230-HC™	Pennzoil	II	42.5/6.4	100
EXCEL 575-HC™	Pennzoil	II	111.0/12.3	100
Yubase 150N™	Yukong	III	37.9/6.6	129
Yubase 240N™	Yukong	III	47.4/7.7	129
Chevron UCBO7R™	Chevron	III	28.2/6.8	137
Shell XHVI™	Shell Chemical	III	47.3/8.2	148

After ref. [33].

One important difference between mineral oil and vegetable oil is that the former consists predominantly of non-polar hydrocarbons, whereas the latter contains polar ester functional groups. As there is a lack of dispersant/inhibitor packages formulated specifically for use with the more polar vegetable oils, only conventional DI packages could be employed for the motor oil blends. In order to solubilise the conventional DI packages in vegetable oil in the required amounts (10–14%), it was necessary to blend the oil with about 10–30% API Group I to III mineral oils (Table 10.7) or API Group IV PAO synthetic oils to lower the polarity. The resulting oils are clear and homogeneous. Note that API Groups II–IV represent environmentally friendly base oils.

As the finishing touch, a pour point depressant was added to lower the pour point of the formulation in view of the presence of the petroleum fraction, whereas an antioxidant was added mainly to protect the vegetable oil fraction. Table 10.8 shows the composition of the developed motor oil blends. In particular, Table 10.9 shows the formulation of a test engine oil composed of a base stock blend consisting of two Group II oils, a modified HOSNO (Cargill Agri-Pure™-560) (see Table 10.22) and a dibase ester; the blend was appropriately VI improved, and DI, PPD and antioxidant additivated. By mixing 0–40 wt% high- and low-viscosity mineral oils to the

Table 10.8 Compositions of environmentally friendly motor oil blends[a]

Component	Range (wt%)
Transesterified triglyceride	40–65
Synthetic ester	15–20
Mineral oil	10–25
Ester type VII	0.5–2
Olefin copolymer type VII	2–4
Dispersant/inhibitor package	5–10
Other additives[b]	1–2

[a] Most preferred range.
[b] Pour point depressant, antioxidant, friction modifier, rust inhibitor, corrosion inhibitor, and antifoaming agent.
After ref. [33].

environmentally friendly lubricant, it was possible to prepare a full range of SAE grade motor oils for gasoline-fuelled and biodiesel-fuelled engines. The SAE grade motor oils that can be achieved include 0W-30, 5W-30, 10W-30 and 10W-40.

The formulation was submitted for ASTM Sequence IIIF engine tests. The Sequence IIIF test is a fired-engine, dynamometer lubricant test for evaluating automotive engine oils for certain high-temperature performance characteristics, including oil thickening, sludge and varnish deposition, oil consumption and engine wear. According to ASTM Sequence III engine tests, the oil formulation of Table 10.9 meets the bench test of the ILSAC GF-3/API SL minimum performance standard; the formulation also passes all API SL bench tests requirements (see Table 10.10).

It was thus shown that by combining polar and non-polar types of viscosity modifiers, a wide range of viscosity grades of motor oils can be produced. Further, when blended with synthetic esters, a motor oil is obtained meeting the desired viscosity, GI specifications, and other specifications needed to make an environmentally friendly engine lubricant. Plant-based automotive lubricants offer opportunities (see also Sections 12.2 and 12.2.1).

10.5.2 Stabilized™ HOBS

Hybrid grown HOBS are modified agricultural vegetable oils which show performance improvement over petroleum-based oils in nearly all categories of lubricant properties (see Section 5.3.1). Monounsaturated triglycerides provide optimum oxidative stability and low-temperature properties. The oxidative stability of high-oleic oils is three to six times greater than

Table 10.9 Formulation and physical properties of a proprietary SAE 5W-30 grade motor oil formulation

Formulation	Description	Contents (wt%)
Lubrizol™7764	VII	1.3
Lubrizol™7075F	VII	1.2
Excel 100-HC™	Group II oil	5.0
Excel 575-HC™	Group II oil	19.0
Cargill™AP560	Modified vegetable oil	40.4
Emkarate™1130	Dibase ester	20.0
Lubrizol™20000A	DI	12.0
Viscoplex™1-3004	PPD	0.1
Irganox™L-64	AO	1.0
Total		100.0

Physical properties	Value
Kinematic viscosity @ 100/40 °C	10.94/59.09 cSt
VI	180
CCS @ −35/−30 °C	12 160/6180 cP
TBS @ 150 °C	3.5
Brookfield temperature @ 30 000/40 000 cP	−31.2/−32 °C
Gelation index	8.3
Gelation temperature	−32 °C
MRV TP-1 @ −35 °C	15 900 cP
MRV YS @ −35 °C	0
Pour point	<−52 °C

After ref. [33].

conventional vegetable oils [34]. Such oils perform adequately for some low operational temperature applications. However, for more demanding applications both the oxidative stability and low-temperature fluidity need further improvement. High monounsaturated oils provide cost-effective, optimal performance raw materials to generate environmentally friendly higher-performance lubricants. The chemical structures of these raw materials may be modified to provide high-performance bio-based synthetic esters. Alternatively, specifically designed additive packages may be used in commercial formulations of high-oleic base stocks. RLI has developed proprietary additives with exceptional performance in HOBS (from renewable and sustainable commodity crops such as soy, corn, canola, rapeseed and sunflower) as lubricant oils. Stabilized™ is RLI's trademark on their proprietary and patented antioxidant, anti-wear and cold-flow technology [24, 25]. The Stabilized™ additive technology allows HOBS to perform as a high-performance formula in high- and low-temperature applications, reducing oil thickening and deposits. Stabilized™ HOBS represent a green chemistry breakthrough. RLI's additive technology provides a low-cost

Table 10.10 Performance testing of a proprietary SAE 5W-30 grade motor oil formulation

Test	Test method	Test result	ILSAC GF-3/API SL limit
H&M (homogeneity and miscibility) test	FTM-3470.1	Non-separation	Non-separation
TEOST MHT 4	ASTM D 6335M		
Total deposit (rod + filter)		20.6 mg	45 mg (max)
Foam test GF-3	ASTM D 892		
Sequence I, foaming/setting		0/0	10 max. /0
Sequence II, foaming/setting		5/0	50 max. /0
Sequence III, foaming/setting		0/0	10 max. /0
High temperature foam test	ASTM D 6082		
Foam tendency		20	100 mL (max)
Foam stability		0	0
EOFT (engine oil filterability test)	GM 9099P		
% Flow reduction		15.07	50 (max)
EOWTT (engine oil filterability/water tolerance test)	GM 9099P		
with 0.6% water		22.12	50 (max)
with 1.0% water		12.17	50 (max)
with 2.0% water		13.90	50 (max)
with 3.0% water		15.63	50 (max)
Gelation index	ASTM D 5133	8.3	12 (max)
NOACK volatility (% wt loss)	ASTM D 972	7.11	15 (max)
Volatility loss @ 371 °C	ASTM D 6417	3.20%	10% (max)
Phosphorus (wt%)	ASTM D 4951	0.093	0.1% (max)
Ball rust test (average Gray value)	ASTM D 6557	134	100 (min)

After ref. [33].

method for stabilising vegetable base lubricants for high- and low-temperature engine, transmission, hydraulic, gear, and grease applications. The low-temperature fluidity of non-stabilised vegetable oil-based fluids is poor compared to Stabilized™ HOBS fluids. Table 10.11 lists additive packages for use in HOBS base stocks for various applications. RLI's Stabilized™ HOBS are additivated with surface additives (notably anti-wear and extreme pressure (EP) agents, bio-rust and corrosion inhibitors and detergents/dispersants), performance enhancing additives (VI improvers, PPDs,

Table 10.11 Special requirements for application of HOBS base stocks

Type of machinery	Additive package	Special requirements
Food processing	–	Safety in case of ingestion
Oil hydraulic	AO, AW, PPD, VII, antirust, antifoam	Minimum viscosity change with *T*, minimum wear of steel/steel
Steam and gas turbines	AO, antirust	Good oxidation stability, ready separation from water
Steam engine cylinders	None or fatty oil	Maintenance of film on hot surfaces; resistence to washing away by wet steam
Air compressor cylinders	AO, antirust	Low deposit formation tendency
Gears (steel/steel)	AO, AW, EP, antifoam	Protection against abrasion and scuffing
Gears (steel/ bronze)	AO, oiliness	Reduce friction, temperature rise, wear and oxidation
Machine tool slideways	Oiliness, tackiness	Smooth sliding at very low speeds; film on vertical surfaces
Diesel engines	AO, AW, CI, detergent, dispersant, acid-neutraliser, antifoam	Additive package engine-type dependent

After ref. [35].

antifoaming agents and demulsifiers) and flow protective additives (oxidation inhibitors, anti-leak additives).

Stabilized™ HOBS represent premium quality base oils with all the required physical properties bio-based lubricant formulations must have in order to meet the highest technical standards defined by industry and government and comply with the objectives for the environment, agriculture and national security set in the 1993–2007 White House Executive Orders (see Section 9.3).

Oxidation in HOBS is a major issue in industry, as they tend to oxidise too readily. The Stabilized™ patented bioprocessing technology, developed by Renewable Lubricants, Inc., has allowed to turn HOBS into a preferred lubricant base oil. Stabilized™ chemistry effectively inhibits the breakdown of HOBS. Contrary to the powerful antioxidation properties of Stabilized™ HOBS, petroleum-based hydrocarbons (mineral oils, synthetics, synthetic blends) and commercial vegetable oil formulations will react readily over time to incorporate oxygens into their structure. This leads to gums, acids and sludge which foul or damage metal parts.

RLI's HOBS formulations are as good as or better than mineral oils in oxidation stability with less volatility. The Penn State Micro-Oxidation Test shows excellent correlation with the SAE engine Sequence III series hot engine test. The PSU micro-oxidation test measures deposits, evaporation and remaining liquid. HOBS 10W-30 formulations perform equal to or better than mineral-based 10W-30 SJ and SG engine oils in deposit formation. The HOBS sunflower formulation shows excellent high-temperature stability with only 1.4% deposits. When comparing deposits, evaporation and remaining liquid, HOBS 10W-30 motor oil outperforms mineral-based 10W-30.

Also in extreme wear and extreme pressure studies, RLI's technology outperformed reference mineral-based oils and formulas. Green chemistry-based Stabilized™ HOBS renewable lubricants technology dramatically lowers emissions and reduces wear while boosting power and fuel economy in heavy-duty diesel vehicles, industrial generators, and motor-driven machinery. Stabilized™ HOBS lubricants (also available through United Bio Lube, Inc., UBL) reduce wear and fatigue in equipment, machines and hardware, thereby saving downtime, maintenance time, and replacement costs.

Stabilized™ HOBS also exhibit low foam tendency. ASTM D 892 and ASTM D 6082 high-temperature foam tests indicate zero foam performance, better than comparative petroleum and synthetic based stocks. Less foam means a better lubricating fluid film and steady oil pressure during extensive operation. This property allows HOBS to perform better in high-performance oil pump systems under high volume and/or pressure.

RLI has formulated a complete line of bio hydraulic fluids with patented Stabilized™ chemistry (see also Section 12.3.1). These fluids meet physical demands of modern hydraulic systems (see Table 12.3). Protection of all hydraulic system components against friction, wear, oxidation, rust and corrosion is achieved through the use of Stabilized™ HOBS compounded with an environmentally friendly zinc-free additive system. RLI's Bio Hydraulic Fluids provide superior lubricity in high-pressure/high-temperature hydraulic systems. The super-high VI (216) of Stabilized™ HOBS formula provides advanced multigrade performance and consolidation of inventory. Bio Ultimax™ AW1000 Hydraulic Fluid-10W40 (RLI) replaces two or more ISO grade mineral oils or synthetic hydraulic fluids. Bio Hydraulic Fluids are compatible with mineral oil-based and most synthetic fluids. RLI's hydraulic fluids are 100% ultimate biodegradable, non-hazardous, non-toxic and free of heavy metals.

Bio hydraulic fluids made from Stabilized™ HOBS are low-volatility fluids. RLI's bio-based lubricants have no known health risks, at variance to petroleum-based fluids. The extremely low volatility of Stabilized™ HOBS increases the flash point and fire safety features in these bio-based hydraulic formulas.

Vegetable formulas can pass the standard pour point ASTM D 97 test at –35 °C, but yet solidify within 24 h at higher temperatures (–10 °C) in the mini-rotary viscometer (MRV) test (ASTM D 4684) or in cold temperature storage tests under long-term static conditions. This may determine poor (failed) performance of vegetable formulations. The problem has been overcome by RLI's Bio Hydraulic Fluid ISO 32.

RBOT tests of RLI's fully formulated HOBS hydraulic fluids showed better performance at 260–400 min (to 25 psi loss at 150 °C) than Pennzoil's SJ 10W-30 motor oil and a commercial mineral oil-based HF-O Hydraulic fluid at 248–262 min. Longer RBOT time corresponds to better performance.

Stabilized™ HOBS bio hydraulic fluids permit water to separate and demulsify rapidly. These fluids are engineered for high-pressure (over 6900 kPa) and high-speed (over 1200 rpm) applications. These applications do not permit formation of full fluid films to protect contacting surfaces (i.e. boundary lubrication). Anti-wear additives minimise wear. RLI's Bio Hydraulic Fluids also provide anti-leak and/or non-drip performance. Anti-leak fluids maintain seal and hose flexibility, reduce oil leaks, extend seal and hose life, and prevent blown or cracked hoses. Leaks rank first in the list of most common problems with hydraulic equipment. Any component leaking in the system is increasing the heat load on the system. Leaks cause the pressure in the system to drop. When pressure drops, heat is generated.

RLI's Bio EP Gear Oils meet and exceed the key objectives for a high-performance gear lubricant (see Table 12.6) and may be applied to prevent premature component failure, assure reliable operation, reduce operating cost and increase service life. Bio EP Gear Oils are specially formulated with cold-flow technology to provide pour points as low as –34 °C and include chemical additives capable of producing a film that provides extreme pressure and anti-scuffing protection. Bio EP Gear Oils are *not* formulated with PTFE or WS_2 as these chemical additives are non-renewable, toxic or hazardous. Renewable Stabilized™ HOBS vegetable oils are replacing petroleum-based mineral and synthetic oils as the far safer, superior performing lubricants of choice for all types of bearing applications.

RLI's Bio Turbine R&O Fluids outperform petroleum-based mineral oils and synthetic esters in terms of superior tribological qualities, properties, and features. Bio Turbine R&O Fluids are additivated with Bio Corrosion Inhibitors (BCI™) to protect bearing surfaces against rust and corrosion.

Bio EP Greases based on Stabilized™ HOBS are engineered to outperform petroleum-based EP greases in terms of natural lubricity, shear stability, oxidation stability and load-carrying capacity. RLI's Bio EP Greases are low-volatility greases; low volatility leads to higher flash points and higher operating temperatures. Stabilized™ HOBS technology prevents oxidation or chemical breakdown of Bio EP Grease formulations, thus avoiding deposit formation. These greases will continue to function under prolonged high-temperature exposure. Bio EP Greases are superior in comparison with petroleum-based greases in terms of oxidation breakdown and very low TANs. Enhanced chemical stability ensures long-lasting performance. Bio EP Greases provide exceptional performance over a wide service temperature range (–30 to 280 °C). Multipurpose Bio EP Greases formulated with calcium sulphonate have inherent EP properties and may be used in H1 food-grade applications.

Owing to excellent antioxidation, anticorrosion and thermal stability properties of Stabilized™ HOBS these bio-based lubricants are fully compatible with seals, rubbers, elastomers, paints and all types of yellow and ferrous metals. The systems compatibility of Stabilized™ HOBS vegetable lubricants and petroleum lubricants for the same seals and filters means that no modification or engineering changes need to be made in equipment in case of lubricant consolidation.

Stabilized™ HOBS are now functional lubricants at extreme temperatures, while meeting or even exceeding physical performance properties and qualities of mineral oils and synthetic lubricants. RLI is producing and marketing a line of over 160 lubricants in some 30 categories based on Stabilized™ technology (Table 10.12). All of these renewable products, which are rated 'bio-based' and 'ultimate biodegradable' by USDA and 'worker friendly' by OSHA, outperform the leading petroleum lubricants in all classes: oils, fluids and greases. RLI's safe biodegradable products are setting new performance, health and environmental standards throughout

Table 10.12 Biolubricant categories based on Stabilized™ HOBS technology[a]

Bio 2-Cycle Engine Oils	Bio Hydraulic Fluids
Bio Air Tool Oils	Bio Metalworking Fluids
Bio Assembly Oils	Bio Mold Release Fluids
Bio Bar & Chain Oils	Bio Motor Oils
Bio Chain & Cable Oils	Bio Oven Lubricants
Bio Corrosion Inhibitors	Bio Penetrating Lubricants
Bio Degreasers–Cleaners	Bio Rock Drill Oils
Bio EP Greases	Bio Saw Guide Oils
Bio Firearm Lubricants	Bio Slideway Oils
Bio Food Grade Gear Oils	Bio Spindle Oils
Bio Food Grade Greases	Bio Tractor Fluids
Bio Food Grade Hydraulic Fluids	Bio Transmission Fluids
Bio Food Grade Lubricants	Bio Turbine R&O Fluids
Bio Gear Oils	Bio Vacuum Pump Oils

[a] From United Bio Lube Product Catalog.

multiple industries, including agriculture, assembly, construction, forestry, government, landscaping, manufacturing, marine, metalworking, mills, military and transportation.

10.5.3 High-stability esters for lubricant applications

'Synthetic esters' represent an extensive class of chemicals with different qualities and prices. Synthetic esters are a group of lubricant base oils with wide variation in structure, which can be used to achieve the desired product characteristics such as thermal and hydrolytic stability, low-temperature properties and seal compatibility (see Section 6.2.1). For the development of environmentally acceptable lubricants, esters have to be selected which fulfil the ecological requirements and have more favourable properties than natural fatty oils. Esters are manufactured chemically from alcohols and acids. In alcohols parameters which may be varied comprise chain length and branching characteristics, as well as the number of OH groups (monoalcohols, diols, polyols). Acids include mono-, di- and polycarboxylic acids with aliphatic, aromatic, straight or branched structures, and may be saturated or partially unsaturated. The raw materials may originate from natural materials and petrochemicals. Chemistry offers a wide range of possibilities in the area of synthetic esters, including transesterification, selective hydrogenation, ozonolysis and dimerisation. The most widely used synthetic esters are probably dicarboxylic acid esters, (non-glycerol) polyol esters and complex esters. Fully synthetic, saturated ester oils based on saturated TMP, polyol and complex esters give excellent technical performance and may be used in high-pressure, high-temperature and otherwise highly stressed hydraulic systems. Partially unsaturated ester oils form the largest group of rapidly biodegradable lubricants. The minimum requirements for ester oils are specified by ISO 15380 [5]. Advances in the biotechnology of vegetable oils (see Section 5.3) and chemical derivatisations of natural esters (see Chapter 6) have resulted in high-performance biolubricants at a lower cost compared to synthetic esters.

In view of their high structural diversity, the main characteristics of synthetic esters, as reported in Tables 6.4 and 6.6, are merely indicative. Using an ester instead of a mineral oil improves the viscosity–temperature behaviour of a lubricant. The higher VI of an ester results in a broader temperature range in application; obviously, the recommended working viscosity should always be adhered to. The higher VI can lead to polymer-free multigrade lubricants with improved shear stability. The low evaporation loss of ester oils compared with vegetable oils has great advantages in regard to emissions from machine tools in metalworking as well as for (particle) emissions of internal combustion engines. In regard to hydrolytic stability, normal polyol esters differ only slightly from RSO. However, oxidation resistance is much improved. Both characteristics are significantly improved with complex esters. Usually, improved hydrolytic stability decreases the biodegradability of the base fluid. However, there are complex esters (medium chain, SFAs on TMP or other poyols) which combine excellent thermo-oxidative characteristics with good hydrolysis resistance and good biodegradability.

Table 10.13 compares the typical compositions and properties of two ester base stocks. The TMP ester exhibits more favourable low-temperature properties than the RSO ester. The shift in fatty acid composition in the TMP ester also has a positive effect on the oxidation stability.

Most synthetic esters display excellent lubricity in boundary lubrication conditions. The high degree of polarity of synthetic esters results in superiority against mineral oil-based lubricants. The friction coefficient of vegetable oils, synthetic esters and glycols is about half that of mineral oils. Some esters have a high degree of polarity which makes them more effective boundary lubricants than non-polar mineral oils and, especially, PAOs. However, as the polar components of the ester tend to compete with surface active anti-wear additives, the effectiveness of the latter is reduced in vegetable oil-based formulations. Esters can be ranked according to their polarity by a non-polarity index (Table 10.14), defined as follows:

$$\text{Non-polarity index (NPI)} = \frac{\text{total number of C atoms} \times \text{MW}}{\text{number of carboxylic groups} \times 100} \qquad [10.1]$$

In general, the higher the NPI, the lower the affinity of the ester to the metal surface and the better it will respond to anti-wear additives. As a rule, the higher the NPI beyond the range of 40 to 200 the better the anti-wear

Table 10.13 Typical compositions and properties of two ester base stocks

Parameter	Value	
	Rapeseed oil	TMP ester
SFA (%)	7	10
C18:1 (%)	50	76
C18:2 (%)	36	13
C18:3 (%)	7	1
Viscosity @ 40 °C (cSt)	32	49
Viscosity index	220	190
Pour point (°C)	–20	–50
Flash point (°C)	260	290
Biodegradability (CEC L-33-A-93) %	95	95

After ref. [36].

Table 10.14 Classification of esters according to polarity

Compound	Non-polarity index
TMP/heptanoic acid ester	40
DOZ diester[a]	52
TMP/pelargonic acid ester	61
Pentaerythritol ester	70
TMP/isostearic acid ester	185
Dimer acid ester[b]	208, 220
Mineral oil	>200
Poly-α-olefin	>300
Emulsifiers	2–7
Detergents	<2

[a] DOZ, di-2-ethylhexyl ester of azelaic (C_9) acid.
[b] Di-2-ethylhexyl ester of dimer (C_{36}) acid.
After ref. [36].

properties of the formulated ester-based lubricant and the less it causes elastomers to swell and soften.

Low-viscosity TMP esters are generally prepared from lower fatty acids, such as hexanoic to decanoic acids. Such low-viscosity esters are inherently polar.

An advantage of synthetic fluids is their potential to molecular design, i.e. modification of the molecular structure to meet the required performance. By appropriate selection of raw materials so as to control molecular properties such as molecular weight, polarity, linearity and structural diversity, ester fluids having a very wide range of physical and chemical properties can be produced. Thermolysis at the ester group by β-elimination can be prevented by substitution of the glycerol moiety of the molecule by polyols such as NPG or TMP. The absence of hydrogen atoms on the β-carbon atom of the alcohol portion of an ester leads to superior thermal stability. The advantage in thermal stability of polar esters over diesters is well documented. Structure–property relationships for ester-based fluids, as shown in Table 10.15, can be used to develop a range of diesters or polyol esters optimised for specific applications.

Research on esters as high-performance lubricants originated in pre-World War II Germany with early work by Zorn (I.G. Farben Industries) on castor oil [38, 39]. It was recognised that the oxidation instability of castor oil – with its outstanding lubricating properties, notwithstanding its gumming propensities – was due to the presence of the glycerol moiety. It was also found that the esterified secondary hydroxyl group of glycerol is hydrolytically quite reactive, and easily leads to acids that catalyse further

Table 10.15 Structure–property relationships of ester base fluids

Property	Molecular weight	No. of ester groups	Branches	Longest linear chain	Structural diversity
Viscosity	+	+	+	+	
VI	+		−	+	
Flash point	+	+	−	+	−
Low-temperature fluidity	−	+	+	−	+
Solvency	−	+	−	−	+
Biodegradability	−	+	−		

+ indicates positive correlation, − indicates negative correlation.
After ref. [37].

degradation. At that time, TME was proposed as an improved substitute to glycerol. Oleic and stearic acid esters of TME combine high thermal stability (needed for high output aircraft engines) with good lubricating properties. During the years 1937–1943 the Leuna laboratories proceeded to prepare, evaluate and manufacture some 3500 esters, diesters and polyesters in order to develop a product with high thermal stability, good viscosity index, low pour point, high flash point, better lubricity than mineral oils and good miscibility with natural and synthetic lubricants [40–42]. For practical applications, esters of dibasic acids and methyl branched alcohols were selected and produced. Additives to prevent piston-ring sticking, oil ageing and corrosion, and for EP lubrication were also developed at Leuna. Furthermore, in Zorn's laboratory, a range of ester-based oils was formulated and produced for a wide variety of applications between 1940 and 1945. At the same time, Zisman (US Naval Research Laboratory) was also synthesising esters, especially diesters [43]. This work has led to an understanding of the performance attributes of diesters (adipates, sebacates, azelates, etc.) and of esters of neopolyols such as PE and TMP [44–47]. Esters of neopolyols became the lubricant base stock for aircraft turbines. Although the thermal and oxidative stability benefits of highly branched acid esters were later recognised by Metro *et al.* [48] and Chao *et al.* [49], these products did not get commercial follow-up because of manufacturing problems. Additionally, the highly branched structures have very low VIs and pour points not compatible with aircraft turbine oil specifications. Successively, the oxidative degradation pathways of neopolyol esters have been established [50], in particular the reaction scheme for autooxidation of pentaerythritol tetraheptanoate [51].

Once other market segments beyond aircraft turbine oils (two-stroke and four-stroke engine oils, refrigerant oils, hydraulic fluids, fibre lubricants, etc.)

had expressed interest in synthetic esters, the spectrum of the required performance attributes has broadened. Some of the properties of industrial interest include a wide range of viscosities, lubricity, fuel economy and biodegradability. Exxon has been particularly active in the development of commercially useful esters [52–58], among others describing preferred compositions with high biodegradability and high viscosity useful as two-stroke engine oils [52–54]. Compositions with high stability built on esters derived from 3,5,5-trimethylhexanol and/or 3,5,5-trimethylhexanoic acid [55, 56] and partially converted polyols [56–58] were disclosed in patents. Schlosberg *et al.* [59] have described esters of TME and TMP showing enhanced thermal and oxidative stability. Partially converted polyol esters (containing unconverted hydroxyl groups) exhibit some advantages over the corresponding full esters [60], as measured by high-pressure differential scanning calorimetry (HPDSC). HPDSC has been used to evaluate the thermo-oxidative stabilities of formulated automotive lubricating oils [60], of synthetic lubricating oils [61] and of polyol ester aviation turbine oils [62].

Polyol esters have greater thermo-oxidative stability than adipates and phthalates of branched alcohols. Polyol esters of shorter chain acids have greater stability than do those of longer chain acids, which is in accordance with work of Chao *et al.* [49]. Polyol esters of certain branched acids (3,5,5-trimethylhexanoic acid and to a lesser extent 2-ethylhexanoic acid) are much more stable in HPDSC than the polyol esters of the corresponding linear acids. Partial polyol esters (especially of certain branched acids, trialkyl acetic acids, or 3,5,5-trimethylhexanoic acid) are more stable than their fully converted analogues. Partial polyol esters based on linear acids also provide thermal/oxidative benefits, although not to the extent of specific branched acid esters. For branched polyol esters, see also Section 10.5.4.

There exists also a strong influence of the hydrocarbon side chain of the polyol on the overall product stability, especially for (partial) polyol esters based solely on iso-octanoic acid [59]. The methyl group of TME is thermally/oxidatively beneficial compared with the ethyl group of TMP even in the iso-octanoate derivatives.

Many polyol esters in current use are built from linear acids. Polyol esters are used in aircraft turbine oils, in full synthetic oils such as Mobil 1 synthetic lubricant and Exxon's ULTRON™ synthetic lubricant, and elsewhere because they provide enhanced stability performance. Mobil 1 synthetic lubricant consists of a blend of PAO and a full ester built on TMP and the C_8–C_{10} acids derived from coconut. Most aircraft turbine oils employ similar FAs Aircraft turbine oil systems require certain viscosity attributes for which the viscosities of esters of branched acids and even more, partial esters of branched systems, are beyond the specification. Within the range of esters available to the formulator further enhancements are possible, such as the use of TME-based neopolyol esters [59]. With carboxylic acids such as 3,5,5-trimethylhexanoic acid excellent performance is provided and outstanding thermal/oxidative stabilities can be achieved. Table 10.16 summarises the special properties of polyol esters.

Uosukainen *et al.* [63] have described the synthesis of biodegradable TMP esters from rapeseed oil methyl ester (RME) via both chemical and enzymatic methods. Approximately 99% total conversion was achieved at 110–120 °C at 3.3 kPa in 10 h over 0.5 wt% $NaOCH_3$. Using 40 wt% *Candida rugosa* lipase only 64% of the TMP was converted to triesters in 24 h at 5.3 kPa and 47 °C. With immobilised *Rhizomucor miehei* (50 wt%), the highest conversion to the TMP triesters of about 90% was achieved in 66 h.

In Eur. Patent No. EP 0,712,834 A1 (to Lubrizol Corp.) Lawate and Lal have disclosed high-oleic (72–95%) polyol esters [64]. In another development, esters of neopentyl polyols were prepared by an esterification reaction between PE and erucic acid at 200 °C catalysed by 4-toluene sulphonic acid (*p*-TSA) in xylene [65, 66]. With PE as backbone of the new esters, four types of esters are obtained: tetra-, tri-, di- and monoesters. Gryglewicz *et al.* [67] have described the basic physicochemical properties of neopentyl glycol (NPG) and TMP esters of rapeseed oil, olive oil and lard (viscosity at 40 °C, 13.5–37.6 cSt; pour point –10.5 to –17.5 °C; VI > 200). Generally, TMP esters of fatty acids have higher viscosity than analogous NPG esters. The synthesised esters show moderate thermo-oxidative stability, exceeding that of native triglycerides. The slow reaction rate of conversion of animal fat (lard) into TMP polyol esters using calcium methoxide with a yield of 85–90% after 20 h of reaction was attributed to steric hindrance [67]. NPG and TMP esters of lard have a much lower pour point (below –10 °C) than natural lard (+33 °C). A two-stage low-crystallisation process was used to improve the pour points.

Table 10.16 Special properties of polyol esters

• High natural VI
• Adequate lubrication at high temperatures
• Very good thermal and chemical stability
• Excellent flow properties at low temperatures
• Long oil life
• Constant thermal conductivity
• Compatibility with sealing materials (NBR, HNBR, EPDM, etc.)[a]
• Excellent solubility in FC and HFC refrigerants[b]
• Ultra-dried before use

[a] NBR, acrylonitrile butadiene rubber; HNBR, hydrogenated nitrile butadiene rubber; EPDM, ethylene–propylene–diene rubber.
[b] FC, fluorocarbon; HFC, hydrofluorocarbon.

Yunus *et al.* [68] synthesised palm oil TMP esters containing 98 wt% triesters in 1 h in appropriate reaction conditions (20 mbar, 120 °C, molar ratio of 3.9:1, 0.9 wt% $NaOCH_3$ catalyst); application of vacuum is vital. They have also carried out a similar study on the transesterification with 98% conversion of palm kernel oil methyl esters (PKOMEs) with TMP using $NaOCH_3$ as a catalyst (20 mbar, 130 °C, molar ratio 3.9:1, 0.9 wt% $NaOCH_3$ catalyst) [69]. The final product consisted mainly (85%) of triesters (TE) and a small fraction (15%) of diesters (DE).

Intentionally, the new synthetic palm-based lubricants should have superior oxidative stability as well as lower pour point than the original palm oil products. The lubrication properties of palm oil and palm kernel oil TMP esters, which include viscosity, VI, pour point, oxidative stability, and wear and friction, have been evaluated by Yunus *et al.* [70]. As shown in Table 10.17, high VI ranges between 170 and 200 were recorded for these base stocks. The presence of partial esters such as ME and DE is expected to affect viscosity and VI of both palm and palm kernel TMP esters. Lower VIs were observed for samples with high percentages of partial esters. Intermolecular hydrogen bonding between —OH groups of partial esters increases oil viscosity. Transformation of PME and PKOME into TMP esters results in only marginal changes in viscosity.

Pour points of the oils and polyols are affected by the FA chain length (mainly C_{12} for PKO and C_{16}–C_{18} in PMO) as well as by the chemical structure [72]. *Cis*-unsaturation levels in the oleic fraction favour decreased PPs in vegetable oils [73]. Also, hydroxyl groups present in monoesters and diesters favour low PP (see castor oil). On the whole, pour points of palm oil and palm kernel oil TMP esters are rather similar (see Tables 10.17 and

Table 10.17 Properties of non-additivated palm-based lubricants

Product[a]	PMOTE	PKOTE	Rapeseed-based TMP ester
Origin[b]	PME	PKOME	RME
Product composition (wt%)			
ME	0.0–0.3	0.0–0.2	n.a.
DE	0.3–4.2	0.3–4.8	n.a.
TE	93.0–97.5	94.5–98.1	n.a.
Viscosity @ 40 °C (cSt)	44.8–52.4	40.5–43.2	34.9
Viscosity @ 100 °C (cSt)	9.1–10.2	8.1–8.4	7.8
Viscosity index	186–189	176–180	210
Pour point (°C)	2–4	3–4	−15

[a] PMOTE, palm oil TMP esters; PKOTE, palm kernel oil TMP esters.
[b] PME, palm oil methyl ester; PKOME, palm kernel oil methyl ester.
n.a., not available.
After refs [63, 69, 70].

10.18), but distinctively higher than for esters derived from conventional and non-conventional vegetable oils. This is mainly due to higher saturation levels in palm oil products.

The low-temperature properties of palm-based TMP esters are inferior compared to other vegetable oil-based lubricants due to a higher level of saturation. Palm oil TMP esters exhibit a better VI and a slightly lower pour point than palm kernel oil TMP esters. Palm oil TMP esters contain more than 53% unsaturated fraction while palm kernel oil TMP esters contain less than 16 wt%. The higher level of unsaturation decreases their oxidative stability. The presence of partial polyol esters is known to improve the oxidative stability, as shown for neopolyol esters [59]. While palm oils usually contain 47% SFA, 44% MUFA and 9% PUFA, palm kernel oils are even far more saturated (84.5% SFA, 15% MUFA, 0.5% PUFA) [74]. The high saturation levels in palm-based oils confer high pour points to the TMP esters (0–4 °C) compared to esters derived from other vegetable oils. To improve the low-temperature properties of palm oil TMP esters, which are well below the requirements of lubricant manufacturers (Table 10.18), high-oleic palm oil TMP esters with PP of –33 °C were synthesised [70].

The lubricating properties of TMP esters depend primarily on the structures of these esters. Wear of palm oil TMP esters is comparable to fully additivated commercial hydraulic fluids. The presence of DE improves wear. Since the viscosity of palm oil TMP esters containing a high percentage of partial esters is similar to castor oil it appears that hydrodynamic lift becomes dominant when viscosity is exceptionally high. For moderate vis-

Table 10.18 Pour points of selected oil samples

Oil sample	Pour point (°C)
Mineral oil (MO)	–21
High-oleic sunflower oil (HOSNO)	–12
Canola oil[a]	–18
Castor oil[a]	–24
RBD palm oil (RBDPO)	25
RBD palm kernel oil (RBDPKO)	23
RBD palm olein (RBDPOo)	9
Palm oil methyl ester (PME)	12
Palm kernel oil methyl ester (PKOME)	–1
Palm kernel oil TMP ester (PKOTE)	0
Palm oil TMP ester (PMOTE)	3
High-oleic palm oil TMP ester (HOPMOTE)	–33
Commercial vegetable oil-based TMP ester	–14

[a] Ref. [71].
After ref. [70].

cosity fluids, the rate of wear will be proportional to viscosity. It has also been noted that the presence of FFAs in vegetable oils, which are polar compounds with —OH bonds, improves wear properties of these oils [75].

TMP esters synthesised from palm oil and palm kernel oil methyl esters show good potential as base stock in biodegradable lubricant formulations. Apart from high PP, other lubrication properties such as viscosity, VI, wear and friction properties are comparable to commercial hydraulic fluids [70]. In particular, their oxidative stability is superior to high-oleic vegetable oil, producing less deposit. The effects of impurities such as partial esters and unconverted methyl esters are positive in relation to lubrication properties, as long as they are maintained at moderate levels.

By careful processing and selection of feedstocks naturally based synthetic esters can be produced without some of the drawbacks of conventional natural esters. Outstanding Unichema synthetic base fluid products are Priolube® 2088 and 2089 oleic-based polyol esters and Priolube® 1421 made from fatty acids with a low level of PUFAs. The esters are prepared from natural oils and fats which are upgraded by hydrogenation, distillation or fractionation and modified by polymerisation. Suitable FAs and dicarboxylic acids are selected from the range for the production of high-quality synthetic esters for use by lubricant formulators. The main products are polyol/hindered esters and complex esters (Table 10.19).

The Unichema manufacturing process (Fig. 10.5) breaks down the various natural vegetable oil and animal fat triglycerides into crude FAs and glycerine. The fatty acids are then separated, refined and modified by a splitting, hydrogenation, distillation, fractionation and dewaxing process to produce pure stearic, oleic and other fatty acids. Additional modification by dimeri-

Table 10.19 Properties of Unichema synthetic esters as lubricant base oils

Property	Value	
	Polyol/hindered esters	Complex esters
Viscosity @ 40 °C (cSt)	17.5–13.8	400–47 000
VI	136–160	up to 175
Pour point (°C)	–3 to –60 °C	down to –30 °C
Nitrile rubber swell (%)	–2 to 22[a]	–1[d]
Biodegradability	>90%[b]	Poor
Evaporation loss (%)	0.8–8.0[c]	0.1–1.0

[a] Cf. mineral oil: –2.
[b] Cf. mineral oil: 30%.
[c] Cf. mineral oil: 9.1.
[d] In 168 h at 150 °C.

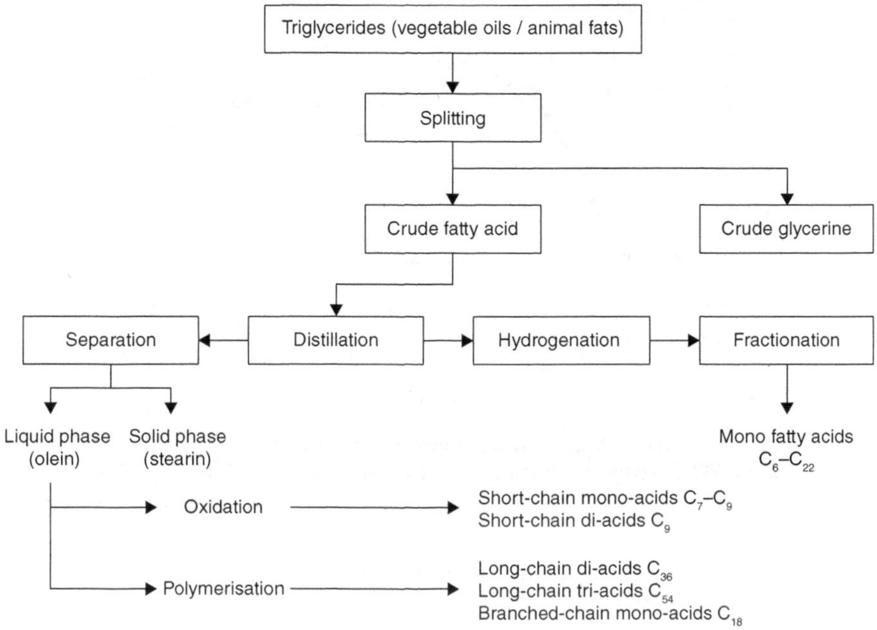

10.5 Unichema International process scheme for biolubricant base stocks. After ref. [36].

sation of oleic acid gives mono- and dicarboxylic acids, and polymerisation yields long-chain di- and tricarboxyl acids. Reaction with alcohols, in particular hindered polyols, produces esters. By selecting the appropriate FAs and alcohols, esters with the desired properties for each application can be made. These mono-, di-, poly- and complex esters vary from water soluble to mineral oil soluble depending on the choice of FA and alcohol. Mineral oil compatible esters are most important for the lubricants industry.

Table 10.19 shows some properties of these high-quality synthetic esters for use by lubricant formulators. Priolube® 2088/2089 are formulated into a biodegradable hydraulic fluid with a wide operating temperature range. Priolube® 1421 has enhanced oxidation stability for longer fluid lifetimes and extended service intervals. Both are WGK 0 rated and classify as Blue Angel hydraulic fluids. Figure 10.6 shows the improved low-temperature viscosity performance of Priolube® 2088/2089 esters with little viscosity increase even after 168 h at –30 °C. By contrast, the viscosity of normal trimethylolpropane oleate ester increases to 14 000 mm^2/s after only 24 h at –30 °C despite a pour point of below –50 °C.

Applications of Unichema's range of Priolube® lubricant base fluids are listed in Table 10.20. Polyol/hindered esters are for use in four- and two-stroke petrol and diesel engines, air compressors, in greases, as hydraulic

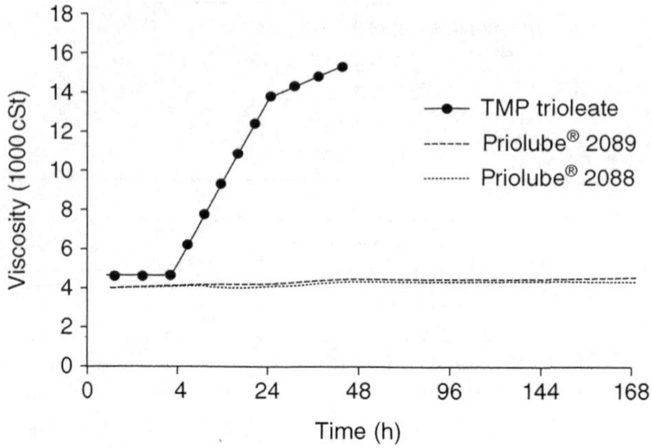

10.6 Viscosity stability at low temperature (–30 °C) of Priolube® 2088/2089 esters in comparison to TMP trioleate. After ref. [36].

Table 10.20 Applications of Unichema's range of Priolube® lubricants

• Hydraulic fluids
• Internal combustion engines
• Air and refrigerant compressors[a]
• Gear oils and automatic transmission fluids
• Metalworking fluids
• Textile machinery
• Food processing[b]
• Biodegradable greases
• Industrial equipment

[a] Compatible with HFC-134a and other HFC refrigerants.
[b] FDA approval required.
After ref. [36].

fluids and in gas turbine lubricants. There is also a range of diesters with similar applications. Complex esters with high viscosities are made as bases for gear oils and in lubricants for high-temperature engines. Unichema esters are used in synthetic 5W-30 and 10W-30 four-stroke engine crankcase lubricants and its polyol esters together with PAO synthetic hydrocarbons form the base for fully synthetic engine oils.

Bio-based polyol esters

The chemical structure of highly monounsaturated oils can be modified chemically to improve the functional properties of the oil. An example is a family of bio-based polyol esters that Cargill, Inc. has recently successfully introduced on the market and that has much improved low-temperature fluidity, is oxidative stable, has very high VI, low volatility and is biodegradable [34, 76]. Kodali has described a new class of bio-based synthetic esters derived from vegetable oils with better performance (in particular oxidative stability and excellent low-temperature flow properties) at a lower cost than synthetic esters [34]. This is possible due to recent advances in the biotechnology of vegetable oils and the chemical modifications that could be applied to convert these natural esters into high-performance biolubricants. A general chemical structure of such bio-based polyol esters in shown in Fig. 10.7. Despite the fact that these molecules contain >50% saturated fatty acyl chains the low-temperature flow properties and oxidative stability of these fluids are superior to the vegetable oils they are derived from.

Low-temperature properties of polyol esters can be improved by:

- changing chain lengths of the acyl/alkyl chains;
- introducing branching in the acyl/alkyl chains;
- changing the polyol backbone structure;
- mixing asymmetry of the backbone/acyl/alkyl chains.

The chemical structures of high monounsaturated vegetable oils modified by these principles produce bio-based polyol esters with superior functional properties (Table 10.21).

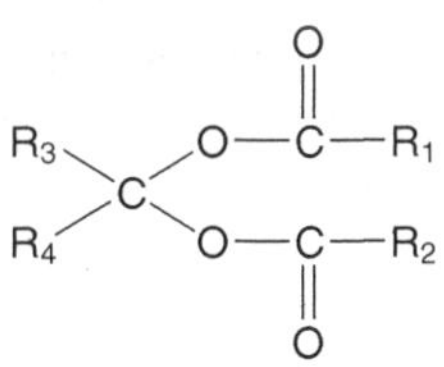

10.7 General structure of bio-based polyol esters. After ref. [76].

Table 10.21 Functional properties of bio-based polyol esters

• Excellent low-temperature fluidity (–25 °C)
• Good oxidative stability
• Very high viscosity index (180–210)
• Range of viscosities (18–46 cSt at 40 °C)
• Low volatility (<7%)
• Good boundary lubrication
• Readily biodegradable
• Cost effective

After refs [34, 76].

In US Patent No. 6,278,006 B1 (to Cargill, Inc.) Kodali and Nivens [76] disclose transesterified oil formulations of bio-based polyol esters obtained by transesterifying short saturated FA esters with triacylglycerol (TAG) containing oils, such as vegetable oils, to obtain oils with improved lubrication properties, in particular oxidative stability and low-temperature properties. Figures 10.8 and 10.9 show typical reaction schemes. The vegetable oil has a high MUFA content (70–90%), whereas the saturated short-chain fatty ester typically contains from six to ten carbons, is normal or branched, and can be a methyl ester or a polyol ester; for example, trimethylolpropane triheptanoate is a useful ester. The transesterification reaction (i.e. the exchange of an acyl group of one ester with that of another ester) randomises the distribution of short and long fatty acid chains among the molecular backbones. Transesterification of a short-chained polyol ester with a vegetable oil results in the short FA chains of the polyol and the long FA chains of the TAG being randomly distributed among both the non-glycerol and TAG polyol backbones. The heterogeneous bio-based (glycerol) polyol ester reaction products show interesting properties (Table 10.21). These polyol esters exhibit lower viscosities, low volatilitics and high VIs compared with vegetable oils. The very high oxidative stability of these glycerol esters is mainly due to very high concentrations of saturated and monounsaturated fatty acyl chains. Asymmetric molecules have a tendency not to crystallise.

Agri-Pure™-560 (AP-560) (Cargill, Inc., Minneapolis, MN) is a representative of the family of bio-based synthetic esters. The chemical, physical and biodegradability properties of unformulated AP-560 are given in Table 10.22. Formulations made from AP-560 are suitable in environmentally

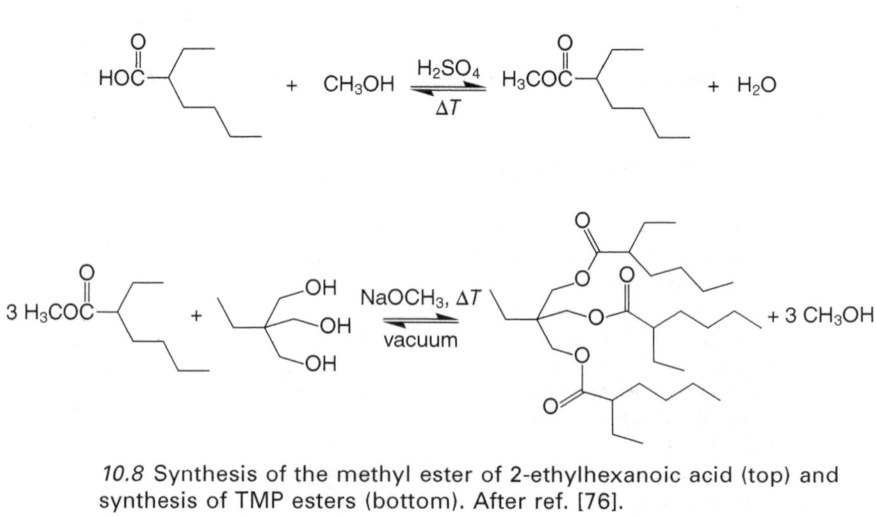

10.8 Synthesis of the methyl ester of 2-ethylhexanoic acid (top) and synthesis of TMP esters (bottom). After ref. [76].

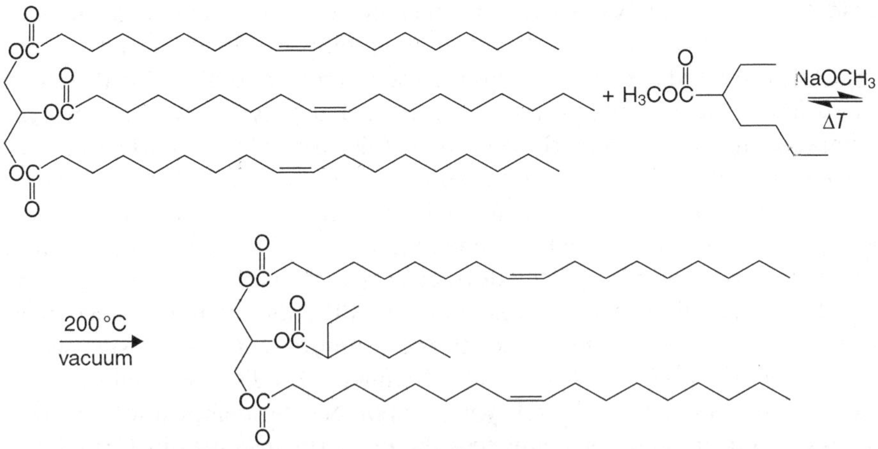

10.9 Transesterification of methyl esters with IMC-130 canola oil (Cargill, Inc.).

Table 10.22 Chemical and physical properties of unformulated Agri-Pure™-560 (Cargill, Inc.)

Property (units)	Result	Test method
Viscosity (mm^2/s)		ASTM D 445
40 °C	28.8	
100 °C	6.5	
–5 °C	230	
Viscosity index	190	ASTM D 2270
Specific gravity (g/cm^3)	0.924	ASTM D 1298
Noack volatility (%)	3.5	ASTM D 6375–99A
Flash point (°C)	275	ASTM D 92
Low-temperature behaviour at –25 °C	Stays fluid for >30 days	–
Oxidative stability (h)	>1500	ASTM D 943 Dry
PDSC at 180 °C (min)	50	ASTM D 6186–98
Biodegradability (%)	>95%	CEC L-33-A-94
Biodegradability (%)	>80%	ASTM D 5864–95

acceptable biodegradable hydraulic fluid applications, and meet the Caterpillar BF-1 specification. This spec includes an oxidative stability requirement of 1250 h by ASTM D 943 (modified method).

Hydroxyl esters

Polyol esters having unconverted hydroxyl groups from the reaction product of a polyol with a branched and/or linear acid exhibit enhanced thermo-oxidative stability, lower friction coefficient, reduced sediment formation, and lower wear compared with conventional fully esterified synthetic esters [58]. Unconverted hydroxyl groups are used to substantially delay the onset of oxidative degradation versus fully esterified polyol esters. Such ester compositions are particularly useful for fully formulated lubricating oils, such as aircraft turbine oils, and require only reduced amounts of antioxidants (e.g. arylamines). Mineral oil is insufficient to accommodate the extreme temperature variations encountered in aircraft turbine engines. Synthetic base stocks have been developed that can withstand the high temperatures and still provide a sufficiently low PP. Other characteristics to be considered include lubricity, viscosity and oxidation-corrosion resistance.

US Patent No. 5,698,502 (to Exxon Chemical Co.) discloses that incorporating somewhat high hydroxyl synthetic polyol ester compositions as base stocks into formulated oils enhances the properties versus oils formulated from fully esterified polyol esters in terms of improved thermo-oxidative stability in the liquid phase, comparable thermo-oxidative stability in the vapour phase, and no corrosion [58]. In particular, polyol ester base stocks with higher hydroxyl values (HV, 10–80 mg KOH/g, corresponding to 2–10% unconverted hydroxyl groups) show an optimum performance, provided that the hydroxyl values remain within critical ranges with respect to sediment formation. Figure 10.10 demonstrates the impact of hydroxyl value on HPDSC induction times for a series of linear acid TMP derivatives.

High hydroxyl esters are typically resistant to high-temperature oxidation. Having —CHROH functional groups bonded to the ester backbone is believed to allow these higher hydroxyl esters to cause the thermal-oxidative stability of the novel polyol ester compositions to increase.

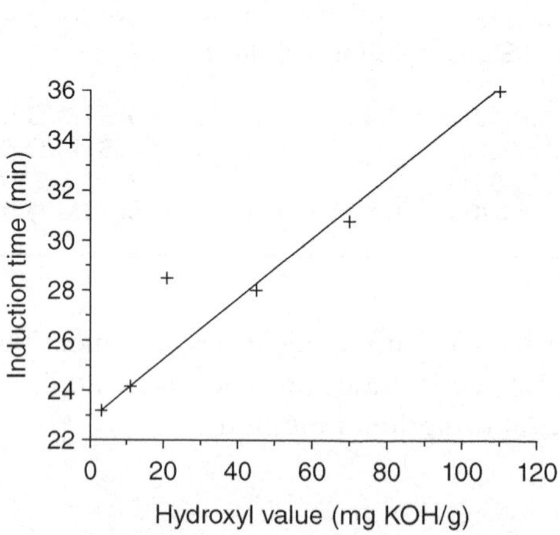

10.10 Effect of hydroxyl value of linear acid TMP compositions with unconverted hydroxyl groups (with antioxidants) on induction times according to the HPDSC test. After ref. [58].

The unconverted hydroxyl groups, which are bonded to the resultant ester moiety, act similarly to an antioxidant such that they transfer a hydrogen atom to the unstable carbon radical which is produced when the ester molecule is under thermal stress, thereby converting the radical to a stable alcohol and oxygen. These unconverted hydroxyl groups, which behave as internal antioxidants, can substantially reduce or even eliminate the need for addition of conventional and costly antioxidants to the polyol ester formulation. Aircraft turbine oil formulations typically comprise about 85–100 wt% of a synthetic polyol ester composition and 0–15 wt% additive package. The fact that polyol esters with unconverted hydroxyl groups also exhibit lower friction coefficients and wear volume than similar fully esterified polyol esters suggests that these polyol esters can also be used as antiwear agents or friction modifiers.

Polyester compositions with unconverted hydroxyl groups can also be used in the formulation of crankcase lubricating oils (i.e. passenger car motor oils, heavy-duty diesel motor oils, and passenger car diesel oils) for spark-ignited and compression-ignited engines.

10.5.4 Branched biodegradable lubricants

General wisdom holds that increased branching causes increased resistance to biodegradation [77–79]. Resistance becomes exceptionally great when quaternary branching occurs. However, Duncan and Meade [52] have shown that non-biodegradability of branched esters is more a function of steric hindrance around the ester linkage – site of enzymatic hydrolysis – than of the inability of microorganisms to break down tertiary and quaternary carbons. By relieving the steric hindrance around the ester linkage(s), biodegradation can more readily occur with branched esters. US Patent No. 5,817,607 (to Exxon Chemical Co.) teaches that biodegradable base stocks with good cold-flow properties, good solubility with dispersants and good lubricity can be formed by incorporating branching along the chains of the acyl and/or alcohol portions of the ester molecule [52]. Branching is needed to build viscosity without increasing molecular weight, and multiple isomers are helpful in attaining low-temperature properties. Furthermore, branched biodegradable lubricants provide the following advantages over all-linear biodegradable lubricants: (i) decreased PP; (ii) increased solubilities of other additives; (iii) increased detergency/dispersancy of the lubricant oil; and (iv) increased oxidative stability. The lubricants are useful in hydraulic fluid and catapult oil applications.

Even when synthesised from linear acids and highly branched alcohols, such as polyol esters of linear acids, high viscosity esters with good low-temperature properties can be difficult to achieve. In addition, PE esters of linear acids exhibit poor solubility with dispersants whereas TMP esters of low-molecular-weight linear acids (<C_{14}) do not provide sufficient lubricity. This lower quality of lubricity is also seen with adipate esters of branched alcohols. However, when both the alcohol and acid moieties of the ester are highly branched, such as with polyol esters of highly branched oxo acids, the resulting molecule tends to exhibit poor biodegradation as measured by the Modified Sturm test (OECD test No. 301B). Additionally, while highly branched acid esters offer outstanding thermo-oxidative properties, their viscosity indexes are frequently very low and their PPs fairly high. Difficulties may also be encountered in synthesis due to slow reaction rates caused by highly sterically hindered components.

In US Patent No. 5,817,607 Duncan and Meade [52] formulate synthetic polyol ester base stocks using mixed monocarboxylic branched and linear mono- or dicarboxylic acids (C_7–C_{10}) with no more than 10% of the branched acids (preferably methyl branches) containing a quaternary carbon. It is desirable to use a branched acid composed of multiple isomers. The proposed lubricants are typically a reaction product of technical grade PE (86–92% mono-PE, 6–12% di-PE and 1–3% tri-PE) with 30–70 mol% C_8–C_{10} linear acids (<20% di-acids) and 30–70% iso-C_8 (Cekanoic 8, Ck8) branched acids. Introduction of linear acids into the molecule relieves the steric crowding around the quaternary carbon of the branched alcohol. By having two branched acids and two linear acids on pentaerythritol, the enzymes have access to the ester linkages, which allows the first stage of biodegradation – hydrolysis of the ester – to occur. NPG can be totally esterified with 2-ethylhexanoic acid or an iso-C_8 acid and still maintain about 90% biodegradation according to the Modified Sturm test. Once two branched acids have been added to a branched polyol, the ester linkages begin to become crowded around the quaternary carbon of the branched alcohol. Additional branched acids added to the branched alcohol begin to lower the biodegradation of the molecule such that by the fourth addition of a branched acid to the branched alcohol the biodegradation of the resulting molecule drops from about 80% to less than 15% biodegradation as measured by the Modified Sturm test.

Table 10.23 compares the properties of a biodegradable branched synthetic polyol ester base stock of ref. [52] with various conventional biodegradable lubricants. The TPE/Ck8/C810 polyol ester is superior to RSO in cold-flow properties and stability, superior to di-tridecyladipate (DTDA) in stability, biodegradation and aquatic toxicity, and superior to TMP/iso-C18 in cold-flow properties, stability and biodegradation. Native RSO is highly biodegradable, exhibits very poor low-temperature properties, does not lubricate very well due to its instability, and breaks down in the engine causing deposit formation, sludge and corrosion problems. Polyol esters of low-molecular-weight linear acids do not provide lubricity, and those of high-molecular-weight linear or semi-linear acids have poor low-temperature

Table 10.23 Properties of biodegradable branched synthetic polyol ester base stock and conventional biodegradable lubricant base stocks[a]

Property	TPE /Ck8/ C810[b]	Rapeseed oil	DTDA[c]	TMP/iC18[d]
Pour point (°C)	–45	0	–54	–20
Flash point (°C)	274	162	221	n.g.
Viscosity (cps) @ –25 °C	3600	Solid	n.g.	358 000
(cSt) @ 40 °C	38.8	47.8	26.9	78.3
(cSt) @ 100 °C	6.7	10.2	5.3	11.9
Viscosity index	128	208	135	147
Oxidation induction time (min)[e]	16.0	2.1	3.9	4.3
Lubricity[f]	Pass	n.g.	Fail	Pass
Biodegradation (%)[g]	85	85	60	65
Toxicity (LC50, ppm)	>5000	>5000	<1000	n.g.
Solubility with dispersant	Soluble	n.g.	Soluble	n.g.
Acid number (mg KOH/g)	0.01	0.35	0.04	1.9
Hydroxyl number (mg KOH/g)	1.91	n.g.	1.49	n.g.

[a] After ref. [52].
[b] TMP, trimethylolpropane; Ck8, Cekanoic-8 acid; C810, mixture of n-C_6 to n-C_{12} acids.
[c] DTDA, di-tridecyladipate.
[d] Tri-ester of trimethylolpropane and isostearic acid.
[e] HPDSC.
[f] Yamaha engine.
[g] Modified Sturm test.
n.g., not given.

properties. Moreover, PE esters of linear acids are not soluble with polyamide dispersants. DTDA is only marginally biodegradable and does not provide lubricity. Polyol esters of TMP or PE and branched oxo acids do not biodegrade easily due to the steric hindrance around the ester linkage.

The biodegradable branched synthetic ester base stocks have been formulated for application as two-cycle engine oils, catapult oils, hydraulic fluids, drilling fluids, water turbine oils, compressor oils and greases. In all cases at least 60% biodegradation (as measured by the Modified Sturm test) has been achieved with branching along the chains of the acyl and/or alcohol portions of the synthetic ester with improved cold-flow properties and dispersant solubility.

It is important to notice that the lubricant compositions of US Patent No. 5,817,607 (to Exxon Chemical Co.) differ from the polyols formulated by Carr and DeGeorge [80] in US Patent No. 4,826,633 (to Hatco Chemical Co.) using iso-nonanoic acid. The latter highly stable formulations (HPDSC, 35–65 min), useful in gas turbine engines, are not biodegradable on account of the high content of quaternary carbons within the 3,5,5-trimethylhexanoic moiety. Conversely, the lubricants according to US Patent No. 5,817,607, in which no more than 10% of branched acids contain a quaternary carbon, are characterised by lower stability (HPDSC, 12–17 min), but are biodegradable.

Exxon Chemical has been particularly active in the introduction of commercially useful biodegradable branched synthetic ester base stocks with improved performance attributes [53, 54]. Compositions with high stability were based on esters derived from 3,5,5-trimethylhexanol and/or 3,5,5-trimethylhexanoic acid [55, 56]. Polyol esters of branched acids such as 3,5,5-trimethylhexanoic acid and 2-ethylhexanoic acid show higher oxidative stability (in HPDSC) than the corresponding polyol esters of straight chained acids [59]. Highly oxidative stable high-molecular-weight lubricants with a plurality of branch groups were also obtained by telomerisation of a complex oleate ester with vegetable oils containing at least one linoleyl moiety of the three groups in the triglyceride [81].

Esters of dicarboxylic acids with branching have been used as lubricants and hydraulic fluids over a wide temperature range [82, 83]. Similarly, other branched esters have been applied in hydraulic oils [84]. Branched C_{18} monoacid (isostearic), C_{36} diacid (dimer acid) and C_{54} triacid (trimer acid) are useful raw materials for the synthesis of high-viscosity fluid esters.

10.5.5 Origin of biodiesel lubricity

Biodiesel is an alternative diesel fuel obtained through transesterification of vegetable oils and animal fats with monohydric alcohols to give the corresponding monoalkyl esters (see also Section 6.2.1). As a result of the transesterification reaction, biodiesel contains small amounts of glycerol, FFAs, partially reacted acylglycerols (mono- and diacylglycerols), as well as residual starting material (triacylglycerols). These contaminating materials are limited by biodiesel specifications (EN 14214 and ASTM D 6751). Although the biodiesel standards do not currently contain lubricity specifications, it is nonetheless a critical fuel property. Fuels with poor lubricity can cause failure of diesel engine parts that rely on lubrication by the fuel, such as fuel pumps and injectors [85]. As such, lubricity specifications are included in petrodiesel standards in the United States (ASTM D 975) and EU (EN 590). Fuels with higher viscosities are more likely to cause formation of engine deposits. High viscosity is the major fuel property explaining why neat vegetable oils have been largely abandoned as alternatives to conventional diesel fuel. Compared with SME, partially hydrogenated methyl esters of soybean oil (PHSME) has better oxidative stability but inferior low-temperature properties, kinematic viscosity and lubricity [86].

Pure biodiesel (fatty acid methyl esters, FAMEs) has excellent lubricating properties, which even improves somewhat with chain length and the presence of double bonds. Biodiesel extends engine life by 25% because of the lubricity and cleaning powers of the fuel itself. Neat biodiesel possesses inherently greater lubricity than low-sulphur petrodiesel [87, 88]. The reason for the poor lubricity of the latter is not so much removal of the sulphur-containing compounds but rather reduction in polar compounds with other heteroatoms such as oxygen and nitrogen [89]. The change to low-sulphur diesel in all on-road applications since 1993 has given rise to recent interest in diesel fuel lubricity. Fatty compounds possess better lubricity than hydrocarbons, because of their polarity imparting O atoms. Significant lubricity improvement to low-sulphur petrodiesel is already achieved by low-volume (1–2%) blends of biodiesel, which seems to imply that the alkyl esters of biodiesel are responsible for this lubricity enhancement. However, individual FAMEs have little effect [90], whereas adding FFAs or other selected oxygenated compounds to low-lubricity petrodiesel at additive levels does enhance lubricity [91]. Neat FFAs, monoacylglycerols and glycerol possess better lubricity than neat esters, because of their free OH groups. An order of oxygenated moieties enhancing lubricity was established as follows: COOH > CHO > OH > $COOCH_3$ > C=O > C—O—C [85].

It is also known that the principal virtue of vegetable oils as lubricants is their superior ability to cling to metal surfaces in the form of very thin films. This property appears to be due largely to the surface activity conferred by the small amount of FFAs occurring in the oils. The boundary lubrication performance of FFAs in vegetable oils has extensively been studied [92–97]; see Section 5.2.3, p. 170. FFAs enhance the boundary lubrication behaviour of SNO formulations [96]. For the tribological performance of stearic acid; see again Section 5.2.3, p. 170. The FFAs are highly polar and tend to become adsorbed in layers of molecular dimensions at the metal–oil interface. It is also of interest to notice that esters of vegetable oils with hydroxylated FAs such as castor and *Lesquerella* oils improve lubricity at lower levels than the esters of non-hydroxylated vegetable oils [98, 99]. Also, oxidised biodiesel shows improved lubricity compared with its non-oxidised counterpart [100]. Wain and Perez have determined optimum blend ratios of oxidised biodiesel in a low sulphur diesel fuel to obtain maximum lubricity [100].

As commercial biodiesel improves lubricity of low-sulphur petrodiesel more than neat fatty esters, it is concluded that other biodiesel components cause lubricity enhancement at low biodiesel blend levels, in particular contaminants such as FFAs and monoacylglycerols [85]. Biodiesel's impact on engine lubrication is not yet completely understood [101].

10.6 Sources of further information and advice

Th. Mang, K. Bobzin and Th. Bartels, *Industrial Tribology: Tribosystems, Friction, Wear and Surface Engineering, Lubrication*, Wiley-VCH, Weinheim (2010).

L.R. Rudnick (ed.), *Lubricant Additives: Chemistry and Applications*, 2nd edn, CRC Press, Boca Raton, FL (2009).

J.S. Marth, *Renewable Lubricants Manual: Biobased Oils, Fluids & Greases*, United Bio Lube, Palo Alto (2007); http://www.renewablelubricants.com.

T.R. Lynch, *Process Chemistry of Lubricant Base Stocks*, CRC Press, Baton Rouge, FL (2007).

Th. Mang and W. Dresel (eds), *Lubricants and Lubrication*, 2nd edn, Wiley-VCH Verlag, Weinheim (2007).

L.R. Rudnick (ed.), *Synthetics, Mineral Oils, and Bio-based Lubricants: Chemistry and Technology*, CRC Press/Taylor & Francis, Boca Raton, FL/ New York, NY (2006).

G.E. Totten, S.R. Westbrook and R.J. Shah (eds), *Fuels and Lubricants Handbook: Technology, Properties, Performance, and Testing*, ASTM International, West Conshohocken, PA (2003).

J. Denis, J. Briant and J.-C. Hipeaux, *Lubricant Properties Analysis and Testing*, Éditions Technip, Paris (2000).

E.R. Booser (ed.), *Tribology Data Handbook*, CRC Press, Boca Raton, FL (1997).

R.C. Gunderson and A.W. Hart (eds), *Synthetic Lubricants*, Reinhold, New York, NY (1962).

10.7 References

[1] A. Hubmann, K. Reiter and W. Baumann, Engine oils with optimized environmental properties, *Mineralöltechnik* **39**(9) (1994) 1–14.

[2] K. Kollmann, T. Gürtler, H. Land, W. Warnecke and H.D. Müller, Extended oil drain intervals – conservation of resources or reduction of engine life (Part II), *SAE Techn. Paper Series* No. 981443 (1998).

[3] C.A. Amann, The stretch for better passenger car fuel economy, *SAE Techn. Paper Series* No. 972658 (1997).

[4] J.A. McGeetam, The pivotal role of crankcase oil in preventing soot wear and extending filter life in low emission diesel engines, *SAE Techn. Paper Series* No. 1999-01-1525 (1999).

[5] ISO 15380, *Lubricants, Industrial Oils and Related Products (Class L) – Family H (Hydraulic Systems) – Specification for Categories HETG, HEPG, HEES and HEPR*, International Organization for Standardization, Geneva (2002).

[6] S. Noll and R. Müller, Laboratory methods for testing lubricants, in *Lubricants and Lubrication* (Th. Mang and W. Dresel, eds), 2nd edn, Wiley-VCH, Weinheim (2007), pp. 715–35.

[7] T. Bartels, Mechanical-dynamic test methods for lubricants, in *Lubricants and Lubrication* (Th. Mang and W. Dresel, eds), 2nd edn, Wiley-VCH, Weinheim (2007), pp. 736–85.

[8] C.G.A. von Eberan-Eberhorst and M.A. Cantab, The Coordinating European Council (CEC) Reference Oil System: development, results, and challenges, *Mineralöltechnik* **45**(4) (2000) 1–27.

[9] E. Choi, K. Akiyama, T. Astida, K. Kado, F. Ueda and H. Otira, Engine testing comparison of the relative oxidation stability performance of two engine oils, *SAE Techn. Paper Series* No. 952530 (1995).
[10] S. Jahanmir, Examination of wear mechanisms in automotive camshafts, *Wear* **108**(3) (1986) 235–54.
[11] L.R. Rudnick, Testing methods for additive/lubricant performance, in *Lubricant Additives: Chemistry and Applications* (L.R. Rudnick, ed.), CRC Press, Boca Raton, FL (2009), pp. 669–83.
[12] L.R. Rudnick, Additives for industrial lubricant applications, in *Lubricant Additives: Chemistry and Applications* (L.R. Rudnick, ed.), CRC Press, Boca Raton, FL (2009), pp. 493–509.
[13] DIN 50322, *Kategorien der Verschleißprüfung*, Beuth Verlag, Berlin.
[14] W. Mach and A. Gervé, *IAVF Information Brochure*, Institut für angewandte Verschleißforschung, Karlsruhe (2000).
[15] M. Scherge, K. Pöhlmann and A. Gervé, Wear measurement using radionuclide technique, *Wear* **254** (2003) 801–17.
[16] M. Harperscheid and J. Omeis, Lubricants for internal combustion engines, in *Lubricants and Lubrication* (Th. Mang and W. Dresel, eds), 2nd edn, Wiley-VCH, Weinheim (2007), pp. 191–229.
[17] P.S. De Caro and A. Gaset, Lubricants based on renewable resources: which future for the development of this value chain?, *Actual. Chim.* **10** (1999) 9–14.
[18] CTVO-Net – Lubricants and Hydraulic Fluids (2000); http://www.dainet.de/fnr/ctvo.
[19] C.A. Holmes, *IENICA – European Summary Report 2000–2005*, QLK5-CT-2000-00111.
[20] A.V. Avellan and C. Alfos, Valorisation non alimentaire des huiles de friture usagées en tant que lubrifiants biodegradables, *Oléag., Corps Gras, Lipides* **12**(4) (2005) 290–4.
[21] REFLAX – Rational engineering of lipid metabolism in flax to produce branched-chain fatty acids, unusual fatty acids with potent biolubricant properties, Proj. QLK3-2000-00349 (2001–2004); http://ec.europa.eu/research/quality-of-life-/cell-factory/volume1/projects/qlk3-2000-00349en.html
[22] T. Foglia, G. Piazza, V. Wyatt, M.J. Haas and W. Marmer, ARS project, Production of value-added lipids, biofuels, and biobased products from fats and oils, Project # 1935-41000-066-00 (2004–2009); http://www.ars.usda.gov/research/projects/projects.htm?ACCN_NO=408597
[23] P.S. Lathi and B. Mattiasson, Green approach for the preparation of biodegradable lubricant base stock from epoxidized vegetable oil, *Appl. Catal. B: Environm.* **69** (2007) 207–12.
[24] W.W. Garmier and A.W. Rotondo (to Renewable Lubricants, Inc.), Biodegradable vegetable oil compositions, US Patent No. 6,383,992 B1 (7 May 2002).
[25] W.W. Garmier and A.W. Rotondo (to Renewable Lubricants, Inc.), Biodegradable vegetable oil compositions, US Patent No. 6,534,454 B1 (18 March 2003).
[26] S.J. Miller (to Chevron), Production of biofuels and biolubricants from a common feedstock, US Patent Appl. No. 2009/0084026 A1 (2 April 2009).
[27] http://www.terresolve.com/product.html

[28] A.E.F. Fontes, W.S. Nogueira, F.A.D. Sanitago and J.R. Gomes (to Petrobas), Process to obtain biolubricants and bioparaffins by hydroprocessing mixtures of wax obtained from renewable resources and waxes of mineral origin, Eur. Patent Appl. No. EP 1,927,642 A2 (4 June 2008).
[29] C. Leray, Waxes, in *Kirk-Othmer Encyclopedia of Chemical Technology*, Vol. 25, John Wiley & Sons, Inc., New York, NY (2006).
[30] A. Igartua, X. Fernández, O. Areitioaurtenaa, R. Luther, C. Seyfert, J. Rausch, I. Illarramendi, M. Berg, H. Schultheiß, B. Duffau, S. Plouseau and M. Woydt, Biolubricants and triboreactive materials for automotive applications, *Tribol. Intl.* **42** (2009) 561–8.
[31] A. Igartua, J. Barriga and A. Aranzabe (eds), *Biodegradable Lubricants,* Virtual Tribology Institute Edition, Eibar (2005).
[32] Frost & Sullivan, *Biolubricants – Global Technology Developments (Technical Insights)*, Palo Alto, CA (30 September 2008).
[33] I.-Ching Chiu, S. Gunsel and P. Lacey (to Pennzoil-Quaker State Co.), Environmentally friendly lubricants, PCT Int. Publ. No. WO 03/027212 A1 (3 April 2003).
[34] D.R. Kodali, High performance ester lubricants from natural oils, *Ind. Lubr. Tribol.* **54**(4) (2002) 165–70.
[35] M.J. Neale, *Lubrification and Reliability Handbook*, Newnes, Oxford (2000).
[36] B. Wilson, Lubricants and functional fluids from renewable sources, *Ind. Lubr. Tribol.* **50**(1) (1998) 6–15.
[37] S.J. Randles, Esters, in *Synthetic Lubricants and High-performance Functional Fluids* (L.R. Rudnick and R.L. Shubkin, eds), Marcel Dekker, Inc., New York, NY (1999), p. 63.
[38] H.L. West, *Major Developments in Synthetic Lubricants and Additives in Germany*, Section B1, Technical Information and Documents Unit, London (1948), pp. 40–60.
[39] H. Zorn, *Esters as Lubricants* (US Air Force translation), Report F-TS-957-RE (1946–47).
[40] E. Jantzen, The origins of synthetic lubricants: The work of Hermann Zorn in Germany. Part 1: Basic studies of lubricants and the polymerisation of olefins, *J. Synth. Lubr.* **12**(4) (1996) 283–301.
[41] H. Zorn, Chemischer Aufbau und physikalische Eigenschaften der Schmierstoffe, *VDI – Berichte* **20** (1957), 47.
[42] E. Jantzen, The origins of synthetic lubricants: The work of Hermann Zorn in Germany. Part 2: Esters and additives for synthetic lubricants, *J. Synth. Lubr.* **13**(2) (1996) 113–28.
[43] W.A. Zisman, Historical review: lubricants and lubrication, in *Synthetic Lubricants* (R.C. Gunderson and A.W. Hart, eds), Reinhold, New York, NY (1962), pp. 6–60.
[44] R.S. Barnes and M.Z. Fainman, Synthetic ester lubricants, *Lubr. Engng.* **13** (1957) 454–7.
[45] C.M. Murphy and W.A. Zisman, Noninflammable hydraulic fluids, *Lubr. Engng.* **5** (1949) 231–5, 264–9.
[46] C.M. Murphy and W.A. Zisman, Structural guides for synthetic lubricant development, *J. I&EC* **42** (1950) 2415–20.

[47] G.E. Bohner, J.A. Krimmel, J.J. Schmidt-Collerus and R.D. Stacy, Properties of polyester fluids with desirable synthetic lubricant characteristics, *J. Chem. Engng. Data* **7**(4) (1962) 547–53.
[48] S.J. Metro, E.M. Hoffman and A.H. Matuszak (to Esso Research & Engineering Co.), Synthetic ester, Can. Patent No. 859,771 (29 December 1970).
[49] T.S. Chao, M. Kjonaas and J. DeJovine, Esters from branched-chain acids and neopentylglycols and phenols as base fluids for synthetic lubricants, *I&EC Prod. Res. Dev.* **22** (1983) 357–62.
[50] H.S. Aldrich, K.E. Edwards, K.L. Ingold, L. Gschwender and C. Snyder, Insights into oxidation kinetics for degradation pathways, *Proc. 9th EUCHEM Conf. Organic Free Radicals*, Grottaferrata (4–9 July 1998).
[51] E.J. Hamilton Jr, S. Korcek, L.R. Mahoney and M. Zinbo, Kinetics and mechanism of the autoxidation of pentaerythrityl tetraheptanoate at 180–220 °C, *Intl. J. Chem. Kin.* **12**(9) (1980) 577–603.
[52] C.B. Duncan and L.K. Meade (to Exxon Chemical Co.), Biodegradable branched synthetic ester base stocks and lubricants formed therefrom, US Patent No. 5,817,607 (6 October 1998).
[53] C.B. Duncan and L.K. Meade (to Exxon Chemical Co.), Biodegradable branched synthetic ester base stocks and lubricants formed therefrom, US Patent No. 5,767,047 (16 June 1998).
[54] C.B. Duncan and L.K. Meade (to Exxon Chemical Co.), Biodegradable branched synthetic ester base stocks and lubricants formed therefrom, US Patent No. 5,681,800 (28 October 1997).
[55] R.H. Schlosberg, D.W. Turner, M.A. Krevalis, W.J. Munley Jr and H.S. Aldrich (to Exxon Chemical Co.), High stability and low metals esters based on 3,5,5-trimethyl-1-hexanol, US Patent No. 5,798,319 (25 August 1998).
[56] R.H. Schlosberg, L.D. Sherwood-Williams, H.S. Aldrich and J.S. Szobota (to Exxon Chemical Co.), Polyol ester compositions with unconverted hydroxyl groups, US Patent No. 5,665,686 (9 September 1997).
[57] R.H. Schlosberg, H.S. Aldrich, L.D. Sherwood-Williams, J.S. Szobota, M.A. Krevalis, D.P. Leta, D.G.L. Holt and F.H. Gordon, Polyol ester compositions with unconverted hydroxyl groups, US Patent No. 5,744,434 (28 April 1998).
[58] B.J. Pafford, J.T. Kim, P.E. Godici, H.S. Aldrich, R.H. Schlosberg and M.A. Krevalis (to Exxon Chemical Co.), Polyol ester compositions with unconverted hydroxyl groups for use as lubricant base stocks, US Patent No. 5,698,502 (16 December 1997).
[59] R.H. Schlosberg, J.W. Chu, G.A. Knudsen, E.N. Suciu and H.S. Aldrich, High stability esters for synthetic lubricant applications, *Lubr. Engng.* **57**(2) (2001) 21–6.
[60] J.A. Walker and W. Tsang, Characterization of lubricating oils by differential scanning calorimetry, *SAE Techn. Paper Series* No. 801383 (1980).
[61] M. Wakakura and T. Sato, High pressure thermal analysis of synthetic lubricating oils. Part 1. Estimation of fire-resistance and oxidation stability, *Sekiyu Gakkaishi (J. Jap. Petr. Inst.)* **24**(6) (1981) 385–92.
[62] A. Zeman, R. Stuwe and K. Koch, The DSC cell – a versatile tool in the study of thermal-oxidative degradation of lubricants and related problems, *Thermochim. Acta* **80**(1) (1984) 1–9.

[63] E. Uosukainen, Y.-Y. Linko, M. Lämsä, T. Tervakangas and P. Linko, Transesterification of trimethylolpropane and rapeseed oil methyl ester to environmentally acceptable lubricants, *J. Am. Oil Chem. Soc.* **75**(11) (1998) 1557–63.

[64] S.S. Lawate and K. Lal (to Lubrizol Corp.), High oleic polyol esters, compositions and lubricants, functional fluids and greases containing the same, Eur. Patent No. EP 0,712,834 A1 (14 November 1995).

[65] V. Eychenne, Z. Mouloungui and A. Gaset, Neopentylpolyol ester-based lubricants, *Oléag., Corps Gras, Lipides* **3** (1996) 57–63.

[66] V. Eychenne, Z. Mouloungui and A. Gaset, Total and partial erucate of pentaerythritol. Infrared spectroscopy study of relationship between structure, reactivity, and thermal properties, *J. Am. Oil Chem. Soc.* **75**(2) (1998) 293–9.

[67] S. Gryglewicz, W. Piechocki and G. Gryglewicz, Preparation of polyol esters based on vegetable and animal fats, *Bioresour. Technol.* **87** (2003) 35–9.

[68] R. Yunus, A. Fakhru'l-Razi, T.L. Ooi, S.E. Iyuke and A. Idris, Development of optimum synthesis method for transesterification of palm oil methyl esters and trimethylolpropane to environmentally acceptable palm oil-based lubricants, *J. Palm Oil Res.* **15**(2) (2003) 35–41.

[69] R. Yunus, A. Fakhru'l-Razi, T.L. Ooi, S.E. Iyuke and A. Idris, Preparation and characterization of trimethylolpropane esters from palm kernel oil methyl esters. *J. Oil Palm. Res.* **15**(2) (2003) 42–9.

[70] R. Yunus, A. Fakhru'l-Razi, T.L. Ooi, S.E. Iyuke and J.M. Perez, Lubrication properties of trimethylolpropane esters based on palm oil and palm kernel oils, *Eur. J. Lipid. Sci. Technol.* **106**(1) (2004) 52–60.

[71] D.E. Weller Jr. and J.M. Perez, A study of the effect of chemical structure on friction and wear. Part 2. Vegetable oils and esters, *Lubr. Engng.* **57**(5) (2001) 20–6.

[72] D.E. Weller Jr. and J.M. Perez, A study of the effect of chemical structure on friction and wear. Part 1. Synthetic ester base fluids, *Lubr. Engng.* **56**(11) (2000) 39–44.

[73] S. Asadauskas and S.Z. Erhan, Depression of pour points of vegetable oils by blending with diluents used for biodegradable lubricants, *J. Am. Oil Chem. Soc.* **76** (1999) 313–16.

[74] J.C.J. Bart, N. Palmeri and S. Cavallaro, *Biodiesel Science and Technology. From Soil. to Oil*, Woodhead Publishing, Ltd/CRC Press, Cambridge/Boca Raton, FL (2010).

[75] D. Swern, Miscellaneous oils and fat products, in *Bailey's Industrial Oil and Fat Products* (D. Swern, ed.), 4th edn, John Wiley & Sons, Inc., New York, NY (1982), Vol. 2, pp. 335–40.

[76] D.R. Kodali and S.C. Nivens (to Cargill, Inc.), Transesterified oils, US Patent No. 6,278,006 B1 (21 August 2001).

[77] S.J. Randles and M. Wright, Environmentally considerate ester lubricants for the automotive and engineering industries, *J. Synth. Lubr.* **9**(2) (1992) 145–61.

[78] R.D. Swisher, *Surfactant Biodegradation*, 2nd edn, Marcel Dekker, Inc., New York, NY (1987), pp. 415–17.

[79] N.S. Battersby, S.E. Pack and R.J. Watkinson, A correlation between the biodegradability of oil products in the CEC L-33-T-82 and Modified Sturm tests, *Chemosphere* **24**(12) (1992) 1989–2000.
[80] D.D. Carr and N. DeGeorge (to Hatco Chemical Corp.), Synthetic lubricant base stock of monopentaerythritol and trimethylolpropane esters, US Patent No. 4,826,633 (2 May 1989).
[81] A.J. O'Lenick Jr. (to Lambent Technologies, Inc.), Telomerized complex ester triglycerides, US Patent No. 5,985,806 A (16 November 1999).
[82] D.C. Kenneth, V.E. Sohns, R.B. Perkins Jr and E.L. Huffman, Brassylic acid: chemical intermediate from high-erucic oils, *I & EC Prod. Res. Dev.* **16** (1977) 95–100.
[83] R.G. Kadesch, Fat-based dibasic acids, *J. Am. Oil Chem. Soc.* **56**(11) (1979) 845A–9A.
[84] S. Busch and J. Röder (to Cognis), Esters and mixtures thereof and their application as lubricants or in hydraulic oils, Eur. Patent No. EP 1,958,932 A1 (20 August 2008).
[85] G. Knothe and K.R. Steidley, Lubricity of components of biodiesel and petrodiesel. The origin of biodiesel lubricity, *En. Fuels* **19** (2005) 1192–1200.
[86] B.R. Moser, M.J. Haas, J.K. Winkler, M.A. Jackson, S.Z. Erhan and G.R. List, Evaluation of partially hydrogenated methyl esters of soybean oil as biodiesel, *Eur. J. Lipid. Sci. Technol.* **109**(1) (2007) 17–24.
[87] A.K. Bhatnagar, S. Kaul, V.K. Chhibber and A.K. Gupta, HFRR studies on methyl esters of nonedible vegetable oils, *En. Fuels.* **20** (2006) 1341–4.
[88] G. Knothe, The history of vegetable oil-based diesel fuels, in *The Biodiesel Handbook* (G. Knothe, J. Van Gerpen and J. Krahl, eds), AOCS Press, Champaign, IL (2005), pp. 4–16.
[89] D. Wei and H.A. Spikes, The lubricity of diesel fuels, *Wear* **111** (1986) 217–35.
[90] D.P. Geller and J.W. Goodrum, Effects of specific fatty acid methyl esters on diesel fuel lubricity, *Fuel.* **83** (2004) 2351–6.
[91] E. Kenesey and A. Ecker, Oxygen compounds for improvement of the lubricity in fuels, *Tribol. Schmierungstech.* **50** (2003) 21–6.
[92] Y. Cao, L. Yu and W. Liu, Study of the tribological behavior of sulfurized fatty acids as additives in rapeseed oil, *Wear* **244** (2000) 126–31.
[93] Y. Cao and L. Yu, Effect of alcohols and fatty acids as additives on the antiwear and extreme pressure properties of rapeseed oil, *Mocaxue Xuebao/Tribol.* **20**(4) (2000) 288–91.
[94] I. Minami, H.-S. Hong and N.C. Mathur, Lubrication performance of model organic compounds in high oleic sunflower oil, *J. Synth. Lubr.* **16** (1999) 3–12.
[95] J. Vižintin, A. Arnšek and T. Ploj, Lubricating properties of rapeseed oils compared to mineral oils under a high-load oscillating movement, *J. Synth. Lubr.* **17**(3) (2000) 201–17.
[96] N.J. Fox, B. Tyrer and G.W. Stachowiak, Boundary lubrication performance of free fatty acids in sunflower oil, *Tribol. Lett.* **16**(4) (2004) 275–81.
[97] I.-ul-Haq, M. Farooq and N. Muhammd, Some studies on the use of vegetable oils as environmentally-friendly lubricants, *Tribol. Online* **6**(5) (2011) 230–4.

[98] D.C. Brown, K. Harper and E. Frame, Screening vegetable oil alcohol esters as fuel lubricity enhancers, *J. Am. Oil Chem. Soc.* **78**(6) (2001) 579–84.
[99] J.W. Goodrum and D.P. Geller, Influence of fatty acid methyl esters from hydroxylated vegetable oils on diesel fuel lubricity, *Bioresour. Technol.* **96**(7) (2005) 851–5.
[100] K.S. Wain and J.M. Perez, Oxidation of biodiesel fuels for improved fuel lubricity, *Proc. Internal Combustion Engine Division* (Lubrication and Friction Committee ASME, Rockford, IL), **38** (2002) 27–34.
[101] G.M. Parsons, Biodiesel and engine lubrication. Part I, *Lubr. Mag.* (October 2007); Part II, *Lubr. Mag.* (December 2007), Chevron Products Company, San Ramon, CA.

11
Environmental life-cycle assessment (LCA) of lubricants

Abstract: Lubricants impact on the environment from raw material extraction to manufacturing, usage and disposal or reuse. Life-cycle assessment (LCA) is utilised for evaluation of the environmental impacts of energy and material flows throughout the product life. Several comprehensive LCAs have compared mineral oil-based and bio-based lubricants, reporting varying conclusions. Environmental savings of biolubricants are more apparent at a local level than at a global scale. Negative impacts mainly stem from their agricultural production and performance in the use phase. LCAs should be used in industrial product design of biolubricants. Sustainable development indicators also consider the socio-economic impact of a product. Optimised lubricants determine significant energy savings.

Key words: life-cycle assessment (LCA), environmental impact, product design, sustainable development, energy efficient lubricants.

11.1 Introduction

The use of alternatives for petro-based mineral oils should improve the quality of a product towards more sustainability; in no case should it worsen the product quality. Qualities to maintain or improve include the prolonged lifespan, health and safety functions during use (lubricants increase safety and reduce wear of machinery), biodegradability, ecotoxicity, reduced material consumption, lower mobility and energy efficiency. The quality of lubricants is co-determined by environmentally aware customers and regulators who demand certain characteristics.

Ecology should be considered on an equal footing with other factors in evaluating products and production methods [1]. In many cases, current manufacturing processes are not adequately clean and produce substances which have the potential to pollute air, water and soil, as shown in Fig. 11.1 for a general manufacturing process. Lubricants should be non-toxic to human health through contact and inhalation during production and usage. Accidental and usage losses are highly damaging to the environment. Approximately 30–50% of all traditional lubricants, mostly not completely biodegradable mineral-based oils with high toxic content compared with bio-based options, are released into the environment (especially water sources) during use, spills and disposal [2, 3]. Several billion dollars have

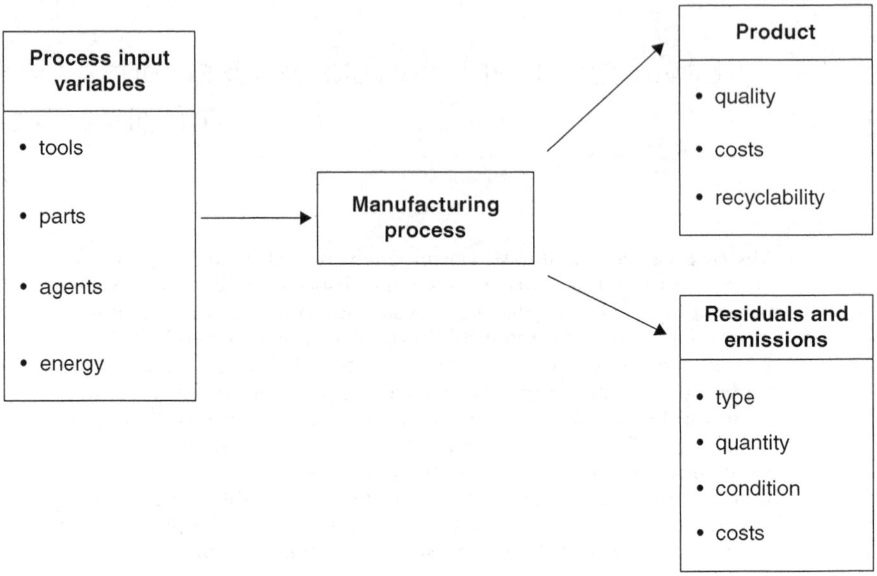

11.1 Balance of a manufacturing process.

been spent in North America alone to clean up accidental lubricant oil spills. A healthy environment is essential to long-term prosperity and quality of life. Today's citizens demand a high level of environmental protection. High environmental standards are an engine for innovation and business opportunities. Reduced emissions are directly related to changes in lubricant specifications and quality. One of the major advantages of recently developed bio-based synthetic esters is better performance at a lower cost compared with synthetic esters [4].

Lubricants impact on the environment at all stages of production, usage and disposal [5]. Awareness and concern over the usage of petroleum-based products and their impact on the environment have created an opportunity to produce environmentally acceptable lubricants (EALs). Lubricant manufacturers operating in the EAL market need to consider total product stewardship, requiring the 'cradle-to-grave' approach to be used in product design [6]. Consideration must therefore be given to aspects such as the origin of the raw materials, and the production, packaging, transportation, application and disposal of the product.

Concern over the usage of petroleum-based lubes in relation to preservation of the environment has been the main driver for the development of a wide variety of environmentally compatible products, mostly derivatives from agricultural feedstocks. Substitution of petroleum-based lubricants with bio-based (i.e. plant-derived) lubricants for industrial applications appears to be an attractive option because biolubricants are biodegradable, relatively non-toxic, more easily extracted and processed than petrolubricants, and politically more independent. There are, however, several aspects of the use of biolubricants that have to be addressed in order to better estimate the societal, environmental, and technological benefits and impacts of widespread use of these substitutes. These aspects fall into three areas: performance (see Chapter 12), regulatory (see Chapter 9) and life cycle (see Section 11.2). It is accepted that potentially more than 90% of all lubricants could be manufactured by harvestable raw materials.

As shown in Chapter 9, in many countries government policies, laws and initiatives aim at promoting the purchase and use of bio-based, environmentally preferable, energy- and water-efficient products that reduce hazards and toxicity. As for lubricants, in the United States the Department of Agriculture (USDA) manages the BioPreferred^{SM} Biobased Products Catalog that includes more than 200 lubricants in the following categories: chain and cable lubricants, food-grade lubricants, forming lubricants, gear lubricants, greases (see Table 13.6), hydraulic fluids penetrating lubricants, metalworking fluids and two-cycle engine oils.

It is often assumed that bio-based products are environmentally preferable to petroleum products due to their renewable nature. To determine the validity of this assumption, a comparative life-cycle assessment (LCA) allows a quantitative comparison of the energy and material flows throughout the stages of each product, from inception to disposal or reuse [7]. Environmental management tools are important non-technical drivers for growth of the EAL market segment. Biolubricants are employed to various degrees in order to minimise the life-cycle environmental impacts compared with mineral-based lubricants. Research on bio-based lubricants has generated varying conclusions regarding the environmental effects of these products. Biolubricants can provoke negative impacts on the environment, caused in part by the agricultural production of their raw materials.

11.2 Life-cycle assessment (LCA)

LCA is one of several tools of environmental management (e.g. risk assessment, environmental impact assessment and environmental auditing). It is a tool used to determine the major sources of environmental impacts caused by a product or process throughout its life cycle. The stages considered in the lifetime of a product are raw material extraction, transportation, manufacturing, use and disposal. Recycling or reuse of materials can also be envisaged. LCA quantifies raw materials inputs; energy inputs; emissions to air, soil, and water; and waste generated at every stage of the life of the product or process. The result of LCA is a detailed strengths, weaknesses, opportunities and threats (SWOT) analysis of the environmental effects. Demand for LCA methodology has come originally (and increasingly so since the early 1990s) from manufacturers in search of ecolabeling. Meanwhile, LCAs have become a matter of vital importance to the chemical industry and are now considered as valuable business planning and marketing tools. LCA presents both opportunities and threats. Products with improved environmental performance have much to gain.

The LCA methodology is standardised by ISO 14040-44 [8–12]. Other organisations, such as the US Environmental Protection Agency (EPA), the Society of Environmental Toxicology and Chemistry (SETAC), and the German Institute of Standardisation (DIN) [13], have also developed criteria to standardise LCA procedures. ADEME (France) has recently worked out a simplified LCA methodology adapted to bioproducts [14]. Basically, LCA is conducted in four steps (see Fig. 11.2):

- definition of goal and scope (sets objectives and system boundaries);
- life-cycle inventory (LCI) analysis (documents material and energy flows);
- life-cycle impact assessment (LCIA) (characterises the environmental effects);
- interpretation and improvement analysis (identifies areas for reduction of environmental impact).

In the first stage of an LCA (goal and scope definition) modelling specifications are selected. The objectives are precisely defined as are the frame of the investigation, time, spatial and physical system boundaries, environmental impacts to be considered, selected procedures, definition of the functional unit, level of detail of the data, and other facets. In any LCA it is essential to select an adequate functional unit, which allows results from

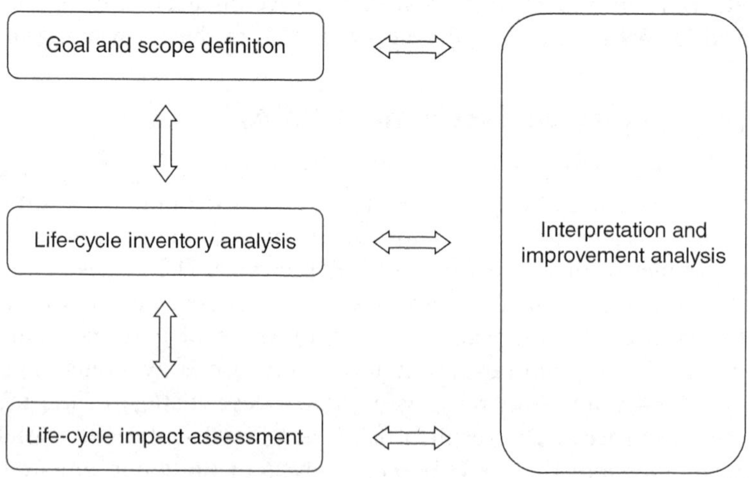

11.2 LCA phases according to ISO 14040.

the different stages and from the different feedstocks to be expressed quantitatively and to be compared. One of the most important elements is to define the life-cycle boundaries. The system boundaries establish which life-cycle stages are included in the LCA. Also natural, geographical, temporal and technical boundaries need to be defined [15]. A geographical boundary is relevant to LCA since various locations may show specific differences in infrastructure (and costs) and sensitivities of ecosystems (legislation). Environmental impacts are evaluated based on impact categories that affect human health, ecosystem health, resource depletion and social welfare. A variety of impact categories may be considered such as global warming, acidification, eutrophication and land use.

Owing to its great complexity and time-consuming nature, an LCA may not include all of the life-cycle stages or all of the aforementioned LCA phases. LCAs may model a system from 'cradle-to-grave', i.e. from raw material extraction to end-of-life (EOL) or disposal to establish the consumption of energy, materials used and emissions [15]. In so-called 'cradle-to-gate' modelling the use and disposal stages are not included in the analysis, which is consequently focused more on the manufacturing or production stage of the system's life cycle. In 'cradle-to-cradle' assessments, products are reused or recycled instead of being disposed of in landfills, making the system essentially waste free [16]. Since all tools for environmental management have their limitations, it is of importance to understand these in a specific LCA study. It is always necessary to keep the focus on the system boundaries and assumptions used when comparing different LCA studies.

LCI is the modelling phase of the LCA and consists of several steps, starting with the development of detailed flow diagrams for the products to be analysed, and is followed by qualitative and quantitative data collection. Some much used LCI databases are US LCI, GREET (Greenhouse Gases, Regulated Emissions, and Energy Use in Transportation), SimaPro, ecoinvent and BUWAL [17–23], which may be used to generate inventory data by calculation of the consumption of total energy, fossil fuels, petroleum, coal and natural gas, greenhouse gas (GHG) and air pollutant emissions. For each individual process, the expenditures (use of energy, equipment and raw materials) and all of the environmental effects (e.g. emissions) are determined and compiled for complete life-cycle comparison. Data used should be validated, as required per ISO 14041 [9]. Inputs and outputs do not only include resource use, energy use, emissions to air and water, but also allocations (to co-products).

In the LCIA stage [10], the LCI results (individual inventory parameters) are merged into impact categories to obtain environmental impacts from the individual entities. There are a host of LCIA software tools available including TRACI (Tool for the Reduction and Assessment of Chemical and

Other Environmental Impacts), Eco-indicator 99 and EDIP/UMIP [24–31], which match the LCI to the impact categories, and calculate the impacts caused by the system being evaluated. These tools vary by the inclusion of different impact categories, characterisation factors, and the use of normative factors. For example, TRACI, developed by the US Environmental Protection Agency (EPA) [24], is a tool which can be used to determine the effects of pollutants on some 12 impact categories (acidification, ecotoxicity, eutrophication, global warming, human health criteria, human health – cancer, human health – non-cancer, ozone depletion, photochemical smog, fossil fuel use, land use and water use).

Interpretation and improvement analysis is the fourth and final step of the LCA (see Fig. 11.2). This requires evaluation of the environmentally relevant parameters which have been determined. As the various products compared will show different environmental aspects (pros and cons) it is frequently not immediately obvious which product or process is best in terms of ecological relevance. Making hard decisions throughout an LCA is a difficult task owing to complexity, uncertainty, multiple conflicting objectives and different perspectives [32]. LCA results can be utilised in developing a framework for decision making. The decision-making process may be influenced by external (socio-economic) factors. For example, preservation of fossil energy sources and the greenhouse effect currently have the greatest political significance in Europe. Decision analysis can possibly result in a recommended course of action.

Performing an LCA is an elaborate process where data is often unavailable or varies significantly due to location, approach and tools utilised. Additionally, allocation is frequently an obstacle in obtaining accurate results. All LCAs have their limitations that could alter the outcome. LCAs are also characterised by a great heterogeneity among the methodologies used, as well as by a lack of transparency in the results obtained. The output from an LCA is a rather self-contained view, valid under well-defined constraints. The impacts associated with the running of machinery can be severe and depend on the application. Spills of oil result in ecological and environmental damage. Plants and animals may be burned or scorched by significant hot spills. Such local impacts are not adequately reflected in any of the standard environmental impact categories used in life-cycle assessments. Also, LCAs do normally not include the economic and social aspects of a product. LCA studies may be combined with cost–benefit analysis (CBA) to describe the environmental and socio-economic impacts of replacing mineral oils with environmentally acceptable lubricants. The use of bio-based lubricants within a system is not a licence to care less about spillage into the environment and minimise maintenance procedures. Although spillage of biodegradable oils may result in a faster recovery, it still causes environmental damage.

LCAs generally concur in indicating that products based on renewable resources usually are more ecologically compatible when compared to petrochemical analogues, but this requires assessment for any new product.

11.2.1 LCAs of biolubricants

The adoption of biolubricants is based on their performance and better environmental credentials. The lubrication performance of biolubricants formulated for various applications must be measured and compared with petro-based counterparts in order to determine such qualifying factors as material wear, lubricant stability and quantitative needs. For describing the ecological (dis)advantages of biolubricants compared with conventional lubricants, LCA, which was developed for comparing industrial products and systems, is especially suitable. The lower ecotoxicity and higher biodegradability combined with lower vapour pressure of the higher molecular weight components of biolubricants suggest that solid waste issues and volatile organic compound (VOC) emissions for biolubricants will be less severe than for petrolubricants. Quantification of the overall environmental impact of lubricants requires a detailed LCA.

Comprehensive lubricant LCAs are complicated for a variety of reasons. Lubricants are typically manufactured as co-products in integrated production networks, based on petroleum refining, oleochemical refining or chemical processing. Consequently, allocation of resource requirements and environmental impacts to lubricants in such networks is difficult. Life-cycle stages considered during biolubricant production include the amount and source of energy used in the biorefinery, the production of chemical and other biorefinery inputs, and the conversion process itself. Moreover, since lubricants differ greatly in performance and in the amount required for any specific application, detailed knowledge is required for the definition of an appropriate functional unit for comparison. As a result, a simple comparison of lubricant types, based only on their resource requirements per standard weight or volume unit, may give misleading results. Finally, many applications do not allow easy EOL tracing of lubricants. For example, whereas the EOL treatment of industrial lubricants used in manufacturing sites is usually well documented, this is not the case of used car engine oils. The actual fate of a large proportion of the overall lubricant production, and therefore the environmental impact, is not well known.

LCA was originally developed for industrial systems, where processes are usually carefully structured and controlled, with known or measurable material and energy flows. Agricultural processes differ from most industrial processes because of inconsistent material fluxes in terms of uncertain input variables (e.g. nitrogen fixation and soil mineralisation), emission profiles, the time scale to produce a product, and an inherently high degree of variability and uncertainly due to agricultural practices, soil type, geography, climate, etc. To reduce variability among parameters when examining agricultural systems, average data are mostly used or the analysis is limited to a relatively small geographical area. Incorporation of data variability in LCA provides a more comprehensive system description. LCA can be used to estimate the many implications of expansion of agricultural production of plants as feedstocks for biolubricants. Agricultural production is itself dependent on the use of petroleum products.

Comprehensive life-cycle comparisons of lubricants are rather few (see Table 11.1) and concern particular application areas such as hydraulic equipment for forestry applications [33–35] and municipal cleaning [33], chainsaws [35], domestic refrigerators [36], metalworking [37, 38], re-refined products [39, 40] and engine oils [41]. LCAs can be used as a basis for improvement of the various stages in development of lubricants. For the development of EALs, LCA methodology is a useful tool in gaining an oversight of environmental impact. LCA has not yet extensively been used in the development of lubricants. Detailed analyses of the environmental impact of various industrial products, carried out by lubricant manufacturers, are not always available in the open literature; see, however, refs [34, 39, 41–45].

The goal of most studies is a comparative LCA of biolubricants derived from plant-based oils (mainly rapeseed oil (RSO) and soybean oil (SBO)) and traditional lubricants made from mineral oil. Figure 11.3 shows simplified flowcharts of the life-cycle stages for RSO and SBO biolubricants, and for production of mineral-based lubricants. The main life-cycle stages of a biolubricant include farming, milling, lubricant manufacture, use and disposal. Renewable feedstock production comprises the production of farm or field inputs, field preparation activities, planting and establishment activities, feedstock harvesting and collection. For traditional lubricants the life-cycle stages include crude oil recovery, crude refining, lubricant refining, use and disposal.

The first step in the LCI is to develop *detailed* flowcharts for the products to be analysed. The detailed life cycle of a biolubricant is shown in Fig. 11.4. In the first stage, farming equipment is used to prepare the land, plant the seeds, apply fertilisers, herbicides and pesticides, irrigate and harvest the oil seeds [33, 37]. Plant residues are assumed to be left on the field. The milling stage consists of drying, heating and rolling the seed oils before the oil extraction process. Hexane is used as a solvent to allow further extraction of the oil, and is then recovered through distillation from the oil produced, and by desolventising the meal by-product [33]. The vegetable oil is refined by removing gums with hot water or steam, neutralising free fatty acids with an alkali solution, removing excess water, adsorbing coloured substances, and deodorising through distillation. The oil extracted from oilseed crops

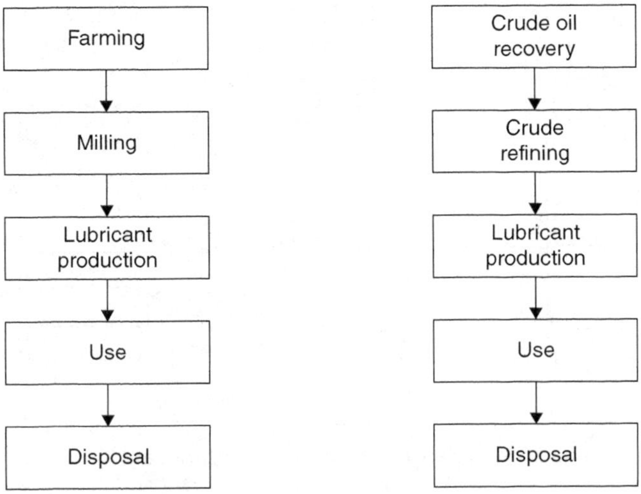

11.3 Life-cycle stages of a biolubricant (left) and a mineral lubricant (right).

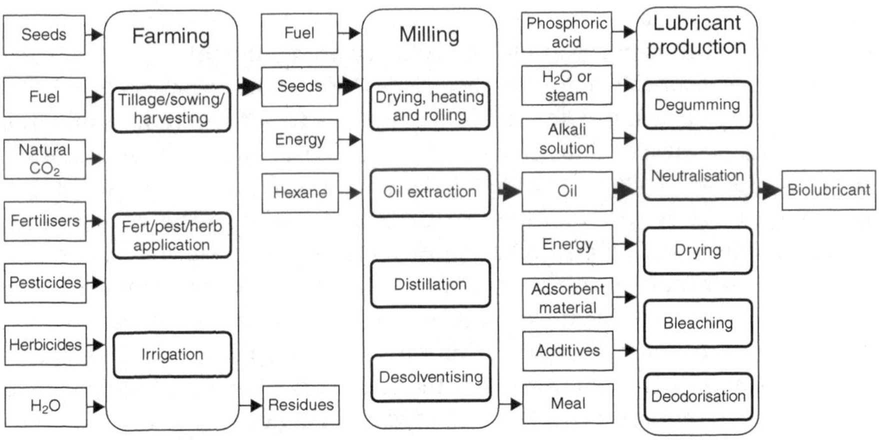

11.4 Detailed life cycle of a biolubricant.

is utilised as base oil for lubricant production [46]. By blending additives with the base oil a lubricant formulation of the desired quality is finally obtained.

The detailed life cycle of a mineral-based lubricant is shown in Fig. 11.5. Crude oil can be recovered in three different phases [47]. Typically, in the primary recovery, 10% of the reservoir's crude oil is obtained by natural pressure, gravity and pumping. In a secondary step, 20–40% is obtained by injecting water or gas to displace the oil. The remaining 30–60% can be recovered at high cost by enhanced oil recovery techniques. Crude oil is then distilled and the vacuum distillates and residue are further refined into

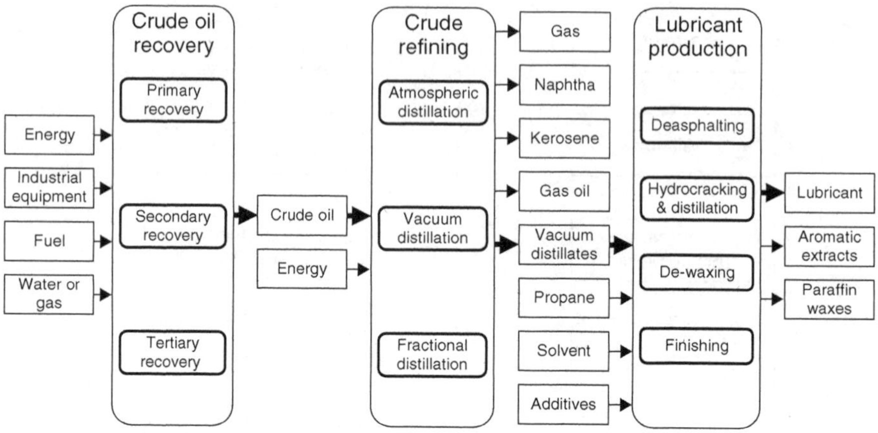

11.5 Detailed life cycle of a mineral-based lubricant.

lubricants by deasphalting, hydrocracking, dewaxing and finishing (see Sections 3.2 and 3.3). Finally, additives are added to the base oil. Crude oil extraction is responsible for the majority of air emissions in mineral oil production, especially for methane, where fugitive emissions from oil extraction and transportation to the refinery are abundant.

Table 11.1 lists the main characteristics of various comparative LCA studies of biolubricants. Detailed flow diagrams for production of mineral, soybean and rapeseed lubricants may be found in the original literature (e.g. refs [33, 34, 37, 48]). Individual balances for different vegetable oils are necessary. LCA results of rapeseed oil are not representative for other vegetable oils [49]. Soybean oil as a raw material for lubricants possesses a significant attribute. Soybean cultivation in corn-soybean rotation mode (US practice) or in wheat–soybean succession (Argentina's second-class soybean) requires a lower application rate of nitrogen fertiliser. This drastically improves GHG emissions and the net energy balance. Median fossil energy consumption for SBO is 5.27 MJ/kg oil, divided as follows: oil processing, 55%; farming operations, 28%; transportation, 11%; and upstream processes, 6% [37].

McManus *et al.* [33] have carried out an LCA on the use of mineral oil and RSO in mobile hydraulic systems (forestry harvester and road sweeper). Despite the fact that the results indicated that mineral oil had the most impact on GHG emissions, the LCA study shows that mobile hydraulic systems running on RSO are not necessarily better for the environment. This is mainly because of the performance characteristics of RSO in use. Rapeseed fluids are more easily degraded and released when subjected to high pressure and temperature than mineral oils, causing the need for more frequent oil change during use. RSO is also less compatible with certain hydraulic components (e.g. rubber seals and hoses), which need more fre-

Table 11.1 Characteristics of comparative LCAs of biolubricants and conventional lubricants

Products	Functional unit	Application	Tool(s)	Country	Impact categories[a]	Reference
Mineral oil Rapeseed oil	1 kg of oil	Mobile hydraulic systems	SimaPro Eco-indicator 95	UK	GHG, OD, AP, EP, HM, CE, WS, SS, SW, energy use, pesticides	[33]
Mineral oil Synthetic ester Rapeseed oil	1 m^3 of hydraulic fluid	Mobile hydraulic systems	LCA inventory Tool (LCAiT) 3.0	Sweden	GWP, AP, CED	[34]
Mineral oil Soybean oil	Area of aluminium rolled	Metalworking	GREET 1.6 TRACI	USA	AP, EP, PS, HH, climate change, fossil energy	[37]
Mineral oil Rapeseed oil ester Palm oil ester Animal fat ester Used cooking oil ester	1000 work pieces produced	Coolant	ISO 14040	Germany	GWP, AP, NP, CED, PS, PM, RD, CE	[38]
Mineral oil Rapeseed oil	Volume of oil used to cut 1000 m^3 of wood[b]	Chainsaw oil	SETAC guidelines	UK	GWP, NP	[35]
Mineral oil Rapeseed oil	1 t of lubricant	Loss lubrication	ISO 14040	Unspecified	GWP, AP, EP, CED	[48]
Rapeseed oil Soybean oil Mineral oil	1 kg of lubricant	Lubricant	GREET 1.8c SimaPro 7.1 ecoinvent	USA Europe	GWP, AP, CE, NCE, RE, EP, OD, ET, PS	[50]

[a] GWP, global warming potential; GHG, greenhouse gases; OD, ozone depletion; AP, acidification potential; EP, euthrophication potential; HM, heavy metals; CE, carcinogenic effects; WS, winter smog; SS, summer smog; PS, photochemical smog; HH, human health; NP, nutrification potential; SW, solid waste; CED, cumulative energy demand; PM, particulate matter; RD, abiotic resource depletion; RE, respiratory effects; NCE, non-carcinogenic effects.
[b] 34 litres for rapeseed chainsaw oil; 56 litres for mineral oil.

quent replacement during the lifetime of the machine. This study recommends improvements in RSO production and in the design of the components within the hydraulic systems to reduce the overall life-cycle impacts of RSO. Continued use of mineral oil was not recommended, since it is derived from non-renewable resources.

In another LCA evaluation of the impact of base oils in the use of a forestry harvester in Swedish conditions, Våg *et al.* [34] assumed that the hydraulic fluids were used in total-loss mode, did not consider the influence of additives and ignored the effect of pesticides in rapeseed production. It was concluded that the production of RSO (Swedish cultivation) has the lowest global warming potential (GWP) and acidification potential (AP) as compared with mineral oil and a synthetic ester lubricant (trimethylol propane (TMP) trioleate). Rapeseed oil also had the lowest cumulative energy demand (CED) of the three base oils studied. No other impact categories were studied. However, it would be erroneous to conclude that RSO is the best base fluid for the formulation of an environmentally adapted hydraulic fluid. In this total-loss application, where fresh fluid was frequently added to the system, lubricant life has not been considered. Differences in useful product life do greatly influence the outcome of an LCA. A much higher consumption rate of RSO in comparison to that of the synthetic ester clearly affects the GWP. The outcome of an LCA study of low-loss systems, where the useful life of the lubricant is included explicitly, might rate the base fluids differently. Clearly, the total-loss end-use in this LCA model is not an acceptable technical solution in the long term. The development of sustainable practices will have to include the design and operation of hydraulic systems that do not waste so much hydraulic fluid.

Miller *et al.* [37, 51] have performed an LCA for use of petroleum and soybean-based lubricants at an aluminium rolling plant. Carbon sequestration and EOL impacts were considered. Unlike most metalworking operations using bio-based lubricants, rolling processes recycle lubricants in a continuous loop. Soybean lubricant has indicated superior performance over the traditional mineral oil with a considerably reduced impact on climate change and fossil fuel use, and improved surface quality of the metal. However, a significant impact on eutrophication of regional water was observed when compared with petroleum-based lubricants. Other impact categories, such as acidification, human health and smog, presented varying results. As 75% less SBO is required in the rolling process to achieve metal production similar to mineral oils, this results in proportionally lower life-cycle emissions. The results of the assessment indicate that the comparative environmental impacts are largely dependent on the performance of SBO in the use phase.

Herrmann *et al.* [38] compared mineral oil (Variocut G500, Castrol), and methyl and 2-ethylhexyl esters of plant oils (PMO, RSO), animal fat, and used cooking oil for use as lubricants in cooling applications by evaluating the technical, ecological and economical aspects involved in production of these materials. Again, the cheap mineral oil has the highest impacts on GWP, carcinogenic effects (CE), CED, photochemical smog (PS) and abiotic resource depletion (RD) (see Table 11.1), and has the potential to cause most harm to the environment. However, rapeseed contributes the most to the acidification and nutrification potential. The palm oil ester system has the highest impact score for particulate matter. Used cooking oil and animal fats have the lowest environmental impacts, but these products are not readily available on the commercial biolubricant market. Figure 11.6 shows a market price/environmental impact portfolio for the various products.

Various other comparative LCA studies of RSO and mineral oil lubricants have been reported [35, 48, 52, 53]. In the use of lubricants for a chainsaw application, mineral oil confirmed the largest impact on GWP, while RSO made the largest contribution to nutrient enrichment potential [35, 52, 53]. The product life cycle for oilseed rape was assumed to start with ploughing of the land in preparation for crop production, and end after the lubricant has completely degraded in the environment. The product life cycle of the mineral oil was taken from oil exploration and production activities up to degradation in the environment.

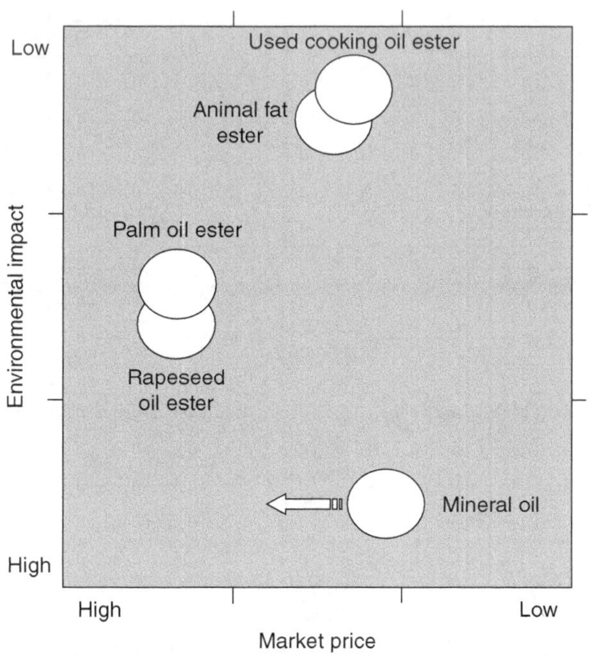

11.6 Market price/environmental impact portfolio. After ref. [38].

Reinhardt *et al.* [48] have made a life-cycle comparison of RSO and a conventional lubricant with different scenarios of post-use treatment (total loss, and energetic recovery as light oil and as natural gas). Figure 11.7 depicts the system boundaries for rapeseed and mineral-based lubricants used in this work. Rapeseed-derived lubricants are advantageous in terms

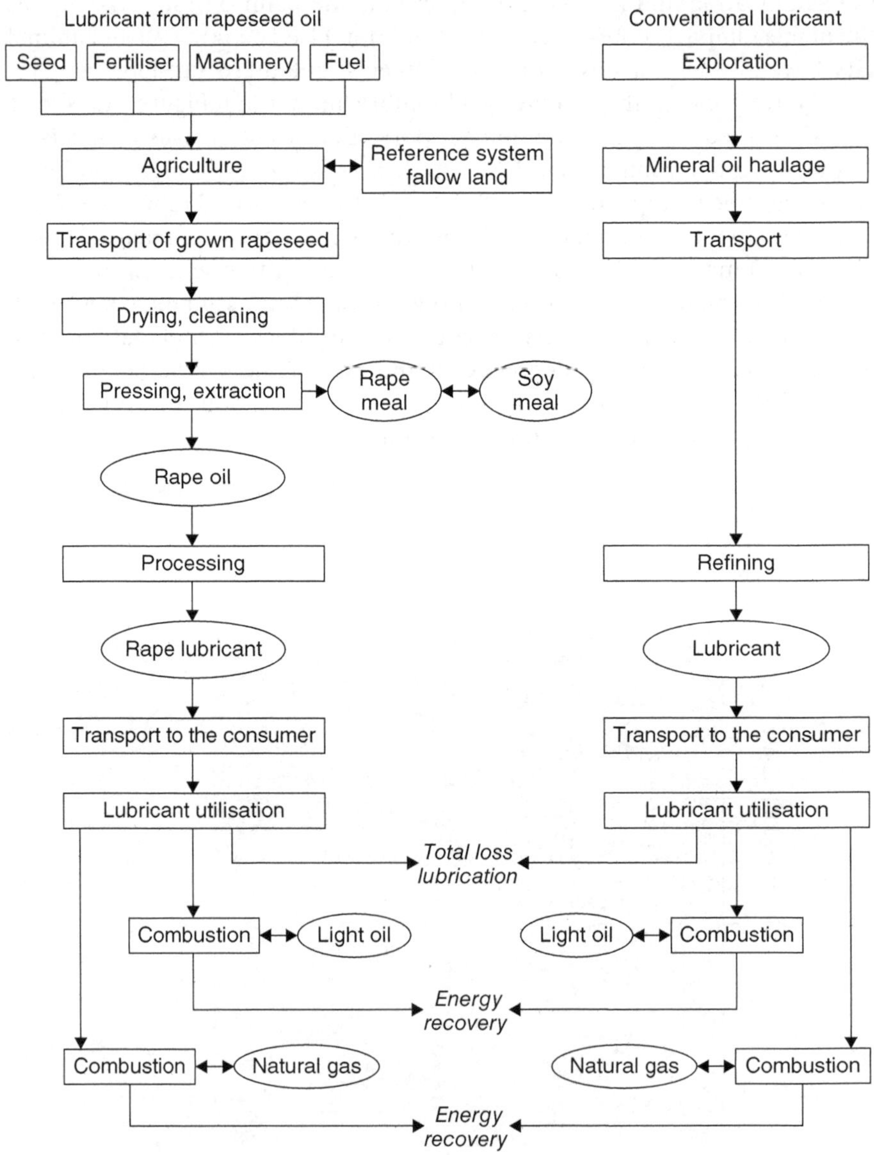

11.7 Life cycles of rapeseed and mineral-based lubricants, including various post-use treatments. After ref. [48].

of saving exhaustible energy resources and in terms of the greenhouse effect. In loss lubrications the biolubricants show even better results than in combustion after use. Acidification, eutrophication and stratospheric ozone depletion are negative influences for rapeseed-derived lubricants. If one considers saving of exhaustible energy resources and diminishing the greenhouse effect as having the highest ecological priority over other impact categories, then the evaluation is in favour of the biolubricant.

Cuevas [50] has recently reported a comparative LCA of rapeseed, soybean and mineral-based lubricants. The system boundary of this study included cultivation of rapeseed and soybean up to the production of biolubricant, and crude oil recovery to lubricant production for the mineral-based lubricant. The LCA thus evaluates the impacts to air and water during extraction and production of these materials. Lubricant use and disposal phases were not included in this study; neither were additives. LCIA results for rapeseed, soybean and mineral lubricants are shown in Table 11.2 and Figs 11.8–11.11 for several impact categories. It appears that rapeseed lubricants dominate the majority of the affected impact categories, notably the AP, CE, non-carcinogenics (NCE), respiratory effects (RE), eutrophication potential (EP) and PS impact categories when compared with soybean and lubricant oils (cf. Figs 11.8 and 11.9). Mineral lubricants dominate the GWP and ozone depletion potential (ODP) categories (cf. Figs 11.10 and 11.11). The farming and fertiliser stages were the largest contributors to rapeseed LCIA results. However, milling was also a significant contributor to GWP and PS. Reduction in fertiliser usage would significantly reduce the overall environmental impacts caused by rapeseed lubricant. Reduction in

Table 11.2 LCIA results for 1 kg of rapeseed, soybean and mineral lubricant

Impact category	Unit	Rapeseed lubricant	Soybean lubricant	Lubricating oil
Global warming	kg CO_2 eq	–0.36	–1.65	1.07
Acidification potential	H^+ moles eq	2.70	0.20	0.46
Carcinogenics	kg benzene eq	0.006	0.001	0.003
Non-carcinogenics	kg toluene eq	33.4	3.3	14.3
Respiratory effects	kg PM2.5 eq	0.005	0.001	0.002
Eutrophication potential	kg N eq	0.076	0.029	0.002
Ozone depletion potential	kg CFC-11 eq	2.8×10^{-7}	4.7×10^{-8}	6.5×10^{-7}
Ecotoxicity	kg 2,4-D eq	1.78	–2.08	1.22
Photochemical smog	kg NO_x eq	0.023	0.004	0.003

After ref. [50].

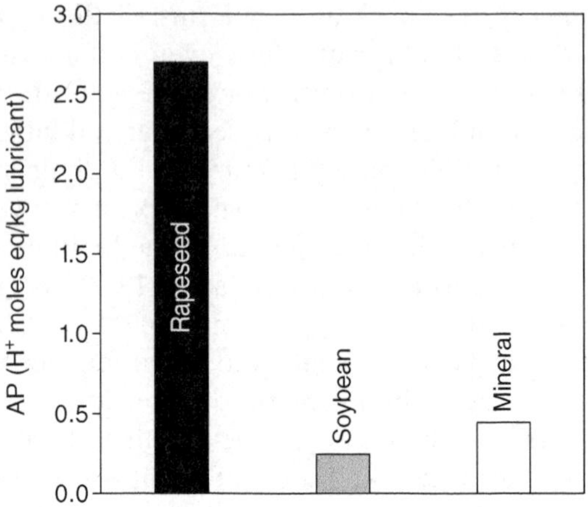

11.8 Acidification potential for rapeseed, soybean and mineral lubricants. After ref. [50].

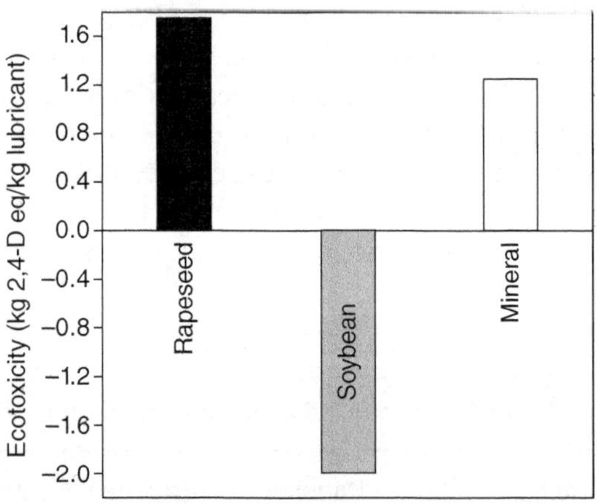

11.9 Ecotoxicity for rapeseed, soybean and mineral lubricants. After ref. [50].

the contributions from the rapeseed farming stage would require a considerable cutback in machinery use. In the GWP category, the rapeseed and soybean lubricants presented negative values due to the assumption that CO_2 is sequestered during the farming stage.

The various studies mentioned in Table 11.1, which all compared biolubricants with traditional mineral-based lubricants, each had their own specific system characteristics with significant differences between goals, system boundaries, applications, functional units, tools and impact categories. This does not facilitate easy comparison of LCA results. Goal and scope set

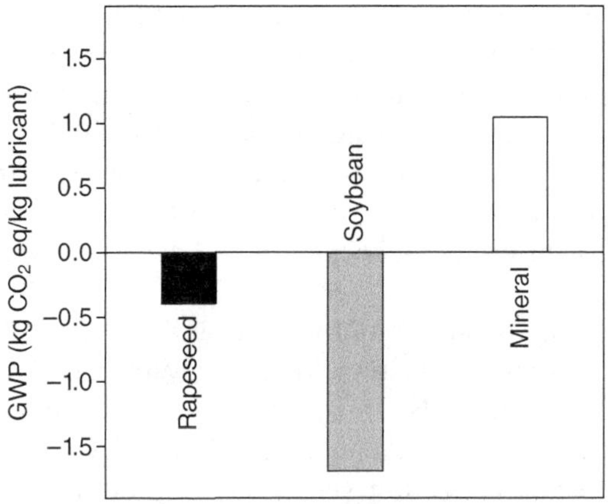

11.10 Global warming potential for rapeseed, soybean and mineral lubricants. After ref. [50].

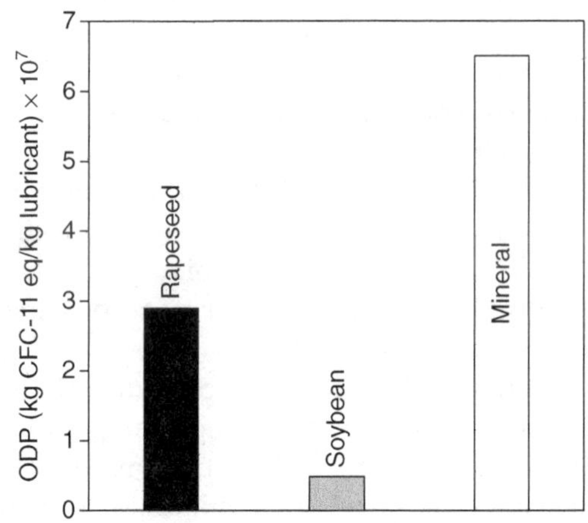

11.11 Ozone depletion potential for rapeseed, soybean and mineral lubricants. After ref. [50].

system boundaries for LCA. Differences may arise from disregard of certain flows or life-cycle stages, co-product allocation, availability of databases, etc. Also, the functional unit selected depends on the definition of goals and scope. In particular, the choice of functional unit is often dictated by products and applications. Whereas McManus *et al.* [33] and Cuevas [50] included the stages from farming up to milling (biolubricant production) and all stages from crude oil extraction up to refining (lube oil production), Våg *et al.* [34] analysed rapeseed and mineral lubricants from farming/crude oil extraction to lubricant use. However, the latter did not include the use of pesticides in their assessment and also did not address additives. Fertilisers and pesticides tend to have significant effects on the impact categories acidification and eutrophication potential. In the aforementioned studies, the most common impact categories evaluated are GWP, ODP, CED, AP and EP, but others may relate to toxicity, human health, smog, particulate matter and land use. Finally, the tools used in any LCA study should be selected carefully. Use of different tools can lead to very different results [28, 54].

Nonetheless, overall comparison of biolubricants and conventional lubricants consistently indicates similar results when analysing GWP, which is highest for mineral-based lubricants. Rapeseed contributes the most to GWP on account of fertiliser production and use [38]. However, at variance with these results, biodiesel production (from soy oil) in Argentina and Brazil leads to a higher GWP than fossil diesel [55]. Results are not always quite as expected: although mineral oil had the most impact on GHG emissions, RSO had greater environmental impacts due to agricultural and performance factors [33]. As to the energy used at the different life-cycle stages, in most cases the largest contributor was mineral oil due to its high embodied energy [34, 37, 38, 48, 55]. However, McManus *et al.* [33] ascribe a higher energy contribution to rapeseed oil due to the energy required in the crushing stage in RSO production.

Compared with mineral-based lubricants, SBO lubricants have a reduced impact on climate change and fossil fuel use, but a significant impact on eutrophication [37], as also noted for RSO [33]. Similarly, Reinhardt *et al.* [48] concluded that AP, EP and ODP are affected more by rapeseed oil lubricant than conventional mineral lubricant.

LCA results may be affected by lubricant performance. Biolubricants have several limitations, including sensitivity to high temperatures, poor low-temperature stability and poor oxidative stability. Consequently, biolubricants need to be replaced more frequently than mineral-based lubricants. This could change LCA results significantly [33].

In addition to environmental impacts, biolubricants also have a significant impact on land use. Biolubricant use can cause changes in cropland. Out of the total US cropland (19% of US total land), the common crops suitable for biolubricant production, soy and rapeseed, account for 9% and 0.2%, respectively. Any change in biolubricant use will typically cause variations in other land categories. Based on the land use evaluation, approximately 32 million acres of land would be required to produce 2 billion gallons per year (Bgy) of US lubricant demand from soybean; only approximately 25 million acres would be needed if the lubricants were produced from rapeseed. However, also other measures can be envisaged to meet biolubricant demand, such as an increase in crop yield [56].

In the LCA community there have been numerous discussions concerning direct and indirect land use. Direct land use change occurs within a specific supply chain for a specific production facility, whereas indirect land use changes (ILUCs) occur in land that is not part of a specific supply chain due to market forces [57]. Converting carbon-rich ecosystems to biolubricant or biofuel production results in large amounts of CO_2 being released for decades following land clearing (so-called 'carbon debts') [58, 59]. GHG emissions depend on the types of land converted (rainforests, peatlands, savannahs or grasslands). Conversely, only biolubricants from perennials grown on degraded or abandoned cropland and from waste biomass avoid land use change, minimise habitat destruction, competition with food production, and carbon debts, and can offer immediate and sustained GHG advantages [60]. In Germany two sustainability regulations exist referring to different kinds of biofuels. They contain specific criteria for sustainability, e.g. sustainable agricultural use and greenhouse gas emission reduction potential (BGB 2009). Recently, serious concerns have been expressed about the effects of ILUC associated with expanded use of biofuels and bioliquids in the EU [61].

LCA takes into account various impact categories, but often not land use. However, as noted above, the use of land resulting from the cultivation of oilcrops needs careful consideration (e.g. promotion of fallow oil plant cultures in Europe, deforestation of tropical rainforests for palm tree plantations in South-East Asia and soy-related deforestation in Argentina and the Amazon) [56]. In Austria and Germany oilseed rape for bio-based use is currently grown only on set-aside fields in rotation.

The increase in bio-based products (e.g. biofuels) has led to many studies on the life-cycle impacts of these products, particularly during the agricultural phase and up to the production of oil. However, as shown before, only few studies have been conducted on the production of lubricants from renewable sources. Various other comparative LCA studies of agricultural products have been reported, as summarised in Table 11.3; see also ref. [83]. Varying conclusions have been drawn regarding the environmental effects of these products. Most studies conclude that the agricultural phase has the highest environmental impact, contributing to acidification, eutrophication and smog impact [33, 34, 37]. Generally, lower impacts are reported in the GWP and climate change categories for the bio-based products compared with the mineral options. In particular, the prime ecological advantage of biodiesel is its net energy yield, whereas the main ecological disadvantage is the impact of energy crop farming [84]. In case of microalgae, 90% of the process energy consumption is dedicated to lipid extraction [80]. Some LCAs of palm oil-based biofuel are flawed because they overlook indirect land use [68, 69].

Table 11.3 LCA studies of bio-based products

Product(s)	Application	Geographical area(s)	Reference(s)
Rapeseed oil Mineral oil	Rape biodiesel	Germany, Europe	[49, 62, 63]
Vegetable oils	Biodiesel	Europe	[64]
Canola, soybean, waste vegetable oils, animal fats	Biodiesel	Canada	[65]
Corn, soy, alfalfa, switchgrass	Biomass	USA	[66]
Soybean, rapeseed, sunflower, coconut oil	Biodiesel	Europe, USA	[67]
Palm oil	Palm biodiesel	Malaysia	[68, 69]
Soybean, palm and rapeseed biodiesel; fossil diesel	Biodiesel	Argentina, Switzerland	[55, 70]
Sugar beet ethanol; rapeseed methyl ester	Biofuels	Europe	[71]
Canola, corn, soy, wheat	Conventional and organic crop production	Canada	[72]
Soybean meal, palm, rapeseed, spring barley	Lifestock protein	Argentina, Netherlands	[73]
Soybean biodiesel, soybean-based renewable diesel, fossil diesel and gasoline, renewable gasoline	Fuels	USA	[74]
Palm oil, rapeseed oil	Vegetable oils, biodiesel	Denmark, Malaysia, Indonesia	[56, 75, 76]
Rape, soy, palm oil Mineral oil	Biodiesel, oil diesel	Europe, USA, Malaysia	[77–79]
Microalgae	Biodiesel	Mediterranean[a], UK[b]	[80, 81]
Next generation feedstocks[c], corn grain, soybean	Next-generation biofuels	USA	[82]

[a] Open raceways.
[b] Raceways, tubular bioreactors.
[c] Corn stover, switchgrass, forest residues.

11.3 Sustainable product design

While the societal benefits of lubricants are enormous in terms of increased productivity, protection of machinery, resource savings and energy efficiency, the environmental benefits are more questionable. Lubricants can improve their environmental impact or reduce negative effects by good biodegradability, low toxicity and efficiency in application (low friction, low treatment, low losses, long duration) [85]. Although lubricants do not contribute directly to improve the quality of the environment they should not be a burden either; they need to be sustainable.

Our reliance on fossil-based materials for continued economic growth is reducing more and more. Resource depletion jeopardises the survival of manufacturing in the present form. Sustainable product design is a key challenge. Sustainable means 'clean enough' to meet the needs of the present without compromising the ability of future generations to meet their own needs [86]. The concept of sustainability dictates that modern conversion technologies and practices should be founded on renewable resources. In the age 'beyond petroleum', for which the world is now preparing, a bio-based industrial future is foreseen based on biomass chemicals. In a true bio-based economy the products are sustainable. However, bio-based products are not necessarily more sustainable than alternatives. This needs to be evaluated on a case-by-case basis (using LCA). Sustainability needs to take into account all potential impact factors, including the effects on deforestation, environmental degradation, soil quality, competition with food production, import trade and costs of distribution. However, production of sufficient food and non-food biomass without increasing emissions and waste and with the need to mitigate the anticipated climate change can no longer be taken for granted. The use of non-sustainable agricultural practices worldwide for bioproduct supply should be discouraged. Intensive single-crop farming for chemical or fuel production (e.g. US corn for bio-ethanol) is environmentally unsustainable. Intensive agricultural systems are largely responsible for the increases in global reactive nitrogen compounds, which are associated with significant environmental impacts. Also, truly sustainable production requires substitute technologies, which are not readily at hand and require development. Consequently, science and technology are faced with enormous challenges to overcome constraints on the future availability of raw materials (both organic and inorganic), energy, fresh water, arable land and renewable resources.

While several decades of serious investment in chemical and agronomical sciences and technology in a few countries, notably Brazil, China and India, has positively affected agricultural production, urgent action and global commitment to more scientific endeavour in long-term R&D are required to create sustainable production of more food and animal proteins for the rapidly increasing world population while at the same time securing abundant non-food biomass as the only practical sustainable source of carbon for our production processes. Resource efficiency should be a declared objective of the international community by 2020. The EC communication '*Roadmap to a Resource Efficient Europe*' holds the vision that by 2050 all resources are sustainably managed, from raw materials to energy, water, air, land and soil.

Environment and sustainability are at the heart of societal challenges. Chemistry is fundamental to understanding the problems humanity is facing, and can provide innovative solutions. Green chemical technology is ideally placed to sustainably support the needs of the growing global population with minimal impact on Earth's natural resources. Actions for sustainable product design by the chemical sciences are listed in Table 11.4. The principles of green chemistry include consideration of product use, and in particular the impact of the product on energy consumption and waste generation. This area may be further optimised.

The environmental impact of a product is determined largely at the design stage. Sustainable substances should either be designed in such a way that they can be fully recycled without loss in quality or fully degrade or mineralise during disposal (no waste in the long run). For full sustainability, there is a major step the lubricant industry has to take, namely the use of renewable resources for base stocks *and* additives. While use of renewable raw materials and material recycling are undoubtedly requirements of a future sustainable technology, they are not necessarily the most important issues to address. For example, water efficiency is becoming ever

Table 11.4 Sustainable product design

Environmental impact:
• Improved LCA tools and metrics
• Standardisation of LCA methods and data gathering
• Methods to assess recycled materials
• Improved understanding of ecotoxicity
• Efficiency in application
Chemistry and chemical engineering
• Structure–property relationships
• Technology for designing biodegradability into finished products
• Green chemistry
• Manufacturing process intensification and optimisation
• Process modelling, analysis and control
• Substitution of toxic substances
• Readily recyclable products
• Improved recovery processes

Source: www.rsc.org/roadmap.

more important. Production and application up to disposal of lubricants (see Chapter 14) must cover the requirements of the best possible protection of the environment. Health hazards may arise directly or indirectly (through our environment). Direct contact between lubricants and people and nature has to be checked by compatibility testing. LCA, which should accompany any product development and should be applied across the entire supply chain, allows identification of the major environmental impact events.

Whereas LCA is meant to study the resource and environmental impacts of a product, sustainable development (SD) indicators concentrate on the environmental, social and economic impact of a product [43, 87–89]. Fundamental to effective indicator development is an understanding of the trade-offs in satisfying conflicting objectives between society and business. The sustainability assessment tool helps in sustainable product development. The SD assessment tool more easily includes local impacts than LCA. LCA studies of lubricants tend to focus on global impacts of the fluids; however, it is equally essential to consider the impact on the local environment because this is often the area where the most significant environmental effects are likely to occur. Making industry more sustainable can provide the means of reducing environmental impacts or even improving the environment and quality of life. Developing a sustainable bio-based economy that uses eco-efficient bio-processes and renewable bioresources is one of the key strategic challenges for the 21st century.

Shell Global Solutions has developed a sustainability assessment tool for biolubricants. The sustainability of an environmentally acceptable hydraulic fluid meeting Swedish Standard SS 155434 has been compared with that of a conventional mineral oil-based product [43]. The environmentally acceptable hydraulic fluid provides a real benefit in terms of environmental impact. This is largely due to the benefits the fluid presents at a local scale. In sensitive environments, the local ecological advantage means the environmentally acceptable fluid is preferable. The environmental savings at a global scale are less distinct, as found in several full LCAs [33–35]. Social benefits are highly susceptible to subjectivity. Environmentally acceptable fluids are inferior to traditional fluids in terms of economic indicators on the use phase of the product. Some bio-based lubricants have a higher price, and the performance of the product may be inferior.

11.3.1 Energy-efficient lubricants

Key performance of lubricants comprises reduction of friction in machinery. Reducing friction implies reduced energy consumption for equivalent output of useful work. Therefore, optimising friction performance may reduce overall consumption of energy and associated emissions to the environment. Minimising wear by efficient lubrication prolongs the useful life of machines, thereby minimising the consumption of non-renewable resources. A third key performance of lubricants is the stability or lifetime of the lubricant itself, which affects the environmental impact. Adoption of high-performance lubricant technologies has led to a significantly lower lubricant consumption.

Lubricants have a particularly important contribution to make in relation to energy conservation. Truly green lubricants are those that not only minimise wear in machinery and maximise service lifetimes but also optimise energy efficiency. The use of optimised lubricants can lead to significant energy savings, i.e. reduced fuel or electricity costs [44, 45]. Significant savings in machinery energy consumption can be achieved by using energy-efficient lubricants. It has been estimated that about 11% of the US total energy consumption in the major areas of transportation, turbomachinery, power generation and industrial processes can be saved through developments in lubrication and tribology [90]. Reported energy savings by the correct selection of high-quality lubricants are generally between 5% and 15%, and typically range from 5–6% for engines, to 6–12% plus for gearboxes, up to 7% for hydraulics and 20–26% for drive chains/conveyors [91]. Chain conveyor systems and gearboxes are prime beneficiaries of improved lubricants. It is well known that gearboxes with worm gears are inefficient energy users.

Lubricants influence energy efficiency mainly through reducing energy losses, which include churning losses and friction losses in the hydrodynamic (HD), elastohydrodynamic (EHD) and boundary lubrication regimes. The main sources of lubricant-related energy losses in mechanical equipment are on account of:

- metal-to-metal contact causing friction;
- churning and windage of mechanical parts through the oil;
- internal lubricant friction (traction) under EHD lubrication conditions.

Energy losses are determined by the initial oil viscosity, chemical composition, temperature, pressure and shear rate dependence. The main physicochemical factors influencing oil energy efficiency comprise viscosity shear-thinning (reversible or irreversible), pressure-thickening effects, compressibility and viscoelasticity. Boundary friction coefficient, high-temperature high-shear viscosity and pressure–viscosity coefficient are the most important oil properties defining oil energy efficiency [92].

Throughout the life of a machine, energy (fuel) is one of the largest components of operating cost. Fuel-efficient lubricants start with optimised base stock. As the base stock is the main component of a lubricant formulation, the energy savings potential is highly dependent on the ability of the base stock to reduce energy losses. A base stock with higher viscosity index (VI) reduces the likelihood of metal-to-metal contact. The higher the oil viscosity, the more energy that is lost moving the mechanical parts through the oil. The effect of oil chemical structure on friction losses was studied by Naruse *et al.* [93]. Hydrocracked and poly-α-olefin (PAO) oils show significant advantages over conventional mineral oils in friction reduction under EHD lubrication conditions. Synthetic base stocks such as PAOs with low traction coefficients offer the ability to save energy, especially when there is a high degree of sliding such as in gears, particularly worm gears. Energy savings of 5–8% were achieved by changing lubricants in compressors from mineral to synthetic diester and other synthetic oils of identical ISO viscosity grades [94, 95]. Fuel-saving benefits obtained with synthetic and hydrotreated engine oils and transmission fluids have been extensively reported [92, 96, 97].

Lubricant designers use a variety of formulation strategies to maximise fuel economy performance. Additives are designed to improve base stock properties such as viscosity and reduce vulnerability to corrosion, friction and wear. Many of these properties, especially viscosity, make a direct contribution to fuel efficiency. Judicious choice of additives is critical. As additives generally tend to thicken the oil, which reduces fuel economy, it is imperative to use fewer – but better performing – additives. Polymeric VI improver additives can boost vehicle fuel economy [98]. Additives have to minimise frictional losses in engine parts that are operating under a broad range of operating conditions and must ensure that other performance aspects of the lubricants such as wear and cleanliness are not compromised in the effort to maximise fuel economy gains.

Lubricants have been helping vehicle fuel efficiency since CAFE (Corporate Average Fuel Economy) standards were enacted in the United States in 1975. However, CAFE ratings are changing more slowly than originally planned, namely from 18 mpg (1978) to 27.5 mpg (1985) and 36 mpg (2015 target). Demands to reduce fuel consumption and improve efficiency will increase. Fuel consumption monitoring programmes, which allow for comparison of different automotive vehicle models, induce to manufacturing of low specific fuel consumption (SFC) and low exhaust emission vehicles. Lubricants that provide higher fuel economy (FE) than a reference lubricant are classified by the American Petroleum Institute (API) as 'Energy Conserving' [99]. Standard Sequence VIA-D engine dynamometer testing (ASTM D 6202) is used to measure a lubricant's ability to improve the fuel economy of passenger cars and light-duty trucks. The Sequence VIA engine oil test evaluates fresh oil fuel economy whereas the more recent Sequence VID engine test checks FE on both fresh and used oil.

Improving the fuel economy of internal combustion engines is a key objective of the automotive industry (see Section 12.2). Engine oils can contribute to improvements in fuel economy in two ways: (i) indirectly, by supporting engine design that allows fuel economy without compromising high performance; and (ii) directly, through new formulations. It has been observed that friction and wear, and fuel consumption are affected by engine cylinder liner surface treatments and finishes [100, 101]. Engine crankcase oils influence automotive fuel efficiency through the energy required to overcome viscous shear and boundary friction. Consequently, the main factors contributing to the fuel economy properties of engine oils are high-temperature high-shear (HTHS) viscosity and the boundary friction coefficient. Suitable viscometric techniques have been developed for the determination of high-shear viscosity (see Section 3.6). Besides the shear thinning effect, VI-improved oils exhibit another aspect of non-Newtonian behaviour, namely viscoelasticity. The viscoelastic effect is considered to be important for energy conservation and friction reduction [102]. Crankcase and transmission lubricants can contribute in improving energy efficiency in many ways (e.g. use of lower-viscosity oils, additive technology).

Engine-based savings can reach a theoretical 8–10% [85]. However, as engine oils cannot totally eliminate frictional losses, savings potentials of 4 to 5% already present a great challenge. In practice, lower values are being reported. Fuel savings in gasoline engines operated in certain conditions were found to reach an optimum of 3.5% for HTHS viscosity values between 2.5 and 2.9 mPa s [103]. The absolute savings figures on car fuel consumption depend largely on the test conditions, base oil formulations and reference oil used. Standardised dynamometer tests, which reflect driving conditions more accurately, provide more realistic values than bench tests, which cannot reproduce all operating conditions. The requirement for fuel economy leads to the use of low-viscosity oils with friction modifier additives to protect the engine cylinder surface against wear [104]. According to original equipment manufactures (OEMs), HTHS values must not be lower than 2.6 mPa s to guarantee safe engine operation. The HTHS viscosity of SAE 0W-20 oil (2.6 mPa s) is at the limit [98]. The European specification for fuel economy oils for gasoline engines spans from 2.9 to 3.5 mPa s. Evonik (Darmstadt) has recently developed an innovative 0W-20 engine oil with VISCOPLEX® and VISCOBASE® components.

According to Smeeth *et al.* [105] Group III base oils do not produce higher fuel economy than Group I base oils in general engine lubrication. Inoue *et al.* [106] verified that the use of low-viscosity oils is an efficient way to increase fuel economy in spark ignition engines. Addition of molybdenum dithiocarbamate (MoDTC) to a SAE 5W-20 low-viscosity oil reduces fuel consumption at high temperature operation and increases energy efficiency by 2.7% in comparison with a conventional SAE 10W-30 lubricant. Devlin *et al.* [107] measured the influence of physical properties of lubricating oils on the fuel economy of general motor vehicles. Reductions in an oil's HTHS, boundary friction coefficient and pressure–viscosity coefficient were found to improve fuel economy. Moore [108] has described test methods for the measurement of boundary friction of automotive engine crankcase oils which correlate well with measured fuel savings. Tseregounis *et al.* [109] have quantified the viscosity and friction modification effects of eight lubricating oils on the fuel economy of gasoline-fuelled engines. Modified Stribeck curves allow insight in the effects of oil formulations on the fuel efficiency [110].

As diesel engines operate under high temperature, lubricant viscosity is increased due to evaporation of light compounds, oil oxidation and nitration, and contamination by soot. Manni *et al.* [111] studied the effects of lubricating oil characteristics on emissions, fuel economy and oil consumption for light-duty diesel engines. Measuring energy savings is very challenging for heavy-duty applications. Fotheringham *et al.* [97] evaluated exhaust emissions and fuel economy for advanced engine oils (very-high viscosity index (VHVI), PAO) in a 12-L heavy-duty diesel truck operating in standardised conditions (European Transient Cycle). SAE 5W-30 lubricants formulated with PAOs produced a significant reduction of particulate matter emissions in comparison with mineral base oils and high-viscosity oils. On average, SAE 5W-30 lubricants showed a 2% higher fuel economy in comparison with SAE 15W-40 lubricants (see Fig. 11.12). For seven different lubricants in a 9.6 L bus engine a fuel economy of 1.6% was observed with differences on hydrocarbon emissions and particulate matter from

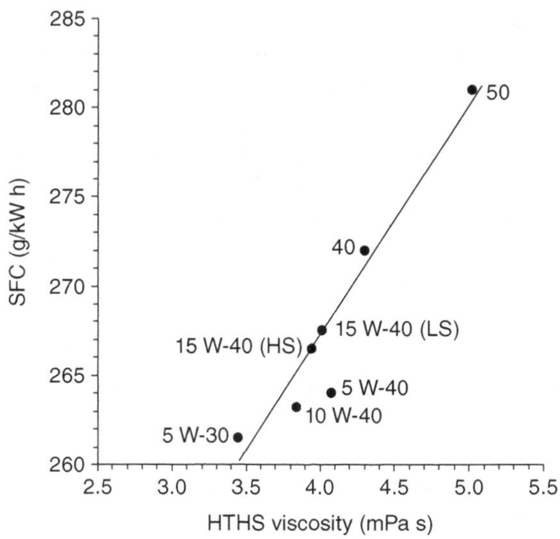

11.12 Variation of specific fuel consumption in a diesel engine as a function of HTHS viscosity for various SAE mono- and multigrade oils. After ref. [98].

0.6% to 22%, respectively [112]. Figure 11.12 shows the SFC of a single-cylinder, four-stroke, direct injection diesel engine operated in standardised conditions as a function of the HTHS viscosity of seven mono- and multi-grade mineral (Group I), synthetic (Groups III, IV) and semi-synthetic (Groups I, III) oils, all formulated with the same concentration of antioxidant and detergent and dispersant additives [98]. In accordance with Smeeth *et al.* [105], at all test conditions the monograde lubricants showed higher SFC than multigrade oils which have VI improver additives in their formulation. The highest SFC reductions compared with the monograde SAE 50 lubricant were of the order of 8.5% to 8.9% for multigrade oils. As shown in Fig. 11.12, the specific fuel consumption reduces linearly with the decrease in lubricant HTHS viscosity from 5.0 to 3.5 mPa s. This is in accordance with similar findings by Devlin *et al.* [107]. The recommended HTHS viscosity for heavy-duty diesel engines is above 3.5 mPa s [104]. This precludes the use of SAE 0W-20 lubricants for that purpose. The shear-thinning effect due to temporary mechanical realignment of a polymeric VI improver has been observed for multigrade engine oils to give better fuel economy than monograde oils [98]. The reversible behaviour reduces the effective viscosity of the lubricant under high-shear operating conditions resulting in a reduction of viscous drag and an improvement in fuel economy. The presence of high-shear stable polymer as an additive to multigrade SAE 15W-40 mineral oil reduces viscosity and fuel consumption in comparison with the use of a low-shear stable polymer in the same oil. For all lubricants tested, the highest specific fuel consumption was observed at low engine speed and load (1500 rev/min and 10 N m).

For passenger cars there has been a general trend toward lighter viscosity grades of lubricants along with friction-modifier additives. The drive to improve vehicle fuel consumption has led to the widespread availability of low-viscosity lubricants (e.g. friction-modified SAE 5W-30 grades compared to SAE 15W-40 grade lubricants) with 2–5% fuel economy improvement. It is imperative that any reduction in lubricant viscosity does not result in decreased durability. The key concern in using low-viscosity oils and oils of thin films in engines is component wear. HTHS minimum limits have been set.

In response to recent (2010) EPA and NHTSA (National Highway Traffic Safety Administration) regulations to reduce greenhouse gases and improve fuel efficiency the lubricant industry has started defining the next generation of HD lubricants. The latest minimum performance standard for US passenger car engine oils (ILSAC GF-5, 2010) calls for improving simultaneously the fuel economy and piston cleanliness, which poses a considerable challenge. The ILSAC GF-5 oil rating for gasoline engines in cars and light trucks requires Sequence VID limits for fuel economy and is roughly equivalent to the API SN ('resource conserving') rating.

Machinery components requiring lubrication that have the greatest effect on fuel efficiency include the crankcase, transmission and gearbox. Energy-efficient lubricants are less widely used in industrial applications but the potential savings may be even greater. Different sources of lubricant-related power losses in industrial systems (hydraulic pumps, gears) have been described [92]. Approximately 5–8% energy savings are achievable by using energy-efficient industrial lubricants. The energy efficiency of most industrial oils containing a standard additive package can be improved by using additional supplements or by replacing a conventional oil with a new, fully formulated, more efficient oil. Lubricant-related energy losses in a hydraulic system are load dependent and consist of churning and friction losses. Churning losses are mostly influenced by oil viscosity, while friction losses also greatly depend on oil chemical composition. In hydraulic systems, friction is dominated by pipe losses and is therefore directly proportional to lubricant dynamic viscosity, whereas in an engine friction is proportional to the square root of viscosity. Hence, changing the ISO viscosity grade, or moving to a higher VI lubricant could lead to significant energy savings under cold-start conditions. Since hydraulic fluids are usually subjected to high pressures, the pressure–viscosity behaviour is one of the most significant fluid properties to be taken into account when selecting a hydraulic fluid. A minimal pressure–viscosity dependence, as shown for PAOs, is the most beneficial for energy conservation as well as for anti-wear performance.

Hydraulic pumps operate at high fluid shear rates. The shear-thinning and viscoelastic effect of viscosity index-improved hydraulic fluids can be utilised for friction reduction and energy conservation. A hydraulic fluid is often simultaneously exposed to temperature, pressure and shear stresses. The energy dissipated in hydraulic fluids may be much higher than in gear oils, in which temperature is usually controlled by friction heat developed in the gear mesh. Energy-saving benefits have been obtained with synthetic industrial gear oils in comparison with paraffinic mineral oil [93, 113]. In moving from mineral base stocks to more highly processed base stocks for gear lubricants, the decrease in the pressure–viscosity coefficient of the lubricant results in both decreased friction *and* decreased temperature rises [45]. Important energy savings have also been reported through appropriate refrigeration compressor base fluid design (see Section 12.7.1) [44].

New lubricant standards eventually will require oils to provide improved fuel economy and include additive chemistry that does not degrade emission system components [110]. The next challenge is the development of additive components and formulation strategies to minimise the internal friction losses in different engine parts operating under different tribological regimes as well as to mitigate engine wear that may occur as the lube viscosity is minimised in an effort to further increase the energy conserving capability of lubrication.

Energy efficient lubricants are generally more expensive than conventional lubricants, and a life-cycle cost analysis, which takes account of operating costs as well as the initial purchase cost of the lubricant, is required to bring about the true benefits of these products. It is reasonable to expect that in future years the life-cycle cost analysis will be weighted even more in favour of energy efficient lubricants.

11.4 Sources of further information and advice

C. Jiménez-González and D.J.C. Constable, *Green Chemistry and Engineering. A Practical Design Approach*, Wiley-Blackwell, Oxford (2011).

Umweltbundesamt (ed.), *Guide on Sustainable Chemicals*, Federal Environment Agency, Dessau-Roβlau (2011).

Nuffield Council on Bioethics, *Biofuels: Ethical Issues*, Nuffield Press, Abingdon (UK) (2011); http://www.nuffieldbioethics.org

Commission of the European Communities, *Roadmap to a Resource Efficient Europe*, Brussels (20 September 2011).

A. Matlack, *Introduction to Green Chemistry*, 2nd edn, CRC Press, Boca Raton, FL (2010).

M. Lancaster, *Green Chemistry. An Introductory Text*, RSC Publishing, Cambridge (2010).

N. Winterton, *Chemistry for Sustainable Technologies. A Foundation*, RSC Publishing, Cambridge (2010).

J. Emsley, *A Healthy, Wealthy, Sustainable World*, RSC Publishing, Cambridge (2010).

National Renewable Energy Laboratory, *U.S. Life-Cycle Inventory Database*, NREL, Golden, CO (2009); http://www.nrel.gov/lci/

R. Höfer (ed.), *Sustainable Solutions for Modern Economies*, RSC Publishing, Cambridge (2009).

G.L. Baldo, M. Marino and S. Rossi, *Analisi del Ciclo di Vita LCA*, Edizione Ambiente Srl, Milan (2008).

S. Takata and Y. Umeda (eds), *Advances in Life Cycle Engineering for Sustainable Manufacturing Business*, Springer, London (2007).

R.A. Sheldon, I. Arends and F. van Rantwijk, *Green Chemistry and Catalysis*, Wiley-VCH, Weinheim (2006).

H. Baumann and A.-M. Tillman, *The Hitch Hiker's Guide to LCA*, Studentlitteratur AB, Lund (2004).

W. McDonough and M. Braungart, *Cradle to Cradle*, North Point Press, New York, NY (2002).

J.B. Guinée (ed.), *Handbook on Life Cycle Assessment*, Kluwer Academic Publishers, Dordrecht (2002).

H. Wenzel, M. Hauschild and L. Alting, *Environmental Assessment of Products. Vol. I. Methodology, Tools, and Case Studies in Product Development*, Kluwer Academic Publishers, Dordrecht (2001).

G.A. Reinhardt and G. Zemanek, *Ökobilanz Bioenergieträger, Basisdaten, Ergebnisse, Bewertungen* (LCA of Bioenergy Sources, Data, Results, Evaluations), Erich Schmidt-Verlag, Berlin (2000).

J. Borken, A. Patyk and G.A. Reinhardt, *Base Data for Ecological Assessments: Transportation, Agriculture and Mining*, Vieweg-Verlag, Braunschweig/Wiesbaden (1999) (in German).

P.T. Anastas and J.C. Warner, *Green Chemistry: Theory and Practice*, Oxford University Press, New York, NY (1998).

B. Baldo, *LCA Istruzioni per l'Uso*, Esculapio, Bologna (1998).

M. Kaltschmitt and G.A. Reinhardt (eds), *Nachwachsende Energieträger: Grundlagen, Verfahren, ökologische Bilanzierung* (Biobased Energy Sources: Basics, Procedures, Ecological Assessment), Vieweg Verlag, Braunschweig/Wiesbaden (1997).

G. Boyle (ed.), *Renewable Energy – Power for a Sustainable Future*, Oxford University Press, Oxford (1996).

M. Curran (ed.), *Environmental Life-cycle Assessment*, John Wiley & Sons, Inc., New York, NY (1996).

G. Rice, *LCA Software Review*, Centre for Environmental Strategy, University of Surrey (UK) (1996).

Canadian Standards Association, *Life Cycle Assessment*, Environmental Technology, Toronto (1994).

SETAC, *Guidelines for Life-cycle Assessment: A Code of Practice*, SETAC, Brussels (1993).

B.W. Vigon, D.A. Tolle, B.W. Cornaby, H.C. Latham, C.L. Harrison, T.L. Boguski, R.G. Hunt and J.D. Sellers, *Life-cycle Assessment: Inventory, Guidelines and Principles*, U.S. Environmental Protection Agency, Battelle and Franklin Associates, Ltd, Cincinnati, OH (1992).

J.B. Guinée, R. Heijungs and G. Huppes, *Manual for the Environmental Life Cycle Analysis of Products; Outline*, Leiden University, The Netherlands (1992).

Society of Environmental Toxicology & Chemistry, *A Technical Framework for Life Cycle Assessments*, SETAC, Washington, DC (1991).

O. Pinkus and B.F. Wilcock, *Strategy for Energy Conservation Through Tribology*, 2nd edn, ASME, New York, NY (1981).

11.5 References

[1] T. Pfeifer, W. Eversheim, W. König and M. Weck, *Manufacturing Excellence: The Competitive Edge*, Chapman & Hall, London (1994).

[2] M.P. Schneider, Plant-oil based lubricants and hydraulic fluids, *J. Sci. Food Agric.* **86**(12) (2006) 1769–80.
[3] W.J. Bartz, Lubricants and the environment, *Tribol. Intl.* **31**(1–3) (1998) 35–47.
[4] D. Kodali, High performance ester lubricants from natural oils, *Ind. Lubr. Tribol.* **54**(4) (2002) 165–70.
[5] P. De Caro and C. Cecutti, Lubrifiants et environnement, *Oléag., Corps Gras, Lipides* **12**(4) (2005) 279–84.
[6] J. Sander, The aspects of designing lubricants using an environmental product assessment, *NLGI Spokesman* **61**(11) (1998) 18.
[7] B.W. Vigon, D.A. Tolle, B.W. Cornaby, H.C. Latham, C.L. Harrison, T.L. Boguski, R.G. Hunt and J.D. Sellers, *Life-cycle Assessment: Inventory, Guidelines and Principles*, US Environmental Protection Agency, Battelle and Franklin Associates, Ltd, Cincinnati, OH (1992).
[8] ISO Standard 14040, *Environmental Management – Life Cycle Assessment – Principles and Framework*, International Organization for Standardization, Geneva (1997).
[9] ISO Standard 14041, *Environmental Management – Life Cycle Assessment – Goal and Scope Definition and Inventory Analysis*, International Organization for Standardization, Geneva (1998).
[10] ISO Standard 14042, *Environmental Management – Life Cycle Assessment – Life Cycle Impact Assessment*, International Organization for Standardization, Geneva (2000).
[11] ISO Standard 14043, *Environmental Management – Life Cycle Assessment – Life Cycle Interpretation*, International Organization for Standardization, Geneva (2000).
[12] ISO Standard 14044, *Environmental Management – Life Cycle Assessment – Requirements and Guidelines*, International Organization for Standardization, Geneva (2006).
[13] (a) SETAC, *Guidelines for Life-Cycle Assessment: A Code of Practice*, SETAC, Brussels (1993); (b) Deutsches Institut für Normung (ed.), DIN EN 14040 ff: *Umweltmanagement – Produkt-Ökobilanz*, Beuth-Verlag, Berlin (1997–99).
[14] H. Bewa (ed.), *Study for a Simplified LCA Methodology Adapted to Bioproducts*, ADEME/Bio Intelligence Service, Angers/Paris (Dec. 2009).
[15] H. Baumann and A.-M. Tillman, *The Hitch Hiker's Guide to LCA*, Studentlitteratur AB, Lund (2004).
[16] W. McDonough and M. Braungart, *Cradle to Cradle: Remaking the Way We Make Things*, North Point Press, New York, NY (2002).
[17] G.A. Norris, *SimaPro Database Manual*, The Franklin US LCI Library, PRé-Consultants, Amersfoort, The Netherlands (2003).
[18] M. Wang, *GREET – Greenhouse Gases, Regulated Emissions, and Energy Use in Transportation*, Argonne National Laboratory, Argonne, IL (1999).
[19] M. Wang, *Overview of GREET Model Development at Argonne, GREET User Workshop*, Argonne, IL (2007).
[20] Argonne National Laboratory, *The Greenhouse Gases, Regulated Emissions, and Energy Use in Transportation (GREET) Model* (2008); http://www.transportation.anl.gov/modeling_simulation/GREET/index.html
[21] R. Frischknecht and N. Jungbluth, *Overview and Methodology: Ecoinvent Data v.2.0*, Swiss Centre for Life Cycle Inventories, Dübendorf (2007).

[22] PRéConsultants, *SimaPro v.7.0*, Amersfoort, The Netherlands (2008).
[23] R. Spriensma, *SimaPro Database Manual: The BUWAL 250 Library*, PRé-Consultants, Amersfoort, The Netherlands (2004).
[24] J.C. Bare, G.A. Norris, D.W. Pennington and T. McKone, TRACI The tool for reduction and assessment of chemical and other environmental impacts, *J. Ind. Ecol.* **6**(3–4) (2003) 49–78.
[25] H. Wenzel, M. Hauschild and L. Alting, *Environmental Assessment of Products, Vol. I. Methodology, Tools, and Case Studies in Product Development*, Kluwer Academic Publishers, Dordrecht (2001).
[26] M. Goedkoop and R. Spriensma, *The Eco-indicator 99 – A Damage Oriented Method for Life Cycle Impact Assessment: Methodological Report*, PRéConsultants, Amersfoort, The Netherlands (2001).
[27] M. Goedkoop, *The Eco-indicator 95*, PRéConsultants, Amersfoort, The Netherlands (2004).
[28] L.C. Dryer, A.L. Niemann, and M.Z. Hauschild, Comparison of three different LCIA methods: EDIP97, CML2001 and Eco-indicator 99. Does it matter which one you choose? *Int. J. LCA* **8**(4) (2003) 191–200.
[29] O. Jolliet, M. Margni, R. Charles, S. Humbert, J. Payet, G. Rebitzer and R. Rosenbaum, IMPACT 2002+: A new life cycle impact assessment methodology, *Int. J. LCA* **8**(6) (2003) 324–30.
[30] N. Itsubo and A. Inaba, A new LCIA method: LIME has been completed, *Int. J. LCA* **8**(5) (2003) 305.
[31] J.C. Bare and T.P. Gloria, Critical analysis of the mathematical relationships and comprehensiveness of life cycle assessment approaches, *Environ. Sci. Technol.* **40**(4) (2006) 1104–13.
[32] R.T. Clemen, *Making Hard Decisions: An Introduction to Decision Analysis*, Duxbury Press, Pacific Grove, CA (1997).
[33] M.C. McManus, G.P. Hammond and C.R. Burrows, Life cycle assessment of mineral and rapeseed oil in mobile hydraulic systems, *J. Ind. Ecol.* **7**(3–4) (2004) 163–77.
[34] C. Våg, A. Marby, M. Kopp, L. Furberg and T. Norrby, A comparative life cycle assessment of the manufacture of base fluids for lubricants, *J. Synth. Lubr.* **19**(1) (2002) 39–57.
[35] P.S. Wightman, R.M. Davis, K.C. Walker, S.E. Batchelor, S.P. Carruthers and E.J. Booth, Life-cycle assessment of chainsaw lubricants made from rapeseed oil or mineral oil, *Proc. 10th Intl. Rapeseed Congress*, Canberra (August 1999).
[36] C. Ciantar and M. Hadfield, Investigating the sustainable development of domestic refrigerating systems working in an HFC-134a environment, in *Proc. Recent Developments in Refrigeration and Heat Pump Technologies* (Professional Engineering Publishing, ed.), London (1999), pp. 47–59.
[37] S.A. Miller, A.E. Landis, T.L. Theis and R.A. Reich, A comparative life cycle assessment of petroleum and soybean-based lubricants, *Environm. Sci. Technol.* **41**(11) (2007) 4143–9.
[38] C. Herrmann, J. Hesselbach, R. Bock, A. Zein, G. Öhlschläger and T. Dettmer, Ecological benign lubricants – Evaluation from a life cycle perspective, *Clean – Soil, Air, Water* **35**(5) (2007) 427–32.
[39] M. Levizzari, M. Voglino and P. Volpi, Environmental and economic impact or re-refined products: a life cycle analysis, in *Lubricants for the Future and Environment, Proc. 6th Intl. LFE Congress*, Brussels (1999).

[40] Taylor Nelson Sofres Consulting, *Critical Review of Existing Studies and Life Cycle Analysis on the Regeneration and Incineration of Waste Oils* (Report 20AW83–5 to European Commission DG Environment), Montrouge (December 2001).
[41] M. Völtz, *Ökobilanz für Schmierstoffe – Grundlagen und Vorgehensweise*, in *Lubricants, Materials and Lubrication Engineering* (W.J. Bartz, ed.), *Proc. 13th Int. Conf.*, Technische Akademie Esslingen (2002), p. 1041.
[42] T. Norrby, Environmentally adapted lubricants – Where are the opportunities?, *Ind. Lubr. Tribol.* **55**(6) (2003) 268–74.
[43] B. Cunningham, N. Battersby, W. Wehrmeyer and C. Fothergill, A sustainability assessment of a biolubricant, *J. Ind. Ecol.* **7**(3–4) (2004) 179–92.
[44] S. Boyde, Green lubricants – Environmental benefits and impacts of lubrication, *Green Chem.* **4** (2002) 293–307.
[45] R.I. Taylor, R.T. Dixon, F.D. Wayne and S. Gunsel, Lubricants and energy efficiency: life-cycle analysis, *Tribol. Interf. Engng. Ser.* **48** (Life Cycle Tribology) (2005) 565–72.
[46] J. Adamczewska and D. Wilson, Development of ecologically responsive lubricants, *Lubric. Sci.* **14**(2) (2006) 129–42.
[47] U.S. Department of Energy (DOE), Enhanced oil recovery/CO_2 injection, *Oil & Natural Gas Supply & Delivery* (2008); http://www.fossil.energy.gov/programs/oilgas/eor/index.html
[48] G.A. Reinhardt, R. Herbener and S.O. Gärtner, Life cycle analysis of lubricants from rape seed oil in comparison to conventional lubricants (2001); http://www.p2ways.org/ref/35/34311.pdf
[49] G. Zemanek and G.A. Reinhardt, Notes on life cycle assessments of vegetable oils, *Fett/Lipid* **101** (1999) 321–7.
[50] P. Cuevas, *Comparative Life Cycle Assessment of Biolubricants and Mineral Based Lubricants*, MSc. Thesis, University of Pittsburgh, Pittsburgh, PA (2010).
[51] S.A. Miller, *A Comparative Life Cycle Assessment of Soybean-Based and Mineral Oil Lubricants in Aluminum Rolling*, University of Illinois, Chicago, IL (2006); *Proc. 59th Ann. Mtg. American Society of Tribologists and Lubrication Engineers (STLE)*, Las Vegas, NV (15 May 2005).
[52] P.S. Wightman, R.M. Eavis, S.E. Batchelor, K.C. Walker, R. Bennet, S.P. Carruthers and R. Tranter, Comparison of rapeseed and mineral oils using life-cycle assessment and cost–benefit analysis, *OCL- Oléag., Corps Gras, Lipides* **6**(5) (1999) 384–8.
[53] P.S. Wightman, R.M. Eavis, K.C. Walker, S.E. Batchelor and S.P. Carruthers, A comparative LCA of hydraulic lubricants made from mineral oil and rapeseed oil, *Proc. 6th Symp. on Renewable Resources for the Chemical Industry*, Bonn (March 1999).
[54] S.A. Miller and T.L. Theis, Comparison of life-cycle inventory databases: a case study using soybean production, *J. Ind. Ecol.* **10**(1–2) (2006) 133–47.
[55] L. Panichelli, A. Dauriat and E. Gnansounou, Life cycle assessment of soybean-based biodiesel in Argentina for export, *Int. J. LCA* **14**(2) (2009) 144–59.
[56] J.H. Schmidt, P. Christensen and T.S. Christensen, Assessing the land use implications of biodiesel use from an LCA perspective, *J. Land Use Sci.* **4**(1–2) (2009) 35–52.

[57] H. Kim, S. Kim and B.E. Dale, Biofuels, land use change, and greenhouse gas emissions: Some unexplored variables, *Environm. Sci. Technol.* **43**(3) (2009) 961–7.

[58] J. Fargione, J. Hill, D. Tilman, S. Polasky and P. Hawthorne, Land clearing and biofuel carbon debt, *Science* **319** (28 Feb. 2008) 1235–8.

[59] T. Searchinger, R. Heimlich, R.A. Houghton, F. Dong, A. Elobeid, J. Fabiosa, S. Tokgoz, D. Hayes and T.-H. Yu, Use of US croplands for biofuels increases greenhouse gases through emissions from land-use change, *Science* **319** (29 February 2008) 1238–40.

[60] D. Tilman, J. Hill and C. Lehman, Carbon-negative biofuels from low-input high-diversity grassland biomass, *Science* **314** (2006) 1598–600.

[61] C. Bowyer, *Anticipated Indirect Land Use Change Associated with Expanded Use of Biofuels and Bioliquids in the EU – An Analysis of National Renewable Energy Action Plan*, Institute for European Environmental Policy (IEEP), London (March 2011).

[62] E. Poitrat, Ecological assessment of rapeseed methyl ester in comparison to mineral fuel (life cycle analysis), *Proc. 8th European Biomass Conference*, Vienna (3–5 October 1994), Vol. 2, pp. 1081–93.

[63] G.A. Reinhardt, J. Borken, A. Patyk and G. Zemanek, *Ressourcen- und Emissionsbilanzen, Räpsol und RME im Vergleich zu Dieselkraftstoff*, IFEU, Heidelberg (1997).

[64] G.A. Reinhardt, Life cycle assessments (LCAs) of liquid biofuels: Latest results, *Proc. 2nd European Motor Biofuels Forum*, Graz (22–25 September 1996), pp. 297–303.

[65] Levelton Engineering and (S&T)^2 Consulting Inc., *Assessment of Biodiesel and Ethanol Diesel Blends. Greenhouse Gas Emissions, Exhaust Emissions, and Policy Issues*, Ottawa, ON (2002).

[66] S. Kim and B.E. Dale, Cumulative energy and global warming impact from the production of biomass for biobased products, *J. Ind. Ecol.* **7**(3–4) (2004) 147–62.

[67] S.O. Gärntner and G.A. Reinhardt, Environmental implications of biodiesel (life-cycle assessment), in *The Biodiesel Handbook* (G. Knothe, J. Van Gerpen and J. Krahl, eds), AOCS Press, Urbana, IL (2005), pp. 219–30.

[68] H. Helms, G.A. Reinhardt and N. Rettenmaier, Bioenergie aus Palmöl: Ökologische Chancen und Risiken, Energiewirtsch. Tagesfragen, *Ztschr. f. Energiewirtsch., Recht, Techn. Umwelt* **56**(11) (2006) 70–3.

[69] N. Rettenmaier, Life-cycle assessment of palm oil biodiesel, *Proc. Intl. Congress on Biodiesel*, Vienna (5–7 November 2007).

[70] L. Panichelli, *Análisis de Ciclo de Vida (ACV) de la Producción de Biodiesel (B100) en Argentina*, Graduate School of Agronomy, University of Buenos Aires, BA (2006).

[71] H. Halleux, S. Lassaux, R. Renzoni and A. Germain, Comparative life cycle assessment of two biofuels: ethanol from sugar beet and rapeseed methyl ester, *Intl. J. LCA* **13**(3) (2008) 184–90.

[72] N. Pelletier, N. Arsenault and P. Tyedmers, Scenario modeling potential eco-efficiency gains from a transition to organic agriculture: life cycle perspectives on Canadian canola, corn, soy, and wheat production, *Environm. Managm.* **42**(6) (2008) 989–1001.

[73] R. Dalgaard, J. Schmidt, N. Halberg, P. Christensen, M. Thrane and W.A. Pengue, LCA of soybean meal, *Intl. J. LCA* **13**(3) (2008) 240–54.
[74] H. Huo, M. Wang, C. Bloyd and V. Putsche, Life-cycle assessment of energy use and greenhouse gas emissions of soybean-derived biodiesel and renewable fuels, *Environm. Sci. Technol.* **43**(3) (2009) 750–6.
[75] J.H. Schmidt, Comparative life cycle assessment of rapeseed oil and palm oil, *Intl. J. LCA* **15**(2) (2010) 183–97.
[76] J.H. Schmidt, *Life Cycle Assessment of Rapeseed and Palm Oil*, Department of Development and Planning, Aalborg University (2007).
[77] N. Jungbluth, M. Chudacoff, A. Dauriat, F. Dinkel, G. Doka, M.F. Emmenegger, E. Gnansounou, N. Kljun and M. Spielmann, *Life Cycle Inventories of Bioenergy*, Final Rept. ecoinvent v.2.0 No. 17, Swiss Centre for Life Cycle Inventories, Dübendorf (2007).
[78] R. Hischier, R. Zah, P. Wäger, M. Lehmann and M. Gauch, *Life Cycle Assessment of Energy Products: Environmental Impact Assessment of Biofuels – Executive Summary* (Ökobilanz von Energieprodukten: ökologische Bewertung von Biotreibstoffen – Schlußbericht), EMPA, St. Gallen (2007).
[79] R. Zah, H. Böni, M. Gauch, R. Hischier, M. Lehmann and P. Wäger, *Life Cycle Assessment of Energy Products: Environmental Assessment of Biofuels*, EMPA, St. Gallen (2007).
[80] L. Lardon, A. Hélias, B. Sialve, J.-P. Steyer and O. Bernard, Life-cycle assessment of biodiesel production from microalgae, *Environm. Sci. Technol.* **43**(17) (2009) 6475–81.
[81] A.L. Stephenson, E. Kazamia, J.S. Dennis, C.J. Howe, S.A. Scott and A.G. Smith, Life-cycle assessment of potential algal biodiesel production in the United Kingdom: A comparison of raceways and air-lift tubular bioreactors, *En. Fuels* **24** (2010) 4062–77.
[82] P.R.D. Williams, D. Inman, A. Aden and G.A. Heath, Environmental and sustainability factors associated with next-generation biofuels in the U.S.: what do we really know?, *Environm. Sci. Technol.* **43**(13) (2009) 4763–75.
[83] J.C.J. Bart, N. Palmeri and S. Cavallaro, *Biodiesel Science and Technology. From Soil to Oil*, Woodhead Publishing, Ltd/CRC Press, Cambridge/Boca Raton, Fl (2010), pp. 630–7.
[84] T.-U. Eden and G.A. Reinhardt, Environmental implications of biodiesel use. An assessment, in *Biomass for Energy, Environment, Agriculture and Industry, Proc. 8th European Biomass Conference* (P. Chartier, A.A.C.M. Beenackers and G. Grassi, eds), Vienna (3–5 October 1994), Elsevier, Oxford (1995), pp. 1141–8.
[85] M. Völtz, Umweltfreundliche Schmierstoffe – Ein Überblick, in *Biologisch schnell abbaubaure Schmierstoffe und Arbeitsflüssigkeiten* (W.J. Bartz, ed.), Expert-Verlag, Ehningen (1993), pp. 136–57.
[86] Brundtland Report A/RES/42/187, United Nations World Commission on Environment and Development, New York, NY (4 August 1987).
[87] A. Azapagic and S. Perdan, Indicators of sustainable development for industry: a general framework, *TransIMechE* **78**(B) (2000) 243–61.
[88] D. Meadows, *Indicators and Information Systems for Sustainable Development*, International Institute for Sustainable Development, Winnipeg, MB (1998).

[89] J. Fiksel, J. McDaniel and D. Spitzley, Measuring product sustainability, *J. Sustain. Prod. Design* **6**(7) (1998) 7–19.

[90] O. Pinkus and B.F. Wilcock, *Strategy for Energy Conservation through Tribology*, 2nd edn, ASME, New York, NY (1981).

[91] J. Van Rensselar, How basestocks and additives are improving energy efficiency, *Tribol. Lubr. Technol.* **67**(10) (2011) 34–46.

[92] L.A. Bronshteyn and J.H. Kreiner, Energy efficiency of industrial oils, *Tribol. Lubr. Technol.* **67**(10) (2010) 48–56, and references therein contained.

[93] C. Naruse, R. Nemoto, S. Haizuka and M. Yoshizaki, Influence of oil viscosity, chemical structure, and chemical additives on friction loss of spur gears. Concerning the influence of synthetic oil and mineral oil, *Tribol. Trans.* **37**(2) (1994) 358–68.

[94] P.J. Douglas, An environmental case for synthetic lubricants, *Lubric. Engng.* **48**(9) (1992) 696–700.

[95] K.A. Hildebrant and J.R. Norgate, Energy savings using synthetic lubricants, *Canadian Institute of Mining, Metallurgy and Petroleum (CIM) Bulletin* **87**(984) (1994) 44–6.

[96] T. Sasaki, I. Ohmori, M. Furumoto, H. Tanaka, K. Komiya, T. Ohsumi and M. Henmi, Development of automotive lubricants based on high-viscosity index base stock, *SAE Techn. Paper Series* No. 951028 (1995).

[97] J. Fotheringham, G. Bleimschein and A. Plomer, Base oil effects on emissions and fuel economy in a heavy-duty vehicle over the European Transient Cycle (ETC), *SAE Techn. Paper Series* No. 2002-01-2768 (2002).

[98] M.J. Souza de Carvalho, P.R. Seidl, C.R. Pereira Belchior and J.R. Sodré, Lubricant viscosity and viscosity improver additive effects on diesel fuel economy. *Tribol. Intl.* **43** (2010) 2298–302.

[99] A. Comfort, An introduction to heavy-duty diesel engine frictional losses and lubricant properties affecting fuel economy. Part I, *SAE Techn. Paper Series* No. 2003-01-3225 (2003).

[100] P.E. Fox, Numerical evaluation of the potential for fuel economy improvement due to boundary friction reduction within heavy-duty diesel engines, *Tribol. Intl.* **38** (2005) 265–75.

[101] E. Tomanik, Friction and wear bench tests of different engine liner surface finishes, *Tribol. Intl.* **41** (2008) 1032–8.

[102] B.P. Williamson and A. Milton, Characterization of the viscoelasticity of engine lubricants at elevated temperatures and shear rates, *SAE Techn. Paper Series* No. 951032 (1995).

[103] M. Nakada, Requirements to lubricants from engine technology, *Proc. 1st Ann. Fuels and Lubes Asia Conference*, Singapore (23–25 January 1995).

[104] R.I. Taylor, R. Mainwaring and R.M. Mortier, Engine lubricant trends since 1990, *J. Engng. Tribol.* **219** (2005) 331–46.

[105] M. Smeeth, H.A. Spikes and S. Gunsel, The formation of various surface films by polymer solutions: boundary or elastohydrodynamic lubrication?, *Tribol. Trans.* **39** (1996) 720–5.

[106] K. Inoue, E. Tominaga, K. Akiyama and T. Ashida, Effects of lubricant composition on fuel efficiency in modern engines, *SAE Techn. Paper Series* No. 951037 (1995).

[107] M.T. Devlin, W.Y. Lam and T.F. McDonnell, Critical oil physical properties that control the fuel economy performance of General Motors vehicles, *SAE Techn. Paper Series* No. 982503 (1998).
[108] A.J. Moore, Fuel efficiency screening tests for automotive engine oils, *SAE Techn. Paper Series* No. 932689 (1993).
[109] S.I. Tseregounis, M.L. McMillan and R. Olree, Engine oil effects on fuel economy in GM vehicles. Separation of viscosity and friction modifier effects. *SAE Techn. Paper Series* No. 982502 (1998).
[110] R. Glyde, Smoothing the ride, *Chem. Brit.* (July 1997) 39–41.
[111] M. Manni, C. Gomellini and G. Sabbioni, Effect of physical characteristics of lubricating oils on emissions, fuel, economy and oil consumption in a light duty diesel engines *SAE Techn. Paper Series* No. 952552 (1995).
[112] T.T. Murtonen and M. Sutton, New crankcase lubricants for heavy-duty diesel engines: effect on fuel consumption and exhaust emissions, *SAE Techn. Paper Series* No. 2005-01-3717 (2005).
[113] K. Michaelis and B.R. Horn, Influence of lubricants on power loss of cylindrical gears, *Tribol. Trans.* **37**(1) (1994) 161–7.

12

Biolubricant product groups and technological applications

Abstract: This chapter focuses on specific applications of biodegradable lubricants, both bio-based and from mineral origin. Biolubricants are attractive in both the industrial and automotive market segments although to a different extent. Consumer acceptance of these materials, ranging from total-loss oils to hydraulic fluids, gear lubrication oils, compressor and refrigeration oils, turbine oils, metalworking and machining fluids, to crankcase oils, food-processing oils and greases, depends largely on how well they perform during use. Minimum performance specifications are outlined per product group. Performance of biolubricants is frequently comparable to conventional products without environmental compatibility. Direct use of high-viscosity native vegetable oils as biolubricants is limited.

Key words: industrial lubricants, automotive lubricants, specialties, biolubricant formulations, technological applications.

12.1 Introduction

Around 2000 BC the Sumerians and Egyptians looked for practical solutions to move axle bearings and potter's wheels smoothly, and to make bore holes, though of course without any theoretical notion of the concept of tribology. The earliest lubricants were oils, fats, water and air. An Egyptian bas-relief, found in Tehuti-Hetep's grave (1880 BC), documents the use of a lubricant (water or oil) to reduce friction between the surface and a sleigh used for transportation of a heavy statue. Nowadays our high-tech world requires a very great variety of specially designed lubricants for numerous different applications. In many cases materials are subjected to highly stressing temperature and pressure or voltage conditions, sliding speeds, and to variations in temperature, pressure or corrosion.

The primary functions of lubricants are controlled friction and wear and other effects, including heat transfer, contaminant suspension, liquid sealing and corrosion protection. Between 5000 and 10 000 different lubricant formulations are necessary to satisfy more than 90% of all lubricant applications. Lubricants (from fluid to solid) are designed for specific uses and applications to ensure they meet system specifications and operating conditions. The following major types of components have lubricant requirements: plain bearings, rolling contact bearings, enclosed gears, steam turbines, open gears, ropes and chains, clock and instrument pivots, and hinges, locks, and latches. Each component demands the use of a particular lubricant. For example, enclosed gears usually utilise oil for lubrication, while rolling contact bearings typically use grease, which prevents lubricant loss and lubricant contamination. However, many others factors influence lubricant selection (see Section 14.2).

Modern lubricants are a product of intensive R&D and generally are robust and durable. Commercial lubricants are quality products designed to provide effective and economical performance when they are used as recommended in the applications for which they are intended. Manufacturer-specific literature is widely available (e.g. refs [1, 2]). Several manufacturers provide helpdesk services (see Section 13.1). The quality of lubricants is to be guaranteed by a quality management (QM) system such as according to the automotive standard ISO/TS 16949 [3]. Every practical precaution in product storage and handling is taken by the manufacturer to ensure that the products are maintained on specification through delivery to the customer [4].

This chapter describes the most important tribosystems. In terms of applications, fluids and lubricants can be classified broadly as automotive (see Section 12.2), industrial, marine (see Section 12.5) and aviation (see Section 12.8.1), and process oils (see Section 12.14). An alternative classification is as engine and non-engine lubricants (see Table 12.1). Industrial oils and greases are widely used in the chemical sector, metalworking industry, paper industry, food processing and elsewhere. Industrial lubricants are usually subdivided in industrial oils and their specialties, the latter being primarily greases, metalworking lubricants and solid lubricant films. Process oils, used as raw materials in processes (e.g. rubber processing), are a special product category in terms of the concept of lubrication.

Lubricating oils find wide application in internal combustion engines (from automotive to marine), general industrial uses (as turbine oils, fire-resistant fluids, etc.), as hydraulic fluids and transmission oils (automatic transmission fluids, tractor universal oils, power steering fluids, brake oils, etc.), and as grease (for automotive and industrial use). Other typical applications include drilling fluids, precoat oils, metalworking oils, cutting fluids, gear lubrication oils, way lubricants, steam cylinder oils, aviation oils, compressor oils, textile lubricants, circulating oils, spindle oils, food-grade and food-processing oils, and much more. Cars, engines, elevators, lawnmowers, chainsaws, railroad tracks, bikes, machinery, pumps, heavy equipment (earthmoving machines, shovels, excavators and tractors) used in road construction, hydraulic engineering and mining, and numerous other types of mobile or static equipment greatly depend on lubricants for smooth operation. The introduction of new engineering materials has further increased the demand for tailor-made lubricants. Performance trends call for base oils

Table 12.1 Classification of lubricants as engine and non-engine lubricants

Engine lubricants
• Gasoline engine oils
• Diesel engine oils
Automotive diesel oils
Stationary diesel oils
Railroad diesel oils
Marine diesel oils
• Natural gas engine oils
• Aviation engine oils
• Two-stroke cycle engine oils
Non-engine lubricants
• Transmission fluids
Automatic transmission fluids
Manual transmission fluids
Power transmission fluids
• Power steering fluids
• Shock absorber fluids
• Gear oils
Automotive gear oils
Industrial gear oils
• Hydraulic fluids
Tractor hydraulic fluids
Industrial hydraulic fluids
• Metalworking fluids
• Greases
• Specialities

with higher viscosity index (VI), lower volatility, zero sulphur, excellent oxidation stability and improved additive response.

12.1.1 Technical application requirements of lubricants

Given the great number of product groups, technical applications and lubricant formulations, in this context only an illustrative rather than comprehensive overview can be presented of the present status of use and future opportunities of biolubricants. For each lubricant product group the general technical application requirements are given with the object of examining use and challenges for bio-based lubricants. There are many characteristics of a (bio)lubricant that need to be considered when matching a formulation to a given application. The requirements to lubricant/grease applications vary considerably per product group (see Table 12.2). Lubricants with very strict and tight specifications serve in technologically advanced areas, such as aeronautics, automotive, turbines, etc. In other technological applications extremely high stability is not required. Oils synthesised from natural prod-

Table 12.2 Selected requirements to various lubricant applications

Lubricant	Requirements
Passenger car motor oils (PCMO)	Biodegradable, non-toxic Reduced metal-organic additives High fuel efficiency Low exhaust emission levels Long drain intervals
Heavy-duty engine oils	Lubrication and cooling Wear protection Soot and particle handling; deposit characteristics
High-temperature lubricants	Thermo-oxidative properties Boundary lubrication properties Deposit forming tendencies
Metalworking lubricants	Biodegradability Safety
Tunnel boring machinery	Biodegradability Superior tackiness and adhesion characteristics Resistance to water spray off Good pumpability Good anti-wear/extreme pressure (AW/EP) properties Fire resistance
Aqueous lubricants	Excellent drag coefficient Compatibility with printing oils Low stress cracking of polymers

Table 12.3 US consumption of lubricants (%)

Crankcase	76.4	Metal cutting	0.25
Hydraulic	14.1	Bar chain	0.15
Marine	3.7	Wire rope	0.13
Turbine	2.8	Rail flange	0.06
Two-cycle	1.3	Dust control	0.06
Drip oil	1.0		

ucts are generally approved of as environmentally friendly functional fluids and can therefore be used widely, for example in small machinery as chainsaws and pumps.

The major oil uses of lubricants (by volume) are for engine oils, transmission oils, gear oils (automotive and industrial), hydraulic oils and greases, where the first three make up about 80%, hydraulic over 10% and grease 3% of the world total use of lubricants. Table 12.3 shows a rough breakdown of US consumption of lubricants per product group.

Mineral oils still take the lion's share of the lubricant market. Currently, the synthetic fluids that are manufactured commercially are mainly poly-α-olefins (PAOs), polyalkylene glycols (PAGs) and synthetic esters (SEs). With the introduction of the more highly biodegradable synthetic esters to a wide variety of market segments other than hydraulic fluids, a broader spectrum of performance attributes is required. For example, use in aircraft turbine oils [5], two- and four-stroke engines [6] and refrigerant oils [7] extends requirements for base oils with new and improved lubrication characteristics. Vegetable oil-based lubricants are slowly emerging as environmentally friendly, high-performance alternatives to petroleum-based lubricants. As early as 1995, Castrol (Malaysia), Ltd had launched Castrol Palmtec 2T and 4T as the world's first palm-based motorcycle and passenger car engine oils [8]. Vegetable oil-based products are now finding increasing use as lubricants for industrial and transportation applications with quite satisfactory performance. Although the use of biodegradable lubricants in diverse applications continues to increase, prospectives for rapid replacement of mineral oil for biolubricants vary greatly by application. Industrial oils are an important focus area for plant-based lubricant manufacturers.

Optimisation of lubricant performance in use in terms of friction, wear and lubricant stability across the very many widely different lubricant applications, each with greatly different performance criteria, requires a wide range of practical solutions and development of increasingly specialised products. Product application specialists may offer good advice. Lubricant developers nowadays do benefit from being able to draw on a wide range of available raw materials and functional additives which permits optimisation of application performance through chemical design at the molecular level.

Price/performance is of great importance to the end-user. Whereas the purchase cost of synthetic lubricants is some three times higher, the oil drain intervals are frequently also three times longer than the service life of mineral oils. An overall cost–benefit analysis of gear lubrication oils, considering also longer life and reduction in service cost, waste oil, stop periods, etc., indicates that use of synthetic base fluids can substantially reduce total lubricant-related system costs.

Lubricating agents are used on a very large scale. Every mechanically operated device requires some form of lubrication if inadvertent material damage is to be avoided. High-performance lubricants, which are usually liquids under operating conditions and are being introduced between two moving surfaces in machinery so as to reduce friction and wear, are meant to ensure a safe and trouble-free power transmission and to improve service life. Oil lubrication is suitable when heat generated from a bearing or heat being transferred to a bearing must be discharged. Some stabilisation problems are solved with blends or combinations of lubricants and additives, making use of synergic effects. In many cases, formulations are being improved and optimised by using existing products with the aim of obtaining better, and above all, cheaper products, thus avoiding the time-consuming and expensive development of entirely new chemicals.

Different engineering designs make different demands on the lubricant used. Lubricating oils need to fulfil both mechanical-dynamic requirements and physicochemical specifications. The most important parameter for determining the lubricity of oils is viscosity. The demands made on today's high-performance machines are constantly growing. Increased utilisation and efficiency lead to higher speeds and greater forces. This results in ever-increasing stress on the lubricant. With improved quality of lubricant base oil/additive formulations, more demanding higher operating temperatures can be met and the output of the equipment is increased. Industrial production processes can thus reach higher levels of efficiency and performance and products are improved through the added value of lubrication. Modern industrial lubricants have excellent long-life characteristics which provide optimum lubricity and minimum wear even under severe loads. In general, fast running machines permit the use of lower viscosities than slow running machines. With the present installation of high-speed equipment, lubricant manufacturers are frequently faced with the challenge of increased heat generation. High bearing loads require higher viscosities than low bearing loads. Higher temperatures and pressures increase the demands made on the lubricant, e.g. for turbine oils. In general, oil lubrication is more suitable for high-speed running applications or high-temperature applications than grease application. Bright stocks are used in applications where lubrication is needed at high temperature or under heavy loads (monograde engine oils, industrial gear oils, marine oils).

12.1.2 Applications and performance of biolubricants

Special technical requirements for lubricants differ according to their function. However, primary and total biodegradability and adequate ecotoxicity are nowadays to be considered as general important assets for a lubricant formulation, in particular for environmentally sensitive applications such as farm tractors and marine dredging. Technical, economic and legislative aspects play a role in the replacement of lubricants of petroleum origin by those based on renewable raw materials. Biolubricants are now available for numerous applications, with performance being comparable to and sometimes even better than that of mineral oil products. Bio-based materials have clear advantages and are already economically competitive in a growing number of industrial sectors but further development work is required to achieve additional cost reduction. The best opportunity for biolubricant usage is in situations where the lubricant can be unintentionally exposed to humans, food or the environment, since in these scenarios the use of high-toxicity products has the potential to cause severe damage [9]. This mainly concerns lubricants used in total-loss applications, systems with leakage, or the food and pharma industry. For biolubricant formulations to be used in these applications also the additives should ideally be biodegradable and have low toxicity characteristics [10]. Bio-based lubricants can be utilised in many applications (Table 12.4).

Recently, lubricant technology has greatly advanced as a result of growing environmental concerns, driven by public demand, industrial R&D (see Section 10.5), legislation and governmental agency policies (see Chapter 9). Sustainable products (both oils and greases) have now been brought to the forefront in many high-profile applications in environmentally sensitive areas and offer a level of performance on a par with their mineral-based lubricant counterparts [12].

Vegetable oils and polyol esters are such highly biodegradable materials. Natural oils, which include vegetable oils, possess the best biodegradability of all the lubricating base stocks. Many factors should be considered when choosing a biodegradable fluid for a given application, including temperature, pressure, fluid life, water interference, seals and elastomers, spill potential and cost. Tables 12.10, 12.15, 12.17 and 12.21 allow comparison of physicochemical properties of biodegradable base fluids and mineral oils. Compared with equiviscous mineral oils, rapidly biodegradable rapeseed oil (RSO)- or ester-based products reduce emissions by up to 90%. High flash points ensure higher safety on the shop floor. Tool life is enhanced because the higher wetting tendency of polar esters leads to friction reduction. Overall cost savings are on account of less maintenance, personnel, storage and disposal cost. Compromised performance and increased cost are frequently no longer critical issues when making the change to environmentally considerate lubricants. Biolubricants can also

Table 12.4 Major applications for biodegradable lubricants

Total-loss lubricants	Sealed/circulating systems
• Chain bar oils	• Hydraulic fluids
• Concrete mould-release oils	• Compressor oils
• Two-stroke oils	• Milking machine lubes
• Chain and conveyor lubes	• Engine oils[a]
• Greases	• Gear and transmission lubes[a]
• Corrosion preventive oils	• Metalworking fluids[a]
• Wire rope lubricants	• Machine tool lubricants[a]

[a] Under development.
After ref. [11].

perform equally well if not better than petroleum lubricants as additives to enhance power output and combustion characteristics of engines running a biofuel [13].

Lubricants have multiple functions in practical applications (see Tables 12.7 and 12.12). Plant-based fluids meet naturally several of these requirements ideally with excellent tribological properties (ester functions sticking to metal surfaces), lower friction coefficients and lower evaporation (up to 20% less) than mineral oil-based fluids. Not surprisingly, therefore, in the beginning of the development of biolubricants the focus was on the formulation of products based on pure vegetable oils. The first biodegradable lubricants were introduced in the 1980s, i.e. as two-stroke engine oils for boating (1980) and chainsaw lubricants (1985) [11], but meanwhile numerous other applications have followed suit both as total-loss lubricants (TLLs), where the lubrication oils are lost to the environment (see Table 12.4), and in circulating systems. Of the various product groups, total-loss applications, chain bar oils, two-stroke oils and hydraulic fluids have undergone the highest conversion into bio-based lubricant formulations. In applications where engine oils and gear and transmission lubes are contained in sealed or circulating systems the environmental problem is less critical. However, the major commercial potential for biolubricants lies in the conversion of automotive and industrial oils that make up the bulk of the total global lubricant consumption.

Native plant oils can only be employed for low-performance applications, at least without suitable additivation. More recently, synthetic ester types are also being used, which may be partly derived from renewable resources. Fully saturated (branched alcohol) diesters generally show good low-temperature performance and high VI. By adjusting the chain length of the dicarboxylic acid the viscosity can be modified. These oils are highly stable towards oxidation. As vegetable oil-based lubricants are prone to thermal degradation, friction heating limits their lifetime. The use of advanced low friction coatings offers a solution to this problem (LUBRICOAT project) [14].

Relative stabilities are established as follows: bio-based esters (straight vegetable oils) < bio-based synthetic esters (e.g. TMPs) < mineral oil-based saturated synthetic esters, mineral oils, PAOs. Bio-based synthetic esters in which the glycerol (GL) moiety has been replaced by trimethylolpropane (TMP), neopentyl glycol (NPG) or pentaerythritol (PE) polyols show improved lifetime, are suitable for higher performance requirements and frequently outperform mineral oil-based fluids. The lower the degree of unsaturation, the longer the lifetime. As oxidative stability increases, biodegradability decreases. Consequently, a compromise between lubrication performance and biodegradability is always necessary (see Section 12.9 for metalworking fluids). Except for total-loss applications, additives against corrosion, attrition (wear and tear) and oxidation are usually needed, regardless of the employed base oil. Also low-temperature performance (viscosity and pour point, PP) may be improved by additivation.

While for loss lubricants the rapidly biodegradable lubricants (native or synthetic base oils) dominate the market with about 80%, for all kinds of lubricants without loss lubricants the biolubricant market share is still disappointingly low (4% in Germany, 2005) in a nearly constant overall market volume [15]. The main deterrent for wider use of biolubricants in other applications is often still the perceived cost. There is a strong need for effective lubricants for general use in internal combustion engines and in industrial machinery, hydraulics and metalworking and in loss applications, as well as for lubricating greases which are renewable, biodegradable, non-toxic to the environment and cost effective to produce and to market. For the last two decades, industry has been trying hard (and successfully) to formulate biodegradable lubricants with satisfactory technical characteristics. The use of environmentally compatible lubricants enhances the image of the lubricant manufacturer and vendor as well as of the equipment manufacturer who employs such a fluid.

Bio-based lubricants are attractive in both the industrial and automotive market segments although to a different extent. Consumer acceptance of biodegradable materials such as total-loss oils, including marine oils, two-stroke engine oils, drip oils and drilling oils, industrial lubricants and greases such as hydraulic fluids, gear lubrication oils, compressor and refrigeration oils, turbine oils, metalworking and machining fluids, as well as biolubricants for internal combustion engines and for the food-processing industry, depends largely on how well they perform during use. Advances in the biotechnology of vegetable oils and chemical derivatisations of natural esters have resulted in high-performance biolubricants partly at a lower cost compared to mineral oils and synthetic lubricants.

Biodegradable lubricants – both bio-based and from mineral origin – are now available for numerous applications. From a technical point of view, more than 90% of all current lubricants can be formulated to be rapidly biodegradable. This sets the total market potential. Important biolubricant product groups are discussed in the following sections in more detail.

Currently, some 20 Mt/yr of fatty products are used for industrial purposes, including biosolvents, printing inks, biodiesel and biolubricants. At present, vegetable oils as the main biodegradable lubricant base stock represent a niche market [16]. Because of their physical properties, refined, chemically unchanged vegetable oils can be proposed as alternatives to petroleum products in a large number of lubricants. Compared with mineral oils, their performance in most applications is limited by their relatively poor ageing resistance.

Vegetable oil-based lubricants and hydraulic fluids differ from mineral oils in their chemical properties. Hence, some machines need adjusting if they are to be run with biolubricants. However, only minor adjustments are necessary in most cases (often in relation to material compatibility), and a number of original equipment manufacturers (OEMs) have already approved their products for biodegradable oils. Occasionally, entire systems have undergone fundamental changes (e.g. two-cycle engines). Entirely new requirements have been formulated in case of the conversion of refrigeration systems to chlorofluorocarbon (CFC)-free refrigerants or the introduction of biodegradable hydraulic fluids.

Some of the rapidly biodegradable lubricants are based on pure, unmodified vegetable oils (typically RSO, sunflower oil (SNO)). Natural triglycerides are very rapidly biodegradable and are highly effective lubricants. However, as their thermal, oxidation and hydrolytic stability is limited, pure vegetable oil formulations are primarily indicated for low-performance demands such as total-loss applications, including rail oils for trains, bar/chain oils for wood cutting (100%), metal cutting oils and concrete mould release oils, hydraulic oils and transmission fluids, those with very low thermal stress, as well as for products with relatively low technical specifications (mould-release oils and some types of greases). Applications of neat oils are restricted to $T < 70\,°C$. Use of vegetable oils (e.g. RSO, SNO, palm oil) has been claimed as substitutes for petroleum-based lubricating oil and synthetic esters [17–19]. RSO has good lubricating and protective (anti-corrosive) properties. Its high VI (210), meaning low variation of viscosity with temperature, is not equalled by any additivated mineral oil. Rapeseed oil is used in hydraulic devices and chainsaw lubrication [20–22]. Canola oil has superior low-temperature stability (cryogenic differential scanning calorimetry (DSC), PP, chronopotentiometry (CP)).

The oxidative stability of vegetable oils has been improved by using genetically modified (GM) oils (such as high-oleic soybean oil (HOSBO)) [23], transesterification of TMP and rapeseed oil methyl ester (RSME) [24], and by selective hydrogenation of polyunsaturated bonds of fatty acids (FA) chains [25], as well as by additivation. A high-oleic sunflower oil (HOSNO)-based lubricant formulation (BIOVIT) has successfully been used in textile and tannery industries, although with lack of economic advantage compared with conventional mineral oils [26]; see Section 12.12. A blend of high-oleic vegetable oil and castor oil (with a 10–15% hydroxy fatty acid (HFA) content) is an excellent feedstock for a variety of lubricant applications, from crankcase oil to hydraulic and two-stroke.

Renewable lubricants have moved into well-established usage [11] and biolubricants with technical performance comparable to or even better than conventional lubricants are already commercially available for the majority of applications. For example, numerous applications are covered by the

FUCHS (Mannheim, Germany) PLANTO product range and the demand for new products drives further development. PLANTO products are available in the following fields:

- total-loss lubricants (e.g. chainsaw oils for forestry use, two-stroke engine oils, concrete mould-release oils, greases, lubricants for railroad applications); and
- circulated lubricants (hydraulic oils, engine oils, gear oils, multipurpose oils, metalworking oils).

Despite the fact that up to now vegetable oil-based lubricants lack a considerable market share, for almost all applications a suited oleochemical ester can be selected from the great variety of existing structures. Some applications primarily use a transesterified vegetable oil, converting the glycerides to a free fatty acid (FFA) form prior to use. Esters derived from vegetable oils have several actual and potential applications as biofuel, lubricants, hydraulic fluids, ingredients for greases and printing inks, solvents, etc. At variance to ultralow-sulphur fuels, which fail to provide adequate lubrication for the fuel injection system of modern diesel engines, virgin vegetable oils and biodiesel (fatty acid methyl esters (FAMEs)) act as a lubricity additive to petroleum diesel fuels [27, 28]; see also Section 10.5.5.

Esters have been used as lubricating oils in various applications for almost 70 years although initially not because of their environmental properties. In the mid-1980s, biodegradable chainsaw oils based on natural esters of RSO were introduced on the market as a result of regulatory pressure. Esters of vegetable oils (TMP esters) are now used as hydraulic oils or for drilling ($T > 70\,°C$). They may be applied in numerous other applications as straight metalworking fluids, food industry lubricants (food-grade or just food-processing), open gear oils, biodegradable grease, hydraulic fluids, marine oils and outboard engine lubricants, oils for water and underground pumps, rail flange lubricants, shock absorber lubricants, tractor oils, agricultural equipment lubricants, elevator oils, mould-release oils, automotive two-stroke engine oils, transmission/gear oils, fuel additives, textile and tannery industries and forestry applications. As early as the 1960s, London Bus and other companies were using castor oil-based fluids in their rear axles, resulting in an excellent low coefficient of friction. Market sectors such as metalworking, gear oils, turbine oils and engine oils require more development effort. Saturated esters of modified vegetable oils with superior performance with respect to mineral oils can be used in automotive or aeronautics or as additives in four-stroke engines, with hydrolysis and oxidation resistance up to 120 °C. In extreme, high-temperature, environmentally sensitive applications readily biodegradable synthetic fluids should be used. SAE low-temperature viscosity requirements have been difficult to meet in vegetable-based oils. Disposal of ester-based lubricants is generally not problematic.

In the past, applications of vegetable oils and derivatives for lubrication have also been focused on the use of these oils as additives to a petroleum base oil. US Patent No. 5,282,989 to International Lubricants, Inc. relates to the use of lubricant additives with antifriction properties that contain vegetable oil and its derivatives (such as wax esters of vegetable oil FAs, sulphurised vegetable oil triglycerides, etc.) [29]. In applications where the amount of pressure applied to the moving part is quite high, lubricants which consist of only hydrocarbon base stock tend to fail. In these cases extreme pressure additives (e.g. sulphurised jojoba oil) are used as lubricant additives to increase the high-pressure performance.

For bio-based lubricants the lubricating base fluid and property-enhancing additives all need to be tailored for an individual application (e.g. extreme pressure and temperature conditions). Engine bio-oils need to fulfil different functions, including lubricating and cooling the system, wear protection, soot and particle handling with less deposit tendency, etc. Thermal degradation of biolubricants by frictional heating limits their utilisation in industrial applications. Insufficient thermal resistance of plant- or animal-based oils for use in gears sets a temperature limit of a maximum of 80 °C, while for some ester-based and polyglycol lubricants vehicle and off-road machinery producers recommend a temperature limit of 100 °C [30]. As shown in this chapter, there is a wide variety of performance levels among biodegradable products. When an environmentally preferable product is required outside the common temperature range, a biodegradable synthetic is usually required. While offering biodegradation, these (significantly more expensive) products can operate at temperatures in excess of 200 °C and still offer long fluid life.

12.2 Automotive lubricants

Many types of lubricants are used in a vehicle. Automotive lubricants, which dominate demand for base oil (>20 Mt/yr), comprise transmission oils, engine oils, gear oils, axle oils, brake oils, greases, etc. On a volume basis, engine oil and transmission fluid provide a major portion of the lubricants in a typical car or truck, and a great number of standard tests are associated with these fluids. The diversity of the oil function strongly influences the choice of chemical composition and physical properties for a given type of lubricant and dictates that a variety of lubricants are required to perform the various lubrication functions in a vehicle. Different vehicle components require different lubricant base stocks, fluid viscosities, and type and quantity of additive treatment for the formulation to function property. Different engine components require substantially different kinds of protection. Thus a variety of specialised lubricants are required for automotive applications (Table 12.5). For example, a system such as a pick-up truck requires the variety of lubricants of Table 12.6. As lubricants are considered as spare parts within the European regulation all technical information must be readily available to motor vehicle operators.

Different automotive components rely upon different modes of lubrication to achieve acceptable performance, and each may experience more than one lubrication regime per cycle. For example, engine bearings are designed to operate in the hydrodynamic lubrication regime. Metal-to-metal contact takes place only at low speeds and high loads and with low-viscosity lubricants. On the other hand, transmission clutch sliding generally takes place under mixed or boundary lubricant conditions. In a single stroke of the piston, the piston ring interface with the cylinder wall may experience boundary, mixed and full-fluid film lubrication [31]. Elastohydrodynamic (EHD) lubrication of piston rings is also feasible in both gasoline and diesel engines. The importance of different lubricating regimes for each engine

Table 12.5 Tribological components of a motor vehicle

Automotive component	Requirements to lubricants
Engine	
Engine bearings	Hydrodynamic lubrication, minimum film thickness (1.0–2.5 μm)
Piston assembly	Large variations of load, speed, temperature; various lubrication regimes
Valve train	Mixed or boundary lubrication, high temperature (300–500 °C) stability, friction modifier, wear resistance, corrosion resistance
Transmission and drive line	
Transmission clutch	High shear forces and compressive loads, wide operating temperature range, mixed or boundary lubrication, friction modifier
Traction drive	High traction coefficient at 140 °C, no degradation by shear or oxidation
Velocity joints	High loads, low entrainment velocity, boundary lubrication
Wheel bearings	Grease
Drive chains	Minimal wear
Axle	Appropriate viscosity, metal and elastomer compatibility, seal performance
Auxiliaries	
Pumps[a]	Protection of pump, compatibility with hoses
Door locks	Grease

[a] Air-conditioning, oil and water pumps.

Table 12.6 Lubrication requirements for a pick-up truck

Biolubricant	Pour point (°C)	Flash point (°C)	Viscosity (VI)
Bio gear oil	–30	280	180
Bio transmission fluid	–20	315	183
Bio bearing grease	–36	279	186
wheel bearing grease			
chassis grease			
Bio engine/motor oil	–18	215	220[a]
Bio steering fluid	–36	271	190
Bio brake fluid	–30	274	190
Bio hydraulic fluids (ISO 32)	–36	278	194
Bio lock oil	–30	145	180

[a] GF-2/GF-3/GF-4.
After http://www.biolubricants.com

component may change with the surface roughness at the interface, wear of critical interacting surfaces and lubricant degradation.

In addition to providing a fluid film and a chemical surface film, automotive lubricants must also inhibit corrosion, since partially burned fuel can be acidic. Automotive lubricants protect against corrosion by virtue of alkaline agents to neutralise acids that form in hot spots. Vehicle, component or lubricant manufacturers designate the lubricant best fit for a particular application.

Some 60% of the total lubricant volume is used in engine oils. Compared with other automotive lubricants, engine oils are subject to accelerated degradation due to combustion gases, fuel, water, external contaminants (airborne materials, dirt, sand) which cause unique filtration problems, and result in relatively short change intervals. For engine oil lubrication blends of lubricating base oils are being used [32]; see also Section 10.5.1. Automotive engine oil must remain effective despite contamination with fuel combustion products and water and must not evaporate or degrade excessively during high-temperature operation. Engine oil needs to protect the engine in various conditions of service: (i) at high engine speed and high oil temperature; (ii) in city service (oil fully warm); and (iii) during extreme short trip service (essentially cold oil). Various types of bearings require the presence of a fluid film. Under highly loaded, high-temperature conditions, fluid-film lubrication may not be sufficient to provide complete wear protection of the piston assembly. In such cases, the lubricant must contain additives (e.g. zinc dialkyldithiophosphate, ZDDP) that interact with rubbing surfaces to form anti-wear chemical surface films. Engine oil transports anti-wear surface-active agents to the nose of the cam on a camshaft, a region in which chemical agents are essential for wear protection when a lifter slides over a cam nose. Once the additives are sufficiently degraded, the engine oil needs to be changed, otherwise engine wear will accelerate. Identification of appropriate test methods for proper evaluation of the performance of engine oils is a challenging task facing automobile and lubricant manufacturers. The reason for the difficulty lies in the particularly harsh environment in which engine oils are called to operate. When bench tests are used in place of engine tests to measure a given performance parameter, great care must be taken to ensure correlation with real-world conditions. For ASTM testing methods for environmentally friendly engine oils, see Table 10.4; for performance testing of motor oil formulations, see Table 10.10. For engine oil performance designations, see ref. [33].

Transmission oils optimise the performance of everything that revolves, rolls or moves. Transmission fluids lubricate a vehicle's transmission. They may be composed of synthetic or mineral oil. Transmission fluids must withstand high temperatures and loads. Synthetic oils show enhanced resistance to deterioration from exposure to heat. Clutch and band friction characteristics, transmission fluid viscosity, fluid oxidation, fluid shear stability and wear properties of friction materials all play a crucial role in the overall performance and durability of the transmission. Transmission fluids contain additives to maintain the desired friction properties, minimise wear and provide corrosion inhibition to the transmission components. Antioxidants are required to prolong the life of transmission fluids. Great care must be taken that an appropriate additive is chosen to ensure stable frictional properties. Wear rates must remain low. Various standard test methods are available to ensure sustained transmission performance. Tests for automatic transmission fluids (ATFs) include measurement of physical and chemical properties: viscosity, flash point, PP corrosion resistance, wear resistance, friction characteristics and resistance to foaming. Transmission technology is now evolving rapidly in response to market driven demands for better performance and smoother operation and regulatory requirements for lower emissions and improved fuel economy [34].

A lubricant typically reduces friction, but in some cases an automotive lubricant is designed to provide a desired amount of traction, as is the case with traction fluids in some CVTs (constantly variable transmissions). Naphthenic base compounds and several types of synthetic fluids have better traction characteristics than paraffinic or aromatic compounds. Traction fluids have a chemical structure that, under stress, forms a three-dimensional network that resists flow.

Gear lubricants must be designed to withstand high loads that are concentrated on gear surfaces. As a consequence, high temperatures may be generated at the point of contact. Despite the harsh environment, the gear lubricant must provide sufficient wear and extreme pressure (EP) protection to prevent material failure, and the gear lubricant must remain functional for extended periods of time. The properties of the gear lubricant must remain stable, including oxidative stability, anti-wear protection, maintenance of appropriate friction characteristics, low-temperature viscosity, corrosion protection, and resistance to foam formation. Various test methods for gear oils are available from CRC (Coordinating Research Council), such as CRC L-37 (gear distress), CRC L-42 (resistance to gear scoring), CRC L-33 (rust test), and CRC L-60–1 (oxidation and deposits).

Requirements for axle lubricants are described in the American Petroleum Institute API-GL-5 requirements for gear lubricants. Desirable properties for axle lubricants include an appropriate viscosity (14–15 cSt at 100 °C). Axle lubricants should not degrade during service, nor damage metal components. In addition, axle lubricants must not damage seals (ASTM D 5662). Desirable axle-lubricant properties include fuel efficiency, thermal stability, long life and appropriate limited slip differential performance.

Brake fluid must remain in place and continue to function even in adverse weather and road conditions. A mixture of lubricant and air-conditioning fluid must have sufficient lubricating properties to protect the air-conditioning pump, but the lubricant/fluid mixture should be compatible with materials composing the system. Automotive air-conditioning systems require a specialised lubricant that can be transported by the air-conditioning medium.

Greases provide a variety of highly specialised functions and are found in many locations within a vehicle (in door locks, gears for seat adjustments and windshield wipers, bearings, electrical contacts, etc.). Greases are also used to lubricate rolling/sliding contacts in various components of automobiles due to their good retaining abilities in the lubricated conjunctions. Test methods for automotive greases focus on such parameters as separation of the thickener from the oil, oxidation resistance, wear protection, and ability to withstand high temperature.

As the component scale of automotive components in lubricated contacts becomes smaller and smaller, nanolubrication needs to be considered, which differs from the unique mechanisms operating in conventional lubrication. Nanotribology and nanolubrication technology are developing fields [35]; see also Section 12.12. For an overview of the lubrication aspects of the automotive powertrain system, including the engine, transmission, drive line and other components, see ref. [33].

A variety of (standard) tests are needed to guarantee that a given lubricant performs as it should. For automotive lubricants these include engine tests, physicochemical characterisation of the used lubricant, evaluation of its corrosion and wear protection and remaining effectiveness of the lubricant's additive package, see also Section 10.5.1. Additional information includes deposit formation tendency, extent of oil oxidation, fuel efficiency and emission system protection. For some automotive lubrication applications, standard test methods are available, while in other cases manufacturer's (or some other) designated tests should be followed [36]. Bench tests (for viscosity, PP volatility, shear stability, wear rate, wear depth, wear mechanism, etc.) are often used to indicate whether certain fundamental lubricant properties are within a desired range. Chemical and physical tests measure the composition and properties of a lubricant, the changes that occur as a result of ageing, and the nature of contaminants and wear debris that may have entered the lubricant. The relevance of bench tests to real-world conditions should be ascertained. Engine tests such as ASTM D 6923 are used to determine whether a proposed oil formulation meets current certification requirements. Before a new oil is introduced it needs to be tested in laboratory engines. The engine is then run under known conditions for a given period (typically up to 480 hours) according to standard requirements and oil samples are taken at intervals for analysis. Engine components are examined after the test. Oils passing the test with satisfactory results are judged reliable when used successfully in field engines. However, such tests are expensive because they require long periods of engine operation. Therefore, tests are usually carried out only for commercial oil investigation. For degradation of lubricating oils (ageing studies), see Section 14.4.

A modified engine rig test with decreased oil sump capacity and accelerated oil degradation has been used to test the performance of several high-stability vegetable oils as crankcase oils [37]. The oils were effective as lubricants with no failures, but did increase in viscosity over time and cause deposits on the bore. The use of vegetable oil as an internal combustion engine lubricant has been investigated since 1993. Caterpillar and General Motors were the first engine manufacturers to approve compounded cranckcase lubricating oils on the basis of satisfactory performance in laboratory engine tests; see ASTM D 6923 [38]. Development and testing of environmentally friendly lubricant compositions for internal combustion engines are illustrated in Section 10.5.1.

The current automotive engine lubricant formulations are designed for cast iron blocks. Recent industrial developments envisage use of lightweight materials such as non-ferrous materials (Al, Mg) for engine and drivetrain materials to replace the current heavy-weight cast iron blocks. Aluminium blocks have poor wear resistance. Several industrial-academic consortia have been formed to develop new generation vehicles utilising advanced powertrains, lighter, stronger materials, and environmentally friendly lubricants.

Automotive tribology may be advanced by developing a better understanding of lubricant chemistry and how additives affect the interactions between lubricants and rubbing surfaces. It is also of interest to gain better insight in the rate of degradation of oil additives. This will allow developing new lubricants that will be longer-lasting, environmentally friendly, capable of handling increased soot and acid loadings, compatible with catalysts and with new, lightweight non-ferrous materials.

Legislative pressure is calling for better fuel economy and reduced emissions. Despite the US Corporate Average Fuel Economy (CAFE) regulations of 1975 (in the wake of the 1973 Arab oil embargo), intended to improve the average fuel economy of cars and light trucks, the on-road vehicle fleets in the United States and Canada have the lowest overall fuel economy among first-world nations: 25 mpg (9.4 L/100 km) in the United States versus 45 mpg (5.2 L/100 km) in the EU and Japan. Improving the fuel efficiency of internal combustion engines is a key objective of the automotive industry. Improvement requires more than just optimised lubricants. With the large number of internal combustion engines in service even the smallest improvements in engine efficiency, emission levels and durability have a major effect on the world fuel economy and environment in the long term. Reducing friction and wear in engine and drive train components in transportation could save the US economy as much as US$ 120 billion per year [39]. Engine oils, transmission fluids and rear axle lubricants influence the vehicle fuel economy. The energy derived from combustion of the fuel is distributed between the engine and the powertrain system. It has been reported that a 10% reduction in mechanical losses would lead to a 1.5% reduction in fuel consumption [40].

Modern downsizing technology aims at smaller but equally powerful gasoline engines with lower weights, lower fuel consumption and CO_2 emissions. Smaller engines with new higher-pressure injection designs (up to 2000 bar) to increase performance (e.g. FIAT's TwinAir bicylinder) are more efficient because less energy is wasted in friction between surfaces and as a result of lower car weight. Crankcase and transmission lubricants can contribute in improving energy efficiency in many ways. Owing to their greatly reduced volatility and good low-temperature performance, API Groups II–IV allow the formulation of lighter automotive viscosity grade oils such as SAE 5W-40, 0W-30 and even 0W-20 to achieve better fuel economy (see also Fig. 11.12). However, the use of thinner base oils increases the risk of engine wear unless appropriate friction modifiers are simultaneously deployed in the formulation. The downside of smaller engines and thinner base oils is also that lubricants experience more stressful conditions. Heat management is more difficult as advanced lubricants are required to operate at higher sump temperatures [41]. Thermo-oxidative stress is increased and favours use of polar fluids having excellent oxidative stability. At the same time, also better rather than more additives are wanted, notably new antioxidant and detergent and dispersant (DD) technology for both lubricants and fuel [42]. For efficient vehicles, additives are arguably even more important in engine lubricants than in fuels, in particular for viscosity control (right viscosity grade, high VI).

To design engine bearings operated in the mixed lubrication regime will be a critical and significant approach to improve engine performance. The design of cleaner, more efficient engines with higher specific power outputs is forcing tribological engine components to be operated with generally thinner oil films. Energy efficiency is improved by use of lower-viscosity oils (SAE 5W-20 or 0W-20 instead of SAE 5W-30) with lower volatility than hydrocarbons. This trend is already noticeable. While this helps to reduce friction losses, it also leads to potential durability problems. In addition, there is a drive to extend engine service intervals. Engine oils are expected to last longer and simultaneously reduce engine losses. The engine has therefore to withstand an increasingly contaminated and degraded lubricant.

The requirements of the automotive industry are growing permanently. Each generation of vehicles poses new challenges for the development of lubricants. Technologically optimised engines and sensitive electronic components ask for tailor-made lubricant solutions. These include life lubrication and fuel economy. New engine oil specifications are requiring oil formulators to develop formulations containing less phosphorus and also reducing engine wear. While the performance specifications have been increased, allowable treat rates for lubricant performance additives have been reduced.

A gasoline-driven internal combustion engine uses only 20% of the energy content of the fuel, the rest is lost as heat. In US Patent No. 7,465,696 B2 (to Chevron Oronite Co.) Roby and Ruelas disclose a lubricating base oil composition composed of a Fischer–Tropsch-derived lubricating base oil and at least one polyol ester with improved fuel economy in an internal combustion engine [43]. The fuel efficiency may also be improved through additive technology. A significant contribution to overall frictional losses in an internal combustion engine arises from contacts which operate under boundary lubrication conditions [44]. Friction loss is the major portion (48%) of the energy consumption within an engine [45]. The other contributions are the acceleration resistance (35%) and cruising resistance (17%). Surface additives can be used to modify the surface chemistry and reduce friction. The presence of additives may reduce the friction coefficient by approximately 20% [46]. There is growing interest in organic friction modifiers based on derivatives of natural long-chain FAs. Development of improved additive systems offering higher activity and improved oxidative stability is an active area of research.

Opportunities galore for plant-based automotive lubricants. Their widespread acceptance depends to some extent on the issue of certification. Currently, automotive engine manufacturers require that motor oils meet standards set by the API. This is a costly process involving testing methodologies that have been developed over the last century for petroleum-derived lubricants. The industry is divided on this issue. Some industry representatives hold the view that every vegetable oil-based lubricant should meet the same testing protocols as petroleum lubricants; establishment of separate standards is seen as detrimental to the plant-based lubricant industry. On the other hand, Agro Management Group, Inc. (AMG) is attempting to establish a new testing protocol for vegetable oil-based motor oil in cooperation with Savant, Inc. (Edison, NJ), an independent testing laboratory that has developed testing methods for API and ASTM. It should certainly be considered that vegetable oils represent completely different chemistry, and in fact vegetable oil-based lubricants have performance characteristics that are not incorporated into the existing petroleum testing protocols. Used automotive lubricants have a high toxicity because of polycyclic aromatic hydrocarbon (PAH) concentration that establishes them as potential carcinogen [47].

12.2.1 Lubricants for internal combustion engines

The term 'internal combustion' describes engines that develop power directly from the gases of combustion. This class of engine includes the reciprocating piston engines, used in a wide variety of applications, as well as most gas turbines. However, closed-system gas turbines are not truly internal combustion systems. Piston engines range in size from small model airplanes (<1 hp) to engines for industrial use and marine propulsion (>50 000 hp). This wide range of engine sizes and types of application presents a variety of lubrication challenges. Nonetheless, certain parameters affecting lubrication are more or less common to all reciprocating engines.

Historically, lubricants were used to ensure the durability of the engine and related components. Engine oils have to perform in different types of vehicles that run on either petrol or diesel, and must operate in particularly harsh environments. Typical engine temperatures are 130–140 °C; diesel engines operate under even higher temperatures. Combustion gases, fuel and water in the oil as well as outside contaminants (dirt, sand and other airborne materials) accelerate oil degradation, cause filtration problems, and result in much shorter change intervals compared with other typical automotive lubricants. Today, the properties and composition of the engine oil are also important in relation to a vehicle's optimal fuel economy and low exhaust emissions. A low-performance engine oil may risk contamination of the emission system of the vehicle as well as deterioration of the catalyst, resulting in increased emissions [33]. Transport accounts for 25% of emissions from fuel consumption and about 15% of man-made emissions.

Table 12.7 Basic functions of an engine oil

• Prevents metal-to-metal contacts • Reduces friction between moving parts • Transfers heat from engine to the cooling system • Protects against corrosion • Keeps the engine clean by acting as a solvent for contaminants • Isolates and transports particles for oil filter trapping • Creates a contaminate seal between parts • Contributes to fuel efficiency • Satisfies minimum performance standards

Engine oils fulfil a wide range of functions in engines (Table 12.7). The primary objectives of lubrication of reciprocating engines are the prevention of wear and maintenance of power-producing ability and efficiency. The tribological task consists of ensuring the functionality of all friction points in all operating conditions. Engine oils should reduce friction and wear both at low-temperature start-ups as well as at high temperatures and pressures. At low temperature the oil should still flow well and additives must not precipitate. At high temperature the oil must offer resistance to thermal and mechanical ageing. The stability of the lubricating film should not diminish by fuel dilution of up to 10%. Additional functions of engine oils range from sealing of piston and cylinder during the combustion process to dissipation of heat, corrosion protection, and transport of sludge, soot, dirt and wear particles to the oil filter. Moreover, any water formed during the combustion process should be emulsified. Engine oils should contribute in keeping the engine clean in order to maintain optimal running conditions. Notably, the UK's fuels boast among the highest fuel detergent levels in the world, causing correspondingly low levels of deposit build up. This contrasts with the United States, where minimum additive levels are set by legal mandate. The internal combustion engine will continue to be the powertrain of choice for the foreseeable future.

Four-stroke engine oils

Base oils used as motor oils are generally classified by the API as being mineral oils (Groups I, II and III) or synthetic oils (Groups IV, V and VI) [48]; see Table 3.6. Over the years the engine market has shifted away from conventional products to semi-synthetic and synthetic formulations with higher ecological demands. In recent years, fully synthetic oils dominate the premium-quality market. Biodegradable motor oils are based on the esters (or hydrocarbon-ester blends). Esters were the first synthetic crankcase motor oils in automotive applications but were later replaced by PAOs due to their lower cost and formulation similarities to mineral oils. However, in fully synthetic motor oils, esters are almost always used in combination with PAOs to balance the effect on seals, additive solubility, volatility reduction, and energy efficiency improvement by enhanced lubricity. Another class of base oils suitable for engine oils are PAGs, which offer zero-ash, bio-no-tox properties and lean burn characteristics [49]. Depending on the base oil used and the required engine performance, engine oils can contain up to 30 different additives whose percentage content can range from 5 to 25% in total. In the oil industry, a differentiation is made between viscosity improvers, flow improvers and performance additives [50]; see Section 7.2. Performance additives used in engine oils comprise antioxidants, anti-wear (AW) agents, DD, friction modifiers and anti-misting agents. Multigrade engine oils are based on oils containing VI improvers or base oils with high natural VI. In the automotive world, the synthetic motor oil Mobil 1 (see Table 12.10), based on AW SuperSyn technology, is known as the absolute leader, but the semi-synthetic Esso Ultra and fully synthetic Esso Ultron multigrade engine oils for cars also enjoy an excellent reputation.

There is no significant pressure from regulatory agencies or environmental groups to force a change from the use of mineral oils as crankcase lubricants. In normal use, the potential contact of crankcase oils with the environment is limited to disposal of used products. Recycling or energy recovery is the preferred alternative to limit environmental exposure (see Section 14.7). The industry continues to adopt new crankcase oil standards for which further decreases in volatility and increases in VI of oil are stressed. These new standards are already causing crankcase oil manufacturers to reformulate engine oils and substitute more expensive API Group II+ and Group III hydrocracked base stocks and API Group IV synthetics for standard API Groups I and II solvent refined oils. API Group III oils produced through hydroprocesses and gas-to-liquid (GTL) base oils are free from sulphur and nitrogen and have a high viscosity index (≥120). Therefore, the engine oil formulation will have to change considerably in the near future. The very high VIs (>200) and extremely low volatility of vegetable oils may prove useful in formulations meeting the new standards if other performance characteristics such as oxidative stability are improved.

Driving factors for improvement of engine oils are emission protection, longevity and fuel economy. Reduction of oil viscosity improves the fuel economy. Lower sulphur and phosphorus contents (0.05 and 0.08%, respectively) have impact on wear. Tetrosyl, Ltd (Bury, UK) produces the Carlube range of low sulphated ash, phosphorus and sulphur (SAPS) fully synthetic engine oils for the automotive aftermarket to meet the requirements of the latest generation of vehicles with exhaust aftertreatment devices. Similarly, ENOC (Dubai) has recently launched two new environmentally friendly fully synthetic engine oils, namely Protec Green 5W40 and Protec Xtreme

5W30 based on SAPS additive technology. A while ago, a niche for environmentally friendly, biodegradable products (esters) was also created in the German-speaking market. Hubmann *et al.* [51] have described test results of a biodegradable SAE 5W-30 engine oil based on synthetic carboxylic acid esters.

While the automotive engine oil market is huge, strict performance requirements and the low price of mineral oil alternatives makes it a difficult market for vegetable oil-based products. The present operating conditions of engine oils cannot be satisfied by chemically unmodified vegetable oils. Vegetable oil formations have restricted lifetime. However, the substitution of circulating oils by TLLs in some applications (e.g. oil-refreshing systems for lubrication of four-stroke engines) offers prospects for chemically unchanged vegetable oils in these new applications. The concept consists in continuously refreshing a vegetable-based engine oil. The use of vegetable oils for engine lubrication depends on the progressive burning of used engine oil with the fuel. The advantage of the oil-refreshing method is that chemically unmodified, low additivated vegetable oils can be used and no waste oil has to be disposed of [52]. UNI-NABL is working on a continuous oil recycling system (CORS) for the use of vegetable oils as diesel engine oil [53].

The performance requirements for crankcase lubricants are severe. In order to provide engine lubricants based on vegetable oils, certain standards should be met, including specifications required by the SAE (Society of Automotive Engineers, Inc.), API (American Petroleum Institute) and ILSAC (International Lubricant Standardisation and Approval Committee) [54]. ILSAC was formed in 1992 by AAMA (American Automobile Manufacturers Association) and JAMA (Japan Automobile Manufacturers Association) to define the need, parameters, licensing and administration of lubricating specifications. Minimum performance standards for an oil for ICE use that satisfies ILSAC GF-3/API SL include a gelation index of less than about 12; high temperature TEOST (thermo-oxidative engine oil simulation) of total deposits of 45 mg maximum; remain homogeneous and miscible when mixed with SAE reference mineral oils; low volatility; phosphorus level of 0.1% maximum; and passes foam, filterability and ball rust tests. ILSAC GF-3 oil meets stringent parameters regarding long-term effects of the oil on the vehicle emission system, improved fuel economy and volatility, deposit control and viscosity performance. The latest minimum performance standard for passenger car engine oils (ILSAC GF-5; ratified December 2009, market introduction October 2010) provides the following improvements: emissions compatibility (Seq. IIIG phosphorus retention), oil robustness (Seq. IIIG WPD, TEOST MHT and 33C, E85 emulsion), and fuel economy (Seq. VI-D) [55]. Improving fuel economy and piston cleanliness simultaneously is a considerable challenge, as these factors largely contradict each other. European cars need to meet emissions targets of an average of 130 g CO_2/km by 2015.

Application of a biodegradable lubricant should not result in any unusual wear on the engine. Stringent performance requirements, especially the need for oxidative stability, have long kept bio-based lubricants out of the crankcase oils market. Major automotive manufacturers have limited interest in bio-based oils as crankcase lubricants unless cost-saving performance benefits result. Yet, some lubricant manufacturers are developing and testing vegetable-based crankcase oils. In particular, bench-scale tests of higher-oleic vegetable oils by several formulators have shown results comparable to current mineral-based oils in oxidative stability (see Sections 10.5.1 and 10.5.2). Table 10.9 shows the formulation and physical properties of a proprietary SAE 5W-30 grade motor oil.

Various companies have developed and are marketing plant-based automotive engine oils. Agro Management Group, Inc. (AMG) (Colorado Springs, CO) has disclosed a general purpose oil formulation for small air-cooled engines (lawn-mowers, etc.) and total-loss lubrication, consisting of an unrefined base oil (i.e. without degumming, bleaching or deodorising), an HFA and wax (see Table 12.8) [56]. Base oils should be composed of at least 65% of C_{16} to C_{22} fatty acids. The preferred source is transgenic *Brassica napus* (rapeseed and canola), eventually blended with soybean, crambe, safflower or sunflower oil. A blend of 83 vol% canola oil and 17 vol% soybean oil (SBO) gives excellent performance. The other components of the formulation, HFAs and waxes, are particularly well suited to bonding to metals, assuring the user of reduced metal-to-metal contact. The very

Table 12.8 Typical biodegradable liquid lubricant formulations (vol%) for internal combustion engines and total-loss applications

Component	Application			
	Bar chain	Rail oil/ cutting oil	Small engines[e]	Automotive engine
Base oil[a]	85	90	82	75
Hydroxy fatty acid[b]	10	8	10	13
Vegetable or animal wax[c]	5	2	5	7
Antioxidants[d]	0	0	3	5

[a] Preferred base oil: canola.
[b] Castor or lesquerella oil.
[c] Jojoba, meadowfoam or lanolin wax.
[d] Natural or synthetic.
[e] Small air-cooled four-cycle engine (3.5 to 20 hp).

reactive hydroxyl groups help preventing breakdown of the oils under extreme (heat and friction) conditions by forming dimers as well as by reacting with metals in contact with the lubricant. Jojoba liquid wax ester is an essential component of the general-purpose oil formulation. Wax esters (C_{38}–C_{46}) tend to bond to metal, thus coating the wear surfaces and reducing wear. AMG 2000 is the first renewable, biodegradable motor oil derived from canola, soy or other vegetable oils as a replacement for petroleum oil in automotive engines (recipient of the World's Best Technology 2004 award). In field testing conducted by AMG and the US Postal Service in Michigan, AMG 2000 performed as well as conventional engine oil and had the added advantages of producing significantly fewer toxic air emissions and achieving greater fuel efficiency.

Environmentally friendly engine lubricant compositions (>60% biodegradable; gelation index, 12) of a full range of SAE viscosities for use in gasoline- and diesel-fuelled internal combustion engines, consisting of a transesterified (high-oleic) triglyceride base oil (40–65%) together with a synthetic ester (e.g. polyol ester, diester) (15–20%), mineral oil (10–25%), and an olefin copolymer VI improver, that meet GF-3/API SL specifications, were recently disclosed by Chiu *et al.* [57], see Section 10.5.1. Renewable Lubricants, Inc. (RLI; Hartville, OH) has developed a crankcase oil using additives aimed at bringing high-oleic vegetable oils up to mineral oil standards. RLI uses canola, sunflower, soy and corn oils. In PCT Int. Publ. No. WO 2006/116502 A1 (to RLI) Garmier [58] discloses a lubricant composition composed of a bio-based synthetic or natural oil (preferably with C18:1 > 80%), eventually blended with another base oil, boron nitride (>2%) and an additive package for EP performance in high-temperature (>500 °C) applications such as combustible engines, ovens, chains, cables, bearings, etc. Boron nitride is listed by the US Department of Agriculture (USDA) and National Sanitation Foundation International (NSF) as food-grade approved and non-toxic. Valvoline has evaluated mid-oleic SBO for potential crankcase application. Green Earth Technologies, Inc. (Stamford, CT) has developed a completely biodegradable tallow-based motor oil (G-Oil™ 5W-30) for gasoline and diesel engines as well as hybrid cars, which has received the API's seal of approval and was launched at the 2007 Automotive Aftermarket Products Expo (AAPEX), Las Vegas, NV [59]. Natoil AG (Immensee, CH) has claimed epoxidised unsaturated fatty esters and glycerides as lubricating oil stabilisers for internal combustion engines [60]. Consumers today are demanding more from their engine oils than ever before. The use of renewable lubricants such as estolides can meet such demands.

General acceptance of plant-based automotive lubricants depends not only on technical performance and cost, but also on certification (not unlike the introduction of biodiesel). As vegetable oil-based lubricants have performance characteristics differing from those in existing petroleum motor oil testing protocols, establishment of separate standards for vegetable oils may be needed. However, this is a very costly and lengthy process. For the time being, every vegetable oil-based lubricant necessarily needs to meet the same testing protocols as petroleum lubricants.

For vegetable-based engine oils, see also refs [61, 62].

Two-stroke engine oils

Two-stroke engines are mainly used in small motorcycles and boats (outboard engines), more particularly when high specific power, low weight and low price are desirable, as in case of mopeds, scooters, go-carts, snowblowers, jet skis, chain saws and gasoline-powered gardening equipment such as hedge trimmers, leaf blowers and soil cultivators. Almost all two-stroke engines use TLL. Often, these motors are not exposed to as wide service temperature ranges as in vehicles, so these oils may be single viscosity oils. In small two-stroke engines, the oil is premixed with the gasoline, often in a rich gasoline:oil ratio (typically 40:1), and burned in use along with the gasoline. Larger two-stroke engines used in boats and motorcycles will have a more economical oil injection system rather than oil pre-mixed with the gasoline. In most simple two-strokes, the engine breathes through classic carburetors. Contrary to four-stroke engines, the fresh fuel/air mixture in a classical two-stroke scavenges the cylinder after combustion. This causes about 30% of the fresh mixture to be exhausted as unburned oil mist. Along with the only partial burning of the oil, two-stroke engines generate high emissions and cause severe odour, smoke and noise pollution.

Two-stroke oil formulations are composed of 85–90% base oils, about 1–5% solvent, with the remainder consisting of additives. While four-stroke engine oils are largely based on polyolefin and hydrocracked oil, two-stroke engines use all common base oils. Lubricants for two-cycle gasoline engines are generally composed of a mineral oil or synthetic base fluid, performance additives and a solvent (usually a relatively low-boiling petroleum distillate) to enhance gasoline/lubricant miscibility. Higher quality two-stroke lubricants often contain various synthetic esters or polyisobutylene and this is particularly the case for biodegradable oils which were specifically developed for marine outboards. The use of biodegradable synthetic ester-based lubricants combines the highest technical performance with improved environmental compatibility. Low-smoke two-stroke oils contain a significant amount (10 to 50%) of polybutenes. The additives in two-stroke oils (DD and AW additives, antioxidants, rust and corrosion inhibitors, defoamers, flow improvers) are matched to the requirements of the engine. Two-stroke oils for outboard engines require non-ash forming additives.

Following up the detection of hydrocarbon compounds in the sediments of Lake Constance (Bodensee) in the 1980s two-stroke outboard motors in the bordering countries (Switzerland, Germany and Austria) were required to use biodegradable oils. Vegetable oils based on RSO were used first. The Swiss government implemented a ban on the use of two-stroke engines above 10 hp on the Bodensee.

Modern two-stroke engines require high-quality oils for reliable operation and long life. The principle criteria for the quality of two-stroke oils are shown in Table 12.9. The ISO classifies two-stroke oils into three categories, ISO-L-EGB, -EGC and -EGD using lubricity, smoke, exhaust deposits, detergent effect and piston cleanliness as test criteria. The performance categories for two-stroke oils for outboard engines were primarily developed by the American National Marine Manufacturers Association (NMMA) and laid down in 'NMMA certification for two-stroke cycle gasoline engine lubricants' of NMMA TCW-3. As a result of their basic design, small gasoline engines show high hydrocarbon emissions, far exceeding those from four-cycle engines. There is growing public concern over these comparatively high emission levels as hydrocarbons do not readily biodegrade. The increasing impact of environmental legislation is also affecting two-stroke oils, especially outboard engine oils. Ecologically optimised oils often have regionally differing classifications which respect local environmental legislation (cf. EU standards, US Environmental Protection Agency (EPA) and Canadian regulations). Consequently, their biodegradability depends on varying minimum requirements. At the international level, the International Council of Marine Industry Associations (ICOMIA) has specified harmonised requirements. In 1997, ICOMIA Standard 27–97 was adopted for environmentally friendly outboard engine oils [63]. These oils are based on fully synthetic components with base oils which offer very low algae, daphnia and fish toxicity and rapid biodegradability. By using correspondingly high quality esters, these products are the very best two-stroke oils and can even be used for lubricating chainsaws. The use of ester-based lubricants combines the highest technical performance with improved environmental compatibility.

Table 12.9 Criteria for the quality of two-stroke engine oils

- Lubricity and anti-wear properties
- Cleaning function (detergent/dispersant properties)
- Avoidance of deposits in the exhaust system
- Low smoke
- Spark plug cleanliness and the avoidance of pre-ignition
- Good fuel miscibility even at low temperatures
- Corrosion protection
- Good flowing properties

After ref. [54].

EC Directives 97/68/EC [64] and 2002/88/EC [65] regulate environmental and performance requirements of small non-road mobile machinery engines equipped with internal combustion, such as two-stroke engines and recreational vehicles including snowmobiles and go-carts [66]. The EPA has restricted the emissions from non-road large spark-ignition engines such as those used in forklifts and airport ground-service equipment and recreational engines (marine and land-based) as from 7 January 2003 [67, 68].

It was long considered that mineral oil-based two-cycle lubricants are a prime example of a product in need of replacement, especially in marine applications. According to the NMMA, the new EPA regulations with strict emission standards for two-cycle engines [69] cannot be met by these engines and mean the end of this traditional carburetion engine type.

Branched synthetic polyol ester base stocks can be used in the formulation of biodegradable two-cycle oils together with selected lubricant additives [6]. Certain other types of ester are also useful as base stocks for air-cooled two-cycle gasoline engine lubricants. US Patent No. 5,912,214 to Henkel Corp. discloses ester base stocks for smokeless and solvent-free two-cycle gasoline lubricant compositions with excellent miscibility with gasoline and a smoke index of at least 75 (JASO M 342–92 test) [70]. The ester base stocks consist of a mixture of esters or of complex esters (such as linear oligoesters and polyol esters). Examples are TMP tristearate and complex polyesters composed of glycerol-adipic acid-nonanoic acid/octanol (mole ratio 1/2/1/2). Some of these base stocks are also biodegradable, as determined by the Coordinating European Council standard test method CEC L-33-A-93 (Biodegradability of Two-Stroke Cycle Outboard Gasoline Engine Oils in Water).

A vegetable oil-based two-cycle engine oil is marketed by Agro Management Group, Inc. (Colorado Springs, CO). Green Earth Technologies, Inc. (GET; Stamford, CT) has developed a completely biodegradable (green) motor oil, G-Oil™ 2-Cycle Oil, based on tallow, with excellent performance in terms of four qualifiers of an oil: VI, sulphated ash, flash point and PP (Table 12.10) [59]. Ash content is particularly low, which ensures low particulate content of the exhaust. Only limited quantities of commercially available additives are needed to make G-Oil™ 2-Cycle Oil meet or exceed the API TC rating.

Coconut oil, which is abundantly available in southern states of India, is being widely used as two-stroke engine lubricant (2T oil) by autorickshaws but effects increased engine wear. The tribological properties of coconut oil have been evaluated using a four-ball tester and a test rig to test the wear on two-stroke engines [71]. Addition of an AW/EP additive brings about considerable reduction in wear with coconut oil as 2T oil.

In the past, jojoba oil has been evaluated as an (expensive) two-stroke engine lubricant, but probably without commercial applications in mind

Table 12.10 Comparison of motor oils

Motor oil	Qualifying parameters			
	VI	Ash[a]	Flash point (°C)	PP (°C)
G-Oil™ 2-Cycle Oil	147	<1	128	–42
Gulf Pride	130	<10	63	–39
Mobile One	154	150	110	–42

[a] Per 1000 parts of oil by weight.

[72]. Formulations of jojoba oil with commercial lubricants displayed good miscibility with petrol, comparable scuffing and deposit formation tendencies and improved wear performance.

In a palm oil methyl ester (PME) blended lubricant in a two-stroke engine PME acts as a lubricating oil additive [73, 74]. Mineral-based lubricating oil exhibits better performance in terms of friction compared with palm oil-based lubricating oil, but the latter is superior in terms of performance in wear [75].

Recent advances in two-stroke engine technology have significantly reduced emissions and improved fuel efficiency. The marine industry has developed a new high-pressure fuel injected engine with injection directly into the combustion cylinder that is more efficient, requires less lubricant and can meet the new EPA emission standards. These E-TEC engines are expected to replace all two-cycle engines of present design as from 2010. Consequently, pressure to use substitute (biodegradable) lubricants in this market segment has greatly been diminished.

Tractor oils

No machinery is so uniquely suited to using vegetable oil-based lubricants as agricultural and construction equipment. Since this equipment is particularly close to the environment, the lubricant can easily come in contact with the soil, ground water and crops. Tractor oils should guarantee long machinery life in all-season conditions as well as extended service intervals and reduced down-time.

In the past, simple engine oils or low-additive gear oils were used for general lubrication as well as for the hydraulic circuits in tractors. Nowadays, the state-of-the-art is complex hydraulic systems, hydrodynamic drive units (retarders/power splitting) or wet (oil) braking systems.

For reasons of simplified servicing, universal oils were developed which satisfy the various functional requirements of tractors. In this application area essentially two different oil technologies are used, namely

Universal Tractor Transmission Oils (UTTO) and Super Tractor Oils Universal (STOU). Both oil types apply to the hydraulic system, gearbox and wet (oil) brakes, and STOU also to engine oils. Technical advances in tractors have determined a significant improvement in operating fluids (not dissimilar from progress in the whole automotive area) and in additive technology. In particular, the ageing stability of fluids has increased dramatically.

All-season use is now a standard requirement for tractor oils (viscosity grades SAE 15W-40, 10W-40 and 5W-40). The use of these fluids in highly-stressed mechanical drive units makes great demands on the wear protection of life of the lubricant. A special challenge to tractor oil formulations is the use of these products in wet brakes. UTTO additive objectives comprise friction and wear control, low-temperature and oxidative stability, corrosion protection and foam inhibition. Special friction modifiers are required. The use of UTTO oils as engine oils requires vastly different additives. Engine oils additionally require a DD package to ensure engine cleanliness and sludge/soot transportation. Tractor engine manufacturers issue approvals for oils after successfully completed tests in the corresponding engines. The latest generation of tractor oils not only includes products with significantly higher performance than in the past, but also oils which offer better environmental compatibility. Such products often contain rapeseed or sunflower base oils or (more expensive) synthetic, rapidly biodegradable esters. The use of environmentally compatible lubricants enhances the image of the lubricant manufacturer and vendor as well as of the equipment manufacturer who employs such a fluid.

Lubrizol Corp. and a leading OEM (John Deere) have co-developed the first successful vegetable oil (canola)-based tractor hydraulic transmission oil (UTTO) in 1993. This product (BIO-HY-GARD), which is suitable for lubricating the transmission, axle and hydraulic system, contains 68% RSO [76]. More recently, Kunz has described in great detail the successful joint product development (again by Lubrizol Corp. and John Deere) of a follow-up environmentally friendly, renewable fluid for vehicles with a common oil tank for transmission and hydraulics [77]. The development goals of this second generation UTTO (Table 12.11) were met except for the metal (zinc) content, albeit mainly due to time constraints. Zn and P levels of the novel BIO-HY-GARD II fluid are 1400 and 400 ppm, respectively (cf. 480 and 650 ppm in BIO-HY-GARD). As a result, the lubricant is only classified as WHC1. Oil lifetime approaches closely that of the mineral oil-based HY-GARD. As expensive synthetic esters were avoided by use of cheap renewable plant oil (75% RSO) as the base stock the attractive price level of the first generation BIO-HY-GARD product could be maintained for BIO-HY-GARD II. In this commercial development the role of additive technology was fundamental. Valvoline STOU is a multifunctional oil for

Table 12.11 Development goals for the universal tractor transmission oil BIO-HY-GARD II[a]

- All-season use
- Improved oxidative stability at higher operating temperature (CEC L-48-A-95)
- Improved flow properties (pour point)
- Heavy metal-free additive package
- Biodegradability (CEC L-33-A-93) >90%
- Not water hazardous (nwh)
- Extended service intervals (1500 h)
- Metal and elastomer compatibility
- Conformity to JDM J20C [78] and Allison C4 quality standards
- Broad suitability for agricultural machinery
- Favourable price

[a] Trademark.
After ref. [77].

all modern tractors, providing excellent lubrication for the engine, transmission, hydraulics, power take-off and wet brakes.

12.2.2 Transmission fluids

Gear lubrication oils for motor vehicles have to meet the specific technical requirements of the gear drives. While gearboxes, which serve to change the gear ratio in passenger cars, are based on manual transmission with synchronisation in Europe, in America and Asia Pacific mostly (semi) automatic transmissions or so-called CVTs are in use. With respect to these automatics or CVTs, the lubricant is responsible for the gear drive lubrication as well as for the function-related operation of wet clutches and wet brakes. It is not possible to meet all requirements of the aforementioned types of transmissions with only one fluid [79]. In this field, tailor-made individual lubricant designs are being developed for specific transmission types. Gear lubrication oils for motor cars are under continuous development with the object of improving reliability, driving comfort and increasing oil-drain intervals. The trend is for fill-for-life transmission fluids (>300 000 km in passenger cars) and for environmentally friendly fluids and disposal. Passenger cars require ever-increasing performance levels which translate into a high torque transfer in smaller gearboxes. The operation of a transmission causes friction losses in all sliding rolling contacts, with resulting heating up. Oil sump temperatures in transmissions in passenger cars or mobile hydraulic systems often reach 130 °C. In these cases the use of high-stability oils (e.g. synthetic lubricants) is recommended. Lubricants are also expected to optimise the total efficiency of the engine. New passenger cars also include transmissions with a higher number of gear stages, as well as new lighter and more solid materials with new friction and wear characteristics. The variety of friction material partners used in synchronisation of passenger cars requires more and more tailor-made solutions with respect to lubrication and determine an increasing number of requirements and approval tests.

New and advanced differential gear transmission systems are being developed by the automotive industry. These new systems often involve high energy requirements. High performance increases the heat generation in the transmissions, which accelerates the ageing of the gear lubrication oils. Therefore, component protection technology must be developed to meet the increasing energy requirements of these advanced systems. The need for use of highly technical testers makes automatic transmission fluid (ATF) development quite expensive.

Transmission oils lubricate an engine's transmission. The most important consumers of functional fluids for automatic transmission and drive systems, the ATFs, are off-highway vehicles and machines and heavy-duty commercial vehicles such as buses. Moreover, light-duty ATFs are also used to fill power steering systems in light trucks and passenger cars. The total estimated ATF volume worldwide amounts to approximately 1.2 Mt (2006). An increasing number of vehicles use automatic transmissions. Manufacturers of vehicles and of transmission systems have issued particular specifications for ATFs. The highly OEM-specific ATFs are all expensive small-volume fluids with lower and higher viscosities.

Automatic transmissions (wet clutches) require hydrodynamic clutches, torque converters and wet brakes. Due to the function-related fluid losses, these always require low-viscosity, mildly additivated gear lubrication with a high oxidation stability. In order to reduce fuel consumption ATFs must have lower viscosities while at the same time maintaining protection for gears and bearings despite the lower lubricant film thickness. Short-term oil temperatures of more than 160 °C during operation of the vehicle are not unusual. A good viscosity–temperature behaviour (VI), corrosion protection, optimal foaming behaviour and air release properties are required. Synthetic ATFs used in power steering units and hydraulic systems thus provide excellent viscosity–temperature behaviour (similar to engine oils), shear stability, corrosion protection, high oxidation stability, hydrolytic stability, minimal foaming, outstanding air release properties and compatibility with elastomers. They have better performance and a wider temperature range of operation than commercial petroleum fluids. Transmissions usually operate under severe conditions. With smaller sump sizes, containing less fluid, new fuel-efficient transmissions will have higher operating temperatures adding extra stress to the ATF.

Base oils used for the lubrication of transmissions comprise mineral oils, synthetic hydrocarbons (PAOs), PAGs, (synthetic) esters and naphthenic oils. Requirements for better fluid durability, as expressed by tougher oxidation tests, have essentially eliminated Group I base stocks from ATF application. To provide the desired shift properties, one of the additives present in the oil is typically a long-chain hydrocarbon that has a polar group on one end. The long-chain hydrocarbon portion permits the additive to dissolve in oil, and the polar group on the end is attracted to metal surfaces and provides the desired amount of friction [33]. The various types of ATF differ in the fluid's friction characteristics. Apart from friction modifiers the same requirements apply for both ATFs and CVTs in motor vehicles, which enable economic operation of a combustion engine. Viscosity and VI for automatic transmissions are very similar to engine oils. Engine oils are frequently used for light transmission applications. To facilitate the distinction between ATFs and engine oils, ATFs are always dyed red. As a result of legislation the share of mineral-based oils is decreasing and environmentally friendly, synthetic and ester-based multigrade oils are nowadays being preferred. Due to better thermal oxidation stability, the ester-based oils are used together with PAOs and hydrocracked base oils.

CVTs in motor vehicles enable operation of a combustion engine while minimising fuel consumption. The same requirements apply for both CVT fluids and ATFs. Viscosity, additives and base oils are very similar; however, the friction modifiers have great importance.

Suitable vegetable oil-based transmission fluids for agricultural and forestry use (including tractors) may be composed of 69 wt% RSO or HOSNO and 31 wt% Lubrizol LZ9999 [80]. High-oleic vegetable oils (C18:1 > 80%) with 1–10% double unsaturated and 0.1–1% triple unsaturated fatty acids and a natural VI ≥ 200 have been claimed as hydraulic fluid and transmission fluid [81]. The product line of International Lubricants, Inc. (ILI, Seattle, WA), a producer of plant-based lubricants (including total-loss lubricants, hydraulic fluids and others), comprises a successful ATF supplement (Lubegard™). The product is based on liquid wax ester technology that links the FA of high-erucic rapeseed oil (HERO) with alcohols to form a linear liquid wax; this substance is similar to sperm whale oil used as a transmission additive in the past (until 1972) and has superior wear and other performance characteristics.

12.3 Hydraulic oils

Hydraulics stands for transfer of energy and pressure through fluids (governed by Pascal's law). Hydraulic oils are incompressible fluids that flow easily and efficiently transmit power throughout the system to drive, control and move. There exists hardly any (stationary or mobile) machinery or equipment which operates without hydraulic components. This omnipresence makes fluids technology a quite essential and rapidly expanding technology. Hydraulics is divided into hydrostatics and hydrodynamics. Hydrostatic systems are characterised by energy transfer requiring high static pressure and low flow rates. In hydrodynamic systems, where the kinetic energy of the flowing fluid is used, pressures are low and flow rates are high. Fluids for hydrodynamic application are known as transmission oils and those for hydrostatic applications as hydraulic oils.

Modern hydraulics is divided into stationary, mobile and aviation hydraulics, each with specific demands on the hydraulic medium. Hydraulic fluids, which may also be in the form of water and oil emulsions, are used in hydraulic machinery, such as brakes, shock absorbers, automatic transmission of motor vehicles, power steering systems, control mechanisms, etc. Examples of stationary and (more demanding) mobile equipment using hydraulic fluids include industrial machinery, excavators, garbage trucks and aircraft flight control systems. Pump lubricants are used to keep irrigation systems working smoothly. Hydraulic pumps operate at high fluid shear rates. Hydraulic systems are also found throughout marine installations – on cranes, winches, life-boat davits and deck hydraulic remote control systems, often located where leakage or spillage can escape into the marine environment. High flow rates, high operating pressures and extensive use of flexible hoses combine to make these systems vulnerable to a rapid loss of fluid caused by chafe or mechanical damage. With application of higher pressures and system operating temperatures, and increased circulation at lower system volumes, stress on the hydraulic fluids is increasing and higher performances are required. Optimum applications reduce wear, increase machine life, reduce maintenance intervals and determine considerable economic savings [46]. Lubricants are crucial in maximising industrial productivity. Sweden's Forestry Engineering Group machinery is required to provide a service life of 20 000 hours and service intervals are to be kept to a minimum. The organisation attributes half of all downtime to faulty hydraulic systems. (For an evaluation of hydraulic fluids in forestry applications, see Section 8.2.2.) Similar demands are made across many industries and all specified objectives and requirements must be taken into account in new system designs.

After automotive lubricants, hydraulic oils are the next most important group of lubricants, accounting for some 13% of total worldwide lubricant consumption. Industrial hydraulic fluids represent a 222 million gallon market in the United States. Mineral-based hydraulic oils account for about 80–85% of all hydraulic oils, biodegradable and fire-resistant fluids have a market share of c. 7% each, and synthetic PAO- or hydrocarbon-based fluids about 1% (Germany, 2005).

Displacement pumps and motors used in hydraulic systems to transfer energy are subject to great hydraulic stress whereas the fluid is subject to constant mechanical stresses. The main functions of hydraulic fluids are to protect drive components and bearings of a hydraulic system from wear and corrosion and reduce friction, thus minimising the accumulation of deposits. The pumping action also creates a substantial build-up of heat in use and hydraulic fluids serve to dissipate heat. Industrial applications often present an environment where the hydraulic fluid is cooled so as to maintain a relatively stable temperature. Ideally, a hydraulic fluid should have no effect on seal materials and elastomers of the hydraulic system. However, interaction between a hydraulic medium and seal material is unavoidable. ISO TC 131/SC7 N343 lists methods describing the influence of fluids on standard elastomeric materials [82]. The properties relevant in seal compatibility include hardness and volume changes, tensile strength and elongation breakage-point. Polar fluids are quite suitable with non-polar elastomers ('like dissolves like' principle).

The main scope of any hydraulic fluid is to:

- Transmit power mechanically through hydraulic power systems. Hydraulic systems will work most efficiently if the hydraulic fluid has low compressibility (i.e. high bulk modulus) and can flow easily.
- Provide lubrication to all parts of a hydraulic system and protect them from corrosion, rust, oxidation, wear and demulsibility (water release). These qualities are usually provided through additives.
- Maintain fluid characteristics during use so as to keep appropriate pressure to operate hydraulic actuators.
- Provide the seal necessary between the low-pressure and high-pressure side of valve ports. The amount of leakage will depend on the viscosity and the tolerance between adjacent surfaces.
- Remove heat generated throughout the system.

The multiple requirements for hydraulic fluids are listed in Table 12.12. Hydraulic fluids are used to transmit power. Although lubrication is not their primary purpose, hydraulic fluids are regarded as a class of industrial lubricants because they must provide adequate lubrication to prevent wear of the pump and actuators, which would impair efficiency. One of the most important properties of hydraulic fluids is oxidation stability. Oils with low values of oxidation stability will oxidise rapidly in the presence of air at elevated temperature. Table 12.13 lists the most important physical properties of hydraulic oils which affect performance. From a rheological and fluid technological standpoint viscosity of a hydraulic fluid should be as low as possible; however, for efficient power transmission, without high viscous losses or internal leakage, the viscosity of hydraulic fluids should normally be in the range of 22–46 cSt at 40 °C. If the viscosity is too low, the fluid film may also be insufficient to prevent excessive wear or possible seizure of moving parts. Ultimately, pump efficiency may decrease and sluggish operation may result. If viscosity is too high, then friction, pressure drop, power

Table 12.12 Functions and requirements of hydraulic fluids

Function	Requirements
Medium for efficient power transfer and control	Low compressibility (high bulk modulus) Fast air release Low foaming tendency Low volatility
Medium for heat dissipation	Good thermal capacity and conductivity
Sealing medium	Adequate viscosity and high VI Shear stability Anti-leak and non-drip protection
Lubrication	Viscosity for film maintenance Low-temperature fluidity (low PP) Thermal and oxidative stability Hydrolytic stability/water tolerance Cleanliness and filterability Good water release Demulsibility/emulsibility Anti-wear characteristics
Pump efficiency	Proper viscosity to minimise leakages High VI
Environmental impact	Low ecotoxicity (fresh and aged) Biodegradability
Functioning life	Long life and consistent performance Material compatibility (metals, elastomers, other hydraulic fluids) Inertness Corrosion protection Good filterability High VI Low vapour pressure
Special functions	Low flammability (fire resistance) Controlled friction Radiation resistance Safety

Table 12.13 Physical properties of hydraulic oils affecting performance

• Viscosity; V–*T* and V–*p* behaviour	• Pour point
• Density	• Corrosion behaviour
• Compressibility	• Water content
• Gas solubility, cavitation	• Ageing stability
• Air release	• Neutralisation number
• Foaming	• Wear protection
• Demulsification	• Shear stability

consumption and heat generation will increase. High thermal and oxidative stability is required so as to maintain viscosity and prevent formation of sludge. The viscosity–temperature (V–T) behaviour is equally of great importance. Normally oils with VI ≈ 100 are used but higher VIs are recommended for special hydraulic systems, at higher temperatures and in mobile applications. Hydraulic oils of high VI enable easy pump start-up, less loss in performance at low ambient temperatures, and improved sealing and wear protection at high operating temperatures. High-VI oils improve the efficiency of a system, increase the life of components subject to wear and enable grade rationalisation in industrial applications. The viscosity–pressure behaviour of a lubricant is principally responsible for the load-carrying properties of a hydraulic oil film. The dynamic viscosity of fluid media increases with pressure. Precise power transmission in high-pressure hydraulic systems requires the fluid viscosity (as well as volume) to be the least sensitive to pressure (low pressure–viscosity coefficient α).

Gas absorption does not necessarily negatively affect the characteristics a hydraulic fluid but foam and high air entrainment may cause erratic operation of servos and contribute to pump cavitation. Systems should be designed to hinder ingress of air into the fluid. Water condensed from entrained air in a hydraulic system causes rust if the metal surfaces are not properly protected. Water that enters a hydraulic system can emulsify and promote the collection of dust and dirt. Free water should rapidly be removed from hydraulic fluids. Demulsification is the capacity of a hydraulic fluid to repel ingressed water. The neutralisation number is an important criterion for evaluation of the condition of a hydraulic fluid. If the neutralisation number of a hydraulic oil increases as a result of ageing, aged products can cause serious system failures.

Though not total-loss lubricants, hydraulic fluids have been classified as 'high risk loss' lubricants. Hydraulic fluids are increasingly being used in mobile equipment for transformation and construction in applications in environmentally sensitive areas where a *potential* total loss could occur, such as tractors, excavators, in agricultural and forestry applications, and in fresh water (groundwater) sensitive areas [83]. They are used in large volumes in equipment that is susceptible to spills. Spills and accidents with closed-system lubricants such as hydraulic oils occur regularly and contribute considerably to the environmental burden posed by the use of lubricants. Modern hydraulic systems operate under high pressure and temperature and leaks or failures from hoses, seals or cylinders can result in large amounts of hydraulic fluid being released into the environment. Industry experts estimate that 70–80% of hydraulic fluids leave systems through leaks, spills, line breakage and fitting failure. US Patent No. 3,816,316 describes an antileak hydraulic oil of gel-thickened type [84].

Typical hydraulic fluids are composed of 95–98% base oils and 2–5% additives. The most important base stocks, accounting for 80–85% of total volume, are mineral oils (paraffinic, naphthenic and white oils, and/or their mixtures; Group I oils). Commonly used petroleum-based hydraulic fluids are chemically saturated or unsaturated, straight-chained, branched or ring-type hydrocarbons [85]. The ageing stability of mineral oil-base hydraulic fluids is determined in accordance with ASTM D 943. Synthetic fluids based on hydrocracked oils (HC or Group III oils), PAOs, polyol ester oils (POEs) and PAGs are mainly used for rapidly biodegradable, fire-resistant or specialty hydraulic fluids (e.g. extreme temperature applications). Brake fluids are hydraulic fluids with high boiling point. Rapidly biodegradable fluids often contain natural vegetable oils such as RSO or SBO. Food-grade hydraulic oils are generally based on special white oils, PAOs and PAGs. The required lubricity of hydraulic fluids is usually provided by special additives. The most important additives for hydraulic fluids are base oil-active additives and surface-active additives (see Table 12.14). Hydraulic fluid additive systems are commonly distinguished in systems containing zinc and ash (70–80% of total volume) and zinc- and ash-free (ZAF) hydraulic oils (20–30% of total volume). Conventional industrial hydraulic fluids for plant indoor applications do not usually contain VI improvers as the fluids are not supposed to be used at sub-zero temperatures.

Conventional hydraulic fluids are satisfactory for low-pressure and low-speed applications, but need AW additives for high-pressure (over 6900 kPa) and high-speed (over 1200 rpm) applications. These applications do not permit the formation of full-fluid film lubrication to protect contacting surfaces – a condition known as boundary lubrication. Use of a hydraulic fluid without the proper AW additives will cause premature wear of the pumps and cause inadequate system pressure. Air entering a hydraulic system through the reservoir or through air leaks within the hydraulic system contributes to surface foaming on the oil and may lead to erratic operation. Foam depressants are added to (bio)hydraulic fluids to expedite foam break-up and release of dissolved air. Water entering a hydraulic

Table 12.14 Most important types of additives in hydraulic fluids

• Antioxidants (L)	• Steel/iron corrosion inhibitors (S)
• Metal (copper) deactivators (L)	• Anti-wear additives (S)
• Antifoam agents (L)	• Extreme pressure additives (S)
• Viscosity index improvers (P)	• Friction modifiers (S)
• Pour point depressants (P)	• Detergents/dispersants (S)
• Dyes	• Tackifiers (S)

L, lubricant-protective additive; P, performance-enhancing additive; S, surface-active additive.

system can emulsify and promote the collection of dirt, which can adversely affect operation of valves and pumps, increase wear and corrosion, promote fluid oxidation, deplete additives, and plug filters. To prevent rust, biohydraulic fluids use bio-corrosion inhibitors (BCIs) that deposit a protective film on metal surfaces. BCI additives exhibit excellent hydrolytic stability in the presence of water to prevent fluid breakdown and the acid formation that causes oxidation.

The variety of technical requirements for hydraulic fluids is vast and so are their formulations. The wide variety of different characteristics required for hydraulic fluids (Table 12.12) requires performance characteristics which cannot easily be satisfied by just one base oil. The Shell industrial lubricants line comprises the Tellus hydraulic fluids with new polymer technology which maintain their viscosity throughout a wide temperature range even under high loads. Optimum fluid selection depends on the application (design of the hydraulic system, type of pump), specific technical requirements (operating temperature range and pressure, etc.), required fluid life, economic and environmental conditions. The proper choice affects the performance, lifetime and cost effectiveness of the machinery and hydraulic equipment. In technical and economic terms, widely varying specifications and application conditions require different hydraulic fluids. These have to satisfy a broad performance spectrum. Table 12.15 compares the properties of various base fluids for lubricants and hydraulic fluids.

The type of base fluid and the types of additives determine the classification of hydraulic oils. Figure 12.1 shows a commonly accepted classification

Table 12.15 Properties of various base fluids for lubricants and hydraulic fluids

Base fluid	Viscosity @ 40 °C (cSt)	Pour point (°C)	Oxidative stability[a]	Biodegradability CEC[b]	Biodegradability OECD[c]
RSO	33	−21	+	100	>70
HOSNO	39	−21	++	100	>70
SBO	31	−9	−	99	>70
TMP trioleate	50	−50	+	90	>60
TMP trioleate, high oleic	47	−45	++	90	>60
PAO-8	47	−50	++	<50	30
Mineral oil (150N)	ISO 32	−12	+++	15	5

[a] +++ excellent; ++ good; + moderate; − poor.
[b] Coordinating European Council (www.cec.org).
[c] OECD 301B (Organisation of Economic Cooperation and Development).

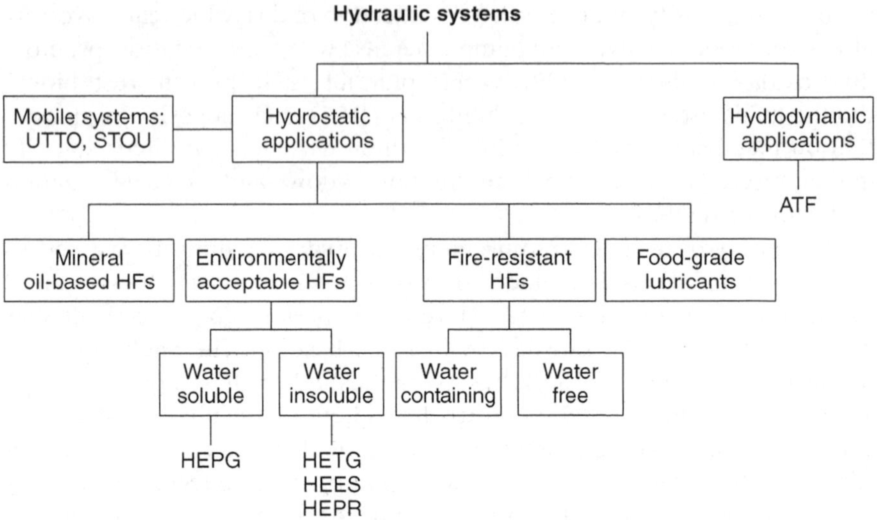

12.1 Classification of hydraulic fluids (HFs). HEPG, hydraulic oil environmental polyalkyline glycol; HETG, hydraulic oil environmental triglyceride; HEES, hydraulic oil environmental ester synthetic; HEPF, hydraulic oil environmental PAO/related hydrocarbon.

of hydraulic systems and fluids. Some 30 hydraulic fluid classes may be distinguished (Fig. 12.1 details only environmentally acceptable hydraulic fluid classes). The main classes of hydraulic fluids commonly used are petroleum-based, water/glycol solutions, water-in-oil emulsions, and oil-in-water emulsions. There are several notable disadvantages to conventional hydraulic fluids (including non-renewability, environmental and health risks, flammability). Water-based hydraulic fluids are used in machinery that is not exposed to freezing. Existing water-based hydraulic fluids, however, often lack the stability or lubricity required to operate at high pressure and temperature. Consequently, there is scope for water-based hydraulic fluids which have the characteristics of desirable shear stability, rheology control, lubricity, stability, volatility, very little swelling effect on (synthetic) rubber and AW properties while reducing potential harm to the environment. These (partially conflicting) characteristics are hard to achieve in a single composition. The main hydraulic fluid groups and their applications are listed in Table 12.16.

Non-renewable, non-sustainable, non-biodegradable petroleum-based hydraulic fluids include mineral oils, synthetic PAOs, synthetic esters and semi-synthetic PAO/mineral oil blends. Petroleum-based fluids are the most commonly used stocks for hydraulic applications where there is no danger of fire, no possibility of leakage that may cause contamination of other products, no wide temperature fluctuations and no environmental impact.

Table 12.16 Most important hydraulic fluid groups and their applications

Category[a]	Composition	Application	Market share (%)
HLP	Wear-protected refined	Universal hydraulic oil	88
HFA	Fire resistant emulsion/ solution	Mining hydraulics	<1
HFC	Aqueous PAG solution	Steel-making industry	4–5
HFDR	Fire resistant phosphoric acid ester	Turbines and aviation systems	<1
HFDU/HEES	Carboxylic acid ester or polyester/synthetic ester	Highly stressed industrial hydraulic systems	<1/2–3
HETG	Triglyceride	Medium-temperature and low-pressure	1–2

[a] According to DIN 51524–2, DIN 51502, ISO 6743/4, ISO 15380 and VDMA 24568.

HLP denotes the dominant group of universal hydraulic oils (refined mineral oil with improved AW properties).

In order to reduce the consequences of accidental discharge biodegradable fluids are desirable when hydraulic equipment is used in environmentally sensitive applications such as agriculture, forestry and water course maintenance. In some applications this requirement can be met by use of natural oils. However, native oils are limited by their relatively low thermo-oxidative stability and poor low-temperature fluidity. Synthetic esters are well suited to meet these performance requirements. Using optimised short-chain saturated polyol esters the high-pressure (HP) DSC induction time at 200 °C can be approximately five times greater than that of the mineral oil formulation, while retaining the biodegradability equivalent to that of the vegetable-based formulation [46].

To compare costs for hydraulic fluids: in 2006 they ranged from US$ 4 to US$ 6 per gallon for petroleum products, while soy versions cost between US$ 7 and US$ 10 per gallon [86]. Nowadays synthetic biodegradable fluids are more likely in the US$ 25–35 per gallon range, against an average of US$ 13 for vegetable-based fluids and US$ 7–8 for high-quality petroleum fluids.

In applications where fire hazards or environmental pollution are a concern, water-based or aqueous fluids offer distinct advantages. Fire-resistant fluids consist of water-glycols (35–60% water) and water–oil fluids with emulsifiers, stabilisers and additives. Oil-in-water fluids consist of very small oil droplets dispersed in a continuous water phase. These fluids have low viscosities, excellent fire resistance and good cooling capability due to the larger proportion of water. Additives must be used to improve their inherently poor lubricity and to protect against rust. The water content of water-in-oil fluids may be as much as 40%. These fluids consist of very small water droplets dispersed in a continuous oil phase. Synthetic fire-resistant fluids are based on phosphate esters or halogenated hydrocarbons.

Fire-resistant fluids should not spontaneously ignite when they contact hot surfaces. Fire-resistant hydraulic fluids with significantly higher ignition temperature offer better fire protection than mineral oils and have been developed for mining, die-casting, steel mill and aviation applications. Perfluoropolyether (PFPE)/Krytox™, developed by Du Pont in the 1960s, was aimed at fire-resistant hydraulic fluid applications in the US Airforce. For safety reasons, fire-resistant fluids are also used in control and governor circuits of power stations. This applies in particular to hydraulic systems in high-temperature zones such as near superheated steam pipes. The fire-resistant hydraulic fluids used in power stations are generally water-free, synthetic fluids based on phosphoric acid esters (HFDR category). Polyol ester or carboxylic acid ester-based HFDU fluids are synthetic, water-free, biodegradable and fire-resistant hydraulic fluids with tribological properties similar to those of mineral oils [87]. They are used in hydrostatic drives and in high-performance industrial hydraulic systems at temperatures up to 70–100 °C and pressures of 250–350 bar.

Base oils for food-grade hydraulic fluids are principally special white oils, or special PAO and PAG grades. Foodstuff-compatible hydraulic oils are classified according to FDA and NSF guidelines (see Section 12.10).

Early hydraulic brake fluids essentially consisted of a solution of castor oil in *n*-butanol or diacetone alcohol. Later, higher-performance fluids were based on ethylene oxide derivatives with a higher boiling temperature (190 °C instead of 140 °C). With more powerful cars and use of automatic transmission (which provides less engine braking) even more efficient braking systems were required. These trends resulted in more energy having to be dissipated as heat in the braking system. In order to avoid vapour locking failures high-boiling-point brake fluids (boiling point of 260–290 °C) were developed. With PEG-based brake fluids water absorption is a problem.

ATFs are functional fluids for automatic transmissions in vehicles and machinery. These products have excellent viscosity–temperature behaviour (high VI), shear stability, high oxidation stability, minimal foaming and outstanding air release. For the requirements for transmission fluids, see Section 12.2.2. For hydraulic fluids in tractors and agricultural machinery, see Section 12.2.1, p. 593. International Lubricants, Inc. (ILI; Seattle, WA)'s Lubegard™ automatic transmission fluid supplement is an extremely successful automotive lubricant that contains plant material. Lubegard™ is based on liquid wax ester technology. ILI produces a full line of plant-based lubricants (including total-loss lubricants, hydraulic fluids and others). Another universal AW biodegradable hydraulic transmission fluid was developed consisting of a compounded base oil of PAO, synthetic ester, API Group II oil, and modified castor oil [88].

Marine and offshore applications require water tolerance. Most bio-based fluids are more or less susceptible to hydrolytic breakdown, which results in acid formation. Terresolve Technology, Ltd (Eastlake, OH) markets a high-performance biodegradable, modified vegetable oil-based hydraulic fluid (XBD) for marine applications with excellent water tolerance and elastomer compatibility.

The principal functions of aircraft hydraulics are controlling the elevator, rudder, ailerons, flaps, landing gear, doors and brakes, etc. Aircraft hydraulics have to work under extreme climatic conditions with great temperature fluctuations and at high pressures (100–280 bar). Aerospace hydraulic systems experience both very low and high temperatures and are required to operate from –54 to 290 °C (MIL-PRF 27601) [89]. This calls for outstanding thermal stability and high VI of the hydraulic fluids which must also be fire resistant and free of particle contamination. A wide variety of hydraulic fluids (based on phosphoric acid esters, PAOs, silicate esters and special mineral oils) are currently being used in aircrafts. Fluid volumes used are relatively small (from 20 to 40 L). Hydraulic fluids have been reviewed previously [90, 91].

12.3.1 Environmentally acceptable hydraulic fluids

Hydraulic fluids present a high need for biodegradability. ASTM D 6006 describes the assessment of biodegradability of hydraulic fluids; for the classification of hydraulic fluids for environmental impact, see ASTM D 6046. Biodegradable hydraulic fluids based on synthetic esters [92, 93], vegetable oils [94–96], and specially treated RSOs [97] are on the market as from the early 1990s. In a typical advanced lubricant market (Germany, 2005) hydraulic oils consist of 80–85% mineral oils, 7% fire-resistant fluids and 7% rapidly biodegradable fluids, and 1% PAO-based fluids. Caterpillar, Inc. estimates that 12% of the European hydraulic fluid market is served by either rapeseed-based lubricants or synthetic fluids.

In many applications where hydraulic fluids are used, in particular in mobile equipment, there is risk of leakage of the lubricant from the system to the environment. This has led to the development of environmental adapted lubricants (EALs) that is a growing segment for vegetable-based oils [11, 98, 99]. Hydraulic systems in construction and forestry machines are vulnerable and pose a big threat to the environment as they have large reservoirs and external lines that if damaged would leak into the soil. Use of bio oils in hydraulic systems of mobile equipment originated in Europe in the late 1980s following up pollution of Lake Constance (Bodensee). Environmentally friendly, rapidly biodegradable hydraulic fluids were originally developed to ensure ecological compatibility. They are currently being used in both stationary and mobile systems. Their market share is growing, at the expense of mineral oil-based hydraulic fluids. Environmentally considerate hydraulic fluids have now been on the market for more than 20 years and a wide range of fluids is available for pump machinery that offer different levels of performance [12]. Many of the historic field problems with the use of environmentally considerate hydraulic fluids, such as short service lives, loss of fluidity at low temperatures and excessive foaming, can be attributed either to the use of a fluid with a poorly balanced technical profile, or cross-contamination with the previously used mineral oil (retrofitting problems).

As shown in Fig. 12.1, the so-called environmentally acceptable (EA) hydraulic fluids are allocated to four product families:

- HETG: triglyceride (vegetable-oil) types (non-water-soluble);
- HEES: synthetic ester types (non-water-soluble);
- HEPG: polyalkylene glycol types (water-soluble);
- HEPR: poly-α-olefins and related hydrocarbon products (non-water-soluble).

Although EA fluids are readily biodegradable, not all are made from non-renewable, non-sustainable petroleum base oils. Minimum requirements for HETG, HEES and HEPG fluids (as formulated end-products) are specified by VDMA guideline 24568 [100] and ISO 6743/4 [101], see Table 12.17. ISO 15380, which is intended particularly for hydraulic fluid power systems, also

Table 12.17 Hydraulic fluids[a]

Property	HV	HETG	HEES	HEPG
	Mineral oils	Vegetable oils	Synthetic esters	Polyglycols
Biodegradability (%)[b]	10–40	70–100	60–100	10–90
Viscosity index	90–100	200–220	140–190	180–200
Pour point, °C	–54 to –15	–20 to 10	–60 to –20	–40 to 20
Compatibility with mineral oils	–	Good	Good	Not miscible
Oxidative stability	Good	Poor to good	Poor to good	Poor
Water solubility	No	No	No	Yes
Relative cost	1	2 to 3	3 to 20	2 to 4

[a] VDMA (Verband Deutscher Maschinen- und Anlagenbau e.V.) designations [100].

[b] According to CEC L-33-T-82.

embraces HEPR [102]. It is noticed that most 'bio' hydraulic fluids are made from non-renewable synthetic POEs. Traditionally, this type of product is designated in industry more appropriately as an EA fluid (see Section 3.5) rather than a bio-based fluid. Reduction of the fraction of very toxic substances that are presently allowed in hydraulic fluids would further improve their environmental impact.

Because of cost and availability, the initial EA hydraulic fluids offered in the market were of vegetable oil (triglyceride) type, designated as HETG. Rapeseed and sunflower oils are mostly used as the raw material for HETG lubricants or for synthetic esters. The selection of the most suitable triglyceride and additive system is crucial to the quality of the HETG lubricant. Such truly renewable biohydraulic fluids, with a very narrow operating temperature range, solidify at lower temperatures and oxidise rapidly at the higher normal service temperatures. Because HETG hydraulic oils withstand lower thermal and oxidative loads than mineral oils, they are used for medium-temperature and low-pressure applications, often in total-loss mode. Vegetable hydraulic fluids are compatible with the working conditions of forestry equipment. Oleic HETG formulations represent a good price/performance compromise [103]. Under demanding pump operations the main degenerative process in vegetable oils is more alike pyrolytic degradation than oxidation.

Synthetic ester-type (HEES) hydraulic fluids extend the operating temperatures of HETG fluids, but with a significant increase in cost. Their advantages include outstanding oxidation stability, good material compatibility and excellent tribological performance. Synthetic esters are a group of substances with wide variation in structure (see Sections 6.2.1, 10.5.3 and 10.5.4). Partially unsaturated ester oils form the largest group of rapidly biodegradable lubricants. Not all HEES fluids survive higher operating temperatures for long times. Fully synthetic, saturated TMP, polyol and complex ester oils give the best technical performance and are used in high pressure and temperature and highly stressed hydraulic systems, e.g. mobile hydraulics and hydrostatic drives under harsh conditions. Such saturated synthetic esters are stable at higher temperatures for prolonged service times (from 2000 up to 6000 h). TMP trioleate ester is widely used in biodegradable hydraulic fluids and cutting fluids where oxidative stability is less critical [104]. Battersby *et al.* [105] examined ageing of saturated and unsaturated HEES-based fluids for applications in mobile hydraulic equipment. Negligible fluid degradation or hydraulic pump wear was seen after 3500–5000 h of service. US Patent No. 5,707,945 (to Unichema) describes a non-hygroscopic biohydraulic base fluid composed of the esterification product of polymerised unsaturated C_{12}–C_{24} fatty acids (such as dimer or trimer acids) with a saturated monohydric alkoxylated alcohol (straight or branched chained, or alkoxy poly(alkylene)glycol) [106]. Such products are marketed by Uniqema (Wilmington, DE) as water-dispersable esterified trimer acids (ETAs) (Priolube® 3951/3952/3955) for hydraulic and metal-working applications [107].

In US Patent No. 5,817,607 (to Exxon Chemical) Duncan and Meade [6] have disclosed highly biodegradable lubricants for hydraulic fluids and catapult oil applications which are based on synthetic ester (polyol) base stocks which incorporate both highly branched acids and linear acids, and selected additives. The additive package may include VII improvers, rust and corrosion inhibitors, wetting agents, water loss improving agents, bactericides and drill bit lubricants. Catapults are instruments used on aircraft carriers at sea to eject the aircraft off the carrier.

Castrol's BioBar is a high VI, low PP hydraulic oil (HEES fluid) which allows start-ups at low temperatures and provides a thicker lubricating film at high temperatures for additional AW protection. Careful choice of the saturated synthetic ester enables the product to operate up to +120 °C system temperature. BioBar high-specification hydraulic oils are intended as replacements for conventional mineral oils. BioBar is recommended for hydraulic systems and hydrostatic transmissions incorporating gear pumps, vane pumps, radial or axial piston pumps and motors where there is a perceived risk of egress into the environment in the event of spillage or leakage. The Shell Naturelle HF-E product line is composed of advanced, readily biodegradable hydraulic fluid esters for use in power transmission and hydraulic systems (heavy-duty hydraulic systems for construction and earth moving equipment, machine tool hydraulic systems, hydrostatic drive gears, general industrial control equipment and hydraulic systems, etc.). Shell Naturelle HF-E46 is approved for Hitachi's heavy earth-moving excavators, is on the German *Positivliste* and has qualified for funding under the 'Market Introduction Programme' (MIP) for biolubricants. Ester-based fluids must be maintained in a dry state to obtain maximum performance.

Environmentally compatible hydraulic fluids may also be PAG-based [108]. HEPG (polyglycol) fluids show better lower temperature properties than HETG (vegetable oil) fluids, but their oxidative stability is poor. HEES (synthetic ester) fluids show a superior operating range. In the past, base fluids on the basis of PAGs have been proposed for hydraulic fluids, but these are hygroscopic and, owing to the water absorption from the environment, the flash point is lowered. Ethylene oxide-based polyglycols are highly water soluble, poorly miscible with mineral oil, and of high polarity. Polyglycols with a high proportion of propylene oxide are not or only slightly water soluble. Biodegradable polyglycol-based hydraulic fluids are primarily used when the application leads to unavoidable contact with ingressed water, such as in the water-supply industry, in canal lock hydraulics and in offshore applications. US Patent No. 2,755,251 describes fluid compositions (containing high-molecular-weight polyglycol ether-esters) suitable for transferring energy in hydraulic systems at very low temperatures, where the ether esters are condensation products of 1,2-propylene oxide with dimerised, unsaturated, long-chain fatty acids (such as linoleic acid) [109].

A fourth category of biodegradable hydraulic fluids is HEPR (poly-α-olefins and related hydrocarbon products). Low-molecular-weight PAOs and the correspondingly derived hydrocarbons are classified as environmentally friendly hydraulic fluids. HEPR fluids are more rapidly biodegradable than mineral oils but significantly less so than most ester oils and natural vegetable oils. Apart from faster biodegradability, the technical properties of these oils are similar to mineral oils. PAO-based fluids can be tailor-made to fit specific requirements. These fluids do not hydrolyse (break down when mixed with water) and, as such, are much more stable in hot, wet conditions.

The majority of vegetable oil-based lubricant applications have concerned hydraulic fluids. The structure of the triglyceride molecule is generally more stable against mechanical stress existing in hydraulic systems than the linear structure of mineral oils. In addition, the ability of the polar triglyceride molecule to generally adhere onto metallic surfaces improves the lubricating properties of these triglycerides. The intended use of natural triglycerides for hydraulic purposes is limited by their ease of oxidation. Also, variations in temperature, in particular high-temperature environments, are known to impact the performance of a vegetable oil-based fluid. Vegetable oil-based hydraulic fluids conforming to the minimum requirements of ISO 15380 have found successful industrial applications. Biodegradable hydraulic fluids have been based on suitably additivated base stocks composed of highly monounsaturated (85–90%) vegetable oils [110]. In Germ. Offenl. No. DE 60,122,400 T (to Nippon Mitsubishi Corp.) Konishi and Kikuchi disclose hydraulic oil formulations which are excellent in oxidative stability, lubricating properties and biodegradability [111]. The hydraulic oils comprise a vegetable oil with low total degree of unsaturation (<0.2, as determined according to JIS K1557–1970), and a specific 0.01–5 wt% antioxidant package (phenolic, aminic or zinc dithiophosphate). Modified ASTM D 943 (Dry TOST) is reliable to evaluate the impact on oxidative stability of various additive packages in vegetable oil hydraulic fluids.

Use of vegetable oil-based hydraulic fluids in the out-of-doors environment, potentially within mobile equipment, presents a much bigger challenge than in indoor applications (e.g. food processing). Success of such fluids has been more limited, even though commercial RSO-based hydraulic fluids have been on offer for these applications. A process of hydrogenation may be necessary to obtain optimum stability needed for such demanding applications. The key characteristics for mobile hydraulic fluid use are indicated in Table 12.18. Stationary (indoor) hydraulic systems may

Table 12.18 Key characteristics for mobile hydraulic fluids

• High oxidation stability
• Viscosity characteristics (high VI, low PP)
• Extreme pressure performance
• Corrosion inhibition properties
• Seal and polymer compatibility
• Foam suppression
• Controlled friction

not require all of the above properties, although most still apply. Agricultural tractors present the harshest conditions for hydraulic fluids. An additional demand placed upon a tractor hydraulic fluid is the requirement that it functions as a transmission fluid, which means that it must withstand greater amounts of heat generated within the operational environment and must have improved frictional qualities. Field trials of fully formulated vegetable oil-based hydraulic/transmission fluids have been positive [22, 112]. For the degradation of hydraulic oils, see also Section 14.4. The increased demand on hydraulic systems has accelerated the development of high-performance biodegradable hydraulic fluids that meet the Caterpillar BF-1 specification [113].

Natural oils (such as RSO) are used as base stocks for fluids where biodegradability and renewable resources are considered important. Fresh RSO-based hydraulic fluids are more readily biodegradable and less toxic than mineral oil-based fluids. RSO-based hydraulic fluids are normally used in lightly stressed systems and are changed at the same intervals as mineral oils. Rapeseed-based lubricants have great potential for use in hydraulic systems, as power transmission oils, and in two-stroke engines. Low-cost RSO-based hydraulic fluids are suitable only up to 70 °C tank temperature. Such lubricating oils can endure 2000 h operation without problems at a working temperature below 80 °C [114]. Long-term testing of an RSO-based Esso PFL hydraulic fluid with 37 agricultural machines has shown the suitability and environmental compatibility of such fluids for this application. High biodegradability and low ecotoxicity were invariable properties during the test period [115]. Hydraulic fluids based on RSO tend to oxidise and hydrolyse above 80 °C. For such more severe conditions more expensive high-performance ester oils are required.

Various manufacturers are now marketing EA hydraulic fluids in the United States. For example, ExxonMobil, Chevron and E.F. Houghton, a supplier of industrial hydraulic fluids, offer rapeseed-based products. Mobil EAL 224H (available in ISO VG 32, 46, 68 grades) is a readily biodegradable and non-toxic hydraulic and circulation oil for hydraulic systems under moderate to severe conditions. It is formulated from high viscosity index (HVI) vegetable oils and additives that satisfy stringent criteria for biodegradability and ecotoxicity. Pennzoil offers a hydraulic fluid made with sunflower oil. G-OIL Hydraulic™ is a non-ash hydraulic oil in the green (G)-branded product line of Green Earth Technologies, Inc. (Stamford, CT) based on an ultimate biodegradable tallow (beef, pork or chicken)-based base stock, which even rivals synthetics for low operating temperatures. G-OIL Hydraulic™ is a green alternative for all engines requiring AW 32, AW 46 and AW 68 hydraulic oils. Other green hydraulic oils use vegetable oils rather than animal fats as a raw material.

US Patent No. 5,451,334 to Henkel [116] describes environmental-friendly basic oils for formulating ester-based hydraulic fluids for hydrostatic power transmission having improved viscosity at elevated temperature and lower PP. Industrial stationary equipment applications such as these require the use of hydraulic oils of water hazard class 0 (now 'nwh'). Formulations consist of purified RSO or SBO as the main component and antioxidants specifically designed to perform satisfactorily in ester oils. As the high solidification point of RSO or SBO presents considerable problems for the practical application of such hydraulic oils at low ambient temperatures, which cannot be solved by the addition of PP depressants, about equal amounts of selected TMP or NPG esters were added to the formulations. Biodegradable hydraulic fluids are frequently constituted of TMP trioleate ester base [117]. The primary focus of US Patent No. 4,783,274 is the effectiveness of RSO as the base component of an hydraulic fluid used within a laboratory environment [118]; other usable triglyceride oils were olive oil, peanut and maize oil. These oily triglycerides are liquids at 15–20 °C with saturated fatty acids (SFA) <20% and viscosities which change on temperature to a lesser extent than those of hydrocarbon basic oils. The VIs of triglycerides exceed those of hydrocarbon oils with no additives. US Patent No. 5,454,965 to International Lubricants, Inc. [119] is mainly concerned with telomerised triglycerides as a performance enhancing additive for use in industrial lubricant applications. Improved lower viscosity range telomerised oils according to US Patent No. 5,567,345 are designed as a thickener additive for high-viscosity oils [120]. Fluid used vegetable oils (UFOs) ('Mixoil': mainly RSO-based, IV 78 g I_2/hg; 4.4% FFA) can be used as raw materials for environmentally friendly polyol ester hydraulic fluids, though with restricted applicability in view of a relatively high PP [121]. Semi-solid UFOs (from tallow and palm oils) do not qualify. Rapidly biodegradable RSO-based hydraulic fluids have been compared with the performance of mineral oils [22].

Soy-based hydraulic fluids work as well as or better than the mineral oil products, particularly in terms of lubricity, biodegradability and reduced flammability. However, although SBO performs well on lubricating pump components, the lack of oxidative stability causes an increase in the viscosity of untreated SBO during use in hydraulic systems [122]. US Patent No. 5,972,855 to Honary [123] describes use of partially hydrogenated SBO (Cargill, Inc. #110 PKS-WOO) in hydraulic systems of mobile outdoor equipment that is operated under widely varying conditions. The base oil has been optimised for improved stability and winterised to remove crystallised fats and improve the PP (–40 °C). Finally, a specially designed hydrocarbon-based performance additive package (Lubrizol LZ9999) for mobile equipment application was used which includes friction and anti-chatter materials for operation of wet clutches and brakes. The formulated hydraulic oil (70 wt% SBO and 30% chemical lubricant additive) has a kinematic viscosity in the range of 45–50 cSt at 40 °C (cf. high-viscosity functional SBO-based lubricants with viscosities of 2000–2500 cSt at 25 °C). The lubrication qualities were comparable to those associated with other vegetable oil or petroleum-based fluids. BioSOY®, the soybean-based hydraulic fluid developed in the framework of the ABIL (Ag-base Industrial Lubricants) research programme (University of Northern Iowa) has been tested in 20 mobile hydraulic machines including tractors, street sweepers, garage trucks, fork lifts, motor graders, front end loaders and others. Other soy-based ABIL development projects are a hydraulic oil for food-processing plants and a grease for preventing wear to railroad tracks. Terresolve Technology, Ltd. (Eastlake, OH) uses a proprietary technology to chemically modify soy and canola oils to produce commercially available hydraulic fluids and two-stroke engine oils. SoyFluid Hydraulic All Season oil is a premium biodegradable soy-based hydraulic fluid developed by Environmental Lubricants Manufacturing (ELM; Grundy Center, IA) with super high VI (214), low PP (–36 °C) and high flash point (208 °C) for increased safety and use in hydraulic systems operating throughout all seasons. A soy-based hydraulic fluid is used in the elevators at the Statue of Liberty.

Renewable Lubricants, Inc.'s (RLI; Hartville, OH) Bio Hydraulic Fluids based on Stabilized™ high-oleic base stocks (HOBS) offer exceptional performance properties designed to improve efficiency and reliability in both production and maintenance (see also Section 10.5.2). Bio Hydraulic Fluids permit water to separate and demulsify readily and are suitable for the toughest applications in extreme operating environments. There are clear performance differences between hydraulic fluid types formulated from Stabilized™ HOBS and those formulated from traditional mineral oils, synthetics or commercial vegetable oils. The natural oily fluid films of these high-viscosity, multigrade products outperform petroleum-based fluids and commercial vegetable oil formulations in terms of lubricity, oxidation stability, thermal shear stability and load-carrying capacity. Bio Hydraulic Fluids render obsolete the use of harmful petroleum-based fluids. Oxidation stability clearly differentiates Stabilized™ HOBS hydraulic

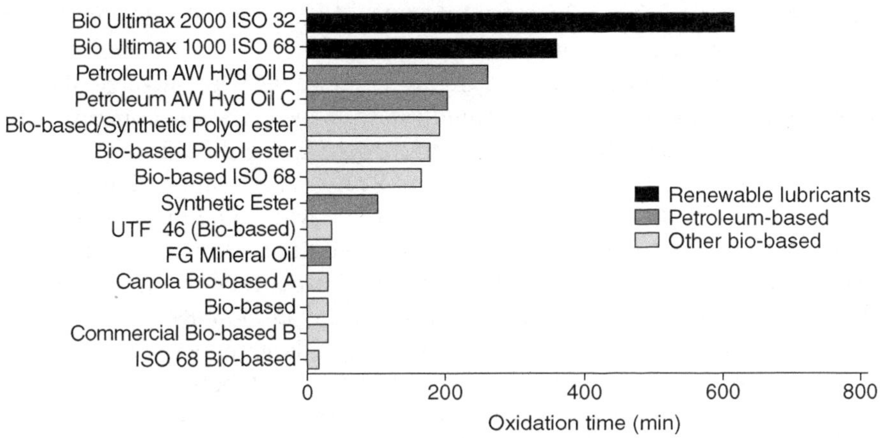

12.2 Comparative Rotating Bomb Oxidation Tests (ASTM D 2272) of hydraulic fluids. After ref. [1]

fluids from competitive offering (mineral oil, synthetic ester, synthetic ester/mineral oil blends, readily biodegradable petroleum-based, ultimately biodegradable and bio-based). As shown in Fig. 12.2, Bio Ultimax™ AW 1000 and 2000 Hydraulic Fluids (RLI) outclass leading brands, including:

- petroleum-based mineral oil hydraulic fluids;
- commercial bio-based hydraulic fluids;
- blended bio-based (sunflower)/30% synthetic polyol esters (trimethylolpropane oleate, TMPO);
- full synthetic POE-based hydraulic fluids.

Full synthetic POE-based hydraulic fluids are considered and promoted in industry as very stable formulations. Apart from superior oxidation stability Bio Ultimax™ AW 1000 and 2000 hydraulic fluids also show the lowest total acid number (TAN, 0.28–0.43 mg KOH/g). Hydraulic systems may suffer acidic attack of internal metal components for TAN >2.0 mg KOH/g.

Bio Hydraulic Fluids can operate in static, mid-dynamic and dynamic areas of hydraulic systems and are compatible with a wide range of materials, such as NBR (medium to high nitrile rubber; Buna N), NBR-L (low nitrile rubber), FPM (fluoroelastomer; Viton), AU (polyurethane; Adiprene, Millathene), PTFE (polytetrafluoroethylene, Teflon) and CR (polychloroprene; Neoprene). Consequently, conversion procedures are easy, as no engineering design changes to filters, seals, hoses and accumulator bladders are necessary. They are also compatible with mineral oil-based fluids and most synthetic based fluids (PAOs, semi-synthetic PAO and mineral oil blends). To convert a system it is sufficient to simply drain and recharge.

For proper biodegradability, at least 90% of the mineral oil-based fluid should be removed from the system.

Bio Hydraulic Fluids are also highly compatible with most SE-based fluids, in particular polyol esters and diesters. However, some compatibility problems may arise with complex esters. Most SE-based fluids show excessive seal swell on NBR-L; NBR, FPM and PTFE are recommended in those cases. Bio Hydraulic Fluids are also compatible with some phosphate ester fluids, but are not compatible with water-based fluids, such as water glycol (HFC) and invert emulsions (HFB). Standard practices for flush and purification of hydraulic systems are found in ASTM D 4174–89.

ISO 32, ISO 46 and ISO 68 apply to biodegradable hydraulic fluids. For biodegradable hydraulic fluids the most important ecolabels are the German Blue Angel UZ-79 and the Swedish Standard SS 155434. Whereas the former is optional, the latter is a legal requirement: hydraulic fluids not fulfilling the criteria of SS 155434 (Version 4, July 2000) are not permitted on the Swedish market. The sustainability of a biolubricant (an 'environmentally acceptable' synthetic ester-based hydraulic fluid meeting Swedish Standard SS 155434) has been compared with that of a conventional mineral oil-based product [124]. The former is more sustainable but more expensive. The European ecolabel 'Marguerite' was introduced in 2005 with EU Directive 2005/360/EC [125]; see Section 9.4. For biodegradable hydraulic fluids it is based on technical specification ISO 15380. Besides the requirement for ultimate aerobic biodegradability of >95%, also sustainability is requested, namely at least 50% of the raw materials must be derived from renewable materials.

Binol X/BioSafe XE range is the first long-life hydraulic oil based on renewable raw material that has been awarded the EU ecolabel. Carl Bechem GmbH's (Hagen, Germany) Hydrostar hydraulic oils are partially labelled with the European ecolabel or German 'Blauer Engel' environmental emblem. Cognis's ProEco HE 801 series of fully formulated next-generation biohydraulic oils (for >90% from renewable feedstocks), available in various viscosity grades, combines excellent performance and compliance with the exacting requirements of the EU ecolabel [126]. These 'green' ester-based products, available in various viscosity grades, overcome the well-known problems of swelling or other interactions with elastomers or sealings (commonly NBRs) that are encountered with conventional esters. Combined with this are superior oxidative and hydraulic stability – important factors for long fluid lifetime, reduced maintenance and reduced sludge formation. The lubricants have a high VI. Even under arctic conditions, the products provide outstanding long-term mechanical protection, due to their excellent flowability.

Environmentally friendly hydraulic fluids have been reviewed [127–131]. For use of bio-oils in hydraulic power, see also refs [132, 133].

12.4 Biodegradable loss lubricants

It is claimed that more than 50% of all lubricating oils sold worldwide pollute the environment by way of leaks, spillage, evaporation, loss applications or otherwise [134]. Other estimates limit this figure to 30% of used lubricants [135]. However, the loss of hydraulic fluids is even as high as 70–80% [134]. Apparently, in the EU alone 13% (or 600 kt, 1998) of all lubricants used disappear uncontrolled, i.e. are lost to the environment [136]. Similarly, in the United States 32% (430 million gallons per year Mgy) of lubricating oils end up in landfills or are dumped. Some 300 kilogallons per year (kgy) of oil are already being discarded along the railway lines within the United States alone [56]. More than 95% of lost materials are currently mineral-oil based. This calls for concerted actions, especially as not all losses can be prevented even in case of proper management practices. Actions required involve government (environmental legislation and control), consumers (environmental consciousness) and lubricant manufacturers (environmentally friendly product compositions).

Loss lubrication refers to lubrication with direct emission to the proximate environment during use. Particularly sensitive environments are found in marine and agroforestry applications. Also major civil engineering projects always require the use of large amounts of lubricant, which are then likely to be released into the environment. Clearly, non-toxic biodegradable fluids are ideal for use in heavy machinery and construction equipment that are exposed to the environment.

Lubricants are designed and produced for systems prone to leakage and total-loss applications. Specifically, such applications include lubricants for brush cutters and hedge trimmers, chainsaw bar oils, two-stroke engine oils (see Section 12.2.1, p. 590), outboard engine lubricants, drilling fluids (see Section 12.4.2), open gear oils, railroad frange oils and greases, hydraulic fluids, drip oils, mould-release oils, chain and conveyor lubes, corrosion preventive oils, wire rope lubricants, dust suppressant and marine lubricants (see Section 12.5). Loss lubricants account for 8% of all lubricants, and the market is divided as follows: two-stroke engine oils, 30%; mould-release agents, 25%; various greases, 17%; protective oils, 10%; chainsaw oils (CSO), 8%. Although the immediate consumption of total-loss oils is relatively small, the cumulative effect on the environment is devastating. Because these types of lubricants pose immediate harm to the surrounding environment, there is increasing regulatory pressure to seek biodegradable and less toxic alternatives. Many European countries enforce the use of biodegradable lubricants in selective applications. EU ecolabel requires ultimate aerobic biodegradability of >95% for TLLs (chainsaw oils, concrete release or demoulding agents, etc.) and >75% for two-stroke engine oils.

Vegetable oil-based lubricants have focused primarily on the sector of industrial oils. Within this category the applications best suited for plant oils are various TLLs and hydraulic fluids. The use of *pure* native plant oils is limited to total-loss applications and those with very low thermal stress. Natural oils are a good option for once-through applications whereby stability is not an issue and ready biodegradability is a requirement. Consequently, vegetable oils such as rapeseed, canola and soybean are widely used in applications like chainsaw lubricants and railroad greases. Apart from their environmental properties vegetable-based chainsaw oils in forestry also contribute significantly to the safety of loggers. TESSOL GmbH (Stuttgart) is a producer of RSO-based (Blue Angel) chainsaw oil. In US Patent No. 5,888,947 (to Agro Management Group, Inc.) Lambert and Johnson [56] describe a vegetable-based lubricant composed of a mono-, di- and triglycerol base oil (from the Crucifereae, Leguminoseae or Compositeae families), a vegetable oil additive containing HFAs (castor or Lesquerella oil), and a liquid vegetable wax (jojoba or lanolin), designed for total-loss applications (see also Table 12.8).

When producing parts by moulding, it is important to control the demoulding step in order to obtain an intact and homogeneous surface. In particular, hydraulic-setting compositions such as concrete tend to react with the surface of the mould, which lowers the quality of the moulded part and the surface life of the mould. Mould-release agents are loss lubricants, formulated as emulsions, which are used at high temperatures (250–350 °C) for mould release of very different materials; they therefore require excellent thermo-oxidative properties. The main mould-release ingredient is commonly made by mixing a mineral oil, animal fat or vegetable oil lubricant with silicon oil, wax or fatty acid to improve thermal resistance and adhesive properties, before being emulsified by a surface active agent. Demoulding agents based on fluid UFOs (mainly composed of RSO) have been developed within the context of the European VALUIL project [121]. US Patent Appl. No. 2010/0107928 A1 (to Lafage) [137] describes a demoulding composition (emulsion) for the production of moulded parts with a homogeneous surface appearance in hydraulic-setting material which typically includes 30 wt% vegetable oil or animal fat (or neopentylpolyol esters of fatty acids) and 70 wt% calcium salt in aqueous solution. The oils make it possible to reduce the friction coefficient and hence facilitate the demoulding. Vegetable demoulding oils available on the market are Rheofinish 201 (Degussa), Decobio VGI 200 (Pieri) and Chryso® Dem Eco 2 (Chryso).

Objectives of the LLINCWA project on Loss Lubricants in Inland and Coastal Water Activities (FP5: IPS-1999–0005) were: (i) reduction of diffuse water pollution with lubricants and greases; (ii) increased use of non-toxic biodegradable lubricants; and (iii) protection of sweet water reserves [138].

In LLINCWA focus was on the direct emissions to surface water, which represents the largest hazard potential.

Hydraulic fluids are of increasing importance for applications in environmentally sensitive areas where a potential total loss could be encountered, such as excavators, earthmoving equipment and tractors, in agricultural and forestry applications, and in fresh water (groundwater) sensitive areas. Triglyceride (vegetable oil-based) HETG hydraulic oils, which withstand lower thermal and oxidation loads than mineral oils, are used for medium-temperature and low-pressure applications, often in total-loss mode. Vegetable oil-based lubricants do not pose threats to the environment which are characteristic of mineral oils. They significantly reduce equipment emissions, and the fluid that is released readily biodegrades and is low in toxicity. Vegetable oils are much safer, both for workers and the environment. See also Section 12.3.1.

Two-cycle (two-stroke) gasoline engines are mostly used when high specific power, low weight and low price are requested. These engines have gained considerable popularity as a power source for outboard motors, snow mobiles, jet skis, motorcycles, mopeds and landscaping equipment (lawnmowers, chainsaws, string trimmers and blowers). Almost all two-stroke engines operate in TLL mode. The oil is not circulated as in case of four-stroke engines but added to the fuel. A considerable fraction (about one-quarter) is not burnt in the combustion process and generates high emissions. Water-cooled outboard motors exhaust directly into the water, giving rise to water pollution, whereas air-cooled two-cycle gasoline engines produce emissions that pose a serious air pollution problem (up to 50 times that of truck engines per horsepower hour). In principle, all common base oils can be used. Apart from hydrocarbon types, higher quality two-stroke lubricants often contain various SEs. The ever-increasing severity of environmental legislation is affecting the development of two-stroke oils, especially marine (outboard) engine oils. Their biodegradability depends on varying minimum requirements of local environmental legislation. At the international level, ICOMIA Standard 27–97 has harmonised requirements for environmentally friendly outboard engine oils [63]. These oils, with very low algae, daphnia and fish toxicity and rapid biodegradability according to ISO and OECD standards are based on fully synthetic components, with the base oils usually being rapidly biodegradable SEs. High-quality esters are the best two-stroke oils and can also be used as chainsaw lubricants. In this application, the use of RSME is favoured over RSO in view of better thermal properties. The use of ester-based lubricants ensures the highest technical performance with improved environmental compatibility. Manufacturers of vegetable-oil based two-cycle engine oils claim increased engine life. See also Section 12.2.1, p. 590.

Chains are supposed to guarantee non-stop smooth operation. Chains working almost continuously and at maximum capacity require high-performance oils for lubrication at lowest maintenance efforts. The oil inside the chain and not the lubricant which is visible outside the chain is responsible for lubrication. Chains are running under mixed friction. The requirements for high-performance chain oils are summarised in Table 12.19. Priolube 1962, Priolube 1963 and Priolube 1965 (Croda Lubricants) are chain-oil base fluids (POEs) capable of operating at high temperatures for extended periods of time with reduced energy consumption. Vegetable oil-based lubricants with their natural adhesive capacity thanks to their polar character guarantee lowest oil consumption and avoid dripping and centrifuging off of the oil in case of correct dosage. Selection of the right viscosity for the application (temperature) is crucial. Duration of the lubricating effect is optimal for oils with a higher viscosity. Bechem GmbH (Hagen, Germany) is a producer of high-performance synthetic chain oils (mainly esters) for universal and special applications (Berusynth® series), including a native/NSF H1 chain oil for bakeries (Bechem Plantfluid®) and a chain oil (polymer + MoS_2) for agricultural machinery.

The method of lubricating chain conveyors is generally governed by speed, environment and accessibility. Some method of continuously or periodically lubricating the chain conveyor in service should always be considered. Manual lubrication is normally sufficient for slow-speed conveyors. Drip lubrication is often required for moderate- to high-speed conveyors. In drip lubrication, oil drips from an oiler provided on the top of a bearing unit should strike the upper edges of the link plate or sidebar in the lower span of the chain. This enables gravity and centrifugal force to carry the lubricant into the critical bearing areas. The amount of oil and the frequency of its application must be adequate. Drip lubrication is quite acceptable in reasonably clean, dry, non-abrasive environments. Total-loss application

Table 12.19 Requirements for high-performance roller chain oils

- Excellent friction and wear reducing properties
- Optimal viscosity (ISO VG 220/320)
- Good wetting and creeping properties (oil penetration)
- Adhesive capacity
- Load-carrying capacity
- Low evaporation loss even at high temperatures (ISO VG 100 to 220/320)
- Cleaning effect
- Low oil consumption (long relubrication intervals)
- Low dust absorption
- Corrosion protection and paint compatibility
- Extended lifetime

drip oils may range from high viscosity, low PP oils made from paraffin distillate stocks to accommodate the most severe of temperature ranges (e.g. Rugged Drip Oil; Colorado Petroleum, Denver, CO) to soy-based products (GEMTEK, Phoenix, AZ). At extremely high or low temperatures, special synthetic lubricants may be required. Similarly, in wet environments, special lubricants may be needed.

Reportedly, in Nebraska alone 700 kgy of drip oil were lost to the environment each year, possibly contaminating groundwater sources. In order to obviate this inconvenience, soy-based drip oil for pump lubrication has been developed [139]. GEMTEK (Phoenix, AZ), which produces a full line of total-loss lubricants and hydraulic fluids, uses vegetable seed oil telomers, other seed esters and rarified waxy substances as additives. Safe Lube™ industrial lubricants comprise a soy-based Bar, Chain and GP Oil suitable for use in chainsaws and other forest-cutting machines.

12.4.1 Biodegradable penetrating lubricants

Penetrating lubricants are liquid compositions that have the ability to penetrate rapidly between metallic surfaces that are in close contact, such as the leaves of springs, hinges, bolts, locks, pipe fittings, etc., and to loosen the adjacent metallic surfaces that somehow have become bound together. Surface tension and viscosity of such liquids are somewhat lower than those of an ordinary lubricant used on rotating parts. Penetrating lubricants are usually composed of petroleum-based oils. The use of vegetable oils for this application has not been thoroughly explored despite the fact that vegetable oils possess many desirable properties (biodegradability, ecotoxicity, good lubrication and viscosity, high flash point, excellent corrosion inhibition). In penetrating lubricants, a balance must be reached between the penetrating and lubricating functions. In PCT Int. Publ. No. WO 03/106599 A1 to Renewable Lubricants, Inc. [140] Garmier describes biodegradable penetrating lubricants based on (chemically or genetically modified) triglycerides and an organic solvent (such as ethyl lactate or methyl ester) and other additives (antioxidant, corrosion inhibitor, viscosity modifier, AW inhibitor and emulsifier). In this product the organic solvent provides the penetrating function. BPL™ (Bio-Penetrating Lubricant, with bio-corrosion inhibitors) is a very successful replacement for hazardous petroleum-based multipurpose penetrants.

Glysol PenTrate is an all-purpose soy-based spray-on lubricant, developed by Workcell Systems, Inc. (Highland Park, IL). Another all-purpose lubricant that has been developed with support from US soybean farmers is Naturelube 700, manufactured by BioPlastic Polymers & Composites LLC (Okemos, MI), which is suitable as a base for rust- and corrosion-preventive formulations. Naturelube 700 has exceptionally low values of the friction coefficient even under heavy loads. The oxidative stability of Naturelube 700 is also similar to high-performance synthetic lubricants.

12.4.2 Biodegradable drilling fluids

Conventional fossil fuels are usually extracted from deposits buried below Earth's surface. When drilling oil and gas wells, it has long been the practice to circulate a drilling fluid or drilling mud past the drilling apparatus and to cool the drill bit, lift cuttings out of the hole, and counterbalance pressure. One of the more important functions of a drilling fluid is to reduce the considerable torque on the rotating drill pipe caused by friction between the outside of the drill pipe and the wall of the well. The drill mud helps lubricate the drill and its pipe as well as hold open the well bore. Drilling fluids are designed to deposit a low permeability layer on the surface of the borehole to limit leakage of the drilling fluid into the oil-bearing formation and to prevent invasion of solids into the oil production zones. Once the well is drilled to the desired depth, the low permeability layer must be removed in order to maximise oil production rates. Traditional drilling fluids are muds – dispersions of clay mineral in water or oil where the clay provides the required viscosity and the oil provides the lubrication. These muds pose two problems: (i) the oil used in their formulation can have negative environmental impacts and requires treatment; and (ii) the strong acid required to remove the low permeability layer is toxic to the environment, corrodes equipment and does not uniformly remove the low permeability layer. Brookfield's TT100 in-line viscometer is an industry standard for drill mud.

In offshore crude oil exploration the demands placed on lubricants (drilling fluids) are particularly high as they are disposed of directly into the sea. In addition to the good lubricating effect biodegradability assumes particular relevance in this application. It is well known that petroleum cuts having a high aromatic compound content are highly toxic to marine life. On the other hand, aromatic-free mineral oil fractions such as white oils are considered as low-polluting oils [141]. However, highly de-aromatised mineral hydrocarbon cuts (especially gas oil) lose much of their lubricating power.

A typical biodegradable drilling fluid is composed of 60–90% base stock, 5–25% hydrocarbon solvents, with the remainder comprising an additive package [142]. In US Patent No. 4,181,617 (to Milchem, Inc.) Elrod and Nance [143] disclose an aqueous drilling fluid lubricant consisting essentially of a reaction product of a fatty vegetable oil (castor, coconut, corn, palm or cottonseed) with 4,4′-thiodiphenol. Petro-Canada Lubricants, Inc. (Mississauga, ON) produces inherently to readily biodegradable Pure-Drill™ drilling mud base fluids.

The US EPA has repeatedly reviewed and adjusted the binding standards for drilling fluids (2001, 2004) thereby significantly tightening the environmental regulations governing oil exploration on the open seas. The objective of the National Pollutant Discharge Elimination System (NPDES) General Permit of EPA is reducing the negative effects of oil and gas exploration in the delicately balanced marine ecosystem. Only non-toxic and biologically degradable drilling fluids are permitted. Very few fluids are compliant with the stringent EPA standards. Most contain high-performance vegetable esters. Both vegetable esters and low-viscosity C_8 esters exhibit an environmental performance superior to that of all the other synthetic-based fluids and are now practically indispensable for environmentally acceptable drilling fluids.

The first ester-based drilling fluids containing vegetable esters have been developed by Henkel chemical division (Cognis) in collaboration with the drilling fluids service company Baroid, Ltd (Halliburton Group) [144, 145]. The first trials of such ester drilling fluids were carried out in Norwegian waters in 1990; since then, these products have been used successfully on drilling rigs all over the world. Ester-based formulations provide drilling fluids with sufficiently low viscosity for the greatest possible equipment safety, even when subjected to high pressure and extreme temperature fluctuations. Specially designed vegetable ester fluids remain stable above 175 °C. Shorter chain length low-viscosity C_8 esters have been developed in response to the increasing importance of viscosity issues as operators move into deeper waters. The specially developed FA ester fluid Petrofree® for efficient drill bit lubrication not only fulfils the tightened environmental regulations but also has a better lubricating effect than mineral oil products. Ester-based formulations contribute to worker safety by posing a much lesser risk to human health than oil-based fluids and diesel. Esters do not smell, are not toxic upon inhalation, do not cause skin irritation and can be tailored to satisfy the most demanding needs. Cognis has introduced the EQ ('Ester Quality') emblem to guarantee greater environmental protection, safety and technical performance of original esters on drilling rigs.

Baroid's Accolade® drilling fluid is a blend composed of several high-performance vegetable esters and an internal olefin that allows excellent suspension and hole cleaning while keeping the fluid extremely shear-sensitive. Low and stable viscosity means low equivalent circulating density (ECD) – the most sought-after property in a deepwater drilling fluid. Accolade® fluid, which has a consistently lower and more stable viscosity than Petrofree® SF, shows good performance in the 10-day *Leptocheirus* test and 62% biodegradation after 275 days, compared with 42% for an internal olefin standard and 73% for a typical ester base fluid. *Leptocheirus plumulosus*, a tiny crustacean, plays an important role in all maritime toxicity tests, as it reacts most sensitively to pollution of the sea water.

Biodegradable compounds such as triglyceride esters of plant or animal fatty acids in invert emulsion drilling fluids suffer from hydrolysis which results in unwanted changes in the viscosity of emulsions. Less sensitive esters of (un)saturated monocarboxylic acids (C_{12}–C_{14}) and monofunctional alcohols are still susceptible to hydrolysis, particularly at high temperatures ($T > 160\,°C$), as is the case in rock drilling and deep offshore drilling [144, 145]. In PCT Int. Publ. No. WO 2001/083640 A1 Total discloses an economic highly lubricating drilling fluid for deep offshore sites composed of approximately 70 wt% highly de-aromatised (hydrogenated) mineral hydrocarbon oil cut (C_{13}–C_{16} with <10 wt% *n*-paraffins and >35 wt% isoparaffins) and 30 wt% FA ester (preferably Finagreen® BMDF, derived from RSO and 2-ethylhexanol; now Radiagreen®) with relatively low flow point (<–30 °C) and kinematic viscosity (<10 mm^2/s) at deep water temperatures (4–5 °C) and a suitable anaerobic biodegradability level of over 30% (according to ISO standard 11734) compared with 6% for a pure mineral cut [146]. It is noticed that the viscosity of offshore drilling fluids needs to be regulated at the deep water temperature of 4–5 °C. The benefits of renewable drilling fluids will never offset the environmental damage due to the disastrous crude oil spill in the Gulf of Mexico caused by the BP Deepwater Horizon oil platform.

Mud additives that were formulated earlier from water-soluble substances with combinations of Si, P, Al and B were replaced by thinner sunflower lubricating agents in Russian drilling operations [147]. These environmentally safe additives both thin the mud and improve the mud inhibition level.

In a very different development, British Petroleum Exploration has been using bio-organic drilling fluids containing mixtures of bio-organic polymers such as xanthan gum, which provides viscosity, and starch or cellulose, which acts as a binder. The formulation also contains an inert solid or bridging agent with a particle size allowing it to bridge pores in the structure of the rock being drilled. This formulation is non-toxic and avoids the problems of conventional drilling muds: (i) there is no oil or other component which requires treatment before release into the environment; and (ii) the enzymes used in removing the low permeability layer not only perform better but also do not corrode equipment or pose environmental hazard. Biotechnology has been used to optimise the characteristics of these enzymes (cellulase, hemicellulase, amylase and pectinase) to work under the conditions found in a borehole.

In US Patent Appl. No. 2008/0207472 A1 Valls describes a well drill lubricant composed of crude glycerol and graphite beads suspended in liquid alcohol with gum as a viscosity improver [148]. The composition is non-toxic and fully biodegradable, and possesses high lubricity. Graphene oxide is an efficient fluid-loss-control additive in water-based drilling fluids [149].

Various cost-effective, environmentally friendly and non-polluting biodegradable hammer oil grades (from 150 to 460 cP) from blends of vegetable oils with excellent EP lubrication qualities and efficient wear protection to moving parts, which maintain optimum viscosity under high temperature and pressure while maximising the performance of the hammer, have been developed by Australian Mud Company (AMC) [150]. Hammer oils are used in all types of air driven down hole hammers as air line lubricants. A recent strategic analysis of the European market for oilfield chemicals has given attention to the drilling chemicals market [151].

12.5 Marine lubricants

The demands for marine lubricants are much higher now than they were several years ago. Naphthenic bright stocks have made in-roads in marine oil formulations. Lubrication quality and its supervision play a decisive role. A most important issue is to be able to guarantee the continuity of the machines in order to avoid unnecessary delays through technical malfunctions. Equally important is environmental responsibility (either voluntary or enforced).

Accidental leakage is a routine occurrence in the maritime industry, especially among older vessels. It is estimated that over 30 million gallons of oil are leaked from ships during normal operations in ports and harbours every year – more than three times the *Exxon Valdez* oil spill. Underlying causes of pollution incidents include failure and misuse of valves, overflows, defects in plate and pipes, hose rupture, contaminated bilges and faulty gauges. As oil leakage can occur even on well-maintained and managed ships it is advisable to use biodegradable oils for equipment where there is risk of accidental spillage or leakage to minimise environmental damage. In these cases, replacement of conventional mineral oils with biodegradable lubricants could eventually even become mandatory. The management and control of marine pollution is recently assuming a higher priority for ship operators as global regulatory authorities are increasingly enforcing legislation that regulates the release of oil and chemicals into the marine environment whereas official bodies focus on environmental infringement and threaten personnel with criminal prosecution. Given the business impact of oil leakage (ranging from a fine to suspension of activities) shipping operators are nowadays demanding higher levels of environmental responsibility from their carriers.

BP-Castrol Marine is one of the major lubricant suppliers that have developed high-performance environmentally responsible lubricants to help shipping companies reduce their environmental footprint and protect the world's oceans. In 2007, Castrol Marine has launched a comprehensive range of biodegradable marine lubricants (also including hydraulic, stern tube, gear oils and greases), which exceed EPA requirements [152].

Castrol CAREMAX™ is a comprehensive, advanced used-oil analysis service to ship operators which helps maximising the value from marine lubricants. Onboard test kits allow quick determination of a wide range of lubricant characteristics.

Marine and offshore applications require water tolerance. Terresolve Technology, Ltd (Eastlake, OH) markets a high-performance biodegradable, modified vegetable oil-based hydraulic fluid (XBD) for marine applications with excellent water tolerance and elastomer compatibility.

12.5.1 Marine diesel engine oils

Marine diesel engine oils are heavily influenced by the fuel characteristics and engine design [153]. Nowadays, marine diesel engines are fuelled by the heaviest crude oil fuel cuts or low-cost fuel residues, such as vacuum residues, propane de-asphalting residues, heavy solvent extracts from lube refining, etc. Bunker oil or heavy fuel oil (HFO) is a residual oil of the crude oil distillation or from crackers with a high viscosity from 300 up to 30 000 cSt at 100 °C. After dilution to the specific viscosity (30–700 cSt at 50 °C) this HFO is used as marine fuel oil (MFO) for marine diesel engines. Marine diesel fuels are characterised by a large fraction of high-molecular-weight substances and high sulphur contents (up to 3%). This results in poor combustion, large amounts of deposits and oil acidification. The sulphur content of marine diesel fuels needs to be regulated.

Cylinder oils in crosshead engines are total-loss products which lubricate the sliding motion of the piston rings. Additives are needed for neutralising corrosive acids and to achieve good dispersant properties. Marine grade lubricants generally will run total base number (TBN) values from 15 to 50 mg KOH/g, but can be as high as 70–80 mg KOH/g (e.g. Exxon's MobileGuard 570 or Castrol's Cyltech 80AW). Oils with TBNs of up to 100 mg KOH/g are by no means an exception. The high TBN level is designed to allow a longer operating time between changes, under harsh operating conditions. High sulphur fuels will decrease the TBN value rapidly. Crankcase oils are hardly contaminated by combustion chamber residues and these oils therefore contain relatively few additives. The neutralising capacity can be relatively low (TBN = 5 mg KOH/g); antioxidants are essential. Contamination of crankcase oil with cylinder oils may lead to an undesired increase in emulsifying capacity. Water separation is therefore another characteristic of crankcase oils. An increase in dispersant and detergent properties by ingress of cylinder oil may cause wear problems. Development of marine diesel engine oils requires time-consuming (>1 year) and expensive on-board trials.

The Cyltech range of marine diesel engine cylinder oils, formulated by Castrol for optimum engine protection in two-stroke crosshead engines, was developed to match engine performance with the demands of lower sulphur fuels (<1.5%). Cyltech 40SX is the first cylinder oil specifically approved by MAN B&W and Wärtsilä (Sulzer) for low sulphur operation. Castrol TLX Plus is an advanced, high-performance marine lubricant developed for medium speed engines to provide high levels of performance and safety in operation. Castrol Bio Range Lubricants and other such products are in increasing demand.

Chevron has launched Delo® 710 LE SAE 20W-40 engine oils formulated for use with low sulphur diesel (LSD) and ultra low sulphur diesel (ULSD) fuels in marine, railroad, stationary power generation and offshore drilling and production where zinc-free, non-chlorinated oils and long oil life are required.

12.6 Gear lubrication oils

Gearboxes are designed to convert and transfer movement and forces. In gear systems, energy is transmitted from a power source to a terminal point, through gears that change speeds, directions and torque. The main functions of a gear lubricant are wear prevention and friction reduction through a lubricating film between the gear mating elements. Apart from the important function of lubricating the sliding rolling contacts, the oil also fulfils the task of cooling and removing the friction heat generated in the sliding rolling contacts. Gear oil performance requirements can vary substantially depending on the amount of sliding and rolling and the load-carrying limits of gear types. High-performance lubricant formulations for efficient gearboxes, which yield higher gear speeds and loads and higher operating temperatures, should have the characteristics listed in Table 12.20.

The main markets are stationary and mobile gears, i.e. gears for industrial and automotive application, respectively. Friction conditions of the various gear types (geometry, load and speed conditions, operating mode, lubrication conditions) are highly variable. The two major gear types are enclosed gears (lubricated by splash) and open gears (lubricated by intermittent spray). For enclosed gears the lubricant has also to dissipate the heat developed during friction. In most gears, rolling action is common. Worm gears are special because they operate under high sliding velocity and moderate loads. The sliding action produces friction that produces higher operating temperatures than those occurring in other gear sets. The sliding action allows fluid-film lubrication to take place. Lubricants for worm gears must be resistant to the thinning due to high temperatures and must provide adequate cooling.

Table 12.20 Requirements for high-performance gear lubrication oils

• Reduction of friction and wear • EP properties • Good surface fatigue operation • Good high-temperature performance and oxidative resistance • Improved heat transfer • Long demulsibility time • Removal of foreign and wear particles • Good foam resistance • Reduction of operating noise • Corrosion prevention • Long machine service life and drain intervals • Environmental compatibility • Biodegradability

When selecting a lubricant for any gear application the following issues must be considered:

- nature and material of gears;
- type of application;
- operating conditions (temperature range; rolling or sliding speed; type of steady load);
- method of lubricant application;
- type of service;
- environment.

The demands on gear lubricants vary widely in view of variations in speed, torque and other operating conditions. Clearly defined industry performance specifications have been defined (API Categories GL-4, GL-5, MT-1, MIL-PRF-2105E, SAE J2360). Lubrication requirements for gear sets are prescribed by the equipment manufacturers, based on the operating characteristics and ambient conditions under which the equipment will operate. In general, gear lubricants are formulated to comply with AGMA 9005-D94 ('Industrial Gear Lubrication Standard') by the American Gear Manufacturers Association, classifying industrial gear oils into nine types. AGMA 9005-D94 provides lubrication guidelines for enclosed and open gearing that is installed in general industrial power transmission applications. It is important that the lubricating gear oil prevents premature wear of gear tooth surfaces. The type of gear application sets requirements on this type of lubricant. Enclosed gears, where the oil is contained in an oil-tight housing, require more additives (e.g. to prevent rust and oxidation) than open gears. For open gears it is vital that the lubricant does not disappear from the tooth surfaces due to the centrifugal force. Consequently, these lubricants need to be highly adhesive as in case of heavy oils, asphalt-based compounds or soft greases [154].

Selection of a gear oil lubricant requires knowledge of the operating conditions (specific loads, switching periods) and environmental conditions of the application. Low operating temperatures, low power losses and high efficiencies are major issues in modern gearboxes. Most gear oils typically operate at temperature ranges from 50 to 55 °C above ambient that result in inherently little oxidation of the oil. Oils operating at high temperature require good viscosity and high resistance to oxidation and foaming. The heat development and, thus, the oil temperature are determined by the type of gear, the transferred torque, the specific load, operating mode (permanent or intermittent), type of application (mobile or stationary) and environmental (climatic) conditions. If the gear case is located in a heated atmosphere and no provision is made for cooling, the oil will break down.

Service life requirements for gears for torque transfer typically range from 100 to 500 h for household machinery (often lubricated with grease instead of oil), to 3000 h for passenger cars, 5000 h for office and data-processing equipment and 100 000 h for paper machines. The latter case requires oil exchange intervals of 2500 h of operation. In general, an increase in the lubricant's viscosity has a service-life extending effect.

Gearboxes in machine tools serve to reduce drive speed to the feed velocity of supports. Industrial gear applications are quite distinct from vehicle transmissions by the larger variety in toothing types and sizes, and widely differing possible operating and ambient conditions. Housing dimensions and lubrication systems differ. Industrial gears can often be lubricated manually (open gears). Gear lubrication oil is an important machine component for gear and transmission. Stress on machine tool gearboxes is relatively small and gear and general lubricating oils are often used. Service life requirements are considerably higher (up to 100 000 h) than for vehicle transmissions (1000–10 000 h).

The viscosity–temperature behaviour of gear lubrication oils is an important issue. Modern gears and hydraulic systems apply considerable force to their fluid. VI improvers may eventually be affected by this force, leading to breaking (shearing) of the VI improvers. Corrosion protection plays a very important role with respect to lubrication oils for industrial gears. Other requirements relate to oxidation stability, flash and pour points, demulsibility and water separation, air release, paint and seal compatibility, foaming, miscibility with mineral oils, and environmental and skin compatibility. Owing to regular draining, the environmental compatibility of industrial gear oils should especially be taken into account. Compared with lubrication oils for vehicle transmissions, the number of specifications for lubricants for industrial applications is rather small.

Various types of lubricant are used in the lubrication of gear and transmissions, mainly mineral oils, synthetic hydrocarbons (PAOs), PAGs (homopolymers) and naphthenic oils. Table 12.21 compares the base fluid data of industrial gear oils. Worm drives are often lubricated with PAG-based gear oils. Synthetic PAO-based oils are used for thermally stressed gearboxes. Synthetic biodegradable esters, RSO and castor oil are also used as industrial gear lubrication oils. Most open gear lubricants are heavy oils, asphalt-based compounds, or soft greases. Compounded gear lubricants are blends of petroleum-based oils with 3–10% fatty or synthetic fatty oils. They are particularly useful in worm gear drives. Gear oils and machine lubes, previously formulated using sulphurised sperm oil, are typically based on low FFA, higher viscosity base oils with higher alkyl ester additives.

Special gear lubricants, so-called multipurpose fluids, are used in agricultural machines such as tractors, harvesters, etc., where special friction, hydraulic and wear requirements have to be met even at low temperatures. Multipurpose oils almost always have a low viscosity and stand out due to a very good viscosity–temperature behaviour (high VI). The presence of water and dirt has a significant impact on these oils, especially in respect to foaming and air release properties. Multipurpose oils for tractor engines are known as UTTOs and oils with engine-related performance are STOUs. Major manufacturers of tractors and agricultural machines have set their own specifications. John Deere (in cooperation with Lubrizol Corp.) has developed an UTTO under the trade-name of BIO-HY-GARD [76], later BIO-HY-GARD II [77]; see also Section 12.2.1, p. 593.

'Mixoil'-type UFOs are useful for lubricant applications as gearbox oil and mould-release agents [121]. The quality of UFOs directly influences the physicochemical characteristics of the lubricant bases (POEs) that are manufactured from them.

Apart from the various base oils, the type and quantity of additives, which depends greatly on the base oil, have a significant influence on the function and service life of gear and transmission. Gear oil additives are EP addi-

Table 12.21 Base fluid data of industrial gear oils

Property	Synthetic ester	PAO	PAG	Mineral oil
Viscosity at 40 °C/100 °C (mm^2/s)	220/28.0	220/25.7	220/34.7	220/18.4
Viscosity index	160	150	200	95
Flash point (°C)	280	260	230	230
Pour point (°C)	–48	–54	–33	–18
Water solubility	No	No	Partly	No

After ref. [2].

tives, friction modifiers, VI improvers and anti-misting additives. Depending on the operating conditions, enclosed gears usually require an oil with various additives, i.e. rust and oxidation (R&O) and foam inhibitors are common. R&O oils are common gear lubricants usually formulated to include chemical additives to enhance their performance qualities. R&O oils have essential application properties for gear and bearings, good lubrication qualities and adequate cooling qualities. Properly formulated biodegradable low-toxicity ester-based oils show favourable gear power loss performance (FZG test rig) in comparison to commercial mineral oils [155].

Inappropriate selection of the lubricant may lead to early failure of gears and transmission by extreme abrasive wear, early endurance failure (fatigue), scuffing and scoring of the friction contacts. Gear failures can be due to mechanical problems or lubricant failure. Lubricant-related failures are usually traced to contamination, oil film collapse, additive depletion or use of improper lubricant for the application. The most common failures are due to particle contamination of the lubricant. Fourier transform infrared (FTIR) testing has been used in case of degradation from contamination. Wear cannot be completely eliminated. Speed, lubricating viscosity and temperature impose practical limits on gear operating conditions. Gears that are highly loaded, operate at slow speeds and rely on boundary lubrication are particularly subject to excessive wear. In case of biotransmission fluids for heavy loaded gearboxes, biolubricants are required to have superior EP properties. The properties can be screened using the four-ball wear tester assembly. This test can be supplemented with a scuffing evaluation of lubricants by increasing the load until scuffing occurs.

Bio EP Gear Oils (Renewable Lubricants, Inc.; Hartville, OH) meet and exceed the key objectives for a high-performance gear lubricant (Table 12.22). Gear oils should have a minimum PP of 5 °C lower than the lowest expected operating temperature. The PP for mineral gear oil is typically –7 °C. Bio EP Gear Oils are specially formulated with cold-flow technology to provide PPs as low as –34 °C. They are characterised by super-high VIs (>200). These oils not only have the advantage of stable application over a

Table 12.22 Characteristics of formulated Bio EP Gear Oils (RLI)

- Good viscosity
- Super high viscosity index (>200)
- Excellent low-temperature properties (PP, –34 °C)
- Good thermo-oxidative stability
- Low volatility
- Non corrosive
- Compatibility with seals and other lubricants

wide temperature range, but also exhibit good oxidation stability at high temperatures, and low volatility. Bio EP Gear Oils may be used for all types of applications at high temperature where other lubricants fail. Another advantage of BIO EP Gear Oils is their compatibility with seals and other lubricants. Using Bio EP Gear Oils ensures that exposed materials will not be damaged; they are non-corrosive to yellow metals, i.e. bronze. Bio EP Gear Oils are specially formulated to include chemical additives capable of producing a film that provides extreme pressure and anti-scuffing protection; they are not formulated with PTFE or WS_2 as these chemical additives are non-renewable, toxic and hazardous. RLI's Bio EP Gear Oils are formulated and applied to prevent premature component failure, assure reliable operation, reduce operating cost and increase service life. The oils with superior lubricating properties are also recommended for worm gears even though the EP additive is usually not required.

Low-viscosity R&O oils, such as turbine oils, are commonly used in high-speed, low-load gear units. For high-speed, low-load gear applications, Bio EP Gear Oils without anti-scuffing and EP agents can be used successfully provided the oil viscosity is capable of maintaining the required film thickness. However, low-speed gears are usually heavily loaded so anti-scuffing and EP agents are necessary to ensure adequate protection.

Castrol's BioTrans gear oils (viscosity grades ISO VG 150 and 220) are environmentally responsible synthetic gear oils. BioTrans oils are formulated from a synthetic ester. AW and corrosion protection additives improve the natural characteristics of the synthetic base oil. BioTrans gear oils are suited for application from –25 °C to +90 °C in spur, helical and planetary gear units, couplings, rolling and sliding bearings.

Castrol BioStat is a marine gear oil with effective lubrication up to 20% water contamination and reduced environmental impact when compared with conventional lubricants. BioBlend (Joliet, IL)'s 99% biodegradable vegetable based food-grade lubricant blended in compliance with 21 CFR §178.3570 from temperature stable high-oleic canola oils is recommended for use with gears, bearings and chains in facilities under the Federal Food Safety Inspection Service [156].

The higher cost of bio gear oils must be set against the savings generated by longer life (compared with mineral oils) and reduction in service cost. Gear lubrication oils have been recently reviewed [79].

12.6.1 Bio bearing oils

Bearings most often found in gearboxes are plain and roller bearings, although plain bearings are seldom used in machine tools. The most common types are ball and cylindrical roller bearings. Some bearings are lubricated by air (e.g. dentist's drill) or water, others by a general or specific gear oil, or grease. Plain bearings depend on a hydrodynamic lubricating oil film to reduce friction. Rolling contact bearings can operate with small amounts of lubricant, are relatively insensitive to lubricant viscosity, have low wear rates and require little maintenance. Traditionally, for plain bearings non-renewable petroleum-based mineral oils have generally been used except in extreme hot and cold temperature applications where petroleum-based synthetics provide superior performance. Oil is used for faster rotational speeds and higher operating temperatures and in maximum loading situations. In general, oils will be the medium to high VI type with R&O inhibitors. Oil additives are not required in plain bearing applications. EP oils are required when operating under heavy loads or shock conditions. Some additives and contaminants may cause corrosion. Grease is used for slower rotational speeds or oscillating movements, lower temperatures, and low to medium loads. It is also used in cases of extreme loading where the bearing operates in boundary lubrication conditions. Today, renewable Stabilized™ HOBS vegetable oils are replacing petroleum-based mineral and synthetic oils as the far safer, superior performing lubricants of choice for all types of bearing applications. Table 12.23 lists some of the important considerations regarding selection of a bearing lubricant.

Table 12.23 Selection of a bearing lubricant

Lubricant	Operating range	Remarks
Petroleum lubricants		
Mineral oils	Moderate conditions of load and speed	Wide range of viscosities Potential corrosion problems (additives)
Synthetic oils	All conditions of load and speed	All viscosities available Good high-low temperature properties Costly
Mineral/synthetic greases	Restricted operating speeds	Use for intermittent motion
High-oleic base stocks		
Bio EP Gear Oils[a]	All conditions of load and speed	All viscosities available Advanced EP protection Excellent high-low temperature properties Safe, affordable
Bio Spindle Oils[a]	Lighty loaded bearings	Low viscosities (ISO 10, 22) Advanced EP protection Safe, affordable
Bio EP Greases[a]	Restricted operating speeds	Use for intermittent motion AW/EP

[a] Renewable Lubricants, Inc. (RLI).

12.7 Compressor oils

Compressors decrease the volume and increase the pressure of air or any gaseous medium in one or more stages and thus transfer energy to the medium. The main functions of compressor oils are to lubricate bearings, pistons, cylinders and valves, thus reducing friction and wear. The lubricant also plays a role in cooling, preventing corrosion, achieving sealing of the compression chambers, and minimising viscosity dilution and reactions with process gases. Compressor oil requirements can vary substantially. Requirements depend on compressor type, the properties of the gas being compressed and discharge pressures and temperatures. A significant portion of compressor oils have a synthetic base. Universal lubrication requirements for compressor oils include:

- low flammability;
- oxidation resistance;
- thermal stability;
- viscosity retention;
- AW capabilities;
- corrosion prevention;
- good demulsibility.

Few lubrication challenges are more difficult than creating an ideal oil for air compressors, particularly for the rotary screw air compressor. The compression process creates high temperatures and the oil, either in thin film or small droplets, intimately mixes with hot air. This mixture creates severe oxidation conditions, promoting rapid degradation of the oil. Consequently, compressor oil needs to be highly stable in the presence of air and water (i.e. exhibit oxidative and hydrolytic stability), provide excellent rust protection, separate easily from air and water, and be biodegradable for easy disposal. Present lubricant options range from traditional mineral oils to synthetic lubricants and more recently to POEs [157, 158]. In severe operating conditions synthetic base stocks (typically PAO or mPAO with a solvent co-base stock such as an ester or alkylated naphthalene to improve additive solvency and seal capability) present a wide range of potential benefits, namely: (i) higher thermal and oxidative stability; (ii) higher VI and lower PP; (iii) potential for energy savings; (iv) enhanced wear protection and cleanliness; and (v) enhanced safety. In rotary air compressors modern fully synthetic oils can extend oil life up to 8000 h compared with 1000–2000 h for mineral oils.

Compressors are manufactured in several types and for a variety of purposes. It is possible to differentiate between air and gas compressors, vacuum pumps and refrigerant compressors. Vacuum pump lubrication requires synthetic oils with low vapour pressure (mostly synthetic ester oils). Enormous stress exerted onto the oil resulting from the high temperatures created when the medium is compressed may cause oxidation and favour deposit formation. Lubricating requirements vary considerably with the type of compressor, the pressures involved, the outlet temperature and gas being compressed. Most compressed air systems operate between 100 and 125 psig. Although air and gas compressors are mechanically similar, the main difference is in the effect of the gas on the lubricant and the compressor components. Compressors whose chambers are lubricated pose particular safety problems if air or aggressive gases contact the lubricant. Aggressive gases from the surroundings can influence the performance of the lubricant in an extremely negative way. The selection criteria for oils for various compressor types differ also greatly in relation to special demands. The latter include low foaming, excellent air release and good demulsibility (separation) of condensed water. Refrigeration and air conditioning compressors also require special consideration because of the recirculation of the refrigerant and mixing with the lubricant. If the medium to be extracted is not air, but a refrigerant, a compatible refrigeration oil should be used. The requirements for lubricants suitable for different types of compressors are given by Johnson [159]. For another detailed overview of compressor oils, see ref. [160].

A wide range of lube oils is available for use in compressors (Table 12.24). Lubricants commonly used are mineral oils (for normal and medium working conditions), hydrocracked oils (for medium to severe working conditions), and diester, polyester, PAG- or PAO-based lubricants (for very severe operating conditions). White oils are used as low-density polyethylene (LDPE) compressor oils. Water resistance, thermal stability, long life,

Table 12.24 Overview of compressor oil types

Oil type[a]	Type of compressor		
	Piston	Screw	Turbo
MO	+	+	
PAO	+	+	
Diester	+		
HC		+	
POE		+	
TDL			+
Synthetic			+

[a] MO, mineral oil; PAO, poly-α-olefin; HC, hydrocracked oil (so-called Group III oils); POE, biodegradable polyol esters; TDL, turbine oils.

resistance to oxidation and resistance to absorption of process gases are important characteristics. The lube's usable lifespan can be reduced considerably by contaminants. Compressor lube oils are formulated to work well and remain stable at high temperatures and pressures. Hydrotreated mineral oils are used for their low gas solubility (1–5%). Synthetic compressor lubricants (PAO, PAG) are used depending on the process and how much gas is present. Petro-Canada Lubricants, Inc. (Mississauga, ON) has recently introduced a synthetic PAO-based Purity^{TM} FG Compressor Fluid for food-grade applications (NSF H1) that protects against wear, oxidation, rust and corrosion. The product's ASTM D 2272 resistance to oxidation (RPVOT, 4554 min) compares favourably with that of competitive synthetic products (1129–2364 min). PAG oils, which do not readily absorb gases, are used in applications where process gases are compressed. Mobile compressors are often lubricated with monograde engine oil (SAE 20 to 40). Because of explosion hazards, oxygen compressors require inert lubricants based on (extremely expensive) perfluoroether oils. Screw compressors require lubricants of moderate viscosity (ISO VG 46 or 68) with excellent oxidation stability and mild/high AW/EP performance; biodegradable polyol ester base oils qualify. Rotary piston compressors are lubricated by total-loss systems or by direct oil injection. SOILCY, an EU-sponsored sustainable compressor oil programme (FP6), has optimised the life cycle of an environmentally friendly polyglycerol ester-based compressor oil derived from the renewable resource glycerol (as a by-product of biodiesel).

Most of the common heat applications in pulp and paper, polymer, textile, agricultural and food industries require heating up to 140 °C, but some food drying and petrochemical processes require temperatures above 200 °C. A theoretical analysis of subcritical/transcritical heat pumps using natural refrigerants for high-temperature heating applications has been reported [161].

Environmentally benign natural refrigerants such as carbon dioxide, ammonia, propane, butane, isobutane and propylene may be used for heat pump applications [162]. Ammonia-based high-temperature heat pumps require special types of cooling and lubrication systems that are cost effective.

The stability of compressor lubricants at high temperatures is a major issue for heat pumps employed in high-temperature heating. Discharge temperature in case of propane, isobutane and CO_2 systems (below 215 °C) does not pose any serious problem. Several lubricants are available for use up to a maximum temperature range of about 250 °C [163–165]. The synthetic lubricants for CO_2 compressors, PAG, POE, PVE (polyvinyl ether) and PC (polycarbonate) have been tested for a maximum temperature of 220 °C [166]. The problem of high discharge temperature is very serious for an ammonia compressor, both for cooling and lubrication. Adequate cooling is required to increase compressor life. Adhvaryu and Erhan [167] proposed epoxidised soybean oil (ESO) as a potential high-temperature lubricant. Lubrication Technology, Inc. (LTI; Franklin Furnace, OH) manufactures high-technology synthetic lubricants based on perfluoropolyethers (PFPE), fluorinated polysiloxanes, esters, PAOs and hydrocarbon co-oligomers under the tradename CHRISTO-LUBE®, which cover a wide range of operating temperatures (from –96 °C to +288 °C) for demanding conditions [168]. Also Krytox fluorinated lubricants (operating range up to 426 °C) [169] may be employed in high-temperature applications.

Branched synthetic polyol ester base stocks can be used in the formulation of biodegradable compressor oils together with selected lubricant additives (oxidation inhibitors, additive solubilisers, rust inhibitors/metal passivators, demulsifying agents and AW agents) [6]. A typical biodegradable compressor oil contains 80–99% base stock, 1–15% solvent, with the remainder comprising the additive package. For additives in (PAO-based) compressor oils, see also ref. [170].

In compressors, lube oil is used to seal the compressor from gas leaks, lubricate moving parts and manage temperature during operation. The condition of the lubricant oil is a critical factor in extending a compressor's bearing life and overall reliability. Hydrocarbons such as ethane and propane are easily soluble in mineral oil. This causes the viscosity of the lubricating oil to fall if mineral oil-based products are used. Ester- or polyglycol-based lubricants with lower hydrocarbon solubility are then recommended. Many factors can affect lube viscosity of a compressor, including oxidation, temperature changes, dilution, contamination and bubble formation. When a lubricant oil is diluted or comes in contact with vapours of methane or light hydrocarbons, the viscosity may break down quickly and unpredictably, increasing the risk of equipment failure. Compressor failure in a single part of a refinery determines important losses in production and revenues. Real-time in-line monitoring of lube viscosity in critical compressors, rather than monthly external lab testing, is the best and most cost-effective way to avoid such costly failures [171]. Viscosity also plays a role in energy efficiency; demand for more efficient compressors is driving the use of lower-viscosity lubricants.

12.7.1 Refrigeration compressor lubricants

Refrigeration compressors deserve special mention in lubricant technology. Refrigerator oils are used in refrigerators, air conditioners, dehumidifiers, cold-storage chests, freezers, automatic vending machines, cooling units in chemical plants, etc., serving as a hydraulic control and functional fluid. In a domestic refrigerator or freezer compressor, motor bearings require lubrication. Refrigeration oils should help to ensure that the compressor and system components receive the proper lubrication for years of trouble-free operation. Since motor and compressor are sealed inside the refrigeration circuit, the lubricant cannot be changed for the lifetime of the appliance, typically 10–20 years. The lubricant is not exposed to atmospheric oxygen and water; consequently, oxidative and hydrolytic stability are of less importance than thermal stability. Refrigeration lubricants must have very good electrical insulation properties, must be stable to electrically induced decomposition and exhibit very good low-temperature fluidity characteristics with no tendency to deposition of waxes. High dielectric strength allows the oil/refrigerant mixture to serve as an insulator between the motor windings and the body in a compressor. The requirements of today's refrigeration and air conditioning compressor lubricants are complex (see Table 12.25).

The principal functions of refrigeration oils are lubrication of pistons or rotors, sealing of valves and dissipation of heat. High-quality refrigeration oils are required in view of the longevity expected of such compressors, interaction with other substances and low and high temperatures. A mixture of lubricant and air-conditioning fluid must have sufficient lubricating properties to protect the air-conditioning pump, but should not be aggressive towards materials making up the system. Refrigeration oils should maintain high film strength even when diluted with refrigerant. High chemical stability minimises the reaction with refrigerants and other materials (e.g. elastomers) that are part of the system. Important factors are the lubricity of the refrigeration oil, interaction with the refrigerant, evaporation behaviour, solubility and mixture behaviour. Refrigerants and lubricants should

Table 12.25 General requirements for refrigeration lubricants

- Lubricity properties
- Excellent thermal stability
- Superior low-temperature performance[a]
- High chemical stability
- Compatibility with refrigerants[b]
- Compatibility with polymeric materials[c] and other materials of construction
- Compatibility with other refrigeration oils[d]
- Controlled viscosity
- High VI
- Long service life
- Freedom from moisture
- High dielectric strength
- Low energy consumption

[a] Much lower PP than mineral oils.
[b] Typically CFCs, HCFCs, HFCs, CO_2, hydrocarbons, ammonia.
[c] Typically BR, NR, NBR, HNBR, nylon-6,6, Teflon, EPDM, Neoprene, etc.
[d] For retrofitting.

be fully miscible and have high mutual solubility (at all temperatures that the system will experience), as in use of hydrofluorocarbon (HFC) refrigerants and POE lubricants. Lubricants with a thermal stability of >175 °C generally work well in refrigeration systems. Ultra-low temperature refrigerants (as low as –100 °C) require specialty lubricants (e.g. EMKARATE RL 'H'). Obviously, for such applications all the components of the lubricant, including additives, need good solubility and low-temperature flow characteristics.

The most important parameter for determining the lubricity of oils or oil–refrigerant mixtures is viscosity. Selection of an optimum refrigeration oil depends on the specifications of the compressor, the system as a whole and the refrigerant. Refrigeration lubricants must have a wide variety of viscosities to achieve optimal effectiveness in the many different types of refrigeration equipment. For conventional home refrigerators, equipped with low power compressors, lubricants with relatively low viscosities at normal operating temperatures are generally satisfactory. Low-viscosity lubricants are preferred for economy of operation. By contrast, automobile air-conditioners and industrial refrigeration systems require relatively higher viscosity lubricants by reason of the more extreme conditions of operation. A high VI is an indication that their effective viscosity will not change drastically with the wide swings in temperature seen by most typical refrigeration and air conditioning systems. This means that energy consumption will be minimised at low temperatures and lubrication will be maximised at elevated temperatures.

Moisture can enter the refrigeration systems by improper evacuation of the system, system leaks, system components, improper handling of the refrigerant or of the lubricants. The moisture content for refrigeration lubricants should be < 100 ppm. Hygroscopicity of refrigeration lubricants ranks (in descending order) as PAGs > PVEs > POEs > alkyl benzenes (ABs) > mineral oils. The rate at which POEs pick up moisture is dependent on temperature, relative humidity, exposure time, and relative surface area. PAG lubricants are typically used in automotive applications and PVE is used sparingly in certain regions of the world.

Refrigeration oils differ according to the refrigerants being compressed (ASHRAE Standard 34–1992) [172]. A refrigeration lubricant must be miscible with the refrigerant gas. Modern refrigerants are halohydrocarbons (e.g. HFCs), such as HFC-134a and HFC-152a, as substitutes for the ozone layer destructive chlorofluorocarbons (CFCs), now in controlled use. Under the terms of the Montreal Protocol CFCs were phased out in 2000; HCFCs should be completely phased out in 2030. Hydrofluorocarbon (HFC) gases, which are currently the most widely used non-flammable, zero ozone depletion refrigerants, are relatively polar and not miscible with mineral oils.

Typical refrigerator oils are generally naphthenic or paraffinic mineral oils, ABs, polyglycolic oils, ester oils or their (additivated) mixtures with kinematic viscosity of 10–200 cSt at 40 °C. Mineral oils are still the most significant group of oils for traditional refrigeration compressors using ammonia along with CFCs and hydrochlorofluorocarbons (HCFCs), but ABs or PAOs are also being used. For instance, Glova [173] has disclosed a refrigeration lubricating oil composition based on branched-chain alkylbenzenes. These oils are not sufficiently soluble in the new chlorine-free substitute refrigerant mixtures (fluorocarbons (FCs) and HFCs) such as R 134a, R 404a and R 507. R 134a is currently the most commonly used new refrigerant. Modern synthetic refrigeration oils should not only have high lubricity, low hygroscopicity and high thermal and chemical stability, but also excellent compatibility with the alternative halohydrocarbons to ensure lubricant suitability. The tribological performance of POE and PVE synthetic lubricants operating with HFC-134a refrigerant has been reported [174]. Biodegradable POEs are the lubricants of choice for use with HFC gases [175].

Honeywell and DuPont are to produce a new eco-friendly refrigerant for use in car air conditioning systems. The refrigerant, dubbed HFO-1234yf (2,3,3,3-tetrafluoropropene), has a global warming potential (GWP) of 4; cf. GWP of 1 for CO_2. The currently used refrigerant HFC-134a (1,1,1,2-tetrafluoroethane; main supplier: Mexichem) has a GWP of 1430, which is above the EU Mobile Air Conditioning Directive (2006/40/EC) that states that by 2011 all new vehicle models must use a refrigerant with a GWP below 150 [176]. Notice that HFC-23 (CHF_3, fluoroform) is an extremely potent greenhouse gas with GWP of 14 800.

In the past, the widespread commercial use of chlorine-free refrigerant heat transfer fluids such as 1,1,1,2-tetrafluoroethane (formerly Klea 134a, ICI) has been hampered by the lack of commercially adequate lubricants. Examples of refrigeration oils that are miscible with HFCs include oxygen-containing compounds such as esters, ethers and carbonates. In particular, ester-based oils now constitute a most important product group. Various recent patents describe esters useful as refrigerator oils (see Table 12.26). Schnur [177] discloses an ester blend, including an ester based on NPG and/or PE and 2-ethylhexanoic acid. As the solubility of NPG and PE esters in non-chlorinated HFCs is often only fair, improved polyol-based esters are required. US Patent No. 6,582,621 to Sasaki *et al.* (to Nippon Mitsubishi Oil) [178] describes the development of mono- and dicarboxylic acid polyol esters with high insulating properties, whereas US Patent Appl. No. 2004/0046146 to Ankner *et al.* (to Neste Chemicals OY) [179] claims polyol esters comprising a sterically hindered diol esterified with mono- and dibasic carboxylic acids. In US Patent No. 6,831,045 [180] and US Patent No. 7,045,490 [181] Shimomura and Takigawa describe alicylic di- and polycarboxylic acid ester compounds. In US Patent No. 6,656,891 B1 (to Idemitsu

Table 12.26 Selected refrigeration oil compositions

Lubricant base oil	Refrigerant	Reference
NPG/PE-2EH polyol ester blend	HFC (R 134a)	[177]
Mono- or dicarboxylic acid polyol[a] ester; epoxy compound	HFCs	[178]
Diol[b] containing polyol esters	HFCs	[179]
Alicyclic dicarboxylic acid esters	HFCs, CO_2, hydrocarbons	[180]
Alicyclic polycarboxylic acid esters	HFCs, CO_2, hydrocarbons	[181]
Diesters, triesters	HFCs (R 134a, R 410a)	[182]
Polyol esters[c]	HFCs (R 290, R 600a)	[183]

[a]EG, NPG, PE, TME, etc.
[b]CHDM, 1,4-dimethylcyclohexane; ETHD, 2-ethyl-1,3-hexanediol.
[c]EMKARATE RL™ product line.

Kosan Co.) Sakanoue *et al.* [184] disclose a refrigerating machine oil composition composed of a blend of PVE or POE with a PAG alkyl ether and an AB, suitable for HFC type, HC type, ether type, CO_2 or ammonia-type refrigerants. Polyether-based lubricant oil compositions for refrigerators have also been disclosed by Enna *et al.* (to Asahi Glass Co.) [185] and Kaimai and Takahashi (to Japan Energy Corp.) [186]. Reflo™ is an inherently biodegradable ammonia-type refrigeration compressor oil (Petro-Canada Lubricants, Inc., Mississauga, ON).

Uniqema (formerly ICI)'s EMKARATE RL™ POE line is specifically designed for use with HFC and HCFC refrigerants and is widely accepted by major OEMs. These POE refrigeration lubricants cover a wide viscosity range from 7 cSt (35–40 SUS) to 220 cSt (1100 SUS). EMKARATE RL™ is designed specifically for use with environmentally friendly HFC refrigerants. The superior performance has been achieved with base fluids that are specifically formulated to deliver optimum performance with minimal additive levels. EMKARATE RL™ lubricants are compatible with all CFCs, HCFCs, HFCs, CO_2 and hydrocarbons (e.g. R 290, R 600a) refrigerants, but not with ammonia (R 717) [183]. HFC refrigerants and the POE lubricant have both polar molecular structures, which attract polar water molecules. The solubility of water in HFCs, such as R 134a, is many times greater than in the CFCs they replace. POEs, such as EMKARATE RL™, are also considerably more hygroscopic than traditionally used mineral oils. Some physical properties of EMKARATE® RL 9HPlus are as follows: viscosity at 40/100 °C, 8.6/2.5 mm^2/s; VI, 120; PP, –49 °C; FP, 195 °C; AV, 0.01 mg KOH/g; no foaming [183].

Branched synthetic POEs have been used extensively in non-biodegradable applications, such as refrigeration lubricants, and have proven to be quite effective if more than 25% 3,5,5-trimethylhexanoic acid is incorporated into the molecule. However, trimethylhexanoic acid is not biodegradable (OECD 301B) and its incorporation into the lubricant base oil would drastically lower the biodegradation of the polyol ester due to the quaternary carbons contained therein. Also incorporation of trialkyl acetic acids (or neo acids) into POEs produces very useful refrigeration lubricants. Again, these acids do not biodegrade and cannot be used to produce POEs for biodegradable applications. POEs of all branched acids can be used in refrigeration oils, but they do not rapidly biodegrade. Although POEs made from linear C_5 and C_{10} acids for refrigeration application are biodegradable under the Modified Sturm test, they would not be adequate as a lubricant in hydraulic or two-cycle engine applications because the viscosities would be too low and wear additives would be needed.

Recently, Chevron Corp. [182] has developed various diester and triester lubricating base oils for refrigeration application with molecular mass from 340 to 780 a.m.u. (Fig. 12.3) having kinematic viscosity of at least 3 cSt at 100 °C and PP below –20 °C. Diester lubricants may be prepared according to Fig. 6.41 in which the C_8–C_{16} olefin used is a reaction product of a Fischer–Tropsch process; the carboxylic acid can be derived from alcohols generated by Fischer–Tropsch and/or it can be a bio-derived FA [187]. Figure 12.4 shows the synthetic strategy for conversion of a mono-unsaturated fatty acid (MUFA; C_{10}–C_{22}) to a triester derivative. The demand for polyol ester-based synthetic refrigeration oils is increasing. It is essential to shield these ester oils from water in the compressor.

In subcritical CO_2-based car air-conditioning systems special synthetic polyglycols (PAGs) or ester oils with AW/EP additives are to be used to guarantee the lifetime of the compressor under the severe conditions of the CO_2 transcritical process.

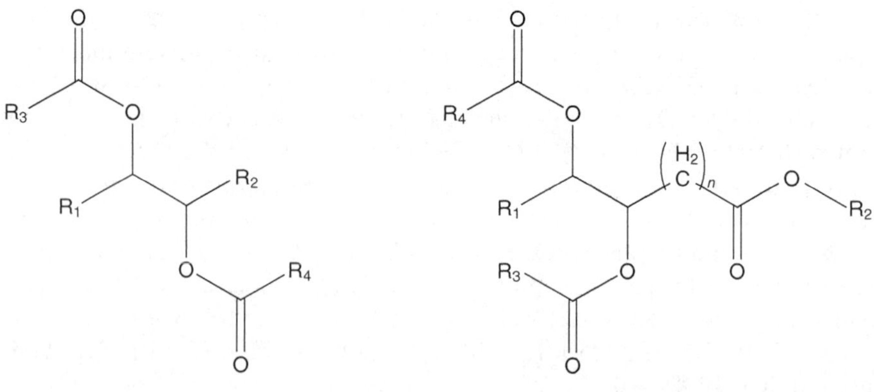

$R_i = C_2$–C_{17} $R_i = C_2$–C_{20}, n = 2–20

12.3 Diester and triester refrigerator base oil compositions. After ref. [182].

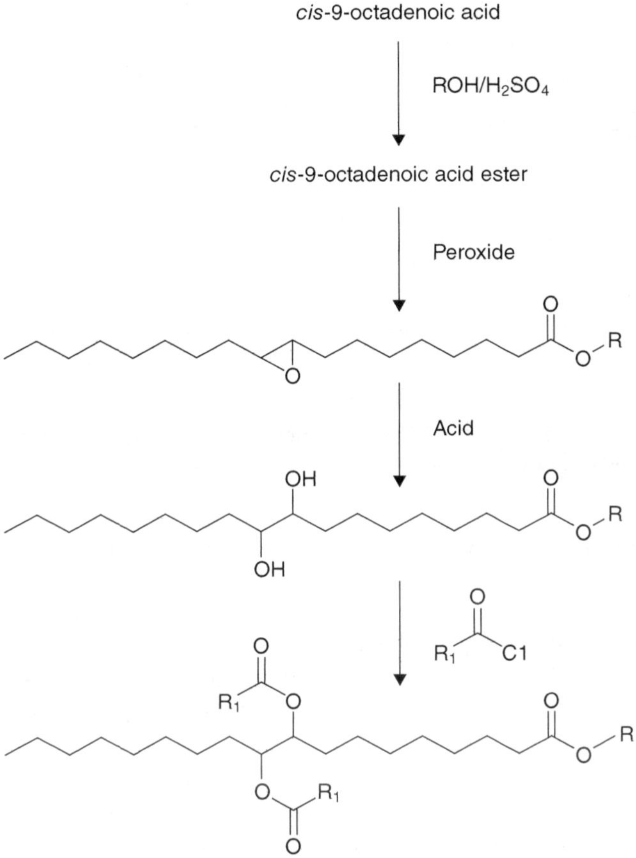

12.4 Synthetic strategy for conversion of oleic acid to triester derivatives. After ref. [182].

Reducing electrical energy consumption through lubricant base fluid design and optimisation of lubricant performance are of considerable economic interest. Some 5–10% of the UK's domestic electricity consumption is used to power refrigerators and freezers (up to 15% worldwide). A similar proportion of non-domestic power consumption is shared by industrial and commercial refrigeration and air conditioning [46].

Small appliance refrigeration compressors are designed to operate under conditions of hydrodynamic lubrication, where metal-to-metal contact is avoided by ensuring that a full-fluid film separates the surfaces. The energy required for shearing the entrained hydrodynamic film increases with the viscosity of the lubricant. Therefore, the energy requirement of an appliance compressor can be reduced by reducing the lubricant viscosity. However, if the lubricant viscosity is reduced too much, a full-fluid film is no longer

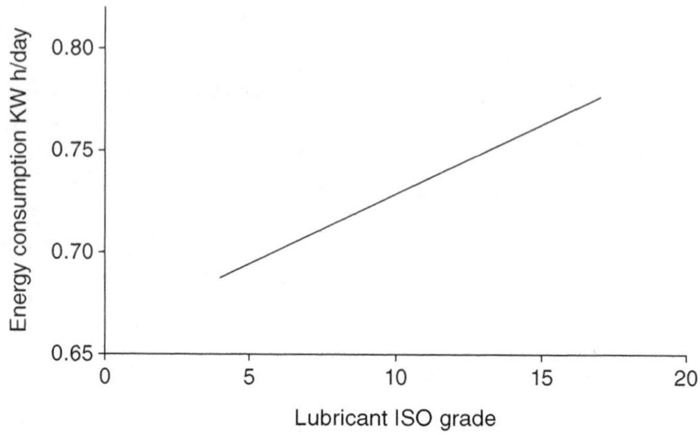

12.5 Dependence of refrigerator energy consumption on lubricant viscosity. After ref. [188].

ensured and the system will move into the boundary lubrication regime. Here, metal-to-metal contact occurs; wear reduces efficiency of the operation and limits the lifetime of the appliance. As friction coefficients in mixed and boundary lubrication greatly exceed those in the hydrodynamic regime, also increased frictional losses occur. Higher friction leads to overheating of the fluid and contacting surfaces, again compromising the expected lifetime. Where possible, a periodic refrigerant analysis is important to detect and control contaminants (moisture, acid, particulate/solids, organic matter – sludge, wax, tars – and non-condensable gases) in the refrigerant, which can result in degradation/failure of the various components, and cause inefficient operation of the unit.

Structure–property relationships (see Table 10.15) have been used for optimisation of the energy efficiency of POEs in refrigeration lubrication applications [188]. Optimisation of POEs has permitted a gradual reduction in the industry standard lubricant viscosity from 18 cSt at 40 °C when polyol esters were first introduced to 10 cSt today. This translates into real benefits in energy consumption (about 10%) for a standard domestic refrigerator, as shown in Fig. 12.5. The new generation optimised POEs are also largely based on renewable materials and readily biodegradable (Fig. 12.6).

12.8 Turbine oils

Major equipment such as turbines, pumps, governors, gates, hoists and gear drives for specific equipment generally encountered at dams, hydroelectric power plants pumping plants and related water conveyance facilities is often custom designed. Environmental concerns are having a growing impact on the development and use of lubricants for such important appli-

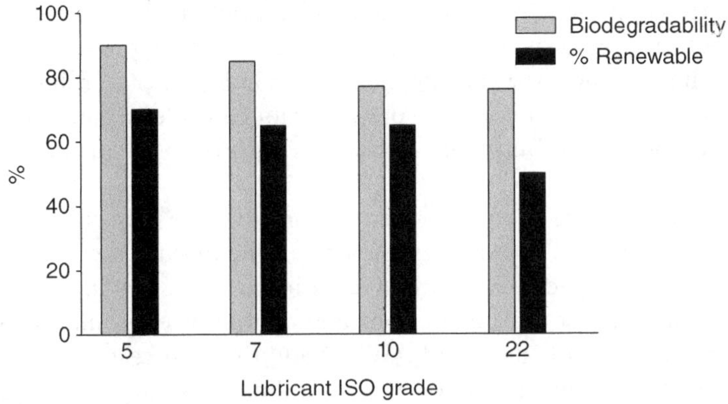

12.6 Biodegradability by OECD 301B and percentage of renewable raw materials for various polyol ester refrigeration lubricant base fluids. After ref. [46].

cations. Operators should exercise caution when evaluating alternative lubricants to ensure that the product selected complies with the manufacturer's requirements. Better lubrication makes power stations, combined heat and power (CHP) plants, and wind turbines more efficient. ExxonMobil is a leading lubricant manufacturer for the world of power generation.

The following performance requirements are typically required for circulating oils, particularly steam and gas turbine oils:

- provide bearing lubrication;
- remove heat through circulation;
- serve as hydraulic fluid for control equipment;
- lubricate reducing gears;
- protect against corrosion;
- allow rapid separation of water;
- resist foaming;
- resist oxidation and sludge formation.

Additives typically used in circulating oils and turbine oils are antioxidants, rust inhibitors, metal deactivators, foam inhibitors and demulsifiers.

Modern turbine lubricants must provide hydrodynamic and EHD film formation, surface protection from metallic contact (or turning gear), seal formation and hydrostatic lift, offer resistance to extreme (high) operating temperatures and chemical and electrokinetic erosion, and control hydraulic actuation. The specific demands on turbine oils are determined by the nature of the turbines and their operating conditions. Apart from mechanical-dynamic requirements, which comprise hydrodynamic lubrication of all bearings and lubrication of gearboxes, heat dissipation, avoidance of friction and wear in gearboxes, and functional fluid for control and safety circuits, various physicochemical specifications (including ageing stability for long operating periods, hydrolytic stability, corrosion prevention – especially in the presence of water, steam and/or condensation –, reliable separation of water, vapour or condensed water, rapid air release and low foaming tendency, good filterability and purity) have to be fulfilled for steam and gas turbines.

Steam turbines are used by power stations to drive electrical power generators. The stringent demands on steam turbine oils are met with carefully selected base oils and special additives. Turbine oils are exposed to severe pressure and temperature conditions and contain ageing inhibitors, corrosion protection agents and mild AW/EP additives (in case of load-bearing capacity). Modern turbine oils offer outstanding reliability and protection of the components from wear. High temperatures (550 °C) and pressures (130–240 bar) of steam turbines are setting increased demands on the lubricants used. It is imperative that the condition of the oil used in steam turbines is monitored and kept within the required specification. Since the level of moisture in the lubricating oil changes over time as a function of environmental and operating conditions, there is a need to rapidly obtain accurate analytical information (see Section 8.6). When the moisture in the lubricating fluid exceeds the allowable specification, corrective action is taken to remove the water in the oil. There are two methods to adjust the moisture content of the turbine oil: (i) the turbine gland steam pressure is manually adjusted if the unit is to operate at a lower than normal load; or (ii) a mechanical device that separates water from oil is used to remove moisture from the turbine main oil tank. ASTM D 2272 is the standard test method for oxidation stability of steam turbine oils by the rotating pressure vessel [189].

Typical turbine oil base stocks today are special paraffinic oils with good viscosity–temperature characteristics (high VI). Water separation and air release properties should be inherent to the base oil because these features cannot subsequently be improved with additives. ISO 6743-5 and DIN 51515 specifications classify turbine lubricants. Different engineering designs make different demands on the lubricant used. The main differences between these standards relate to the ageing and/or oxidation tests and AW/EP performance. Turbine oils also use small amounts of (silicone-free) antifoams and PP depressants. All additives should be ash-free.

The pressurised oil circuits (10–40 bar) play an important role in the lubrication of power station turbines; oil tank temperatures range from 40 to 60 °C. Oil circuits are usually fitted with filtering systems to remove contaminants, aged by-products and sludge. Oil ageing is also influenced by the oil pumping velocity. In normal circumstances, oil monitoring intervals of one year (evaluation in the oil manufacturer's laboratory) are quite acceptable. Oil life of 100 000 h is not uncommon in large steam turbines. The life of turbine oil depends heavily on its quality, operating conditions (pressure and temperature), oil circulation speed, filtering, maintenance and the amount of oil topped-up (to restore the gradually decaying antioxidant level). During operation, turbine oils are exposed to considerable oxygen enrichment. Oil tanks serve the purpose of releasing any trapped air and settling solid contaminants and water.

Gas turbine oils are used in stationary turbines which produce either electricity or heat. Successfully operated gas turbine plants for industrial use date from just before World War II. Common mineral oil-based steam turbine oils are used for conventional gas turbines. The oil temperature is usually 50–55 °C. However, as the temperature of some bearings in gas turbines is higher than in steam turbines (locally up to 200–280 °C), the volume of gas turbine oils generally smaller and circulation more rapid, premature oil ageing can be expected with shorter service life (from 20 000 to 80 000 h). Common mineral oil-based steam turbine oils with additives are used for conventional gas turbines. Semi-synthetic oils (special hydrotreated base oils, Group III), or fully synthetic turbine oils based on synthetic PAOs are recommended for more demanding conditions.

In US Patent No. 4,826,633 (to Hatco Chemical Corp.) Carr and DeGeorge [190] describe a synthetic ester lubricant based on TMP or PE with a mixture of aliphatic monocarboxylic acids (C_5–C_{10}), preferably iso-nonanoic acid (i.e. 3,5,5-trimethylhexanoic acid). The quaternary carbons of the iso-nonanoic moiety inhibit biodegradation of the polyol ester product. A highly stable formulated composition (HPDSC, 35–65 min; viscosity of 5.0 cSt at 99 °C and PP of –54 °C) is particularly useful in gas turbines. In recent years, turbine OEMs have modified their designs to improve operating efficiency. The changes are pronounced for gas turbine applications with improvements in fuel-to-power efficiency and elevated firing temperatures. These changes place increasing demands on turbine oil formulations, particularly with regard to temperature-specific performance parameters (thermal stability, oxidative stability, operating viscosity). Turbine lubricants are to survive in extreme environments. Reformulations have concerned changes in base oil and additive technologies. In the gas turbine industry varnish has become a high-profile issue [191].

Hydroelectric power plants need to pay particular attention to the handling of potentially water-polluting substances such as lubricants. In water turbines there is usually a large amount of lubricating oil in the bearing oil system. Most hydroturbines are connected to a plant oil system that has a centrally located oil filtration and moisture removal system.

Lubricants used in hydroelectric power stations must display good water and air release, low foaming, good corrosion properties, high wear protection, good ageing resistance and compatibility with standard elastomers. There are no established standards for water turbine oils and product specifications for general turbine oils are usually adopted. Viscosity is probably the most important quality of a turbine oil. Hydro generator manufacturers specify the viscosities acceptable for various operating conditions and will include recommendations concerning AW additives, if necessary. Turbine oil ISO 32 is the most common viscosity in use in hydroturbine bearings. Water turbine oils are subject to little thermal stress, and as oil tank volumes are high, service life of water turbine oils is very long. It is common for the lubricating oil in a hydropower unit to be kept in service for 20 years or more. Even though the operating temperatures are moderate, the oil is exposed to air continuously and it is necessary to include a high-performance antioxidation package. The oil must pass ASTM D 943 (Turbine Oil Oxidation Test) and should be over 3500 hours to reach an acid neutralisation number of 2.0. In the bearing tubes, turbine oil splashes and entrains air. The lubricating oil should release entrained air quickly. Antifoam additives increase the air release rate. Lubricants with antifoam additives should meet and exceed ASTM D 892 (Foam Test, Sequences I, II and III).

The turbine's lubricating oil circuit should be sealed from possible water ingress. As the oil in hydroturbines often becomes contaminated with water it is important that oil and water do not remain in emulsion as this affects the oil's film strength and causes increased oxidation and corrosion rates. Applicable standards that must be passed by the oil are ASTM D 665 B (rust test) and ASTM D 130 1B (copper strip corrosion test). Demulsibility properties should meet and exceed ASTM D 1401 (emulsion test). A primary function of hydroturbine oil is to protect against rust on steel bearing surfaces and corrosion in close tolerance oil passageways, since hydroturbine oils are naturally susceptible to water contamination. Table 12.27 lists the main considerations for selecting turbine oils.

Lubricants with or without additives have been used in hydroelectric power stations to lubricate the bearings and gearboxes of machinery and for hydraulic functions in control equipment. Traditionally, petroleum-based turbine R&O oils have been the type of lubricating oil selected for use in hydroturbines. In recent years, rapidly biodegradable water turbine

Table 12.27 Considerations for selecting bio turbine oils

Property	Requirement(s)
• Viscosity	Manufacturer's specification
• Rust and corrosion inhibition	ASTM D 665 B ASTM D 130 1B
• Oxidation inhibition	ASTM D 943
• Antifoaming	ASTM D 892
• Water release (demulsibility)	ASTM D 1401
• Anti-wear	Manufacturer's specification

oils based on saturated esters, allocated in a lower water pollution category than mineral oils, have proven successful in practice. Branched synthetic POE base stocks can be used in the formulation of biodegradable water turbine oils with a conventional water turbine oil additive package [6]. The additive package must include VI improvers, corrosion and oxidation inhibitors, thickeners, dispersants, anti-emulsifying agents, colour stabilisers, detergents and rust inhibitors, and PP depressants. The additives are typically used in amounts so as to provide their normal attendant functions.

Recently biodegradable lubricants for turbines based on mixed esters composed of TMP and/or PE with mixed FAs have been disclosed [192]. McCarthy *et al.* [193] examined ageing of unsaturated and saturated ester-based turbine oils used in the hydropower industry. Measurement of acid number was taken as a simple means of evaluation of the oxidative state of the fluids. Samples with high acid numbers (>2 mg KOH/g) were tested to assess extreme ageing. Friction is reduced for aged samples in comparison to fresh fluids. Nevertheless, even aged samples with acid numbers of 4 mg KOH/g continued to display acceptable wear performance. This suggests a limited risk for catastrophic failure of the equipment during normal operational use. The coefficient of friction appeared to reduce with fluid age. EA lubricating oils should not be used in hydroelectric turbine applications, such as bearing oil, runner hub oil, or governor oil, until extensive testing has been carried out.

Bio Turbine R&O Fluids (Renewable Lubricants, Inc.; Hartville, OH) outperform petroleum-based mineral oils and synthetic esters in terms of superior tribological qualities, properties and features. Bio Turbine R&O Oil-ISO 32 is the ideal biodegradable lubricant for the application, which is fully compatible with gear materials and seals. Bio Turbine R&O Fluids are additivated with Bio Corrosion Inhibitors (BCI™) to protect bearing surfaces against rust and corrosion. For safety reasons, fire-resistant fluids are used in circuits which are exposed to ignition and fire hazards (e.g. hydraulic systems in high-temperature zones in power stations).

Turbine oils have four different sources of possible degradation; viscosity breakdown, particulate contamination, additive breakdown and water contamination; see Section 14.5.1. Power stations generally have an on-site lubrication monitoring program in place. A portable FTIR analyser (e.g. iPAL FTIR, A2 Technologies, Danbury, CT) allows measuring the level of moisture in a lubricating fluid virtually in real time [194]. For physical performance and chemical changes in turbine oils from oxidation, see ref. [195]. Gatto *et al.* [196] have faced the question of modifying current bench tests to adequately mimic oxidation processes in modern turbine lubricant formulations.

A relatively new application of lubricants is in wind turbine installations, where there are conflicting demands of long drain interval (due to remoteness of installations and difficulty of access) and the need for optimal performance over lubricant life (due to the high costs associated with equipment failure). Lubricant performance is here a critical element [197]. A windmill biolubricant has recently been developed within the framework of the Spanish BIOVESIN programme (2006–2012). A wind energy unit requires 500 to 600 litres of high-value lubricants. For the evolution of turbine base oil technology, see ref. [198].

12.8.1 Aircraft turbine oils

Turbojet-propelled aircrafts were developed towards the end of World War II. Turbojet design for aircraft propulsion is an extension of the industrial gas turbine. Aircraft engines used for propulsion in civil and military aviation need very specialised lubricants. All relevant civilian and military product specifications must be fulfilled. Aviation turbine oils (ATOs) have to be able to withstand the extremely low temperatures that prevail in the stratosphere, ranging from –50° to –70 °C. One of the requirements is therefore that the lubricant contains no water at all. At these low temperatures it would freeze and seize the hydraulic systems. ATOs represent one of the most demanding lubricant applications in terms of thermal and oxidative requirements. Aviation turbine oils containing special naphthenic base oils (ISO VG 15–32) with good temperature characteristics are in use. Because of the high temperatures encountered, special low-viscosity (ISO VG 10, 22) synthetic oils based on saturated esters (e.g. polyol ester oils) are also being applied in aviation engines or turbines [199]. These high VI synthetic esters (neat or additivated) show good thermal stability, oxidation resistance and excellent low-temperature characteristics with PPs of –50 to –70 °C.

SE lubricating oils have widely been used in those areas where the oils are subjected to extreme temperature variations such as in aircraft engines. These ester base oils do not, however, inherently exhibit high load-carrying ability and cannot be used, without modification, where a high degree of lubricity is required. Also, the SE oil compositions are subject to oxidative degradation and cannot be used, without further modification, for long periods of time under oxidising conditions. Dioctyl diphenyl amine is a standard antioxidant in the aircraft turbine oil industry.

In US Patent No. 4,440,657 (to Exxon) Metro and Carr [200] describe a turbine oil useful in high-temperature applications such as in the operation of aircraft engines comprising a selected synthetic (complex) ester base oil and an additive package consisting of a combination of diphenyl *t*-butylphenyl phosphate (as a metal passivator, load carrying and extreme pressure additive) and a long-chain alkyl amine for hydrolytic stabilisation. Aircraft turbine oil formulations have also been based on POE compositions with partially unconverted hydroxyl groups [201]; see Section 10.5.3, p. 513. Many gas turbine engines with maximum operating temperature of about 200 °C are experiencing oil system coking and other problems related to the limited thermal and oxidative stability of the conventional, ester-based gas turbine engine oils. Target requirements for advanced high-temperature air force ester-based gas turbine engine oils are a temperature range of –40 °C to 232 °C [202]. MIL-PRF-27601 sets the operational temperature range for aerospace lubricants (gas turbine engine oils and greases, hydraulic fluids) even at –54 to 290 °C [89]. For space applications the space environment poses other unique and challenging demands. Customary additives are usually not adequate for aerospace applications, and more development is needed.

12.8.2 Transformer insulating fluids

The electrical industry uses a variety of cost-effective (non-inflammable) insulating fluids. Examples are polychlorinated biphenyl (PCB; now banned), silicone fluid, fluorocarbon-based oils and polyol esters, used in transformers, power cables and capacitors; see, for example, ref. [203]. Many of these fluids are not biodegradable. Vegetable oils are fully degradable, but most oils available in the market are not electrical grade. A few vegetable oils such as RSO and castor oil have been used in limited quantities, mostly in capacitors, but these are not oleic esters. There is a need for a fully biodegradable fluid and for a method of processing vegetable oil to electrical grade.

Various patents describe electrical insulating fluids. US Patent No. 5,958,851 A to Cannon and Honary [204] discloses a biodegradable (hydrogenated and winterised) soybean oil-based fluid. A special soybean oil grade may act as a non-toxic, non-flammable, readily biodegradable transformer oil. US Patent No. 6,312,623 B1 to ABB Power T&D Company, Inc. [205] claims high-oleic acid oil compositions (C18:1 > 75%, C_{16}–C_{22} diunsaturates <10%, C_{16}–C_{22} triunsaturates <3%, C_{16}–C_{22} saturates <8%) for electrical insulation fluids. The process for preparing an electrical insulating high-oleic acid triglyceride composition comprises the steps of refining, bleaching and deodorising (RBD) for removal of all non-oily compounds (gums, phospholipids, pigments, etc.), winterisation (removal of saturates), stabilisation (using non-toxic additives) and purification (neutral clay treatment) for removal of trace polar impurities and acidic materials. Electrical tests showed that such purified oils have properties similar to currently used high-temperature fluids in distribution transformers (Table 12.28). Typical additives for electrical insulation base fluids comprise oxidation inhibitors, copper deactivators and PP depressants. High-oleic acid oils with favourable oxidation stability and gassing ten-

Table 12.28 Comparison of purified vegetable oil with high-temperature fluids used in transformers

Property	Test method	High-oleic vegetable oil[a]	High-temperature mineral oil[b]	Synthetic fluid ester[c]
Dielectric strength (KV/100 mil gap)	ASTM D 877	42.4	40–45	50
Dissipation factor (% at 25 °C)	ASTM D 924	0.02	0.01	0.1
Neutralisation number (mg KOH/g)	ASTM D 974	0.05	–	0.03
Electrical conductivity (pS/m, 25 °C)	ASTM D 2624	0.25–1.0	0.1–10[d]	5.0[d]
Flash point (°C)	ASTM D 92	328	275–300	257
Pour point (°C)	ASTM D 97	–28	–24	–48

[a] For purified oil without additives.
[b] RTEemp (Cooper Power Fluid Systems; Waukesha, WI).
[c] MIDEL 7131 and REOLEC 138.
[d] Deduced from resistivity.
After ref. [205].

dency, specific heat, thermal expansion and thermal conductivity properties make them particularly suitable for use in transformer insulation fluids. Also Garmier [58] has described high-temperature boron nitride containing bio-based synthetic or natural oil (>80% C18:1) lubricant compositions which provide high dielectric strength that is useful in electrical insulation fluids.

The first SBO-based transformers were retrofitted units and were put into service on the Waverly Light and Power (WLP) distribution system in 1997. The oil product, later called BioTrans™, was further developed within the framework of the American Public Power Association's DEED R&D programme (1999). In 2002 Cargill, Inc. began manufacturing BioTrans™ transformer oils in concert with ERMCO, Inc. (Indianapolis, IN). In another development, Cooper Power Systems (CPS; Houston, TX) has been actively interested in a vegetable-based transformer oil since the early 1990s. A successive alliance between CPS and Cargill, Inc. has resulted in the production and distribution of Envirotemp® FR3™ insulating fluid (based on food-grade SBO) [206, 207]. Envirotemp® FR dielectric fluids are fire-resistant dielectric coolants developed for applications requiring high performance under extreme temperature conditions. With their low freezing and melting points these low-viscosity fluids with excellent dielectric, thermal and mechanical properties are well suited for forced cooling operations such as in traction transformers. Envirotemp® FR3™ offers considerable advantages over traditional dielectric fluids (non-conductive fluids used to insulate and cool electrical distribution products such as transformers), including better performance and increased insulation life. The fluid is commercially available worldwide.

Luminol™ grades are Petro-Canada's inherently biodegradable electrical insulating fluids. The Brazilian electric power utilities have interest in developing an insulating vegetable oil based on castor oil [208].

12.9 Metalworking fluids (MWFs) and metal-forming lubricants

Machine tools, which are most important in the metalworking industry, are used for a variety of operations including forming, cutting and bending. For lubrication a machine tool can be divided into a number of major elements: hydraulic unit, gearbox, spindle, slideway, linear system, plain and roller bearings, and cutting zone. In general, a different lubricant is required for every component. Machine tools require lubricating oils, hydraulic fluids and greases. Apart from cutting fluids, hydraulic oils are volumetrically the most significant group of machine lubricants, followed by slideway oils and gear oils.

Metalworking fluids (MWFs) are fluids used in the machining and working of metals. MWFs currently contribute 17% of the total US industrial lubricant market. Market forces favour semi-synthetic and synthetic formulations. Lubricant manufacturers continue to reduce the use of mineral oil as a component in their formulations because environmentally friendly synthetic and semi-synthetic formulations with longer useful lives decrease the amount of fluid for disposal and are therefore more cost-effective. New lubricant formulations focus on environmental issues, process compatibility, and worker health and safety. Biological control of fluids, in use and in disposal, also draws considerable attention [209].

Friction and lubrication are important aspects of metalworking processes [210]. MWFs for machining operations (milling, turning, drilling, boring, reaming, grinding, punching, honing, lapping) are used to reduce heat when cutting or forming metal. They also lubricate the cutting area, remove contaminants and prevent corrosion. Cutting fluids, which are among the most commonly used process materials in the metalworking industry, have been introduced into the cutting process with the purpose of improving the characteristics of the tribological processes which occur on the contact surfaces between tool and workpiece. Cutting fluids act as a medium between machine, workpiece and tool and must possess a number of different characteristics enabling it to satisfy a wide variety of requirements. By providing lubrication, cooling and corrosion control to help machine or form parts at the highest rate of speed, they maximise tool life, minimise downtime with the fewest possible rejected parts and improve the efficiency of the production system, all while maintaining dimensional accuracy and surface finish requirements within prescribed limits. There are several types of MWFs that may be used to carry out such tasks. The demands MWFs are required to meet vary widely depending on the operation concerned and determine their characteristics. Synthetic MWFs are typically used in applications where cooling is the primary criterion, whereas semi-synthetic and oil soluble fluids are used in applications where lubricity is more important. Cooling and lubrication properties of MWFs help in decreasing tool wear and extending the life cycle. These properties also give the desired size, shape and finish to the workpiece. The cooling characteristics are meant to lower the risk of dimensional changes by maintaining the machined material at ambient temperature. If enough lubrication is present, less coolant is required to absorb the heat generated at the cutting interface as there is little friction to be dissipated. A decrease in wear leads to improved surface finish, which improves the quality of the finished component. In addition to technical requirements, aspects of toxicology and industrial medicine exert an important influence on the formulation of coolants.

The final machining process must yield a quality acceptable part that meets required tolerances. Since every machinery process has its own set of variables, the lubricant used must be formulated to compensate for these variables, by balancing performance criteria which accommodate the range of applications. Metalworking cutting and grinding share both similarities and differences relevant to performance specifications. Grinding involves metal-removal machinery. In the grinding process, which generates a considerable amount of heat and calls for effective cooling by cutting with a geometrically defined cutting edge, the priorities are different. The most important lubricant conditions for grinding are heat transfer and rust protection. Lubricity is minimised in the design because it can interfere with surface finish quality if present at too high levels. In grinding coolants wetting is formulated at the proper level to keep metal fines flushed away from the contact point so as not to interfere with the surface finish. High load-carrying chemistry is not necessary for most grinding applications, unlike cutting fluids. Generally, grinding coolants do not make acceptable cutting fluids due to lack of lubricity, which allows heat (friction) to build up. Cutting requires lubricity. Cutting forces are usually much greater than grinding and tend to generate more friction build-up [211]. The trend in MWFs is towards lower viscosities for high-speed cutting, which requires good heat dissipation.

Metal-forming (non-cutting: drawing, stamping, pressing, welding, rolling, forging, extrusion) requires a lubricant design which accommodates deformation of the metal. Starting as flat sheets or coil stock, the lubricant must form a barrier between the die and workpiece to alter the metal to 'stretch' as desired without breakage, scoring, welding to die or out-of-specified tolerance. Heavy-duty metal-forming lubes must withstand tremendous forces over a wide surface area of metal. Such fluids need to be 'superwetters' to achieve maximum load-carrying capabilities. Typical metal-forming pressures exerted on metal during formation are usually 500 t or more. The lubricant film must remain intact and continuous to meet these demands. Apart from providing effective lubrication and cooling of the workpiece/working elements, rolling oils must also be capable of providing a continuous surface coating of a minimum required thickness which is maintained at high pressures. Moreover, the lubricant needs to give corrosion protection to the rolled strip and should burn off cleanly during the annealing operation.

Although metalworking and metal-forming lubricants both facilitate motion, the complexity of their design differs significantly. Whether grinding, cutting or metal-forming, greatest efficiency requires the proper lubricant for the job. Bio-based lubricants are currently used as substitutes for hydraulic fluids and once-through operations such as cutting and stamping. They have yet to make inroads in continuous-loop operations such as aluminium rolling; however, such potential does exist [212].

Today's MWFs are sophisticated materials. MWFs used for metal-forming or -shaping are usually very viscous because such high-viscosity materials give best performance. The viscosity of most metalworking oils is between 2 and 46 mm^2/s at 40 °C. Low-viscosity oils offer better heat dissipation. The material composition of cutting fluids is quite complex. Users of cutting fluids usually find it difficult to obtain information about the composition and health-related risks. In metalworking operations, such as cutting, threading, broaching, drawing and stamping the presence of free fatty acids (as in UFOs) is acceptable.

Lubricants are working materials which affect the workplace. Industrial hygiene aspects need to be considered, particularly in the field of metal-working. Machine operators are easily affected by contact with various substances within the cutting fluids. Exposure to MWFs (mineral oils) has revealed evidence for human health problems such as rectal cancer, as observed in the automobile manufacturing industry [213]. Metalworking lubricants, which are necessary production aids to the engineering industry, must allow the manufacture of performing products at competitive cost without causing harm to operators or the environment. The use of cutting fluids is potentially dangerous. When formulating MWFs safety aspects relate to skin contact, inhalation and toxicity. In low-emission metalworking applications the lubricants should be toxicologically and dermatologically safe, with favourable lubrication properties and high thermal stress capacity. Moreover, cutting fluids change during use in machining by formation of reaction products and inclusion of foreign bodies and microorganisms. The microbiology of MWFs and its impact on performance and workers' health and safety has greatly advanced in the last decade [214].

Lubricants and coolants are well known to contain problem contaminants, dissolved organic compounds such as amines, esters, surfactants, FAs, petroleum compounds and biocides. These products are specially formulated to be both highly soluble and stable during use, which makes them difficult to break down and remove from wastewater. If not treated or removed, wastewater contamination levels can be significantly raised, leading to substantial oxygen depletion and nutrient loading posing environmental and health risk (see Section 8.2.1). Zeller+Gmelin GmbH & Co (Eislingen, Germany) have developed the concept of re-using neat MWFs [215].

One of the most significant problems in the development and application of coolants is their microbial susceptibility. Microorganisms use the organic material of the coolant as their main source of energy. Microorganisms find very favourable conditions in the continuously re-supplied coolant systems. In practice, microorganisms can attack and decompose the essential organic components of a lubricant formulation, such as emulsifiers and corrosion inhibitors. Users of traditional metal cutting fluids often experience problems with bacteria growth and foam buildup. Bacteria growth generates odours, degrades the lubricant and can cause operator health and safety concerns. Henkel Corp. (Rocky Hill, CT) has recently introduced low-foaming and *bioresistant* semi-synthetic water-soluble metal removal fluids for machining and grinding ferrous and non-ferrous alloys (Multan® B236 and B414), which can be used continuously for years with minimal biological degradation. Also the readily biodegradable and REACH compliant CORRGUARD™ EXT Amino Alcohol (Angus Chemical Co., Buffalo Grove, IL), based on advanced primary amine technology, resists biodegradation at the levels present in working dilutions.

Microorganisms do not only constitute a technical problem but also a hygienic and toxicological risk for machine personnel. EU implementation of the Biocidal Products Directive (BPD) [216] and Registration, Evaluation, Authorisation and Restriction of Chemicals (REACH) [217] have regulated the use of MWF microbicides. BPD is similar to the US EPA's FIFRA-based regulations [218]. ASTM Standard E 2169 is common practice for selecting antimicrobial pesticides for use in water-miscible metalworking fluids [219]. Although some 80 active ingredients are currently approved by the EPA for use in MWFs, in practice fewer than a dozen account for more than 90% of the total volume of MWF microbicide use. Hexahydro-1,3,5-tris(2-hydroxyethyl)-*S*-triazine (HTHT) is still most widely used in the US. Other MWF microbicide chemicals in current use are isothiazolines (5-chloro-2-methyl-4-isothiazolin-3-one (CIT), 2-methyl-3(2H)-isothiazolin-3-one (MIT)). Increasingly, formulators are being asked to develop fluids with non-formaldehyde releasing biocides. Benz-isothiazolinone (ROCIMA™ BT 2S Antimicrobial) does not contain or release formaldehyde. The universal biodynamic cutting and grinding fluids TRIM® E720 and E909 (Master Chemical Europe, Ltd; Needham Market, UK), suitable for machining of aluminium and steel alloys and for yellow metals (E720), contain no aggressive biocides and are free from boron, amines and formaldehyde donor biocides. It is noticed that formaldehyde and boric acid are included in the candidate Substance of Very High Concern (SVHC) List (REACH). Taking into account the small MWF biocide market, high regulatory costs and shrinking R&D efforts of microbicide manufacturers, it is unlikely that new active ingredients will be developed specifically for use in MWFs [220]. For microbicides, see also Section 7.2.3.

Owing to the multiplicity of negative effects that cutting fluid wastes have on people and the environment, nowadays there is an increasing attention to carefully select cutting fluids that are both efficient and environmentally friendly. Apart from the usual tribological requirements, novel cutting fluids have to satisfy the requirements of environmental protection set either by in-house regulations, governmental directives or international ISO 14000 standards. Products which are highly poisonous, irritating or corrosive, flammable, easily ignitable or explosive are all regulated.

Apart from causing harm in the immediate working environment, cutting fluids also give rise to certain problems in waste disposal [221] (see Section 14.7). Cooling lubricants (such as cutting fluids in machining processes) are potentially great polluters. The increasing number of laws and directives governing industrial safety and environmental protection means that nowadays the use of cutting fluids is putting considerable economic pressure on manufacturing [222]. Rising costs for waste disposal are triggering manufacturers to implement improved cooling lubricant strategies. This constitutes a technological challenge. Ecological parameters of the cutting fluids are usually determining new constraints on machinability parameters. Short-term measures for improved cutting fluids management and maintenance are reduction of the quantity of cutting fluids used (minimal quantity lubrication, MQL) or re-use and modification of the cutting fluid medium (substitution of emulsion with oil). This requires detailed understanding of the complex interrelationships between process, workpiece and machine tool.

Conventional metalworking (cutting and grinding) oils fall into two general categories: neat and water based [223]. Water-soluble oils are used by diluting lubricant components and surfactants with water. Neat cutting fluids products or water-insoluble (immiscible) oils are not mixed with water prior to use and are used directly as stock solutions composed mainly of mineral oils. Neat MWFs are classified according to ISO 6743/7 on the basis of additives used (antifriction and EP agents). They offer both advantages and disadvantages over water-miscible fluids. Generally speaking, water-insoluble cutting and grinding oils exhibit superior lubricating performance (longer tool and oil life), while water-soluble or water-dispersible oils exhibit superior cooling performance. The three major types of water-extendable MWFs are soluble oil, synthetic and semi-synthetic systems; selection depends upon the specific application. In the USA most metalworking is done with water-extendable fluids, which contain numerous components (e.g. base oil, water, emulsifier, corrosion inhibitor, biocide). Aqueous cutting fluids (soluble, semi-synthetic and synthetic) are less resistant to microorganisms such as bacteria, yeasts and fungi. It is desirable to use a metalworking composition that is biostable without the need for biocides.

In relation to the requirements of the various application methods, cutting fluids have a wide variety of compositions (base fluids and additives). The choice of the lubricant depends on its function. If lubrication is more important than cooling, a neat oil will be used. If cooling is more important, water-soluble emulsions will be used, because water can take up three times more heat than oils. The high process temperatures also require high flash points and the lubricant should not cause staining of the processed metal. Cutting and grinding oils that are effective for improving working efficiency are frequently environment unfriendly. Whether water-insoluble or water-soluble, oils undergo gradual degradation with use and eventually become unusable. With growth of microorganisms in water-soluble oils the solution stability is lost, resulting in separation of the components. On the other hand, progressive oxidation of water-insoluble oils produces acidic components that corrode metallic materials and produce significant changes in viscosity. In both cases, the degraded oils must be disposed of as chemical waste. Oils with excellent biodegradability are preferred from the viewpoint of waste treatment and working environment. Cutting fluids containing chloroparaffins as EP additives are difficult to dispose of, are classified after use as hazardous waste, and should no longer be used. Sokovićand Mijanoviĉ [224] have described a new semi-synthetic cutting fluid (Teolin AIK) as a replacement of oils with chlorinated hydrocarbons. TRIM® E950 (Master Chemical Europe, Ltd; Needham Market, UK) is a biostable cutting and grinding emulsion based on a proprietary blend of new vegetable-based technology and premium, non-chlorinated and non-sulphurised EP additives [225]. Perfad™ 8100 (Croda Lubricants; Cowick Hall, UK) is a bio-based additive that replaces chlorinated paraffins in neat mineral oil and ester-based forming fluids.

Early lubrication technology was entirely based on straight petroleum oil designs, which served as base stock. Use of animal fats was common to impart 'wetting' or surface coating. Emulsifier technology advanced to accommodate use of chlorinated paraffins and sulphurised fats and oils into lubricant designs in pursuit of versatile performance. These designs offered both boundary lubrication and EP performance. During the evolutionary period of lubricant technology, metal parts produced were commonly cleaned with chlorinated solvent such as 1,1,1-trichloroethane. Such cleaning solvents are now considered as being 'ozone depleters', which has led to regulatory scrutiny and their ultimate demise in industry. Regulatory restrictions regarding workplace volatile organic compounds (VOCs), use and disposal of chlorinated solvents and paraffins have induced industry to look for alternatives.

Neat MWFs without additives are limited in use to simple machining operations on easy-to-machine materials such as cast iron and non-ferrous metals. Most MWFs based on petroleum hydrocarbons contain additives to improve lubricity, wear protection and other additives to influence characteristics such as foaming, misting, dispersion, wetting, flushing, oxidation stability and corrosion protection. Table 12.29 lists the main ingredients of MWFs. Greater amounts of additives normally increase the cutting performance of a metalworking oil. Neat oils are more skin-compatible than emulsions. They do not contain biocides and fungicides. Industrial hygiene and dermatological problems occur when machine personnel is exposed to repeated skin contact with metalworking oils. Oil misting is still a problem. Replacing conventional emulsions with neat oils enhances safety-at-work at superior technical performance and lower cost.

Most neat MWFs are petroleum hydrocarbon base oils (preferably low-aromatic, paraffinic solvent extracted or hydrocracked products). Oils used in metalworking [226] typically contain mono- and diethanolamine-based

Table 12.29 Main ingredients of metalworking fluids

Mineral oil hydrocarbons (85–95%)
Synthetic hydrocarbons (PAOs), synthetic esters, fatty oils
Additives:
Emulsifiers
Corrosion inhibitors
Stabilisers, coupling agents
Extreme pressure (EP) additives
Anti-wear additives
Lubricating film improvers
Antifoam agents
Microbiocides
Anti-mist additives
Complexing agents

FA amines, which are suspected carcinogens, and/or petroleum-based hydrocarbons, which can damage the environment.

Substitution of water-immiscible lubricating oils for metal processing for more renewable and less polluting oils is a valuable objective. In Germany alone, the consumption of these lubricants amounts to 40 kt/yr. Mineral oil fractions applied as lubricant base oils can be replaced by FA ester oils, bearing technical and ecological advantages. Recently, fluids based on vegetable, animal or synthetic sources (in particular esters) have successfully been introduced, initially in Europe. Recent trends in Germany focus heavily on SEs and PAOs. However, so far ester oils are only parsimoniously applied in industrial practice mainly since they are more expensive than mineral oil products and also because of a lack of user acceptance.

Design preference may be governed by raw material availability, environmental regulations and global market trends. Endorsement of rapeseed-based technology in Germany to improve environmental capability on MWFs is an example of market trend setting. Similarly, Europe shows a strong preference for chlorine-free metalworking lubricants to satisfy various waste disposal and regulatory trends. Non-petroleum, non-EP product designs are favoured to insure regulatory compatibility.

Going green in cutting tool engineering has an important specific advantage [227]. The less chemically active petroleum oil molecules adhere more loosely to the metal surfaces, and therefore are unable to withstand the shearing stresses exerted in the metalworking process. Metalworking lubricants based on natural triglycerides can be utilised throughout the industry for a variety of processes including rolling, stamping, drawing, pickling, cutting and extruding. Aqueous formulations of triglycerides are used in cold rolling of steel to cool the rolls. Vegetable oil cutting fluids are more environmentally friendly, less of a health hazard and superior in performance, and could become the lubricants of choice. Vegetable oil-based emulsion products are inherently more subject to attack by bacteria and fungi than comparable petroleum oil products. The very low evaporation values for ester oils are particularly interesting. Ester oils also display tribological advantages due to their superior friction behaviour [228]. Even without any additives, they also offer improved friction and wear characteristics due to their polarity. This advantage as well as high VI, low water pollution hazards and good biodegradability make this group of base oils an excellent choice for use in machine tools as cutting and machine lubricants.

In view of variations in the metals being worked, the different operating conditions and application methods employed, numerous metalworking oils based on natural fats and oils have been developed [229–236]. Drawbacks of vegetable oils as MWFs are emulsification, hydrolysis and oxidation. Natural oils are more susceptible to hydrolysis than petro-based lubricants, a condition that is accelerated by the alkaline nature of MWFs. Metal ions introduced to the fluid during machining speed up the process of oxidation. Most of the formulations have involved the use of different triglyceride sources, partial replacement with a petroleum product, additivation, and advanced emulsifier technology. Natural oils and fats have also been chemically modified to form ester products (e.g. ethylhexyl ester of RSO) [237]. Mixed esters (reaction products of a triglyceride, a polyoxyethylene glycol and a short-chain mono- or dibasic acid) have significantly improved water solubility [238]. Among vegetable-based metalworking lubricant additives FAs have the best grinding performance of all oiliness agents, friction modifiers and EP agents [239, 240]. Elektrionized™ vegetable oils are highly efficient as lubricity and oiliness additives in formulation of metalworking lubricants (typically 5–15% in cutting and 10–20% in forming applications [236], see also Sections 7.1 and 7.3). Perfad™ 8400 (Croda Lubricants) is a lubricity additive for neat MWFs (mineral oil and ester-based) that reduces energy consumption and extends tool life. Low environmental impact additives such as Polartech® LA8005 and Polartech® LA8330 (Afton Chemical Corp.) provide superior MWF lubrication performance. Boundary lubrication encompasses a significant portion of lubrication phenomena in metalworking conditions.

The potential for application of bio-based ester oils for use as lubricants in sustainable metalworking production processes has been described by Wichmann and Bahadir [233]. 2-Ethylhexanol esters from cheap (winterised) renewable waste plant oils and technical animal fats perform very well as cooling lubricants. Jain and Bisht [234] have formulated oil-in-water (O/W) metalworking emulsions by blending mineral oils with non-edible industrial vegetable oils like high-erucic rapeseed and karanja oils. The performance of such metalworking formulations (5% O/W), consisting of vegetable oil, mineral oil (N150 and N500 paraffinic lubestock), an emulsifier and commercial additives (Lubrizol® 5620, ZDDP, silicone defoamer and Fynol AC28), was tested in tribotesters which can simulate the conditions prevailing in rolling mills. The formulated O/W emulsions were close to commercial reference oils in lubricity, wear scar diameter (WSD), weld load (WL) and particle size. The blend formulation consisting of 66 vol% mineral oil, 17 vol% karanja oil and 17 vol% rapeseed oil performed best in the low saponification region. According to US Patent Appl. No. 2007/191240 a metalworking oil comprising a triester of FAs and glycerin, with an oleic acid content of 40–98 wt% on the total amount of FAs can achieve a high level of machining performance without a chlorine-based EP agent [241]. In US Patent Appl. No. 2008/026967 A1 to Nippon Oil Corp. [235] Suda *et al.* disclose a natural or SE oil with moisture content of 200–20000 ppm and a kinematic viscosity of 1–20 mm^2/s at 40 °C as a water-insoluble oil for non-ferrous metalworking and for cutting, grinding or rolling supplied as an oil mist in an MQL system. An SE is preferred from the viewpoint of stability of the obtained lubricating oil and a natural ester from the standpoint of effects on the environment. The physical properties of vegetable oils set some limits to their performance as metal cutting fluids [242].

PCT Int. Publ. No. WO 2008/001933 discloses an oil composition for use in metalworking composed of a base oil derived from natural fats or an SE, and a sorbitan FA ester and phospholipid [243]. This oil composition is excellent in lubricity and rust prevention and is suitable for processing a metallic material such as cast iron, (stainless) steel, or non-ferrous material with minimal quantity lubrication. Feng [244] has described a modified vegetable oil-based water-soluble antirust lubricity additive for use in a water-based metal cutting fluid.

Conventional MWF formulations use anionic emulsifiers which can be depleted by calcium build up from the water hardness. This can lead to the separating out of insoluble soaps, poor wettability, reduced lubricity and loss of fluid cutting performance. Certain lipid solutions, called self-emulsifying systems, by virtue of their ingredients, have the capacity to undergo spontaneous emulsification when introduced into an aqueous phase following gentle agitation, yielding O/W emulsions. Self-emulsification of vegetable oil-nonionic surfactant mixtures was described by Wakerly *et al.* [245]. Self-emulsifiable esters (SEEs) were developed by Unichema [246] to overcome the problems caused by calcium soap formation, as can occur with MWFs. The self-emulsifiable SE base oils are unaffected by water hardness, allowing a large reduction in components in the MWF formulation and longer emulsion lifetimes. SEEs have excellent oxidation resistance, good hydrolytic stability but poor biodegradability. SEEs contain anionic functional groups which have to be activated by neutralisation with an alkanol amine or KOH to form very stable emulsions. Mineral oils and vegetable oils are often used as components of formulated aqueous metal-working compositions. In these instances, an emulsifying agent must be used to hold these water insoluble oils in uniform suspension in the water matrix, which adds significant cost to the commercial formulations. In PCT Int. Publ. No. WO 2005/071050 A1 Lange *et al.* (to Lubrizol Corp.) have disclosed a succinated triglyceride oil derived from maleating an unsaturated vegetable oil for use as a self-emulsifying agent for MWFs, replacing alkyl-benzene petroleum sulphonates as emulsifiers [247].

Esterified trimer acids (ETAs) provide certain desirable properties for metal treatment fluid systems, such as biostability, and can be disposed of using standard techniques. Depending on the specific derivative, ETAs can be water immiscible or water dispersable. Water immiscible ETAs are (nearly) fully esterified with alkyl alcohols whereas water dispersable ETAs are only partially esterified with alkylalkoxy alcohols. US Patent No. 6,060,438 discloses the use of water immiscible ETAs, such as Priolube®

3953 (Uniqema, DE) for use in semi-synthetic O/W emulsions [248]. On the other hand, US Patents No. 5,707,945 and 5,688,750 disclose water dispersable ETAs, such as Priolube® 3951/3952/3955 (Uniqema, DE) [106, 249]. However, ETA technology does not provide a synthetic (water-based) emulsion system that exhibits good lubricity and transparency, and is low foaming, biostable and waste treatable. Therefore, there remains a need for such a synthetic (water-based) emulsion system. In US Patent Appl. No. 2005/0037933 A1 Bingeman describes such compositions comprising: (i) a water immiscible component (e.g. polyhydrostearic acid); (ii) a water dispersable esterified dimer or trimer acid (ETA); and (iii) a low emulsifier (e.g. a C_8–C_{20} straight chain or branched organic acid) [250]. The dimer/trimer acids are composed of polymerised products of unsaturated C_{16}–C_{20} fatty acids (SBO, tall oil, etc.). Dimer and trimer acids are commercially available, such as Empol® 1008 (Henkel Corporation/Emery Group) and Pripol® 1004 (Uniqema).

Today both oils and greases as well as esters and fatty alcohols are in use, apart from conventional mineral oils and water-miscible coolants. Ester oils are used with preference for all machining operations in which the lubrication effect is paramount. It is important to make available low emission, dermatologically and toxicologically acceptable media with high flash point. SE oils and fatty alcohols, with low evaporation characteristics and high flash point are toxicologically harmless, and have proven themselves in practice [91]. Inolex Chemical Co. (Philadelphia, PA) offers cost-effective high-temperature base stocks based on extremely stable POEs and low-volatility ester compounds. Lexolube® IPP and IPO are isopropyl palmitate and oleate base metalworking additives, respectively, whereas the Lexolube® CPE-series constitutes extremely stable POE formula metalworking additives. Ester oils and fatty alcohols with additives with low toxicological values score high in terms of safety aspects. Ester oils and fatty alcohols should be given preference in the case of processes where high heat is developed to avoid residues on workpieces and machines. Table 12.30 shows the differences between esters and fatty alcohols [251]. Equiviscous fatty alcohols have a lower flash point than ester oils. They evaporate relatively

Table 12.30 Differences between esters and fatty alcohols as coolants

Property	Ester	Fatty alcohol
Evaporation	Very low	High
Lubrication	Very good	Low
Flash point	High	Low
Water pollution category	–/1	–/1

easily and as a result have a low cooling effect. Compared with ester oils their lubricating effect is relatively low. Fatty alcohols are preferred for cutting operations where the lubricating effect takes second place to the cooling effect.

Much technical testing is required to determine which applications are most suitable for ester-based cutting oils. Success is assured when the ester product performs better, has a universal use, and is less expensive than competitive products. Belluco and De Chiffre [252] have evaluated the performance of a range of mineral and vegetable oil-based cutting fluids in machining operations. Vegetable-based oil formulations displayed equal or better performance than the reference commercial mineral oil in all operations. In comparison to mineral oil-based coolants causing ongoing stability problems in production, vegetable oil-based coolants for metalworking machines performing processes such as broaching, mill cutting and grinding improve tool life, increase system life, eliminate bacterial infections, swarf buildup and unpleasant working conditions, and reduce overall fluids costs [253].

As many different oils (hydraulic, gear, slideway and spindle oils) and greases are used in machine tooling mutual contamination of hydraulic and cutting fluid circuits may easily occur. In metalworking applications lubricant compatibility ('universal fluid') has a high priority. Within the framework of the German UNICUT project a low-viscosity (10 mm^2/s) non-water-hazardous multifunctional ester oil (Unifluid 10, FUCHS GmbH) has been developed which serves both as a cutting fluid, slideway oil and hydraulic fluid [254]. There are some special oils in use which, as alternatives to pure ester oils, are mixtures of hydrocracked oils and ester oils and thus combine excellent tribological properties with the somewhat lower price of hydrocracked products.

Although plant oils are environmentally friendly alternatives their negligible water solubility limits their use for water-based applications as in metalworking. However, transesterification of triethanolamine (TEA) or ethoxylated triethanolamine (ETEA) and a triacylglycerol (TAG) oil to form partial esters improves the water dispersability resulting in a more suitable oil for metalworking [255]. Other drawbacks of vegetable-oil based or esterified lower molecular weight polyglycerol-based compounds as lubricants are their low viscosity profile, which limits their usefulness and insolubility in functional fluids. The viscosity of vegetable oils may be increased in a variety of ways including blending with synthetic oils or polyol esters. Reaction products of polyols and natural fat- and oil-based products (e.g. canola oil) are well known as metalworking lubricants. For example, US Patent No. 4,885,104 [256] discloses metalworking lubricants composed of a reaction product of natural fat or oil, a C_5–C_{15} polyol and a C_2–C_{36} dicarboxylic acid. Although biodegradable, such esters are still relatively low in viscosity and therefore of limited utility. Higher viscosities may be achieved by using dibasic acids to increase molecular weight, but at the expense of biodegradability.

The most important function of a mould lubricant in continuous casting of steel is to prevent sticking. Lubricants used to date have been HERO (*Crambe*), fish oil (sulphurised sperm oil, SSO), paraffin wax, etc. Sperm whale oil has been banned from industrial use since 1971. Butterfat or lard can also be used in MWFs. State-of-the-art plant-based products are available for metalworking applications. GEMTEK (Phoenix, AZ) has developed an extensive line of industrial lubricants and MWFs (Safe Lube™) based on FA esters mixed with degummed soy or canola base oils. Ester-based metalworking oils typically sell for twice the price of conventional oils. CIMCOOL Global Fluids (Cincinnati, OH) has developed CIMFREE Botanical Fluids, a line of nine environmentally friendly vegetable-based products for high-speed machining and metalworking operations. TRIM® E850 (Master Chemical Corp.; Perrysburg, OH), an emulsion metalworking (metal-removal) fluid for machining high-tech alloys, is a proprietary formulation based on new 'green' vegetable-based technology and premium traditional lubricity additives.

There are very many processes in sheet metal working [257]. Not surprisingly, in no other field of engineering as many different lubricants are used as in metal-forming operations. To select the best lubricant, detailed knowledge is necessary, ranging from how the lubricant is applied through to any subsequent operations performed on the pressed part. Information about the tools and materials involved ultimately lead to tailor-made, process-compatible forming lubricants. All metal-forming liquid lubricants follow three basic designs:

- Straight oils: Use neat (100%) oils. Water is viewed as a contaminant in straight oil systems.
- Soluble oils: Consist of oil-based concentrates, designed to be diluted with water to form milky emulsions.
- Lube fluids: 100% water-based containing no petroleum oil. True plant-based chemical solutions.

Quenchants (e.g. UCON™, Dow Chemical Co.) are non-flammable, aqueous solutions containing special polymers (PAGs) and corrosion inhibitors for quenching ferrous and non-ferrous metals. All lubricants for sheet metal-forming must be removable with commonly available cleaners.

Sheet metal-forming processes are among the most important metalworking operations for which large quantities of liquid (and solid) lubricants are used. Hot and cold flat-rolling operations used to produce aluminium sheets in continuous casting operations normally rely on mineral oil-based MWFs. The fluid's chief function is to keep the metal and the heavy steel rolls used to flatten it from welding together. Cutting down on VOC emissions is a driving force behind switching from petroleum-based fluids to vegetable oils. A cost-effective, oxidation resistant, bio-based equivalent, based on modified soy oil and antioxidants, developed by ARS, has successfully been evaluated by Alcoa [258]. Another benefit of soy-based fluids is less worker exposure to synthetic fluids and chemical additives they may contain, such as biocides and corrosion inhibitors. The efficacy of any formulated cold rolling oil is ideally finally evaluated by conducting long duration tests on an actual production rolling mill. Such field testing is expensive. Selected locally produced (Nigeria) vegetable oils (palm, palm kemel and sheabutter oils) were evaluated for use as lubricants in metal-forming processes [259].

Xianyan [260] has disclosed an aluminium–magnesium alloy draw lubricant composed of 85–94% base oil (vegetable oil or animal grease of low unsaturation degree), 3–8% polyurethane and 1–2% antirust agent. The product has excellent thermal, oxidation and hydrolytic stability, slows ageing and prolongs service life.

The pressing of car body parts is one of the most important sheet metal drawing processes. Current solid lubricants are used for protecting sheet metal and/or as sheet metal-forming lubricants. The most obvious requirement for a lubricant is that it be present as a homogeneous layer of adequate thickness in the zone of deformation. All dry-lubes have in common the solid state, or at least a waxy consistency, at ambient temperature. Solids or waxes are difficult to apply and are usually thinned before application [261]. For optimum versatility, dry-lubes should be suitable for all sheet-metal qualities (uncoated steel, aluminium, etc.). The application of solid films in the finishing lines of rolling mills presents major obstacles. In stamping operations, wet (mineral oil-based) or dry-film lubricants are used to protect the surfaces of the sheet blanks, reduce friction during deep drawing, and minimise required surface treatments before assembly [262]. Dry-film lubes are either water-soluble (but incompatible with most adhesives used in automotive body construction) or water-free (so-called hotmelts, adhesive compatible). Hotmelt dry-lubes are favoured in the sheet metal-forming industry. The benefits of dry lubricants in sheet metal-forming are given in Table 3.8. Starch–oil composites are preferred ingredients for formulating solid lubricants because they are based on renewable and abundantly available raw materials and have superior environmental and health characteristics. Starch–oil composites based on purified food-grade corn starch (PFGS) and SBO were evaluated in water-based solid lubricant formulations for a one-time sheet metal-forming application [263].

Apart from lubricating properties, extrusion oils must have high thermal stability in order to cope with local overheating on the extrusion bodies and tools. Extrusion oils also have an essential cooling function, especially in the case of high-speed automatic machines. Further demands include low evaporation, little odour nuisance and oil misting tendency. The preferred viscosity range of extrusion oils is between 35 and 65 mm^2/s at 40 °C. In general, the difference between extrusion oils and non-water miscible oils used for cutting operations lies in high additive content. Extrusion oils are classified into four classes by type and amount of polar and EP additives (on S, P or Cl basis). For environmental reasons chlorine containing additives are no longer used in Europe. Caminaga *et al.* [264] have compared lubricants (mineral oil, chlorine-free semi-synthetic oil, powder soap and wheat flour) and lubrication methods (immersion and hydrostatic) for cold extrusion of steel shafts with the object of producing industrially acceptable results in terms of extrusion load and product quality. Wheat flour was found promising.

Sulphurised mineral oils, sulphurised fatty oils and polysulphides play a role as sulphur carriers. Apart from the sulphurised fatty oils there are also other products in the polar additive sector such as thermally stable synthetic esters (also as base oils). Products based on low-cost animal fats (generically called 'lard oils') and their derivatives have extensively been used as lubricant additives for metalworking and industrial use. For example, 'methyl lardate' has found metalworking application both as is and in sulphurised form. Relatively small quantities of sulphurised vegetable oil products are being used [265]. Unlike products from the wax ester jojoba oil, sulphurised vegetable triglycerides alone (from soybean, sunflower, cottonseed, high-erucic rapeseed, canola) have physical properties (oxidation and thermal stability) generally undesirable for lubricant additives. *Limnanthes* (meadowfoam)-containing products generally give the best performance but cannot compete with low-cost animal fat-based products. Thermally stable, biodegradable ester oils with excellent tribological characteristics as extrusion oils will gain more significance in the future.

FUCHS GmbH (Mannheim) is a leading global manufacturer of specialty lubricants for the metalworking industries with ANTICORIT Prelube oils in car body manufacturing and RENOFORM process-compatible deep drawing oils. RENOFORM products are designed for hot and cold extrusion, section, tube and wire drawing. Uniflex® 164PT (Arizona Chemical, Jacksonville, FL) is an *n*-butyl tallate ester developed as a lubricant ester for metal-forming applications. Biodegradable Radia®-branched products (Oleon; formerly Atofina Oleochemicals) meet the most stringent requirements of the lubricants industry. Radia® fatty esters of methanol, propanol, butanol, octanol, etc., are lubricity improvers for steel, copper and aluminium rolling; they are widely used in neat oils and in emulsions for cutting, grinding and drilling operations. Bio-based MWFs are being distributed by Performance Biolubes LLC (Morton Grove, IL), a spin-off of

Environmental Lubricants Manufacturing, Inc. (Cedar Falls, IA). Utilisation of vegetable oils in metalworking fluids has been reviewed [266].

12.10 Lubricants in the food-processing industry

The equipment used in the food-processing industry varies by segment, with the leading segments comprising meat and poultry, beverages, snack foods, vegetables and dairy. The moving parts in food-processing equipment such as bearings, gears and slide mechanisms are similar and usually require lubrication. The lubricants most often used in these applications include hydraulic, refrigeration and gear oils as well as oven, chain and cable lubricants, mould-release agents and all-purpose greases. 'Food grade' lubricants are designated for use in food-processing plants. They are designed for light-duty applications where the fluid does not enter in contact with food. Regulations allow contamination of up to 10 ppm with food-grade lubricants compared with zero ppm contamination allowed with non-food-grade lubricants.

Lubricants for use in the food-processing and pharmaceutical industries need to be internationally approved, neutral in task, odour and colour, as well as in good microbiological status (human health protection), and must meet more stringent standards than other industrial lubricants. In countries of the industrial world, their applications are usually controlled by laws and regulations. Such lubricants need to be approved by government agencies. In the United States the food industry must comply with the (mandatory) rules and regulations set forth by the USDA. Foodstuff-compatible oils are classified according to FDA (Food and Drug Administration, Silver Spring, MD) and NSF (National Sanitation Foundation, Ann Arbor, MI) guidelines [267]. The FDA lists ingredients permitted for use in food-grade formulations (Code of Federal Regulations: 21 CFR §178.3570); white oils are listed in 21 CFR §178.3620). According to the Food Safety and Inspection Service (FSIS) lubricants and other substances that are susceptible to incidental food contact are considered as indirect food additives under USDA regulations. Therefore, these lubricants, classified as either H1 or H2, are required to be approved before use in food-processing plants. Inspection by FSIS of non-food compounds used in federally inspected plants is required. The most stringent requirement, NSF H1 approval, applies to food-grade lubricants for which incidental food contact is technically unavoidable. More precisely, the relevant category codes for 'food-grade' products are H1 for finished oils, and HX1 for additives and base oils used in these oils. The *NSF White Book* lists the brand names and manufacturers of registered food-grade products. The base oils for such products are principally special white oils, special PAO grades and polyglycols. The manufacturing processes employed for making H1 lubricants should likewise be certified by NSF according to ISO 21469 [268], to ensure compliance with the strict hygienic requirements of food and pharmaceutical production. The selection of lubricant additives for the food-processing industry is of special concern [269]. The NSF H2 classification is for lubricants for general use in food-processing technology, without the possibility of direct food contact, and ensures that no known poisons or carcinogens are used in the lubricant. Such products can be standard lubricants if special precautions are taken. Compliance with hygienic regulations in these key industries (food and pharma) requires H1 lubricants with high-performance capabilities, enabling MQL, combined with precise lubricant metering. The USDA is no longer issuing any new approvals; NSF has taken over the registration activities as from 1998.

In Europe, Council Directive 93/43/EEC of 14 June 1993 (Hygiene Directive) lays down the general rules of hygiene for foodstuffs and requires adequate safety procedures in food processing, including protection against contamination by equipment [270]. Conveyances used for transporting foodstuffs must protect foodstuffs from contamination. Various decisions of the European Commission aim at improving food safety [271, 272]. HACCP-9000® is an international programme that integrates HACCP (Hazard Analysis and Critical Control Points), ISO 9000 and food hygiene practices into one management system to ensure food safety and quality. Clearly, a large number of material certifications and approvals have been put in place for lubricants in the food-processing industry, depending on the application: H1 registration, ISO 21469 certification, drinking water approval, compliance with BSE guideline EMEA/410/01 [273], halal and kosher certifications, neutrality towards beer froth and the like. Several other relevant initiatives deserve mentioning. The purpose of the European Hygienic Engineering & Design Group (Frankfurt/Main) is to spread the best practice and eliminate poor lubrication and hygiene practices in food manufacturing and its supply claim, whereas the European Lubricating Grease Institute (ELGI, Amsterdam) has activated a Food Grade Working Group.

In addition to meeting the requirements for safety set by regulatory agencies, the product must also be an effective lubricant. Lubricating oils for food-processing plants should lubricate machine parts, resist viscosity change and oxidation, protect against rusting and corrosion, provide wear protection, and resist the formation of deposits and sludge in service. The product should perform effectively at various lubrication regimes, ranging from hydrodynamic thick film to boundary thin film regimes. In particular, there is a need for a lubricant that exhibits excellent lubricity, extreme pressure and AW resistance. Croda offers a wide range of NSF H1 registered base fluids (including PAGs, fatty acids and Priolube™ esters of varying viscosity grades) and an anti-corrosion additive for use in food-grade lubricants. FUCHS Lubritech GmbH (Kaiserslautern) CASSIDA and FM food-grade oils and greases are certified to ISO 21469 standard, which ensures a complete audit trail.

Some specific requirements for conveyor lubricants for lubricating transport operations of glass, plastic, plastic impregnated or coated cardboard or metal beverage containers on a conveyor of stainless steel or nylon are listed in Table 12.31 [274]. In commercial container filling or packaging operations, containers are moved by a conveying system at very high rates of speed. Usually copious amounts of aqueous dilute lubricant solutions (obtained by on-site dilution of concentrated lubricant) are applied to the conveyor or containers using spray or pumping equipment. Aqueous dilute lubricant compositions are suitable for conveyor lubrication in the beverage or food industries. Diluted lubricants are advantageous because they are an effective way of lubricating conveyor surfaces while using less of the concentrated lubricant composition. Disadvantages of this approach are the wasteful use of large amounts of water, environmental unfriendliness, cost, promotion of growth of microbes, error-prone lubricant dilution procedures, and risk of environmental stress cracking (ESC) of thermoplastic containers. For example, dilution on site of concentrated lubricant solutions can cause ESC of poly(ethylene terephthalate) (PET) bottles. 'Dry lubes' (lubricant compositions with less than 50% water) are a solution to the disadvantages of dilute aqueous solutions. The application of 'dry lubes' without dilution requires special dispensing equipment. Silicone materials have been the most popular 'dry lube'. However, as these are not equally effective at lubricating all types of containers (plastic, glass, metal), a plant has to stock more than one type of conveyor lubricant. In US Patent Appl. No. 2010/0286005 A1 (to Ecolab, Inc.) Valencia Sil *et al.* disclose a spray-applicable lubricant composition having greater than 50% water and comprising a water-miscible silicone material (such as polydimethylsiloxane, PDMS) and a water-miscible lubricant (such as an FA) [275]. In contrast to dilute aqueous lubricants, the aforementioned system, applicable in low amounts, provides drier lubrication of the conveyors and containers. For practical application of 'dry lube' technology with PET bottles, acceptable PET compatibility of the lubricant composition is required [275]. Some silicone emulsions do not induce stress cracking in PET containers [276].

Table 12.31 Specific requirements for conveyor lubricants in the food and beverage industries

- Excellent drag coefficient
- Compatibility with printing inks
- Low stress cracking of polymeric materials
- Control of microbiological growth
- Detergency for removal of spillages

In the past, in beverage plants in the food industry conveyor chain lubricants have been used which were mostly based on FAs in the form of their water-soluble alkali or alkanolamine salts or on fatty amines, preferably as their (in)organic salts. While both classes of substances can be used without any problems in splash lubrication, they display disadvantages in the central chain lubrication systems that are conventionally used today. In US Patent Appl. No. 2008/0108532 A1 (to Ecolab, Inc.) Küpper *et al.* [277] disclose oil-in-water (O/W) emulsions for lubricating conveyor belt systems in the food industry, which are stable in storage. The disclosed O/W emulsions consist of up to 50 wt% of at least one wax ester $R^1CO–OR^2$ with R^1CO being a saturated or unsaturated acyl residue (C_6–C_{22}) and R^2 an alkyl or alkenyl residue (C_6–C_{22}) and other components such as triglycerides, partial glycerides and/or fatty alcohol polyglycol ethers.

Drive chains combine the flexibility of a belt drive with the positive action of a gear drive. Roller chains have several moving parts and require periodic lubrication. The most important considerations in chain lubrication are boundary lubrication and corrosion. Chain life can be extended through the proper selection and application of lubricant for the operating conditions. Requirements to chain lubricants are good adhesion to the chains, good lubricating properties and, if necessary, easy removal from the chains of films formed. Chain lubricants should be made available in formulations stable on storage. Typical chain lubricants include light general-purpose oils, turbine oils, gear oils and penetrating fluids. When the potential for environmental contamination or pollution is a major concern, a food-grade lubricant may be required to prevent contamination of ground, water and air. Chain lubricants may require special formulation or incorporation of multiple lubricants to cope with severe operating conditions including wet, dusty and gritty environments. Most chains, such as those used on conveyors, transporters and hoists, are accessible and easily lubricated while in service. Lubrication of these chains can be accomplished by various methods, including oil baths, brushing, dipping, pump, slinger, oil mist or spray application. The preferred method depends on the operating conditions such as load, speed and size, and whether the components are exposed or enclosed. Non-toxic H1 food-grade Lexolube® (Inolex Chemical Co., Philadelphia, PA) is a high-performance POE-based oven chain lube for temperatures up to 260 °C. Bechem Plantfluid® (Carl Bechem GmbH, Hagen) is a food-grade (NSF H1) native chain oil for bakeries, suitable for oven temperatures up to 290 °C.

Food manufacturing and service applications perfectly suit biolubricants based on vegetable oil chemistry, which are natural, refined or modified renewable products. For these lubricants incidental contact with food is less of a problem. Consequently, quite a large number of lubricant base oils meet regulators' requirements for food-contact applications. Details can be obtained from lubricant suppliers. Cold-temperature performance is also of little concern as most of the food processing is carried out indoors. Biodegradable food-grade palm-derived base fluids are designed for the formulation of a wide range of lubricants for different applications [278]. Purity™ FG food-grade lubricants (Petro-Canada Lubricants, Inc.; Mississauga, ON) are inherently readily biodegradable. Q8Oils (Kuwait Petroleum) offers 16 food-grade lubricants.

Food-grade lubricants have been developed for the lubrication, load transmission or corrosion protection of machines for the production and treatment of food, food-related material, tobacco manufacture, cane sugar production and animal food. Several patents describe food-grade lubricant compositions for equipment in the food service industry, usually based on chemically or genetically modified vegetable oils [58, 279]. The reaction between FAs and glycerol or propylene glycol results in lubricants recommended for food-contact applications. In US Patent No. 5,185,091 to Nippon Oil Company, Ltd [280] Ogake *et al.* describe a food-grade lubricant for food-processing machines, composed of a mixture of an FA ester of polyglycerol (preferably decaglyceryl monostearate) as a hydrophilic surface active agent, glycerol and medium-chain (C_6–C_{10}) saturated triglycerides (MCTs) to ensure smoothness. The maximum degree of polymerisation of glycerol in polyglycerol esters of FAs permitted by law as food additives is ten. The hydrophilic product is substantially insoluble in triglyceride oils. The most important characteristic of the greasy composition is that it does not flow down during the service period (dropping point of 80 °C or more).

US Patent No. 5,380,469 to Calgene Chemical, Inc. [281] teaches that the viscosity profile and solubility of biodegradable functional fluids may be increased by reacting polyglycerols with FAs (or their methyl esters), triglycerides (preferably canola oil), and/or blending the resulting esterified polyglycerols with natural fats and oils. A functional fluid is any material that may be used in any working device, hydraulic, crankcase or lubricating fluid. Polyglycerols (for $n \geq 8$) are highly viscous and highly polar, and therefore oil insoluble. Esterification with FAs or transesterification with methyl esters of FAs or triglycerides alters their polarity, hydrophilic–lipophilic characteristics and improve solubility of the polyglycerols in oil. As the polyglycerol chain length increases, viscosity increases and the hydroxyl value decreases. Controlled polymerisation is a means of increasing the viscosity of the fluid (from 1000 to 3000 cSt at 25 °C) and changes oil solubility. Esterification can take place at any or all of the hydroxy groups on the polyglycerol chain (usually at least 25%) depending upon the reaction conditions. The saponification value of the reaction products typically varies between 100 and 170 mg KOH/g. The reaction products can be used as high-viscosity lubricants or blended with triglyceride-based functional fluids as completely biodegradable products for use in mud drilling, as low-temperature lubricants, cutting fluids, chainsaw fluids, hydraulic fluids, H1 food-grade lubricants or as additives for other functional fluids.

In US Patent No. 5,538,654 to Lubrizol Corp. [279] Lawate *et al.* describe environmentally friendly food-grade lubricants from edible GM triglycerides containing FDA approved additives, useful as hydraulic oil, gear oil and compressor oil. A 99% biodegradable BioBlend (Joliet, IL) vegetable-based food-grade lubricant blended in compliance with 21 CFR §178.3570 from temperature stable high-oleic canola oil is recommended for use with gears, bearings and chains in facilities under the federal FSIS [156].

Bio food-grade lubricants are a specialty market for soy-based lubricants. The use of soy oil is exceptionally useful in industrial equipment applications that may come into direct contact with food. Renewable Lubricants, Inc. (Hartville, OH) produces several food-grade lubricants for food-safe use in applications such as hydraulic fluid and gear oil. Methyl soyate is an FDA-accepted incidental food-contact product when used in rolling metal foils. Greases meeting food-grade standards do not necessarily meet any of the standards for biodegradability or toxicity.

Whereas polymer-based products have traditionally provided adequate lubrication in sugar mill rolling, Faci *et al.* [282] have recently described a cost-effective bio-based lubricant with excellent physical and performance characteristics in a severe application such as that in cane sugar mills.

A standard practice used by cooks to promote the release of baked products from cookware is to apply a coating of a solid or liquid fat or oil to the inside of the pan, and then to dust the coating with flour. Formulas and products have been developed for applying pre-mixed compositions by aerosol or pump-type sprays. In US Patent No. 3,821,007 Carey describes a non-aqueous sprayable pan release material based on vegetable oil, lecithin, soft wheat flour and high fat soyflour [283]. Also freeze–thaw stable W/O lecithin emulsions have been proposed for use as cooking lubricants [284]. US Patent No. 4,425,164 discloses an aerosol cookware lubricant composition, which comprises a vegetable oil solution of an emulsifier (e.g. lecithin) in admixture with a hydrocarbon propellant (such as isobutane) [285].

12.11 Biodegradable lubricating greases

Lubricating greases represent about 9% of total industrial lubricants and 3% of all lubricant sales. Grease is a solid to semifluid product of dispersion of a thickening agent in a liquid lubricant [286, 287]. Other ingredients (additives) imparting special properties may be included. Lubricating greases owe their consistency to a gel-forming network. The liquid medium is usually a lubricating oil that provides lubricant features to the system. The thickener is primarily a bodying or gelling agent. The function of a thickener is to constitute a kind of three-dimensional network that will hold the fluid in place in order for it to execute its lubricating properties. The fluid can be mineral, synthetic or vegetable oil. The most common gelling agents are FA soaps of lithium, calcium, sodium and aluminium. These FAs are generally a mixture of saturated, mono- and polyunsaturated moieties.

The function of grease is to remain in contact with and lubricate moving surfaces without leaking out under gravity or centrifugal action nor be squeezed out under pressure. Its major practical requirement is that it must retain its properties under shear at temperatures to which it may be subjected to during use. In grease application, proper grease gun practices are important. A grease gun can deliver pressures up to 15 000 psi, or about 30 times what a typical bearing seal can handle. Grease must be able to flow into the bearing through grease guns and from spot to spot in the lubricated machinery as needed, but must not add significantly to the power required to operate the machine, particularly at start-up. These conditions represent an exacting set of rheological requirements. Grease has been described as a temperature-regulated feeding medium: when the lubricant film between wearing surfaces thins, the resulting heat softens the adjacent grease, which expands and releases oil to restore film thickness.

The base oil selected in formulating a grease should have the same characteristics as if the equipment is to be lubricated by oil. For instance, lower-viscosity base oils are used for grease applications at lower temperatures or high speeds and light loads. Pre-industrial grease-like lubricants in the form of mutton or beef fat (tallow), vegetable oils (such as olive oil) and their reaction products with lime, were used already as early as 1400 BC by the Egyptians to reduce axle friction in chariots. These ecologically compatible lubricants were rapidly biodegradable. The semi-solid consistency of fat allowed it to remain in place longer than liquids such as vegetable oils. Wool grease (lanolin) has been used as a lubricating grease, as a concrete mould lubricant and in rust prevention. Waxes of animal origin tend to be replaced by vegetal waxes and these in turn by much cheaper synthetic waxes. Since oil discovery, mineral oils and synthetic base oils have been used to formulate greases. For petroleum oils in general, naphthenic oils tend to chemically mix better with soaps and additives and form stronger structures than paraffinic oils. The mineral oil in traditional mineral oil greases evaporates at temperatures above 177 °C (350°F), leading to hardening. Higher evaporation rates determine the need for more frequent relubrication. Synthetic oils are more effective than mineral oils in high- and low-temperature operations.

Greases generally contain from 65 to 95 wt% base oils, from 5 to 35 wt% thickeners and up to 10 wt% additives. For greases usually oils with kinematic viscosities from 15 to 1500 mm^2/s at 40 °C are used. High-molecular-weight esters made from diols or polyols and dicarboxylic acids with viscosities > 2000 mm^2/s at 40 °C are used as base oils in greases that must be solvent resistant. Dispersed in the base oil, a soap thickener gives grease its physical character. Soap thickeners provide consistency to grease, and affect properties such as water and heat resistance and pumpability. The soap influences how a grease will flow, change shape and age as it is mechanically worked. The principal ingredients of a soap are the FA (as derived from soybean, corn, canola/rapeseed, sunflower, olive, castor, palm or peanut oils) and an alkali (lithium, sodium, calcium and aluminium). Incompatible grease mixtures should be avoided. Bio-based greases may be processed using microwave technology [53].

Lithium greases are the most widely used but calcium, aluminium, polyurea, and sodium-calcium are also used. Traditional petroleum-based lithium soap grease contains lithium 12-hydroxystearate (dropping point 204 °C, max. operating temperature 135 °C). The high temperatures generated by modern equipment necessitate an increase in the heat resistance of normal soap-thickened greases. As a result, complex soap greases have been developed. A complexing agent made from a salt of the named metal is the additional ingredient in forming a complex grease. A complex soap is formed by the reaction of an FA and alkali to form a soap, and the simultaneous reaction of the alkali with a short-chain organic or inorganic acid to form a metallic salt, the complexing agent. Common organic acids are acetic or lactic acid, and common inorganic acids are carbonates or chlorides. Lithium complex grease (dropping point 260 °C) is true multipurpose grade.

In order to improve characteristics such as anti-corrosion, anti-wear, etc., in greases, chemical additives (VI improvers, rust and oxidation inhibitors, AW/EP agents, detergents, PP depressants, metal deactivators) are commonly used [286, 287]. It is important that the additives are compatible with the thickener. Surface protecting and performance enhancing additives can effectively improve the overall performance of a grease. Solid lubricants (e.g. MoS_2, graphite) are added to grease in certain applications for high-temperature and high-pressure applications. The use of MoS_2 as the EP additive can be minimised by developing new grease chemistry based on the synergistic interaction between an organothiophosphate (OTP) and fluoropolymers to address the wear and EP behaviour under intermediate and extreme loading conditions [288]. Alcell™ organosolv lignin (Repap Enterprises, Inc.; Stamford, CT) is a multifunctional grease additive (antioxidant and AW/EP functions) [289]. Use of additives either in base oils or in thickeners also offers several disadvantages. Thus, there is interest in developing additive-free (eco-friendly) greases that should have at least equal (or better) tribological results compared with commercially available greases containing additives [290]. The following types of greases are commonly made from either petroleum or renewable vegetable base oils:

calcium, sodium, aluminium, polyurea and organo-clay greases. Polyurea is the most important organic non-soap thickener; organo-clay is the most commonly used inorganic thickener. Shell Stamina® are polyurea greases.

Greases (multipurpose, food-grade, rail-track, chassis, truck, etc.) are qualitatively classified by penetration number (NLGI Grade) (see Table 3.7) and the type of soap or other thickener. Viscosity of the oil included in the grease also needs to be considered in selecting greases for certain applications.

Grease structure can best be visualised as a 3D network of soap fibres. The structure will flow under an applied stress, the magnitude of which depends on the rigidity of the soap fibre network, which in turn is governed by the forces holding the fibres together [291]. Soap thickeners provide rigidity to grease and affect properties such as water, heat resistance and pumpability. They lower the coefficient of friction over that of the base oil alone [292]. Composition and processing conditions of greases greatly affect their microstructures and rheology. The evolution of good tribological characteristics in greases strongly depends on the process variables, viz. preparation temperature, cooling temperature, and homogenisation treatment, etc., during grease synthesis.

The functional properties of a grease include holding solid lubricants in suspension and functioning as a sealant to minimise leakage and keep out contaminants. Greases are easier to contain than oil, with less costly retention devices. There is generally no need for control and monitoring of fluid levels.

Table 12.32 lists the main characteristics of greases. It is necessary to distinguish viscosity of grease and oil. Viscosity of grease, more appropri-

Table 12.32 Grease characteristics

- Grease classification (NLGI)
- Consistency, penetration (ASTM D 217, ASTM D 1403)
- Apparent viscosity (ASTM D 1092)
- Pumpability, feedability
- Bleeding (migration, syneresis)
- Corrosion and rust resistance
- Dropping point
- Volatility
- Frictional wear
- Oxidation stability
- Shear stability
- High temperature effects (oxidation, softening, bleeding)
- Low temperature effects
- Texture
- Water resistance

ately referred to as apparent viscosity, holds only for the shear rate and the temperature at which it was determined. At start-up, grease has a resistance to motion (i.e. high viscosity). However, as grease is sheared between wearing surfaces and moves faster, its resistance to flow diminishes. In other words, the viscosity decreases as the shear rate increases. By contrast, an oil at constant temperature would have the same viscosity at start-up and at full motion.

The most important feature of a grease is its rigidity, or consistency (resistance to deformation by an applied force). Consistency depends on the apparent viscosity of the base oil and the type and amount of thickener used. The measure of consistency is called penetration (ASTM D 217, ASTM D 1403). NLGI has established consistency numbers or grade numbers, corresponding to specified ranges of penetration numbers (see Table 3.7). Pumpability is the ability of a grease to be pumped or pushed through a system. Feedability refers to the ability of a grease to be drawn or sucked into a pump. Fibrous textured greases tend to have good feedability, but poor pumpability, while butter textured greases tend to have good pumpability but poor feedability.

Bleeding is a condition when the liquid lubricant separates from the thickener. This may occur at high temperatures or after long storage periods. Migration is a form of bleeding under certain circumstances (e.g. pumping through a pipe). Syneresis is a special form of bleeding caused by shrinking or structural rearrangement due to physicochemical changes in the thickener. The dropping point is an indicator of the heat resistance of greases and denotes the temperature at which a grease becomes fluid enough to drip. High temperatures harm greases more than they harm oils as greases cannot dissipate heat by convection like a circulating oil. Effective grease lubrication under high-temperature operation is a function of the grease consistency. The base oil's PP is considered the low temperature limit of a grease. The natural corrosion and rust resistance of greases depend upon the thickener type and can be extended by additives.

The lubricating properties of greases are significantly affected by the base oil and type of thickeners used. Greases are applied when it is not practical or convenient to use oil. They are used instead of oils because they stay put, with or without seals. They reduce dripping and spattering, require less frequent application, resist the washing action of water, and seal out contaminants. They can prolong the life of worn parts and provide a better mechanical cushion for extreme conditions (shock loading, reversing operations, low and high speeds). Because of the semi-solid nature of grease, it does not perform the cooling and cleaning functions of a fluid lubricant. Grease is generally applied for machinery that is used intermittently or in storage for long periods. Further, greases provide good lubrication for machinery that is not easily accessible for frequent lubrication and for those that operate under severe conditions such as high temperatures, high pressures or low speed. Grease can maintain thicker films in clearances enlarged by wear and thereby provide an opportunity to extend the lifespan of such equipment [154]. Greases used to lubricate open gear sets or other slow-moving applications cannot be used in speed reducers. Speed reducer gears run at relatively high speeds and require an oil bath to ensure proper lubrication of the fatty contacting gear tooth surfaces. Greases designed to extend bearing life offer the advantages of reduced grease consumption and waste, less frequent re-greasing and lower environmental impact. Some disadvantages of greases are: (i) poor heat dissipation by convection; (ii) resistance to motion (at start-up); and (iii) difficult handling (dispensing, draining, refilling and metering). Greases are not appropriate for low-torque, high-speed operations.

A grease's environmental acceptance has high priority. The biodegradability of greases essentially reflects that of their base oils. Based on biocompatible fluids, lubricating greases can be formulated covering the requirements regarding a fast biodegradability. Reduction of the toxic potential of greases requires elimination of polycyclic hydrocarbons from base oils and chlorine substitution. Most of the thickeners and some of the additional compounds added to impart other desired properties have remained ecologically acceptable.

A grease is called synthetic when the base oil is not a mineral oil but a synthetic product. The term fully synthetic grease refers to a product in which also the thickener is synthetic. Thickeners not only transform liquid lubricants into consistent lubricants, but also change the properties of these formulations. In general, a maximum thickening effect is achieved with C_{18} carboxylic acid derivatives, such as soaps from vegetable-derived 12-hydroxystearic acid or its esters [293]. Hydroxystearic acid, the hydrolysis product of hydrogenated castor oil, is used in the manufacture of multipurpose lithium and calcium greases [294, 295]. RSO has been introduced in Central Europe as a base oil for greases on a technical scale, mainly for loss-lubrication applications [296]. Initially calcium 12-hydroxystearate was chosen as the most suitable thickener [297]. Greases based on RSO and calcium 12-hydroxystearate show biodegradability of close to 100%. The limitations of vegetable oil-based greases have led to the introduction of transesterified vegetable oils [298] and synthetic esters. Soybean-based greases improved to withstand high reaction temperatures (>200 °C) have been developed [299]. Conventional grease performance can be exceeded at approximately the same price as petroleum-based greases.

French Patent No. FR 2,873,712 to Grignou *et al.* (to Christol Grease S.p.A.) [300] describes a lubricating grease comprising an epoxidised vegetable oil in admixture to a mineral, vegetable or synthetic oil. The use of FAs from soapstock as a thickener for multigrade lubricating grease has been investigated by El-Adly [301]. Low-cost cottonseed soapstock (IV, 60 g I_2/hg; SN, 198 mg KOH/g; C16:0, 27%; C18:0, 2%; C18:1, 29%; C18:2, 42%) and unrefined animal bone fat (IV, 45 g I_2/hg; SN, 180 mg KOH/g; C14:0, 9%; C16:0, 23%; C18:0, 15%; C18:1, 48%; C18:2, 4%) are good thickeners for the preparation of moderate and light-duty greases, respectively. Split soapstock sunflower oleins have also been used for estolide formation [302]; see Section 6.2.2. Oleogels, prepared by dispersing sorbitan monostearate (SMS) in castor oil, are potentially suitable as biodegradable lubricating greases [303]. Such oleogel samples exhibit poor mechanical stability after sample working in a rolling element, but completely recover their initial consistency after a moderate resting time, also exhibiting remarkable lubricant properties. However, their friction coefficients are inferior to those of model lithium lubricating greases.

Experiences with biodegradable greases are promising [304]. Branched synthetic polyol ester base stocks can be used in the formulation of biodegradable greases together with selected lubricant additives [6]. A biodegradable grease composition with excellent EP and thermal oxidation stability properties has been described, consisting of at least a polyol ester or complex ester as a base oil, and a lithium soap based thickener [305]. Application of greases in both an environmentally responsible and economical way requires extension of lifetime and minimisation of the amount of lubricant necessary. BIOMON (FP6 programme) has aimed at developing long-life biolubricants based on high-performance native esters (TMP, PE) and high-temperature and oxidation-resistant biodegradable greases based on polyurea thickeners (methyl diisocyanate types) mixed with native esters. Two additive-free calcium stearate greases, prepared by using dry technology [306] and developed for the requirements of lubrication in automotive components, show comparable tribological behaviour with commercial greases containing additives [290].

Dresel has described properties of vegetable oil (mainly RSO)-based lubricating greases (biodegradability, lubricity and oxidative stability), comparison with standard and high-performance lubricating greases, fields of application and markets [297]. Lubricating greases have major impact in the transportation (automotive, railroad), rugged industry (e.g. steel mill), farming, aircraft and military sectors. The main consumer market for grease is the automotive market, with about 50%. Greases find applications in automotive wheel bearings, in ball and roller bearings in electric motors, and in machine tools. An SNO-based grease for heavy-duty applications (earth moving equipment) with better properties than mineral oil-based grease has been developed [307]. Soybean-based greases tend to oxidise rapidly along with viscosity increase. The SBO can be stabilised through use of antioxidants or can be genetically improved to withstand high reaction temperatures (>200 °C) required for common thickeners. Stabilised SBO can produce greases for long shelf-life and stable use [299]. NABL (University of Northern Iowa) has developed a soy-based 'lost-in-use' grease that lubricates the slide contact between the rail flange and railroad track. The product exceeds performance of most conventional greases at economical prices approaching petroleum-based greases. NABL has also developed a soy-based Fifth Wheel Grease for lubrication of the contact plate between a trailer and a semi-truck, which exceeds conventional grease performance at the same price of petroleum-based greases. Vibroacoustics Solutions, Inc. (Ames, IA) offers a truck lubricant made from corn, soybeans and graphite, called 5th Wheel LubePaste. Another (minor) application is a soy-based cotton picker spindle grease (NABL).

Among industrial grease consumers, the roller bearing industry is important [308]. Greases are also used for the lubrication of small gear drives and many slow-speed sliding applications. An environmentally friendly palm grease has been formulated from epoxy RBDPO (refined, bleached, deodorised palm oil) as base oil and lithium soap as thickener [309]. This palm grease for general use in bearings, gears and other mechanical systems requiring grease lubrication in environmentally sensitive areas such as agriculture, forestry and coastal marine, recreation areas and others, provides better anti-wear properties than mineral (HVI 160S) grease. This has been ascribed to the relatively polar groups in the base oil (hydroxide, ester, oxirane). US Patent Appl. No. 2009/029881 A1 to Okamura *et al.* [310] claims a heat-resistant noise-reduction lithium grease composition comprising 5–20 wt% of a C_{26}–C_{32} long-chain fatty acid lithium soap as a thickener and 80–95 wt% of a comb shaped polymer ester oil as the base oil.

Compared with lubricating oils, greases are not the right choice when heat transfer is a problem; they also have a lower speed limit in view of their higher effective viscosity. Good multipurpose properties, e.g. high dropping point, good water resistance and good shear stability – which has been related to hydrogen-bonding of the hydroxyl groups – and good response to additives favour use of 12-hydroxystearate-based greases (as lithium soaps). Such greases are found in a wide range of applications as EP greases for heavy loads (oil basis with kinematic viscosities of about 200 to 1000 mm^2/s at 40 °C), multipurpose greases for bearings (mineral oil basis with kinematic viscosities from 60 to 120 mm^2/s at 40 °C) and greases for high speeds (diester or PAO oil basis with kinematic viscosities of 15 to 30 mm^2/s at 40 °C). The upper temperature limit of application lies between 120 and 150 °C. The performance of lubricating greases is described by ISO 6743–9, and for automotive greases by ASTM D 4950.

Biodegradability is vitally important to the rail industry. Locomotive gear oils or gear greases, wheel flange lubrication, switch lubrication and rail track-side lubrication cause environmental problems. For these applications biodegradable greases based on esters have better wear protection and consumption performance [304]. Biodegradable (bear-friendly) grease for advanced rail track-side lubrication in Canada proved its superior load-carrying and wear-protection performance [20]. R.S. Clare & Co., Ltd (Liverpool, UK), a European market leader for the supply of advanced rail track grease, has a range of approved biodegradable lubricants for all track applications and mobile and portable plants. The B1099 rail lubricant is a biodegradable grease for track-side. In Germany, products used in switches have been completely replaced by ester-based biodegradable greases. Trials with biodegradable wheel-flange greases have been successful [311]. For the specific application in railroad grease, vegetable-based products cost less (US$ 1.50/lb) than petroleum-based grease (US$ 1.75/lb) [86]. However, vegetable-based lubricants cost much more for other applications. Aircraft and military applications, which are beyond the capabilities of petroleum-based greases, utilise synthetic greases, which can sustain extreme temperatures (from –55 to 315 °C).

Several EU-funded and co-financed research projects such as IBIOLAB, VOSULUB and BIOVESIN have aimed at developing biogreases from renewable resources. BIOGREASE (FP5 programme, 2001–2004) has formulated new environmentally friendly multifunctional greases based on renewable resources and advanced thickener technology for use in loss lubrication earth-moving equipment. Agro Management Group, Inc. (Colorado Springs, CO) has 18 bio-based lubricants in its product line, ranging from hydraulic fluids to food-grade grease. Castrol's BioTac EP2 grease, based on an SE and a lithium/calcium soap as a thickener, is a multipurpose lubricant formulated with a readily biodegradable base oil. The product provides excellent lubrication over a wide range of working conditions (from –35 °C to 120 °C). Its biodegradability is of particular benefit where there is a risk of environmental contamination due to leakage in total-loss lubrication systems. BioTac EP2 grease, with very good EP protection, is suitable for both rolling and plain bearings of certain engine room equipment and can also be used for wire-rope lubrication. BioTac EP2 offers excellent resistance against water spray off, reducing the need for frequent application.

Bio EP Greases (RLI; Hartville, OH) are mixtures of a base oil (HOBS), a thickener (commonly soaps and organic or inorganic non-soap thickeners) and additives. The corrosion resistance of Bio EP Greases is enhanced by bio-corrosion inhibitors (BCI™). RLI's biodegradable greases are low-volatility greases, i.e. having higher flash points and allowing higher operating temperatures. Multipurpose Bio EP Greases formulated with calcium sulphonate have inherent EP properties and may be used in H1 food-grade applications. The NLGI classifications for Bio Multi-Purpose High Temp EP Grease and Bio Graphite EP Grease are NLGI 2 and NLGI 1, respectively.

RLI's Bio EP Greases work efficiently in extreme hot and cold applications. The high VI of the HOBS base oil naturally lowers heat build up by reducing friction and wear. Bio EP Greases provide exceptional performance over a wide temperature range of –30 to 280 °C. Stabilized™ HOBS technology prevents oxidation or chemical breakdown of Bio EP Grease formulations, thus avoiding deposit formation; the greases will continue to function under prolonged high-temperature exposure. Bio EP Greases are engineered to outperform petroleum-based EP greases and synthetics in terms of natural lubricity, shear stability, oxidation stability, TAN and load-carrying capacity, and have the added advantage of lower cost. Enhanced chemical stability ensures long-lasting performance.

Bio EP Greases are generally used for:

- machinery that runs intermittently or is in storage for extended periods of time;
- machinery that is not easily accessible for frequent lubrication;
- machinery operating under extreme conditions such as high $p, T,$ shock loads, or slow speed under heavy load;
- worn components.

High quality Bio EP Greases can lubricate isolated or relatively inaccessible components for extended periods of time without frequent replenishing. Because Bio EP Greases remain in place, a lubricating film can form instantly. Under extreme conditions oil films can be too thin and can rupture, whereas Bio EP Greases provide thicker film cushions. Bio EP Greases can extend the life of worn parts that were previously oil lubricated. Thicker grease films also provide noise insulation.

NABL (National Ag-based Lubricants Center, Cedar Falls, IA) has developed a food-grade soy grease based on a biodegradable aluminium complex. Bio-Food Grade EP Greases (RLI) are premium-grade aluminium complex bio-based food-grade greases designed to meet the more stringent demands for food, beverage, pharmaceutical, and cosmetic processing plants. These high-performance greases are engineered to withstand the moisture, high temperature (280 °C), and high speeds of advanced processing machinery. The maximum usable temperature is almost 100 °C higher than aluminium soap greases. A calcium complex grease, prepared by adding calcium acetate, provides a product with extreme pressure characteristics without using an additive (dropping point 280 °C, maximum usable temperature 177 °C). As yet, the majority of greases on the market are composed of mineral oil blended with a soap thickener. These greases lack many of the advanced performance properties and qualities inherent in Bio EP Greases. Table 13.6 is a bio grease database for the US market.

As engine and equipment performance requirements are constantly changing and pushing boundaries Shell Lubricants has recently launched an improved range of its Gadus multipurpose greases for extreme conditions (of temperatures, pressures, load-carrying capabilities and sliding speeds) [312]. Shell is also marketing Shell Naturelle Grease EP conforming to the highest environmental standards [12]. Lubrizol® 5288 is a multifunctional grease additive based on less harmful chemistry that meets or exceeds the NLGI GC-LB automotive service grease specification and the DIN FE9 High Temperature Bearing Grease Oxidation Life Test. Lubricating grease is a critical design component in the lubrication of rolling element bearings for both industrial and automotive applications.

Important development topics in the European grease market are railway lubricants, bio-based greases, grease and environment, food-industry lubrication, aerospace greases, grease shelf-life, grease cleanliness, test methods and REACH. Grease life testing may be carried out according to SKF's R0F+ methodology [313]. More efforts should be addressed to biodegradability of thickening systems in greases. ELGI (European Lubricating Grease Institute, Amsterdam) is exploring the formation of a Biobased Performance Working Group to coordinate resources for increasing knowledge and awareness of the performance of bio-based greases. The tribological behaviour of bio-greases, mainly in terms of rolling bearing wear and power loss, requires more detailed study [314]. For current developments and marketing opportunities of bio-based greases, see refs [315, 316].

12.12 Specialty lubricants

Besides the usual industrial and automotive lubricants various specialty products are on the market or are under development (Table 12.33). With the introduction of new highly performing materials for civilian and military needs lubrication science is regularly faced with new challenges. New tech-

Table 12.33 Specialty lubricants

- Low- and high-temperature lubricants (<–50 °C, >250 °C)
- Lubricants for reactive light metals and alloys
- Electrical contact lubricants
- Household appliance lubricants
- Lubrication of microsystems
- Valve lubricants
- Textile-processing lubricants
- Dry lubricants (e.g. anti-friction coatings)
- Noise-damping lubricants
- Vibration absorbers
- Plastics lubrication
- Food-grade lubricants
- Gas-expanded lubricants

nologies are being developed to solve problems ranging from traditional lubricant applications for classical materials to emerging uses in microelectromechanical systems [317] and in the disc drive industry [318].

Nanotribology and nanolubrication are developing fields [35]. The nanolubrication strategy is based on the direct feeding of the sliding interface with nanoparticles of tribo-active phases (e.g. layered WS_2 compounds, metal soap micelles, encapsulated particles) in stable dispersions of lubricant base oils [319–321]. Using composite or functionalised nanoparticles in dispersion in lubricating oils or greases the new lubrication strategies confer the needed functions to the lubricants. Nanoparticles are well adapted for a perfect feeding of the sliding interface and allow the possibility of design of composite colloidal particles with multifunctional properties (e.g. AW, corrosion inhibitor, CI). Colloidal physics and chemistry offer interesting possibilities for reducing friction and wear in various metallurgical contacts. The nanolubricant approach is to overcome the disadvantages of conventional AW and friction reduction additives related to their need of chemical reactions with substrates and the resulting induction period to produce the tribo-film on sliding surfaces.

Special lubricants may be needed in optical, fine mechanical and medical applications, in electrical contact lubrication of micro-switches, in miniature or high-precision bearings. Miniaturised devices and micro/nanoelectrical mechanical systems (MEMS/NEMS) with large surface-to-volume ratios, which are operated under very small normal loads, require thin lubrication films with thickness of a few tenth of nanometres. Ionic liquids (ILs) are a class of advanced lubricants which are being proposed for use in nanotribological MEMS/NEMS applications (see Chapter 15). The properties of ILs that make the difference with conventional lubricants are their excellent thermal and electrical conductivities. Electrical contact lubricants also meet the extreme requirements of household appliance technology and high voltages.

Most conventional lubricants are developed to meet specific applications but their properties cannot be altered if changes occur in the operation unless the lubricant is removed. Gas-expanded lubricants (GELs), binary mixtures of a lubricant and a dissolved gas (typically CO_2), provide the end-user with the ability to tune the physical properties of a lubricant in real time to meet changes in operating conditions [322]. In particular, slight changes in the CO_2 content can lead to huge changes in the viscosity of a gas-expanded lube. Synthetic base stocks such as PAGs, PAOs and POEs can be used to prepare GELs with CO_2, but mineral oils are not compatible with CO_2. Lubricants are necessary in rotating machinery to ensure separation between solid surfaces but they are also responsible for undesirable power losses that can occur through a bearing in power turbines. These losses can be reduced by the use of a binary mixture of a synthetic lubricant and dissolved liquid-phase CO_2 [323]. GELs increase the energy efficiency in power turbines. A 25% reduction in power loss is achievable using gas-expanded lubricants [322].

Commercial lubricants are not suitable for all materials in all conditions. Effective lubricants are under development for reactive light metals and alloys. Special lubricants are required for ultra-low-temperature systems (−100 °C). ILs may find application for high-temperature applications.

The task of sealing process plants tends to culminate in a highly complex tribological system. Apart from low friction and minimal wear the effect of the lubricant on its chemical environment (process media and seal material) is most important. In plastics lubrication above all compatibility with the used materials is crucial. Valve lubricants require special sealing performance or excellent sliding capacity. Lube&Seal is an initiative of the German Freudenberg Process Seals (Weinheim) and tribology expert Klüber Lubrication (Munich) to offer optimally integrated solutions for lubrication and sealing tasks in hygienic processes [324]. As a result, the reliability and service life of machines and plants is substantially improved. The relevance of the tribological systems for the function and service life of a production line is obvious.

Because lubricating greases could contaminate textiles, they are commonly formulated with white oils, and greases are thickened with water-soluble sodium soaps, which are thus removable by washing. HOSNO lubricant formulations (BIOVIT, BIOlubrificanti Vegetali per l'Industria Toscana) have potential of substituting mineral lubricants (mainly alkylbenzenes) which at present are widely used in textile processing (spinning phase of wool) and tannery processing [26]. Although economic analysis has indicated a higher production cost and price of such vegetable oil lubricants they offer prospects due to a lower utilisation rate. Inolex Chemical Company (Philadelphia, PA) markets various adipate and polyol ester textile lubricants (Lexolube® series). Lexolube® NBS, T-110 and B-109 are *n*-butyl stearate, 2-ethylhexyl stearate and tridecylstearate textile lubricants, respectively, whereas Lexolube® 3G-310 is a trimethylolpropane trioleate-based fire-resistant fluid. Also monoglycerides are used as lubricants in the textile industry.

12.13 Solid biodegradable lubricants

Solid lubricants are typically composed of a solid, a binder and additives such as corrosion inhibitors or solvents. Solid lubricants, which are used where it is important for the lubricant to stay in place, typically have a temperature range over which they are effective. Above the optimum range they may degrade chemically or physically. Organic biolubricants such as the cereal lubricants wheat flour or starch degrade at fairly low temperatures whereas graphite can withstand 650 °C and moderate loads. ASTM D 2714 provides a test method for determination of the friction coefficient of a solid lubricant.

Starch-based lubricants are well known (see Sections 6.4 and 6.4.1). An early (1875) US patent made reference to improvements to starch lubricants [325]. Lubricants based on cereal food materials (30% lipids, 65% carbohydrates, 3% proteins and 0.6% $CaCO_3$) have potential for practical applications and are non-polluting [326]. In US Patent Appl. No. 2006/0058583 A1 Matsumoto and Ito [327] disclose a solid lubricant for endoscopes consisting of a porous carbonaceous material formed by carbonisation of bran or similar plant material (e.g. rice bran, wheat bran, rice hulls, soybean husks, etc.). Yoshimura *et al.* [328] have disclosed a grain-type lubricant useful for press forming or extruding of metal strips composed of dextrin, glutens and grain powders containing protein and carbohydrates from wheat flour, sweet corn powder, rice flour or soybean flour in 1–30% aqueous suspension.

Old lubricants for coating wires for drawing operations consisted of wheat or rye flour paste or lime paste [329]. A method for processing of metals, cutting, grinding, drilling extruding, drawing, etc., comprises coating the material surface with lubricants containing grain powder composed of carbohydrates and protein as major components, e.g. flours of wheat, corn, rice, soybean [330]. Caminaga *et al.* [264] compared four lubricants (mineral oil, semi-synthetic oil, powder soap and wheat flour) and two lubrication methods (immersion and hydrostatic) for cold extrusion of steel shafts. For one deformation stage, wheat flour was the best lubricant.

Liquid suspensions containing 0.3–30% grain powder, such as wheat, corn, rice and soybean, have been applied as machine lubricants to decrease friction [331]. The lubrication by various cereal flours, such as wheat or rice flour in aqueous suspension (followed by coating and drying), as non-polluting and recyclable lubricants was studied for sheet metal-forming [332–334]. A deep drawing test showed that sheet coated with wheat flour has a higher deformability than a sheet coated with press oil. The lubrication effect was ascribed to wheat flour components such as gluten. Yoshimura and Okamua have also disclosed an environmentally friendly, aqueous lubricant containing sugar, wheat flour, corn flour, rice flour and/or potato starch for application on a (non)metal material in machining [335].

Starch has been applied as a binder in drilling muds; see Section 12.4.2. Refractory shutters have been lubricated with aqueous lubricants containing 30 wt% powdered graphite and 10 wt% starch glue (as a binder) [336]. Water-based lubricants for finishing cotton fibres and automobile interior sheet fabrics were disclosed, composed of wax (paraffin, animal or plant-derived) and composites of cationic starch and anionic polymers [337, 338]. Starch–oil composites may be used as ingredients for formulating solid lubricants based on renewable and abundantly available raw materials, which also have superior environmental and health characteristics [263].

Sizing mixtures for glass fibres, which give perfect coating with high adhesion to the fibre, commonly consist at least of binders (proteins, starch), lubricants and emulsifiers [339]. Composition, application methods and properties of aqueous starch-based lubricants for glass fibres were reviewed [340]. Sizing agents for glass fibres were described which contain starches, lubricants, softening agents and antiseptics, where the starch is a mixture comprising 30–70 wt% starch containing ≥ 60 wt% amylase and the balance cross-linking etherified starch containing ≤ 30 wt% amylase [341]. Starch-oil sizing compositions for glass fibres, giving improved fibre processability, contain hydroxypropylated high-amylose corn starch, highly cross-linked low-amylose corn starch, lubricants, coupling agents, and additives [342]. The presence of a lubricant (prepared from paraffin wax, tallow and a non-ionic emulsifier) in a corn starch sizing of a cotton–polyester blend yarn decreases the cohesive force of the size but does not change the chemical structure of the starch film [343]. Magnesium stearate dihydrate lubricant compositions are used to lubricate a variety of solid industrial or consumer products [344], in particular for pharmaceutical and nutraceutical applications.

12.14 Process oils

Process oils, often included in lubricant statistics, are oils which are used as raw materials in the manufacturing of diverse products like adhesives, explosives, fertilisers, PVC, but above all as plasticisers or extenders for the rubber industry. For these productions, manufacturers usually make use of the cheapest base oil on the market. Just like most lubricants, process oils are mineral oil products resulting from the refinery of base oils. Recently, the industry, for environmental reasons among others, has shifted from aromatic base oils to using naphthenic base oils for process oils. The demand for process oils corresponds to about 10% of the total volume of all lubricating oils, i.e. to a substantial volume of 3.6 Mt.

Process oils are applications for which vegetable-based oils could expand into by replacing existing mineral-based oils. If tyres were to be made using vegetable-based oils, this would result in great environmental gains due to the shift to a bio-based renewable source.

12.14.1 Lubricants for plastics processing

The macromolecular structure of polymers results in very high melt viscosities. Polymer melt is very complex. In practice, processing of plastics requires the appropriate selection of the combination of machine and mould as well as of lubricants that affect the rheology of the polymer melt in the desired way. The major function of lubricants in polymers is to decrease internal and external friction. This results in:

- reduced shear, or lower temperature and reduced tendency for polymer degradation;
- reduced equipment wear;
- increased production rates;
- reduced energy consumption.

Substances commonly used as lubricants not only affect melt rheology, but also surface structure and interfacial properties of fillers and pigments. Most plastics lubricants (slip agents, flow aids, plasticisers) are natural or (semi) synthetic waxes; most of the others are fat derivatives [345].

In polymer processing, the term lubricant is not used in a standard way. High-polymeric processing aids for PVC ('flow aids') overlap with lubricants in their effects, but are to be distinguished from them by a different mode of action. For polyolefin processing the term means an additive ('slip agent') that reduces the frictional resistance between the surfaces of finished parts at service temperature.

Different terms are used in connection with lubricants for polymer processing: internal lubricants are incorporated into the polymer, whereas external lubricants are applied directly onto the metal surfaces of the processing machine or the mould to prevent the polymer melt from sticking. Internal lubricants are mostly soluble in the polymer. They work 'internally' by facilitating ('lubricating') the movement of polymeric molecular chains past one another during processing. External (insoluble) lubricants work 'externally' by migrating to the surface of the polymer melt during processing providing a sliding film, a boundary layer, between the melt and the metal wall of the processing equipment. Most lubricants display some combination of both external and internal characteristics. The lubricant's stability is determined by molecular structure and its polarity in relation to the polymer. Total solubility in the polymer is undesirable. The fact that essentially all lubricants have both external and internal characteristics makes it impractical to describe lubricants by external vs internal classification. Instead, lubricants for polymer processing are best grouped by chemical class.

The effect of a lubricant depends on its concentration. This can be explained by the fact that even internal lubricants have a finite solubility limit beyond which undissolved portions may act as an external lubricant. The saturation concentration is temperature dependent.

Important lubricant product groups used in the plastics processing industry are listed in Table 12.34 in order of decreasing polarity [345]. The price of polymer processing lubricants is determined by the relatively low cost of

Table 12.34 Lubricant product groups for plastics processing

Chemical class	Structural characteristics[a,b]	Typical representatives
Alcohols	$R = C_{12} - C_{34}$	Fatty alcohols
Fatty acids	$R = C_{12} - C_{34}$	Stearic acid, tall oil fatty acids
Metallic soaps	$R = C_{12} - C_{34}$; M^{n+} ($n = 1 - 3$)	Calcium stearate, montan wax acid soaps
Amides ('amide waxes')	$R = C_{12} - C_{22}$	Oleamide, erucamide, ethylenebisstearamide (EBS)
Fatty acid esters	$R = C_{12} - C_{22}$	Oleic acid derivatives, glycerol monostearate (GMS)
Polyol esters	$R = C_{12} - C_{34}$; $n = 1 - 6$	Polyglyceryl esters, TMP stearic acid esters
Complex esters	$R = C_{12} - C_{34}$	Oligomeric fatty acid esters
Waxes	$R = C_{28} - C_{32}$; $R' = C_{28} - C_{32}$	Montan waxes
Polar waxes	$R = C_{35} - C_{70}$	Oxidised PE waxes
Polyolefin waxes	$R = C_{120} - C_{700}$	PP waxes
Paraffin waxes	$R = C_{20} - C_{70}$	Fischer–Tropsch (FT) waxes

[a] R, R′ = hydrocarbon chain.
[b] M = Al, Ca, Pb, Na, Sn, Zn.

the raw materials used: FAs originating from animal fats and vegetable oils, petroleum wax from crude oil, and montan wax from lignite. The main lubricant product groups are FA esters, hydrocarbon waxes, amide waxes, ester waxes and metal stearates, with an overriding demand of the PVC processing industry.

Natural waxes, such as carnauba wax and beeswax, generally are no longer of importance in the plastics industry. The raw materials for producing high-molecular-weight lubricants (e.g. polyolefin waxes) and montan waxes are obtained only from fossil resources. FAs and their derivatives are mostly produced from animal resources such as beef tallow. Spermaceti (from whales) has long been banned for industrial use.

Except for hydrocarbon waxes, all lubricants are interfacially active substances. Their polar centres determine their affinity, not only for polar macromolecules, but also for fillers and reinforcing materials, pigments, and metal surfaces. The lubricating effect actually originates from the non-polar, longer-chain aliphatic groups of a molecule (alkyl groups with a chain length $< C_{12}$ are usually ineffective). Paraffinic hydrocarbon waxes are classified as external lubricants in PVC processing. In non-polar polymers, the situation is reversed. Hydrocarbon waxes are readily soluble in PE, while polar esters are incompatible and hence, should be referred to as external lubricants.

Reduction of the adhesive force between polymer melts and solid surfaces is called the release effect. Mould-release effects are attributed to the demoulding forces required to eject a solidified part from the mould. The specific reduction of the demoulding force is a high priority target in many applications. Mould-release agents help to separate the moulding, i.e. the product, from its mould when it has been made. Internal release agents are used in moulding compounds, including injection moulded plastics. Lubricants with a positive release effect are usually chemically characterised by polar groups in the molecule. Zinc stearate is one of the commonest. Hydrocarbon waxes, which are incompatible with polar plastics and are classified as typical external lubricants, have only a moderate release effect. External mould-release agents are for use with thermosetting resins, mainly in open mould and low throughput operations. Hydrocarbon waxes and poly(vinylalcohol) can also be used.

Montan wax is a naturally occurring fossil vegetable wax which is found in lignite deposits and consists of esters of montanic acid (C_{28}–C_{32}) with long-chain aliphatic alcohols (C_{28}–C_{32}). Montan wax acid is the most important wax acid for lubricant production. PEG esters of fatty acids have been disclosed as lubricants in the processing of thermoplastics [346]. Rigid PVC (rPVC) processing is impossible without lubricants [347].

All lubricants added to plastic materials intended to come into contact with food are given special attention by government regulatory agencies (see Section 12.10). Conformity to EU Directive 2002/72/EC, which requires that the white oil must be a medicinal quality paraffinic oil, with a viscosity of not less than 8.5 cSt at 100 °C and an average molecular weight exceeding 480 Da, makes such oils acceptable as a process aid in the production of polymers for food packaging. Large volumes of white oils are used in the plastics processing industry, especially as polystyrene extender oil and in low-density polyethylene (LDPE) manufacturing [348]. Small volumes are also used as a die lubricant for extruding spaghetti.

12.15 Sources of further information and advice

K. Chipasa, *Best Practice Guide for the Disposal of Water-mix Metalworking Fluids*, UKLA, Berkhamsted (2011); http://www.ukla.org.uk

Carlube, *Commercial Vehicle Oil Guide*, Tetrosyl Ltd, Bury (UK) (2011); http://www.carlube.co.uk

Th. Mang, K. Bobzin and Th. Bartels, *Industrial Tribology: Tribosystems, Wear and Surface Engineering, Lubrication*, Wiley-VCH, Weinheim (2010).

R.M. Mortier, M.F. Fox and S.T. Orszulik (eds), *Chemistry and Technology of Lubricants*, 3rd edn, Springer, Heidelberg (2009).

J.P. Byers (ed.) *Metalworking Fluids*, 2nd edn, STLE/CRC Press, Park Ridge, IL/Boca Raton, FL (2009).

H.P. Bloch, *Practical Lubrication for Industrial Facilities*, The Fairmont Press, Inc., Lilburn, GA (2009).

J.S. Marth, *Renewable Lubricants Manual: Biobased Oils, Fluids & Greases*, United Bio Lube, Palo Alto, CA (2007); http://www.renewablelubricants.com

Th. Mang and W. Dresel (eds), *Lubricants and Lubrication*, 2nd edn, Wiley-VCH, Weinheim (2007).

T. Dettmer, *Nichtwassermischbare Kühlschmierstoffe auf Basis nachwachsender Rohstoffe*, Vulkan-Verlag, Essen (2006).

L.R. Rudnick (ed.), *Synthetics, Mineral Oils, and Bio-based Lubricants: Chemistry and Technology*, CRC Press/Taylor & Francis, Boca Raton, FL/New York, NY (2006).

G.E. Totten, *Handbook of Lubrication and Tribology: Application and Maintenance*, CRC Press, Boca Raton, FL (2006).

National Association of Lubricating Grease Manufacturers, *Lubricating Grease*, NLGI, Kansas City, MO (2006).

A. Igartua, J. Barriga and A. Aranzabe (eds), *Biodegradable Lubricants*, Virtual Tribology Institute Edition, Eibar (2005).

R.F. Haycock, A.J. Caines and J.E. Hillier, *Automotive Lubricants Reference Book*, 2nd edn, John Wiley & Sons, Ltd, Chichester (2005).

S.Z. Erhan (ed.), *Industrial Uses of Vegetable Oils*, AOCS Press, Urbana, IL (2005).

J.-L. Ligier, *Avaries en Lubrification,* Éditions Technip/IFP, Paris/Rueil-Malmaison (2004).

S.Z. Erhan and J.M. Perez (eds), *Biobased Industrial Fluids and Lubricant*s, AOCS Press, Champaign, IL (2003).

G.E. Totten, S.R. Westbrook and R.J. Shah (eds), *Fuels and Lubricants Handbook: Technology, Properties, Performance, and Testing*, ASTM International, West Conshohocken, PA (2003).

V. Štěpina and V. Veselý, *Lubricants and Special Fluids,* Elsevier, Amsterdam (2002).

W.R. Herguth and T.M. Warne, *Turbine Lubrication in the 21st Century,* ASTM International, West Conshohocken, PA (2001).

J. Denis, J. Briant and J.C. Hipeaux, *Lubricant Properties Analysis and Testing*, Éditions Technip, Paris (2000).

L.R. Rudnick and R.L. Shubkin (eds), *Synthetic Lubricants and High-performance Functional Fluids,* 2nd edn, Marcel Dekker, Inc., New York, NY (1999).

M.J. Neale, *Lubrication, A Tribology Handbook,* Butterworth-Heinemann, Ltd, Oxford (1999).

Fachagentur Nachwachsende Rohstoffe (ed.), *Chemische Nutzung heimischer Pflanzenöle*, Landwirtschaftsverlag, Münster (1998).

E.R. Booser (ed.), *Tribology Data Handbook,* CRC Press, Boca Raton, FL (1997).

A.J. Caines and R.F. Haycock, *Automotive Lubricants Reference Book,* 2nd edn, SAE, Inc., Warrendale, PA (1996).

K.E. Bannister, *Lubrication for Industry*, Industrial Press, Inc., New York, NY (1996).

Schmierstoffe und Steuerflüssigkeiten, VDEW Ölbuch, VDEW Verlag, Frankfurt a. Main (1996), Band I.

W.J. Bartz *et al., Hydraulikflüssigkeiten – Eigenschaften, Normung und Prüfung, Anwendung*, Expert-Verlag GmbH, Renningen (1995).

A.R. Lansdown, *High Temperature Lubrication*, Mechanical Engineering Publications, London (1994).

W.J. Bartz (ed.), *Biologically Fast Degradable Lubricants and Operational Fluids*, Expert-Verlag GmbH, Ehningen (1993).

O. Pigors, *Werkstoffe in der Tribotechnik*, Deutscher Verlag für Grundstoffindustrie, Leipzig-Stuttgart (1993).

U.J. Möller and U. Boor, *Schmierstoffe im Betrieb,* VDI-Verlag GmbH, Düsseldorf (1986).

E.S. Nachtman and S. Kalpakjian, *Lubricants and Lubrication in Metalworking Operations*, CRC Press, Boca Raton, FL (1985).

A. Dorinson and K.C. Ludema, *Mechanics and Chemistry in Lubrication*, Elsevier, Amsterdam (1985).

W.L. Robertson and W.S. Robertson, *Lubrication in Practice*, Marcel Dekker, Inc., New York, NY (1984).

J.A. Schey, *Tribology in Metalworking Friction, Lubrication and Wear*, American Society for Metals, Metals Park, OH (1983).

D. Klamann, *Schmierstoffe und verwandte Produkte: Herstellung, Eigenschaften, Anwendung,* Verlag Chemie, Weinheim (1982).

C.J. Boner, *Modern Lubricating Greases*, Scientific Publications, Broseley (1976).

J.M. McDermott, *Drilling Mud and Fluid Additives,* Noyes Data Corp., London (1973).

Noria Corporation, *Lubricating Grease Guide*, Noria Corporation, Tulsa, OK (n.d.); http://www.noria.com

12.16 References

[1] J.S. Marth, *Renewable Lubricants Manual: Biobased Oils, Fluids & Greases*, United Bio Lube, Palo Alto, CA (2007); http://www.renewablelubricants.com

[2] W. Bock, *High-performance Gear Oil, Industrial Oils*, FUCHS Brochure, FUCHS Petrolub AG, Mannheim (2004).

[3] ISO/TS 16949, *Quality Management Systems – Particular Requirements for the Application of ISO 9001:2008 for Automotive Production and Relevant Service Part Organizations*, International Organization for Standardization, Geneva (2009).
[4] D.M. Pirro and A.A. Wessol (eds), Handling, storing and dispersion lubricants, in *Lubrication Fundamentals*, 2nd edn, CRC Press, Boca Raton, FL (2001), Chapter 18.
[5] R.H. Schlosberg, J.N. Chu, G.A. Knudsen, E.N. Suciu and H.S. Aldrich, High stability esters for synthetic lubricant applications, *Lubric. Engng.* **57**(2) (2001) 21–6.
[6] C.B. Duncan and L.K. Meade (to Exxon Chemical Co.), Biodegradable branched synthetic ester base stocks and lubricants formed therefrom, US Patent No. 5,817,607 (6 October 1998).
[7] http://www.uniqema.com/news
[8] Anon., Motorcycle palm oil in market now, *The Malay Mail* (24 November 1995), p.6.
[9] M. Pal and S. Singhal, Environmentally adapted lubricants. Part I. An overview, *J. Synth. Lubr.* **17**(2) (2000) 135–43.
[10] IENICA, *Biolubricants – Market Data Sheet*, IENICA, Central Science Laboratory (UK), pp. 1–11.
[11] C.W. Lea, European development of lubricants derived from renewable resources, *Ind. Lubr. Tribol.* **54**(6) (2002) 268–74.
[12] F. Farshchi, Why invest in environmentally considerate lubricants?, *World Pumps* **496** (2008) 36–7.
[13] S.K. Lo, Y.M. Choo, A.N. Ma and Y. Basiron, Additives for petroleum fuels and lubricants, biofuels and biolubricants – a review, *Oil Palm Bull.* **40** (2000) 1–5.
[14] LUBRICOAT, Environmentally friendly lubricants and low friction coatings a route towards sustainable products and production processes, CORDIS FP5 (2001–2004).
[15] W. Klose, Significance of rapidly biodegradable lubricants in the German lubricant market, *Tribol. Schmierungstechn.* **50** (2003) 45–9.
[16] L.R. Rudnick and S.Z. Erhan, Natural oil as lubricants, in *Synthetics, Mineral Oils, and Bio-based Lubricants: Chemistry and Technology* (L.R. Rudnick, ed.), CRC Press/Taylor & Francis, Boca Raton, FL/New York, NY (2006), pp. 353–60.
[17] S. Asadauskas, J.M. Perez and J.L. Duda, Oxidative stability and antiwear properties of high oleic vegetable oils, *Lubric. Engng.* **52** (1996) 877–82.
[18] S.J. Randles and M. Wright, Environmentally considerate ester lubrication for the automotive and engineering industries, *J. Synth. Lubr.* **9** (1992) 145–61.
[19] X. Wu, X. Zhang, S. Yang, H. Chen and D. Wang, The study of epoxidized rapeseed oil used as a potential biodegradable lubricant, *J. Am. Oil Chem. Soc.* **77**(5) (2000) 561–3.
[20] R.L. Goyan, R.E. Melley, P.A. Wissner and W.C. Ong, Biodegradable lubricants, *Lubric. Engng.* **54**(7) (1998) 10–17.
[21] O. Glavati and L. Glavati, Lubricating oils of plant origin, *Tribol. Schmierungstechn.* **42** (2000) 17–18.

[22] R. Šraj, J. Vižintin, M. Svoljšak and M. Feldin, Rapidly biodegradable hydraulic fluids on the basis of rapeseed oil, *Lubric. Engng.* **54**(4) (2000) 34–9.
[23] G. Biresaw, A. Adhvaryu, S.Z. Erhan and C.J. Carriere, Friction and adsorption properties of normal and high-oleic soybean oils, *J. Am. Oil Chem. Soc.* **79** (2002) 53–8.
[24] E. Uosukainen, Y.-Y. Linko, M. Lämsä, T. Tervakangas and P. Linko, Transesterification of trimethylol propane and rapeseed oil methyl ester to environmentally acceptable lubricants, *J. Am. Oil Chem. Soc.* **75** (1998) 1557–63.
[25] L.E. Johansson and S.T. Lundin, Copper catalyst in the selective hydrogenation of soybean and rapeseed oils: I. The activity of the copper chromite catalyst, *J. Am. Oil Chem. Soc.* **56** (1979) 974–80.
[26] L. Lazzeri, M. Mazzoncini, A. Rossi, E. Balducci, G. Bartolini, L. Giovannelli, R. Pedriali, R. Petroselli, G. Patalano, G. Agnoletti, A. Borgioli, B. Croce and L. D'Avino, Biolubricants for the texile and tannery industries as an alternative to conventional minerals oils: an application experience in the Tuscany province, *Ind. Crops Prod.* **24** (2006) 280–91.
[27] L. Schumacher, Biodiesel lubricity, in *The Biodiesel Handbook* (G. Knothe, J. Van Gerpen and J. Krahl, eds), AOCS Press, Urbana, IL (2005), pp. 137–44.
[28] J. Van Gerpen, S. Soylu and D. Chang, *Evaluation of Lubricity of Soybean Oil-based Additives in Diesel Fuel*, Iowa State University, Ames, IA (1998).
[29] F.L. Erickson, R.E. Anderson and P.S. Landis (to International Lubricants, Inc.), Vegetable oil derivatives as lubricant additives, US Patent No. 5,282,989 (1 February 1994).
[30] M. Vogel, Biolubricants in gears, *Tribol. Schmierungstechn.* **45** (1998) 25–8.
[31] B.L. Ruddy, D. Dowson and P.N. Economou, A review of studies of piston ring lubrication, *Proc. 9th Leeds–Lyon Symp. on Tribology: Tribology of Reciprocating Engines*, Leeds (7–10 September 1982), pp. 109–21.
[32] K. Haniyuda and E. Nagatomi (to Showa Shell Sekiyu), Lubricating oil composition for engine, Jap. Patent No. JP 2008/031289 (14 February 2008).
[33] S.C. Tung and R.L. McMillan, Automotive tribology overview of current advances and challenges for the future, *Tribol. Intl.* **37** (2004) 517–36.
[34] P.A. Willermet, Topics in transmission tribology, *Lubric. Engng.* **55**(2) (1999) 39–43.
[35] S.M. Hsu and Z.C. Ying (eds), *Nanotribology: Critical Assessment and Research Needs* (based on the *Nanotribology Workshop*, Gaithersburg, MD, 13–15 March 2000) (2003).
[36] S.E. Schwartz, S.C. Tung and M.L. McMillan, Automotive lubricants, in *Fuels and Lubricants Handbook: Technology, Properties, Performance and Testing* (G.E. Totten, ed.), ASTM International, West Conshohocken, PA (2003).
[37] A. Permsuwan, D.J. Picken, K.D.R. Seare and M.F. Fox, Engine development and test for vegetable oil lubricant, *Int. J. Ambient Energy* **17**(3) (1996) 157–61.
[38] ASTM D 6923–10a, Standard Test Method for Evaluation of Engine Oils in a High Speed, Single-cylinder Diesel Engine – Caterpillar 1R Test Procedure, in *ASTM Annual Book of Standards*, American Society for Testing and Materials, West Conshohocken, PA (2010).
[39] R.R. Fessler and G.R. Fenske (eds), *Multiyear Program Plan: Reducing Friction and Wear in Heavy Vehicles*, Argonne National Laboratory (13 December 1999).

[40] B.S. Andersson, Company perspectives in vehicle tribology – Volvo, in *Vehicle Tribology* (D. Dowson, C.M. Taylor and M. Godet, eds), Elsevier, Ltd, Oxford (1991), pp. 503–6.
[41] S. Korcek, J. Sorab, M.D. Johnson and R.K. Jensen, Automotive lubricants for the next millennium, *Ind. Lubr. Tribol.* **52** (2000) 209.
[42] A. Extance, Engines of innovation, *Chem. World* (October 2011) 48–51.
[43] S.H. Roby and S.G. Ruelas (to Chevron Oronite Co.), Lubricating base oil compositions and methods for improving fuel economy in an internal combustion engine using same, US Patent No. 7,465,696 B2 (16 December 2008).
[44] R.C. Coy, Practical applications of lubrication models in engines, in *New Directions in Tribology* (I.M. Hutchings, ed.), IMechE, London (1997), p. 197.
[45] M. Nakasa, Engine friction overview, *Proc. Intl Tribology Conference*, Yokohama (29 October – 2 November 1995), no pp. given.
[46] S. Boyde, Green lubricants – environmental benefits and impacts of lubrication, *Green Chem.* **4** (2002) 293–307.
[47] P.R.M. Lopes, R.N. Montagnolli, R. de F. Domingues and E.D. Bidoia, Toxicity and biodegradation in sandy soil contaminated by lubricant oils, *Bull. Environm. Contam. Toxicol.* **84**(4) (2010) 454–8.
[48] American Petroleum Institute, *Engine Oil Licensing and Certification Systems (EOLCS)*, Publ. No. 1509, 15th edn, API, Washington, DC (2002).
[49] M. Woydt, No/low SAP and alternative engine oil development and testing, *J. ASTM Intl.* **4**(10) (2007), online.
[50] W. Grossmann, Additives for lubricants, *Erdöl Erdgas Kohle* **107** (1991) 417–21.
[51] A. Hubmann, K. Reiter and W. Baumann, Engine oils with optimized environmental properties, *Mineralöltechn.* **39**(9) (1994) 1–14.
[52] R. Luther, Lubricants in the environment, in *Lubricants and Lubrication* (Th. Mang and W. Dresel, eds), 2nd edn, Wiley-VCH, Weinheim (2007), pp. 119–82.
[53] L.A.T. Honary, unpublished results (2010).
[54] M. Harperscheid and J. Omeis, Lubricants for internal combustion engines, in *Lubricants and Lubrication* (Th. Mang and W. Dresel, eds), 2nd edn, Wiley-VCH, Weinheim (2007), pp. 191–229.
[55] http://www.delfinusa.com/articles/item/279
[56] J.W. Lambert and D.L. Johnson (to Agro Management Group, Inc.), Vegetable oil lubricants for internal combustion engines and total loss lubrication, US Patent No. 5,888,947 (30 March 1999).
[57] I.-Ching Chiu, S. Gunsel and P. Lacey (to Pennzoil-Quaker State Co.), Environmentally friendly lubricants, PCT Int. Publ. No. WO 03/027212 A1 (3 April 2003).
[58] W.W. Garmier (to Renewable Lubricants, Inc.), High temperature biobased lubricant compositions comprising boron nitride, PCT Int. Publ. No. WO 2006/116502 A1 (2 November 2006).
[59] http://www.GETgreenearth.com; http://www.getg.com
[60] V. Schäfer and O.H. Botz (to Natoil AG), Stabilisator- und Additivzusammensetzung für Verbrennungsmotoren, Eur. Patent No. EP 2,138,556 A1 (30 December 2009).

[61] B.N. Rhodes and D. Johnson, Vegetable-based motor oils, in *Biobased Industrial Fluids and Lubricants* (S.Z. Erhan and J.M. Perez, eds), AOCS Press, Champaign, IL (2003), pp. 85–109.
[62] J.M. Perez, Vegetable oil-based engine oils, in *Industrial Uses of Vegetable Oils* (S.Z. Erhan, ed.), CRC Press, Boca Raton, FL (2005), Chapter 3.
[63] International Council of Marine Industry Associations (ICOMIA) Standard 27–97, Egham, UK (1997).
[64] EU Directive 97/68/EC (16 December 1997), Measures against the emissions of gaseous and particulate pollutants from internal combustion engines to be installed in non-road mobile machinery, *Off. J. Eur. Comm.* **L59** (1998) 1–9.
[65] EU Directive 2002/88/EC (9 December 2002), amending Directive 97/68/EC on the approximation of the laws of the Member States relating to measures against the emissions of gaseous and particulate pollutants from internal combustion engines to be installed in non-road mobile machinery, *Off. J. Eur. Comm.* **L35** (2003) 28–33.
[66] B. Van Zeebroeck, F. Vanhove and L. Franckx, *Impact Assessment – Reviewing Directive 97/68/EC – Emissions from non-road mobile machinery*, Arcadis, Louvain (30 January 2009).
[67] *Control of Emissions from Nonroad Large Spark-Ignition Engines, and Recreational Engines (Marine and Land-Based)*, Federal Register (8 November 2002), Vol. 67, Number 217.
[68] http://www.epa.gov/EPA-AIR/2002/November/Day-08/a23801.htm
[69] http://www.epa.gov/oms/regs/nonroad/marinesiequipld/420f07032.htm
[70] E.R. Zehler and B.J. Beimesch (to Henkel Corp.), Smokeless two-cycle engine lubricants, US Patent No. 5,912,214 (15 June 1999).
[71] N.H. Jayadas, K.P. Nair and G. Ajithkumar, Tribological evaluation of coconut oil as an environment-friendly lubricant, *Tribol. Int.* **40**(2) (2007) 350–4.
[72] G.A. Sivasankaran, R.P.S. Bisht, V.K. Jain, M. Gupta, A. Sethumariah and V.K. Bhatia, Jojoba oil-based two-stroke gasoline engine lubricant, *Tribol. Int.* **21**(6) (1988) 327–33.
[73] H.H. Masjuki and M.A. Maleque, Wear, performance and emissions of a two-stroke engine running on palm oil methyl ester blended lubricant, *Proc. IMechE, Part J: J. Engng. Tribol.* **210** (1997) 213–19.
[74] H.H. Masjuki and M.A. Maleque, Investigation of the anti-wear characteristics of palm oil methyl ester using a four-ball tribometer test, *Wear* **206** (1997) 179–86.
[75] H.H. Masjuki, M.A. Maleque, A. Kubo and T. Nonaka, Palm oil and mineral oil based lubricants – their tribological and emission performance, *Tribol. Int.* **32** (1999) 305–14.
[76] R.E. Gapinsky, I.E. Joseph and B.D. Layzell, A vegetable oil based tractor lubricant, *SAE Techn. Paper Series* No. 941758, SAE, Warrendale, PA (1994).
[77] A. Kunz, Development of a universal tractor transmission oil (UTTO) based on renewable raw materials, *Mineralöltechn.* **51** (2006) 1–24; *Mat.-wiss. u. Werkstofftech.* **37**(2) (2006) 191–201.
[78] Anon., *JDM J20C – Specification for Anti-brake Chatter Transmission/ Hydraulic Fluids*, Deere & Company, Moline, IL (September 1992).
[79] T. Bartels, Gear lubrication oils, in *Lubricants and Lubrication* (Th. Mang and W. Dresel, eds), 2nd edn, Wiley-VCH, Weinheim (2007), pp. 230–73.

[80] S. Castellucci, *Utilizzazione di biolubrificanti di origine vegetale nelle machine e nelle apparecchiature dei settori agricoli forestale*, Ph.D Thesis, Università degli Studi della Tuscia, Viterbo (2006).
[81] O.H. Botz (to Venyard GmbH), Vegetable oil-based hydraulic fluid and transmission fluid, US Patent Appl. No. 2011/0195885 A1 (11 August 2011).
[82] ISO TC 131/SC7 N343, *Hydraulic Fluid Power – Compatibility Between Fluids and Standard Elastomeric Materials*, International Organization for Standardization, Geneva (1999) (Draft).
[83] A. Willing, Lubricants based on renewable resources: an environmentally comparative alternative to mineral oil products, *Chemosphere* **43** (2001) 89–98.
[84] J.Q. Griffith III, E.S. Williams and W.H. Reiland Jr (to Sun Oil Co. of Pennsylvania), Soap thickening hydraulic oil composition, US Patent No. 3,816,316 (11 June 1974).
[85] W.A. Givens and P.A. Michael, Hydraulic fluids, in *Fuels and Lubricants Handbook: Technology, Properties, Performance, and Testing* (G. Totten, ed.), ASTM International, West Conshohocken, PA (2003), pp. 353–82.
[86] P. Patel-Predd, Lubes from vegetable oils become affordable, *Environm. Sci. Technol. Online, Technol. News*, ACS Publications (24 January 2007).
[87] W.J. Bartz *et al.*, *Hydraulikflüssigkeiten*, Band 475, Expert Verlag, Renningen (1995).
[88] F. Huang, Y. Zhang, H. Wang and Z. Zeng, Study on the universal biodegradable heavy-load hydraulic transmission fluid, *Runhua Yu Mifeng* **33**(12) (2008) 82–5.
[89] C.E. Snyder, L.J. Gschwender and S.K. Sharma, Long-term additive trends in aerospace applications, in *Lubricant Additives: Chemistry and Applications* (L.R. Rudnick, ed.), 2nd edn, CRC Press, Boca Raton, FL (2009), pp. 637–46.
[90] A.G. Papay and L.R. Rudnick, Hydraulics, in *Synthetic Lubricants and High-performance Functional Fluids* (L.R. Rudnick and R.L. Shubkin, eds), Marcel Dekker, Inc., New York, NY (1999), pp. 595–624.
[91] W. Bock, Hydraulic fluids, in *Lubricants and Lubrication* (Th. Mang and W. Dresel, eds), 2nd edn, Wiley-VCH, Weinheim (2007), pp. 274–337.
[92] Anon., Texaco biodegradable lubricants, *Ind. Lubr. Tribol.* **44**(4) (1992) 28.
[93] Anon., New biodegradable hydraulic fluid, *Ind. Lubr. Tribol.* **45**(5) (1993) 16.
[94] V. Cheng, A.A. Wessol, P. Bandovin, M.T. BenKinney and N.J. Norrick, Biodegradable and nontoxic hydraulic oils, *SAE Techn. Paper Series* No. 910964, SAE, Warrendale, PA (1991).
[95] F.J. Flider, Use of rapeseed oil in lubricants, *INFORM* **6**(9) (1995) 1031–5.
[96] Anon., Silkolene environmentally acceptable lubricants, *Ind. Lubr. Tribol.* **45**(3) (1993) 16.
[97] Anon., The environmental 1990s, *Ind. Lubr. Tribol.* **45**(3) (1993) 15.
[98] S.Z. Erhan and S. Asadauskas, Lubricant basestocks from vegetable oils, *Ind. Crops Prod.* **11** (2000) 277–82.
[99] T. Norrby, Environmentally adapted lubricants – where are the opportunities?, *Ind. Lubr. Tribol.* **55**(6) (2003) 268–74.
[100] VDMA Blätter 24568/24569, *Fluidtechnik biologisch schnell abbaubarer Druckflüssigkeiten, Technische Mindest-anfordeungen, Umstellungsrichtlinien*,

Verband Deutscher Maschinen-und Anlagenbau e.V., Beuth Verlag GmbH, Berlin.
[101] ISO 6743/4, *Lubricants, Industrial Oils and Related Products (Class L) – Classification – Part 4: Family H (Hydraulic Systems)* (with ISO 11158), International Organization for Standardization, Geneva (1997).
[102] ISO 15380, *Lubricants, Industrial Oils and Related Products (Class L) – Family H (Hydraulic Systems) – Specification for Categories HETG, HEPG, HEES and HEPR*, International Organization for Standardization, Geneva (2002).
[103] P.S. De Caro, A. Gaset and N.N. The, Vegetable based lubricants in forestry: results from a test campaign, *Agro-Food Ind. Hi-Tech.* **12** (2000) 26–30.
[104] S.J. Randles, Esters, in *Synthetic Lubricants and High Performance Functional Fluids* (L.R. Rudnick and R.L. Shubkin, eds), Marcel Dekker, Inc., New York, NY (1999), p. 63.
[105] N. Battersby, S. Greenall and G. Gustafsson, Environmentally acceptable hydraulic fluids. Field performance in mobile equipment, *Proc. 14th Intl. Colloquium on Tribology*, Esslingen (2004), Vol. II, pp. 759–68.
[106] N.A.J. Cooban and H.L. Rieffe (to Unichema Chemie), Base fluids, US Patent No. 5,707,945 (13 January 1998).
[107] Anon., Uniquema launches new range of Priolube biohydraulic basefluids, *Ind. Lubr. Technol.* **53**(3) (2001) 97.
[108] D. Da Costa (to Enbio Ind., Inc.), Environmentally compatible hydraulic fluid, Eur. Patent No. EP 1,899,442 (19 March 2008).
[109] G.E. Barker (to Atlas Powder Co.), Hydraulic fluid composition, US Patent No. 2,755,251 (17 July 1956).
[110] D.R. Kodali, Z. Fan and L.R. DeBonte (to Cargill, Inc.), Plants, seeds, and oils having an elevated total monounsaturated fatty acid content, PCT Int. Publ. No. WO 00/007433 (17 February 2000).
[111] T. Konishi and Y. Kikuchi (to Nippon Mitsubishi Oil Corp.), Hydraulic oil composition with improved biodegradable properties, Germ. Patent No. DE 60,122,400 T (8 November 2007).
[112] J.Z. Adamczewska and D. Wilson, Development of ecologically responsive lubricants, *Synth. Lubr.* **14**(2) (1997) 129–42.
[113] D.R. Kodali, High performance ester lubricants from natural oils, *Ind. Lubr. Tribol.* **54**(4) (2002) 165–70.
[114] J. Vižintin, A. Arnšek and T. Ploj, Lubricating properties of rapeseed oils compared to mineral oils under a high load oscillating movement, *J. Synth. Lubr.* **17**(3) (2000) 201–17.
[115] E. Remmele and B. Widmann, Suitability and environmental compatibility of rapeseed based hydraulic fluids for agricultural machinery, *J. Synth. Lubr.* **16**(2) (1999) 129–45.
[116] F. Bongardt, K. Schmid and R. Wüst (to Henkel KaA), Environment-friendly basic oil for formulating hydraulic fluids, US Patent No. 5,451,334 (19 September 1995).
[117] A. Zeman, Biodegradable metalworking and hydraulic fluids, *Tribol. Schmierungstechn.* **44** (1997) 160–5.
[118] K.V.J. Jokinen, H.K. Kerkkonen, E.A. Leppämaki and E.I. Piirilä (to Oy Kasviöljy-Växtolje Ab.), Hydraulic fluids, US Patent No. 4,783,274 (8 November 1988).

[119] A. Shanahan and P.S. Landis, Telomerized triglyceride oil product (to International Lubricants, Inc.), US Patent No. 5,454,965 (3 October 1995).
[120] P.S. Landis (to International Lubricants, Inc.), Lower viscosity telomer oil, US Patent No. 5,567,345 (22 October 1996).
[121] A.V. Avellan and C. Alfos, Valorisation non alimentaire des huiles de friture usagées en tant que lubrifiants biodégradables, *Oléag., Corps Gras, Lipides* **12**(4) (2005) 290–4.
[122] L.A.T. Honary, An investigation of the use of soybean oil in hydraulic systems, *Bioresour. Technol.* **56** (1996) 41–7.
[123] L.A.T. Honary, Soybean based hydraulic fluid, US Patent No. 5,972,855 A (26 October 1999).
[124] B. Cunningham, N. Battersby, W. Wehrmeyer and C. Fothergill, A sustainability assessment of a biolubricant, *J. Ind. Ecol.* **7** (2004) 179–92.
[125] Commission Directive 2005/360/EC of 26 April 2005 establishing ecological criteria and the related assessment and verification requirements for the award of the Community eco-label to lubricants, *Off. J. Eur. Comm.* **L118** (5 May 2005) 26–34.
[126] http://www.cognis.com/company/Press+and+Media/Press+Release/2008/080131_EN_FP.htm
[127] G.H. Kling and J.M. Perez, Worldwide use of bio oils for hydraulic fluids, in *Biobased Industrial Fluids and Lubricants* (S.Z. Erhan and J.M. Perez, eds), AOCS Press, Champaign, IL (2003), pp. 120–7.
[128] S. Lawate, Environmentally friendly hydraulic fluids, in *Biobased Industrial Fluids and Lubricants* (S.Z. Erhan and J.M. Perez, eds), AOCS Press, Champaign, IL (2003), pp. 35–46.
[129] M. Pal and S. Singhal, Environmentally adapted lubricants. Part II. Hydraulic fluid, *Synth. Lubr.* **17**(3) (2000) 219–24.
[130] S. Lawate, Environmentally friendly hydraulic fluids, in *Synthetics, Mineral Oils and Bio-Based Lubricants: Chemistry and Technology* (L.R. Rudnick, ed.), CRC Press/Taylor & Francis, Boca Raton, FL/New York, NY (2006), pp. 541–50.
[131] H.F. Eichenberger, Biodegradable hydraulic lubricants. An overview of current developments in Central Europe, *SAE Techn. Paper Series* No. 910962, SAE, Warrendale, PA (1991).
[132] H. Theissen, Application of bio-oil in hydraulics, *o+p Ölhydr. Pneumatik* **49**(1) (2005) 47–52 (in German).
[133] H. Theissen, Experiences with bio hydraulic oil, *o+p Ölhydr. Pneumatik* **50**(3) (2006) 132–7.
[134] M. Schneider, Review – Plant-oil-based lubricants and hydraulic fluids, *J. Sci. Food Agric.* **86** (2006) 1769–80.
[135] P.S. De Caro and A. Gaset, Lubricants based on renewable resources: which future for the development of this value chain?, *Actual. Chim.* **10** (1999) 9–14.
[136] B. Wilson, Lubricants and functional fluids from renewable sources, *Ind. Lubr. Technol.* **50**(1) (1998) 6–15.
[137] S. Venon, G. Chanvillard, I. Duvois-Brugger, C. Levy and V. Teissier (to Lafage), Demolding composition, US Patent Appl. No. 2010/0107928 A1 (6 May 2010).

[138] P. van Broekhuizen, D. Theodori, K. Le Blansch and S. Ullmer (eds), *Lubrication in Inland and Coastal Water Activities*, LLINCWA Final Report, IVAM UvA BV, Amsterdam (2003).
[139] J. Nelson, Harvesting lubricants, *Carbohydr. Econ.* **3**(1) (2000), no pp. given.
[140] W.W. Garmier (to Renewable Lubricants, Inc.), Biodegradable penetrating lubricant, US Patent Appl. No. 2003/0069146 A1 (10 April 2003); PCT Int. Publ. No. WO 03/106599 A1 (24 December 2003).
[141] P.A. Boyd, D.L. Whitfill, T.S. Carter and J.P. Allamon, New base oil used in low-toxicity oil muds, *J. Petrol. Technol.* **37** (1985) 137–42.
[142] T.O. Walker, D.W. Brownawell and A. Gutierrez (to Exxon Research & Engineering Co.), Drilling fluids containing an additive composition, US Patent No. 4,382,002 (3 May 1983).
[143] S.H. Elrod and W.B. Nance (to Milchem, Inc.), Aqueous drilling fluid and lubricant composition, US Patent No. 4,181,617 (1 January 1980).
[144] H. Müller, C.-P. Herold, S. von Tapavicza, D.J. Grimes, J.-M. Braun and S.P.T. Smith (to Henkel K.A./Baroid, Ltd.), Verwendung ausgewählter Esteröle in Bohrspülungen insbesondere zur off-shore-Erschliessung von Erdöl- bzw. Erdgasvorkomnen, Eur. Patent Nos. EP 0374,671/0374,672 (11 December 1989).
[145] H. Müller, C.-P. Herold, S. von Tapavicza, M. Neuss, W. Zöllner and F. Burbach (to Henkel K.A.), Ester von Carbonsauren mittlerer Kettenlänge als Bestandteil der Ölphase in Invert-Bohrspülschlämmer, Eur. Patent No. EP 0,386,636 (1 March 1990).
[146] B. Genuyt, M. Janssen, R. Reguerre, J. Cassiers and F. Breye (to Total Raffinage Distribution), Biodegradable lubricating composition and its use in a drilling fluid, PCT Int. Publ. No. WO 2001/083640 A1 (8 November 2001).
[147] R.A. Padavich and L. Honary, A market research and analysis report on vegetable-based industrial lubricants, *Trans. SAE J. Fuel Lube* **104**(4) (1995) 845–54.
[148] M. Valls, Drilling mud lubricant and shale stabilizer, US Patent Appl. No. 2008/0207472 A1 (28 Aug. 2008).
[149] D.V. Kosynkin, G. Ceriotti, K.C. Wilson, J.R. Lomeda, J.T. Scorsone, A.D. Patel, J.E. Friedheim and J.M. Tour, Graphene oxide as a high-performance fluid-loss-control additive in water-based drilling fluids, *ACS Appl. Mater. Interfaces* **4**(1) (2012) 222–7.
[150] http://www.ausmud.com/pds.php?id = 78
[151] Frost & Sullivan, Inc., *Strategic Analysis of the European Market for Oilfield Chemicals,* San Antonio, TX (24 October 2008).
[152] http://www.castrol.com/castrol/genericarticle.do?categoryld=4504740&contentld=7038530
[153] B.H. Carter, Marine lubricants, in *Chemistry and Technology of Lubricants* (R.M. Mortier and S.T. Orszulik, eds), VCH, Weinheim (1992).
[154] US Army Corps of Engineers, *Engineering and Design, Lubricants and Hydraulic Fluids, Engineer Manual*, 1110-2-1424, Washington, DC (1999).
[155] R. Martins, N. Cardoso and J. Seabra, Gear power loss performance of biodegradable low-toxicity ester-based oils, *Proc. IMechE Part J: J. Engng Tribol.* **222**(3) (2008) 431–40.

[156] http://www.bioblend.com/component/page.shop.browse/category_id.7/option.com_virtuemart/Itemid.69
[157] A. Jacobson, Air compressor lubricants: the next generation, *Tribol. Lubr. Technol.* **60**(7) (2004) 30–2, 34–6, 38.
[158] Anon. Synthetic base stocks provide benefits for air compressor lubricants, *Lube Mag.* **106** (2011) 19–21.
[159] M. Johnson, Positive displacement compressors, selecting the correct lubricant, *Tribol. Lubr. Technol.* **64**(12) (2008) 20–5.
[160] W. Bock and G. Lingg, Compressor oils, in *Lubricants and Lubrication* (Th. Mang and W. Dresel, eds), 2nd edn, Wiley-VCH, Weinheim (2007), pp. 338–52.
[161] J. Sarkar, S. Bhattacharyya and M.R. Gopal, Natural refrigerant-based subcritical and transcritical cycles for high temperature heating, *Int. J. Refrig.* **30** (2006) 3–10.
[162] S.F. Pearson, Natural refrigerants for heat pumps, *IEA Heat Pump Center Newsl.* **20**(1) (2004) 13–16.
[163] http://www.prod.exxonmobil.com/lubes/marine/products/mn_synthetic_lubricants.html
[164] http://www.cromptoncorp.com/lubricants-grease.html
[165] http://www.ecompressedair.com/lubricant.shtml
[166] H. Ikeda, J. Yagi and Y. Kawaguchi, Evaluation of various compressor lubricants for a carbon dioxide heat pump system, in *Proc. Sixth IIR G. Lorentzen Natural Working Fluids Conference*, Glasgow (29 August – 1 September 2004).
[167] A. Adhvaryu and S.Z. Erhan, Epoxidized soybean oil as a potential source of high-temperature lubricants, *Ind. Crops Prod.* **15** (2002) 247–54.
[168] http://www.lubricationtechnology.com/products.htm
[169] http://www.miller-stephenson.com
[170] D.H. Culpon Jr (to Texaco, Inc.), High temperature compressor oil, US Patent No. 5,156,759 (20 October 1992).
[171] R. Kasameyer, Lubricant viscosity control for gas compressors, *Petro Ind. News* **11**(6) (2011) 22.
[172] American Society of Heating, Refrigerating and Air-conditioning Engineers (ASHRAE), ASHRAE Standard ANST/ASHRAE 34–1992, Atlanta, GA (1992).
[173] D.J. Glova (to Chevron Research Co.), Refrigeration lubricating oil containing dialkyl sulfosuccinate, US Patent No. 4,556,496 (3 December 1985).
[174] C. Ciantar, M. Hadfield, A. Swallow and A. Smith, The influence of POE and PVE lubricant blends within hermetic refrigerating compressors with HFC-134a refrigerant, *Wear* **241** (2000) 53–64.
[175] S.J. Randles, Refrigeration lubes, in *Synthetic Lubricants and High Performance Functional Fluids* (L.R. Rudnick and R.L. Shubkin, eds), Marcel Dekker, Inc., New York, NY (1999), p. 563.
[176] Directive 2006/40/EC, Emissions from air-conditioning systems in vehicles and amending Council Directive 70/156/EEC, *Off. J. Eur. Comm.* **L161/**12 (14 June 2006).
[177] N.E. Schnur (to Cognis Corp.), Blended polyol ester lubricants for refrigerant heat transfer fluids, US Patent No. 6,551,523 B1 (22 April 2003).

[178] U. Sasaki, T. Ishikawa, H. Hasegawa and N. Ishida (to Nippon Mitsubishi Oil Corp.), Refrigerator oils for use with chlorine-free hydrocarbon refrigerants, US Patent No. 6,582,621 (24 June 2003).

[179] K. Ankner, E. Glad, M. Aalto and V. Saranpaa (to Neste Chemicals OY), Complex esters for use with fluorinated refrigerants, US Patent Appl. No. 2004/0046146 (11 March 2004).

[180] Y. Shimomura and K. Takigawa, Refrigerating machine oil composition, US Patent No. 6,831,045 (7 August 2003).

[181] Y. Shimomura and K. Takigawa, Refrigerating machine oil composition, US Patent No. 7,045,490 (4 September 2003).

[182] D.C. Kramer, R. Shah, S.J. Miller and S.A. Elomari (to Chevron Corp.), Refrigeration oil from gas-to-liquid derived and bio-derived diesters, US Patent Appl. No. 2009/0159837 A1 (25 June 2009).

[183] S.J. Randles and S.J. McTavish (to ICI), Lubricant composition, PCT Int. Publ. No. WO 2004/005441 A2 (15 January 2004).

[184] S. Sakanoue, M.M. Takesue and M. Takagi (to Idemitsu Kosan Co.), Refrigerating machine oil composition, US Patent No. 6,656,891 B1 (2 December 2003).

[185] G. Enna, H. Fukuda and Y. Ohtsuka (to Asahi Glass Co.), Lubricant oil compositions for freezer, PCT Int. Publ. No. WO 98/037170 A1 (27 August 1998).

[186] T. Kaimai and H. Takahashi (to Japan Energy Corp.), Lubricating oil for compression refrigerators, working fluid for refrigerators, and method for lubrificating refrigeration system, PCT Int. Publ. No. WO 98/036040 A1 (20 August 1998).

[187] S.J. Miller, Z. Zhou and S.A. Elomari (to Chevron U.S.A., Inc.), Synthesis of diester-based biolubricants from epoxides, US Patent Appl. No. 2009/198075 A1 (6 August 2009).

[188] S. Boyde, P. Gibb, S.J. Randles, S. Corr, P. Dowdle and A. McNicol, Effects of lubricant properties on efficiency of refrigeration compressors, *Proc. 2000 Intl Compressor Engineering Conference* (W. Soedel, ed.), West Lafayette, IN (2000), p. 311.

[189] ASTM D 2272–09, Standard Test Method for Oxidation Stability of Steam Turbine Oils by Rotating Pressure Vessel, *Annual Book of ASTM Standards*, American Society for Testing and Materials, West Conshohocken, PA (2009), Vol. 05–01.

[190] D.D. Carr and N. DeGeorge (to Hatco Chemical Corp.), Synthetic lubricant base stock of monopentaerythritol and trimethylolpropane esters, US Patent No. 4,826,633 (2 May 1989).

[191] G. Livingstone, Varnish: looking ahead, *Turbomach. Intl.* (1 September 2008), no pp. given.

[192] T. Kaimai and Y. Morishima (to Japan Energy Corp.), Biodegradable lubricant for turbines, Jpn. Kokai Tokkyo Koho JP 2004/315553 A (11 November 2004).

[193] D.M.C. McCarthy, S.B. Glavatskih, M. Pach and A. Dmitriev, Performance evaluation for aged environmentally adapted lubricants in bronze-steel sliding contacts, *J. Synth. Lubr.* **25**(3) (2008) 115–29.

[194] R.C.J. Wilson, iPAL FTIR Analyser for moisture determination in lubricating oils – An advancement over Karl Fischer titration, http://www.a2technologies.

com/downloads/ferrybridgedownload.pdf; http://www.mt-online.com/component/content/article/276-mtsept 2010 (1 September 2010).
[195] G.J. Livingstone, B.T. Thompson and M.E. Okazaki, Physical, performance and chemical changes in turbine oils from oxidation, *J. ASTM Int.* **4**(1) (2007) JAI 100465.
[196] V.J. Gatto, W.E. Moehle, T.W. Cobb and E.R. Schneller, Oxidation fundamentals and its application to turbine oil testing, *J. ASTM Int.* **3**(4) (2006) JAI 13498.
[197] Kline & Company, *Lubricants for Wind Turbines 2012. Global Market Analysis and Opportunities*, http://www.klinegroup.com
[198] D.C. Kramer, B.K. Lok and R.R. Krug, The evolution of base oil technology, in *Turbine Lubrication in the 21st Century*, ASTM STP#1407 (W.R. Herguth and T. M. Warne, eds), ASTM Intl., West Conshohocken, PA (2001).
[199] E.V. Zaretsky, *Tribology for Aerospace Applications*, STLE Publication SP – 37, STLE, Park Ridge, IL (1997).
[200] S.J. Metro and D.D. Carr (to Exxon Research & Engineering Co.), Synthetic ester lubricating oil composition containing particular *t*-butylphenyl substituted phosphates and stabilized hydrolytically with particular long chain alkyl amines, US Patent No. 4,440,657 (3 April 1984).
[201] B.J. Pafford, J.T. Kim, P.E. Godici, H.S. Aldrich, R.H. Schlosberg and M.A. Krevalis (to Exxon Chemical Co.), Polyol ester compositions with unconverted hydroxyl groups for use as lubricant base stocks, US Patent No. 5,698,502 (16 December 1997).
[202] L. Gschwender, C.E. Snyder Jr., L. Nelson, L. Carswell, G.W. Fultz and C. Saba, Research and development of advanced high-temperature air force turbine engine oil, *Lubric. Engng.* **56/**5 (2000) 20–4.
[203] Y. Nakagami (to Fuji Electric Co.), Stationary induction apparatus containing uninflammable insulating liquid, US Patent No. 5,336,847 (9 August 1994).
[204] G.S. Cannon and L.A.T. Honary, Soybean based transformer oil and transmission line fluid, US Patent No. 5,958,851 A (28 September 1999).
[205] T.V. Oommen and C.C. Claiborne (to ABB Power T&D Company, Inc.), High oleic acid compositions and methods of making and electrical insulation fluids and devices comprising the same, US Patent No. 6,312,623 B1 (6 November 2001).
[206] Cargill, Inc., *Development of Soybean Oil-based Transformer Fluid* (20 February 2004); Presented at United Soybean Technical Advisory Panel on Lubricants (22 September 2004).
[207] Anon., The evolution of soybean based transformer oil at Waverly Light and Power, http://www.waverlyia.com
[208] J.R. Marinho, E.G. Sampaio and M.M. Monteiro, Castor oil as an insulating fluid, *CIGRÉ-500–06. Symposium 05–87*, Vienna (1987).
[209] D.S. Shukla and V.K. Jahn, Water-based lubricants for metalworking, CHEMTECH **27**(5) (1997) 32–6.
[210] J.A. Schey, *Tribology in Metalworking: Friction, Lubrication and Wear*, American Society of Metals, Metals Park, OH (1983), pp. 27–130.
[211] Th. Mang, C. Freiler and D. Hörner, Metalworking fluids, in *Lubricants and Lubrication* (Th. Mang and W. Dresel, eds), 2nd edn, Wiley-VCH, Weinheim (2007), pp. 384–521.

[212] S.A. Miller, A.E. Landis, T.L. Theis and R.A. Reich, A comparative life cycle assessment of petroleum and soybean based lubricants, *Environm. Sci. Technol.* **41** (2007) 4143–9.

[213] E.J. Malloy, K.L. Miller and E.A. Eisen, Rectal cancer and exposure to metalworking fluids in the automobile manufacturing industry, *Occup. Environ. Med.* **64** (2007) 244–9.

[214] J.P. Byers (ed.), *Metalworking Fluids*, 2nd edn, STLE/CRC Press, Park Ridge, IL/Boca Raton, FL (2009).

[215] http://www.zeller-gmelin.de

[216] Directive 98/8/EC of the European Parliament and of the Council of 16 February 1998 concerning the placing of biocidal products on the market, *Off. J. Eur. Comm.* **L 123** (24 April 1998)

[217] Regulation 06/1907/EC (18 Dec. 2006), Registration, Evaluation, Authorisation and Restriction of Chemicals (REACH), establishing a European Chemicals Agency, amending Directive 1999/45/EC and repealing Council Regulation (EEC) No. 793/93 and Commission Regulation (EC) No. 1488/94 as well as Council Directive 76/769/EEC and Commission Directives 91/155/EEC, 93/67/EEC, 93/105/EEC and 2002/21/EC, *Off. J. Eur. Comm.* **L 396** (2006).

[218] Federal Insecticide, Fungicide, and Rodenticide Act (FIFRA), http://www.epa.gov/agriculture/lfra. html

[219] ASTM Standard E 2169, Practice for Selecting Antimicrobial Pesticides for Use in Water Miscible Metalworking Fluids, *Annual Book of ASTM Standards*, American Society for Testing and Materials, West Conshohocken, PA (2007).

[220] F.J. Passman, Current trends in MWF microbicides, *Tribol. Lubr. Technol.* **66**(5) (2010) 30–9.

[221] Th. Mang, Disposal of used lubricating oils, in *Lubricants and Lubrication* (Th. Mang and W. Dresel, eds), 2nd edn, Wiley-VCH, Weinheim (2007), pp. 183–90.

[222] T. Pfeifer, W. Eversheim, W. König and M. Weck, *Manufacturing Excellence: The Competitive Edge*, Chapman & Hall, London (1994).

[223] DIN 51385, *Kühlschmierstoffe Begriffe*, Hrsg. Deutscher Normenausschluß (June 1991).

[224] M. Sokovič and K. Mijanovič, Ecological aspects of the cutting fluids and its influence on quantifiable parameters of the cutting processes, *J. Mater. Proc. Technol.* **109** (2001) 181–9.

[225] http://www.masterchemical.com

[226] E.S. Nachtman and S.Kalpakjian, *Lubricants and Lubrication in Metalworking Operations*, CRC Press, Boca Raton, FL (1985).

[227] S. Woods, Going green, *Cutting Tool Engng.* **57**(2) (2005) 1–3.

[228] K. Michaelis and B.-R. Höhn, Reibungsverhalten biologisch leicht abbaubarer Schmierstoffe, *9. Internationales Kolloquium Tribologie*, Esslingen (11–13 January 1994).

[229] P.V. Joseph, S. Depak and D.K. Sharma, Study of some non-edible vegetable oils of Indian origin for lubricant application, *J. Synth. Lubr.* **24** (2007) 181–97.

[230] E. Durak and F. Karaosmanoğlu, Using of cottonseed oil as an environmentally accepted lubricant additive, *En. Sources* **26**(7) (2004) 611–25.

[231] A.K. Singh and A.K. Gupta, Metalworking fluids from vegetable oils, *J. Synth. Lubr.* **23**(4) (2006) 167–76.
[232] D.V. Lightcap Jr (to Midwest Biologicals, Inc.), Water-dispersible metal working fluid, US Patent No. 6,204,225 B1 (20 March 2001).
[233] H. Wichmann and M. Bahadir, Bio-based ester oils for use as lubricants in metal working, *Clean: Soil, Air, Water* **35** (2007) 49–51.
[234] A. Jain and R.P.S. Bisht, Metalworking emulsions from industrial vegetable oils, *J. Synth. Lubr.* **25**(3) (2008) 87–94.
[235] S. Suda, H. Yokota and M. Ibi (to Nippon Oil Corp.), Metal working fluid, US Patent Appl. No. 2008/026967 A1 (31 January 2008).
[236] M. Roegiers, H. Zhang and B. Zhmud, Elektrionized™ vegetable oils as lubricity components in metalworking lubricants, *Proc. 6th Intl. Conf. on Tribology Balkantrib '08,* Sozopol (12–14 June 2008), Paper BT – 97; *FME Transactions* **36** (2008) 133–8.
[237] A. Zeman, Biodegradable lubricants – analysis of metalworking oils and hydraulic fluids by mass spectroscopy *J. Synth. Lubric.* **13**(2) (1996) 149–61.
[238] R.J. Sturworld and F.O. Barrett (to Emery Industries, Inc.), Water soluble triglyceride compositions and method for their preparation, US Patent No. 3,928,401 (23 December 1975).
[239] Y. Yamanaka, M. Hayama, T. Oi, J. Imai and M. Satoh, Development of new grinding fluid for CBN grinding wheels. Part III, *Lubric. Engng.* **54**(8) (1998) 24–30.
[240] Y. Yamanaka, T. Oi, H. Nanao and M. Satoh, Development of new grinding fluids for CBN grinding wheel. Part V. A study on the grinding performance of various types of carboxylic acids, *Lubric. Engng.* **56**(3) (2000) 25–31.
[241] S. Suda, Metal working fluid, US Patent Appl. No. 2007/191240 (16 August 2007).
[242] H.S. Abdalla and S. Patel, The performance and oxidation stability of sustainable metalworking fluid derived from vegetable extracts, *Proc. Inst. Mech. Engineers, Part B – J. Engng. Manuf.* **220**(12) (2006) 2027–40.
[243] K. Goto, K. Takeda and E. Niwa (to Kyodo Yushi Co. Ltd.), Metal working oil composition, method of metal working, and product of metal working, PCT Int. Publ. No. WO 2008/001933 (3 January 2008).
[244] L. Feng (to Changzhou Xialei Chemical Co.), Method for preparing water-soluble anti-rust additive, Chin. Patent No. CN 101265435 (17 September 2008).
[245] M.G. Wakerly, C.W. Pouton, B.J. Meakin and F.S. Morton, Self-emulsification of vegetable oil-nonionic surfactant mixtures, *ACS Symp. Ser.* **311** (1986) 242–55.
[246] B. Wilson, Lubricants and functional fluids from renewable sources, *Ind. Lubr. Tribol.* **50**(1) (1998) 6–15.
[247] R.M. Lange, L.S. Bartley, C.G. Ollinger and J.M. Hogan (to Lubrizol Corp.), Maleated vegetable oils and derivatives, as self-emulsifying lubricants in metalworking, PCT Int. Publ. No. WO 2005/071050 A1 (4 August 2005).
[248] T.P. Oleksiak (to D.A. Stuart), Emulsion for the hot rolling of non-ferrous metals, US Patent No. 6,060,438 (9 May 2000).
[249] N.A.J. Cooban and H.L. Rieffe (to Unichema Chemie), Base fluid, US Patent No. 5,688,750 (18 November 1997).

[250] R.E. Bingeman, Low foaming, lubricating, water based emulsions, US Patent Appl. No. 2005/0037933 A1 (17 February 2005).

[251] H. Sefrin, Neue Technologien zur Kühlschmierung, *Seminarniederschrift TAE Esslingen*, (July 1998).

[252] W. Belluco and L. De Chiffre, Testing of vegetable-based cutting fluids by hole making operations, *Lubric. Engng.* **57**(1) (2001) 12–16.

[253] R. Garcia, Veggie coolants help Eaton slim down, *Lubes 'n' Greases* (1 June 2008), p. 26.

[254] Projektträger Fachagentur Nachwachsende Rohstoffe e. V., *1. Zwischenbericht zum Verbundvorhaben Entwicklung und praktische Erprobung eines Mehrzweck-Prozeß- und Betriebsöls auf Basis nachwachsender Rohstoffe für die spanende Metallbearbeitung (UNICUT-Projekt) gefördert von Bundesministerium für Ernährung, Landwirtschaft und Forsten* (1998).

[255] D.R. Kodali and S.C. Nivens (to Cargill, Inc.), Water soluble vegetable oil esters for industrial applications, PCT Int. Publ. No. WO 2001/0424405 A2 (14 June 2001).

[256] R.J. Sturwold (to Cincinnati-Vulcan Co.), Metalworking lubricants derived from natural fats and oils, US Patent No. 4,885,104 (5 December 1989).

[257] Th. Mang, F. Kubicki, A. Losch and W. Buss, Forming lubricants, in *Lubricants and Lubrication* (Th. Mang and W. Dresel, eds), 2nd edn, Wiley-VCH, Weinheim (2007), pp. 522–647.

[258] J. Suszkiw, Metalworking lube from soy oil, http://www.ars.usda.gov.is/AR/archive/may06/lube0506.htm (8 May 2006).

[259] B.L. Abdulquadir and M.B. Adeyemis, Evaluations of vegetable oil-based lubricants for metal-forming processes, *Ind. Lubr. Tribol.* **60**(5) (2008) 242–8.

[260] Z. Xianyan, Aluminum–magnesium alloy wiredrawing lubricant, Chin. Patent No. CN 101205499 (25 June 2008).

[261] A. Losch, Drylubes for sheet metal forming, *Abstracts 15th TAE Intl. Colloquium Tribology Automotive and Industrial Lubrication*, Esslingen (January 2006).

[262] H. Kim and T. Alt'an, Evaluation of dry film lubricants for automotive applications, Part I. A general review, *Stamping J.* (9) (2005) 38.

[263] G. Biresaw and S.M. Erhan, Solid lubricant formulations containing starch–soybean oil composites, *J. Am. Oil Chem. Soc.* **79** (2002) 291–6.

[264] C. Caminaga, F.O. Neves, F.C. Gentile and S.T. Button, Study of alternative lubricants to the cold extrusion of steel shafts, *J. Mat. Proc. Tech.* **182**(1–3) (2007) 432–9.

[265] K.P. Kammann, Jr and A.I. Phillips, Sulfurized vegetable oil products as lubricant additives, *J. Am. Oil Chem. Soc.* **62** (1985) 917–23.

[266] S.J. Asadauskas and J.M. Perez, Utilization of vegetable oils in metalworking fluids, in *Biobased Industrial Fluids and Lubricants* (S.Z. Erhan and J.M. Perez, eds), AOCS Press, Champaign, IL (2003), pp. 59–64.

[267] United States Department of Agriculture (USDA), *List of Proprietary Substances and NonFood Compounds*, Washington, DC (1996), No. 1419, 01/1996; together with National Sanitation Foundation (NSF), *Standard for 'food grade lubricants'*, Ann Arbor, MI.

[268] ISO 21469, *Safety of Machinery – Lubricants with Incidental Product Contact – Hygiene Requirements*, International Organization for Standardization, Geneva (2006).
[269] S. Lawate, Formulation components for incidental food-contact lubricants, in *Lubricant Additives: Chemistry and Application*s (L.R. Rudnick, ed.), 2nd edn, CRC Press, Boca Raton, FL (2009), pp. 511–22.
[270] Council Directive 93/49/EEC of 14 June 1993 on the hygiene of foodstuffs, Luxemburg (14 June 1993).
[271] Commission Decision 97/579/EC (23 July 1997), Scientific Committee in the field of consumer health and food safety, *Off. J. Eur. Comm.* **L 237** (28 August 1997) 18–23.
[272] EC Regulation No. 178/2002 (28 January 2002), General principles and requirements of food law, *Off. J. Eur. Comm.* **L 31** (1 February 2002) 1–24.
[273] Note for guidance on minimizing the risk of transmitting animal spongiform encephalopathy agents via human and veterinary medicinal products (EMEA/410/01 Rev. 2 – October 2003) adopted by the Committee for Proprietary Medicinal Products (CPMP) and by the Committee for Veterinary Medicinal Products (CVMP), *Off. J. Eur. Un.* **C24/**6 (28 January 2004).
[274] J. Taylour and C.T. Wright (to LHS), Aqueous lubricant compositions, Eur. Patent Appl. No. EP 1,001,005 A1 (17 May 2000).
[275] A.S. Valencia Sil, L.A. Grab, B.E. Schmidt, D.A. Halsrud, G.-J.J. Wei, E.D. Morrison and H.R. Dibenedetto (to Ecolab, Inc.), Dry lubricants for conveying containers, US Patent Appl. No. 2010/0286005 A1 (11 November 2010).
[276] A.S. Valencia Sil, L.A. Grab, B.E. Schmidt, D.A. Halsrud, G.J-.J. Wei, E.D. Morrison, H.R. Dibenedetto, J.L. Anacker, R.D. Johnson, M.W. Malvey, S.C. Bennett and S. Seemeyer (to Ecolab, Inc.), Lubricant for conveying containers, US Patent Appl. No. 2006/211582 A1 (21 September 2006).
[277] S. Küpper, C. Kohlstedde and M. Schneider (to Ecolab, Inc.), Use of O/W emulsions for chain lubrication, US Patent Appl. No. 2008/108532 A1 (8 May 2008).
[278] Loh Soh Kheang and Choo Yuen May, Food-grade palm-based lubricant base fluids, *MPOB Information Series* No. 325 (June 2006).
[279] S.S. Lawate, P.C. Naegely and V.A. Carrick (to Lubrizol Corp.), Environmental friendly food grade lubricants from edible triglycerides containing FDA approved additives, US Patent No. 5,538,654 (23 July 1996).
[280] R. Ogake, M. Okada, H. Takashima, T. Shimizu, N. Yamamoto and H. Uenoyama (to Nippon Oil Co., Ltd), Greasy oil and fat composition for food processing machines, US Patent No. 5,185,091 (9 February 1993).
[281] F.J. Flider (to Calgene Chemical, Inc.), Polyglycerol esters as functional fluids and functional fluid modifiers, US Patent No. 5,380,469 (10 January 1995).
[282] H. Faci, B. Cisler and C. Barrett, Biodegradable sugar mill lubricant: what makes it suitable for a sweet but aggressive environment, *NLGI Spokesman* **71**(9) (2007) 7–16.
[283] W.C. Carey, Spray-type pan release and method of making the same, US Patent No. 3,821,007 (28 June 1974).
[284] C. Doumani, Freeze–thaw stable emulsion cookware lubricant (to Blue Cross Laboratories, Inc.), US Patent No. 4,073,411 (14 Febuary 1978).

[285] J. B. Bliznak and A. Algas (to Alberto-Culver Co.), Aerosol cookware lubricant composition, US Patent No. 4,425,164 (10 January 1984).
[286] C.J. Boner, *Modern Lubricating Greases*, Scientific Publications, Broseley (1976), Chp. 5.
[287] W. Dresel and R.-P. Heckler, Lubricating greases, in *Lubricants and Lubrication* (Th. Mang and W. Dresel, eds), 2nd edn, Wiley-VCH, Weinheim (2007), pp. 648–714.
[288] A. Suresh and P. Aswath, Development of a high performance grease using design of experiments, *Tribol. Lubr. Technol.* **66**(4) (2010) 16–7.
[289] E.K. Pye and J. H. Lora, The Alcell™ process. A proven alternative to Kraft pulping, *Tappi J.* (March 1991) 113–17.
[290] N. Anandan, C.R. Jagga and R.K. Pandey, Tribological behaviour of additive free calcium stearate greases, *Tribol. Online* **2** (2007) 34–9.
[291] M.A. Delgado, M.C. Sánchez, C. Valencia, J.M. Franco and C. Gallegos, Relationship among microstructure, rheology and processing of a lithium lubricating grease, *Trans. IChemE, Part A, Chem. Eng. Res. Design* **83**(A9) (2005) 1085–92.
[292] H.B. Silver and I.R. Stanley, The effect of the thickener on the efficiency of load-carrying additives in greases, *Tribol. Int.* **7**(3) (1974) 113–18.
[293] J. Bhatia, R.L. Dhawan and R.L. Sovani, Tailor-made fatty acids based on castor oil for processing of high performance greases, *NLGI Spokesman* **56** (1992) 265–72.
[294] Y.L. Ishchuk, L.P. Ishchuk, A.K. Maskaev and A.A. Kotok, Influence of adding oleic and ricinoleic acids to 12-hydroxystearic acid on the structure and properties of lithium greases, *Chem. Technol. Fuels Oils* **10** (1974) 833–5.
[295] Y.L. Ishchuk, L.N. Dugina, M.E. Krasnokutskaya and B.A. Godun, Influence of composition of technical 12-hydroxystearic acid on properties of anhydrous calcium greases, *Chem. Technol. Fuels Oils* **22** (1986) 402–5.
[296] A. Hubmann, Rapsöl – ein alternatives Basisöl fur Schmierstoffe, *Mineralöltechn.* **34** (1989) 9.
[297] W.H. Dresel, Biologically degradable lubricating greases based on industrial crops, *Ind. Crops Prod.* **2** (1994) 281–8.
[298] H.-G. Schmidt, Modified natural base oils – their properties and use for lubricating greases, *ELGI Preprint* (1993); Komplexester aus pflanzlichen Ölen, *9th Int. Colloquium Esslingen* (1994), Vol. I, pp. 1–29.
[299] L.A.T. Honary, Performances and characteristics of soybean-based greases thickened with clay, aluminum complex and lithium complex, *NLGI Spokesman* **65** (2001) 18–27.
[300] H. Grignou, Hoang Le Chien and L. Moigner (to Christol Grease S.p.A.) Process for producing a lubricating grease from a vegetable oil and grease obtained, Fr. Patent No. FR 2,873,712 (3 February 2006).
[301] R.A. El-Adly, Producing multigrade lubricating greases from animal and vegetable fat by-products, *J. Synth. Lubr.* **16**(4) (2000) 323–32.
[302] L.A. García-Zapateiro, M.A. Delgado, J.M. Franco, C. Valencia, M.V. Ruiz-Méndez, R. Garcés and C. Gallegos, Oleins as a source of estolides for biolubricant applications, *Grasas Aceites* **61**(2) (2010) 171–4.
[303] R. Sánchez, J.M. Franco, M.A. Delgado, C. Valencia and C. Gallegos, Effect of thermo-mechanical processing on the rheology of oleogels potentially

applicable as biodegradable lubricating greases, *Chem. Engng. Res. Design* **86**(10A) (2008) 1073–82.

[304] P.P. Heckler, Application experience with biodegradable greases, *Proc. 1st Lubricating Grease Conference (Greasetech India)*, Naina Tal (February 1999), Vol. II, pp. 5–9.

[305] K. Hayashi and S. Tsuji (to Cosmo Sekiyu Lubricants KK), Biodegradable grease composition, Jap. Patent No. JP 2008/208240 (11 September 2008).

[306] P.A. Bessette, Manufacturing grease using dry technology, *NLGI Spokesman* **65** (2004) 14–17.

[307] J. Barriga, A. Aranzabe, P. Galda, M. Kruse, S. Jumaine and M. Andret, Sunflower based grease for heavy duty applications, *Riv. APAET – Mec. Exp.* **13** (2006) 129–33.

[308] P.M. Lugt, A review on grease lubrication in rolling bearings, *Tribol. Trans.* **52**(4) (2009) 470–80; *Tribol. Lubr. Technol.* **66**(7) (2010) 44–8, 50–6.

[309] R.F. Sukirno, S. Bismo and M. Nasikin, Biogrease based on palm oil and lithium soap thickener: evaluation of antiwear property, *World Appl. Sci. J.* **6**(3) (2009) 401–7.

[310] S. Okamura, M. Akiyama, S. Hashimoto and H. Iwamatsu, Heat resistant lithium grease composition and a small motor reduced noise bearing, US Patent Appl. No. 2009/029881 A1 (29 January 2009).

[311] W.-D. Abel, Biologisch schnell abbaubare Weichenschmierfette (Prüfung und Anwendung), *11th Int. Colloquium*, Esslingen (1998), Vol. 1, 225–36.

[312] Anon., Gadus greases for extreme conditions, *Lube Mag.* **104** (2011) 10–11.

[313] P.M. Lugt, A.v.d. Kommer, H. Lindgren and C. Roth, The R0F+ methodology for grease life testing, *ELGI 23rd Ann. General Mtg.*, Paris (30 April–3 May 2011).

[314] B.M. Graça, A.J.V. Campos and J.H.O. Seabra, Taper roller bearing lubricated with bio-grease, *Rev. Ass. Port. Anal. Exp. Tens.* (Mec. Exp.) **17** (2009) 117–28.

[315] A. Adhvaryu and B.K.Sharma, Current developments of biodegradable grease, in *Industrial Uses of Vegetable Oils* (S.Z. Erhan, ed.), CRC Press, Boca Raton, FL (2005), Chapter 2.

[316] NLGI, Marketing opportunities for biobased greases, National Lubricating Grease Institute, Kansas City, MO (2006), Paper# 0604.

[317] N. Canter, Tribology: the science of thinking small, *Tribol. Lubr. Technol.* **60**(6) (2004) 43–9.

[318] T.E. Karis, Lubricants for the disk drive industry, in *Lubricant Additives: Chemistry and Applications* (L.R. Rudnick, ed.), 2nd edn, CRC Press, Boca Raton, FL (2009), pp. 523–84.

[319] R. Tenne, M. Homyonfer and Y. Feldman, Nanoparticles of layered compounds with hollow cage structures, *Chem. Mater.* **10**(11) (1998) 3225.

[320] L. Rapoport, Y. Bilik, Y. Feldman, M. Homyonfer, S.R. Cohen and R. Tenne, Hollow nano-particles of WS_2 as potential solid state lubricants, *Nature* **387** (1997) 791.

[321] J.L. Mansot, Y. Bercion, L. Romana and J.M. Martin, Nanolubrication, *Braz. J. Phys.* **39**(1A) (2009) 186–95.

[322] N. Canter, Gas-expanded lubricants, *Tribol. Lubr. Technol.* **66**(5) (2010) 12–13.

[323] S. Wang and A. Clarens, Feasibility of gas-expanded lubricants for increased energy efficiency of power turbines, *Preprints ACS Div. Fuel Chem.* **55**(1) (2010) 255–6.

[324] C. Riebesell, J. Fett-Schudnagis and H. Thomas, Pas de deux of lubricant and seal, *Chem. Plants Proc.* **1** (2011) 44–5.

[325] M.E. Robinson, Improvement in starch lubricants, US Patent No. 163,809 A (1875).

[326] H. Yoshimura, D. Yamada, T. Tanaka, J. Nakahigashi, T. Nonishi and N. Inouchi, Application of cereal food materials as lubricants, in *Aerospace Materials and Manufacturing: Emerging Materials, Processes, and Repair Techniques* (M. Jahaz, M. Elboujdani and P.C. Patnaik, eds), *Proc. Intl. Symp. Aerospace Materials and Manufacturing: Emerging Materials, Processes, and Repair Techniques*, Montreal, QC (1–4 October 2006), pp. 439–45.

[327] J. Matsumoto and H. Ito, Lubricant for endoscope and endoscope, US Patent Appl. No. 2006/0058583 A1 (16 March 2006).

[328] H. Yoshimura, T. Nonishi, T. Komaki and N. Inouchi, Grain-type lubricant for metal processing, Jpn. Kokai Tokkyo Koho JP 2003/096484 A (3 April 2003).

[329] V.D.P. Kenerson and C.H. Morgan, Inprovement in processes of wire-drawing, US Patent No. 203,346 A (7 May 1878).

[330] H. Yoshimura and E. Fuwa, Material processing method using lubricants containing grain powder as major component, Jpn. Kokai Tokkyo Koho JP 11140481 A (25 May 1999).

[331] H. Yoshimura and E. Fuwa, Method for using liquid suspension mainly containing grain powder as machine lubricant, Jpn. Kokai Tokkyo Koho JP 11270786 A (5 October 1999).

[332] H. Yoshimura, T. Hayashi, J. Katayama and H. Fuwa, Improvement of Erichsen values of titanium and stainless steel sheets by wheat flour lubricant, *J. Jap. Soc. Technol. Plast.* **40**(457) (1999) 134–8.

[333] H. Yoshimura, T. Nishihara, T. Nonishi and N. Inouchi, Improvement of Erichsen values of titanium sheets using various cereal flour lubricants, *Tetsi to Hagane* **87**(7) (2001) 474–9.

[334] H. Yoshimura, S. Torikai, T. Nishihara, T. Nonishi and N. Inouchi, Application of wheat flour lubricants to the press-forming process, *J. Mat. Proc. Technol.* **125–6** (2002) 375–8.

[335] H. Yoshimura and K. Okamura, Sugar-containing liquid lubricant, Jpn. Kokai Tokkyo Koho JP 2010/168519 A (5 August 2010).

[336] Anon., Lubricants for refractory shutters, Jpn. Kokai Tokkyo Koho JP 57121097 A (28 July 1982).

[337] J. Takahashi, Water-based lubricants for finishing automobile interior sheet fabrics, Jpn. Kokai Tokkyo Koho JP 05331766 A (14 December 1993).

[338] J. Takahashi, S. Nakane and M. Okamura, Cationic lubricants for fibers and their manufacture, Jpn. Kokai Tokkyo Koho JP 05257845 A (24 September 1993).

[339] M. Svoboda and A. Pohorska, Sizing mixture for glass fibers, Czech Patent No. CS 204825 B1 (30 April 1981).

[340] V. Sauer and K. Rothschein, Indirect starch lubricants, *Sklar a Keramik* **34**(2) (1984) 30–3.

[341] T. Maeda and S. Sakaguchi, Sizing agents for glass fibers and seizing agent-coated glass fibers, Jpn. Kokai Tokkyo Koho JP 01242444 A (27 September 1989).

[342] W.B. Rice, Starch–oil sizing compositions for glass fibers, PCT Int. Publ. No. WO 9315140 A1 (5 August 1993).

[343] H.Y. Tin, C.A. Lin and J.J. Lin, Studies on the applicability of lab-prepared lubricant formulations for the sizing of polyester cotton blends, *Jiemian Kexue* **20** (1983) 30–7.

[344] S.H. Wu, B.K. Cheng, G.A. Nichols and J.H. Park (to Mallinckrodt, Inc.), Use of magnesium stearate dihydrate for lubrication of solid industrial or consumer products, PCT Int. Publ. No. WO 2009/114227 A1 (17 September 2009).

[345] E. Richter, Lubricants, in *Plastics Additives Handbook* (H. Zweifel, ed.), 5th edn., Hanser Verlag, Munich (2000), pp. 511–52.

[346] E.-U. Brand and P. Daute (to Cognis Oleochemicals), Use of polyethylene glycol esters of fatty acids as lubricants for plastics, Can. Patent Appl. No. CA 2,594,153 (20 January 2008).

[347] J.B. Williams, J.A. Falter and K.S. Geick, Lubricating system for rigid PVC, in *Plastic Additives. An A–Z Reference* (G. Pritchard, ed.), Chapman & Hall, London (1998), pp. 450–8.

[348] M. Peters, White mineral oils, *Lube Mag.* **106** (2011) 6–7; *Lube Mag.* **107** (2012) 26–8.

13
Markets for biolubricants

Abstract: The highly diversified global lubricant market is heavily dominated by the mineral oil-based products of (multi)national oil companies. The market share of biodegradable lubricants of fossil origin and biogenous oils is less than 2% and best in Europe. In Germany, which has actively pursued biolubricant market introduction, mobile hydraulic applications show the highest share of biogenous oils. Major constraints limiting the biolubricant market penetration are perceived cost and general lack of awareness of technical potential. The market outlook for biolubricants is influenced by many factors (political, economic, demographic, environmental, regulatory). Growth rates for biolubricants are higher than for the overall lubricant market.

Key words: biodegradable and biogenous oils, biolubricant markets, German Market Introduction Programme (MIP), market share, market mechanisms.

13.1 Introduction

Various recent market research studies are available for the global lubricants industry in general and for biolubricants in particular [1–8], with forecasts up to 2017 [1, 7, 9]. The world annual lubricant production of about 36 Mt is less than 1% of total refinery throughput of crude oil (global capacity of 4103 Mt/yr of 723 petroleum refineries in 2004). World lubricant use amounted to 37.4 Mt in 2004, nearly 38 Mt in 2005 with a projected increase of 1.2% up to 2015 [10], but down to 35.1 Mt in 2011 [11]. The rate of consumption of total lubricants is almost flat, with a modest 2.2% growth for synthetic lubricants [12]. A most recent (optimistic) forecast holds that the global lubricant demand is to reach 40.5 Mt in 2012 [1], which is a downward adjustment in comparison to an earlier forecast of 41.8 Mt in 2010 [3]. More highly industrialised and motorised regions increasingly go for high-performance lifetime lubrication, with the effect of lower volatility rates, and thus reduced specific lube demand. As a result of higher-quality lubricants and improved materials fewer lubricants are needed per unit of economic output in the course of time. On the other hand, strongest volume growth is expected in the emerging markets of Asia/Pacific, Africa/Mideast and Latin American regions due to rapidly advancing industrialisation and rising car ownership. For detailed historical lubricants demand for the years 1997, 2002 and 2007, and forecasts for 2012 and 2017 by formulation (e.g. petroleum, synthetic, re-refined, vegetable-based), products (e.g. engine oils, process oils, hydraulic fluids, metalworking fluids), and markets, see ref. [1]. Engine oils will remain as the largest segment while process oils grow the fastest.

Figure 13.1 shows the world lubricant demand by region, with a predominance for Asia/Pacific, which has overtaken North America and Europe at 25.8% or 9.3 Mt in 2008. The present global ranking of the top-ten lube markets is as follows: the United States (6500 kt), China, Japan, Russia, Brazil, India, Germany, South Korea, the United Kingdom and Indonesia (in this order). Differences in regional pro-capita lubricant consumption are considerable: 20 kg in North America, 11.4 kg in Europe vs 3 kg for the BRIC region (Brazil, Russia, India and China).

Mid-2009 the world had 157 virgin base oil refineries (to be unchanged up to 2015) with a total nameplate capacity of 46.5 Mt, to increase shortly to 50 Mt; Europe holds about 25% of global capacity volumes. The number of lube manufacturers has decreased from about 1700 players in the mid-1990s to about 700 manufacturers at the end of 2005. Companies and cooperatives engaged in manufacturing finished lubricants as their core business range from big petroleum multinationals to small lube companies serving niche markets. The 20 main lubricant manufacturers worldwide are as follows, in decreasing order of capacity: Royal Dutch Shell, ExxonMobil, BP, Chevron, PetroChina, Sinopec, Lukoil, Total, FUCHS, Nippon Oil, Idemitsu, Valvoline, ConocoPhillips, Petronas, CPC, Petramina, PDVSA, Repsol, SK Corp. and Indian Oil (subject to change). Major players are multinational and national oil companies for which lubricants account only for a very small part of their business. These large, vertically integrated

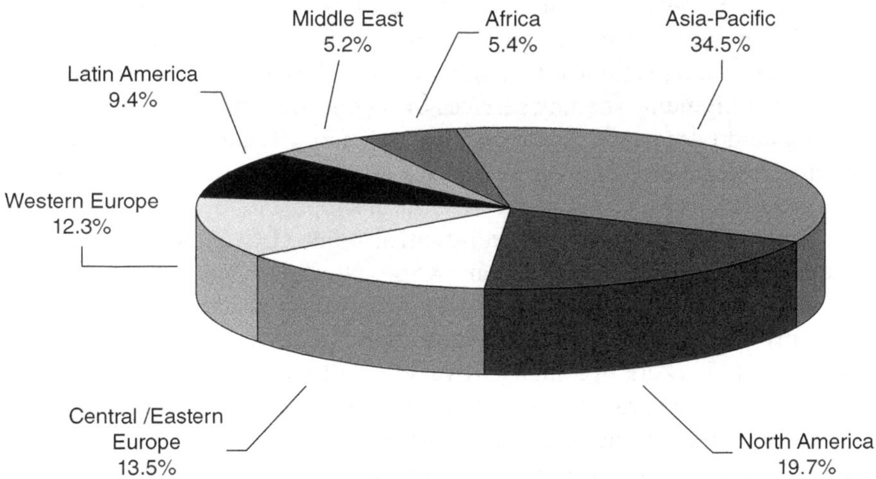

13.1 Global lubricants demand 2008 (36.2 Mt). After ref. [11].

companies (about 180 worldwide) mainly focus on high-volume lubricants such as engine, hydraulic and gear oils. The top three lubricant producers (Royal Dutch Shell, ExxonMobil and BP Amoco) hold over 50% of the total lubricant market. On the other hand, hundreds of independent lubricant companies, for which lubricants is their core business, concentrate on manufacturing and marketing of specialties and tailor-made lubricants. Both these independent lube companies and the petroleum industry are lately experiencing continuing concentration in order to maintain critical mass. The dedicated producers generally do not operate base oil refineries but purchase their mineral base oils from the large petroleum companies and other raw materials from the chemical and oleochemical industry. Specialties often require the use of chemical processing, such as saponification (for greases), esterification (for ester base oils or additives), amidation (for metalworking lubricants), etc. The many thousands of lube applications require a diverse number of lubricant systems, such as oils, emulsions (oil-in-water, water-in-oil, oil-in-oil), water-based solutions, greases, solid lubricant suspensions or solid lubricants.

Historically, marketers have promoted their products using lubricant product catalogues and other recommendation manuals. Manufacturer-specific literature is widely available (e.g. refs. [13, 14]) and easily accessible (see Appendix A.4). OATS, Ltd (Swindon, UK) is a lubricants database provider. Innovative digital portals allow access to on-line marketing support and can be used to maximise lubricant profitability. For instance, Shell Lubricants operates a Product Plus web-based interactive mode service to assist customers to get the most suitable (Shell) product for a particular application and to answer important lubricant questions. The Shell Lubricants portfolio includes LubeAdvisor, LubeAnalyst, LubeClinic, LubeCoach, LubeExpert, LubeMatch, LubeReclaim and LubeVideoCheck [15]. The on-line service, which has already seen over 2.5 million visitors worldwide, claims significant savings for its customers and real improvements to equipment. The key services bring on-line lubrication recommendations, field-based technical expertise and support, and practical tools to monitor performance of lubricants whilst in use. Also other lubricant manufacturers, such as FUCHS, provide technical support to manufacturing sites, assistance with lubrication critical control point (LCCP) surveys and oil condition monitoring (see Section 14.5.1). Nowadays there is also a burgeoning number of applications (apps) designed for social networking sites which provide relevant product information [16, 17].

Lubricant markets are highly diversified (Table 13.1). The volume of lubricants used in various automotive and industrial applications, especially engine oils and hydraulic fluids, is large (about 30 Mt/yr). In 2004, the breakdown of the 37.4 Mt (total) lubricant market (with a global growth rate of 0.5% p.a.) was as follows: 53% automotive lubricants (engine oils,

Table 13.1 Lubricant markets

Commercial/industrial
• Agriculture
• Construction equipment
• Mining
• Petroleum and natural gas
• Power generation
Durable goods manufacturing
• Transportation equipment
• Machinery
• Metals
Light vehicle aftermarket
• Light vehicle lubricants
• Light vehicle engine oils
Transportation equipment aftermarket
• Heavy-duty trucks and buses
• Railroad equipment
• Marine equipment
• Aerospace equipment
Non-durable goods manufacturing
• Chemicals
• Cosmetics and toiletries
• Food and beverages
• Paper and textiles
• Plastics and rubber
• Printing inks

transmission fluids), 32% industrial lubricants, 5% marine oils and 10% process oils (actually often not comprised in lubricant statistics). The industrial lubricant sector consists of 37% hydraulic oils, 7% gear oils, 16% metalworking fluids, 9% lubricating greases and 31% other industrial oils (such as turbine and compressor oils, demoulding oils, chainsaw oils, etc.) [18]. The global economic importance of lubricants can be illustrated by an estimate of the US Department of the Environment (DOE) that reducing friction and wear in engine and train components could save the US economy up to US$ 120 billion per year [19]. Upgraded lubricants reduce energy bills, maintenance costs, replacement costs and pollution.

The complicated lubricant market is made up of three main product segments. Petroleum-based lubricants (or mineral oils), derived from crude oil, make up the main share. Synthetic lubricants, synthesised from intermediate petrochemicals, possess high-performance attributes, often including biodegradability. Ag-based lubricants, derived from agricultural products such as vegetable oils and therefore inherently biodegradable and low in toxicity, occupy a niche market segment. Although environmentally friendly, biodegradable alternatives are technically available for almost all mineral oil-based lubricants, penetration of biolubricants within the market is very low (less than 2% of total market) and the market share of small and medium enterprises (SMEs) is poor. Incentives are needed to help an immature market to compete. It is the aim of various measures taken by legislators and regulatory authorities to increase the market share. Production and use of biodegradable lubricants illustrate diligence by the lubricant manufacturer, original equipment manufacturer (OEM) and the consumer in contributing to a cleaner environment. Two major constraints limiting the biolubricant penetration to the market are perceived cost and performance compared with current mineral-based lubricants. Market value for an application can be achieved by creating the desired functionality through an in-depth understanding of the molecular origins of the structure–properties–functionality relationship (see Fig. 5.3). Mineral and biogenous oils are based on completely different chemical products.

The market of lubricants is still dominated by mineral oil-based products, the ecological characteristics of which are not satisfactory. Today, market demands for lubricants are primarily superior performance (including longer lifetime) but also increased environmental compatibility, reduced emissions, higher energy efficiency and greater occupational safety. The pollution problem is so severe that approximately 50% of all lubricants sold worldwide end up in the environment via accidental spillage, non-recoverable usage, volatility, industrial and municipal waste, urban runoff, refinery processes, and total-loss applications [20]. The oils contaminate soil, air and drinking water and pose a threat to animal and plant life. For historic field problems with mineral oil-based hydraulic fluids, see Section 12.3.1.

Various factors make it difficult to obtain reliable estimates on the actual consumption of vegetable oils in the lubricants sector. As there is no universally accepted definition of biolubricants the term may be used to denote either biodegradable products only or biodegradable *and* renewable materials (see Section 3.4). However, it is generally accepted that biolubricants must fulfil the OECD 301 test for biodegradability [21]. Not all biolubricants are completely vegetable oil-based. In some countries, getting an ecolabel only requires that 50% of the oil is renewable; synthetic esters or even petroleum oils can be used in formulations. While the application areas of lubricants are very diverse there is also a disturbing lack of consistency with the use of the term 'biolubricant' by different product manufacturers and customers. 'Readily biodegradable' lubricants are not necessarily manufactured using renewable raw materials. For example, according to the classification of hydraulic oils, final products encoded HEES (hydraulic oil environmental ester synthetic) stand for readily biodegradable synthetic ester oils that can be obtained by appropriate chemical modifications of both mineral oil and vegetable oil. As the terminology 'biolubricant' includes both biodegradable lubricants of fossil origin and lubricants of vegetable origin it is often difficult to know their respective market shares. Most biolubricants are composed of natural or synthetic esters or mixtures, and additives. The biolubricant market is estimated to be based on synthetic products for 80% and 20% on non-modified vegetable oils [22].

Vegetable oils and animal fats served as lubricants before the domination of petroleum-based lubricants introduced into the market since the rise of petroleum in the early 1900s. The renaissance of ag-based lubricants is determined by dwindling petroleum reserves, greater environmental conscience and increasing pressure from legislators and regulators. Consequently, high-performance vegetable oil-based lubricants have recently been developed and marketed. Both plant-based and synthetic lubricants have potential to capture the market of biodegradable lubricants.

Out of the total global oils and fats production (153 Mt, 2006/07) about 14% or 21.4 Mt/yr provides the basis of the oleochemical industry (soaps, candles, detergents, toiletries, cosmetics, surfactants, emulsifiers, perfumes, solvents, etc.), and lately also of new applications such as biodiesel and biolubricants. The global biodiesel production has increased rapidly up to 12.7 Mt (2008) with an overoptimistic installed nameplate capacity of approximately 37.9 Mt/yr (2009) [23]. Even though the present requirements for vegetable oils supply for bio-based lubricants are very modest (probably not exceeding about 0.2 Mt/yr), this adds to a presently already tight market and sets limits to growth.

Plant-based lubricant manufacturers are faced with several crucial issues, including concerns with technical potential, performance standards, testing protocols, certification and governmental policy. Introduction of biodegradable lubricants into a market dominated by Big Oil is a huge challenge. The decisive factor is that the oleochemical products (usually esters), which are proposed as replacements for mineral oils, do not only exhibit ecologically compatible properties but also show equivalent or better performance than traditional products. That this is possible has very clearly been demonstrated for the various product groups where biodegradable lubricants have already proved themselves. Specially designed fatty acid esters are available as valid environmentally friendly alternatives for almost all mineral oil-based products in a wide range of application areas, for example outboard motor oils, gearbox oils, hydraulic oils, and lubricants for oil drilling. The enhanced risks of contamination of soil, water and air, and damage to crops connected with direct exposure of cultivations to agricultural and forestry machinery determines the special interest of these sectors for the use of biolubes. While vegetable oil-based biolubricants are generally less suited for use in motor oils, they show excellent performance as hydraulic oils and transmission fluids. Biolubricants can also be used in (two-stroke) chainsaw engines.

The market outlook is influenced by many factors including global economy and population, manufacturing prospects (in particular motor vehicles), world oil industry, and environmental and regulatory trends. Success of biodegradable lubricants greatly depends on statutory regulations which should govern the use of environmentally friendly products [24]. Investor confidence also has a major impact on biolubricant development, which is not favoured by the volatile nature of renewable resource governance. For a recent market analysis of 77 biolubricant manufacturers, see ref. [9].

13.2 European biolubricant markets

Biolubricants have been in use in Europe for over 20 years. Their use has been driven by regulations in some countries and perhaps a 'greener' view of the environment and the need for renewable materials in some other countries. Already in the 1990s regulatory and environmental pressures in Europe required that biodegradable oils be used in areas where petroleum products were deemed unacceptable (see Section 9.3).

The Western European lubricants (export) market is approximately 5.8 Mt/yr (2006) with Germany being the largest national market in Europe. Lubricant consumption (Europe, 2004) is roughly divided as follows: automotive, 47%; industrial, 32% (including hydraulic oils, which account for approximately 15% of total lubricant applications); process oils, 11.3%; and marine and aviation, 9.4%. According to Frost & Sullivan 2007, the European biolubricants market 2006 amounted to 127 kt or about 2% [5], to be compared with 100 kt (2001) [25] and 110 kt in 2002. While volume growth is small, revenue growth is larger because of the higher price of the biolubes. Various estimates of the biolubricant share of the total lubricant usage have been given, ranging from an optimistic 3.2% to 1% (FUCHS Oil), which can partly be traced back to the differences in accounting for and the definition of a 'biolubricant' (see Section 13.1).

Various European countries, notably Germany, Switzerland, Austria and those in Scandinavia, are leading the world to advance biolubricants to the markets through development and legislation. These countries are more bio-oriented than others, such as France, Spain and the United Kingdom. The major vegetable oil in use in Europe for industrial products is rapeseed. European biogenous lubricants are for the most part (70%) rapeseed oil (RSO)-based. Processes for modification of rapeseed oil to improve stability in lubricant use have been developed. These developments and the current political support and infrastructure will make it difficult to replace rapeseed oil in the European lubricant market.

Market share of biolubricants in the EU market is in the order hydraulic fluids (67%), chainsaw oils (14%), gear oils/two-stroke engine oils and cutting oils (12%) and concrete release agents (7%). Estimated growth rates are highest for hydraulic fluids.

As technically over 90% of all lubricants could be replaced by biolubricants this gives a present potential European biolubricant market of at most 5 Mt/yr. However, the long-term biolubricant potential in Europe is estimated being only 10–20% of the total market (500–1000 kt/yr), but with widely different prospectives per product group (Table 13.2) [26]. The increase in oilcrop production in Europe has led to overoptimistic expectations of the growth of the biolubricant market. Similarly, optimistic marketing expectations for soybean-based industrial lubricants in the United States have also not yet come true.

The European biolubricant market is worth about €0.11 bn (2007). While the total lubricant market is rather static, the bio-segment grows by ~3–5%/yr. In Europe at least 500 different lubricating agents are being marketed as being 'non-toxic' and 'biodegradable'. In Germany and France vegetable oil-based biolubricants are available for all applications; in Italy (2004) access to market is limited to chainsaw oils. The French bio-chainsaw oil market (2007) is about 200 000 litres. Although all major petrochemical companies are able to provide biolubricants the European market is dominated by some 70–80 independent manufacturers supporting 80% of the biolubricants markets. European competitors are Cognis (now BASF), Uniqema/Croda, Oleon (now Sofiprotéol), BINOL (dominant in the Nordic area), and various family-owned companies in Germany, some long established. For environmentally adapted lubricants in the Nordic marketplace, see ref. [27].

Table 13.2 Long-term European market expectations for biodegradable biolubricants[a]

Product group	Biolubricant market volume (kt/yr)	Market share (%)
Automotive oils	250	11
Industrial gear oils	10	5
Hydraulic oils	200	27
Compressor oils	25	38
Turbine oils	20	10
Metalworking fluids	10	2
Demoulding oils	110	100
Chainsaw oils	60	100
Lubricating greases	100	100
Process oils	200	30

[a] According to ref. [26].

Table 13.3 Bio-oil consumption (kt) in Germany (2005)

Lubricants	Biogenous oils	Other bio-oils	Total
Biolubricants (total)	7.1	16.7	23.8
Bio-hydraulic fluids	4.1	6.6	10.8
Bio-metalworking oils	1.4	1.9	3.3
Bio-greases	0.4	1.9	2.4

After ref. [28].

At present, Germany holds the largest share of the total European biolubricants market. The domestic sales volumes for lubricants amounted to 1066 kt in 2003, 1040 kt in 2005, and 1109 kt in 2008, thus experiencing only minimal growth. No comparable statistics are available for biolubricants. An estimated share of bio-oils (i.e. lubricants with minimal 50% renewable components) in 2005 amounted to 23.8 kt or 2.3% (incomplete data for total-loss lubricants) (Table 13.3). Out of these only 7.1 kt or 0.7% were on account of biogenous oils, mainly composed of mobile hydraulic fluids (including universal tractor transmission oil, UTTO, and super tractor oil universal, STOU) [28]. Use of biogenous hydraulic fluids doubled since the introduction of the German Market Introduction Programme (MIP) in 2000. In 2005 the market share of total bio-oils respectively biogenous oils amounted to 16%/6% for mobile hydraulic fluids, 3.9%/1.6% for metalworking oils, and 7.5%/1.2% for greases. Bio-hydraulic fluids for mobile equipment (excavators, wheel loaders, etc.) score high in the biolubricants market. Hydraulic systems operating with pressures of several hundred bar represent a high loss risk. Typically, a forest harvester may lose several hundred litres of hydraulic oil per year. Partly as a result of the MIP the market share of bio-oils in mobile hydraulic applications is growing fast in Germany [29–31]. Expectations are that the market for mobile bio-hydraulic fluids in Germany might reach 30% by 2010 [28].

It should be stressed that other volume estimates have been made. Kaup [32] indicates a German biolubricant consumption of the order of 40 kt/yr, if full account is taken of total-loss lubricants. Fachagentur Nachwachsende Rohstoffe (FNR)'s 2005/2006 'Market Analysis for Renewable Resources' confirms an overall market for biolubricants in Germany of 46.4 kt or about 4% (including readily biodegradable mineral oil-based lubricants). However, the biogenous portion in the individual products on the market varies greatly (with products containing less than 50% biogenous components). The overall market for biolubricants with more than 50% biogenous components (in accordance with the German MIP funding requirements) amounted to about 7.1 kt, which represents less than 1% of the overall lubricant market in Germany. The market for hydraulic oils is the largest, followed by metalworking oils, chainsaw oils, penetrating oils, gear oils and motor oils [33].

The aforementioned estimate of 7.1 kt for the total volume of lubricants made from renewable materials in Germany seems rather low. A more recent study performed by the nova-Institut estimates the current overall market for biolubricants manufactured from renewable raw materials (RSO, sunflower oil (SNO) and animal fats) in Germany at approximately 24.0 kt for 2008/2009 [34]. FNR assumes that the actual consumption is around 35 kt or about 3% of domestic sales for lubricants [35].

In an attempt to stimulate national markets temporarily various incentives have been offered to biolubricant users (e.g. German MIP [36, 37] and Dutch VAMIL scheme [38]). The German government has supported the conversion of existing or new equipment from mineral oil to bio-based oil through financial grants, technical advice and publicity. The MIP 'Bio-based Fuels and Lubricants' [39], launched in 2000 by the German Federal Ministry of Agriculture (BMELV), and administered by the Agency for Renewable Raw Materials (FNR), has been operating until 2008 with an initial yearly budget of 10 million Euros, reduced stepwise. Aims of the German MIP for bio-based oils were:

- increased usage of vegetable oils for lubrication
- promotion of environmentally friendly technologies
- use of renewable resources (CO_2-neutral production)
- support for agricultural producers

through reimbursement of changeover cost from mineral oil lubricants to biolubricants. Motivations for users to change over are: (i) public pressure to use biolubricants in certain applications; (ii) legislation (federal, state or local authorities); and/or (iii) corporate 'green' image. Financial subsidies were intended to promote the changeover of petrochemical to bio-based lubricants (except for chainsaw and concrete mould-release applications). Only lubricants from the *Positivliste* did qualify [37]. Table 13.4 shows the minimum requirements of the German *Positivliste*. In November 2008 a

Table 13.4 Minimum requirements of the German *Positivliste*

• Renewable mass content >50%
• German WGK class 1
• Biodegradability 80% (CEC L-33-A-93 standard) or 60% (OECD 301B standard)
• Technical information (supplied by manufacturer)
• Motor, gear, hydraulic and other lubricants admitted
• Chainsaw oils and mould-release agents excluded

Positive List of acceptable fluids contained 225 oil products, namely 130 lubricants (59 vegetable oils and 71 synthetic esters) and 95 hydraulic fluids (80 HEES, 13 hydraulic oil environmental triglyceride (HETG) and 2 hydraulic oil environmental poly-α-olefin/related hydrocarbon (HEPR)), supplied by 40 vendors [40]. Products on the positive list contain a minimum of 50% renewable carbon, are rapidly biologically degradable and little or not hazardous to water. These requirements conform to the EU Euromarguerite label. A radiocarbon analysis method may be used for detection of renewable raw material (RRM) in MIP-admitted products (see Fig. 8.4). For a quick check on biodegradable oil in a vehicle a portable sensor has been developed which detects the presence of esters in a lubricant [41].

Work within the framework of MIP has demonstrated that there are only a few problems related with the use of bio-oil, especially if the proper precautions are taken. One of the major obstacles in market conversion is the sensitivity of bio-based ester fluids or biodegradable fluids to contamination with mineral oils. Petroleum-based fluids must never be mixed with bio-oils! Such contamination may generate reaction products and deposits. Incomplete flushing during conversion from mineral oil to bio-oil may also lead to problems such as foaming, poor air release, or filter clogging, connected with the contamination level. The degree of contamination is related to the amount of metals introduced through the mineral oil additives. The effects are specific for different bio-oil products [42]. Göhler and Rüschkamp [43] have examined the compatibility of bio-oils (HEES 46, HOSNO 32/68 and HEPR 32/68) and zinc-containing mineral oils (HLP 46, HLPD 46, HVLP 46 and SAE 15W-40). Incompatibility of some mixtures was observed. Consequently, during changeover important technical precautions are required to avoid contamination with mineral oils. Both ISO 15380 and VDMA for changeover [44] recommend a maximum mineral oil contamination level of 2%. Complete draining of the old mineral oil fill is therefore an important issue. With mobile equipment typically having about 60% of its oil volume in the tank and 40% distributed over the hydraulic system, just draining of the tank is obviously inadequate. Several flushing cycles and large quantities of oil may be required. Flushing is time consuming and expensive. Future bio-oil development should aim at enhancing the tolerance against components present in mineral oils. Equally important is the use of bio-compatible materials (seals, flexible hoses, etc.). Increased service life of the expensive bio-oils is guaranteed only by dry and clean operation. Most hydraulic bio-oils are esters which are sensitive to water. Improved maintenance practices like bypass microfiltering and regular laboratory checks are important for high levels of oils cleanliness (see also Section 14.5). This is especially meaningful when more expensive lubricants are being used. Extension of oil life is also influenced by efficient temperature control.

In the 2000–2008 period some 25 000 changeovers have been carried out within the 'Bio-based Fuels and Lubricants' MIP; 95% of the conversions concerned mobile hydraulic equipment (e.g. agricultural, forestry, construction), hydraulic waterway installations, wind turbine gear oils, and loss lubrication greases [45]. Mobile hydraulic machines are especially interesting applications because of the high loss risk of the fluid, given the high pressures, extended piping, and the large quantity of fluid present on the vehicle. Since 2002 about 10 kt of biohydraulic oil were sold in Germany. MIP has supported an annual quantity of 1 ML biolubricant and has contributed to reducing market barriers for biolubricants through:

- communication of high technical performance level;
- compensation for higher product price and changeover cost;
- long-term increased volume;
- long-term reduced price difference.

Bio-based lubricants have successfully been introduced in Germany with a low degree of technical problems and a high percentage (94%) of consumer satisfaction. Taking care of some special features helps maintaining cost-effective operation. A combined offer of oil, advice and technical service to the consumer is the most successful option. Success is largely dependent on the sales force of biolubricant dealers. For a real breakthrough, the German market is probably too small, and an initiative at European level is needed.

In the context of the Dutch sustainable purchasing policy criteria have been developed for 80 product groups, including lubricants. The Dutch VAMIL regulation (tax break) allows for accelerated depreciation of investments in equipment in which environmentally friendly lubricants are used [38]. In 2008, the VAMIL programme set the EU ecolabel as the official standard for lubricants and Blue Angel for greases (due to a lack of European Ecolabel for Lubricants (EEL) products). The Dutch Water Management Authority (RWS) has introduced the use of biolubricants for hydraulic applications in waterways. In France, lubricants awarded with the EEL are exempted from TGAP (general tax on polluting activities), which amounted to €44.02/t in 2008.

Despite mandates and the aforementioned incentive programmes the relatively modest growth in market share for biolubricants over recent years suggests, however, that more action is needed if market penetration of these products is to be successful. For example, the Swedish forestry industry is driven by environmental concerns from German timber consumers [27]. In the United States a mandatory procurement of bio-based products by government agencies is operating. New European regulations (461/2010, 330/2010, 595/2009, 715/2007) can have impact on the lubricant business.

While recent statistics show that environmentally considerate products account for only 2% of lubricant sales in Europe, significant growth in market share is likely as application techniques and ultimate benefits of newly developed products become more widely understood. As part of Royal Dutch Shell's commitment to sustainable development, the company has developed and promotes the use of lubricants designed to reduce the impact on the environment. All of the environmentally considerate lubricants produced by Shell are formulated using carefully selected unsaturated and saturated synthetic esters, or advanced technology RSO. Optimised additive packages are used to give good extreme pressure (EP) and anti-wear (AW) performance, corrosion protection and oxidative stability, while maintaining environmental acceptability. Shell's line of environmentally considerate products includes Shell Naturelle HF-E 46 for hydraulic systems and Shell Naturelle Grease EP. These products conform to the highest environmental standards for biodegradability and low ecotoxicology. Shell Naturelle HF-E 46 is on the German *Positivliste*. Shell Naturelle HF-E 32 and 46 both meet the environmental requirements of the Swedish Standard SS 155434. Hitachi has approved Shell Naturelle HF-E 46 for use in all types of its heavy earth-moving excavators and is capable of more than 2000 h of use in the field without fluid change [46]. For the same lubricant, used in dam gates at a hydroelectric power station operated by the Electric Corporation of New Zealand (ECNZ), a service life of at least 10–15 years is reasonably expected.

At the European level, various biolubricant product development initiatives should be mentioned (see Section 10.5). For example, Table 13.5 lists

Table 13.5 Top ten IBIOLAB biolubricants

Product group	Product name	Application(s)
Metalworking fluid and emulsion	BAS SML	Metal cutting (steel and non-ferrous metals)
Metalworking fluid	Ibiolab 2009	Multipurpose neat oil
Metalworking fluid emulsion and release agent	EVL 28 8A and EVL 28 8B	Rail drilling; cost-effective Bioemulsion concentrate
Hydraulic fluid	HF BSA-1	Biodegradable fluid for hydraulic systems
Hydraulic fluid	L 0907/1	Hydraulic fluid
Two-stroke engine oil	L 0906/1	Two-stroke engine oil
Grease	Ibiolab XP4a	Curved rail grease
Grease	Ibiogrease Cali-2	Low-cost multipurpose grease
Grease	Ibiogrease Li XP-2	Multipurpose high-performance grease

bio base stocks and emulsifiers developed within the European IBIOLAB (Improvement of Biolubricant Manufacturing and Development thanks to the Obtaining of Ecolabels in a wide Range of Industrial Sectors) project, a collective research effort supported by EU FP6 contract no. 030307 (September 2006 to September 2009). The products conform to EU Ecolabel 2005/360/EC.

In December 2007, the European Commission has launched the Lead Market Initiative (LMI) for Europe to foster the emergence of six highly innovative lead markets with high economic and societal value, namely bio-based products, eHealth, sustainable construction, protective textiles, recycling, and renewable energy [47]. The LMI is a comprehensive effort at EU level for a coordinated demand-side innovation policy approach using various policy instruments (regulation, public procurement, standardisation and complementary activities) to facilitate the uptake of new innovative products and services in the market. As reflected in the actions of the LMI, a strong role is implied for the government as a lead buyer. About 45% of the EU's GDP is dedicated to public expenditure. A key action in the LMI is promoting networking and cooperation among public procurers. Governments can also apply their regulatory prowess by promoting targeted legislation to catalyse a market transformation process (e.g. in environmental regulation). In the corresponding 3–5 year action plan, the lack of product quality standards is quoted as a factor limiting the demand for bio-based products and standards are seen as essential elements in aggregating initial demand. In October 2008, the EC has also issued two standardisation mandates to the European standardisation bodies (CEN). Within mandate M 429 for a standardisation programme for bio-based products, BT/WG 209 is requested to prepare a definition of the term 'bio-based'. Mandate M 430, accepted on 19 February 2009 by the CEN BT/TCMG (BT Technical Committee Management Group), concerns development of CEN technical specifications and European standards for biolubricants. The standardisation will be prepared by CEN TC 19 'Gaseous and liquid fuels, lubricants and related products of petroleum, synthetic and biological origin'. The CEN technical specification will deal with biodegradability, product functionality, impact on greenhouse gas (GHG) emissions and amount of different renewable materials. Although the sectorial importance of the mandate is broad, there is not a large number of SMEs in industry sectors related to biolubricants. The potential impacts on SMEs of the standardisation activities of M 430 are not considered to be very great.

In another EU initiative, within the Seventh Framework Programme (FP7), a call targets sustainable biorefineries for developing technologies to make biorefining production cost-effective and coordinating better existing biorefinery related research in Europe [48]. Moreover, a cross-disciplinary expert group ('Ad-hoc Advisory Group for Bio-based Products') has been set up at European level for matters concerning renewable raw materials and bio-based products. The LMI is still in an early phase of implementation. In the short term, the activities of the LMI can stimulate the demand side in member states. However, real impact can only be expected after 2014. EU public procurement rules are to be modernised to enhance the green public procurement aspect.

13.3 Biolubricant markets in the United States

North America is the second major consumer of lubricants, which are mostly manufactured for use in automotive, industrial, marine and aviation applications. Most of the lubricants used in the United States are mineral based, mainly due to the inherent tendency to design components for use with mineral-based lubricants. According to a Freedonia 2007 report [3] sales of all lubricants in the United States amounted to 3.6 billion gallons per year (Bgy) in 2006. This contrasts with about 2.0 Bgy according to the Energy Information Administration (EIA) in 2008 [49] and with the 2.5 Bgy stated in a National Petroleum Refineries Association (NPRA) report 2006, which is probably more accurate. Demand for lubricants in the United States is mainly on account of process oils and high-performance lubricants. Most of these products (e.g. general industrial oils, engine oils, transmission and hydraulic fluids, gear oils and greases) are based on mineral oils. The lubricant market (2006) is segmented as follows: automotive, 56.1%; industrial, 21.2%; process oils, 18.1%; metalworking, 2.1%; and greases, 2.4% [50]. Hydraulic oils make up about 225 Mgy. In spite of the increase in vehicles and industrial machinery requiring lubrication, the expected low growth rate in volume (about 1%/yr) is attributed to the development of higher-performance oils and longer oil change intervals. ExxonMobil is the absolute market leader in industrial oils and greases. Biolubricant manufacturers/distributors are numerous (see Table 14.1). Cuevas [51] has given an extensive list of some 300 commercial biolubricant products, manufacturers/distributors (US market), applications and websites. Table 13.6 shows the database for bio grease.

There is far less market information available on lubricants in general and biolubes in particular in the US market than for the EU. Perhaps this is due to the higher level of environmental awareness in Europe. Biolubricant consumption (2006) amounted to 8.2 million gallons per year (Mgy) or about 0.3% of the total market volume [50]. Commercial estimates (2002) indicate that over 75% of bio-based lubricants are hydraulic fluids [52]. In 1997 there were no substantial volumes of vegetable oil-based lubricants sold commercially in the United States. At that time, a few hydraulic oils based on rapeseed oil were being imported from Europe and some test work on soybean oil (SBO) looked promising in small volume

Table 13.6 Bio grease database (US market)

Commercial product name	Manufacturer/Distributor	BioPreferred[a]
Bio-Food Grade E.P. Grease NLGI #0, #1, #2 (High Temperature)	Renewable Lubricants, Inc.	+
Bio-High Temp 180 E.P. Grease NLGI #2 (Multipurpose Lithium Complex)	Renewable Lubricants, Inc.	+
Bio-Graphite E.P. Grease NLGI #1	Renewable Lubricants, Inc.	+
Bio Grease EP2	BioBlend	+
Tru Grease AP2	BioBlend	–
Bio Grease DR3	BioBlend	–
BioMultis SEP 2	Total	–
BioMerkan RS	Total	–
Mobil SHC Grease 100 EAL Series	Mobil	–
5th Wheel Biobased Grease	Creative Composites, Ltd.	+
5th Wheel Trailer Grease	Plews/Edelmann	+
Disc/Drum Wheel Bearing Grease	Plews/Edelmann	+
Eco Line Food Machinery Grease	Cortec Corporation	+
Eco Line Heavy Duty Grease	Cortec Corporation	+
ELM Cotton Picker Spindle Grease	Environmental Lubricants Manufacturing, Inc.	+
ELM Textile Grease	Environmental Lubricants Manufacturing, Inc.	+
EP Lithium Grease	Environmental Lubricants Manufacturing, Inc.	+
EP Plus Grease	SoyClean	+
Grease EP	RyDol Products, Inc.	+
LMX TM Red Grease	Plews/Edelmann	+
Moly EP Grease	Plews/Edelmann	+
Multi-Purpose Grease	Plews/Edelmann	+
SoyGrease Food Machinery (NLGI Grade 1 & 2)	Environmental Lubricants Manufacturing, Inc.	+
SoyGrease Heavy Duty Truck	Environmental Lubricants Manufacturing, Inc.	+
SoyGrease Multi-Purpose EP	Bio-Gem Services, Inc.	+
SoyGrease Semi-Truck Fifth Wheel	Environmental Lubricants Manufacturing, Inc.	+
SoyGrease EP Plus	Environmental Lubricants Manufacturing, Inc.	+
SoyGrease EP Premium	Environmental Lubricants Manufacturing, Inc.	+

Table 13.6 Continued

Commercial product name	Manufacturer/Distributor	BioPreferred[a]
SoyGrease™ Multipurpose Equipment Grease, TF 0-100 Winter	Environmental Lubricants Manufacturing, Inc.	+
SoyGrease™ Multipurpose Equipment Grease, TF 35-160 Summer	Environmental Lubricants Manufacturing, Inc.	+
SoyGrease™ Multipurpose Equipment Grease, TFHD 35-160 Summer w/moly	Environmental Lubricants Manufacturing, Inc.	+
SoyGrease™ Multipurpose Equipment Grease, TFHD Winter 0-100 w/moly	Environmental Lubricants Manufacturing, Inc.	+
SoyTrak Arctic Blend Grease	Bio-Gem Services, Inc.	+
SoyTrak Multi-Season Grease	Bio-Gem Services, Inc.	+
SoyTrak Summer Grease	Bio-Gem Services, Inc.	+
SoyTrak Winter Grease	Bio-Gem Services, Inc.	+
TempFlex™ 0-100 Rail Curve Lubricant	Environmental Lubricants Manufacturing, Inc.	+
TempFlex™ 0-100 w/MoS2 Rail Curve Lubricant	Environmental Lubricants Manufacturing, Inc.	+
TempFlex™ 35-160 Rail Curve Lubricant	Environmental Lubricants Manufacturing, Inc.	+
TempFlex™ 35-160 w/MoS2 Rail Curve Lubricant	Environmental Lubricants Manufacturing, Inc.	+
ThermaPlex Bio Green Bearing Grease	LPS Laboratories	+
Vane Spindle Grease (VSG)	Eco Fluid Center, Ltd.	+
White Lithium Grease	Plews/Edelmann	+

[a] Included (+) and excluded (–) from BioPreferred℠ catalogue.

potential areas, such as drip oils and wire rope lubricants. The attitude in most market segments was that there was little incentive to switch from mineral oil-based to vegetable oil-based lubricants. There were no regulations and little environmental pressure forcing this change. Moreover, natural vegetable oil products present several disadvantages (less oxidatively stable, low-temperature problems, more expensive) compared with mineral oil products. Even up to November 2008 there were still no regulations in the United States to mandate adoption of vegetable oils in environmentally sensitive application areas to replace mineral oils, as is occurring increasingly in Europe. The present lack of US regulations mandating the use of biolubricants in environmentally sensitive areas is a major hurdle to overcome. However, 'bio-based' and 'buy-bio' are nowadays the new buzz words in the US government and some relevant legislative initiatives have been taken. The Food, Conservation and Energy Act of 2008, the Farm Security and Rural Investment Act (FSIRA) of 2002 (art. IX), various Presidential Executive Orders (Nos 13101, 13123, 13134), Public Laws (106–224), the Federal Acquisition Regulation (FAR) and other legislative actions all require US federal agencies to develop procedures favouring products derived from bio-based, renewable, sustainable and low environmental impact resources. The measures aim at promoting the market development infrastructure necessary to induce greater use of such resources in commercial non-food products by giving preference to the purchase of bio-based products in place of petroleum-based products when reasonably competitive and suited to the application. For this purpose, the US Department of Agriculture (USDA) has developed a product list (so called BioPreferred^{SM} products); see Chapter 9. Table 9.8 shows the BioPreferred minimum bio-based content for lubricant product groups while Table 13.6 lists bio-greases available on the US market and their BioPreferred status.

In contrast to environmental regulations, US federal procurement regulations thus do provide for a distinction between petroleum-based and bio-based products (including vegetable oil-based lubricants and fluids). The requirements of the 2002 Farm Bill and subsequent legislative initiatives should become more effective in the near future. In addition, the USDA is developing a 'USDA Certified Biobased Product' label that can be used by manufacturers and vendors to identify and validate their products in the commercial and consumer marketplace as well as the federal marketplace. Many hope that increased federal procurement will allow plant-based companies to more successfully move into the consumer market. On the state level, Iowa and Michigan have implemented similar policies. Rather than restrict the use of petroleum-based lubricants, these states, recognising the value of products derived from renewable resources, have implemented procurement mandates for plant-based lubricants. Iowa Senate File 2249 encourages state agencies to use soybean-based lubricants, whereas Michigan House Concurrent Resolution 34 mandates the use of vegetable oil-based engine oil.

Despite improvements in performance, the US market for biolubricants is slow to develop. The main reasons for this are perceived cost and the lack of regulatory pressures to change. Bremmer and Plonsker [50] have recently published a market opportunity study for soybean as a base lubricant in the US market.

The biolubricant database categories (Table 13.7) are illustrative of the current status of commercial development in the United States. The absence

Table 13.7 Biolubricant database categories

Biolubricant	No. entries	Biolubricant	No. entries
Air tools	cf. Rock drill oils	Hydraulic fluids	74
Assembly oils	1	Metalworking fluids	61
Bar and chain oils	1	Motor oils	13
Cable and saw guides	2	Mould-release fluids	1
Chain and cable oils	8	Penetrating lubricants	13
Compressor oils	1	Refrigeration compressor oils	1
Corrosion inhibitors	8	Rock drill oils	5
Degreasers-cleaners	4	Rust and oxidation	1
Food-grade gear oils	1	Saw guide oils	cf. Cable oils
Food-grade greases	3	Slideway oils	cf. Gear oils
Food-grade hydraulic fluids	1	Specialty products	9
Food-grade lubricants	8	Spindle oils	cf. Gear oils
Forming lubricants	2	Transmission fluids	6
Gas and diesel fuel conditioners	8	Two-cycle engine oils	6
Gear oils	18	UTTO transmission oils	2
Greases	43		

After ref. [51].

of biolubricant products from various major US lubricant suppliers suggests that their commercial interest in this area is still so limited as not to warrant active participation. Nevertheless, the US biolubricant market is bound to grow faster in the next years than in the past because:

- commercial quantities of high oleics (especially high-oleic sunflower oil, HOSNO) allow more stable and better performing products in all lubricant classes; and
- the federal government BioPreferred^{SM} programme will promote bio-based products in all areas of the government as well as in the private sector.

A growth forecast of bio-based lubricant products and fluids of 7–10% per year has been given [53].

A very successful soy-based replacement product for mineral oil-derived lubricants is transformer oil. In the past, transformer dielectric fluid formulations have undergone significant changes. In the 1970s, polychlorinated biphenyls or PCBs (fire retardant, toxic and non-biodegradable) were replaced by naphthenic mineral oils, high-molecular-weight hydrocarbons, synthetic esters and silicone fluids (Dow Corning). A soybean-based transformer fluid (BioTrans 1000) was initially introduced by ABIL (University of Northern Iowa), ERMCO (Indianapolis, IN) and Cargill (Minneapolis, MN) in the 1990s and later further developed by Cargill and United Soybean Board (USB) [54]. In 2004, Cargill ceased production of BioTrans 1000 in favour of a premium soybean-based transformer oil EnviroTemp® FR-3™, a co-production of Cooper Power Systems (Waukesha, WI) and Cargill Industrial Oils and Lubricants (Minneapolis, MN).

The US market is also served by a vegetable oil-based product under the BIOTEMP® brand name. BIOTEMP® is an advanced biodegradable high-performance electrical insulating vegetable oil for high-voltage (HV) power transformers and apparatus, developed by ABB (Jefferson City, MO) [55]. The first worldwide application of BIOTEMP® vegetable oil as an insulating and cooling fluid in a HV (145 kV) power transformer has been made by CEMIG (Belo Horizonte, MG) [56]. The fluid has excellent dielectric characteristics with high-temperature stability and superior flash and fire resistance (see Table 13.8). It is an inhibited oil, has been approved according to both ASTM D 2440 (Standard Test Method for Oxidation Stability for Mineral Insulating Oil) and ASTM D 3487 type II (Standard Specification for Mineral Insulating Oil Used in Electrical Apparatus) and does not contain any PCBs [57]. The high acceptability of the present generation of vegetable oil-based transformer fluids is on account of both an economic advantage and superior performance (fire point >300 °C, increased service life of the transformer, biodegradability, low toxicity and much lower spill remediation costs) over conventional mineral oil-based products (non-biodegradable, toxic, fire point ~145 °C). The total size of the US transformer oil market is estimated at about 60 Mgy. Transformer oils are expected to show growth [9].

Crankcase oils are by far the largest segment of the global lubricants market, representing a demand in excess of 1 Bgy in the United States alone. For this potential biolubricant market only quite recently are development products getting close to being qualified to meet industry standards (see Section 10.5.1). This market segment is extremely demanding in performance requirements, particularly in long-term oxidative stability where vegetable oils face inherent barriers not yet completely surmounted. The extremely high viscosity index and very low volatility of SBO (and other biofluids) are of great interest and allow savings in viscosity modifier treatment necessary to obtain a given viscosity grade. However, SBO requires additivation for stabilisation at high and low temperatures. Valvoline (Lexington, KY) is committed to biodegradable engine oils for high mileage cars. Valvoline (with United Soybean Board support) has conducted extensive

Table 13.8 Typical transformer insulating fluid properties

Property	Method	BIOTEMP®	Mineral oil	H.T.H.[a]
Electrical:				
Dielectric strength (kV)	ASTM D 877	45	30	40
Physical:				
Viscosity @ 100/40/0 °C (cSt)	ASTM D 445	10/45/300	3/12/76	11.5/110/2200
Flash point (°C)	ASTM D 92	330	145	285
Fire point (°C)	ASTM D 92	360	160	308
Pour point (°C)	ASTM D 97	−15 to −25	−40	−24
Specific heat (cal/g/°C)	ASTM D 2766	0.47	0.43	0.45
Expansion coefficient (per °C)	ASTM D 1903	6.88×10^{-4}	7.55×10^{-4}	7.3×10^{-4}
Specific gravity (g/cm³)	ASTM D 1298	0.91	0.91	0.87
Environmental:				
Aquatic biodegradation rate (%)	CEC L-33-A-93/94[b]	97.0	25.2	27.1

[a] High-temperature hydrocarbon fluid (ASTM D 5222).
[b] 21-day test.

engine and bench testing in the quest for a formulation that fully meets the latest industry standards (API SM, API Energy Conserving and ILSAC GF-4). Of the SBO varieties available for product development, those with oleic content at the level found in conventional SBO (up to 27%) were not considered because of known deficiencies in high-temperature stability. Stability at high temperature and maintaining good flow properties at low temperature were also found to be main hurdles for mid-oleic (>53% C18:1) soybean oil (MOSBO). The low-temperature flow problems of MOSBO were resolved by using selected (proprietary) cold-flow improver/pour point depressant additives. Bearing corrosion, as measured by the Sequence VIII engine test, has then led to the selection of a variety of corrosion and oxidation inhibitors that fully protect copper-lead bearings in engine service. Sequence IIIG (oxidation stability) and Sequence VIB (fuel economy) engine test issues were the most difficult to resolve, with the Sequence IIIG as the most critical hurdle. Although a reduction in SBO content from 20% to 5% and a modification of the additive package improved Sequence IIIG performance, this was insufficient to obtain a test pass. It turned out to be necessary to use a 5% high-oleic soybean oil (HOSBO). SBO thus only makes up a small percentage of this total formulated oil for which field testing is still to be carried out. The Valvoline engine oil product development programme illustrates the following key points:

- simple replacement of a mineral oil base stock in a lubricant with a soybean (or other biofluid) component, even at low levels, compromises performance; and
- additive selection can regain performance lost by addition of a biofluid.

Using canola, sunflower, soy and corn oils, Renewable Lubricants, Inc. (RLI; Hartville, OH) has also developed a crankcase oil using additives aimed at bringing higher-oleic vegetable oils up to mineral-oil standards (see also Section 10.5.1). An Agro Management Group (AMG) product (AMG 2000) is derived from canola, soy and various other vegetable oils; see Section 12.2.1, p. 585. International Lubricants, Inc. (ILI; Seattle, WA) produces another extremely successful lubricant (Lubegard™). However, this biodegradable automatic transmission fluid supplement is based on liquid wax ester technology (see Section 5.4).

After crankcase oil, the hydraulic fluids segment represents the most important product line in the North American market (225 Mgy). Not surprisingly, therefore, most biolubricant suppliers have a range of products to meet the needs of this industry. Various (joint) projects have been developed to produce biodegradable hydraulic fluids, following different approaches:

- chemically modified SBO (Cargill/PSU/USB/Caterpillar);
- additivation of SBO (RLI/USB; University of Northern Iowa) [58];
- HOSBO (RLI; DuPont).

The biolube share in this market segment is still very small (about 1% or 2 Mgy). Growth rate in the hydraulic fluid area has been small. Bio-based hydraulic fluids are finding increased use at military bases, national laboratories and national parks. The segment that has achieved most significant success is elevator oil. An SBO-based elevator hydraulic fluid, developed by the Agricultural Research Service (USDA) and improved by Bunge/AgriTech (White Plains, NY), has successfully been introduced. Hydro Safe Oil Division, Inc. (Dewitt, MI) has developed a commercial canola oil-based elevator hydraulic fluid [59]. Hydro Safe® can be used as a superior replacement for conventional petroleum oil. The totally biodegradable fluid exhibits excellent thermal and oxidative properties and AW performance. Its high viscosity index (about 200) allows outdoor operations in winter and summer without the need for fluid changes or equipment modifications. The market for elevator oils is substantial (approximately 20 Mgy for 600 000 hydraulic elevators operated in North America). Because most elevator fluids are indoors, pour point is often not an issue, but price is. A realistic hurdle in this market are the substantially higher prices for the bio-based products; there are no regulations to mandate adoption.

Back in 1997 it was considered that mineral oil-base two-cycle lubricants were a prime example of a product in bad need of replacement, especially in marine applications. According to the National Marine Manufacturers Associations (NMMA), the new Environmental Protection Agency (EPA) regulations with strict emission standards for two-cycle engines [60] cannot be met by these engines and mean the end of this traditional carburetion engine type. Meanwhile, however, the marine industry has developed a new high-pressure fuel injected engine with injection directly into the combustion cylinder that is more efficient, requires less lubricant and can meet the new EPA emission standards. These E-TEC engines are to replace all two-cycle engines of present design as from 2010. Consequently, pressure to use substitute lubricants in this market segment has diminished. The North Amcrican bio two-cycle oil market is poised to deliver strong growth (>12%) through 2017 [9].

Another successful application of a soybean lubricant is in the rolling oil area of metalworking. Several metalworking fluids have been introduced, many of which based on methyl soyate (used by the Food and Drug Administration (FDA) as an incidental food-contact product when rolling metal foils). USDA/Alcoa have developed a bio-based rolling lubricant which is being used in various aluminium rolling mill operations worldwide (100 kgy). Miller *et al.* [61–63] have reported a comparative life-cycle assessment of soybean-based and mineral oil lubricants in aluminium rolling (see Section 11.2.1). Bio-based fluids are also being used in metal cutting and casting (mould release). Vegetable oil-based cooling and cutting oils have met with success [64]. An E.F. Houghton (Valley Forge, PA) biolubricant, based on an unidentified vegetable oil, shows substantial savings in coolant waste stream treatment and disposal, biocide use, tool savings, and increased grinding throughput. CIMCOOL Global Fluids (Cincinnati, OH) has developed CIMFREE Botanical Fluids, a line of nine environmentally friendly vegetable-based products for high-speed machining and metalworking operations.

Little progress has been made in the total-loss lubricant segment with little growth for bar and chain oils, despite the fact that biolubricants are available for this application. The US bar and chain oil market is small (2–3 Mgy). Although it makes sense to replace mineral oil products with biolubricants the present US demand for biolubes in this total oil loss application is very limited. There are biolubes available, at higher price than the mineral oil counterparts. Without regulations, US consumers usually go for the cheaper mineral oil-based product. Only about 5% of the bar and chain oils purchased in the United States is a biodegradable product (100 kgy).

There are two types of rail-related lubricants that are total loss-type products, namely top-of-rail lubricants and gauge face greases. Only the latter application allows for the use of bio-based product (market potential for these greases <10 Mgy).

Wire rope oils constitute one of the few regulated areas, with coast guard regulations in effect. However, the biolubricant volume in this application is small and without growth. The US market for biodegradable wire rope grease is small (<1 Mgy). A previously commercialised (GEMTEK) SBO product for drip oil application (pump lubricant for irrigation systems) has been withdrawn from the market after build-up problems developed.

Cargill Industrial Oils and Lubricants (Minneapolis, MN) has formulated various cost-effective vegetable oil-based products, such as Agri-Pure® base oils (polyol esters for high-performance industrial lubrication application), Agri-Pure® fatty acids and esters (solvents, diluents) and Agri-Pure Gold™ vegetable oils (metalworking lubricants, rolling oils, process oils). Improved SBO basestocks have also been investigated by DuPont/Pioneer (GM soybean with high-oleic levels), ADM (increased C18:1 by low-cost modification process to improve high-temperature stability) and the USDA (non-transgenic modification to produce more stable oils with higher C18:1 and lower saturate levels). BioBlend (Joliet, IL) has combined highly stable vegetable base oils with unique additive technologies, thus creating a product line of over 40 renewable bio-based products (USDA standards) with excellent thermal and oxidative stability while delivering superior anti-wear and anti-corrosion protection.

Biolubricants in the US have been increasingly successful in displacing mineral oil products in the areas of transformer fluids, elevator and other hydraulic fluids and metal rolling oils. The amounts of biolubricants used per year in these areas are estimated at: transformer fluids, 6 Mgy; hydraulic fluids, 2 Mgy; bar and chain oils, 100 kgy; railroad greases, 50 kgy; wire rope, 20 kgy; and two-cycle oils, 20 kgy. The US market value (US$ 13.4 billion, 2007) for lubricants is expected to rise 4% annually through 2012, driven by the impact of high base oil costs and shifts toward higher-quality formulations [2]. The key engine oil segment will grow the fastest based on increasing motor vehicles in use. Environmental concerns will favour bio-based lubricants and re-refined base stocks. Market forecasts up to 2015 are given by lubricant base oil, product and market. An annual growth of 5–8% for environmentally favourable lubricants is expected on the US market over the next few years compared with a rate of only 2% for the overall lubricant market [50].

13.3.1 Soy-based lubricants

While soy oil has recently been overtaken by palm oil as the main global vegetable oil (Table 5.2), soybean (*Glycine max*) remains a most important oilcrop in the United States. World soybean production has increased from 125 Mt in 1995 to over 257.5 Mt in 2009/10 with the USA, Argentina and Brazil as the top three producers. The US soybean production capacity (in excess of 3 billion bushels on about 74 M acres in 2006) accounts for almost 40% of the world soybean output and for 50% of the total available bio-based oil for industrial applications [65]. In the past, the soybean market has been driven by demand for meal, but more recently – after the advent of biodiesel [66] – has moved towards an oil-driven market. Global soy oil production (2009/10) is of the order of 38 Mt. Soy, especially enhanced varieties, shows the ability to compete in significant segments of the lubricant market, dominated by petroleum- or mineral-based oils. The market for SBO in lubricants is driven by a combination of environmental concerns, economics and performance issues.

Whereas it is traditionally considered that under boundary lubrication conditions the reduction in friction and wear is mostly dependent on EP additives rather than the base stock, vegetable oils also contribute to the lubricity under this regime. Lubricating characteristics of SBO and paraffinic mineral oil (150 N) of similar viscosity and four additives were compared under boundary lubrication conditions [67]. The results suggest that soy oil might provide cost effective routes to industrial fluids, gear lubricants and possibly engine oils. However, Section 5.2.1 has pointed out the general need for enhancing the properties of SBO for use in lubricant applications.

SBO competes for a share of the emerging environmentally sensitive and renewable lubricant markets with other vegetable oils and with synthetic lubricants. Vegetable oils, including soy, cost less than synthetics and will be the product of choice when they meet customer performance requirements. Moreover, soy oil is cheaper than RSO and can have an advantage if the oil can be modified to improve stability. Wider commercialisation of soy-based lubricants depends on the development of a cost-effective, commercially viable source of economical and stable base stock. Market penetration will be highest in niches where environmental and safety concerns are greatest.

Growers' Associations such as the United Soybean Board (USB), American Soybean Association and Iowa Soybean Promotion Board are committed to expanding the market for crops grown in the United States by funding applied research and development leading toward technology transfer and product commercialisation of new industrial uses for soybeans as a replacement of petrochemicals, including environmentally friendly, bio-based lubricants as one of the target areas. Alternative industrial products have seen a heightened demand due to increasing US government regulations. Advances in formulations (chemical modifications and additives) and development of genetically modified, and more oxidatively stable SBOs, have created opportunities for bio-based lubricants and greases [68]. The USB has investigated nine areas to determine the potential acceptance of SBO as a commercial alternative to mineral-based oil (Table 13.9). The potential soybean market share is an estimate of the potential penetration once a technically and economically competitive product is developed and introduced. Providing performance superior to rapeseed-based oils will facilitate penetration into the European market and is a priority for SBO-lubricant formulators. SBO is significantly cheaper in the United States than RSO or canola oil. This economic advantage makes it possible for SBO to capture a large share of an emerging US market for renewable lubricants if performance can be enhanced to meet or exceed the rapeseed standard. RSO or canola oil sets the target for lubricant performance of SBO.

Higher performance and cost advantages spur demand for soy-based lubricants. Some of the more proven lubricant product areas incorporating SBO are listed in Table 13.10. A selection of industrial developments of lubrication products from soy oils is given in Table 13.11; others are mentioned throughout this text. Development of SBO-based lubricants and hydraulic fluids has been among Battelle's research programmes. The Industrial Uses of Soybean Oil Award 2010 by the American Oil Chemists' Society has been attributed to Benecke (Battelle Memorial Institute, Columbus, OH). The market-focused approach of the University of Northern Iowa Ag-Based Industrial Lubricants (ABIL) Research Center (Waverly, IA) has led to over 30 formulated industrial lubricants, greases

Table 13.9 Opportunities of soy-based lubricants

Product group	Potential acceptance (%)	Market share potential (%)	Market share (MBu)
Hydraulic fluid	40	5	8.0
Two-cycle	20	10	1.5
Bar/chain	60	50	1.0
Crankcase	10/80[a]	10	87.3
Drip oil	65	80	0.8
Rail and flange	55	50	0.4
Wire rope	70	70	1.1
Metal cutting	30	10	2.8
Dedusting	50	50	0.4

[a] Automotive water-cooled-engine use/air-cooled-engine use.
After ref. [69].

Table 13.10 Commercially available soy-based lubricants and greases

- All-purpose (household) lubricants
- Tractor transmission hydraulic fluids
- Industrial hydraulic fluids for process and machinery applications
- Metalworking oils and coolants
- Food-grade hydraulic fluids
- Chainsaw bar oil
- Gear lubes
- Compressor oil
- Transformer and transmission line cooling fluids
- Greases including automotive, machinery, rail curve, track and food grade

After ref. [70].

and base oils from high-oleic (genetically modified) soybeans, each licensed to the market. Various patented and proprietary technologies are based on chemical modification of soy oil. ABIL's programme uses hydrogenation and other chemical processes to increase the oxidative stability. ABIL has developed various successful hydraulic fluid formulations targeted for applications ranging from food-grade industrial hydraulic fluid to elevator hydraulic oil, universal tractor oil, and ISO 15 and ISO 22 low-temperature hydraulic fluids. A group of farmers profiting from plant-based lubricant production is West Central Cooperative, a 3500-member farmer co-op in Ralston, IA.

The outlook for soy-based hydraulic fluids is positive if soy can meet performance specifications and any emerging regulatory requirements, while remaining lower in cost. BioSoy® hydraulic oil, the first soybean-based tractor hydraulic fluid, developed by the Ag-based Industrial Lubricants (ABIL) Research Program, was launched in 1997 [71]. The product, based on Cargill, Inc. (Minneapolis, MN) SBO and Lubrizol Corp. (Cleveland, OH) additives, is marketed by ITEC (West Des Moines, IA), and priced at 2–2.5 times the cost of similar petroleum-based products. BioSoy® can be used both indoors and outdoors and is suitable for mobile and stationary machinery. Environmental Lubricants Manufacturing, Inc. (ELM; Grundy Center, IA) has produced numerous soy-based lubricants, including the premium biodegradable bio-based SoyFluid Hydraulic All Seasons oil (viscosity index, 214; pour point, –36 °C) formulated for use in hydraulic systems operating throughout all seasons. Soylube™ and CanolaLube™ are household soy- and canola-based lubricants developed by UNI-NABL and offered by ELM. A biodegradable soy-based hydraulic fluid, developed by ARS (Peoria, IL) and produced by Agri-Lube, Inc. (Defiance, OH), has been used to operate the elevator system in the Statue of Liberty since November 2002 [72]. Terresolve Technologies (Eastlake, OH) chemically

Table 13.11 Selection of soy-based lubricants and greases

Product group	Development[a]	Trade-name	Reference(s)
All-purpose lubricants	Workcell Systems, Inc.	Glycol Pen Trate	–
	Bioplastics Polymers and Composites, LLC	Naturelube 700	–
	UNI-NABL/ELM, Inc.	Soylube™	–
Hydraulic fluids	ABIL/ITEC	BioSoy®[b]	[71]
	ELM, Inc.	SoyFluid Hydraulic All Seasons	[71]
	ARS/Bunge/Agri-Lube	(Elevator oil)	[72]
	Terresolve Technologies	XBO	–
	UNI-NABL	Cargill Inc. #110 PKS-WOO[c]	[73, 74]
Pump lubricant	UNI-NABL/GEMTEK	Soy Bio-Drip	[75]
Bar and chain oils	UNI-NABL	(Chainsaw)	[74]
	GEMTEK Products	SafeLube™[d]	–
	Cortec® Corp.	EcoLine™	–
Transformer oils	ABIL/ERMCO/Cargill; Cargill/USB	BioTrans 1000[e]	[54, 76]
	Cargill/Cooper Power Systems	EnviroTemp® FR-3™[f]	[54]
Metal-forming oils[g]	ARS/Alcoa	–	[77]
Greases	ABIL	–	[78]
Heavy-duty truck	UNI-NABL/ELM, Inc.	SoyGrease Heavy Duty Truck	[74, 79]
Truck lubricant	Vibroacoustics Solutions, Inc.	5th Wheel LubePaste[h]	–
Railroad track	UNI-NABL/ELM, Inc.	TempFlex™	[74]
Cotton picker spindle	UNI-NABL/ELM, Inc.	ELM Cotton Picker Spindle Grease	[74]
Food-grade processing	UNI-NABL/ELM, Inc.	Soy Grease Food Machinery	[74]
	RLI, Inc.	Bio-Food Grade EP Grease	[80]

[a] R&D, production, marketing.
[b] Tractor hydraulic fluid.
[c] Multigrade hydraulic fluid, partially hydrogenated SBO-based.
[d] Degummed soy- and canola-based.
[e] Hydrogenated and winterised SBO.
[f] Food-grade SBO-based.
[g] Hot-rolling.
[h] Based on corn, soybeans and graphite.

modified soy and canola oils to produce commercially available hydraulic fluids and two-stroke engine oils.

EcoLine™ bearing, chain and roller lubricant (Cortec® Corp.; St. Paul, MN) is a bio-based, rust preventative lubricant formulated with SBO which offers superior lubricity over most conventional lubes also on high-speed systems. It contains a corrosion inhibitor that forms a thin, tight, layer on the metal surface, completely sealing off the surface against moisture or air.

USB-sponsored research has also focused on formulating SBO or methyl soyate for technically more challenging areas, such as extreme-pressure metalworking. Several new metalworking fluids have been introduced, many of which are based on methyl soyate. Methyl soyate has many advantages, including use by the FDA as an incidental food-contact product when used in rolling metal foils. Another technology is soy-based hot rolling lubricants that are being used by the large aluminium manufacturer Alcoa for hot and cold flat-rolling operations, which produce aluminium sheets for applications ranging from aircraft-wing panels to beer cans [77]. Agro Management Group (AMG; Colorado Springs, CO) has developed a soy-based oil that is usable in air-cooled engines. Valvoline (Lexington, KY) is evaluating HOSBO formulations for potential crankcase applications.

13.4 Market opportunities for bio-based lubricants

A sustainable decision on the manufacture and use of biolubricants must consider the economic consequences. It is normally not possible to immediately phase out the production or use of a less sustainable chemical. However, product sustainability should be a high long-term priority. The alternatives for non-sustainable substances (biolubricants instead of mineral oils) have to be applicable in existing technology. Modern manufacturing processes are often very fast (e.g. newspaper printing) or very complex (e.g. electronic devices) and production machines are fine-tuned to the use of particular chemicals. In substitution processes, the technical feasibility and requirements of the entire production chain need to be assessed. Substitution of lubricants is connected to economic risks: the scarce information on proven performance of substitutes in particular applications, the practical and technical implementation of substitution, as well as the risk of lower acceptance of new products on the market.

A fast and significant market penetration of plant-based lubes has been impeded by the original choice of (high-erucic) RSO as a base component. Despite the fact that the limitations of such RSO-based liquids were sufficiently well known, most lubricants initially produced on the basis of this oil were formulated with additives originally developed for non-polar mineral oils. Their resulting disappointing performance has damaged the reputation of biolubes. Another reason for the weak market penetration of biolubes is the high price of saturated esters. Basically, today three types of base oil are used predominantly for biolubes: triglycerides (high oleics), unsaturated and saturated esters.

Biogenous lubricants in today's market are mostly made from RSO, SBO, and SNO or a mix of these vegetable oils. While vegetable oils have captured a small but significant portion of the lubricant market, they present good potential for further substantial growth due to several market drivers (Table 13.12) [81]. The market for biolubricants is driven by environmental concerns, as well as by economics and performance issues. The biggest driver is environmental concern (Cancun Agreement 2010) and follow-up legislation. CO_2 emission targets are being tightened (in Europe from 140 to 95 g/km by 2020) and fuel economy targets (in USA from about 25 to 35 mpg for passenger vehicles by 2020). Both challenges affect fluid formulators. Economic, regulatory and political issues are highly influential on the demand and supply dynamics of the global lubricants market. Although the

Table 13.12 Market drivers and barriers for agro-based lubricant market development

Drivers:
- Environmental awareness of consumers
- Legislative initiatives
- Market conversion programmes
- Environmental regulations and fiscal incentives
- Ecolabelling schemes
- Influence of non-governmental organisations (NGOs)
- Increased use of life-cycle analysis
- Economic, ecological and social sustainability
- Chemical notification policies
- Rising petroleum base oil prices
- Improved fuel economy
- Penalties at accidental release
- End-of-life costs

Barriers:
- Specifications
- Lack of quality standards for bio-based products
- Quality/price
- Feedstock supply
- Development of non-food oil crops
- Technical awareness
- Public opinion (acceptability)
- Low market share
- Changeover costs
- Warranties for machinery

After ref. [81].

volume growth is moderate, modernisation in manufacturing machinery and increase in motor vehicle fleets calls for superior quality lubricants. China offers a huge market for lubricant suppliers. Economic factors in favour of biolubes include the increasing cost of mineral oil, tax incentives to promote bio-based products, the cost of recycling and penalties applied in case of accidental release. There are also performance requirements and environmental regulations demanding biodegradability and low toxicity in sensitive areas. Nowadays, end-users are increasingly prepared to pay a surcharge on lubricants meeting environmental criteria. Ecolabelling schemes (non-mandatory) and non-governmental organisations (environmental NGOs), such as Greenpeace, Friends of the Earth and Bellona, have a strong influence on society towards sustainable and environmentally friendly options.

The two most significant opportunities for biolubricants are high-risk lubricants (applications with a high probability of accidental exposure of sensitive environments to the lubricants) and total-loss lubricants (where by design of the equipment or application the lubricant ends up almost entirely in the environment). The market interest in bio-oils is characterised by the desire for immediate protection of the environment, i.e. minimising the contamination effect in case of a severe loss of fluid. Most users do not care if the bio-oil originates from petrochemical or renewable resources. Bio-based products are now successfully being applied in metalworking and industrial lubricants, greases, tractor oils and hydraulic fluids. The use of bio-based automotive lubricants is so far very limited because of performance issues. However, some bio-based products are highly efficient as lubricity additives in formulations of fuel-economy engine oils and transmission fluids.

Vegetable oil-based lubricants and fluids can take advantage of increasing environmental awareness among industrial and commercial users if life-cycle analysis (LCA) can show reduced GHG impacts compared with petroleum-based lubricants and fluids. Increased use is being made of LCA in manufacturing processes of new products (see Section 11.2). Environmental sustainability indexes influence where investments are made. Shell Global Solutions has developed a sustainability assessment tool for biolubricants. The sustainability of an environmentally acceptable hydraulic fluid (Shell Naturelle HF-E 32) meeting Swedish Standard SS 155434 has been compared with that of a conventional mineral oil-based product [82].

Other drivers are legislative regulations and mandatory procurement of bio-based products by government agencies. Government-sponsored market conversion programmes for lubricants based on renewable resources have been enacted in several countries (Germany, The Netherlands, France). Numerous governmental and international organisations are exercising environmental policies and legislation. At an equally advanced technological level, early legislation in Europe has helped to gain locally a greater market share for biolubricants than in the United States. The plant-based lubricant industry is young, particularly in the United States. Growing regulatory impacts on lubricants should result in more use of bio-based lubricants in the United States during the next 5 or 10 years. US government regulatory action already requires a minimum renewable content for different product categories for preferred procurement (see Table 9.8). The US Air Force supports bio-based products from crops as an essentially strategic approach to national security. Finally, chemical registration policies (in the EU, the United States and China) may also act as incentives toward increased usage of bio-based lubricants. Further challenges and restraints will be the emissions legislation and the continuously increasing oil change intervals that will further reduce unit volumes. Historically, refined vegetable oils have held substantially higher market price levels than their petroleum base stock counterparts. However, this price spread has narrowed in recent years. In the future, the rising price of petroleum base oils will be an important driver rendering vegetable-based base oils more competitive in lubricant formulations.

Despite considerable ecological and economical advantages, the present market share of biolubricants is relatively small (about 2% worldwide, with an estimated annual growth rate of 5–10%) [83, 84]. Table 13.12 mentions some significant barriers to increased use of agro-based raw materials for lubricants. The main barriers to the introduction of biogenous oils are the widely diffused insufficient awareness of their performance characteristics (technical know-how), and fear for technical problems (contamination, damage), even more than price. There still exists a persistent prejudice that bio-oils have lower performance. In changing over to bio-oils mineral oil contaminations should be kept below 2% (see DIN ISO 15380 and ref. [85]); see also Section 13.2. To increase the market share, the product acceptability must be improved. This can be performed by overcoming the inherent disadvantages of vegetable oils.

The purchase cost of biodegradable lubricants is frequently higher than that of comparable mineral oil-based products (see Fig. 13.2 for the US market) and also depends on the corresponding additivation. Relative costs (in brackets) of the main classes of synthetic base oils in comparison to mineral oils are as follows: alkylated aromatics (2–3), polybutenes (3–5), dibasic acid esters (4–15), poly-α-olefins (4–15), poly(alkylene glycols) (6–15), polyol esters (10–20), polysiloxanes (silicone oils) (25–200) and perfluoroalkyl polyethers (350–800) [87]. The cost varies with the specific application (see Table 13.13 for the US market), but in general the purchase price of formulated vegetable oil-based lubricants is frequently twice as high as mineral oils, and half as much as the most common synthetic lubricants (see Fig. 13.2). At variance to hydraulic fluids, some vegetable oil-

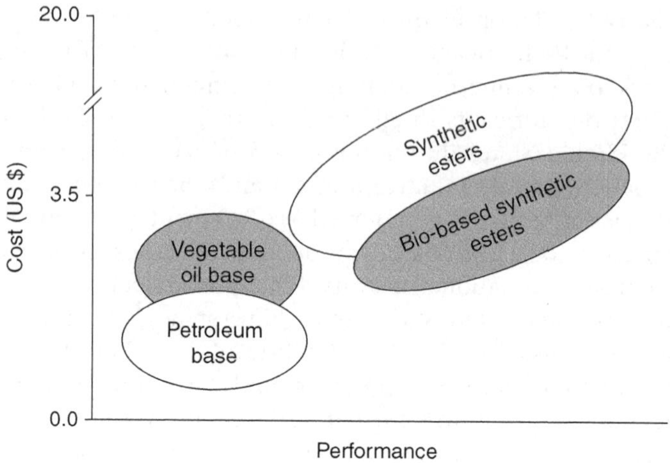

13.2 Cost vs performance for various lubricant base oils.

Table 13.13 Price comparison between commercial bio-based and conventional lubricants[a]

- Hydraulic tractor fluid (1.5–2×)
- Food-grade hydraulic fluid (1–1.5×)
- Industrial hydraulic fluid (1–2×)
- Chainsaw bar lubricants (1–1.5×)
- Multipurpose truck grease (1×)
- Soy- and cotton oil-based greases (1–1.5×)
- Rail curve grease (0.9–1.2×)
- Metalworking fluids (1–1.5×)
- Gear lube (1.5–2×)
- Retail soy lubes (1–1.2×)

[a] Ratio biolubricant/petrolubricant.
After ref. [86].

based products such as railroad flange oils and cutting oils come much closer to mineral oil lubricants in price. The price differential between bio-oils and mineral distillates reflects both the more expensive materials cost (base oils and additives) and the development costs to be sustained by a small market volume. Vegetable oil prices and thus also oil ester (derivative) prices depend on the world market, such as the size of the soya harvest and the price of soybeans. The effect of EU agricultural subsidies on the manufacturing costs of biolubricants is relatively small. As to additives, according to Terresolve Technologies (Eastlake, OH), their cost in vegetable oils is approximately five times that of those used for petroleum-based lubricants. As the share of raw materials on the overall lubricant cost is much smaller than for fuels, increasing petroleum prices have a relatively small effect on the lubricant market. The price of biolubricants is a major hurdle for their commercial development. Consequently, the market share of biodegradable lubricants is small, but opportunities for global expansion are promising [88, 89]. Although current price differences between bio-based and mineral oil-based fluids are expected to become less relevant in the near future [90], use of food material for non-food applications poses similar problems as for biodiesel with highly uncertain markets and volatility. Growth prospects therefore remain uncertain.

Even more important than purchase cost is the cost-to-benefit ratio of a product. The overall cost of a lubricant is influenced by various parameters (Table 13.14). As vegetable oil-based lubricants evaporate less quickly and adhere better to metal surfaces, end-users often use less of a product per application. Higher prices are acceptable by users who benefit from longer product life. The higher initial price for natural or synthetic products as compared with mineral oils can be compensated by savings generated by significantly longer drain intervals, reduction of oil sump temperature, reduction of stop periods, parts wear, overall service costs and amount of waste oil. Other cost benefits relate to energy (fuel) efficiency, lower disposal fees, reduction in environmental and safety penalties in case of accidental spills or less expense during clean-up. If all these aspects are considered, use of synthetic base fluids can substantially reduce total lubricant-based system costs. Consequently, one might find that the overall costs of plant-based lubricants are not too different from the cost of mineral oils.

Table 13.14 Parameters influencing the overall cost structure of a lubricant

• Initial purchase price
• Ageing and temperature stability
• Machine compatibility
• Machine investments
• Maintenance measures
• Oil change intervals
• Operating costs
• Clean-up costs after spillage
• Disposal costs
• Reconditioning possibilities
• Cost of preventive health and safety measures
• Measures for emission control
• Storage costs (multifunctional oils)
• Risk minimisation
• Plant licence procurement
• Subsidies

After ref. [91].

In the field of metalworking, ecologically friendly cutting fluids can significantly reduce the cost of washing operations between machining steps, downtime, maintenance, compliance with water laws and disposal. Advanced condition monitoring of biolubes is expected to allow an increase of 10–20% of the speed limits and performance (20% reduction of wear and life improvement) of actual lubricants (FP6 BIOMON project, 2004–2006). Together with a 10% reduction of the quantity of used lubricant the maintenance costs could be decreased by 30%. Finally, the use of non-toxic biodegradable lubricants brings with it a 'green image' and constitutes an important marketing instrument with positive business economical impact.

Cost-effective operation is therefore a better criterion of choice than litre purchasing price. The most important market for biolubricants is the mobile hydraulics market. Hydraulic equipment with high oil volumes on vehicle (100 L) at high pressures (up to 400 bar) constitute a very high environmental risk. Biohydraulic oils (mostly HEES type, very little HETG) are more expensive per litre than mineral oil. However, if good filtering and maintenance is provided, an oil fill can last very long, and the oil-related cost per operating hour of mobile equipment decreases. Figure 13.3 reveals a much reduced difference in operating cost for a bio-hydraulic fluid (HEES) at extended oil change intervals (>3000 h) than at the usual short fluid change intervals of 1000 h for a petroleum hydraulic oil (HLP), as illustrated for the case of an excavator 300 L system volume [45]. As one of the advantages of bio-oils is their high viscosity index, which allows to use low-viscosity oils over a wide range of temperature and reduce wear and consumption, HLP 46 should actually be replaced by HEES 32 (rather than HEES 46). Other factors are equally important, such as higher

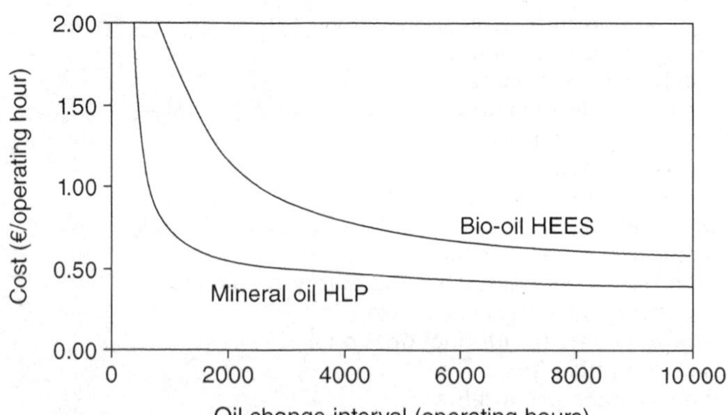

13.3 Cost comparison bio-oil vs. mineral oil for the mobile hydraulics market. After ref. [45].

reliability due to improved maintenance, lower cost of a potential oil accident, or the possibility to use the equipment free of restrictions also in environmentally sensitive areas. However, water is a major risk for many bio-oils (esters). In the next few years, Germany expects a five-fold increase in the use of biogenous mobile hydraulic oils up to reaching a market share of over 30% [28]. Growth indications for the hydraulic fluids and cutting oil segments are strong.

Quality and purchase price are not always valued the same way. Quality-sensitive customers prefer the premium segment for new engines. Synthetic esters, if well maintained, have a very long life expectation. However, when an engine gets old and oil consumption increases, customers become more price sensitive. Also in the vehicle segment many customers are not even conscious of the quality of a lubricant brand as they rely totally on their authorised or local mechanics.

The high biogenous oil price represents more a psychological than a real technical barrier. Considerable quantities of biodegradable mineral-based lubricants are on the market which are only marginally cheaper than biogenous oils. Bio-oils guarantee better lubricity, longer life time, lower energy consumption and reduced environmental impact. A marketing problem is that biogenous oils are not easily differentiated from mineral-based bio-oils. Ecolabels can play an important role in this respect. Many consumers do not differentiate between bio-oils and biogenous oils.

Bio-based lubricant markets have the most potential where

- the application is total loss;
- the application is indoors (no need for low pour point);
- the application is in an environmentally sensitive area (biodegradability required in case of leakage and spills);
- the application is used in/near food-processing equipment;
- preferred purchase is required (e.g. federal regulations).

A common problem to SMEs of the biolubricant sector, who represent 80% of the European producers, is the difficulty for marketing and integration of bioproducts. The IBIOLAB research consortium (FP6 Contract no. 030307; 2006–2009) has addressed several needs to enhance the market share of biolubricants, and thus reinforcing the competitiveness of SMEs, namely:

- increasing the performance of biolubricants through the development of new manufacturing processes for improved base stocks;
- decreasing the overall costs of biolubricants;
- procuring ecolabels for biolubricants;
- enhancing the knowledge base on best practices in biolubricant base stocks, bioadditives and formulations; and

- promoting the use of biolubricants by a communication campaign involving all the actors and potential end-users.

The acceptability of biolubricants must be improved. The inherent disadvantages of vegetable oils need to be overcome. The USB has investigated the potential acceptance of SBO as a commercial alternative to mineral-based oil (Table 13.9). Whereas the technical advantages of RSO and current political support and infrastructure will make it difficult to replace RSO by SBO in the European lubricant market, SBO is significantly cheaper in the United States than RSO or canola oil. Caterpillar, Inc. estimates that environmentally acceptable hydraulic fluids constitute 12% of the European hydraulic fluids market. In the USA, ExxonMobil, Chevron and E.F. Houghton supply RSO-based hydraulic fluids. Pennzoil offers an SNO-based hydraulic fluid, whereas AMG has developed a canola, sunflower and soy-based oil that is usable in air-cooled engines (lawnmowers, chain saws).

A major obstacle for the introduction of biolubricants is further their present negligible market share (application dependent), which does neither act as an incentive to manufacturers to make necessary changes nor to customers to put at risk their warranties for machinery. As to market conversion, the usual strong resistance of users towards the introduction of new unknown products eventually requiring changes in working procedures needs to be overcome. In recent years growth in market share for biolubricants has been modest and below expectations. In order to stimulate national markets various incentive schemes have been adopted such as the German MIP, Dutch VAMIL, EU LMI and others (see Section 13.2). Significant growth in market share is likely as application techniques and benefits become more widely understood.

Other barriers to increased use are feedstock supply and the development of new and novel non-food oilcrops. As European biolubricant producers are well aware, both the political climate and negative public sentiments can interfere with successfully bringing bio-based products to the market. In particular, genetic modification and the food vs fuel debate are sensitive issues. There are also limitations in producing biolubricants. Cuevas [51] has considered the consequences of direct land use for the production of 2 Bgy of lubricants from bio-based instead of mineral-based sources. At a more modest level a limitation is obtaining enough high-oleic oil from sources such as canola and safflower. These oilseed crops have more lucrative outlets in biodiesel and nutritional markets [92]. Another limitation is the need for certification by the Petroleum Institute. While environmental soundness, ecolabels and certifications are increasingly becoming selling points for manufacturers and users in the industrial lubricants market, clearly, new and improved natural-based lubricant products should meet the minimum performance features of conventional, less sustainable products. For instance, Cognis's ProEco HE 801 series of fully formulated biohydraulic oils combines excellent performance and compliance with the EU ecolabel. Production costs for vegetable oil derived base oils are not yet competitive with mineral-based oils.

Mineral-base oil prices tripled in the 2004–2008 period. The impact of recession on the lubricants industry has been felt, in particular in the manufacturing sector for metalworking fluids and less so in the maintenance market for the transportation engine oil segment [9]. Demand in the developed and mature markets is stagnant or declining. The global lubricant market volume level of 2008 will not be reached in the short term. Long term (5–10 years), FUCHS expects only a moderately increasing lube market on a global basis, driven by growing quality standards, which will compensate the volume growth in emerging markets [11]. In a tight domestic market Group I base oil prices in Europe are recently (mid-2011) at levels just below the peak of 2008 but such high prices are deemed unsustainable. Group III base oils are very tight globally but new production came on stream in the Middle East by end of 2011.

Future growth in different market segments worldwide will be a gradual process based on slow replacement and country-specific legislative norms. Industrial oils are an important focus for vegetable oil lubricant manufacturers. The increasing number of motor vehicles and the concomitant need for oil change remain some of the main growth opportunities for (bio) lubricants. The long-term potential of biodegradable lubricants is estimated to be 10–20% of the total market [93].

13.5 Sources of further information and advice

Global Industry Analysts, Inc., *Biolubricants – A North American & European Market Report*, San José, CA (April 2011); http://www.researchandmarkets.com

Global Industry Analysts, Inc., *Lubricating Oils and Greases: A Global Strategic Business Report*, San José, CA (10 February 2011); http://www.researchandmarkets.com

Union Indépendante de l'Industrie Européenne des Lubrifiants, *Purchase and Use of Lubricants within the European Union*, UEIL, Brussels (2011); http://www.ueil.org

Freedonia Group, Inc., *World Lubricants to 2012 – Market Research, Market Share, Market Size, Sales, Demand Forecast, Market Leaders, Company Profiles, Industry Trends*, Freedonia Group Study # 2454, Cleveland, OH (February 2009); http://www.freedoniagroup.com/Lubricants.html

Frost & Sullivan, *Biolubricants Market in North America*, San Francisco, CA (11 February 2009).

Frost & Sullivan, *European Biolubricants Market*, San Francisco, CA (2007).

13.6 References

[1] Freedonia Group, Inc., *World Lubricants to 2012 – Market Research, Market Share, Market Size, Sales, Demand Forecast, Market Leaders, Company Profiles, Industry Trends*, Freedonia Group Study # 2454, Cleveland, OH (February 2009).

[2] Freedonia Group, Inc., *Lubricants to 2012 – Demand on Sales Forecasts, Market Share, Market Size, Market Leaders*, Freedonia Group Study # 2384, Akron, OH (September 2008).

[3] Freedonia Group, Inc., *World Lubricants to 2010*, Freedonia Group Study # 2182, Cleveland, OH (1 April 2007).

[4] Frost & Sullivan, *Biolubricants Market in North America*, San Francisco, CA (11 February 2009).

[5] Frost & Sullivan, *European Biolubricants Market*, San Francisco, CA (2007).

[6] Global Industry Analysts, Inc., *Lubricating Oils and Greases: A Global Strategie Business Report*, San José, CA (10 February 2011); http://www.researchandmarkets.com

[7] Kline Europe S.A., *Competitive Intelligence for the Global Lubricants Industry, 2006–2016*, Brussels (2008).

[8] Kline & Company, Inc., *Lubricants for Wind Turbines 2012. Global Market Analysis and Opportunities*; http://www.klinegroup.com

[9] Global Industry Analysts, Inc., *Biolubricants – Global Strategic Business Report*, San José, CA (April 2011); http://www.researchandmarkets.com

[10] Kline & Company, Inc., *Competitive Intelligence for the Global Lubricants Industry, 2004–2014*, Little Falls, NJ (2006).

[11] A. Gosalia, The European base oils and lubricants industry, Trends in a global context, *Lube Mag.* **95** (2010) 8–16; Sustainability and the global industry, *Lube Mag.* **109** (2012) 20–5.

[12] L. Tocci, Rethinking hydraulic fluids, *Lubes 'n' Greases* **12** (2006) 26.

[13] J.S. Marth, *Renewable Lubricants Manual: Biobased Oils, Fluids & Greases*, United Bio Lube, Palo Alto, CA (2007); http://www.renewablelubricants.com

[14] Carlube, *Commercial Vehicle Oil Guide*, Tetrosyl, Ltd, Bury (2011); http://carlube.co.uk

[15] http://www.shell.com/lubeadvisor; http://www.shell.com/lubematch, etc.

[16] S. Crawshaw, The rapid development of lubricants data includes targeting consumers through a wider range of media, *Lube Mag.* **105** (2011) 6–8; http://www.oats.co.uk

[17] H. Oosting and E. Oster, Social media and apps and their impact on the global lubricant market, *Lube Mag.* **105** (2011) 16–20.

[18] Th. Mang, Lubricants and their market, in *Lubricants and Lubrication* (Th. Mang and W. Dresel, eds), 2nd edn, Wiley-VCH, Weinheim (2007), pp. 1–6.

[19] S.C. Tung and M.L. McMillan, Automotive tribology overview of current advances and challenges for the future, *Tribol. Intl.* **37** (2004) 517–36.

[20] S.Z. Erhan, B.K. Sharma, Z. Liu and A. Adhvaryu, Lubricant base stock potential of chemically modified vegetable oils, *J. Agric. Food Chem.* **56** (2008) 8919–25.

[21] OECD 301, *Guideline for Testing of Chemicals*, Organisation for Economic Co-operation and Development, Paris (1992).

[22] J. Legrand, Performing lubricants and solvents based on oleochemicals, *Proc. Eur. Conf. Renewable Raw Materials*, Gmunden (6–8 October 1998).
[23] J.C.J. Bart, N. Palmeri and S. Cavallaro, *Biodiesel Science and Technology. From Soil to Oil*, Woodhead Publishing, Ltd/CRC Press, Cambridge/Boca Raton, FL (2010), p. 678.
[24] Bundesministerium für Ernährung, Landwirtschaft und Forsten (ed.), *Bericht über den Einsatz biologisch schnell abbaubarer Schmierstoffe und Hydraulikflüssigkeiten und Maßnahmen der Bundesregierung*, Berlin (1996).
[25] Environmental Data Services, Climate change action could boost plant-based products, *ENDS Rept. 323*, London (2001), p. 33.
[26] J. Legrand and K. Dürr, High-performing lubricants based on renewable resources, *Agro-Food Industry Hi-Tech* **9**(5) (1998) 16.
[27] M. Torbacke, T. Norrby and M. Kopp, Environmentally adapted lubricants in the Nordic marketplace: recent developments, in *Proc. 13th Intl. Colloquium Tribology: Lubricants, Materials, and Lubrication Engineering*, Technische Akademie Esslingen (15–17 January 2002), Vol. 1, pp. 293–9.
[28] H. Theissen, *The Market Situation of Biodegradable and Biobased Lubricants in Germany 2006*, IFAS/RWTH, Aachen (2006); http://www.bioschmierstoffe.info (in German).
[29] H. Theissen, Application of bio-oil in hydraulics, *o+p Ölhydraulik u. Pneumatik* **49**(1) (2005) 47–52 (in German).
[30] H. Theissen, Experiences with bio hydraulic oil, *o+p Ölhydraulik u. Pneumatik* **50**(3) (2006) 132–7.
[31] W. Klose, Significance of rapidly biodegradable lubricants in the German lubricant market, *Tribol. Schmierungstechn.* **50** (2003) 45–9.
[32] M. Kaup, Entwicklungs- und Erfolgsfaktoren für Produkte aus nachwachsenden Rohstoffe in Deutschland und der EU im Spannungsfeld zwischen Ökonomie und Ökologie, Thesis, University of Cologne (2002), p. 146.
[33] méo Consulting Team, Siegsdorf (Germany) (2005).
[34] nova–Institut für politische und ökologische Innovation GmbH (nova–Institute for Political and Ecological Innovation), Hürth/Cologne (Germany).
[35] D. Peters, N. Holst, B. Herrmann, S. Lulies and H. Stolte, *Renewable Resources in Industry,* 3rd edn, Fachagentur Nachwachsende Rohstoffe e.V., Gülzow (2010).
[36] H. Theissen, Experiences with the German Market Introduction Program for biolubricants, in *Proc. European VTI-LLINCWA Workshop on Biolubricants* (The Virtual Tribology Institute/Mol and Fundacion Tekniker/Eibar, eds), Mol (15 January 2003).
[37] H. Murrenhoff, *Markteinführungs programm Biogene Treib- und Schmierstoffe – Positivliste*, RWTH, Aachen (2006).
[38] *VAMIL (Accelerated Depreciation of Environmental Investments Measure)*, Dutch Ministry of Housing, Spatial Planning and Environmental Management (VROM), The Hague (2003); http://www.vrom.nl/pagina.html?id=9575
[39] German Market Introduction Programme, *Bio-based Fuels and Lubricants*; http://www.bioschmierstoffe.info (in German).
[40] Fachagentur Nachwachsende Rohstoffe e.V. (FNR), *Positivliste. Hersteller- und Produktverzeichnis für auf nachwachsenden Rohstoffe basierende Schmierstoffe und Hydraulikflüssigkeiten*, Gülzow (15 November 2008).

[41] T. Meindorf and H. Theissen, Fast recognition of bio-oil, *o+p Ölhydraulik u. Pneumatik* **53**(9) (2009) 390–4 (in German).

[42] H. Theissen, Effects of contamination of biobased hydraulic fluids with mineral oil, *J. ASTM Intl.* **6** (2009) 1–9.

[43] C. Göhler and H. Rüschkamp, The problem of mixture: Contamination of biogenous oils with mineral oils; http://www.ifas.rwth-aachen.de/Main/Institut/IFK%20Paper/2_10_rueschkamp_paper.pdf

[44] VDMA Blatt 24589, *Richtlinie für die Umstellung von Druckflüssigkeiten auf Mineralölbasis nach DIN 51524 auf biologisch schnell abbaubare Druckflüssigkeiten und erforderliche Maßnahmen für den Betrieb*, Beuth Verlag GmbH, Berlin.

[45] H. Theissen, The German Market Introduction Program for biobased lubricants, *Tribol. Online* **5**(5) (2010) 225–9.

[46] F. Farshchi, Why invest in environmentally considerate lubricants?, *World Pumps* **496** (2008) 36–7.

[47] EC Commission Communication, *A Lead Market Initiative for Europe*, Brussels, COM (2007) 806 (final 21. December 2007); http://ec.europe.eu/enterprise/leadmarket/leadmarket.htm

[48] FP7-2009-BIOREFINERY_CP

[49] US Energy Information Administration (EIA), *Annual Energy Review*, Department of Energy, Washington, DC (2010); http://www.eia.doe.gov/

[50] B.J. Bremmer and L. Plonsker, *Bio-Based Lubricants: A Market Opportunity Study Update*, United Soybean Board, Chesterfield, MO (November 2008).

[51] P. Cuevas, *Comparative Life Cycle Assessment of Biolubricants and Mineral Based Lubricants*, M.Sc. Thesis, University of Pittsburgh (2010).

[52] S. Miller, C. Scharf and M. Miller, Utilising new crops to grow the biobased market, in *Trends in New Crops and New Uses* (J. Janick and A. Whipkey, eds), ASHS Press, Alexandria, VA (2000), pp. 26–8.

[53] D. De Guzman, Vegetable oil in lubricants and additives slated for strong growth, *Chem. Market Reporter* **261**(23) (10 June 2002) 11.

[54] Cargill Industrial Oils and Lubricants, Inc., *Development of Soybean Oil-based Transformer Fluid* (20 February 2004); presented at United Soybean Technical Advisory Panel on Lubricants (22 September 2004).

[55] T.V. Oommen and C.C. Claiborne (to ABB Power T&D Company, Inc.), High oleic acid compositions and methods of making and electrical insulating fluids and devices comprising the same, US Patent No. 6,312,623 B (6 November 2001).

[56] J.C. Mendes, A.S.G. Reis, E.C. Nogawa, C. Ferra, A.J.A.L. Martins and A.C. Passos, Oiled and ready to go, *ABB Review* **3** (2007) 53–7.

[57] T.V. Oommen and C.C. Claiborne, Biodegradable insulating fluid from high oleic vegetable oils, *CIGRÉ Symposium*, Paris (1998) 15–302.

[58] L.A.T. Honary, Potential utilization of soybean oil as an industrial hydraulic oil, *SAE Techn. Paper Series* No. 941760 (1994).

[59] http://www.hydrosafe.com

[60] http://www.epa.gov/oms/regs/nonroad/marinesiequipld/420f07032.htm

[61] S.A. Miller, *A Comparative Life Cycle Assessment of Soybean-Based and Mineral Oil Lubricants in Aluminum Rolling*, Ph.D Thesis, University of Illi-

nois, Chicago, IL (2006); *Proc. 59th Ann. Mtg. American Society of Tribologists and Lubrication Engineers (STLE)*, Las Vegas, NV (15 May 2005).
[62] S.A. Miller, A.E. Landis, T.L. Theis and R.A. Reich, A comparative life cycle assessment of petroleum and soybean-based lubricants, *Environ. Sci. Technol.* **41** (2007) 4143–9.
[63] S.A. Miller, A.E. Landis and T.L. Theis, Use of Monte Carlo analysis to characterize nitrogen fluxes in agroecosystems, *Environ. Sci. Technol.* **40** (2006) 2324–32.
[64] R. Garcia, Veggie coolants help Eaton slim down, *Lubes 'n' Greases* (1 June 2008) 26.
[65] A.J. Kinney and T.E. Clemente, Modifying soybean oil for enhanced performance in biodiesel blends, *Fuel Proc. Technol.* **86** (2005) 1137–47.
[66] J.C.J. Bart, N. Palmeri and S. Cavallaro, *Biodiesel Science and Technology. From Soil to Oil*, Woodhead Publishing, Ltd/CRC Press, Cambridge/Boca Raton, FL (2010).
[67] S.J. Asadauskas, T. McClure and G. Biresaw, Comparison of extreme pressure additive treat rates in soybean and mineral oils under boundary lubrication conditions, *Abstr. 100th AOCS Mtg*, Orlando, FL (3–6 May 2009), p. 15.
[68] L.A.T. Honary, A status report on promoting the use of biobased lubricants made of commodity and enhanced vegetable oils, *NLGI Spokesman* **66** (2002) 14–19.
[69] United Soybean Board, http://www.soynewuses.org
[70] J. Van Rensselar, Biobased lubricants: gearing up for a green world, *Tribol. Lubr. Technol.* **66**(1) (2010) 32–4, 36–41.
[71] http://www.cns.uni.edu/CONNECTIONS/FALL97/page1.html
[72] J. Suszkiw, *Statue of Liberty Goes Green With Soy-based Elevator Fluid*, Agricultural Research Service, USDA (Oct. 2004); http://www.ars.usda.gov/is/pr/2004/041019.html
[73] L.A.T. Honary, Soybean based hydraulic fluid, US Patent No. 5,972,855 A (26 October 1999).
[74] http://www.uni.edu/nabl/applications.htm
[75] J. Nelson, Harvesting lubricants, *Carbohydr. Econ.* **3**(1) (2000), no pp. given.
[76] G.S. Cannon and L.A.T. Honary, Soybean-based transformer oil and transmission line fluid, US Patent No. 5,958,851 A (28 September 1999).
[77] J. Suszkiw, *Biodegradable Lube for Metalworking Operations*, Agricultural Research Service, USDA (8 May 2006); http://www.ars.usda.gov/is/pr/2006/060508.htm
[78] L.A.T. Honary, Performance characteristics of soybean-based greases thickened with clay, aluminum complex and lithium, *NLGI Spokesman* **65** (2001) 18–27.
[79] L.A.T. Honary, Fluid test results of soybean-based greases developed by UNI-ABIL research program, *NLGI Spokesman* **64** (2000) 22–8.
[80] J.S. Marth, *Renewable Lubricants Manual: Biobased Oils, Fluids & Greases*, United Bio Lube, Palo Alto, CA (2007); http://www.renewablelubricants.com
[81] T. Norrby, Environmentally adapted lubricants – where are the opportunities?, *Ind. Lubr. Tribol.* **55**(6) (2003) 268–74.
[82] A. Cunningham, N. Battersby, W. Wehrmeyer and C. Fothergill, A sustainability assessment of a biolubricant, *J. Ind. Ecol.* **7** (2004) 179–92.

[83] R.D. Whitby, Will regulations help improve the bio-lubricant market?, *Tribol. Lubr. Technol.* **60**(3) (2004) 64.
[84] R.D. Whitby, Market share of biolubricants in Europe and the USA, *Lipid Technol.* **16** (2004) 125–9.
[85] H. Rüschkamp, *Untersuchung des Verhaltens von Bioölen mit Restantanteilen von Mineralöl in hydraulischen Anlagen*, IFAS/RWTH, Aachen (2006); FNR FKZ 22,014,803.
[86] L.A.T. Honary, *Biobased Greases and Lubricants: From Research to Commercialization*, Biomass R&D Technical Advisory Committee, National Ag-Based Lubricants (NABL) Center, University of Northern Iowa (2007); http://www.bdisolutions.com/SiteDocs/.../P4_BRDA27_2007.pdf
[87] R.L. Shubkin (ed.), *Synthetic Lubricants and High-performance Functional Fluids*, Marcel Dekker, Inc., New York, NY (1993).
[88] Frost & Sullivan, *European Biolubricants Market*, San Francisco, CA (2000).
[89] Conseil de Recherches Agro-alimentaires du Canada (CRAC), in *Biolubrifiants. Rapport sur l'évaluation des possibilités et des défis d'une économie axée sur la biotechnologie pour la recherche agroalimentaire au Canada* (2003), pp. 13–146.
[90] M. Schneider, Plant-oil-based lubricants and hydraulic fluids, *J. Sci. Food Agric.* **86** (2006) 1769–80.
[91] R. Luther, Lubricants in the environment, in *Lubricants and Lubrication* (Th. Mang and W. Dresel, eds), 2nd edn, Wiley-VCH, Weinheim (2007), pp. 119–82.
[92] D. Johnson, Future crops for bio-diesel and bio-lubricants, http://www.harvestcleanenergy.org/conference/HCE5_PPTs/Johnson.pdf
[93] Th. Mang, Environmental criteria for the use of vegetable oils and their derivatives in lubricants, *Fett/Lipid* **100** (1998) 524–7.

14
Lubricant use and disposal

Abstract: Criteria are defined for optimum machine-specific selection of conventional, high-performance and specialty lubricants. Lubrication consolidation is indicated as a means of rationalisation of inventories. Intended use of lubricants may be compromised by oxidation, water and air contamination, additive depletion and accumulation of contaminants, including wear debris, and biological degradation. Strategic oil analysis is described from simple in-shop sensory inspections to primary on-site standard testing and more comprehensive secondary testing methods as an operational maintenance tool for machine and lubricant condition monitoring to estimate remaining lubricant life time and prevent premature machine failure. The disposal of spent lubricants, including waste oil legislation and management, and re-refining technologies, are discussed.

Key words: lubricant selection, consolidation, monitoring, disposal, operational maintenance.

14.1 Introduction

Increasing economic development is putting pressure on world's capacity to sustain demands for resources or to absorb pollution. Society must work towards decoupling environmental impacts and degradation from economic growth. Consumption patterns have to become more sustainable. Stakeholders and consumers must take more ownership of efforts to protect the environment. Industry must operate in a more eco-efficient way, i.e. producing the same, better or more products with less input and less waste.

In previous chapters we have examined extensively the industrial development of biolubricant base stocks and the formulation of biolubricants for various applications. This chapter deals with the last two stages of the life cycle of a lubricant, namely use and final disposal (see Figs 11.3 and 11.7). Lubricant use comprises all stages from oil selection and consolidation (Sections 14.2 and 14.3) to procurement, storage, handling and application (see Chapter 12). In-use lubricant testing stands in close relation to operational maintenance of the machinery (Section 14.5).

After use, the life cycle of products should take into account the waste phase up to end-of-waste. In the lubricant sector, various waste prevention and reduction measures have been taken in the past:

- research and development (either on a proprietary basis or within the framework of joint actions) into producing cleaner and less wasteful products and technologies;
- eco-design (integration of environmental aspects into product design with the aim of improving the environmental performance of the product throughout its whole life cycle);
- creditable ecolabels and evaluation of their effectiveness;
- green procurement;
- reuse of waste oils through collection, recycling and re-refining;
- remediation;
- voluntary commitments.

A recycling society with a high level of resource efficiency seeks to avoid waste generation and uses waste as a resource.

The Sixth Environmental Action Programme of the European Community (2002–2012) calls for measures aimed at ensuring source separation, collection and recycling of priority waste streams (plastics, paper, glass, metal, oils) [1]. Used vegetable oils are to be considered as biodegradable waste or 'bio-waste' [2] and require separate collection from industrial oil wastes (see Section 14.7).

Principal objectives of any waste policy are to minimise the negative effects of the generation of waste on human health and the environment, to reduce the use of resources and to improve the efficiency of such use. It is imperative to recognise the potential benefits to the environment and human health of using waste as a resource. The used lubricant waste producer and waste holder should manage the waste in a way that guarantees a high level of protection of the environment and human health. Within a 'name, shame and fame' strategy, the 'polluter pays' principle is a guidance at international level.

The quantities of waste oils and emulsions have increased. Recycling of waste oils may be conducive to a fuel supply policy. Waste has value as a resource. Waste ceases to be waste when it has undergone a recovery operation, including recycling, into a useful commercial product meeting specific legislative and environmental criteria (end-of-waste status).

Recently, the European Commission has unveiled plans on how to achieve a resource-efficient society. Among others, the EU *Roadmap to a Resource Efficient Europe* (20 September 2011) aims at ensuring that waste is treated and used as a resource and fed back into the economy as a raw material. Measures to be taken include revision of recycling targets in EU waste legislation, replacing all environmentally harmful incentives with environmentally friendly ones, stimulating demand for recycled materials through appropriate economic incentives and development of EU end-of-waste criteria, creating stakeholders working groups to develop EU resource efficiency targets and resource productivity indicators.

14.2 Selection of (renewable) oil lubricants and greases

For lubrication of a machine tool, typically composed of a hydraulic unit, gearbox, spindle, slideway, linear system, plain and roller bearings, and cutting zone, a variety of different types of lubricants may be required. The choice of a lubricant for a specific application is determined by matching the machinery design and operating conditions with desired lubricant characteristics. In order to ensure reliable and long-term machine service life, selection of the most appropriate lubricant in relation to other machine components is of crucial importance and can significantly improve one's operations.

Lubricants are formulated and produced for numerous types of applications, machines, equipment, etc. (see Chaper 12). Various considerations must be taken into account when selecting the correct lubricant for each machine application. In order to make a systematic choice of a suitable lubricant it is absolutely necessary to understand the relationship between friction, wear and lubrication, and the interaction between the elements of the tribological system and the specific properties of each element. Special attention must be given to the different stress factors and the structure of the tribological system. It is equally important to accurately define the conditions of use before comparing the lubricating regime of the application to producer's product data specifications. Minimising wear by efficient lubrication prolongs the useful life of machinery, thereby minimising the consumption of non-renewable resources (metals), even more so when biolubricants are being employed. Optimum lubricant procurement can lead to improvements in energy efficiency of up to 10% (see Section 12.7.1). Adoption of high-performance lubricants (see Section 14.2.1) also leads to a significantly lower lubricant consumption.

It is advisable to adopt standardised lubricant selection criteria. Selection should be driven by quality and specifications, not by other concerns (brand, margin, price, consolidation, product availability within the brand, etc.). Lubricant viscosity and additive-type specifications deserve greatest attention. Regardless of how much information may be implied by a commercial brand name, it is insufficient to select a lubricant. Intended use designations can be misleading. A user must always refer to the producer's product information brochures, performance data sheets and material safety data sheets (MSDSs) to determine the intended use, additives and specifications.

Lubricant selection can be a very difficult process due to the great variety of manufacturers, vendors, and products in today's market (see Tables 10.2, 13.6 and 14.1). Lubricant selection is dictated by the application (see Chapter 12). A lubricant has many properties that need evaluation prior to selection to ensure that it is compatible with the system where it is to be used, and that it continues to provide adequate lubrication under operating conditions. In selection of a mineral oil for lubrication, viscosity and viscosity index are the main characteristics. If a synthetic ester (SE)-based lubricant is considered for an application, many additional properties need to be evaluated. For example, polarity and thermal conductivity are properties which differ more between different SEs than between mineral oils. Viscosity index (VI) is crucial in selecting a lubricant for a given application, and is especially critical in extremely hot or cold climates. While a main property of a lubricant is its viscosity, other important factors used to select a lubricant are temperature stability, chemical stability, compatibility, corrosiveness, flammability, toxicity, environmental effects, availability and cost [3]. These properties were considered by Bartz [4] in a decision matrix (DM) for different base oils (Table 14.2). Physical and chemical data (such as viscosity, flash point, solubility, corrosion properties, compatibility with

Table 14.1 Biolubricant manufacturers/distributors in the US market

Acuity Specialty Products, Inc.	Green Earth Technologies
Agro Management Group, Inc.	Hill and Griffith Company
American Chemical Technologies	E.F. Houghton & Co.
Arizona Chemical	Hydro Safe Oil Division, Inc.
BioBlend Renewable Resources, LLC	International Lubricants, Inc.
Bio-Gem Services, Inc.	LPS Laboratories
Bi-O-Kleen Industries, Inc.	Lubrizol Corporation
BioPlastic Polymers and Composites, LLC	McNovick, Inc.
Bunge Oils	Milacron LLC
Calgene, Inc.	NATOIL AG
Cargill Industrial Oils & Lubricants	Nutek LLC
Castrol	Panolin America, Inc.
Cognis Corporation	Plews/Edelmann
Cortec Corporation	Renewable Lubricants, Inc.
Creative Composites, Ltd	RS Farm and Harvest Supply, Inc.
Desilube Technology, Inc.	RyDol Products, Inc.
DSI Ventures, Inc.	SoyClean
Eco Fluid Center, Inc.	Starbrite Distributing
Environmental Lubricants Manufacturing, Inc.	Sunnen Products Company
FUCHS Lubricants Co.	Terresolve Technologies
G-C Lubricants	The Dow Chemical Company
GEMTEK Products	UNIST, Inc.

Table 14.2 Decision matrix for different base oils

Evaluation: 1 excellent, 2 very good, 3 good, 4 moderate, 5 poor.	Mineral oils	Polyalphaolefins	Polyalkylene glycols	Dicarboxylic acid esters	Neopentyl polyesters	Rapeseed oils
Viscosity–temperature behaviour (VI)	4	2	2	2	2	2
Low temperature behaviour (pour point)	5	1	3	1	2	3
Liquid range	4	2	3	2	2	3
Oxidation stability (ageing)	4	2	3	2/3	2	5
Thermal stability	4	4	3	3	2	4
Evaporation loss, volatility	4	2	3	1	1	3
Fire resistance, flash temperature	5	5	4	4	4	5
Hydrolytic stability	1	1	3	4	4	5
Corrosion protection properties	1	1	3	4	4	1
Seal material compatibility	3	2	3	4	4	4
Paint and lacquer compatibility	1	1	4	4	4	4
Miscibility with mineral oil	–	1	5	2	2	1
Solubility of additives	1	2	4	2	2	3
Lubricant properties, load-carrying capacity	3	3	2	2	2	1
Toxicity	3	1	3	3	3	1
Biodegradability	4	3/4	1/2	1/2	1/2	1
Price relation against mineral oil	1	3–5	6–10	4–10	4–10	2–3

After ref [4].

other materials and evaporation loss), that can be utilised as DM inputs, can easily be obtained from MSD sheets and other technical data sheets and used for evaluation. Even physical and chemical properties are not enough to select the best lubricant for an engine. Complex and expensive practical and bench engine tests are performed to test and understand the performance of a lubricant (see Section 10.5.1). These requirements reappear in delivery conditions, in-house standards and general specifications. Other properties, such as biodegradability, toxicity, water hazard impact, and cost information should also be valued for a specific application.

Cost (or preferably cost-to-benefit; see Section 13.4) is one of the factors to be considered when selecting lubricants. This is especially true when making substitutions such as using renewable bio-based lubricants, fluids and greases in place of mineral oils and synthetics. Bio-based lubricants are more expensive than mineral lubricants. Therefore, justification for their use must be based on operating requirements for which suitable mineral and synthetic lubricants are not available. It is unlikely that expensive high-performance lubricants are the best choice in total-loss applications.

Selection tools also comprise recommendations of the equipment manufacturer and lubricant producer. An informed selection or substitution of a lubricant depends on understanding the lubricating regime (i.e. film, mixed, boundary), established conventions of classifications, and the ability to interpret and apply the producer's product data specifications to the equipment. Prime considerations in selecting a lubricant for a given application are film thickness and wear. Although film thickness can be approached by calculation, wear assessment requires testing.

Lubricant manufacturers provide product information in brochures, pamphlets, handbooks (e.g. refs. [5, 6]), on the web, as well as on the product container or packaging. This allows the intended use, oil type, additives (anti-wear (AW), extreme pressure (EP), rust and oxidation (R&O), etc.), and the specifications to be established. Some producers identify the product by its usage classification (see Section 3.3) or machinery class. Producers do not usually specify additives. Product literature also states the base oil types, i.e. bio-based: soy, high-oleic base stocks (HOBS); petroleum-based: paraffinic, naphthenic, residual compounded; or synthetic. This is important when distinguishing between renewable biolubricants, non-renewable mineral oils and synthetic fluids. Environmentally acceptable (EA) lubricating oils should not be used in certain applications, such as runner hub oil, or governor oil, until extensive testing has been carried out.

Machine manufacturers use a combination of testing and in-house experience to determine which classes of lubricants are likely to perform well in their equipment. For this purpose also established standard specifications and classifications for lubricants, as developed by various professional societies and based on collective experience, are useful. Original equipment manufacturers (OEMs) normally supply lubricant recommendation tables. However, OEM recommendations might not necessarily reflect the best solution. Individual manufacturers may express different opinions based on their proper experience, equipment design and products. Definition of the 'best' lubricant for a given application is much a matter of opinion. Nonetheless, the manufacturer is expected to be in the best position to recommend a lubricant or indicate alternative lubricants. The recommendation should always be followed, especially when the equipment is still under warranty, unless the lubricant fails to perform satisfactorily. Lubricant recommendations should be updated every few years to take advantage of new lube developments. Machine manufacturers also often refer to the lubricant recommendations issued by component manufacturers. These recommendations issued by the manufacturers of hydraulic components, gearboxes, slideways, etc. must be observed.

On the basis of an equipment manufacturer's product description, operating characteristics, special operating requirements, and lubricant specification, product engineers of lubricant producers are other experts who can advise on lubricants meeting the OEM specifications (see Section 13.1). Unless unrealistic, viscosity should be of the equipment manufacturer's recommended grade. Products proposed by different lubricant manufacturers are likely to vary in performance, quality and cost.

It is finally the responsibility of the user to ensure that all applicable criteria are met in practice. Generally, the user should follow the manufacturer's specification. If no manufacturer's specifications are available, it is useful to determine what lubricant is currently in use and select a product with equal or better specifications. In any case, selection should respect the class recommended by the machinery manufacturer (EP, AW, R&O, etc.) and within the same base stock category (paraffinic, naphthenic or synthetic). Physical and chemical properties should be equal or exceeding those specified by the equipment manufacturer. It is also useful having clear ideas about the relative importance of lubricant properties in a particular class to facilitate rating. Finally, additional special requirements may be considered that can be beneficial when selecting lubricants (see Table 10.11). Vendor and product quality are important, but even the best quality products do not deliver acceptable results if they are abused during the application practice.

Selection of a lubricant needs to consider various variables such as load, speed, temperature and location. Load is the amount of weight or pressure to which the lubricant is subjected. The smaller the load, the lighter the lubricant can be. Equipment that operates at higher speeds generally has less pressure than slower speeds and can use a lighter lubricant. However, higher speeds determine more wear and therefore the lubricant must have the proper viscosity to reduce wear. The flow properties of a lubricant have a direct relationship to temperature. Lower temperatures need additives to improve flow, whereas higher temperatures require additives to stabilise the lubricant. Both for interior and exterior locations the atmospheric conditions of moisture, heat and cold need to be considered as well as dust and chemical or corrosive environments.

When selecting a lubricant one must look at more than physical properties alone, since the impacts caused by these products not only affect the environment but also socio-economic outcomes. In particular, life-cycle analysis (LCA) criteria (see Section 11.2.1) are helpful in terms of lubricant selection. While LCA results are useful in evaluation of bioproducts, they are obviously not the only metrics to be used in a lubricant selection process, as physical properties of a lubricant, life cycle, cost and other selected criteria should be included in the overall decision matrix framework. Factors that influence selection of bio-based lubricants include: operating conditions, viscosity, VI, pour point (PP), EP properties, oxidation inhibitors, rust inhibitors, and detergent/dispersant additives.

Cunningham *et al.* [7] developed a sustainability or decision matrix to evaluate environmental, social and economic impacts of a biolubricant (hydraulic fluid). The purpose of the tool was to quicken and reduce the costs of performing a full LCA by focusing on specific criteria of the life cycle of the product. Several LCA criteria such as energy use, CO_2 emissions, air emissions, impact of water supplies, and others were considered in the sustainability matrix. More recently, Cuevas [8] has considered the criteria utilised by Bartz [4] and Cunningham *et al.* [7] in combination with full LCA results in the development of a DM framework for use in lubricant selection. The resulting decision making tool may be utilised to evaluate mineral-based lubricants and bio-based lubricants derived from plant oils, such as rapeseed in Europe and soybean in the United States In particular, Cuevas [8] has included normalised results from nine life-cycle impact assessment (LCIA) categories – global warming potential (GWP), acidification potential (AP), carcinogenics, non-carcinogenics, respiratory effects (RE), eutrophication potential (EP), ozone depletion (OD), ecotoxicity and photochemical smog (PS) (see Table 11.1) – in the decision matrix for screening and evaluation of potential lubricants.

The DM developer establishes the relevant criteria needed to evaluate a certain product, e.g. lubricants. The products are then screened against the criteria and scored (e.g. 1 for excellent and 5 for poor). The criteria can eventually also be weighed to express relevance [9]. Product ranking is achieved on the basis of a total score. The DM framework can be customised by the user wishing to evaluate different products. For example, one might consider that inclusion of a lubricant in the US Department of Agriculture (USDA) BioPreferred^{SM} programme is another valuable selection criterion. This procurement programme aims at increasing the use of renewable, environmentally friendly bio-based products by providing a database of bioproducts and bioproduct manufacturers that can be used by US federal and contractor personnel to find products that meet federal regulations regarding green purchasing [10]. The BioPreferred^{SM} programme has established a minimum bio-based content percentage for more than 40 items including greases, hydraulic fluids and multiple types of lubricants. Several lubricant products and their minimum bio-based content according to the BioPreferred^{SM} programme are listed in Table 9.7. The decision matrix could profitably take into account inclusion of a product in the BioPreferred^{SM} catalogue. It is obvious that the bio-based content is not necessarily the sole criterion for selecting lubricants; an organisation may justify the purchase of non-biobased lubricants. As to lubricant selection, it is the case to notice that HOBS lubricants are compatible with the same seals and filters as are petroleum lubricants.

It is through a decision matrix that all relevant information for selection can easily be combined and evaluated to ensure an informed decision is made regarding the appropriate product to be utilised. Table 14.3 shows a partially compiled lubricant decision matrix according to Cuevas [8]. According to the reported value on the basis of the LCIA categories alone, in this case soybean would result as the better option. However, as explained above, obviously also other lubricant properties should duly be taken into account.

While LCA and decision matrix development are complex and time consuming, these methods can improve how decisions are made by choosing the most appropriate products that will have the least negative impact on the environment, economy and society. Mineral and bio-based lubricants have different environmental impacts. For example, rapeseed lubricants have a higher AP than mineral-based lubricants, whereas the opposite holds for ozone depletion potential (ODP). Consequently, an increase in biolubricant use would minimise ODP impacts, but also cause an increase in AP emissions. Knowing the environmental impacts caused by these products can be useful in determining how to regulate the lubricant market. Inclusion of additional parameters, such as the impacts of land use, could further improve the lubricant selection process.

Table 14.3 Lubricant decision matrix

Criteria	Rapeseed	Soybean	Mineral
LCIA categories			
GWP	–0.33	–0.64	1
AP	1	0.07	0.17
Carcinogenics	1	0.17	0.43
Non-carcinogenics	1	0.06	0.26
RE	1	0.14	0.47
EP	1	0.38	0.03
ODP	0.44	0.07	1
Ecotoxicity	1	–0.40	0.24
PS	1	0.17	0.14
Lubricant properties			
Viscosity			
Flash point			
Evaporation loss			
Biodegradability			
Other			
BioPreferred^{SM}	0	0	1
Cost			
Sample total score	7.11	0.02	4.74

After ref. [8].

Greases are classified by penetration number and type of soap or other thickener. Although these are important factors, they are not the only ones of interest. The final selection of a grease must be based on other information provided in the producer's specifications. Viscosity of the oil included in a grease must be considered. Producers provide information and specifications for grease in brochures, pamphlets, on the web or on the product container or packaging. Grease specifications normally consist of soap base, penetration, included oil viscosity, pumpability and dropping point. The producer may also specify ASTM test information on wear, loading, lubrication life, water washout, corrosion, oil separation and leakage. Grease additives are not usually stated except for solid additives.

14.2.1 High-performance (HP) and specialty lubricants

Substantial differences exist between conventional, high-performance (HP) and specialty lubricants, even when the viscometric properties are similar. Differences in the expected longevity, rate of deposit formation, film strength and wear resistance are of interest for comparison sake. *A priori* quantification of these differences is not always easy. Despite higher prices, HP products can actually result in a lower net overall use cost. Improvement is based on evidence such as reduced heat, wear debris, energy consumption and product consumption.

HP products are developed to thoroughly exceed the minimum allowable technical standard for a given application. The materials, manufactured with specialised raw materials, are designed to offer enhanced performance capability in the form of improved oxidation and deposit resistance, improved surface protection (wear resistance, nanoscale surface enhancement), improved long-term viability (greases), etc., vs conventional product performance. Common HP product examples include gear oils and greases incorporating solid film agents (synergistic metallic-molybdenum blends, graphite, borates), hydraulic oils manufactured with severely hydrotreated base stocks and ashless anti-wear agents, turbine oils manufactured with hydrocracked stocks and deposit-resistant oxidation inhibitors. These products are designed to be physically and chemically stable in the production environment. These materials are required to operate with minimal degradation in temperature extremes, under vacuum and exposure to contaminants that would quickly destroy other lubricants. Some examples include:

- radiation-resistant lubricants;
- lubricants operating in the space vacuum (–200 °C);
- lubricants operating in a chemical-rich environment.

Examples of the latter are lubricants operating in oxygen or CO_2 compressor conditions. Machines that pump natural gas and light hydrocarbons require lubricants that will not mix with (absorb) the gas during transmission. Only about 1% of plant applications require a specialty lubricant with unique properties for optimum machine performance. The price difference between the specialty and commodity product can be as high as 100:1.

Performance lubricants are often not considered for use because of price objections. Although cost control is always legitimate, limiting the use of a lubricant material because of price is shortsighted. Lubricant purchases represent only 1–3% of maintenance expenses. When lubricants are used as a process fluid (for rolling, cutting, grinding, etc.) the value is about 3%. Where lubricants are only used to protect load-bearing surfaces and provide hydromechanical functions, the value is about 1% [11]. Cost control is a better criterion than the product's litre price. With HP products there is a potential for significant benefits in terms of both reduced deposit formation and extended lubricant life cycle. Other benefits may include reduced power consumption, reduced downtime from change events, extended machine life cycle/reduced repairs. Use of HPs requires a value-based selection. A more expensive choice can well be the less costly choice. The biggest threat to loss of value for HP products comes from the risk of leakage (arising from incompatibility with seals, o-rings, etc.). It is important to decide which machine applications best fit a HP product line. Over-application of HPs should be avoided.

Extended lubricant life cycles, reduced deposits, improved machine performance (less leakage, fewer valve repairs, less wear debris, better repeatability, etc.) and optimised use cost can all be achieved with improved handling and contamination control strategies, regardless of the relative quality of the product.

14.3 Lubricant consolidation

Manufacturers of machinery give guidelines on the operation, lubrication and maintenance required for their particular equipment. In complex industrial facilities or service organisations many different types of unrelated equipment, often supplied by many different manufacturers, may be operating, each requiring lubrication. This is an error-prone situation. Consequently, there may be a need to select as many similar lubricants as possible and to use related machine techniques. Moreover, lubrication of older machines, usually operating at slow speeds and light loads, is less critical than that of modern, high-speed machines operating under heavier loads and with tighter mechanical tolerances. Servicing both older and newer versions of similar equipment means maintaining inventories of several types of lubricant. Consolidation is then a rational approach to handling the lubrication requirements at a facility while reducing the total number of lubricants in the inventory. Lubrication consolidation aims at achieving operating advantages by increasing the reliability of the processes and the productivity of production plants, and cost savings by reduction of inventories, storage requirements, worker safety and health hazards, environmental hazards and lubricant costs. This can be achieved by using only a limited number of lubricant types, optimising the relubrication intervals and choosing high-performance lubricants. By also making use of lubrication charts and cutting the costs for energy, it is often possible to bring down manufacturing costs considerably. Neale [12] provides overall guidance on how to coordinate the recommendations of the various equipment manufacturers.

Lubricants are frequently purchased in large quantities and must be stored safely. The amount of stored material should be minimised to reduce the potential for contamination, deterioration, and health and explosion hazards associated with lubricant storage. Typically, non-renewable hydraulic systems are susceptible to explosion hazards. Synthetic fluids are less flammable than mineral oils. Use of fire-resistant bio-hydraulic fluids (made from low-volatility Stabilized™ HOBS) significantly reduces the risk of an explosion.

In general, lubricants specified by the equipment manufacturer should be used until expiration of the warranty. Voiding the terms of a warranty is not normally advisable. However, if more highly performing lubricants have become commercially available in the meantime (e.g. Stabilized™ HOBS), some equipment owners might be willing to trade off their warranty protection for other advantages such as advanced equipment protection, improvement in worker health and safety, compliance with environmental regulations, production line efficiencies, lower energy expenditure and other cost savings.

A consolidation process requires careful analysis and matching of the equipment operating characteristics, lubricating requirements and lubricant properties. The equipment should be sorted by type of lubricant (oil, hydraulic fluid, synthetics, biodegradable, grease) required. For each type, the essential properties of each lubricant should be listed, e.g. oil viscosity, AW/EP requirements, rust and oxidation inhibitors, detergent-dispersant requirements for oils, or NLGI grade, viscosity of the oil component, pumpability, etc. for greases. The goal of the exercise is to figure out if a single lubricant can span the range. If so, consolidation is possible.

Several precautions must be followed when consolidating lubricants. Consideration should be given to the more severe requirements of any of the original and consolidated lubricants. Lubricants with vastly different viscosity requirements must not be consolidated. The same holds for applications with very specific lubricant requirements. Consideration should also be given to the fact that some lubricant additives may show material incompatibility. Generally, the user will make a selection to meet the equipment manufacturer's specifications or to substitute a new brand for one previously in use. Product selection could start by using substitution lists available from most lubricant producers. However, users should be aware of the fact that such substitution lists may not specify classes of lubricants. Charts of interchangeable industrial lubricants are available for some lubricant categories. Information brochures provided by the lubricant producers must be examined to verify viscosity, intended use, class of lubricant, performance specifications and additives.

It is not uncommon that different hydraulic fluids (such as Warren R&O, NAPA AW-46 and AW-315, Conoco MV-22) are in use for various pieces of hydraulic equipment – bucket trucks, cranes, backhoes, rollback wreckers, dump trucks, scissors lifts, etc. It is useful to know the compatibility and interchangeability of the various products. The equipment owner of Table 14.4 operates with both R&O and AW (mineral) hydraulic oils in various viscosity grades, one of which (Conoco MV-22) is VI improved. A single hydraulic oil, such as ISO HM-VG 46, could possibly cover the entire fleet [13]. This requires individual assessment for each piece of equipment by considering:

- starting viscosity at worst-case cold start;
- maximum expected operating temperature;
- permissible and optimum viscosity range for individual hydraulic components in the system.

In changing equipment over to the 'consolidated' oil it is essential to check the compatibility of seal materials and miscibility of oils (to avoid additive 'drop out'); see also Section 13.2.

Consolidation may easily result in the use of higher-grade lubricants that are capable of meeting the requirements of different machinery. Although the cost of such high-grade lubricants is greater, the disadvantage may be offset by several benefits, consisting of reduction in the number of different lubricants needed and in inventory management requirements, and price

Table 14.4 Consolidation of a hydraulic oil inventory

Hydraulic oil	Grade	ISO classification[a]	Viscosity grade (VG)
Warren R&O	Rust and oxidation	HL	n.a.
Conoco MV-22	Multi-viscosity	HR or HV	VG 22
NAPA AW-46	Anti-wear	HM	VG 46
NAPA AW-315	Anti-wear	HM	VG 68

[a] ISO 6743/4.
n.a., not available.

discounts for purchasing lubricants in greater quantity. Prospective consolidated lubricants may also be multipurpose or other general-purpose (bio-based) lubricants which can be applied with a wide range of equipment and help reduce the number of lubricants required. For instance, super universal tractor oils (STOUs), multifunctional oils for tractors which provide lubrication for engine, transmission, hydraulics, power take-off and wet-brakes (see Section 12.2.1, p. 593), contribute to reducing stock holding and operational costs. Other examples of lubricant consolidation are the UNICUT universal metalworking fluid (MWF) (see Section 12.9) and conveyer lubes in the food-processing industry (see Section 12.10).

Plant operators and OEMs considering a changeover of conventional to biodegradable EA lubricants and hydraulic fluids should be aware of the differences between these products. In particular, as outlined in Section 3.5, EA fluids are not necessarily equal to one another. It is therefore important to assess thoroughly the requirements of the specific application to determine the feasibility of the proposed substitution without compromising quality or performance. Table 14.5 lists the most important considerations, measures or adaptations to the system required for switching to EA. It is important that the change to environmentally considerate lubricants is carried out in a carefully controlled way to avoid any cross-contamination with mineral oil. Some redesign of the installation might be needed (e.g. to deal with foaming problems, materials compatibility, use of stainless steel components). Change of performance characteristics at low temperature and accelerated fluid degradation at high temperature should be carefully investigated. The oxidation rate of vegetable-based EA lubricants increases markedly above 80 °C and lengthy exposure at low temperature can cause gelling of some products. RLI's Bio Ultimax™ AW1000 Hydraulic Fluid – 10W40 replaces two or more ISO grade mineral oil or synthetic hydraulic fluids. Also the new metal-free multipurpose additive package Additin® RC 9330 (Rhein Chemie, Mannheim) for both hydraulic oils and industrial gear oils reduces inventory costs and simplifies formulation.

Table 14.5 Assessment of consequences of substitution of conventional for EA lubricants

- Need for redesign
- Materials compatibility
- Need for extensive testing
- Temperature controls
- Change of performance characteristics
- Rate of fluid degradation
- Filter changes
- Need for moisture scavengers
- Disposal of used fluids

14.4 Degradation of lubricating oils and hydraulic fluids

Predominating machine operating conditions can induce property changes in a lubricant that can affect its performance. For instance, the frequency with which the oil is pumped through a circuit influences oil ageing. If an oil is pumped too fast, excessive amounts of air are either dispersed or dissolved, leading to premature ageing. Oil tank foaming can also occur. Engineering design measures, such as fitting baffles to a tank, can positively influence air release and tank foaming. These have the effect of prolonging the time in which air, water and solid contaminants can be released from the oil.

A lubricating oil may become unsuitable for its intended use as a result of one or several processes, notably oxidation, water and air contamination (foaming, air release, demulsibility), partial evaporation, additive depletion, accumulation of contaminants, including wear debris, thermal or biological degradation. Oxidation is the factor that most often limits the lubricant's useful life. Oxidation can lead to various undesirable consequences such as acidity, corrosion, sludge formation, volatility and/or changes in viscosity. Ageing studies of lubricants are several [14–20]. Fox [21] provides an overview of degradation of lubricants in service use. Methods of investigating lubricant degradation may consist of:

- simple tests (density, flash point, foaming, system corrosion/rusting, demulsibility, interfacial tension) using instrumental analytical techniques;
- lab tests for lubricant deterioration (viscosity and VI improvers, trace metals, particulates and ash in lubricants, acidity and base reserve, water contents, antioxidant reserve);
- field tests (direct observation of lubricant condition).

Sepčic *et al.* have determined the principal volatile compounds in degraded engine oil using headspace gas chromatography mass spectrometry (HS-GC-MS) [22]. Zeman *et al.* [23] reported quality control of used oils by means of pressure differential scanning calorimetry (PDSC).

The reason we need to change an oil is often that the additive package has worn out. The buffers get used up neutralising acids and/or detergents and dispersants get consumed by clinging to particles too small for the oil filters. Frequently also VI improver molecules have been broken down, and the oil has therefore thinned out enormously. Antioxidant depletion is another frequent occurrence.

Fairly little is known about the influence of ageing on the performance of environmentally adapted lubricants (EALs) in mixed and boundary lubricated tribological contacts. McCarthy *et al.* [14] have examined ageing of unsaturated and saturated ester-based turbine oils. Measurement of acid number was taken as a simple means of evaluation of the oxidative state of the fluids. Samples with high acid numbers (>2 mg KOH/g) were tested to assess extreme ageing. Friction is reduced for aged samples in comparison to fresh fluids. Nevertheless, even aged samples with acid numbers of 4 mg KOH/g continued to display acceptable wear performance. This suggests a limited risk for catastrophic failure of the equipment during normal operational use. The coefficient of friction appeared to reduce with fluid age. For the specific application investigated, the EALs continue to function well in terms of their friction and wear performances even at high acid numbers. Battersby *et al.* [15] have examined ageing of saturated and unsaturated SE-based fluids for applications in mobile hydraulic equipment. Negligible fluid degradation or hydraulic pump wear was seen after 3500–5000 h of service.

Water may be dissolved or emulsified in oil. Water contamination is a problem for machine health. Water affects viscosity, promotes oil degradation and equipment corrosion, and interferes with lubrication. More precisely, water contamination (from leaking heat exchangers and seals, condensation, etc.) in fluid systems may cause fluid breakdown (such as additive precipitation and oil extraction), reduced fluid viscosity and lubricating film thickness, accelerated metal surface fatigue, galvanic corrosion, loss of dielectric strength in insulating oils, etc. Even low concentrations of moisture (<100 ppm) can have a meaningful negative impact on machine components such as bearings [24]. In particular, it has been observed that the fatigue life of bearings can be extended considerably by reducing the amount of water in petroleum-based lubricants. Typical water saturation levels are 30–50 ppm for transformer oils and 100–1000 ppm for hydraulic fluids, depending on the specific grade. Below the saturation level water will completely dissolve in the oil. The saturation level increases at higher operating temperatures. Water contamination is a common occurrence for hydraulic fluids because of their hygroscopic nature. Hydraulic system operation may be compromised when water contamination reaches 1–2%. The separation of small droplets of water can be assisted using centrifuges or coalescers. Dissolved air (and other gases) in oil systems may cause foaming, slow system response with erratic operation, higher fluid temperatures, pump damage due to cavitation, inability to reach full system pressure, and acceleration of oil oxidation.

Oxidation is a general term used to denote chemical degradation of lubricants. Oxidation is the process through which lubricants change chemically by incorporating oxygen. Oxidation accelerators include air, moisture, catalytic metals and heat. Debris caused by wear acts as a catalyst to induce degradation. Group I base oils are characterised by 10% aromatics and more than 300 ppm sulphur. Sulphur compounds and aromatic molecule structures are highly reactive and tend to react rapidly with oxygen to form various soluble and insoluble degradation by-products. The high percentage of aromatics generally means a higher risk of oxidation within the lubricant sump and short lubricant life cycles. In hydrocracked Group II oils the aromatics fraction and sulphur and nitrogen components are nearly eliminated from the stock. As a consequence, the useful life of the finished lubricant product is extended. Group III base oils, which are even more severely hydrotreated, perform on a par with some synthetic Group IV oils. Group II/III-based products are inherently more stable, more resistant to oxidation, experience less temperature-induced viscosity change and produce less sludge than their Group I counterparts. A drawback is their reduced solvency (ability to dissolve or suspend polar additives and/or the polar by-products of lubricant degradation). Lubricant stocks with high solvency (such as vegetable oils) are able to soak up polar additive compounds or the polar by-products of the oxidation process.

Oxidation has an impact on the lubrication performance of vegetable oils (see Table 14.6) and quite possibly also plays a role in the lubrication process itself. Correlating studies of oxidation with lubrication performance will lead to greater understanding into the role of oxidation in the lubrication mechanism of vegetable oils [25]. Controlling the consequences of oxidation degradation (deposits, varnish and residue formation) begins with controlling the oxidative processes within the system. Additives are used to

Table 14.6 Impact of vegetable oil oxidation compounds on lubrication

Compound	Lubrication impact	Reference(s)
Hydroperoxides[a]	Increased wear	[25–28]
Volatiles[b]	Virtually negligible	
Non-volatiles[b]		
Epoxides	High viscosity, greater oxidation stability, similar boundary lubrication, lower deposit forming tendencies	[22, 29, 30]
Other compounds[c]	Uncertain effects	
High-molecular-weight compounds[b,d]	Increased viscosity Formation of deposits Uncertain impact on lubrication	
Free fatty acids[b]	Improved boundary lubrication properties	[20, 31–33]
	Lower oxidation stability	[34, 35]

[a] Primary oxidation compounds.
[b] Secondary oxidation compounds.
[c] Many, often with short lifetimes.
[d] Products of cyclisation and polymerisation reactions.

help stop, or at least postpone, oxidation. As antioxidants are meant to be gradually consumed by chemical reaction with oxygen, this determines reduced oxidation protection and stability with time. Once antioxidant additives are consumed, oxidation can continue unabated. Additive depletion is thus a cause of degradation of lubricants. Also water may react with additives. For example, zinc dialkyldithiophosphate (ZDDP), an AW additive commonly used for boundary lubrication applications in high-pressure pumps, gears and bearings, is destroyed by chemical reaction with water at operating temperatures above 60 °C. The end result is premature component failure due to metal fatigue. The effects of additive depletion may also be illustrated for turbine lubricating oil where R&O-inhibiting agents are essential. Water may deplete such inhibitors and increase the rate of rusting.

Lubricating oils can equally become unfit for further service by accumulation of contaminants, for instance MWFs. Internal sources of contamination are rust, wear and sealing products. Dirt is an external contaminant. Solid contaminants may also enter the system during maintenance and repair.

Lubricating oils and hydraulic fluids are susceptible to microbiological deterioration in certain conditions. Growth of microbes such as bacteria, algae, yeast and fungi can occur in hydraulic systems contaminated with water. This can lead to clogging of the system, production of significant waste products and acids capable of corrosion of metals, and component failure. Procedures for preventing bio-contamination include addition of biocides, frequent draining of moisture from the system, cleaning and sterilisation. Biodegradable lubricants are inherently more susceptible to biological deterioration and their concentrations should be monitored more frequently.

MWFs play a significant role in machining operations and have a substantial impact on tool life, shop productivity and the quality of the workpiece. An important aspect of water-soluble MWFs is the influence of microbial effects on working results in machining. Microbial degradation leads to drastic changes in the physical and chemical properties of the MWF during service life [36, 37]. Besides the demands on the efficiency and economy of cutting processes, environmental friendliness becomes a crucial issue.

Adenosine triphosphate (ATP) assays are used to quantify bioburdens (biomass) in low-organic-compound content fluids (freshwater, seawater, cooling tower water and similar fluids). A new test protocol for ATP in emulsifiable oil, semisynthetic and synthetic MWFs at end-use dilutions is available [38]. Comparison of field tests of ATP with other MWF condition monitoring data demonstrates the applicability of the new protocol.

The loss and effect of mineral, synthetic and vegetable oil-based lubricants from agricultural machinery into soil have been investigated [17]. A range of synthetic lubricants applied to field plots was found to degrade more rapidly and extensively than mineral oil-based lubricants, but not as rapidly or extensively as natural vegetable oils. Loss of the oils was enhanced by the addition of nitrogen; the greatest losses were observed in the warmer summer months. The extent and rate of biodegradation of the oils were much lower than that found previously in laboratory-based studies, but the relative biodegradability of the oils was respected. Performance, biodegradability and ecotoxicity of a commercially available rapeseed oil (RSO)-based hydraulic fluid were also monitored during field trials in a variety of agricultural machinery over a period of 6 years (operation for 35 000 h and 21 000 km) [18]. The hydraulic fluid caused no damage and there was no more leakage than with mineral oil. On average 60% of the oil reserve was replaced during the 6-year field experiment. The fluid remained biodegradable with low toxicity throughout its lifetime. Used petroleum-based automotive lubricants, contaminating sandy soils, were found to be more biodegradable than fresh oils, but more toxic (as a result of high polycyclic aromatic hydrocarbon (PAH) concentrations) [39]. Used lubricants represent a problem for the environment (see Section 14.7).

14.5 Operational maintenance

Competitiveness on the global markets requires optimisation of production facilities and plant equipment. The criteria are long service life, minimal downtime and low maintenance costs. The lifespan of machinery depends on the lube. The quality and performance of the lube become critical to long-term operation of engines and industrial systems alike. It appears that 50% of all bearing failures are the result of poor lubrication or contamination [40]. Viscosity loss, oxidation stress and contamination are reliability concerns for the machine. A proactive maintenance culture is essential. Machine reliability management starts with lubricant selection, purchase, in-plant receipt, storage and handling (safety and health hazards), establishes operational maintenance strategies and machine-specific routine lubrication programme practices, defines oil operating conditions and assesses machine and lubricating oil health on a regular and timely basis. It is the case to distinguish between machine wear problems and lubricant health and controlled degradation.

For the highest quality and performance due attention should be given to various aspects related to lubricants: vendor and product selection (see Section 14.2), lubricant storage and handling (see ref. [41]), in-use lubricant testing (from sampling and method development to laboratory quality verification and alarms limits development) (see Section 14.5.1). Recognising the potential for container conditions and human error to influence the final cleanliness and quality of the lubricant delivered at a plant site, plant reliability and lubrication managers should apply systematic quality control measures to all incoming lubricants [42]. Typical in-plant lubricant handling and storage problems that render the products unacceptable for use are:

- use of poorly maintained transfer delivery containers;
- transfer of incoming lubricants into contaminated intermediate storage tanks.

Lubricants that have become contaminated by previously delivered bulk products should not be allowed in the consumption stream. Wherever the concentration of solid contaminants in the lubricant exceeds the quality control target limits for allowable contaminants, the lubricant should be filtered before storage or use. A storage plan should be operating.

A comprehensive lubrication programme should include appropriate steps to ensure proper handling, storage and dispensing of selected quality oils and greases to avoid untimely degradation, as well as proper handling and disposal of used lubricants, in the interest of the plant, safety of the personnel and of the environment. A good lubrication scheduling programme should be inherently reliability-centred, and machine-specific. Lubrication plans, which cover all machine components in need of lubrication, should describe:

- the precise location of all lubrication points;
- the lubricant specifications (type of lubrication and lubricant required);
- the lubrication timetable.

There should be an up-to-date plant record of lubricant viscosity and additive-type specifications for each machine sump. The purpose of a lubrication plan as part of routine servicing is to ensure that a sufficient quantity of the correct lubricant is applied to the right spot at the right time. OEMs normally supply lubricant recommendation tables. Technical similar lubricants can be grouped to enable some lubricant rationalisation (see Section 14.3).

Johnson [43] has described the prospects of operator-based lubrication practices rather than mechanic- or lubrication technician-based programmes. The former is usual practice in metalworking operations but is not widely adopted in continuous process industries such as petroleum refining, chemicals production, cement and paper manufacturing. On the basis of the information supplied by the machine and lubricant manufacturers a lubrication maintenance programme can be set up for every machine which details the type of lubricant and lubrication interval (cf. Shell LubePlanner).

Significant economic and environmental pressure exists to extend the service lifetime of lubricants, to allow engines to optimise combustion conditions, and to detect impending failures. In a recent series of papers Johnson and Spurlock have addressed the practical challenge of measuring and managing lubricant vitality [44–48]. Strategic oil analysis comprises estimating remaining lubricant life [44], on-site oil analysis [45, 46] and rate-of-change alarms for machine and lubricant condition [47]. For benchmarking of a lubrication programme, see ref. [49]. Table 14.7 lists the main categories for lubrication programme development.

Oil analysis should provide information about the state of the lubricant condition, the cleanliness of the sump and the condition of the machine. The added value of oil analysis or any other condition monitoring technique is high (Table 14.8). Machine condition monitoring based on oil analysis has become an accepted practice in maintenance management. Oil-based condition monitoring techniques provide benefits in terms of machine health and cost savings by failure avoidance and productive capacity improvement [40]. Plants can achieve important savings in equipment repair and replacement costs through a well-managed operational and preventive maintenance (PM) programme. Clearly defined lubrication practices help improving productivity (machine reliability) and quality (machine repeatability). Productive capacity improvement is an even more important reason

Table 14.7 Lubrication programme benchmark categories

- Vendor selection process
- Lubricant delivery, storage and handling processes
- Lubricant technical selection practices
- Lubricant application practices
- Oil analysis programme practices
- Lubricant contamination and condition control practices
- Programme management and personnel development
- Lubrication practices standardisation (SOPs)
- Health, safety and environmental (HSE) practices

After ref. [49].

Table 14.8 Benefits of a machine maintenance programme

- Increased productivity (better use of equipment)
- Reduced equipment downtime
- Increased equipment life
- Reduced safety hazards
- Better utilisation of in-shop maintenance (less emergency work)
- Improved control of spare-part inventory and reduced parts usage
- Environmental benefits (reduced spill potential)
- Reduced downtime related to environmental release
- Reduced fines and clean-up costs due to environmental release
- Savings on lubes
- Documented warranty claims

to adopt an oil analysis or any other form of condition assessment programme than reducing maintenance costs. Considerable savings attributable to oil analyses have been reported for a great many common production processes, as diverse as an automotive stamping plant (by wear debris analysis), a catalytic reformer hydrogen recycle compressor (by gas contamination analysis) and an hydraulic mining shovel (by contamination and degradation monitoring and control) [48, 50].

Machine maintenance strategies are invariably intended to result in extended lifetime of the equipment, increased productivity, cost savings and environmental benefits. PM programmes schedule maintenance and/or replacement of parts and equipment based on manufacturer's suggestions. A PM programme relies on elapsed time, i.e. established service intervals without regard to the actual operating conditions of the equipment, as in the predictive maintenance (PdM) programme where trend analysis is key. Predictive maintenance is a proven, cost-effective tool for certain types of machinery and industries, such as the power generation (particularly wind turbine) and aggregates industry. PM programmes often result in excess downtime and premature replacement of equipment. Professional maintenance organisations (e.g. Fluid Life; Bloomington, MN) may improve PM practices, which ultimately contributes to healthier operations.

Modern maintenance schedules or computer maintenance management systems (CMMSs) are essential to maintenance activities for complex manufacturing facilities. The primary goals of CMMSs include scheduling resources, optimising resource availability and reducing the cost of production, labour, materials and tools. The most effective CMMSs are integrated with a PdM programme which relies on condition monitoring of machines to help determine when maintenance or replacement is necessary. Condition monitoring periodically records the machine's condition by trend spotting. The most commonly measured characteristic is vibration, but other tests may include lubricant analysis, thermography and ultrasonics. Vibration and oil analysis are complementary: oil condition can confirm certain problems detected through vibration. A PdM strategy uses analysis to identify an incipient failure condition and to monitor that condition by a maintenance action in order to derive maximum component life. Wear particle analysis is a PdM action. It is forecasting machine stoppage due to progressive failure of mechanical components. Trend analysis is used to help predict failure of a particular machine component and to schedule maintenance and order parts. The obvious advantage of condition monitoring is that failure can often be predicted, repairs planned, and downtime and costs reduced. Condition-based oil analysis will invariably save money compared with calendar-based oil analysis for industrial equipment. Condition-based oil change and contamination monitoring are proactive maintenance actions [51]. A properly executed analysis plan provides strong support of a proactive strategy. The proactive maintenance strategy seeks to achieve machine lifetime extension through the elimination of root causes of potential failure.

Even with the best maintenance programme, there is still potential for a hydraulic fluid or lubricant spill. In the US large spills will result in a Resource Conservation and Recovery Act (RCRA) clean-up (and inherent costs). All mineral oil-based hydraulic fluid spills are reportable events. While spilling large quantities of biodegradable fluid is still considered to be a reportable event under the RCRA, agencies are required to evaluate bio-based oils differently from petroleum-based oils. As awareness of biodegradable fluids increases, federal and state agencies are becoming more lenient regarding clean-up and fines [52].

Table 14.9 lists the main hazards threatening the life and performance of machinery [53]. Contamination has the single largest impact on the life of both the equipment and lubricant. A lubricant contaminant is 'any amount of material or energy which constitutes a chemically reactive hazard or that mechanically impairs the proper functioning of the equipment and the performance of the system' [54]. Contamination control of sumps is a vital requirement to preserve machine function and component health for all mechanical systems. It is good practice to pre-filter a lubricant before use.

Air-to-oil contact cannot generally be avoided. Air in a tribosystem may be either free, dissolved or entrained (as small or large bubbles). Free air may become trapped in a supply line. Dissolution of air in a lubricant depends on the nature of the lube, liquid density and pressure, air/gas pressure and temperature. The oxidative effect is noticeable (as smell and colour change) especially at temperatures above 60 °C, which are common sump temperatures. Oxidation enhances acidity, produces oil thickening (through formation of high-molecular-weight components), and favours sludge deposits. Air and water contaminants can disrupt the fluid film required for surface separation.

Water is a potentially debilitating contaminant that is present in many production environments. Water may enter lubricated systems from many different sources, such as operator washdown activities, rain, runoff, supersaturated production areas, high atmospheric relative humidity, long-term

Table 14.9 Most frequently occurring contaminants at manufacturing facilities

• Air	• Process chemicals
• Moisture	• Microbes
• Atmospheric particulate	• Heat
• Machine particulate (metals and oxides)	• Stored energy

sump condensate build-up, lubricant storage, product handling methods, etc. It is common for moisture from air to gradually accumulate in circulating systems and static sumps with normal heating/cooling cycles. Water in the tribosystem may be either free, dissolved or emulsified (entrained small and large droplets). Free water allows for microbe growth and poses a chemical threat to the lubricant if not removed. It poses only a minimal threat to lubricated components as long as the water does not pass through a load zone within the machine. Dissolved water cannot be observed visually but can be measured with common oil analysis techniques. At increasing concentrations, microscopic water droplets are formed which cause cloudiness. Suspension or precipitation of these agglomerates depends on: (i) type of base stock; (ii) type and concentration of additives; (iii) presence of demulsifier additives; (iv) temperature; (v) system turbulence; and (vi) change in surface tension. Polar base fluids, polar additives, higher temperatures and moving oil all favour water suspension. As water can have a dramatic negative influence on lubricant performance and life cycle it is important to identify even small amounts of water during ongoing production activities. Large water concentrations can create stable emulsions and form sludge. Emulsified water (oil-in-water) conditions are particularly threatening to mechanical components. Water can be highly damaging to both machine parts and lubricant. Since water has no load-supporting capacity, a water film formed in the load zone collapses readily and allows surface contact, crashing surface asperities, thereby forming wear debris, burning the oil, vaporising the water, and forming oxidation by-products. Water accelerates machine component corrosion (rust) and base oil degradation/oxidation, degrades and depletes additives, and contributes to formation of bacterial growth. In modern conventional power stations, the overall condition of the fluids that lubricate large, high-value machinery is critical to the successful and economic operation of the plant. In particular, the amount of moisture present in the oil can affect the lubricating fluid's performance since water can wash out critical antioxidative compounds, contribute to lubricant oxidation and subsequent loss of lubricant performance.

Solid particles are a threat to the component surfaces in all tribosystems. Microscopic particle intrusion into an oil sump promotes micropitting, fatigue wear, and two- and three-body abrasive wear (for various types of wear, see ref. [55]). Particle shape, hardness, size and concentration influence the wear mechanisms that particles produce. Environmental and process chemicals cause surface abrasion, deformation and corrosion, i.e. lost surface integrity and accelerated wear. A correlation exists between the concentration of environmental particles (many with hardness exceeding machine steel) and component wear. Catalytic wear metals increase the rate of oxidation and lubricant degradation and degrade the machine surface profiles. Machine wear debris also contributes to increased particle concentrations. There are several methods useful to gauge particle concentrations. When establishing an oil analysis programme that is proactive in controlling contamination, particle counting (ISO 11171:2010, 16232:2007) – typically developed to measure residual particles resulting from manufacturing processes and from the environment – is a vital component to the routine testing set. The ISO fluid cleanliness code (ISO 4406:99) – as it relates to solid contaminants – is a three-part system which expresses concentrations of particles with sizes equal to or exceeding 4, 6 and 14 μm. ISO 4406:99 is not the only method by which the cleanliness of an oil sample can be reported. Other standards include NAS 1638 and MIL-STD 1246 C. Atmospheric dust can change the surface profile of a machine component in particular if the hardness of the particles exceeds that of the component construction material. Particles that are 1–3 μm are frequently particularly tough and destructive to machine surfaces.

Microbial growth is most common to process sumps (for metalworking applications) but can also occur in machine lubricant sumps. Poor thermal conditions also represent a hazard. The ideal sump temperature is 50–60 °C. Stored energy from potential, kinetic, pressure and tension/strain sources is another threat for structural change.

Oil analysis (OA) is a powerful tool in the lubricated machine condition monitoring toolbox; it has an important machine care function [56]. Oil analysis testing constitutes a mainstay of a plant or facility reliability programme and is equally important in over-the-road (OTR) and construction equipment engine and systems monitoring. Table 14.10 shows the various stages of a quality oil analysis programme and setting up a testing programme. The first step in the process is selection of the equipment for review. Essential considerations are criticality and failure modes. Machine criticality is a key factor for determining which machines justify development of reliability-centred lubrication and oil practices [57]. Criticality assessments take into account operations and maintenance priorities, and considerations for worker safety and the environment. High criticality machines operating in highly stressful production environments with narrow ageing limits should be screened and lab-tested quite regularly. Low criticality machines may warrant analysis just to determine oil change requirements only.

Table 14.10 Oil analysis programme development

• Machine selection
• Development of sample collection methods
• Selection of primary and secondary testing methods
• Development of alarm settings
• Data review
• Corrective actions

Understanding the potential failure modes of a component being tested will help determine whether or not oil analysis is a suitable technology. Common machine failure modes are fatigue, two-body and three-body abrasive wear (resulting in very small particles), aggressive wear (with formation of large particles > 14 μm), environmental contamination (from dirt or process debris), fluid degradation (by oxidation, moisture, heat), inadvertent use of incorrect lubricant and cross-contamination (fuel, coolant, lubricant). An appropriate lubricant-identification system should be in place. Inadvertent mixing of a dissimilar lubricant viscosity grade is a common lubrication problem. It is important to determine the predictive technologies to be applied for condition monitoring purposes. It may not always make good economic sense to include oil analysis in an overall predictive tool kit.

The progressions to failure of most lubricated components are predictable. Lost lubricant health and excessively high contaminant levels lead to abrasion wear, fatigue wear and adhesion wear. Each wear mode causes increased component surface degradation up to a recognisable state which is often beyond the stage of having the opportunity for corrective actions. A consistent analysis programme may identify distinct wear mode signatures in the wear-development profile. The metals of primary concern in routine wear debris analysis are those known to be used in component construction, typically Fe, Cu, Sn, Pb, Ni, Cr and Al. Less common machine component construction metals include Ag, Ti, V and Mo. Component composition information is available from the OEM. There are also metals (Ca, K, Na, B, Si, Zn, P and Sb) that are typically associated with the additive package in use and the contaminant types likely to be found in the production environment. There are some useful (average) default wear metal limits provided by OEMs and industry associations. For example, the American Gear Manufacturers Association (Alexandria, VA) has guidelines for gear wear limits. Also many OTR engine manufacturers have guidelines for wear as do industrial equipment and component manufacturers. While many OEMs have established minimum guidelines, these may not be the most appropriate for individual plant reliability objectives. Established wear metal limits set by the user should be respected.

Vehicular fleets, such as the aircraft, ships, tanks and trucks of the US Department of Defence and of major freight transport and construction equipment operators, have adopted a periodic oil sampling system in which samples are taken from all equipment and sent to oil analysis laboratories for evaluation. This system is costly and unwieldy. Clearly, there is a need for a ubiquitous and user-transparent sensor technology that is capable of direct integration into a vehicle with the ability to analyse the lubricants and hydraulic fluids *in situ* [58]; see also Section 14.5.1.

Oil analysis has become an extremely successful predictive maintenance tool applied by the commercial industry and military to greatly improve safety and reduce maintenance costs. The Joint Oil Analysis Program (JOAP) uses oil analysis as a maintenance diagnostic tool to:

- determine the internal condition of aeronautical and non-aeronautical engines, transmissions, and gearboxes, and their oil-wetted components through the analysis of used lubricating oils, grease, and fluids. Its goal is flight safety, enhanced equipment readiness, reduced maintenance costs, and the extension of component life;
- determine the suitability of lubricants and fluids for continued use, resulting in savings and early detection of harmful conditions that, if not corrected, could promote premature component failure.

Following selection and execution of an oil testing programme, the reliability manager must apply an alarm or ageing limit structure for each test [59]. Armed with the information gained regarding the state of oil health the maintenance team can then take appropriate actions to remove root causes of failure before damage can occur (oil change, seal repair, etc.) and to monitor the progressive degradation of a specific mechanical component, thereby predicting when it will fail. Heat, contaminants and moisture all contribute markedly to oxidation and shortened life cycles. More frequent filter changes may be needed. Moisture alarm limits for lubricating oils are customarily set in ppm. The challenge of driving moisture out of a system begins once the target alarm levels have been reached. Vacuum dehydration is an effective and preferred method because it allows for rapid water evaporation at a relatively low temperature [53]. Vacuum dehydrators are used in large systems and systems with high water contamination risk, such as paper machines, turbines and hydraulic systems operating in humid environments. Less complex air-stripping technology is an effective alternative. Smaller quantities of water may be removed by using super-absorbent materials in filter elements. Moisture scavengers may be necessary to ensure low water content in the lubricant.

When the sumps are large enough to justify cost vs frequent oil changes, specially designed electrophoresis and electrophysical separation techniques allow for removal of oil waste and extension of lubricant lifetime without risk to the production machinery. Electrophoresis uses charge attractions to trap high-molecular-weight insoluble compounds. Electrophysical filtration, based on ion exchange, is designed to attract charged soft particles. For fluid decontamination, see also ref. [60]. Benchmarking is an excellent way of achieving excellence in maintenance management and is a highly beneficial tool for lubrication programmes [49, 61].

14.5.1 Lubricant monitoring

In PM mode, typically a fixed operating mileage, time or an algorithm based on engine operating conditions is used as an indication that the oil is approaching the end of its useful life and should be changed. These interval indicators are usually specified to underestimate the remaining oil life so as to avoid damage to the engine or machine. There is room for a more accurate method for determining the remaining useful life of lubrication oil in such a system. Techniques for determining when lubricating oil, in use in an engine or other machine, has reached the end of its useful life are fairly advanced, but less sophisticated than the technology used for their development. Condition-based oil change enables the useful life of the lubricant to be utilised without discarding the material too early, and avoids using the fluid too long to the detriment of the host machine or engine.

Lubricant analysis programmes are tests used to determine whether a lubricant remains effective. The three critical areas of industrial oils are oil chemistry, system contamination and machine wear. Strategic oil analysis is a properly developed, scheduled and executed program intended to keep 'normal' machine wear to a minimum [44, 56]. Oil analysis has an important machine care function. Lubricant-based machine condition analysis is the control tool that reliability engineers use to grade the effectiveness of lubrication practices and activities, in their efforts to maintain machine performance.

Lubricated machine surfaces do wear, degrade and fail. Accordingly, a properly developed, scheduled and executed oil analysis condition monitoring practice is useful to avoid downtime and production losses related to lubricated mechanical components. Oil analysis practices provide insight into machine operation by focusing on lubricant health, cleanliness of the sump and changing machine conditions. Degraded lubricant health and contaminated sumps accelerate failure of many lubricated components. As the contaminant loads increase, lubricant health further declines. A multitude of tests is required to deliver the desired information.

Systematic used oil analysis (for mobile equipment) originated in the US railroad industry after World War II, was subsequently used to monitor military jet engines (now within the framework of the JOAP; see Section 8.8) and was later extended to general bench-top testing and plant floor screening. In the United States, alone more than 250 used oil analysis laboratories are operating, including some commercial service providers. For example, CSI (Emerson Process Management; Knoxville, TN) and CMI (Condition Monitoring International; Miami, FL) are advanced industrial oil analysis laboratory and consultants in fluid analysis, respectively. Similarly, Oelcheck GmbH (Brannenburg) is an important German used-oil analytical laboratory. PerkinElmer 0.1 Express™ is a recent oil condition monitoring system. ERASPEC OIL (Eralytics, Vienna) allows for portable FTIR oil analysis.

In-use lubricant testing starts with analysis sampling, comprises test slate development, alarm limits setting and laboratory quality verification. The benefits and drawbacks of a properly executed lubricant analysis programme are listed in Table 14.11. Key lubricant performance criteria for measurement include viscosity, contaminants, moisture, silt, solids and wear debris, oil chemistry, oxidation resistance, sludging and residue formation, and chemical integrity. This preferably calls for a multiparameter instrument approach. On-site minilab oil testing options shorten the turnaround time in comparison to off-site lab oil analysis.

The most frequently encountered lubricant failure modes are:

- contaminated sumps (dust, water, machine wear debris, air, heat, other lubricants; see Table 14.9);
- inadvertent mixing of dissimilar lubricant (viscosity grade, additive package);
- incorrect sump volume.

In practice, lubricants are replaced either because the lubricant is dirty or worn out. Both conditions can be defined and measured. At any point during its use, if the lubricant is no longer capable of protecting the machine from wear, it must be changed. The ageing process is dominated by chemical changes, generally involving oxidation. High concentrations of oxidation/degradation by-products and/or low concentrations of oxidation inhibitors indicate a looming threat to a lubricant and a machine. The pro-oxidants are well known: heat, water, air, metals and soluble gases. Eliminating any

Table 14.11 Benefits and drawbacks of a lubricant analysis programme

Benefits:
• Reduces the frequency of oil changes
• Decreases consumption, purchase and storage of virgin oil
• Reduces the generation of waste oil
• Provides valuable diagnostic information
• Reduces maintenance costs
• Reduces plant production outage
Drawbacks:
• Data must be collected over time and analysed to determine trends
• Know-how is required to take representative samples and perform diagnostic tests
• Oil analysers must be calibrated to the type and manufacturer of the oil being used
• Results are subject to interpretation
• Capital and operational costs

of these contributors adds appreciably to the long-term health of the lubricant. For hot machines, such as combustion and gas turbine engine applications, heat is often the primary pro-oxidant. In case of limited heat-removal options short lubricant change intervals are expected. Additionally, sludge and varnish deposits are an expected outgrowth of degradation from specific types of oxidation inhibitors and degradation of the base oil itself. As lubricants age they also become more acidic. Even lubricants that are highly alkaline in nature (engine oils) age towards an acidic condition. As oil ages its rate of degradation often increases. If the oil is hot, wet and/or contaminated with iron or copper wear debris, the rate of decay could accelerate.

In hydroturbines there is usually a large amount of oil in the bearing oil system. In this case it is more appropriate to consider maintenance rather than change intervals. Such facilities usually run a testing laboratory for periodic oil quality tests. Regular sampling and testing can determine the timing and effectiveness of filtration, and can help identify unexpected problems. The lubricants used present four main causes for degradation: viscosity breakdown, particulate contamination, additive depletion and water contamination:

- *Viscosity breakdown*: The ability of a turbine oil to maintain separation between the surfaces and in the bearing depends on its film strength, which is related to viscosity. A loss in viscosity is mostly on account of shearing stresses in the bearing. An increase in viscosity usually indicates that the oil temperature is high enough for boiling off the lighter oil fraction. An increased viscosity can determine a higher bearing operating temperature.
- *Particulate contamination*: Unless bearing surfaces are actually contacting, the major cause of wear is through contamination by particles. Internal contamination may derive from additive breakdown; external contamination may be due to dust and dirt introduced through vents, or poor filters.
- *Additive breakdown*: By performing their intended functions, some additives are used up. Depletion of additives may increase wear by allowing corrosion to create oxide particles that can damage bearings. Filtration may remove components of the additive system.
- *Water contamination*: This form of degradation can sometimes be observed visually, usually by the oil taking on a cloudy or milky cast. However, by the time enough water is mixed in the oil to be visible, the oil's strength has already severely been decreased. Thus testing for water should preferably be performed by other tests even if the oil visually does not appear to be contaminated with water.

For gas and steam turbines, typically oil monitoring intervals in the laboratory of one year are perfectly acceptable, apart from weekly visual inspections. The control of used fire-resistant fluids, used lubricants and turbine oils is usually performed by the laboratory of the oil supplier. Effective contamination control for hydraulic and circulation systems improves sump life cycles.

In terms of simplicity and costs lubricant health programmes evolve from simple sensory inspection to more extensive in-shop, on-site or off-site oil analysis monitoring. Sensory inspection of oil-charged equipment is a low-cost, high-value on-site practice for both machine and lubricant assessment and represents the first stage of a more extensive oil analysis and machine maintenance programme [45]. The overriding question for spotting of looming problems is: What changed? Table 14.12 shows the possible outcome of simple sensory inspections (visual, audible, olfactory, tactile) at the machine. Oxidation, water contamination, lubricant cross-contamination, sludge formation, high wear debris and loss of dispersancy are all patterns that can be recognised by sensory inspections. Dropping sump levels may be indicative of either internal or external leakage. Some large circulating systems, such as those found in metal-rolling mills or paper machines, are prone to leakage because of their (poor) design; on the other hand, large engines may consume large volumes of oil with normal operation. High sump levels may be on account of air entrainment, internal leakage or water ingression. Air ingression could be the result of degradation of seals. As air has no load-bearing capability and creates considerable heat when compressed (eventually causing surface wear) air contamination warrants immediate action. Haziness or cloudiness are indicative of a rise in concentration of either air or moisture.

Table 14.12 Sensory inspection of oil-charged equipment

Observation[a–d]	Diagnosis
• Changes in oil level[a]	Oil losses or gains
• Large bubbles, tight foam[a]	Air intrusion
• Cloudy or opaque appearance[a]	Air or moisture entrainment
• Foul odours[c]	Chemical changes
• Vapour emissions[a,b,c]	Elevated temperatures
• Oil darkening[a]	Lubricant degradation
• Sediment[a]	Lubricant damage, solid contamination and wear debris
• Used filters[a]	Solid contamination and wear debris
• Hot surfaces[d]	Elevated temperatures

[a] Visual.
[b] Audible.
[c] Olfactory.
[d] Tactile.

Common methods for measuring moisture loads require collection of a lubricant sample to report a value. Several options for continuous monitoring measure water concentration as a percentage of fluid saturation at a given temperature [62]. Water intrusion in a lubricant is easily spotted by a simple (pass/fail) on-site crackle/hotplate test, where a 1000 ppm-plus of water concentration in a sample produces an audible sound. Other commonly accepted methods for moisture measurement in lubricating oil are calcium hydroxide and Fourier transform infrared (FTIR; quantitative, low sensitivity), ASTM D 95 (quantitative, distillation method) and various Karl Fischer (KF) procedures. If water is identified by simple on-site testing, different levels of maintenance action should be considered depending on the water contamination.

Foul odours denote chemical changes. For example, water entrainment in EP and AW-type lubricants can lead to conversion of zinc–sulphur–phosphorus-based chemistries into odorous sulphur-based acids and gases. Oxidation is another cause of odour formation. Vapour emissions from normally hot machinery (gas turbines, combustion engines) could indicate either lubricating oil evaporation or water vapour release. Some machines, such as gas turbines and synthetic fibre manufacturing machines can reach temperatures high enough to drive off light-ends of the base oil. As lubricants darken upon ageing and some lubricants darken even simply by exposure to light, a colour change alone does not necessarily mean the lubricant is damaged. However, alarm might be warranted depending upon the extent of colour change. Sediments should be observed under the microscope for further identification of sludge or wear debris. Used filter inspections may reveal important information.

For oil health measurements properly devised sample collection procedures are the first consideration. Sampling is obviously the process of taking a sample. Not all machines are designed for easy and accurate sample collection. It may be necessary to retrofit the equipment with fixed oil sampling ports. Sample collection from oil drain lines, from the top of a reservoir or from a circulated oil system all present particular challenges. The same holds for sampling splash-, collar- or ring-lubricated systems and for safe, effective high-pressure sampling from hydraulic systems. The sampling technique should guarantee that the fluid sample to be analysed fully represents the actual conditions for the entire system, is relevant and repeatable. Although the liquid lubricant is largely homogeneous, contamination and wear debris are not. Consequently, oil analysis might deliver greatly different results depending on where the sample is pulled. There are two types of samples – static and dynamic. Static samples are drawn from a fluid body at rest and may suffer from concentration gradients and particle segregation by size and density. Dynamic samples are taken from a fluid in motion. The basic types of dynamic sampling are laminar and turbulent flow sampling. Laminar flow sampling has no value for particulate contaminant analysis. Fluid sampling can create a major source of error in particulate contaminant analysis [63].

Next, the physicochemical parameters to be monitored should be determined. Tests that best reflect real failure conditions and root causes are to be selected preferentially. The selection of the oil testing methods is driven by machine criticality, environmental conditions, strictness of the ageing alarm set and the type of components under surveillance. High criticality sumps require ample testing to clearly delineate lubricant degradation and contamination conditions beyond such simple routine measurements as crackle testing, particle count or FTIR. With a properly designed and executed set of tests, oil analysis is capable to diagnose machine health, lubricant condition and contaminant level. Setting alarm limits (absolute or relative values) is used to draw attention to changes in condition well before failure and damage.

Samples can be analysed in-shop using field-portable analysers, on-site using laboratory analysers, or off-site using contracted lab services or in combination. In the manufacturing sector, laboratory approaches are unsuitable due to the lag-time between sampling and results reporting. Therefore, the continuous measurement approach is considered the most ideal, yielding instantaneous results reporting and quality control feedback.

In-shop tools are the first line of instrumental testing. In-shop analysis has the advantage of giving immediate feedback to machine operators. The strength of the approach is both its low cost and the motivation of finding and fixing machine problems. Effective use of in-shop lube analysis tools is to monitor the non-critical applications in a plant that often make up about 70–90% of the plant equipment. On the other hand, the most critical equipment in a plant will commonly be analysed more comprehensively relying on on-site or contracted lab services for more in-depth analysis of selected samples. However, also in this case in-shop analysers may be used to monitor the equipment. Monitoring of the dielectric constant of a lubricating oil, which requires a relatively low capital investment and is therefore useful at shop-level, can be used as a field test which indicates oxidation of lubricant molecules. Moderate dielectric increase indicates contamination due to fuel soot, dirt, oxidation or acid build-up. Severe dielectric increase indicates water, antifreeze or metal particles, and calls for immediate action to avoid potentially serious equipment damage. For in-shop lube analysis also OilView® Model 5100 (CSI; Knoxville, TN) can be used [40]. In-shop oil analysis, often in conjunction with vibration analysis programmes and laboratory oil analysis, is intended to result in cost savings through extended oil exchange intervals by condition monitoring of the lube, maintenance cost avoidance and plant production uptime. Garvey [40] has described several case histories and cost savings using in-shop and contracted lab services for oil analysis in industrial applications. Castrol's CAREMAX™ Onboard is a range of on-board test kits which allow quick determination of a wide range of marine lube characteristics [64].

Selection of testing methods provides users of oil analysis with a narrow set of options [44, 65, 66]. The objective in test selection is to maximise the information gained from a relatively narrow and cost-effective set of primary routine (or standard) tests, eventually followed by secondary non-routine tests (for the asking) providing more comprehensive details. Primary tests should provide a broad view of health indicators (oil health, contaminant load, machine health) with relatively few tests. Secondary tests should provide much more specific information that closely pertains to the type of machine being scrutinised. As shown in Table 14.13, relatively few generic test methods are commonly employed as used machine and engine oil

Table 14.13 Instrumental analytical tools for generic used-oil analysis

Condition monitoring	Test profile	
	Primary tests[a]	Scondary tests[b]
Machine condition analysis		
Wear debris measurement	ICP, AES, PC	PQ, DRF, FPT, A/F, SEM/EDX
Lubricant fluid analysis		
Viscosity	(ASTM D 445)	EA, FTIR, TAN/TBN, KF
Acidity/basicity	TAN/TBN	
Oxidation	FTIR	RPVOT, LSV
Process contaminants	FTIR	
Additive depletion	FTIR, ICP, AES	
Contaminant monitoring		
Environmental particle contamination	PC	AES
Air contamination	Visual	
Soot	AES, FTIR	Pentane insolubles, MPC
Water	Hot plate, FTIR	KF
Glycol	FTIR	
Volatiles	FP	GC

[a] ICP, inductively coupled plasma; AES, atomic emission spectrometry; PC, particle counting; TAN/TBN, total acid number/total base number; FTIR, Fourier transform infrared; FP, flash point.
[b] PQ, particle quantifier; DRF, direct read ferrography; FPT, filter patch test; A/F, analytical ferrography; SEM/EDX, scanning election microscopy/energy dispersive X-rays; EA, elemental analysis; FTIR, Fourier transform infrared; TAN/TBN, total acid number/total base number; KF, Karl Fischer; RPVOT, rotary pressure vessel oxidation test; LSV, linear-sweep voltammetry; AES, atomic emission spectrometry; MPC, membrane patch colorimetry; GC, gas chromatography.

evaluation techniques. Monitoring lubricant health has a high return-on-investment (ROI) potential as proper use of oil analysis can easily result in largely extended oil change intervals, avoid premature machine failure and damage, and optimise productivity.

Oil inevitably degrades on use. The value of oil condition monitoring programmes is clearly recognised. Infrared analysis is a rapid, inexpensive test that can profitably provide specific information about the chemical condition of the oil, also allowing inferences about the state of the machine component from which the sample was taken. Infrared spectroscopy plays a key role as a tool for the rapid, sensitive measurement of oil degradation, contamination and additive depletion. As mentioned elsewhere (Section 8.8) various ASTM standard practices cover in-service monitoring of lubricating oils (ASTM D 6224, D 7412, D 7414, D 7415, D 7418, D 7686, E 2412). Trend data, gathered by regular sampling to monitor engine performance and operating conditions, allows oil change at optimal intervals. The benefits of oil condition monitoring can be further increased by complementing routine measurements at a centralised laboratory with more rapid on-site analysis using a portable analyser [67]. Early detection of wear is essential for timely replacement, avoiding unnecessary maintenance while reducing the risk of catastrophic breakdown.

For generating actionable knowledge about the machine under review it is necessary to replace generic oil analysis test methods with machine-specific primary and secondary lubricant testing methods and alarm sets. The selection of the tests for any given machine depends on the machine owner's strategy (proactive or predictive) and should include test methods that effectively measure for the common failure modes for the machine under consideration [65, 66]. Appropriate application-specific parameters need to be monitored. Johnson and Spurlock [66] have discussed machine-specific tests and alarm sets for some of the most common component types, including industrial gearboxes, hydraulic systems, turbine generators, pumps and compressors. Testing programmes for any given machine should include analytical methods that effectively measure for the common failure modes for the machine under investigation and take into account the machine owners strategy as to use of the analytical data (in proactive or predictive mode). Table 14.14 shows the recommended test methodology for machine-specific test methods.

There are several tests that measure oil and machine health, some overlapping. Most tests were developed to measure fresh oil properties and performance, but work well to grade used oil and machine health when adapted accordingly. Oil analysis should provide information about the state of the lubricant condition, the cleanliness of the sump and the condition of the machine. Strategic oil analysis estimates the remaining lubricant life [44]. A variety of tests may be used to deliver this type of information.

Table 14.14 Recommended machine-specific test methods

Test method	Test methodology
Elemental analysis[a]	ASTM D 5185 or D 6595
Kinematic viscosity @ 40/100 °C	ASTM D 445
Water	Crackle test
ISO particle count	ISO 11500
Acid number	ASTM D 664[b], ASTM D 9744
FTIR	ASTM D 7686
LSV	ASTM D 6971
RPVOT	ASTM D 2272
Flash point	ASTM D 92, D 93
Varnish tendency testing[c]	ASTM WK 13070
Water[d]	ASTM D 1744 or D 6304
Ferrous density	None
Analytical ferrography	None
Demulsibility	ASTM D 1401

[a] Wear metals, contaminants, additives.
[b] Preferred.
[c] MCP, membrane patch colorimetry.
[d] Karl Fischer.

Some common test parameters that are indicative of lubricant health include viscosity, chemical state (total acid number/total base number, TAN/TBN), additive levels (typically Ca, Zn, Mg, Ba, Sb, P, B), dielectric constant and rotary pressure vessel oxidation test (RPVOT). It is useful to set ageing limits to these parameters [59]. Contamination alarms include: solids (wear and dirt) via particle count (ISO Cleanliness Code), and chemical contamination (water, glycol and fuel). Some common water limits based on component type are as follows: rotary screw and centrifugal compressors, 500 ppm; turbine and components using circulating oils, 200 ppm; hydraulic systems, 100 ppm; electrohydraulic control systems, 100 ppm; industrial gearboxes, 400 ppm [59].

Viscosity is only one of several oil quality parameters. For optimal condition-based monitoring the measurement of viscosity is considered essential. The presence of water, fuel and other contaminants causes the viscosity of the oil to decrease. The build-up of soot and sludge in the oil causes the viscosity to increase. Therefore, monitoring oil viscosity can provide timely warnings of deteriorating oil quality. For industry to transit from scheduled preventive maintenance to predictive maintenance sensor technology is a requirement. Requirements of real-time viscosity sensors for industrial/automotive applications are reliable, affordable and meaningful *in situ* oil condition monitoring, long lifetimes (10 years) without recalibration, oil condition measurements at shear rates relevant to its function, operation in a wide temperature range (–50 to 150 °C) and insensitivity to high vibration, solvent exposure, corrosive chemicals, and abrasive conditions. Modern intelligent sensor methods, which employ microelectromechanical system (MEMS) devices with monolithic piezoelectric sensor (MPS) technology, provide the technical means to enable affordable embedded condition monitoring of oil and hydraulic fluids of critical significance to industrial process control, and military, commercial and consumer transportation engines. Highly sensitive solid-state viscosity sensors in an on-line configuration are now available for use in challenging process control environments and for OEM applications.

Properly implemented thickness shear mode devices [68] appear to offer the best reliability, are rugged in corrosive and abrasive environments, and measure properties most relevant to the actual conditions of use of the lubricant and rated as such by most oil viscosity engineers [69]. In such devices, viscosity is measured by placing a piezoelectric thickness shear mode resonator (sensing element) in contact with oil. The top surface of the sensing element interacts with the oil, forming a microns thin fluid layer that moves with the vibrating surface. The oil's viscosity determines the thickness of the fluid layer that is hydrodynamically coupled to the surface of the resonator. The shear wave interaction with the fluid changes with variations in viscosity and density, which can be measured by the electrical properties of the piezoelectric sensing element. The ViSmart™ in-line/on-line solid-state viscosity sensor (SenGenuity; Hudson, NH), which provides *in situ*, temperature-compensated viscosity, is ideally suited for measuring oil quality. Simultaneous real-time measurement of viscosity and density is possible [70, 71].

Considerable efforts have been devoted by the lubrication and automotive industries to the seeking of an on-line sensor capable of determining the chemical condition and degree of degradation of industrial and automotive lubricants in order to signal the need for an oil change. However, the variety of oil formulations has complicated significantly the task of implementing such a device [58, 72–74]. Wang and coworkers [72, 73] have described on-board electrochemical oil-condition microsensors sensitive to the TAN and glycol contamination of engine oils. Iridium oxide chronopotentiometric (CP) sensors offer prospects for the detection of acidity and basicity levels in lubricants over other types of acidity sensors, such as glass electrode and ion-selective field effect transistors [74].

Many car manufacturers use on-board engine computer calculations to estimate the oil's condition. Some systems, known as oil life monitors, use an optical sensor for determining the clarity of the oil in the engine. General Motors Corporation has developed a compact impedance sensor for monitoring electrical properties of hydrocarbon oils in operating automotive vehicles [58]. These same electrical sensors developed for mobile engines, which measure physical oil property data (resistivity ρ and permittivity ε), can also be used in assessing the remaining useful life of high-resistivity lubricant oils operating in virtually any stationary equipment. The resistivity ρ of an oil is proportional to the viscosity η of the medium, and inversely proportional to the number density of free charge carriers (ions) N_i moving through the medium under the influence of an applied electric field, i.e.

$$\rho \sim \eta / N_i \quad [14.1]$$

When the number of free charge carriers is constant (i.e. in the absence of chemical reactions), the resistivity is a measure of the fluid viscosity. Variations in the resistivity of the oil as a function of engine operating time reflect physical and chemical changes in the oil, and its time dependence is indicative of oil quality. The permittivity ε is a measure of the reorientation of molecular dipoles in the lubricant under the influence of an applied electric field. The permittivity of the oil will change through chemical reactions or through the addition of polar liquids such as water, ethanol or glycol to the oil. Engine oils are fairly non-polar dielectric liquids, with relative electrical permittivities ranging from 2 to 3. Increases in oil permittivity are to be expected throughout the life of the oil, since the combustion process will cause both oxidation and nitration of the molecular components of an engine oil, and water may also be introduced into the oil.

Figure 14.1 shows the typical behaviour of the electrical resistivity ρ for a passenger car motor oil with operating time. After an initial steady increase ρ decreases to a minimum value and then increases until the useful

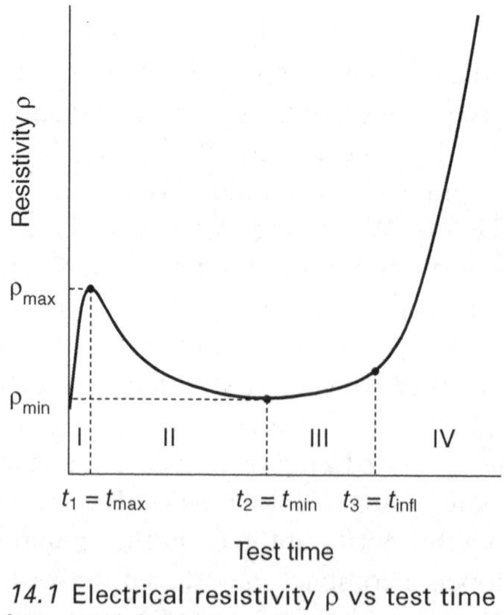

14.1 Electrical resistivity ρ vs test time *t* for a representative engine oil in a spark ignition engine operated under steady-state conditions. After ref. [58].

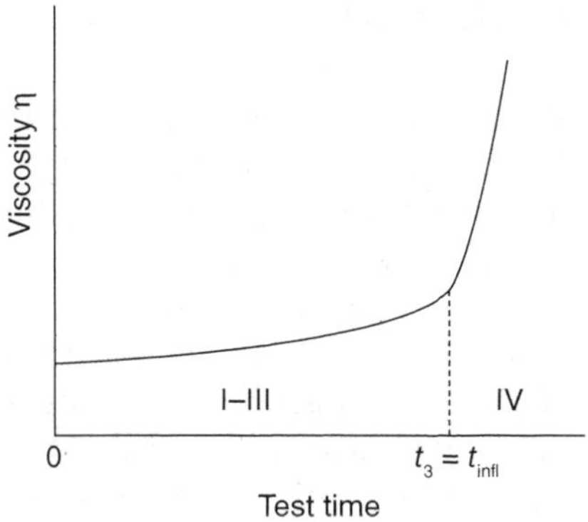

14.2 Viscosity η vs test time *t* for the engine oil of Fig. 14.1 in an engine operated under steady-state conditions. After ref. [58].

life of the oil has been depleted (Region III). The end of useful lubricant life is characterised by a very rapid increase in resistivity values at t_3 (inflection point): $(d\rho/dt)_{t<t_3} \ll (d\rho/dt)_{t>t_3}$.

The viscosity of the oil increases steadily during most of its useful lifetime in the engine, except for a short period of time (up to 10–15 h) at the beginning of its lifetime when it may decrease due to shear thinning. On the whole, the viscosity exhibits a monotonic change over the whole time interval but exhibits a drastic change in Region IV (see Fig. 14.2), signalling imminent engine failure due to improper lubrication.

Note that the time axes of Figs 14.1 and 14.2 are not drawn to scale. The time dependence of the electrical resistivity in Region I (1–20 h) is dominated by fast chemical processes that deplete charge carriers. The time dependence in Region II (typically 40–150 h) is on account of a slow chemical process that releases free charge carriers into the oil. The increase in resistivity in Regions III and IV is entirely due to the increase in viscosity.

On-site lubricant analysis

On-site lubricant screening, as opposed to offsite lab analysis, is especially useful in one of the following situations: (i) the rate of change from heat, water or chemical degradation is high; (ii) the site exercises frequent wash-down activities of critical production equipment; (iii) specific production machinery with tight alarm limits for parameters subject to rapid change (moisture, solid contaminants, chemical contamination); or (iv) no readily available cost-effective and timely lab response. On-site sensory inspection (see Table 14.12) and testing are particularly useful for machines with very short time-cycles between incipient and manifest failure. On-site lubricant analysis as a means of evaluating machine reliability is a viable option for many plant environments.

Clean environments, such as those found in electronics manufacturing sites, are much less prone to lubricant degradation than harsh wet or chemical-laden environments. Production processes that produce highly aggressive environments are most likely to derive added benefit from both quality and timeliness of rendered information for sump management. For example, in cement manufacturing environments, hot and heavily laden with highly abrasive solid particles, the rate of change of machine lubricants is particularly high. On-site particle counting is then a valuable asset; in the plants, lubricant viscosity changes are slow. On the other hand, food-processing plants are characterised by low hard-particulate loads but often exhibit extremely aggravating water and chemical contamination conditions. High-pressure hot water and wash chemicals enhance the risk of chemical degradation of lubricants. Also in this case the value of on-site screening is high.

Physical parameters and contamination levels are measured fairly easily with relatively inexpensive on-site instruments. Chemical properties measurements are more difficult to accomplish on-site; the cost of instrumentation is also appreciably higher. On-site wear debris analysis (WDA) permits immediate identification of the type and source of contaminants or wear debris particles through visual examination. Often, this information is the key to finding and eliminating root causes of recurring problems that shorten machine life. On-site or contracted oil analysis offers more sophisticated tools than in-shop testing, and involves greater knowledge of lube oil technology. The maximum benefit is obtained by guided information transfer between laboratory and maintenance personnel regarding lube analysis results and the consequences for machinery.

Any suspected conditions arising from sensory inspection should lead to follow-up actions by the reliability management. An obvious second step is instrument-based on-site lubricant analysis [46]. There are multiple instruments with which to monitor oil health. Some provide only pass/fail level identification while others provide detailed output as to the nature of the observed changes. The VISGAGE® (Louis C. Eitzen Co., Glenwood Springs, CO) pocket viscosity comparator is an option for simple on-site viscosity testing [75]. On-site infrared analysis may provide important information about a variety of lubricant indicators, including TOG (total oil and grease), acid number, base number, oxidation, additive depletion, water, and soot. See also Sections 8.7.1 and 8.8.1. The effects of lubricant soot content on engine wear are complex and can even be beneficial.

When properly developed, near-infrared spectroscopy (NIRS) is an excellent off-, at- and on-line quality control tool which allows rapid assessment of a variety of chemical, physical and ecotoxicological properties (see Table 8.12). Linear-sweep cyclic voltammetry (LSV) according to ASTM D

6971 is a measure of the remaining useful antioxidants in lubricants and can be used proactively to determine proper oil change intervals or extend intervals through timely sweetening of the fluid to replenish antioxidant levels.

Comprehensive all-in-one testing systems allow several parameters of a lubricant to be measured. The CSI 5200 Industrial Oil Analyser (CSI Technology, Inc., Knoxville, TN) is part of a complete desktop-sized minilab system that provides oil analysis results concerning oil chemistry (kinematic viscosity, dielectric constant and lube degradation), system contamination (water, large non-ferrous particles > 60 μm), laser particle counting (ISO 4406 and particle distribution), machine wear (magnetic differentiation of wear particles, ferrous index value), and wear debris analysis (small and large ferrous and non-ferrous particles). The rapid multifunctional CSI OilView® 5200 analyser, covered by ASTM D 7416-08 Standard Practice for Analysis of In-Service Lubricants, is capable of detecting most lubricant-related problems in gears, pumps, compressors, turbines, engines, hydraulics and process machinery. On-Site Analysis, Inc. (OSA; Palm Beach Gardens, FL) provides on-site fluid analysis equipment for wear metals (Al, Fe, Cr, Sn, Pb, Mo, Cu, Si, Na, K), extended metals (Ni, Ti, Mn, Ca, Mg, Ba, Zn, P, V, B), physical properties, viscometry and particle count. Other on-site analysers are tuned towards the need of fleet owners, where focus is on engine oil analysis, and the failure mode reflects a high incidence of corrosive and chemical wear [76]. Analysed results may trigger recommended follow-on actions which might include conducting more precise standard measurements at a laboratory.

Metal levels in engine oils can vary depending on such factors as:

- engine metallurgy;
- types of engine lubricants and additives;
- oil/lubricant consumption and replacement;
- dispersion characteristics of the oil's additive package;
- filtration efficiency.

The predominant threat to long-term machine performance is surface wear. Small particle wear debris measurement is commonly performed by elemental spectrographic analysis by means of inductively coupled plasma (ICP) or atomic emission spectrometry (AES). However, both techniques are particle-size limited and less useful to characterise wear debris from components in escalating wear conditions where high concentrations of large particles are being produced. The accuracy of ICP is significantly compromised for particles above 8 μm; similarly, AES for wear particles above 10 μm. Consequently, the effectiveness of ICP and AES at diagnosing aggressive wear patterns producing wear debris particles above 10 μm is limited.

Tests such as wear metals are highly specific to machine-condition assessment. Common WDA tests are particle quantifier (PQ), direct read ferrography (DRF), and analytical ferrography (A/F), but ferrous particle count and patch analysis may also be used. Particle counting is considered by many to be the single most important test for used oil analysis. Many large industrial systems generate ferrous wear particles when they are in conditions of an abnormal wear mode or impending machinery failure. If the equipment being tested is prone to ferrous wear, some type of ferrous density testing may be beneficial. In industrial gearbox and pump applications, where gears and element bearings are the primary wearing subcomponents, ferrous density analysis should be part of a standard test set. Conversely, if sampling from a sleeve bearing housing where non-ferrous debris is the major wear component, ferrous density testing would not be required. PQ is a ferrous density device using the Hall effect to determine the ferromagnetic particle concentration of an oil sample [77]. PQ analysis (as expressed by the PQ index) can be used as an early warning system to quickly identify the presence of large ferrous particles (>5–10 μm) [78]. The method is inexpensive, highly repeatable and not dependent on particle size. Table 14.15 compares results of PQ index values to spectrographic analysis for some typical samples. Low PQ index numbers indicate a low concentration and larger numbers represent high metal content. The method is most useful to test differential, transmission and gearbox oils. Kittiwake Developments, Ltd (Littlehampton, UK) has recently launched the ANALEX-pqL Ferrous Debris Monitor for on-site wear debris testing at ppm level in lubricating oils, hydraulic fluids and greases.

The PQ and DRF methods are appropriate for both standard and non-routine testing to identify the onset of aggressive wear. DRF, commonly referred to as the ferrous density reading, measures the amount of ferromagnetic wear debris in an oil sample using a high-powered magnet [77]. Ferrous density measurement is a useful secondary test method in case of

Table 14.15 Identification of wear of ferrous components

Sample	Analytical result		Test result
	PQ index	Spectrographic analysis (ppm)	
No. 1	60	286	Corrosive wear
No. 2	66	64	Small wear particles
No. 3	82	12	Large ferrous wear particles; approaching catastrophic failure

After ref. [78].

evidence of abnormal wear (increasing wear metal concentrations and particle counts). The quantitative analysis of ferrous particles is generally given in terms of DL and DS, representing large (>5 μm) and small (<5 μm) particles, respectively. DRF data may be used to derive total wear particle concentration (WPC) and percentage of large particles (PLP). The method will not detect non-ferrous wear or contamination particles. The filter patch test (FPT) is another common method for making a detailed determination of wear occurrence, especially of non-ferrous materials. LaserNet Fines-C (LNF-C) (Spectro, Inc.; Littleton, MA) is a particle counter and particle shape classifier.

A/F is among the most powerful diagnostic tools in oil analysis today. It is biased to ferrous material, but other non-ferrous wear particles and contaminants are also captured and identified. A/F provides a root cause for wear mechanisms (two-body adhesive wear, three-body cutting wear) based on the morphology and composition of the particles [79]. It is, however, time consuming, and therefore costly.

Ferrographic analysis is a predictive method for determining equipment condition long before signs of wear are detected. The size, number, composition and type of particles will indicate the severity of the wear. Although total wear debris measurement is most common, it does not reflect the state of ongoing machine change and machine health. The rate of wear generation is a more insightful metric than total wear [47]. Statistically derived wear debris limits should be developed around individual machines.

As shown in Table 14.13, parameters for condition monitoring of lubricant health are viscosity, acidity/basicity, oxidation, process contaminants and additive concentration. As an industrial oil ages, viscosity, oxidation value, acid number (AN) and process contamination are all likely to increase, whereas additives are generally being depleted. Lubricant viscosities (ASTM D 445) change over time. Viscosity alone cannot be used to determine the health of a lubricant, as different parameters (oxidation, contamination) can cause changes in viscosity beyond the normal ageing process. When a change in viscosity is observed, it is imperative that the cause of change be identified and corrected. Tests that can provide insight in the nature of viscosity changes are elemental analysis (indicator of incorrect oil; additive metals), IR spectroscopy (oxidation, thermal degradation, chemical contamination, soot loading), TAN/TBN (oxidation, chemical degradation), KF (moisture) and flash point (fuel contamination).

The growth of strong and weak acids is routinely measured through a neutralisation technique. High TAN values generally mean that the sump should be changed. However, oils with AW/EP additives and some corrosion inhibitors are intrinsically high TAN oils. In such cases, the value of TAN decreases slightly over time with additive depletion and then rises with the accumulation of degradation products. TAN values above 4.0 mg

KOH/g are considered potentially corrosive. Measurement of AN according to ASTM D 664/D 974 is insensitive to weak organic acids routinely produced by lubricant degradation. TAN and TBN measurements are carried out for industrial oils and engine oils, respectively. TBN (ASTM D 4739, preferred for used oil analysis) measures the reserve alkalinity of engine oil. Engine oils are equipped with additives that create an overbase condition intended to neutralise acids produced during normal operation. Once this reserve has been depleted, engine oils can become corrosive. It is good practice to change the oil when TBN has dropped to 20% of its initial value. Rapid changes in TBN may be caused by high sulphur fuel, incomplete combustion, coolant contamination (glycol), etc.

FTIR spectroscopy provides a means to monitor multiple oil parameters simultaneously: degradation (products of oxidation, sulphation, other oil waste), contaminants (water, glycol, soot, chemicals, additives). Also oxidation alone is not a total indicator of oil age. In lubricating oil, common wear metals act as catalysts to accelerate oxidation. Process contaminants and water also catalytically accelerate the rate of oxidation. RPVOT should be used in collaboration with other oil stability tests, and not serve as a stand-alone indicator. Some antioxidants can greatly extend RPVOT values but produce high quantities of oil waste upon degradation. When the oil reaches an oxidation threshold (limit), it must be replaced. Cooling machine sumps increases lubricant life cycles. Any sump with oil temperatures above about 70 °C is prone to rapid lubricant degradation.

Maintaining the additive package is a prime reason for using a lubricant maintenance programme offered by the manufacturer of the oil. Determining whether an additive package is depleted is difficult largely because additive packages vary from one manufacturer to another. In addition, most available analytical tools do not directly measure the concentration of the additive package. Chemical fingerprints of antioxidants may be traced with commonly used oil analysis techniques, including FTIR spectroscopy (ASTM E 2412-04) and linear-sweep voltammetry (LSV; ASTM D 6971). FTIR and LSV methods can thus be used to identify additive depletion. These techniques can help the programme manager assess the concentration of oxidation-limiting chemicals left in the lubricant for the purpose of projecting either lubricant change schedules or other activities to manage the remaining lubricant life cycle.

As oils age in operation, lubricant degradation products are formed. Many of these products are highly acidic, and negatively affect the acid level of the oil. Other degradation products have high molecular weights, are insoluble in the oil and form deposits. Lubricating oils have some degree of solvency that allows for the suspension of these insoluble products for a period of time. As the level of insolubles builds up in the oil these products eventually drop out and harmful varnish and sludge forms on components.

Growing levels of pentane insolubles (ASTM D 893) point to lubricant degradation. The various oxidation by-products, generically called varnish and sludge, can also be measured – but not always routinely – by acid build-up (AN or TAN), ultracentrifuge (UC) and varnish potential (membrane patch colorimetry, MPC) techniques. UC (no ASTM method) for sediment and sludge concentration is a subjective test method. Using visible light, MPC (ASTM Draft) is a generic, in-plant test that measures colour bodies of soft insoluble contaminants in lubricants, primarily circulation oils (such as in-service turbine oils), and is used to identify the oil's potential for harmful varnish and sludge build-up. In industrial equipment such as turbines, injection moulding machines, and large hydraulic systems the oil may be in use for many years before a complete oil change. Such equipment runs the risk of harmful varnish and sludge build-up if not properly monitored.

Numerous methods exist for determining water, both titrimetically (KF) and spectroscopically (FTIR, NIR spectroscopy). KF titration [80] for at-line on-site water determination according to various ASTM procedures (D 1533, D 1744, D 4377, D 4928, D 6304) is considered the most accurate and reproducible of all methods. It determines not only free water, but also surface-adsorbed and crystal water [81]. Moisture analysis by various KF procedures is influenced by sulphur-containing AW/EP agents and some rust inhibitors. The co-distillation method (ASTM D 6304) is probably the most dependable for additive-enhanced lubricating oils (without additive interferences). This (time-consuming) method distils moisture and then titrates the condensate and is accurate into single-digits ppm. Recently, various continuous moisture sensors (Vaisala MMT317, Pall Corp. WS10, Kittiwake ANALEXrs, EESiFLO, EASZ-1, Serveron TM8, Elektronik GmbH EE36) have been marketed that are designed to measure water as a relative humidity or as a percentage of water saturation in the fluid for a given temperature [62].

Glycol contamination (ASTM D 2982, D 4291) through seal failure or a leak in the cooling system is a problem associated with combustion engines. Analytical methods are colorimetry (Schiff's reagent, ASTM D 2982) for screening, elemental analysis and FTIR (low sensitivity). Flash point test (ASTM D 92) identifies the presence of fuel or other volatile organics.

Rapid fluctuations of oil properties, such as contaminant metal concentrations, are more important than strict adherence to the published ranges of criteria in determining whether an oil fails. Rapid changes in oil properties may indicate faulty equipment, severe operating conditions or insufficient maintenance activities.

Interpretation of used-oil analysis data is an even bigger challenge than just obtaining accurate and reliable analytical data. This is especially true for an external oil analysis supplier. In order to create a truthful sample context, the testing lab needs to consider the correct background details of

Table 14.16 Effects of electrostatic filtration of a hydraulic fluid

Test method	Electrostatic filtration	
	Before	After
AN (ASTM D 664), mg KOH/g	0.67	0.59
Particles >4 μm (per mL)	44 621	3833
Particles >6 μm (per mL)	8799	675
Particles >14 μm (per mL)	368	26
ISO Code 4406:1999	23/20/16	19/17/12
MPC (varnish potential)[a]	52	6

[a] Rating scale.
After ref. [82].

the sample in terms of the lubricant's mechanical function, application (food-processing plant, cement plant, steel mill, gas turbine, etc.) and operating conditions. This includes information about the machine's operating state, machine criticality, lube brand and grade, oil analysis history, oil age, extent of filtration, etc.

The most difficult aspect of effective oil analysis is to translate the measured data in corrective actions to protect machine capacity and plant productivity in either a predictive or proactive fashion. Table 14.16 shows the effect of several days of electrostatic filtration of an oil with marginally high AN value which had been in use in a hydraulic system for over 10 years. After filtration the particle count had dropped four ISO cleanliness codes and the varnish potential was reduced from a 52 (critical) range to 5 (none) [82].

Extension of drain intervals is an expected outcome of improvements in lubrication practices. Mineral oil life cycles should be extendable by a factor of three to five if the lubricant is maintained in a cool, clean (no atmospheric contaminants and/or wear debris) and dry state. Disposal of the used fluids should be in accordance with environmental regulations and procedures (see Section 14.7.1).

14.6 Essential properties of used oil

Monitoring a fingerprint of an oil using a scheduled oil sampling (SOS) used analysis programme may ensure that the oil has not exceeded maximum allowable limits for viscosity change, TAN and water. Used oil must retain certain critical properties to ensure continued service, such as viscosity, oxidation stability and corrosion protection. The recommended oil operating temperature range for a particular application is usually specified by the equipment manufacturer. Exceeding these values may reduce oil viscosity and impair lubrication. Subjecting oil to high temperature also increases the oxidation rate (for every 10° above 66 °C the oil's oxidation rate doubles and oil life is halved).

In perspective of waste reduction and cost, lubricating oils are demanded to have more long-drain capabilities. Longevity is especially critical in some applications, e.g. for turbines in hydroelectric generating units where the oil life expectancy is several years. New turbine oils (such as ISO-VG 68 or ISO-VG 100, based on highly refined lubricant base stocks) are highly stable in the presence of air or oxygen. In service, oxidation is gradually accelerated by the presence of metals (acting as catalysts) and by the depletion of antioxidant additives. When oxidation stability decreases, the oil will undergo complex reactions that eventually end up in the formation of insoluble sludge. This sludge may interfere with the lubrication and cooling functions of oil. It is very important to carefully monitor the oxidation of oils (ASTM D 2272) and to determine the remaining useful life of the oil (oxidation stability reserve).

Corrosion protection provided by a lubricant is of great importance for turbine systems. Rust inhibitor additives may be depleted by normal usage, removal with water in the oil, absorption on wear particles and debris, or chemical reaction with contaminants. For this purpose, ASTM D 665 provides the analytical tool for monitoring.

Other important properties of used lubricating oils are colour, foaming characteristics, water content, inhibitor content, wear and contaminant metals. Significant colour changes in a short time may indicate that the oil is contaminated or excessively degraded. Foaming can result in poor system performance and can cause serious mechanical damage. Foaming characteristics are measured by the ASTM D 892 test method. Most lubricants contain antifoam additive to break up foams. Oil containing large amounts of water does not have the same viscosity and lubricating effect of clean oil. Water may be removed after settling or by means of purification systems. The water separability characteristics of an oil are adequately measured using the ASTM D 1401 test procedure.

The presence of abrasive solids in the oil cannot be tolerated since they will cause serious problems to the system. Once abrasive solids have been detected, they must be removed by filtration or centrifugation. Quantitative spectrographic analysis of used oil samples may be used to detect trace metals and silica.

As pointed out in Section 8.2.1, used oils present a complex toxicological picture with potential hazards. Used automotive lubricant oils are more biodegradable than fresh mineral oils, but also the most toxic [39]. Their high toxicity is on account of PAH concentrations that establishes them as potential carcinogens.

14.7 Disposal of spent lubricants

In addressing the waste phase, the lubricant manufacturer or user has to identify the likely disposal pathways of the product and assess potential problems in relation to the substance properties. Used oils are toxic waste with harmful repercussions on the environment. Each year a large part of the lubricants in use enters the environment by way of total-loss applications, leaks, evaporation and other routes. Waste oils that are found in rivers, lakes and streams threaten aquatic life. A litre of waste oil can contaminate a million litres of water.

Out of the lubricants in the market only about 49% (W. Europe, 1996) is deemed collectable and only an even smaller part (28%) is actually collected [83, 84]. According to other estimates, out of the 2006 EU lubricant consumption (5.8 Mt) some 52% (3.0 Mt) was consumed in use, whereas 48% (2.8 Mt) was generated as waste oils of which 25% (0.7 Mt) was unaccounted for and presumably dumped into the environment. The remaining 75% was collected of which 36% (0.76 Mt) was re-refined into lubricating oils and 64% (1.34 Mt) burned as fuel. There is at least 2.1 Mt waste oil to be managed every year in the EU. Despite the fact that intense gathering of collectable oils is being pursued, only a small share of consumed lubricants is recovered through waste-oil collection programmes and reprocessed or used for the generation of energy. A substantial amount is still being released into the environment.

As most other industrial products, lubricants lose intrinsic value on use, which cannot easily be restored. Zeller+Gmelin GmbH & Co. (Eislingen, Germany) has recently introduced the concept of quality-controlled recovery and recycling of used neat MWFs such as cutting, grinding and quenching oils from oil/water mixtures resulting at the end of industrial metalworking production processes [85]. Otherwise, the value of the used mineral lube drops approximately to the level of heating oils or fuels.

Used lubricants are created when their required performance characteristics can no longer be maintained at the desired level. This is typically the case when chemical additives have been depleted or important ageing by-products have built up in the oil. The notion that the original performance can easily be restored is often unrealistic or at least largely uneconomical. This picture does not change with legislative provisions or subsidies. Ageing of lubricant formulations is not dissimilar from that of other complex materials such as plastics (polymer + additives). Used lubricants are eventually only slightly better off as the molecular weight of the base oil is not as greatly affected as that of polymer chains. Re-refining can restore the quality of the base oil but not the equally essential additive package (see Section 14.7.3).

Used lubricants represent a threat to the environment. Their ecologically compatible use is an important protection measure. Disposal of waste oils should not cause harmful effects to the environment. 'Disposal' means the processing or destruction of waste oils as well as their storage and tipping above or under ground; processing stands for all operations designed to permit the reuse of waste oils (regeneration, combustion). Disposal costs for EA oils are greater than for mineral oils because traditional recyclers are reluctant to accept them. Nevertheless, development and enhanced use of environmentally acceptable, renewable and fully biodegradable lubricants constitute important steps towards improvement of the ecological compatibility of these important industrial products.

Severe soil contamination can result from waste oils being left on the ground. (Bio)remediation of waste lubricant oil contaminated sites is an important topic, but beyond the scope of the present text. Various innovative techniques for remediation by enhanced biodegradation have been developed. *In situ* chemical oxidation can be used for degrading hazardous wastes for the remediation of soil at spill sites. Some full-scale treatments using *in situ* chemical oxidation (mainly using hydrogen peroxide and ozone) for soil remediation are known such as the patented injection processes Clean-Ox®, C-Sparge™, Geo-Cleanse®, ISOTEC™, etc. Bioremediation via bioaugmentation or biostimulation or both is considered a feasible method for treating petroleum hydrocarbon-contaminated soils [86–88]. Also genetically engineered systems can be useful in bioremediation [89]. Combined chemical (Fenton-like and ozonation) and biological treatment for the remediation of transformer oil contaminated soil has also been studied [90] as well as the application of (bio)surfactants [91, 92]. Park *et al.* [92] have considered alkaline-enhanced surfactant washing of waste lubricant oils from contaminated railway soil. Although the removal efficiency is up to 52%, the residual concentration still exceeds the local (Korean) regulatory level. Lopes *et al.* [39] have examined toxicity and biodegradation of sandy soil contaminated by automotive lubricant oils. Research is ongoing to enhance the recovery of waste lubricants from contaminated soil [93].

14.7.1 Waste oil legislation

'Waste oils' (WOs) have variously been defined as any semi-liquid or liquid used product totally or partially consisting of mineral or synthetic oil, including the oily residues from tanks, oil–water mixtures and emulsions [94], or as any mineral or synthetic lubrication or industrial oils which have become unfit for the use for which they were originally intended, such as used combustion engine oils and gearbox oils, lubricating oils, oils for turbines and hydraulic oils [95]. In the EU, classification of waste as hazardous waste is based on the Community legislation on chemicals [95]. 'Used oil' is defined by the EPA as 'any oil that has been refined from crude oil, or any synthetic oil, that has been used and as a result of such use is contaminated by physical or chemical impurities'. Animal or vegetable oils (including when used as a lubricant) are excluded from the EPA's definition of used oil. In several US states (e.g. California) used oils have been classified as hazardous waste.

Objectives of the recycling of used oils concern the respect of the environment and the economy of raw materials. When technical and/or economical conditions do not allow it, incineration in cement factories or waste treatment centres with energy recovering remains an acceptable solution. Disposal of used oils and waste fluids should be carried out in conformity with local, state and (supra)national regulations. In this respect plant-derived oils are at an advantage over mineral-based ones as they are deemed to be more environmentally friendly and sustainable. In accordance with ASTM D 7066-4, portable infrared analysers can be used for on-site determinations of oil and grease concentration levels in plant discharges for compliance with regulatory limits (see Section 8.8.1).

Legislation has influenced waste oil collection and reconditioning. The EU Waste Oil Directive 75/439/EEC [94] is designed to create a harmonised system for the collection, storage, recovery and disposal of waste oils, such as lubricant oils from vehicles, gearboxes and engines, and hydraulic oils. The directive prohibits discharge of waste oils into inland surface waters, ground water, coastal waters and drainage systems. Processing of waste oils should not cause air pollution beyond prescribed levels. Council Directive 87/101/EEC contains a recommendation concerning the regeneration of used oils insofar compatible with economic, technical and organisational conditions [96]. The European waste oil legislation has significantly been influenced by the PCB/PCT (perchlorinated bi- and terphenyls) problem. This problem has also led to a division of waste oils into two categories. Used oils containing more than 0.2% Cl cannot be re-refined and are subject to expensive disposal procedures. In turn, this has promoted the development of chlorine-free lubricants (base oils and additives). Emission thresholds for incineration plants (<3 MW) make burning of untreated oils difficult.

Burning lubricants for disposal produces airborne pollutants and heavy metal-containing waste [97]. Under certain conditions, EU member states may ban the combustion of waste oils on their territory. Legislation permits incineration in high-temperature furnaces and by the cement manufacturing industry.

Waste and hazardous waste have been classified in accordance with the list of types of waste as established by Commission Decision 2000/532/EC

[98]. Recently, Directives 2006/12/EC and 2008/98/EC have established the latest legislative framework on waste in the EU and have reiterated the obligation to handle waste in a way that does not have a negative impact on the environment or human health [95, 99]. Directive 2006/12/EC defines key concepts such as waste, recovery and disposal and puts in place the essential requirements for the management of waste [99]. Waste prevention is the first priority of waste management; reuse and material recycling are preferred to energy recovery from waste, insofar as they are the best ecological options. Directives 75/439/EEC and 2006/12/EC have been repealed and replaced by the revised Waste Framework Directive (WFD) 2008/98/EC (in particular Article 21) with effect from 12 December 2010 in all separate member states [95]. Directive 2008/98/EC also regulates waste oil management and moves the EU towards a recycling society, seeking to avoid waste generation and to use waste as a resource. Waste oils should be collected separately where this is technically feasible, should be treated in accordance with a certain waste hierarchy and with a high level of environmental protection. Waste oils of different characteristics should not be mixed, and they should also not be mixed with other kinds of waste or substances if such mixing impedes their treatment. The waste hierarchy (from most to least desirable) according to the directive was given as follows: prevention, preparation for reuse, recycling, other recovery (e.g. energy recovery), and disposal (intended as any operation which is not recovery). The EU envisages that by 2020 energy recovery should be limited to non-recyclable materials and landfilling should be virtually eliminated. Higher priority will be given to recycling and reuse in all relevant EU policies. The waste hierarchy lays down a priority order of what constitutes the best overall environmental option in waste legislation and policy. However, member states do not favour regeneration of WOs, but on the contrary are widely using waste oils as fuel in industrial installations [100]. In view of studies showing that regeneration of waste oils is environmentally and economically no more beneficial than combustion, the revised WFD does not give an absolute priority to regeneration but allows member states which want to promote regeneration to prescribe that waste oils should continue to be treated in this way. It is expected that the EU Waste Directive will result in a collection of more and better quality used oil and increasing volumes of re-refined base oils, including premium quality stocks. The EU facilitates the separate collection and proper treatment of bio-waste. Legislative measures for the management of bio-waste are forthcoming.

In Germany various regulations of the law apply for the disposal of used lubricants [4]. Regulations covered within the Waste Material Act comprise:

- Waste Allocation Regulation
- Waste and Remainder Material Monitoring Regulation
- Used Oil Regulation
- Technical Regulation Waste.

These regulations are not without discrepancies. According to the Used Oil Regulation RSOs are used oils not to be regarded as special material, whereas the Technical Regulation Waste establishes that RSOs are special waste materials requiring special control. It appears from the framework of laws and regulations that the evaluation of environmental impairment in Germany is very complex and not unequivocal. For the UK legislation governing disposal and discharge of waste-mix MWFs, see ref. [101].

In the United States, the White House Executive Order #13101 (1998) deals with 'Greening the Government' through waste prevention, recycling and federal acquisition. US regulations are becoming more restrictive with regard to the contents, use and disposal of lubricants. Four regulations are of particular interest at the Federal level:

- Toxic Substance Control Act (TSCA) – governs the manufacturing, importation, distribution, and processing of all toxic chemicals (require EPA approval).
- Resource Conservation and Recovery Act (RCRA) – regulates hazardous waste and solid waste.
- Comprehensive Environmental Response Compensation and Liability Act (CERCLA) – imposes liability for cleaning up contamination caused by hazardous substances.
- Superfund Amendments and Reauthorization Act (SARA) – amends and extends CERCLA to include toxicological profiles of hazardous substances.

Globally, some quite different situations are found in practice. Some countries do not regulate the collection and disposal of used oil, which consequently is generally not collected. Other countries (notably Germany and the USA) show high collection and disposal rates. While national and EU regulations give the priority to regeneration, France (2000) still realised elimination of collected oils by incineration for 65 wt%. In France operates SRRHU (Société de Ramassage pour la Régénération des Huiles Usagées). In Italy, CONOE (Consorzio Obbligatorio Nazionale Oli Esausti) is charged with collection and disposal of waste vegetable oils. Out of the 280 kt/yr of vegetable oils and animal fats currently used in Italy (domestic use, 50%; restaurants, 20%; industrial, 30%) in 2009 only 16% (45 kt) was collected and reused, mainly (50%) as a feedstock for biodiesel. CONOE has a recovery target of 100 kt by 2012. In some countries, fuels and heating oils reclaimed from used lubricants are not taxed. Re-refined used oils are subsidised in Italy. In Germany, the legal requirement to properly dispose of waste oils has been transferred from lubricant manufacturers and distributors to collection organisations with financing by the lubricant consumers.

14.7.2 Waste oil management

Waste oils are only partly biodegradable and should not be dispersed in the environment. The environment must be protected against pollution by lubricants and hydraulic fluids based on mineral oils. This is best done by preventing undesirable losses and by reclaiming and reusing lubricants. Strategic objectives are improved gathering of collectable oils and improvement in the environmental compatibility of lubricants. The preferred method of waste treatment is through biodegradation. Traditionally, biostable compositions often complicate disposal and suffer a disadvantage from this perspective.

Used oils are commercially distinguished in three categories, so-called clear and dark industrial oils, and used cooking oils. Waste mineral oil-based lubricants and waste vegetable oils require separate storage and collection systems, and completely different recycling processes. It is important to take notice of the disposal best practices. The disposal of used lubricating oils has been discussed by Mang [102]. A generally accepted waste hierarchy consists of prevention, recycling/combustion (with or without energy recovery) and landfill, in this order of preference. Disposal of used vegetable-based industrial oils is less problematic and less expensive since ester-based lubricants present low pollution hazards.

Clear oils with a low degree of deterioration can easily be regenerated by purification processes (filtration or centrifugation). Dark oils, which are mainly derived from industrial or automotive lubrication and have been subject to severe thermal and mechanical conditions, usually contain metallic contaminants (from wear) and residues of combustion and oxidation. Depending on their characteristics disposal of dark waste oils may take place in the form of regeneration, reuse, combustion, physical and/or chemical treatment, and thermodestruction as waste, in decreasing order of ecological sustainability.

Regeneration of waste oils is intended as any high-quality recycling operation whereby base oils can be produced by refining waste oils, in particular by removing the contaminants, the oxidation products and the additives contained in such oils. Regeneration consists in obtaining new base oils with characteristics similar to the original mineral-based oils. Regeneration of waste oils containing PCBs or PCTs may be permitted if the regeneration processes make it possible either to destroy the PCBs or PCTs or to reduce them so that the regenerated oils do not contain PCB/PCT beyond a maximum limit (50 ppm) [96]. When untreated waste oils do not satisfy the requirements for regeneration but are still re-utilisable they can be used as flux oils for bitumen or for the manufacture of secondary feeds in sec-feed plants for catalytic crackers and as blending stock for high-sulphur fuels [102, 103].

Material recycling and reuse are preferred to energy recovery by combustion, generally in cement plants [104]. The cement industry has great interest in replacing fossil fuels with alternative fuels (waste and industrial by-products). The calorific value of waste lubricant oils is 39.7 MJ/kg (or 9500 kcal/kg), which is similar to that of combustion oils. Combustion of waste oils should be carried out under environmentally acceptable conditions [96]. When even combustion is no longer acceptable because some parameters of the waste oil are beyond specification, physical and/or chemical pretreatment is required before regeneration or combustion can again be considered. In all other circumstances (i.e. for about 0.5% of total waste oil volume), the product is referred to thermodestruction plants, which are subject to stringent emission control. As far as possible, the disposal of waste oils should be carried out by recycling (regeneration and/or combustion other than for destruction).

The European Commission is considering a mandatory requirement for incorporation of a minimum percentage of recycled fluids into all new lubricant products. This would seriously impair and constrain the development of optimised products tailored to the requirements of particular applications. Incorporation of significant levels of highly standardised re-refined mineral oil would preclude development of biodegradable lubricants for use in environmentally sensitive areas, or high stability fluids with extended service lifetime. The lubricant industry regards fuel use as a more practical and equally acceptable option, as in case of other product groups (polymers, paper products). The vision that the highest priority should be given to collection of used oil for disposal by whichever route [105] is questionable. There has been much debate over whether it is environmentally preferable for used oils to be regenerated for reuse as lubricants or used as fuel. This largely depends on product type and application.

At the US federal level, 'used oil' is managed under a set of standards (40 CFR 279) developed by the EPA under the RCRA. US federal environmental regulations governing oil spill prevention, spill notification and clean-up make little distinction between petroleum oil and vegetable oil. By contrast, used oil waste management regulations exempt vegetable oil products. Waste management requirements for vegetable oil-based lubricant and fluid products require careful attention. Used lubricant and fluid products containing a mixture of vegetable oil and petroleum or synthetic oil can be 'recycled' under US federal 'used oil' management standards, which is a less costly approach than managing the used oil as a hazardous waste. If a formulated product contains no petroleum or synthetic base oils, it would not be subjected to the 'used oil' management standards. However, the vegetable oil-based product would still need to be assessed after its use to determine if it contains any contaminants that would classify it as a hazardous waste. If the product contains no such contaminants, it can be managed as a solid waste, a more flexible and less costly management option. Certain states have adopted used oil management standards that are more stringent than the federal standards.

Oils refined from crude oil, or any synthetic oil, that have been used as lubricants, hydraulic fluids, heat transfer fluids, buoyants or for similar purposes (e.g. engine oil, transmission fluid, metalworking fluids and oils, electrical insulating oil, and industrial process oils) *and* that are sent for 'recycling' in accordance with the EPA's used oil regulations are excluded from the federal definition of 'hazardous waste'. Used oil containing more than 50 ppm of PCBs cannot be managed under the RCRA used oil standards but instead must be managed under the Toxic Substances Control Act regulations. Used oil that is not recycled in accordance with federal standards is subject to being managed as a hazardous waste if it meets the regulatory definition of hazardous waste (e.g. contains certain EPA listed metals and chemicals above certain limits).

The EPA's used oil regulations allow for four types of recycling:

- reconditioning the oil on site to remove impurities;
- using the oil as a feedstock for a petroleum refinery;
- re-refining the oil into a new base stock; or
- processing and burning the oil for energy recovery.

Specialist disposal of waste oils and emulsions is becoming increasingly more important and various options are available; see Fig. 14.3. Significant amounts of water-mixed cutting fluids need to be disposed of. The water and oil phases are separated first. For waste treatment of MWFs, see also Byers [106]. Waste oils must be handled with care. The main concern is skin adsorption. Use of proper personal protective equipment is required.

Recycled used cooking oils can be used for industrial applications such as production of lubricants, biodiesel (by direct acid transesterification), surfactants and soaps. As to a second life of (partially additive depleted) used vegetable oil-based lubricants, used vegetable oils are known to contain high amounts of free fatty acids, water and oxidised components. It is not at all evident that these by-products can easily and economically be removed to create a virgin biolube base stock. Moreover in a performing biolubricant formulation additives should be replenished in the correct amounts (requiring costly analyses of the remaining fractions). It might be more worthwhile investigating transformation of used vegetable oil-based lubricants into biodiesel as low-value feedstocks in a robust acid-catalysed transesterification process.

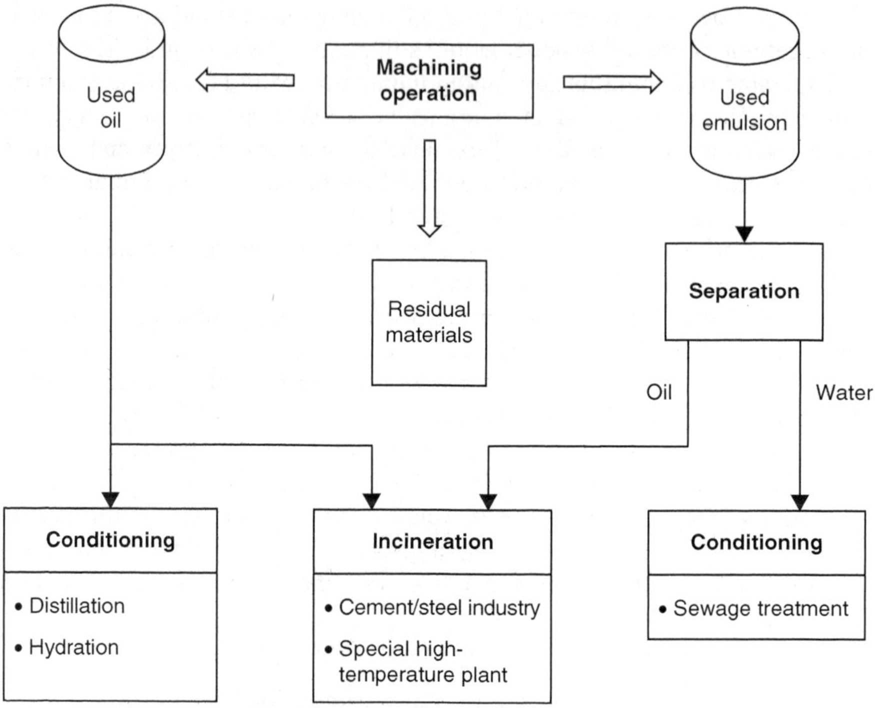

14.3 Disposal of cutting fluids in cutting technology.

14.7.3 Re-refining technologies

Re-refined oils are the exact opposite of total-loss oils which are used once only, and then discarded in the proximate environment. The US EPA defines re-refined products as containing at least 25% re-refined base stock [107], but other standards are significantly higher, e.g. 70% in California [108]. Numerous re-refining technologies have been patented but only few have developed into large-scale application. Some of the more well-known technologies are the Meinken process, propane extraction (IFP, Snamprogetti), KTI (Kinetics Technology International) process, CEP-Mohawk Process, PROP process (Phillips Petroleum Company), Safety-Kleen process, DEA and Revivoil™ technologies [102, 109]. Rincon *et al.* [110] have described regeneration of used lubricant oil by propane extraction. A range of different refining processes have been commercialised, including the traditional (now obsolete) acid/clay treatment, solvent extraction, distillation/clay treatment and distillation/hydrotreatment.

Top re-refiners are Safety-Kleen (USA) with 15% global capacity, Puralube (Germany) and Viscolube (Italy). More than half of the global re-refining volume lies in Europe. The re-refining nameplate capacity of W.

Europe amounts to 1.1 Mt and will shortly reach 1.3 Mt. With some 58 re-refining plants (65 by 2015) and a nameplate capacity of 2.5 Mt (up to 3 Mt in the near future), the average plant capacity per re-refinery is 7× lower than that of a virgin base oil refinery [111].

In the preparation of re-refined oil, oil undergoes an extensive processing to remove contaminants to produce fresh base oil. Process stages for re-refining of used oils generally comprise:

- separation of solid impurities, usually by sedimentation;
- separation of water and volatiles (fuel residues in engine oils, solvents and low boiling point lubricant components), normally by atmospheric distillation;
- separation of additives and ageing by-products, normally by acid refining, solvent (propane) extraction, vacuum distillation or hydrogenation;
- separation of remaining impurities (additives, ageing by-products, refining reaction products), commonly by hydrofinishing or selective solvent extraction.

The re-refining industry generally differentiates only between two base oil qualities: 71% Group I (50% Europe, 25% Americas, 25% rest of the world) and 29% Group II (50% Europe, 50% N. America). Expectations are a shift towards 75% Group I and 25% Group II re-refined oils by 2015 [111].

The case for recycling synthetic lubricant base fluids is stronger than that for mineral oils, because of the relatively higher energy requirements in the original manufacture of the synthetics. Additionally, oxygen-containing base fluids such as esters, vegetable oils and derivatives, and polyalkylene glycols have significantly lower calorific value and therefore lower value as fuels than hydrocarbons. Where such oils can readily be segregated and handled in large volumes, reprocessing of specific high grade oils can be a realistic option, e.g. for the polyol esters used for aviation jet engine lubricants [112].

The difficulty of re-refining used lubricating oils depends on the severity of the application, and the degree of control on recovery and collection. For some industrial oils such as hydraulic fluids and transformer fluids, which are subject to regular condition monitoring, it is common practice to periodically restore the oils by filtration, ion exchange and additive replenishment to the standard specification [113]. Recycling or energy recovery is the preferred alternative for crankcase oils to limit environmental exposure. In general, lubricant additives are not recyclable due to degradation in service use [114]. They are therefore removed as sludge during lubricant re-refining and contribute to waste generation. Most additives are combustible.

It is worthwhile to optimise the performance of the existing processes of regeneration or to develop new processes that are competitive from an environmental and economical point of view. Recycling technology continues to be an active area of research, with recent developments including separation of unoxidised base oil components using supercritical fluid extraction (SFE) [115], separation of additives [116], and separation of mineral oils from higher value synthetic base fluids [117]. Regeneration of used oils by means of SFE has first been reported in 1993 [118]. The yield is dependent on the solubility of the oil in the supercritical fluid. Gourgouillon *et al.* [119] have described an environmentally friendly process for the regeneration of used oils based on separation by ultra-filtration membranes using supercritical CO_2 ($scCO_2$) as a viscoreductor agent to improve the membrane permeability. The feasibility study was carried out with PEG 400 (as model compound), mineral oils, namely low-viscosity S3699 (IFP, 69 mPa s at 40 °C) and high-viscosity S3701 (IFP, 462 mPa s at 40 °C), and a used motor oil (S5010). It is possible to regenerate 96 wt% of the used oil and to confine the majority of the metals in an extremely viscous black gum.

Regeneration of Italian waste oils takes advantage of three advanced technologies, all based on various dehydration, deasphaltation/fractionation and refining processes. Figure 14.4 shows a flowsheet of the Revivoil™ process. In another, similar process removal of water and light hydrocarbons takes place in two units in series (preflash 1 and 2); the dehydrated

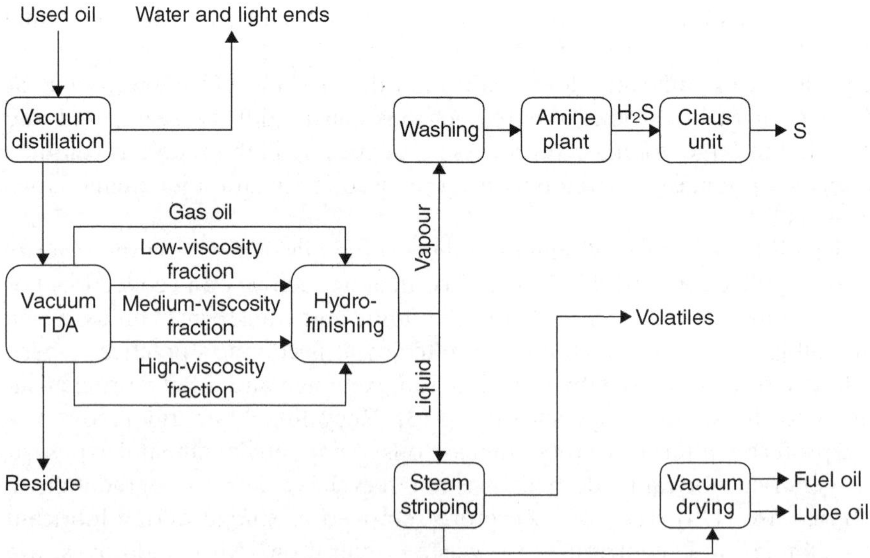

14.4 Flowsheet of the Revivoil™ process.

oil then undergoes deasphaltation with propane (PDA I) for removal of heavy fractions (for use as bitumen). The deasphalted oil is distilled into a gas oil, four side fractions and a bottom fraction; the latter is subjected to a second propane deasphaltation stage (PDA II) for removal of residual heavy fractions. The four side fractions and the PDA II product finally undergo hydrofinishing. A third process for re-refining of waste oils in Italy is based on distillation and chemical refining. The process starts with thin-film evaporation (TFE) and conventional distillation. Fractionation of the used oil yields various products: water, light hydrocarbons, gas oil, medium and heavy distillates, and a bituminous residue. The medium and heavy distillates are then refined chemically by acidification, decantation, neutralisation and discoloration.

Viscolube S.p.A. (Pieve Fissiraga, Italy) is one of the European companies specialising in waste oil re-refining which is able to produce lubricating base oils with properties equal to or even better than virgin base oils currently on the market. In 2009 Viscolube processed 130 kt of waste oil, producing 85 kt of lubricating bases in its Pieve Fissiraga (Lodi) and Ceccano (Frosinone) plants, both based on Revivoil™ technology.

Re-refining of used oils to lube base oils has been practised since 1935. Re-refined oils are usually admixed with virgin oils. However, their market share is still quite limited. Only a small fraction of lubricants is being recycled. In Western Europe only 8% of base oil demand was satisfied by re-refined products in 2010 [111]. The re-refinery capacity utilisation in the region is currently only 30%. Forecasts hold a considerable increase in re-refined base oil production (up to 23% in W. Europe). The economic and environmental impact of re-refined products has been evaluated [100, 120]. Reasons for limited acceptance of re-refining technologies are high process costs, inadequate removal of carcinogenic polycyclic aromatics, the increasing complexity of base oil blends in engine and industrial lubricants, and scepticism of users that the lubricant maintains its quality. According to a recent public opinion survey in Italy, 51% of the respondents (out of 1000) consider re-refined oils as eco-sustainable whereas 26% express serious concern about the quality of such oils [121]. Some 47% is in favour of purchasing regenerated oil, or even up to 78% in case of a price discount compared with virgin oil. It is to be mentioned here that Italian lubricant oil producers do not usually state the percentage of re-refined oil on their product labels, as opposed to similar good practice in the United States. The survey also indicated that only 59% of the Italian public shows positive environmental awareness in general.

There is disagreement as to just how recyclable bio-based lubricants are. According to some, once the oil is used it is difficult to restore to its original state, and recycling is even more challenging when it is mixed with mineral or synthetic fluids. Others see no fundamental reason why mixed fluids cannot be processed and reused. Factors influencing the choice for environmental fate of vegetable oil-based lubricants include:

- availability of a facility to re-refine used oil;
- transportation distance to a re-refining oil facility;
- availability of an industrial furnace (with appropriate air pollution control);
- internal uses for used oil;
- economics for the life-cycle phases of oil production, use and reuse.

There are no facilities dedicated to re-refining of bio-based lubricants due to their relatively low volumes. Most collection practices do not segregate bio-based oils from petroleum oils and lubes. If any used oil collection segregation is employed, it is usually for oils that contain emulsified or oil/water mixtures.

Current re-refining facilities for used petroleum oils are not equipped to separate and recover the vegetable oil lube fraction from a mixture of oils. Petroleum re-refiners are reluctant to accept greater than 2% bio-based lubricants in their incoming mixtures due to concern regarding degrading the oxidation stability of the base oil they produce. If a vegetable oil-based lubricant is present in an incoming mixture, it could be separated from the desired petroleum fraction owing to the higher boiling point. This higher boiling fraction would include higher boiling petroleum fractions and be used as an industrial fuel oil feedstock. Testing by Safety Kleen has shown that soybean-based lubricants can be disposed and re-refined along with conventional mineral oil-based lubricants. Despite some efforts to recycle bio-based lubricants, it is currently not profitable.

When re-refining facilities are too distant for economical transport, used oils are commonly used as a low-cost fuel feedstock for industrial furnaces (asphalt plants, cement kilns, etc.). These furnaces can accommodate a high ratio of vegetable oil lubricants in the fuel feedstock. Oil re-refining reduces greenhouse gases and heavy metal emissions when compared to burning used oil as fuel.

Based on economics and ease of handling, the most common end-of-life option for used vegetable oil lubricant products is to utilise them as a supplemental fuel source for industrial furnaces. This option is competitive with used petroleum-based lubricant environmental fate options. In terms of the US preferred procurement practices (see Section 13.3) re-refined products have been given a priority over bio-based products [122].

Revivoil™ technology (Viscolube, Axens)

This technology, developed at the end of the 1990s jointly by the Italian Viscolube and French Axens/IFP, enables production of re-refined base oils with API Group II characteristics by high-pressure catalytic hydrogenation. The first-rate products with low environmental impact, higher viscosity index and clearer colour than virgin base oils have a low sulphur and unsaturates content and very low aromatics content [123].

The Revivoil™ process (Fig. 14.4) combines preflashing (dehydration) with thermal de-asphalting (TDA)/fractionation and hydrofinishing (refining). In the preflash step the used oil is heated up to 140 °C and then vacuum distilled separating water and light hydrocarbons. Subsequently, the dehydrated oil is distilled at 360 °C in a vacuum TDA column. The process leads to asphaltic and bituminous products (bottom), three side fractions of different viscosity, and intermediate gas oil (top). The three side fractions and the gas oil are hydrofinished batchwise in a high-pressure (100 bar) catalytic plant at 300 °C, where the reactor effluent is separated into two phases, a vapour and a liquid phase. The vapour phase is washed with water to remove chlorine and sulphur compounds whereas the liquid phase is stripped with steam to eliminate the most volatile compounds and restore the flash point. The sulphur-containing streams are post-treated in an amine

Table 14.17 Physicochemical characteristics of re-refined Viscolube grades

Property	Method	Limit	Value		
			80 N	150 N	400 N
Colour	ASTM D 1500	max	0.5	1.0	2.0
Flash point (°C)	ASTM D 92	min	180	215	235
Kinematic viscosity (mm^2/s)	ASTM D 445	mean	15.5	30.0	58.0
Viscosity index	ASTM D 2270	min	97	105	105
Pour point (°C)	ASTM D 97	max	–6	–9	–9
Cold cranking simulator (CCS) viscosity (cP)	ASTM D 5293	max	2400/1300[a]	3850/2150[b]	6500/3300[c]
Noack (wt%)	CEC L-40-A-93	max	35	13	10
Sulphur (wt%)	ASTM D 5453	max	0.08	0.1	0.2
Aromatic carbon (mol%)	IP 451	max	7	5	6

[a] At –30/–25 °C.
[b] At –25/–20 °C.
[c] At –20/–15 °C.

plant to separate H_2S from the other compounds. A Claus unit transforms H_2S to pure liquid sulphur. The final product is a clear lubricating base oil with very low sulphur and polynuclear aromatics content. The process is characterised by high yields: 100 kg waste oil are converted into 60 kg re-refined lube oil and 20–25 kg gas oil/fuel oil.

Viscolube's re-refined lubricating bases, obtained through high-pressure catalytic hydrogenation, are suitable for use in a variety of lubrication applications, in both the automotive and industrial sector. Table 14.17 lists the physicochemical characteristics of various Viscolube grades (80 N, 150 N and 400 N). Revivoil™ is equal in quality to lubricating oils currently on the market and is considered as a green oil by Lega Ambiente, Italy's main environmental association.

14.8 Sources of further information and advice

Union Indépendante de l'Industrie Européenne des Lubrifiants, *Purchase and Use of Lubricants within the European Union*, UEIL, Brussels (2011); http://www.ueil.org

K. Chipasa, *Best Practice Guide for the Disposal of Water-mix Metalworking Fluids*, UKLA, Berkhamsted (2011); http://www.ukla.org.uk

Commission of the European Communities, *Roadmap to a Resource Efficient Europe*, Brussels (20 Sept. 2011).

H.P. Bloch, *Practical Lubrication for Industrial Facilities,* The Fairmont Press, Inc., Lilburn, GA (2009).

E.C. Fitch, *Fluid Contamination Control*, FES, Inc., Stillwater, OK (2008).

J.S. Marth, *Renewable Lubricants Manual: Biobased Oils, Fluids & Greases*, United Bio Lube, Palo Alto, CA (2007); http://www.renewablelubricants.com

Noria Corporation, *Fundamentals of Machine Lubrication*, Noria Corporation, Tulsa, OK; http://www.noria.com

A. Lansdown, *Lubrication and Lubricant Selection: A Practical Guide*, ASME Press, Champaign, IL (2004).

T. Wireman, *Benchmarking Best Practices in Maintenance Management*, Industrial Press, Inc., New York, NY (2004).

D.M. Pirro and A.A. Wessol (eds), *Lubrication Fundamentals*, 2nd edn, CRC Press, Inc., Boca Raton, FL (2001).

D. Nyman and J. Levitt, *Maintenance Planning, Scheduling and Coordination*, Industrial Press, Inc., New York, NY (2001).

M.J. Neale, *Lubrication and Reliability Handbook*, Newnes, Oxford (2000).

M. Alexander, *Biodegradation and Bioremediation*, Academic Press, San Diego, CA (1999).

E.R. Booser (ed.), *Tribology Data Handbook: An Excellent Friction, Lubrication, and Wear Resource*, CRC Press, Boca Raton, FL (1997).

T. Pfeifer, W. Eversheim, W. König and M. Weck, *Manufacturing Excellence: The Competitive Edge*, Chapman & Hall, London (1994).
E.C. Fitch, *Proactive Maintenance for Mechanical Systems*, Elsevier Advanced Technology, Oxford (1992).

14.9 References

[1] Commission of the European Communities, *Proposal for a Sixth Environment Action Programme of the European Community. Environment 2010: Our Future, Our Choice*, Brussels, COM (2001) 31 final.
[2] Directive 2008/98/EC of the European Parliament and of the Council of 19 November 2008 on waste and repealing certain Directives, *Off. J. Eur. Comm.* **L 312** (22 November 2008), pp. 3–30.
[3] A. Lansdown, *Lubrication and Lubricant Selection: A Practical Guide*, 3rd edn, ASME Press, Champaign, IL (2004).
[4] W.J. Bartz, Lubricants and the environment, *Tribol. Int.* **31**(1–3) (1998) 35–47.
[5] J.S. Marth, *Renewable Lubricants Manual: Biobased Oils, Fluids & Greases*, United Bio Lube, Palo Alto, CA (2007); http://www.renewablelubricants.com
[6] W. Bock, *High-performance Gear Oils Industrial Oils*, FUCHS Brochure, FUCHS Petrolub AG, Mannheim (2004).
[7] B. Cunningham, N. Battersby, W. Wehrmeyer and C. Fothergill, A sustainability assessment of a biolubricant, *J. Ind. Ecol.* **7**(3–4) (2003) 179–92.
[8] P. Cuevas, *Comparative Life Cycle Assessment of Biolubricants and Mineral Based Lubricants*, M.Sc. Thesis, University of Pittsburgh, PA (2010).
[9] A.A. Mullur, C.A. Mattson and A. Messac, New decision matrix based approach for concept selection using linear physical programming, *Proc. 44th AIAA/ASME/ASCE/AHS Structures, Structural Dynamics, and Materials Conference*, Norfolk, VA (7–10 April 2003), Paper No. AIAA-2003-1446.
[10] USDA, BioPreferred (2010); http://www.biopreferred.gov
[11] M. Johnson, High-performance lubricants: cost vs. performance, *Tribol. Lubr. Technol.* **66**(5) (2010) 24–9; The business case for high-performance lubricants, *Tribol. Lubr. Technol.* **67**(12) (2011) 28–32.
[12] M.J. Neale, *Lubrication and Reliability Handbook*, Newnes, Oxford (2000).
[13] B. Casey, Hydraulic oil – how to consolidate your inventory, *Machinery Lubr.* **5** (2009); http://www.machinerylubrication.com/Articles/Print/2153
[14] D.M.C. McCarthy, S.B. Glavatskih, M. Pach and A. Dmitriev, Performance evaluation for aged environmentally adapted lubricants in bronze-steel sliding contacts, *J. Synth. Lubr.* **25**(3) (2008) 115–29.
[15] N. Battersby, S. Greenall and G. Gustafsson, Environmentally acceptable hydraulic fluids – field performance in mobile equipment, Shell Global Solutions, *Proc. 14th Int. Colloquium on Tribology*, Esslingen (2004), Vol. II, pp. 759–68.
[16] M. Pach, Å. Byheden, T. Norrby, D. McCarthy, S.B. Glavatskih and E. Kassfeldt, Aged environmentally adapted lubricants. I. Procedures, techniques and affected properties for aged oils, *Proc. 15th Intl. Colloquium on Tribology, Automotive and Industrial Lubrication,* Technische Akademie Esslingen (17–19 January 2006).

[17] S.D. Haigh, Fate and effects of synthetic lubricants in soil: biodegradation and effect on crops in field studies, *Sci. Total Environm.* **168**(1) (1995) 71–83.
[18] E. Remmele and B. Widmann, Suitability and environmental compatibility of rapeseed oil based hydraulic fluids for agricultural machinery, *J. Synth. Lubr.* **16**(2) (1999) 129–45.
[19] K. Hosonuma, K. Yoshida and A. Matsunaga, The decomposition products of zinc dialkyldithiophosphate in an engine and their interaction with diesel soot. *Wear* **103** (1985) 297–309.
[20] J. Vižintin, A. Arnšek and T. Ploj, Lubricating properties of rapeseed oils compared to mineral oils under a high load oscillating movement, *J. Synth. Lubr.* **17**(3) (2000) 201–18.
[21] M.F. Fox, The degradation of lubricants in service use, in *Handbook of Lubrication and Tribology: Application and Maintenance* (G.E. Totten, ed.), 2nd edn, CRC Press, Boca Raton, FL (2006), Vol. I, pp. 29/3–29/40.
[22] K. Sepčic, M. Josowicz and J. Janata, Determination of the principal volatile compounds in degraded engine oil, *Tribol. Lubr. Technol.* **60**(3) (2004) 40–7.
[23] A. Zeman, R. Stuwe and K. Koch, The DSC cell – a versatile tool in the study of thermal-oxidative degradation of lubricants and related problems, *Thermochim. Acta* **80**(1) (1984) 1–9.
[24] R.E. Cantley, The effect of water in lubricating oil on bearing fatigue life. *ASLE Trans.* **20**(3) (1977) 244–8.
[25] N.J. Fox and G.W. Stachowiak, Vegetable oil-based lubricants – a review of oxidation, *Tribol. Int.* **40** (2007) 1035–46.
[26] R.A. Newly, H.A. Spikes and P.B. Macpherson, Oxidative wear in lubricated contact, *J. Lubr. Technol.* **102** (1980) 539–44.
[27] J.J. Habeeb and W.H. Stover, The role of hydroperoxides in engine wear and the effect of ZDDPs, *ASLE Trans.* **30**(4) (1987) 419.
[28] F. Rounds, Effects of hydroperoxides on wear as measured in four-ball wear tests, *Trib. Trans.* **36**(2) (1993) 297–303.
[29] X. Wu, X. Zhang, S. Yang, H. Chen and D. Wang, The study of epoxidised rapeseed oil used as a potential biodegradable lubricant, *J. Am. Oil Chem. Soc.* **77**(5) (2000) 561.
[30] A. Adhvaryu and S.Z. Erhan, Epoxidized soybean oil as a potential source of high-temperature lubricants, *Ind. Crops. Prod.* **15** (2002) 247–54.
[31] Y. Cao and L. Yu, Effect of alcohols and fatty acids as additives on the anti-wear and extreme pressure properties of rapeseed oil, *Mocaxue Xuebae/ Tribol.* **20**(4) (2000) 288–91.
[32] I. Minami, H.-S. Hong and N.C. Mathur, Lubrication performance of model organic compounds in high oleic sunflower oil, *J. Synth. Lubr.* **16**(1) (1999) 3–12.
[33] N.J. Fox, B. Tyrer and G.W. Stachowiak, Boundary lubrication performance of free fatty acids in sunflower oil, *Tribol. Lett.* **16**(4) (2004) 275–81.
[34] K. Miyashita and T. Takagi, Study on the oxidative rate and prooxidant activity of free fatty acids, *J. Am. Oil Chem. Soc.* **63**(10) (1986) 1380.
[35] N. Frega, M. Mozzon and G. Lercker, Effects of free fatty acids on oxidative stability of vegetable oil, *J. Am. Oil Chem. Soc.* **76**(3) (1999) 325–9.
[36] E. Brinksmeier, M. Garbrecht, C. Heinzel, Th. Koch and J. Eckebrecht, Current approaches in design and supply of metalworking fluids, *Tribol. Lubr. Technol.* **66**(3) (2010) 36–8, 40–2, 44–9.

[37] F.J. Passman, Current trends in MWF microbicides, *Tribol. Lubr. Technol.* **66**(5) (2010) 30–9.
[38] F.J. Passman, G.L. Egger, S. Hallahan, B.W. Skinner and M. Deschepper, Real-time testing of bioburdens in metalworking fluids using adenosine triphosphate as a biomass indicator, *Tribol. Lubr. Technol.* **66**(5) (2010) 40–5.
[39] P.R.M. Lopes, R.N. Montagnolli, R. de F. Domingues and E.D. Bidoia, Toxicity and biodegradation in sandy soil contaminated by lubricant oils, *Bull. Environm. Contam. Toxicol.* **84**(4) (2010) 454–8.
[40] R. Garvey, *Case Histories and Cost Savings Using In-shop Oil Analysis for Industrial Plant Applications*, CSI Industry Report; http://www.mhm.assetweb.com
[41] D.M. Pirro and A.A. Wessol, Handling, storing and dispensing lubricants, in *Lubrication Fundamentals*, CRC Press, Boca Raton, FL (2001), Chapter 18.
[42] M. Johnson, Proper storage and handling of lubricants, *Tribol. Lubr. Technol.* **64**(2) (2008) 22–30.
[43] M. Johnson, Operator-based lubrication practices, *Tribol. Lubr. Technol.* **66**(3) (2010) 20–5.
[44] M. Johnson and M. Spurlock, Strategic oil analysis: estimating remaining lubricant life, *Tribol. Lubr. Technol.* **66**(1) (2010) 22–4, 26–30.
[45] M. Johnson, Strategic oil analysis: Onsite practices using sensory inspection, *Tribol. Lubr. Technol.* **66**(4) (2010) 18–23.
[46] M. Johnson and M. Spurlock, Strategic oil analysis: instrument-based on-site lubricant analysis, *Tribol. Lubr. Technol.* **66**(7) (2010) 26–33.
[47] M. Johnson, Strategic oil analysis: time-dependent alarms for extended lubricant lifecycles, *Tribol. Lubr. Technol.* **66**(8) (2010) 26–9.
[48] M. Johnson, Selecting an oil analysis laboratory, *Tribol. Lubr. Technol.* **66**(12) (2010) 22; Oil analysis: finding economic value, *Tribol. Lubr. Technol.* **66**(7) (2011) 24–8.
[49] M. Johnson, Measuring lubrication program health, *Tribol. Lubr. Technol.* **66**(2) (2010) 24–31.
[50] J. Van Rensselar, Used-oil analyses for predictive maintenance, *Tribol. Lubr. Technol.* **68**(1) (2012) 34–45.
[51] E.C. Fitch, *Proactive Maintenance for Mechanical Systems*, Elsevier Advanced Technology, Oxford (1992).
[52] M. Johnson and M. Miller, Eco-friendly fluids for the lubricants industry, *Tribol. Lubr. Technol.* **66**(10) (2010) 28–30, 32–4.
[53] M. Johnson, Industrial fluid contaminants and their effect, *Tribol. Lubr. Technol.* **65**(9) (2009) 26–8, 30–1.
[54] E.C. Fitch, *Proactive Maintenance for Mechanical Systems*, Elsevier Advanced Technology, Oxford (1992), p. 19.
[55] E.C. Fitch, *Proactive Maintenance for Mechanical Systems*, Elsevier Advanced Technology, Oxford (1992), Chapter 8.
[56] M. Johnson and M. Spurlock, Strategic oil analysis: systems, tools and tactics, *Tribol. Lubr. Technol.* **65**(4) (2009) 29–33.
[57] M. Johnson, Lubrication program development and scheduling, in *Handbook of Lubrication and Technology: Application and Maintenance* (G.E. Totten, ed.), 2nd edn, CRC Press, Boca Raton, FL (2006), Vol. I, pp. 33/1–33/30.

[58] I.C. Halalay, E.W. Schneider and M.W. Rogers (to General Motors Corporation), Determining quality of lubricating oils in use, US Patent Appl. No. 2006/0232267 A1 (19 October 2006).
[59] M. Johnson and M. Spurlock, Strategic oil analysis: Selecting alarms, setting limits, *Tribol. Lubr. Technol.* **65**(11) (2009) 18–20, 22–3.
[60] E.C. Fitch, *Fluid Contamination Control*, FES, Inc., Stillwater, OK (2008).
[61] T. Wireman, *Benchmarking Best Practices in Maintenance Management*, Industrial Press, Inc., New York, NY (2004).
[62] M. Johnson, Integrating online sensors and filtration for moisture management, *Tribol. Lubr. Technol.* **67**(10) (2011) 28–33.
[63] E.C. Fitch, *Fluid Contamination Control*, FES, Inc., Stillwater, OK (2008), p. 46.
[64] http://www.castrol.com/castrol/genericarticle.de?categoryld=4504740&contentld=7038530 (14 November 2007).
[65] M. Johnson and M. Spurlock, Strategic oil analysis: setting the test slate, *Tribol. Lubr. Technol.* **65**(5) (2009) 20–2, 24–7.
[66] M. Johnson and M. Spurlock, Strategic oil analysis: developing the test slates, *Tribol. Lubr. Technol.* **65**(7) (2009) 28–30, 32–3.
[67] B. Perston, Raising the bar in FT-IR getting the most out of on-site oil monitoring, *Lube Mag.* **104** (2011) 16–20.
[68] R. Hughes, A. Ricco, M. Butler and S. Martin, Chemical microsensors, *Science* **254** (1991) 74–80.
[69] K. Durdag and J. Andle, Real-time viscosity measurement for condition-based monitoring using solid-state viscosity sensor, *Tribol. Lubr. Technol.* **66**(2) (2010) 40–7.
[70] F. Herrmann, D. Hahn and G. Flik (to Robert Bosch GmbH), Sensor array and method for determining the density and viscosity of a liquid, US Patent No. 6,543,274 B1 (8 April 2003).
[71] J. Andle (to Biode, Inc.), Measurements of density and viscoelasticity with a single acoustic wave sensor, PCT Int. Publ. No. WO 2005/114138 A2 (1 December 2005).
[72] S.S. Wang and H.-S. Lee, The development of *in-situ* electrochemical oil-condition sensors, *Sens. Actuators* **B17** (1994) 179–85.
[73] H.-S. Lee, S.-S. Wang, D.J. Smolenski, M.B. Viola and E. Klusendorf, *In-situ* monitoring of high-temperature degraded oil condition with microsensors, *Sens. Actuators* **B20** (1994) 49–54.
[74] M.F. Smiechowski and V.F. Lvovich, Iridium oxide sensors for acidity and basicity detection in industrial lubricants, *Sens. Actuators* **B96** (2003) 261–7.
[75] http://www.visgage.com
[76] http://www.on-siteanalysis.com
[77] M. Spurlock, Particle counting or ferrous density . . . or both?, *Practic. Oil Anal.* 11 (2006); http://www.machinerylubrication.com/Read/955/particle-counting-ferr
[78] WearCheck Application Note WC-PQ-Analysis-2008-04-25, *PQ Analysis*, WearCheck Canada, Inc., Burlington, ON (2008); http://www.wearcheck.com
[79] M. Barrett and M. McMahon, Analytical ferrography – make it work for you, *Practic. Oil Anal.* 10 (2000); http://www.machinerylubrication.com/Read/5/analytical-ferrography
[80] K. Fischer, Neues Verfahren zur maßanalytischen Bestimmung des Wassergehaltes von Flüssigkeiten und festen Körpern, *Angew. Chem.* **48** (1935) 394–6.

[81] F. Portala, M. Burkhard, N. Geil and A. Steinbach, On-site water determination, *Chem. Plants Proc.* (2) (2009) 76–8.

[82] WearCheck Application Note WC-MPC-2009-04-23, *Membrane Patch Colorimetry*, WearCheck Canada, Inc., Burlington, ON (2009); http://www.wearcheck.com

[83] The Oil Companies' Organisation for the Conservation of Clean Air and Water in Europe (CONCAWE), *Collection and Disposal of Used Lubricating Oil*, CONCAWE Report 5, Brussels (1996).

[84] P. de Caro and C. Cecutti, Lubrifiants et environnement, *Oléag., Corps Gras, Lipides* **12**(4) (2005) 279–84.

[85] http://www.zeller-gmelin.de

[86] T.-C. Lin, P.-T. Pan and S.-S. Cheng, Ex situ bioremediation of oil contaminated soil, *J. Hazard. Mater.* **176** (2010) 27–34.

[87] M. Alexander, *Biodegradation and Bioremediation*, Academic Press, San Diego, CA (1999).

[88] E.J.M. Thomassin-Lacroix, M. Eriksson, K.J. Reimer and W.W. Mohn, Biostimulation and bioaugmentation for on-site treatment of weathered diesel fuel in Arctic soil, *Appl. Microbiol. Biotechnol.* **59** (2002) 551–6.

[89] G.S. Sayler and S. Ripp, Field applications of genetically engineered microorganisms for bioremediation processes, *Curr. Opin. Biotechnol.* **11** (2000) 286–9.

[90] A. Goi, N. Kulik and M. Trapido, Combined chemical biological treatment of oil contaminated soil, *Chemosphere* **63** (2006) 1754–63.

[91] L.-M. Whang, P.-W.G. Liu, C.-C. Ma and S.-S. Cheng, Application of biosurfactants, rhamnolipid, and surfaction, for enhanced biodegradation of diesel-contaminated water and soil, *J. Hazard. Mater.* **151** (2008) 155–63.

[92] S.-W. Park, J.-Y. Lee, K.-J. Kim, J.-S. Yang and K. Baek, Alkaline enhanced-separation of waste lubricant oils for railway contaminated soil, *Sep. Sci. Technol.* **45** (2010) 1988–93.

[93] J.L. Assunção Filho, L.G. Moraes de Moura and A.C. da Silva Ramos, Polycyclic aromatic hydrocarbons (PAHs) adsorption on solid surfaces applied to waste lubricant oils recovery process, *Can. J. Chem. Engng.* **88**(3) (2010) 411–16.

[94] Council Directive 75/439/EEC (16.06.1975) on disposal of waste oils, *Off. J. Eur. Comm.* **L 194** (25 July 1975), 23.

[95] Directive 2008/98/EC of the European Parliament and the Council of 19 November 2008 on waste and repealing certain Directives, *Off. J. Eur. Comm.* **L 312** (22 November 2008), 3–30.

[96] Council Directive 87/101/EEC of 22 December 1986 amending Directive 75/439/EEC on the disposal of waste oils, *Off. J. Eur. Comm.* **L 042** (12 February 1987), 43–7.

[97] I. Gulyurtlu, H. Lopes and I. Cabrita, The determination of emissions of pollutants from burning waste oils, *Fuel* **75**(8) (1996) 940–4.

[98] Commission Decision 2000/532/EC of 3 May 2000 replacing Decision 94/3/EC establishing a list of wastes pursuant to Article 1(a) of Council Directive 75/442/EEC on waste and Council Directive 94/904/EC establishing a list of hazardous waste pursuant to Article 1(4) of Council Directive 91/689/EEC on hazardous waste, *Off. J. Eur. Comm.* **L 226** (6 September 2000), 30.

[99] Directive 2006/12/EC of the European Parliament and of the Council of 5 April 2006 on waste, *Off. J. Eur. Comm.* **L 114** (27 April 2006), 9.

[100] Taylor Nelson Sufres Consulting, *Critical Review of Existing Studies and Life Cycle Analysis on the Regeneration and Incineration of Waste Oils*, Rept. 20 AW83-5 to European Commission DG Environment, Montrouge (France) (December 2001).

[101] K. Chipasa, *Best Practice Guide for the Disposal of Water-mix Metalworking Fluids*, UKLA, Berkhamsted (2011); http://www.ukla.org.uk

[102] Th. Mang, Disposal of used lubricating oils, in *Lubricants and Lubrication* (Th. Mang and W. Dresel, eds), 2nd edn, Wiley-VCH, Weinheim (2007), pp. 183–90.

[103] B. Krishna, Recovery, re-refining and disposal of used oil of steel plants under Steel Authority of India Limited, *Proc. 15th Intl. Colloquium Tribology, Automotive and Industrial Lubrication*, Technische Akademie Esslingen (TAE), Esslingen (January 2006).

[104] http://www.holcim.it/oliesausti

[105] The Oil Companies' Organisation for the Conservation of Clean Air and Water in Europe (CONCAWE), Disposing of used lubricating oils, *Concawe Review* **9** (2000) 21.

[106] J.P. Byers (ed.), *Metalworking Fluids*, 2nd edn, STLE/CRC Press, Park Ridge, IL/Boca Raton, FL (2009).

[107] US EPA Comprehensive Procurement Guideline, *Re-refined Lubricating Oil*; http://www.epa.gov/osw/conserve/tools/cpg/products/lubricant.htm

[108] California State Contract Code 12209; http://law.onecle.com/california/public-contract/12209.html

[109] C. Kajdas, Used oil re-refining: Overview of current technologies used, *Proc. 3rd European Congress on Re-refining*, Lyon (1996).

[110] J. Rincon, P. Canizares, M.T. Garcia and I. Gracia, Regeneration of used lubricant oil by propane extraction, *I&EC Res.* **42**(20) (2003) 4867–73.

[111] A. Gosalia, The European base oils and lubricants industry. Trends in a global context, *Lube Mag.* **95** (2010) 8–16.

[112] G. Glasgow and R.J. Bruns, *Reclamation of Synthetic Turbine Engine Oil Mixtures*, USAF Report UFAPL-TR-78-50, MRC-DA-780 (1979), from *Gov. Rep. Announce Index* **79** (1979) 167; *C.A.* **92**, 200605n (1980).

[113] W.D. Philips and D.I. Sutton, Improved maintenance and life extension of phosphate esters using ion-exchange treatment, *Synth. Lubr.* **13** (1996) 225.

[114] The Technical Committee of Petroleum Additive Manufacturers in Europe (ATC), *Lubricant Additives and the Environment*, ATC Document 49, CEFIC, Brussels (1993).

[115] L. Schrive, S. Sarrade and D. Gourgouillon (to Commissariat á l'Energie Atomique), Method for treating an oil using a liquid in a supercritical state, PCT Int. Publ. No. WO 2000/052118 A (8 September 2000).

[116] C.W. Harrison, A.G. Gorneau, R.M. Steinberg and B.R. Bond (to Texaco, Inc.), Used lubricating oil reclaiming, US Patent No. 5,306,419 (26 April 1994).

[117] R. Deutsch, Separation and recovery of synthetic oils from used motor sump oils, Eur. Patent Appl. No. EP 1,028,158 A (16 August 2008).

[118] C. Jian, F. Yaohua and Z. Yufu, Regeneration of used oils by supercritical fluid extraction, *Proc. First Industrial Chemical Technology Topical Conference*, St. Louis, MO (7–12 November 1993), pp. 262–4.
[119] D. Gourgouillon, L. Schrive, S. Sarrade and G.M. Rios, An environmentally friendly process for the regeneration of used oils, *Environm. Sci. Technol.* **34** (2000) 3469–73.
[120] M. Levizzari, M. Voglino and P. Volpi, Environmental and economic impact of re-refined products: a life cycle analysis, in *Lubricants for the Future and Environment, Proc. 6th Intl. LFE Congress*, Brussels (1999).
[121] Istituto per gli Studi sulla Pubblica Opinione (ISPO), *L'atteggiamento della popolazione italiana verso gli oli rigenerati*, Milan (May–June 2008).
[122] B.J. Bremmer and L. Plonsker, *Bio-based Lubricants: A Market Opportunity Study Update*, United Soybean Board, Chesterfield, MO (November 2008).
[123] http://www.viscolube.it

15
Advanced lubricant fluids

Abstract: This chapter outlines the current status of research, development and applications of (room-temperature) ionic liquids as lubricants. Ionic liquids (ILs) possess unique properties such as non-volatility, non-flammability and high thermo-oxidative stability. The potential use of ILs as very promising high-performance lubricants was first proposed in 2001 and since then has attracted increasing attention in tribology on account of the excellent friction and wear capabilities in comparison to conventional lubricants in general use. IL lubricants may find engineering applications as base oils, additives or thin films. Lubrication mechanism, IL additive technology to further improve lubricity, current challenges and outlook for commercialisation are discussed.

Key words: ionic liquids (ILs), designer fluids, synthetic lubricants, additives, tribological properties.

15.1 Introduction

The lubricants currently used in industry (mainly mineral oils) are restricted in their application. Commercial lubricants derived from petroleum present environmental problems and are not suitable for all materials in all conditions. Despite considerable technological and marketing efforts the commercial impact of biolubricants is still disappointingly low. These materials also present limitations in various demanding applications. Improvements in lubricant performance are urgently needed (and guarantee a rapid return on investment) considering that in the transportation sector 10–15% of the energy generated in automotive engines is lost to friction. Also, industrial oils – currently limited by low thermal stability above 250 °C – are one of the barriers to increase efficiency for engines by operating at higher combustion temperatures. Moreover, oils contribute to hydrocarbon exhaust emissions due to oil blow-by and burn out. Thus, there are civilian and military needs for new, higher-performing lubricant materials.

A major goal in tribology is to develop effective lubricants for reactive light metals and alloys such as magnesium, aluminium and titanium, which are gradually substituting conventional iron-based materials. Among the light alloys, the most studied are aluminium alloys owing to their wide range of application in sliding components, particularly in the automotive industry. Development of advanced lubricants is highly desirable.

15.2 Ionic liquids (ILs)

The term ionic liquid (IL) refers to thermally stable molten salts that melt below 100 °C [1]. Room-temperature ionic liquids (RTILs) present melting points lower than ambient temperature. The seemingly divergent fields of ILs and molten salts are two branches of the same field that are separated only by working temperatures [2].

Ionic liquids are charged fluids which are usually composed of unique ion pairs consisting of bulky, asymmetric cations – typically containing nitrogen and phosphorus – and weakly coordinating, more highly symmetric or almost spherical anions. ILs are invariably organic salts possessing a high degree of asymmetry that frustrates packing and thus inhibits crystallisation. The possible choices of cation or anion result in numerous ILs. Some common cationic species are imidazolium, quaternary ammonium, *N,N*-dialkylpyrrolidinium, *N*-alkylpyridinium and *P,P,P,P*-tetra-alkylphosphonium (see Fig. 15.1). The anions are categorised into hydrophobic anions with weak hydration capacity (such as PF_6^-, $(CF_3SO_2)_2N^-$ or Tf_2N^-, $(CF_3CF_2SO_2)_2N^-$ or BETI, $(CF_3CF_2)_3PF_3^-$ or FAP), and hydrophilic anions (such as Cl^-, Br^-, NO_3^-, ClO_4^-, BF_4^-, $CF_3CO_2^-$, $CH_3CO_2^-$, $CF_3SO_3^-$). Combinations of these cations and anions may not generally be liquid at room temperature. Popular ILs are dialkylimidazolium salts. ILs come in two main categories, namely simple salts (composed of a single cation and anion) and binary ILs (salts where an equilibrium is involved).

The first room-temperature IL, $[EtNH_3][NO_3]$ (m.p. 12 °C), was discovered in 1914 by the Latvian-German chemist Paul Walden [3]. Much later, in the 1975–2004 period outstanding contributions to IL chemistry were made by Osteryoung and co-workers. Design of IL solvents requires fundamental knowledge on how the nature of the cation, anion and substituents affects chemical and physical properties. ILs are a flexible technology:

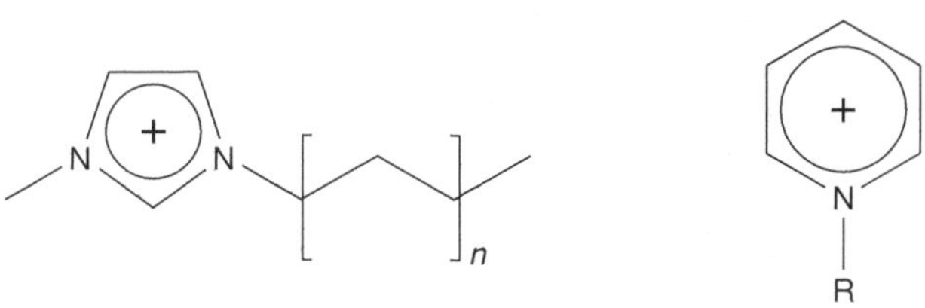

Imidazolium Pyridinium

$[R_4N^+]$ Ammonium

$[R_4P^+]$ Phosphonium

15.1 Some common ionic liquid cations used in lubrication.

the properties can be tailored. Table 15.1 summarises the main characteristics of ionic fluids.

A key feature of ILs is that their chemical and physical properties (such as hydrophobicity, melting point, viscosity, density, lubricity) can be varied virtually at will and can be tailored by judicious selection of cation, anion and substituents [4–6]. Not surprisingly, ILs are therefore often referred to as designer liquids. Up to now fewer than 10^3 ILs have been investigated, whereas the number of combinations of suitable cations and anions is staggering (estimated as $>10^{18}$).

The melting point and viscosity of ILs, important parameters for lubrication oils, are strongly dependent on their molecular structures, the cations, the nature and length of alkyl chains, and the anions [7]. For ILs with the same anion and alkyl chain on cation, the viscosities increase in the order imidazolium < pyridinium < pyrrolidinium. Longer alkyl chains in imidazo-

Table 15.1 Characteristics of ionic liquids

- Designer fluids
- Highly asymmetric organic salts with low melting point (<100 °C[a])
- Very low pour points (below –50 °C)
- Large liquidus range (from below 25 to over 200 °C)
- Allowance for tailoring of physicochemical properties
- High versatility
- Ultra-low volatility
- Non flammability
- High thermo-oxidative stability
- High polarity
- Controlled miscibility with organic compounds
- Water soluble
- Wetting properties
- Mostly hygroscopic
- Newtonian fluids
- Very large viscosity range (up to 1000 cP at 25 °C)
- High viscosity index (typically 100–150)
- Excellent tribological properties
- Strong boundary film formation capability
- High thermal conductivity
- Molar conductivity (<10 S cm^2/mol)
- Good specific electrical conductivity (<20 mS/cm)
- Dielectric constant (≤30)
- Wide electrochemical window (2.0–4.5 V)
- Unfavourable cost/performance ratio
- Corrosive[b]
- Recyclable
- Unexplored disposal methods

[a] RTILs, m.p. < 25 °C.
[b] Structure dependent.

lium cations increase the viscosity and diminish the viscosity index (VI) of ILs. The viscosity increases with anion types, as follows: Tf_2N^- < FAP^- < $BETI^-$ < BF_4^- < PF_6^- < Cl^- < Br^-. Because of the highly symmetric or almost spherical anions, ILs are highly viscous.

Components in ILs strongly interact with each other through the coulomb force, whereas normal molecular liquids are bonded through van der Waals forces and/or hydrogen bonding forces. Therefore, they have non-volatility, non-flammability, high thermal stability and other outstanding properties [5]. The cation–anion interactions in ILs determine high viscosity, reaction dynamics and the nucleophilicity of the anion [8].

As RTILs are liquid in their pure state at room temperature, they have essentially a negligible vapour pressure and therefore do not evaporate (quite at variance to volatile organic compounds, VOCs). Common molecular solvents are characterised by a relatively narrow liquidus region (range of temperatures between freezing and boiling point) and are relatively volatile at many process conditions. An estimated 20 million tonnes of VOCs are discharged into the atmosphere each year as a result of industrial operating conditions [9]. Use of ILs can overcome this problem, although most ILs are water soluble and can easily enter the biosphere this way.

Other unique characteristics of ILs are their high polarity, high chemical stability, high solvent power (miscibility with water and with organic solvents) and interesting electrochemical properties. Ionic liquids act much like good organic solvents, dissolving both polar and non-polar species. In many cases they have been found to perform much better than commonly used molecular solvents. The highly chemically stable ILs are described as green solvents (which is open to debate) and can be recycled into a process almost indefinitely. With the exception of those forming liquid crystals, ILs are Newtonian fluids [10]. While most ILs are non-flammable, some decomposition products formed on heating are combustible.

Depending on the application, the unique properties of ILs may be either desirable or undesirable. The drawback of flexibility is considerable complexity of the research as the possible formulations are almost infinite. Properties such as melting point, viscosity, polarity, acidity and hydrophobicity can be adjusted by changes to the cation–anion structure. Some ILs are highly miscible with water, while others are hydrophobic. The drawback of water miscibility can be decomposition. Some cations decompose irreversibly. Hydrophobic anions create more stability than hydrophilic anions.

There is an enormous potential of IL designer fluids. ILs are of interest to synthetic chemists [11], electrochemists [12, 13] as well as to chemical engineers [9]. Potential applications of ionic liquids are numerous, some already of commercial use (see Table 15.2). ILs are innovative fluids which are finding a variety of technological applications, such as eco-friendly solvents [14] and in clean separation technology [15], as sophisticated chemical

Table 15.2 Potential applications of ionic liquids

• Phase transfer catalysis
• Extraction processes
• Green chemical synthesis
• Reaction solvents
• Solvents for cleaning operations
• Gas separations
• Liquid separations
• Electrochemistry
• Electrolytes/batteries/fuel cells
• Lubricants
• Heat-transfer fluids
• Composite materials
• Surface engineering
• Nanotechnology
• Embalming fluids
• Analysis

reaction media (different from molecular solvents) and as green chemical engineering fluids [9, 16, 17]. The bulk of investigations on ILs has focused on their use as solvents for reactions. They have been used successfully as solvents in a variety of chemical reactions, including hydrogenations, hydroformylations, isomerisations, alkylations and others, with reaction rates and selectivities as good as or better than conventional organic solvents.

The use of ILs is expanding rapidly in many branches of science. Applications of ILs in carbohydrate chemistry are developing [18], in particular for dissolution and functionalisation of sugars, cellulose, etc., which bring about attractive prospects for solution processing (dissolution and regeneration) of (ligno)cellulosic and woody materials [19, 20]. A class of water-stable and miscible ILs, 1-alkyl(R)-3-methylimidazolium chloride salts, [RMIM]Cl, can solubilise cellulose, which is insoluble in common organic solvents or water [21]. In a recent advance in cellulosic fuel technology, gentle hydrolysis of cellulose in wood and in edible plant material has been achieved by combining ionic liquids (e.g. BMIMCl) and solid catalysts (Amberlyst® 15 Dry), thereby easing the crucial first stage in converting waste biomass [22]. The reaction products, cellulose oligomers and sugars, can then be transformed enzymatically into chemical feedstocks and fuels. Cellulose dissolved in certain other ILs has recently been converted under relatively mild conditions (<140 °C, 1 atm) into water-soluble sugars with high yield (97%), or into hydroxymethylfurfural (HMF) in high conversion (89%) in the presence of $CrCl_2$ [23].

Because ILs are easy to alter, this is beneficial for solvent extractions or product separations where the relative solubility of the ionic and extraction phase can be adjusted to facilitate separation and recycling. Many ILs are hygroscopic, efficiently removing water vapour from gas mixtures [24]. Since the choice of cation, anion and substituents provides tremendous opportunity to tailor the solubility of various compounds in the IL, these new compounds have applications as extractants in the selective separation of liquids [25]. ILs are attractive in cleaning operations as they are remarkably good solvents for a wide variety of both polar and non-polar compounds [26].

RTILs were initially developed by electrochemists for use as electrolytes in batteries or for metal electrodeposition. ILs have replaced sulphuric and phosphoric acids in electropolishing. The large liquidus range and excellent thermal stability are desirable characteristics for both good heat-transfer fluids and lubricants (see Section 15.2.1). Other emerging fields of application are surface engineering and nanotechnology. The unique properties of ILs offer considerable challenges to chemical engineers, allowing for new processing options.

ILs do not charge the environment with volatile organics because of their extremely low vapour pressure. There is also a growing interest in the use of less toxic cations (such as ammonium salts) rather than the more toxic halogenated compounds, and towards new anions such as bistriflimide. There are numerous studies on toxicity of ILs. They can often be recycled, with considerable environmental and cost benefits.

Commercialisation barriers to the introduction of ILs have been addressed [27]. Barriers include cost reduction, assessment of environmental impacts and breakdown products, reduced toxicity, increased lifetime, reproducible production of high purity and high-quality ILs, increased contamination tolerability and economic benefit analysis. Development of ILs from cheap starting materials is in progress [28]. There remains much to discover about them. The emerging field of ILs has been reviewed by several authors [5, 29–32].

15.2.1 IL lubricants

IL lubricant research is very new. An article on the use of a molten salt (melting point 460 °C) as a lubricant in a high-temperature bearing test at 650–815 °C was published in 1961 [33]. In 1992 Wilkes and Zaworotko formulated an air and water-stable IL with 1-*n*-alkyl-3-methylimidazolium cation and BF_4^- anion [21]. Such stable alkylimidazolium tetrafluoroborates were later selected as lubricants in a wide range of contact materials. RTILs were recognised as potentially very promising high-performance lubricating fluids in 2001 [34] and since then have attracted considerable attention in the field of tribology because of their remarkable lubrication and antiwear (AW) capabilities compared with conventional lubrication oils. The number of scientific papers and patents pertaining to the use of ILs in tribology has increased steadily to over 40 per year by 2009 (Thomson ISI Web of Knowledge database). Most of the work is currently being carried out in China, Japan and Europe; there is very little IL tribological research in the United States. During the last decade, ILs have established themselves as promising advanced lubricants due to many of the same properties that make them useful as solvents and engineering fluids.

Highly polar ILs possess a combination of excellent and unique properties, including ultra-low vapour pressure, non-flammability, extremely high thermal stability, low melting point, low-temperature fluidity, wide viscosity range, high ionic conductivity, high load-carrying capability and miscibility with organic compounds (see Table 15.1). ILs are characterised by a large liquidus range and tend to wet metal, polymeric and inorganic surfaces. Their viscosity index is similar to other viscous fluids. Viscosities of RTILs are 1–3 orders of magnitude higher than those of conventional solvents [35], which is considered to be a drawback in certain applications (e.g. electrochemistry). Viscosities for some representative RTILs at 20 °C are given by Hapiot and Lagrost [12] and range from 32 mPa s at 20 °C in [EMIM]Tf_2N to 860 mPa s in [OMIM]PF_6. When the molecular order is high within the ionic liquid, the viscosity is increased. Electrical conduction is a most distinctive property that distinguishes ILs from conventional lubricants. These characteristics make them ideal candidates for advanced high-performance lubricants where conventional oils and greases or solid lubricants fail. Such conditions include ultra-high vacuum and extreme temperatures. Very harsh friction conditions require lubricating oils to have high thermal stability and chemical inertness. The decomposition temperatures of imidazolium ILs are generally above 350 °C (even up to 480 °C), with low-temperature fluidity (pour points) below –50 °C (even –100 °C). Consequently, ILs can function in a wide temperature range. As ILs do not evaporate like most other liquids they hold much promise as lubricants.

From the viewpoint of energy and resource conservation, development of low viscous base stocks is of interest. However, such molecular fluids are inherently volatile. ILs may act as novel functional fluids that can exhibit both low viscosity and low volatility.

The most notable characteristic that distinguishes ILs from other synthetic lubrication oils is the high polarity which stands in contrast to the nature of the normally used synthetic lubricants which are generally non-ionisable and even less polar. The high polarity of ILs allows them to form a very strong effective absorption film and ready tribochemical reactions, which contribute to their outstanding AW capability. ILs may be considered novel lubricating fluids as they exhibit AW properties under boundary conditions. The tribological properties of various ILs, derived from cations such as imidazolium, pyridinium and ammonium, and anions such as BF_4^- and $TFSA^-$, have been investigated [34, 36–38]. It was found that ILs undergo tribochemical reactions under boundary conditions. The tribological properties of many ionic fluids such as BMI.BF_4 are comparable to or better than those of conventional lubricants such as synthetic esters or fluorinated ethers. The tribological properties of ILs are at least partially attributable to the anionic moiety of the molecule. ILs have been verified as versatile lubricants for different sliding pairs and exhibit high friction reduction, AW performance and high load-carrying capacity. They can also be used in the form of ultrathin films.

The benefits of using IL lubricants include:

- reduced energy loss by reducing friction;
- extended service life and maintenance cycles due to wear reduction;
- expanded high-temperature lubricant usage on account of high thermal stability;
- safe transport and storage because of non-flammability;
- essential non-volatility.

IL lubricants are fundamentally very different from existing lubricants, both chemically and in terms of performance. Expectations are high in view of potential friction and wear reductions of up to 50%.

Imidazolium ionic liquids constitute the more extensively studied family of IL neat lubricants and lubricant additives [36, 39, 40], but trialkylammonium ILs have also demonstrated promising lubricating properties as neat lubricants or lubricant additives, particularly for use with difficult to lubricate metals like aluminium [41, 42]. The thermal stability of imidazole is beneficial for lubricants [43]. Alkylimidazolium ILs provide effective lubrication under high vacuum [36, 40] and could thus qualify for space applications. Also aromatic amine and phosphonium cations were investigated. Compared with alkylimidazolium ILs, phosphonium ILs contain phosphorus that is considered as an essential element in lubricant additives, such as zinc dialkyldithiophosphate (ZDDP), or in synthetic lubricant base oils, such as phosphazenes or phosphate esters.

The choice of cation and anion in an IL as well as the design of ion side chains determines the fundamental properties of ILs, which permits creating tailor-made lubricants and lubricant additives [7]. The study of the influence of molecular structure on thermophysical properties of ILs is essential for their use in lubrication. Structure–performance relationships of ILs in tribology are highly desired as they are recognised as designer fluids.

A large majority of the IL cations examined are 1,3-dialkylimidazolium derivatives, with a higher alkyl substituent being beneficial for good lubrication but detrimental to thermo-oxidative stability. As shown in Fig. 15.2, the friction coefficient of 1-alkyl-3-methylimidazolium bis(trifluoromethanesulphonyl)amide (TFSA) salts decreases with an increase in chain length of the alkyl group. Generally, the increment of alkyl

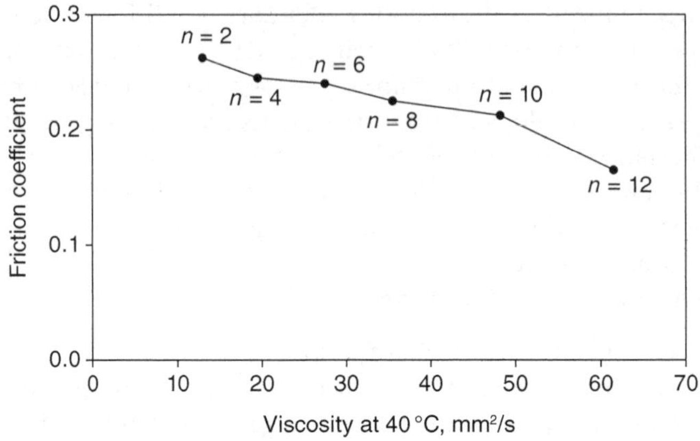

15.2 Effect of chain length in 1-alkyl-3-methylimidazolium TFSA salts on friction. After ref. [44].

chain length reduces friction and wear of sliding pairs in elastohydrodynamic lubrication (EHL) as a consequence of increased viscosity [45, 46].

The major anions reported for ionic liquids are trifluoroborate (BF_4^-) and hexafluorophosphate (PF_6^-). However, this does not imply that these are also the most suitable anions for lubricants. Anions have a more complicated effect on tribological properties in that they cannot only change the viscosity but also affect the formation of the boundary film (in terms of physical adsorption or chemical reaction film). For identical cations, the AW properties of Tf_2N^- and FAP^- are usually better than those of BF_4^- and PF_6^- [47]; FAP^- is usually better than Tf_2N^- [48].

Figure 15.3 shows the relationship between molecular structure of ILs and their performance. Ideally, the anionic moieties should be hydrophobic to improve the tribological properties and thermo-oxidative stability. Hydrophobic anions exhibit better tribological properties than BF_4^- and PF_6^- for steel–steel contacts. Higher alkyl groups in imidazolium improve the tribological properties at the expense of thermo-oxidative stability. AW properties of ILs can be improved by means of additive technology (see Section 15.2.3). Friction reduction and corrosion inhibition can further be improved. Halogen-free anions are highly desirable. Complex cations containing multifunctional groups are of considerable interest.

The question arises why neat ILs exhibit superior tribological behaviour. The view is held that ILs with their unique dipolar structure establish a tribolayer that is physically adsorbed onto and/or chemically reacted with the sliding surfaces of frictional pairs. As a result, friction and wear under boundary lubrication are effectively reduced [34]. However, ILs are not always as effective as reference hydrocarbon lubricants in reducing wear of steel/steel surfaces by tribocorrosion [49].

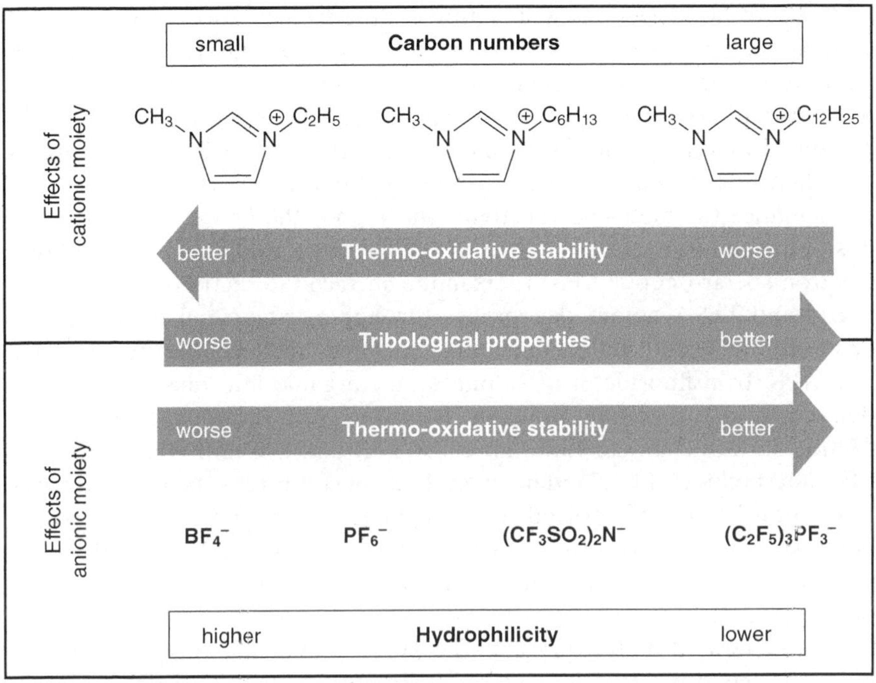

15.3 Relation between structure and lubricant properties of room-temperature ionic liquids. After ref. [44].

Boundary film generation studies of various phosphonium ILs on metal and ceramic surfaces have indicated that SiO_2 and $Mg(OH)_2$ interact strongly with IL anions and cations, whereas Al_2O_3, MgO, ZnO and ZrO_2 appear less active because they have fewer surface hydroxyl groups [50]. Stronger interacting ILs provide better lubrication capability. It has been suggested that under medium–high pressure ILs do not adsorb on sliding surfaces as single layers but rather in multilayered mode (as for the solid lubricants graphite and MoS_2), which should then account for the good lubricity of ILs under relatively low shear strength [51]. This also explains why longer alkyl chains, which are more likely to form densely packed structures, exhibit significantly better tribological properties.

Under harsh sliding conditions a single adsorption layer might be damaged and IL decomposition initiated, with formation of a reaction film. Tribo-products can prevent a fresh metal surface from further wear. When the two contact surfaces are reactive to each other, neat ILs may not be appropriate because of their electrical conductivity.

ILs are stable compounds. However, tribochemical reactions of ILs have been observed [34, 40, 47]. Quite complicated tribochemical reactions take place during friction, mainly on account of reactions of anions with the surface, but also reactions with cationic moieties have been reported [52]. Complex surface chemistry takes place at the material–IL interface, particularly in the case of ceramic and metal lubrication with imidazolium hexafluorophosphates and tetrafluoroborates (including oxidation, dissociation of anions, formation of metallic phosphates, oxides and fluorides, etc.). During the wear process, nascent metallic surfaces are produced which can enhance the metal–IL reactivity and favour the process of IL decomposition. The lubrication potential of ILs has been studied by X-ray photoelectron spectroscopy (XPS) to examine surface interactions.

BF_4 and TFSA anions decompose through a tribochemical reaction to generate metal fluoride on the rubbed surface, as evidenced by ToFSIMS and XPS. Iron fluoride, iron sulphate and organic fluorides are typically derived from the anionic moieties. Reaction products identified by post-friction surface characterisation are also FeB and B_2O_3 for BF_4^-, $FePO_4$ for PF_6^-, and FeS for Tf_2N^-. Under severe friction conditions the tetrafluoroborate anion has been reported to decompose to form anti-scratch components such as fluoride, B_2O_3 and BN, as shown by XPS [34]. Reaction products and chemisorbed cations affect the tribological behaviour of ILs [45]. These surface compounds are important in preventing seizure in freshly generated metal surfaces. The view is held that this is one of the reasons accounting for the superior tribological characteristics of ILs in comparison with conventional lubricants (without these anions) [53].

The surface interactions and tribological reactions of ILs with the sliding counterparts need to be thoroughly examined in order to ensure the formation of protecting tribofilms and to avoid serious corrosion of the substrate. Many ILs contain halogen atoms and may release toxic and corrosive hydrogen halides to the environment [54, 55]. In particular, the fluorine-containing anions such as BF_4^- and PF_6^- occasionally cause corrosion of steel under humid conditions by HF generation. The high polarity of ILs enhances their water-absorbing capacity, thus favouring hydrolysis. Replacing conventional anions with fluorine-free or more hydrophobic anions represents a way to alleviate the corrosion problem. Hydrophobic anions are less corrosive and exhibit good tribological properties [56]. New halogen-free ILs are being developed to avoid corrosion and toxicity [57], but little toxicological and/or ecotoxicological data is available for ILs at present. In order to avoid serious tribocorrosion, particularly in case of fluorine-containing ILs, use has been made of reactive ILs as lubricant additives rather than as neat lubricants (see Section 15.2.2), and corrosion inhibitor additives have been added to IL lubricants (see Section 15.2.3).

Minami *et al.* [58] have studied the thermo-oxidation and volatility of ILs in detail. The thermal stability of ILs is generally much greater than that of conventional synthetic lubricants, as shown in Fig. 15.4. This is particularly

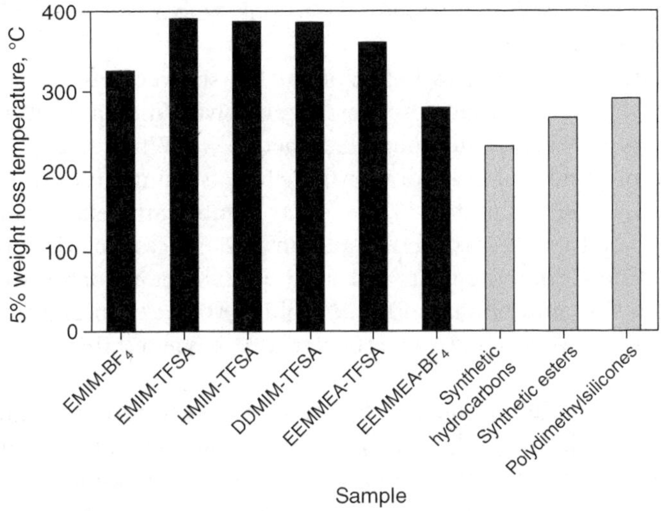

15.4 Thermogravimetric analysis: temperature for 5% weight loss. For acronyms of samples, see Appendix A.1. After ref. [44].

true for alkylimidazolium-based ILs. The thermal stability depends both on the anion and cation. Generally, imidazolium-based dicationic liquids show higher degradation temperatures (>400 °C) than the triazolium analogues. Dicationic ILs have excellent tribological properties at 300 °C and therefore have potential as high-temperature lubricants. The thermo-oxidative stability of IL lubricants may further be improved by means of additives. Oxidation mechanisms and products require further investigation. Existing problems with IL lubricants can profitably be tackled by molecular design [53].

IL-lubricated sliding contacts investigated have mainly concerned steel; Fe, Al, Cu and Ni alloys; Si_3N_4; and Si, SiO_2 [44]. ILs were evaluated for steel–steel contacts [34, 40, 47, 52, 59–61], but also for steel–aluminium [39, 62–64]. Lightweight materials such as aluminium alloys and silicon are difficult to lubricate. The most commonly used aluminium-on-steel couple is an extremely difficult system to be lubricated even at a modest load. Lubrication of aluminium is rendered difficult because of its hardness and susceptibility to attack by tribo-products such as HF. Aluminium requires lubricants with stronger film formation capability. Aluminium alloys are extensively being used for tribological applications, such as in internal combustion engines, plain bearings, compressors and refrigerators. However, their poor resistance to seizure and difficulty to be lubricated make them vulnerable in the sliding process. RTILs are currently attracting interest as high-performance multifunctional lubricants for reactive light alloys [65].

For lubrication of aluminium alloys phosphonyl-containing imidazolium ILs are suitable [65–67].

At all testing temperatures imidazolium ILs showed lower friction and wear values than the synthetic ester propylene glycol dioleate (PGDO) and a mineral oil for steel–aluminium contacts [64]. When compared with 15W-40 diesel engine oil in laboratory bench tests ammonium ILs show up to 35% less friction and up to 55% less wear in lubricating steel–aluminium contacts, and up to 55% less friction and up to 21% less wear in lubricating Cr-plated piston rings against cast iron [68]. Jiménez and Bermúdez [69] have studied ammonium and imidazolium ILs as titanium–steel lubricants. Very little work has been reported on polymer lubrication by ILs [70, 71].

The results reported so far indicate that ILs are a family of new high-performance lubricants [53], which may find commercial and technological applications where other lubricants are not suitable or fail to prevent friction, wear and surface damage of materials. IL lubricants are used in three different ways: as base oils, additives (see Section 15.2.2) and thin films. Imidazolium derivatives are the most widely studied family of IL lubricants. ILs as lubrication oils generally exhibit good tribological performance for variable sliding pairs such as steel/steel, ceramic/ceramic or steel/aluminium. They have low friction coefficients as compared to competing synthetic oils (such as perfluoropolyether, PFPE). The load-carrying capacity is also much higher [40]. IL lubricants can be used in a wide temperature range (from below –50 up to 400 °C). The use of these advanced fluids could contribute to solve some difficult tribological problems such as the lubrication of steel–aluminium contacts [39, 65, 67]. In order to find adequate IL lubricants for use in special environments (high vacuum, cryogenic and high temperatures, high pressures) a large number of ILs should be tested under variable conditions. Low vapour pressure and high thermal stability are inherent IL properties of considerable interest under vacuum conditions, and in particular for aerospace and spacecraft applications, with tribological performance and stability exceeding that of multiply-alkylated cyclopentanes (MACs) [53]. The non-flammability makes IL lubricants useful in combustion engines and machining processes, in (petro)chemical and aircraft industries and in high-temperature applications where fire resistance is significant.

Miniaturised devices or micro/nanoelectrical mechanical systems (MEMS/NEMS) with large surface-to-volume ratios, which are operated under very small normal loads, cannot be lubricated with lubrication oils, but use a thin lubrication film with thickness below a few tens of nanometres. The potential of ILs in thin film lubrication has been exploited as a replacement for PFPEs [72, 73]. The properties of ILs used in the nanotribological MEMS/NEMS applications that are different from conventional lubricants are their excellent thermal and electrical conductivity, which allows minimisation of contact resistance between sliding surfaces.

The study of IL lubricants is rapidly expanding mainly because they are highly versatile. Issues that need to be addressed are shown in Table 15.3. It is hoped that these designer fluids will eventually solve some of the current most difficult industrial lubricating problems. Neat ILs exhibit remarkably better lubrication and AW properties than conventional synthetic oils and are considered promising alternatives with considerably lower development costs [29]. However, although ionic liquids have many advantages, their cost/performance is still prohibitive. At present, the uncertainty of the cost of the liquid and availability makes ILs less attractive to industrial users. The current market for ILs is very small, and product costs are still too great for high-volume industrial applications such as metal machining. However, this cost aspect is quite different for use of ILs as additives in base lubrication oils. Some attempts are also already being made to develop ILs from cheap starting materials [28].

It should be stressed that IL lubricants are not yet in commercial use mainly because they are so new. Commercialisation takes time. For the acceleration of IL commercialisation, see ref. [27]. Synthetic lubricants have been in existence for about 50 years but even today have captured only a small share of the total lubricant market. Biolubricants are a repeating story.

Apart from cost, there are as yet also a number of technical challenges associated with turning ionic liquids into practical fluids for industrial use, namely lack of physical properties and corrosion data, and absence of data on ecotoxicity and ecological impact. Not all ionic liquids are inflammable, not all decompose at $T > 400\,^{\circ}\mathrm{C}$ or have a negligible vapour pressure [74]. While ILs are considered as (light) green solvents for the future [1] not all ILs are ecocompatible; certain classes of ionic liquids are toxic [75]. Most ILs studied so far contain halogen atoms and may release toxic and corrosive hydrogen halides to the environment [54, 55]. Along with toxic and non-ecocompatible compounds the vast class of ILs certainly comprises

Table 15.3 Development issues of ionic liquid lubricants

• Health, safety and environmental aspects
• Surface interactions and tribochemical reactions between ILs and sliding counterparts
• Synergistic effect and long-term stability of tribocorrosion-reducing IL additives
• Feasibility of IL lubricants in contact with cryogenic or high-temperature materials
• Economic benefits

After ref. [68].

others with low environmental impact. New halogen-free ILs are being developed to avoid corrosion and toxicity [57], but little (eco)toxicological data is currently available for ILs. The challenge is to develop ILs with desirable technological properties and low environmental impact. Work is needed to determine which ILs have the least negative impacts on aquatic environments. Mammalian toxicity studies of ILs are still rare [76].

In use as lubrication oils, corrosion and thermo-oxidation of ILs must be overcome. As additives, compatibility with base oils is of concern. The high polarity of ILs and the low polarity of conventional oils results in very poor miscibility. These problems can be overcome by molecular design. Miscibility with lubrication oils can be promoted by making combinatorial molecules having components of conventional oils [29, 71], typically poly-α-olefin (PAO)-ILs or similar. This might further improve the lubrication properties of ILs and the tribological properties of base oils. As additives, ILs can be used in greases where the miscibility issue is less of a problem. ILs will not be able to compete in the marketplace until stability, toxicity and cost issues have been solved [68]. There are many reviews of ILs in tribology [44, 53, 68, 77, 78].

15.2.2 ILs as lubricity additives

With their strong boundary film formation capability and good tribological properties, small amounts of ILs may act as additives to conventional lubricant base oils such as polyether, polyester or mineral oils to enhance their tribo-performance. Practical application of this concept requires: (i) compatibility of IL to the base oil (i.e. solubility); (ii) better performance than traditional additives; and (iii) lower cost. Crucial to their use as additives in low concentration is the alleviated tribocorrosion compared with neat ILs.

In view of their miscibility with polar and non-polar solvents, ionic liquids have been used as advanced lubricating additives both in water and in lubricating base oils [39, 63, 64, 79, 80]. ILs were firstly proposed as additives for water in lubricating silicon nitride ceramic sliding contacts [79]. When added to water, ILs reduce the initial period of high friction. However, ILs as additives to water might not be the most appropriate application due to hydrolysis of some commonly used anions, in particular BF_4^- and PF_6^-. When added to grease, base or fully formulated, ILs substantially improve performance, denoting synergy with the additives present in the formulated grease [49].

There is general consensus that ILs as additives show better tribo-performance than neat IL lubricants or conventional base oils. However, the lubricating ability of ILs as oil additives cannot be derived directly from that of neat ILs. The formation of effective lubricating films depends on the chain length in cations, on the anion composition as well as on the miscibility with base oil and the presence of atmospheric moisture.

IL additives to mineral oil were used to lubricate steel/aluminium alloy sliding pairs [41]. More than 30% friction reduction was achieved compared with the neat mineral oil. For 1 wt% imidazolium IL additives in a paraffinic–naphthenic mineral base oil for SAE 52100 steel – ASTM 2011 aluminium contacts at 25 and 100 °C lubrication and AW properties all improved over neat ILs for relatively large sliding velocities (0.15–0.20 m/s) [39]. The solubility of imidazolium ILs in base oils is very low.

Bermúdez and coworkers [64] have evaluated the RTIL 1-*n*-alkyl-3-methylimidazolium salts with different side chain lengths on the cations and different fluorine-containing anions and 1-butyl-3-methylpyridinium TFSA as 1 wt% mineral base oil additives in steel–aluminium contacts. For the same anion [BF_4^-], the more effective neat IL lubricants are those containing cations with longer alkyl chains: octyl [$OMIM^+$] > hexyl [$HMIM^+$] > ethyl [$EMIM^+$]. The opposite is true when used as additives to mineral oil, as reported in Fig. 15.5 which shows the wear rate (material volume loss by unit of sliding distance) of aluminium at room temperature: the more effective wear-reducing mineral oil IL additives are those with shorter alkyl chains. For the same cation ([$EMIM^+$]) and different anions, the wear rate of aluminium decreases in the order $MeC_6H_4SO_3^-$ > $CF_3SO_3^-$ > BF_4^-. Friction coefficients for IL additives are similar or lower than for neat ILs, while wear rates for 1 wt% ILs can be several orders of magnitude lower than those for neat ILs. 1-*n*-Alkyl-3-methylimidazolium ILs have also been studied as 1 wt% additives of the synthetic ester propylene glycol dioleate

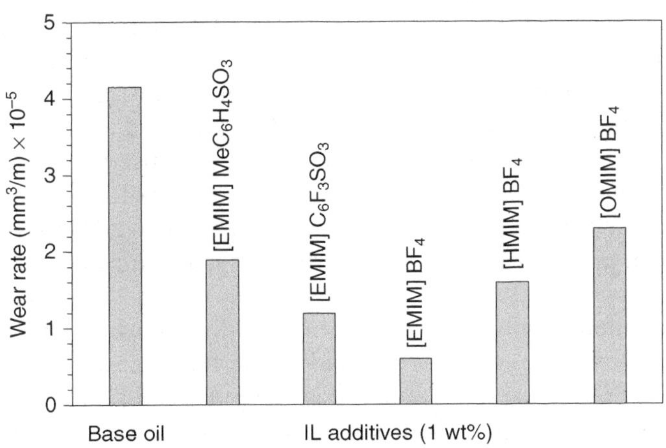

15.5 Ionic liquids as additives in steel-aluminium lubrication. For acronyms, see Appendix A.1. After ref. [78].

(PGDO) in steel–aluminium contacts at 25 and 100 °C [63]. At 100 °C, all additives reduce both friction and wear with respect to the base oil.

The excellent tribological properties of ILs as additives are due to the:

- formation of physically adsorbed films, similar to a friction modifier; and
- formation of tribochemical products during friction, creating an anti-wear boundary film.

15.2.3 Additive technology of ILs

Similar to mineral and synthetic fluids, IL lubricating oils usually have to be optimised by additive technology in order to meet the requirements for practical applications. There are various ways to reduce IL lubricant friction, but additives are key [44]. Additive-enhanced ILs show great promise as lubricants.

Additives for ILs further enhance the AW properties of the neat products. The conventional lubricant additives tricresylphosphate (TCP) and dibenzyldisulphide (DBDS) are effective in ILs and improve the AW properties [38]. Addition of 1% TCP helps to establish a tribofilm very rapidly at high temperature [49]. These additives also prevent the tribochemical decomposition of ILs. Observed boundary film was composed of elements derived from the additives rather than the IL. Therefore, the mechanism of these additives is considered to provide boundary film with AW properties through tribochemical reactions. The additive response of ILs was superior to that for conventional lubricants. However, AW additives are not always necessary because of the ease of formation of surface boundary films. Neat ILs contain tribologically interested elements (N, P, B and F) for AW and extreme pressure (EP) use.

The more reactive ILs – especially those containing fluorine anions – require addition of wear additives as well as anticorrosion agents. It has been found that benzotriazole, a frequently used anticorrosion additive, shows good miscibility with imidazolium ILs and could alleviate their corrosion [81]. Unfortunately, benzotriazole cannot be used at elevated temperatures or under reduced pressures. However, triazole-ILs are expected to exhibit considerable anticorrosion capability (see Fig. 15.6). Innovative additives that inhibit corrosion and prevent ILs from being oxidised require further development.

Most commercially available lubricant additives were developed for mineral base oils (see Section 7.2), and hence they hardly dissolve in ILs [37]. For example, non-polar compounds and polar compounds with higher alkyl groups dissolve poorly in ILs (classified as polar solvents). Generally, the saturated aliphatics do not mix well with ILs, olefins perform better and aldehydes are soluble. Simple compounds such as carboxylic acids and

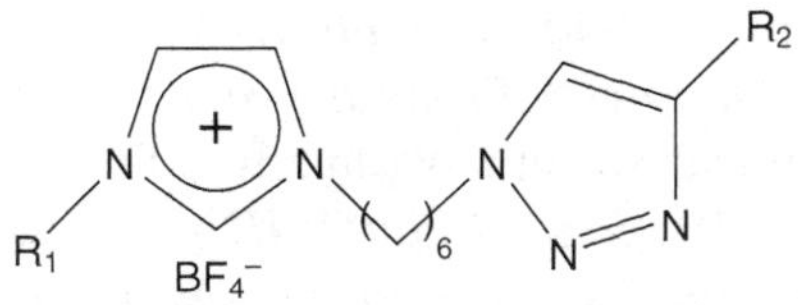

15.6 Structure of triazole containing imidazolium ILs.

benzotriazole (BTA), considered as potential additives for ionic liquids, are actually unsuitable in view of their thermal stability and volatility. Amino-acid-derived salts were developed as additives for ILs [82]. Tetra-alkylammonium and tetra-alkylphosphonium salts of *N*-protected aspartic acid reduce the friction and wear properties of 1-alkyl-3-methylimidazolium bis(trifluoromethylsulphonyl)imide [44]. Addition of aspartic acid derivatives to [BMIM]TFSA significantly reduces friction and wear by 20% and 40%, respectively. The carboxyl group in the additive has an affinity towards steel surfaces. This functional group anchors the molecules to the surface. The relationship between molecular structure and tribological properties suggests the formation of a protective film comprising both additive and base oil molecules.

When it comes to improving wear and friction properties of ILs with additives, the purity of the base ILs is very important. The purity of ILs has significant influence to the tribological properties by additive technology. The dependence of additive response on the refining of base oil is well known [83]. Highly purified additivated ILs reduce friction much better than reagent-grade ILs. Highly purified [BMIM]TFSA provides significant lubricity, while reagent-grade [BMIM]TFSA only provides acceptable lubricity [44]. The subject deserves further investigation.

15.3 Sources of further information and advice

M. Gaune-Escard and K.R. Seddon (eds), *Molten Salts and Ionic Liquids – Never the Twain?*, John Wiley & Sons, Ltd, Chichester (2010).

M. Freemantle, *Introduction to Ionic Liquids*, RSC Publishing, Cambridge (2009).

B. Kirchner (ed.), *Ionic Liquids*, Springer Verlag, Berlin (2009); *Topics in Current Chemistry*, Vol. 290.

P. Wasserscheid and T. Welton (eds), *Ionic Liquids in Synthesis*, 2nd edn, Wiley-VCH, Weinheim (2008).

J.F. Brennecke, R.D. Rogers and K.R. Seddon (eds), *Ionic Liquids IV. Not Just Solvents Anymore*, ACS Symp. Ser. 975, American Chemical Society, Washington, DC (2007).

R.D. Rogers and K.R. Seddon (eds), *Ionic Liquids IIIA: Fundamentals, Progress, Challenges, and Opportunities – Properties and Structure*, ACS Symp. Ser. 901, American Chemical Society, Washington, DC (2005).

R.D. Rogers and K.R. Seddon (eds), *Ionic Liquids IIIB: Fundamentals, Processes, Challenges, and Opportunities – Transformations and Processes*, ACS Symp. Ser. 902, American Chemical Society, Washington, DC (2005).

H. Ohno (ed.), *Electrochemical Aspects of Ionic Liquids*, Wiley-Interscience, Hoboken, NJ (2005).

R.D. Rogers and K.R. Seddon (eds), *Ionic Liquids as Green Solvents: Progress and Prospects*, ACS Symp. Ser. 856, American Chemical Society, Washington, DC (2003).

R.D. Rogers and K.R. Seddon (eds), *Ionic Liquids: Industrial Applications for Green Chemistry*, ACS Symp. Ser. 818, American Chemical Society, Washington, DC (2002).

R.D. Rogers, K.R. Seddon and S. Volkov (eds), *Green Industrial Applications of Ionic Liquids*, Kluwer, Dordrecht (2002).

15.4 References

[1] M.J. Earle and K.R. Seddon, Ionic liquids. Green solvents for the future, *Pure Appl. Chem.* **72** (2000) 1391–8.

[2] M. Gaune-Escard and K.R. Seddon (eds), *Molten Salts and Ionic Liquids – Never the Twain?*, John Wiley & Sons, Ltd, Chichester (2010).

[3] P. Walden, Molecular weights and electrical conductivity of several fused salts, *Bull. Acad. Imper. Sci.* (St. Petersburg) **8** (1914) 405–23.

[4] K.R. Seddon, A. Stark and M.J. Torres, Influence of chloride, water and organic solvents on the physical properties of ionic liquids, *Pure Appl. Chem.* **72** (2000) 2275.

[5] T. Welton, Room-temperature ionic liquids. Solvents for synthesis and catalysis, *Chem. Rev.* **99** (1999) 2071–83.

[6] K.R. Seddon, A. Stark and M.-J. Torres, Viscosity and density of 1-alkyl-3-imidazolium ionic liquids, *ACS Symp. Ser.* **819** (2001) 34–49.

[7] A.S. Pensado, M.J.P. Comuñas and J. Fernández, The pressure-viscosity coefficient of several ionic liquids, *Tribol. Lett.* **31**(2) (2008) 107–18.

[8] R. Bini, C. Chiappe, E. Marmugi and D. Pieraccini, The 'non-nucleophilic' anion $[Tf_2N]^-$competes with the nucleophilic Br^-: an unexpected trapping in the dediazoniation reaction in ionic liquids, *Chem. Commun.* (2006) 897–9.

[9] D.T. Allen and D.R. Shonnard, *Green Engineering*, Prentice Hall, Upper Saddle River, NJ (2002).

[10] J.F. Brennecke, L.A. Blanchard, J.L. Anthony, Z. Gu, I. Zarraga and D.T. Leighton, Separation of species from ionic liquids, *ACS Symp. Ser.* **819** (2002) 82–96.

[11] P. Wasserscheid and T. Welton (eds), *Ionic Liquids in Synthesis*, 2nd edn., Wiley-VCH, Weinheim (2008).

[12] P. Hapiot and C. Lagrost, Electrochemical reactivity in room-temperature ionic liquids, *Chem. Rev.* **108** (2008) 2238–64.

[13] H. Ohno (ed.), *Electrochemical Aspects of Ionic Liquids*, Wiley-Interscience, Hoboken, NJ (2005).
[14] Y. Chauvin and H. Olivier-Bourbigou, Nonaqueous ionic liquids as reaction solvents, *ChemTech* **26** (Sept. 1995) 26–30.
[15] M. Freemantle, Ionic liquids show promise for clean separation technology, *Chem. Engng. News* **76** (24 Aug. 1998) 12.
[16] J.F. Brennecke and E.J. Maginn, Ionic liquids: innovative fluids for chemical processing, *AIChE J.* **47**(11) (2001) 2384–9.
[17] R. Renner, An environmental solution, *Sci. Amer.* **285**(2) (Aug. 2001) 19.
[18] O.A. El Seoud, A. Koschella, L.C. Fidale, S. Dorn and T. Heinze, Applications of ionic liquids in carbohydrate chemistry: a window of opportunities, *Biomacromol.* **8** (2007) 2629–47.
[19] J. Zakzeski, P.C.A. Bruijnincx, A.L. Jongerius and B.M. Weckhuijsen, The catalytic valorization of lignin for the production of renewable chemicals, *Chem. Rev.* **110** (2010) 3552–99.
[20] M. Zavrel, D. Bross, M. Funke, J. Büchs and A.C. Spiess, High-throughput screening for ionic liquids dissolving (ligno)-cellulose, *Bioresour. Technol.* **100** (2009) 2580–7.
[21] J.S. Wilkes and M.J. Zaworotko, Air and water stable 1-ethyl-3-methylimidazolium based ionic liquids, *Chem. Commun.* (1992) 965–7.
[22] R. Rinaldi, R. Palkovits and F. Schüth, Depolymerization of cellulose using solid catalysts in ionic liquid, *Angew. Chem. Int. Ed.* **47** (2008) 8047–50.
[23] Y. Zhang, H. Du, X. Qian and E.Y.-X. Chen, Ionic liquid–water mixtures: enhanced K_w for efficient cellulose biomass conversion, *En. Fuels* **24** (2010) 2410–17.
[24] J.L. Anthony, E.J. Maginn and J.F. Brennecke, Solution thermodynamics of imidazolium-based ionic liquids and water, *J. Phys. Chem. B* **105** (2001) 10942–9.
[25] A.G. Fadeev and M.M. Meagher, Opportunities for the ionic liquids in recovery of biofuels, *Chem. Commun.* (2001) 295–6.
[26] L.A. Branchard and J.F. Brennecke, Recovery of organic products from ionic liquids using supercritical carbon dioxide, *I&EC Res.* **40** (2001) 287–92.
[27] BCS, Inc, *Accelerating Ionic Liquids Commercialization*, Laurel, MD (2004); http://www.chemicalvision2020.org /pdfs/ionicliquid_commercialization.pdf
[28] C. Evans, New ionic liquids from cheap non-toxic sources – saccharine solution to a toxic problem, *Chem. Commun.* (2004) C27.
[29] N.V. Plechkova and K.R. Seddon, Applications of ionic liquids in the chemical industry, *Chem. Soc. Rev.* **37** (2008) 123–50.
[30] J.B. Holbrey and K.R. Seddon, Ionic liquids, *Clean Prod. Proc.* **1**(4) (1999) 223–36.
[31] K.R. Seddon, Ionic liquids for clean technology, *J. Chem. Tech. Biotech.* **68** (1997) 351–6.
[32] A. Stark and K.R. Seddon, Ionic liquids, in *Kirk-Othmer Encyclopedia of Chemical Technology*, John Wiley & Sons, Inc., New York, NY (2007), Vol. 26, pp. 836–920.
[33] P.G. Smith, High temperature molten-salt lubricated hydrodynamic journal bearings, *ASLE Trans.* **4** (1961) 263–74.

[34] C.F. Ye, W.M. Liu, Y.X. Chen and L.G. Yu, Room-temperature ionic liquids: a novel versatile lubricant, *Chem. Commun.* (2001) 2244–5; highlighted by *C&EN* (12 November 2001), 23.
[35] S. Zhang, N. Sun, X. He, X. Lu and X. Zhang, Physical properties of ionic liquids: database and evaluation, *J. Phys. Chem. Ref. Data* **35** (2006) 1475.
[36] H. Wang, Q. Lu, C. Ye, W. Liu and Z. Cui, Friction and wear behaviors of ionic liquid of alkylimidazolium hexafluorophosphates as lubricants for steel/steel contact, *Wear* **256** (2004) 44–8.
[37] R.A Reich, P.A. Stewart, J. Bohaychick and J.A. Urbanski, Base oil properties of ionic liquids, *Lubr. Engng.* **59**(7) (2003) 16–21.
[38] H. Kamimura, T. Kubo, I. Minami and S. Mori, Effect and mechanism of additives for ionic liquids as new lubricants, *Tribol. Int.* **40** (2007) 620–5.
[39] A.E. Jiménez, M.D. Bermúdez, P. Iglesias, F.J. Carrión and G. Martínez-Nicolás, 1-*N*-alkyl-3-methylimidazolium ionic liquids as neat lubricants and lubricant additives in steel–aluminium contacts, *Wear* **260** (2006) 766–82.
[40] W.M. Liu, C.F. Ye, Q.Y. Gong, H.Z. Wang and P. Wang, Tribological performance of room-temperature ionic liquids as lubricants, *Tribol. Lett.* **13** (2002) 81–5.
[41] J. Qu, J.J. Truhan, S. Dai, H. Luo and P.J. Blau, Ionic liquids with ammonium cations as lubricants or additives, *Tribol. Lett.* **22** (2006) 207–14.
[42] J. Qu, P.J. Blau, D. Sheng, L. Huimin, J.M. Meyer and J.J. Truhan, Tribological characteristics of aluminium alloys sliding against steel lubricated by ammonium and imidazolium ionic liquids, *Wear* **267** (2009) 1226–31.
[43] H. Ohtani, S. Ishimura and M. Kumai, Thermal decomposition behaviors of imidazolium-type ionic liquids studied by pyrolysis-gas chromatography, *Anal. Sci.* **24** (2008) 1335–40.
[44] I. Minami, Ionic liquids in tribology, *Molecules* **14** (2009) 2286–2305.
[45] H. Kamimura, T. Chiba, T. Kubo, H. Nanao, I. Minami and S. Mori, Relationship between structure and tribological properties of ionic liquids composed of imidazolium cations, *Toraiborojisuto* **51**(11) (2006) 826–34; *Japanese J. Tribol.* **51** (2006) 675–87.
[46] H. Kamimura, T. Chiba, N. Watanabe, T. Kubo, H. Nanao, I. Minami and S. Mori, Effects of carboxylic acids on wear and friction reducing properties of alkylimidazolium-derived ionic liquids, *Tribol. Online* **1** (2006) 40–3.
[47] Q. Lu, H. Wang, C. Ye, W.M. Liu and Q. Xue, Room temperature ionic liquid 1-ethyl-3-hexylimidazoliumbis(trifluoromethylsulfonyl)-imide as lubricant for steel/steel contact, *Tribol. Int.* **37** (2004) 547–52.
[48] I. Minami, M. Kita, T. Kubo, H. Nanao and S. Mori, The tribological properties of trifluorotris (pentafluoroethyl) phosphate derived ionic liquids, *Tribol. Lett.* **30** (2008) 215–23.
[49] M.F. Fox and M. Priest, Tribological properties of ionic liquids as lubricants and additives. Part 1: Synergistic tribofilm formation between ionic liquids and tricresyl phosphate, *PIME Part J J. Engng. Tribol.* **222** (2008) 291.
[50] M. Forsyth, T.F. Kemp, P.C. Howlett, J. Sun and M.E. Smith, A potential novel rapid screening NMR approach to boundary film formation at solid interfaces in contact with ionic liquids, *J. Phys. Chem. C* **112** (2008) 13801–4.
[51] X.Q. Liu, F. Zhou, Y.M. Liang and W.M. Liu, Tribological performance of phosphonium based ionic liquids for an aluminum-on-steel system and opinions on lubrication mechanism, *Wear* **261** (2006) 1174–9.

[52] B.S. Phillips, G. John and J.S. Zabinski, Surface chemistry of fluorine containing ionic liquids on steel substrates at elevated temperature using Mössbauer spectroscopy, *Tribol. Lett.* **26** (2007) 85–91.
[53] F. Zhou, Y. Liang and W. Liu, Ionic liquid lubricants: designed chemistry for engineering applications, *Chem. Soc. Rev.* **38** (2009) 2590–9.
[54] R.P. Swatloski, J.D. Holbrey and R.D. Rogers, Ionic liquids are not always green: hydrolysis of 1-butyl-3-methylimidazolium hexafluorophosphate, *Green Chem.* **5** (2003) 361–3.
[55] B. Jastorff, R. Stormann, J. Ranke, K. Molter, F. Stock, B. Oberheitmann, W. Hoffmann, J. Hoffmann, M. Nuchter, B. Ondruschka and J. Filser, How hazardous are ionic liquids? Structure activity relationships and biological testing as important elements for sustainability evaluation, *Green Chem.* **5** (2003) 136–42.
[56] T. Itoh, N. Watanabe, K. Inada, A. Ishioka, S. Hayase, M. Kawatsura, I. Minami and S. Mori, Design of alkylsulphate ionic liquids for lubricants, *Chem. Lett.* **38**(1) (2009) 64–5.
[57] P. Wasserscheid, R. van Hal and A. Bosmann, 1-*n*-Butyl-3-methylimidazolium ([bmim])octylsulfate – an ever 'greener' ionic liquid, *Green Chem.* **4** (2002) 400–4.
[58] I. Minami, H. Kamimura and S. Mori, Thermo-oxidative stability of ionic liquids as lubricating fluids, *J. Synth. Lubr.* **24** (2007) 135–47.
[59] M. Yao, Y. Liang, Y. Xia and F. Zhou, Bisimidazolium ionic liquids as the high-performance antiwear additives in poly(ethylene glycol) for steel–steel contacts, *ACS Appl. Mater. Interf.* **1** (2009) 467–71.
[60] A.H. Battez, R. Gonzalez, J.L. Viesca, D. Blanco, E. Asedegbega and A. Osorio, Tribological behaviour of two imidazolium ionic liquids as lubricant additives for steel/steel contacts, *Wear* **266** (2009) 1224–8.
[61] L.J. Weng, X.Q. Liu, Y.M. Liang and Q.J. Xue, Effect of tetraalkylphosphonium based ionic liquids as lubricants on the tribological performance of steel-on-steel system, *Tribol. Lett.* **26** (2007) 11–17.
[62] A.E. Jiménez and M.D. Bermúdez, Ionic liquids on lubricants for steel-aluminium contacts at low and elevated temperatures, *Tribol. Lett.* **26** (2007) 53–60.
[63] A.E. Jiménez and M.D. Bermúdez, Imidazolium ionic liquids as additives of the synthetic ester propylene glycol dioleate in aluminium–steel lubrication, *Wear* **265** (2008) 787–98.
[64] A.E. Jiménez, M.D. Bermúdez, J.F. Carrión and G. Martínez-Nicolás, Room temperature ionic liquids as lubricant additives in steel–aluminium contacts: influence of sliding velocity, normal load and temperature, *Wear* **261** (2006) 347–59.
[65] Z.G. Mu, F. Zhou, S.X. Zhang, Y.M. Liang and W.M. Liu, Effect of the functional groups in ionic liquid molecules on the friction and wear behaviour of aluminum alloy in lubricated aluminum-on-steel contact, *Tribol. Lett.* **38** (2005) 725–31.
[66] Z.G. Mu, F. Zhou, S.X. Zhang, Y.M. Liang and W.M. Liu, Preparation and characterization of new phosphonyl-substituted imidazolium ionic liquids, *Helv. Chim. Acta* **87** (2004) 2549.
[67] Z.G. Mu, W.M. Liu, S.X. Zhang and F. Zhou, Functional room-temperature ionic liquids as lubricants for aluminum-on-steel system, *Chem. Lett.* **33** (2004) 524–5.

[68] J. Van Rensselar, Unleashing the potential of ionic liquids, *Tribol. Lubr. Technol.* **66**(4) (2010) 24–31.
[69] A.E. Jiménez and M.D. Bermúdez, Ionic liquids as lubricants of titanium–steel contacts, *Tribol. Lett.* **33** (2009) 111–26.
[70] J. Sanes, J.F. Carrión, M.D. Bermúdez and G. Martínez-Nicolás, Ionic liquids as lubricants of polystyrene and polyamide 6–steel contacts. Preparation and properties of new-polymer-ionic liquid dispersions, *Tribol. Lett.* **21** (2006) 121–33.
[71] H. Kondo, Protic ionic liquids with ammonium salts as lubricants for magnetic thin film media, *Tribol. Lett.* **31** (2008) 211–18.
[72] Y. Mo, W. Zhao, M. Zhu and M. Bai, Nano/microtribological properties of ultrathin functionalized imidazolium wear-resistant ionic liquid films on single crystal silicon, *Tribol. Lett.* **32** (2008) 143–51.
[73] B. Bhushan, M. Palacio and B. Kinzig, AFM-based nano-tribological and electrical characterization of ultrathin wear-resistant ionic liquid films, *J. Colloid Interf. Sci.* **317** (2008) 275–87.
[74] M.J. Earle, J.M.S.S. Esperança, M.A. Gilea, J.N. Canongia Lopes, L.P.N. Rebelo, J.W. Magee, K.R. Seddon and J.A. Widegren, The distillation and volatility of ionic liquids, *Nature* **1439** (2006) 831–4.
[75] C. Chiappe, Ionic liquids. A world to disclose beyond their 'green solvents' properties, *Chim. Ind.* (Milano) **88**(9) (2006) 52–5.
[76] J. Pernak, A. Czepukowicz and R. Pozniak, New ionic liquids and their antielectrostatic properties, *I&EC Res.* **40** (2001) 2379.
[77] Z. Zeng, F. Huang, Y. Zhang, H. Wang and K. Yi, Development brief review of green lubricants, *Hecheng Runhua Cailiao* **35**(4) (2008) 17–20.
[78] M.D. Bermúdez, A.E. Jiménez, J. Sanes and F.J. Carrión, Ionic liquids as advanced lubricant fluids, *Molecules* **14** (2009) 2888–908.
[79] B.S. Phillips and J.S. Zabinski, Ionic liquid lubrication effects on ceramics in a water environment, *Tribol. Lett.* **17** (2004) 533–41.
[80] B.A. Omotowa, B.S. Phillips, J.S. Zabinski and J.M. Shreeve, Phosphazene-based ionic liquids: synthesis, temperature-dependent viscosity, and effect of additives in water lubrication of silicon nitride ceramics, *Inorg. Chem.* **43** (2004) 5466–71.
[81] B. Yu, F. Zhou, C. Pang, B. Wang, Y.M. Liang and W.M. Liu, Tribological evaluation of α-ω-diimidazoliumalkylene hexafluorophosphate ionic liquid and benzotriazole as additive, *Tribol. Lett.* **41** (2008) 797–801.
[82] I. Minami, N. Watanabe, H. Nanao, S. Mori, K. Fukumoto and H. Ohno, Aspartic acid-derived wear-preventing and frictioning-reducing agents for ionic liquids, *Chem. Lett.* **37** (2008) 300–1.
[83] F.G. Rounds, Some effects of nonhydrocarbon base oil constituents on the friction and surface coating formation obtained with three additives, *ASLE Trans.* **11** (1968) 19–30.

16
Epilogue: the outlook for biolubricant science and technology

Abstract: Biolubricants are a reality but not products without challenges: sustainable feedstock sources and regularity of supplies, international quality standards per product group, oxidative stability, low-temperature flow, regulatory issues, production costs, marketing, industrial and public awareness/acceptance, etc. The physicochemical, tribological and environmental properties of vegetable oils lend them for use as lubricants. Technology advances are based on physical blending, chemical transformation, advanced additive technology, non-transgenic and genetic modification. Progressive change-over of the chemical industry and transportation to renewable and sustainable products is an inevitable necessity, but market penetration of high-performance biolubricants is slow. Drivers and barriers are indicated and estimates of biolubricant market potential are given.

Key words: status of green lubricants, technological advances, future lubricant targets, drivers and barriers, market potential.

16.1 Introduction

The use of natural oils and fats dates back to antiquity, their usefulness being determined by their chemical structure, while their chemical and physical properties lend them for use as lubricants. In more modern times, among others, the family-owned company Carl Bechem GmbH (Hagen), established in 1834 as the first German oil factory, initially used animal fats and vegetable oils for the production of lubricants. Prior to the mid-19th century, almost all lubricants were derived from natural sources such as vegetable oils, animal lard or whale oil. However, in the 1880s and 1890s these raw materials were largely being replaced by mineral oil products, more harmful to the biosphere. The rise of petroleum, with its low crude oil prices and extensive transportation infrastructure, gave economic advantage to the chemically quite different petrolubricants, which quickly dominated the commercial markets and still do, not least because of the abundancy of feedstock and high operating life (long drain intervals). However, this trend is shifting, but slowly. Attention was already refocused on vegetable oils during wartime (World Wars I and II) and in oil shortage situations (oil embargo of 1973). After the oil crisis in the 1970s vegetable base oils were again used increasingly for various environmentally friendly developments. Within a short time, these vegetable lubricant products have become a completely new generation of semi- and fully synthetic biodegradable lubricants which meet the standards of conventional formulations. Renewed interest in vegetable oil-based lubricants has recently occurred as environmental awareness has increased. The harmful effects of petroleum on environmental and human health are becoming ever more obvious. At the same time, high-performance vegetable oil-based lubricants have been developed and are being marketed. Continuing its long-standing tradition, in 2012 Carl Bechem GmbH is still an active producer of a high-performance range of environmentally friendly lubricants and additives (circulation oils, greases, industrial lubricants, rail lubricants, chain oils, machining oils and food-processing oils).

In order to be successful the chemical industry must take notice of important societal developments and consumer needs – megatrends – such as increased ageing, health and environmental hazards (greenhouse gases (GHGs), CO_2, ozone hole, acid rain, climate change, etc.) and resource management (food, water, energy, raw material feedstocks). Chemistry plays an essential role as a provider of solutions for society's problems. Sustainable development of (limited) resources is a strategic goal, which imposes new approaches to the chemical industry. Technological advances can create an industry of maximum innovation and minimal environmental impact.

The progressive change-over of the chemical industry to renewable and sustainable feedstocks is an inevitable necessity. The terrestrial biomass is composed of low- and high-molecular weight products, such as sugars, hydroxy and amino acids, lipids and a multitude of biopolymers (hemicellulose, cellulose, starch, lignin, chitin, and proteins), together being considerably more complex than fossil raw materials. Partial readjustment of the chemical industry to such renewable raw materials will require completely new approaches in research, development and production. The biological sciences will have a particularly powerful impact on both raw materials and processing technologies underlying the bio-based economy. Chemical and biological sciences will meet in the building of 21st century industries. The bio-based economy is expected to grow rapidly.

Sustained economic growth depends on a secure supply of raw material inputs. The rapid increase in price of petroleum products in recent years, combined with the declining rate of production from older oil fields, the increased dependence on offshore sources, the decrease in the rate of discovering of new oil reserves, continuing demographic development and rapid changes in consumer demands are pressing reasons for finding replacements for petroleum products where practical. There is an urgent need to exploit renewable resources for industrial production and energy needs in a sustainable way. In the light of current debates on energy and environmental policy fast and partly radical changes are to be expected in the fuel and lubricants market.

To ease the transition of fossil to bio huge investments in the development of renewable energy techniques are necessary as well as in carbon capture and storage (CCS) in order to maintain the much needed stability of supply (see also Chapter 4). As liquid fossil fuels with their high energy concentration are easy to use and their physical, social and legislative infrastructures are well established, they will not cease to be used in the short or medium term, even if finding new deposits of light oil and gas is increasingly more difficult. A sustainable future demands that the world switches to renewable energy sources as soon as possible. In the meantime, it will be necessary to minimise emissions of GHGs, including development of technologies avoiding or reducing CO_2 emissions [1, 2].

In many countries bio-based products are now being recognised as critical to the economy, national security and environment, and are being promoted at national, federal and state levels, and by farmers, the industry and the government. Bio-based products add value to commodities and offer a great potential for revitalising rural communities. Bio-based products are likely to dramatically alter our world. Refining biomass will largely decentralise fuel and chemical production, will reduce the release of petroleum-originated GHGs into the atmosphere, and will reduce the dependence on oil imports. Bio-based products, which are being marketed on the basis of economy and use, are already building a strong performance record, exceeding minimum quality specifications and becoming cost competitive. Increase in the price of petroleum due to increased worldwide demand at decreasing reserves offers opportunities for investment in bio-based products.

Various regulations were placed in the way of petroleum products in Europe during the 1990s, necessitating the use of biodegradable lubricants. At the same time, many US companies also began developing biodegradable products. Examples are Mobil's environmental awareness lubricants (EALs) line of hydraulic fluids and the development of sunflower oil-based lubricants and lubricant additives by the Lubrizol Corporation. However, regulatory mandates in the United States have been slow in following up technological progress. Meanwhile, as the era of low-cost oil has finished, biodegradable oils are moving into a more competitive position. Due to rising oil prices, the tight supplies of natural resources, global climatic change and increased environmental sensitivity, various alternatives to petroleum-based lubricants are now being explored. The current emphasis on conservation has resulted in renewed interest in the use of (modified) natural oils for non-edible (industrial) purposes. The last decade has witnessed important progress in the use of biodegradable lubes and greases, arising from:

- advances in biodegradable lubricants and additive technology;
- research on non-food uses of crops;
- genetic enhancement to seed oils;
- active promotion of the use of environmentally friendly products in various countries.

Interest in biolubricants, derived from renewable sources such as vegetable oils, waxes and carbohydrates, is growing due to their biodegradability, low toxicity and environmentally benign nature. It is estimated that in certain applications as much as 50% of all the oil ends up in the environment. This represents another strong incentive to provide lubricants that are biodegradable. Growth of the biolubricant industry is based on these pressures, environmental concerns and sustainability.

In particular in the last two decades natural fatty oils and their oleochemical derivatives have experienced a renaissance as lubricant base oils because of their technical characteristics, ecotoxicity and rapid biodegradability. The modern plant-based lubricant industry is still young, especially in North America.

16.2 Biodegradable lubricants for a sustainable life

Plant-based renewable and sustainable resources are strategic options for meeting the growing needs for industrial building blocks. Compared with sugar and starch-based agricultural products, biomass at large has a much higher potential to become a renewable energy and chemical product source of the future. If we consider plant biomass composition (75% carbohydrates, 20% lignin and 5% other natural compounds), it appears that traditionally or even more recently only readily extractable components (vegetable oils, wax esters, and starches) have been employed as lubricating materials, either directly (as is) or in some modified form. Industrial lubricants are not being synthesised *ab initio* from platform building blocks (sugar, cellulose). Use of biomass *residues* is to be preferred over transformation of biomass suitable for food production. So far, most biolubricant chemistry developed is based on biomass for food consumption (e.g. edible vegetable oils, starch) rather than on non-edible biomass (e.g. cellulose, castor oil, wax esters) and crop wastes. The cellulose platform is still all to be developed.

In the lubricant sector there is a significant potential for new product development based on renewable resources. These advances are possible due to the progress in both vegetable oil biotechnology and chemical modifications, which can be used to turn natural products into components of high-performance biolubricant formulations (see Chapter 6 and Section 10.5). Only a close integration of the agronomical and chemical sciences can ensure a successful future for biolubricants.

The use of vegetable oils is an obvious choice among environmentally friendly lubricants, because of their excellent biological degradability and non-toxicity, and the fact that they can be readily extracted from plant matter. In the last two decades, natural fatty oils and their oleochemical derivatives have experienced a renaissance as base oils because of their technical characteristics and rapid biodegradability. Owing to growing environmental concerns, plant oils are increasingly finding their way into lubricants for industrial and transportation purposes. In particular, industrial groups have been pressured by environmentalists to use bio-based lubricants instead of petro-based lubes in situations which are potentially highly dangerous to the environment (offshore drilling, agricultural and forestry machinery, etc.). Polar vegetable oils are also more effective lubricants than non-polar petroleum-based oils. This can translate into improved part quality and surface finish, optimised tool life and potentially lower total fluid consumption, reducing costs and the amount of fluid introduced into the environment.

Industrial use of vegetable oils aligns perfectly with the principles of responsible care and is an important part of green chemistry and sustainability. Vegetable oils have a number of inherent qualities that give them advantages over petroleum oils as the feedstock for lubricants. Because vegetable oils are derived from a renewable resource, their use avoids the upstream pollution associated with petroleum extraction and refining. Also from a worker safety perspective, plant-based lubricants are more attractive than their petroleum counterparts. In addition to their physical advantages, plant-based lubricants hold great potential for rural economic development.

Biolubricants formulated from plant oils also offer several important performance advantages derived from the chemistry of the base stock. The excellent lubrication ability, high viscosity indices, and superior anticorrosion properties are due to the higher affinity of plant oils to metal surfaces. Higher lubricity leads to lower friction losses, yielding more power, and better fuel economy. In addition, the high flash points (>300 °C) classify plant oils as non-flammable liquids. Lower volatility results in decreased exhaust emissions. Plant oils show high shear stability, high dispersancy and high detergency, limiting the need for detergent additives. Plant oils are also superior to petrolubricants in terms of biodegradability and hence constitute decreased environmental/toxicological hazards. Because an important fraction of all lubricants is expelled into the air or ground during normal use, the main environmental advantage of bio-based lubricants is the biodegradability of the exhaust and leakage rather than their recyclability or ultimate disposal. As non-toxic and biodegradable plant oils allow clean and renewable processing, and are easily disposed of, they are an attractive alternative to petro-based lubricants. The application of plant oils and animal fats for industrial purposes has been in practice for many years. At the same time, however, they also present several drawbacks. The fatty acid compositions of most of the readily available and inexpensive natural vegetable oils are not directly suitable for most lubrication applications. Bio-based lubricants are fully biodegradable because they oxidise easily, but rapid oxidation also determines short drain intervals. Moreover, the initial cost per volume is generally two to three times that of a petro-based lubricant. Bio-based lubricants can be applied over only a moderate range of temperatures and require specialised crops (rather than food by-products) as raw materials for higher-quality lubricants. Chemical modification of vegetable oils is an attractive option for solving these problems by improving certain characteristics, notably the sensitivity to oxidative and hydrolytic attacks, and poor low-temperature behaviour (see Chapter 6).

The improvement of plant biomass characteristics for biolubricant production is in the initial stages. Oilcrops for use as biolubricant raw materials have been identified and are at various levels of domestication and cultivar selection, while genetic and genomic resources for these species are being developed. When it comes to lubricant production, vegetable oils are not all created equal. The most commonly used vegetable oils are rapeseed (or canola), castor, soy and sunflower. Rapeseed is the primary type of vegetable oil used for lubricants in the European market. This oil (as canola) has also made the most headway into the slowly emerging North American market. Frost & Sullivan estimates that out of the vegetable oil used for lubricant production in the US 85% is canola oil and 15% soy and other oils such as sunflower. A major limitation of these products is insufficient ageing resistance. Consequently, they are less suitable for circulation lubricant systems or in more critical and severe applications such as engine oils.

Despite their obvious advantages, of the 70 or so commercially available vegetable oils, only a handful are currently used in technical applications, such as hydraulics, gears, bar and chain, transformer and metalworking oils or greases (see Table 13.11). However, other oils have technical potential yet to be exploited in these applications. Wheat germs, olive, peanut, sesame and soybean oils can all be considered as viable alternatives to rapeseed oil, which is already used extensively in technical applications. In particular, wheat germ, palm and sesame oils show good oxidative stability. The industrial suitability of these oils should further be evaluated, especially as to hydrolytic stability and compatibility with additives. Researchers also continue to experiment with crops such as safflower and castor bean.

Demand for improved technical performance and stringent environmental regulations are the main drivers for new technological developments aiming at replacements for fossil oil products. Biolubricants based on vegetable oils have to overcome their inherent instability based on the presence of polyunsaturates in the natural oil to compete with products based on mineral oils. Optimisation of natural oils refers to specific technical performance such as hydrolytic stability, viscosity, evaporation, oxidation stability, low-temperature characteristics, etc. Functional properties of vegetable oils such as stability and viscosity can be adjusted by chemical modification. With an adequate understanding of structure–property relationships almost any synthetic ester can now be designed and tailor-made with lubrication performance as good as or better than mineral oils. From a chemical point of view, success in this area also greatly depends on more selective and cost-effective (bio)catalytic processes for the modification of vegetable oils. This leaves ample scope to improved catalyst design. Plant oil fatty acid compositions can also be modified through biotechnology and breeding efforts to make value-added products useful as lubricants. Major breakthroughs on the understanding of lipid metabolism and plant cell wall biosynthesis and structure are still needed to overcome low oil yields and the resistance of lignocellulose, respectively, for efficient and cost-competitive conversion to biolubricants. Among the most promising vegetable oil base stocks developed for biolubricants we mention high-oleic acid sunflower (HOSNO) and low-erucic acid rapeseed (LEAR) oil. Demand for bioproducts accelerates the advances in seed oil technology.

The specific chemical structure of vegetable oils and derivatives (such as esters) means that they show particularly good tribological properties, combining good boundary friction lubricity and general wear protection with stable viscosity–temperature behaviour and very low evaporation. In view of their easy biodegradability they are excellent raw materials for the formulation of ecologically friendly lubricants of the highest technical performance as functional elements in engineering.

Natural wax esters (oxo esters of long-chain fatty acids and long-chain fatty alcohols), such as jojoba oil, exhibit native lubrication properties that are superior to ordinary vegetable oils due to their high oxidation stability and resistance to hydrolysis. However, in view of their high cost, these products are confined to niche applications. The non-food oil crop *Crambe* has been proposed as a promising agricultural raw material for the biotechnological production of liquid wax esters (EPOBIO project, see Section 5.4.2). Results are not to be expected before 2015–2020.

Starch (a carbohydrate) is one of the most abundantly available agricultural products. The ability of starch to adjust the viscosity of solutions and pastes is used in lubrication. The great advantage of starch products is their low cost and their sustainability because of biodegradability. Formulation of starch-based lubricants requires solubilising starch and various lubricant ingredients in water. Aqueous starch–oil composites are used as dry-film lubricants (niche application). Production of another bioproduct (bioethanol) from starch-containing raw materials, e.g. maize and wheat, is booming.

Recently, Boyde has critically examined the green aspects of lubricants [3]. Truly green lubricants are those that optimise energy efficiency and minimise wear in machinery which they lubricate, and which have maximised service lifetimes in order to reduce the amount of lubricant required. Table 16.1 illustrates the application of green chemistry principles to biolubricants. Lubricants have a particularly important contribution to make in relation to energy conservation, minimisation of waste and development of durable products.

Manufacturing of biolubricants formulated from plant oils is characterised by high mass efficiencies, and relatively low process energy requirements. The most important issues to consider in assessing the environmental impact of lubricants are the performance parameters related to friction reduction, wear reduction and lubricant stability. Lubricants have a unique capability to deliver environmental benefits in use, mainly through the role in reducing friction in machinery, which improves energy efficiency, conserves fuel resources and reduces emissions of CO_2 and acid gases. The energy efficiencies potentially accessible through lubricant optimisation exceed the energy requirements for lubricant manufacture. Reduction of

Table 16.1 Green chemistry metrics for biolubricants

Principle	Biolubricant practice
• Feedstocks	Renewables: edible and non-edible vegetable oils and animal fats, wax esters, starch. Low cost.
• Chemistry	Molecular design principles.
• Manufacturing mass efficiency	High mass efficiency for base fluid production.
• Resource consumption for lubricant manufacture	Extraction; relatively low process energy requirements.
• Synthesis	(Bio)catalysis for renewables.
• Industrial application	Consumer choice, ecolabelling schemes.
• Performance in use	Energy efficiency and environmental benefits by lubricant selection; resource conservation; degradation in use.
• Environmental impact	Environmentally friendly, few emissions; application-dependent lubricant lifetime.
• Health issues	Inherently safe; complex toxicology for used oils.
• Analysis for pollution prevention	Condition monitoring.
• End-of-life	Reclamation, reconditioning, recycling technology[a]; waste disposal regulations.
• Sustainability	Optimised lubricants solutions, economic incentives, renewable resources.

[a] Re-refining facilities available mainly for mineral oils.

wear extends the useful life of machinery and contributes to non-renewable resource conservation. Lubricant consumption can be minimised by further optimisation of the stability of lubricant base fluids.

As a result of their poor thermo-oxidative stability the operational lifetime of (neat) vegetable oil products is shorter than that of mineral oils. Consequently, large volumes of biolubricants need to be manufactured and disposed of. Moreover, where a closely specified viscosity is required for optimum energy efficiency in a given application, the narrow range of viscosities available from natural oils may not be optimal, thus easily resulting in energy inefficiencies. The performance limitations of vegetable oils mean that their use as lubricants is not necessarily the environmentally preferred option. Only renewable resources as raw materials for high-performance lubricants and additives with extended lifetimes and reduced material requirements will make a significant contribution to sustainability. At present, synthetic ester lubricants having optimised performance are not manufactured from renewable resources only.

Renewable resources are not necessarily sustainable. True sustainability requires economic, social and environmental sustainability. Agricultural practices need to be sustainable. Exclusive use of renewable resources is not a sufficient condition for sustainable technology. There is potential for counterproductive effects of legislative or economic incentives.

Biolubricants from vegetable oils are not a completely green product. The principles of Green Chemistry may be applied to further development of more sustainable lubricant technology [3]. However, prices for niche products with improved sustainability will often be (much) higher than for vegetable oils and significantly higher than petroleum-based lubricants, thus limiting the desired substantial increase in market share.

16.3 Technology advances

Use of renewable materials requires new concepts of alternative processing. Developments in separation technology, pyrolysis and gasification technology, and the use of microorganisms and enzymes are means of extraction of value-added chemicals and intermediates from plant-based materials at competitive cost. In particular, it is expected that biotechnology processes will contribute substantially to manufacturing of bulk chemicals, polymers and specialty chemicals. Strategic partnerships between the chemical industry, food, textiles and agricultural sectors are a natural consequence and will accelerate growth of the emerging bioproducts industry. Successful progress depends on integrated and multidisciplinary research. Chemical process knowledge, to be applied to bio-based systems, is already partly available.

EALs have been a slowly growing segment of the lubricants business since the early 1970s [4]. Over the years the evolution of environmental thinking has led to a change of focus on lubricants, from biodegradability to renewability and sustainability, fuel economy and lower emissions. This influences the design of such products. EALs have been repeatedly heralded as one of the few future growth segments of the lubricants business, hence the relatively large increase in R&D activity over the last 15 years. Development of high-performance biolubricants requires expertise in tribology, (bio)chemistry, lubricant manufacturing and maintenance.

In addressing a number of technical concerns with the performance of native vegetable oil-based lubricants, directly related to market penetration, namely operational temperature limitations and oxidative instability, without impairing the already excellent tribological and environmentally relevant properties, industry has developed a number of plant-based lubricants that meet or exceed the performance expectations of petroleum lubricants. The key to lubricant acceptance of vegetable oils is the development of a commercial-scale, economical and more stable vegetable oil source. There are several ways for improving the stability of a vegetable oil base stock. Technology advances (Fig. 16.1) have mainly been achieved by physical, chemical or biotechnological means:

- physical blending (see Section 7.1);
- chemical transformation (see Section 6.2);
- advanced additive technology (see Sections 7.2 and 7.3);
- non-transgenic and genetic modification (see Sections 5.3 and 5.4.2).

A combination of several of these approaches offers the best opportunity for introducing significant levels of vegetable oil in finished lubricant formulations. All of these means, which aim at improving the stability and performance of vegetable oils, come at some additional cost beyond the feedstock oil itself.

Many patented and proprietary technologies are based on the approach of chemical modification (molecular design) of vegetable base oils by transesterification, partial hydrogenation, branching, etc. Using structure–property relationships as a basis (see Section 5.2.3, p. 170), synthetic ester properties can be tailor-made to fit an application. Molecular modelling of seed oil derivatives, understanding of the additive effects of structural moieties, etc., are ways to select the optimum structure for each application. From a chemical point of view, success in this area also greatly depends on more selective and cost-effective catalytic processes for the modification of vegetable oils. Apart from the well-known use and applications of synthetic esters few new groups of lubricants based on chemical modification of vegetable oils appear to have been proposed to the market recently, but work is in progress (see Fig. 16.1). In recent years, research has improved the physicochemical properties of bio-based lubricants (see Chapter 6). Techniques to increase the oxidative stability of vegetable oils include

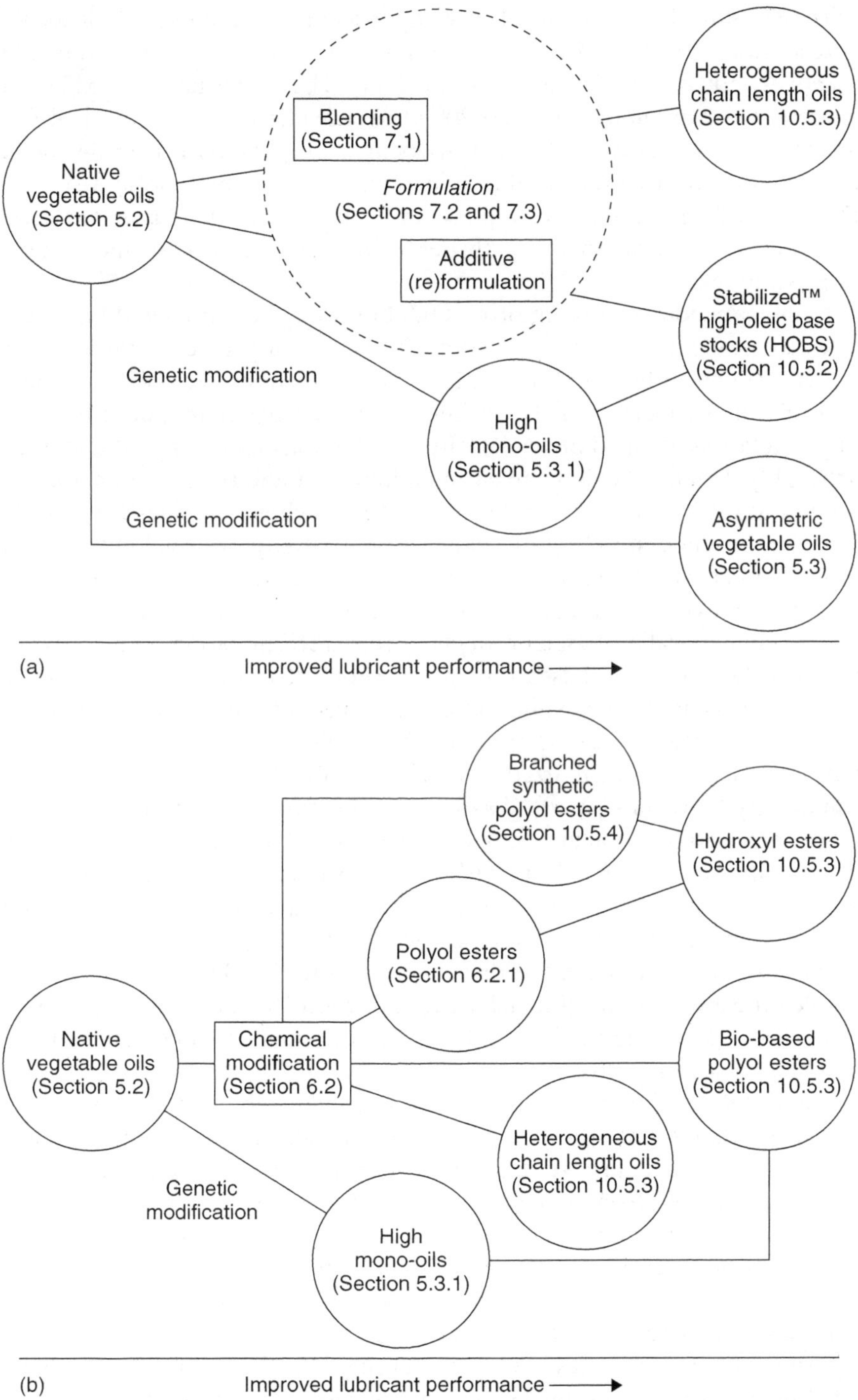

16.1 (a, b) Effects of technology advances on biolubricant oil properties.

(partial) saturation through selective hydrogenation and epoxidation. The epoxidation process is of considerable interest as the oxirane ring is highly reactive and can be further processed [5]. The Prileshajev epoxidation process, which annually generates 200 kt of epoxidised soybean oil (ESBO), increases the oxidative stability of soybean oil [6]. Amberlyst-15 has been used as a catalyst to reduce the pour point of vegetable oils, and Archer Daniels Midland (ADM) Company and others have increased the oleic content by modification of oil through low-cost chemical or mechanical processing.

Chemical additives that improve stability offer the most rapid and cost-effective route to commercialisation. It is necessary to develop a better understanding of how additives affect the interactions between lubricants and rubbing surfaces. It is also of interest to gain better insight in the rate of degradation of oil additives. Additive technology and lubricant tribology are highly dynamic fields. Conventional lubricant additives for mineral oils are not always effective for polar base stocks such as esters and ethers [7]. This can be understood by the balance of solubility of an additive into a base oil and adsorptivity of an additive on a surface [8]. Generally, additives are easy to dissolve in polar solvents. Bio-based lubricants, which are more polar than mineral oil-based products, require reformulation of the conventional additive packages used for hydrocarbon substrates. Once developed, such products have successfully been introduced into the market (e.g. RLI's Stabilized™ HOBS), see Section 10.5.2. Roegiers *et al.* (E-ION, Brussels) have developed Elektrionized™ vegetable oils as lubricity additives for top-quality lubricants [9, 10]. Wax esters constitute a target lubricant base oil but can also function as additives to vegetable-base oils. Wax esters from genetically modified (GM) *Crambe* are expected by 2015–2020. Further advances still need to be made in the area of compatibility of additives and property enhancers (and biodegradable additives in particular) with more polar base fluids. For example, dispersant/inhibitor packages for polar vegetable oil base fluids are still lacking (see Section 10.5.1). Progress in development work is hampered by the expected small commercial product volumes of such additives.

The rapid development of new plant molecular biology tools during the last two decades makes it now possible to produce novel chemicals tailor-made for technical purposes from agricultural crops that can compete with products obtained from mineral sources. Most research on bio-based lubricants has aimed at improving their physical properties through bioengineering. A typical research product is the development of high-oleic vegetable oils (see Section 5.3.1). Non-transgenic modification has resulted in new varieties with superior oil traits using genes from wild and commercial soy varieties. New soy varieties have been developed with higher C18:1 contents and lower levels of saturates. Genetic modification of oilseed crops has been directed to the production of higher mono-unsaturated (C18:1) oils (see Section 10.5.2), branched and asymmetric vegetable oils and wax esters (see Sections 5.3 and 5.4.2). High-oleic base stocks (HOBS) are now found in many commercial biolubricant formulations (RLI product line, see Sections 5.3.1 and 10.5.2; Pennzoil, see Section 10.5.1; Cargill, see Section 5.3.1; and others). The generation of transgenic plants with high-oleic acid contents (in particular HOSNO) represents an important contribution to the biolubricant industry. Vegetable oils with high-oleic acid contents yield stable lubricants that oxidise much more slowly [11]. Du Pont/Pioneer have developed a genetically modified high-oleic soybean oil (HOSBO). While conventional soybeans contain only approximately 20% C18:1, (more expensive) bioengineered soybean seeds reach 85% of oleic acid. The derived oil is about 30 times more oxidatively stable than conventional soybean oil in hydraulic pump tests [12]. Biolubricant technology is transferable to other crop oils (soy, rapeseed/canola, sunflower, palm, etc.), native to other geographic areas. Pricing and supply issues have prevented high-oleic canola oil from gaining wide market acceptance in lubricants.

Branching is the object of the EPOBIO wax ester project based on genetic modification of crambe oil (see Section 5.4.2). Cargill, Inc. is actively pursueing plant oil modification to obtain asymmetric oil compositions (see Section 5.3). Biotechnology will be ever more important in this field.

As illustrated by the first German oil factory, now in its sixth generation (see Section 16.1), lubricant technology can look back at a long industrial tradition. In the meantime, recent advances in lubricant technology have brought to the forefront environmentally considerate products in many high-profile applications in environmentally sensitive areas. For instance, Cargill, Inc. has successfully developed and introduced on the market a family of environmentally acceptable bio-based high-performance polyol ester lubricants that exhibit excellent lubrication properties – low viscosity, high viscosity index, low volatility, high flash point and high oxidative stability – and are cost effective [13, 14].

Scientific focus on lubricants and lubrication technology is relatively recent. While synthetic chemists are needed to develop base oil compositions, devise chemical processes and suggest lubricant formulations, physical chemistry is to contribute with structure–property relationships and (bio) catalytic expertise is required to ensure high-yield and selective lube oil manufacturing processes at low cost. Tribological research taking place at materials science and engineering departments of academic institutions usually concentrates on materials science aspects and application engineering. However, without taking into account the interdisciplinarity of the field, development of optimal performing lubricants for specific applications is hardly possible. As illustrated in Chapter 10, joint research projects in collaboration with lubricant manufacturers are nowadays considered as the best guarantee to lay the basis for successful product development. Lubricants should always be considered in relation to their specific application and engineering criteria. There is no single answer to the question: What is the best (bio)lubricant technology? A product simply needs to meet the minimum requirements for the application, at the lowest possible cost. Excess performance is never being rewarded: good is good enough! Lubricating base oil technology is changing continuously and has changed also during the last decade. API Group III oils produced through hydroprocesses and gas-to-liquid (GTL) base oils are free from sulphur and nitrogen and have a very high viscosity index (see Section 3.3).

Lubricants have become a design element of future original equipment manufacturer (OEM) developments, rather than, as in the past, a less than perfect retrofitted component. Lubricants have to be kept on the newest technological level because of the continuous development of tribological systems (see also Sections 12.2.1, 12.12 and 15.1). Table 16.2 lists some current challenges, including the extended operating range of high-temperature ester-based aircraft gas turbine engines from –40 to 232 °C (previously 200 °C) and the higher temperatures and pressures of steam turbines. Today's industry systems are getting smaller but performance demands are not. In compact equipment reservoirs, which contain less oil, fluids circulate more rapidly with shorter cycle times and less allowance for cooling and settling of contaminants. Harsher environments can also result in rising operating temperatures, which contribute to rapid formation of varnish. The tougher operating environment leads in particular to more frequent problems in higher-performance systems. The need for higher stability lubricants is increasing. Smaller automotive engines and higher-pressure injection systems to increase performance require new additives for both fuel and oil. Higher-stability lubricants are also needed when longer drain intervals and decreased maintenance are desired. New cleaning technology (e.g. Lubrizol® 6786) extends equipment life and reduces downtime. Extending equipment life and reducing downtime are major needs of the end-user. Other continuously challenging projects in the field of tribology comprise the improvement of fuel economy of internal combustion engines (see

Table 16.2 Selection of advanced lubrication targets

- High-temperature aircraft engines
- High-performance steam turbine
- Increased internal combustion engine operating temperatures and pressures
- Development of low viscous base stocks (fuel economy)
- New engine types (e.g. automotive, two-cycle)
- New engineering materials
- Biodegradable engine oil formulations (base stocks *and* additives)

Section 12.2) and other equipment (see Section 12.7.1). Ongoing projects from the viewpoint of saving energy and resources aim at the development of low-viscosity base stocks. Ester oils can reduce fuel consumption by 5%, compared with conventional motor oils, as a consequence of their better lubrication properties at lower viscosity [15]. Low-viscosity lubricating fluids, which exhibit low fluid friction in hydrodynamic lubrication conditions, are usually unsatisfactory under boundary conditions and therefore require effective anti-wear additives and friction modifiers. Lubricant developers will also be challenged to meet the requirements of new engines or fuel cells using hydrogen as a prospected clean fuel, where there would be no soot, no sulphur and no carbon deposits. Also, the introduction of new engineering materials has increased the demand for tailor-made lubricants. Engine oil formulations are to change considerably in the near future with increased use of biodegradable lubricants mainly based on vegetable oils or genetically modified vegetable oils.

Modern lubricants are targeted to add value in the following areas:

- achieve sustainability objectives;
- reduce emissions;
- improve fuel economy;
- extend oil drain and mantainance intervals;
- increase production efficiency;
- improve equipment durability.

Performance requirements for lubricants are becoming ever more demanding. The substantial need for enhanced stability manifests itself in a number of ways: the desire for longer drain intervals for engine oils, the need to slow viscosity growth of lubricants, control of inlet valve and combustion chamber deposits, etc. There is a need to fully understand all of the structural elements of high-performance lubricants to best address the specific requirements for high stability in demanding end-uses.

16.4 Biolubricants: drivers and barriers

During the last 20 years the chemical industry has developed a large number of new lubricants made partly or totally from renewable raw materials. In the long run these will be indispensable as a replacement of petrochemical products. After some initial technical difficulties with lubricants made from simple rapeseed oil, today we have a range of chemically processed oils and greases for almost all technical purposes. These mature and environmentally friendly lubricants, which often even have technical advantages over conventional mineral oil-based products, have not yet been generally accepted because of a lack of consumer awareness and higher litre price. Bio-based lubricants are renewable products but for certain applications still lack some of the desirable physical properties that petro-based lubricants possess. More research is needed to improve these properties. Factors affecting the success of bio-based products are technological advancement, investment by farmers (availability of resources), demand for petroleum and political initiatives.

Performance and price, OEM acceptance and institutional support are key to encouraging consumers to use biolubricants and are among the main drivers and barriers to an increased market share of biolubricants (Table 16.3). Essential to growth of bioproducts is the acceptance and awareness by the public, business enterprises and governments that biomass is the only realistic alternative to global resource problems. Biolubricant development cannot do without OEMs guaranteeing their engines (warranties). Various national governments (e.g. Germany, France, Brazil) and supranational institutions (EU) are actively promoting and developing the use of renewables. Biolubricant markets have been stimulated by market introduction measures, such as the German Market Introduction Programme (MIP), Dutch VAMIL, EU Lead Market Initiative (LMI), and the US bio-based product preferred procurement scheme (BioPreferred^{SM}). The Dutch chemical industry aims at sourcing 50% of its raw materials from biomass by 2030; various other chemical companies elsewhere are following the same line (see Section 4.1). The EU is actively stimulating the emergence of sustainable biorefineries (FP7 call). Resource-efficient product substitution is a main (medium-to-long term) target.

Table 16.3 Biolubricants: drivers and limitations

Drivers	Limitations
• General acceptance (OEMs, end-users)	• Purchasing and changeover costs
• Need for innovation in chemical industry	• Performance
• Economic security and benefits	• Sustainable raw material supply
• High fossil oil price	• Feedstock quality
• Environmental benefits	• (Bio)catalytic conversion of biomass
• Technical development	• Investments (farmers, research and technology development)
• Advances in biotechnology	• Limited additive technology
• Legislation, regulations, mandates	• International product quality standards
• Ecolabels	• Testing protocols
• Cost efficiency	• Strong market position for mineral oils
• Market conversion programmes	• After-use
• Rural development	• Negative product quality perception

The need and eagerness to reduce the dependence on imported oil are important economic drivers for the development and use of locally available renewable resources coming from rapeseed (canola), soy, corn, sunflower, palm, starch and other biomaterials for lubricant production. In the future, economics are expected to lean more favourably to the direction of renewable feedstocks. The main factors for (economically) successful application of bioproducts are a high fossil oil price, low prices for agricultural raw materials (oilseeds, cereals and sugar) and a rapid (bio)technological development. Cost, technical performance and environmental factors are major driving forces for using plant oils and their derivatives as lubricants. The worldwide trend to the use of more environmentally friendly lubricants as replacements for fossil oil products is strongly motivated by strict legislation to reduce the environmental impact of lubricants. Environmental regulations do affect the quantity and/or quality of lubricants chosen by the end-user. In this respect (but not technologically), the almost non-regulated US market is lagging behind Europe. The impetus for bio-based product substitution requires favourable government policies. For example, biodiesel production is centred in regions where state or local governments have established mandates. Similarly, biogenous fluids are only slowly being introduced where not legally required. Manufacturing incentives are beneficial to improve efficiency, reduce waste and extend the service life of machines. Minimum industry standards in respect to energy savings and environmental impact are useful.

Governments, industry and pressure groups are seeking to accelerate a transition to more sustainable products and technologies by use of regulations (Registration, Evaluation, Authorisation and Restriction of Chemicals (REACH), Globally Harmonised System (GHS), Classification, Labelling and Packaging (CLP), etc.), economic incentives and by harnessing the power of consumer choice through ecolabelling schemes. Ecolabels have been developed for identifying lubricants with reduced environmental impact and high-performance criteria. The European ecolabel represents an important unification effort of national ecolabels developed in recent years. The US Department of Agriculture (USDA) is developing a 'USDA Certified Biobased Product' label, which is expected to provide a strong third party validation of bio-based products in the United States. Existing ecolabelling schemes with their focus on toxicity, ecotoxicity and biodegradability overlook the potential environmental benefits of optimised performance in use. This could lead to misleading conclusions. Lifecycle assessments (LCAs) provide more objective and quantitative information.

Technical development drivers include the availability of suitable base stocks (such as HOBS) and (bio)additives for lubricants formulation and the adaptation of technical standards, OEM specifications and ecolabels.

Important non-technical development drivers include environmental management tools. Environmental policy, and procurement guidelines for government organisations and city councils, clearly have a large impact.

Industrial use of vegetable crops favours rural development and generates new income for farmers and foresters. Development of a large bioindustry requires a reliable material supply, new processing, distribution and logistics, and new service industries. Consequently, the biolubricant industry faces several challenges. Compared to another bio-based product, biodiesel, the necessary oilcrop volumes required for biolubricants are rather small. However, biolubricant production volumes could be limited by the availability of high-oleic oils, which are presently being produced only in rather small volumes and mainly for use in the food industry [16]. This could give rise to undesirable competition. In the future, careful allocation of crops to chemicals manufacturing is needed, in particular in view of the limited availability of arable land. This requires consensus by government, agriculture and industry. In order to make the chemical industry more sustainable and competitive, access to renewable feedstocks should be improved.

Cost is among the various aspects of the bio-based lubricant economy that needs to be considered for it to be an effective alternative. Commercial users will choose bio-based lubricants as substitutes for petrolubricants only if they are cost effective. While the initial acquisition cost may deter potential lubricant buyers, this is frequently offset by reduced energy costs resulting from the high lubricity of bio-based lubricants. The high lubricity makes bio-based lubricants attractive alternatives to use in various high-temperature applications such as injection moulding equipment or heated presses [17]. Indicative energy savings have been indicated as follows: hydraulics, 1–3%; plastics injection moulding equipment, 2–3%; compressors, 2–7%; worm gears, 15–30%; spur gears, 1–5%. The increased cost of the lubricant is negligible when compared with the savings in energy and maintenance costs. Consequently, higher purchasing prices of plant-oil based lubricants as compared to mineral oils do not seem to be a deterrent of market expansion as these are frequently being compensated by overall cost benefits. More important is the fact that there are no political, physical and social infrastructures that favour an economy that prefers bio-based lubricants.

From a technological point of view biolubricant formulation is challenged by the need for development of more highly performing (bio)catalysts for renewables [18] and the reformulation of (renewable) additive packages differing from those applicable for hydrocarbon base stocks. These needs derive from the different nature of the raw materials: wet, polar biomass vs hydrophobic, apolar petroleum fractions.

More tailor-made analytical methods for biolubricants should be developed to allow a better understanding of the modifications occurring during use and to identify the best strategies for their performance and condition monitoring. While certified reference materials (CRMs) are not readily available for the enormous variety of lubricants, there is an almost complete lack of CRMs for biolubricants for use in quality assurance programmes. Also, the literature is rather scarce in relation to the use of spectroscopic methods coupled with multivariate analysis as an analytical tool to evaluate biolubricant characteristics. In particular, quantitative at-line/on-line process control using near-infrared (NIR) spectroscopy (e.g. for quantitative analysis of blend oil components) is still underdeveloped.

Continued collaboration between lubricant manufacturers, OEMs, end-users and testing institutions is required to assure performance standards and testing protocols. Lack of product quality standards for biolubricants is a factor limiting the demand for bio-based products. Standards are essential elements for triggering initial demand. In particular, currently existing testing protocols for lubricants (including ASTM standards) were made to fit petroleum-based oils, are not equally well suited for the chemically quite different plant-based biolubricants, and put vegetable oil-based lubricants at a disadvantage. Uniform standards for environmental attributes such as biodegradability, toxicity and bioaccumulation must be clearly established, and incorporated in regulatory action. There exists no meaningful standardised oil ageing protocol in order to generate reproducible and representative samples of aged oil (e.g. for the evaluation of ecotoxicity and biodegradability). The after-use of bio-based lubricants is also more of a problem than for mineral oils: mineral oil re-refining installations are reluctant to accept biolubricants. Recycling of used lubricants is debatable. It has been argued that mandatory enforcement of minimum recycled components into new lubricant products would lead to sub-optimal lubrication for given applications and compromises the development of biodegradable lubricants for use in environmentally sensitive areas [3]. The better solution is reduction of the amount of waste oil at the source by promoting the use of high-performance lubricants having improved stability and extended drain intervals, followed by reuse for fuel value by combustion. Governmental policy should reward the range of benefits offered by plant-based lubricants.

The negative preconceptions associated with early and less developed technologies have gradually been surmounted as the industry has proved that plant-based lubricants meet the required performance. The technical and political barriers that once slowed the use of bio-lubes in the United States and Europe are steadily being overcome.

16.5 Current and future potential for biolubricants

The current perspectives for use of biodegradable lubricants and greases in a wide variety of applications with very different technical requirements, environmental, and health and safety concerns have been illustrated (see Chapter 12). As economic health begins to align with environmental health, the lubricants and greases market is expected to witness increasingly a focus on reducing the environmental footprint of lubricant consumption. Technology development in the industry in the medium to long term will be driven by the concept of sustainability to minimise pollution and maximise resource utilisation. Commodity lubricants will be replaced by new generation lubricants. R&D activities of lubricant producers are focused on improving product properties to meet the increased requirements of new applications. Vegetable oils, especially enhanced varieties, show the ability to compete in significant segments of the lubricant market and exceed the technical performance of mineral oils, not only for less demanding applications. The majority of companies manufacturing plant-based lubricants have focused primarily on industrial oils. Within this category, certain applications are better suited for plant-based lubricants. Biolubricants are especially important in agricultural, forestry and offshore applications in view of their direct contact with the environment (crops, soil, air, surface water) and contribute to creating a green market image. These applications include, but are not limited to, mobile hydraulic fluids and total-loss lubricants (two-stroke engine oils, chainsaw bar and chain oil, railroad flange oils and greases, drip oils, wire rope lubricants, dust suppressant and marine lubricants). Because the latter types of lubricants pose immediate harm to the surrounding environment, there is increasing regulatory pressure to seek biodegradable and less toxic alternatives. Many European countries require biodegradable lubricants to be used in such selected applications.

It is distressing to realise that yearly millions of dollars are wasted as a result of: (i) premature equipment failure caused by poor lubrication practices; (ii) energy loss due to friction and wear; (iii) lost productivity due to equipment downtime; (iv) loss due to environmental issues; and (v) misspent funds purchasing the cheapest *vs.* the best value lubricants. Industry should improve its ways. While the overall lubricants market is predicted to remain fairly steady (expected annual consumption of 40.5 Mt in 2012 [19]) a growing market for biolubricants and bio-oils is forecasted, owing to their environmental and technical advantages. From a technical standpoint more than 90% of all lubricant products could be made from biogenous materials, according to estimates of the lubricant industry (FUCHS, 2009). At present, biolubricants constitute a slowly growing, small segment of the lubricants business (about 2%). Frost & Sullivan's forecast that the biodegradable lubricants market will increase by 6.8% in 2000 has not come true, even if we consider that both plant-based and synthetic lubricants have the potential to capture this market. According to another previous estimate (2002) of the potential of biomass to replace petrochemical-based

Table 16.4 Estimated EU potential of major biomass-based products

Market sector	Total consumption market (1998) (kt)	Renewable consumption market (1998) (kt)	Potential of renewables in 2010 (kt)	Potential share in 2010 (%)
Polymers	33000	25	500	1.5
Lubricants	4240	100	200	5
Solvents	4000	60	235	12.5
Surfactants	2260	1180	1450	52

After ref. [20].

products in the areas of polymers, lubricants, solvents and surfactants (Table 16.4) also the potential share of 5% for biolubricants in 2010 has not been realised. Some of the other applications, especially solvents and surfactants, can be achieved by direct extraction of the components from biomass without additional (bio)conversion steps.

Although in principle biolubricants can share a global market in relation to mineral-based products for many applications, their market share is still disappointingly low. The main reason for this modest market share (3% in Germany 2008/2009) is that, although many biogenous lubricants have been developed and are ready for market introduction, purchase costs of these products are still higher than those of established, mineral oil-based products (due to higher raw material costs of both base oils and reformulated additives). The lubricant performance–cost relationship is complicated and depends on many variables (such as energy efficiency, temperature, wear) which are difficult to quantify in an easy way. However, taking properly into account various costs benefits associated with vegetable oils (lower product use per application, lower penalties in case of spills, reduced parts wear, maintenance costs and disposal fees), then plant-based lubricants are cost competitive with mineral-based and synthetic lubricants. Increased expenditures on better lubricants pay off [21]. In this respect, it is useful to perform life-cycle costing (LCC) exercises. Moreover, biolubricants can offer comparative and sometimes superior qualities which enhance the effectiveness of machinery. The technical advantages of biolubricants allow users to save considerably in operating and maintenance costs over the entire usage period of the product, thus outperforming conventional, mineral oil-based lubricants. As chemical products derived from fossil raw material will inevitably increase in price as depletion of these resources progresses, bio-based products will become even more competitive within the next two or three decades at the latest.

At present, commercial biodegradable lubricants are mainly confined to synthetic esters and vegetable-oil based products. For further advancements, it is imperative that carbohydrates and other biomass components are systematically exploited in efficient, environmentally benign and economical processes for their large-scale conversion into industrially viable products. While conceptual formulations towards this goal are readily available [22–25], practical implementation is currently (too) slow. The main reason is lack of sufficient support from funding institutions and the chemical industry. At present, the funding for research is not as high as it once was (see Chapter 10.5). Incentives are required to companies willing to innovate the production of new lubricants in order to meet the present and future challenges.

Introduction of a new product into a market dominated by big petroleum companies is a daunting task. Successful market introduction of biolubricants still needs to overcome several considerable hurdles, notably:

- perceived cost competiveness with mineral oils;
- lack of economy of scale (for cost optimisation);
- increased consumer awareness of the technical capabilities of biolubricants;
- limited regulation; and
- tight raw materials market and scattered marketing activities.

In order to ensure success, plant-based lubricant manufacturers must address a number of existing concerns, including (disappearing) negative industry preconceptions about plant-based lubricants, unresolved questions about performance standards and testing protocols and the direction of regulatory actions. Before converting vegetable-based lubricants into an attractive alternative to mineral oils, industry needs robust governmental policy, price stability, detailed technical specifications (in particular regarding degradation over time) and appropriate instruments for waste management. Adequate end-of-life scenarios should be established. Another problem is lack of acceptance by potential users, owing to inadequate information and practical experience with these products. Biolube product services lack far behind those for the long-established petro-based products (see Section 13.1). For a good future development, political hurdles must also be overcome. There is a need to further convince policymakers of the benefits of plant-based lubricants and consumer expectations. This situation has not changed much since 2003 [4].

Vegetable oils are envisioned of becoming a routine component in lubricants and green technologies [22]. Vegetable oils are competitive in particular in areas where regulations are in place requiring expensive clean-up and disposal (e.g. drilling fluids). The market for natural esters is increasing slowly but strong efforts are required from the ester base oil, additive and lubricant industries to develop improved products which will be more generally accepted by end-users, toxicologists and legislators. This requires in particular self-regulation of lubricant suppliers and increased transparency. Environmental methodologies and quality control methods should be standardised. An increase in the number of users is needed to achieve the critical mass necessary for market acceptance of biolubricants despite an almost stagnant overall lubricant market demand. Ecolabels, national incentives and mandates for specific applications in protected areas (e.g. waterways) are useful instruments for an increased market share. A partnership between the public and private sectors is required, similar to the present accelerated development of electrical cars.

Many outstanding contributions in lubrication science and technology have been made to improve the performance of bio-based lubricants. In the future, the tribological issues industry is facing will be more complex than ever. Stringent legislation is calling for better economy and reduced emissions. In addition, in qualifying applications such as automotive, design and legislative pressures for cleaner, more efficient engines with higher specific power outputs are forcing tribological engine components to be operated with generally thinner oil films. This has direct effects on wear resistance and durability of such tribological systems. Development of light-weight, non-ferrous (Al, Mg) materials for engine and drivetrain materials as replacements of the current heavy-weight cast iron blocks requires extensive automotive lubricant development. Biolubricants are works in progress.

16.6 Sources of further information and advice

R.B. Laughlin, *Powering the Future*, Basic Books, New York, NY (2011).

Global Industry Analysts, Inc., *Lubricating Oils and Greases – A Global Strategic Business Report*, San José, CA (10 February 2011).

S.P. Srivastava, *Advances in Lubricant Additives and Tribology*, CRC Press, Boca Raton, FL (2010).

16.7 References

[1] C. Tsouris and D. Aaron, Do we really need carbon capture and storage?, *Chem. World* (September 2010), 40.

[2] J. Gluyas and S. Daniels, Do we really need carbon capture and storage?, *Chem. World* (September 2010), 41.

[3] S. Boyde, Green lubricants. Environmental benefits and impacts of lubrication, *Green Chem.* **4** (2002) 293–307.

[4] T. Norrby, Environmentally adapted lubricants – where are the opportunities?, *Ind. Lubr. Tribol.* **55**(6) (2003) 268–74.

[5] P.S. Lathi and B. Mattiasson, Green approach for the preparation of biodegradable lubricant base stock from epoxidised vegetable oil, *Appl. Catal. B: Environm.* **69** (2007) 207–12.
[6] S.C. Maurer and R.D. Schmid, Biocatalysts for the epoxidation and hydroxylation of fatty acids and fatty alcohols, in *Handbook of Industrial Biocatalysis* (C.T. Hou, ed.), CRC Press, Boca Raton, FL (2005) pp. 4/1–4/25; http://elib.uni-stuttgart.de/opus/volltexte/2006/2655/pdf/Maurer_2005b.pdf
[7] I. Minami, S. Kikuta and H. Okabe, Anti-wear and friction reducing additives composed of *ortho*-phenylene phosphate-amine salts for polyether type base stocks, *Tribol. Intl.* **31**(6) (1998) 305–12.
[8] I. Minami and S. Mori, Anti-wear additives for ester oils, *J. Synth. Lubr.* **22**(2) (2005) 105–21.
[9] M. Roegiers, H.L. Zhang and B. Zhmud, Elektrionized vegetable oils as lubricity components in metalworking lubricants, *Proc. 6th Intl. Conference on Tribology, Balkantrib'08*, Sozopol (12–14 June 2008), Paper BT-97; *FME Transactions* **36** (2008) 133–8.
[10] M. Roegiers and B. Zhmud, Tribological performance of ionised vegetable oils as lubricity and fatty oiliness additives in lubricants and fuels, *Lubr. Sci.* **21**(5) (2009) 169–82.
[11] W. Castro, J.M. Perez, S.Z. Erhan and F. Caputo, A study of the oxidation and wear properties of vegetable oils: soybean oil without additives, *J. Am. Oil Chem. Soc.* **83**(1) (2006) 47–52.
[12] L.A.T. Honary, Biodegradable/biobased lubricants and greases, *Mach. Lubr. Mag.* **9** (2001) 240.
[13] D.R. Kodali and S.C. Nivens (to Cargill, Inc.), Transesterified oils, US Patent No. 6,278,006 B1 (21 August 2001).
[14] D.R. Kodali, High performance ester lubricants from natural oils, *Ind. Lubr. Tribol.* **54**(4) (2002) 165–70.
[15] R. Luther, Bioöle aus Sicht eines Experten – Stand der Wissenschaft und Blick in die Zukunft, *Presented at the FNR Seminar Biolubricants*, Ulm (2 February 2004).
[16] D. Johnson, Future crops for bio-diesel and bio-lubricants, *Proc. Harvesting Clean Energy Conf. V*, Great Falls, MT (20–21 January 2005); http://www.harvestcleanenergy.org/conference/HCE5/HCE5_PPTs/Johnson.pdf
[17] L. Cliff, Energy savings through use of advanced biodegradable lubricants, *Ind. Lubr. Tribol.* **59**(3) (2007) 132–6.
[18] G. Centi and R.A. van Santen (eds), *Catalysis for Renewables*, Wiley-VCH, Weinheim (2007).
[19] TransWorldNews, World lubricant demand to reach 40.5 million metric tons in 2012, *TransWorldNews* (2009); http://www.transworldnews.com/NewsStory.aspx?id=81121&cat=1
[20] J. Ehrenberg, *Current Situation and Future Prospects of EU Industry Using Renewable Raw Materials*, European Renewable Resources and Materials Association (ERRMA), Brussels (2002).
[21] L.A. Bronshteyn and J.H. Kreiner, Energy efficiency of industrial oils, *Tribol. Lubr. Tribol.* **67**(10) (2011) 48–56.
[22] Committee on Biobased Industrial Products, *Bio-based Industrial Products: Priorities for Research and Commercialization* (National Research Council

USA, ed.), National Academies Press, Washington, DC (2000); http://www.nap.edu/openbook.php?isbn=0309053927
[23] Biomass Research & Development Technical Advisory Committee, *Roadmap for Biomass Technologies in the United States*, Washington, DC (December 2002), http://www1.eere.energy.gov/biomass/publications.html
[24] Biomass Research & Development Technical Advisory Committee, *Vision for Bioenergy & Biobased Products in the United States*, Washington, DC (October 2002); http://www.bioproducts-bioenergy.gov/pdfs/BioVision_03_web.pdf
[25] M. Eissen, J.O. Metzger, E. Schmidt and U. Schneidewind, Concepts on the contribution of chemistry to a sustainable development, *Angew. Chem.* **114** (2002) 402–24; *Angew. Chem. Intl. Ed.* **41** (2002) 414–36.

Appendix

A.1 (Bio)chemical and physical nomenclature

AAS	Atomic absorption spectrometry
AB	Alkyl benzene
ACP	Acyl carrier protein
AD	Ashless dispersant
ADI	Animal-derived ingredient
AES	Atomic emission spectrometry
AF	Antifoaming
A/F	Analytical ferrography
AL	Anti-leak
AN	(1) Acid number; (2) alkylated naphthalene
AO	Antioxidant
AOM	Active oxygen method
APT	Attached proton test (NMR)
ASP	Aspartic acid
ATF	Automatic transmission fluid
ATO	Aircraft turbine oil
ATP	Adenosine triphosphate
AV	Acid value
AW	Anti-wear
BCFA	Branched-chain fatty acid
BCI	Bio corrosion inhibitor
BD	Biodiesel
BDHAS	Butyl-10,12-dihydroxy-9-acyloxystearate
BDHBS	Butyl-10,12-dihydroxy-9-behenoxystearate
BDHOS	Butyl-10,12-dihydroxy-9-octyloxystearate
BDO	1,4-Butanediol
BETI	Bis(pentafluoroethylsulphonyl)imide (anion)
BMIM	1-Butyl-3-methylimidazolium
BMIMCl	1-Butyl-3-methylimidazolium chloride
BN	Base number
BO	Butylene oxide
BOCLE	Ball-on-cylinder lubricity evaluator

BR	Butadiene rubber, polybutadiene
BRI	Bio rust inhibitor
Bt	*Bacillus thuringiensis* (toxin)
BTA	Benzotriazole
BTL	Biomass-to-liquid
BTX	Benzene, toluene, xylene
CAS	Castor bean oil
CCS	(1) Cold cranking simulator; (2) carbon capture and storage
CD	Conjugated diene
CFC	Chlorofluorocarbon
CI	Corrosion inhibitor
CID	Charged-injection device
CIT	5-Chloro-2-methyl-4-isothiazolin-3-one
CJO	Crude jojoba oil
CL	Chemiluminescence
CMFO	Crude meadowfoam oil
CMR	Carcinogenic, mutagenic or toxic to reproduction
CMSO	Chemical modified soybean oil
CNO	Coconut oil
CO	Canola oil
CoA	Coenzyme A
COD	Cyclooctadiene
CP	Chronopotentiometry
CPD	Cyclopentadienyl
CPR	Catalysed pressure reduction
CRM	Certified reference material
CRO	Corn oil
CSO	(1) Cottonseed oil; (2) chainsaw oil
CT	Conjugated triene
DA	Diels–Alder
DAB	Diaminobutane
DAG	Diacylglycerol
DBDS	Dibenzyldisulphide
DBPC	2,6-Di-*t*-butyl-*p*-cresol
DCPD	Dicyclopentadiene
DD	Detergent and dispersant
DDMIM	1-Dodecyl-3-methylimidazolium
DDT	Dichlorodiphenyltrichloroethane
DE	Diester
DEPT	Distortionless enhancement of polarisation transfer (NMR)
DGAT	Diacylglycerolacyltransferase

DHASA	10,12-Dihydroxy-9-acyloxystearic acid
DHBSA	10,12-Dihydroxy-9-behenoxystearic acid
DHOSA	10,12-Dihydroxy-9-octyloxystearic acid
DI	Dispersant/inhibitor
DIDA	Diisodecyl adipate
DJO	Deodorised jojoba oil
DMC	Double-metal cyanide
DMSO	Dimethylsulphoxide
DOA	Di-2-ethylhexyl adipate
DOS	Di-2-ethylhexyl sebacate
DPE	Dipentaerythritol
DRF	Direct read ferrography
DSC	Differential scanning calorimetry
DTA	Differential thermal analysis
DTDA	Di-tridecyladipate
E2EHS	Epoxidised 2-ethylhexyl soyate
EA	Elemental analysis
EAL	Environmentally adapted lubricant
EBS	Ethylene bisstearamide
ECD	Equivalent circulating density
EDAX®, EDX	Energy dispersive X-ray spectrometry
EDXRF	Energy dispersive X-ray fluorescence
EEMMEA	Diethylmethyl-(2-methoxyethyl)amine
EH	Ethylhexyl
EHC	Electrohydraulic control
EHD	Elastohydrodynamic
EHL	Elastohydrodynamic lubrication
EHS	Di(2-ethylhexyl)sebacate
EMIM	1-Ethyl-3-methylimidazolium
EMLEN	Epoxidised methyl linolenate
EMLO	Epoxidised methyl linoleate
EMO	Epoxidised methyl oleate
EN	Estolide number
EO	Ethylene oxide
EOE	1,3-Dierucoyl 2-oleoylglycerol
EOFT	Engine oil filterability test
EOWTT	Engine oil filterability/water tolerance test
EP	Extreme pressure
EPDM	Ethylene–propylene–diene rubber
EPR	Electron paramagnetic resonance
ERA	Epoxidised ricinoleic acid
ERSO	Epoxidised rapeseed oil
ESBO, ESO	Epoxidised soybean oil

ESC	Environmental stress cracking
ETA	Esterified trimer acid
ETBE	Ethyl *tert*-butyl ether
ETEA	Ethoxylated triethanolamine
EVA	Ethylene–vinylacetate copolymer
FA	Fatty acid
FAAE	Fatty acid alkyl ester
FAME	Fatty acid methyl ester
FAP	Tris(pentafluoroethyl)trifluorophosphate
FAR	Fatty acyl-CoA reductase
FAS	Fatty acid synthase
FC	Fluorocarbon
FCC	Fluid catalytic cracking
FDA	Filter debris analysis
FDCA	2,5-Furandicarboxylic acid
FE	Fuel economy
FFA	Free fatty acid
FM	Friction modifier
FOF	Finish-on-fibre
FOG	Fats, oils and grease
FPT	Filter patch test
FR	(1) Flame retardant; (2) fire resistant
FT	Fischer-Tropsch
FTIR	Fourier transform infrared
FTNMR	Fourier transform nuclear magnetic resonance
FU	Functional unit
GBSS	Granule-bound starch synthase
GEL	Gas-expanded lubricant
GHG	Greenhouse gas
GL	Glycerol
GLU	Glutamic acid
GMS	Glycerol monostearate
GTL	Gas-to-liquid
3-HBL	3-Hydroxybutyrolactone
HC	(1) Hydrocracked oil; (2) hydrocarbon
HCE	Hydrocarbon ester blend
HCFC	Hydrochlorofluorocarbon
HCO	(1) Hydrogenated castor oil; (2) heavy cycle oil
HD	Hydrodynamic
HDMO	Heavy-duty motor oil
HDT	Hydrotreatment
HEAR	High-erucic acid rapeseed
HEES	Hydraulic oil environmental ester synthetic

	(water-insoluble)
HEPG	Hydraulic oil environmental polyalkylene glycol (water-soluble)
HEPR	Hydraulic oil environmental poly-α-olefin/related hydrocarbon (water-insoluble)
HERO	High-erucic acid rapeseed oil
HETG	Hydraulic oil environmental triglyceride (water-insoluble)
HF	(1) Hydrofinishing; (2) hydraulic fluid
HFA	Hydroxy fatty acid
HFB	Water-in-oil emulsion (mineral oil-based)
HFC	Hydrofluorocarbon
HFCS	High-fructose corn syrup
HFDU	Water-free, fire-resistant hydraulic fluid based on carboxylic acid ester or polyol ester
HFO	Heavy fuel oil
HFRR	High-frequency reciprocating rig
HIDW	Hydroisodewaxing
HISM	Hydroxy isobutoxy stearic methyl ester
HLB	Hydrophilic–lipophilic balance
HLSFO	High-linoleic safflower oil
HM	Heavy metal
H&M	Homogeneity and miscibility (test)
HMF	5-Hydroxymethylfurfural
HMIM	1-Hexyl-3-methylimidazolium
HMW	High molecular weight
HNBR	Hydrogenated nitrile butadiene rubber
HO	High oleic
HOBS	High-oleic base stock
HOCAS	High-oleic castor oil
HOCSO	High-oleic cottonseed oil
HOLL	High oleic, low linolenic
HOPMOTE	High-oleic palm oil trimethylolpropane ester
HOPNO	High-oleic peanut oil
HORSO	High-oleic rapeseed oil
HOSBO	High-oleic soybean oil
HOSFO	High-oleic safflower oil
HOSME	High-oleic sunflower methyl ester
HOSNO	High-oleic sunflower oil
HP	High performance
HPA	Hydroxypropionic acid
HPDSC	High-pressure differential scanning calorimetry
HPKO	Hydrogenated palm kernel oil

HPV	High production volume (chemicals)
HS	High shear
HSGC, HS-GC	Headspace gas chromatography
HS-GC-MS	Headspace gas chromatography mass spectrometry
HTHS	High temperature high shear
HTHT	Hexahydro-1,3,5-tris(2-hydroxyethyl)-*S*-triazine
HV	Hydroxyl value
HVI	High viscosity index
HVO	Hydrotreated vegetable oil
ICP-OES	Inductively coupled plasma-optical emission spectrometry
IL	Ionic liquid
IS	Internal standard
IV	Iodine value
JCO	*Jatropha curcas* oil
JME	Jatropha methyl ester
JO	Jojoba oil
KF	Karl Fischer (coulometry)
LAO	Linear alpha olefin
LBC	Load-bearing capacity
LCF	Lignocellulosic feedstock
LCO	Light cycle oil
LDA	Lithium diisopropylamide
LDPE	Low-density polyethylene
LEAR	Low erucic-acid rapeseed oil
LL	Low linolenic
LMW	Low molecular weight
LOX	Lipoxygenase
LR-NMR	Low-resolution nuclear magnetic resonance
LS	Low shear
LSD	Low-sulphur diesel
LSV	Linear-sweep voltammetry
LVFA	Low-velocity friction apparatus
MAC	Multiply-alkylated cyclopentanes
MC	Methylcellulose
MCD	Magnetic chip detector
MCFA	Medium-chain fatty acid
MCT	Medium-chain triglyceride
MDTC	Metal dialkyldithiocarbamate
ME	Monoester
MEMS	Microelectromechanic system
MFO	(1) Meadowfoam oil; (2) multifunctional oil; (3) marine fuel oil

MgSt	Magnesium stearate
MgSt-D	Magnesium stearate dihydrate
MIT	2-Methyl-3(2H)-isothiazolin-3-one
MLEN	Methyl linolenate
MLO	Methyl linoleate
MO	(1) Mineral oil; (2) methyl oleate
MoDTC	Molybdenum dialkyldithiocarbamate
MoDTP	Molybdenum dithiophosphate
MOSBO	Mid-oleic soybean oil
mPAO	Metallocene poly-α-olefin
MPC	Membrane patch colorimetry
MPS	Monolithic piezoelectric sensor
MQL	Minimal quantity lubrication
MRV	Mini-rotary viscometer
MS	Mass spectrometry
MSO	Maleated soybean oil
MSW	Municipal solid waste
MUFA	Monounsaturated fatty acid
MW	Molecular weight
MWF	Metalworking fluid
NBR	Acrylonitrile–butadiene rubber, nitrile rubber
NEMS	Nanoelectromechanic system
NIR(S)	Near-infrared (spectroscopy)
NMR	Nuclear magnetic resonance
NPG	Neopentyl glycol
NPI	Non-polarity index
NR	Natural rubber, polyisoprene
OA	Oil analysis
OBS	Over-based detergent (Ca/Mg sulphonate or phenate)
OCP	Olefin copolymer type (VII)
OIT	Oxidation induction time
OMIM	1-Octyl-3-methylimidazolium
OO	Olive oil
OSA	On-site analysis
OSI	(1) Oxidation stability index; (2) oil stability index
OSP	Oil-soluble polyalkylene glycol
OTP	Organo thiophosphate
O/W	Oil-in-water (emulsion)
PAD	Programmable array detector
PAG	Polyalkylene glycol
PAH	Polycyclic aromatic hydrocarbon
PAO	Poly-α-olefin

PBS	Polybutylsuccinate
PBT	Persistent, bioaccumulative and toxic substance
PC	Polycarbonate
PCA	Polycyclic aromatic
PCB	Perchlorinated biphenyl
PCMO	Passenger car motor oil
PCT	Perchlorinated terphenyl
PDA	Propane deasphaltation
PDMS	Polydimethylsiloxane
PDO	Propanediol
PDSC	Pressure differential scanning calorimetry
PE	Pentaerythritol
PEG	Polyethylene glycol
PET	Poly(ethylene terephtalate)
PFC	Perfluorocarbon
PFGS	Purified food-grade corn starch
PFPE	Perfluoropolyether
PGDO	Propylene glycol dioleate
PGSE	Pulsed-gradient spin-echo (NMR)
PHSME	Partially hydrogenated soybean oil methyl ester
PI	Polarity index
PIB	Polyisobutylene
PIBS	Polyisobutylene succinimide
PIO	Poly(internal olefin)
PJO	Cold pressed jojoba oil
PKO	Palm kernel oil
PKOME	Palm kernel oil methyl ester
PKOTE	Palm kernel oil trimethylolpropane ester
PLA	Polylactic acid
PM	Particulate matter
PMA	Poly(methacrylate)
PME, PMME	Palm oil methyl ester
PMO	Palm oil
PMOTE	Palm oil trimethylolpropane ester
PNA	Polynuclear aromatics
PNO	Peanut oil
PO	Propylene oxide
POE	Polyol ester
POo	Palm olein
POP	Persistent organic pollutant
PPD	Pour point depressant
PPG	Polypropylene glycol
PQ	Particle quantifier

PTFE	Poly(tetrafluoro-ethylene)
PTP	Performance testing programme
p-TSA	*p*-Toluenesulphonic acid
PUFA	Polyunsaturated fatty acid
PV	Peroxide value
PVA	Poly(vinyl alcohol)
PVC	Poly(vinyl chloride)
PVE	Polyvinyl ether
QCL-IR	Quantum cascade laser infrared
RA	Ricinoleic acid
RBD	Refined, bleached and deodorised
RBDPKO	Refined, bleached, deodorised palm kernel oil
RBDPO	Refined, bleached, deodorised palm oil
RBDPOo	Refined, bleached, deodorised palm olein
RBOT	Rotating bomb oxidation test (cf. RPVOT)
RBX	Rice bran wax
RCAS	Refined castor oil
RHC	Raffinate hydroconversion
RJO	Refined jojoba oil
RME, RSME	Rapeseed oil methyl ester
RMFO	Refined meadowfoam oil
RNT	Radionuclide technique
R&O	Rust and oxidation
rPVC	Rigid polyvinyl chloride
RPVOT	Rotary pressure vessel oxidation test (formerly RBOT)
RR	Renewable resource
RRM	Renewable raw material
RSBO	Refined soybean oil
RSME, RME	Rapeseed oil methyl ester
RSO	Rapeseed oil
RTIL	Room-temperature ionic liquid
SAPL	Surface-active phospholipid
SAPS	Sulphated ash, phosphorus and sulphur
SBE	Starch branching enzyme
SBME, SME	Soybean oil methyl ester
SBO	Soybean oil
SBS	Styrene hydrogenated butadiene–styrene terpolymer
SE	Synthetic ester
SEC	Size-exclusion chromatography
SEE	Self-emulsifiable ester
SEM	Scanning electron microscopy
SEOF	Self-emulsifying oily formulation

SFA	Saturated fatty acid
SFE	Supercritical fluid extraction
SFO	Safflower oil
SJO	Sulphurised jojoba oil
SL	Superlubricity
SME, SBME	Soybean oil methyl ester
SMS	Sorbitan monostearate
SN	(1) Saponification number; (2) solvent neutral
SNO	Sunflower oil
SOP	Standard operating procedure
SOS	Scheduled oil sampling
SRMO	Super-refined mineral oil
SSO	Sulphurised sperm oil
SSS	Soybean starch synthase
STOU	Super Tractor Oil Universal
SVHC	Substance of very high concern
SVO	Straight vegetable oil
TAG	Triacylglycerol
TAN	Total acid number
TBA	Thiobarbituric acid
TBARS	2-Thiobarbituric acid reactive substances
TBN	Total base number
TCP	Tricresylphosphate
TDA	Thermal deasphalting
TE	Triester
TEA	Triethanolamine
TEOST	Thermo-oxidation engine oil simulation test
Tf	Trifluoromethanesulphonyl
TFE	Thin-film evaporation
TFMO	Thin-film micro-oxidation
Tf_2N	Bis(trifluoromethanesulphonyl)amide
TfO	Triflate
TFSA	Bis(trifluoromethanesulphonyl)amide (cf. Tf_2N^-)
TG	Triglyceride
TGA	Thermogravimetric analysis
TLC	Thin-layer chromatography
TLL	Total-loss lubricant/lubrication
TLW	Tallow
TME	Trimethylolethane
TMP	Trimethylolpropane
TMPO	Trimethylolpropane oleate
TMPTH	Trimethylolpropane triheptanoate
ToF-SIMS	Time-of-flight secondary ion mass spectrometry

TOG	Total oil and grease
TON	Turnover number
TOST	Turbine oxidation stability test
TPH	Total petroleum hydrocarbons
TPP	Tetraphenylporphyrin
TPPT	Triphenylphosphorothionate
TSM	Thickness shear mode (resonator)
UC	Ultra-centrifuge
UCBO	Unconventional base oil
UCG	Underground coal gasification
UCO	Used cooking oil
UFO	Used frying oil
UHMWPE	Ultra-high-molecular-weight polyethylene
ULSD	Ultra-low sulphur diesel
UTTO	Universal Tractor Transmission Oil
VG	Viscosity grade
VHVI	Very-high-viscosity index
VI	Viscosity index
VII	Viscosity index improver
VLCFA	Very-long-chain fatty acid
VO	Vegetable oil
VOC	Volatile organic compound
vPvB	Very persistent and very bioaccumulative substance
WDA	Wear debris analysis
WDT	Wax disappearance temperature
WDXRF	Wavelength dispersive X-ray fluorescence
WO	Waste oil
WS	Wax ester synthase
XHVI	Extra-high-viscosity index
XPS	X-ray photoelectron spectroscopy
XRD	X-ray diffraction
XRF	X-ray fluorescence
ZAF	Zinc- and ash-free
ZDDP, ZDTP	Zinc dialkyldithiophosphate
ZDTC	Zinc dialkyldithiocarbamate

A.2 Physicochemical and mathematical symbols

ac	Acre
a.c.	Alternating current
Bbbl	Billion barrels
bbl	Barrel; 1 bbl = 42 gal
Bgy	Billion gallons per year

bn	Billion
b.p.	Boiling point
Bpd	Barrels per day
Bt	Billion tons
BTU, Btu	British thermal unit; 1 Btu = 0.252 kcal
Bty	Billion tons per year
bu	Bushel
°C	Degrees centigrade
C, c	Concentration or molar concentration
CN	Cetane number
CoF	Coefficient of friction
CP	Cloud point
cP	centiPoise (dynamic viscosity)
cSt	centiStokes (mm^2/s) (kinematic viscosity)
d	Day
d	(1) Diameter, thickness; (2) density
Da	Dalton or atomic mass unit
E	Energy (in eV)
EC	Effect concentration
EC50	Median effect concentration
EJ	Exajoule (10^{12} joule)
FIP	Fire point
FP	Flash point
FT	Fourier transform
ft	Foot
G	Free enthalpy (Gibbs free energy)
g	Gram
GI	Gelation index
gpy	Gallons per year
GtC	Gigatonnes (10^{15} g) carbon
H	Enthalpy
h	Hour
hp	Horse power
HV	High voltage
in.	Inch
J	Joule, a unit of energy
K	Kelvin
k	Kilo (10^3)
kg	Kilogram
kgd	Kilogallons per day
kgoe	Kilogram oil equivalent
kgy	Kilogallons per year
kPa	kiloPascal

KV	Kinematic viscosity
L	Litre
L, *l*	Length (column length; chain length)
lb	Pound; 0.453 kg
LC50	Lethal concentration
LLD	Lower limit of detection
M	(1) Molarity (mol/L); (2) mega (10^6)
M	Atomic or molecular weight
M_n	Number average molecular weight
m	(1) Milli; (2) metre
Mbd	Million barrels per day
Mbu	Million bushels
MCA	Monte Carlo analysis
mg, mmol, mL	Milligram, millimole, millilitre (10^{-3})
Mgy	Million gallons per year
MHz	megaHertz
MMgy	Million metric gallons per year
MMLy	Million metric litres per year
MMT	Million metric tons
m.p.	Melting point
MPa	megaPascal
mpg	Miles per gallon
Mt	Million tonnes
Mtoe	Million tonnes of oil equivalent
MW	(1) Molecular weight; (2) megawatt
Nd, n.d.	Not detectable
ng, nm, nmol	Nanogram, nanometre, nanomole (10^{-9})
OT	Onset temperature of oxidation
p	Pressure
Pa	Pascal
PC	Particle counting
PCA	Principal component analysis
PCR	Principal component regression
PDA	Principal discriminant analysis
PLP	Percent of large particles (direct read ferrography)
PLS(R)	Partial least squares (regression)
pmt	Per metric tonne
PP	Pour point
ppb	Parts per billion
ppm	Parts per million
r	(1) Reaction rate; (2) radius
r.t.	Room temperature
s, sec	Second

SFC	Specific fuel consumption
SKE	Steinkohleneinheit (1 kg SKE = 0.984 kg bituminous coal = 29 307.6 kJ)
SMT	Signal maxima temperature
S/N	Signal-to-noise ratio
SUS	Saybolt universal second
T	Absolute temperature (K)
T_c	Critical temperature
T_m	Melting temperature
t	Tonne
t	Time
TEP	Tonne of petroleum equivalent; 1 TEP = 45.217 GJ
TOE, toe	Tonne of oil equivalent; 1 toe = 42 GJ
V	Volt
V	Volume; molar volume
v	Velocity
v/v	Volume/volume (solution concentration)
W	Watt
W	Wear volume
WL	Weld load
WPC	Wear particle concentration (direct read ferrography)
WSD	Wear scar diameter
w/w	Weight/weight (solution concentration)
x,y,z	Cartesian coordinates
Y%	Extraction yield
Z	Atomic number
α	Pressure–viscosity coefficient
δ	Hildebrand solubility parameter ($MPa^{1/2}$)
δ_d, δ_p, δ_h	Solubility parameter based on dispersive, polar and hydrogen bonding interactions
ε	Permittivity
ε_0	Relative permittivity
η	Dynamic viscosity of medium
μ	(1) Micro (10^{-6}); (2) absorption coefficient; (3) friction coefficient
μm	Micrometre
ν	Kinematic viscosity of medium
ρ	(1) Density (g/cm^3); (2) resistivity

A.3 General abbreviations

AACC	American Association of General Chemists International (St. Paul, MN)

ABIL	Ag-based Industrial Lubricants (University of Northern Iowa, Cedar Falls, IA)
ACEA	Association des Constructeurs Européens d'Automobiles, European Automobile Manufacturers' Association (Brussels, Belgium)
ADEME	Agence de l'Environnement et de la Maitrise de l'Energie, French Environment and Energy Management Agency (Paris, France)
ADM	Archer Daniels Midland Co. (Decatur, IL)
AF	Air Force (USA)
AFI	Air Force Instruction (USA)
AFNOR	Association Française de Normalisation (Saint Denis, France)
AGMA	American Gear Manufacturers Association (Alexandria, VA)
AIAA	American Institute of Aeronautics and Astronautics (Reston, VA)
AIChE	American Institute of Chemical Engineers (New York, NY)
AMG	Agro Management Group (Colorado Springs, CO)
ANSI	American National Standards Institute (Washington, DC)
AOAC	Association of Official Analytical Chemists International (Arlington, VA)
AOAP	Army Oil Analysis Program (USA)
AOCS	American Oil Chemists' Society (Urbana, IL)
AP	Acidification potential
APAET	Associação Portuguesa de Análise Experimental de Tensões (Lisbon, Portugal)
API	American Petroleum Institute (Washington, DC)
ARS	Agricultural Research Service (USDA)
ASA	American Society of Agronomy (Madison, WI)
ASHRAE	American Society of Heating, Refrigerating and Air-Conditioning Engineers (Atlanta, GA)
ASLE	American Society of Lubrication Engineers (now STLE)
ASM	American Society of Metals Intl. (Russell Township, OH)
ASME	American Society of Mechanical Engineers (New York, NY)
ASTM	ASTM International, formerly American Society for Testing and Materials (West Conshohocken, PA)

ATC	Additive Technical Committee of Petroleum Additive Manufacturers in Europe (Brussels, Belgium)
ATIEL	Association Technique de l'Industrie Européenne des Lubrifiants, Technical Association of the European Lubricant Industry (Auderghem, Belgium)
BBE	Bio-based economy
BFPA	British Fluid Power Association (Chipping Norton, UK)
BMELV	Bundesministerium für Ernährung, Landwirtschaft und Verbraucherschutz, Federal Ministry of Food, Agriculture and Consumer Protection (Germany)
BPD	Biocidal Products Directive (EU)
BSE	Bovine spongiform encephalopathy ('Mad cow disease')
BWM	Bio-Wetenschappen en Maatschappij (NWO) (The Hague, The Netherlands)
CAFE	Corporate Average Fuel Economy regulations (USA)
CAP	Common Agricultural Policy (EU)
CAS	Chemical Abstracts Service (Columbus, OH)
CBA	Cost–benefit analysis
CBI	Confidential business information
CE	Carcinogenic effect
CEC	Coordinating European Council for the Development of Performance Tests for Lubricants and Engine Fuels (Brussels, Belgium)
CED	Cumulative energy demand
CEFIC	Conseil Européen des Federations de l'Industrie Chimique, European Chemical Industry Council (Brussels, Belgium)
CEN	Comité Européen de Normalisation, European Committee for Standardisation (Brussels, Belgium)
CEPA	Canadian Environmental Protection Act (1999)
CERCLA	Comprehensive Environmental Response Compensation and Liability Act (USA)
CFR	*Code of Federal Regulations* (USA)
CIS	Commonwealth of Independent States
CLP	Classification, Labelling and Packaging Regulation (EC)
CMI	Condition Monitoring International (Miami, FL)
CMMS	Computer Maintenance Management System
CNPC	China National Petroleum Corporation (Beijing, P.R. China)

CONCAWE	The Oil Companies' Organisation for the Conservation of Clean Air and Water in Europe (Brussels, Belgium)
CONOE	Consorzio Obbligatorio Nazionale Oli Esausti, Italian Agency for Waste Vegetable Oil Collection and Disposal (Milan, Italy)
CORS	Continuous Oil Recycling System
COSHH	Control of Substances Hazardous to Health (UK)
CPS	Cooper Power Systems (Houston, TX)
CRADA	Cooperative Research and Development Agreement
CRC	Coordinating Research Council (Alphretta, GA)
CRMMA	Chemical Reference Material Manufacturers Association (USA)
CRP	Cancer risk potential
CVT	Constantly variable transmission
CWA	Clean Water Act (USA)
DA	US Department of the Army (Washington, DC)
DIN	(1) Deutsches Institut für Normung, German Institute of Standardisation (Berlin, Germany); (2) Deutsche Industrie Normen, German Industrial Standards
DLO	Stichting Dienst Landbouwkundig Onderzoek, Dutch Agricultural Research Foundation (Wageningen, The Netherlands)
DM	Decision matrix
DoD	US Department of Defense (Washington, DC)
DOE	US Department of Energy (Washington, DC)
DOI	US Department of Interior (Washington, DC)
DPD	Dangerous Preparations Directive (EC)
DSD	Dangerous Substances Directive (EC)
DSL	Domestic Substances List (CEPA, Canada)
DWC	Dow Wolff Cellulosics (Midland, MI)
EA	Environmentally acceptable/adapted/aware
EAL	Environmentally adapted lubricant
EC	European Community
ECHA	European Chemicals Agency (Helsinki, Finland)
ECNZ	Electricity Corporation of New Zealand, Ltd (New Zealand)
EDGE	Efficiency Directed Genome Engineering
E&E	Electrical and electronic
EEC	European Economic Community
EEE	Electrical and Electronic Equipment
EEL	European Ecolabel for Lubricants (EU)

EFSA	European Food Safety Authority (Parma, Italy)
EHEDG	European Hygenic Engineering & Design Group (Frankfurt/Main, Germany)
EIA	Energy Information Administration (Washington, DC)
ELGI	European Lubricating Grease Institute (Amsterdam, The Netherlands)
ELM	Environmental Lubricants Manufacturing, Inc. (Grundy Center, IA)
EMA	Engine Manufacturers Association (Chicago, IL)
EMPA	Eidgenössische Materialprüfungs- und Forschungsanstalt, Swiss Federal Laboratories for Materials Testing and Research (St. Gallen, Suitzerland)
ENDS	Environmental Data Services (London, UK)
ENOC	Emirates National Oil Company (Dubai, UAE)
EO	Executive Order (White House)
EOL	End of life
EP	Eutrophication potential
EPA	Environmental Protection Agency (USA)
ERRMA	European Renewable Resources and Materials Association (Brussels, Belgium)
ESA	European Society for Agronomy (Montpellier, France)
ETC	European Transient Cycle
EU	European Union
EUEB	EU Eco-labelling Board (Brussels, Belgium)
FAO	Food and Agriculture Organization of the United Nations (Rome, Italy)
FAR	Federal Aquisition Regulations (USA)
FAS	Foreign Agricultural Service (USA)
FDA	Food and Drug Administration (Silver Spring, MD)
FEDER	Fonds Européen de Développement Régional, European Fund for Regional Development (EU)
FIFRA	Federal Insecticide, Fungicide, and Rodenticide Act (USA)
FNR	Fachagentur Nachwachsende Rohstoffe e.V., Agency for Renewable Resources (Gülzow, Germany)
FP	Framework Programme (EU)
FSIS	Food Safety and Inspection Service (Washington, DC)
FSRIA	Farm Security and Rural Investment Act 2002 (USA)

FTMS	Federal Test Methods Standardisation (USA)
FZG	Forschungsstelle für Zahnräder und Getriebebau, Research Centre for Toothed Wheel and Gearing Engineering (Garching, Germany)
GAP	Good Agricultural Practice
GDP	Gross domestic product
GEIR	Groupement Européen de l'Industrie de la Régéneration, European Waste Oils Regeneration Industry (Brussels, Belgium)
GEN	Global Ecolabelling Network
GET	Green Earth Technologies, Inc. (Stamford, CT)
GHS	Globally Harmonised System (EU)
GIA	Global Industry Analysts, Inc. (San José, CA)
GLWQI	Great Lakes Water Quality Initiative (USA)
GM(O)	Genetically modified (organism)
GNP	Gross national product
GOM	Gulf of Mexico
GP	General purpose
GRAS	Generally recognised as safe
GREET	Greenhouse Gases, Regulated Emissions, and Energy Use in Transportation
GWP	Global warming potential
HACCP	Hazard Analysis and Critical Control Points
HGCA	Home-Grown Cereals Authority (Kenilworth, UK)
HH	Human health
HMSO	Her Majesty's Stationery Office (UK)
HSE	Health, Safety and Environment
IAR	Industries & Agro-Resources (Laon, France)
IAVF	Institut für angewandte Verschleißforschung GmbH (Karlsruhe, Germany)
ICE	Internal combustion engine
ICML	International Council for Machinery Lubrication (Broken Arrow, OK)
ICOMIA	International Council of Marine Industry Associations (Egham, UK)
IEEE	Institute of Electrical and Electronics Engineers, Inc. (Washington, DC)
IEEP	Institute for European Environmental Policy (London, UK)
IFAS	Institut für fluidtechnische Antriebe und Steuerungen, Institute for Fluid Power Drives and Controls (Aachen, Germany)

IFEU	Institut für Energie- und Umweltforschung, Institute for Energy and Environmental Research (Heidelberg, Germany)
IGC	International Grains Council (London, UK)
IJEC	International Jojoba Export Council (Scottsdale, AZ)
ILI	International Lubricants, Inc. (Seattle, WA)
ILSAC	International Lubricants Standardization and Approval Committee
ILUC	Indirect land use change
IP	(1) Institute of Petroleum (London, UK); (2) identity preserved
IS	Indian Standard
ISAAA	International Service for the Acquisition of Agri-Biotechnology Applications
ISI	Fraunhofer-Institut für System- und Innovationsforschung, Fraunhofer Institute for Systems and Innovation Research (Karlsruhe, Germany)
ISO	International Organization for Standardization (Geneva, Switzerland)
ISP	International Specialty Products (now Ashland Specialty Ingredients; Wayne, NJ)
ISPO	Istituto per gli Studi sulla Pubblica Opinione, Institute of Public Opinion Studies (Milan, Italy).
ITEC	Industrial and Transportation Equipment Company (Des Moines, IA)
IUPAC	International Union of Pure and Applied Chemistry (Research Triangle Park, NC)
JAMA	Japan Automobile Manufacturers Association (Tokyo, Japan)
JETOC	Japan Chemical Industry Ecology – Toxicology & Information Centre (Tokyo, Japan)
JGPSSI	Japan Green Procurement Survey Standardization Initiative
JOAP	Joint Oil Analysis Program (USA)
JPI	Japan Petroleum Institute (Tokyo, Japan)
jv	Joint venture
KECI	Korea Existing Chemicals Inventory
KTI	Kinetics Technology International (Cheltenham, Australia)
LCA	Life-cycle assessment
LCC	Life-cycle costing

LCCP	Lubrication critical control point
LCI	Life-cycle inventory
LCIA	Life-cycle impact assessment
LMI	Lead Market Initiative (EC)
LTI	Lubrication Technology, Inc. (Franklin Furnace, OH)
LU	Land use
MDG	Millennium Development Goal
MEMS	Microelectromechanic system
MEP	Ministry of Environmental Protection (P.R. China)
METI	Ministry of Economy, Trade and Industry (Japan)
MIL	Military Standard (USA)
MIP	Market Introduction Programme (Germany)
MISTRA	Swedish Strategic Fund for Environmental Research (Stockholm, Sweden)
MIT	Massachusetts Institute of Technology (Boston, MA)
MPOB	Malaysian Palm Oil Board (Kuala Lumpur, Malaysia)
MSDS	Material Safety Data Sheet
MWV	Mineralölwirtschaftsverband, Association of the German Petroleum Industry (Berlin, Germany)
NABL	National Ag-based Lubricants Center (University of Northern Iowa, Cedar Falls, IA)
NAL	North American Lubricants, Co. (Scottsdale, AZ)
NAS	National Academy of Sciences (USA)
NASS	National Agricultural Statistical Service (Washington, DC)
NATO	North Atlantic Treaty Organization (Brussels, Belgium)
NAVAIR	Navy Air (USA)
NCE	Non-carcinogenic effect
NCGA	National Corn Growers Association (Chesterfield, MO)
NCNS	New Chemical Notification Scheme (OECD)
NEMS	Nanoelectromechanic system
NGVA	European Association for Bio/Natural Gas Vehicles (Madrid, Spain)
NHTSA	National Highway Traffic Safety Administration (USA)
NIST	National Institute of Standards and Technology (Gaithersburg, MD)
NLGI	National Lubricating Grease Institute (Kansas City, MO)

NMMA	National Marine Manufacturers Association (Chicago, IL)
NNFCC	National Non-Food Crops Centre (Heslington, UK)
NOEC	No observed effect concentration
NP	Nutrification potential
NPDES	National Pollutant Discharge Elimination System (EPA)
NPRA	National Petrochemical & Refiners Association (Washington, DC)
NRL	US Naval Research Laboratory (Washington, DC)
NSF	(1) National Sanitation Foundation Intl. (Ann Arbor, MI); (2) National Science Foundation (Arlington, VA)
NWO	Nederlandse Organisatie voor Wetenschappelijk Onderzoek, Dutch Organisation for Scientific Research (The Hague, The Netherlands)
OAP	Oil Analysis Program (USA)
OCNS	Offshore Chemical Notification Scheme (NL, UK)
OD(P)	Ozone depletion (potential)
OECD	Organisation for Economic Co-operation and Development (Paris, France)
OEM	Original equipment manufacturer
OGL	*Oléagineux, Corps Gras, Lipides*
OLM	Oil life monitor
ONIDOL	National Interprofessional Organization of Oil Producers (France)
OSA	On-Site Analysis, Inc. (Palm Beach Gardens, FL)
OSHA	Occupational Safety and Health Administration (USA)
OSPAR	Oslo and Paris Convention for the Protection of the Marine Environment of the North East Atlantic
OSR	Oil Storage Regulations (UK)
OTR	Over-the-road
OVID	Verband der Ölsaatenverarbeitenden Industrie in Deutschland e.V., Association of the Oilseed Processing Industry in Germany (Berlin, Germany)
PBL	Performance BioLubes, LLC (Morton Grove, IL)
PCT	Patent Cooperation Treaty
PdM	Predictive maintenance
PM	Preventive maintenance
P&P	Paper and pulp
PQ	Product quality
PS	Photochemical smog

PSU	Pennsylvania State University (University Park, PA)
QA	Quality assurance
QC	Quality control
QM	Quality management
®	Registered
RAL	Deutsches Institut für Gütesicherung und Kennzeichnung e.V., German Institute for Quality Assurance and Labelling (Sankt Augustin, Germany)
RCRA	Resource Conservation and Recovery Act (USA)
RD	Resource depletion
R&D	Research and development
RE	Respiratory effects
REACH	Registration, Evaluation, Authorisation and Restriction of Chemicals (EU)
RED	Renewable Energy Directive (EU)
RLI	Renewable Lubricants, Inc. (Hartville, OH)
RLM	Reichsluftfahrtministerium (Berlin, Germany)
RoHS	Restrictions on Hazardous Substances (EU)
ROI	Return-on-investment
RSC	Royal Society of Chemistry (London, UK)
RTD	Research & technology development
RWS	Rijkswaterstaat, Dutch Water Management Authority (The Netherlands)
RWTH	Rheinisch-Westfälische Technische Hochschude (Aachen, Germany)
SAE	Society of Automotive Engineers, Inc. (Warrendale, PA)
SARA	Superfund Amendments and Reauthorization Act (USA)
SCF	Scientific Committee on Food (CEC)
SEPA	State Environmental Protection Agency (P.R. China)
SETAC	Society for Environmental Toxicology and Chemistry (Pensacola, FL)
SM	Service mark
SME	Small- and medium-sized enterprise
SOP	Standard operating procedure
SPCC	Spill prevention control and countermeasures (40 CFR 112)
SRRHU	Société de Ramassage pour la Régeneration des Huiles Usagées (Brazey-en-Plaine, France)
SS	Summer smog

STLE	Society of Tribologists and Lubrication Engineers (Park Ridge, IL)
SusChem	European Technology Platform for Sustainable Chemistry (*see* CEFIC)
SW	Solid waste
SWOT	Strengths, weaknesses, opportunities, threats
TAE	Technische Akademie Esslingen (Ostfildern, Germany)
TBK	Thész, Boros, Király
TBS	Tapered bearing simulator
TGAP	Taxe Générale sur les Activitées Polluantes, General Tax on Polluting Activities (France)
TM	Trade mark
TRACI	Tool for the Reduction and Assessment of Chemical and Other Environmental Impacts
TSC	Technical support centre
TSCA	Toxic Substance Control Act (USA)
TSE	Transmissible spongiform encephalopathy
TURA	Toxic Use Reduction Act (MA)
UBA	Umweltbundesamt, Federal Environmental Agency (Berlin, Germany)
UBL	United Bio Lube, Inc. (Palo Alto, CA)
UEIL	Union Indépendante de l'Industrie Européenne des Lubrifiants, European Lubricants Industry Organization (Brussels, Belgium)
UK	United Kingdom
UKLA	United Kingdom Lubricants Association, Ltd (Berkhamsted, Herts, UK)
UKWAS	UK Woodland Assurance Standard
UN	United Nations
UNC	University of North Carolina (Chapel Hill, NC)
US, USA	United States of America
USAF	United States Air Force
USB	United Soybean Board (Chesterfield, MO)
USDA	US Dept. of Agriculture (Washington, DC)
USP	United States Pharmacopœia
USTC	University of Science and Technology of China (Hefei, PRC)
UvA	Universiteit van Amsterdam, Amsterdam University (The Netherlands)
VCI	Verband der Chemischen Industrie e.V., German Chemical Industry Association (Frankfurt/Main, Germany)

VDMA	Verband Deutscher Maschinen- und Anlagenbau e.V., German Machinery and Plant Manufacturers' Association (Frankfurt/Main, Germany)
VROM	Ministry of Housing, Spatial Planning and the Environment (The Hague, The Netherlands)
WAOB	World Agricultural Outlook Board (Washington, DC)
WCC/RRC	Western Canadian Canola/Rapeseed Recommending Committee, Inc. (Winnipeg, MB)
WFD	Waste Framework Directive (EU)
WGK	Wassergefährdungsklasse, water hazard class
WHC	Water Hazard Classification
WLP	Waverly Light and Power (Waverly, IA)
WP	Workpackage
WS	Winter smog
WTC	Wetenschappelijke en Technische Commissie, Scientific Technical Committee (The Netherlands)

A.4 Main Internet sources of further information and advice

http://lubricants.petro-canada.ca
http://pubs.acs.org
http://www.aak.com
http://www.abiluni.org
http://www.acieu.net
http://www.aftonchemical.com
http://www.aimpoints.hq.af.mil
http://www.alibaba.com
http://www.anderol-europe.com
http://www.applied.com
http://www.ars.usda.gov
http://www.ashland.com
http://www.atiel.org/codeprac.htm
http://www.ausmud.com
http://www.bechem.com
http://www.bechem.de
http://www.bio2cycleengineoils.com
http://www.bioassemblyoils.com
http://www.biobarchainoils.com
http://www.biobasedbusiness.com
http://www.biobased.oce.usda.gov
http://www.biobaseurope.com

http://www.bioblend.com
http://www.biochainoils.com
http://www.biocorrosioninhibitors.com
http://www.biofoodgradegearoils.com
http://www.biofoodgradegreases.com
http://www.biofoodgradehydraulicfluids.com
http://www.biofoodgradelubes.com
http://www.biogreases.com
http://www.biolubricants.com
http://www.biopreferred.gov
http://www.bioschmierstoffe.info
http://www.blauer-engel.de
http://www.bunge.com
http://www.cargill.com
http://www.carlube.co.uk
http://www.cassida-lubricants.com
http://www.castrol.com
http://www.catalog.biopreferred.gov
http://www.chemtura.com
http://www.chevron.com
http://www.chevronbaseoils.com
http://www.chevronlubricants.com
http://www.cognis.com
http://www.connekt.nl
http://www.cooperpower.com
http://www.CortecVCI.com
http://www.creativecoatingsco.com
http://www.crodalubricants.com
http://www.curtis-systems.de
http://www.dev.hgca.com
http://www.dsifluids.com
http://www.ec.europe.eu/environment/ecolabel/index_en.htm
http://www.eco-label.com
http://www.eion-additives.com
http://www.elgi.org
http://www.elmusa.com
http://www.environment-agency.gov.uk/osr
http://www.epobio.net
http://www.exxonmobil.com
http://www.fnr.de
http://www.forestry.gov.uk
http://www.freedonia.ecnext.com
http://www.freedoniagroup.com

http://www.freepatentsonline.com
http://www.fuchslubricants.com
http://www.fuchs-oil.com
http://www.gannonoils.co.uk
http://www.getg.com
http://www.GETgreenearth.com
http://www.getgreenearthsolutions.com
http://www.goliath.ecnest.com
http://www.govexee.com
http://www.identitypreserved.com
http://www.ijec.net
http://www.imakenews.com/lubritec
http://www.imcdgroup.com
http://www.ineosoligomers.com
http://www.inolex.com
http://www.ispcorp.com
http://www.ispmetal.com
http://www.italmatch.it
http://www.itctribology.org
http://www.jergensinc.com
http://www.klinegroup.com
http://www.lca.jrc.ec.europa.eu
http://www.linnaeus.net
http://www.lubline.com
http://www.lubricants.elf.com
http://www.lubricants.total.com
http://www.lubrizol.com
http://www.machinerylubrication.com
http://www.masterchemical.com
http://www.MetalformingFacts.com
http://www.mobil.com
http://www.mobilindustrial.com
http://www.mooreballiewoil.com
http://www.motul.com
http://www.mpob.gov.my
http://www.mscdirect.com
http://www.nesteoil.com/baseoil
http://www.nlgi.com
http://www.nlgi.org
http://www.noco.com
http://www.oats.co.uk
http://www.oem.bmj.com
http://www.olyslager.nl

http://www.onf.fr
http://www.orisonllc.com
http://www.orisonmarketing.com
http://www.oronitegoes.com
http://www.pacsciinst.com
http://www.perkinelmer.com
http://www.petronas.nl
http://www.portal.isiknowledge.com
http://www.predictusa.com
http://www.processonline.com.au
http://www.profoodnetwork.com
http://www.q8oils.co.uk
http://www.ral.de
http://www.ram-pac.com
http://www.renewablelube.com
http://www.renewablelubricants.com
http://www.repsol.com
http://www.researchandmarkets.com
http://www.rheinchemie.com
http://www.rsclare.com
http://www.sae.org
http://www.sciencedirect.com
http://www.sealandchem.co.uk
http://www.sevia-onyx.com
http://www.shamrock.com.cy
http://www.shell.com/lubematch/
http://www.shell.com/lubricants/
http://www.shell.com/naturelle
http://www.shimadzu.co.uk
http://www.soybiobased.org/lifecyclechecklist
http://www.soyclean.biz
http://www.soynewuses.org
http://www.soypower.net
http://www.springerlink.com
http://www.statoillubricants.com
http://www.stle.org
http://www.svanen.nu
http://www.syntheticwarehouse.com
http://www.tae.de/tribology
http://www.techoils.cargill.com
http://www.terrabond.net
http://www.terresolve.com
http://www.thomasnet.com

http://www.thomasregister.com
http://www.tritechlubricants.com
http://www.uclan.ac.uk
http://www.ueil.org
http://www.ukla.org.uk
http://www.uni.edu/nabl/applications.htm
http://www.unitedbiolube.com
http://www.unitedsoybean.org
http://www.vickers-oil.com
http://www.viscoinc.com
http://www.vsi-schmierstoffe.de
http://www.yubase.com
http://www.zeller-gmelin.de
http://www1.mscdirect.com

Index